Precalculus
Mathematics for Calculus

Fourth Edition

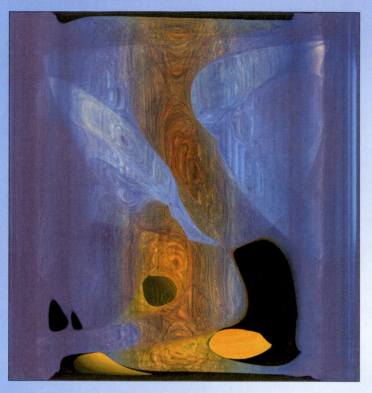

About the Cover

The art on the cover was created by Bill Ralph, a mathematician who uses modern mathematics to produce visual representations of "dynamical systems." Examples of dynamical systems in nature include the weather, blood pressure, the motions of the planets, and other phenomena that involve continual change. Such systems, which tend to be unpredictable and even chaotic at times, are modeled mathematically using the concepts of composition and iteration of functions (see Section 2.8 and the Laboratory Project on page 228). The basic idea is to start with a particular function and evaluate it at some point in its domain, yielding a new number. The function is then evaluated at the new number. Repeating this process produces a sequence of numbers called *iterates* of the function. The original domain is "painted" by assigning a color to each starting point; the color is determined by certain properties of its sequence of iterates and the mathematical concept of "dimension." The result is a picture that reveals the complex patterns of the dynamical system. In a sense, these pictures allow us to look, through the lens of mathematics, at exotic little universes that have never been seen before.

Professor Ralph teaches at Brock University in Canada. He can be contacted by email at bralph@spartan.ac.brocku.ca.

About the Authors

James Stewart was educated at the University of Toronto and Stanford University, did research at the University of London, and now teaches at McMaster University. His research field is harmonic analysis.

He is the author of a best-selling calculus textbook series published by Brooks/Cole, including *Calculus, 4th Ed., Calculus: Early Transcendentals, 4th Ed.,* and *Calculus: Concepts and Contexts, 2nd Ed.,* as well as a series of high-school mathematics textbooks.

Lothar Redlin grew up on Vancouver Island, received a Bachelor of Science degree from the University of Victoria, and a Ph.D. from McMaster University in 1978. He subsequently did research and taught at the University of Washington, the University of Waterloo, and California State University, Long Beach.

He is currently Professor of Mathematics at The Pennsylvania State University, Abington College. His research field is topology.

Saleem Watson received his Bachelor of Science degree from Andrews University in Michigan. He did graduate studies at Dalhousie University and McMaster University, where he received his Ph.D. in 1978. He subsequently did research at the Mathematics Institute of the University of Warsaw in Poland. He also taught at The Pennsylvania State University.

He is currently Professor of Mathematics at California State University, Long Beach. His research field is functional analysis.

The authors have also published *College Algebra, Third Edition* (Brooks/Cole, 2000), and *Algebra and Trigonometry* (Brooks/Cole, 2001).

Precalculus
Mathematics for Calculus

Fourth Edition

James Stewart
McMaster University

Lothar Redlin
The Pennsylvania State University

Saleem Watson
California State University, Long Beach

BROOKS/COLE

THOMSON LEARNING™

Australia • Canada • Mexico • Singapore • Spain • United Kingdom • United States

BROOKS/COLE

THOMSON LEARNING

Publisher: *Robert W. Pirtle*
Marketing Team: *Karin Sandberg, Samantha Cabaluna, and Darcie Pool*
Assistant Editor: *Carol Ann Benedict*
Editorial Assistant: *Molly Nance*
Production Editor: *Tom Novack*
Production Service: *Luana Richards*
Manuscript Editor: *Luana Richards*
Permissions Editor: *Sue Ewing*
Interior Design: *John Edeen*

Cover Design: *Roy R. Neuhaus*
Cover Illustration: *Bill Ralph*
Interior Illustration: *Brian Betsill*
Photo Researcher: *Kathleen Olson*
Print Buyer: *Vena Dyer*
Typesetting: *Better Graphics, Inc.*
Cover Printing: *Phoenix Color Corporation*
Printing and Binding: *Quebecor World, Taunton*

For more information about this or any other Brooks/Cole products, contact:
BROOKS/COLE
511 Forest Lodge Road
Pacific Grove, CA 93950 USA
www.brookscole.com
1-800-423-0563 (Thomson Learning Academic Resource Center)

For permission to use material from this work, contact us by
www.thomsonrights.com
fax: 1-800-730-2215
phone: 1-800-730-2214

Printed in United States of America

10   9   8   7   6   5   4

Library of Congress Cataloging-in-Publication Data

Stewart, James, [date]
    Precalculus : mathematics for calculus / James Stewart, Lothar Redlin, Saleem
Watson.—4th ed.
        p.   cm.
    Includes index.
    ISBN 0-534-38541-9
    1. Mathematics.   I. Redlin, L.   II. Watson, Saleem.   III. Title.

QA39.3.S74 2002
510—dc21   2001043621

To Viola and David Panman,
for their encouragement and support

Preface

The art of teaching is the art of assisting discovery.
MARK VAN DOREN

What do students really need to know to be prepared for calculus? What tools do instructors need to assist their students in preparing for calculus? These two questions have motivated the writing of this book.

To be prepared for calculus a student needs not only technical skill but also a clear understanding of concepts. Indeed, *conceptual understanding* and *technical skill* go hand in hand, each reinforcing the other; both are necessary for success, not only in calculus, but also in subsequent mathematics and science courses. A student also needs an appreciation that mathematics contains great ideas that can be used to model real-world situations. A precalculus course presents many opportunities for students to discover this practical power of mathematics.

To prepare students for calculus, instructors use many different methods. Good teaching comes in different forms, and each instructor brings unique strengths to the classroom; but certainly all instructors are committed to the goal of encouraging conceptual understanding. To implement this goal some instructors use *technology* to help students become active learners; others use the *rule of four*, "topics should be presented geometrically, numerically, algebraically, and verbally," as well as an expanded emphasis on *applications* to promote conceptual reasoning; still others use *group learning*, *extended projects*, or *writing exercises* as a way of encouraging students to explore their own understanding of a given concept; and all present mathematics as a *problem-solving* endeavor. In this book we have used all these methods as enhancements to a central core of fundamental skills. These methods are tools to be utilized by instructors and their students to navigate their own course of action toward the goal of conceptual understanding.

In writing this fourth edition, our purpose was to further sharpen the clarity of the exposition while retaining the main features that have contributed to the success of this book. We have reorganized and rewritten several of the chapters (as described below) with a view to making the book usable in a wider variety of instructional approaches. We have included more use of the powerful capabilities of the graphing calculator throughout the book. We have also included a new chapter on limits, designed to lead into an easy transition to Stewart's calculus texts.

Many of the features of this book are the result of our own experience in teaching precalculus. We have benefited from the insightful comments of our colleagues who have used the text and critiqued it section by section. We have also benefited from the sharp insight of our reviewers.

■ Special Features

Exercise Sets

The most important way to foster conceptual understanding is through the problems that the instructor assigns. To that end we have provided a wide selection of exercises. Each exercise set is carefully graded, progressing from basic conceptual exercises and skill-development problems to more challenging problems requiring synthesis of previously learned material with new concepts.

A Complete Review Chapter

Chapter 1 is a review chapter; it contains the fundamental concepts a student needs to begin a precalculus course. We have included this chapter primarily as a handy reference for the student for basic concepts in algebra and analytic geometry. As much or as little of this chapter as necessary can be covered in class, depending on the background of the students. The test at the end of this chapter is designed as a diagnostic test for determining what parts of this review chapter need to be emphasized.

Graphing Calculators and Computers

Calculator and computer technology extends in a powerful way our ability to calculate and visualize mathematics. These capabilities affect not only how a topic is taught but also what is emphasized. We have integrated the use of the graphing calculator throughout the text—to graph and analyze functions, families of functions, and sequences, to calculate and graph regression curves, to perform matrix algebra, to graph linear inequalities, and other powerful uses. We also exploit the programming capabilities of the graphing calculator to provide simple programs that model real-life situations (see, for instance, pages 686, 903, and 905). The graphing calculator sections, subsections, examples, and exercises, all marked with the special symbol 📈, are optional and may be omitted without loss of continuity.

The availability of graphing calculators makes it not less important, but far more important to understand the concepts that underlie what the calculator produces. Accordingly, all our calculator-oriented subsections are preceded by sections in which students must graph or calculate by hand, so that they can understand precisely what the calculator is doing when they later use it to simplify the routine, mechanical part of their work.

Interactive Video Skillbuilder CD-ROM and Video

Every copy of the text now includes the Interactive Video Skillbuilder CD-ROM, which contains over eight hours of interactive video instruction. The problems in each video lesson are listed beside the video screen so that students can try them before watching the solution. The icon 💻 points students to material contained in the CD-ROM. To help students evaluate their progress, each section is linked to a 10-question web quiz, the results of which can be e-mailed to the instructor.

A tutorial series on VHS videotape, which provides additional examples and explanations for each chapter, is also available to departments that adopt the text.

Flexible Approach to Trigonometry

One way to teach trigonometry is to begin with the right triangle approach (Chapter 6), building on the foundation of a conventional high-school course in trigonometry. Another way is to begin with the unit circle approach (Chapter 5), thus emphasizing that the trigonometric functions are functions of real numbers, just like polynomial, exponential, or logarithmic functions. Each way of teaching trigonometry has its merits, and the trigonometry chapters of this text have been written so that either approach (Chapter 5 or Chapter 6) may be studied first. Moreover, putting these two approaches in different chapters, each with its applications, helps clarify the purpose of each approach. Another way to teach trigonometry is to intertwine the two approaches: Sections 5.1, 5.2, 6.1, 6.2, 6.3, 5.3, 5.4, 6.4, and 6.5. Our organization makes it easy to do this without obscuring the fact that the two approaches involve distinct representations of the same functions.

Focus on Modeling

In addition to many applied problems where students are given a model to analyze, we have included several sections and subsections in which students are required to *construct* models of real-life situations. In addition, we have concluded several of the chapters with sections entitled *Focus on Modeling*. The first such section, after Chapter 2, introduces the basic idea of modeling a real-life situation by fitting lines to data (linear regression). Other sections present ways in which polynomial, exponential, logarithmic, and trigonometric functions, systems of inequalities, and probability can all be used to model familiar phenomena from the sciences and from everyday life.

Focus on Problem Solving

We are committed to helping students develop mathematical thinking rather than "memorizing all the rules," an endless and pointless task that many students are not taught to avoid. Thus, an emphasis on problem solving is integrated throughout the text. In addition, we have concluded several chapters with sections entitled *Focus on Problem Solving*, each of which highlights a particular problem-solving technique. The first such section, following Chapter 1, gives a general introduction to the principles of problem solving. Each *Focus* section includes problems of a varied nature that encourage students to apply their problem-solving skills. In selecting these problems, we have kept in mind the following advice from David Hilbert: "A mathematical problem should be difficult in order to entice us, yet not inaccessible lest it mock our efforts."

Projects

One way to engage students and make them active learners is to have them work (perhaps in groups) on extended projects that give a feeling of substantial accomplishment when completed. We have included four types of projects. *Discovery Projects* encourage discovery through algebraic and geometric pattern recognition; the one on page 35 involves connecting factoring formulas to areas and volumes. *Laboratory Projects* involve technology; the one on page 672 uses repeated matrix multiplication to predict population distribution within a species. The *Writing Project* on page 76 compares modern algebraic methods to ancient ones and the *Applied Project* on page 934 shows how tangents can be used to design a roller coaster.

Mathematical Vignettes

Throughout the book we make use of the margins to provide short biographies of interesting mathematicians as well as applications of mathematics to the "real world." The biographies often include a key insight that the mathematician discovered and which is relevant to precalculus. (See, for instance, the vignettes on Viète, page 51; Salt Lake City, page 95; and radiocarbon dating, page 368). They serve to enliven the material and show that mathematics is an important, vital activity, and that even at this elementary level it is fundamental to everyday life. A series of vignettes, entitled *Mathematics in the Modern World*, emphasizes the central role of mathematics in current advances in technology and the sciences (see pages 16, 192, and 626, for example).

Real-World Applications

We have included substantial applied problems that we believe will capture the attention of students. These are integrated throughout the text in both examples and exercises. Applications from engineering, physics, chemistry, business, biology, environmental studies, and other fields show how mathematics is used to model real-world situations. We have carefully chosen applications that show the relevance of precalculus to our daily lives and its power as a problem-solving tool. (See, for instance, the examples and exercises on computer graphics, pages 668–672; energy expended in bird flight, page 217; terminal velocity of a sky diver, page 345; establishing time of death, page 388, Exercise 24; and finding the height of a mountain, page 511, Exercise 29.) We have used real-world data where appropriate.

Discovery, Writing, and Group Learning

Each exercise set ends with a block of exercises labeled *Discovery • Discussion*. These exercises are designed to encourage the students to experiment, preferably in groups, with the concepts developed in the section, and then to write out what they have learned, rather than simply look for "the answer."

Check Your Answer

The *Check Your Answer* feature is used wherever possible to emphasize the importance of looking back to check whether an answer is reasonable. (See, for instance, pages 54–55.)

Review Sections and Chapter Tests

Each chapter ends with an extensive review section, including a *Chapter Test* designed to help the students gauge their progress. Brief answers to odd-numbered exercises in each section (including the review exercises), and to all questions in the Chapter Tests, are given in the back of the book.

The review material in each chapter begins with a *Concept Check*, designed to get the students to think about and explain in their own words the ideas presented in the chapter. These can be used as writing exercises, in a classroom discussion setting, or for personal study.

■ Major Changes for the Fourth Edition

- More than 20% of the exercises in each chapter are new. New exercises have been chosen to provide more practice with basic concepts, as well as to explore ideas that we do not have space to cover in the discussion and examples in the text itself.

- The *Chapter 1 Test*, concluding an extensive chapter that thoroughly reviews algebra basics, has been redesigned to serve as a diagnostic test to help students gauge exactly what algebra topics they need to review, rather than waste time on topics they are already comfortable with.

- A graphic calculator can do much more than just graph functions. We have included many more of the capabilities of the graphing calculator throughout the text. In particular, we show how the `TABLE`, `Intersect`, `Zero`, `Maximum`, and `Mimimum` commands can be used to analyze graphs of functions. We utilize all the available regression commands (`LinReg`. `QuadReg`, `PwrReg`, `ExpReg`, `LnReg`, `Logistic`, and `SinReg`) to show how to find functions that model real-life data. We also use other capabilities of the graphing calculator to analyze matrices, sequences, and inequalities. In addition, some simple programs are used to model computer graphics, Monte Carlo simulations, and other applications of mathematics. However, we continue to make the book completely usable by instructors who prefer not to use the graphing calculator by clearly marking graphing calculator subsections, examples, and exercises with the special symbol ⬚.

- Each chapter now includes one or more *Projects*, designed to get students to work in groups on extended projects that require them to apply and write about, in a substantial way, the mathematical knowledge they have just acquired.

- In addition to the existing mathematical "vignettes" (a popular feature of earlier editions), which give snapshot views of interesting mathematical people and topics, we have added a series of new vignettes called *Mathematics in the Modern World*. These discuss the many ways that mathematics has been used to develop such current innovations as cellular telephones, compact discs, CAT scans, global positioning systems, and the like. They also present the impact of mathematics on modern psychology, sociology, and biology; for instance, one vignette discusses what mathematics has to say about fair voting systems in elections.

- The *Focus on Modeling* sections have been rewritten to simplify their structure and make them more usable. We have added more real-world examples to these sections.

- The "modeling" theme has been used throughout to unify and clarify the many applications of precalculus. There are numerous applied problems throughout the book where students are given a model to analyze (see, for instance, page 152, Exercises 35–42; page 323, Exercises 67–72; and page 373, Exercises 65–68). But the material on "modeling," where students are required to *construct* mathematical models for themselves, has been organized into clearly defined sections and subsections (see for example, Sections 1.6, 2.7, and 4.5). We have made a special effort, in these modeling sections and subsections, to clarify the essential process of translating problems from English into the language of mathematics.

- In Chapter 1 we have incorporated more extensive discussion of graphical solutions of equations (Section 1.9).

- In Chapter 2 we have added a new section on average rates of change. We have also completely rewritten and reorganized the material on modeling with functions.

■ In Chapter 3 we have expanded the first section on graphs of polynomial functions. We have also simplified the treatment of rational functions, including some material on using the function transformations from Chapter 2 to help graph rational functions.

■ In Chapter 4 we have combined the treatment of arbitrary exponential functions with the treatment of the natural exponential function. We have also clarified and reorganized the sections that describe how exponential and logarithmic functions are used to model real-life processes.

■ In Chapters 5 and 6 we continue to use our flexible approach to trigonometry, presenting the unit circle and triangle approaches in such a way that either approach may be treated first, depending on the preference of the instructor. We give new enhanced explanations of why each approach is important and how they are related to each other, regardless of which approach the instructor chooses to teach first. We have also incorporated more use of the graphing calculator (in the Chapter 5 *Focus on Modeling* section) to provide trigonometric models for data.

■ In Chapter 7 we have completely rewritten Section 7.5 on trigonometric equations to organize the topic more clearly. We have also added a section on the dot product of vectors, at the request of users of the book.

■ In Chapter 8 we have reorganized the first three sections so that they can be used as stand-alone sections by those instructors who prefer not to use matrices when teaching systems of equations.

■ In Chapter 9 we have reorganized the first four sections, providing more summary boxes and simpler explanations.

■ In Chapters 10 and 11 we have incorporated more use of the graphing calculator, primarily to enhance the students' visual understanding of the nature of sequences and probability.

■ Chapter 12, an introduction to limits and the calculus, has been added at the request of many users. It is designed to lead into an easy transition to Stewart's Calculus books.

SECTION 2.6   Extreme Values of Functions   **201**

Expressing a quadratic function in standard form helps us sketch its graph as well as find its maximum or minimum value. If we are interested only in finding the maximum or minimum value, then a formula is available for doing so. This formula is obtained by completing the square for the general quadratic function as follows:

$$f(x) = ax^2 + bx + c$$

$$= a\left(x^2 + \frac{b}{a}\,x\right) + c \qquad \text{Factor } a \text{ from the } x\text{-terms}$$

$$= a\left(x^2 + \frac{b}{a}\,x + \frac{b^2}{4a^2}\right) + c - \frac{b^2}{4a} \qquad \begin{array}{l}\text{Complete the square:} \\ \text{Add } \frac{b^2}{4a^2} \text{ inside parentheses,} \\ \text{subtract } a\left(\frac{b^2}{4a^2}\right) \text{ outside}\end{array}$$

$$= a\left(x + \frac{b}{2a}\right)^2 + c - \frac{b^2}{4a}$$

This equation is in standard form with $h = -b/(2a)$ and $k = c - b^2/(4a)$. Since the maximum or minimum value occurs at $x = h$, we have the following result.

> **MAXIMUM OR MINIMUM VALUE OF A QUADRATIC FUNCTION**
>
> The maximum or minimum value of a quadratic function
> $f(x) = ax^2 + bx + c$ occurs at
> $$x = -\frac{b}{2a}$$
> If $a > 0$, then the **minimum value** is $f\left(-\dfrac{b}{2a}\right)$.
>
> If $a < 0$, then the **maximum value** is $f\left(-\dfrac{b}{2a}\right)$.

EXAMPLE 4 ■ Finding Maximum and Minimum Valu

Find the maximum or minimum value of each quadratic

(a) $f(x) = x^2 + 4x$ (b) $g(x) = $

SOLUTION

(a) This is a quadratic function with $a = 1$ and $b = 4$. minimum value occurs at

$$x = -\frac{b}{2a} = -\frac{4}{2 \cdot 1} = -2$$

MATHEMATICS IN THE MODERN WORLD

Weather Prediction

Modern meteorologists do much more than predict tomorrow's weather. They research long-term weather patterns, depletion of the ozone layer, global warming, and other effects of human activity on the weather. But daily weather prediction is still a major part of meteorology; its value is measured by the innumerable human lives saved each year through accurate prediction of hurricanes, blizzards, and other catastrophic weather phenomena. At the turn of the century mathematicians proposed to model weather with equations that used the current values of hundreds of atmospheric variables. Although this model worked in principle, it was impossible to predict future weather patterns with it because of the difficulty of measuring all the variables accurately and solving all the equations. Today, new mathematical models combined with high-speed computer simulations have vastly improved weather prediction. As a result, many human as well as economic disasters have been averted. Mathematicians at the National Oceanographic and Atmospheric Administration (NOAA) are continually researching better methods of weather prediction.

◀ Mathematical vignettes provide applications of mathematics to the real world or give short biographies of mathematicians, contemporary as well as historical. A series of vignettes entitled *Mathematics in the Modern World* emphasize the central role of mathematics in current advances in technology and the sciences.

◀ Summary boxes organize and clarify topics and highlight key ideas.

SECTION 3.1   Polynomial Functions and Their Graphs   **269**

With a graphing calculator we can quickly draw the graphs of many functions at once, on the same viewing screen. This allows us to see how changing a value in the definition of the functions affects the shape of its graph. In the next example we apply this principle to a family of third-degree polynomials.

EXAMPLE 8 ■ A Family of Polynomials

Sketch the family of polynomials $P(x) = x^3 - cx^2$ for $c = 0, 1, 2,$ and 3. How does changing the value of c affect the graph?

SOLUTION

The polynomials

$$P_0(x) = x^3 \qquad\qquad P_1(x) = x^3 - x^2$$
$$P_2(x) = x^3 - 2x^2 \qquad\qquad P_3(x) = x^3 - 3x^2$$

are graphed in Figure 13. We see that increasing the value of c causes the graph to develop an increasingly deep "valley" to the right of the y-axis, creating a local maximum at the origin and a local minimum at a point in quadrant IV. This local minimum moves lower and farther to the right as c increases. To see why this happens, factor $P(x) = x^2(x - c)$. The polynomial P has zeros at 0 and c, and the larger c gets, the farther to the right the minimum between 0 and c will be. ■

FIGURE 13
A family of polynomials
$P(x) = x^3 - cx^2$

◀ Graphing calculator examples, exercises, and subsections are identified with the graphing calculator icon.

◀ Families of curves can be studied using the graphing calculator, so that students can see how varying a parameter in an equation affects the graph.

3.1   EXERCISES

1–10 ■ Sketch the graph of the function by transforming the graph of an appropriate function of the form $y = x^n$. Indicate all x- and y-intercepts on each graph.

1. $P(x) = x^3 - 27$ 2. $P(x) = -x^3 + 8$
3. $P(x) = -(x + 2)^3$ 4. $P(x) = 2(x - 1)^3$
5. $P(x) = 2x^4 + 8$ 6. $P(x) = 3x^4 - 27$
7. $P(x) = -(x - 1)^4 + 16$ 8. $P(x) = (x + 2)^4 - 1$
9. $P(x) = 2(x + 3)^5 - 64$ 10. $P(x) = -\frac{1}{2}(x - 2)^5 + 16$

11–36 ■ Sketch the graph of the function. Make sure your graph shows all intercepts and exhibits the proper end behavior.

11. $P(x) = (x - 1)(x + 2)$
12. $P(x) = (x - 1)(x + 1)(x - 2)$
13. $P(x) = x(x - 3)(x + 2)$
14. $P(x) = (2x - 1)(x + 1)(x + 3)$
15. $P(x) = (x - 3)(x + 2)(3x - 2)$
16. $P(x) = \frac{1}{5}x(x - 5)^2$

17. $P(x) = (x - 1)^2(x - 3)$
18. $P(x) = \frac{1}{4}(x + 1)^3(x - 3)$
19. $P(x) = \frac{1}{12}(x + 2)^2(x - 3)^2$
20. $P(x) = (x - 1)^2(x + 2)^3$
21. $P(x) = x^3(x + 2)(x - 3)^2$ 22. $P(x) = (x^2 - 2x - 3)^2$
23. $P(x) = x^3 - x^2 - 6x$ 24. $P(x) = x^3 + 2x^2 - 8x$
25. $P(x) = -x^3 + x^2 + 12x$ 26. $P(x) = -2x^3 - x^2 + x$
27. $P(x) = x^4 - 3x^3 + 2x^2$ 28. $P(x) = x^5 - 9x^3$
29. $P(x) = x^3 + x^2 - x - 1$
30. $P(x) = x^3 + 3x^2 - 4x - 12$
31. $P(x) = 2x^3 - x^2 - 18x + 9$
32. $P(x) = \frac{1}{8}(2x^4 + 3x^3 - 16x - 24))^2$
33. $P(x) = x^4 - 2x^3 - 8x + 16$
34. $P(x) = x^4 - 2x^3 + 8x - 16$
35. $P(x) = x^4 - 3x^2 - 4$ 36. $P(x) = x^6 - 2x^3 + 1$

68   CHAPTER 1   Fundamentals

EXAMPLE 8 ■ Dividing a Lottery Jackpot

A group of people come forward to claim a $1,000,000 lottery jackpot, which the winners are to share equally. Before the jackpot is divided, three more winning ticket holders show up. As a result, each person's share is reduced by $75,000. How many winners were in the original group?

SOLUTION

We are asked for the number of people in the original group. So let

Identify the variable

$$x = \text{number of winners in the original group}$$

We translate the information in the problem as follows:

Express all unknown quantities in terms of the variable

In Words	In Algebra
Number of winners in original group	x
Number of winners in final group	$x + 3$
Winnings per person, originally	$\dfrac{1,000,000}{x}$
Winnings per person, finally	$\dfrac{1,000,000}{x + 3}$

Now we set up the model.

Set up the model

$$\boxed{\begin{array}{c}\text{winnings per}\\\text{person, originally}\end{array}} - \boxed{\$75,000} = \boxed{\begin{array}{c}\text{winnings per}\\\text{person, finally}\end{array}}$$

$$\frac{1,000,000}{x} - 75,000 = \frac{1,000,000}{x + 3}$$

Solve

$$1,000,000(x + 3) - 75,000x(x + 3) = 1,000,000x$$ Multiply by LCD $x(x + 3)$

$$40(x + 3) - 3x(x + 3) = 40x$$ Divide by 25,000

$$x^2 + 3x - 40 = 0$$ Expand, simplify, and divide by 3

$$(x + 8)(x - 5) = 0$$

$$x + 8 = 0 \quad \text{or} \quad x - 5 = 0$$

$$x = -8 \qquad x = 5$$

Since we can't have a negative number of people, we co[...] winners in the original group.

CHECK YOUR ANSWER

winnings per person, originally = $\dfrac{\$1,000,00}{5}$

winnings per person, finally = $\dfrac{\$1,000,00}{8}$

$$\$200,000 - \$75,000 = \$125,000$$

◀ Example titles clarify the purpose of examples.

◀ The process of translating a word problem into an algebraic model is clearly explained, step by step.

◀ Author notes provide step-by-step comments on solutions.

Check Your Answer emphasizes the importance of looking back at your work to check whether an answer is reasonable. ▲

Real-world applications show the relevance of precalculus to everyday life and indicate its remarkable problem-solving power. ▶

Discovery·Discussion exercises encourage students to experiment with, discuss, and write about the concepts they have learned. ▶

SECTION 1.5   Equations   **59**

of miles. How far from the center of the earth is the "dead spot" where no net gravitational force acts upon the object? (Express your answer to the nearest thousand miles.)

87. One method for determining the depth of a well is to drop a stone into it and then measure the time it takes until the splash is heard. If d is the depth of the well (in feet) and t_1 the time (in seconds) it takes for the stone to fall, then $d = 16t_1^2$, so $t_1 = \sqrt{d}/4$. Now if t_2 is the time it takes for the sound to travel back up, then $d = 1090t_2$ because the speed of sound is 1090 ft/s. So $t_2 = d/1090$. Thus, the total time elapsed between dropping the stone and hearing the splash is $t_1 + t_2 = \sqrt{d}/4 + d/1090$. How deep is the well if this total time is 3 s?

Time stone falls: $t_1 = \dfrac{\sqrt{d}}{4}$

Time sound rises: $t_2 = \dfrac{d}{1090}$

🟠 **DISCOVERY · DISCUSSION**

88. A Family of Equations The equation

$$3x + k - 5 = kx - k + 1$$

is really a **family of equations**, because for each value of k, we get a different equation with the unknown x. The letter k is called a **parameter** for this family. What value should

we pick for k to make the given value of x a solution of the resulting equation?
(a) $x = 0$ (b) $x = 1$ (c) $x = 2$

89. Proof That 0 = 1? The following steps appear to give equivalent equations, which seem to prove that 1 = 0. Find the error.

$x = 1$	Given
$x^2 = x$	Multiply by x
$x^2 - x = 0$	Subtract x
$x(x - 1) = 0$	Factor
$\dfrac{x(x - 1)}{x - 1} = \dfrac{0}{x - 1}$	Divide by $x - 1$
$x = 0$	Simplify
$1 = 0$	Given $x = 1$

90. Volumes of Solids The sphere, cylinder, and cone shown here all have the same radius r and the same volume V. Express the heights of the cylinder and cone in terms of r only, not V. (Use the volume formulas given on the inside front cover of this book.)

91. Relationship between Roots and Coefficients The quadratic formula gives us the roots of a quadratic equation from its coefficients. We can also obtain the coefficients from the roots. For example, find the roots of the equation $x^2 - 9x + 20 = 0$ and show that the product of the roots is the constant term 20 and the sum of the roots is 9, the negative of the coefficient of x. Show that the same relationship between roots and coefficients holds for the following equations:

$$x^2 - 2x - 8 = 0$$
$$x^2 + 4x + 2 = 0$$

Use the quadratic formula to prove that in general, if the equation $x^2 + bx + c = 0$ has roots r_1 and r_2, then $c = r_1r_2$ and $b = -(r_1 + r_2)$.

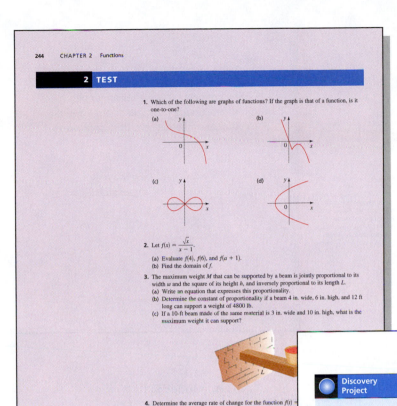

244   CHAPTER 2   Functions

2 TEST

1. Which of the following are graphs of functions? If the graph is that of a function, is it one-to-one?

    (a)   (b)

    (c)   (d)

2. Let $f(x) = \dfrac{\sqrt{x}}{x-1}$.

    (a) Evaluate $f(4)$, $f(6)$, and $f(a+1)$.
    (b) Find the domain of f.

3. The maximum weight M that can be supported by a beam is jointly proportional to its width w and the square of its height h, and inversely proportional to its length L.
    (a) Write an equation that expresses this proportionality.
    (b) Determine the constant of proportionality if a beam 4 in. wide, 6 in. high, and 12 ft long can support a weight of 4800 lb.
    (c) If a 10-ft beam made of the same material is 3 in. wide and 10 in. high, what is the maximum weight it can support?

4. Determine the average rate of change for the function $f(t) =$
    $t = 2 + h$.

5. (a) Sketch the graph of the function $f(x) = x^3$.
    (b) Use part (a) to graph the function $g(x) = (x-1)^3 - 2$.

◄ Each chapter ends with a review section containing a *Concept Check*, extensive review exercises, and a *Chapter Test*.

Discovery, Laboratory, Writing, and ▶
Applied Projects are designed to help students become active learners by providing them with an extended problem, often involving experimentation and a written report, that requires them to use their mathematics skills in a substantial fashion.

Discovery Project

Where to Sit at the Movies

Everyone knows that the apparent size of an object depends on its distance from the viewer. The farther away an object, the smaller its apparent size. The apparent size is determined by the angle the object subtends at the eye of the viewer.

If you are looking at a painting hanging on a wall, how far away should you stand to get the maximum view? If the painting is hung above eye level, then the following figures show that the angle subtended at the eye is small if you are too close or too far away. The same situation occurs when choosing where to sit in a movie theatre.

Small θ Large θ Small θ

1. The screen in a theater is 22 ft high and is positioned 10 ft above the floor, which is flat. The first row of seats is 7 ft from the screen and the rows are 3 ft apart. You decide to sit in the row where you get the maximum view, that is, where the angle θ subtended by the screen at your eyes is a maximum. Suppose your eyes are 4 ft above the floor, as in the figure, and you sit at a distance x from the screen.

    (a) Show that $\theta = \tan^{-1}\left(\dfrac{28}{x}\right) - \tan^{-1}\left(\dfrac{6}{x}\right)$.

    (b) Use the subtraction formula for tangent to show that
    $$\theta = \tan^{-1}\left(\frac{22x}{x^2 + 168}\right)$$

    (c) Use a graphing device to graph θ as a function of x. What value of x maximizes θ? In which row should you sit? What is the viewing angle in this row?

2. Now suppose that, starting with the first row of seats, the floor of the seating area is inclined at an angle of $\alpha = 25°$ above the horizontal, and the distance that you sit up the incline is x, as shown in the figure.

    (a) Use the Law of Cosines to show that
    $$\theta = \cos^{-1}\left(\frac{a^2 + b^2 - 484}{2ab}\right)$$

Focus on Problem Solving
Adding Something Extra

It is often necessary to combine different problem-solving principles. The princi-ples of **drawing a diagram** and **introducing something extra** are used in the two problems that follow. Euler's solution of the "Königsberg bridge problem" is so brilliant that it is the basis for a whole field of mathematics called *network theory*, which has practical applications to electric circuits and economics.

The Königsberg Bridge Problem

The old city of Königsberg, now called Kaliningrad, is situated on both banks of the Pregel River and on two islands in the river. Seven bridges connect the parts of the city, as shown in Figure 1. After the last bridge was built, the Königsberg townspeople posed the problem of finding a route through the city that would cross all seven bridges in a continuous walk, without recrossing any bridge. No one was able to find such a route. When Euler heard of the problem, he realized that an interesting mathematical principle must be at work. To discover the principle, Euler first stripped away the nonessential parts of the problem by drawing a simpler dia-gram of the city in which he represented the land masses as points and the bridges as lines connecting the points (Figure 2).

FIGURE 1

The problem now was to draw the diagram without lift... retracing any line. To show that this was impossible, Eul... extra by labeling each point with the number of line segm... (Figure 3). He then argued that every pass through a poi... ends (one entering and one leaving). Thus for the walk to... except possibly the starting and ending points, must be lab... Since more than two points are labeled by odd numbers, th...

The beauty of Euler's solution is that it is completely... guage a diagram consisting of points and connecting lines... point is a *vertex*, each line an *edge*, and the number of ed... the *order* of the vertex. A network is *traversable* if it can...

FIGURE 3

◀ *Focus on Problem Solving* sections include many problems that are related to the problem-solving principle discussed in the section or to the material of the preceding chapters.

Focus on Modeling sections show ▶ how precalculus mathematics is used to model important phenomena from the sciences and from everyday life.

Mathematical vignettes provide short ▶ biographies of interesting mathematicians, contemporary as well as historical, or describe applications of mathematics to the real world.

Focus on Modeling
The Path of a Projectile

Galileo Galilei (1564–1642) was born in Pisa, Italy. He studied med-icine, but later abandoned this in favor of science and mathematics. At the age of 25 he demonstrated that light objects fall at the same rate as heavier ones, by dropping cannonballs of various sizes from the Leaning Tower of Pisa. This contradicted the then-accepted view of Aristotle that heavier objects fall more quickly. He also showed that the distance an object falls is proportional to the square of the time it has been falling, and from this was able to prove that the path of a projectile is a parabola.
Galileo constructed the first tele-scope, and using it, discovered the moons of Jupiter. His advocacy of the Copernican view that the earth revolves around the sun (rather than being stationary) led to his being called before the Inquisition. By then an old man, he was forced to recant his views, but he is said to have muttered under his breath "the earth nevertheless does move." Galileo revolutionized science by expressing scientific principles in the language of mathematics. He said, "The great book of nature is written in mathe-matical symbols."

Modeling motion is one of the most important ideas in both classical and modern physics. Much of Isaac Newton's work dealt with creating a mathematical model for how objects move and interact—this was the main reason for his invention of calculus (see page 921). Albert Einstein developed his Special Theory of Relativity in the early 1900s to refine Newton's laws of motion.
In this section we use coordinate geometry to model the motion of a projectile, such as a ball thrown upward into the air, a bullet fired from a gun, or any other sort of missile. A similar model was created by Galileo (see the margin), but we have the advantage of using our modern mathematical notation to make describing the model much easier than it was for Galileo!
Suppose that we fire a projectile into the air from ground level, with an initial speed v_0 and at an angle θ upward from the ground. The initial *velocity* of the pro-jectile is a vector (see Section 7.7) with horizontal component $v_0 \cos \theta$ and vertical component $v_0 \sin \theta$, as shown in Figure 1.

FIGURE 1

If there were no gravity (and no air resistance), the projectile would just keep moving indefinitely at the same speed and in the same direction. Since distance = speed × time, the projectile's position at time t would therefore be given by the fol-lowing parametric equations (assuming the origin of our coordinate system is placed at the initial location of the projectile):

$$x = (v_0 \cos \theta)t \qquad y = (v_0 \sin \theta)t \qquad \text{No gravity}$$

But, of course, we know that gravity will pull the projectile back to ground level. Using calculus, it can be shown that the effect of gravity can be accounted for by subtracting $\frac{1}{2} gt^2$ from the vertical position of the projectile. In this ex-pression, g is the gravitational acceleration: $g \approx 32$ ft/s² ≈ 9.8 m/s².

Acknowledgments

We thank the following reviewers for their thoughtful and constructive comments.

Reviewers for the Third Edition

Edward Dixon, *Tennessee Technological University;* Richard Dodge, *Jackson Community College;* Floyd Downs, *Arizona State University at Tempe;* Marjorie Kreienbrink, *University of Wisconsin–Waukesha;* Donna Krichiver, *Johnson Community College;* Wayne Lewis, *University of Hawaii* and *Honolulu Community College;* Adam Lutoborski, *Syracuse University;* Keith Oberlander, *Pasadena City College;* Christine Panoff, *University of Michigan at Flint;* Susan Piliero, *Cornell University;* Gregory St. George, *University of Montana;* Gary Stoudt, *Indiana University of Pennsylvania;* Arnold Vobach, *University of Houston;* Tom Walsh, *City College of San Francisco;* Muserref Wiggins, *University of Akron;* Diane Williams, *Northern Kentucky University;* Suzette Wright, *University of California at Davis;* and Yisong Yang, *Polytech University*

Reviewers for the Fourth Edition

Michelle Benedict, *Augusta State University;* Linda Crawford, *Augusta State University;* Vivian G. Kostyk, *Inver Hills Community College;* Heather C. McGilvray, *Seattle University.*

We have benefited greatly from the suggestions and comments of our colleagues who have used our books in previous editions. We extend special thanks in this regard to Linda Byun, Bruce Chaderjian, David Gau, Daniel Hernandez, YongHee Kim-Park, Daniel Martinez, David McKay, Robert Mena, Kent Merryfield, Viet Ngo, Marilyn Oba, Robert Valentini, and Derming Wang, from California State University, Long Beach; to Karen Gold, Betsy Huttenlock, Cecilia McVoy, Mike McVoy, Samir Ouzomgi, and Ralph Rush, of the Pennsylvania State University, Abington College; to Gloria Dion, of Educational Testing Service, Princeton, New Jersey; and to Fred Safier, of the City College of San Francisco.

We especially thank Luana Richards, manager of our production service, for her excellent work and her tireless attention to quality and detail. Her energy, devotion, experience, and intelligence were essential components in the creation of this book. At TECH·arts we thank Brian Betsill for his elegant graphics. At Brooks/Cole, our thanks go to Editorial Production Supervisor Tom Novack, Assistant Editor Carol Ann Benedict, Editorial Assistant Molly Nance, Marketing Manager Karin Sandberg, Senior Advertising Project Manager Samantha Cabaluna, National Sales Manager Lucas R. Tomasso, and Art Director Vernon Boes. They have all done an outstanding job.

We are particularly grateful to our editor Bob Pirtle for guiding this book through every stage of writing and production. His support and editorial insight when crucial decisions had to be made were invaluable.

ANCILLARIES FOR PRECALCULUS, MATHEMATICS FOR CALCULUS, FOURTH EDITION

For the Instructor

PRINTED

Instructor's Solutions Manual
by John Banks, San Jose City College and Evergreen Valley College
0-534-38542-7

- Solutions to all even-numbered text exercises

Printed Test Items
by Andrew Bulman-Fleming
0-534-38546-X

- Text specific
- Contains six sample tests per chapter

TESTING SOFTWARE

BCA Testing
0-534-38547-8
Available on CD-ROM or over the Web, this text-specific software provides algorithmically generated problems for every section of the text, enabling you to create and print multiple versions of the same test. You can also post tests, quizzes, and practice problems for your students on your Web site or local campus network.

BCA Course Management
All results from BCA Testing and BCA Tutorial flow directly to the BCA Gradebook. This powerful function gives you the tools to create a Web course, automatically enroll a class, track student progress, and weight assignments as a portion of total grade. A feature that will enable direct communication with individual students, sections, or entire courses is coming soon.

Technical Support
- Toll-free technical support:
  (800) 423-0563 or e-mail: support@kdc.com

VIDEO

Text-Specific Video Tutorial Series
0-534-38543-5

- Provides additional examples and explanations for each chapter of the text

For the Student

PRINTED

Student Solutions Manual
by John Banks, San Jose City College and Evergreen Valley College
0-534-38544-3

- Solutions to all odd-numbered text exercises

Study Guide
by John Banks, San Jose City College and Evergreen Valley College
0-534-38545-1

- Detailed explanations
- Worked-out practice problems

TUTORIAL SOFTWARE

Interactive Video Skillbuilder CD-ROM

This effective and engaging CD-ROM contains over eight hours of video instruction. The problems worked during each video lesson are listed next to the viewing screen, so that students can try working them before watching the solutions. In order to help students evaluate their progress, each section contains a 10-question Web quiz (the results of which can be e-mailed to the instructor) and each chapter contains a chapter test, with answers to each problem on each test.

Technical Support
- Toll-free technical support:
  (800) 423-0563 or e-mail: support@kdc.com

RELATED PRODUCTS

Mastering Mathematics: How to Be a Great Math Student, 3/e © 1998
by Richard Manning Smith, Bryant College
0-534-34947-1

This practical guide will help you
- Avoid mental blocks during math exams
- Identify and improve areas of weakness
- Get the most out of class time
- Study more effectively
- Overcome a perceived low math ability
- Be successful on math tests
- Get back on track when feeling lost

To the Student

This textbook was written for you to use as a guide to mastering precalculus mathematics. Here are some suggestions to help you get the most out of your course.

First of all, you should read the appropriate section of text *before* you attempt your homework problems. Reading a mathematics text is quite different from reading a novel, a newspaper, or even another textbook. You may find that you have to reread a passage several times before you understand it. Pay special attention to the examples, and work them out yourself with pencil and paper as you read. With this kind of preparation you will be able to do your homework much more quickly and with more understanding.

Don't make the mistake of trying to memorize every single rule or fact you may come across. Mathematics doesn't consist simply of memorization. Mathematics is a *problem-solving art*, not just a collection of facts. To master the subject you must solve problems—lots of problems. Do as many of the exercises as you can. Be sure to write your solutions in a logical, step-by-step fashion. Don't give up on a problem if you can't solve it right away. Try to understand the problem more clearly—reread it thoughtfully and relate it to what you have learned from your teacher and from the examples in the text. Struggle with it until you solve it. Once you have done this a few times you will begin to understand what mathematics is really all about.

Answers to the odd-numbered exercises, as well as all the answers to each chapter test, appear at the back of the book. If your answer differs from the one given, don't immediately assume that you are wrong. There may be a calculation that connects the two answers and makes both correct. For example, if you get $1/(\sqrt{2} - 1)$ but the answer given is $1 + \sqrt{2}$, your answer *is* correct, because you can multiply both numerator and denominator of your answer by $\sqrt{2} + 1$ to change it to the given answer.

The symbol ⊘ is used to warn against committing an error. We have placed this symbol in the margin to point out situations where we have found that many of our students make the same mistake.

The Interactive Video Skillbuilder CD-ROM bound into the cover of this book contains video instruction designed to further help you understand the material of this course. The symbol ▣ points to topics for which additional examples and explanations can be found on the CD-ROM.

Calculators and Calculations

Calculators are essential in most mathematics and science subjects. They free us from performing routine tasks, so we can focus more clearly on the concepts we are studying. Calculators are powerful tools but their results need to be interpreted with care. In what follows, we describe the features that a calculator suitable for a pre-calculus course should have, and we give guidelines for interpreting the results of its calculations.

■ Scientific and Graphing Calculators

For this course you will need a *scientific* calculator—one that has, as a minimum, the usual arithmetic operations ($+$, $-$, $\times$, $\div$) and exponential, logarithmic, and trigonometric functions (e^x, 10^x, $\ln x$, $\log$, $\sin$, $\cos$, $\tan$). In addition, a memory and at least some degree of programmability will be useful.

Your instructor may recommend or require that you purchase a *graphing* calculator. This book has optional subsections and exercises that require the use of a graphing calculator or a computer with graphing software. These special subsections and exercises are indicated by the symbol ▰. Besides graphing functions, graphing calculators can also be used to find functions that model real-life data, solve equations, perform matrix calculations (which are studied in Chapter 8), and help you perform other mathematical operations. All these uses are discussed in this book.

It is important to realize that, because of limited resolution, a graphing calculator gives only an *approximation* to the graph of a function. It plots only a finite number of points and then connects them to form a *representation* of the graph. In Section 1.9, we give guidelines for using a graphing calculator and interpreting the graphs that it produces.

■ Calculations and Significant Figures

Most of the applied examples and exercises in this book involve approximate values. For example, one exercise states that the moon has a radius of 1074 miles. This does not mean that the moon's radius is exactly 1074 miles but simply that this is the radius rounded to the nearest mile.

One simple method for specifying the accuracy of a number is to state how many **significant digits** it has. The significant digits in a number are the ones from the first nonzero digit to the last nonzero digit (reading from left to right). Thus, 1074 has four significant digits, 1070 has three, 1100 has two, and 1000 has one significant digit. This rule may sometimes lead to ambiguities. For example, if a distance is 200 km to the nearest kilometer, then the number 200 really has three significant digits, not just one. This ambiguity is avoided if we use scientific notation—that is, if we express the number as a multiple of a power of 10:

$$2.00 \times 10^2$$

When working with approximate values, students often make the mistake of giving a final answer with *more* significant digits than the original data. This is incorrect because you cannot "create" precision by using a calculator. The final result can be no more accurate than the measurements given in the problem. For example, suppose we are told that the two shorter sides of a right triangle are measured to be 1.25 and 2.33 inches long. By the Pythagorean Theorem, we find, using a calculator, that the hypotenuse has length

$$\sqrt{1.25^2 + 2.33^2} \approx 2.644125564 \text{ in.}$$

But since the given lengths were expressed to three significant digits, the answer cannot be any more accurate. We can therefore say only that the hypotenuse is 2.64 in. long, rounding to the nearest hundredth.

In general, the final answer should be expressed with the same accuracy as the *least*-accurate measurement given in the statement of the problem. The following rules make this principle more precise.

RULES FOR WORKING WITH APPROXIMATE DATA

1. When multiplying or dividing, round off the final result so that it has as many *significant digits* as the given value with the fewest number of significant digits.

2. When adding or subtracting, round off the final result so that it has its last significant digit in the *decimal place* in which the least-accurate given value has its last significant digit.

3. When taking powers or roots, round off the final result so that it has the same number of *significant digits* as the given value.

As an example, suppose that a rectangular table top is measured to be 122.64 in. by 37.3 in. We express its area and perimeter as follows:

Area = length × width = 122.64 × 37.3 ≈ 4570 in^2 *Three significant digits*

Perimeter = 2(length + width) = 2(122.64 + 37.3) ≈ 319.9 in.   *Tenths digit*

⊘ Note that in the formula for the perimeter, the value 2 is an exact value, not an approximate measurement. It therefore does not affect the accuracy of the final result. In general, if a problem involves only exact values, we may express the final answer with as many significant digits as we wish.

Note also that to make the final result as accurate as possible, *you should wait until the last step to round off your answer.* If necessary, use the memory feature of your calculator to retain the results of intermediate calculations.

Abbreviations

cm	centimeter	MHz	megahertz
dB	decibel	mi	mile
F	farad	min	minute
ft	foot	mL	milliliter
g	gram	mm	millimeter
gal	gallon	N	Newton
h	hour	qt	quart
H	henry	oz	ounce
Hz	Hertz	s	second
in.	inch	Ω	ohm
J	Joule	V	volt
kcal	kilocalorie	W	watt
kg	kilogram	yd	yard
km	kilometer	yr	year
kPa	kilopascal	°C	degree Celsius
L	liter	°F	degree Fahrenheit
lb	pound	K	Kelvin
M	mole of solute per liter of solution	⇒	implies
m	meter	⇔	is equivalent to

Mathematical Vignettes

MATHEMATICS IN THE MODERN WORLD

Contents

9   Topics in Analytic Geometry   722

10   Sequences and Series   806

Counting and Probability   866

Limits: A Preview of Calculus   906

Precalculus
Mathematics for Calculus

Fourth Edition

1 Fundamentals

Algebra has been an essential element of scientific progress in every age, from the construction of the pyramids to the development of the space shuttle.

In this first chapter we review the basic ideas from algebra and coordinate geometry that will be needed throughout this book.

1.1 REAL NUMBERS

The different types of real numbers were invented to meet specific needs. For example, natural numbers are needed for counting, negative numbers for describing debt or below-zero temperatures, rational numbers for concepts like "half a gallon of milk," and irrational numbers for measuring certain distances, like the diagonal of a square.

Let's recall the types of numbers that make up the real number system. We start with the **natural numbers**:

$$1, 2, 3, 4, \ldots$$

The **integers** consist of the natural numbers together with their negatives and 0:

$$\ldots, -3, -2, -1, 0, 1, 2, 3, 4, \ldots$$

We construct the **rational numbers** by taking ratios of integers. Thus, any rational number r can be expressed as

$$r = \frac{m}{n} \qquad \text{where } m \text{ and } n \text{ are integers and } n \neq 0$$

Examples are: $\quad \frac{1}{2} \qquad -\frac{3}{7} \qquad 46 = \frac{46}{1} \qquad 0.17 = \frac{17}{100}$

(Recall that division by 0 is always ruled out, so expressions like $\frac{3}{0}$ and $\frac{0}{0}$ are undefined.) There are also real numbers, such as $\sqrt{2}$, that cannot be expressed as a ratio of integers and are therefore called **irrational numbers**. It can be shown, with varying degrees of difficulty, that these numbers are also irrational:

$$\sqrt{3} \qquad \sqrt{5} \qquad \sqrt[3]{2} \qquad \pi \qquad \frac{3}{\pi^2}$$

The set of all real numbers is usually denoted by the symbol $\mathbb{R}$. When we use the word *number* without qualification, we will mean "real number." Figure 1 is a diagram of the types of real numbers that we work with in this book.

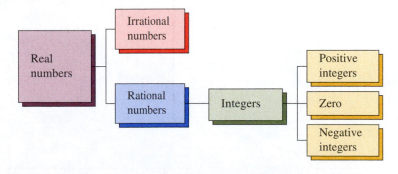

FIGURE 1

The real number system

Every real number has a decimal representation. If the number is rational, then its corresponding decimal is repeating. For example,

$$\frac{1}{2} = 0.5000\ldots = 0.5\overline{0} \qquad\qquad \frac{2}{3} = 0.66666\ldots = 0.\overline{6}$$

$$\frac{157}{495} = 0.3171717\ldots = 0.3\overline{17} \qquad \frac{9}{7} = 1.285714285714\ldots = 1.\overline{285714}$$

A repeating decimal such as

$$x = 3.5474747\ldots$$

is a rational number. To convert it to a ratio of two integers, we write

$$1000x = 3547.47474747\ldots$$

$$\underline{10x = \quad 35.47474747\ldots}$$

$$990x = 3512.0$$

Thus, $x = \frac{3512}{990}$. (The idea is to multiply x by appropriate powers of 10, and then subtract to eliminate the repeating part.)

(The bar indicates that the sequence of digits repeats forever.) If the number is irrational, the decimal representation is nonrepeating:

$$\sqrt{2} = 1.414213562373095\ldots \qquad \pi = 3.141592653589793\ldots$$

If we stop the decimal expansion of any number at a certain place, we get an approximation to the number. For instance, we can write

$$\pi \approx 3.14159265$$

where the symbol $\approx$ is read "is approximately equal to." The more decimal places we retain, the better the approximation we get.

■ Properties of Real Numbers

In combining real numbers using the familiar operations of addition and multiplication, we use the following properties of real numbers.

PROPERTIES OF REAL NUMBERS

Property	Example	Description
Commutative Properties		
$a + b = b + a$	$7 + 3 = 3 + 7$	When we add two numbers, order doesn't matter.
$ab = ba$	$3 \cdot 5 = 5 \cdot 3$	When we multiply two numbers, order doesn't matter.
Associative Properties		
$(a + b) + c = a + (b + c)$	$(2 + 4) + 7 = 2 + (4 + 7)$	When we add three numbers, it doesn't matter which two we add first.
$(ab)c = a(bc)$	$(3 \cdot 7) \cdot 5 = 3 \cdot (7 \cdot 5)$	When we multiply three numbers, it doesn't matter which two we multiply first.
Distributive Property		
$a(b + c) = ab + ac$	$2 \cdot (3 + 5) = 2 \cdot 3 + 2 \cdot 5$	When we multiply a number by a sum of two numbers, we get the same result as multiplying the number by each of the terms and then adding the results.
$(b + c)a = ab + ac$	$(3 + 5) \cdot 2 = 2 \cdot 3 + 2 \cdot 5$	

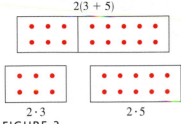

2(3 + 5)

2·3   2·5

FIGURE 2

The Distributive Property

The Distributive Property is crucial because it describes the way addition and multiplication interact with each other.

From our experience with numbers we know intuitively that these properties are valid. To reacquaint yourself with these properties, calculate the expressions on both sides of the equal sign in each example in the table. The operations in the parentheses are to be performed first. Thus,

$$(2 + 4) + 7 = 6 + 7 = 13 \quad \text{and} \quad 2 + (4 + 7) = 2 + 11 = 13$$

The Distributive Property applies whenever we multiply a number by a sum. Figure 2 explains why this property works for the case in which all the numbers are positive integers, but the property is true for any real numbers a, b, and c.

EXAMPLE 1 ■ Using the Properties of the Real Numbers

Let x, y, z, and w be real numbers.

(a) $(x + y)(2zw) = (2zw)(x + y)$ Commutative Property for multiplication

(b) $(x + y)(z + w) = (x + y)z + (x + y)w$ Distributive Property (with $a = x + y$)

$$= (zx + zy) + (wx + wy) \quad \text{\color{red}Distributive Property}$$

$$= zx + zy + wx + wy \quad \text{\color{red}Associative Property of addition}$$

In the last step we removed the parentheses because, according to the Associative Property, the order of addition doesn't matter.   ■

The number 0 is special for addition; it is called the **additive identity** because $a + 0 = a$ for any real number a. Every real number a has a **negative**, $-a$, that satisfies $a + (-a) = 0$. **Subtraction** is the operation that undoes addition; to subtract a number from another, we simply add the negative of that number. By definition

$$a - b = a + (-b)$$

To combine real numbers involving negatives, we use the following properties.

🚫 Don't assume that $-a$ is a negative number. Whether $-a$ is negative or positive depends on the value of a. For example, if $a = 5$, then $-a = -5$, a negative number, but if $a = -5$, then $-a = -(-5) = 5$ (Property 2), a positive number.

PROPERTIES OF NEGATIVES	
Property	**Example**
1. $(-1)a = -a$	$(-1)5 = -5$
2. $-(-a) = a$	$-(-5) = 5$
3. $(-a)b = a(-b) = -(ab)$	$(-5)7 = 5(-7) = -(5 \cdot 7)$
4. $(-a)(-b) = ab$	$(-4)(-3) = 4 \cdot 3$
5. $-(a + b) = -a - b$	$-(3 + 5) = -3 - 5$
6. $-(a - b) = b - a$	$-(5 - 8) = 8 - 5$

Property 6 states the intuitive fact that $a - b$ and $b - a$ are negatives of each other. Property 5 is often used with more than two terms:

$$-(a + b + c) = -a - b - c$$

The word *algebra* comes from the ninth century Arabic book *Hisâb al-Jabr w'al-Muqabala*, written by al-Khowarizmi. The title refers to transposing and combining terms, two processes used in solving equations. In Latin translations the title was shortened to *Aljabr*, from which we get the word *algebra*. The author's name itself made its way into the English language in the form of our word *algorithm*.

EXAMPLE 2 ■ **Using Properties of Negatives**

Let x, y, and z be real numbers.

(a) $-(x + 2) = -x - 2$ Property 5

(b) $-(x + y - z) = -x - y - (-z)$ Property 5

$\qquad\qquad\quad = -x - y + z$ Property 2 ■

The number 1 is special for multiplication; it is called the **multiplicative identity** because $a \cdot 1 = a$ for any real number a. Every nonzero real number a has an **inverse**, $1/a$, that satisfies $a \cdot (1/a) = 1$. **Division** is the operation that undoes multiplication; to divide by a number, we multiply by the inverse of that number. If $b \neq 0$, then, by definition.

$$a \div b = a \cdot \frac{1}{b}$$

We write $a \cdot (1/b)$ as simply a/b. We refer to a/b as the **quotient** of a and b or as the **fraction** a over b; a is the **numerator** and b is the **denominator** (or **divisor**). To combine real numbers using the operation of division, we use the following properties.

PROPERTIES OF FRACTIONS

Property	Example	Description
1. $\dfrac{a}{b} \cdot \dfrac{c}{d} = \dfrac{ac}{bd}$	$\dfrac{2}{3} \cdot \dfrac{5}{7} = \dfrac{2 \cdot 5}{3 \cdot 7} = \dfrac{10}{21}$	To multiply fractions, multiply numerators and denominators.
2. $\dfrac{a}{b} \div \dfrac{c}{d} = \dfrac{a}{b} \cdot \dfrac{d}{c}$	$\dfrac{2}{3} \div \dfrac{5}{7} = \dfrac{2}{3} \cdot \dfrac{7}{5} = \dfrac{14}{15}$	To divide fractions, invert the divisor and multiply.
3. $\dfrac{a}{c} + \dfrac{b}{c} = \dfrac{a + b}{c}$	$\dfrac{2}{5} + \dfrac{7}{5} = \dfrac{2 + 7}{5} = \dfrac{9}{5}$	To add fractions with the same denominator, add the numerators.
4. $\dfrac{a}{b} + \dfrac{c}{d} = \dfrac{ad + bc}{bd}$	$\dfrac{2}{5} + \dfrac{3}{7} = \dfrac{2 \cdot 7 + 3 \cdot 5}{35} = \dfrac{29}{35}$	To add fractions with different denominators, find a common denominator. Then add the numerators.
5. $\dfrac{ac}{bc} = \dfrac{a}{b}$	$\dfrac{2 \cdot 5}{3 \cdot 5} = \dfrac{2}{3}$	Cancel numbers that are common factors in numerator and denominator.
6. If $\dfrac{a}{b} = \dfrac{c}{d}$, then $ad = bc$.	$\dfrac{2}{3} = \dfrac{6}{9}$, so $2 \cdot 9 = 3 \cdot 6$	Cross multiply.

When adding fractions with different denominators, we don't usually use Property 4. Instead we rewrite the fractions so that they have the smallest possible common denominator (often smaller than the product of the denominators), and then we use Property 3. This denominator is the **Least Common Denominator (LCD)** described in the next example.

EXAMPLE 3 ■ Using the LCD to Add Fractions

Evaluate: $\dfrac{5}{36} + \dfrac{7}{120}$

SOLUTION

Factoring each denominator into prime factors gives

$$36 = 2^2 \cdot 3^2 \qquad \text{and} \qquad 120 = 2^3 \cdot 3 \cdot 5$$

We find the least common denominator (LCD) by forming the product of all the factors that occur in these factorizations, using the highest power of each factor. Thus, the LCD is $2^3 \cdot 3^2 \cdot 5 = 360$. So

$$\frac{5}{36} + \frac{7}{120} = \frac{5 \cdot 10}{36 \cdot 10} + \frac{7 \cdot 3}{120 \cdot 3} \qquad \textcolor{red}{\text{Property 5}}$$

$$= \frac{50}{360} + \frac{21}{360}$$

$$= \frac{71}{360} \qquad \textcolor{red}{\text{Property 3}} \qquad ■$$

The Real Line

The real numbers can be represented by points on a line, as shown in Figure 3. The positive direction (toward the right) is indicated by an arrow. We choose an arbitrary reference point O, called the **origin**, which corresponds to the real number 0. Given any convenient unit of measurement, each positive number x is represented by the point on the line a distance of x units to the right of the origin, and each negative number $-x$ is represented by the point x units to the left of the origin. Thus, every real number is represented by a point on the line, and every point P on the line corresponds to exactly one real number. The number associated with the point P is called the coordinate of P, and the line is then called a **coordinate line**, or a **real number line**, or simply a **real line**. Often we identify the point with its coordinate and think of a number as being a point on the real line.

FIGURE 3
The real line

The real numbers are *ordered*. We say that ***a* is less than *b*** and write $a < b$ if $b - a$ is a positive number. Geometrically, this means that a lies to the left of b on the number line. (Equivalently, we can say that ***b* is greater than *a*** and write $b > a$.) The symbol $a \leq b$ (or $b \geq a$) means that either $a < b$ or $a = b$ and is read "*a* is less than or equal to *b*." For instance, the following are true inequalities (see Figure 4):

$$7 < 7.4 < 7.5 \qquad -\pi < -3 \qquad \sqrt{2} < 2 \qquad 2 \leq 2$$

FIGURE 4

Sets and Intervals

In the discussion that follows, we need to use set notation. A **set** is a collection of objects, and these objects are called the **elements** of the set. If S is a set, the notation $a \in S$ means that a is an element of S, and $b \notin S$ means that b is not an element of S. For example, if Z represents the set of integers, then $-3 \in Z$ but $\pi \notin Z$.

Some sets can be described by listing their elements within braces. For instance, the set A that consists of all positive integers less than 7 can be written as

$$A = \{1, 2, 3, 4, 5, 6\}$$

We could also write A in **set-builder notation** as

$$A = \{x \mid x \text{ is an integer and } 0 < x < 7\}$$

which is read "A is the set of all x such that x is an integer and $0 < x < 7$."

If S and T are sets, then their **union** $S \cup T$ is the set that consists of all elements that are in S *or* T (or in both). The **intersection** of S and T is the set $S \cap T$ consisting of all elements that are in both S *and* T. In other words, $S \cap T$ is the common part of S and T. The **empty set**, denoted by $\varnothing$, is the set that contains no element.

EXAMPLE 4 ■ Union and Intersection of Sets

If $S = \{1, 2, 3, 4, 5\}$, $T = \{4, 5, 6, 7\}$, and $V = \{6, 7, 8\}$, find the sets $S \cup T$, $S \cap T$, and $S \cap V$.

SOLUTION

$$S \cup T = \{1, 2, 3, 4, 5, 6, 7\} \qquad \text{All elements in } S \text{ or } T$$

$$S \cap T = \{4, 5\} \qquad \text{Elements common to both } S \text{ and } T$$

$$S \cap V = \varnothing \qquad \text{S and V have no element in common} \qquad ■$$

Certain sets of real numbers, called **intervals**, occur frequently in calculus and correspond geometrically to line segments. For example, if $a < b$, then the **open interval** from a to b consists of all numbers between a and b and is denoted by the symbol (a, b). Using set-builder notation, we can write

$$(a, b) = \{x \mid a < x < b\}$$

FIGURE 5
The open interval (a, b)

FIGURE 6
The closed interval $[a, b]$

Note that the endpoints, a and b, are excluded from this interval. This fact is indicated by the parentheses ( ) in the interval notation and the open circles on the graph of the interval in Figure 5.

The **closed interval** from a to b is the set

$$[a, b] = \{x \mid a \le x \le b\}$$

Here the endpoints of the interval are included. This is indicated by the square brackets [ ] in the interval notation and the solid circles on the graph of the interval in Figure 6. It is also possible to include only one endpoint in an interval, as shown in the table of intervals below.

We also need to consider infinite intervals, such as

$$(a, \infty) = \{x \mid x > a\}$$

This does not mean that ∞ ("infinity") is a number. The notation (a, ∞) stands for the set of all numbers that are greater than a, so the symbol ∞ simply indicates that the interval extends indefinitely far in the positive direction.

The following table lists the nine possible types of intervals. When these intervals are discussed, we will always assume that $a < b$.

Notation	Set description	Graph
(a, b)	$\{x \mid a < x < b\}$	
$[a, b]$	$\{x \mid a \le x \le b\}$	
$[a, b)$	$\{x \mid a \le x < b\}$	
$(a, b]$	$\{x \mid a < x \le b\}$	
(a, ∞)	$\{x \mid a < x\}$	
$[a, \infty)$	$\{x \mid a \le x\}$	
$(-\infty, b)$	$\{x \mid x < b\}$	
$(-\infty, b]$	$\{x \mid x \le b\}$	
$(-\infty, \infty)$	$\mathbb{R}$ (set of all real numbers)	

EXAMPLE 5 ■ Graphing Intervals

Express each interval in terms of inequalities, and then graph the interval.

(a) $[-1, 2) = \{x \mid -1 \le x < 2\}$

(b) $[1.5, 4] = \{x \mid 1.5 \le x \le 4\}$

(c) $(-3, \infty) = \{x \mid -3 < x\}$

EXAMPLE 6 ■ Finding Unions and Intersections of Intervals

Graph each set.

(a) $(1, 3) \cap [2, 7]$ \hspace{3cm} (b) $(1, 3) \cup [2, 7]$

SOLUTION

(a) The intersection of two intervals consists of the numbers that are in both intervals. Therefore

$$(1, 3) \cap [2, 7] = \{x \mid 1 < x < 3 \text{ and } 2 \leq x \leq 7\}$$
$$= \{x \mid 2 \leq x < 3\}$$
$$= [2, 3)$$

This set is illustrated in Figure 7.

(b) The union of two intervals consists of the numbers that are in either one interval or the other (or both). Therefore

$$(1, 3) \cup [2, 7] = \{x \mid 1 < x < 3 \text{ or } 2 \leq x \leq 7\}$$
$$= \{x \mid 1 < x \leq 7\}$$
$$= (1, 7]$$

This set is illustrated in Figure 8.

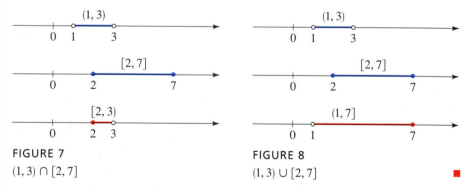

FIGURE 7
$(1, 3) \cap [2, 7]$

FIGURE 8
$(1, 3) \cup [2, 7]$ ■

Absolute Value and Distance

The **absolute value** of a number a, denoted by $|a|$, is the distance from a to 0 on
the real number line (see Figure 9). Distance is always positive or zero, so we have
$|a| \geq 0$ for every number a. Remembering that $-a$ is positive when a is negative,
we have the following definition.

FIGURE 9

DEFINITION OF ABSOLUTE VALUE
If a is a real number, then the **absolute value** of a is $$|a| = \begin{cases} a & \text{if } a \geq 0 \\ -a & \text{if } a < 0 \end{cases}$$

EXAMPLE 7 ■ Evaluating Absolute Values of Numbers

(a) $|3| = 3$

(b) $|-3| = -(-3) = 3$

(c) $|0| = 0$

(d) $|\sqrt{2} - 1| = \sqrt{2} - 1$ (since $\sqrt{2} > 1 \Rightarrow \sqrt{2} - 1 > 0$)

(e) $|3 - \pi| = -(3 - \pi) = \pi - 3$ (since $\pi > 3 \Rightarrow 3 - \pi < 0$)   ■

When working with absolute values, we use the following properties.

PROPERTIES OF ABSOLUTE VALUE

Property	Example	Description
1. $\|a\| \geq 0$	$\|-3\| = 3 \geq 0$	The absolute value of a number is always positive or zero.
2. $\|a\| = \|-a\|$	$\|5\| = \|-5\|$	A number and its negative have the same absolute value.
3. $\|ab\| = \|a\|\|b\|$	$\|-2 \cdot 5\| = \|-2\|\|5\|$	The absolute value of a product is the product of the absolute values.
4. $\left\|\dfrac{a}{b}\right\| = \dfrac{\|a\|}{\|b\|}$	$\left\|\dfrac{12}{-3}\right\| = \dfrac{\|12\|}{\|-3\|}$	The absolute value of a quotient is the quotient of the absolute values.

FIGURE 10

What is the distance on the real line between the numbers -2 and 11? From Figure 10 we see that the distance is 13. We arrive at this by finding either $|11 - (-2)| = 13$ or $|(-2) - 11| = 13$. From this observation we make the following definition (see Figure 11).

DISTANCE BETWEEN POINTS ON THE REAL LINE

If a and b are real numbers, then the **distance** between the points a and b on the real line is

$$d(a, b) = |b - a|$$

FIGURE 11
Length of a line segment $= |b - a|$

From Property 6 of negatives it follows that $|b - a| = |a - b|$. This confirms that, as we would expect, the distance from a to b is the same as the distance from b to a.

EXAMPLE 8 ■ Distance between Points on the Real Line

The distance between the numbers -8 and 2 is

$$d(a, b) = |-8 - 2| = |-10| = 10$$

FIGURE 12

We can check this calculation geometrically, as shown in Figure 12.   ■

1.1 EXERCISES

1–6 ■ State the property of real numbers being used.

1. $3x + 4y = 4y + 3x$

2. $c(a + b) = (a + b)c$

3. $(x + 2y) + 3z = x + (2y + 3z)$

4. $2(A + B) = 2A + 2B$

5. $(5x + 1)3 = 15x + 3$

6. $(x + a)(x + b) = (x + a)x + (x + a)b$

7–12 ■ Use properties of real numbers to write the expression without parentheses.

7. $3(x + y)$

8. $(a - b)8$

9. $4(2m)$

10. $\frac{4}{3}(-6y)$

11. $-\frac{5}{2}(2x - 4y)$

12. $(3a)(b + c - 2d)$

13–16 ■ Perform the indicated operations.

13. (a) $\dfrac{4}{13} + \dfrac{3}{13}$

(b) $\dfrac{3}{10} + \dfrac{7}{15}$

14. (a) $\dfrac{7}{45} + \dfrac{2}{25}$

(b) $\dfrac{5}{14} - \dfrac{1}{21} + 1$

15. (a) $\dfrac{2}{5} \div \dfrac{9}{10}$

(b) $\left(4 \div \dfrac{1}{2}\right) - \dfrac{1}{2}$

16. (a) $\left(\dfrac{1}{8} - \dfrac{1}{9}\right) \div \dfrac{1}{72}$

(b) $\left(2 \div \dfrac{2}{3}\right) - \left(\dfrac{2}{3} \div 2\right)$

17–20 ■ State whether each inequality is true or false.

17. (a) $-6 < -10$

(b) $\sqrt{2} > 1.41$

18. (a) $\dfrac{10}{11} < \dfrac{12}{13}$

(b) $-\dfrac{1}{2} < -1$

19. (a) $-\pi > -3$

(b) $8 \le 9$

20. (a) $1.1 > 1.\overline{1}$

(b) $8 \le 8$

21–22 ■ Write each statement in terms of inequalities.

21. (a) x is positive
(b) t is less than 4
(c) a is greater than or equal to π
(d) x is less than $\frac{1}{3}$ and is greater than -5
(e) The distance from p to 3 is at most 5

22. (a) y is negative
(b) z is greater than 1

(c) b is at most 8
(d) w is positive and is less than or equal to 17
(e) y is at least 2 units from π

23–26 ■ Find the indicated set if

$$A = \{1, 2, 3, 4, 5, 6, 7\} \qquad B = \{2, 4, 6, 8\}$$

$$C = \{7, 8, 9, 10\}$$

23. (a) $A \cup B$

(b) $A \cap B$

24. (a) $B \cup C$

(b) $B \cap C$

25. (a) $A \cup C$

(b) $A \cap C$

26. (a) $A \cup B \cup C$

(b) $A \cap B \cap C$

27–28 ■ Find the indicated set if

$$A = \{x \mid x \ge -2\} \qquad B = \{x \mid x < 4\}$$

$$C = \{x \mid -1 < x \le 5\}$$

27. (a) $B \cup C$

(b) $B \cap C$

28. (a) $A \cap C$

(b) $A \cap B$

29–34 ■ Express the interval in terms of inequalities, and then graph the interval.

29. $(-3, 0)$

30. $(2, 8]$

31. $[2, 8)$

32. $[-6, -\frac{1}{2}]$

33. $[2, \infty)$

34. $(-\infty, 1)$

35–40 ■ Express the inequality in interval notation, and then graph the corresponding interval.

35. $x \le 1$

36. $1 \le x \le 2$

37. $-2 < x \le 1$

38. $x \ge -5$

39. $x > -1$

40. $-5 < x < 2$

41–46 ■ Graph the set.

41. $(-2, 0) \cup (-1, 1)$

42. $(-2, 0) \cap (-1, 1)$

43. $[-4, 6] \cap [0, 8)$

44. $[-4, 6) \cup [0, 8)$

45. $(-\infty, -4) \cup (4, \infty)$

46. $(-\infty, 6] \cap (2, 10)$

47–52 ■ Evaluate each expression.

47. (a) $|100|$ (b) $|-73|$

48. (a) $|\sqrt{5} - 5|$ (b) $|10 - \pi|$

49. (a) $||-6| - |-4||$ (b) $\dfrac{-1}{|-1|}$

50. (a) $|2 - |-12||$ (b) $-1 - |1 - |-1||$

51. (a) $|(-2) \cdot 6|$ (b) $\left|\left(-\tfrac{1}{3}\right)(-15)\right|$

52. (a) $\left|\dfrac{-6}{24}\right|$ (b) $\left|\dfrac{7 - 12}{12 - 7}\right|$

53–54 ■ Find the distance between the given numbers.

53. (a) 2 and 17
   (b) -3 and 21
   (c) $\tfrac{11}{8}$ and $-\tfrac{3}{10}$

54. (a) $\tfrac{7}{15}$ and $-\tfrac{1}{21}$
   (b) -38 and -57
   (c) -2.6 and -1.8

55–56 ■ Express each repeating decimal as a fraction. (See the margin note on page 4.)

55. (a) $0.\overline{7}$ (b) $0.2\overline{8}$ (c) $0.\overline{57}$

56. (a) $5.\overline{23}$ (b) $1.3\overline{7}$ (c) $2.1\overline{35}$

57. Suppose an automobile's fuel consumption is 28 mi/gal in city driving and 34 mi/gal in highway driving. If x denotes the number of city miles and y the number of highway miles, then the total miles this car can travel on a 15-gallon tank of fuel must satisfy the inequality

$$\tfrac{1}{28}x + \tfrac{1}{34}y \le 15$$

Use this inequality to answer the following questions. (Assume the car has a full tank of fuel.)
(a) Can the car travel 165 city miles and 230 highway miles before running out of gas?
(b) If the car has been driven 280 miles in the city, how many highway miles can it be driven before running out of fuel?

58. Sums and Products of Rational and Irrational Numbers Explain why the sum, the difference, and the product of two rational numbers are rational numbers. Is the product of two irrational numbers necessarily irrational? What about the sum?

59. Combining Rational Numbers with Irrational Numbers Is $\tfrac{1}{2} + \sqrt{2}$ rational or irrational? Is $\tfrac{1}{2} \cdot \sqrt{2}$ rational or irrational? In general, what can you say about the sum of a rational and an irrational number? What about the product?

60. Limiting Behavior of Reciprocals Complete the tables. What happens to the size of the fraction $1/x$ as x gets large? As x gets small?

x	$1/x$	x	$1/x$
1		1.0	
2		0.5	
10		0.1	
100		0.01	
1000		0.001	

61. Irrational Numbers and Geometry Using the following figure, explain how to locate the point $\sqrt{2}$ on a number line. Can you locate $\sqrt{5}$ by a similar method? What about $\sqrt{6}$? List some other irrational numbers that can be located this way.

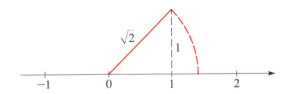

1.2 EXPONENTS AND RADICALS

In this section we give meaning to expressions such as $a^{m/n}$ in which the exponent m/n is a rational number. To do this, we need to recall some facts about integer exponents, radicals, and nth roots.

Integer Exponents

A product of identical numbers is usually written in exponential notation. For example, $5 \cdot 5 \cdot 5$ is written as 5^3. In general, we have the following definition.

EXPONENTIAL NOTATION

If a is any real number and n is a positive integer, then the **nth power** of a is

$$a^n = \underbrace{a \cdot a \cdot \cdots \cdot a}_{n \text{ factors}}$$

The number a is called the **base** and n is called the **exponent**.

EXAMPLE 1 ■ Exponential Notation

(a) $\left(\frac{1}{2}\right)^5 = \left(\frac{1}{2}\right)\left(\frac{1}{2}\right)\left(\frac{1}{2}\right)\left(\frac{1}{2}\right)\left(\frac{1}{2}\right) = \frac{1}{32}$

(b) $(-3)^4 = (-3) \cdot (-3) \cdot (-3) \cdot (-3) = 81$

(c) $-3^4 = -(3 \cdot 3 \cdot 3 \cdot 3) = -81$ ■

> Note the distinction between $(-3)^4$ and -3^4. In $(-3)^4$ the exponent applies to -3, but in -3^4 the exponent applies only to 3.

We can state several useful rules for working with exponential notation. To discover the rule for multiplication, we multiply 5^4 by 5^2:

$$5^4 \cdot 5^2 = \underbrace{(5 \cdot 5 \cdot 5 \cdot 5)}_{4 \text{ factors}}\underbrace{(5 \cdot 5)}_{2 \text{ factors}} = \underbrace{5 \cdot 5 \cdot 5 \cdot 5 \cdot 5 \cdot 5}_{6 \text{ factors}} = 5^6 = 5^{4+2}$$

It appears that *to multiply two powers of the same base, we add their exponents*. In general, we can confirm this by considering any real number a and any positive integers m and n:

$$a^m a^n = \underbrace{(a \cdot a \cdot \cdots \cdot a)}_{m \text{ factors}}\underbrace{(a \cdot a \cdot \cdots \cdot a)}_{n \text{ factors}} = \underbrace{a \cdot a \cdot a \cdot \cdots \cdot a}_{m + n \text{ factors}} = a^{m+n}$$

Thus, we have shown that

$$a^m a^n = a^{m+n}$$

whenever m and n are positive integers.

We would like this rule to be true even when m and n are 0 or negative integers. Thus, for instance, we must have

$$2^0 \cdot 2^3 = 2^{0+3} = 2^3$$

But this can happen only if $2^0 = 1$. Likewise, we want to have

$$5^4 \cdot 5^{-4} = 5^{4+(-4)} = 5^{4-4} = 5^0 = 1$$

and this will be true if $5^{-4} = 1/5^4$. These observations lead to the following definition.

ZERO AND NEGATIVE EXPONENTS

If $a \neq 0$ is any real number and n is a positive integer, then

$$a^0 = 1 \qquad \text{and} \qquad a^{-n} = \frac{1}{a^n}$$

EXAMPLE 2 ■ **Zero and Negative Exponents**

(a) $\left(\frac{4}{7}\right)^0 = 1$

(b) $x^{-1} = \frac{1}{x^1} = \frac{1}{x}$

(c) $(-2)^{-3} = \frac{1}{(-2)^3} = \frac{1}{-8} = -\frac{1}{8}$ ■

Familiarity with the following rules is essential for our work with exponents and bases. In the table the bases a and b are real numbers, and the exponents m and n are integers.

LAWS OF EXPONENTS

Law	Description
1. $a^m a^n = a^{m+n}$	To multiply two powers of the same number, add the exponents.
2. $\dfrac{a^m}{a^n} = a^{m-n}$	To divide two powers of the same number, subtract the exponents.
3. $(a^m)^n = a^{mn}$	To raise a power to a new power, multiply the exponents.
4. $(ab)^n = a^n b^n$	To raise a product to a power, raise each factor to the power.
5. $\left(\dfrac{a}{b}\right)^n = \dfrac{a^n}{b^n}$	To raise a quotient to a power, raise both numerator and denominator to the power.

We have seen how to prove Law 1 whenever m and n are positive integers, but it can also be proved for any integer exponents by using the definition of negative exponents. We now give the proofs of Laws 3 and 4. (You are asked to prove Laws 2 and 5 in Exercise 88.)

■ **Proof of Law 3**   If m and n are positive integers, we have

$$(a^m)^n = \underbrace{(a \cdot a \cdots a)}_{m \text{ factors}}{}^n$$

$$= \underbrace{(a \cdot a \cdots a)}_{m \text{ factors}}\underbrace{(a \cdot a \cdots a)}_{m \text{ factors}} \cdots \underbrace{(a \cdot a \cdots a)}_{m \text{ factors}}$$

$$\underbrace{}_{n \text{ groups of factors}}$$

$$= \underbrace{a \cdot a \cdots a}_{mn \text{ factors}} = a^{mn}$$

The cases for which $m \le 0$ or $n \le 0$ can be proved using the definition of negative exponents.   □

■ **Proof of Law 4**   If m and n are positive integers, we have

$$(ab)^n = \underbrace{(ab)(ab) \cdots (ab)}_{n \text{ factors}} = \underbrace{(a \cdot a \cdots a)}_{n \text{ factors}} \cdot \underbrace{(b \cdot b \cdots b)}_{n \text{ factors}} = a^n b^n$$

Here we have used the Commutative and Associative Properties repeatedly. If $m \le 0$ or $n \le 0$, Law 4 can be proved using the definition of negative exponents.   □

EXAMPLE 3 ■ Using Laws of Exponents

(a) $x^4 x^7 = x^{4+7} = x^{11}$ Law 1

(b) $y^4 y^{-7} = y^{4-7} = y^{-3} = \dfrac{1}{y^3}$ Law 1

(c) $\dfrac{c^9}{c^5} = c^{9-5} = c^4$ Law 2

(d) $(b^4)^5 = b^{4 \cdot 5} = b^{20}$ Law 3

(e) $(3x)^3 = 3^3 x^3 = 27x^3$ Law 4

(f) $\left(\dfrac{x}{2}\right)^5 = \dfrac{x^5}{2^5} = \dfrac{x^5}{32}$ Law 5

EXAMPLE 4 ■ Simplifying Expressions with Exponents

Simplify: (a) $(2a^3b^2)(3ab^4)^3$ (b) $\left(\dfrac{x}{y}\right)^3\left(\dfrac{y^2 x}{z}\right)^4$

SOLUTION

(a) $(2a^3b^2)(3ab^4)^3 = (2a^3b^2)[3^3 a^3 (b^4)^3]$ Law 4

$= (2a^3b^2)(27a^3b^{12})$ Law 3

$= (2)(27)a^3 a^3 b^2 b^{12}$ Group factors with the same base

$= 54a^6 b^{14}$ Law 1

most exciting of these just within the past decade.

   In other *Mathematics in the Modern World*, we will describe in more detail how mathematics affects us in our everyday activities.

(b) $\left(\dfrac{x}{y}\right)^3\left(\dfrac{y^2x}{z}\right)^4 = \dfrac{x^3}{y^3}\dfrac{(y^2)^4 x^4}{z^4}$ Laws 5 and 4

$= \dfrac{x^3}{y^3}\dfrac{y^8 x^4}{z^4}$ Law 3

$= (x^3 x^4)\left(\dfrac{y^8}{y^3}\right)\dfrac{1}{z^4}$ Group factors with the same base

$= \dfrac{x^7 y^5}{z^4}$ Laws 1 and 2   ■

When simplifying an expression, you will find that many different methods will lead to the same result; you should feel free to use any of the rules of exponents to arrive at your own method. We now give two additional laws that are useful in simplifying expressions with negative exponents.

LAWS OF EXPONENTS

6. $\left(\dfrac{a}{b}\right)^{-n} = \left(\dfrac{b}{a}\right)^{n}$ To raise a fraction to a negative power, invert the fraction and change the sign of the exponent.

7. $\dfrac{a^{-n}}{b^{-m}} = \dfrac{b^m}{a^n}$ To move a number raised to a power from numerator to denominator or from denominator to numerator, change the sign of the exponent.

   ■ **Proof of Law 7**   Using the definition of negative exponents and then Property 2 of fractions, we have

$$\frac{a^{-n}}{b^{-m}} = \frac{1/a^n}{1/b^m} = \frac{1}{a^n}\cdot\frac{b^m}{1} = \frac{b^m}{a^n}$$ □

(You are asked to prove Law 6 in Exercise 88.)

EXAMPLE 5 ■ Simplifying Expressions with Negative Exponents

Eliminate negative exponents and simplify each expression.

(a) $\dfrac{6st^{-4}}{2s^{-2}t^2}$ (b) $\left(\dfrac{y}{3z^2}\right)^{-2}$

SOLUTION

(a) We use Law 7, which allows us to move a number raised to a power from the numerator to the denominator (or vice versa) by changing the sign of the exponent.

$$\frac{6st^{-4}}{2s^{-2}t^2} = \frac{6ss^2}{2t^4 t^2}$$ Law 7

$$= \frac{3s^3}{t^6}$$ Law 1

(b) We use Law 6, which allows us to change the sign of the exponent of a fraction by inverting the fraction.

$$\left(\frac{y}{3z^2}\right)^{-2} = \left(\frac{3z^2}{y}\right)^2 \qquad \text{Law 6}$$

$$= \frac{9z^4}{y^2} \qquad \text{Laws 5 and 4}$$

■ Scientific Notation

Exponential notation is used by scientists as a compact way of writing very large numbers and very small numbers. For example, the nearest star beyond the sun, Proxima Centauri, is approximately 40,000,000,000,000 km away. The mass of a hydrogen atom is about 0.00000000000000000000000166 g. Scientists usually write such numbers in a more convenient way, called *scientific notation*. A positive number x is said to be written in **scientific notation** if it is expressed as follows:

$$x = a \times 10^n \qquad \text{where } 1 \le a < 10 \text{ and } n \text{ is an integer}$$

For instance, when we state that the distance to the star Proxima Centauri is 4×10^{13} km, the positive exponent 13 indicates that the decimal point should be moved 13 places to the *right*:

$$4 \times 10^{13} = 40{,}000{,}000{,}000{,}000$$

$$\underbrace{\qquad\qquad\qquad}_{13 \text{ places}}$$

When we state that the mass of a hydrogen atom is 1.66×10^{-24} g, the exponent -24 indicates that the decimal point should be moved 24 places to the *left*:

$$1.66 \times 10^{-24} = 0.00000000000000000000000166$$

$$\underbrace{\qquad\qquad\qquad}_{24 \text{ places}}$$

EXAMPLE 6 ■ Writing Numbers in Scientific Notation

(a) $56920 = 5.692 \times 10^4$ \qquad\qquad (b) $0.000093 = 9.3 \times 10^{-5}$ ■

To use scientific notation on a calculator, use a key labeled $\boxed{\text{EE}}$ or $\boxed{\text{EXP}}$ or $\boxed{\text{EEX}}$ to enter the exponent. For example, to enter the number 3.629×10^{15} on a TI-83 calculator, we enter

$$3.629 \;\boxed{\text{2ND}}\; \boxed{\text{EE}}\; 15$$

and the display reads

$$\boxed{\text{3.629E15}}$$

Scientific notation is often used on a calculator to display a very large or very small number. For instance, if we use a calculator to square the number 1,111,111, the display panel may show (depending on the calculator model) the approximation

$$\boxed{\text{1.234568 \quad 12}} \quad \text{or} \quad \boxed{\text{1.234568 \quad E12}}$$

Here the final digits indicate the power of 10, and we interpret the result as

$$1.234568 \times 10^{12}$$

EXAMPLE 7 ■ Using Scientific Notation

If $a \approx 0.00046$, $b \approx 1.697 \times 10^{22}$, and $c \approx 2.91 \times 10^{-18}$, use a calculator to approximate the quotient ab/c.

SOLUTION

We could enter the data using scientific notation, or we could use laws of exponents as follows:

$$\frac{ab}{c} \approx \frac{(4.6 \times 10^{-4})(1.697 \times 10^{22})}{2.91 \times 10^{-18}}$$

$$= \frac{(4.6)(1.697)}{2.91} \times 10^{-4+22+18}$$

$$\approx 2.7 \times 10^{36}$$

We state the answer correct to two significant figures because the least accurate of the given numbers is stated to two significant figures.   ■

■ Radicals

We know what 2^n means whenever n is an integer. To give meaning to a power, such as $2^{4/5}$, whose exponent is a rational number, we need to discuss radicals.

The symbol $\sqrt{}$ means "the positive square root of." Thus

$$\sqrt{a} = b \qquad \text{means} \qquad b^2 = a \quad \text{and} \quad b \geq 0$$

Since $a = b^2 \geq 0$, the symbol $\sqrt{a}$ makes sense only when $a \geq 0$. For instance,

$$\sqrt{9} = 3 \qquad \text{because} \qquad 3^2 = 9 \quad \text{and} \quad 3 \geq 0$$

Square roots are special cases of nth roots. The nth root of x is the number that, when raised to the nth power, gives x.

> It is true that the number 9 has two square roots, 3 and -3, but the notation $\sqrt{9}$ is reserved for the *positive* square root of 9 (sometimes called the *principal square root* of 9). If we want the negative root, we *must* write $-\sqrt{9}$, which is -3.

DEFINITION OF nth ROOT

If n is any positive integer, then the **principal nth root** of a is defined as follows:

$$\sqrt[n]{a} = b \qquad \text{means} \qquad b^n = a$$

If n is even, we must have $a \geq 0$ and $b \geq 0$.

Thus

$$\sqrt[4]{81} = 3 \qquad \text{because} \qquad 3^4 = 81 \quad \text{and} \quad 3 \geq 0$$

$$\sqrt[3]{-8} = -2 \qquad \text{because} \qquad (-2)^3 = -8$$

But $\sqrt{-8}$, $\sqrt[4]{-8}$, and $\sqrt[6]{-8}$ are not defined. (For instance, $\sqrt{-8}$ is not defined because the square of every real number is nonnegative.)

Notice that

$$\sqrt{4^2} = \sqrt{16} = 4 \qquad \text{but} \qquad \sqrt{(-4)^2} = \sqrt{16} = 4 = |-4|$$

Thus, the equation $\sqrt{a^2} = a$ is not always true; it is true only when $a \geq 0$. However, we can always write $\sqrt{a^2} = |a|$. This last equation is true not only for square roots, but for any even root. This and other rules used in working with nth roots are listed in the following box. In each property we assume that all the given roots exist.

PROPERTIES OF nth ROOTS

Property	Example
1. $\sqrt[n]{ab} = \sqrt[n]{a}\,\sqrt[n]{b}$	$\sqrt[3]{-8 \cdot 27} = \sqrt[3]{-8}\,\sqrt[3]{27} = (-2)(3) = -6$
2. $\sqrt[n]{\dfrac{a}{b}} = \dfrac{\sqrt[n]{a}}{\sqrt[n]{b}}$	$\sqrt[4]{\dfrac{16}{81}} = \dfrac{\sqrt[4]{16}}{\sqrt[4]{81}} = \dfrac{2}{3}$
3. $\sqrt[m]{\sqrt[n]{a}} = \sqrt[mn]{a}$	$\sqrt{\sqrt[3]{729}} = \sqrt[6]{729} = 3$
4. $\sqrt[n]{a^n} = a$ if n is odd	$\sqrt[3]{(-5)^3} = -5,\ \sqrt[5]{2^5} = 2$
5. $\sqrt[n]{a^n} = |a|$ if n is even	$\sqrt[4]{(-3)^4} = |-3| = 3$

EXAMPLE 8 ■ Simplifying Expressions Involving nth Roots

(a) $\sqrt[3]{x^4} = \sqrt[3]{x^3 x}$ Factor out the largest cube

$\qquad = \sqrt[3]{x^3}\,\sqrt[3]{x}$ Property 1

$\qquad = x\,\sqrt[3]{x}$ Property 4

(b) $\sqrt[4]{81x^8y^4} = \sqrt[4]{81}\,\sqrt[4]{x^8}\,\sqrt[4]{y^4}$ Property 1

$\qquad = 3\,\sqrt[4]{(x^2)^4}\,|y|$ Property 5

$\qquad = 3x^2\,|y|$ Property 5   ■

It is frequently useful to combine like radicals in an expression such as $2\sqrt{3} + 5\sqrt{3}$. This can be done by using the Distributive Property. Thus

$$2\sqrt{3} + 5\sqrt{3} = (2 + 5)\sqrt{3} = 7\sqrt{3}$$

The next example further illustrates this process.

⊘ Avoid making the following error:

$$\sqrt{a + b} \ne \sqrt{a} + \sqrt{b}$$

For instance, if we let $a = 9$ and $b = 16$, then we see the error:

$$\sqrt{9 + 16} \overset{?}{=} \sqrt{9} + \sqrt{16}$$

$$\sqrt{25} \overset{?}{=} 3 + 4$$

$$5 \overset{?}{=} 7 \quad \text{Wrong!}$$

EXAMPLE 9 ■ Combining Radicals

(a) $\sqrt{32} + \sqrt{200} = \sqrt{16 \cdot 2} + \sqrt{100 \cdot 2}$ Factor out the largest squares

$\qquad = \sqrt{16}\,\sqrt{2} + \sqrt{100}\,\sqrt{2}$ Property 1

$\qquad = 4\sqrt{2} + 10\sqrt{2} = 14\sqrt{2}$ Distributive Property

(b) If $b > 0$, then

$$\sqrt{25b} - \sqrt{b^3} = \sqrt{25}\sqrt{b} - \sqrt{b^2}\sqrt{b} \qquad \text{Property 1}$$

$$= 5\sqrt{b} - b\sqrt{b} \qquad \text{Property 5, } b > 0$$

$$= (5 - b)\sqrt{b} \qquad \text{Distributive Property} \qquad \blacksquare$$

Rational Exponents

To define what is meant by a *rational exponent* or, equivalently, a *fractional exponent* such as $a^{1/3}$, we need to use radicals. In order to give meaning to the symbol $a^{1/n}$ in a way that is consistent with the Laws of Exponents, we would have to have

$$(a^{1/n})^n = a^{(1/n)n} = a^1 = a$$

So, by the definition of nth root,

$$\boxed{a^{1/n} = \sqrt[n]{a}}$$

In general, we define rational exponents as follows.

DEFINITION OF RATIONAL EXPONENTS

For any rational exponent m/n in lowest terms, where m and n are integers and $n > 0$, we define

$$a^{m/n} = \left(\sqrt[n]{a}\right)^m$$

or equivalently

$$a^{m/n} = \sqrt[n]{a^m}$$

If n is even, then we require that $a \geqslant 0$.

With this definition it can be proved that *the Laws of Exponents also hold for rational exponents.*

EXAMPLE 10 ■ Using the Definition of Rational Exponents

(a) $4^{1/2} = \sqrt{4} = 2$

(b) $8^{2/3} = (\sqrt[3]{8})^2 = 2^2 = 4$ Alternative solution: $8^{2/3} = \sqrt[3]{8^2} = \sqrt[3]{64} = 4$

(c) $(125)^{-1/3} = \dfrac{1}{125^{1/3}} = \dfrac{1}{\sqrt[3]{125}} = \dfrac{1}{5}$

(d) $\dfrac{1}{\sqrt[3]{x^4}} = \dfrac{1}{x^{4/3}} = x^{-4/3}$ $\blacksquare$

EXAMPLE 11 ■ **Using the Laws of Exponents with Rational Exponents**

(a) $a^{1/3}a^{7/3} = a^{8/3}$ Law 1

(b) $\dfrac{a^{2/5}a^{7/5}}{a^{3/5}} = a^{2/5+7/5-3/5} = a^{6/5}$ Laws 1 and 2

(c) $(2a^3b^4)^{3/2} = 2^{3/2}(a^3)^{3/2}(b^4)^{3/2}$ Law 4

$\qquad\qquad = (\sqrt{2})^3 a^{3(3/2)}b^{4(3/2)}$ Law 3

$\qquad\qquad = 2\sqrt{2}\,a^{9/2}b^6$

(d) $\left(\dfrac{2x^{3/4}}{y^{1/3}}\right)^3\left(\dfrac{y^4}{x^{-1/2}}\right) = \dfrac{2^3(x^{3/4})^3}{(y^{1/3})^3} \cdot (y^4x^{1/2})$ Laws 5, 4, and 7

$\qquad\qquad = \dfrac{8x^{9/4}}{y} \cdot y^4x^{1/2}$ Law 3

$\qquad\qquad = 8x^{11/4}y^3$ Laws 1 and 2   ■

EXAMPLE 12 ■ **Simplifying by Writing Radicals as Rational Exponents**

(a) $(2\sqrt{x})(3\sqrt[3]{x}) = (2x^{1/2})(3x^{1/3})$ Definition of rational exponents

$\qquad\qquad = 6x^{1/2+1/3} = 6x^{5/6}$ Law 1

(b) $\sqrt{x\sqrt{x}} = (xx^{1/2})^{1/2}$ Definition of rational exponents

$\qquad\qquad = (x^{3/2})^{1/2}$ Law 1

$\qquad\qquad = x^{3/4}$ Law 3   ■

■ **Rationalizing the Denominator**

It is often useful to eliminate the radical in a denominator by multiplying both numerator and denominator by an appropriate expression. This procedure is called **rationalizing the denominator**. If the denominator is of the form $\sqrt{a}$, we multiply numerator and denominator by $\sqrt{a}$. In doing so we multiply the given quantity by 1, so we do not change its value. For instance,

$$\frac{1}{\sqrt{a}} = \frac{1}{\sqrt{a}} \cdot 1 = \frac{1}{\sqrt{a}} \cdot \frac{\sqrt{a}}{\sqrt{a}} = \frac{\sqrt{a}}{a}$$

Note that the denominator in the last fraction contains no radical. In general, if the denominator is of the form $\sqrt[n]{a^m}$ with $m < n$, then multiplying numerator and denominator by $\sqrt[n]{a^{n-m}}$ will rationalize the denominator, because (for $a > 0$)

$$\sqrt[n]{a^m}\,\sqrt[n]{a^{n-m}} = \sqrt[n]{a^{m+n-m}} = \sqrt[n]{a^n} = a$$

EXAMPLE 13 ■ Rationalizing Denominators

(a) $\dfrac{2}{\sqrt{3}} = \dfrac{2}{\sqrt{3}} \cdot \dfrac{\sqrt{3}}{\sqrt{3}} = \dfrac{2\sqrt{3}}{3}$

(b) $\dfrac{1}{\sqrt[3]{x^2}} = \dfrac{1}{\sqrt[3]{x^2}} \dfrac{\sqrt[3]{x}}{\sqrt[3]{x}} = \dfrac{\sqrt[3]{x}}{\sqrt[3]{x^3}} = \dfrac{\sqrt[3]{x}}{x}$

(c) $\sqrt[7]{\dfrac{1}{a^2}} = \dfrac{1}{\sqrt[7]{a^2}} = \dfrac{1}{\sqrt[7]{a^2}} \dfrac{\sqrt[7]{a^5}}{\sqrt[7]{a^5}} = \dfrac{\sqrt[7]{a^5}}{\sqrt[7]{a^7}} = \dfrac{\sqrt[7]{a^5}}{a}$ ■

1.2 EXERCISES

1–6 ■ Write each radical expression using exponents, and each exponential expression using radicals.

1. $\dfrac{1}{\sqrt{17}}$

2. $\sqrt[4]{7^3}$

3. $4^{2/3}$

4. $\sqrt[3]{b^5}$

5. $a^{3/5}$

6. $w^{-3/2}$

7–14 ■ Evaluate each number.

7. (a) $(-2)^4$ (b) -2^4 (c) $(-2)^0$

8. (a) $\left(\tfrac{1}{2}\right)^4 4^{-2}$ (b) $\left(\tfrac{1}{4}\right)^{-2}$ (c) $\left(\tfrac{1}{4}\right)^0 2^{-1}$

9. (a) $2^4 5^{-2}$ (b) $\dfrac{10^7}{10^4}$ (c) $(2^3 \cdot 2^2)^2$

10. (a) $\sqrt{64}$ (b) $\sqrt[3]{-64}$ (c) $\sqrt[5]{-32}$

11. (a) $\sqrt{\tfrac{4}{9}}$ (b) $\sqrt[4]{256}$ (c) $\sqrt[6]{\tfrac{1}{64}}$

12. (a) $\sqrt{7}\,\sqrt{28}$ (b) $\dfrac{\sqrt{48}}{\sqrt{3}}$ (c) $\sqrt[4]{24}\,\sqrt[4]{54}$

13. (a) $\left(\tfrac{4}{9}\right)^{-1/2}$ (b) $(-32)^{2/5}$ (c) $(-125)^{-1/3}$

14. (a) $1024^{-0.1}$ (b) $\left(-\tfrac{27}{8}\right)^{2/3}$ (c) $\left(\tfrac{25}{64}\right)^{3/2}$

15–18 ■ Evaluate the expression using $x = 3$, $y = 4$, and $z = -1$.

15. $\sqrt{x^2 + y^2}$

16. $\sqrt[4]{x^3 + 14y + 2z}$

17. $(9x)^{2/3} + (2y)^{2/3} + z^{2/3}$

18. $(xy)^{2z}$

19–22 ■ Simplify the expression.

19. $\sqrt[3]{108} - \sqrt[3]{32}$

20. $\sqrt{8} + \sqrt{50}$

21. $\sqrt{245} - \sqrt{125}$

22. $\sqrt[3]{54} - \sqrt[3]{16}$

23–40 ■ Simplify the expression and eliminate any negative exponent(s).

23. $a^9 a^{-5}$

24. $(3y^2)(4y^5)$

25. $(12x^2y^4)\left(\tfrac{1}{2}x^5y\right)$

26. $(6y)^3$

27. $\dfrac{x^9(2x)^4}{x^3}$

28. $\dfrac{a^{-3}b^4}{a^{-5}b^5}$

29. $b^4\left(\tfrac{1}{3}b^2\right)(12b^{-8})$

30. $(2s^3 t^{-1})\left(\tfrac{1}{4}s^6\right)(16t^4)$

31. $(rs)^3(2s)^{-2}(4r)^4$

32. $(2u^2v^3)^3(3u^3v)^{-2}$

33. $\dfrac{(6y^3)^4}{2y^5}$

34. $\dfrac{(2x^3)^2(3x^4)}{(x^3)^4}$

35. $\dfrac{(x^2y^3)^4(xy^4)^{-3}}{x^2y}$

36. $\left(\dfrac{c^4d^3}{cd^2}\right)\left(\dfrac{d^2}{c^3}\right)^3$

37. $\dfrac{(xy^2z^3)^4}{(x^3y^2z)^3}$

38. $\left(\dfrac{xy^{-2}z^{-3}}{x^2y^3z^{-4}}\right)^{-3}$

39. $\left(\dfrac{q^{-1}rs^{-2}}{r^{-5}sq^{-8}}\right)^{-1}$

40. $(3ab^2c)\left(\dfrac{2a^2b}{c^3}\right)^{-2}$

41–56 ■ Simplify the expression and eliminate any negative exponent(s). Assume that all letters denote positive numbers.

41. $x^{2/3}x^{1/5}$

42. $(-2a^{3/4})(5a^{3/2})$

43. $(4b)^{1/2}(8b^{2/5})$

44. $(8x^6)^{-2/3}$

45. $(c^2d^3)^{-1/3}$

46. $(4x^6y^8)^{3/2}$

47. $(y^{3/4})^{2/3}$

48. $(a^{2/5})^{-3/4}$

49. $(2x^4y^{-4/5})^3(8y^2)^{2/3}$

50. $(x^{-5}y^3z^{10})^{-3/5}$

51. $\left(\dfrac{x^6y}{y^4}\right)^{5/2}$

52. $\left(\dfrac{-2x^{1/3}}{y^{1/2}z^{1/6}}\right)^4$

53. $\left(\dfrac{3a^{-2}}{4b^{-1/3}}\right)^{-1}$

54. $\dfrac{(y^{10}z^{-5})^{1/5}}{(y^{-2}z^3)^{1/3}}$

55. $\dfrac{(9st)^{3/2}}{(27s^3t^{-4})^{2/3}}$

56. $\left(\dfrac{a^2b^{-3}}{x^{-1}y^2}\right)^3\left(\dfrac{x^{-2}b^{-1}}{a^{3/2}y^{1/3}}\right)$

57–64 ■ Simplify the expression. Assume the letters denote any real numbers.

57. $\sqrt[4]{x^4}$

58. $\sqrt[3]{x^3y^6}$

59. $\sqrt[3]{x^3y}$

60. $\sqrt{x^4y^4}$

61. $\sqrt[5]{a^6b^7}$

62. $\sqrt[3]{a^2b}\,\sqrt[3]{a^4b}$

63. $\sqrt[3]{\sqrt{64x^6}}$

64. $\sqrt[4]{x^4y^2z^2}$

65–68 ■ Rationalize the denominator.

65. (a) $\dfrac{1}{\sqrt{6}}$ (b) $\sqrt{\dfrac{x}{3y}}$ (c) $\sqrt{\dfrac{3}{20}}$

66. (a) $\sqrt{\dfrac{x^5}{2}}$ (b) $\sqrt{\dfrac{2}{3}}$ (c) $\sqrt{\dfrac{1}{2x^3y^5}}$

67. (a) $\dfrac{1}{\sqrt[3]{x}}$ (b) $\dfrac{1}{\sqrt[5]{x^2}}$ (c) $\dfrac{1}{\sqrt[7]{x^3}}$

68. (a) $\dfrac{1}{\sqrt[3]{x^2}}$ (b) $\dfrac{1}{\sqrt[4]{x^3}}$ (c) $\dfrac{1}{\sqrt[3]{x^4}}$

69–70 ■ Write each number in scientific notation.

69. (a) 69,300,000
(b) 0.000028536
(c) 129,540,000

70. (a) 7,259,000,000
(b) 0.0000000014
(c) 0.0007029

71–72 ■ Write each number in ordinary decimal notation.

71. (a) 3.19×10^5
(b) 2.670×10^{-8}
(c) 7.1×10^{14}

72. (a) 8.55×10^{-3}
(b) 6×10^{12}
(c) 6.257×10^{-10}

73–74 ■ Write the number indicated in each statement in scientific notation.

73. (a) A light-year, the distance that light travels in one year, is about 5,900,000,000,000 mi.
(b) The diameter of an electron is about 0.0000000000004 cm.

(c) A drop of water contains more than 33 billion billion molecules.

74. (a) The distance from the earth to the sun is about 93 million miles.
(b) The mass of an oxygen molecule is about 0.000000000000000000000053 g.
(c) The mass of the earth is about 5,970,000,000,000,000,000,000,000 kg.

75–80 ■ Use scientific notation, the Laws of Exponents, and a calculator to perform the indicated operations. State your answer correct to the number of significant digits indicated by the given data.

75. $(7.2 \times 10^{-9})(1.806 \times 10^{-12})$

76. $(1.062 \times 10^{24})(8.61 \times 10^{19})$

77. $\dfrac{1.295643 \times 10^9}{(3.610 \times 10^{-17})(2.511 \times 10^6)}$

78. $\dfrac{(73.1)(1.6341 \times 10^{28})}{0.0000000019}$

79. $\dfrac{(0.0000162)(0.01582)}{(594621000)(0.0058)}$ **80.** $\dfrac{(3.542 \times 10^{-6})^9}{(5.05 \times 10^4)^{12}}$

81. Proxima Centauri, the star nearest to our solar system, is 4.3 light-years away. Use the information in Exercise 73(a) to express this distance in miles.

82. The speed of light is about 186,000 mi/s. Use the information in Exercise 74(a) to find how long it takes for a light ray from the sun to reach the earth.

83. The average ocean depth is 3.7×10^3 m, and the area of the oceans is 3.6×10^{14} m^2. What is the total volume of the ocean in liters? (One cubic meter contains 1000 liters.)

84. As of January 2001, the population of the United States was 2.83×10^8, and the national debt was 5.736×10^{12} dollars. How much was each person's share of the debt?

85. A sealed room in a hospital, measuring 5 m wide, 10 m long, and 3 m high, is filled with pure oxygen. One cubic meter contains 1000 L, and 22.4 L of any gas contains 6.02×10^{23} molecules (Avogadro's number). How many molecules of oxygen are there in the room?

86. Police use the formula $s = \sqrt{30fd}$ to estimate the speed s (in mi/h) at which a car is traveling if it skids d feet after the brakes are applied suddenly. The number f is the coefficient of friction of the road, which is a measure of the "slipperiness" of the road. The table gives some typical estimates for f.

	Tar	Concrete	Gravel
Dry	1.0	0.8	0.2
Wet	0.5	0.4	0.1

(a) If a car skids 65 ft on wet concrete, how fast was it moving when the brakes were applied?
(b) If a car is traveling at 50 mi/h, how far will it skid on wet tar?

87. Due to the curvature of the earth, the maximum distance D that you can see from the top of a tall building of height h is estimated by the formula

$$D = \sqrt{2rh + h^2}$$

where $r = 3960$ mi is the radius of the earth and D and h are also measured in miles. How far can you see from the observation deck of the Toronto CN Tower, 1135 ft above the ground?

CN Tower

r

88. Prove the given Laws of Exponents for the case in which m and n are positive integers and $m > n$.

(a) Law 2   (b) Law 5   (c) Law 6

DISCOVERY · DISCUSSION

89. Limiting Behavior of Powers Complete the following tables. What happens to the nth root of 2 as n gets large? What about the nth root of $\frac{1}{2}$?

n	$2^{1/n}$
1	
2	
5	
10	
100	

n	$\left(\frac{1}{2}\right)^{1/n}$
1	
2	
5	
10	
100	

Construct a similar table for $n^{1/n}$. What happens to the nth root of n as n gets large?

90. Comparing Roots Without using a calculator, determine which number is larger in each pair.

(a) $2^{1/2}$ or $2^{1/3}$
(b) $\left(\frac{1}{2}\right)^{1/2}$ or $\left(\frac{1}{2}\right)^{1/3}$
(c) $7^{1/4}$ or $4^{1/3}$
(d) $\sqrt[3]{5}$ or $\sqrt{3}$

91. Easy Powers That Look Hard Calculate these expressions in your head. Use the Laws of Exponents to help you.

(a) $\dfrac{18^5}{9^5}$
(b) $20^6 \cdot (0.5)^6$

92. Distances between Powers Which pair of numbers is closer together?

$$10^{10} \text{ and } 10^{50} \quad \text{or} \quad 10^{100} \text{ and } 10^{101}$$

93. Relativity The theory of relativity states that as an object travels with velocity v, its rest mass m_0 changes to a mass m given by the formula

$$m = \frac{m_0}{\sqrt{1 - \dfrac{v^2}{c^2}}}$$

where $c \approx 3.0 \times 10^8$ m/s is the speed of light. By what factor is the rest mass of a spaceship multiplied if the ship travels at one-tenth the speed of light? At one-half the speed of light? At 90% of the speed of light? How does the mass of the spaceship change as it travels at a speed very close to the speed of light? Do we need to know the actual value of the speed of light to answer these questions?

1.3   ALGEBRAIC EXPRESSIONS

Algebraic expressions such as

$$2x^2 - 3x + 4 \qquad ax + b$$

$$\frac{y-1}{y^2+2} \qquad \frac{cx^2y + dy^2z}{\sqrt{x^2 + y^2 + z^2}}$$

are obtained by starting with variables such as x, y, and z and constants such as 2, -3, a, b, c, and d, and combining them using addition, subtraction, multiplication, division, and roots. A **variable** is a letter that can represent any number in a given set of numbers, whereas a **constant** represents a fixed (or specific) number. The **domain** of a variable is the set of numbers that the variable is permitted to have. For instance, in the expression $\sqrt{x}$ the domain of x is $\{x \mid x \geq 0\}$, whereas in the expression $2/(x-3)$ the domain of x is $\{x \mid x \neq 3\}$.

The simplest types of algebraic expressions use only addition, subtraction, and multiplication. Such expressions are called **polynomials**. The general form of a polynomial of degree n (where n is a nonnegative integer) in the variable x is

Monomials:

$$5, \quad 6x, \quad ax^2, \quad -10ab$$

Binomials:

$$2 + 3x, \quad 6x^2 - 5, \quad ax + b$$

Trinomials:

$$x^2 + x + 1, \quad ax + by + cz,$$
$$x^5 + 2x^3 + 3x$$

$$a_n x^n + a_{n-1} x^{n-1} + \cdots + a_1 x + a_0$$

where $a_0, a_1 \ldots, a_n$ are constants and $a_n \neq 0$. The **degree** of a polynomial is the highest power of the variable. Any polynomial is a sum of **terms** of the form ax^k, called **monomials**, where a is a constant and k is a nonnegative integer. A **binomial** is a sum of two monomials, a **trinomial** is the sum of three monomials, and so on. Thus, $2x^2 - 3x + 4$, $ax + b$, and $x^4 + 2x^3$ are polynomials of degree 2, 1, and 4, respectively; the first is a trinomial, the other two are binomials.

▪ Combining Algebraic Expressions

We **add** and **subtract** polynomials using the properties of real numbers that were discussed in Section 1.1. The idea is to combine **like terms** (that is, terms with the same variables raised to the same powers) using the Distributive Property. For instance,

Distributive Property

$$ac + bc = (a + b)c$$

$$5x^7 + 3x^7 = (5 + 3)x^7 = 8x^7$$

🚫 In subtracting polynomials we have to remember that if a minus sign precedes an expression in parentheses, then the sign of every term within the parentheses is changed when we remove the parentheses:

$$-(b + c) = -b - c$$

[This is simply a case of the Distributive Property, $a(b + c) = ab + ac$, with $a = -1$.]

EXAMPLE 1 ■ Adding and Subtracting Polynomials

(a) Find the sum $(x^3 - 6x^2 + 2x + 4) + (x^3 + 5x^2 - 7x)$.

(b) Find the difference $(x^3 - 6x^2 + 2x + 4) - (x^3 + 5x^2 - 7x)$.

SOLUTION

(a) $(x^3 - 6x^2 + 2x + 4) + (x^3 + 5x^2 - 7x)$

$= (x^3 + x^3) + (-6x^2 + 5x^2) + (2x - 7x) + 4$ Group like terms

$= 2x^3 - x^2 - 5x + 4$ Combine like terms

(b) $(x^3 - 6x^2 + 2x + 4) - (x^3 + 5x^2 - 7x)$

$= x^3 - 6x^2 + 2x + 4 - x^3 - 5x^2 + 7x$ Distributive Property

$= (x^3 - x^3) + (-6x^2 - 5x^2) + (2x + 7x) + 4$ Group like terms

$= -11x^2 + 9x + 4$ Combine like terms   ■

To find the **product** of polynomials or other algebraic expressions, we need to use the Distributive Property repeatedly. In particular, using it three times on the product of two binomials, we get

$$(a + b)(c + d) = a(c + d) + b(c + d) = ac + ad + bc + bd$$

This says that we multiply the two factors by multiplying each term in one factor by each term in the other factor and adding these products. Schematically we have

The acronym **FOIL** helps us remember that the product of two binomials is the sum of the products of the **F**irst terms, the **O**uter terms, the **I**nner terms, and the **L**ast terms.

$$(a + b)(c + d) = ac + ad + bc + bd$$
$$\qquad\qquad\quad \uparrow \quad \uparrow \quad \uparrow \quad \uparrow$$
$$\qquad\qquad\quad F \quad O \quad I \quad L$$

In general, we can multiply two algebraic expressions by using the Distributive Property and the Laws of Exponents.

EXAMPLE 2 ■ Multiplying Algebraic Expressions

(a) $(2x + 1)(3x - 5) = 6x^2 - 10x + 3x - 5$ Distributive Property

$= 6x^2 - 7x - 5$ Combine like terms

(b) $(x^2 - 3)(x^3 + 2x + 1) = x^2(x^3 + 2x + 1) - 3(x^3 + 2x + 1)$ Distributive Property

$= x^5 + 2x^3 + x^2 - 3x^3 - 6x - 3$ Distributive Property

$= x^5 - x^3 + x^2 - 6x - 3$ Combine like terms

(c) $(1 + \sqrt{x})(2 - 3\sqrt{x}) = 2 - 3\sqrt{x} + 2\sqrt{x} - 3(\sqrt{x})^2$ Distributive Property

$= 2 - \sqrt{x} - 3x$ Combine like terms   ■

Certain types of products occur so frequently that you should memorize them. You can verify the following formulas by performing the multiplications.

See the Discovery Project on page 35 for a geometric interpretation of some of these formulas.

SPECIAL PRODUCT FORMULAS

1. $(A - B)(A + B) = A^2 - B^2$

2. $(A + B)^2 = A^2 + 2AB + B^2$

3. $(A - B)^2 = A^2 - 2AB + B^2$

4. $(A + B)^3 = A^3 + 3A^2B + 3AB^2 + B^3$

5. $(A - B)^3 = A^3 - 3A^2B + 3AB^2 - B^3$

The key idea in using these formulas (or any other formula in algebra) is the **Principle of Substitution**: We may substitute any algebraic expression for any letter in a formula. For example, to find $(x^2 + y^3)^2$ we use Product Formula 2:

$$(A + B)^2 = A^2 + 2AB + B^2$$

We substitute x^2 for A and y^3 for B to get

$$(x^2 + y^3)^2 = (x^2)^2 + 2(x^2)(y^3) + (y^3)^2$$

This type of substitution is valid because every algebraic expression (in this case, x^2 or y^3) represents a number.

EXAMPLE 3 ■ **Using the Special Product Formulas**

Use the Special Product Formulas to find each product.

(a) $(3x + 5)^2$ (b) $(2x - \sqrt{y})(2x + \sqrt{y})$ (c) $(x^2 - 2)^3$

SOLUTION

(a) Product Formula 2, with $A = 3x$ and $B = 5$, gives

$$(3x + 5)^2 = (3x)^2 + 2(3x)(5) + 5^2 = 9x^2 + 30x + 25$$

(b) Using Product Formula 1 with $A = 2x$ and $B = \sqrt{y}$, we have

$$(2x - \sqrt{y})(2x + \sqrt{y}) = (2x)^2 - (\sqrt{y})^2 = 4x^2 - y$$

(c) Substituting $A = x^2$ and $B = 2$ in Product Formula 5, we get

$$(x^2 - 2)^3 = (x^2)^3 - 3(x^2)^2(2) + 3(x^2)(2)^2 - 2^3$$
$$= x^6 - 6x^4 + 12x^2 - 8$$ ■

Factoring

We used the Distributive Property to expand algebraic expressions. We sometimes need to reverse this process (again using the Distributive Property) by **factoring** an expression as a product of simpler ones. For example, we can write

$$x^2 - 4 = (x - 2)(x + 2)$$

◀ **EXPANDING** ▬

We say that $x - 2$ and $x + 2$ are **factors** of $x^2 - 4$. The easiest type of factoring occurs when the terms have a common factor.

EXAMPLE 4 ■ Factoring Out Common Factors

Factor each expression.

(a) $3x^2 - 6x$ (b) $8x^4y^2 + 6x^3y^3 - 2xy^4$

SOLUTION

(a) The greatest common factor of the terms $3x^2$ and $-6x$ is $3x$, so we have

$$3x^2 - 6x = 3x(x - 2)$$

(b) We note that

8, 6, and -2 have the greatest common factor 2

x^4, x^3, and x have the greatest common factor x

y^2, y^3, and y^4 have the greatest common factor y^2

So the greatest common factor of the three terms in the polynomial is $2xy^2$, and we have

$$8x^4y^2 + 6x^3y^3 - 2xy^4 = (2xy^2)(4x^3) + (2xy^2)(3x^2y) + (2xy^2)(-y^2)$$
$$= 2xy^2(4x^3 + 3x^2y - y^2)$$ ■

To factor a second-degree polynomial, or **quadratic**, of the form $x^2 + bx + c$, we note that

$$(x + r)(x + s) = x^2 + (r + s)x + rs$$

so we need to choose numbers r and s so that $r + s = b$ and $rs = c$.

EXAMPLE 5 ■ Factoring $x^2 + bx + c$ by Trial and Error

Factor: $x^2 + 7x + 12$

SOLUTION

We need to find two integers whose product is 12 and whose sum is 7. By trial and error we find that the two integers are 3 and 4. Thus, the factorization is

$$x^2 + 7x + 12 = (x + 3)(x + 4)$$ ■

CHECK YOUR ANSWER

Multiplying gives

$$3x(x - 2) = 3x^2 - 6x$$ ✓

CHECK YOUR ANSWER

Multiplying gives

$$2xy^2(4x^3 + 3x^2y - y^2) =$$
$$8x^4y^2 + 6x^3y^3 - 2xy^4$$ ✓

CHECK YOUR ANSWER

Multiplying gives

$$(x + 3)(x + 4) = x^2 + 7x + 12$$ ✓

To factor a quadratic of the form $ax^2 + bx + c$ with $a \neq 1$, we look for factors of the form $px + r$ and $qx + s$:

$$ax^2 + bx + c = (px + r)(qx + s) = pqx^2 + (ps + qr)x + rs$$

factors of a

$$ax^2 + bx + c = (px + r)(qx + s)$$

factors of c

Therefore, we try to find numbers p, q, r, and s such that $pq = a$, $rs = c$, $ps + qr = b$. If these numbers are all integers, then we will have a limited number of possibilities to try for p, q, r, and s.

EXAMPLE 6 ■ Factoring $ax^2 + bx + c$ by Trial and Error

Factor: $6x^2 + 7x - 5$

SOLUTION

We can factor 6 as $6 \cdot 1$ or $3 \cdot 2$, and -5 as $-5 \cdot 1$ or $5 \cdot (-1)$. By trying these possibilities, we arrive at the factorization

$$6x^2 + 7x - 5 = (3x + 5)(2x - 1)$$ ■

CHECK YOUR ANSWER

Multiplying gives

$$(3x + 5)(2x - 1) = 6x^2 + 7x - 5 \quad \checkmark$$

Some special algebraic expressions can be factored using the following formulas. The first three are simply the Special Product Formulas written backward.

FACTORING FORMULAS	
Formula	**Name**
1. $A^2 - B^2 = (A - B)(A + B)$	Difference of squares
2. $A^2 + 2AB + B^2 = (A + B)^2$	Perfect square
3. $A^2 - 2AB + B^2 = (A - B)^2$	Perfect square
4. $A^3 - B^3 = (A - B)(A^2 + AB + B^2)$	Difference of cubes
5. $A^3 + B^3 = (A + B)(A^2 - AB + B^2)$	Sum of cubes

EXAMPLE 7 ■ Factoring Differences of Squares

Factor each polynomial.

(a) $4x^2 - 25$ (b) $(x + y)^2 - z^2$

SOLUTION

(a) Using the formula for a difference of squares with $A = 2x$ and $B = 5$, we have

$$4x^2 - 25 = (2x)^2 - 5^2$$
$$= (2x - 5)(2x + 5)$$

(b) We use the formula for the difference of squares with $A = x + y$ and $B = z$.

$$(x + y)^2 - z^2 = (x + y - z)(x + y + z)$$ ■

EXAMPLE 8 ■ Factoring Differences and Sums of Cubes

Factor each polynomial: (a) $27x^3 - 1$ (b) $x^6 + 8$

SOLUTION

(a) Using the formula for a difference of cubes with $A = 3x$ and $B = 1$, we get

$$27x^3 - 1 = (3x)^3 - 1^3 = (3x - 1)[(3x)^2 + (3x)(1) + 1^2]$$
$$= (3x - 1)(9x^2 + 3x + 1)$$

(b) Using the formula for a sum of cubes with $A = x^2$ and $B = 2$, we have

$$x^6 + 8 = (x^2)^3 + 2^3 = (x^2 + 2)(x^4 - 2x^2 + 4)$$ ■

A trinomial is a perfect square if it is of the form

$$A^2 + 2AB + B^2 \quad \text{or} \quad A^2 - 2AB + B^2$$

So, we **recognize a perfect square** if the middle term ($2AB$ or $-2AB$) is plus or minus twice the product of the square roots of the outer two terms.

EXAMPLE 9 ■ Recognizing Perfect Squares

Factor each trinomial: (a) $x^2 + 6x + 9$ (b) $4x^2 - 4xy + y^2$

SOLUTION

(a) Here $A = x$ and $B = 3$, so $2AB = 2 \cdot x \cdot 3 = 6x$. Since the middle term is $6x$, the trinomial is a perfect square. By Formula 2 we have

$$x^2 + 6x + 9 = (x + 3)^2$$

(b) Here $A = 2x$ and $B = y$, so $2AB = 2 \cdot 2x \cdot y = 4xy$. Since the middle term is $-4xy$, the trinomial is a perfect square. By Formula 2 we have

$$4x^2 - 4xy + y^2 = (2x - y)^2$$ ■

When we factor an expression, the result can sometimes be factored further. In general, we first factor out common factors, then inspect the result to see if it can be factored by any of the other methods of this section. We repeat this process until we have factored the expression completely.

EXAMPLE 10 ■ Factoring an Expression Completely

Factor each expression completely.

(a) $2x^4 - 8x^2$ (b) $x^5y^2 - xy^6$

SOLUTION

(a) We first factor out the power of x with the smallest exponent.

$$2x^4 - 8x^2 = 2x^2(x^2 - 4)$$ Common factor is $2x^2$
$$= 2x^2(x - 2)(x + 2)$$ Factor $x^2 - 4$ as a difference of squares

(b) We first factor out the powers of x and y with the smallest exponents.

$$x^5y^2 - xy^6 = xy^2(x^4 - y^4) \qquad \text{Common factor is } xy^2$$

$$= xy^2(x^2 + y^2)(x^2 - y^2) \qquad \text{Factor } x^4 - y^4 \text{ as a difference of squares}$$

$$= xy^2(x^2 + y^2)(x + y)(x - y) \qquad \text{Factor } x^2 - y^2 \text{ as a difference of squares} \quad \blacksquare$$

In the next example we factor out variables with fractional exponents. This type of factoring occurs in calculus.

EXAMPLE 11 ■ Factoring Expressions with Fractional Exponents

Factor each expression.

(a) $3x^{3/2} - 9x^{1/2} + 6x^{-1/2}$ (b) $(1 + x)^{-2/3}x + (1 + x)^{1/3}$

SOLUTION

(a) Factor out the power of x with the *smallest exponent*, that is, $x^{-1/2}$.

$$3x^{3/2} - 9x^{1/2} + 6x^{-1/2} = 3x^{-1/2}(x^2 - 3x + 2) \qquad \text{Factor out } 3x^{-1/2}$$

$$= 3x^{-1/2}(x - 1)(x - 2) \qquad \text{Factor the quadratic } x^2 - 3x + 2$$

(b) Factor out the power of $1 + x$ with the *smallest exponent*, that is, $(1 + x)^{-2/3}$.

$$(1 + x)^{-2/3}x + (1 + x)^{1/3} = (1 + x)^{-2/3}[x + (1 + x)] \qquad \text{Factor out } (1 + x)^{-2/3}$$

$$= (1 + x)^{-2/3}(1 + 2x) \qquad \blacksquare$$

CHECK YOUR ANSWER

To see that you have factored correctly, multiply using the Laws of Exponents.
(a) $3x^{-1/2}(x^2 - 3x + 2)$
    $= 3x^{3/2} - 9x^{1/2} + 6x^{-1/2}$ ✓
(b) $(1 + x)^{-2/3}[x + (1 + x)]$
    $= (1 + x)^{-2/3}x + (1 + x)^{1/3}$ ✓

Polynomials with at least four terms can sometimes be factored by grouping terms. The following example illustrates the idea.

EXAMPLE 12 ■ Factoring by Grouping

Factor each polynomial.

(a) $x^3 + x^2 + 4x + 4$ (b) $x^3 - 2x^2 - 3x + 6$

SOLUTION

(a) $x^3 + x^2 + 4x + 4 = (x^3 + x^2) + (4x + 4)$ Group terms

$$= x^2(x + 1) + 4(x + 1) \qquad \text{Factor out common factors}$$

$$= (x^2 + 4)(x + 1) \qquad \text{Factor out } x + 1 \text{ from each term}$$

(b) $x^3 - 2x^2 - 3x + 6 = (x^3 - 2x^2) - (3x - 6)$ Group terms

$$= x^2(x - 2) - 3(x - 2) \qquad \text{Factor out common factors}$$

$$= (x^2 - 3)(x - 2) \qquad \text{Factor out } x - 2 \text{ from each term} \quad \blacksquare$$

1.3 EXERCISES

1–34 ■ Perform the indicated operations and simplify.

1. $(3x^2 + x + 1) + (2x^2 - 3x - 5)$

2. $(3x^2 + x + 1) - (2x^2 - 3x - 5)$

3. $(x^3 + 6x^2 - 4x + 7) - (3x^2 + 2x - 4)$

4. $3(x - 1) + 4(x + 2)$

5. $8(2x + 5) - 7(x - 9)$

6. $4(x^2 - 3x + 5) - 3(x^2 - 2x + 1)$

7. $2(2 - 5t) + t^2(t - 1) - (t^4 - 1)$

8. $5(3t - 4) - (t^2 + 2) - 2t(t - 3)$

9. $\sqrt{x}\,(x - \sqrt{x})$ **10.** $x^{3/2}\bigl(\sqrt{x} - 1/\sqrt{x}\bigr)$

11. $y^{1/3}\,(y^2 - 1)$ **12.** $y^{1/4}\,(y^{1/2} + 2y^{3/4})$

13. $(3t - 2)(7t - 5)$ **14.** $(4x - 1)(3x + 7)$

15. $(x + 2y)(3x - y)$ **16.** $(4x - 3y)(2x + 5y)$

17. $(1 - 2y)^2$ **18.** $(3x + 4)^2$

19. $(2x^2 + 3y^2)^2$ **20.** $\left(c + \dfrac{1}{c}\right)^2$

21. $(2x - 5)(x^2 - x + 1)$ **22.** $(1 + 2x)(x^2 - 3x + 1)$

23. $(x^2 - a^2)(x^2 + a^2)$ **24.** $(x^{1/2} + y^{1/2})(x^{1/2} - y^{1/2})$

25. $\left(\sqrt{a} - \dfrac{1}{b}\right)\left(\sqrt{a} + \dfrac{1}{b}\right)$

26. $\left(\sqrt{h^2 + 1} + 1\right)\left(\sqrt{h^2 + 1} - 1\right)$

27. $(1 + a^3)^3$ **28.** $(1 - 2y)^3$

29. $(x^2 + x - 2)(x^3 - x + 1)$

30. $(1 + x + x^2)(1 - x + x^2)$

31. $(1 + x^{4/3})(1 - x^{2/3})$

32. $(1 - b)^2(1 + b)^2$

33. $(3x^2y + 7xy^2)(x^2y^3 - 2y^2)$

34. $(x^4y - y^5)(x^2 + xy + y^2)$

35–88 ■ Factor the expression completely.

35. $12x^3 + 18x$ **36.** $30x^3 + 15x^4$

37. $6y^4 - 15y^3$ **38.** $5ab - 8abc$

39. $x^2 - 2x - 8$ **40.** $x^2 - 14x + 48$

41. $y^2 - 8y + 15$ **42.** $z^2 + 6z - 16$

43. $2x^2 + 5x + 3$ **44.** $2x^2 + 7x - 4$

45. $9x^2 - 36x - 45$ **46.** $8x^2 + 10x + 3$

47. $6x^2 - 5x - 6$ **48.** $6 + 5t - 6t^2$

49. $4t^2 - 12t + 9$ **50.** $4x^2 + 4xy + y^2$

51. $r^2 - 6rs + 9s^2$ **52.** $25s^2 - 10st + t^2$

53. $x^2 - 36$ **54.** $4x^2 - 25$

55. $49 - 4y^2$ **56.** $4t^2 - 9s^2$

57. $(a + b)^2 - (a - b)^2$ **58.** $\left(1 + \dfrac{1}{x}\right)^2 - \left(1 - \dfrac{1}{x}\right)^2$

59. $x^2(x^2 - 1) - 9(x^2 - 1)$ **60.** $(a^2 - 1)b^2 - 4(a^2 - 1)$

61. $t^3 + 1$ **62.** $x^3 - 27$

63. $8x^3 - 125$ **64.** $x^6 + 64$

65. $x^6 - 8y^3$ **66.** $27a^3 + b^6$

67. $x^3 + 2x^2 + x$ **68.** $3x^3 - 27x$

69. $x^4 + 2x^3 - 3x^2$ **70.** $x^3 + 3x^2 - x - 3$

71. $y^3 - 3y^2 - 4y + 12$ **72.** $y^3 - y^2 + y - 1$

73. $2x^3 + 4x^2 + x + 2$ **74.** $3x^3 + 5x^2 - 6x - 10$

75. $(x - 1)(x + 2)^2 - (x - 1)^2(x + 2)$

76. $(x + 1)^3x - 2(x + 1)^2x^2 + x^3(x + 1)$

77. $y^4(y + 2)^3 + y^5(y + 2)^4$

78. $n(x - y) + (n - 1)(y - x)$

79. $x^{5/2} - x^{1/2}$ **80.** $3x^{-1/2} + 4x^{1/2} + x^{3/2}$

81. $x^{-3/2} + 2x^{-1/2} + x^{1/2}$ **82.** $(x - 1)^{7/2} - (x - 1)^{3/2}$

83. $(x^2 + 1)^{1/2} + 2(x^2 + 1)^{-1/2}$

84. $x^{-1/2}(x + 1)^{1/2} + x^{1/2}(x + 1)^{-1/2}$

85. $(a^2 + 1)^2 - 7(a^2 + 1) + 10$

86. $(a^2 + 2a)^2 - 2(a^2 + 2a) - 3$

87–90 ■ Factor the expression completely. (This type of expression arises in calculus when using the "product rule.")

87. $3x^2(4x - 12)^2 + x^3(2)(4x - 12)(4)$

88. $5(x^2 + 4)^4(2x)(x - 2)^4 + (x^2 + 4)^5(4)(x - 2)^3$

89. $3(2x - 1)^2(2)(x + 3)^{1/2} + (2x - 1)^3\bigl(\tfrac{1}{2}\bigr)(x + 3)^{-1/2}$

90. $\tfrac{1}{3}(x + 6)^{-2/3}(2x - 3)^2 + (x + 6)^{1/3}(2)(2x - 3)(2)$

91. (a) Show that $ab = \frac{1}{2}[(a+b)^2 - (a^2 + b^2)]$.
   (b) Show that $(a^2 + b^2)^2 - (a^2 - b^2)^2 = 4a^2b^2$.
   (c) Show that

$$(a^2 + b^2)(c^2 + d^2) = (ac + bd)^2 + (ad - bc)^2$$

   (d) Factor completely: $4a^2c^2 - (a^2 - b^2 + c^2)^2$.

92. Verify each formula algebraically.
   (a) Special Product Formulas 1 and 2
   (b) Special Product Formulas 3 and 4
   (c) Factoring Formula 4
   (d) Factoring Formula 5

 DISCOVERY · DISCUSSION

93. The Power of Algebraic Formulas   Use the Difference of Squares Formula to factor $17^2 - 16^2$. Notice that it is easy to calculate the factored form in your head, but not so easy to calculate the original form in this way. Evaluate each expression in your head:
   (a) $528^2 - 527^2$
   (b) $122^2 - 120^2$
   (c) $1020^2 - 1010^2$

Now use Special Product Formula 1 to evaluate these products in your head:
   (d) $49 \cdot 51$
   (e) $998 \cdot 1002$

94. Volume of a Cylindrical Shell   Using the formula for the volume of a cylinder given on the inside back cover of this book, explain why the volume of the cylindrical shell shown in the figure is

$$V = \pi R^2 h - \pi r^2 h$$

Factor this to show that

$$V = 2\pi \cdot \text{average radius} \cdot \text{height} \cdot \text{thickness}$$

Use the "unrolled" diagram to explain why this makes sense geometrically.

95. Differences of Even Powers
   (a) Factor the expressions completely: $A^4 - B^4$ and $A^6 - B^6$.
   (b) Verify that $18{,}335 = 12^4 - 7^4$ and that $2{,}868{,}335 = 12^6 - 7^6$.
   (c) Use the results of parts (a) and (b) to factor the integers 18,335 and 2,868,335. Show that in both of these factorizations, all the factors are prime numbers.

96. Factoring $A^n - 1$ Verify the factoring formulas in the list by expanding and simplifying the right-hand side in each case.

$$A^2 - 1 = (A - 1)(A + 1)$$

$$A^3 - 1 = (A - 1)(A^2 + A + 1)$$

$$A^4 - 1 = (A - 1)(A^3 + A^2 + A + 1)$$

Based on the pattern displayed in this list, how do you think $A^5 - 1$ would factor? Verify your conjecture. Now generalize the pattern you have observed to obtain a factorization formula for $A^n - 1$, where n is a positive integer.

97. Factoring $x^4 + ax^2 + b$ A trinomial of the form $x^4 + ax^2 + b$ can sometimes be factored easily. For example, $x^4 + 3x^2 - 4 = (x^2 + 4)(x^2 - 1)$. But $x^4 + 3x^2 + 4$ cannot be factored in this way. Instead, we can use the following method.

$$\begin{aligned} x^4 + 3x^2 + 4 &= (x^4 + 4x^2 + 4) - x^2 &&\text{\color{red}Add and subtract } x^2 \\ &= (x^2 + 2)^2 - x^2 &&\text{\color{red}Factor perfect square} \\ &= [(x^2 + 2) - x][(x^2 + 2) + x] &&\text{\color{red}Difference of squares} \\ &= (x^2 - x + 2)(x^2 + x + 2) \end{aligned}$$

Factor the following using whichever method is appropriate.
   (a) $x^4 + x^2 - 2$
   (b) $x^4 + 2x^2 + 9$
   (c) $x^4 + 4x^2 + 16$
   (d) $x^4 + 2x^2 + 1$

Visualizing a Formula

Many of the Special Product Formulas that we learned in this section can be "seen" as geometrical facts about length, area, and volume. For example, the figure shows how the formula for the square of a binomial can be interpreted as a fact about areas of squares and rectangles.

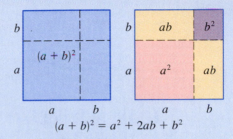

$$(a + b)^2 = a^2 + 2ab + b^2$$

In the figure, a and b represent lengths, a^2, b^2, ab, and $(a + b)^2$ represent areas. The ancient Greeks always interpreted algebraic formulas in terms of geometric figures as we have done here.

1. Explain how the figure verifies the formula $a^2 - b^2 = (a + b)(a - b)$.

2. Find a figure that verifies the formula $(a - b)^2 = a^2 - 2ab + b^2$.

3. Explain how the figure verifies the formula
$(a + b)^3 = a^3 + 3a^2b + 3ab^2 + b^3$.

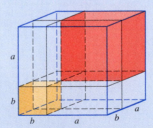

4. Is it possible to draw a geometric figure that verifies the formula for $(a + b)^4$? Explain.

5. (a) Expand $(a + b + c)^2$.
(b) Make a geometric figure that verifies the formula you found in part (a).

1.4 FRACTIONAL EXPRESSIONS

A quotient of two algebraic expressions is called a **fractional expression**. We assume that all fractions are defined; that is, *we deal only with values of the variables such that the denominators are not zero.*

A common type of fractional expression occurs when both the numerator and denominator are polynomials. This is called a **rational expression**. For instance,

$$\frac{4x^3 + 2x + 5}{x + 3}$$

is a rational expression whose denominator is 0 when $x = -3$. So, in dealing with this expression, we implicitly assume that $x \neq -3$.

■ Simplifying Fractional Expressions

To **simplify fractional expressions**, we factor both numerator and denominator and use the following property of fractions:

$$\boxed{\frac{AC}{BC} = \frac{A}{B}}$$

This says that we can **cancel** common factors from numerator and denominator.

EXAMPLE 1 ■ Simplifying Fractional Expressions by Cancellation

Simplify: $\dfrac{x^2 - 1}{x^2 + x - 2}$

SOLUTION

⊘ We can't cancel the x^2's in
$\dfrac{x^2 - 1}{x^2 + x - 2}$ because x^2 is not a factor.

$$\frac{x^2 - 1}{x^2 + x - 2} = \frac{(x - 1)(x + 1)}{(x - 1)(x + 2)} \qquad \text{Factor}$$

$$= \frac{x + 1}{x + 2} \qquad \text{Cancel common factors} \qquad ■$$

■ Multiplying and Dividing Fractional Expressions

To **multiply fractional expressions**, we use the following property of fractions:

$$\boxed{\frac{A}{B} \cdot \frac{C}{D} = \frac{AC}{BD}}$$

This says that to multiply two fractions we multiply their numerators and multiply their denominators.

EXAMPLE 2 ■ Multiplying Fractional Expressions

Perform the indicated multiplication and simplify: $\dfrac{x^2 + 2x - 3}{x^2 + 8x + 16} \cdot \dfrac{3x + 12}{x - 1}$

SOLUTION

We first factor.

$$\frac{x^2 + 2x - 3}{x^2 + 8x + 16} \cdot \frac{3x + 12}{x - 1} = \frac{(x - 1)(x + 3)}{(x + 4)^2} \cdot \frac{3(x + 4)}{x - 1} \qquad \text{Factor}$$

$$= \frac{3(x - 1)(x + 3)(x + 4)}{(x - 1)(x + 4)^2} \qquad \text{Property of fractions}$$

$$= \frac{3(x + 3)}{x + 4} \qquad \text{Cancel common factors} \qquad ■$$

To **divide fractional expressions**, we use the following property of fractions:

$$\frac{A}{B} \div \frac{C}{D} = \frac{A}{B} \cdot \frac{D}{C}$$

This says that to divide a fraction by another fraction we invert the divisor and multiply.

EXAMPLE 3 ■ Dividing Fractional Expressions

Perform the indicated division and simplify: $\dfrac{x - 4}{x^2 - 4} \div \dfrac{x^2 - 3x - 4}{x^2 + 5x + 6}$

SOLUTION

$$\frac{x - 4}{x^2 - 4} \div \frac{x^2 - 3x - 4}{x^2 + 5x + 6} = \frac{x - 4}{x^2 - 4} \cdot \frac{x^2 + 5x + 6}{x^2 - 3x - 4} \qquad \text{Invert and multiply}$$

$$= \frac{(x - 4)(x + 2)(x + 3)}{(x - 2)(x + 2)(x - 4)(x + 1)} \qquad \text{Factor}$$

$$= \frac{x + 3}{(x - 2)(x + 1)} \qquad \text{Cancel common factors} \qquad ■$$

■ Adding and Subtracting Fractional Expressions

To **add or subtract fractional expressions**, we first find a common denominator and then use the following property of fractions:

$$\frac{A}{C} + \frac{B}{C} = \frac{A + B}{C}$$

🚫 Avoid making the following error:

$$\frac{A}{B + C} \ \cancel{\times} \ \frac{A}{B} + \frac{A}{C}$$

For instance, if we let $A = 2$, $B = 1$, and $C = 1$, then we see the error:

$$\frac{2}{1 + 1} \overset{?}{=} \frac{2}{1} + \frac{2}{1}$$

$$\frac{2}{2} \overset{?}{=} 2 + 2$$

$$1 \overset{?}{=} 4 \quad \textbf{Wrong!}$$

Although any common denominator will work, it is best to use the **least common denominator** (LCD) as explained in Section 1.1. The LCD is found by factoring each denominator and taking the product of the distinct factors, using the highest power that appears in any of the factors.

EXAMPLE 4 ■ Adding and Subtracting Fractional Expressions

Perform the indicated operations and simplify.

(a) $\dfrac{3}{x - 1} + \dfrac{x}{x + 2}$

(b) $\dfrac{1}{x^2 - 1} - \dfrac{2}{(x + 1)^2}$

SOLUTION

(a) Here the LCD is simply the product $(x - 1)(x + 2)$, so we have

$$\frac{3}{x - 1} + \frac{x}{x + 2} = \frac{3(x + 2)}{(x - 1)(x + 2)} + \frac{x(x - 1)}{(x - 1)(x + 2)} \qquad \text{Write fractions using LCD}$$

$$= \frac{3x + 6 + x^2 - x}{(x - 1)(x + 2)} \qquad \text{Add fractions}$$

$$= \frac{x^2 + 2x + 6}{(x - 1)(x + 2)} \qquad \text{Combine terms in numerator}$$

(b) The LCD of $x^2 - 1 = (x - 1)(x + 1)$ and $(x + 1)^2$ is $(x - 1)(x + 1)^2$, so we have

$$\frac{1}{x^2 - 1} - \frac{2}{(x + 1)^2} = \frac{1}{(x - 1)(x + 1)} - \frac{2}{(x + 1)^2} \qquad \text{Factor}$$

$$= \frac{(x + 1) - 2(x - 1)}{(x - 1)(x + 1)^2} \qquad \text{Combine fractions using LCD}$$

$$= \frac{x + 1 - 2x + 2}{(x - 1)(x + 1)^2} \qquad \text{Distributive Property}$$

$$= \frac{3 - x}{(x - 1)(x + 1)^2} \qquad \text{Combine terms in numerator}$$

■

■ Compound Fractions

A **compound fraction** is a fraction in which the numerator, the denominator, or both, are themselves fractional expressions.

EXAMPLE 5 ■ Simplifying a Compound Fraction

Simplify: $\dfrac{\dfrac{x}{y} + 1}{1 - \dfrac{y}{x}}$

SOLUTION 1

We combine the terms in the numerator into a single fraction. We do the same in the denominator. Then we invert and multiply.

$$\frac{\dfrac{x}{y} + 1}{1 - \dfrac{y}{x}} = \frac{\dfrac{x+y}{y}}{\dfrac{x-y}{x}} = \frac{x+y}{y} \cdot \frac{x}{x-y}$$

$$= \frac{x(x+y)}{y(x-y)}$$

SOLUTION 2

We find the LCD of all the fractions in the expression, then multiply numerator and denominator by it. In this example the LCD of all the fractions is xy. Thus

$$\frac{\dfrac{x}{y} + 1}{1 - \dfrac{y}{x}} = \frac{\dfrac{x}{y} + 1}{1 - \dfrac{y}{x}} \cdot \frac{xy}{xy} \qquad \text{Multiply numerator and denominator by } xy$$

$$= \frac{x^2 + xy}{xy - y^2} \qquad \text{Simplify}$$

$$= \frac{x(x+y)}{y(x-y)} \qquad \text{Factor} \qquad ■$$

The next two examples show situations in calculus that require the ability to work with fractional expressions.

EXAMPLE 6 ■ Simplifying a Compound Fraction

Simplify: $\dfrac{\dfrac{1}{a+h} - \dfrac{1}{a}}{h}$

SOLUTION

We begin by combining the fractions in the numerator using a common denominator.

$$\frac{\dfrac{1}{a+h} - \dfrac{1}{a}}{h} = \frac{\dfrac{a - (a+h)}{a(a+h)}}{h} \qquad \text{Combine fractions in the numerator}$$

$$= \frac{a - (a+h)}{a(a+h)} \cdot \frac{1}{h} \qquad \begin{array}{l}\text{Property 2 of fractions} \\ \text{(invert divisor and multiply)}\end{array}$$

$$= \frac{a - a - h}{a(a+h)} \cdot \frac{1}{h} \qquad \text{Distributive Property}$$

$$= \frac{-h}{a(a+h)} \cdot \frac{1}{h} \qquad \text{Simplify}$$

$$= \frac{-1}{a(a+h)} \qquad \begin{array}{l}\text{Property 5 of fractions} \\ \text{(cancel common factors)}\end{array}$$

■

EXAMPLE 7 ■ Simplifying a Compound Fraction

Simplify: $\quad \dfrac{(1+x^2)^{1/2} - x^2(1+x^2)^{-1/2}}{1+x^2}$

SOLUTION 1

Factor out the power of $1 + x^2$ with the *smallest* exponent, in this case $(1+x^2)^{-1/2}$.

Factor $(1+x^2)^{-1/2}$ from the numerator.

$$\frac{(1+x^2)^{1/2} - x^2(1+x^2)^{-1/2}}{1+x^2} = \frac{(1+x^2)^{-1/2}[(1+x^2) - x^2]}{1+x^2}$$

$$= \frac{(1+x^2)^{-1/2}}{1+x^2} = \frac{1}{(1+x^2)^{3/2}}$$

SOLUTION 2

Since $(1+x^2)^{-1/2} = 1/(1+x^2)^{1/2}$ is a fraction, we may clear all fractions by multiplying numerator and denominator by $(1+x^2)^{1/2}$.

$$\frac{(1+x^2)^{1/2} - x^2(1+x^2)^{-1/2}}{1+x^2} = \frac{(1+x^2)^{1/2} - x^2(1+x^2)^{-1/2}}{1+x^2} \cdot \frac{(1+x^2)^{1/2}}{(1+x^2)^{1/2}}$$

$$= \frac{(1+x^2) - x^2}{(1+x^2)^{3/2}} = \frac{1}{(1+x^2)^{3/2}}$$

■

Rationalizing the Denominator or the Numerator

If a fraction has a denominator of the form $A + B\sqrt{C}$, we may rationalize the denominator by multiplying numerator and denominator by the **conjugate radical** $A - B\sqrt{C}$. This is effective because, by Product Formula 1 in Section 1.3, the product of the denominator and its conjugate radical does not contain a radical:

$$\left(A + B\sqrt{C}\right)\left(A - B\sqrt{C}\right) = A^2 - B^2C$$

EXAMPLE 8 ■ Rationalizing the Denominator

Rationalize the denominator: $\dfrac{1}{1 + \sqrt{2}}$

SOLUTION

We multiply both the numerator and the denominator by the conjugate radical of $1 + \sqrt{2}$, which is $1 - \sqrt{2}$.

Product Formula 1

$$(a + b)(a - b) = a^2 - b^2$$

$$\frac{1}{1 + \sqrt{2}} = \frac{1}{1 + \sqrt{2}} \cdot \frac{1 - \sqrt{2}}{1 - \sqrt{2}}$$

Multiply numerator and denominator by the conjugate radical

$$= \frac{1 - \sqrt{2}}{1^2 - \left(\sqrt{2}\right)^2}$$

Product Formula 1

$$= \frac{1 - \sqrt{2}}{1 - 2} = \frac{1 - \sqrt{2}}{-1} = \sqrt{2} - 1$$

■

EXAMPLE 9 ■ Rationalizing the Numerator

Rationalize the numerator: $\dfrac{\sqrt{4 + h} - 2}{h}$

SOLUTION

We multiply numerator and denominator by the conjugate radical $\sqrt{4 + h} + 2$.

$$\frac{\sqrt{4 + h} - 2}{h} = \frac{\sqrt{4 + h} - 2}{h} \cdot \frac{\sqrt{4 + h} + 2}{\sqrt{4 + h} + 2}$$

Multiply numerator and denominator by the conjugate radical

$$= \frac{\left(\sqrt{4 + h}\right)^2 - 2^2}{h\left(\sqrt{4 + h} + 2\right)}$$

Product Formula 1

$$= \frac{4 + h - 4}{h\left(\sqrt{4 + h} + 2\right)}$$

$$= \frac{h}{h\left(\sqrt{4 + h} + 2\right)} = \frac{1}{\sqrt{4 + h} + 2}$$

Property 5 of fractions (cancel common factors)

■

■ Avoiding Common Errors

 Don't make the mistake of applying properties of multiplication to the operation of addition. Many of the common errors in algebra involve doing just that. The following table states several properties of multiplication and illustrates the error in applying them to addition.

Correct multiplication property	Common error with addition
$(a \cdot b)^2 = a^2 \cdot b^2$	$(a + b)^2 \ne a^2 + b^2$
$\sqrt{a \cdot b} = \sqrt{a}\,\sqrt{b} \quad (a, b \ge 0)$	$\sqrt{a + b} \ne \sqrt{a} + \sqrt{b}$
$\sqrt{a^2 \cdot b^2} = a \cdot b \quad (a, b \ge 0)$	$\sqrt{a^2 + b^2} \ne a + b$
$\dfrac{1}{a} \cdot \dfrac{1}{b} = \dfrac{1}{a \cdot b}$	$\dfrac{1}{a} + \dfrac{1}{b} \ne \dfrac{1}{a + b}$
$\dfrac{ab}{a} = b$	$\dfrac{a + b}{a} \ne b$
$a^{-1} \cdot b^{-1} = (a \cdot b)^{-1}$	$a^{-1} + b^{-1} \ne (a + b)^{-1}$

To verify that the formulas in the right-hand column are wrong, simply substitute numbers for a and b and calculate each side. For example, if we take $a = 2$ and $b = 2$ in the fourth property, we find that the left-hand side is

$$\frac{1}{a} + \frac{1}{b} = \frac{1}{2} + \frac{1}{2} = 1$$

whereas the right-hand side is

$$\frac{1}{a + b} = \frac{1}{2 + 2} = \frac{1}{4}$$

Since $1 \ne \frac{1}{4}$, the stated formula is wrong. You should similarly convince yourself of the error in each of the other formulas. (See Exercise 82.)

1.4 EXERCISES

1–52 ■ Simplify the expression.

1. $\dfrac{x - 2}{x^2 - 4}$

2. $\dfrac{x^2 - x - 2}{x^2 - 1}$

3. $\dfrac{x^2 + 6x + 8}{x^2 + 5x + 4}$

4. $\dfrac{x^2 - x - 12}{x^2 + 5x + 6}$

5. $\dfrac{y^2 + y}{y^2 - 1}$

6. $\dfrac{y^2 - 3y - 18}{2y^2 + 5y + 3}$

7. $\dfrac{2x^3 - x^2 - 6x}{2x^2 - 7x + 6}$

8. $\dfrac{1 - x^2}{x^3 - 1}$

9. $\dfrac{t - 3}{t^2 + 9} \cdot \dfrac{t + 3}{t^2 - 9}$

10. $\dfrac{x^2 - x - 6}{x^2 + 2x} \cdot \dfrac{x^3 + x^2}{x^2 - 2x - 3}$

11. $\dfrac{x^2 + 7x + 12}{x^2 + 3x + 2} \cdot \dfrac{x^2 + 5x + 6}{x^2 + 6x + 9}$

12. $\dfrac{x^2 + 2xy + y^2}{x^2 - y^2} \cdot \dfrac{2x^2 - xy - y^2}{x^2 - xy - 2y^2}$

13. $\dfrac{2x^2 + 3x + 1}{x^2 + 2x - 15} \div \dfrac{x^2 + 6x + 5}{2x^2 - 7x + 3}$

14. $\dfrac{4y^2 - 9}{2y^2 + 9y - 18} \div \dfrac{2y^2 + y - 3}{y^2 + 5y - 6}$

15. $\dfrac{\dfrac{x^3}{x + 1}}{\dfrac{x}{x^2 + 2x + 1}}$

16. $\dfrac{\dfrac{2x^2 - 3x - 2}{x^2 - 1}}{\dfrac{2x^2 + 5x + 2}{x^2 + x - 2}}$

17. $\dfrac{x/y}{z}$

18. $\dfrac{x}{y/z}$

19. $\dfrac{1}{x + 5} + \dfrac{2}{x - 3}$

20. $\dfrac{1}{x + 1} + \dfrac{1}{x - 1}$

21. $\dfrac{1}{x + 1} - \dfrac{1}{x + 2}$

22. $\dfrac{x}{x - 4} - \dfrac{3}{x + 6}$

23. $\dfrac{x}{(x + 1)^2} + \dfrac{2}{x + 1}$

24. $\dfrac{5}{2x - 3} - \dfrac{3}{(2x - 3)^2}$

25. $u + 1 + \dfrac{u}{u + 1}$

26. $\dfrac{2}{a^2} - \dfrac{3}{ab} + \dfrac{4}{b^2}$

27. $\dfrac{1}{x^2} + \dfrac{1}{x^2 + x}$

28. $\dfrac{1}{x} + \dfrac{1}{x^2} + \dfrac{1}{x^3}$

29. $\dfrac{2}{x + 3} - \dfrac{1}{x^2 + 7x + 12}$

30. $\dfrac{x}{x^2 - 4} + \dfrac{1}{x - 2}$

31. $\dfrac{1}{x + 3} + \dfrac{1}{x^2 - 9}$

32. $\dfrac{x}{x^2 + x - 2} - \dfrac{2}{x^2 - 5x + 4}$

33. $\dfrac{2}{x} + \dfrac{3}{x - 1} - \dfrac{4}{x^2 - x}$

34. $\dfrac{x}{x^2 - x - 6} - \dfrac{1}{x + 2} - \dfrac{2}{x - 3}$

35. $\dfrac{1}{x^2 + 3x + 2} - \dfrac{1}{x^2 - 2x - 3}$

36. $\dfrac{1}{x + 1} - \dfrac{2}{(x + 1)^2} + \dfrac{3}{x^2 - 1}$

37. $\dfrac{\dfrac{x}{y} - \dfrac{y}{x}}{\dfrac{1}{x^2} - \dfrac{1}{y^2}}$

38. $x - \dfrac{y}{\dfrac{x}{y} + \dfrac{y}{x}}$

39. $\dfrac{1 + \dfrac{1}{c - 1}}{1 - \dfrac{1}{c - 1}}$

40. $1 + \dfrac{1}{1 + \dfrac{1}{1 + x}}$

41. $\dfrac{\dfrac{5}{x - 1} - \dfrac{2}{x + 1}}{\dfrac{x}{x - 1} + \dfrac{1}{x + 1}}$

42. $\dfrac{\dfrac{a - b}{a} - \dfrac{a + b}{b}}{\dfrac{a - b}{b} + \dfrac{a + b}{a}}$

43. $\dfrac{x^{-2} - y^{-2}}{x^{-1} + y^{-1}}$

44. $\dfrac{x^{-1} + y^{-1}}{(x + y)^{-1}}$

45. $\dfrac{1}{1 + a^n} + \dfrac{1}{1 + a^{-n}}$

46. $\dfrac{\left(a + \dfrac{1}{b}\right)^m \left(a - \dfrac{1}{b}\right)^n}{\left(b + \dfrac{1}{a}\right)^m \left(b - \dfrac{1}{a}\right)^n}$

47. $\dfrac{\dfrac{1}{a + h} - \dfrac{1}{a}}{h}$

48. $\dfrac{(x + h)^{-3} - x^{-3}}{h}$

49. $\dfrac{\dfrac{1 - (x + h)}{2 + (x + h)} - \dfrac{1 - x}{2 + x}}{h}$

50. $\dfrac{(x + h)^3 - 7(x + h) - (x^3 - 7x)}{h}$

51. $\sqrt{1 + \left(\dfrac{x}{\sqrt{1 - x^2}}\right)^2}$

52. $\sqrt{1 + \left(x^3 - \dfrac{1}{4x^3}\right)^2}$

53–58 ■ Simplify the expression. (This type of expression arises in calculus when using the "quotient rule.")

53. $\dfrac{3(x + 2)^2(x - 3)^2 - (x + 2)^3(2)(x - 3)}{(x - 3)^4}$

54. $\dfrac{2x(x + 6)^4 - x^2(4)(x + 6)^3}{(x + 6)^8}$

55. $\dfrac{2(1 + x)^{1/2} - x(1 + x)^{-1/2}}{x + 1}$

56. $\dfrac{(1 - x^2)^{1/2} + x^2(1 - x^2)^{-1/2}}{1 - x^2}$

57. $\dfrac{3(1 + x)^{1/3} - x(1 + x)^{-2/3}}{(1 + x)^{2/3}}$

58. $\dfrac{(7 - 3x)^{1/2} + \frac{3}{2}x(7 - 3x)^{-1/2}}{7 - 3x}$

59–62 ■ Rationalize the denominator.

59. $\dfrac{2}{3 + \sqrt{5}}$

60. $\dfrac{1}{\sqrt{x} + 1}$

61. $\dfrac{2}{\sqrt{2} + \sqrt{7}}$

62. $\dfrac{y}{\sqrt{3} + \sqrt{y}}$

63–68 ■ Rationalize the numerator.

63. $\dfrac{1 - \sqrt{5}}{3}$

64. $\dfrac{\sqrt{3} + \sqrt{5}}{2}$

65. $\dfrac{\sqrt{r} + \sqrt{2}}{5}$

66. $\dfrac{\sqrt{x} - \sqrt{x + h}}{h\sqrt{x}\sqrt{x + h}}$

67. $\sqrt{x^2 + 1} - x$

68. $\sqrt{x + 1} - \sqrt{x}$

69–78 ■ State whether the given equation is true for all values of the variables. (Disregard any value that makes a denominator zero.)

69. $\dfrac{16 + a}{16} = 1 + \dfrac{a}{16}$

70. $\dfrac{b}{b - c} = 1 - \dfrac{b}{c}$

71. $\dfrac{2}{4 + x} = \dfrac{1}{2} + \dfrac{2}{x}$

72. $\dfrac{x + 1}{y + 1} = \dfrac{x}{y}$

73. $\dfrac{x}{x + y} = \dfrac{1}{1 + y}$

74. $2\left(\dfrac{a}{b}\right) = \dfrac{2a}{2b}$

75. $\dfrac{-a}{b} = -\dfrac{a}{b}$

76. $\dfrac{1 + x + x^2}{x} = \dfrac{1}{x} + 1 + x$

77. $\dfrac{x^2 + 1}{x^2 + x - 1} = \dfrac{1}{x - 1}$

78. $\dfrac{x^2 - 1}{x - 1} = x + 1$

79. If two electrical resistors with resistances R_1 and R_2 are connected in parallel (see the figure), then the total resistance R is given by

$$R = \dfrac{1}{\dfrac{1}{R_1} + \dfrac{1}{R_2}}$$

(a) Simplify the expression for R.
(b) If $R_1 = 10$ ohms and $R_2 = 20$ ohms, what is the total resistance R?

 DISCOVERY · DISCUSSION

80. Limiting Behavior of a Rational Expression   The rational expression

$$\dfrac{x^2 - 9}{x - 3}$$

is not defined for $x = 3$. Complete the tables and determine what value the expression approaches as x gets closer and closer to 3. Why is this reasonable? Factor the numerator of the expression and simplify to see why.

x	$\dfrac{x^2 - 9}{x - 3}$	x	$\dfrac{x^2 - 9}{x - 3}$
2.80		3.20	
2.90		3.10	
2.95		3.05	
2.99		3.01	
2.999		3.001	

81. Is This Rationalization?   In the expression

$$\dfrac{2}{\sqrt{x}}$$

we would eliminate the radical if we were to square both numerator and denominator. Is this the same thing as rationalizing the denominator?

82. Algebraic Errors   The left-hand column in the table lists some common algebraic errors. In each case, give an example using numbers that shows that the formula is not valid. An example of this type, which shows that a statement is false, is called a *counterexample*.

Algebraic error	Counterexample
$\dfrac{1}{a} + \dfrac{1}{b} \neq \dfrac{1}{a + b}$	$\dfrac{1}{2} + \dfrac{1}{2} \neq \dfrac{1}{2 + 2}$
$(a + b)^2 \neq a^2 + b^2$	
$\sqrt{a^2 + b^2} \neq a + b$	
$\dfrac{a + b}{a} \neq b$	
$(a^3 + b^3)^{1/3} \neq a + b$	
$a^m / a^n \neq a^{m/n}$	
$a^{-1/n} \neq \dfrac{1}{a^n}$	

83. The Form of an Algebraic Expression An algebraic expression may look complicated, but its "form" is always simple; it must be a sum, a product, a quotient, or a power. For example, consider the following expressions:

$$(1 + x^2)^2 + \left(\frac{x + 2}{x + 1}\right)^3 \qquad (1 + x)\left(1 + \frac{x + 5}{1 + x^4}\right)$$

$$\frac{(5 - x^3)}{1 + \sqrt{1 + x^2}} \qquad \sqrt{\frac{1 + x}{1 - x}}$$

With appropriate choices for A and B, the first has the form $A + B$, the second AB, the third A/B, and the fourth $A^{1/2}$. Recognizing the form of an expression helps us expand, simplify, or factor it correctly. Find the form of the following algebraic expressions.

(a) $x + \sqrt{1 + \dfrac{1}{x}}$ (b) $(1 + x^2)(1 + x)^3$

(c) $\sqrt[3]{x^4(4x^2 + 1)}$ (d) $\dfrac{1 - 2\sqrt{1 + x}}{1 + \sqrt{1 + x^2}}$

1.5 EQUATIONS

An equation is a statement that two mathematical expressions are equal. For example,

$$3 + 5 = 8$$

is an equation. Most equations that we study in algebra contain variables, which are symbols (usually letters) that stand for numbers. In the equation

$$4x + 7 = 19$$

the letter x is the variable. We think of x as the "unknown" in the equation, and our goal is to find the value of x that makes the equation true. The values of the unknown that make the equation true are called the **solutions** or **roots** of the equation, and the process of finding the solutions is called **solving the equation**.

Two equations with exactly the same solutions are called **equivalent equations**. To solve an equation, we try to find a simpler, equivalent equation in which the variable stands alone on one side of the "equal" sign. Here are the properties that we use to solve an equation. (In these properties, A, B, and C stand for any algebraic expressions, and the symbol $\Leftrightarrow$ means "is equivalent to.")

PROPERTIES OF EQUALITY

Property	Description
1. $A = B \Leftrightarrow A + C = B + C$	Adding the same quantity to both sides of an equation gives an equivalent equation.
2. $A = B \Leftrightarrow CA = CB \quad (C \neq 0)$	Multiplying both sides of an equation by the same nonzero quantity gives an equivalent equation.

These properties require that you *perform the same operation on both sides of an equation* when solving it. Thus, if we say "*add* -7" when solving an equation, that is just a short way of saying "*add* -7 to each side of the equation."

This is how we use the properties of equality to solve the equation $4x + 7 = 19$:

$$4x + 7 + (-7) = 19 + (-7) \qquad \text{Add } -7$$
$$4x = 12 \qquad \text{Simplify}$$
$$\tfrac{1}{4} \cdot 4x = \tfrac{1}{4} \cdot 12 \qquad \text{Multiply by } \tfrac{1}{4}$$
$$x = 3 \qquad \text{Simplify}$$

So the solution of this equation is $x = 3$. To verify this, we check our answer by subsituting $x = 3$ to make sure that this value of x does indeed make the equation true:

$$\boxed{x = 3}$$
$$\downarrow$$
$$4(3) + 7 \stackrel{?}{=} 19$$
$$19 = 19 \qquad \text{Correct!}$$

Because it is so important to CHECK YOUR ANSWER, we do this in many of the examples throughout the book. In these checks, LHS stands for "left-hand side" of the original equation, and RHS stands for "right-hand side."

■ Linear Equations

The simplest type of equation is a **linear equation**, or first-degree equation, which is an equation in which each term is either a constant or a nonzero multiple of the variable. This means that it is equivalent to an equation of the form $ax + b = 0$. Here a and b represent real numbers with $a \neq 0$, and x is the unknown variable that we are solving for. The equation in the following example is linear.

EXAMPLE 1 ■ Solving a Linear Equation

Solve the equation $7x - 4 = 3x + 8$.

SOLUTION

We solve this by changing it to an equivalent equation with all terms that have the variable x on one side and all constant terms on the other.

$$7x - 4 = 3x + 8$$
$$7x = 3x + 12 \qquad \text{Add 4}$$
$$4x = 12 \qquad \text{Subtract } 3x$$
$$x = 3 \qquad \text{Multiply by } \tfrac{1}{4}$$

CHECK YOUR ANSWER

$x = 3$:   LHS $= 7(3) - 4$ RHS $= 3(3) + 8$
                                              $= 17$ $= 17$

LHS = RHS   ✔

■

Many formulas in the sciences involve several variables, and it is often necessary to express one of the variables in terms of the others. In the next example we solve for a variable in Newton's Law of Gravity.

EXAMPLE 2 ■ Solving for One Variable in Terms of Others

Solve for the variable M in the equation

$$F = G\frac{mM}{r^2}$$

This is Newton's Law of Gravity. It gives the gravitational force F between two masses m and M that are a distance r apart. The constant G is the universal gravitational constant.

SOLUTION

Although this equation involves more than one variable, we solve it as usual by isolating M on one side and treating the other variables as we would numbers.

$$F = \left(\frac{Gm}{r^2}\right)M \qquad \text{Factor } M \text{ from RHS}$$

$$\left(\frac{r^2}{Gm}\right)F = \left(\frac{r^2}{Gm}\right)\left(\frac{Gm}{r^2}\right)M \qquad \text{Multiply by reciprocal of } \frac{Gm}{r^2}$$

$$\frac{r^2F}{Gm} = M \qquad \text{Simplify}$$

The solution is $M = \dfrac{r^2F}{Gm}$. ■

EXAMPLE 3 ■ Solving for One Variable in Terms of Others

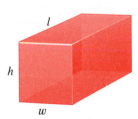

FIGURE 1
A closed rectangular box

The surface area A of the closed rectangular box shown in Figure 1 can be calculated from the length l, the width w, and the height h according to the formula

$$A = 2lw + 2wh + 2lh$$

Solve for w in terms of the other variables in this equation.

SOLUTION

Although this equation involves more than one variable, we solve it as usual by isolating w on one side, treating the other variables as we would numbers.

$$A = (2lw + 2wh) + 2lh \qquad \text{Collect terms involving } w$$

$$A - 2lh = 2lw + 2wh \qquad \text{Subtract } 2lh$$

$$A - 2lh = (2l + 2h)w \qquad \text{Factor } w \text{ from RHS}$$

$$\frac{A - 2lh}{2l + 2h} = w \qquad \text{Divide by } 2l + 2h$$

The solution is $w = \dfrac{A - 2lh}{2l + 2h}$. ■

■ Quadratic Equations

Linear equations are first-degree equations of the form $ax + b = 0$. Quadratic equations are second-degree equations; they contain an additional term involving the square of the variable.

> ### QUADRATIC EQUATIONS
>
> A **quadratic equation** is an equation of the form
>
> $$ax^2 + bx + c = 0$$
>
> where a, b, and c are real numbers with $a \neq 0$.

Some quadratic equations can be solved by factoring and using the following basic property of real numbers.

> ### ZERO-PRODUCT PROPERTY
>
> $$AB = 0 \quad \text{if and only if} \quad A = 0 \ \text{ or } \ B = 0$$

This means that if we can factor the left-hand side of a quadratic (or other) equation, then we can solve it by setting each factor equal to 0 in turn. This method works only when the right-hand side of the equation is 0.

EXAMPLE 4 ■ Solving a Quadratic Equation by Factoring

Solve the equation $x^2 + 5x = 24$.

SOLUTION

We must first rewrite the equation so that the right-hand side is 0.

$$x^2 + 5x = 24$$

$$x^2 + 5x - 24 = 0 \qquad \text{Subtract 24}$$

$$(x - 3)(x + 8) = 0 \qquad \text{Factor}$$

$$x - 3 = 0 \quad \text{ or } \quad x + 8 = 0 \qquad \text{Zero-Product Property}$$

$$x = 3 \qquad\qquad x = -8 \qquad \text{Solve}$$

The solutions are $x = 3$ and $x = -8$. ∎

CHECK YOUR ANSWERS

$x = 3$:

$\quad (3)^2 + 5(3) = 9 + 15 = 24$ ✓

$x = -8$:

$\quad (-8)^2 + 5(-8) = 64 - 40 = 24$ ✓

Do you see why one side of the equation must be 0 in Example 4? Factoring the equation as $x(x + 5) = 24$ does not help us find the solutions, since 24 can be factored in infinitely many ways, such as $6 \cdot 4, \frac{1}{2} \cdot 48, \left(-\frac{2}{3}\right) \cdot (-60)$, and so on.

A quadratic equation of the form $x^2 - c = 0$, where c is a positive constant, factors as $(x - \sqrt{c})(x + \sqrt{c}) = 0$, and so the solutions are $x = \sqrt{c}$ and $x = -\sqrt{c}$. We often abbreviate this as $x = \pm\sqrt{c}$.

SOLVING A SIMPLE QUADRATIC EQUATION

The solutions of the equation $x^2 = c$ are $x = \sqrt{c}$ and $x = -\sqrt{c}$.

EXAMPLE 5 ■ Solving Simple Quadratics

Solve each equation: (a) $x^2 = 5$ (b) $(x - 4)^2 = 5$

SOLUTION

(a) From the principle in the preceding box, we get $x = \pm\sqrt{5}$.

(b) We can take the square root of each side of this equation as well.

$$(x - 4)^2 = 5$$

$$x - 4 = \pm\sqrt{5} \qquad \text{Take the square root}$$

$$x = 4 \pm \sqrt{5} \qquad \text{Add 4}$$

The solutions are $x = 4 + \sqrt{5}$ and $x = 4 - \sqrt{5}$.   ■

See page 31 for how to recognize when a quadratic expression is a perfect square.

As we saw in Example 5, if a quadratic equation is of the form $(x \pm a)^2 = c$, then we can solve it by taking the square root of each side. In an equation of this form the left-hand side is a *perfect square*: the square of a linear expression in x. So, if a quadratic equation does not factor readily, then we can solve it using the technique of **completing the square**. This means that we add a constant to an expression to make it a perfect square. For example, to make $x^2 - 6x$ a perfect square we must add 9, since $x^2 - 6x + 9 = (x - 3)^2$.

Completing the Square

Area of blue region is

$$x^2 + 2\left(\frac{b}{2}\right)x = x^2 + bx$$

Add a small square of area $(b/2)^2$ to "complete" the square.

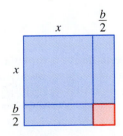

COMPLETING THE SQUARE

To make $x^2 + bx$ a perfect square, add $\left(\frac{b}{2}\right)^2$, the square of half the coefficient of x. This gives the perfect square

$$x^2 + bx + \left(\frac{b}{2}\right)^2 = \left(x + \frac{b}{2}\right)^2$$

EXAMPLE 6 ■ Solving Quadratic Equations by Completing the Square

Solve each equation.

(a) $x^2 - 8x + 13 = 0$ (b) $3x^2 - 12x + 6 = 0$

SOLUTION

(a) $x^2 - 8x + 13 = 0$

$$x^2 - 8x = -13 \qquad \text{Subtract 13}$$

$$x^2 - 8x + 16 = -13 + 16 \qquad \text{Complete the square: add } \left(\frac{-8}{2}\right)^2 = 16$$

$$(x - 4)^2 = 3 \qquad \text{Perfect square}$$

$$x - 4 = \pm\sqrt{3} \qquad \text{Take square root}$$

$$x = 4 \pm \sqrt{3} \qquad \text{Add 4}$$

⊘ When completing the square, make sure the coefficient of x^2 is 1. If it isn't, you must factor this coefficient from both terms that contain x:

$$ax^2 + bx = a\left(x^2 + \frac{b}{a}x \qquad \right)$$

Then complete the square inside the parentheses. Remember that the term added inside the parentheses is multiplied by a.

(b) After subtracting 6 from each side of the equation, we must factor the coefficient of x^2 (the 3) from the left side to put the equation in the correct form for completing the square.

$$3x^2 - 12x + 6 = 0$$

$$3x^2 - 12x = -6 \qquad \text{Subtract 6}$$

$$3(x^2 - 4x \quad) = -6 \qquad \text{Factor 3 from LHS}$$

Now we complete the square by adding $(-2)^2 = 4$ inside the parentheses. Since everything inside the parentheses is multiplied by 3, this means that we are actually adding $3 \cdot 4 = 12$ to the left side of the equation. Thus, we must add 12 to the right side as well.

$$3(x^2 - 4x + 4) = -6 + 3 \cdot 4 \qquad \text{Complete the square: add 4}$$

$$3(x - 2)^2 = 6 \qquad \text{Perfect square}$$

$$(x - 2)^2 = 2 \qquad \text{Divide by 3}$$

$$x - 2 = \pm\sqrt{2} \qquad \text{Take square root}$$

$$x = 2 \pm \sqrt{2} \qquad \text{Add 2}$$ ■

We can use the technique of completing the square to derive a formula for the roots of the general quadratic equation $ax^2 + bx + c = 0$. First, we divide each side of the equation by a and move the constant to the right side, giving

$$x^2 + \frac{b}{a}x = -\frac{c}{a}$$

François Viète (1540–1603) had a successful political career before taking up mathematics late in life. He became one of the most famous French mathematicians of the 16th century. Viète introduced a new level of abstraction in algebra by using letters to stand for *known* quantities in an equation. Before Viète's time, each equation had to be solved on its own. For instance, the quadratic equations

$$3x^2 + 2x + 8 = 0$$

and $$5x^2 - 6x + 4 = 0$$

had to be solved separately for the unknown x. Viète's idea was to consider all quadratic equations at once by writing

$$ax^2 + bx + c = 0$$

where a, b, and c are known quantities. Thus, he made it possible to write a *formula* (in this case, the quadratic formula) involving a, b, and c that can be used to solve all such equations in one fell swoop. Viète's mathematical genius proved valuable during a war between France and Spain. To communicate with their troops, the Spaniards used a complicated code which Viète managed to decipher. Unaware of Viète's accomplishment, King Philip II of Spain protested to the Pope, claiming that the French were using witchcraft to read his messages.

We now complete the square by adding $(b/2a)^2$ to each side of the equation:

$$x^2 + \frac{b}{a}x + \left(\frac{b}{2a}\right)^2 = -\frac{c}{a} + \left(\frac{b}{2a}\right)^2 \qquad \begin{array}{l}\text{Complete the square:}\\ \text{Add } \left(\dfrac{b}{2a}\right)^2\end{array}$$

$$\left(x + \frac{b}{2a}\right)^2 = \frac{-4ac + b^2}{4a^2} \qquad \text{Perfect square}$$

$$x + \frac{b}{2a} = \pm\sqrt{\frac{-4ac + b^2}{4a^2}} = \pm\frac{\sqrt{b^2 - 4ac}}{2a} \qquad \text{Take square root}$$

$$x = \frac{-b \pm \sqrt{b^2 - 4ac}}{2a} \qquad \text{Subtract } \frac{b}{2a}$$

This is the quadratic formula.

> ### THE QUADRATIC FORMULA
>
> The roots of the quadratic equation $ax^2 + bx + c = 0$, where $a \neq 0$, are
>
> $$x = \frac{-b \pm \sqrt{b^2 - 4ac}}{2a}$$

The quadratic formula could be used to solve the equations in Example 6. You should carry out the details of these calculations.

EXAMPLE 7 ■ Using the Quadratic Formula

Find all solutions of each equation.

(a) $3x^2 - 5x - 1 = 0$ (b) $4x^2 + 12x + 9 = 0$ (c) $x^2 + 2x + 2 = 0$

SOLUTION

(a) Using the quadratic formula with $a = 3$, $b = -5$, and $c = -1$, we get

$$x = \frac{-(-5) \pm \sqrt{(-5)^2 - 4(3)(-1)}}{2(3)} = \frac{5 \pm \sqrt{37}}{6}$$

If approximations are desired, we use a calculator and obtain

$$x = \frac{5 + \sqrt{37}}{6} \approx 1.8471 \qquad \text{and} \qquad x = \frac{5 - \sqrt{37}}{6} \approx -0.1805$$

(b) Using the quadratic formula with $a = 4$, $b = 12$, and $c = 9$ gives

$$x = \frac{-12 \pm \sqrt{(12)^2 - 4 \cdot 4 \cdot 9}}{2 \cdot 4} = \frac{-12 \pm 0}{8} = -\frac{3}{2}$$

This equation has only one solution, $x = -\frac{3}{2}$.

(c) Using the quadratic formula with $a = 1$, $b = 2$, and $c = 2$ gives

$$x = \frac{-2 \pm \sqrt{2^2 - 4 \cdot 2}}{2} = \frac{-2 \pm \sqrt{-4}}{2} = \frac{-2 \pm 2\sqrt{-1}}{2} = -1 \pm \sqrt{-1}$$

Since the square of any real number is nonnegative, $\sqrt{-1}$ is undefined in the real number system. The equation has no real solution. ■

In Section 3.4 we study the complex number system, in which the square roots of negative numbers do exist. The equation in Example 7(c) does have solutions in the complex number system.

The quantity $b^2 - 4ac$ that appears under the square root sign in the quadratic formula is called the *discriminant* of the equation $ax^2 + bx + c = 0$ and is given the symbol D. If $D < 0$, then $\sqrt{b^2 - 4ac}$ is undefined, and the quadratic equation has no real solution, as in Example 7(c). If $D = 0$, then the equation has only one real solution, as in Example 7(b). Finally, if $D > 0$, then the equation has two distinct real solutions, as in Example 7(a). The following box summarizes these observations.

THE DISCRIMINANT

The **discriminant** of the general quadratic $ax^2 + bx + c = 0$ $(a \neq 0)$ is $D = b^2 - 4ac$.

1. If $D > 0$, then the equation has two distinct real solutions.

2. If $D = 0$, then the equation has exactly one real solution.

3. If $D < 0$, then the equation has no real solution.

EXAMPLE 8 ■ Using the Discriminant

Use the discriminant to determine how many real solutions each equation has.

(a) $x^2 + 4x - 1 = 0$ (b) $4x^2 - 12x + 9 = 0$ (c) $\frac{1}{3}x^2 - 2x + 4 = 0$

SOLUTION

(a) The discriminant is $D = 4^2 - 4(1)(-1) = 20 > 0$, so the equation has two distinct real solutions.

(b) The discriminant is $D = (-12)^2 - 4 \cdot 4 \cdot 9 = 0$, so the equation has exactly one real solution.

(c) The discriminant is $D = (-2)^2 - 4\left(\frac{1}{3}\right)4 = -\frac{4}{3} < 0$, so the equation has no real solution. ■

Now let's consider a real-life situation that is modeled by a quadratic equation.

This formula depends on the fact that acceleration due to gravity is constant near the earth's surface. Here we neglect the effect of air resistance.

FIGURE 2

6400 ft

EXAMPLE 9 ■ The Path of a Projectile

An object thrown or fired straight upward at an initial speed of v_0 ft/s will reach a height of h feet after t seconds, where h and t are related by the formula

$$h = -16t^2 + v_0 t$$

Suppose that a bullet is shot straight upward with an initial speed of 800 ft/s. Its path is shown in Figure 2.

(a) When does the bullet fall back to ground level?

(b) When does it reach a height of 6400 ft?

(c) When does it reach a height of 2 mi?

(d) How high is the highest point the bullet reaches?

SOLUTION

Since the initial speed in this case is $v_0 = 800$ ft/s, the formula is

$$h = -16t^2 + 800t$$

(a) Ground level corresponds to $h = 0$, so we must solve the equation

$$0 = -16t^2 + 800t$$

$$= -16t(t - 50)$$

Thus, $t = 0$ or $t = 50$. This means the bullet starts $(t = 0)$ at ground level and returns to ground level after 50 s.

(b) Setting $h = 6400$ gives the equation

$$6400 = -16t^2 + 800t$$

$$16t^2 - 800t + 6400 = 0$$

$$t^2 - 50t + 400 = 0 \qquad \text{Divide by 16}$$

$$(t - 10)(t - 40) = 0 \qquad \text{Factor}$$

$$t = 10 \qquad \text{or} \qquad t = 40$$

The bullet reaches 6400 ft after 10 s (on its ascent) and again after 40 s (on its descent to earth).

(c) Two miles is $2 \times 5280 = 10{,}560$ ft.

$$10{,}560 = -16t^2 + 800t$$

$$16t^2 - 800t + 10{,}560 = 0$$

$$t^2 - 50t + 660 = 0 \qquad \text{Divide by 16}$$

2 mi

10,000 ft

The discriminant of this equation is $D = (-50)^2 - 4(660) = -140$, which is negative. Thus, the equation has no real solution. The bullet never reaches a height of 2 mi.

(d) Each height the bullet reaches is attained twice, once on its ascent and once on its descent. The only exception is the highest point of its path, which is reached only once. This means that for the highest value of h, the equation

$$h = -16t^2 + 800t$$

or

$$16t^2 - 800t + h = 0$$

has only one solution for t. This in turn means that the discriminant of the equation is 0, and so

$$D = (-800)^2 - 4(16)h = 0$$

$$640,000 - 64h = 0$$

$$h = 10,000$$

The maximum height reached is 10,000 ft.   ■

Other Equations

So far we have learned how to solve linear and quadratic equations. Now we study other types of equations, including those that involve fractional expressions, radicals, and higher powers.

EXAMPLE 10 ■ An Equation Involving Fractional Expressions

Solve the equation $\dfrac{3}{x} + \dfrac{5}{x+2} = 2$.

SOLUTION

To simplify the equation, we multiply each side by the common denominator.

$$\left(\frac{3}{x} + \frac{5}{x+2}\right)x(x+2) = 2x(x+2) \qquad \text{Multiply by LCD } x(x+2)$$

$$3(x+2) + 5x = 2x^2 + 4x \qquad \text{Expand}$$

$$8x + 6 = 2x^2 + 4x \qquad \text{Expand LHS}$$

$$0 = 2x^2 - 4x - 6 \qquad \text{Subtract } 8x + 6$$

$$0 = x^2 - 2x - 3 \qquad \text{Divide both sides by 2}$$

$$0 = (x-3)(x+1) \qquad \text{Factor}$$

$$x - 3 = 0 \quad \text{or} \quad x + 1 = 0 \qquad \text{Zero-Product Property}$$

$$x = 3 \qquad\qquad x = -1 \qquad \text{Solve}$$

CHECK YOUR ANSWERS

$x = 3$:

$\text{LHS} = \dfrac{3}{3} + \dfrac{5}{3+2}$

$\quad = 1 + 1 = 2$

$\text{RHS} = 2$

$\text{LHS} = \text{RHS}$ ✓

$x = -1$:

$\text{LHS} = \dfrac{3}{-1} + \dfrac{5}{-1+2}$

$\quad = -3 + 5 = 2$

$\text{RHS} = 2$

$\text{LHS} = \text{RHS}$ ✓

We must check our answers because multiplying by an expression that contains the variable can introduce extraneous solutions (see the *Warning* below). From *Check Your Answers* we see that the solutions are $x = 3$ and -1.   ■

When you solve an equation that involves radicals, you must be especially careful to check your final answers. The next example demonstrates why.

EXAMPLE 11 ■ An Equation Involving a Radical

Solve the equation $2x = 1 - \sqrt{2 - x}$.

SOLUTION

To eliminate the square root, we first isolate it on one side of the equal sign, then square.

$$2x - 1 = -\sqrt{2 - x} \qquad \text{Subtract 1}$$

$$(2x - 1)^2 = 2 - x \qquad \text{Square each side}$$

$$4x^2 - 4x + 1 = 2 - x \qquad \text{Expand LHS}$$

$$4x^2 - 3x - 1 = 0 \qquad \text{Add } -2 + x$$

$$(4x + 1)(x - 1) = 0 \qquad \text{Factor}$$

$$4x + 1 = 0 \quad \text{or} \quad x - 1 = 0 \qquad \text{Zero-Product Property}$$

$$x = -\tfrac{1}{4} \qquad\qquad x = 1 \qquad \text{Solve}$$

The values $x = -\frac{1}{4}$ and $x = 1$ are only potential solutions. We must check them to see if they satisfy the original equation. From *Check Your Answers* we see that $x = -\frac{1}{4}$ is a solution but $x = 1$ is not. The only solution is

$$x = -\tfrac{1}{4} \qquad\qquad ■$$

CHECK YOUR ANSWERS

$x = -\frac{1}{4}$:

$\quad$ LHS $= 2\left(-\frac{1}{4}\right) = -\frac{1}{2}$

$\quad$ RHS $= 1 - \sqrt{2 - \left(-\frac{1}{4}\right)}$

$\qquad = 1 - \sqrt{\frac{9}{4}}$

$\qquad = 1 - \frac{3}{2} = -\frac{1}{2}$

$\quad$ LHS $=$ RHS $\qquad$ ✔

$x = 1$:

$\quad$ LHS $= 2(1) = 2$

$\quad$ RHS $= 1 - \sqrt{2 - 1}$

$\qquad = 1 - 1 = 0$

$\quad$ LHS $\neq$ RHS $\qquad$ ✘

When we solve an equation, we may end up with one or more **extraneous solutions**, that is, potential solutions that do not satisfy the original equation. In Example 11 the value $x = 1$ is an extraneous solution. Extraneous solutions may be introduced when we square each side of an equation because the operation of squaring can turn a false equation into a true one. For example, $-1 \neq 1$, but $(-1)^2 = 1^2$. Thus, the squared equation may be true for more values of the variable than the original equation. That is why you must always check your answers to make sure that each satisfies the original equation.

An equation of the form $aw^2 + bw + c = 0$, where w is an algebraic expression, is an equation of **quadratic type**. We solve equations of quadratic type by substituting for the algebraic expression, as we see in the next two examples.

EXAMPLE 12 ■ A Fourth-Degree Equation of Quadratic Type

Find all solutions of the equation $x^4 - 8x^2 + 8 = 0$.

SOLUTION

If we set $w = x^2$, then we get a quadratic equation in the new variable w:

$$(x^2)^2 - 8x^2 + 8 = 0 \qquad \text{Write } x^4 \text{ as } (x^2)^2$$

$$w^2 - 8w + 8 = 0 \qquad \text{Let } w = x^2$$

$$w = \frac{-(-8) \pm \sqrt{(-8)^2 - 4 \cdot 8}}{2} = 4 \pm 2\sqrt{2} \qquad \text{Quadratic formula}$$

$$x^2 = 4 \pm 2\sqrt{2} \qquad w = x^2$$

$$x = \pm\sqrt{4 \pm 2\sqrt{2}} \qquad \text{Take square roots}$$

So, there are four solutions:

$$\sqrt{4 + 2\sqrt{2}}, \qquad \sqrt{4 - 2\sqrt{2}}, \qquad -\sqrt{4 + 2\sqrt{2}}, \qquad -\sqrt{4 - 2\sqrt{2}}$$

Using a calculator, we obtain the approximations $x \approx 2.61, 1.08, -2.61, -1.08$. ∎

EXAMPLE 13 ■ An Equation Involving Fractional Powers

Find all solutions of the equation $x^{1/3} + x^{1/6} - 2 = 0$.

SOLUTION

This equation is of quadratic type because if we let $w = x^{1/6}$, then $w^2 = x^{1/3}$.

$$x^{1/3} + x^{1/6} - 2 = 0$$

$$w^2 + w - 2 = 0 \qquad \text{Let } w = x^{1/6}$$

$$(w - 1)(w + 2) = 0 \qquad \text{Factor}$$

$$w - 1 = 0 \quad \text{or} \quad w + 2 = 0 \qquad \text{Zero-Product Property}$$

$$w = 1 \qquad\qquad w = -2 \qquad \text{Solve}$$

$$x^{1/6} = 1 \qquad\qquad x^{1/6} = -2 \qquad w = x^{1/6}$$

$$x = 1^6 = 1 \qquad\qquad x = (-2)^6 = 64 \qquad \text{Take the 6th power}$$

From *Check Your Answers* we see that $x = 1$ is a solution but $x = 64$ is not. The only solution is $x = 1$.

CHECK YOUR ANSWERS

$x = 1$:

    LHS $= 1^{1/3} + 1^{1/6} - 2 = 0$

    RHS $= 0$

    LHS $=$ RHS

$x = 64$:

    LHS $= 64^{1/3} + 64^{1/6} - 2$

            $= 4 + 2 - 2 = 4$

    RHS $= 0$

    LHS $\neq$ RHS

EXAMPLE 14 ■ An Absolute Value Equation

Solve the equation $|2x - 5| = 3$.

SOLUTION

By the definition of absolute vaue, $|2x - 5| = 3$ is equivalent to

$$2x - 5 = 3 \quad \text{or} \quad 2x - 5 = -3$$
$$2x = 8 \qquad\qquad 2x = 2$$
$$x = 4 \qquad\qquad x = 1$$

The solutions are $x = 1$, $x = 4$. ■

1.5 EXERCISES

1–4 ■ Determine whether the given value is a solution of the equation.

1. $3x + 7 = 5x - 1$
   (a) $x = 4$ (b) $x = \frac{3}{2}$

2. $\dfrac{1}{x} - \dfrac{1}{x + 3} = \dfrac{1}{6}$
   (a) $x = -3$ (b) $x = 3$

3. $1 - [2 - (3 - x)] = 4x - (6 + x)$
   (a) $x = 2$ (b) $x = 22$

4. $ax - 2b = 0$ $(a \neq 0, b \neq 0)$
   (a) $x = 0$ (b) $x = \dfrac{2b}{a}$

5–16 ■ Solve the linear equation.

5. $3x - 5 = 7$ **6.** $4x + 12 = 28$

7. $x - 3 = 2x + 6$ **8.** $4x + 7 = 9x - 13$

9. $-7w = 15 - 2w$ **10.** $5t - 13 = 12 - 5t$

11. $\frac{1}{2}y - 2 = \frac{1}{3}y$ **12.** $\dfrac{z}{5} = \dfrac{3}{10}z + 7$

13. $2(1 - x) = 3(1 + 2x) + 5$

14. $5(x + 3) + 9 = -2(x - 2) - 1$

15. $4\left(y - \frac{1}{2}\right) - y = 6(5 - y)$ **16.** $\frac{2}{3}y + \frac{1}{2}(y - 3) = \dfrac{y + 1}{4}$

17–22 ■ Solve the equation by factoring.

17. $x^2 + 2x - 8 = 0$ **18.** $x^2 + 6x + 8 = 0$

19. $2y^2 + 7y + 3 = 0$ **20.** $4w^2 = 4w + 3$

21. $6x^2 + 5x = 4$ **22.** $x^2 = 5(x + 100)$

23–28 ■ Solve the equation by completing the square.

23. $x^2 + 2x - 2 = 0$ **24.** $x^2 - 6x - 9 = 0$

25. $x^2 + x - \frac{3}{4} = 0$ **26.** $x^2 + 22x + 21 = 0$

27. $2x^2 + 8x + 1 = 0$ **28.** $3x^2 - 6x - 1 = 0$

29–38 ■ Find all real solutions of the quadratic equation. Use any method.

29. $x^2 + 12x - 27 = 0$ **30.** $8x^2 - 6x - 9 = 0$

31. $3x^2 + 6x - 5 = 0$ **32.** $x^2 - 6x + 1 = 0$

33. $2y^2 - y - \frac{1}{2} = 0$ **34.** $\theta^2 - \frac{3}{2}\theta + \frac{9}{16} = 0$

35. $10y^2 - 16y + 5 = 0$ **36.** $25x^2 + 70x + 49 = 0$

37. $3x^2 + 2x + 2 = 0$ **38.** $5x^2 - 7x + 5 = 0$

39–62 ■ Find all real solutions of the equation.

39. $\dfrac{1}{x - 1} + \dfrac{1}{x + 2} = \dfrac{5}{4}$ **40.** $\dfrac{10}{x} - \dfrac{12}{x - 3} + 4 = 0$

41. $\dfrac{x^2}{x + 100} = 50$ **42.** $\dfrac{1}{x - 1} - \dfrac{2}{x^2} = 0$

43. $\dfrac{x + 5}{x - 2} = \dfrac{5}{x + 2} + \dfrac{28}{x^2 - 4}$

44. $\dfrac{x}{2x + 7} - \dfrac{x + 1}{x + 3} = 1$

45. $\sqrt{2x + 1} + 1 = x$ **46.** $\sqrt{5 - x} + 1 = x - 2$

47. $2x + \sqrt{x + 1} = 8$ **48.** $\sqrt{\sqrt{x - 5} + x} = 5$

49. $x^4 - 13x^2 + 40 = 0$ **50.** $x^4 - 5x^2 + 4 = 0$

51. $2x^4 + 4x^2 + 1 = 0$ **52.** $x^6 - 2x^3 - 3 = 0$

53. $x^{4/3} - 5x^{2/3} + 6 = 0$ **54.** $\sqrt{x} - 3\sqrt[4]{x} - 4 = 0$

55. $4(x + 1)^{1/2} - 5(x + 1)^{3/2} + (x + 1)^{5/2} = 0$

56. $x^{1/2} + 3x^{-1/2} = 10x^{-3/2}$

57. $x^{1/2} - 3x^{1/3} = 3x^{1/6} - 9$ **58.** $x - 5\sqrt{x} + 6 = 0$

59. $|2x| = 3$ **60.** $|3x + 5| = 1$

61. $|x - 4| = 0.01$ **62.** $|x - 6| = -1$

63–74 ■ Solve the equation for the indicated variable.

63. $PV = nRT$;   for R **64.** $F = G\dfrac{mM}{r^2}$;   for m

65. $\dfrac{1}{R} = \dfrac{1}{R_1} + \dfrac{1}{R_2}$;   for R_1 **66.** $P = 2l + 2w$;   for w

67. $\dfrac{ax + b}{cx + d} = 2$; for x

68. $a - 2[b - 3(c - x)] = 6$;   for x

69. $a^2x + (a - 1) = (a + 1)x$;   for x

70. $\dfrac{a + 1}{b} = \dfrac{a - 1}{b} + \dfrac{b + 1}{a}$;   for a

71. $V = \frac{1}{3}\pi r^2 h$;   for r **72.** $F = G\dfrac{mM}{r^2}$;   for r

73. $a^2 + b^2 = c^2$;   for b **74.** $A = P\left(1 + \dfrac{i}{100}\right)^2$;   for i

75–78 ■ Use the discriminant to determine the number of real solutions of the equation. Do not solve the equation.

75. $x^2 - 6x + 1 = 0$ **76.** $3x^2 = 6x - 9$

77. $x^2 + 2.20x + 1.21 = 0$ **78.** $x^2 + rx - s = 0$ $(s > 0)$

79–80 ■ Suppose an object is dropped from a height h_0 above the ground. Then its height after t seconds is given by $h = -16t^2 + h_0$, where h is measured in feet. Use this information to solve the problem.

79. If a ball is dropped from 288 ft above the ground, how long does it take to reach ground level?

80. A ball is dropped from the top of a building 96 ft tall.
   (a) How long will it take to fall half the distance to ground level?
   (b) How long will it take to fall to ground level?

81–82 ■ Use the formula $h = -16t^2 + v_0 t$ discussed in Example 9.

81. A ball is thrown straight upward at an initial speed of 40 ft/s.
   (a) When does the ball reach a height of 24 ft?
   (b) When does it reach a height of 48 ft?
   (c) What is the greatest height reached by the ball?
   (d) When does the ball reach the highest point of its path?
   (e) When does the ball hit the ground?

82. How fast would a ball have to be thrown upward to reach a maximum height of 100 ft?   [*Hint:* Use the discriminant of the equation $16t^2 - v_0 t + h = 0$.]

83. The fish population in a certain lake rises and falls according to the formula

$$F = 1000(30 + 17t - t^2)$$

Here F is the number of fish at time t, where t is measured in years since January 1, 1992, when the fish population was first estimated.
   (a) On what date will the fish population again be the same as on January 1, 1992?
   (b) By what date will all the fish in the lake have died?

84. A small-appliance manufacturer finds that the profit P (in dollars) generated by producing x microwave ovens per week is given by the formula $P = \frac{1}{10}x(300 - x)$ provided that $0 \leq x \leq 200$. How many ovens must be manufactured in a given week to generate a profit of $1250?

85. If F is the focal length of a convex lens and an object is placed at a distance x from the lens, then its image will be at a distance y from the lens, where F, x, and y are related by the *lens equation*

$$\frac{1}{F} = \frac{1}{x} + \frac{1}{y}$$

Suppose that a lens has a focal length of 4.8 cm, and that the image of an object is 4 cm closer to the lens than the object itself. How far from the lens is the object?

86. If an imaginary line segment is drawn between the centers of the earth and the moon, then the net gravitational force F acting on an object situated on this line segment is

$$F = \frac{-K}{x^2} + \frac{0.012K}{(239 - x)^2}$$

where $K > 0$ is a constant and x is the distance of the object from the center of the earth, measured in thousands of miles. How far from the center of the earth is the "dead spot" where no net gravitational force acts upon the object? (Express your answer to the nearest thousand miles.)

87. One method for determining the depth of a well is to drop a stone into it and then measure the time it takes until the splash is heard. If d is the depth of the well (in feet) and t_1 the time (in seconds) it takes for the stone to fall, then $d = 16t_1^2$, so $t_1 = \sqrt{d}/4$. Now if t_2 is the time it takes for the sound to travel back up, then $d = 1090t_2$ because the speed of sound is 1090 ft/s. So $t_2 = d/1090$. Thus, the total time elapsed between dropping the stone and hearing the splash is $t_1 + t_2 = \sqrt{d}/4 + d/1090$. How deep is the well if this total time is 3 s?

Time stone falls:
$$t_1 = \frac{\sqrt{d}}{4}$$

Time sound rises:
$$t_2 = \frac{d}{1090}$$

 DISCOVERY · DISCUSSION

88. A Family of Equations   The equation

$$3x + k - 5 = kx - k + 1$$

is really a **family of equations**, because for each value of k, we get a different equation with the unknown x. The letter k is called a **parameter** for this family. What value should

we pick for k to make the given value of x a solution of the resulting equation?
(a) $x = 0$ (b) $x = 1$ (c) $x = 2$

89. Proof That $0 = 1$?   The following steps appear to give equivalent equations, which seem to prove that $1 = 0$. Find the error.

$x = 1$	Given
$x^2 = x$	Multiply by x
$x^2 - x = 0$	Subtract x
$x(x - 1) = 0$	Factor
$\dfrac{x(x - 1)}{x - 1} = \dfrac{0}{x - 1}$	Divide by $x - 1$
$x = 0$	Simplify
$1 = 0$	Given $x = 1$

90. Volumes of Solids   The sphere, cylinder, and cone shown here all have the same radius r and the same volume V. Express the heights of the cylinder and cone in terms of r only, not V. (Use the volume formulas given on the inside front cover of this book.)

91. Relationship between Roots and Coefficients   The quadratic formula gives us the roots of a quadratic equation from its coefficients. We can also obtain the coefficients from the roots. For example, find the roots of the equation $x^2 - 9x + 20 = 0$ and show that the product of the roots is the constant term 20 and the sum of the roots is 9, the negative of the coefficient of x. Show that the same relationship between roots and coefficients holds for the following equations:

$$x^2 - 2x - 8 = 0$$
$$x^2 + 4x + 2 = 0$$

Use the quadratic formula to prove that in general, if the equation $x^2 + bx + c = 0$ has roots r_1 and r_2, then $c = r_1 r_2$ and $b = -(r_1 + r_2)$.

92. Solving an Equation in Different Ways   We have learned several different ways to solve an equation in this section. Some equations can be tackled by more than one method. For example, the equation $x - \sqrt{x} - 2 = 0$ is of quadratic type: We can solve it by letting $\sqrt{x} = u$ and $x = u^2$, and factoring. Or we could solve for $\sqrt{x}$, square, and then solve the resulting quadratic equation.

Solve the following equations using both methods indicated, and show that you get the same final answers.

(a) $x - \sqrt{x} - 2 = 0$ quadratic type; solve for the radical and square

(b) $\dfrac{12}{(x-3)^2} + \dfrac{10}{x-3} + 1 = 0$ quadratic type; multiply by LCD

1.6   MODELING WITH EQUATIONS

Many problems in the sciences, economics, finance, medicine, and numerous other fields can be translated into algebra problems; this is one reason that algebra is so useful. In this section we use equations as mathematical models to solve real-life problems.

EXAMPLE 1 ■ Interest on an Investment

Mary inherits $100,000 and invests it in two certificates of deposit. One certificate pays 6% and the other pays $4\frac{1}{2}$% simple interest annually. If Mary's total interest is $5025 per year, how much money is invested at each rate?

SOLUTION

The problem asks for the amount she has invested at each rate. So let

Identify the variable

$$x = \text{the amount invested at 6\%}$$

Since Mary's total inheritance is $100,000, it follows that she invested $100,000 - x$ at $4\frac{1}{2}$%. We translate all the information given into the language of algebra.

Express all unknown quantities in terms of the variable

In Words	In Algebra
Amount invested at 6%	x
Amount invested at $4\frac{1}{2}$%	$100,000 - x$
Interest earned at 6%	$0.06x$
Interest earned at $4\frac{1}{2}$%	$0.045(100,000 - x)$

Now we use the fact that Mary's total interest is $5025 to set up the model.

Set up the model

$$\text{interest at 6\%} + \text{interest at } 4\tfrac{1}{2}\% = \text{total interest}$$

Solve

$$0.06x + 0.045(100,000 - x) = 5025$$

$$0.06x + 4500 - 0.045x = 5025 \qquad \text{Multiply}$$

$$0.015x + 4500 = 5025 \qquad \text{Combine the } x\text{-terms}$$

$$0.015x = 525 \qquad \text{Subtract 4500}$$

$$x = \frac{525}{0.015} = 35,000 \qquad \text{Divide by 0.015}$$

So Mary has invested $35,000 at 6% and the remaining $65,000 at $4\frac{1}{2}$%.

CHECK YOUR ANSWER

total interest $= 6\%$ of $\$35,000 + 4\frac{1}{2}\%$ of $\$65,000$

$= \$2100 + \$2925 = \$5025$ ✓

We now adapt the problem-solving principles given on pages 134–137 to the process of modeling real-life problems with equations. The following guidelines should help you set up the equation that expresses the English statement of a problem in the language of algebra. These guidelines are referred to in the margin notes for the examples in this section.

GUIDELINES FOR MODELING WITH EQUATIONS

1. IDENTIFY THE VARIABLE. Identify the quantity that the problem asks you to find. This quantity can usually be determined by a careful reading of the question posed at the end of the problem. **Introduce notation** for the variable (call it x or some other letter).

2. EXPRESS ALL UNKNOWN QUANTITIES IN TERMS OF THE VARIABLE. Read each sentence in the problem again, and express all the quantities mentioned in the problem in terms of the variable you defined in Step 1. To organize this information, it is sometimes helpful to **draw a diagram** or **make a table** (see Examples 2, 3, 4, and 7 in this section).

3. SET UP THE MODEL. Find the crucial fact in the problem that gives a relationship between the expressions you listed in Step 2. **Set up an equation** (or **model**) that expresses this relationship.

4. SOLVE THE EQUATION AND CHECK YOUR ANSWER. Solve the equation, check your answer, and express it as a sentence that answers the question posed in the problem.

EXAMPLE 2 ■ Dimensions of a Poster

A poster has a rectangular printed area 100 cm by 140 cm, and a blank strip of uniform width around the four edges. The perimeter of the poster is $1\frac{1}{2}$ times the perimeter of the printed area. What is the width of the blank strip, and what are the dimensions of the poster?

In a problem such as this, which involves geometry, it is essential to draw a diagram like the one shown in Figure 1.

FIGURE 1

SOLUTION

We are asked to find the width of the blank strip. So let

Identify the variable

$$x = \text{the width of the blank strip}$$

Then we translate the information in Figure 1 into the language of algebra:

Express all unknown quantities in terms of the variable

In Words	In Algebra
Width of blank strip	x
Perimeter of printed area	$2(100) + 2(140) = 480$
Width of poster	$100 + 2x$
Length of poster	$140 + 2x$
Perimeter of poster	$2(100 + 2x) + 2(140 + 2x)$

Now we use the fact that the perimeter of the poster is $1\frac{1}{2}$ times the perimeter of the printed area to set up the model.

Set up the model

$$\text{perimeter of poster} = \tfrac{3}{2} \cdot \text{perimeter of printed area}$$

$$2(100 + 2x) + 2(140 + 2x) = \tfrac{3}{2} \cdot 480$$

Solve

$$480 + 8x = 720 \qquad \text{Expand and combine like terms on LHS}$$

$$8x = 240 \qquad \text{Subtract 480}$$

$$x = 30 \qquad \text{Divide by 8}$$

The blank strip is 30 cm wide, so the dimensions of the poster are

$$100 + 30 + 30 = 160 \text{ cm wide}$$

by

$$140 + 30 + 30 = 200 \text{ cm long} \qquad \blacksquare$$

EXAMPLE 3 ■ Dimensions of a Building Lot

A rectangular building lot is 8 ft longer than it is wide and has an area of 2900 ft^2. Find the dimensions of the lot.

SOLUTION

We are asked to find the width and length of the lot. So let

Identify the variable

$$w = \text{width of lot}$$

Then we translate the information given in the problem into the language of algebra (see Figure 2).

w

$w + 8$

FIGURE 2

Express all unknown quantities in terms of the variable

In Words	In Algebra
Width of lot	w
Length of lot	$w + 8$

Now we set up the model.

Set up the model

$$\left.\begin{array}{c} \text{width} \\ \text{of lot} \end{array}\right. \cdot \left.\begin{array}{c} \text{length} \\ \text{of lot} \end{array}\right. = \left.\begin{array}{c} \text{area} \\ \text{of lot} \end{array}\right.$$

$$w(w + 8) = 2900$$

Solve

$$w^2 + 8w = 2900 \qquad \text{Expand}$$

$$w^2 + 8w - 2900 = 0 \qquad \text{Subtract 2900}$$

$$(w - 50)(w + 58) = 0 \qquad \text{Factor}$$

$$w = 50 \quad \text{or} \quad w = -58 \qquad \text{Zero-Product Property}$$

Since the width of the lot must be a positive number, we conclude that $w = 50$ ft. The length of the lot is $w + 8 = 50 + 8 = 58$ ft.   ■

EXAMPLE 4 ■ Determining the Height of a Building Using Similar Triangles

A man 6 ft tall wishes to find the height of a certain four-story building. He measures its shadow and finds it to be 28 ft long, while his own shadow is $3\frac{1}{2}$ ft long. How tall is the building?

SOLUTION

The problem asks for the height of the building. So let

Identify the variable

$$h = \text{the height of the building}$$

We use the fact that the triangles in Figure 3 are similar. Recall that for any pair of similar triangles the ratios of corresponding sides are equal. Now we translate these observations into the language of algebra.

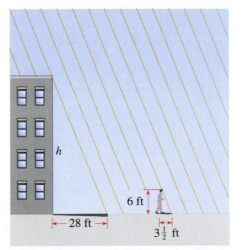

FIGURE 3

In Words	In Algebra
Height of building	h
Ratio of height to base in large triangle	$\frac{h}{28}$
Ratio of height to base in small triangle	$\frac{6}{3.5}$

Express all unknown quantities in terms of the variable

Since the large and small triangles are similar, we get the equation

$$\boxed{\text{ratio of height to base in large triangle}} = \boxed{\text{ratio of height to base in small triangle}}$$

Set up the model

$$\frac{h}{28} = \frac{6}{3.5}$$

Solve

$$h = \frac{6 \cdot 28}{3.5} = 48$$

The building is 48 ft tall.   ■

EXAMPLE 5 ■ Mixtures and Concentration

A manufacturer of soft drinks advertises their orange soda as "naturally flavored," although it contains only 5% orange juice. A new federal regulation stipulates that to be called "natural" a drink must contain at least 10% fruit juice. How much pure orange juice must this manufacturer add to 900 gal of orange soda to conform to the new regulation?

SOLUTION

The problem asks for the amount of pure orange juice to be added. So let

Identify the variable

x = the amount (in gallons) of pure orange juice to be added

In any problem of this type—in which two different substances are to be mixed—drawing a diagram helps us organize the given information (see Figure 4). We then translate the information in the figure into the language of algebra.

In Words	In Algebra
Amount of orange juice to be added	x
Amount of the mixture	$900 + x$
Amount of orange juice in the first vat	$0.05(900) = 45$
Amount of orange juice in the second vat	$1 \cdot x = x$
Amount of orange juice in the mixture	$0.10(900 + x)$

Express all quantities in terms of the variable

To set up the model, we use the fact that the total amount of orange juice in the mixture is equal to the orange juice in the first two vats.

65-80

Volume	900 gallons	x gallons	$900 + x$ gallons
Amount of orange juice	5% of 900 gallons = 45 gallons	100% of x gallons = x gallons	10% of $900 + x$ gallons = $0.1(900 + x)$ gallons

FIGURE 4

Set up the model

amount of orange juice in the first vat	$+$	amount of orange juice in the second vat	$=$	amount of orange juice in the mixture

Solve

$$45 + x = 0.1(900 + x) \qquad \text{From Figure 4}$$

$$45 + x = 90 + 0.1x \qquad \text{Multiply}$$

$$0.9x = 45 \qquad \text{Subtract } 0.1x \text{ and } 45$$

$$x = \frac{45}{0.9} = 50 \qquad \text{Divide by } 0.9$$

The manufacturer should add 50 gal of pure orange juice to the soda.

CHECK YOUR ANSWER

$$\text{amount of juice before mixing} = 5\% \text{ of } 900 \text{ gal} + 50 \text{ gal pure juice}$$
$$= 45 \text{ gal} + 50 \text{ gal} = 95 \text{ gal}$$
$$\text{amount of juice after mixing} = 10\% \text{ of } 950 \text{ gal} = 95 \text{ gal}$$

Amounts are equal. ✔

EXAMPLE 6 ■ Time Needed to Do a Job

Because of an anticipated heavy rainstorm, the water level in a reservoir must be lowered by 1 ft. Opening spillway A lowers the level by this amount in 4 hours, whereas opening the smaller spillway B does the job in 6 hours. How long will it take to lower the water level by 1 ft if both spillways are opened?

SOLUTION

We are asked to find the time needed to lower the level by 1 ft if both spillways are open. So let

Identify the variable

$$x = \text{the time (in hours) it takes to lower the water level}$$
$$\text{by 1 ft if both spillways are open}$$

Finding an equation relating x to the other quantities in this problem is not easy. Certainly x is not simply $4 + 6$, because that would mean that together the two spillways require longer to lower the water level than either spillway alone. Instead, *we look at the fraction of the job that can be done in one hour by each spillway.*

In Words	In Algebra
Time it takes to lower level 1 ft with A and B together	x h
Distance A lowers level in 1 h	$\frac{1}{4}$ ft
Distance B lowers level in 1 h	$\frac{1}{6}$ ft
Distance A and B together lower levels in 1 h	$\frac{1}{x}$ ft

Express all unknown quantities in terms of the variable

Now we set up the model.

Set up the model

$$\boxed{\text{fraction done by A}} + \boxed{\text{fraction done by B}} = \boxed{\text{fraction done by both}}$$

$$\frac{1}{4} + \frac{1}{6} = \frac{1}{x}$$

Solve

$$3x + 2x = 12 \qquad \text{Multiply by the LCD, } 12x$$

$$5x = 12 \qquad \text{Add}$$

$$x = \frac{12}{5} \qquad \text{Divide by 5}$$

It will take $2\frac{2}{5}$ hours, or 2 h 24 min to lower the water level by 1 ft if both spillways are open. ■

The next example deals with distance, rate (speed), and time. The formula to keep in mind here is

$$\boxed{\text{distance} = \text{rate} \times \text{time}}$$

where the rate is either the constant speed or average speed of a moving object. For example, driving at 60 mi/h for 4 hours takes you a distance of $60 \cdot 4 = 240$ mi.

EXAMPLE 7 ■ A Distance-Rate-Time Problem

A jet flew from New York to Los Angeles, a distance of 4200 km. The speed for the return trip was 100 km/h faster than the outbound speed. If the total trip took 13 hours, what was the speed from New York to Los Angeles?

SOLUTION

We are asked for the speed of the jet from New York to Los Angeles. So let

Identify the variable

$$s = \text{speed from New York to Los Angeles}$$

Then $\qquad\qquad s + 100 = $ speed from Los Angeles to New York

Now we organize the information in a table. We fill in the "Distance" column first, since we know that the cities are 4200 km apart. Then we fill in the "Rate" column, since we have expressed both speeds (rates) in terms of the variable s. Finally, we calculate the entries for the "Time" column, using

$$\text{time} = \frac{\text{distance}}{\text{rate}}$$

Express all unknown quantities in terms of the variable

	Distance (km)	Rate (km/h)	Time (h)
N.Y. to L.A.	4200	s	$\dfrac{4200}{s}$
L.A. to N.Y.	4200	$s + 100$	$\dfrac{4200}{s + 100}$

The total trip took 13 hours, so we have the model

Set up the model

$$\begin{array}{c}\text{time from} \\ \text{N.Y. to L.A.}\end{array} + \begin{array}{c}\text{time from} \\ \text{L.A. to N.Y.}\end{array} = \begin{array}{c}\text{total} \\ \text{time}\end{array}$$

$$\frac{4200}{s} + \frac{4200}{s + 100} = 13$$

Multiplying by the common denominator, $s(s + 100)$, we get

$$4200(s + 100) + 4200s = 13s(s + 100)$$

$$8400s + 420,000 = 13s^2 + 1300s$$

$$0 = 13s^2 - 7100s - 420,000$$

Although this equation does factor, with numbers this large it is probably quickest to use the quadratic formula and a calculator.

Solve

$$s = \frac{7100 \pm \sqrt{(7100)^2 - 4(13)(-420,000)}}{2(13)}$$

$$= \frac{7100 \pm 8500}{26}$$

$$s = 600 \qquad \text{or} \qquad s = \frac{-700}{13} \approx -53.8$$

Since s represents speed, we reject the negative answer and conclude that the jet's speed from New York to Los Angeles was 600 km/h. ■

EXAMPLE 8 ■ Dividing a Lottery Jackpot

A group of people come forward to claim a $1,000,000 lottery jackpot, which the winners are to share equally. Before the jackpot is divided, three more winning ticket holders show up. As a result, each person's share is reduced by $75,000. How many winners were in the original group?

SOLUTION

We are asked for the number of people in the original group. So let

Identify the variable

$$x = \text{number of winners in the original group}$$

We translate the information in the problem as follows:

In Words	In Algebra
Number of winners in original group	x
Number of winners in final group	$x + 3$
Winnings per person, originally	$\dfrac{1{,}000{,}000}{x}$
Winnings per person, finally	$\dfrac{1{,}000{,}000}{x + 3}$

Express all unknown quantities in terms of the variable

Now we set up the model.

Set up the model

$$\boxed{\begin{array}{c}\text{winnings per}\\\text{person, originally}\end{array}} - \boxed{\$75{,}000} = \boxed{\begin{array}{c}\text{winnings per}\\\text{person, finally}\end{array}}$$

$$\frac{1{,}000{,}000}{x} - 75{,}000 = \frac{1{,}000{,}000}{x + 3}$$

Solve

$$1{,}000{,}000(x + 3) - 75{,}000x(x + 3) = 1{,}000{,}000x \qquad \text{Multiply by LCD } x(x+3)$$

$$40(x + 3) - 3x(x + 3) = 40x \qquad \text{Divide by 25,000}$$

$$x^2 + 3x - 40 = 0 \qquad \text{Expand, simplify, and divide by 3}$$

$$(x + 8)(x - 5) = 0 \qquad \text{Factor}$$

$$x + 8 = 0 \quad \text{or} \quad x - 5 = 0 \qquad \text{Zero-Product Property}$$

$$x = -8 \qquad\qquad x = 5 \qquad \text{Solve}$$

Since we can't have a negative number of people, we conclude that there were five winners in the original group.

CHECK YOUR ANSWER

$$\text{winnings per person, originally} = \frac{\$1{,}000{,}000}{5} = \$200{,}000$$

$$\text{winnings per person, finally} \;\; = \frac{\$1{,}000{,}000}{8} = \$125{,}000$$

$$\$200{,}000 - \$75{,}000 = \$125{,}000 \quad \checkmark$$

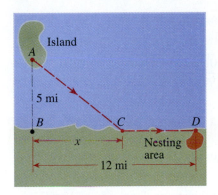

Island

A

5 mi

B C D

x Nesting area

12 mi

FIGURE 5

EXAMPLE 9 ■ Energy Expended in Bird Flight

Ornithologists have determined that some species of birds tend to avoid flights over large bodies of water during daylight hours, because air generally rises over land and falls over water in the daytime, so flying over water requires more energy. A bird is released from point A on an island, 5 mi from B, the nearest point on a straight shoreline. The bird flies to a point C on the shoreline and then flies along the shoreline to its nesting area D, as shown in Figure 5. Suppose the bird has 170 kcal of energy reserves. It uses 10 kcal/mi flying over land and 14 kcal/mi flying over water.

(a) Where should the point C be located so that the bird uses exactly 170 kcal of energy during its flight?

(b) Does the bird have enough energy reserves to fly directly from A to D?

SOLUTION

(a) We are asked to find the location of C. So let

Identify the variable

$$x = \text{the distance from } B \text{ to } C$$

From the figure, and from the fact that

$$\text{energy used} = \text{energy per mile} \times \text{miles flown}$$

we determine the following:

In Words	In Algebra
Distance from B to C	x
Distance flown over water (from A to C)	$\sqrt{x^2 + 25}$ Pythagorean Theorem
Distance flown over land (from B to C)	$12 - x$
Energy used over water	$14\sqrt{x^2 + 25}$
Energy used over land	$10(12 - x)$

Express all unknown quantities in terms of the variable

Now we set up the model.

Set up the model

$$\begin{array}{c} \text{total energy} \\ \text{used} \end{array} = \begin{array}{c} \text{energy used} \\ \text{over water} \end{array} + \begin{array}{c} \text{energy used} \\ \text{over land} \end{array}$$

$$170 = 14\sqrt{x^2 + 25} + 10(12 - x)$$

To solve this equation, we eliminate the square root by first bringing all other terms to the left of the equal sign and then squaring each side.

Solve

$$170 - 10(12 - x) = 14\sqrt{x^2 + 25} \qquad \text{Isolate the square-root term on RHS}$$

$$50 + 10x = 14\sqrt{x^2 + 25} \qquad \text{Simplify LHS}$$

$$(50 + 10x)^2 = (14)^2(x^2 + 25) \qquad \text{Square each side}$$

$$2500 + 1000x + 100x^2 = 196x^2 + 4900 \qquad \text{Expand}$$

$$0 = 96x^2 - 1000x + 2400 \qquad \text{Move all terms to RHS}$$

This equation could be factored, but because the numbers are so large it is easier to use the quadratic formula:

$$x = \frac{1000 \pm \sqrt{(-1000)^2 - 4(96)(2400)}}{2(96)}$$

$$= \frac{1000 \pm 280}{192} = 6\frac{2}{3} \quad \text{or} \quad 3\frac{3}{4}$$

Point C should be either $6\frac{2}{3}$ mi or $3\frac{3}{4}$ mi from B so that the bird uses exactly 170 kcal of energy during its flight.

(b) By the Pythagorean Theorem, the length of the route directly from A to D is $\sqrt{5^2 + 12^2} = 13$ mi, so the energy the bird requires for that route is $14 \times 13 = 182$ kcal. This is more energy than the bird has available, so it can't use this route.   ∎

1.6 EXERCISES

1–8 ■ Express the given quantity in terms of the indicated variable.

1. The sum of three consecutive integers; $n =$ the first integer of the three

2. The sum of three consecutive integers; $n =$ the middle integer of the three

3. The interest obtained after 2 years on an investment at 7% simple interest per year; $x =$ the number of dollars invested.

4. The average of three test scores if the first two scores are 78 and 82; $s =$ the third test score

5. The perimeter (in cm) of a rectangle that is 5 cm longer than it is wide; $w =$ width of the rectangle (in cm)

6. The time (in hours) it takes to travel a given distance at 55 mi/h; $d =$ the given distance (in mi)

7. The concentration (in oz/gal) of salt in a mixture of 3 gal of brine containing 25 oz of salt, to which has been added some pure water; $x =$ volume of pure water added (in gal)

8. The value (in cents) of the change in a purse that contains twice as many nickels as pennies, four more dimes than nickels, and as many quarters as dimes and nickels combined; $p =$ number of pennies

9–79 ■ Use the principles described in this section to answer the question posed.

9. Find three consecutive integers whose sum is 336.

10. Find four consecutive odd integers whose sum is 272.

11. Find two numbers whose sum is 55 and whose product is 684.

12. The sum of the squares of two consecutive even integers is 1252. Find the integers.

13. Phyllis invested $12,000, a portion earning a simple interest rate of $4\frac{1}{2}$% per year and the rest earning a rate of 4% per year. After one year the total interest earned on these investments was $525. How much money did she invest at each rate?

14. If Ben invests $4000 at 4% interest per year, how much additional money must he invest at $5\frac{1}{2}$% annual interest to ensure that the interest he receives each year is $4\frac{1}{2}$% of the total amount invested?

15. What annual rate of interest would you have to earn on an investment of $3500 to ensure receiving $262.50 interest after one year?

16. Jack invests $1000 at a certain annual interest rate, and he invests another $2000 at an annual rate that is one-half percent higher. If he receives a total of $190 interest in one year, at what rate is the $1000 invested?

17. An executive in an engineering firm earns a monthly salary plus a Christmas bonus of $8500. If she earns a total of $97,300 per year, what is her monthly salary?

18. A woman earns 15% more than her husband. Together they make $69,875 per year. What is the husband's annual salary?

19. Craig is saving to buy a vacation home. He inherits some money from a wealthy uncle, then combines this with the $22,000 he has already saved and doubles the total in a lucky investment. He ends up with $134,000, just enough to buy a cabin on the lake. How much did he inherit?

20. Helen earns $7.50 per hour at her job, but if she works more than 35 hours in a week she is paid $1\frac{1}{2}$ times her regular salary for the overtime hours worked. One week her gross pay was $352.50. How many overtime hours did she work that week?

21. A plumber and his assistant work together to replace the pipes in an old house. The plumber charges $45 per hour for his own labor and $25 per hour for his assistant's labor. The plumber works twice as long as his assistant on this job, and the labor charge on the final bill is $4025. How long did the plumber and his assistant work on this job?

22. A father is four times as old as his daughter. In 6 years, he will be three times as old as she is. How old is the daughter now?

23. A movie star, unwilling to give his age, posed the following riddle to a gossip columnist. "Seven years ago, I was eleven times as old as my daughter. Now I am four times as old as she is." How old is the star?

24. During his major league career, Hank Aaron hit 31 more home runs than Babe Ruth hit during his career. Together they hit 1459 home runs. How many home runs did Babe Ruth hit?

25. A change purse contains an equal number of pennies, nickels, and dimes. The total value of the coins is $1.44. How many coins of each type does the purse contain?

26. Mary has $3.00 in nickels, dimes, and quarters. If she has twice as many dimes as quarters and five more nickels than dimes, how many coins of each type does she have?

27. A rectangular garden is 25 ft wide. If its area is 1125 ft², what is the length of the garden?

28. A pasture is twice as long as it is wide. Its area is 115,200 ft². How wide is the pasture?

29. A square plot of land has a building 60 ft long and 40 ft wide at one corner. The rest of the land outside the building forms a parking lot. If the parking lot has area 12,000 ft², what are the dimensions of the entire plot of land?

30. Find the length y in the figure, if the shaded area is 120 in².

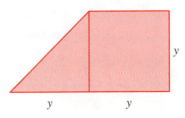

31. Find the length x in the figure, if the shaded area is 144 cm².

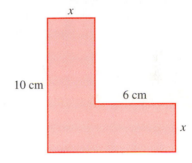

32. Find the length x if the shaded area is 160 in².

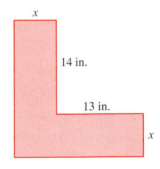

33. Find the length x if the shaded area is 1200 cm².

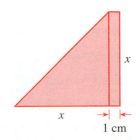

34. A rectangular bedroom is 7 ft longer than it is wide. Its area is 228 ft². What is the width of the room?

35. A farmer has a rectangular garden plot surrounded by 200 ft of fence. Find the length and width of the garden if its area is 2400 ft².

perimeter = 200 ft

36. A box with a square base and no top is to be made from a square piece of cardboard by cutting 4-in. squares from each corner and folding up the sides, as shown in the figure. The box is to hold 100 in³. How big a piece of cardboard is needed?

4 in.

4 in.

37. A large plywood box has a volume of 180 ft³. Its length is 9 ft greater than its height, and its width is 4 ft less than its height. What are the dimensions of the box?

$x + 9$

x

$x - 4$

38. A cylindrical can has a volume of 40π cm³ and is 10 cm tall. What is its diameter? [*Hint:* Use the volume formula listed on the inside front cover of this book.]

10 cm

39. Al paints with watercolors on a sheet of paper 20 in. wide by 15 in. high. He then places this sheet on a mat so that a uniformly wide strip of the mat shows all around the picture. The perimeter of the mat is 102 in. How wide is the strip of the mat showing around the picture?

x

15 in.

20 in.

40. What quantity of a 60% acid solution must be mixed with a 30% solution to produce 300 mL of a 50% solution?

41. A jeweler has five rings, each weighing 18 g, made of an alloy of 10% silver and 90% gold. He decides to melt down the rings and add enough silver to reduce the gold content to 75%. How much silver should he add?

42. A pot contains 6 L of brine at a concentration of 120 g/L. How much of the water should be boiled off to increase the concentration to 200 g/L?

43. The radiator in a car is filled with a solution of 60% antifreeze and 40% water. The manufacturer of the antifreeze suggests that, for summer driving, optimal cooling of the engine is obtained with only 50% antifreeze. If the capacity of the radiator is 3.6 L, how much coolant should be drained and replaced with water to reduce the antifreeze concentration to the recommended level?

44. A health clinic uses a solution of bleach to sterilize petri dishes in which cultures are grown. The sterilization tank contains 100 gal of a solution of 2% ordinary household bleach mixed with pure distilled water. New research indicates that the concentration of bleach should be 5% for complete sterilization. How much of the solution should be drained and replaced with bleach to increase the bleach content to the recommended level?

45. A bottle contains 750 mL of fruit punch with a concentration of 50% pure fruit juice. Jill drinks 100 mL of the punch and then refills the bottle with an equal amount of a cheaper brand of punch. If the concentration of juice in the bottle is now reduced to 48%, what was the concentration in the punch that Jill added?

46. A merchant blends tea that sells for $3.00 a pound with tea that sells for $2.75 a pound to produce 80 lb of a mixture that sells for $2.90 a pound. How many pounds of each type of tea does the merchant use in the blend?

47. A man is walking away from a lamppost with a light source 6 m above the ground. The man is 2 m tall. How far from the lamppost is the man when his shadow is 5 m long? [*Hint:* Use similar triangles.]

48. A woodcutter determines the height of a tall tree by first measuring a smaller one 125 ft away, then moving so that his eyes are in the line of sight along the tops of the trees, and measuring how far he is standing from the small tree (see the figure). Suppose the small tree is 20 ft tall, the man is 25 ft from the small tree, and his eye level is 5 ft above the ground. How tall is the taller tree?

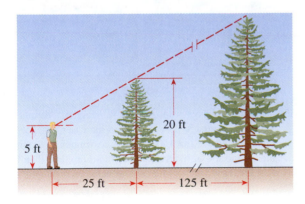

49. A parcel of land is 6 ft longer than it is wide. Each diagonal from one corner to the opposite corner is 174 ft long. What are the dimensions of the parcel?

50. A flagpole is secured on opposite sides by two guy wires, each of which is 5 ft longer than the pole. The distance between the points where the wires are fixed to the ground is equal to the length of one guy wire. How tall is the flagpole (to the nearest inch)?

51. The figure shows a lever system, similar to a seesaw that you might find in a children's playground. For the system to balance the product of the weight and its distance from the fulcrum must be the same on each side; that is

$$w_1 x_1 = w_2 x_2$$

This equation is called the **law of the lever**, and was first discovered by Archimedes (see page 728).

A woman and her son are playing on a seesaw. The boy is at one end, 8 ft from the fulcrum. If the son weighs 100 lb and the mother weighs 125 lb, where should the woman sit so that the seesaw is balanced?

52. A plank 30 ft long rests on top of a flat-roofed building, with 5 ft of the plank projecting over the edge, as shown in the figure. A worker weighing 240 lb sits on one end of the plank. What is the largest weight that can be hung on the projecting end of the plank if it is to remain in balance? (Use the law of the lever stated in Exercise 51.)

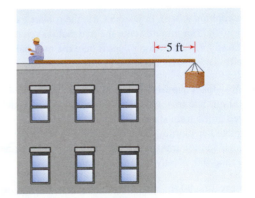

53. Candy and Tim share a paper route. It takes Candy 70 min to deliver all the papers, whereas Tim takes 80 min. How long does it take them when they work together?

54. Stan and Hilda can mow the lawn in 40 min if they work together. If Hilda works twice as fast as Stan, how long would it take Stan to mow the lawn alone?

55. Betty and Karen have been hired to paint the houses in a new development. Working together the women can paint a house in two-thirds the time that it takes Karen working alone. Betty takes 6 h to paint a house alone. How long does it take Karen to paint a house working alone?

56. Next-door neighbors Bob and Jim use hoses from both houses to fill Bob's swimming pool. They know it takes 18 h using both hoses. They also know that Bob's hose, used alone, takes 20% less time than Jim's hose alone. How much time is required to fill the pool by each hose alone?

57. Henry and Irene working together can wash all the windows of their house in 1 h 48 min. Working alone, it takes Henry $1\frac{1}{2}$ h more than Irene to do the job. How long does it take each person working alone to wash all the windows?

58. Jack, Kay, and Lynn deliver advertising flyers in a small town. If each person works alone, it takes Jack 4 h to deliver all the flyers, and it takes Lynn 1 h longer than it takes Kay. Working together, they can deliver all the flyers in 40% of the time it takes Kay working alone. How long does it take Kay to deliver all the flyers alone?

59. A commercial jet took off from Kansas City for San Francisco, a distance of 2550 km, at a speed of 800 km/h. At the same time a private jet, traveling at 900 km/h, left San Francisco for Kansas City. How long after takeoff will the jets pass each other?

60. After robbing a bank in Dodge City, the robber gallops off at 14 mi/h. Ten minutes later the marshal leaves in hot pursuit at 16 mi/h. How long does it take the marshal to catch up with the bank robber?

61. Wendy took a trip from Davenport to Omaha, a distance of 300 mi. She traveled part of the way by bus, which arrived at the train station just in time for Wendy to complete her journey by train. The bus averaged 40 mi/h and the train 60 mi/h. The entire trip took $5\frac{1}{2}$ h. How long did Wendy spend on the train?

62. Two cyclists, 90 mi apart, start riding toward each other at the same time. One cycles twice as fast as the other. If

they meet 2 h later, at what average speed is each cyclist traveling?

63. A pilot flew a jet from Montreal to Los Angeles, a distance of 2500 mi. On the return trip the average speed was 20% faster than the outbound speed. The round-trip took 9 h 10 min. What was the speed from Montreal to Los Angeles?

64. A woman driving a car 14 ft long is passing a truck 30 ft long. The truck is traveling at 50 mi/h. How fast must the woman drive her car so that she can pass the truck completely in 6 s, from the position shown in figure (a) to the position shown in figure (b)?   [*Hint:* Use feet and seconds instead of miles and hours.]

(a)

(b)

65. A salesman drives from Ajax to Barrington, a distance of 120 mi, at a steady speed. He then increases his speed by 10 mi/h to drive the 150 mi from Barrington to Collins. If the second leg of his trip took 6 min more time than the first leg, how fast was he driving between Ajax and Barrington?

66. Kiran drove from Tortula to Cactus, a distance of 250 mi. She increased her speed by 10 mi/h for the 360-mi trip from Cactus to Dry Junction. If the total trip took 11 h, what was her speed from Tortula to Cactus?

67. A social club charters a bus at a cost of $900 to take a group of members on an excursion to Atlantic City. At the last minute, five people in the group decide not to go. This raises the transportation cost per person by $2. How many people originally intended to take the trip?

68. A group of friends decides to buy a vacation home for $120,000, sharing the cost equally. If they could find one more person to join them, each person's contribution would drop by $6000. How many people are in the group?

69. The town of Foxton lies 10 mi north of an abandoned east-west road that runs through Grimley, as shown in the figure. The point on the abandoned road closest to Foxton is 40 mi from Grimley. County officials are about to build a new road connecting the two towns. They have determined that restoring the old road would cost $100,000 per mile, while building a new road would cost $200,000 per mile. How much of the abandoned road should be used (as indicated in the figure) if the officials intend to spend exactly $6.8 million? Would it cost less than this amount to build a new road connecting the towns directly?

70. A boardwalk is parallel to and 210 ft inland from a straight shoreline. A sandy beach lies between the boardwalk and the shoreline. A man is standing on the boardwalk, exactly 750 ft across the sand from his beach umbrella, which is right at the shoreline. The man walks 4 ft/s on the boardwalk and 2 ft/s on the sand. How far should he walk on the boardwalk before veering off onto the sand if he wishes to reach his umbrella in exactly 4 min 45 s?

71. Grain is falling from a chute onto the ground, forming a conical pile whose diameter is always three times its height. How high is the pile (to the nearest hundredth of a foot) when it contains 1000 ft^3 of grain?

72. A spherical tank has a capacity of 750 gallons. Using the fact that one gallon is about 0.1337 ft^3, find the radius of the tank (to the nearest hundredth of a foot).

73. A rectangular parcel of land is 50 ft wide. The length of a diagonal between opposite corners is 10 ft more than the length of the parcel. What is the length of the parcel?

74. Two television monitors sitting beside each other on a shelf in an appliance store have the same screen height. One has a conventional screen, which is 5 in. wider than it is high. The other has a wider, high-definition screen, which is 1.8 times as wide as it is high. The diagonal measure of the wider screen is 14 in. more than the diagonal measure of the smaller. What is the height of the screens, correct to the nearest 0.1 in.?

75. A running track has the shape shown in the figure, with straight sides and semicircular ends. If the length of the track is 440 yd and the two straight parts are each 110 yd long, what is the radius of the semicircular parts (to the nearest yard)?

76. A storage bin for corn consists of a cylindrical section made of wire mesh, surmounted by a conical tin roof, as shown in the figure. The height of the roof is one-third the entire height of the structure. If the total volume of the structure is 1400π ft^3 and its radius is 10 ft, what is its height? [*Hint:* Use the volume formulas listed on the inside front cover of this book.]

77. A wire 360 in. long is cut into two pieces. One piece is formed into a square and the other into a circle. If the two figures have the same area, what are the lengths of the two pieces of wire (to the nearest tenth of an inch)?

78. This problem is taken from a Chinese mathematics text-book called *Chui-chang suan-shu*, or *Nine Chapters on the Mathematical Art*, which was written about 250 B.C.

A 10-ft-long stem of bamboo is broken in such a way that its tip touches the ground 3 ft from the base of the stem, as shown in the figure. What is the height of the break?

[*Hint:* Use the Pythagorean Theorem.]

3 ft

79. The ancient Babylonians knew how to solve quadratic equations. Here is a problem from a cuneiform tablet found in a Babylonian school dating back to about 2000 B.C.

I have a reed, I know not its length. I broke from it one cubit, and it fit 60 times along the length of my field. I restored to the reed what I had broken off, and it fit 30 times along the width of my field. The area of my field is 375 square nindas. What was the original length of the reed?

Solve this problem, using the fact that 1 ninda = 12 cubits.

⬤ **DISCOVERY · DISCUSSION**

80. Historical Research   Read the biographical notes on Pythagoras (page 210), Euclid (page 46), and Thales (page 487). Choose one of these mathematicians and find out more about him from the library. Write a short essay on your findings. Include both biographical information and a description of the mathematics for which he is famous.

⬤ **Writing Project**

Equations through the Ages

Equations have been used to solve problems throughout recorded history, in every civilization. (See, for example, Exercises 78 and 79, above.) Here is a problem from ancient Babylon (ca. 2000 B.C.).

I found a stone but did not weigh it. After I added a seventh, and then added an eleventh of the result, I weighed it and found it weighed 1 mina. What was the original weight of the stone?

The answer given on the cuneiform tablet was $\frac{2}{3}$ mina, 8 sheqel, and $22\frac{1}{2}$ se, where 1 mina = 60 sheqel, and 1 sheqel = 180 se.

In ancient Egypt, knowing how to solve word problems was a highly prized secret. The Rhind Papyrus (ca. 1850 B.C.) contains many such problems (see page 705). Problem 32 in the Papyrus states

A quantity, its third, its quarter, added together become 2. What is the quantity?

The answer in Egyptian notation is $1 + \overline{4} + \overline{76}$, where the bar indicates "reciprocal," much like our notation 4^{-1}.

The Greek mathematician Diophantus (ca. 250 A.D., see page 37) wrote the book *Arithmetica*, which contains many word problems and equations. The Indian mathematician Bhaskara (12th century A.D., see page 533) and the Chinese mathematician Chang Ch'iu-Chien (6th century A.D.) also studied and wrote about equations. Of course, equations continue to be important today.

1. Solve the Babylonian problem and show that their answer was correct.

2. Solve the Egyptian problem and show that their answer was correct.

3. The ancient Egyptians and Babylonians used equations to solve practical problems. From the examples given here, do you think that they may have enjoyed posing and solving word problems just for fun?

4. Consider the following problem from 6th century China.

   If a rooster is worth 5 coins, a hen 3 coins, and three chicks together one coin, how many roosters, hens, and chicks, totalling 100, can be bought for 100 coins?

   This problem has several answers. Use trial and error to find at least one answer. Is this a practical problem or more of a riddle? Write a short essay to support your opinion.

5. Write a short essay explaining how equations affect your own life in today's world.

1.7 INEQUALITIES

Some problems in algebra lead to **inequalities** instead of equations. An inequality looks just like an equation, except that in the place of the equal sign is one of the symbols, $<$, $>$, $\leq$, or $\geq$. Here is an example of an inequality:

$$4x + 7 \leq 19$$

The table in the margin shows that some numbers satisfy the inequality and some numbers don't.

To **solve** an inequality that contains a variable means to find all values of the variable that make the inequality true. Unlike an equation, an inequality generally has infinitely many solutions, which form an interval or a union of intervals on the

x	$4x + 7 \leq 19$	
1	$11 \leq 19$	✓
2	$15 \leq 19$	✓
3	$19 \leq 19$	✓
4	$23 \leq 19$	✗
5	$27 \leq 19$	✗

real line. The following illustration shows how an inequality differs from its corresponding equation:

		Solution	Graph
Equation:	$4x + 7 = 19$	$x = 3$	
Inequality:	$4x + 7 \le 19$	$x \le 3$	

To solve inequalities, we use the following rules to isolate the variable on one side of the inequality sign. These rules tell us when two inequalities are *equivalent* (the symbol $\Leftrightarrow$ means "is equivalent to"). In these rules the symbols A, B, and C stand for real numbers or algebraic expressions. Here we state the rules for inequalities involving the symbol $\le$, but they apply to all four inequality symbols.

RULES FOR INEQUALITIES

Rule	Description
1. $A \le B \iff A + C \le B + C$	**Adding** the same quantity to each side of an inequality gives an equivalent inequality.
2. $A \le B \iff A - C \le B - C$	**Subtracting** the same quantity from each side of an inequality gives an equivalent inequality.
3. If $C > 0$, then $A \le B \iff CA \le CB$	**Multiplying** each side of an inequality by the same *positive* quantity gives an equivalent inequality.
4. If $C < 0$, then $A \le B \iff CA \ge CB$	**Multiplying** each side of an inequality by the same *negative* quantity *reverses the direction* of the inequality.
5. If $A > 0$ and $B > 0$, then $A \le B \iff \dfrac{1}{A} \ge \dfrac{1}{B}$	**Taking reciprocals** of each side of an inequality involving *positive* quantities *reverses the direction* of the inequality.
6. If $A \le B$ and $C \le D$, then $A + C \le B + D$	Inequalities can be added.

Pay special attention to Rules 3 and 4. Rule 3 says that we can multiply (or divide) each side of an inequality by a *positive* number, but Rule 4 says that if we multiply each side of an inequality by a *negative* number, then we reverse the direction of the inequality. For example, if we start with the inequality

$$3 < 5$$

and multiply by 2, we get

$$6 < 10$$

but if we multiply by -2, we get

$$-6 > -10$$

Linear Inequalities

An inequality is **linear** if each term is constant or a multiple of the variable.

EXAMPLE 1 ■ Solving a Linear Inequality

Solve the inequality $3x < 9x + 4$ and sketch the solution set.

SOLUTION

$$3x < 9x + 4$$

$$3x - 9x < 9x + 4 - 9x \qquad \text{Subtract } 9x$$

$$-6x < 4 \qquad \text{Simplify}$$

$$\left(-\tfrac{1}{6}\right)(-6x) > -\tfrac{1}{6}(4) \qquad \text{Multiply by } -\tfrac{1}{6} \text{ (or divide by } -6)$$

$$x > -\tfrac{2}{3} \qquad \text{Simplify}$$

The solution set consists of all numbers greater than $-\tfrac{2}{3}$. In other words the solution of the inequality is the interval $\left(-\tfrac{2}{3}, \infty\right)$. It is graphed in Figure 1. ■

FIGURE 1

EXAMPLE 2 ■ Solving a Pair of Simultaneous Inequalities

Solve the inequalities $4 \le 3x - 2 < 13$.

SOLUTION

The solution set consists of all values of x that satisfy both inequalities. Using Rules 1 and 3, we see that the following inequalities are equivalent:

$$4 \le 3x - 2 < 13$$

$$6 \le 3x < 15 \qquad \text{Add 2}$$

$$2 \le x < 5 \qquad \text{Divide by 3}$$

FIGURE 2

Therefore, the solution set is $[2, 5)$, as shown in Figure 2. ■

Nonlinear Inequalities

To solve inequalities involving squares and other powers of the variable, we use factoring, together with the following principle.

> **THE SIGN OF A PRODUCT OR QUOTIENT**
>
> If a product or a quotient has an *even* number of *negative* factors, then its value is *positive*.
>
> If a product or a quotient has an *odd* number of *negative* factors, then its value is *negative*.

EXAMPLE 3 ■ A Quadratic Inequality

Solve the inequality $x^2 - 5x + 6 \le 0$.

SOLUTION

First we factor the left side.

FIGURE 3

$$(x - 2)(x - 3) \leq 0$$

We know that the corresponding equation $(x - 2)(x - 3) = 0$ has the solutions 2 and 3. As shown in Figure 3, the numbers 2 and 3 divide the real line into three intervals: $(-\infty, 2)$, $(2, 3)$, and $(3, \infty)$. On each of these intervals we determine the signs of the factors using **test values**. We choose a number inside each interval and check the sign of the factors $x - 2$ and $x - 3$ at the value selected. For instance, if we use the test value $x = 1$ for the interval $(-\infty, 2)$ shown in Figure 4, then substitution in the factors $x - 2$ and $x - 3$ gives

$$x - 2 = 1 - 2 = -1 < 0$$

and

$$x - 3 = 1 - 3 = -2 < 0$$

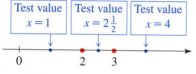

FIGURE 4

So both factors are negative on this interval. (The factors $x - 2$ and $x - 3$ change sign only at 2 and 3, respectively, so they maintain their signs over the length of each interval. That is why using a single test value on each interval is sufficient.)

Using the test values $x = 2\frac{1}{2}$ and $x = 4$ for the intervals $(2, 3)$ and $(3, \infty)$ (see Figure 4), respectively, we construct the following sign table. The final row of the table is obtained from the fact that the expression in the last row is the product of the two factors.

Interval	$(-\infty, 2)$	$(2, 3)$	$(3, \infty)$
Sign of $x - 2$ **Sign of $x - 3$**	$-$ $-$	$+$ $-$	$+$ $+$
Sign of $(x - 2)(x - 3)$	$+$	$-$	$+$

If you prefer, you can represent this information on a real number line, as in the following sign diagram. The vertical lines indicate the points at which the real line is divided into intervals:

	2	3	
Sign of $x - 2$	$-$	$+$	$+$
Sign of $x - 3$	$-$	$-$	$+$
Sign of $(x - 2)(x - 3)$	$+$	$-$	$+$

We read from the table or the diagram that $(x - 2)(x - 3)$ is negative on the interval $(2, 3)$. Thus, the solution of the inequality $(x - 2)(x - 3) \leq 0$ is

$$\{x \mid 2 \leq x \leq 3\} = [2, 3]$$

FIGURE 5

We have included the endpoints 2 and 3 because we seek values of x such that the product is either less than *or equal to* zero. The solution is illustrated in Figure 5.

We use the following guidelines to solve an inequality that can be factored.

GUIDELINES FOR SOLVING NONLINEAR INEQUALITIES

1. MOVE ALL TERMS TO ONE SIDE. If necessary, rewrite the inequality so that all nonzero terms appear on one side of the inequality sign. If the nonzero side of the inequality involves quotients, bring them to a common denominator.

2. FACTOR. Factor the nonzero side of the inequality.

3. FIND THE INTERVALS. Use the factorization to find all solutions of the equation corresponding to the given inequality. These numbers will divide the real line into intervals. List the intervals determined by these numbers.

4. MAKE A TABLE OR DIAGRAM. Use test values to make a table or diagram of the signs of each factor on each interval. In the last row of the table determine the sign of the product (or quotient) of these factors.

5. SOLVE. Determine the solution of the inequality from the last row of the sign table. Be sure to check whether the inequality is satisfied by some or all of the endpoints of the intervals (this may happen if the inequality involves $\leq$ or $\geq$).

 The factoring technique described in these guidelines works only if all nonzero terms appear on one side of the inequality symbol. If the inequality is not written in this form, first rewrite it, as indicated in Step 1. This technique is illustrated in the examples that follow.

EXAMPLE 4 ■ An Inequality Involving a Quotient

Solve $\dfrac{1+x}{1-x} \geq 1$.

SOLUTION

First we move all nonzero terms to the left side, and then we simplify using a common denominator.

$$\frac{1+x}{1-x} \geq 1$$

$$\frac{1+x}{1-x} - 1 \geq 0 \qquad \text{Subtract 1}$$

$$\frac{1+x}{1-x} - \frac{1-x}{1-x} \geq 0 \qquad \text{Common denominator } 1-x$$

$$\frac{1+x-1+x}{1-x} \geq 0 \qquad \text{Combine the fractions}$$

$$\frac{2x}{1-x} \geq 0 \qquad \text{Simplify}$$

The numerator is zero when $x = 0$ and the denominator is zero when $x = 1$, so we construct the following sign diagram using these values to define intervals on the real line.

From the diagram we see that the solution set is $\{x \mid 0 \leq x < 1\} = [0, 1)$. We include the endpoint 0 because the original inequality requires the quotient to be greater than *or equal to* 1. However, we do not include the other endpoint 1, since the quotient in the inequality is not defined at 1. Always check the endpoints of solution intervals to determine whether they satisfy the original inequality.

The solution set $[0, 1)$ is illustrated in Figure 6. ■

FIGURE 6

EXAMPLE 5 ■ Solving an Inequality with Three Factors

Solve the inequality $x < \dfrac{2}{x - 1}$.

SOLUTION

After moving all nonzero terms to one side of the inequality, we use a common denominator to combine the terms.

$$x - \frac{2}{x - 1} < 0 \qquad \text{Subtract } \tfrac{2}{x-1}$$

$$\frac{x(x - 1)}{x - 1} - \frac{2}{x - 1} < 0 \qquad \text{Common denominator } x - 1$$

$$\frac{x^2 - x - 2}{x - 1} < 0 \qquad \text{Combine the fractions}$$

$$\frac{(x + 1)(x - 2)}{x - 1} < 0 \qquad \text{Factor the numerator}$$

The factors in this quotient change sign at -1, 1, and 2, so we must examine the intervals $(-\infty, -1)$, $(-1, 1)$, $(1, 2)$, and $(2, \infty)$. Using test values, we get the following sign diagram.

	-1		1	2	
Sign of $x + 1$	$-$	$+$		$+$	$+$
Sign of $x - 2$	$-$	$-$		$-$	$+$
Sign of $x - 1$	$-$	$-$		$+$	$+$
Sign of $\dfrac{(x + 1)(x - 2)}{x - 1}$	$-$	$+$		$-$	$+$

Since the quotient must be negative, the solution is

$$(-\infty, -1) \cup (1, 2)$$

FIGURE 7

as illustrated in Figure 7.

Absolute Value Inequalities

We use the following properties to solve inequalities that involve absolute value.

PROPERTIES OF ABSOLUTE VALUE INEQUALITIES

Inequality	Equivalent form	Graph
1. $\lvert x \rvert < c$	$-c < x < c$	
2. $\lvert x \rvert \leq c$	$-c \leq x \leq c$	
3. $\lvert x \rvert > c$	$x < -c$ or $c < x$	
4. $\lvert x \rvert \geq c$	$x \leq -c$ or $c \leq x$	

FIGURE 8

These properties can be proved using the definition of absolute value. To prove Property 1, for example, we see that the inequality $\lvert x \rvert < c$ says that the distance from x to 0 is less than c, and from Figure 8 you can see that this is true if and only if x is between c and $-c$.

EXAMPLE 6 ■ Solving an Absolute Value Inequality

Solve the inequality $\lvert x - 5 \rvert < 2$.

SOLUTION 1

The inequality $\lvert x - 5 \rvert < 2$ is equivalent to

$$-2 < x - 5 < 2 \qquad \text{Property 1}$$

$$3 < x < 7 \qquad \text{Add 5}$$

The solution set is the open interval $(3, 7)$.

SOLUTION 2

Geometrically, the solution set consists of all numbers x whose distance from 5 is less than 2. From Figure 9 we see that this is the interval $(3, 7)$.

FIGURE 9

EXAMPLE 7 ■ Solving an Absolute Value Inequality

Solve the inequality $|3x + 2| \geq 4$.

SOLUTION

By Property 4 the inequality $|3x + 2| \geq 4$ is equivalent to

$$3x + 2 \geq 4 \qquad \text{or} \qquad 3x + 2 \leq -4$$

$$3x \geq 2 \qquad\qquad\qquad 3x \leq -6 \qquad \text{Subtract 2}$$

$$x \geq \tfrac{2}{3} \qquad\qquad\qquad x \leq -2 \qquad \text{Divide by 3}$$

So the solution set is

$$\{x \mid x \leq -2 \quad \text{or} \quad x \geq \tfrac{2}{3}\} = (-\infty, -2] \cup [\tfrac{2}{3}, \infty)$$

The set is graphed in Figure 10.

FIGURE 10

Modeling with Inequalities

Modeling real-life problems frequently leads to inequalities because we are often interested in determining when one quantity is more (or less) than another.

EXAMPLE 8 ■ Carnival Tickets

A carnival has two plans for tickets.

> Plan A: $5 entrance fee and 25¢ each ride
>
> Plan B: $2 entrance fee and 50¢ each ride

How many rides would you have to take for plan A to be less expensive than plan B?

SOLUTION

We are asked for the number of rides for which plan A is less expensive than plan B. So let

> Identify the variable

$$x = \text{number of rides}$$

The information in the problem may be organized as follows

> Express all unknown quantities in terms of the variable

In Words	In Algebra
Number of rides	x
Cost with Plan A	$5 + 0.25x$
Cost with Plan B	$2 + 0.50x$

Now we set up the model.

Set up the model

$$\begin{array}{c|c}\text{cost with}\\\text{plan A}\end{array} < \begin{array}{c}\text{cost with}\\\text{plan B}\end{array}$$

$$5 + 0.25x < 2 + 0.50x$$

Solve

$$3 + 0.25x < 0.50x \qquad \text{Subtract 2}$$

$$3 < 0.25x \qquad \text{Subtract } 0.25x$$

$$12 < x \qquad \text{Divide by } 0.25$$

So if you plan to take *more than* 12 rides, plan A is less expensive. ■

EXAMPLE 9 ■ Relationship between Fahrenheit and Celsius Scales

The instructions on a box of film indicate that the box should be stored at a temperature between 5°C and 30°C. What range of temperatures does this correspond to on the Fahrenheit scale?

SOLUTION

The relationship between degrees Celsius (C) and degrees Fahrenheit (F) is given by the equation $C = \frac{5}{9}(F - 32)$. Expressing the statement on the box in terms of inequalities, we have

$$5 < C < 30$$

so the corresponding Fahrenheit temperatures satisfy the inequalities

$$5 < \tfrac{5}{9}(F - 32) < 30$$

$$\tfrac{9}{5} \cdot 5 < F - 32 < \tfrac{9}{5} \cdot 30 \qquad \text{Multiply by } \tfrac{9}{5}$$

$$9 < F - 32 < 54 \qquad \text{Simplify}$$

$$9 + 32 < F < 54 + 32 \qquad \text{Add 32}$$

$$41 < F < 86 \qquad \text{Simplify}$$

The film should be stored at a temperature between 41°F and 86°F. ■

EXAMPLE 10 ■ Concert Tickets

A group of students decide to attend a concert. The cost of chartering a bus to take them to the concert is $450, which is to be shared equally among the students. The concert promoters offer discounts to groups arriving by bus. Tickets normally cost $50 each but are reduced by 10¢ per ticket for each person in the group (up to the maximum capacity of the bus). How many students must be in the group for the total cost per student to be less than $54?

SOLUTION

We are asked for the number of students in the group. So let

Identify the variable

$$x = \text{number of students in the group}$$

The information in the problem may be organized as follows.

In Words	In Algebra
Number of students in group	x
Bus cost per student	$\dfrac{450}{x}$
Ticket cost per student	$50 - 0.10x$

Express all unknown quantities in terms of the variable

Now we set up the model.

Set up the model

$$\boxed{\begin{array}{c}\text{bus cost}\\\text{per student}\end{array}} + \boxed{\begin{array}{c}\text{ticket cost}\\\text{per student}\end{array}} < \boxed{54}$$

$$\frac{450}{x} + (50 - 0.10x) < 54$$

Solve

$$\frac{450}{x} - 4 - 0.10x < 0 \qquad \text{Subtract 54}$$

$$\frac{450 - 4x - 0.10x^2}{x} < 0 \qquad \text{Common denominator}$$

$$\frac{4500 - 40x - x^2}{x} < 0 \qquad \text{Multiply by 10}$$

$$\frac{(90 + x)(50 - x)}{x} < 0 \qquad \text{Factor numerator}$$

	-90		0		50	
Sign of $90 + x$	$-$		$+$		$+$	$+$
Sign of $50 - x$	$+$		$+$		$+$	$-$
Sign of x	$-$		$-$		$+$	$+$
Sign of $\dfrac{(90 + x)(50 - x)}{x}$	$+$		$-$		$+$	$-$

The sign diagram shows that the solution of the inequality is the set $(-90, 0) \cup (50, \infty)$. Because we cannot have a negative number of students, it follows that the group must have more than 50 students for the total cost per person to be less than \$54.

■

1.7 EXERCISES

1–4 ■ Let $S = \{-1, 0, \frac{1}{2}, \sqrt{2}, 2\}$. Use substitution to determine which elements of S satisfy the inequality.

1. $2x + 10 > 8$

2. $4x + 1 \leq 2x$

3. $\dfrac{1}{x} \leq \dfrac{1}{2}$

4. $x^2 + 2 < 4$

5–24 ■ Solve the linear inequality. Express the solution using interval notation and graph the solution set.

5. $2x \leq 7$

6. $-4x \geq 10$

7. $2x - 5 > 3$

8. $3x + 11 < 5$

9. $7 - x \geq 5$

10. $5 - 3x \leq -16$

11. $2x + 1 < 0$

12. $0 < 5 - 2x$

13. $3x + 11 \leq 6x + 8$

14. $6 - x \geq 2x + 9$

15. $\frac{1}{2}x - \frac{2}{3} > 2$

16. $\frac{2}{5}x + 1 < \frac{1}{5} - 2x$

17. $4 - 3x \leq -(1 + 8x)$

18. $2(7x - 3) \leq 12x + 16$

19. $2 \leq x + 5 < 4$

20. $5 \leq 3x - 4 \leq 14$

21. $-1 < 2x - 5 < 7$

22. $1 < 3x + 4 \leq 16$

23. $-2 < 8 - 2x \leq -1$

24. $-3 \leq 3x + 7 \leq \frac{1}{2}$

25–54 ■ Solve the nonlinear inequality. Express the solution using interval notation and graph the solution set.

25. $(x - 2)(x - 5) > 0$

26. $(3x + 1)(x - 1) \geq 0$

27. $x^2 - 3x - 18 \leq 0$

28. $x^2 + 5x + 6 > 0$

29. $2x^2 + x \geq 1$

30. $x^2 < x + 2$

31. $3x^2 - 3x < 2x^2 + 4$

32. $5x^2 + 3x \geq 3x^2 + 2$

33. $x^2 > 3(x + 6)$

34. $x^2 + 2x > 3$

35. $x^2 < 4$

36. $x^2 \geq 9$

37. $-2x^2 \leq 4$

38. $(x + 2)(x - 1)(x - 3) \leq 0$

39. $\dfrac{x - 3}{x + 1} \geq 0$

40. $\dfrac{2x + 6}{x - 2} < 0$

41. $\dfrac{4x}{2x + 3} > 2$

42. $-2 < \dfrac{x + 1}{x - 3}$

43. $\dfrac{2x + 1}{x - 5} \leq 3$

44. $\dfrac{3 + x}{3 - x} \geq 1$

45. $\dfrac{4}{x} < x$

46. $\dfrac{x}{x + 1} > 3x$

47. $1 + \dfrac{2}{x + 1} \leq \dfrac{2}{x}$

48. $\dfrac{3}{x - 1} - \dfrac{4}{x} \geq 1$

49. $\dfrac{6}{x - 1} - \dfrac{6}{x} \geq 1$

50. $\dfrac{x}{2} \geq \dfrac{5}{x + 1} + 4$

51. $\dfrac{x + 2}{x + 3} < \dfrac{x - 1}{x - 2}$

52. $\dfrac{1}{x + 1} + \dfrac{1}{x + 2} \leq 0$

53. $x^4 > x^2$

54. $x^5 > x^2$

55–66 ■ Solve the inequality. Express the solution using interval notation and graph the solution set.

55. $|x| < 7$

56. $|x| \geq 3$

57. $|x - 5| \leq 3$

58. $|x - 9| > 9$

59. $|x + 5| < 2$

60. $|x + 1| \geq 3$

61. $|2x - 3| \leq 0.4$

62. $|5x - 2| < 6$

63. $\left| \dfrac{x - 2}{3} \right| < 2$

64. $\left| \dfrac{x + 1}{2} \right| \geq 4$

65. $4|x + 2| - 3 < 13$

66. $3 - |2x + 4| \leq 1$

67. Use the relationship between C and F given in Example 9 to find the interval on the Fahrenheit scale corresponding to the temperature range $20 \leq C \leq 30$.

68. What interval on the Celsius scale corresponds to the temperature range $50 \leq F \leq 95$?

69. Using calculus it can be shown that if a ball is thrown upward with an initial velocity of 16 ft/s from the top of a building 128 ft high, then its height h above the ground t seconds later will be

$$h = 128 + 16t - 16t^2$$

During what time interval will the ball be at least 32 ft above the ground?

70. The gas mileage g (measured in mi/gal) for a particular vehicle, driven at v mi/h, is given by the formula $g = 10 + 0.9v - 0.01v^2$, as long as v is between 10 mi/h and 75 mi/h. For what range of speeds is the vehicle's mileage 30 mi/gal or better?

71. For a certain model of car the distance d required to stop the vehicle if it is traveling at v mi/h is given by the following formula:

$$d = v + \frac{v^2}{20}$$

where d is measured in feet. Kerry wants her stopping distance not to exceed 240 ft. At what range of speeds can she travel?

$\longleftarrow$ 240 ft $\longrightarrow$

72. If a manufacturer sells x units of a certain product, his revenue R and cost C (in dollars) are given by:

$$R = 20x$$

$$C = 2000 + 8x + 0.0025x^2$$

Use the fact that

$$\text{profit} = \text{revenue} - \text{cost}$$

to determine how many units he should sell to enjoy a profit of at least $2400.

73. A car rental company offers two plans for renting a car.
　　Plan A:　$30 per day and 10¢ per mile
　　Plan B:　$50 per day with free unlimited mileage
For what range of miles will plan B save you money?

74. A telephone company offers two long-distance plans.
　　Plan A:　$25 per month and 5¢ per minute
　　Plan B:　$5 per month and 12¢ per minute

For how many minutes of long-distance calls would plan B be financially advantageous?

75. A charter airline finds that on its Saturday flights from Philadelphia to London, all 120 seats will be sold if the ticket price is $200. However, for each $3 increase in ticket price, the number of seats sold decreases by one.
(a) Find a formula for the number of seats sold if the ticket price is P dollars.
(b) Over a certain period, the number of seats sold for this flight ranged between 90 and 115. What was the corresponding range of ticket prices?

76. As dry air moves upward, it expands and in so doing cools at a rate of about 1°C for each 100 m rise, up to about 12 km.
(a) If the ground temperature is 20°C, write a formula for the temperature at height h.
(b) What range of temperatures can be expected if a plane takes off and reaches a maximum height of 5 km?

77. A coffee merchant sells a customer 3 lb of Hawaiian Kona at $6.50 per pound. His scale is accurate to within ±0.03 lb. By how much could the customer have been overcharged or undercharged because of possible inaccuracy in the scale?

78. A riverboat theater offers bus tours to groups on the following basis. Hiring the bus costs the group $360, to be shared equally by the group members. Theater tickets, normally $30 each, are discounted by 25¢ times the number of people in the group. How many members must be in the group so that the cost of the theater tour (bus fare plus theater ticket) is less than $39 per person?

79. A gardener has 120 ft of deer-resistant fence. She wants to enclose a rectangular vegetable garden in her backyard, and she wants the area enclosed to be at least 800 ft². What range of values is possible for the length of her garden?

80. Solve the inequality for x, assuming that a, b, and c are positive constants.
(a) $a(bx - c) \geq bc$　　　(b) $a \leq bx + c < 2a$

81–84 ■ Determine the values of the variable for which the expression is defined as a real number.

81. $\sqrt{16 - 9x^2}$　　　　**82.** $\sqrt{3x^2 - 5x + 2}$

83. $\left(\dfrac{1}{x^2 - 5x - 14}\right)^{1/2}$　　**84.** $\sqrt[4]{\dfrac{1 - x}{2 + x}}$

85. Suppose that a, b, c, and d are positive numbers such that

$$\frac{a}{b} < \frac{c}{d}$$

Show that

$$\frac{a}{b} < \frac{a + c}{b + d} < \frac{c}{d}$$

 DISCOVERY · DISCUSSION

86. Do Powers Preserve Order? If $a < b$, is $a^2 < b^2$? (Check both positive and negative values for a and b.) If $a < b$, is $a^3 < b^3$? Based on your observations, state a general rule about the relationship between a^n and b^n when $a < b$ and n is a positive integer.

87. What's Wrong Here?   It is tempting to try to solve an inequality like an equation. For instance, we might try to solve $1 < 3/x$ by multiplying both sides by x, to get $x < 3$, so the solution would be $(-\infty, 3)$. But that's wrong; for example, $x = -1$ lies in this interval but does not satisfy the original inequality. Explain why this method doesn't work (think about the *sign* of x). Then solve the inequality correctly.

88. Using Distances to Solve Absolute Value Inequalities   Recall that $|a - b|$ is the distance between a and b on the number line. For any number x, what do $|x - 1|$ and $|x - 3|$ represent? Use this interpretation to solve the inequality $|x - 1| < |x - 3|$ geometrically. In general, if $a < b$, what is the solution of the inequality $|x - a| < |x - b|$?

89. Ordering the Points in the Cartesian Plane   Given any two different real numbers x and y, we know that either $x < y$ or $y < x$. We can think of the real numbers as being *totally ordered* by the relation $<$, in the sense that we can always tell which number in any pair is smaller—that is, which comes earlier on the real line. The words in a dictionary are also totally ordered. Given two words, the one whose first letter comes earlier in the alphabet will be earlier in the dictionary; if the first letter is the same, we apply the alphabetic rule to the second letter, and so on. This is called the *lexicographic* order. Thus, we can say that cat < dog, man < mountain, or pantry < pastry. A similar **lexicographic order** can be defined for the points in the coordinate plane. We'll say that $(a, b) < (c, d)$ if $a < c$, or if $a = c$ and $b < d$. Thus, $(1, 4) < (3, 2)$ and $(5, 0) < (5, 1)$.

(a) List the following points in lexicographic order, from smallest to largest.

$$(2, 5) \quad (4, -2) \quad (2, 0) \quad (0, 2) \quad (2, -3) \quad (-1, 4) \quad (-2, 5)$$

(b) Graph the following sets of points in the coordinate plane.

$$S = \{(x, y) \mid (0, 3) < (x, y)\}$$
$$T = \{(x, y) \mid (-2, 2) < (x, y) \leq (4, 0)\}$$

1.8   COORDINATE GEOMETRY

In this section we introduce the coordinate plane. In the coordinate plane, we can draw a graph of an equation, thus allowing us to "see" the relationship of the variables in the equation. We can also reverse this process and describe a geometrical figure (such as a circle or a line) by an algebraic equation. Thus, the coordinate plane is the link between algebra and geometry that enables us to use the techniques of algebra to solve geometric problems and the techniques of geometry to solve algebraic problems.

■ The Coordinate Plane

The Cartesian plane is named in honor of the French mathematician René Descartes (1596–1650), although another Frenchman, Pierre Fermat (1601–1665), also invented the principles of coordinate geometry at the same time. (See their biographies on pages 90 and 532.)

Although the notation for a point (a, b) is the same as the notation for an open interval (a, b), the context should make clear which meaning is intended.

Just as points on a line can be identified with real numbers to form the coordinate line, points in a plane can be identified with ordered pairs of numbers to form the **coordinate plane** or **Cartesian plane**. To do this, we draw two perpendicular real lines that intersect at 0 on each line. Usually one line is horizontal with positive direction to the right and is called the **x-axis**; the other line is vertical with positive direction upward and is called the **y-axis**. The point of intersection of the x-axis and the y-axis is the **origin O**, and the two axes divide the plane into four **quadrants**, labeled I, II, III, and IV in Figure 1 on page 90. (The points *on* the coordinate axes are not assigned to any quadrant.)

Any point P in the coordinate plane can be located by a unique **ordered pair** of numbers (a, b), as shown in Figure 1. The first number a is called the **x-coordinate** of P; the second number b is called the **y-coordinate** of P. We can think of the coordinates of P as its "address," because they specify its location in the plane. Several points are labeled with their coordinates in Figure 2.

FIGURE 1   FIGURE 2

Any point P in the coordinate plane can be located by a unique **ordered pair** of numbers (a, b), as shown in Figure 1. The first number a is called the **x-coordinate** of P; the second number b is called the **y-coordinate** of P. We can think of the coordinates of P as its "address," because they specify its location in the plane. Several points are labeled with their coordinates in Figure 2.

EXAMPLE 1 ■ Graphing Regions in the Coordinate Plane

Describe and sketch the regions given by each set.

(a) $\{(x, y) \mid x \geq 0\}$ (b) $\{(x, y) \mid y = 1\}$ (c) $\{(x, y) \mid \mid y \mid < 1\}$

SOLUTION

(a) The points whose x-coordinates are 0 or positive lie on the y-axis or to the right of it, as shown in Figure 3(a).

(b) The set of all points with y-coordinate 1 is a horizontal line one unit above the x-axis, as in Figure 3(b).

(c) Recall from Section 1.7 that

$$| y | < 1 \qquad \text{if and only if} \qquad -1 < y < 1$$

The given region consists of those points in the plane whose y-coordinates lie between -1 and 1. Thus, the region consists of all points that lie between (but not on) the horizontal lines $y = 1$ and $y = -1$. These lines are shown as broken lines in Figure 3(c) to indicate that the points on these lines do not lie in the set.

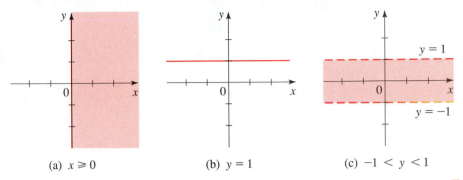

(a) $x \geqslant 0$ (b) $y = 1$ (c) $-1 < y < 1$

FIGURE 3

We now find a formula for the distance $d(A, B)$ between two points $A(x_1, y_1)$ and $B(x_2, y_2)$ in the plane. Recall from Section 1.1 that the distance between points a and b on a number line is $d(a, b) = |b - a|$. So, from Figure 4 we see that the distance between the points $A(x_1, y_1)$ and $C(x_2, y_1)$ on a horizontal line must be $|x_2 - x_1|$, and the distance between $B(x_2, y_2)$ and $C(x_2, y_1)$ on a vertical line must be $|y_2 - y_1|$. Since triangle ABC is a right triangle, the Pythagorean Theorem gives

$$d(A, B) = \sqrt{|x_2 - x_1|^2 + |y_2 - y_1|^2}$$

$$= \sqrt{(x_2 - x_1)^2 + (y_2 - y_1)^2}$$

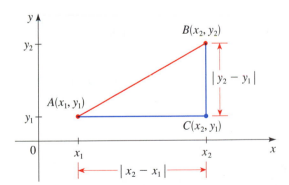

FIGURE 4

DISTANCE FORMULA

The distance between the points $A(x_1, y_1)$ and $B(x_2, y_2)$ in the plane is

$$d(A, B) = \sqrt{(x_2 - x_1)^2 + (y_2 - y_1)^2}$$

FIGURE 5

EXAMPLE 2 ■ Applying the Distance Formula

Which of the points $P(1, -2)$ or $Q(8, 9)$ is closer to the point $A(5, 3)$?

SOLUTION

By the distance formula, we have

$$d(P, A) = \sqrt{(5 - 1)^2 + [3 - (-2)]^2} = \sqrt{4^2 + 5^2} = \sqrt{41}$$

$$d(Q, A) = \sqrt{(5 - 8)^2 + (3 - 9)^2} = \sqrt{(-3)^2 + (-6)^2} = \sqrt{45}$$

This shows that $d(P, A) < d(Q, A)$, so P is closer to A (see Figure 5). ■

Now let's find the coordinates (x, y) of the midpoint M of the line segment that joins the point $A(x_1, y_1)$ to the point $B(x_2, y_2)$. In Figure 6 notice that triangles APM and MQB are congruent because $d(A, M) = d(M, B)$ and the corresponding angles are equal.

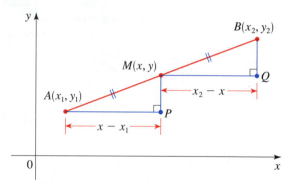

FIGURE 6

It follows that $d(A, P) = d(M, Q)$ and so

$$x - x_1 = x_2 - x$$

Solving this equation for x, we get

$$2x = x_1 + x_2$$

$$x = \frac{x_1 + x_2}{2}$$

Similarly,

$$y = \frac{y_1 + y_2}{2}$$

MIDPOINT FORMULA

The midpoint of the line segment from $A(x_1, y_1)$ to $B(x_2, y_2)$ is

$$\left(\frac{x_1 + x_2}{2}, \frac{y_1 + y_2}{2}\right)$$

EXAMPLE 3 ■ Applying the Midpoint Formula

Show that the quadrilateral with vertices $P(1, 2)$, $Q(4, 4)$, $R(5, 9)$, and $S(2, 7)$ is a parallelogram by proving that its two diagonals bisect each other.

SOLUTION

If the two diagonals have the same midpoint, then they must bisect each other. The midpoint of the diagonal PR is

$$\left(\frac{1 + 5}{2}, \frac{2 + 9}{2}\right) = \left(3, \frac{11}{2}\right)$$

and the midpoint of the diagonal QS is

$$\left(\frac{4 + 2}{2}, \frac{4 + 7}{2}\right) = \left(3, \frac{11}{2}\right)$$

so each diagonal bisects the other, as shown in Figure 7. (A theorem from elementary geometry states that the quadrilateral is therefore a parallelogram.) ■

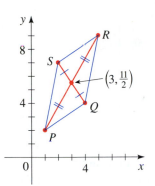

FIGURE 7

Graphs of Equations

Suppose we have an equation involving the variables x and y, such as

$$x^2 + y^2 = 25 \qquad \text{or} \qquad x = y^2 \qquad \text{or} \qquad y = \frac{2}{x}$$

A point (x, y) **satisfies** the equation if the equation is true when the coordinates of the point are substituted into the equation. For example, the point $(3, 4)$ satisfies the first equation, since $3^2 + 4^2 = 25$, but the point $(2, -3)$ does not, since $2^2 + (-3)^2 = 13 \neq 25$.

Fundamental Principle of Analytic Geometry

A point (x, y) lies on the graph of an equation if and only if its coordinates satisfy the equation.

> ### THE GRAPH OF AN EQUATION
>
> The **graph** of an equation in x and y is the set of all points (x, y) in the coordinate plane that satisfy the equation.

The graphs of most of the equations that we will encounter are curves. For example, we will see later in this section that the first equation, $x^2 + y^2 = 25$, represents a circle; we will determine the shapes represented by the other equations later in this chapter. Graphs help us understand equations because they give us visual representations of the equations.

EXAMPLE 4 ■ Sketching a Graph by Plotting Points

Sketch the graph of the equation $2x - y = 3$.

FIGURE 8

SOLUTION

We first solve the given equation for y to get

$$y = 2x - 3$$

This helps us calculate the y-coordinates in the following table.

x	$y = 2x - 3$	(x, y)
-1	-5	$(-1, -5)$
0	-3	$(0, -3)$
1	-1	$(1, -1)$
2	1	$(2, 1)$
3	3	$(3, 3)$
4	5	$(4, 5)$

Of course, there are infinitely many points on the graph, and it is impossible to plot all of them. But the more points we plot, the better we can imagine what the graph represented by the equation looks like. We plot the points we found in Figure 8; they appear to lie on a line. So, we complete the graph by joining the points by a line. (In Section 1.10 we verify that the graph of this equation is indeed a line.) ∎

A detailed discussion of parabolas and their geometric properties is presented in Chapter 9.

EXAMPLE 5 ■ Sketching a Graph by Plotting Points

Sketch the graph of the equation $y = x^2 - 2$.

SOLUTION

We find some of the points that satisfy the equation in the following table. In Figure 9 we plot these points and then connect them by a smooth curve. A curve with this shape is called a *parabola*.

FIGURE 9

x	$y = x^2 - 2$	(x, y)
-3	7	$(-3, 7)$
-2	2	$(-2, 2)$
-1	-1	$(-1, -1)$
0	-2	$(0, -2)$
1	-1	$(1, -1)$
2	2	$(2, 2)$
3	7	$(3, 7)$

EXAMPLE 6 ■ Sketching the Graph of an Equation Involving Absolute Value

Sketch the graph of the equation $y = |x|$.

The coordinates of a point in the xy-plane uniquely determine its location. We can think of the coordinates as the "address" of the point. In Salt Lake City, Utah, the addresses of most buildings are in fact expressed as coordinates. The city is divided into quadrants with Main Street as the vertical (North-South) axis and S. Temple Street as the horizontal (East-West) axis. An address such as

1760 W 2100 S

indicates a location 17.6 blocks west of Main Street and 21 blocks south of S. Temple Street. (This is the address of the main post office in Salt Lake City.) With this logical system it is possible for someone unfamiliar with the city to locate any address immediately, as easily as one locates a point in the coordinate plane.

SOLUTION

We make a table of values:

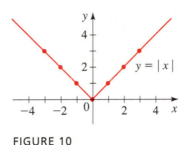

FIGURE 10

x	$y = \lvert x \rvert$	(x, y)
-3	3	$(-3, 3)$
-2	2	$(-2, 2)$
-1	1	$(-1, 1)$
0	0	$(0, 0)$
1	1	$(1, 1)$
2	2	$(2, 2)$
3	3	$(3, 3)$

In Figure 10 we plot these points and use them to sketch the graph of the equation.

■

The x-coordinates of the points where a graph intersects the x-axis are called the **x-intercepts** of the graph and are obtained by setting $y = 0$ in the equation of the graph. The y-coordinates of the points where a graph intersects the y-axis are called the **y-intercepts** of the graph and are obtained by setting $x = 0$ in the equation of the graph.

DEFINITION OF INTERCEPTS

Intercepts	How to find them	Where they are on the graph
x-intercepts: The x-coordinates of points where the graph of an equation intersects the x-axis	Set $y = 0$ and solve for x	
y-intercepts: The y-coordinates of points where the graph of an equation intersects the y-axis	Set $x = 0$ and solve for y	

EXAMPLE 7 ■ Finding Intercepts

Find the x- and y-intercepts of the graph of the equation $y = x^2 - 2$.

SOLUTION

To find the x-intercepts, we set $y = 0$ and solve for x. Thus

$$0 = x^2 - 2 \qquad \text{Set } y = 0$$

$$x^2 = 2 \qquad \text{Add 2 to each side}$$

$$x = \pm\sqrt{2} \qquad \text{Take the square root}$$

The x-intercepts are $\sqrt{2}$ and $-\sqrt{2}$.

To find the y-intercepts, we set $x = 0$ and solve for y. Thus

$$y = 0^2 - 2 \qquad \text{Set } x = 0$$

$$y = -2$$

The y-intercept is -2.

The graph of this equation was sketched in Example 5. It is repeated in Figure 11 with the x- and y-intercepts labeled. ∎

FIGURE 11

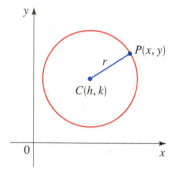

FIGURE 12

Circles

So far we have discussed how to find the graph of an equation in x and y. The converse problem is to find an equation of a graph, that is, an equation that represents a given curve in the xy-plane. Such an equation is satisfied by the coordinates of the points on the curve and by no other point. This is the other half of the fundamental principle of analytic geometry as formulated by Descartes and Fermat. The idea is that if a geometric curve can be represented by an algebraic equation, then the rules of algebra can be used to analyze the curve.

As an example of this type of problem, let's find the equation of a circle with radius r and center (h, k). By definition, the circle is the set of all points $P(x, y)$ whose distance from the center $C(h, k)$ is r (see Figure 12). Thus, P is on the circle if and only if $d(P, C) = r$. From the distance formula we have

$$\sqrt{(x - h)^2 + (y - k)^2} = r$$

$$(x - h)^2 + (y - k)^2 = r^2 \qquad \text{Square each side}$$

This is the desired equation.

EQUATION OF A CIRCLE

An equation of the circle with center (h, k) and radius r is

$$(x - h)^2 + (y - k)^2 = r^2$$

This is called the **standard form** for the equation of the circle.
If the center of the circle is the origin $(0, 0)$, then the equation is

$$x^2 + y^2 = r^2$$

EXAMPLE 8 ■ Graphing a Circle

Graph each equation.

(a) $x^2 + y^2 = 25$ (b) $(x - 2)^2 + (y + 1)^2 = 25$

SOLUTION

(a) Rewriting the equation as $x^2 + y^2 = 5^2$, we see that this is an equation of the circle of radius 5 centered at the origin. Its graph is shown in Figure 13.

(b) Rewriting the equation as $(x - 2)^2 + (y + 1)^2 = 5^2$, we see that this is an equation of the circle of radius 5 centered at $(2, -1)$. Its graph is shown in Figure 14.

FIGURE 13

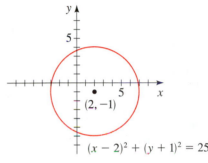

FIGURE 14

EXAMPLE 9 ■ Finding an Equation of a Circle

(a) Find an equation of the circle with radius 3 and center $(2, -5)$.

(b) Find an equation of the circle that has the points $P(1, 8)$ and $Q(5, -6)$ as the endpoints of a diameter.

SOLUTION

(a) Using the equation of a circle with $r = 3$, $h = 2$, and $k = -5$, we obtain

$$(x - 2)^2 + (y + 5)^2 = 9$$

Its graph is shown in Figure 15.

(b) We first observe that the center is the midpoint of the diameter PQ, so by the Midpoint Formula the center is

$$\left(\frac{1 + 5}{2}, \frac{8 - 6}{2}\right) = (3, 1)$$

The radius r is the distance from P to the center, so by the Distance Formula

$$r^2 = (3 - 1)^2 + (1 - 8)^2 = 2^2 + (-7)^2 = 53$$

FIGURE 15

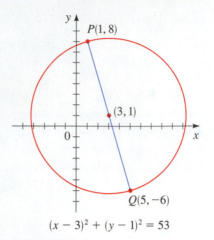

$$(x - 3)^2 + (y - 1)^2 = 53$$

FIGURE 16

Completing the square is discussed on page 49.

⊘ We must add the same numbers to *each side* to maintain equality.

Therefore, the equation of the circle is

$$(x - 3)^2 + (y - 1)^2 = 53$$

Its graph is shown in Figure 16.   ■

EXAMPLE 10 ■ Identifying an Equation of a Circle

Sketch the graph of the equation $x^2 + y^2 + 2x - 6y + 7 = 0$ by first showing that it represents a circle and then finding its center and radius.

SOLUTION

We first group the x-terms and y-terms. Then we complete the square within each grouping. That is, we complete the square for $x^2 + 2x$ by adding $\left(\frac{1}{2} \cdot 2\right)^2 = 1$, and we complete the square for $y^2 - 6y$ by adding $\left[\frac{1}{2} \cdot (-6)\right]^2 = 9$.

$$(x^2 + 2x \qquad) + (y^2 - 6y \qquad) = -7 \qquad \text{Group terms}$$

$$(x^2 + 2x + 1) + (y^2 - 6y + 9) = -7 + 1 + 9 \qquad \text{Complete the square by adding 1 and 9 to each side}$$

$$(x + 1)^2 + (y - 3)^2 = 3$$

Comparing this equation with the standard equation of a circle, we see that $h = -1$, $k = 3$, and $r = \sqrt{3}$, so the given equation represents a circle with center $(-1, 3)$ and radius $\sqrt{3}$. The circle is sketched in Figure 17.

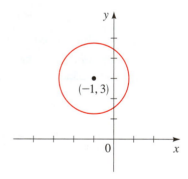

FIGURE 17
$x^2 + y^2 + 2x - 6y + 7 = 0$

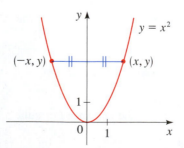

FIGURE 18

■ Symmetry

Figure 18 shows the graph of $y = x^2$. Notice that the part of the graph to the left of the y-axis is the mirror image of the part to the right of the y-axis. The reason is that if the point (x, y) is on the graph, then so is $(-x, y)$, and these points are reflections of each other about the y-axis. In this situation we say the graph is **symmetric with respect to the y-axis**. Similarly, we say a graph is **symmetric with respect to the x-axis** if whenever the point (x, y) is on the graph, then so is $(x, -y)$. A graph is **symmetric with respect to the origin** if whenever (x, y) is on the graph, so is $(-x, -y)$.

DEFINITION OF SYMMETRY

Type of symmetry	How to test for symmetry	What the graph looks like (figures in this section)	Geometric meaning
Symmetry with respect to the x-axis	The equation is unchanged when y is replaced by $-y$	(Figures 13, 20)	Graph is unchanged when reflected in the x-axis
Symmetry with respect to the y-axis	The equation is unchanged when x is replaced by $-x$	(Figures 9, 10, 11, 13, 18, 20)	Graph is unchanged when reflected in the y-axis
Symmetry with respect to the origin	The equation is unchanged when x is replaced by $-x$ and y by $-y$	(Figures 13, 19, 20)	Graph is unchanged when rotated 180° about the origin

The remaining examples in this section show how symmetry helps us sketch the graphs of equations.

EXAMPLE 11 ■ Using Symmetry to Sketch a Graph

Test the equation $y = x^3 - 9x$ for symmetry and sketch its graph.

SOLUTION

If we replace x by $-x$ and y by $-y$ in the equation, we get

$$-y = (-x)^3 - 9(-x)$$

$$-y = -x^3 + 9x$$

$$y = x^3 - 9x$$

$y = x^3 - 9x$

FIGURE 19

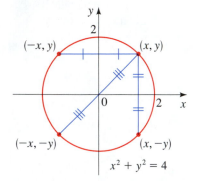

$(-x, y)$ (x, y)

$(-x, -y)$ $(x, -y)$

$x^2 + y^2 = 4$

FIGURE 20

and so the equation is unchanged. This means that the graph is symmetric with respect to the origin. We sketch it by first plotting points for $x > 0$ and then using symmetry about the origin (see Figure 19).

x	$y = x^3 - 9x$	(x, y)
0	0	$(0, 0)$
1	-8	$(1, -8)$
1.5	-10.125	$(1.5, -10.125)$
2	-10	$(2, -10)$
2.5	-6.875	$(2.5 \, -6.875)$
3	0	$(3, 0)$
4	28	$(4, 28)$

EXAMPLE 12 ■ A Circle That Has All Three Types of Symmetry

Test the equation of the circle $x^2 + y^2 = 4$ for symmetry.

SOLUTION

The equation $x^2 + y^2 = 4$ remains unchanged when x is replaced by $-x$ and y is replaced by $-y$, since $(-x)^2 = x^2$ and $(-y)^2 = y^2$, so the circle exhibits all three types of symmetry. It is symmetric with respect to the x-axis, the y-axis, and the origin, as shown in Figure 20.   ■

1.8 EXERCISES

1. Plot the given points in a coordinate plane:

$(2, 3), (-2, 3), (4, 5), (4, -5), (-4, 5), (-4, -5)$

2. Find the coordinates of the points shown in the figure.

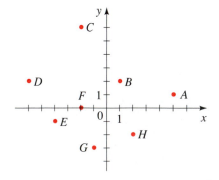

3–8 ■ A pair of points is given.
(a) Plot the points in a coordinate plane.
(b) Find the distance between them.
(c) Find the midpoint of the segment that joins them.

3. $(2, 3), \quad (5, 2)$

4. $(2, -1), \quad (4, 3)$

5. $(6, -2), \quad (-1, 3)$

6. $(1, -6), \quad (-1, -3)$

7. $(3, 4), \quad (-3, -4)$

8. $(5, 0), \quad (0, 6)$

9. Draw the rectangle with vertices $A(1, 3)$, $B(5, 3)$, $C(1, -3)$, and $D(5, -3)$ in a coordinate plane. Find the area of the rectangle.

10. Draw the parallelogram with vertices $A(1, 2)$, $B(5, 2)$, $C(3, 6)$, and $D(7, 6)$ in a coordinate plane. Find the area of the parallelogram.

11. Plot the points $A(1, 0)$, $B(5, 0)$, $C(4, 3)$, and $D(2, 3)$ in a coordinate plane. Draw the segments AB, BC, CD, and DA. What kind of quadrilateral is $ABCD$, and what is its area?

12. Plot the points $P(5, 1)$, $Q(0, 6)$, and $R(-5, 1)$ in a coordinate plane. Where must the point S be located so that the quadrilateral $PQRS$ is a square? Find the area of this square.

13–22 ■ Sketch the region given by the set.

13. $\{(x, y) \mid x \leqslant 0\}$ **14.** $\{(x, y) \mid y \geqslant 0\}$

15. $\{(x, y) \mid x = 3\}$ **16.** $\{(x, y) \mid y = -2\}$

17. $\{(x, y) \mid 1 < x < 2\}$ **18.** $\{(x, y) \mid 0 \leqslant y \leqslant 4\}$

19. $\{(x, y) \mid x \geqslant 1 \text{ and } y < 3\}$

20. $\{(x, y) \mid |y| > 1\}$

21. $\{(x, y) \mid |x| \leqslant 2\}$

22. $\{(x, y) \mid |x| < 3 \text{ and } |y| \leqslant 5\}$

23. Which of the points $A(6, 7)$ or $B(-5, 8)$ is closer to the origin?

24. Which of the points $C(-6, 3)$ or $D(3, 0)$ is closer to the point $E(-2, 1)$?

25. Which of the points $P(3, 1)$ or $Q(-1, 3)$ is closer to the point $R(-1, -1)$?

26. (a) Show that the points $(7, 3)$ and $(3, 7)$ are the same distance from the origin.
   (b) Show that the points (a, b) and (b, a) are the same distance from the origin.

27. Show that the triangle with vertices $A(0, 2)$, $B(-3, -1)$, and $C(-4, 3)$ is isosceles.

28. Find the area of the triangle shown in the figure.

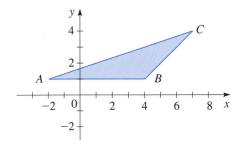

29. Refer to triangle ABC in the figure.
   (a) Show that triangle ABC is a right triangle by using the converse of the Pythagorean Theorem.
   (b) Find the area of triangle ABC.

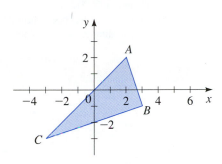

30. Show that the triangle with vertices $A(6, -7)$, $B(11, -3)$, and $C(2, -2)$ is a right triangle by using the converse of the Pythagorean Theorem. Find the area of the triangle.

31. Show that the points $A(-1, 3)$, $B(3, 11)$, and $C(5, 15)$ are collinear by showing that $d(A, B) + d(B, C) = d(A, C)$.

32. Find a point on the y-axis that is equidistant from the points $(5, -5)$ and $(1, 1)$.

33. Find the lengths of the medians of the triangle with vertices $A(1, 0)$, $B(3, 6)$, and $C(8, 2)$. (A *median* is a line segment from a vertex to the midpoint of the opposite side.)

34. Find the point that is one-fourth of the distance from the point $P(-1, 3)$ to the point $Q(7, 5)$ along the segment PQ.

35. (a) Sketch the parallelogram with vertices $A(-2, -1)$, $B(4, 2)$, $C(7, 7)$, and $D(1, 4)$.
   (b) Find the midpoints of the diagonals of this parallelogram.
   (c) From part (b) show that the diagonals bisect each other.

36. The point M in the figure is the midpoint of the line segment AB. Show that M is equidistant from the vertices of triangle ABC.

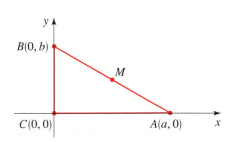

37. Suppose that each point in the coordinate plane is shifted 3 units to the right and 2 units upward.
   (a) The point $(5, 3)$ is shifted to what new point?
   (b) The point (a, b) is shifted to what new point?
   (c) What point is shifted to $(3, 4)$?
   (d) Triangle ABC in the figure has been shifted to triangle $A'B'C'$. Find the coordinates of the points A', B', and C'.

38. Suppose that the y-axis acts as a mirror that reflects each point to the right of it into a point to the left of it.
   (a) The point $(3, 7)$ is reflected to what point?
   (b) The point (a, b) is reflected to what point?
   (c) What point is reflected to $(-4, -1)$?
   (d) Triangle ABC in the figure is reflected to triangle $A'B'C'$. Find the coordinates of the points A', B', and C'.

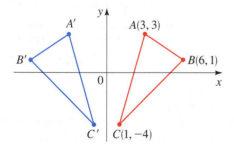

39–42 ■ Determine whether the given points are on the graph of the equation.

39. $y = 2x + 3$; $(0, 0)$, $\left(\frac{1}{2}, 4\right)$, $(1, 4)$

40. $y = \sqrt{x + 1}$; $(1, 0)$, $(0, 1)$, $(3, 2)$

41. $2y - x + 1 = 0$; $(0, 0)$, $(1, 0)$, $(-1, -1)$

42. $y(x^2 + 1) = 1$; $(1, 1)$, $\left(1, \frac{1}{2}\right)$, $\left(-1, \frac{1}{2}\right)$

43–62 ■ Make a table of values and sketch the graph of the equation. Find x- and y-intercepts and test for symmetry.

43. $y = x$ **44.** $y = -x$

45. $y = x - 1$ **46.** $y = 2x + 5$

47. $3x - y = 5$ **48.** $x + y = 3$

49. $y = 1 - x^2$ **50.** $y = x^2 + 2$

51. $4y = x^2$ **52.** $8y = x^3$

53. $y = x^2 - 9$ **54.** $y = 9 - x^2$

55. $y = \sqrt{x}$ **56.** $x + y^2 = 4$

57. $y = |x|$ **58.** $x = |y|$

59. $y = 4 - |x|$ **60.** $y = |4 - x|$

61. $y = x^4$ **62.** $y = x^3 - 1$

63–68 ■ Test the equation for symmetry.

63. $y = x^4 + x^2$

64. $x = y^4 - y^2$

65. $x^2y^2 + xy = 1$

66. $x^4y^4 + x^2y^2 = 1$

67. $y = x^3 + 10x$

68. $y = x^2 + |x|$

69–72 ■ Complete the graph using the given symmetry property.

69. Symmetric with respect to the y-axis

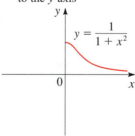

70. Symmetric with respect to the x-axis

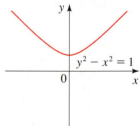

71. Symmetric with respect to the origin

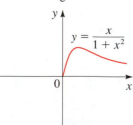

72. Symmetric with respect to the origin

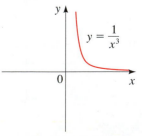

73–78 ■ Find an equation of the circle that satisfies the given conditions.

73. Center $(2, -1)$;   radius 3

74. Center at the origin;   passes through $(4, 7)$

75. Center $(-1, 5)$;   passes through $(-4, -6)$

76. Endpoints of a diameter are $P(-1, 3)$ and $Q(7, -5)$

77. Center $(7, -3)$;   tangent to the x-axis

78. Circle lies in the first quadrant, tangent to both x- and y-axes;   radius 5

79–80 ■ Find the equation of the circle shown in the figure.

79.   **80.**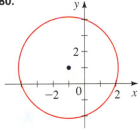

81–86 ■ Show that the equation represents a circle, and find the center and radius of the circle.

81. $x^2 + y^2 - 2x + 4y + 1 = 0$

82. $x^2 + y^2 - 4x + 10y + 13 = 0$

83. $x^2 + y^2 + 6y + 2 = 0$

84. $x^2 + y^2 + x = 0$

85. $x^2 + y^2 + 2x + y + 1 = 0$

86. $x^2 + y^2 - \frac{1}{2}x + \frac{1}{2}y = \frac{1}{8}$

87–88 ■ Sketch the region given by the set.

87. $\{(x, y) \mid x^2 + y^2 \le 1\}$ **88.** $\{(x, y) \mid x^2 + y^2 > 4\}$

89. Find the area of the region that lies outside the circle $x^2 + y^2 = 4$ but inside the circle

$$x^2 + y^2 - 4y - 12 = 0$$

90. Sketch the region in the coordinate plane that satisfies both the inequalities $x^2 + y^2 \le 9$ and $y \ge |x|$. What is the area of this region?

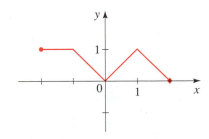

▲ DISCOVERY • DISCUSSION

91. Completing a Line Segment   Plot the points $M(6, 8)$ and $A(2, 3)$ on a coordinate plane. If M is the midpoint of the line segment AB, find the coordinates of B. Write a brief description of the steps you took in finding B, and your reasons for taking them.

92. Completing a Parallelogram   Plot the points $P(-1, -4)$, $Q(1, 1)$, and $R(4, 2)$ on a coordinate plane. Where should the point S be located so that the figure $PQRS$ is a parallelogram? Write a brief description of the steps you took and your reasons for taking them.

93. Circle, Point, or Empty Set?   Complete the squares in the general equation $x^2 + ax + y^2 + by + c = 0$ and simplify the result as much as possible. Under what conditions on the coefficients a, b, and c does this equation represent a circle? a single point? the empty set? In the case that the equation does represent a circle, find its center and radius.

94. Do the Circles Intersect?
(a) Find the radius of each circle in the pair, and the distance between their centers; then use this information to determine whether the circles intersect.
(i) $(x - 2)^2 + (y - 1)^2 = 9$; $(x - 6)^2 + (y - 4)^2 = 16$
(ii) $x^2 + (y - 2)^2 = 4$; $(x - 5)^2 + (y - 14)^2 = 9$
(iii) $(x - 3)^2 + (y + 1)^2 = 1$; $(x - 2)^2 + (y - 2)^2 = 25$
(b) How can you tell, just by knowing the radii of two circles and the distance between their centers, whether the circles intersect? Write a short paragraph describing how you would decide this and draw graphs to illustrate your answer.

95. Making a Graph Symmetric   The graph shown in the figure is not symmetric about the x-axis, the y-axis, or the origin. Add more line segments to the graph so that it exhibits the indicated symmetry. In each case, add as little as possible.
(a) Symmetry about the x-axis.
(b) Symmetry about the y-axis.
(c) Symmetry about the origin.

96. Geometry Versus Coordinate Geometry   What are the differences between geometry and coordinate geometry? Consider a circle drawn in a plane (geometry) and in a coordinate plane (coordinate geometry); they are described as follows.

In geometry: The circle is the set of points whose distance from P is 5.

In coordinate geometry: The circle is the set of points satisfying $x^2 + y^2 = 25$.

(a) The circle and line to the right are drawn first in a plane and then in a coordinate plane. Use the figures to give a description of the line and the points A and B in geometry and in coordinate geometry. Which description gives more information about the figure?

(b) Sketch the graph of the equation $y - x^3 = 0$. Explain why there is no easy geometric description (like that for the circle) for the shape of this graph.

(c) Explain why coordinate geometry allows us to easily describe many more curves than in geometry without coordinates.

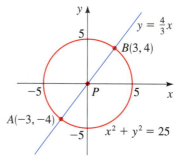

1.9   SOLVING EQUATIONS AND INEQUALITIES GRAPHICALLY

In Sections 1.5 and 1.7 we solved equations and inequalities algebraically. In the preceding section we learned how to sketch the graph of an equation in a coordinate plane. In this section we use graphs to solve equations and inequalities. To do this, we must first draw a graph using a graphing device. So, we begin by giving a few guidelines to help us use graphing devices effectively.

■ Using a Graphing Calculator

A graphing calculator or computer displays a rectangular portion of the graph of an equation in a display window or viewing screen, which we call a **viewing rectangle**. The default screen often gives an incomplete or misleading picture, so it is important to choose the viewing rectangle with care. If we choose the x-values to range from a minimum value of $\text{Xmin} = a$ to a maximum value of $\text{Xmax} = b$ and the y-values to range from a minimum value of $\text{Ymin} = c$ to a maximum value of $\text{Ymax} = d$, then the displayed portion of the graph lies in the rectangle

$$[a, b] \times [c, d] = \{(x, y) \mid a \leq x \leq b, c \leq x \leq d\}$$

as shown in Figure 1. We refer to this as the $[a, b]$ by $[c, d]$ viewing rectangle.

FIGURE 1

The viewing rectangle $[a, b]$ by $[c, d]$

The graphing device draws the graph of an equation much as you would. It plots points of the form (x, y) for a certain number of values of x, equally spaced between a and b. If the equation is not defined for an x-value, or if the corresponding y-value lies outside the viewing rectangle, the device ignores this value and moves on to the next x-value. The machine connects each point to the preceding plotted point to form a representation of the graph of the equation.

EXAMPLE 1 ■ Choosing an Appropriate Viewing Rectangle

Graph the equation $y = x^2 + 3$ in an appropriate viewing rectangle.

SOLUTION

Let's experiment with different viewing rectangles. We'll start with the viewing rectangle $[-2, 2]$ by $[-2, 2]$, so we set

$$\mathsf{Xmin} = -2 \qquad \mathsf{Ymin} = -2$$

$$\mathsf{Xmax} = 2 \qquad \mathsf{Ymax} = 2$$

The resulting graph in Figure 2(a) is blank! This is because $x^2 \geqslant 0$, so $x^2 + 3 \geqslant 3$ for all x. Thus, the graph lies entirely above the viewing rectangle, so this viewing rectangle is not appropriate. If we enlarge the viewing rectangle to $[-4, 4]$ by $[-4, 4]$, as in Figure 2(b), we begin to see a portion of the graph.

Now let's try the viewing rectangle $[-10, 10]$ by $[-5, 30]$. The graph in Figure 2(c) seems to give a more complete view of the graph. If we enlarge the viewing rectangle even further, as in Figure 2(d), the graph doesn't show clearly that the y-intercept is 3.

So, the viewing rectangle $[-10, 10]$ by $[-5, 30]$ gives an appropriate representation of the graph.

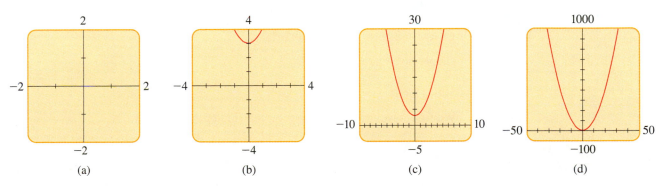

(a)   (b)   (c)   (d)

FIGURE 2

Graphs of $y = x^2 + 3$

Alan Turing (1912–1954) was at the center of two pivotal events of the 20th century—World War II and the invention of computers. At the age of 23 Turing made his mark on mathematics by solving an important problem in the foundations of mathematics that was posed by David Hilbert at the 1928 International Congress of Mathematicians (see page 689). In this research he invented a theoretical machine, now called a Turing machine, which was the inspiration for modern digital computers. During World War II Turing was in charge of the British effort to decipher secret German codes. His complete success in this endeavor played a decisive role in the Allies' victory. To carry out the numerous logical steps required to break a coded message, Turing developed decision procedures similar to modern computer programs. After the war he helped develop the first electronic computers in Britain. He also did pioneering work on artificial intelligence and computer models of biological processes. At the age of 42 Turing died of cyanide poisoning under mysterious circumstances.

EXAMPLE 2 ■ Two Graphs on the Same Screen

Graph the equations $y = 3x^2 - 6x + 1$ and $y = 0.23x - 2.25$ together in the viewing rectangle $[-1, 3]$ by $[-2.5, 1.5]$. Do the graphs intersect in this viewing rectangle?

SOLUTION

Figure 3(a) shows the essential features of both graphs. One is a parabola and the other is a line. It looks as if the graphs intersect near the point $(1, -2)$. However, if we zoom in on the area around this point as shown in Figure 3(b), we see that although the graphs almost touch, they don't actually intersect.

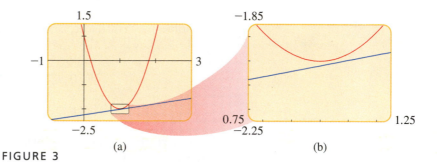

FIGURE 3   (a)   (b)   ■

You can see from Examples 1 and 2 that the choice of a viewing rectangle makes a big difference in the appearance of a graph. If you want an overview of the essential features of a graph, you must choose a relatively large viewing rectangle to obtain a global view of the graph. If you want to investigate the details of a graph, you must zoom in to a small viewing rectangle that shows just the feature of interest.

Most graphing calculators can only graph equations in which y is isolated on one side of the equal sign. The next example shows how to graph equations that don't have this property.

EXAMPLE 3 ■ Graphing a Circle

Graph the circle $x^2 + y^2 = 1$.

SOLUTION

We first solve for y, to isolate it on one side of the equal sign.

$$y^2 = 1 - x^2$$
$$y = \pm\sqrt{1 - x^2}$$

Therefore, the circle is described by the graphs of *two* equations:

$$y = \sqrt{1 - x^2} \qquad \text{and} \qquad y = -\sqrt{1 - x^2}$$

The first equation represents the top half of the circle (because $y \geq 0$), and the second represents the bottom half of the circle (because $y \leq 0$). If we graph the first equation in the viewing rectangle $[-2, 2]$ by $[-2, 2]$, we get the semicircle shown in Figure 4(a). The graph of the second equation is the semicircle in Figure 4(b). Graphing these semicircles together on the same viewing screen, we get the full circle in Figure 4(c).

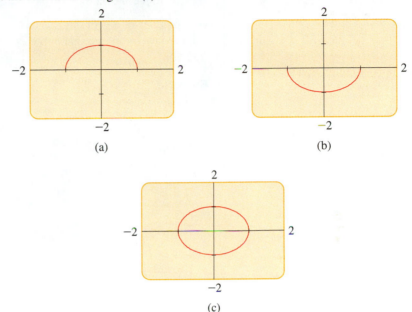

(a)   (b)

The graph in Figure 4(c) looks somewhat flattened. Most graphing calculators allow you to set the scales on the axes so that circles really look like circles. On the TI-82 and TI-83, from the ZOOM menu, choose **ZSquare** to set the scales appropriately. (On the TI-85 the command is **Zsq**.)

(c)

FIGURE 4

Graphing the equation $x^2 + y^2 = 1$

■ Solving Equations Graphically

In Section 1.5 we learned how to solve equations. To solve an equation like

$$3x - 5 = 0$$

"Algebra is a merry science," Uncle Jakob would say. "We go hunting for a little animal whose name we don't know, so we call it x. When we bag our game we pounce on it and give it its right name."

ALBERT EINSTEIN

we used the **algebraic method**. This means we used the rules of algebra to isolate x on one side of the equation. We view x as an *unknown* and we use the rules of algebra to hunt it down. Here are the steps in the solution:

$$3x - 5 = 0$$
$$3x = 5 \qquad \text{Add 5}$$
$$x = \tfrac{5}{3} \qquad \text{Divide by 3}$$

So the solution is $x = \tfrac{5}{3}$.

We can also solve this equation by the **graphical method**. In this method we view x as a *variable* and sketch the graph of the equation

$$y = 3x - 5$$

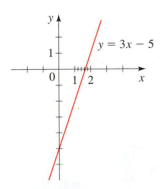

FIGURE 5

Different values for x give different values for y. Our goal is to find the value of x for which $y = 0$. From the graph in Figure 5 we see that $y = 0$ when $x \approx 1.7$. Thus, the solution is $x \approx 1.7$. Note that from the graph we obtain an approximate solution.

We summarize these methods in the following box.

SOLVING AN EQUATION

Algebraic Method	Graphical Method
Use the rules of algebra to isolate the unknown x on one side of the equation.	Move all terms to one side and set equal to y. Sketch the graph to find the value of x where $y = 0$.

Example: $2x = 6 - x$
$$3x = 6 \quad \text{Add } x$$
$$x = 2 \quad \text{Divide by 3}$$

The solution is $x = 2$.

Example: $2x = 6 - x$
$$0 = 6 - 3x.$$
Set $y = 6 - 3x$ and graph.

From the graph the solution is $x \approx 2$.

A numerical method for solving equations is described in the Discovery Project on page 291.

The advantage of the algebraic method is that it gives exact answers. Also, the process of unraveling the equation to arrive at the answer helps us understand the algebraic structure of the equation. On the other hand, for many equations it is difficult or impossible to isolate x.

The graphical method gives a numerical approximation to the answer. This is an advantage when a numerical answer is desired. (For example, an engineer might find an answer expressed as $x \approx 2.6$ more immediately useful than $x = \sqrt{7}$.) Also, graphing an equation helps us visualize how the solution is related to other values of the variable.

EXAMPLE 4 ■ Solving an Equation Algebraically and Graphically

Solve the equations algebraically and graphically.

(a) $x^2 - 4x + 2 = 0$ (b) $x^2 - 4x + 4 = 0$ (c) $x^2 - 4x + 6 = 0$

SOLUTION 1: ALGEBRAIC

We use the quadratic formula to solve each equation.

The quadratic formula is discussed on page 51.

(a) $x = \dfrac{-(-4) \pm \sqrt{(-4)^2 - 4 \cdot 1 \cdot 2}}{2} = \dfrac{4 \pm \sqrt{8}}{2} = 2 \pm \sqrt{2}$

There are two solutions, $x = 2 + \sqrt{2}$ and $x = 2 - \sqrt{2}$.

(b) $x = \dfrac{-(-4) \pm \sqrt{(-4)^2 - 4 \cdot 1 \cdot 4}}{2} = \dfrac{4 \pm \sqrt{0}}{2} = 2$

There is just one solution, $x = 2$.

(c) $x = \dfrac{-(-4) \pm \sqrt{(-4)^2 - 4 \cdot 1 \cdot 6}}{2} = \dfrac{4 \pm \sqrt{-8}}{2}$

There is no real solution.

SOLUTION 2: GRAPHICAL

We graph the equations $y = x^2 - 4x + 2$, $y = x^2 - 4x + 4$, and $y = x^2 - 4x + 6$ in Figure 6. By determining the x-intercepts of the graphs, we find the following solutions.

(a) $x \approx 0.6$ and $x \approx 3.4$

(b) $x = 2$

(c) There is no x-intercept, so the equation has no solution.

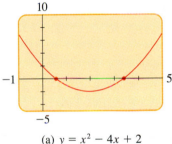

(a) $y = x^2 - 4x + 2$

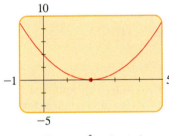

(b) $y = x^2 - 4x + 4$

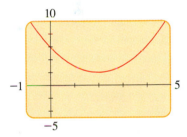

(c) $y = x^2 - 4x + 6$

FIGURE 6

The graphs in Figure 6 show visually why a quadratic equation may have two solutions, one solution, or no real solution. We proved this fact algebraically in Section 1.5 when we studied the discriminant.

EXAMPLE 5 ■ Another Graphical Method

Solve the equation algebraically and graphically: $5 - 3x = 8x - 20$

SOLUTION 1: ALGEBRAIC

$$5 - 3x = 8x - 20$$

$$-3x = 8x - 25 \qquad \text{Subtract 5}$$

$$-11x = -25 \qquad \text{Subtract } 8x$$

$$x = \frac{-25}{-11} = 2\tfrac{3}{11} \qquad \text{Divide by } -11 \text{ and simplify}$$

FIGURE 7

SOLUTION 2: GRAPHICAL

We could move all terms to one side of the equal sign, set the result equal to y, and graph the resulting equation. But to avoid all this algebra, we graph two equations instead:

$$y_1 = 5 - 3x \qquad \text{and} \qquad y_2 = 8x - 20$$

The solution of the original equation will be the value of x that makes y_1 equal to y_2; that is, the solution is the x-coordinate of the intersection point of the two graphs. Using the $\boxed{\text{TRACE}}$ feature or the `intersect` command on a graphing calculator, we see from Figure 7 that the solution is $x \approx 2.27$.   ■

In the next example we use the graphical method to solve an equation that is extremely difficult to solve algebraically.

EXAMPLE 6 ■ Solving an Equation in an Interval

Solve the equation

$$x^3 - 6x^2 + 9x = \sqrt{x}$$

in the interval $[1, 6]$.

SOLUTION

We are asked to find all solutions x that satisfy $1 \le x \le 6$, so we will graph the equation in a viewing rectangle for which the x-values are restricted to this interval.

$$x^3 - 6x^2 + 9x = \sqrt{x}$$

$$x^3 - 6x^2 + 9x - \sqrt{x} = 0 \qquad \text{\textcolor{red}{Subtract } } \sqrt{x}$$

We can also use the `zero` command to find the solutions, as shown in Figures 8(a) and 8(b).

Figure 8 shows the graph of the equation $y = x^3 - 6x^2 + 9x - \sqrt{x}$ in the viewing rectangle $[1, 6]$ by $[-5, 5]$. There are two x-intercepts in this viewing rectangle; zooming in we see that the solutions are $x \approx 2.18$ and $x \approx 3.72$.

FIGURE 8   (a)   (b)   ■

The equation in Example 6 actually has four solutions. You are asked to find the other two in Exercise 57.

Solving Inequalities Graphically

Inequalities can be solved graphically. To describe the method we solve

$$x^2 - 5x + 6 \leq 0$$

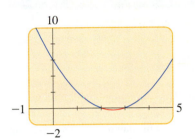

FIGURE 9
$x^2 - 5x + 6 \leq 0$

This inequality was solved algebraically in Section 1.7, Example 3. To solve the inequality graphically, we draw the graph of

$$y = x^2 - 5x + 6$$

Our goal is to find those values of x for which $y \leq 0$. These are simply the x-values for which the graph lies below the x-axis. From Figure 9 we see that the solution of the inequality is the interval $[2, 3]$.

EXAMPLE 7 ■ Solving an Inequality Graphically

Solve the inequality $3.7x^2 + 1.3x - 1.9 \leq 2.0 - 1.4x$.

SOLUTION

We graph the equations

$$y_1 = 3.7x^2 + 1.3x - 1.9$$

$$y_2 = 2.0 - 1.4x$$

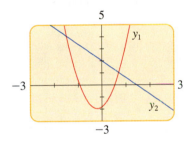

FIGURE 10
$y_1 = 3.7x^2 + 1.3x - 1.9$
$y_2 = 2.0 - 1.4x$

in the same viewing rectangle in Figure 10. We are interested in those values of x for which $y_1 \leq y_2$; these are points for which the graph of y_2 lies on or above the graph of y_1. To determine the appropriate interval, we look for the x-coordinates of points where the graphs intersect. We conclude that the solution is (approximately) the interval $[-1.45, 0.72]$. ■

EXAMPLE 8 ■ Solving an Inequality Graphically

Solve the inequality $x^3 - 5x^2 \geq -8$.

SOLUTION

We write the inequality as

$$x^3 - 5x^2 + 8 \geq 0$$

and then graph the equation

$$y = x^3 - 5x^2 + 8$$

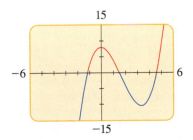

FIGURE 11
$x^3 - 5x^2 + 8 \geq 0$

in the viewing rectangle $[-6, 6]$ by $[-15, 15]$, as shown in Figure 11. The solution of the inequality consists of those intervals on which the graph lies on or above the x-axis. By moving the cursor to the x-intercepts we find that, correct to one decimal place, the solution is $[-1.1, 1.5] \cup [4.6, \infty)$. ■

1.9 EXERCISES

1–6 ■ Use a graphing calculator or computer to decide which viewing rectangle (a)–(d) produces the most appropriate graph of the equation.

1. $y = x^4 + 2$
    (a) $[-2, 2]$ by $[-2, 2]$
    (b) $[0, 4]$ by $[0, 4]$
    (c) $[-8, 8]$ by $[-4, 40]$
    (d) $[-40, 40]$ by $[-80, 800]$

2. $y = x^2 + 7x + 6$
    (a) $[-5, 5]$ by $[-5, 5]$
    (b) $[0, 10]$ by $[-20, 100]$
    (c) $[-15, 8]$ by $[-20, 100]$
    (d) $[-10, 3]$ by $[-100, 20]$

3. $y = 100 - x^2$
    (a) $[-4, 4]$ by $[-4, 4]$
    (b) $[-10, 10]$ by $[-10, 10]$
    (c) $[-15, 15]$ by $[-30, 110]$
    (d) $[-4, 4]$ by $[-30, 110]$

4. $y = 2x^2 - 1000$
    (a) $[-10, 10]$ by $[-10, 10]$
    (b) $[-10, 10]$ by $[-100, 100]$
    (c) $[-10, 10]$ by $[-1000, 1000]$
    (d) $[-25, 25]$ by $[-1200, 200]$

5. $y = 10 + 25x - x^3$
    (a) $[-4, 4]$ by $[-4, 4]$
    (b) $[-10, 10]$ by $[-10, 10]$
    (c) $[-20, 20]$ by $[-100, 100]$
    (d) $[-100, 100]$ by $[-200, 200]$

6. $y = \sqrt{8x - x^2}$
    (a) $[-4, 4]$ by $[-4, 4]$
    (b) $[-5, 5]$ by $[0, 100]$
    (c) $[-10, 10]$ by $[-10, 40]$
    (d) $[-2, 10]$ by $[-2, 6]$

7–18 ■ Determine an appropriate viewing rectangle for the equation and use it to draw the graph.

7. $y = 100x^2$ **8.** $y = -100x^2$

9. $y = 4 + 6x - x^2$ **10.** $y = 0.3x^2 + 1.7x - 3$

11. $y = \sqrt[4]{256 - x^2}$

12. $y = \sqrt{12x - 17}$

13. $y = 0.01x^3 - x^2 + 5$

14. $y = x(x + 6)(x - 9)$

15. $y = x^4 - 4x^3$

16. $y = \dfrac{x}{x^2 + 25}$

17. $y = 1 + |x - 1|$

18. $y = 2x - |x^2 - 5|$

19. Graph the circle $x^2 + y^2 = 9$ by solving for y and graphing two equations as in Example 3.

20. Graph the circle $(y - 1)^2 + x^2 = 1$ by solving for y and graphing two equations as in Example 3.

21. Graph the equation $4x^2 + 2y^2 = 1$ by solving for y and graphing two equations corresponding to the negative and positive square roots. (This graph is called an *ellipse*.)

22. Graph the equation $y^2 - 9x^2 = 1$ by solving for y and graphing the two equations corresponding to the positive and negative square roots. (This graph is called a *hyperbola*.)

23–26 ■ Do the graphs intersect in the given viewing rectangle? If they do, how many points of intersection are there?

23. $y = -3x^2 + 6x - \frac{1}{2}$, $y = \sqrt{7 - \frac{7}{12}x^2}$;
    $[-4, 4]$ by $[-1, 3]$

24. $y = \sqrt{49 - x^2}$, $y = \frac{1}{5}(41 - 3x)$; $[-8, 8]$ by $[-1, 8]$

25. $y = 6 - 4x - x^2$, $y = 3x + 18$; $[-6, 2]$ by $[-5, 20]$

26. $y = x^3 - 4x$, $y = x + 5$; $[-4, 4]$ by $[-15, 15]$

27–36 ■ Solve the equation both algebraically and graphically.

27. $x - 4 = 5x + 12$ **28.** $\frac{1}{2}x - 3 = 6 + 2x$

29. $\dfrac{2}{x} + \dfrac{1}{2x} = 7$

30. $\dfrac{4}{x + 2} - \dfrac{6}{2x} = \dfrac{5}{2x + 4}$

31. $x^2 - 32 = 0$ **32.** $x^3 + 16 = 0$

33. $16x^4 = 625$ **34.** $2x^5 - 243 = 0$

35. $(x - 5)^4 - 80 = 0$ **36.** $6(x + 2)^5 = 64$

37–44 ■ Solve the equation graphically in the given interval. State each answer correct to two decimals.

37. $x^2 - 7x + 12 = 0$; $[0, 6]$

38. $x^2 - 0.75x + 0.125 = 0$; $[-2, 2]$

39. $x^3 - 6x^2 + 11x - 6 = 0$; $[-1, 4]$

40. $16x^3 + 16x^2 = x + 1$; $[-2, 2]$

41. $x - \sqrt{x + 1} = 0$; $[-1, 5]$

42. $1 + \sqrt{x} = \sqrt{1 + x^2}$; $[-1, 5]$

43. $x^{1/3} - x = 0$; $[-3, 3]$

44. $x^{1/2} + x^{1/3} - x = 0$; $[-1, 5]$

45–48 ■ Find all real solutions of the equation, correct to two decimals.

45. $x^3 - 2x^2 - x - 1 = 0$ **46.** $x^4 - 8x^2 + 2 = 0$

47. $x(x - 1)(x + 2) = \frac{1}{6}x$ **48.** $x^4 = 16 - x^3$

49–56 ■ Find the solutions of the inequality by drawing appropriate graphs. State each answer correct to two decimals.

49. $x^2 - 3x - 10 \leqslant 0$ **50.** $0.5x^2 + 0.875x \leqslant 0.25$

51. $x^3 + 11x \leqslant 6x^2 + 6$ **52.** $16x^3 + 24x^2 > -9x - 1$

53. $x^{1/3} < x$ **54.** $\sqrt{0.5x^2 + 1} \leqslant 2|x|$

55. $(x + 1)^2 < (x - 1)^2$ **56.** $(x + 1)^2 \leqslant x^3$

57. In Example 6 we found two solutions of the equation $x^3 - 6x^2 + 9x = \sqrt{x}$, the solutions that lie between 1 and 6. Find two more solutions, correct to two decimals.

 DISCOVERY • DISCUSSION

58. Equation Notation on Graphing Calculators When you enter the following equations into your calculator, how does what you see on the screen differ from the usual way of writing the equations? (Check your user's manual if you're not sure.)

(a) $y = |x|$

(b) $y = \sqrt[5]{x}$

(c) $y = \dfrac{x}{x - 1}$

(d) $y = x^3 + \sqrt[3]{x + 2}$

59. Enter Equations Carefully A student wishes to graph the equations

$$y = x^{1/3} \quad \text{and} \quad y = \frac{x}{x + 4}$$

on the same screen, so he enters the following information into his calculator:

$$Y_1 = X^1/3 \qquad Y_2 = X/X + 4$$

The calculator graphs two lines instead of the equations he wanted. What went wrong? Why are the graphs lines?

60. Algebraic and Graphical Solution Methods Write a short essay comparing the algebraic and graphical methods for solving equations. Make up your own examples to illustrate the advantages and disadvantages of each method.

61. How Many Solutions? This exercise deals with the family of equations $x^3 - 3x = k$.

(a) Draw the graphs of $y_1 = x^3 - 3x$ and $y_2 = k$ in the same viewing rectangle, in the cases $k = -4, -2, 0, 2,$ and 4. How many solutions of the equation $x^3 - 3x = k$ are there in each case? Find the solutions correct to two decimals.

(b) For what ranges of values of k does the equation have one solution? two solutions? three solutions?

1.10   LINES

In this section we find equations for straight lines lying in a coordinate plane. We first need a way to measure the "steepness" of a line, or how quickly it rises (or falls) as we move from left to right. We define *run* to be the distance we move to the right and *rise* to be the corresponding distance that the line rises (or falls). The *slope* of a line is the ratio of rise to run:

$$\text{slope} = \frac{\text{rise}}{\text{run}}$$

Figure 1 shows situations where slope is important. Carpenters use the term *pitch* for the slope of a roof or a staircase; the term *grade* is used for the slope of a road.

Slope of a ramp

Slope $= \frac{1}{12}$

Pitch of a roof

Slope $= \frac{1}{3}$

Grade of a road

Slope $= \frac{8}{100}$

FIGURE 1

If a line lies in a coordinate plane, then the **run** is the change in the *x*-coordinate and the **rise** is the corresponding change in the *y*-coordinate between any two points on the line (see Figure 2). This gives us the following definition of slope.

FIGURE 2

SLOPE OF A LINE

The **slope** m of a nonvertical line that passes through the points $A(x_1, y_1)$ and $B(x_2, y_2)$ is

$$m = \frac{\text{rise}}{\text{run}} = \frac{y_2 - y_1}{x_2 - x_1}$$

The slope of a vertical line is not defined.

The slope is independent of which two points are chosen on the line. We can see that this is true from the similar triangles in Figure 3:

$$\frac{y_2 - y_1}{x_2 - x_1} = \frac{y_2' - y_1'}{x_2' - x_1'}$$

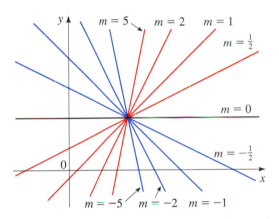

FIGURE 3

Figure 4 shows several lines labeled with their slopes. Notice that lines with positive slope slant upward to the right, whereas lines with negative slope slant downward to the right. The steepest lines are those for which the absolute value of the slope is the largest; a horizontal line has slope zero.

FIGURE 4

Lines with various slopes

EXAMPLE 1 ■ Finding the Slope of a Line through Two Points

Find the slope of the line that passes through the points $P(2, 1)$ and $Q(8, 5)$.

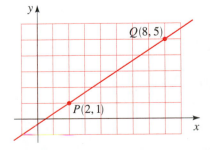

FIGURE 5

SOLUTION

Since any two different points determine a line, only one line passes through these two points. From the definition, the slope is

$$m = \frac{y_2 - y_1}{x_2 - x_1} = \frac{5 - 1}{8 - 2} = \frac{4}{6} = \frac{2}{3}$$

This says that for every 3 units we move to the right, the line rises 2 units. The line is drawn in Figure 5.   ■

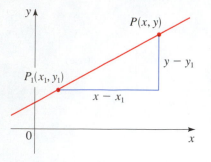

FIGURE 6

Now let's find the equation of the line that passes through a given point $P(x_1, y_1)$ and has slope m. A point $P(x, y)$ with $x \neq x_1$ lies on this line if and only if the slope of the line through P_1 and P is equal to m (see Figure 6), that is,

$$\frac{y - y_1}{x - x_1} = m$$

This equation can be rewritten in the form $y - y_1 = m(x - x_1)$; note that the equation is also satisfied when $x = x_1$ and $y = y_1$. Therefore, it is an equation of the given line.

POINT-SLOPE FORM OF THE EQUATION OF A LINE

An equation of the line that passes through the point (x_1, y_1) and has slope m is

$$y - y_1 = m(x - x_1)$$

EXAMPLE 2 ■ Finding the Equation of a Line with Given Point and Slope

(a) Find an equation of the line through $(1, -3)$ with slope $-\frac{1}{2}$.

(b) Sketch the line.

SOLUTION

(a) Using the point-slope form with $m = -\frac{1}{2}$, $x_1 = 1$, and $y_1 = -3$, we obtain an equation of the line as

$$y + 3 = -\frac{1}{2}(x - 1) \qquad \text{From point-slope equation}$$

$$2y + 6 = -x + 1 \qquad \text{Multiply by 2}$$

$$x + 2y + 5 = 0 \qquad \text{Rearrange}$$

(b) The fact that the slope is $-\frac{1}{2}$ tells us that when we move to the right 2 units, the line drops 1 unit. This enables us to sketch the line in Figure 7. ∎

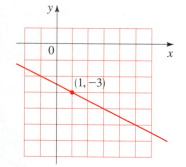

FIGURE 7

EXAMPLE 3 ■ Finding the Equation of a Line through Two Given Points

Find an equation of the line through the points $(-1, 2)$ and $(3, -4)$.

SOLUTION

The slope of the line is

$$m = \frac{-4 - 2}{3 - (-1)} = -\frac{6}{4} = -\frac{3}{2}$$

Using the point-slope form with $x_1 = -1$ and $y_1 = 2$, we obtain

$$y - 2 = -\frac{3}{2}(x + 1) \qquad \text{From point-slope equation}$$

$$2y - 4 = -3x - 3 \qquad \text{Multiply by 2}$$

$$3x + 2y - 1 = 0 \qquad \text{Rearrange}$$

∎

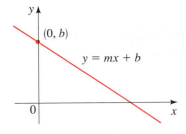

FIGURE 8

Suppose a nonvertical line has slope m and y-intercept b (see Figure 8). This means the line intersects the y-axis at the point $(0, b)$, so the point-slope form of the equation of the line, with $x = 0$ and $y = b$, becomes

$$y - b = m(x - 0)$$

This simplifies to $y = mx + b$, which is called the **slope-intercept form** of the equation of a line.

SLOPE-INTERCEPT FORM OF THE EQUATION OF A LINE

An equation of the line that has slope m and y-intercept b is

$$y = mx + b$$

EXAMPLE 4 ■ Lines in Slope-Intercept Form

(a) Find the equation of the line with slope 3 and y-intercept -2.

(b) Find the slope and y-intercept of the line $3y - 2x = 1$.

SOLUTION

(a) Since $m = 3$ and $b = 2$, from the slope-intercept form of the equation of a line we get

$$y = 3x - 2$$

(b) We first write the equation in the form $y = mx + b$:

$$3y - 2x = 1$$

$$3y = 2x + 1 \qquad \text{Add } 2x$$

$$y = \tfrac{2}{3}x + \tfrac{1}{3} \qquad \text{Divide by 3}$$

From the slope-intercept form of the equation of a line, we see that the slope is $m = \tfrac{2}{3}$ and the y-intercept is $b = \tfrac{1}{3}$. ■

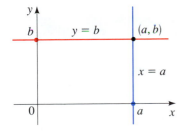

FIGURE 9

If a line is horizontal, its slope is $m = 0$, so its equation is $y = b$, where b is the y-intercept (see Figure 9). A vertical line does not have a slope, but we can write its equation as $x = a$, where a is the x-intercept, because the x-coordinate of every point on the line is a.

VERTICAL AND HORIZONTAL LINES

An equation of the vertical line through (a, b) is $x = a$.

An equation of the horizontal line through (a, b) is $y = b$.

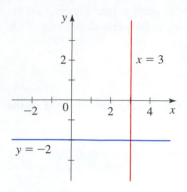

FIGURE 10

EXAMPLE 5 ■ Vertical and Horizontal Lines

(a) The graph of the equation $x = 3$ is a vertical line with x-intercept 3.

(b) The graph of the equation $y = -2$ is a horizontal line with y-intercept -2.

The lines are graphed in Figure 10.   ■

A **linear equation** is an equation of the form

$$Ax + By + C = 0$$

where A, B, and C are constants and A and B are not both 0. The equation of a line is a linear equation:

■ A nonvertical line has the equation $y = mx + b$ or $-mx + y - b = 0$, which is a linear equation with $A = -m$, $B = 1$, and $C = -b$.

■ A vertical line has the equation $x = a$ or $x - a = 0$, which is a linear equation with $A = 1$, $B = 0$, and $C = -a$.

Conversely, the graph of a linear equation is a line:

■ If $B \neq 0$, the equation becomes

$$y = -\frac{A}{B}x - \frac{C}{B}$$

and this is the slope-intercept form of the equation of a line (with $m = -A/B$ and $b = -C/B$).

■ If $B = 0$, the equation becomes $Ax + C = 0$, or $x = -C/A$, which represents a vertical line.

We have proved the following.

GENERAL EQUATION OF A LINE

The graph of every **linear equation**

$$Ax + By + C = 0 \qquad (A, B \text{ not both zero})$$

is a line. Conversely, every line is the graph of a linear equation.

EXAMPLE 6 ■ Graphing a Linear Equation

Sketch the graph of the equation $2x - 3y - 12 = 0$.

SOLUTION 1

Since the equation is linear, its graph is a line. To draw the graph, it is enough to find any two points on the line. The intercepts are the easiest points to find.

x-intercept: Substitute $y = 0$, to get $2x - 12 = 0$, so $x = 6$

y-intercept: Substitute $x = 0$, to get $-3y - 12 = 0$, so $y = -4$

With these points we can sketch the graph in Figure 11.

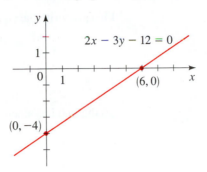

$2x - 3y - 12 = 0$

FIGURE 11

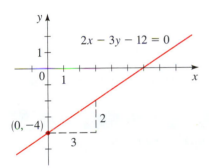

FIGURE 12

SOLUTION 2

We write the equation in slope-intercept form:

$$2x - 3y - 12 = 0$$

$$2x - 3y = 12 \qquad \text{Add 12}$$

$$-3y = -2x + 12 \qquad \text{Subtract } 2x$$

$$y = \tfrac{2}{3}x - 4 \qquad \text{Divide by } -3$$

This equation is in the form $y = mx + b$, so the slope is $m = \tfrac{2}{3}$ and the y-intercept is $b = -4$. To sketch the graph, we plot the y-intercept, and then move 3 units to the right and 2 units up as shown in Figure 12.   ■

Parallel and Perpendicular Lines

Since slope measures the steepness of a line, it seems reasonable that parallel lines should have the same slope. In fact, we can prove this.

PARALLEL LINES

Two nonvertical lines are parallel if and only if they have the same slope.

■ **Proof** Let the lines l_1 and l_2 in Figure 13 have slopes m_1 and m_2. If the lines are parallel, then the right triangles ABC and DEF are similar, so

$$m_1 = \frac{d(B, C)}{d(A, C)} = \frac{d(E, F)}{d(D, F)} = m_2$$

Conversely, if the slopes are equal, then the triangles will be similar, so $\angle BAC = \angle EDF$ and the lines are parallel.   □

EXAMPLE 7 ■ Finding the Equation of a Line Parallel to a Given Line

Find an equation of the line through the point $(5, 2)$ that is parallel to the line $4x + 6y + 5 = 0$.

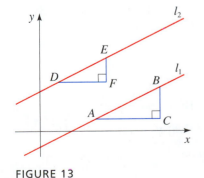

FIGURE 13

SOLUTION

First we write the equation of the given line in slope-intercept form.

$$4x + 6y + 5 = 0$$

$$6y = -4x - 5 \qquad \text{Subtract } 4x + 5$$

$$y = -\tfrac{2}{3}x - \tfrac{5}{6} \qquad \text{Divide by 6}$$

So the line has slope $m = -\tfrac{2}{3}$. Since the required line is parallel to the given line, it also has slope $m = -\tfrac{2}{3}$. From the point-slope form of the equation of a line, we get

$$y - 2 = -\tfrac{2}{3}(x - 5) \qquad \text{Slope } m = -\tfrac{2}{3}, \text{ point } (5, 2)$$

$$3y - 6 = -2x + 10 \qquad \text{Multiply by 3}$$

$$2x + 3y - 16 = 0 \qquad \text{Rearrange}$$

Thus, the equation of the required line is $2x + 3y - 16 = 0$. ■

The condition for perpendicular lines is not as obvious as that for parallel lines.

PERPENDICULAR LINES

Two lines with slopes m_1 and m_2 are perpendicular if and only if $m_1m_2 = -1$, that is, their slopes are negative reciprocals:

$$m_2 = -\frac{1}{m_1}$$

Also, a horizontal line (slope 0) is perpendicular to a vertical line (no slope).

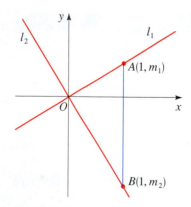

FIGURE 14

■ **Proof**   In Figure 14 we show two lines intersecting at the origin. (If the lines intersect at some other point, we consider lines parallel to these that intersect at the origin. These lines have the same slopes as the original lines.)

If the lines l_1 and l_2 have slopes m_1 and m_2, then their equations are $y = m_1x$ and $y = m_2x$. Notice that $A(1, m_1)$ lies on l_1 and $B(1, m_2)$ lies on l_2. By the Pythagorean Theorem and its converse, $OA \perp OB$ if and only if

$$[d(O, A)]^2 + [d(O, B)]^2 = [d(A, B)]^2$$

By the Distance Formula, this becomes

$$(1^2 + m_1^2) + (1^2 + m_2^2) = (1 - 1)^2 + (m_2 - m_1)^2$$

$$2 + m_1^2 + m_2^2 = m_2^2 - 2m_1m_2 + m_1^2$$

$$2 = -2m_1m_2$$

$$m_1m_2 = -1 \qquad \square$$

EXAMPLE 8 ■ Perpendicular Lines

Show that the points $P(3, 3)$, $Q(8, 17)$, and $R(11, 5)$ are the vertices of a right triangle.

SOLUTION

The slopes of the lines containing PR and QR are, respectively,

$$m_1 = \frac{5 - 3}{11 - 3} = \frac{1}{4} \quad \text{and} \quad m_2 = \frac{5 - 17}{11 - 8} = -4$$

Since $m_1 m_2 = -1$, these lines are perpendicular and so PQR is a right triangle. It is sketched in Figure 15. ■

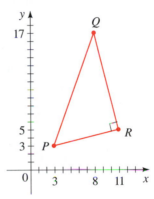

FIGURE 15

EXAMPLE 9 ■ Finding an Equation of a Line Perpendicular to a Given Line

Find an equation of the line that is perpendicular to the line $4x + 6y + 5 = 0$ and passes through the origin.

SOLUTION

In Example 7 we found that the slope of the line $4x + 6y + 5 = 0$ is $-\frac{2}{3}$. Thus, the slope of a perpendicular line is the negative reciprocal, that is, $\frac{3}{2}$. Since the required line passes through $(0, 0)$, the point-slope form gives

$$y - 0 = \tfrac{3}{2}(x - 0)$$

$$y = \tfrac{3}{2}x$$

■

■ Applications of Linear Equations; Rates of Change

Linear equations and their graphs are often used in applications to model the relationship between two quantities. We have seen that the slope measures the steepness of a line. In the next two examples we see that the slope in a linear model can be interpreted as a **rate of change**.

EXAMPLE 10 ■ Slope as Rate of Change

A dam is built on a river to create a reservoir. The water level w in the reservoir is given by the equation

$$w = 4.5t + 28$$

where t is the number of years since the dam was constructed, and w is measured in feet.

(a) Sketch a graph of this equation.

(b) What do the slope and w-intercept of this graph represent?

SOLUTION

(a) This equation is linear, so its graph is a line. Since two points determine a line, we plot two points that lie on the graph and draw a line through them.

When $t = 0$, then $w = 4.5(0) + 28 = 28$, so $(0, 28)$ is on the line.

When $t = 2$, then $w = 4.5(2) + 28 = 37$, so $(2, 37)$ is on the line.

The line determined by these points is shown in Figure 16.

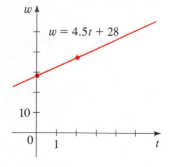

FIGURE 16

(b) The slope is $m = 4.5$; it represents the rate of change of water level with respect to time. This means that the water level *increases* 4.5 ft per year. The w-intercept is 28, and occurs when $t = 0$, so it represents the water level when the dam was constructed. ■

EXAMPLE 11 ■ Linear Relationship between Temperature and Elevation

(a) As dry air moves upward, it expands and cools. If the ground temperature is 20°C and the temperature at a height of 1 km is 10°C, express the temperature T (in °C) in terms of the height h (in kilometers). (Assume that the relationship between T and h is linear.)

(b) Draw the graph of the linear equation. What does its slope represent?

(c) What is the temperature at a height of 2.5 km?

SOLUTION

(a) Because we are assuming a linear relationship between T and h, the equation must be of the form

$$T = mh + b$$

where m and b are constants. When $h = 0$, we are given that $T = 20$, so

$$20 = m(0) + b$$

$$b = 20$$

Thus, we have

$$T = mh + 20$$

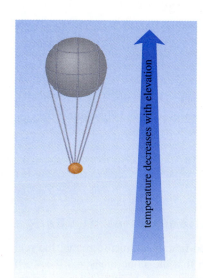

Weather balloons are used to measure temperature at high elevations.

temperature decreases with elevation

$T = -10h + 20$

FIGURE 17

In the supply and demand equations, most economists express p, the price, in terms of y, the amount produced or sold. For clarity, we have chosen to express y in terms of p.

When $h = 1$, we have $T = 10$ and so

$$10 = m(1) + 20$$
$$m = 10 - 20 = -10$$

The required expression is

$$T = -10h + 20$$

(b) The graph is sketched in Figure 17. The slope is $m = -10°\text{C}/\text{km}$, and this represents the rate of change of temperature with respect to distance above the ground.

(c) At a height of $h = 2.5$ km, the temperature is

$$T = -10(2.5) + 20 = -25 + 20 = 5°\text{C}$$ ∎

Economists model supply and demand for a commodity by linear equations. For example, for a certain commodity we might have

Supply equation: $y = 8p - 10$

Demand equation: $y = -3p + 15$

where p is the price of the commodity. In the supply equation, y (the *amount produced*) increases as the price increases because if the price is high, more people will produce the commodity. The demand equation indicates that y (the *amount sold*) decreases as the price increases. The **equilibrium point** is the point of intersection of the graphs of the supply and demand equations; at that point the amount produced equals the amount sold.

 EXAMPLE 12 ■ **Supply and Demand for Wheat**

An economist models the market for wheat by the following equations.

Supply: $y = 8.33p - 14.58$

Demand: $y = -1.39p + 23.35$

Here p is the price per bushel (in dollars), and y is the number of bushels produced and sold (in millions).

(a) At what point is the price so low that no wheat is produced?

(b) At what point is the price so high that no wheat is sold?

(c) Draw the graph of the supply and demand lines in the same viewing rectangle and find the equilibrium point. Estimate the equilibrium price and the quantities produced and sold at equilibrium.

SOLUTION

(a) If no wheat is produced, then $y = 0$ in the supply equation.

$$0 = 8.33p - 14.58 \qquad \text{Set } y = 0 \text{ in the supply equation}$$

$$p = 1.75 \qquad \text{Solve for } p$$

So, at the low price of $1.75 per bushel, the production of wheat halts completely.

(b) If no wheat is sold, then $y = 0$ in the demand equation.

$$0 = -1.39p + 23.35 \qquad \text{Set } y = 0 \text{ in the demand equation}$$

$$p = 16.80 \qquad \text{Solve for } p$$

So, at the high price of $16.80 per bushel, no wheat is sold.

(c) Figure 18 shows the graphs of the supply and demand equations in the same viewing rectangle. By moving the cursor to their point of intersection, we find that the lines intersect (approximately) at the point (3.9, 17.9). Thus, the equilibrium point occurs when the price is $3.90 per bushel and 17.9 million bushels are produced and sold. ∎

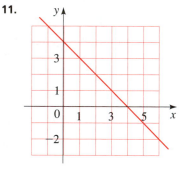

FIGURE 18

1.10 EXERCISES

1–8 ■ Find the slope of the line through P and Q.

1. $P(0, 0)$, $Q(2, 4)$

2. $P(0, 0)$, $Q(2, -6)$

3. $P(2, 2)$, $Q(10, 0)$

4. $P(1, 2)$, $Q(3, 3)$

5. $P(2, 4)$, $Q(4, 12)$

6. $P(2, -5)$, $Q(-4, 3)$

7. $P(1, -3)$, $Q(-1, 6)$

8. $P(-1, -4)$, $Q(6, 0)$

9. Find the slopes of the lines l_1, l_2, l_3, and l_4 in the figure below.

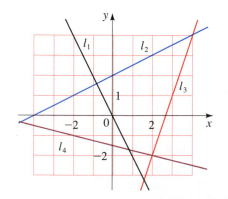

10. (a) Sketch lines through $(0, 0)$ with slopes 1, 0, $\frac{1}{2}$, 2, and -1.

(b) Sketch lines through $(0, 0)$ with slopes $\frac{1}{3}$, $\frac{1}{2}$, $-\frac{1}{3}$, and 3.

11–14 ■ Find an equation for the line whose graph is sketched.

11.

12.

13.

14.

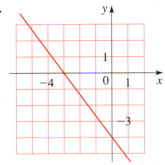

15–34 ■ Find an equation of the line that satisfies the given conditions.

15. Through $(2, 3)$;   slope 1

16. Through $(-2, 4)$;   slope -1

17. Through $(1, 7)$;   slope $\frac{2}{3}$

18. Through $(-3, -5)$;   slope $-\frac{7}{2}$

19. Through $(2, 1)$ and $(1, 6)$

20. Through $(-1, -2)$ and $(4, 3)$

21. Slope 3; y-intercept -2

22. Slope $\frac{2}{5}$; y-intercept 4

23. x-intercept 1; y-intercept -3

24. x-intercept -8; y-intercept 6

25. Through $(4, 5)$;   parallel to the x-axis

26. Through $(4, 5)$;   parallel to the y-axis

27. Through $(1, -6)$;   parallel to the line $x + 2y = 6$

28. y-intercept 6;   parallel to the line $2x + 3y + 4 = 0$

29. Through $(-1, 2)$;   parallel to the line $x = 5$

30. Through $(2, 6)$;   perpendicular to the line $y = 1$

31. Through $(-1, -2)$;   perpendicular to the line
$2x + 5y + 8 = 0$

32. Through $\left(\frac{1}{2}, -\frac{2}{3}\right)$;   perpendicular to the line
$4x - 8y = 1$

33. Through $(1, 7)$;   parallel to the line passing through $(2, 5)$ and $(-2, 1)$

34. Through $(-2, -11)$;   perpendicular to the line passing through $(1, 1)$ and $(5, -1)$

35. (a) Sketch the line with slope $\frac{3}{2}$ that passes through the point $(-2, 1)$.
    (b) Find an equation for this line.

36. (a) Sketch the line with slope -2 that passes through the point $(4, -1)$.
    (b) Find an equation for this line.

37–38 ■ Use a graphing device to graph the given family of lines in the same viewing rectangle. What do the lines have in common?

37. $y = 2 + m(x - 3)$ for $m = 0, \pm 0.5, \pm 1, \pm 2, \pm 10$

38. $y = 1.3x + b$ for $b = 0, \pm 1, \pm 2.8$

39–50 ■ Find the slope and y-intercept of the line and draw its graph.

39. $x + y = 3$

40. $3x - 2y = 12$

41. $x + 3y = 0$

42. $2x - 5y = 0$

43. $\frac{1}{2}x - \frac{1}{3}y + 1 = 0$

44. $-3x - 5y + 30 = 0$

45. $y = 4$

46. $4y + 8 = 0$

47. $3x - 4y = 12$

48. $x = -5$

49. $3x + 4y - 1 = 0$

50. $4x + 5y = 10$

51. Show that $A(1, 1)$, $B(7, 4)$, $C(5, 10)$, and $D(-1, 7)$ are vertices of a parallelogram.

52. Show that $A(-3, -1)$, $B(3, 3)$, and $C(-9, 8)$ are vertices of a right triangle.

53. Show that $A(1, 1)$, $B(11, 3)$, $C(10, 8)$, and $D(0, 6)$ are vertices of a rectangle.

54. Use slopes to determine whether the given points are collinear (lie on a line).
    (a) $(1, 1)$, $(3, 9)$, $(6, 21)$
    (b) $(-1, 3)$, $(1, 7)$, $(4, 15)$

55. Find an equation of the perpendicular bisector of the line segment joining the points $A(1, 4)$ and $B(7, -2)$.

56. Find the area of the triangle formed by the coordinate axes and the line $2y + 3x - 6 = 0$.

57. (a) Show that if the x- and y-intercepts of a line are nonzero numbers a and b, then the equation of the line can be written in the form

$$\frac{x}{a} + \frac{y}{b} = 1$$

This is called the **two-intercept form** of the equation of a line.

(b) Use part (a) to find an equation of the line whose x-intercept is 6 and whose y-intercept is -8.

58. (a) Find an equation for the line tangent to the circle $x^2 + y^2 = 25$ at the point $(3, -4)$. (See the figure.)

(b) At what other point on the circle will a tangent line be parallel to the tangent line in part (a)?

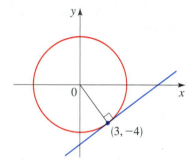

59. West of Albuquerque, New Mexico, Route 40 eastbound is straight and makes a steep descent towards the city. The highway has a 6% grade, which means that its slope is $-\frac{6}{100}$. Driving on this road you notice from elevation signs that you have descended a distance of 1000 ft. What is the change in your horizontal distance?

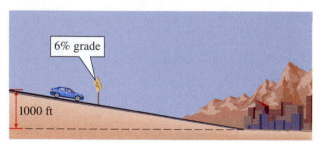

60. Some scientists believe that the average surface temperature of the world has been rising steadily. The average surface temperature is given by

$$T = 0.02t + 8.50$$

where T is temperature in °C and t is years since 1900.

(a) What do the slope and T-intercept represent?

(b) Use the equation to predict the average global surface temperature in 2100.

61. If the recommended adult dosage for a drug is D (in mg), then to determine the appropriate dosage c for a child of age a, pharmacists use the equation

$$c = 0.0417D(a + 1)$$

Suppose the dosage for an adult is 200 mg.

(a) Find the slope. What does it represent?

(b) What is the dosage for a newborn?

62. The manager of a weekend flea market knows from past experience that if she charges x dollars for a rental space at the flea market, then the number y of spaces she can rent is given by the equation $y = 200 - 4x$.

(a) Sketch a graph of this linear equation. (Remember that the rental charge per space and the number of spaces rented must both be nonnegative quantities.)

(b) What do the slope, the y-intercept, and the x-intercept of the graph represent?

63. A small-appliance manufacturer finds that if he produces x toaster ovens in a month his production cost is given by the equation $y = 6x + 3000$ (where y is measured in dollars).

(a) Sketch a graph of this linear equation.

(b) What do the slope and y-intercept of the graph represent?

64. The relationship between the Fahrenheit (F) and Celsius (C) temperature scales is given by the equation $F = \frac{9}{5}C + 32$.

(a) Complete the table to compare the two scales at the given values.

(b) Find the temperature at which the scales agree. [*Hint:* Suppose that a is the temperature at which the scales agree. Set $F = a$ and $C = a$. Then solve for a.]

C	F
$-30°$	
$-20°$	
$-10°$	
$0°$	
	$50°$
	$68°$
	$86°$

65. Biologists have observed that the chirping rate of crickets of a certain species is related to temperature, and the relationship appears to be very nearly linear. A cricket produces 120 chirps per minute at 70°F and 168 chirps per minute at 80°F.

(a) Find the linear equation that relates the temperature t and the number of chirps per minute n.

(b) If the crickets are chirping at 150 chirps per minute, estimate the temperature.

66. A small business buys a computer for $4000. After 4 years the value of the computer is expected to be $200. For accounting purposes, the business uses *linear depreciation* to assess the value of the computer at a given time. This means that if V is the value of the computer at time t, then a linear equation is used to relate V and t.
(a) Find a linear equation that relates V and t.
(b) Find the depreciated value of the computer 3 years from the date of purchase.

67. At the surface of the ocean, the water pressure is the same as the air pressure above the water, 15 lb/in^2. Below the surface, the water pressure increases by 4.34 lb/in^2 for every 10 ft of descent.
(a) Find an equation for the relationship between pressure and depth below the ocean surface.
(b) At what depth is the pressure 100 lb/in^2?

water pressure increases with depth

68. Jason and Debbie leave Detroit at 2:00 P.M. and drive at a constant speed, traveling west on I-90. They pass Ann Arbor, 40 mi from Detroit, at 2:50 P.M.
(a) Express the distance traveled in terms of the time elapsed.
(b) Draw the graph of the equation in part (a).
(c) What is the slope of this line? What does it represent?

69. The monthly cost of driving a car depends on the number of miles driven. Lynn found that in May her driving cost was $380 for 480 mi and in June her cost was $460 for 800 mi.
(a) Express the monthly cost C in terms of the distance driven d, assuming that a linear relationship gives a suitable model.
(b) Use part (a) to predict the cost of driving 1500 mi per month.
(c) Draw the graph of the linear equation. What does the slope of the line represent?
(d) What does the y-intercept of the graph represent?
(e) Why is a linear relationship a suitable model for this situation?

70. The manager of a furniture factory finds that it costs $2200 to manufacture 100 chairs in one day and $4800 to produce 300 chairs in one day.
(a) Assuming that the relationship between cost and the number of chairs produced is linear, find an equation that expresses this relationship. Then graph the equation.
(b) What is the slope of the line in part (a), and what does it represent?
(c) What is the y-intercept of this line, and what does it represent?

 71–72 ■ Equations for supply and demand are given.
(a) Draw the graphs of the two equations in an appropriate viewing rectangle.
(b) Estimate the equilibrium point from the graph.
(c) Estimate the price and the amount of the commodity produced and sold at equilibrium.

71. Supply: $y = 0.45p + 4$
     Demand: $y = -0.65p + 28$

72. Supply: $y = 8.5p + 45$
     Demand: $y = -0.6p + 300$

 DISCOVERY · DISCUSSION

73. What Does the Slope Mean?   Suppose that the graph of the outdoor temperature over a certain period of time is a line. How is the weather changing if the slope of the line is positive? If it's negative? If it's zero?

74. Collinear Points   Suppose you are given the coordinates of three points in the plane, and you want to see whether they lie on the same line. How can you do this using slopes? Using the Distance Formula? Can you think of another method?

1 REVIEW

CONCEPT CHECK

1. Define each term in your own words. (Check by referring to the definition in the text.)
   (a) An integer   (b) A rational number
   (c) An irrational number (d) A real number

2. State each of these properties of real numbers.
   (a) Commutative Property
   (b) Associative Property
   (c) Distributive Property

3. What is an open interval? What is a closed interval? What notation is used for these intervals?

4. What is the absolute value of a number?

5. (a) In the expression a^x, which is the base and which is the exponent?
   (b) What does a^x mean if $x = n$, a positive integer?
   (c) What if $x = 0$?
   (d) What if x is a negative integer: $x = -n$, where n is a positive integer?
   (e) What if $x = m/n$, a rational number?
   (f) State the Laws of Exponents.

6. (a) What does $\sqrt[n]{a} = b$ mean?
   (b) Why is $\sqrt{a^2} = |a|$?
   (c) How many real nth roots does a positive real number have if n is odd? If n is even?

7. Explain how the procedure of rationalizing the denominator works.

8. State the Special Product Formulas for $(a + b)^2$, $(a - b)^2$, $(a + b)^3$, and $(a - b)^3$.

9. State each factoring formula.
   (a) Difference of squares   (b) Difference of cubes
   (c) Sum of cubes

10. What is a solution of an equation?

11. How do you solve an equation involving radicals? Why is it important to check your answers when solving equations of this type?

12. How do you solve an equation
    (a) algebraically?   (b) graphically?

13. Write the general form of each type of equation.
    (a) A linear equation   (b) A quadratic equation

14. What are the three ways to solve a quadratic equation?

15. State the Zero-Product Property.

16. Describe the process of completing the square.

17. State the quadratic formula.

18. What is the discriminant of a quadratic equation?

19. State the rules for working with inequalities.

20. How do you solve
    (a) a linear inequality?
    (b) a nonlinear inequality?

21. (a) How do you solve an equation involving an absolute value?
    (b) How do you solve an inequality involving an absolute value?

22. (a) Describe the coordinate plane.
    (b) How do you locate points in the coordinate plane?

23. State each formula.
    (a) The Distance Formula
    (b) The Midpoint Formula

24. Given an equation, what is its graph?

25. How do you find the x-intercepts and y-intercepts of a graph?

26. Write an equation of the circle with center (h, k) and radius r.

27. Explain the meaning of each type of symmetry. How do you test for it?
    (a) Symmetry with respect to the x-axis
    (b) Symmetry with respect to the y-axis
    (c) Symmetry with respect to the origin

28. Define the slope of a line.

29. Write each form of the equation of a line.
    (a) The point-slope form
    (b) The slope-intercept form

30. (a) What is the equation of a vertical line?
    (b) What is the equation of a horizontal line?

31. What is the general equation of a line?

32. Given lines with slopes m_1 and m_2, explain how you can tell if the lines are
    (a) parallel   (b) perpendicular

EXERCISES

1–4 ■ State the property of real numbers being used.

1. $3 + 2x = 2x + 3$

2. $(a + b)(a - b) = (a - b)(a + b)$

3. $A(x + y) = Ax + Ay$

4. $(A + 1)(x + y) = (A + 1)x + (A + 1)y$

5–6 ■ Express the interval in terms of inequalities, and then graph the interval.

5. $[-2, 6)$

6. $(-\infty, 4]$

7–8 ■ Express the inequality in interval notation, and then graph the corresponding interval.

7. $x \geq 5$

8. $-1 < x \leq 5$

9–18 ■ Evaluate the expression.

9. $\big|3 - |-9|\big|$

10. $1 - \big|1 - |-1|\big|$

11. $2^{-3} - 3^{-2}$

12. $\sqrt[3]{-125}$

13. $216^{-1/3}$

14. $64^{2/3}$

15. $\dfrac{\sqrt{242}}{\sqrt{2}}$

16. $\sqrt[4]{4}\,\sqrt[4]{324}$

17. $2^{1/2}8^{1/2}$

18. $\sqrt{2}\,\sqrt{50}$

19–28 ■ Simplify the expression.

19. $\dfrac{x^2(2x)^4}{x^3}$

20. $(a^2)^{-3}(a^3b)^2(b^3)^4$

21. $(3xy^2)^3(\tfrac{2}{3}x^{-1}y)^2$

22. $\left(\dfrac{r^2s^{4/3}}{r^{1/3}s}\right)^6$

23. $\sqrt[3]{(x^3y)^2y^4}$

24. $\sqrt{x^2y^4}$

25. $\dfrac{x}{2 + \sqrt{x}}$

26. $\dfrac{\sqrt{x} + 1}{\sqrt{x} - 1}$

27. $\dfrac{8r^{1/2}s^{-3}}{2r^{-2}s^4}$

28 $\left(\dfrac{ab^2c^{-3}}{2a^3b^{-4}}\right)^{-2}$

29. Write the number 78,250,000,000 in scientific notation.

30. Write the number 2.08×10^{-8} in ordinary decimal notation.

31. If $a \approx 0.00000293$, $b \approx 1.582 \times 10^{-14}$, and $c \approx 2.8064 \times 10^{12}$, use a calculator to approximate the number ab/c.

32. If your heart beats 80 times per minute and you live to be 90 years old, estimate the number of times your heart beats during your lifetime. State your answer in scientific notation.

33–48 ■ Factor the expression completely.

33. $12x^2y^4 - 3xy^5 + 9x^3y^2$

34. $x^2 - 9x + 18$

35. $x^2 + 3x - 10$

36. $6x^2 + x - 12$

37. $4t^2 - 13t - 12$

38. $x^4 - 2x^2 + 1$

39. $25 - 16t^2$

40. $2y^6 - 32y^2$

41. $x^6 - 1$

42. $y^3 - 2y^2 - y + 2$

43. $x^{-1/2} - 2x^{1/2} + x^{3/2}$

44. $a^4b^2 + ab^5$

45. $4x^3 - 8x^2 + 3x - 6$

46. $8x^3 + y^6$

47. $(x^2 + 2)^{5/2} + 2x(x^2 + 2)^{3/2} + x^2\sqrt{x^2 + 2}$

48. $3x^3 - 2x^2 + 18x - 12$

49–64 ■ Perform the indicated operations and simplify.

49. $(2x + 1)(3x - 2) - 5(4x - 1)$

50. $(2y - 7)(2y + 7)$

51. $(1 + x)(2 - x) - (3 - x)(3 + x)$

52. $\sqrt{x}\left(\sqrt{x} + 1\right)\left(2\sqrt{x} - 1\right)$

53. $x^2(x - 2) + x(x - 2)^2$

54. $\dfrac{x^2 - 2x - 3}{2x^2 + 5x + 3}$

55. $\dfrac{x^2 + 2x - 3}{x^2 + 8x + 16} \cdot \dfrac{3x + 12}{x - 1}$

56. $\dfrac{t^3 - 1}{t^2 - 1}$

57. $\dfrac{x^2 - 2x - 15}{x^2 - 6x + 5} \div \dfrac{x^2 - x - 12}{x^2 - 1}$

58. $\dfrac{2}{x} + \dfrac{1}{x - 2} + \dfrac{3}{(x - 2)^2}$

59. $\dfrac{1}{x - 1} - \dfrac{2}{x^2 - 1}$

60. $\dfrac{1}{x + 2} + \dfrac{1}{x^2 - 4} - \dfrac{2}{x^2 - x - 2}$

61. $\dfrac{\dfrac{1}{x} - \dfrac{1}{2}}{x - 2}$

62. $\dfrac{\dfrac{1}{x} - \dfrac{1}{x + 1}}{\dfrac{1}{x} + \dfrac{1}{x + 1}}$

63. $\dfrac{\sqrt{6}}{\sqrt{3} + \sqrt{2}}$ (rationalize the denominator)

64. $\dfrac{\sqrt{x + h} - \sqrt{x}}{h}$ (rationalize the numerator)

65–80 ■ Find all real solutions of the equation.

65. $7x - 6 = 4x + 9$

66. $8 - 2x = 14 + x$

67. $\dfrac{x + 1}{x - 1} = \dfrac{3x}{3x - 6}$

68. $(x + 2)^2 = (x - 4)^2$

69. $x^2 - 9x + 14 = 0$

70. $x^2 + 24x + 144 = 0$

71. $2x^2 + x = 1$

72. $3x^2 + 5x - 2 = 0$

73. $4x^3 - 25x = 0$

74. $x^3 - 2x^2 - 5x + 10 = 0$

75. $3x^2 + 4x - 1 = 0$

76. $\dfrac{1}{x} + \dfrac{2}{x - 1} = 3$

77. $\dfrac{x}{x - 2} + \dfrac{1}{x + 2} = \dfrac{8}{x^2 - 4}$

78. $x^4 - 8x^2 - 9 = 0$

79. $|x - 7| = 4$

80. $|2x - 5| = 9$

81. The owner of a store sells raisins for \$3.20 per pound and nuts for \$2.40 per pound. He decides to mix the raisins and nuts and sell 50 lb of the mixture for \$2.72 per pound. What quantities of raisins and nuts should he use?

82. Anthony leaves Kingstown at 2:00 P.M. and drives to Queensville, 160 mi distant, at 45 mi/h. At 2:15 P.M. Helen leaves Queensville and drives to Kingstown at 40 mi/h. At what time do they pass each other on the road?

83. A woman cycles 8 mi/h faster than she runs. Every morning she cycles 4 mi and runs $2\frac{1}{2}$ mi, for a total of one hour of exercise. How fast does she run?

84. The hypotenuse of a right triangle has length 20 cm. The sum of the lengths of the other two sides is 28 cm. Find the lengths of the other two sides of the triangle.

85. Abbie paints twice as fast as Beth and three times as fast as Cathie. If it takes them 60 min to paint a living room with all three working together, how long would it take Abbie if she works alone?

86. A homeowner wishes to fence in three adjoining garden plots, one for each of her children, as shown in the figure. If each plot is to be 80 ft^2 in area, and she has 88 ft of fencing material at hand, what dimensions should each plot have?

87–96 ■ Solve the inequality. Express the solution using interval notation and graph the solution set on the real number line.

87. $3x - 2 > -11$

88. $-1 < 2x + 5 \le 3$

89. $x^2 + 4x - 12 > 0$

90. $x^2 \le 1$

91. $\dfrac{x - 4}{x^2 - 4} \le 0$

92. $\dfrac{5}{x^3 - x^2 - 4x + 4} < 0$

93. $|x - 5| \le 3$

94. $|x - 4| < 0.02$

95–98 ■ Solve the equation or inequality graphically.

95. $x^2 - 4x = 2x + 7$

96. $\sqrt{x + 4} = x^2 - 5$

97. $4x - 3 \ge x^2$

98. $x^3 - 4x^2 - 5x > 2$

99–100 ■ Two points P and Q are given.
(a) Plot P and Q on a coordinate plane.
(b) Find the distance from P to Q.
(c) Find the midpoint of the segment PQ.
(d) Sketch the line determined by P and Q, and find its equation in slope-intercept form.
(e) Sketch the circle that passes through Q and has center P, and find the equation of this circle.

99. $P(2, 0)$, $Q(-5, 12)$

100. $P(7, -1)$, $Q(2, -11)$

101–102 ■ Sketch the region given by the set.

101. $\{(x, y) \mid -4 < x < 4 \text{ and } -2 < y < 2\}$

102. $\{(x, y) \mid x \ge 4 \text{ or } y \ge 2\}$

103. Which of the points $A(4, 4)$ or $B(5, 3)$ is closer to the point $C(-1, -3)$?

104. Find an equation of the circle that has center $(2, -5)$ and radius $\sqrt{2}$.

105. Find an equation of the circle that has center $(-5, -1)$ and passes through the origin.

106. Find an equation of the circle that contains the points $P(2, 3)$ and $Q(-1, 8)$ and has the midpoint of the segment PQ as its center.

107–110 ■ Determine whether the equation represents a circle, a point, or has no graph. If the equation is that of a circle, find its center and radius.

107. $x^2 + y^2 + 2x - 6y + 9 = 0$

108. $2x^2 + 2y^2 - 2x + 8y = \frac{1}{2}$

109 $x^2 + y^2 + 72 = 12x$

110. $x^2 + y^2 - 6x - 10y + 34 = 0$

111–118 ■ Test the equation for symmetry and sketch its graph.

111. $y = 2 - 3x$

112. $2x - y + 1 = 0$

113. $x + 3y = 21$

114. $x = 2y + 12$

115. $y = 16 - x^2$

116. $8x + y^2 = 0$

117. $x = \sqrt{y}$

118. $y = -\sqrt{1 - x^2}$

 119–122 ■ Use a graphing device to graph the equation in an appropriate viewing rectangle.

119. $y = x^2 - 6x$

120. $y = \sqrt{5 - x}$

121. $y = x^3 - 4x^2 - 5x$

122. $\dfrac{x^2}{4} + y^2 = 1$

123. Find an equation for the line that passes through the points $(-1, -6)$ and $(2, -4)$.

124. Find an equation for the line that passes through the point $(6, -3)$ and has slope $-\frac{1}{2}$.

125. Find an equation for the line that has x-intercept 4 and y-intercept 12.

126. Find an equation for the line that passes through the point $(1, 7)$ and is perpendicular to the line $x - 3y + 16 = 0$.

127. Find an equation for the line that passes through the origin and is parallel to the line $3x + 15y = 22$.

128. Find an equation for the line that passes through the point $(5, 2)$ and is parallel to the line passing through $(-1, -3)$ and $(3, 2)$.

129–130 ■ Find equations for the circle and the line in the figure.

129.

130.

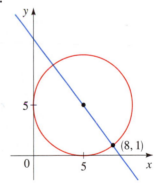

1   TEST

1. (a) Graph the intervals $[-2, 5)$ and $(7, \infty)$ on the real number line.
   (b) Express the inequalities $x \leq 3$ and $0 < x \leq 8$ in interval notation.
   (c) Find the distance between -12 and 39 on the real number line.

2. Evaluate each expression.
   (a) $(-3)^4$ (b) -3^4 (c) 3^{-4} (d) $\dfrac{5^{23}}{5^{21}}$ (e) $\left(\dfrac{2}{3}\right)^{-2}$ (f) $16^{-3/4}$

3. Write each number in scientific notation.
   (a) $186{,}000{,}000{,}000$ (b) 0.0000003965

4. Simplify each expression. Write your final answer without negative exponents.

   (a) $\sqrt{200} - \sqrt{32}$ (b) $(3a^3b^3)(4ab^2)^2$ (c) $\left(\dfrac{3x^{3/2}y^3}{x^2y^{-1/2}}\right)^{-2}$

   (d) $\dfrac{x^2 + 3x + 2}{x^2 - x - 2}$ (e) $\dfrac{x^2}{x^2 - 4} - \dfrac{x + 1}{x + 2}$ (f) $\dfrac{\dfrac{y}{x} - \dfrac{x}{y}}{\dfrac{1}{y} - \dfrac{1}{x}}$

5. Rationalize the denominator and simplify: $\dfrac{\sqrt{10}}{\sqrt{5} - 2}$

6. Perform the indicated operations and simplify.
   (a) $3(x + 6) + 4(2x - 5)$ (b) $(x + 3)(4x - 5)$ (c) $\left(\sqrt{a} + \sqrt{b}\right)\left(\sqrt{a} - \sqrt{b}\right)$
   (d) $(2x + 3)^2$ (e) $(x + 2)^3$

7. Factor each expression completely.
   (a) $4x^2 - 25$ (b) $2x^2 + 5x - 12$ (c) $x^3 - 3x^2 - 4x + 12$
   (d) $x^4 + 27x$ (e) $3x^{3/2} - 9x^{1/2} + 6x^{-1/2}$ (f) $x^3y - 4xy$

8. Find all real solutions.

   (a) $x + 5 = 14 - \frac{1}{2}x$ (b) $\dfrac{2x}{x + 1} = \dfrac{2x - 1}{x}$ (c) $x^2 - x - 12 = 0$

   (d) $2x^2 + 4x + 1 = 0$ (e) $\sqrt{3 - \sqrt{x + 5}} = 2$ (f) $x^4 - 3x^2 + 2 = 0$

   (g) $3|x - 4| = 10$

9. Mary drove from Amity to Belleville at a speed of 50 mi/h. On the way back, she drove at 60 mi/h. The total trip took $4\frac{2}{5}$ h of driving time. Find the distance between these two cities.

10. A rectangular parcel of land is 70 ft longer than it is wide. Each diagonal between opposite corners is 130 ft. What are the dimensions of the parcel?

11. Solve each inequality. Write the answer using interval notation, and sketch the solution on the real number line.
    (a) $-4 < 5 - 3x \leq 17$ (b) $x(x - 1)(x + 2) > 0$

    (c) $|x - 4| < 3$ (d) $\dfrac{2x - 3}{x + 1} \leq 1$

12. A bottle of medicine is to be stored at a temperature between 5°C and 10°C. What range does this correspond to on the Fahrenheit scale? [*Note:* Fahrenheit (F) and Celsius (C) temperatures satisfy the relation $C = \frac{5}{9}(F - 32)$.]

13. For what values of x is the expression $\sqrt{6x - x^2}$ defined as a real number?

14. Solve the equation and the inequality graphically.

(a) $x^3 - 9x - 1 = 0$ 　　　　　(b) $x^2 - 1 \leq |x + 1|$

15. (a) Plot the points $P(0, 3)$, $Q(3, 0)$, and $R(6, 3)$ in the coordinate plane. Where must the point S be located so that $PQRS$ is a square?
(b) Find the area of $PQRS$.

16. (a) Sketch the graph of $y = x^2 - 4$.
(b) Find the x- and y-intercepts of the graph.
(c) Is the graph symmetric about the x-axis, the y-axis, or the origin?

17. Let $A(-7, 4)$ and $B(5, -12)$ be points in the plane.
(a) Find the length of the segment AB.
(b) Find the midpoint of the segment AB.
(c) Find an equation of the line that passes through the points A and B.
(d) Find an equation of the perpendicular bisector of AB.
(e) Find an equation of the circle with center B and radius 13.
(f) Find an equation of the circle for which AB is a diameter.

18. Find the center and radius of each circle and sketch its graph.
(a) $x^2 + (y - 4)^2 = 9$ 　　　　　(b) $x^2 + y^2 - 6x + 10y + 9 = 0$

19. Write the linear equation

$$3x - 4y + 24 = 0$$

in slope-intercept form. What are the slope and y-intercept? Sketch the graph.

20. Find an equation for the line with the given property.
(a) Passing through $(-2, 3)$ and parallel to the line $2x + 6y = 17$
(b) Having x-intercept -3 and y-intercept 12

21. A manufacturer finds that the cost C of producing x blenders is

$$C = 7x + 17,500$$

(a) How much does it cost to produce 2500 blenders?
(b) What do the slope and C-intercept of the cost equation represent?

If you had difficulty with any of these problems, you may wish to review the section of this chapter indicated below.

If you had trouble with this test problem	Review this section
1	Section 1.1
2, 3, 4(a), 4(b)	Section 1.2
4(c), 4(d), 4(e), 4(f), 5	Section 1.4
6, 7	Section 1.3
8	Section 1.5
9, 10	Section 1.6
11, 12, 13	Section 1.7
14	Section 1.9
15, 16, 17(a), 17(b)	Section 1.8
17(c), 17(d)	Section 1.10
17(e), 17(f), 18	Section 1.8
19, 20, 21	Section 1.10

Focus on Problem Solving
General Principles

There are no hard and fast rules that will ensure success in solving problems. However, it is possible to outline some general steps in the problem-solving process and to give principles that are useful in solving certain problems. These steps and principles are just common sense made explicit. They have been adapted from George Polya's insightful book *How To Solve It*.

1 Understand the Problem

The first step is to read the problem and make sure that you understand it. Ask yourself the following questions:

> *What is the unknown?*
> *What are the given quantities?*
> *What are the given conditions?*

For many problems it is useful to

> *draw a diagram*

and identify the given and required quantities on the diagram.
  Usually it is necessary to

> *introduce suitable notation*

In choosing symbols for the unknown quantities, we often use letters such as a, b, c, m, n, x, and y, but in some cases it helps to use initials as suggestive symbols, for instance, V for volume or t for time.

2 Think of a Plan

Find a connection between the given information and the unknown that enables you to calculate the unknown. It often helps to ask yourself explicitly: "How can I relate the given to the unknown?" If you don't see a connection immediately, the following ideas may be helpful in devising a plan.

■ **Try to recognize something familiar**

Relate the given situation to previous knowledge. Look at the unknown and try to recall a more familiar problem that has a similar unknown.

■ **Try to recognize patterns**

Certain problems are solved by recognizing that some kind of pattern is occurring. The pattern could be geometric, or numerical, or algebraic. If you can see regularity or repetition in a problem, then you might be able to guess what the pattern is and then prove it.

■ Use analogy

Try to think of an analogous problem, that is, a similar or related problem, but one that is easier than the original. If you can solve the similar, simpler problem, then it might give you the clues you need to solve the original, more difficult one. For instance, if a problem involves very large numbers, you could first try a similar problem with smaller numbers. Or if the problem is in three-dimensional geometry, you could look for something similar in two-dimensional geometry. Or if the problem you start with is a general one, you could first try a special case.

■ Introduce something extra

You may sometimes need to introduce something new—an auxiliary aid—to make the connection between the given and the unknown. For instance, in a problem for which a diagram is useful, the auxiliary aid could be a new line drawn in the diagram. In a more algebraic problem the aid could be a new unknown that relates to the original unknown.

■ Take cases

You may sometimes have to split a problem into several cases and give a different argument for each case. For instance, we often have to use this strategy in dealing with absolute value.

■ Work backward

Sometimes it is useful to imagine that your problem is solved and work backward, step by step, until you arrive at the given data. Then you may be able to reverse your steps and thereby construct a solution to the original problem. This procedure is commonly used in solving equations. For instance, in solving the equation $3x - 5 = 7$, we suppose that x is a number that satisfies $3x - 5 = 7$ and work backward. We add 5 to each side of the equation and then divide each side by 3 to get $x = 4$. Since each of these steps can be reversed, we have solved the problem.

■ Establish subgoals

In a complex problem it is often useful to set subgoals (in which the desired situation is only partially fulfilled). If you can attain or accomplish these subgoals, then you may be able to build on them to reach your final goal.

■ Indirect reasoning

Sometimes it is appropriate to attack a problem indirectly. In using **proof by contradiction** to prove that P implies Q, we assume that P is true and Q is false and try to see why this cannot happen. Somehow we have to use this information and arrive at a contradiction to what we absolutely know is true.

George Polya (1887–1985) is famous among mathematicians for his ideas on problem solving. His lectures on problem solving at Stanford University attracted overflow crowds whom he held on the edges of their seats, leading them to discover solutions for themselves. He was able to do this because of his deep insight into the psychology of problem solving. His well-known book *How To Solve It* has been translated into 15 languages. He said that Euler (see page 293) was unique among great mathematicians because he explained *how* he found his results. Polya often said to his students and colleagues, "Yes, I see that your proof is correct, but how did you discover it?" In the preface to *How To Solve It*, Polya writes, "A great discovery solves a great problem but there is a grain of discovery in the solution of any problem. Your problem may be modest; but if it challenges your curiosity and brings into play your inventive faculties, and if you solve it by your own means, you may experience the tension and enjoy the triumph of discovery."

■ **Mathematical induction**

In proving statements that involve a positive integer n, it is frequently helpful to use the Principle of Mathematical Induction, which is discussed in Section 10.5.

3 Carry Out the Plan

In Step 2, a plan was devised. In carrying out that plan, you must check each stage of the plan and write the details that prove each stage is correct.

4 Look Back

Having completed your solution, it is wise to look back over it, partly to see if any errors have been made and partly to see if you can discover an easier way to solve the problem. Looking back also familiarizes you with the method of solution, and this may be useful for solving a future problem. Descartes said, "Every problem that I solved became a rule which served afterwards to solve other problems."

   We illustrate some of these principles of problem solving with an example. Further illustrations of these principles will be presented at the end of selected chapters.

PROBLEM ■ **Average Speed**

A driver sets out on a journey. For the first half of the distance she drives at the leisurely pace of 30 mi/h; during the second half she drives 60 mi/h. What is her average speed on this trip?

■ **Thinking about the problem**

It is tempting to take the average of the speeds and say that the average speed for the entire trip is

$$\frac{30 + 60}{2} = 45 \text{ mi/h}$$

But is this simple-minded approach really correct?

   Let's look at an easily calculated special case. Suppose that the total distance traveled is 120 mi. Since the first 60 mi is traveled at 30 mi/h, it takes 2 h. The second 60 mi is traveled at 60 mi/h, so it takes one hour. Thus, the total time is $2 + 1 = 3$ hours and the average speed is

$$\frac{120}{3} = 40 \text{ mi/h}$$

So our guess of 45 mi/h was wrong.

Try a special case

SOLUTION

Understand the problem

We need to look more carefully at the meaning of average speed. It is defined as

$$\text{average speed} = \frac{\text{distance traveled}}{\text{time elapsed}}$$

Introduce notation

State what is given

Let d be the distance traveled on each half of the trip. Let t_1 and t_2 be the times taken for the first and second halves of the trip. Now we can write down the information we have been given. For the first half of the trip, we have

(1)
$$30 = \frac{d}{t_1}$$

and for the second half, we have

(2)
$$60 = \frac{d}{t_2}$$

Identify the unknown

Now we identify the quantity we are asked to find:

$$\text{average speed for entire trip} = \frac{\text{total distance}}{\text{total time}} = \frac{2d}{t_1 + t_2}$$

Connect the given with the unknown

To calculate this quantity, we need to know t_1 and t_2, so we solve Equations 1 and 2 for these times:

$$t_1 = \frac{d}{30} \qquad t_2 = \frac{d}{60}$$

Now we have the ingredients needed to calculate the desired quantity:

$$\text{average speed} = \frac{2d}{t_1 + t_1} = \frac{2d}{\dfrac{d}{30} + \dfrac{d}{60}}$$

$$= \frac{60(2d)}{60\left(\dfrac{d}{30} + \dfrac{d}{60}\right)} \qquad \text{\color{red}Multiply numerator and denominator by 60}$$

$$= \frac{120d}{2d + d} = \frac{120d}{3d} = 40$$

So, the average speed for the entire trip is 40 mi/h. ∎

Problems

Don't feel bad if you don't solve these problems right away. Problems 2 and 7 were sent to Albert Einstein by his friend Wertheimer. Einstein (and his friend Bucky) enjoyed the problems and wrote back to Wertheimer. Here is part of his reply:

Your letter gave us a lot of amusement. The first intelligence test fooled both of us (Bucky and me). Only on working it out did I notice that no time is available for the downhill run! Mr. Bucky was also taken in by the second example, but I was not. Such drolleries show us how stupid we are!

(See *Mathematical Intelligencer*, Spring 1990, page 41)

1. A man drives from home to work at a speed of 50 mi/h. The return trip from work to home is traveled at the more leisurely pace of 30 mi/h. What is the man's average speed for the round-trip?

2. An old car has to travel a 2-mile route, uphill and down. Because it is so old, the car can climb the first mile—the ascent—no faster than an average speed of 15 mi/h. How fast does the car have to travel the second mile—on the descent it can go faster, of course—in order to achieve an average speed of 30 mi/h for the trip?

3. A car and a van are parked 120 mi apart on a straight road. The drivers start driving toward each other at noon, each at a speed of 40 mi/h. A fly starts from the front bumper of the van at noon and flies to the bumper of the car, then immediately back to the bumper of the van, back to the car, and so on, until the car and the van meet. If the fly flies at a speed of 100 mi/h, what is the total distance it travels?

4. Which price is better for the buyer, a 40% discount or two successive discounts of 20%?

5. Use a calculator to find the value of the expression

$$\sqrt{3 + 2\sqrt{2}} - \sqrt{3 - 2\sqrt{2}}$$

The number looks very simple. Show that the calculated value is correct.

6. Use a calculator to evaluate

$$\frac{\sqrt{2} + \sqrt{6}}{\sqrt{2 + \sqrt{3}}}$$

Show that the calculated value is correct.

7. An amoeba propagates by simple division; each split takes 3 minutes to complete. When such an amoeba is put into a glass container with a nutrient fluid, the container is full of amoebas in one hour. How long would it take for the container to be filled if we start with not one amoeba, but two?

8. Two runners start running laps at the same time, from the same starting position. George runs a lap in 50 s; Sue runs a lap in 30 s. When will the runners next be side by side?

9. Player A has a higher batting average than player B for the first half of the baseball season. Player A also has a higher batting average than player B for the second half of the season. Is it necessarily true that player A has a higher batting average than player B for the entire season?

10. A woman starts at a point P on the earth's surface and walks 1 mi south, then 1 mi east, then 1 mi north, and finds herself back at P, the starting point. Describe all points P for which this is possible (there are infinitely many).

11. A spoonful of cream is taken from a pitcher of cream and put into a cup of coffee. The coffee is stirred. Then a spoonful of this mixture is put into the pitcher of cream. Is there now more cream in the coffee cup or more coffee in the pitcher of cream?

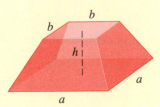

12. An ice cube is floating in a cup of water, full to the brim, as shown in the sketch. As the ice melts, what happens? Does the cup overflow, or does the water level drop, or does it remain the same? (You need to know Archimedes' principle: A floating object displaces a volume of water whose weight equals the weight of the object.)

13. An extended family consists of the members

{Baby, Son, Daughter, Mother, Father, Uncle, Aunt, Grandma, Grandpa}

listed in increasing order of age. If x and y are members of this family, define $x + y$ to be the older of x and y, and define $x \cdot y$ to be the younger of x and y.
   (a) Find (Baby + Uncle) + Mother.
   (b) Find Father $\cdot$ (Grandpa + Aunt).
   (c) Show that the operations + and $\cdot$ that we have defined satisfy the same Commutative, Associative, and Distributive Properties as do the real numbers with regular addition and multiplication.
   (d) Show that the operations also satisfy the property

$$x + (y \cdot z) = (x + y) \cdot (x + z)$$

   Do the operations of regular addition and multiplication with real numbers also satisfy this property?

14. The ancient Egyptians, as a result of their pyramid-building, knew that the volume of a pyramid with height h and square base of side length a is $V = \frac{1}{3}ha^2$. They were able to use this fact to prove that the volume of a truncated pyramid is $V = \frac{1}{3}h(a^2 + ab + b^2)$, where h is the height and b and a are the lengths of the sides of the square top and bottom, as shown in the figure. Prove the truncated pyramid volume formula.

15. The ancient Babylonians developed the following process for finding the square root of a number N. First they made a guess at the square root—let's call this first guess r_1. Noting that

$$r_1 \cdot \left(\frac{N}{r_1}\right) = N$$

they concluded that the actual square root must be somewhere between r_1 and N/r_1, so their next guess for the square root, r_2, was the average of these two numbers:

$$r_2 = \frac{1}{2}\left(r_1 + \frac{N}{r_1}\right)$$

Continuing in this way, their next approximation was given by

$$r_3 = \frac{1}{2}\left(r_2 + \frac{N}{r_2}\right)$$

and so on. In general, once we have the nth approximation to the square root of N, we find the $(n + 1)$st using

$$r_{n+1} = \frac{1}{2}\left(r_n + \frac{N}{r_n}\right)$$

Use this procedure to find $\sqrt{72}$, correct to two decimal places.

16. This problem is adapted from a tenth century Arab manuscript. In his will, a nobleman leaves half of his horses to his oldest son, a third to his second son, and a ninth to his youngest. At his death he has 17 horses, so none of these bequests results in a whole number of horses for any of the sons. The executor of the will solves this dilemma by adding one of his own horses to the estate, making a total of 18. Now, according to the provisions of the will, he gives 9 to the oldest son, 6 to the second, and 2 to the third. Thus, each son inherits a little more than he was originally entitled to, and one horse is left over, which the executor takes back again. Everybody is happy, but something seems wrong. What, in fact, *is* wrong?

17. A piece of wire is bent as shown in the figure. You can see that one cut through the wire produces four pieces and two parallel cuts produce seven pieces. How many pieces will be produced by 142 parallel cuts? Write a formula for the number of pieces produced by *n* parallel cuts.

18. Find the last digit in the number 3^{459}. [*Hint:* Calculate the first few powers of 3, and look for a pattern.]

19. How many digits does the number $8^{15} \cdot 5^{37}$ have? [*Hint:* Express the number as a multiple of a power of 10.]

20. Use the techniques of solving a simpler problem and looking for a pattern to evaluate the number

$$3999999999999^2$$

21. Suppose $x + y = 1$ and $x^2 + y^2 = 4$. Find $x^3 + y^3$.

22. A red ribbon is tied tightly around the earth at the equator. How much more ribbon would you need if you raised the ribbon 1 ft above the equator everywhere? (You don't need to know the radius of the earth to solve this problem.)

23. Prove that it's possible to raise an irrational number to an irrational power and get a rational result. [*Hint:* The number $a = \sqrt{2}^{\sqrt{2}}$ is either rational or irrational. If a is rational, we are done. If a is irrational, consider $a^{\sqrt{2}}$.]

24. Suppose that each point in the coordinate plane is colored either red or blue. Show that there must always be two points of the same color that are exactly one unit apart.

25. Suppose that each point (x, y) in the plane, both of whose coordinates are rational numbers, represents a tree. If you are standing at the point $(0, 0)$, how far could you see in this forest?

26. A thousand points are graphed in the coordinate plane. Explain why it is possible to draw a straight line in the plane so that half of the points are on one side of the line and half are on the other. [*Hint:* Consider the slopes of the lines determined by each *pair* of points.]

27. Sketch the region in the plane consisting of all points (x, y) such that

$$|x| + |y| \leq 1$$

28. Graph the equation

$$x^2y - y^3 - 5x^2 + 5y^2 = 0$$

   [*Hint:* Factor.]

2 Functions

A function is a rule that describes how one quantity depends on another. For example, the distance traveled by a moving object is a function of time; the population of a species is a function of the available habitat.

That flower of modern mathematical thought—the notion of a function.

THOMAS J. McCORMACK

One of the most basic and important ideas in all of mathematics is the concept of a *function*. In this chapter we study functions, their graphs, and some of their applications.

2.1 WHAT IS A FUNCTION?

In this section we explore the idea of a function and then give the mathematical definition of function.

Functions All Around Us

In nearly every physical phenomenon, we observe that one quantity depends on another. For example, your height depends on your age, the temperature depends on the date, the cost of mailing a package depends on its weight (see Figure 1). We use the term *function* to describe this dependence of one quantity on another. That is, we say the following:

- Height is a function of age
- Temperature is a function of date
- Cost of mailing a package is a function of weight

The U.S. Post Office uses a simple rule to determine the cost of mailing a package based on its weight. But it's not so easy to describe the rule that relates height to age or temperature to date.

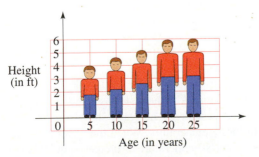

Height is a function of age.

Temperature is a function of date.

Daily high temperature
Columbia, MO, April 1995

w (ounces)	Postage (dollars)
$0 < w \leq 1$	0.34
$1 < w \leq 2$	0.57
$2 < w \leq 3$	0.80
$3 < w \leq 4$	1.03
$4 < w \leq 5$	1.26
$5 < w \leq 6$	1.49

Postage is a function of weight.

FIGURE 1

Can you think of other functions? Here are some more examples:

- The area of a circle is a function of its radius
- The number of bacteria in a culture is a function of time
- The weight of an astronaut is a function of her elevation
- The price of a commodity is a function of the demand for that commodity

The area A of a circle depends on its radius r. The rule that describes this dependence is given by the formula $A = \pi r^2$. The number N of bacteria in a culture depends on the time t (see Table 1). The rule that connects N and t in this case is given by the formula $N = 50 \cdot 2^t$. The weight w of an astronaut depends on her elevation h. Physicists use the rule $w = w_0 R^2 / (R + h)^2$, where R is the radius of the earth. The price p of a commodity (wheat, for instance) depends on the demand y for that commodity (see Example 12, Section 1.10).

t (min)	N (bacteria)
0	50
1	100
2	200
3	400
4	800
5	1600
6	3200

TABLE 1

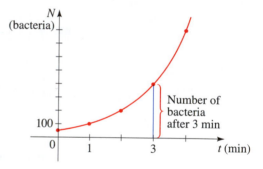

FIGURE 2

We can visualize a function by sketching its *graph*. For instance, to see how the number N of bacteria in a culture depends on the time t, we graph the data given in Table 1. The height of the graph above the horizontal axis represents the number of bacteria at a given time t—the higher the graph, the more bacteria. The shape of the graph in Figure 2 indicates how quickly the bacteria population grows.

Even when a precise rule or formula describing a function is not available, we can still describe the function by a graph. For example, when you turn on a hot water faucet, the temperature of the water depends on how long the water has been running. So we can say

- Temperature of water from the faucet is a function of time

Figure 3 shows a rough graph of the temperature T of the water as a function of the time t that has elapsed since the faucet was turned on. The graph shows that the initial temperature of the water is close to room temperature. When the water from the hot water tank reaches the faucet, the water's temperature T increases quickly. In the next phase, T is constant at the temperature of the water in the tank. When the tank is drained, T decreases to the temperature of the cold water supply.

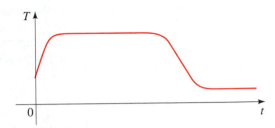

FIGURE 3

Graph of water temperature T
as a function of time t

In the *Focus on Modeling*, pages
246–255, we learn how to find functions that model real-life data.

You can see why scientists routinely use functions in their work. For instance, a biologist observes that the number of bacteria in a culture increases with time. To predict the size of the bacteria culture in the future, the biologist tries to find the rule or function that relates the number of bacteria to the time. A physicist observes that the weight of an astronaut depends on her elevation. The physicist then tries to discover the rule or function that relates these quantities. An economist observes that the price of a certain commodity depends on the demand. The economist then tries to find the rule or function that relates these quantities. The process of finding a function to describe a real-world phenomenon is called *modeling*.

▪ Definition of Function

Before we give the formal definition of a function, let's examine one more example. In the following list the numbers on the right are related to those on the left.

$$1 \rightarrow 1$$
$$2 \rightarrow 4$$
$$3 \rightarrow 9$$
$$4 \rightarrow 16$$

Can you discover the rule that relates these numbers? The rule is "square the number." Thus, if x is any number, we have

$$x \rightarrow x^2$$

In order to talk about this rule, we need to give it a name. Let's call it f. So, in this example, f is the rule "square the number." When we apply the rule f to the number x, we get x^2. We express this mathematically as

$$f(x) = x^2$$

When we write $f(2)$, we mean "apply the rule f to the number 2." Applying the rule gives $f(2) = 2^2 = 4$. Similarly, $f(3) = 3^2 = 9$ and $f(4) = 4^2 = 16$.

The following definition concisely captures these ideas.

DEFINITION OF FUNCTION

A **function** f is a rule that assigns to each element x in a set A exactly one element, called $f(x)$, in a set B.

We have previously used letters to stand for numbers. Here we do something quite different. We use letters to represent *rules*. If the letter *f* represents a function, then the notation $f(x)$ means "apply the rule *f* to the number *x*."

We usually consider functions for which the sets *A* and *B* are sets of real numbers. The symbol $f(x)$ is read "*f* of *x*" or "*f* at *x*" and is called the **value of *f* at *x*,** or the **image of *x* under *f*.** The set *A* is called the **domain** of the function. The **range** of *f* is the set of all possible values of $f(x)$ as *x* varies throughout the domain, that is,

$$\{f(x) \mid x \in A\}$$

The symbol that represents an arbitrary number in the domain of a function *f* is called an **independent variable**. The symbol that represents a number in the range of *f* is called a **dependent variable**. For instance, in the bacteria example, *t* is the independent variable and *N* is the dependent variable.

The $\sqrt{}$ key on your calculator is a good example of a function as a machine. First you input *x* into the display. Then you press the key labeled $\sqrt{}$. If $x < 0$, then *x* is not in the domain of this function; that is, *x* is not an acceptable input and the calculator will indicate an error. If $x \geq 0$, then an approximation to $\sqrt{x}$ appears in the display, correct to a certain number of decimal places. [Thus, the $\sqrt{}$ key on your calculator is not quite the same as the exact mathematical function *f* defined by $f(x) = \sqrt{x}$.]

It's helpful to think of a function as a **machine** (see Figure 4). If *x* is in the domain of the function *f*, then when *x* enters the machine, it is accepted as an input and the machine produces an output $f(x)$ according to the rule of the function. Thus, we can think of the domain as the set of all possible inputs and the range as the set of all possible outputs.

FIGURE 4   Machine diagram of *f*

Another way to picture a function is by an **arrow diagram** as in Figure 5. Each arrow connects an element of *A* to an element of *B*. The arrow indicates that $f(x)$ is associated with *x*, $f(a)$ is associated with *a*, and so on.

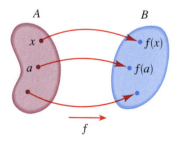

FIGURE 5

Arrow diagram of *f*

EXAMPLE 1 ■ The Squaring Function

The squaring function assigns to each real number *x* its square x^2. It is defined by

$$f(x) = x^2$$

(a) Evaluate $f(3)$, $f(-2)$, and $f(\sqrt{5})$.
(b) Find the domain and range of *f*.
(c) Draw a machine diagram for *f*.

SOLUTION

(a) The values of *f* are found by substituting for *x* in $f(x) = x^2$.

$$f(3) = 3^2 = 9 \qquad f(-2) = (-2)^2 = 4 \qquad f(\sqrt{5}) = (\sqrt{5})^2 = 5$$

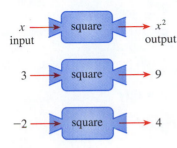

FIGURE 6

Machine diagram

(b) The domain of f is the set $\mathbb{R}$ of all real numbers. The range of f consists of all values of $f(x)$, that is, all numbers of the form x^2. Since $x^2 \geqslant 0$ for all real numbers x, we can see that the range of f is $\{y \mid y \geqslant 0\} = [0, \infty)$.

(c) A machine diagram for this function is shown in Figure 6.   ◼

Evaluating a Function

In the definition of a function the independent variable x plays the role of a "place-holder." For example, the function $f(x) = 3x^2 + x - 5$ can be thought of as

$$f(\ \blacksquare\ ) = 3 \cdot \blacksquare^2 + \blacksquare - 5$$

To evaluate f at a number, we substitute the number for the placeholder.

EXAMPLE 2 ◼ Evaluating a Function

Let $f(x) = 3x^2 + x - 5$. Evaluate each function value.

(a) $f(-2)$ (b) $f(0)$ (c) $f(4)$ (d) $f\left(\frac{1}{2}\right)$

SOLUTION

To evaluate f at a number, we substitute the number for x in the definition of f.

(a) $f(-2) = 3 \cdot (-2)^2 + (-2) - 5 = 5$

(b) $f(0) = 3 \cdot 0^2 + 0 - 5 = -5$

(c) $f(4) = 3 \cdot 4^2 + 4 - 5 = 47$

(d) $f\left(\frac{1}{2}\right) = 3 \cdot \left(\frac{1}{2}\right)^2 + \frac{1}{2} - 5 = -\frac{15}{4}$ ◼

EXAMPLE 3 ◼ A Piecewise Defined Function

Evaluate the following function at $x = -2, 1, 2,$ and 3.

$$f(x) = \begin{cases} 1 - x & \text{if } x \leqslant 1 \\ x^2 & \text{if } x > 1 \end{cases}$$

SOLUTION

Remember that a function is a rule. Here is how we apply the rule for this particular function: First look at the value of the input x. If it happens that $x \leqslant 1$, then the value of $f(x)$ is $1 - x$. On the other hand, if $x > 1$, then the value of $f(x)$ is x^2.

Since $-2 \leqslant 1$, we have $f(-2) = 1 - (-2) = 3$.

Since $1 \leqslant 1$, we have $f(1) = 1 - 1 = 0$.

Since $2 > 1$, we have $f(2) = 2^2 = 4$.

Since $3 > 1$, we have $f(3) = 3^2 = 9$.   ◼

Expressions like the one in part (d) of Example 4 occur frequently in calculus; they are called *difference quotients*, and they represent the average change in the value of f between $x = a$ and $x = a + h$.

EXAMPLE 4 ■ Evaluating a Function

If $f(x) = 2x^2 + 3x - 1$, evaluate the following.

(a) $f(a)$ (b) $f(-a)$

(c) $f(a + h)$ (d) $\dfrac{f(a + h) - f(a)}{h}, \quad h \neq 0$

SOLUTION

(a) $f(a) = 2a^2 + 3a - 1$

(b) $f(-a) = 2(-a)^2 + 3(-a) - 1 = 2a^2 - 3a - 1$

(c) $f(a + h) = 2(a + h)^2 + 3(a + h) - 1$

$$= 2(a^2 + 2ah + h^2) + 3(a + h) - 1$$

$$= 2a^2 + 4ah + 2h^2 + 3a + 3h - 1$$

(d) Using the results from parts (c) and (a), we have

$$\frac{f(a + h) - f(a)}{h} = \frac{(2a^2 + 4ah + 2h^2 + 3a + 3h - 1) - (2a^2 + 3a - 1)}{h}$$

$$= \frac{4ah + 2h^2 + 3h}{h} = 4a + 2h + 3$$

■

The weight of an object on or near the earth is the gravitational force that the earth exerts on it. When in orbit around the earth, an astronaut experiences the sensation of "weightlessness" because the centrifugal force that keeps her in orbit is exactly the same as the gravitational pull of the earth.

EXAMPLE 5 ■ The Weight of an Astronaut

If an astronaut weighs 130 pounds on the surface of the earth, then her weight when she is h miles above the earth is given by the function

$$w(h) = 130\left(\frac{3960}{3960 + h}\right)^2$$

(a) What is her weight when she is 100 mi above the earth?

(b) Construct a table of values for the function w that gives her weight at heights from 0 to 500 mi. What do you conclude from the table?

SOLUTION

(a) We want the value of the function w when $h = 100$; that is, we must calculate $w(100)$.

$$w(100) = 130\left(\frac{3960}{3960 + 100}\right)^2 \approx 123.67$$

So at a height of 100 mi, she weighs about 124 lb.

(b) The table gives the astronaut's weight, rounded to the nearest pound, at 100-mile increments. The values in the table are calculated as in part (a).

h	$w(h)$
0	130
100	124
200	118
300	112
400	107
500	102

The table indicates that the higher the astronaut travels, the less she weighs. ■

▪ The Domain of a Function

The domain of a function may be stated explicitly. For example, if we write

$$f(x) = x^2, \qquad 0 \le x \le 5$$

then the domain is the set of all real numbers x for which $0 \le x \le 5$. If the function is given by an algebraic expression and the domain is not stated explicitly, then by convention *the domain is the set of all real numbers for which the expression is defined as a real number.* For example, the function

$$f(x) = \frac{1}{x - 4}$$

is not defined at $x = 4$, so its domain is $\{x \mid x \ne 4\}$. The function

$$f(x) = \sqrt{x}$$

is not defined for negative x, so its domain is $\{x \mid x \ge 0\}$.

EXAMPLE 6 ■ Finding Domains of Functions

Find the domain of each function.

(a) $f(x) = \dfrac{1}{x^2 - x}$ (b) $g(x) = \sqrt{9 - x^2}$ (c) $h(t) = \dfrac{t}{\sqrt{t + 1}}$

SOLUTION

(a) The function is not defined when the denominator is 0. Since

$$f(x) = \frac{1}{x^2 - x} = \frac{1}{x(x - 1)}$$

we see that $f(x)$ is not defined when $x = 0$ or $x = 1$. Thus, the domain of f is

$$\{x \mid x \neq 0, x \neq 1\}$$

The domain may also be written in interval notation as

$$(\infty, 0) \cup (0, 1) \cup (1, \infty)$$

(b) We can't take the square root of a negative number, so we must have $9 - x^2 \geq 0$. Using the methods of Section 1.7, we can solve this inequality to find that $-3 \leq x \leq 3$. Thus, the domain of g is

$$\{x \mid -3 \leq x \leq 3\} = [-3, 3]$$

(c) We can't take the square root of a negative number, and we can't divide by 0, so we must have $t + 1 > 0$, that is, $t > -1$. So the domain of h is

$$\{t \mid t > -1\} = (-1, \infty)\qquad\blacksquare$$

Four Ways to Represent a Function

To help us understand what a function is, we have used machine and arrow diagrams. To describe a specific function, we can use the following four ways:

- verbally (by a description in words)
- algebraically (by an explicit formula)
- visually (by a graph)
- numerically (by a table of values)

   A single function may be represented in all four ways, and it is often useful to go from one representation to another to gain insight into the function. However, certain functions are described more naturally by one method than by the other. An example of a verbal description is

$$P(t) \qquad \text{is} \qquad \text{"the population of the world at time } t\text{"}$$

The function P can also be described numerically by giving a table of values (see Table 1 on page 394). A useful representation of the area of a circle as a function of its radius is the algebraic formula

$$A(r) = \pi r^2$$

The graph produced by a seismograph (see the box) is a visual representation of the vertical acceleration function $a(t)$ of the ground during an earthquake. As a final example, consider the function $C(w)$, which is described verbally as "the cost of mailing a first-class letter with weight w." The most convenient way of describing this function is numerically—that is, using a table of values.

We will be using all four representations of functions throughout this book. We summarize them in the following box.

FOUR WAYS TO REPRESENT A FUNCTION

Verbal

Using words:

  $P(t)$ is "the population of the world at time t"

Relation of population P and time t

Algebraic

Using a formula:

$$A(r) = \pi r^2$$

Area of a circle

Visual

Using a graph:

Source: Calif. Dept. of Mines and Geology

Vertical acceleration during an earthquake

Numerical

Using a table of values:

w (ounces)	$C(w)$ (dollars)
$0 < w \leqslant 1$	0.34
$1 < w \leqslant 2$	0.57
$2 < w \leqslant 3$	0.80
$3 < w \leqslant 4$	1.03
$4 < w \leqslant 5$	1.26
$\vdots$	$\vdots$

Cost of mailing a first-class letter

2.1 EXERCISES

1–4 ■ Express the rule in function notation. [For example, the rule "square, then subtract 5" is expressed as the function $f(x) = x^2 - 5$.]

1. Multiply by 7, then add 2

2. Add 14, then divide by 7

3. Subtract 4, then square

4. Square, add 9, then take the square root

5–8 ■ Express the function (or rule) in words.

5. $f(x) = \dfrac{x}{2} + 7$

6. $g(x) = \dfrac{x + 7}{2}$

7. $h(x) = 3x^2 - 2$

8. $k(x) = \sqrt{3x - 2}$

9–10 ■ Draw a machine diagram for the function.

9. $f(x) = \sqrt{x}$

10. $f(x) = \dfrac{2}{x}$

11–12 ■ Complete the table.

11. $f(x) = 2x^2 + 1$ **12.** $g(x) = |2x - 3|$

x	$f(x)$
-1	
0	
1	
2	
3	

x	$g(x)$
-2	
0	
1	
3	
5	

13–20 ■ Evaluate the function at the indicated values.

13. $f(x) = 2x + 1$;

   $f(1)$, $f(-2)$, $f\left(\frac{1}{2}\right)$, $f(a)$, $f(-a)$, $f(a + b)$

14. $f(x) = x^2 + 2x$;

   $f(0)$, $f(3)$, $f(-3)$, $f(a)$, $f(-x)$, $f\left(\dfrac{1}{a}\right)$

15. $g(x) = \dfrac{1 - x}{1 + x}$;

   $g(2)$, $g(-2)$, $g\left(\frac{1}{2}\right)$, $g(a)$, $g(a - 1)$, $g(-1)$

16. $h(t) = t + \dfrac{1}{t}$;

   $h(1)$, $h(-1)$, $h(2)$, $h\left(\frac{1}{2}\right)$, $h(x)$, $h\left(\dfrac{1}{x}\right)$

17. $f(x) = 2x^2 + 3x - 4$;

   $f(0)$, $f(2)$, $f(-2)$, $f(\sqrt{2})$, $f(x + 1)$, $f(-x)$

18. $f(x) = x^3 - 4x^2$;

   $f(0)$, $f(1)$, $f(-1)$, $f\left(\frac{3}{2}\right)$, $f\left(\dfrac{x}{2}\right)$, $f(x^2)$

19. $f(x) = 2|x - 1|$;

   $f(-2)$, $f(0)$, $f\left(\frac{1}{2}\right)$, $f(2)$, $f(x + 1)$, $f(x^2 + 2)$

20. $f(x) = \dfrac{|x|}{x}$;

   $f(-2)$, $f(-1)$, $f(0)$, $f(5)$, $f(x^2)$, $f\left(\dfrac{1}{x}\right)$

21–24 ■ Evaluate the piecewise defined function at the indicated values.

21. $f(x) = \begin{cases} x^2 & \text{if } x < 0 \\ x + 1 & \text{if } x \geq 0 \end{cases}$

   $f(-2)$, $f(-1)$, $f(0)$, $f(1)$, $f(2)$

22. $f(x) = \begin{cases} 5 & \text{if } x \leq 2 \\ 2x - 3 & \text{if } x > 2 \end{cases}$

   $f(-3)$, $f(0)$, $f(2)$, $f(3)$, $f(5)$

23. $f(x) = \begin{cases} x^2 + 2x & \text{if } x \leq -1 \\ x & \text{if } x > -1 \end{cases}$

   $f(-4)$, $f\left(-\frac{3}{2}\right)$, $f(-1)$, $f(0)$, $f(1)$

24. $f(x) = \begin{cases} 3x & \text{if } x < 0 \\ x + 1 & \text{if } 0 \leq x \leq 2 \\ (x - 2)^2 & \text{if } x > 2 \end{cases}$

   $f(-5)$, $f(0)$, $f(1)$, $f(2)$, $f(5)$

25–28 ■ Use the function to evaluate the indicated expressions and simplify.

25. $f(x) = x^2 + 1$; $f(x + 2)$, $f(x) + f(2)$

26. $f(x) = 3x - 1$; $f(2x)$, $2f(x)$

27. $f(x) = x + 4$; $f(x^2)$, $(f(x))^2$

28. $f(x) = 6x - 18$; $f\left(\dfrac{x}{3}\right)$, $\dfrac{f(x)}{3}$

29–34 ■ Find $f(a)$, $f(a + h)$, and $\dfrac{f(a + h) - f(a)}{h}$, where $h \neq 0$.

29. $f(x) = 3x + 2$ **30.** $f(x) = x^2 + 1$

31. $f(x) = 5$ **32.** $f(x) = \dfrac{1}{x + 1}$

33. $f(x) = 3 - 5x + 4x^2$ **34.** $f(x) = x^3$

35. The cost C in dollars of producing x yards of a certain fabric is given by the function

$$C(x) = 1500 + 3x + 0.02x^2 + 0.0001x^3$$

   (a) Find $C(10)$ and $C(100)$.
   (b) What do your answers in part (a) represent?
   (c) Find $C(0)$. (This number represents the *fixed costs*.)

36. The surface area S of a sphere is a function of its radius r given by

$$S(r) = 4\pi r^2$$

   (a) Find $S(2)$ and $S(3)$.
   (b) What do your answers in part (a) represent?

37. Due to the curvature of the earth, the maximum distance D that you can see from the top of a tall building or from an airplane at height h is given by the function

$$D(h) = \sqrt{2rh + h^2}$$

where $r = 3960$ mi is the radius of the earth and D and h are measured in miles.
(a) Find $D(0.1)$ and $D(0.2)$.
(b) How far can you see from the observation deck of Toronto's CN Tower, 1135 ft above the ground?
(c) Commercial aircraft fly at an altitude of about 7 mi. How far can the pilot see?

38. A tank holds 50 gallons of water, which drains from a leak at the bottom, causing the tank to empty in 20 minutes. The tank drains faster when it is nearly full because the pressure on the leak is greater. **Torricelli's Law** gives the volume of water remaining in the tank after t minutes as

$$V(t) = 50\left(1 - \frac{t}{20}\right)^2 \qquad 0 \le t \le 20$$

(a) Find $V(0)$ and $V(20)$.
(b) What do your answers to part (a) represent?
(c) Make a table of values for $V(t)$ for $t = 0, 5, 10, 15, 20$.

39. As blood moves through a vein or an artery, its velocity v is greatest along the central axis and decreases as the distance r from the central axis increases (see the figure). The formula that gives v as a function of r is called the **law of laminar flow.** For an artery with a radius 0.5 cm, we have

$$v(r) = 18{,}500(0.25 - r^2) \qquad 0 \le r \le 0.5$$

(a) Find $v(0.1)$ and $v(0.4)$.
(b) What do your answers to part (a) tell you about the flow of blood in this artery?

(c) Make a table of values of $v(r)$ for $r = 0, 0.1, 0.2, 0.3, 0.4, 0.5$.

40. When the brightness x of a light source is increased, the eye reacts by decreasing the radius R of the pupil. The dependence of R on x is given by the function

$$R(x) = \sqrt{\frac{13 + 7x^{0.4}}{1 + 4x^{0.4}}}$$

(a) Find $R(1)$, $R(10)$, and $R(100)$.
(b) Make a table of values of $R(x)$.

41. In a certain country, income tax is assessed according to the following function:

$$T(x) = \begin{cases} 0 & \text{if } 0 \le x \le 10{,}000 \\ 0.08x & \text{if } 10{,}000 < x \le 20{,}000 \\ 1600 + 0.15x & \text{if } 20{,}000 < x \end{cases}$$

(a) Find $T(5{,}000)$, $T(12{,}000)$, and $T(25{,}000)$.
(b) What do your answers in part (a) represent?

42. According to the theory of relativity, the length L of an object is a function of its velocity v with respect to an observer. For an object whose length at rest is 10 m, the function is given by

$$L(v) = 10\sqrt{1 - \frac{v^2}{c^2}}$$

where c is the speed of light.
(a) Find $L(0.5c)$, $L(0.75c)$, and $L(0.9c)$.
(b) How does the length of an object change as its velocity increases?

43–60 ■ Find the domain of the function.

43. $f(x) = 2x$

44. $f(x) = x^2 + 1$

45. $f(x) = 2x, \quad -1 \leqslant x \leqslant 5$

46. $f(x) = x^2 + 1, \quad 0 \leqslant x \leqslant 5$

47. $f(x) = \dfrac{1}{x - 3}$

48. $f(x) = \dfrac{1}{3x - 6}$

49. $f(x) = \dfrac{x + 2}{x^2 - 1}$

50. $f(x) = \dfrac{x^4}{x^2 + x - 6}$

51. $f(x) = \sqrt{x - 5}$

52. $f(x) = \sqrt[4]{x + 9}$

53. $f(t) = \sqrt[3]{t - 1}$

54. $g(x) = \sqrt{7 - 3x}$

55. $h(x) = \sqrt{2x - 5}$

56. $G(x) = \sqrt{x^2 - 9}$

57. $g(x) = \dfrac{\sqrt{2 + x}}{3 - x}$

58. $g(x) = \dfrac{\sqrt{x}}{2x^2 + x - 1}$

59. $g(x) = \sqrt[4]{x^2 - 6x}$

60. $g(x) = \sqrt{x^2 - 2x - 8}$

61. The graph gives the weight of a certain person as a function of age. Describe in words how this person's weight has varied over time. What do you think happened when this person was 30 years old?

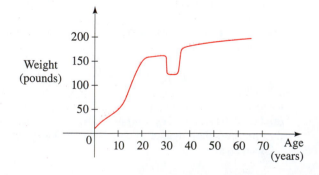

62. The graph gives a salesman's distance from his home as a function of time on a certain day. Describe in words what the graph indicates about his travels on this day.

63. You put some ice cubes in a glass, fill the glass with cold water, and then let the glass sit on a table. Sketch a rough graph of the temperature of the water as a function of the elapsed time.

64. A home owner mows the lawn every Wednesday afternoon. Sketch a rough graph of the height of the grass as a function of time over the course of a four-week period beginning on a Sunday.

65. Three runners compete in a 100-meter hurdle race. The graph depicts the distance run as a function of time for each runner. Describe in words what the graph tells you about this race. Who won the race? Did each runner finish the race? What do you think happened to runner B?

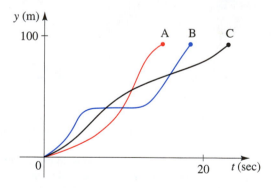

66. Sketch a rough graph of the number of hours of daylight as a function of the time of year in the Northern Hemisphere.

67. The number of Christmas cards sold by a greeting-card store depends on the time of year. Sketch a rough graph of the number of Christmas cards sold as a function of the time of year.

68. You place a frozen pie in an oven and bake it for an hour. Then you take it out and let it cool before eating it. Sketch a rough graph of the temperature of the pie as a function of time.

69. The figure shows the power consumption in San Francisco for September 19, 1996 (P is measured in megawatts; t is measured in hours starting at midnight).
   (a) What was the power consumption at 6 A.M.? At 6 P.M.?
   (b) When was the power consumption the lowest?
   (c) When was the power consumption the highest?

Source: Pacific Gas & Electric

70. The graph shows the vertical acceleration of the ground from the 1994 Northridge earthquake in Los Angeles, as measured by a seismograph. (Here t represents the time in seconds.)
   (a) At what time t did the earthquake first make noticeable movements of the earth?
   (b) At what time t did the earthquake seem to end?
   (c) At what time t was the maximum intensity of the earthquake reached?

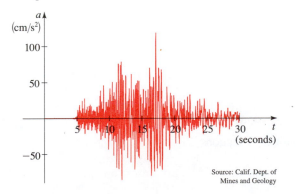

Source: Calif. Dept. of Mines and Geology

71. Temperature readings T (in °F) were recorded every 2 hours from midnight to noon in Atlanta, Georgia, on March 18, 1996. The time t was measured in hours from midnight. Sketch a rough graph of T as a function of t.

t	T
0	58
2	57
4	53
6	50
8	51
10	57
12	61

72. The population P (in thousands) of San Jose, California, from 1984 to 1994 is shown in the table. (Midyear estimates are given.) Draw a rough graph of P as a function of time t.

t	P
1984	695
1986	716
1988	733
1990	782
1992	800
1994	817

 DISCOVERY · DISCUSSION

73. Examples of Functions   At the beginning of this section we discussed three examples of everyday, ordinary functions: Height is a function of age, temperature is a function of date, and postage cost is a function of weight. Give three other examples of functions from everyday life.

74. Four Ways to Represent a Function   In the table on page 151 we represented four different functions verbally, algebraically, visually, and numerically. Think of a function that can be represented all four ways, and write the four representations.

2.2   GRAPHS OF FUNCTIONS

The most important way to visualize a function is through its graph. In this section we investigate in more detail the concept of graphing functions.

▮ Graphing Functions

> **THE GRAPH OF A FUNCTION**
>
> If f is a function with domain A, then the **graph** of f is the set of ordered pairs
>
> $$\{(x, f(x)) \mid x \in A\}$$
>
> In other words, the graph of f is the set of all points (x, y) such that $y = f(x)$; that is, the graph of f is the graph of the equation $y = f(x)$.

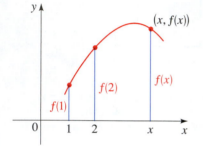

FIGURE 1

The graph of f

The height of the graph above the point x is the value of $f(x)$.

The graph of a function f gives us a useful picture of the behavior or "life history" of a function. Since the y-coordinate of any point (x, y) on the graph is $y = f(x)$, we can read the value of $f(x)$ from the graph as being the height of the graph above the point x (see Figure 1).

A function f of the form $f(x) = mx + b$ is called a **linear function** because its graph is the graph of the equation $y = mx + b$, which represents a line with slope m and y-intercept b. A special case of a linear function occurs when the slope is $m = 0$. The function $f(x) = b$, where b is a given number, is called a **constant function** because all its values are the same number, namely, b. Its graph is the horizontal line $y = b$. Figure 2 shows the graphs of the constant function $f(x) = 3$ and the linear function $f(x) = 2x + 1$.

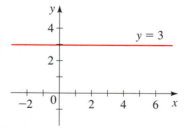

FIGURE 2

The constant function $f(x) = 3$ The linear function $f(x) = 2x + 1$

EXAMPLE 1 ▪ Graphing Functions

Sketch the graphs of the following functions.

(a) $f(x) = x^2$ (b) $g(x) = x^3$ (c) $h(x) = \sqrt{x}$

SOLUTION

We first make a table of values. Then we plot the points given by the table and join them by a smooth curve to obtain the graph. The graphs are sketched in Figure 3.

x	$f(x) = x^2$
0	0
$\pm\frac{1}{2}$	$\frac{1}{4}$
± 1	1
± 2	4
± 3	9

x	$g(x) = x^3$
0	0
$\frac{1}{2}$	$\frac{1}{8}$
1	1
2	8
$-\frac{1}{2}$	$-\frac{1}{8}$
-1	-1
-2	-8

x	$h(x) = \sqrt{x}$
0	0
1	1
2	$\sqrt{2}$
3	$\sqrt{3}$
4	2
5	$\sqrt{5}$

(a) $f(x) = x^2$

(b) $g(x) = x^3$

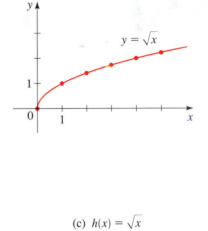

(c) $h(x) = \sqrt{x}$

FIGURE 3

EXAMPLE 2 ■ Graph of a Piecewise Defined Function

Sketch the graph of the function

$$f(x) = \begin{cases} x^2 & \text{if } x \leq 1 \\ 2x + 1 & \text{if } x > 1 \end{cases}$$

SOLUTION

If $x \leq 1$, then $f(x) = x^2$, so the part of the graph to the left of $x = 1$ coincides with the graph of $y = x^2$, which we sketched in Figure 3. If $x > 1$, then $f(x) = 2x + 1$,

On many graphing calculators the graph in Figure 4 can be produced by using the logical functions in the calculator. For example, on the TI-83 the following equation gives the required graph:

$$y = (x \leqslant 1)x^{\wedge}2 + (x > 1)(2x + 1)$$

so the part of the graph to the right of $x = 1$ coincides with the line $y = 2x + 1$, which we graphed in Figure 2. This enables us to sketch the graph in Figure 4.

The solid dot at $(1, 1)$ indicates that this point is included in the graph; the open dot at $(1, 3)$ indicates that this point is excluded from the graph.

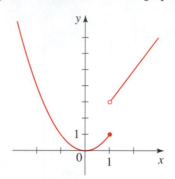

FIGURE 4

$$f(x) = \begin{cases} x^2 & \text{if } x \leqslant 1 \\ 2x + 1 & \text{if } x > 1 \end{cases}$$

■

EXAMPLE 3 ■ Graph of the Absolute Value Function

Sketch the graph of the absolute value function $f(x) = |x|$.

SOLUTION

Recall that

$$|x| = \begin{cases} x & \text{if } x \geqslant 0 \\ -x & \text{if } x < 0 \end{cases}$$

Using the same method as in Example 2, we note that the graph of f coincides with the line $y = x$ to the right of the y-axis and coincides with the line $y = -x$ to the left of the y-axis (see Figure 5).

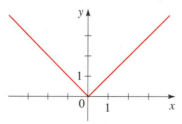

FIGURE 5

Graph of $f(x) = |x|$

■

The **greatest integer function** is defined by

$$[\![x]\!] = \text{greatest integer less than or equal to } x$$

For example, $[\![2]\!] = 2$, $[\![2.3]\!] = 2$, $[\![1.999]\!] = 1$, $[\![0.002]\!] = 0$, $[\![-3.5]\!] = -4$, $[\![-0.5]\!] = -1$.

EXAMPLE 4 ■ Graph of the Greatest Integer Function

Sketch the graph of $f(x) = [\![x]\!]$.

SOLUTION

The table shows the values of f for some values of x. Note that $f(x)$ is constant between consecutive integers so the graph between integers is a horizontal line segment as shown in Figure 6.

x	$[\![x]\!]$
$\vdots$	$\vdots$
$-2 \leqslant x < -1$	-2
$-1 \leqslant x < 0$	-1
$0 \leqslant x < 1$	0
$1 \leqslant x < 2$	1
$2 \leqslant x < 3$	2
$\vdots$	$\vdots$

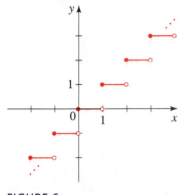

FIGURE 6

The greatest integer function, $y = [\![x]\!]$

The greatest integer function is an example of a **step function**. The next example gives a real-world example of a step function.

EXAMPLE 5 ■ The Cost Function for Long-Distance Phone Calls

The cost of a long-distance daytime phone call from Toronto to New York City is 69 cents for the first minute and 58 cents for each additional minute (or part of a minute). Draw the graph of the cost C (in dollars) of the phone call as a function of time t (in minutes).

SOLUTION

Let $C(t)$ be the cost for t minutes. Since $t > 0$, the domain of the function is $(0, \infty)$. From the given information, we have

$$C(t) = 0.69 \qquad\qquad\qquad\quad \text{if } 0 < t \leqslant 1$$

$$C(t) = 0.69 + 0.58 = 1.27 \qquad \text{if } 1 < t \leqslant 2$$

$$C(t) = 0.69 + 2(0.58) = 1.85 \quad \text{if } 2 < t \leqslant 3$$

$$C(t) = 0.69 + 3(0.58) = 2.43 \quad \text{if } 3 < t \leqslant 4$$

and so on. The graph is shown in Figure 7.

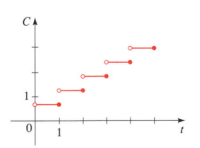

FIGURE 7

Cost of a long-distance call

To understand how the equation of a function relates to its graph, it's helpful to graph a **family of functions**, that is, a collection of functions whose equations are related. Usually, the equations differ only in a single constant called a *parameter*.

In the next example, we graph a family of **power functions** $f(x) = x^n$, when n is a positive integer. In this case, n is the parameter.

EXAMPLE 6 ■ A Family of Power Functions

(a) Graph the functions $f(x) = x^n$ for $n = 2, 4,$ and 6 in the viewing rectangle $[-2, 2]$ by $[-1, 3]$.
(b) Graph the functions $f(x) = x^n$ for $n = 1, 3,$ and 5 in the viewing rectangle $[-2, 2]$ by $[-2, 2]$.
(c) What conclusions can you draw from these graphs?

SOLUTION

The graphs for parts (a) and (b) are shown in Figure 8.

(c) We see that the general shape of the graph of $f(x) = x^n$ depends on whether n is even or odd.

   If n is even, the graph of $f(x) = x^n$ is similar to the parabola $y = x^2$.
   If n is odd, the graph of $f(x) = x^n$ is similar to that of $y = x^3$.   ■

Notice from part (c) that as n increases the graph of $y = x^n$ becomes flatter near 0 and steeper when $x > 1$. When $0 < x < 1$, the lower powers of x are the "bigger" functions. But when $x > 1$, the higher powers of x are the dominant functions.

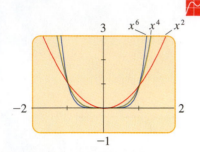

(a) Even powers of x

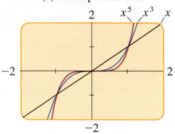

(b) Odd powers of x

FIGURE 8

A family of power functions $f(x) = x^n$

Finding Domain and Range from the Graph

The graph of a function helps us picture the domain and range of the function on the x-axis and y-axis as shown in Figure 9.

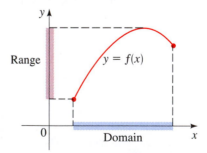

FIGURE 9

Domain and range of f

EXAMPLE 7 ■ **Finding the Domain and Range from a Graph**

(a) Sketch the graph of $f(x) = \sqrt{4 - x^2}$.

(b) Find the domain and range of f.

SOLUTION

(a) We must graph the equation $y = \sqrt{4 - x^2}$. Because we are taking the positive square root, we know that $y \geq 0$. Squaring each side of the equation, we get

$$y^2 = 4 - x^2 \qquad \text{Square both sides}$$

$$x^2 + y^2 = 4 \qquad \text{Add } x^2$$

which we recognize as the equation of a circle with center the origin and radius 2. But, since $y \geq 0$, the graph of f consists of just the upper half of this circle. See Figure 10.

(b) From the graph in Figure 10 we see that the domain is the closed interval $[-2, 2]$ and the range is $[0, 2]$. ■

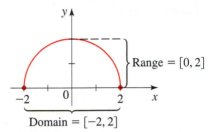

FIGURE 10

Graph of $f(x) = \sqrt{4 - x^2}$

Domain $= [-2, 2]$

Range $= [0, 2]$

The Vertical Line Test

The graph of a function is a curve in the xy-plane. But the question arises: Which curves in the xy-plane are graphs of functions? This is answered by the following test.

THE VERTICAL LINE TEST

A curve in the coordinate plane is the graph of a function if and only if no vertical line intersects the curve more than once.

We can see from Figure 11 why the Vertical Line Test is true. If each vertical line $x = a$ intersects a curve only once at (a, b), then exactly one functional value is

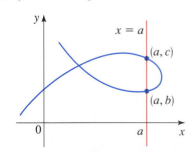

FIGURE 11

Vertical Line Test

Graph of a function

Not a graph of a function

defined by $f(a) = b$. But if a line $x = a$ intersects the curve twice, at (a, b) and at (a, c), then the curve can't represent a function because a function cannot assign two different values to a.

EXAMPLE 8 ■ Using the Vertical Line Test

Using the Vertical Line Test, we see that the curves in parts (b) and (c) of Figure 12 represent functions, whereas those in parts (a) and (d) do not.

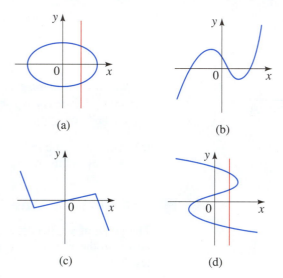

(a)

(b)

FIGURE 12

(c)

(d)

Equations That Define Functions

Any equation in the variables x and y defines a relationship between these variables. For example, the equation

$$y - x^2 = 0$$

defines a relationship between y and x. Does this equation define y as a *function* of x? To find out, we solve for y and get

$$y = x^2$$

We see that the equation defines a rule, or function, that gives one value of y for each value of x. We can express this rule in function notation as

$$f(x) = x^2$$

But not every equation defines y as a function of x, as the following example shows.

EXAMPLE 9 ■ Equations That Define Functions

Does the equation define y as a function of x?

(a) $y - x^2 = 2$ (b) $x^2 + y^2 = 4$

SOLUTION

(a) Solving for y in terms of x gives

$$y - x^2 = 2$$
$$y = x^2 + 2 \qquad \text{Add } x^2$$

The last equation is a rule that gives one value of y for each value of x, so it defines y as a function of x. We can write the function as $f(x) = x^2 + 2$.

(b) We try to solve for y in terms of x:

$$x^2 + y^2 = 4$$
$$y^2 = 4 - x^2 \qquad \text{Subtract } x^2$$
$$y = \pm\sqrt{4 - x^2} \qquad \text{Take square roots}$$

The last equation gives two values of y for a given value of x. Thus, the equation does not define y as a function of x. ■

The graphs of the equations in Example 9 are shown in Figure 13. The Vertical Line Test shows graphically that the equation in Example 9(a) defines a function but the equation in Example 9(b) does not.

FIGURE 13   (a)   (b)

Donald Knuth was born in Milwaukee in 1938 and is now Professor of Computer Science at Stanford University. While still a graduate student at Caltech, he started writing a monumental series of books entitled *The Art of Computer Programming*. President Carter awarded him the National Medal of Science in 1979. When Knuth was a high school student, he became fascinated with graphs of functions and laboriously drew many hundreds of them because he wanted to see the behavior of a great variety of functions. (Now, with progress in computers and programming, it is far easier to use computers and graphing calculators to do this.) Knuth is also famous for his invention of T_EX, a system of computer-assisted typesetting. This system was used in the preparation of the manuscript for this textbook. He has also written a novel entitled *Surreal Numbers: How Two Ex-Students Turned On to Pure Mathematics and Found Total Happiness*.

The following table shows the graphs of some functions that you will see frequently in this book.

SOME FUNCTIONS AND THEIR GRAPHS

Linear functions

$f(x) = mx + b$

$f(x) = b$

$f(x) = mx + b$

Power functions

$f(x) = x^n$

$f(x) = x^2$

$f(x) = x^3$

$f(x) = x^4$

$f(x) = x^5$

Root functions

$f(x) = \sqrt[n]{x}$

$f(x) = \sqrt{x}$

$f(x) = \sqrt[3]{x}$

$f(x) = \sqrt[4]{x}$

$f(x) = \sqrt[5]{x}$

Reciprocal functions

$f(x) = 1/x^n$

$f(x) = \dfrac{1}{x}$

$f(x) = \dfrac{1}{x^2}$

Absolute value function

$f(x) = |x|$

$f(x) = |x|$

Greatest integer function

$f(x) = [\![x]\!]$

$f(x) = [\![x]\!]$

2.2   EXERCISES

1. The graph of a function h is given.
   (a) State the values of $h(-2)$, $h(0)$, $h(2)$, and $h(3)$.
   (b) State the domain and range of h.

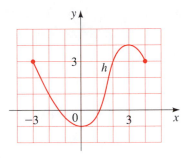

2. The graph of a function g is given.
   (a) State the values of $g(-4)$, $g(-2)$, $g(0)$, $g(2)$, and $g(4)$.
   (b) State the domain and range of g.

Relative function

3. Graphs of the functions f and g are given.
   (a) Which is larger, $f(0)$ or $g(0)$?
   (b) Which is larger, $f(-3)$ or $g(-3)$?
   (c) For which values of x is $f(x) = g(x)$?

Math relation a function

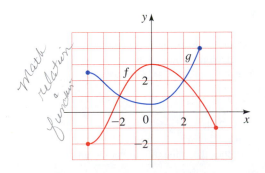

4. The graph of a function f is given.
   (a) Estimate $f(0.5)$ to the nearest tenth.
   (b) Estimate $f(3)$ to the nearest tenth.
   (c) Find all the numbers x in the domain of f so that
      $f(x) = 1$.

Relation funct but not 1 to 1

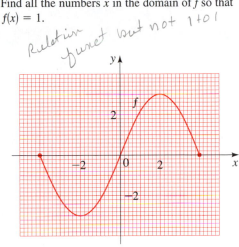

5–6 ■ Determine whether the curve is the graph of a function of x.

5. (a)

(b)

Relation not function

(c)

Relation Function & 1 to 1

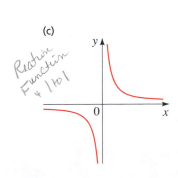

(d) *Relation not F not 1+1*

6. (a)

(b)

(c)

(d)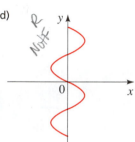

7–10 ■ Determine whether the curve is the graph of a function x. If it is, state the domain and range of the function.

7.

8.

9.

10.

11–18 ■ A function f is given.
(a) Sketch the graph of f.
(b) Find the domain and range of f from the graph, as in Example 7.

11. $f(x) = x - 1$ **12.** $f(x) = 2(x + 1)$

13. $f(x) = x^2 - 4$ **14.** $f(x) = x^2 + 4$

15. $f(x) = \sqrt{16 - x^2}$ **16.** $f(x) = -\sqrt{25 - x^2}$

17. $f(x) = \sqrt{x - 1}$ **18.** $f(x) = \sqrt{x + 2}$

19–40 ■ Sketch the graph of the function by first making a table of values, as in Example 1.

19. $f(x) = 3$ **20.** $f(x) = -5$

21. $f(x) = 2x + 3$ **22.** $f(x) = 6 - 3x$

23. $f(x) = -x + 4, \quad -1 \leqslant x \leqslant 4$

24. $f(x) = \dfrac{x + 3}{2}, \quad -2 \leqslant x \leqslant 2$

25. $f(x) = -x^2$ **26.** $f(x) = x^2 - 4$

27. $g(x) = x^3 - 8$ **28.** $g(x) = 4x^2 - x^4$

29. $g(x) = \sqrt{-x}$ **30.** $g(x) = \sqrt{x + 4}$

31. $F(x) = \dfrac{1}{x}$ **32.** $F(x) = \dfrac{1}{x + 4}$

33. $H(x) = |2x|$ **34.** $H(x) = |x + 1|$

35. $G(x) = |x| + x$ **36.** $G(x) = |x| - x$

37. $f(x) = |2x - 2|$ **38.** $f(x) = \dfrac{x}{|x|}$

39. $g(x) = \dfrac{2}{x^2}$ **40.** $g(x) = \dfrac{|x|}{x^2}$

41–52 ■ Determine whether the equation defines y as a function of x. (See Example 9.)

41. $x^2 + 2y = 4$ **42.** $3x + 7y = 21$

43. $x = y^2$ **44.** $x^2 + (y - 1)^2 = 4$

45. $x + y^2 = 9$ **46.** $x^2 + y = 9$

47. $x^2y + y = 1$ **48.** $\sqrt{x} + y = 12$

49. $2|x| + y = 0$ **50.** $2x + |y| = 0$

51. $x = y^3$ **52.** $x = y^4$

53–58 ■ A family of functions is given. In parts (a) and (b) graph all the given members of the family in the viewing rectangle indicated. In part (c) state the conclusions you can make from your graphs.

53. $f(x) = x^2 + c$
   (a) $c = 0, 2, 4, 6; \quad [-5, 5]$ by $[-10, 10]$
   (b) $c = 0, -2, -4, -6; \quad [-5, 5]$ by $[-10, 10]$
   (c) How does the value of c affect the graph?

54. $f(x) = (x - c)^2$
(a) $c = 0, 1, 2, 3$; $[-5, 5]$ by $[-10, 10]$
(b) $c = 0, -1, -2, -3$; $[-5, 5]$ by $[-10, 10]$
(c) How does the value of c affect the graph?

55. $f(x) = (x - c)^3$
(a) $c = 0, 2, 4, 6$; $[-10, 10]$ by $[-10, 10]$
(b) $c = 0, -2, -4, -6$; $[-10, 10]$ by $[-10, 10]$
(c) How does the value of c affect the graph?

56. $f(x) = cx^2$
(a) $c = 1, \frac{1}{2}, 2, 4$; $[-5, 5]$ by $[-10, 10]$
(b) $c = 1, -1, -\frac{1}{2}, -2$; $[-5, 5]$ by $[-10, 10]$
(c) How does the value of c affect the graph?

57. $f(x) = x^c$
(a) $c = \frac{1}{2}, \frac{1}{4}, \frac{1}{6}$; $[-1, 4]$ by $[-1, 3]$
(b) $c = 1, \frac{1}{3}, \frac{1}{5}$; $[-3, 3]$ by $[-2, 2]$
(c) How does the value of c affect the graph?

58. $f(x) = 1/x^n$
(a) $n = 1, 3$; $[-3, 3]$ by $[-3, 3]$
(b) $n = 2, 4$; $[-3, 3]$ by $[-3, 3]$
(c) How does the value of n affect the graph?

59–72 ■ Sketch the graph of the piecewise defined function.

59. $f(x) = \begin{cases} 0 & \text{if } x < 2 \\ 1 & \text{if } x \geq 2 \end{cases}$

60. $f(x) = \begin{cases} 1 & \text{if } x \leq 1 \\ x + 1 & \text{if } x > 1 \end{cases}$

61. $f(x) = \begin{cases} 3 & \text{if } x < 2 \\ x - 1 & \text{if } x \geq 2 \end{cases}$

62. $f(x) = \begin{cases} 1 - x & \text{if } x < -2 \\ 5 & \text{if } x \geq -2 \end{cases}$

63. $f(x) = \begin{cases} x & \text{if } x \leq 0 \\ x + 1 & \text{if } x > 0 \end{cases}$

64. $f(x) = \begin{cases} 2x + 3 & \text{if } x < -1 \\ 3 - x & \text{if } x \geq -1 \end{cases}$

65. $f(x) = \begin{cases} -1 & \text{if } x < -1 \\ 1 & \text{if } -1 \leq x \leq 1 \\ -1 & \text{if } x > 1 \end{cases}$

66. $f(x) = \begin{cases} -1 & \text{if } x < -1 \\ x & \text{if } -1 \leq x \leq 1 \\ 1 & \text{if } x > 1 \end{cases}$

67. $f(x) = \begin{cases} 2 & \text{if } x \leq -1 \\ x^2 & \text{if } x > -1 \end{cases}$

68. $f(x) = \begin{cases} 1 - x^2 & \text{if } x \leq 2 \\ x & \text{if } x > 2 \end{cases}$

69. $f(x) = \begin{cases} 0 & \text{if } |x| \leq 2 \\ 3 & \text{if } |x| > 2 \end{cases}$

70. $f(x) = \begin{cases} x^2 & \text{if } |x| \leq 1 \\ 1 & \text{if } |x| > 1 \end{cases}$

71. $f(x) = \begin{cases} 4 & \text{if } x < -2 \\ x^2 & \text{if } -2 \leq x \leq 2 \\ -x + 6 & \text{if } x > 2 \end{cases}$

72. $f(x) = \begin{cases} -x & \text{if } x \leq 0 \\ 9 - x^2 & \text{if } 0 < x \leq 3 \\ x - 3 & \text{if } x > 3 \end{cases}$

73–74 ■ Use a graphing device to draw the graph of the piecewise defined function. (See the margin note on page 158.)

73. $f(x) = \begin{cases} x + 2 & \text{if } x \leq -1 \\ x^2 & \text{if } x > -1 \end{cases}$

74. $f(x) = \begin{cases} 2x - x^2 & \text{if } x > 2 \\ (x - 1)^3 & \text{if } x \leq 2 \end{cases}$

75. A taxi company charges $2.00 for the first mile (or part of a mile) and 20 cents for each succeeding tenth of a mile (or part). Express the cost C (in dollars) of a ride as a function of the distance x traveled (in miles) for $0 < x < 2$, and sketch the graph of this function.

76. The domestic postage rate for first-class letters weighing 12 oz or less is 34 cents for a letter weighing 1 oz or less and 23 cents for each additional ounce (or part of an ounce). Express the postage P as a function of the weight x of a letter, with $0 < x \leq 12$.

77–80 ■ Find a function whose graph is the given curve.

77. The line segment joining the points $(-2, 1)$ and $(4, -6)$

78. The line segment joining the points $(-3, -2)$ and $(6, 3)$

79. The bottom half of the parabola $x + (y - 1)^2 = 0$

80. The top half of the circle $(x - 1)^2 + y^2 = 1$

DISCOVERY · DISCUSSION

81. When Does a Graph Represent a Function? For every integer n, the graph of the equation $y = x^n$ is the graph of a function, namely $f(x) = x^n$. Explain why the graph of $x = y^2$ is *not* the graph of a function of x. Is the graph of $x = y^3$ the graph of a function of x? If so, of what function of x is it the graph? Determine for what integers n the graph of $x = y^n$ is the graph of a function of x.

82. Step Functions   In Example 5 and Exercises 75 and 76 we are given functions whose graphs consist of horizontal line segments. Such functions are often called *step functions*, because their graphs look like stairs. Give some other examples of step functions that arise in everyday life.

83. Stretched Step Functions   Sketch graphs of the functions $f(x) = [\![x]\!]$, $g(x) = [\![2x]\!]$, and $h(x) = [\![3x]\!]$ on separate graphs. How are the graphs related? If n is a positive integer, what does the graph of $k(x) = [\![nx]\!]$ look like?

 84. Graph of the Absolute Value of a Function

(a) Draw the graphs of the functions $f(x) = x^2 + x - 6$ and $g(x) = |x^2 + x - 6|$. How are the graphs of f and g related?

(b) Draw the graphs of the functions $f(x) = x^4 - 6x^2$ and $g(x) = |x^4 - 6x^2|$. How are the graphs of f and g related?

(c) In general, if $g(x) = |f(x)|$, how are the graphs of f and g related? Draw graphs to illustrate your answer.

2.3 APPLIED FUNCTIONS: VARIATION

Mathematical models are discussed in more detail in *Focus on Modeling*, which begins on page 246.

When scientists or applied mathematicians talk about a *mathematical model* for a real-world phenomenon, they often mean a function that describes, at least approximately, the dependence of one physical quantity on another. For instance, the model may describe the population of an animal species as a function of time, or the pressure of a gas as a function of its volume. In this section we study a kind of modeling that occurs frequently in the sciences, called *variation*.

Direct Variation

Two types of mathematical models occur so often in various sciences that they are given special names. The first is called *direct variation* and occurs when one quantity is a constant multiple of another, so we use a function of the form $f(x) = kx$ to model this dependence.

DIRECT VARIATION

If the quantities x and y are related by an equation

$$y = kx$$

for some constant $k \neq 0$, we say that y **varies directly as** x, or y **is directly proportional to** x, or simply y **is proportional to** x. The constant k is called the **constant of proportionality**.

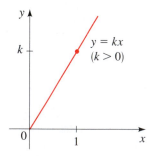

FIGURE 1
Direct variation

Recall that a linear function is a function of the form $f(x) = mx + b$ and its graph is a line with slope m and y-intercept b. So if y varies directly as x, then the equation $y = kx$ or, equivalently, $f(x) = kx$, shows that y is a linear function of x. The graph of this function is a line with slope k, the constant of proportionality (see Figure 1). The y-intercept is $b = 0$, so this line passes through the origin.

EXAMPLE 1 ■ Direct Variation

During a thunderstorm you see the lightning before you hear the thunder because light travels much faster than sound. The distance between you and the storm varies directly as the time interval between the lightning and the thunder.

(a) Suppose that the thunder from a storm 5400 ft away takes 5 s to reach you. Determine the constant of proportionality and write the equation for the variation.

(b) Sketch the graph of this equation. What does the constant of proportionality represent?

(c) If the time interval between the lightning and thunder is now 8 s, how far away is the storm?

SOLUTION

(a) Let d be the distance from you to the storm and let t be the length of the time interval. We are given that d varies directly as t, so

$$d = kt$$

where k is a constant. To find k, we use the fact that $t = 5$ when $d = 5400$. Substituting these values in the equation, we get

$$5400 = k(5) \qquad \text{Substitute}$$

$$k = \frac{5400}{5} = 1080 \qquad \text{Solve for } k$$

Substituting this value of k in the equation for d, we obtain

$$d = 1080t$$

as the equation for d as a function of t.

(b) The graph of the equation $d = 1080t$ is a line through the origin with slope 1080 and is shown in Figure 2. The constant $k = 1080$ is the approximate speed of sound (in ft/s).

(c) When $t = 8$, we have

$$d = 1080 \cdot 8 = 8640$$

So, the storm is 8640 ft ≈ 1.6 mi away. ■

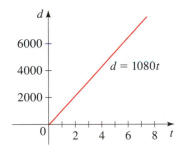

FIGURE 2

Inverse Variation

Another function that is frequently used in mathematical modeling is the function $f(x) = k/x$, where k is a constant.

FIGURE 3
Inverse variation

INVERSE VARIATION

If the quantities x and y are related by the equation

$$y = \frac{k}{x}$$

for some constant $k \neq 0$, we say that y **is inversely proportional to** x, or y **varies inversely as** x.

The graph of the function $f(x) = k/x$ for $x > 0$ is shown in Figure 3 for the case $k > 0$. It gives a picture of what happens when y is inversely proportional to x.

EXAMPLE 2 ■ Inverse Variation

Boyle's law states that when a sample of gas is compressed at a constant temperature, the pressure of the gas is inversely proportional to the volume of the gas.

(a) Suppose the pressure of a sample of air that occupies 0.106 m³ at 25°C is 50 kPa. Find the constant of proportionality, and write the equation that expresses the inverse proportionality.
(b) If the sample expands to a volume of 0.3 m³, find the new pressure.

SOLUTION

(a) Let P be the pressure of the sample of gas and let V be its volume. Then, by the definition of inverse proportionality, we have

$$P = \frac{k}{V}$$

where k is a constant. To find k we use the fact that $P = 50$ when $V = 0.106$. Substituting these values in the equation, we get

$$50 = \frac{k}{0.106} \qquad \text{Substitute}$$

$$k = (50)(0.106) = 5.3 \qquad \text{Solve for } k$$

Putting this value of k in the equation for P, we have

$$P = \frac{5.3}{V}$$

(b) When $V = 0.3$, we have

$$P = \frac{5.3}{0.3} \approx 17.7$$

So, the new pressure is about 17.7 kPa.   ■

Joint Variation

A physical quantity often depends on more than one other quantity. If one quantity is proportional to two or more other quantities, we call this relationship *joint variation*.

JOINT VARIATION

If the quantities x, y, and z are related by the equation

$$z = kxy$$

where k is a nonzero constant, we say that z **varies jointly as** x and y, or z **is jointly proportional to** x and y.

In the sciences, functions of three or more variables are common, and any combination of the different types of proportionality that we have discussed is possible. For example, if

$$z = k\frac{x}{y}$$

we say that z **is proportional to** x and **inversely proportional to** y.

EXAMPLE 3 ■ Newton's Law of Gravitation

Newton's Law of Gravitation says that two objects with masses m_1 and m_2 attract each other with a force F that is jointly proportional to their masses and inversely proportional to the square of the distance r between the objects. Express Newton's Law of Gravitation as an equation.

SOLUTION

Using the definitions of joint and inverse proportionality, and the traditional notation G for the constant of proportionality, we have

$$F = G\frac{m_1 m_2}{r^2}$$

■

If m_1 and m_2 are fixed masses, then the gravitational force between them is $F = C/r^2$ (where $C = Gm_1 m_2$ is a constant). Figure 4 shows the graph of this function for $r > 0$ with $C = 1$. Observe how the gravitational attraction decreases with increasing distance.

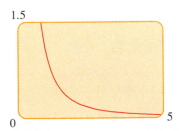

FIGURE 4

Graph of $F = \dfrac{1}{r^2}$

2.3   EXERCISES

1–12 ■ Write an equation that expresses the statement.

1. T varies directly as x.

2. P is directly proportional to w.

3. v is inversely proportional to z.

4. w is jointly proportional to m and n.

5. y is proportional to s and inversely proportional to t.

6. P varies inversely as T.

7. z is proportional to the square root of y.

8. A is proportional to the square of t and inversely proportional to the cube of x.

9. V is jointly proportional to l, w, and h.

10. S is jointly proportional to the squares of r and θ.

11. R is proportional to i and inversely proportional to P and t.

12. A is jointly proportional to the square roots of x and y.

13–22 ■ Express the statement as a formula. Use the given information to find the constant of proportionality.

13. y is directly proportional to x. If $x = 6$, then $y = 42$.

14. z varies inversely as t. If $t = 3$, then $z = 5$.

15. M varies directly as x and inversely as y. If $x = 2$ and $y = 6$, then $M = 5$.

16. S varies jointly as p and q. If $p = 4$ and $q = 5$, then $S = 180$.

17. W is inversely proportional to the square of r. If $r = 6$, then $W = 10$.

18. t is jointly proportional to x and y and inversely proportional to r. If $x = 2$, $y = 3$, and $r = 12$, then $t = 25$.

19. C is jointly proportional to l, w, and h. If $l = w = h = 2$, then $C = 128$.

20. H is jointly proportional to the squares of l and w. If $l = 2$ and $w = \frac{1}{3}$, then $H = 36$.

21. s is inversely proportional to the square root of t. If $s = 100$, then $t = 25$.

22. M is jointly proportional to a, b, and c, and inversely proportional to d. If a and d have the same value, and if b and c are both 2, then $M = 128$.

23. Hooke's law states that the force needed to keep a spring stretched x units beyond its natural length is directly proportional to x. Here the constant of proportionality is called the **spring constant**.
   (a) Write Hooke's law as an equation.
   (b) If a spring has a natural length of 10 cm and a force of 40 N is required to maintain the spring stretched to a length of 15 cm, find the spring constant.
   (c) What force is needed to keep the spring stretched to a length of 14 cm?

24. The period of a pendulum (the time elapsed during one complete swing of the pendulum) varies directly with the square root of the length of the pendulum.
   (a) Express this relationship by writing an equation.
   (b) In order to double the period, how would we have to change the length l?

25. The cost C of printing a magazine is jointly proportional to the number of pages p in the magazine and the number of magazines printed m.
   (a) Write an equation that expresses this joint variation.
   (b) Find the constant of proportionality if the printing cost is $60,000 for 4000 copies of a 120-page magazine.
   (c) How much would the printing cost be for 5000 copies of a 92-page magazine?

26. The pressure P of a sample of gas is directly proportional to the temperature T and inversely proportional to the volume V.
   (a) Write an equation that expresses this variation.
   (b) Find the constant of proportionality if 100 L of gas exerts a pressure of 33.2 kPa at a temperature of 400 K (absolute temperature measured on the Kelvin scale).
   (c) If the temperature is increased to 500 K and the volume is decreased to 80 L, what is the pressure of the gas?

27. The resistance R of a wire varies directly as its length L and inversely as the square of its diameter d.
   (a) Write an equation that expresses this joint variation.
   (b) Find the constant of proportionality if a wire 1.2 m long and 0.005 m in diameter has a resistance of 140 ohms.
   (c) Find the resistance of a wire made of the same material that is 3 m long and has a diameter of 0.008 m.

28. Kepler's Third Law of planetary motion states that the square of the period T of a planet (the time it takes for the planet to make a complete revolution about the sun) is directly proportional to the cube of its average distance d from the sun.
   (a) Express Kepler's Third Law as an equation.
   (b) Find the constant of proportionality by using the fact that for our planet the period is about 365 days and the average distance is about 93 million miles.
   (c) The planet Neptune is about 2.79×10^9 mi from the sun. Find the period of Neptune.

29. The cost of a sheet of gold foil is proportional to its area. If a rectangular sheet measuring 15 cm by 20 cm costs $75, how much would a 3 cm by 5 cm sheet cost?

30. In the short growing season of the Canadian arctic territory of Nunavut, some gardeners find it possible to grow gigantic cabbages in the midnight sun. Assume that the final size of a cabbage is proportional to the amount of nutrients it receives, and inversely proportional to the number of other cabbages surrounding it. A cabbage that received 20 oz of nutrients and had 12 other cabbages around it grew to 30 lb. What size would it grow to if it received 10 oz of nutrients and had only 5 cabbage "neighbors"?

31. The value of a building lot on Galiano Island is jointly proportional to its area and the quantity of water produced by a well on the property. A 200 ft by 300 ft lot has a well pro-

ducing 10 gallons of water per minute, and is valued at $48,000. What is the value of a 400 ft by 400 ft lot if the well on the lot produces 4 gallons of water per minute?

32. The heat experienced by a hiker at a campfire is proportional to the amount of wood on the fire, and inversely proportional to the cube of his distance from the fire. If he is 20 ft from the fire, and someone doubles the amount of wood burning, how far from the fire would he have to be so that he feels the same heat as before?

33. The frequency f of vibration of a violin string is inversely proportional to its length L. The constant of proportionality k is positive and depends on the tension and density of the string.
   (a) Write an equation that represents this variation.
   (b) What effect does doubling the length of the string have on the frequency of its vibration?

34. The rate r at which a disease spreads in a population of size P is jointly proportional to the number x of infected people and the number $P - x$ who are not infected. An infection erupts in a small town with population $P = 5000$.
   (a) Write an equation that expresses r as a function of x.
   (b) Compare the rate of spread of this infection when 10 people are infected to the rate of spread when 1000 people are infected. Which rate is larger? By what factor?
   (c) Calculate the rate of spread when the entire population is infected. Why does this answer make intuitive sense?

▲ **DISCOVERY · DISCUSSION**

35. Is Proportionality Everything?   A great many laws of physics and chemistry are expressible as proportionalities. Give at least one example of a function that occurs in the sciences that is *not* a proportionality.

2.4 AVERAGE RATE OF CHANGE: INCREASING AND DECREASING FUNCTIONS

Functions are often used to model changing quantities. In this section we learn how to determine the rate at which the values of a function change as the variable changes.

Average Rate of Change

We are all familiar with the concept of speed: If you drive a distance of 120 miles in 2 hours, then your average speed, or rate of travel, is $\frac{120 \text{ mi}}{2 \text{ h}} = 60$ mi/h.

Now suppose you take a car trip and record the distance that you travel every few minutes. The distance s you have traveled is a function of the time t:

$$s(t) = \text{total distance traveled at time } t$$

We graph the function s as shown in Figure 1. The graph shows that you have traveled a total of 50 miles after 1 hour, 75 miles after 2 hours, 140 miles after 3 hours, and so on. To find your *average* speed between any two points on the trip, we divide the distance traveled by the time elapsed. Let's calculate your average speed between 1:00 P.M. and 4:00 P.M. The time elapsed is $4 - 1 = 3$ hours. To find the distance you traveled, we subtract the distance at 1:00 P.M. from the distance at 4:00 P.M., that is, $200 - 50 = 150$ mi. Thus, your average speed is

$$\text{average speed} = \frac{\text{distance traveled}}{\text{time elapsed}} = \frac{150 \text{ mi}}{3 \text{ h}} = 50 \text{ mi/h}$$

The average speed we have just calculated can be expressed using function notation:

$$\text{average speed} = \frac{s(4) - s(1)}{4 - 1} = \frac{200 - 50}{3} = 50 \text{ mi/h}$$

Note that the average speed is different over different time intervals. For example, between 2:00 P.M. and 3:00 P.M. we find that

$$\text{average speed} = \frac{s(3) - s(2)}{3 - 2} = \frac{140 - 75}{1} = 65 \text{ mi/h}$$

FIGURE 1

Average speed

Finding average rates of change is important in many contexts. For instance, we may be interested in knowing how quickly the air temperature is dropping as a storm approaches, or how fast revenues are increasing from the sale of a new product. So we need to know how to determine the average rate of change of the functions that model these quantities. In fact, the concept of average rate of change can be defined for any function.

AVERAGE RATE OF CHANGE

The **average rate of change** of the function $y = f(x)$ between $x = a$ and $x = b$ is

$$\text{average rate of change} = \frac{\text{change in } y}{\text{change in } x} = \frac{f(b) - f(a)}{b - a}$$

The average rate of change is the slope of the **secant line** between $x = a$ and $x = b$ on the graph of f, that is, the line that passes through $(a, f(a))$ and $(b, f(b))$.

EXAMPLE 1 ■ Calculating the Average Rate of Change

For the function $f(x) = x^2 + 4$, find the average rate of change of the function between the following points:

(a) $x = 2$ and $x = 6$
(b) $x = 5$ and $x = 10$
(c) $x = a$ and $x = a + h$ $(h \neq 0)$

SOLUTION

(a) Average rate of change $= \dfrac{f(6) - f(2)}{6 - 2}$ Definition

$$= \frac{(6^2 + 4) - (2^2 + 4)}{6 - 2} \qquad \text{\color{red}{Use } } f(x) = x^2 + 4$$

$$= \frac{40 - 8}{4} = \frac{32}{4} \qquad \text{\color{red}{Simplify}}$$

$$= 8$$

(b) Average rate of change $= \dfrac{f(10) - f(5)}{10 - 5}$ Definition

$= \dfrac{(10^2 + 4) - (5^2 + 4)}{10 - 5}$ Use $f(x) = x^2 + 4$

$= \dfrac{104 - 29}{5} = \dfrac{75}{5}$ Simplify

$= 15$

(c) Average rate of change $= \dfrac{f(a + h) - f(a)}{(a + h) - a}$

$= \dfrac{[(a + h)^2 + 4] - (a^2 + 4)}{h}$ Use $f(x) = x^2 + 4$

$= \dfrac{[a^2 + 2ah + h^2 + 4] - (a^2 + 4)}{h}$ Expand

$= \dfrac{2ah + h^2}{h}$ Simplify numerator

$= \dfrac{(2a + h)h}{h}$ Factor

$= 2a + h$ Cancel h ∎

The average rate of change calculated in Example 1(c) is known as a *difference quotient*. In calculus we use difference quotients to calculate instantaneous rates of change. (See Section 12.2.)

EXAMPLE 2 ■ Average Speed of a Falling Object

If an object is dropped from a height of 3000 ft, its distance above the ground (in feet) after t seconds is given by $h(t) = 3000 - 16t^2$. Find the object's average speed for the following times.

(a) Between 1 s and 2 s   (b) Between 4 s and 5 s

SOLUTION

(a) Average rate of change $= \dfrac{h(2) - h(1)}{2 - 1}$ Definition of average rate

$= \dfrac{(3000 - 16(2)^2) - (3000 - 16(1)^2)}{2 - 1}$ Use $h(t) = 3000 - 16t^2$

$= \dfrac{-48}{1}$ Simplify

$= -48$ ft/s

The negative sign indicates that height is decreasing; this makes sense because the object is falling.

(b) Average rate of change $= \dfrac{h(5) - h(4)}{5 - 4}$ Definition of average rate

$$= \dfrac{(3000 - 16(5)^2) - (3000 - 16(4)^2)}{5 - 4}$$ Use $h(t) = 3000 - 16t^2$

$$= \dfrac{-144}{1}$$ Simplify

$$= -144 \text{ ft/s}$$

The object is falling much faster during the 4- to 5-s period than the 1- to 2-s period. ■

Time	Temperature (°F)
8:00 A.M.	38
9:00 A.M.	40
10:00 A.M.	44
11:00 A.M.	50
12:00 NOON	56
1:00 P.M.	62
2:00 P.M.	66
3:00 P.M.	67
4:00 P.M.	64
5:00 P.M.	58
6:00 P.M.	55
7:00 P.M.	51

EXAMPLE 3 ■ Average Rate of Temperature Change

The table gives the outdoor temperatures observed by a science student on a spring day. Draw a graph of the data, and find the average rate of change of temperature between the following times:

(a) 8:00 A.M. and 9:00 A.M.   (b) 1:00 P.M. and 3:00 P.M.
(c) 4:00 P.M. and 7:00 P.M.

SOLUTION

A graph of the temperature data is shown in Figure 2. Let t represent time, measured in hours since midnight (so that 2:00 P.M., for example, corresponds to $t = 14$). Define the function F by

$$F(t) = \text{temperature at time } t$$

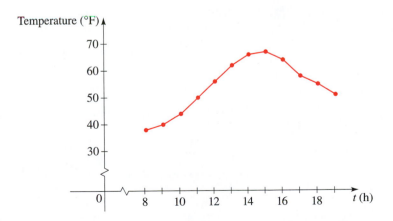

FIGURE 2

(a) Average rate of change $= \dfrac{\text{temperature at 9 A.M.} - \text{temperature at 8 A.M.}}{9 - 8}$

$$= \frac{F(9) - F(8)}{9 - 8}$$

$$= \frac{40 - 38}{9 - 8} = 2$$

The average rate of change was 2°F per hour.

(b) Average rate of change $= \dfrac{\text{temperature at 3 P.M.} - \text{temperature at 1 P.M.}}{15 - 13}$

$$= \frac{F(15) - F(13)}{15 - 13}$$

$$= \frac{67 - 62}{2} = 2.5$$

The average rate of change was 2.5°F per hour.

(c) Average rate of change $= \dfrac{\text{temperature at 7 P.M.} - \text{temperature at 4 P.M.}}{19 - 16}$

$$= \frac{F(19) - F(16)}{19 - 16}$$

$$= \frac{51 - 64}{3} \approx -4.3$$

The average rate of change was about −4.3°F per hour during this time period. The negative sign indicates that the temperature was dropping. ■

EXAMPLE 4 ■ A Function with Constant Rate of Change

Let $f(x) = 3x - 5$. Find the average rate of change of f between the following points.

(a) $x = 0$ and $x = 1$ (b) $x = 3$ and $x = 7$ (c) $x = a$ and $x = a + h$

What conclusion can you draw from your answers?

SOLUTION

(a) Average rate of change $= \dfrac{f(1) - f(0)}{1 - 0} = \dfrac{(3 \cdot 1 - 5) - (3 \cdot 0 - 5)}{1}$

$$= \frac{(-2) - (-5)}{1} = 3$$

(b) Average rate of change $= \dfrac{f(7) - f(3)}{7 - 3}$

$$= \dfrac{(3 \cdot 7 - 5) - (3 \cdot 3 - 5)}{4}$$

$$= \dfrac{16 - 4}{4} = 3$$

(c) Average rate of change $= \dfrac{f(a + h) - f(a)}{(a + h) - a}$

$$= \dfrac{[3(a + h) - 5] - [3a - 5]}{h}$$

$$= \dfrac{3a + 3h - 5 - 3a + 5}{h}$$

$$= \dfrac{3h}{h} = 3$$

It appears that the average rate of change is always 3 for this function. In fact, part (c) proves that the rate of change between any two arbitrary points $x = a$ and $x = a + h$ is 3. ■

As Example 4 indicates, for a linear function $f(x) = mx + b$, the average rate of change between any two points is the slope m of the line. Why is this true? (Think about the definitions of the slope and the average rate of change.)

Increasing and Decreasing Functions

It is very useful to know where the graph of a function rises and where it falls. The graph shown in Figure 3 rises, falls, then rises again as we move from left to right: It rises from A to B, falls from B to C, and rises again from C to D. The function f is said to be *increasing* when its graph rises and *decreasing* when its graph falls. We have the following definition.

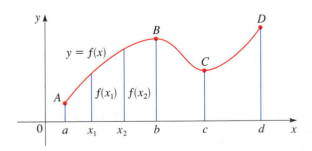

FIGURE 3

DEFINITION OF INCREASING AND DECREASING FUNCTIONS

Definition	What the graph looks like

f is **increasing** on an interval I
if $f(x_1) < f(x_2)$ whenever
$x_1 < x_2$ in I

f is **decreasing** on an interval I
if $f(x_1) > f(x_2)$ whenever
$x_1 < x_2$ in I

EXAMPLE 5 ■ Intervals on which a Function Increases or Decreases

State the intervals on which the function whose graph is shown in Figure 4 is increasing or decreasing.

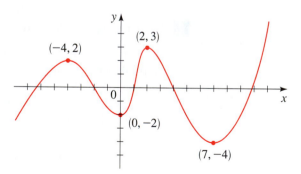

FIGURE 4

SOLUTION

The function is increasing on $(-\infty, -4]$, $[0, 2]$, and $[7, \infty)$. It is decreasing on $[-4, 0]$ and $[2, 7]$.

■

EXAMPLE 6 ■ Using a Graph to Find Intervals where a Function Increases and Decreases

(a) Sketch the graph of the function $f(x) = x^{2/3}$.

(b) Find the domain and range of the function.

(c) Find the intervals on which f increases and decreases.

Some graphing calculators, such as the TI-82 do not evaluate $x^{2/3}$ [entered as $x^{\wedge}(2/3)$] for negative x. To graph a function like $f(x) = x^{2/3}$, we enter it as $y_1 = (x^{\wedge}(1/3))^{\wedge}2$ because these calculators correctly evaluate powers of the form $x^{\wedge}(1/n)$. Newer calculators, such as the TI-83 and TI-86, do not have this problem.

SOLUTION

(a) We use a graphing calculator to sketch the graph in Figure 5.

(b) From the graph we observe that the domain of f is $\mathbb{R}$ and the range is $[0, \infty)$.

(c) From the graph we see that f is decreasing on $(-\infty, 0]$ and increasing on $[0, \infty)$.

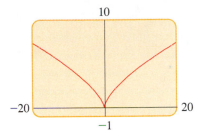

FIGURE 5
Graph of $f(x) = x^{2/3}$

The graphs in Figure 6 show that if a function is increasing on an interval, then the average rate of change between any two points is positive, whereas if a function is decreasing on an interval, then the average rate of change between any two points is negative.

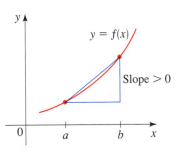

f increasing
Average rate of change positive

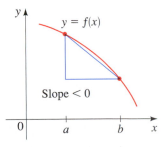

f decreasing
Average rate of change negative

FIGURE 6

2.4 EXERCISES

1–12 ■ A function is given. Determine the average rate of change of the function between the given values of the variable.

1. $f(x) = 3x - 2$; $x = 2, x = 3$

2. $g(x) = 5 + \frac{1}{2}x$; $x = 1, x = 5$

3. $h(t) = t^2 + 2t$; $t = -1, t = 4$

4. $f(z) = 1 - 3z^2$; $z = -2, z = 0$

5. $f(x) = x^3 - 4x^2$; $x = 0, x = 10$

6. $f(x) = x + x^4$; $x = -1, x = 3$

7. $f(x) = 3x^2$; $x = 2, x = 2 + h$

8. $f(x) = 4 - x^2$; $x = 1, x = 1 + h$

9. $g(x) = \dfrac{1}{x}$; $x = 1, x = a$

10. $g(x) = \dfrac{2}{x + 1}$; $x = 0, x = h$

11. $f(t) = \dfrac{2}{t}$; $t = a, t = a + h$

12. $f(t) = \sqrt{t}$; $t = a, t = a + h$

13–14 ■ A linear function is given. (a) Find the average rate of change of the function between $x = a$ and $x = a + h$. (b) Show that the average rate of change is the same as the slope of the line.

13. $f(x) = \frac{1}{2}x + 3$ **14.** $g(x) = -4x + 2$

15–18 ■ The graph of a function is given. Determine the average rate of change of the function between the indicated values of the variable.

15.   **16.**

17.

18.

19. The table gives the population in a small coastal community for the period 1990–1999. Figures shown are for January 1 in each year.
(a) What was the average rate of change of population between 1991 and 1994?
(b) What was the average rate of change of population between 1995 and 1997?
(c) For what period of time was the population increasing?
(d) For what period of time was the population decreasing?

Year	Population
1990	624
1991	856
1992	1,336
1993	1,578
1994	1,591
1995	1,483
1996	994
1997	826
1998	801
1999	745

20. A man is running around a circular track 200 m in circumference. An observer uses a stopwatch to time each lap, obtaining the data in the following table.

(a) What was the man's average speed (rate) between 68 s and 152 s?

(b) What was the man's average speed between 263 s and 412 s?

(c) Calculate the man's speed for each lap. Is he slowing down, speeding up, or neither?

Time (s)	Distance (m)
32	200
68	400
108	600
152	800
203	1000
263	1200
335	1400
412	1600

21. The table shows the number of CD players sold in a small electronics store in the years 1989–1999.

(a) What was the average rate of change of sales between 1989 and 1999?

(b) What was the average rate of change of sales between 1989 and 1990?

(c) What was the average rate of change of sales between 1990 and 1992?

(d) Between which two successive years did CD player sales *increase* most quickly? *decrease* most quickly?

Year	CD players sold
1989	512
1990	520
1991	413
1992	410
1993	468
1994	510
1995	590
1996	607
1997	732
1998	612
1999	584

22. Between 1980 and 2000, a rare book collector purchased books for his collection at the rate of 40 books per year. Use this information to complete the following table. (Note that not every year is given in the table.)

Year	Number of books
1980	420
1981	460
1982	
1985	
1990	
1992	
1995	
1997	
1998	
1999	
2000	1220

23–26 ■ The graph of a function is given. Determine the intervals on which the function is increasing and on which it is decreasing.

23.

24.

25.

26.

 27–34 ■ A function f is given.

(a) Use a graphing device to draw the graph of f.

(b) State approximately the intervals on which f is increasing and on which f is decreasing.

27. $f(x) = x^{2/5}$ **28.** $f(x) = 4 - x^{2/3}$

29. $f(x) = x^2 - 5x$ **30.** $f(x) = x^3 - 4x$

31. $f(x) = 2x^3 - 3x^2 - 12x$

32. $f(x) = x^4 - 16x^2$

33. $f(x) = x^3 + 2x^2 - x - 2$

34. $f(x) = x^4 - 4x^3 + 2x^2 + 4x - 3$

DISCOVERY · DISCUSSION

35. Changing Rates of Change: Concavity   The two tables and graphs give the distances traveled by a racing car during two different 10-s portions of a race. In each case, calculate the average speed at which the car is traveling between the observed data points. Is the speed increasing or decreasing? In other words, is the car *accelerating* or *decelerating* on each of these intervals? How does the shape of the graph tell you whether the car is accelerating or decelerating? (The first graph is said to be *concave up* and the second graph *concave down*.)

(b)

Time (s)	Distance (ft)
30	5208
32	5734
34	6022
36	6204
38	6352
40	6448

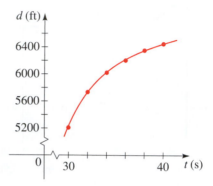

(a)

Time (s)	Distance (ft)
0	0
2	34
4	70
6	196
8	490
10	964

36. Functions That Are Always Increasing or Decreasing   Sketch rough graphs of functions that are defined for all real numbers and that exhibit the indicated behavior (or explain why it is impossible).

(a) f is always increasing, and $f(x) > 0$ for all x

(b) f is always decreasing, and $f(x) > 0$ for all x

(c) f is always increasing, and $f(x) < 0$ for all x

(d) f is always decreasing, and $f(x) < 0$ for all x

2.5   TRANSFORMATIONS OF FUNCTIONS

In this section we study how certain transformations of a function affect its graph. This will give us a better understanding of how to graph functions. The transformations we study are shifting, reflecting, and stretching.

Vertical Shifting

Adding a constant to a function shifts its graph vertically: upward if the constant is positive and downward if it is negative.

EXAMPLE 1 ■ Vertical Shifts of Graphs

Sketch the graph of each function.

(a) $f(x) = x^2 + 3$ (b) $g(x) = x^2 - 2$

SOLUTION

Recall that the graph of the function f is the same as the graph of the equation $y = f(x)$. It is often convenient to express a function in equation form, particularly when discussing its graph. For instance, we may refer to the function $f(x) = x^2$ by the equation $y = x^2$, as we do in Example 1.

(a) We start with the graph of the function

$$y = x^2$$

(from Example 1(a) in Section 2.2). The equation

$$y = x^2 + 3$$

indicates that the y-coordinate of a point on the graph of f is 3 more than the y-coordinate of the corresponding point on the graph of $y = x^2$. This means that we obtain the graph of $f(x) = x^2 + 3$ simply by shifting the graph of $y = x^2$ upward 3 units, as shown in Figure 1.

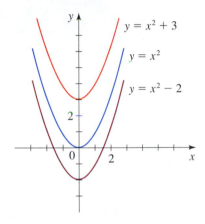

FIGURE 1

(b) Similarly, we get the graph of $g(x) = x^2 - 2$ by shifting the parabola $y = x^2$ downward 2 units (see Figure 1). ■

In general, suppose we know the graph of $y = f(x)$. How do we obtain from it the graphs of

$$y = f(x) + c \qquad \text{and} \qquad y = f(x) - c \qquad (c > 0)$$

The equation $y = f(x) + c$ tells us that the y-coordinate of each point on its graph is c units above the y-coordinate of the corresponding point on the graph of $y = f(x)$. So, we obtain the graph of $y = f(x) + c$ simply by shifting the graph of $y = f(x)$ upward c units. Similarly, we obtain the graph of $y = f(x) - c$ by shifting the graph of $y = f(x)$ downward c units.

We summarize these observations in the following box.

VERTICAL SHIFTS OF GRAPHS

Equation	How to obtain the graph	What the graph looks like
$y = f(x) + c$ $(c > 0)$	Shift graph of $y = f(x)$ upward c units	
$y = f(x) - c$ $(c > 0)$	Shift graph of $y = f(x)$ downward c units	

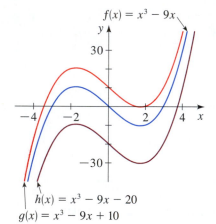

$f(x) = x^3 - 9x$

$h(x) = x^3 - 9x - 20$
$g(x) = x^3 - 9x + 10$

FIGURE 2

EXAMPLE 2 ■ Vertical Shifts of Graphs

Sketch the graph of each function.

(a) $f(x) = x^3 - 9x$ (b) $g(x) = x^3 - 9x + 10$

(c) $h(x) = x^3 - 9x - 20$

SOLUTION

The graphs of f, g, and h are sketched in Figure 2.

(a) The graph of f was sketched in Example 11 in Section 1.8. It is sketched again in Figure 2.

(b) To obtain the graph of g, we shift the graph of f upward 10 units.

(c) To obtain the graph of h, we shift the graph of f downward 20 units. ■

Horizontal Shifting

Now we consider transformations of functions that shift the graph horizontally.

Suppose we know the graph of $y = f(x)$. How do we use it to obtain the graphs of

$$y = f(x + c) \qquad \text{and} \qquad y = f(x - c) \qquad (c > 0)$$

The value of $f(x - c)$ at x is the same as the value of $f(x)$ at $x - c$. Since $x - c$ is c units to the left of x, it follows that the graph of $y = f(x - c)$ is just the graph of

Vertical & Horizontal Shifts

$y = f(x)$ shifted to the right c units. Similar reasoning shows that the graph of $y = f(x + c)$ is the graph of $y = f(x)$ shifted to the left c units. The following box summarizes these facts.

HORIZONTAL SHIFTS OF GRAPHS

Equation	How to obtain the graph	What the graph looks like
$y = f(x - c)$ $(c > 0)$	Shift graph of $y = f(x)$ to the right c units	
$y = f(x + c)$ $(c > 0)$	Shift graph of $y = f(x)$ to the left c units	

EXAMPLE 3 ■ Horizontal Shifts of Graphs

Sketch the graph of each function.

(a) $f(x) = (x + 4)^2$ (b) $g(x) = (x - 2)^2$

SOLUTION

We start with the graph of $y = x^2$, then move it to the left 4 units to get the graph of f and to the right 2 units to obtain the graph of g. See Figure 3.

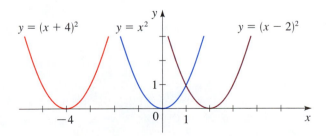

FIGURE 3

EXAMPLE 4 ■ Combining Horizontal and Vertical Shifts

Sketch the graph of the function $f(x) = \sqrt{x - 3} + 4$.

SOLUTION

We start with the graph of the square root function $y = \sqrt{x}$ (Example 1(c) in Section 2.2). We move it to the right 3 units to get the graph of $y = \sqrt{x - 3}$. Then we move the resulting graph upward 4 units to obtain the graph of $f(x) = \sqrt{x - 3} + 4$ shown in Figure 4.

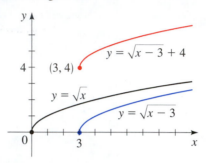

FIGURE 4

Reflecting Graphs

Suppose we know the graph of $y = f(x)$. How do we use it to obtain the graphs of $y = -f(x)$ and $y = f(-x)$? The y-coordinate of each point on the graph of $y = -f(x)$ is simply the negative of the y-coordinate of the corresponding point on the graph of $y = f(x)$. So the desired graph is the reflection of the graph of $y = f(x)$ in the x-axis. On the other hand, the value of $y = f(-x)$ at x is the same as the value of $y = f(x)$ at $-x$ and so the desired graph here is the reflection of the graph of $y = f(x)$ in the y-axis. The following box summarizes these observations.

REFLECTING GRAPHS

Equation	How to obtain the graph	What the graph looks like
$y = -f(x)$	Reflect the graph of $y = f(x)$ in the x-axis	
$y = f(-x)$	Reflect the graph of $y = f(x)$ in the y-axis	

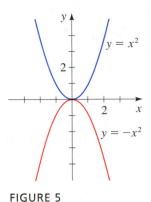

FIGURE 5

EXAMPLE 5 ■ Reflecting Graphs

Sketch the graph of each function.

(a) $f(x) = -x^2$ (b) $g(x) = \sqrt{-x}$

SOLUTION

(a) We start with the graph of $y = x^2$. The graph of $f(x) = -x^2$ is the graph of $y = x^2$ reflected in the x-axis (see Figure 5).

(b) We start with the graph of $y = \sqrt{x}$ (Example 1(c) in Section 2.2). The graph of $g(x) = \sqrt{-x}$ is the graph of $y = \sqrt{x}$ reflected in the y-axis (see Figure 6). Note that the domain of the function $g(x) = \sqrt{-x}$ is $\{x \mid x \leq 0\}$.

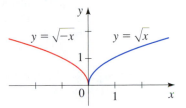

FIGURE 6

■

Vertical Stretching and Shrinking

Suppose we know the graph of $y = f(x)$. How do we use it to obtain the graph of $y = af(x)$? The y-coordinate of $y = af(x)$ at x is the same as the corresponding y-coordinate of $y = f(x)$ multiplied by a. Multiplying the y-coordinates by a has the effect of vertically stretching or shrinking the graph by a factor of a.

VERTICAL STRETCHING AND SHRINKING OF GRAPHS		
Equation	How to obtain the graph	What the graph looks like
$y = af(x)$ $(a > 1)$	Stretch the graph of $y = f(x)$ vertically by a factor of a	
$y = af(x)$ $(0 < a < 1)$	Shrink the graph of $y = f(x)$ vertically by a factor of a	

EXAMPLE 6 ■ Vertical Stretching and Shrinking of Graphs

Sketch the graph of each function.

(a) $f(x) = 3x^2$ (b) $g(x) = \frac{1}{3}x^2$

SOLUTION

FIGURE 7

(a) We start with the graph of $y = x^2$. The graph of $f(x) = 3x^2$ is the graph of $y = x^2$ stretched vertically by a factor of 3. The result is the narrower parabola in Figure 7. The graph is obtained by multiplying the y-coordinate of each point on the graph of $y = x^2$ by 3.

(b) The graph of $g(x) = \frac{1}{3}x^2$ is the graph of $y = x^2$ shrunk vertically by a factor of $\frac{1}{3}$. The result is the wider parabola in Figure 7. The graph is obtained by multiplying the y-coordinate of each point on the graph of $y = x^2$ by $\frac{1}{3}$.   ■

We illustrate the effect of combining shifts, reflections, and stretching in the following example.

EXAMPLE 7 ■ Combining Shifting, Stretching, and Reflecting

Sketch the graph of the function $f(x) = 1 - 2(x - 3)^2$.

SOLUTION

Starting with the graph of $y = x^2$, we first shift to the right 3 units to get the graph of $y = (x - 3)^2$. Then we reflect in the x-axis and stretch by a factor of 2 to get the graph of $y = -2(x - 3)^2$. Finally, we shift upward 1 unit to get the graph of $f(x) = 1 - 2(x - 3)^2$ shown in Figure 8.

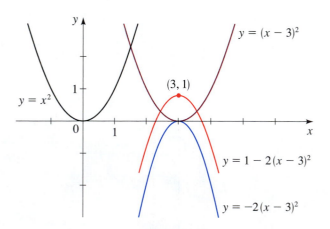

FIGURE 8   ■

Horizontal Stretching and Shrinking

Now we consider horizontal shrinking and stretching of graphs. If we know the graph of $y = f(x)$, then how is the graph of $y = f(ax)$ related to it? The y-coordinate of $y = f(ax)$ at x is the same as the y-coordinate of $y = f(x)$ at ax. Thus, the x-coordinates in the graph of $y = f(x)$ correspond to the x-coordinates in the graph of $y = f(ax)$ multiplied by a. Looking at this the other way around, we see that the x-coordinates in the graph of $y = f(ax)$ are the x-coordinates in the graph of $y = f(x)$ multiplied by $1/a$. In other words, to change the graph of $y = f(x)$ to the graph of $y = f(ax)$, we must shrink (or stretch) the graph horizontally by a factor of $1/a$, as summarized in the following box.

Sonya Kovalevsky (1850–1891) is considered the most important woman mathematician of the 19th century. She was born in Moscow to an aristocratic family. While a child, she was exposed to the principles of calculus in a very unusual fashion—her bedroom was temporarily wallpapered with the pages of a calculus book. She later wrote that she "spent many hours in front of that wall, trying to understand it." Since Russian law forbade women from studying in universities, she entered a marriage of convenience, which allowed her to travel to Germany and obtain a doctorate in mathematics from the University of Göttingen. She eventually was awarded a full professorship at the University of Stockholm, where she taught for eight years before dying in an influenza epidemic at the age of 41. Her research was instrumental in helping put the ideas and applications of functions and calculus on a sound and logical foundation. She received many accolades and prizes for her research work.

HORIZONTAL SHRINKING AND STRETCHING OF GRAPHS

Equation	How to obtain the graph	What the graph looks like
$y = f(ax)$ ($a > 1$)	Shrink the graph of $y = f(x)$ horizontally by a factor of $\dfrac{1}{a}$	
$y = f(ax)$ ($0 < a < 1$)	Stretch the graph of $y = f(x)$ horizontally by a factor of $\dfrac{1}{a}$	

EXAMPLE 8 ■ Horizontal Stretching and Shrinking of Graphs

The graph of $y = f(x)$ is shown in Figure 9. Sketch the graph of each function.

(a) $y = f(2x)$ \qquad\qquad (b) $y = f\left(\tfrac{1}{2}x\right)$

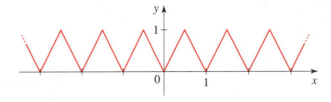

FIGURE 9
$y = f(x)$

SOLUTION

Using the principles described in the preceding box, we obtain the graphs shown in Figures 10 and 11.

FIGURE 10
$y = f(2x)$

FIGURE 11
$y = f\left(\frac{1}{2}x\right)$

Even and Odd Functions

If a function f satisfies $f(-x) = f(x)$ for every number x in its domain, then f is called an **even function**. For instance, the function $f(x) = x^2$ is even because

$$f(-x) = (-x)^2 = (-1)^2 x^2 = x^2 = f(x)$$

The geometric significance of an even function is that its graph is symmetric with respect to the y-axis (see Figure 12). This means that if we have plotted the graph of f for $x \geq 0$, then we can obtain the entire graph simply by reflecting this portion in the y-axis.

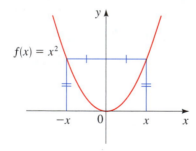

FIGURE 12
$f(x) = x^2$ is an even function.

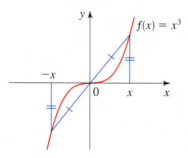

FIGURE 13
$f(x) = x^3$ is an odd function.

If f satisfies $f(-x) = -f(x)$ for every number x in its domain, then f is called an **odd function**. For example, the function $f(x) = x^3$ is odd because

$$f(-x) = (-x)^3 = (-1)^3 x^3 = -x^3 = -f(x)$$

The graph of an odd function is symmetric about the origin (see Figure 13). If we have plotted the graph of f for $x \geq 0$, then we can obtain the entire graph by rotating this portion through 180° about the origin. (This is equivalent to reflecting first in the x-axis and then in the y-axis.)

it in blocks of a million digits. If the first block is mostly 1's, he concludes that you are probably trying to transmit a 1, and so on. To say that this code is not efficient is a bit of an understatement; it requires sending a million times more data than the original message. Another method inserts "check digits." For example, for each block of eight digits insert a ninth digit; the inserted digit is 0 if there is an even number of 1's in the block and 1 if there is an odd number. So, if a single digit is wrong (a 0 changed to a 1, or vice versa), the check digits allow us to recognize that an error has occurred. This method does not tell us where the error is, so we can't correct it. Modern error-correcting codes use interesting mathematical algorithms that require inserting relatively few digits but which allow the receiver to not only recognize, but also correct, errors. The first error-correcting code was developed in the 1940s by Richard Hamming at MIT. It is interesting to note that the English language has a built-in error-correcting mechanism; to test it, try reading this error-laden sentence: Gve mo libty ox giv ne deth.

EVEN AND ODD FUNCTIONS

Definition	Symmetry of graph of f	What the graph looks like
f is **even** if $$f(-x) = f(x)$$ for all x in the domain of f	Graph of f is symmetric with respect to the y-axis	
f is **odd** if $$f(-x) = -f(x)$$ for all x in the domain of f	Graph of f is symmetric with respect to the origin	

EXAMPLE 9 ■ Even and Odd Functions

Determine whether the functions are even, odd, or neither even nor odd.

(a) $f(x) = x^5 + x$

(b) $g(x) = 1 - x^4$

(c) $h(x) = 2x - x^2$

SOLUTION

(a) $f(-x) = (-x)^5 + (-x)$

$\qquad = -x^5 - x = -(x^5 + x)$

$\qquad = -f(x)$

Therefore, f is an odd function.

(b) $g(-x) = 1 - (-x)^4 = 1 - x^4 = g(x)$

So g is even.

(c) $h(-x) = 2(-x) - (-x)^2 = -2x - x^2$

Since $h(-x) \neq h(x)$ and $h(-x) \neq -h(x)$, we conclude that h is neither even nor odd. ∎

The graphs of the functions in Example 9 are shown in Figure 14. The graph of f is symmetric about the origin, and the graph of g is symmetric about the y-axis. The graph of h is not symmetric either about the y-axis or the origin.

(a)

(b)

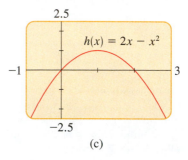

(c)

FIGURE 14

2.5 EXERCISES

1–10 ■ Suppose the graph of f is given. Describe how the graph of each function can be obtained from the graph of f.

1. (a) $y = f(x) - 3$ (b) $y = f(x - 3)$

2. (a) $y = f(x + 4)$ (b) $y = f(x) + 4$

3. (a) $y = 3f(x)$ (b) $y = \frac{1}{3}f(x)$

4. (a) $y = -f(x)$ (b) $y = f(-x)$

5. (a) $y = -2f(x)$ (b) $y = -\frac{1}{2}f(x)$

6. (a) $y = -f(x) + 5$ (b) $y = 3f(x) - 5$

7. (a) $y = f(x - 4) + \frac{3}{4}$ (b) $y = f(x + 4) - \frac{3}{4}$

8. (a) $y = 2f(x + 2) - 2$ (b) $y = 2f(x - 2) + 2$

9. (a) $y = f(4x)$ (b) $y = f(\frac{1}{4}x)$

10. (a) $y = -f(2x)$ (b) $y = f(2x) - 1$

11. The graph of f is given. Sketch the graphs of the following functions.

(a) $y = f(x - 2)$

(b) $y = f(x) - 2$

(c) $y = 2f(x)$

(d) $y = -f(x) + 3$

(e) $y = f(-x)$

(f) $y = \frac{1}{2}f(x - 1)$

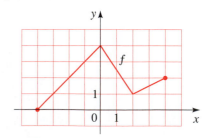

12. The graph of $y = f(x)$ is given. Match each equation with its graph.
  (a) $y = f(x - 4)$
  (b) $y = f(x) + 3$
  (c) $y = \frac{1}{3}f(x)$
  (d) $y = -f(x + 4)$
  (e) $y = 2f(x + 6)$

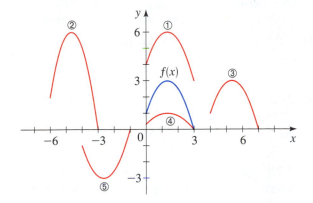

13. (a) Sketch the graph of $f(x) = \dfrac{1}{x}$ by plotting points.
  (b) Use the graph of f to sketch the graphs of the following functions.

  (i) $y = -\dfrac{1}{x}$ (ii) $y = \dfrac{1}{x - 1}$

  (iii) $y = \dfrac{2}{x + 2}$ (iv) $y = 1 + \dfrac{1}{x - 3}$

14. (a) Sketch the graph of $g(x) = \sqrt[3]{x}$ by plotting points.
  (b) Use the graph of g to sketch the graphs of the following functions.

  (i) $y = \sqrt[3]{x - 2}$ (ii) $y = \sqrt[3]{x + 2} + 2$
  (iii) $y = 1 - \sqrt[3]{x}$

15–18 ■ Explain how the graph of g is obtained from the graph of f.

15. (a) $f(x) = x^2,\quad g(x) = (x + 2)^2$
  (b) $f(x) = x^2,\quad g(x) = x^2 + 2$

16. (a) $f(x) = x^3,\quad g(x) = (x - 4)^3$
  (b) $f(x) = x^3,\quad g(x) = x^3 - 4$

17. (a) $f(x) = \sqrt{x},\quad g(x) = 2\sqrt{x}$
  (b) $f(x) = \sqrt{x},\quad g(x) = \frac{1}{2}\sqrt{x - 2}$

18. (a) $f(x) = |x|,\quad g(x) = 3|x| + 1$
  (b) $f(x) = |x|,\quad g(x) = -|x + 1|$

19–24 ■ A function f is given, and the indicated transformations are applied to its graph (in the given order). Write the equation for the final transformed graph.

19. $f(x) = x^2$; shift upward 3 units and shift 2 units to the right

20. $f(x) = x^3$; shift downward 1 unit and shift 4 units to the left

21. $f(x) = \sqrt{x}$; shift 3 units to the left, stretch vertically by a factor of 5, and reflect in the x-axis

22. $f(x) = \sqrt[3]{x}$; reflect in the y-axis, shrink vertically by a factor of $\frac{1}{2}$, and shift upward $\frac{3}{5}$ unit

23. $f(x) = |x|$; shift to the right $\frac{1}{2}$ unit, shrink vertically by a factor of 0.1, and shift downward 2 units

24. $f(x) = |x|$; shift to the left 1 unit, stretch vertically by a factor of 3, and shift upward 10 units

25–40 ■ Sketch the graph of the function, not by plotting points, but by starting with the graph of a standard function and applying transformations.

25. $f(x) = (x - 2)^2$

26. $f(x) = (x + 7)^2$

27. $f(x) = -(x + 1)^2$

28. $f(x) = 1 - x^2$

29. $f(x) = x^3 + 2$

30. $f(x) = -x^3$

31. $y = 1 + \sqrt{x}$

32. $y = 2 - \sqrt{x + 1}$

33. $y = \frac{1}{2}\sqrt{x + 4} - 3$

34. $y = 3 - 2(x - 1)^2$

35. $y = 5 + (x + 3)^2$

36. $y = \frac{1}{3}x^3 - 1$

37. $y = |x| - 1$

38. $y = |x - 1|$

39. $y = |x + 2| + 2$

40. $y = 2 - |x|$

41–44 ■ Graph the functions on the same screen using the given viewing rectangle. How is each graph related to the graph in part (a)?

41. Viewing rectangle $[-8, 8]$ by $[-2, 8]$
  (a) $y = \sqrt[4]{x}$ (b) $y = \sqrt[4]{x + 5}$
  (c) $y = 2\sqrt[4]{x + 5}$ (d) $y = 4 + 2\sqrt[4]{x + 5}$

42. Viewing rectangle $[-8, 8]$ by $[-6, 6]$
(a) $y = |x|$
(b) $y = -|x|$
(c) $y = -3|x|$
(d) $y = -3|x - 5|$

43. Viewing rectangle $[-4, 6]$ by $[-4, 4]$
(a) $y = x^6$
(b) $y = \frac{1}{3}x^6$
(c) $y = -\frac{1}{3}x^6$
(d) $y = -\frac{1}{3}(x - 4)^6$

44. Viewing rectangle $[-6, 6]$ by $[-4, 4]$
(a) $y = \dfrac{1}{\sqrt{x}}$
(b) $y = \dfrac{1}{\sqrt{x + 3}}$
(c) $y = \dfrac{1}{2\sqrt{x + 3}}$
(d) $y = \dfrac{1}{2\sqrt{x + 3}} - 3$

45. The graph of g is given. Use it to graph each of the following functions.
(a) $y = g(2x)$
(b) $y = g\left(\frac{1}{2}x\right)$

46. The graph of h is given. Use it to graph each of the following functions.
(a) $y = h(3x)$
(b) $y = h\left(\frac{1}{3}x\right)$

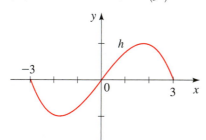

47–48 ■ Using the graph of $f(x) = \llbracket x \rrbracket$ described in Section 1.8, graph the indicated function.

47. $y = \llbracket 2x \rrbracket$
48. $y = \llbracket \frac{1}{4}x \rrbracket$

 49. If $f(x) = \sqrt{2x - x^2}$, graph the following functions in the viewing rectangle $[-5, 5]$ by $[-4, 4]$. How is each graph related to the graph in part (a)?
(a) $y = f(x)$
(b) $y = f(2x)$
(c) $y = f\left(\frac{1}{2}x\right)$

 50. If $f(x) = \sqrt{2x - x^2}$, graph the following functions in the viewing rectangle $[-5, 5]$ by $[-4, 4]$. How is each graph related to the graph in part (a)?
(a) $y = f(x)$
(b) $y = f(-x)$
(c) $y = -f(-x)$
(d) $y = f(-2x)$
(e) $y = f\left(-\frac{1}{2}x\right)$

51–58 ■ Determine whether the function f is even, odd, or neither. If f is even or odd, use symmetry to sketch its graph.

51. $f(x) = x^{-2}$
52. $f(x) = x^{-3}$

53. $f(x) = x^2 + x$
54. $f(x) = x^4 - 4x^2$

55. $f(x) = x^3 - x$
56. $f(x) = 3x^3 + 2x^2 + 1$

57. $f(x) = 1 - \sqrt[3]{x}$
58. $f(x) = x + \dfrac{1}{x}$

59. The graphs of $f(x) = x^2 - 4$ and $g(x) = |x^2 - 4|$ are shown. Explain how the graph of g is obtained from the graph of f.

$f(x) = x^2 - 4$

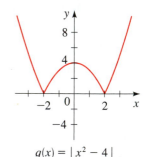
$g(x) = |x^2 - 4|$

60. The graph of $f(x) = x^4 - 4x^2$ is shown. Use this graph to sketch the graph of $g(x) = |x^4 - 4x^2|$.

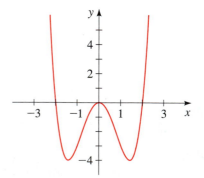

61–62 ■ Sketch the graph of each function.

61. (a) $f(x) = 4x - x^2$ (b) $g(x) = |4x - x^2|$

62. (a) $f(x) = x^3$ (b) $g(x) = |x^3|$

 DISCOVERY · DISCUSSION

63. Sums of Even and Odd Functions   If f and g are both even functions, is $f + g$ necessarily even? If both are odd, is their sum necessarily odd? What can you say about the sum if one is odd and one is even? In each case, prove your answer.

64. Products of Even and Odd Functions   Answer the same questions as in Exercise 63, except this time consider the *product* of f and g instead of the sum.

65. Even and Odd Power Functions   What must be true about the integer n if the function

$$f(x) = x^n$$

is an even function? If it is an odd function? Why do you think the names "even" and "odd" were chosen for these function properties?

2.6   EXTREME VALUES OF FUNCTIONS

An extreme value of a function is the largest or smallest value of the function on some interval. Finding points where functions achieve extreme values is important in many applications. For example, if the function represents the profit in a business, we would be interested in the maximum values; if a function represents the amount of material to be used in a manufacturing process, we would be interested in the minimum values. We begin by giving an algebraic method for finding extreme values of quadratic functions and then use graphing devices to find extreme values of other functions.

Extreme Values of Quadratic Functions

A **quadratic function** is a function f of the form

$$f(x) = ax^2 + bx + c$$

where a, b, and c are real numbers and $a \neq 0$.

In particular, if we take $a = 1$ and $b = c = 0$, we get the simple quadratic function $f(x) = x^2$ whose graph is the parabola that we drew in Example 1 of Section 2.2. In fact, the graph of any quadratic function is a **parabola**; it can be obtained from the graph of $y = x^2$ by the transformations given in Section 2.5.

Completing the square is discussed in Section 1.5.

To graph a quadratic function, we first complete the square to put the function in a more convenient form for graphing.

EXAMPLE 1 ■ **Graphing a Quadratic Function by Completing the Square**

Sketch the graph of the quadratic function $f(x) = 2x^2 - 12x + 23$.

SOLUTION

Since the coefficient of x^2 is not 1, we must factor this coefficient from the terms involving x before we complete the square.

$$f(x) = 2x^2 - 12x + 23$$

$$= 2(x^2 - 6x) + 23 \qquad \text{Factor 2 from the } x\text{-terms}$$

$$= 2(x^2 - 6x + 9) + 23 - 2 \cdot 9 \qquad \text{Complete the square: Add 9 inside parentheses, subtract } 2 \cdot 9 \text{ outside}$$

$$= 2(x - 3)^2 + 5$$

This form of the function tells us that we get the graph of f by taking the parabola $y = x^2$, shifting it to the right 3 units, stretching it by a factor of 2, and moving it upward 5 units. Notice that the lowest point of the parabola is at $(3, 5)$ and the parabola opens upward. We sketch the graph in Figure 1 after noting that the y-intercept is $f(0) = 23$.

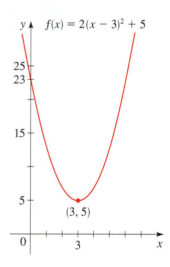

FIGURE 1

If we start with a general quadratic function $f(x) = ax^2 + bx + c$ and complete the square as in Example 1, we arrive at an expression in the standard form $f(x) = a(x - h)^2 + k$. We know from Section 2.5 that the graph of this function is a parabola obtained from the graph of $y = x^2$ by a horizontal shift, a vertical stretch, and then a vertical shift. The *vertex*, that is, the highest or lowest point of the parabola, is the point (h, k). If $a > 0$, the parabola opens upward. In this case, the vertex (h, k) is the lowest point on the parabola, so the *minimum* value of the function is $f(h) = k$. If $a < 0$, the parabola opens downward. In this case, the vertex (h, k) is the highest point on the parabola, so the *maximum* value of the function is $f(h) = k$.

STANDARD FORM OF A QUADRATIC FUNCTION

A quadratic function $f(x) = ax^2 + bx + c$ can be expressed in the **standard form**

$$f(x) = a(x - h)^2 + k$$

by completing the square. The graph of f is a parabola with **vertex** (h, k); the parabola opens upward if $a > 0$ or downward if $a < 0$.

If $a > 0$, then the **minimum value** of f occurs at $x = h$ and this value is $f(h) = k$

If $a < 0$, then the **maximum value** of f occurs at $x = h$ and this value is $f(h) = k$

$f(x) = a(x - h)^2 + k,\ a > 0$

$f(x) = a(x - h)^2 + k,\ a < 0$

EXAMPLE 2 ■ Expressing a Quadratic Function in Standard Form

Consider the quadratic function $f(x) = 5x^2 - 30x + 49$.

(a) Express f in standard form.
(b) Sketch the graph of f.
(c) Find the minimum value of f.

SOLUTION

(a) To express this quadratic function in standard form, we complete the square.

$$f(x) = 5x^2 - 30x + 49$$

$$= 5(x^2 - 6x) + 49 \qquad \text{Factor 5 from the } x\text{-terms}$$

$$= 5(x^2 - 6x + 9) + 49 - 5 \cdot 9 \qquad \begin{array}{l}\text{Complete the square: Add 9 inside}\\ \text{parentheses, subtract } 5 \cdot 9 \text{ outside}\end{array}$$

$$= 5(x - 3)^2 + 4$$

(b) The graph is a parabola that has its vertex at $(3, 4)$ and opens upward, as sketched in Figure 2.

(c) Since the coefficient of x^2 is positive, f has a minimum value. The minimum value is $f(3) = 4$. ■

FIGURE 2

EXAMPLE 3 ■ Expressing a Quadratic Function in Standard Form

Consider the quadratic function $f(x) = -x^2 + x + 2$.

(a) Express f in standard form.
(b) Sketch the graph of f.
(c) Find the maximum value of f.

SOLUTION

(a) To express this quadratic function in standard form, we complete the square.

$$y = -x^2 + x + 2$$

$$= -(x^2 - x) + 2 \qquad \text{Factor } -1 \text{ from the } x\text{-terms}$$

$$= -\left(x^2 - x + \tfrac{1}{4}\right) + 2 - (-1)\tfrac{1}{4} \qquad \begin{array}{l}\text{Complete the square: Add } \tfrac{1}{4} \text{ inside} \\ \text{parentheses, subtract } (-1)\tfrac{1}{4} \text{ outside}\end{array}$$

$$= -\left(x - \tfrac{1}{2}\right)^2 + \tfrac{9}{4}$$

(b) From the standard form we see that the graph is a parabola that opens downward and has vertex $\left(\tfrac{1}{2}, \tfrac{9}{4}\right)$. As an aid to sketching the graph, we find the intercepts. The y-intercept is $f(0) = 2$. To find the x-intercepts, we set $f(x) = 0$ and factor the resulting equation.

$$-x^2 + x + 2 = 0$$

$$-(x^2 - x - 2) = 0$$

$$-(x - 2)(x + 1) = 0$$

Thus, the x-intercepts are $x = 2$ and $x = -1$. The graph of f is sketched in Figure 3.

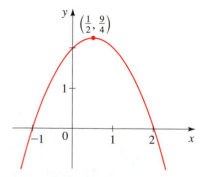

FIGURE 3
Graph of $f(x) = -x^2 + x + 2$

(c) Since the coefficient of x^2 is negative, f has a maximum value, which is $f\left(\tfrac{1}{2}\right) = \tfrac{9}{4}$.

■

Expressing a quadratic function in standard form helps us sketch its graph as well as find its maximum or minimum value. If we are interested only in finding the maximum or minimum value, then a formula is available for doing so. This formula is obtained by completing the square for the general quadratic function as follows:

$$f(x) = ax^2 + bx + c$$

$$= a\left(x^2 + \frac{b}{a}x\right) + c \qquad \text{Factor } a \text{ from the } x\text{-terms}$$

$$= a\left(x^2 + \frac{b}{a}x + \frac{b^2}{4a^2}\right) + c - \frac{b^2}{4a} \qquad \begin{array}{l}\text{Complete the square:} \\ \text{Add } \dfrac{b^2}{4a^2} \text{ inside parentheses,} \\ \text{subtract } a\!\left(\dfrac{b^2}{4a^2}\right) \text{ outside}\end{array}$$

$$= a\left(x + \frac{b}{2a}\right)^2 + c - \frac{b^2}{4a}$$

This equation is in standard form with $h = -b/(2a)$ and $k = c - b^2/(4a)$. Since the maximum or minimum value occurs at $x = h$, we have the following result.

MAXIMUM OR MINIMUM VALUE OF A QUADRATIC FUNCTION

The maximum or minimum value of a quadratic function
$f(x) = ax^2 + bx + c$ occurs at

$$x = -\frac{b}{2a}$$

If $a > 0$, then the **minimum value** is $f\left(-\dfrac{b}{2a}\right)$.

If $a < 0$, then the **maximum value** is $f\left(-\dfrac{b}{2a}\right)$.

EXAMPLE 4 ■ Finding Maximum and Minimum Values of Quadratic Functions

Find the maximum or minimum value of each quadratic function.

(a) $f(x) = x^2 + 4x$
(b) $g(x) = -2x^2 + 4x - 5$

SOLUTION

(a) This is a quadratic function with $a = 1$ and $b = 4$. Thus, the maximum or minimum value occurs at

$$x = -\frac{b}{2a} = -\frac{4}{2 \cdot 1} = -2$$

Since $a > 0$, the function has the minimum value

$$f(-2) = (-2)^2 + 4(-2) = -4$$

(b) This is a quadratic function with $a = -2$ and $b = 4$. Thus, the maximum or minimum value occurs at

$$x = -\frac{b}{2a} = -\frac{4}{2 \cdot (-2)} = 1$$

Since $a < 0$, the function has the maximum value

$$f(1) = -2(1)^2 + 4(1) - 5 = -3 \qquad \blacksquare$$

Many real-world problems involve finding a maximum or minimum value for a function that models a given situation. In the next example we find the maximum value of a quadratic function that models the gas mileage for a car.

EXAMPLE 5 ■ Maximum Gas Mileage for a Car

Most cars get their best gas mileage when traveling at a relatively modest speed. The gas mileage M for a certain new car is modeled by the function

$$M(s) = -\frac{1}{28}s^2 + 3s - 31, \qquad 15 \leqslant s \leqslant 70$$

where s is the speed in mi/h and M is measured in mi/gal. What is the car's best gas mileage, and at what speed is it attained?

SOLUTION

The function M is a quadratic function with $a = -\frac{1}{28}$ and $b = 3$. Thus its maximum value occurs when

$$s = -\frac{b}{2a} = -\frac{3}{-2\left(\frac{1}{28}\right)} = 42$$

The maximum is $M(42) = -\frac{1}{28}(42)^2 + 3(42) - 31 = 32$. Thus, the car's best gas mileage is 32 mi/gal, when it is traveling at 42 mi/h. $\qquad \blacksquare$

Using Graphing Devices to Find Extreme Values

The methods we have discussed apply to finding extreme values of quadratic functions only. We now show how to locate extreme values of any function that can be graphed with a calculator or computer.

If there is a viewing rectangle such that the point $(a, f(a))$ is the highest point on the graph of f *within* the viewing rectangle (not on the edge), then the number $f(a)$ is called a **local maximum value** of f (see Figure 4). Notice that $f(a) \geqslant f(x)$ for all numbers x that are close to a.

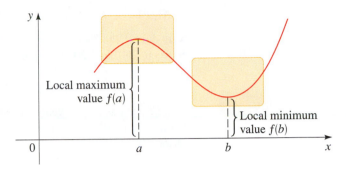

FIGURE 4

Similarly, if there is a viewing rectangle such that the point $(b, f(b))$ is the lowest point on the graph of f within the viewing rectangle, then the number $f(b)$ is called a **local minimum value** of f. In this case, $f(b) \leq f(x)$ for all numbers x that are close to b.

EXAMPLE 6 ■ Finding Local Maxima and Minima from a Graph

Find the local maximum and minimum values of the function $f(x) = x^3 - 8x + 1$, correct to three decimals.

SOLUTION

The graph of f is shown in Figure 5. There appears to be one local maximum between $x = -2$ and $x = -1$, and one local minimum between $x = 1$ and $x = 2$.

Let's find the coordinates of the local maximum point first. We zoom in to enlarge the area near this point, as shown in Figure 6. Using the TRACE feature on the graphing device, we move the cursor along the curve and observe how the y-coordinates change. The local maximum value of y is 9.709, and this value occurs when x is -1.633, correct to three decimals.

We locate the minimum value in a similar fashion. By zooming in to the viewing rectangle shown in Figure 7, we find that the local minimum value is -7.709, and this value occurs when $x \approx 1.633$.

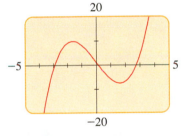

FIGURE 5
Graph of $f(x) = x^3 - 8x + 1$

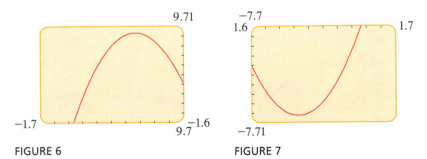

FIGURE 6

FIGURE 7

The `maximum` and `minimum` commands on a TI-82 or TI-83 calculator provide another method for finding extreme values of functions. We use this method in the next example.

EXAMPLE 7 ■ A Model for the Food Price Index

A model for the food price index (the price of a representative "basket" of foods) between 1990 and 2000 is given by the function

$$I(t) = -0.0113t^3 + 0.0681t^2 + 0.198t + 99.1$$

where t is measured in years since midyear 1990, so $0 \le t \le 10$, and $I(t)$ is scaled so that $I(3) = 100$. Estimate the time when food was most expensive during the period 1990–2000.

SOLUTION

The graph of I as a function of t is shown in Figure 8(a). There appears to be a maximum between $t = 4$ and $t = 7$. Using the **maximum** command, as shown in Figure 8(b), we see that the maximum value of I is about 100.38, and it occurs when $t \approx 5.15$, which corresponds to August 1995.

FIGURE 8   (a)   (b)   ■

2.6 EXERCISES

1–14 ■ Sketch the graph of the given parabola and state the coordinates of its vertex and its intercepts.

1. $y = x^2 - 8x$

2. $y = x^2 + 6x$

3. $y = 2x^2 - 6x$

4. $y = -x^2 + 10x$

5. $y = x^2 + 4x + 1$

6. $y = x^2 - 2x + 2$

7. $y = x^2 + 6x + 8$

8. $y = -x^2 - 4x + 4$

9. $y = 2x^2 + 4x + 3$

10. $y = -3x^2 + 6x - 2$

11. $y = 2x^2 - 20x + 57$

12. $y = 2x^2 + x - 6$

13. $y = -4x^2 - 16x + 3$

14. $y = 6x^2 + 12x - 5$

15–24 ■ A quadratic function is given.
(a) Express the quadratic function in standard form.
(b) Sketch its graph.
(c) Find its maximum or minimum value.

15. $f(x) = 2x - x^2$

16. $f(x) = x + x^2$

17. $f(x) = x^2 + 2x - 1$

18. $f(x) = x^2 - 8x + 8$

19. $f(x) = -x^2 - 3x + 3$

20. $f(x) = 1 - 6x - x^2$

21. $g(x) = 3x^2 - 12x + 13$

22. $g(x) = 2x^2 + 8x + 11$

23. $h(x) = 1 - x - x^2$

24. $h(x) = 3 - 4x - 4x^2$

25–34 ■ Find the maximum or minimum value of the function.

25. $f(x) = x^2 + x + 1$

26. $f(x) = 1 + 3x - x^2$

27. $f(t) = 100 - 49t - 7t^2$

28. $f(t) = 10t^2 + 40t + 113$

29. $f(s) = s^2 - 1.2s + 16$

30. $g(x) = 100x^2 - 1500x$

31. $h(x) = \frac{1}{2}x^2 + 2x - 6$

32. $f(x) = -\frac{x^2}{3} + 2x + 7$

33. $f(x) = 3 - x - \frac{1}{2}x^2$

34. $g(x) = 2x(x - 4) + 7$

35. Find a function whose graph is a parabola with vertex $(1, -2)$ and that passes through the point $(4, 16)$.

36. Find a function whose graph is a parabola with vertex $(3, 4)$ and that passes through the point $(1, -8)$.

37–38 ■ Find the domain and range of the function.

37. $f(x) = -x^2 + 4x - 3$

38. $f(x) = x^2 - 2x - 3$

39. If a ball is thrown directly upward with a velocity of 40 ft/s, its height (in feet) after t seconds is given by $y = 40t - 16t^2$. What is the maximum height attained by the ball?

40. A ball is thrown across a playing field. Its path is given by the equation $y = -0.005x^2 + x + 5$, where x is the distance the ball has traveled horizontally, and y is its height above ground level, both measured in feet.
(a) What is the maximum height attained by the ball?
(b) How far has it traveled horizontally when it hits the ground?

41. A manufacturer finds that the revenue generated by selling x units of a certain commodity is given by the function $R(x) = 80x - 0.4x^2$, where the revenue $R(x)$ is measured in dollars. What is the maximum revenue, and how many units should be manufactured to obtain this maximum?

42. A soft-drink vendor at a popular beach analyzes his sales records, and finds that if he sells x cans of soda pop in one day, his profit (in dollars) is given by

$$P(x) = -0.001x^2 + 3x - 1800$$

What is his maximum profit per day, and how many cans must he sell for maximum profit?

43. The effectiveness of a television commercial depends on how many times a viewer watches it. After some experiments an advertising agency found that if the effectiveness E is measured on a scale of 0 to 10, then

$$E(n) = \tfrac{2}{3}n - \tfrac{1}{90}n^2$$

where n is the number of times a viewer watches a given commercial. For a commercial to have maximum effectiveness, how many times should a viewer watch it?

44. When a certain drug is taken orally, the concentration of the drug in the patient's bloodstream after t minutes is given by $C(t) = 0.06t - 0.0002t^2$, where $0 \le t \le 240$ and the concentration is measured in mg/L. When is the maximum serum concentration reached, and what is that maximum concentration?

 45–46 ■ A quadratic function is given.
(a) Use a graphing device to find the maximum or minimum value of the quadratic function f, correct to two decimal places.
(b) Find the exact maximum or minimum value of f, and compare with your answer to part (a).

45. $f(x) = x^2 + 1.79x - 3.21$

46. $f(x) = 1 + x - \sqrt{2}\,x^2$

47–54 ■ Find the local maximum and minimum values of the function and the value of x at which each occurs. State each answer correct to two decimal places.

47. $f(x) = x^3 - x$

48. $f(x) = 3 + x + x^2 - x^3$

49. $g(x) = x^4 - 2x^3 - 11x^2$

50. $g(x) = x^5 - 8x^3 + 20x$

51. $U(x) = x\sqrt{6 - x}$

52. $U(x) = x\sqrt{x - x^2}$

53. $V(x) = \dfrac{1 - x^2}{x^3}$

54. $V(x) = \dfrac{1}{x^2 + x + 1}$

 55. A fish swims at a speed v relative to the water, against a current of 5 mi/h. Using a mathematical model of energy expenditure, it can be shown that the total energy E required to swim a distance of 10 mi is given by

$$E(v) = 2.73v^3 \frac{10}{v - 5}$$

Biologists believe that migrating fish try to minimize the total energy required to swim a fixed distance. Find the value of v that minimizes energy required.

   NOTE This result has been verified; migrating fish swim against a current at a speed 50% greater than the speed of the current.

56. A highway engineer wants to estimate the maximum number of cars that can safely travel a particular highway at a given speed. She assumes that each car is 17 ft long, travels at a speed s, and follows the car in front of it at the "safe following distance" for that speed. She finds that the number N of cars that can pass a given point per minute is modeled by the function

$$N(s) = \frac{88s}{17 + 17\left(\dfrac{s}{20}\right)^2}$$

At what speed can the greatest number of cars travel the highway safely?

57. Between 0°C and 30°C, the volume V (in cubic centimeters) of 1 kg of water at a temperature T is given by the formula

$$V = 999.87 - 0.06426T + 0.0085043T^2 - 0.0000679T^3$$

Find the temperature at which the volume of 1 kg of water is a minimum.

58. When a foreign object lodged in the trachea (windpipe) forces a person to cough, the diaphragm thrusts upward causing an increase in pressure in the lungs. At the same time, the trachea contracts, causing the expelled air to move faster and increasing the pressure on the foreign object. According to a mathematical model of coughing, the velocity v of the airstream through an average-sized person's trachea is related to the radius r of the trachea (in centimeters) by the function

$$v(r) = 3.2(1 - r)r^2 \qquad \tfrac{1}{2} \le r \le 1$$

Determine the value of r for which v is a maximum.

59. Find the maximum value of the function
$f(x) = 3 + 4x^2 - x^4$.   [*Hint:* Let $t = x^2$.]

DISCOVERY · DISCUSSION

60. Maxima and Minima   In Example 5 we saw a real-world situation in which the maximum value of a function is important. Name several other everyday situations in which a maximum or minimum value is important.

61. Minimizing a Distance   When we seek a minimum or maximum value of a function, it is sometimes easier to work with a simpler function instead.
  (a) Suppose $g(x) = \sqrt{f(x)}$, where $f(x) \ge 0$ for all x. Explain why the local minima and maxima of f and g occur at the same values of x.
  (b) Let $g(x)$ be the distance between the point $(3, 0)$ and the point (x, x^2) on the graph of the parabola $y = x^2$. Express g as a function of x.
  (c) Find the minimum value of the function g that you found in part (b). Use the principle described in part (a) to simplify your work.

2.7   MODELING WITH FUNCTIONS

Many of the processes studied in the physical and social sciences involve understanding how one quantity varies with respect to another. Finding a function that describes the dependence of one quantity on another is called *modeling*. For example, a biologist observes that the number of bacteria in a certain culture increases with time. He tries to model this phenomenon by finding the precise function (or rule) that relates the bacteria population to the elapsed time.

In this section we will learn how to find models that can be constructed using geometric or algebraic properties of the object under study. (Finding models from data is studied in the *Focus on Modeling* at the end of this chapter.) Once the model is found, we use it to analyze and predict properties of the object or process being studied.

Modeling with Functions

We begin with a simple real-life situation that illustrates the modeling process.

EXAMPLE 1 ■ **Modeling the Volume of a Box**

A breakfast cereal company manufactures boxes to package their product. For aesthetic reasons, the box must have the following proportions: Its width is 3 times its depth and its height is 5 times its depth.

(a) Find a function that models the volume of the box in terms of its depth.

(b) Find the volume of the box if the depth is 1.5 in.

(c) For what depth is the volume 90 in^3?

(d) For what depth is the volume greater than 60 in^3?

■ **Thinking about the problem.**

Let's experiment with the problem. If the depth is 1 in., then the width is 3 in. and the height is 5 in. So in this case, the volume is $V = 1 \times 3 \times 5 = 15$ in^3. The table gives other values. Notice that all the boxes have the same shape, and the greater the depth the greater the volume.

Depth	Volume
1	$1 \times 3 \times 5 = 15$
2	$2 \times 6 \times 10 = 120$
3	$3 \times 9 \times 15 = 405$
4	$4 \times 12 \times 20 = 960$

SOLUTION

(a) To find the function that models the volume of the box, we use the following steps.

■ **Express the model in words.**

We know that the volume of a rectangular box is

$$\text{volume} = \text{depth} \times \text{width} \times \text{height}$$

■ **Choose the variable.**

There are three varying quantities—width, depth, and height. Since the function we want depends on the depth, we let

$$x = \text{depth of the box}$$

Then we express the other dimensions of the box in terms of x.

In Words	In Algebra
Depth	x
Width	$3x$
Height	$5x$

■ **Set up the model.**

The model is the function V that gives the volume of the box in terms of the depth x.

400

0   3

FIGURE 1

$$\boxed{\text{volume}} = \boxed{\text{depth}} \times \boxed{\text{width}} \times \boxed{\text{height}}$$

$$V(x) = x \cdot 3x \cdot 5x$$

$$V(x) = 15x^3$$

The volume of the box is modeled by the function $V(x) = 15x^3$. The function V is graphed in Figure 1.

■ **Use the model.**

We use the model to answer the questions in parts (b), (c), and (d).

(b) If the depth is 1.5 in. the volume is $V(1.5) = 15(1.5)^3 = 50.625$ in^3.

(c) We need to solve the equation $V(x) = 90$ or

$$15x^3 = 90$$

$$x^3 = 6$$

$$x = \sqrt[3]{6} \approx 1.82 \text{ in}$$

The volume is 90 in^3 when the depth is about 1.82 in.

(d) We need to solve the inequality $V(x) \geq 60$ or

$$15x^3 \geq 60$$

$$x^3 \geq 4$$

$$x \geq \sqrt[3]{4} \approx 1.59$$

The volume will be greater than 60 in^3 if the depth is greater than 1.59 in.   ■

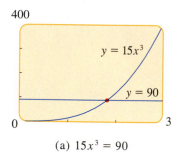

400

0   3

(a) $15x^3 = 90$

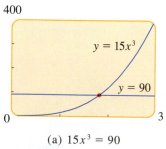

400

0   3

(a) $15x^3 = 90$

FIGURE 2

The graph in Figure 1 gives a visual representation of the model. We can use it to solve parts (c) and (d) graphically, as shown in Figure 2.

The steps in Example 1 are typical of how we model with functions. They are summarized in the following box.

GUIDELINES FOR MODELING WITH FUNCTIONS

1. EXPRESS THE MODEL IN WORDS. Identify the quantity you want to model and express it, in words, as a function of the other quantities in the problem.

2. CHOOSE THE VARIABLE. Identify all the variables used to express the function in Step 1. Assign a symbol, such as x, to one variable and express the other variables in terms of this symbol.

3. SET UP THE MODEL. Express the function in the language of algebra by writing it as a function of the single variable chosen in Step 2.

4. USE THE MODEL. Use the function to answer the questions posed in the problem. (To find a maximum or a minimum, use the algebraic or graphical methods described in Section 2.6.)

EXAMPLE 2 ■ Fencing a Garden

A gardener has 140 feet of fencing to fence in a rectangular vegetable garden.

(a) Find a function that models the area of the garden she can fence.
(b) For what range of widths is the area greater than 825 ft^2?
(c) Can she fence a garden with area 1250 ft^2?
(d) Find the dimensions of the largest area she can fence.

■   **Thinking about the problem.**

If the gardener fences a plot with width 10 ft, then the length must be 60 ft, because $10 + 10 + 60 + 60 = 140$. So the area is

$$A = wl = 10 \cdot 60 = 600 \text{ ft}^2$$

The table shows various choices for fencing the garden (Figure 3). We see that as the width increases, the fenced area increases, then decreases.

Width	Length	Area
10	60	600
20	50	1000
30	40	1200
40	30	1200
50	20	1000
60	10	600

FIGURE 3

Pythagoras (circa 580–500 B.C.) founded a school in Croton in southern Italy, devoted to the study of arithmetic, geometry, music, and astronomy. The Pythagoreans, as they were called, were a secret society with peculiar rules and initiation rites. They wrote nothing down, and were not to reveal to anyone what they had learned from the Master. Although women were barred by law from attending public meetings, Pythagoras allowed women in his school, and his most famous student was Theano (whom he later married).

According to Aristotle, the Pythagoreans were convinced that "the principles of mathematics are the principles of all things." Their motto was "Everything is Number," by which they meant *whole* numbers. The outstanding contribution of Pythagoras is the theorem that bears his name: In a right triangle the area of the square on the hypotenuse is equal to the sum of the areas of the square on the other two sides.

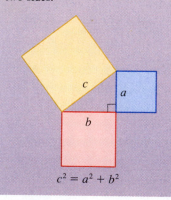

$$c^2 = a^2 + b^2$$

SOLUTION

(a) The model we want is a function that gives the area she can fence.

■ **Express the model in words.**

We know that the area of a rectangular garden is

$$\text{area} = \text{width} \times \text{length}$$

■ **Choose the variable.**

There are two varying quantities—width and length. Since the function we want depends on only one variable, we let

$$x = \text{width of the garden}$$

Then we must express the length in terms of x. The perimeter is fixed at 140 ft, so the length is determined once we choose the width. If we let the length be l as in Figure 3, then $2x + 2l = 140$, so $l = 70 - x$. We summarize these facts.

In Words	In Algebra
Width	x
Length	$70 - x$

■ **Set up the model.**

The model is the function A that gives the area of the garden for any width x.

$$\text{area} = \text{width} \times \text{length}$$

$$A(x) = x(70 - x)$$

$$A(x) = 70x - x^2$$

The area she can fence is modeled by the function $A(x) = 70x - x^2$.

■ **Use the model.**

We use the model to answer the questions in parts (b)–(d).

(b) We need to solve the inequality $A(x) \geq 825$. To solve graphically, we graph $y = 70x - x^2$ and $y = 825$ in the same viewing rectangle (see Figure 4). We see that $15 \leq x \leq 55$.

(c) From Figure 5 we see that the graph of $A(x)$ always lies below the line $y = 1250$, so an area of 1250 ft² is never attained.

(d) We need to find the maximum value of the function $A(x) = 70x - x^2$. Since this is a quadratic function with $a = -1$ and $b = 70$, the maximum occurs at

$$x = -\frac{b}{2a} = -\frac{70}{2(-1)} = 35$$

So the maximum area that she can fence has width 35 ft and length $70 - 35 = 35$ ft.

FIGURE 4

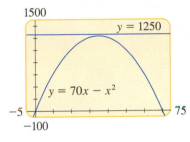

FIGURE 5

EXAMPLE 3 ■ Maximizing Revenue from Ticket Sales

A hockey team plays in an arena with a seating capacity of 15,000 spectators. With the ticket price set at $14, average attendance at recent games has been 9500. A market survey indicates that for each dollar the ticket price is lowered, the average attendance increases by 1000.

(a) Find a function that models the revenue in terms of ticket price.
(b) What ticket price is so high that no one attends, and hence no revenue is generated?
(c) Find the price that maximizes revenue from ticket sales.

■   **Thinking about the problem.**

With a ticket price of $14, the revenue is $9500 \times \$14 = \$133,000$. If the ticket price is lowered to $13, attendance increases to $9500 + 1000 = 10,500$, so the revenue becomes $10,500 \times \$13 = \$136,500$. The table shows the revenue for several ticket prices. Note that if the ticket price is lowered, revenue increases, but if the ticket price is lowered too much, revenue decreases.

Price	Attendance	Revenue
$15	8,500	$127,500
$14	9,500	$133,500
$13	10,500	$136,500
$12	11,500	$138,500
$11	12,500	$137,500
$10	13,500	$135,500

SOLUTION

(a) The model we want is a function that gives the revenue for any ticket price.

■ **Express the model in words.**

We know that

$$\text{revenue} = \text{ticket price} \times \text{attendance}$$

■ **Choose the variable.**

There are two varying quantities—ticket price and attendance. Since the function we want depends on only one variable, we let

$$x = \text{ticket price}$$

Next, we must express the attendance in terms of x.

In Words	In Algebra
Ticket price	x
Amount ticket price is lowered	$14 - x$
Increase in attendance	$1000(14 - x)$
Attendance	$9500 + 1000(14 - x) = 23{,}500 - 1000x$

■ **Set up the model.**

The model is the function R that gives the revenue for a given ticket price x.

$$\text{revenue} = \text{ticket price} \times \text{attendance}$$

$$R(x) = x(23{,}500 - 1000x)$$

$$R(x) = 23{,}500x - 1000x^2$$

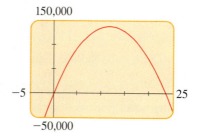

FIGURE 6

■ **Use the model.**

We use the model to answer the questions in parts (b) and (c).

(b) We want to find the ticket price x for which $R(x) = 23{,}500x - 1000x^2 = 0$. We can solve this quadratic equation algebraically or graphically. From the graph in Figure 6 we see that $R(x) = 0$ when $x = 0$ or $x = 23.5$. So, according to our model, the revenue would drop to zero if the ticket price is \$23.50 or higher. (Of course, revenue is also zero if the ticket price is zero!)

(c) Since $R(x) = 23{,}500x - 1000x^2$ is a quadratic function with $a = -1000$ and $b = 23{,}500$, the maximum occurs at

See page 201 for the maximum point of a quadratic function.

$$x = -\frac{b}{2a} = -\frac{23{,}500}{2(-1000)} = 11.75$$

So a ticket price of \$11.75 yields the maximum revenue. At this price the revenue is

$$R(11.75) = 23{,}500(11.75) - 1000(11.75)^2 = \$138{,}062.50$$

■

EXAMPLE 4 ■ Minimizing the Metal in a Can

A manufacturer makes a metal can that holds 1 L (liter) of oil. What radius minimizes the amount of metal in the can?

FIGURE 7

■ **Thinking about the problem.**

To use the least amount of metal, we must minimize the surface area of the can, that is, the area of the top, bottom, and the sides. The area of the top and bottom is $2\pi r^2$ and the area of the sides is $2\pi rh$ (see Figure 7), so the surface area of the can is

$$S = 2\pi r^2 + 2\pi rh$$

The radius and height of the can must be chosen so that the volume is exactly 1 L, or 1000 cm³. If we want a small radius, say $r = 3$, then the height must be just tall enough to make the total volume 1000 cm³. In other words, we must have

$$\pi(3)^2 h = 1000 \qquad \text{Volume of the can is } \pi r^2 h$$

$$h = \frac{1000}{9\pi} \approx 35.4 \text{ cm} \qquad \text{Solve for } h$$

Now that we know the radius and height, we can find the surface area of the can:

$$\text{surface area} = 2\pi(3)^2 + 2\pi(3)(35.4) \approx 729.1 \text{ cm}^3$$

If we want a different radius, we can find the corresponding height and surface area in a similar fashion.

SOLUTION

The model we want is a function that gives the surface area of the can.

■ **Express the model in words.**

We know that for a cylindrical can

$$\text{surface area} \;=\; \text{area of top and bottom} \;+\; \text{area of sides}$$

■ **Choose the variable.**

There are two varying quantities—radius and height. Since the function we want can only depend on one variable, we let

$$r = \text{radius of can}$$

Next, we must express the height in terms of the radius r. Since the volume of a cylindrical can is $V = \pi r^2 h$ and the volume must be 1000 cm³, we have

$$\pi r^2 h = 1000 \qquad \text{Volume of can is 1000 cm}^3$$

$$h = \frac{1000}{\pi r^2} \qquad \text{Solve for } h$$

We can now express the areas of the top, bottom, and sides in terms of r only.

In Words	In Algebra
Radius of can	r
Height of can	$\dfrac{1000}{\pi r^2}$
Area of top and bottom	$2\pi r^2$
Area of sides ($2\pi rh$)	$2\pi r\left(\dfrac{1000}{\pi r^2}\right)$

■ **Set up the model.**

The model is the function S that gives the surface area of the can as a function of the radius r.

surface area = area of top and bottom + area of sides

$$S(r) = 2\pi r^2 + 2\pi r\left(\frac{1000}{\pi r^2}\right)$$

$$S(r) = 2\pi r^2 + \frac{2000}{r}$$

■ **Use the model.**

We use the model to find the minimum surface area of the can. We graph S in Figure 8 and zoom in on the minimum point to find that the minimum value of S is about 554 cm² and occurs when the radius is about 5.4 cm.   ■

FIGURE 8
$S = 2\pi r^2 + \dfrac{2000}{r}$

2.7 EXERCISES

1–18 ■ In these exercises you are asked to find a function that models a real-life situation. Use the principles of modeling described in the text to help you.

1. A rectangular building lot is three times as long as it is wide. Find a function that models its area A in terms of its width w.

2. A poster is 10 inches longer than it is wide. Find a function that models its area A in terms of its width w.

3. A rectangular box has a square base. Its height is half the width of the base. Find a function that models its volume V in terms of its width w.

4. The height of a cylinder is four times its radius. Find a function that models the volume V of the cylinder in terms of its radius r.

5. A rectangle has a perimeter of 20 ft. Find a function that models its area A in terms of the length x of one of its sides.

6. A rectangle has an area of 16 m². Find a function that models its perimeter P in terms of the length x of one of its sides.

7. Find a function that models the area A of an equilateral triangle in terms of the length x of one of its sides.

8. Find a function that models the surface area S of a cube in terms of its volume V.

9. Find a function that models the radius r of a circle in terms of its area A.

10. Find a function that models the area A of a circle in terms of its circumference C.

11. A rectangular box with a volume of 60 ft³ has a square base. Find a function that models its surface area S in terms of the length x of one side of its base.

12. A woman 5 ft tall is standing near a street lamp that is 12 ft tall, as shown in the figure. Find a function that models the length L of her shadow in terms of her distance d from the base of the lamp.

13. Two ships leave port at the same time. One sails south at 15 mi/h and the other sails east at 20 mi/h. Find a function that models the distance D between the ships in terms of the time t (in hours) elapsed since their departure.

14. The sum of two positive numbers is 60. Find a function that models their product P in terms of x, one of the numbers.

15. An isosceles triangle has a perimeter of 8 cm. Find a function that models its area A in terms of the length of its base b.

16. A right triangle has one leg twice as long as the other. Find a function that models its perimeter P in terms of the length x of the shorter leg.

17. A rectangle is inscribed in a semicircle of radius 10, as shown in the figure. Find a function that models the area A of the rectangle in terms of its height h.

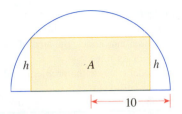

18. The volume of a cone is 100 in³. Find a function that models the height h of the cone in terms of its radius r.

19–36 ■ In these exercises you are asked to find a function that models a real-life situation, and then use the model to answer questions about the situation. Use the guidelines on page 209 to help you.

19. Consider the following problem: Find two numbers whose sum is 19 and whose product is as large as possible.
  (a) Experiment with the problem by making a table like the one below, showing the product of different pairs of numbers that add up to 19. Based on the evidence in your table, estimate the answer to the problem.

First number	Second number	Product
1	18	18
2	17	34
3	16	48
⋮	⋮	⋮

  (b) Find a function that models the product in terms of one of the two numbers.
  (c) Use your model to solve the problem, and compare with your answer to part (a).

20. Find two positive numbers whose sum is 100 and the sum of whose squares is a minimum.

21. Find two numbers whose sum is -24 and whose product is a maximum.

22. Among all rectangles that have a perimeter of 20 ft, find the dimensions of the one with the largest area.

23. Consider the following problem: A farmer has 2400 ft of fencing and wants to fence off a rectangular field that borders a straight river. He does not need a fence along the river. (See the figure on the following page.) What are the dimensions of the field of largest area that he can fence?
  (a) Experiment with the problem by drawing several diagrams illustrating the situation. Calculate the area of

each configuration, and use your results to estimate the dimensions of the largest possible field.

(b) Find a function that models the area of the field in terms of one of its sides.

(c) Use your model to solve the problem, and compare with your answer to part (a).

24. A rancher with 750 ft of fencing wants to enclose a rectangular area and then divide it into four pens with fencing parallel to one side of the rectangle (see the figure).

(a) Find a function that models the total area of the four pens.

(b) Find the largest possible total area of the four pens.

25. A property owner wants to fence a garden plot adjacent to a road, as shown in the figure. The fencing next to the road must be sturdier and costs $5 per foot, whereas the other fencing costs just $3 per foot. The garden is to have an area of 1200 ft².

(a) Find a function that models the cost of fencing the garden.

(b) Find the garden dimensions that minimize the cost of fencing.

(c) If the owner has at most $600 to spend on fencing, find the range of lengths he can fence along the road.

26. A wire 10 cm long is cut into two pieces, one of length x and the other of length $10 - x$, as shown in the figure. Each piece is bent into the shape of a square. Find the value of x that minimizes the total area.

(a) Find a function that models the total area enclosed by the two squares.

(b) Find the value of x that minimizes the total area of the two squares.

27. A baseball team plays in a stadium that holds 55,000 spectators. With the ticket price at $10, the average attendance at recent games has been 27,000. A market survey indicates that for every dollar the ticket price is lowered, attendance increases by 3000.

(a) Find a function that models the revenue in terms of ticket price.

(b) What ticket price is so high that no revenue is generated?

(c) Find the price that maximizes revenue from ticket sales.

28. A community bird-watching society makes and sells simple bird feeders to raise money for its conservation activities. The materials for each feeder cost $6, and they sell an average of 20 per week at a price of $10 each. They have been considering raising the price, so they conduct a survey and find that for every dollar increase they lose 2 sales per week.

(a) Find a function that models weekly profit in terms of price per feeder.

(b) What price should they charge per feeder to maximize profits? What is the maximum profit?

29. A Norman window has the shape of a rectangle surmounted by a semicircle, as shown in the figure at the top of page 217. A Norman window with perimeter 30 ft is to be constructed.

(a) Find a function that models the area of the window.

(b) Find the dimensions of the window that admits the greatest amount of light.

30. A box with an open top is to be constructed from a rectangular piece of cardboard with dimensions 12 in. by 20 in. by cutting out equal squares of side x at each corner and then folding up the sides (see the figure).
(a) Find a function that models the volume of the box.
(b) Find the values of x for which the volume is greater than 200 in³.
(c) Find the largest volume that such a box can have.

31. An open box with a square base is to have a volume of 12 ft³.
(a) Find a function that models the surface area of the box.
(b) Find the box dimensions that minimize the amount of material used.

32. Find the dimensions that give the largest area for the rectangle shown in the figure. Its base is on the x-axis and its other two vertices are above the x-axis, lying on the parabola $y = 8 - x^2$.

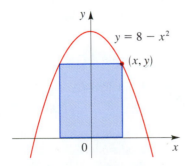

33. A rancher wants to build a rectangular pen with an area of 100 m².
(a) Find a function that models the length of fencing required.
(b) Find the pen dimensions that require the minimum amount of fencing.

34. A man stands at a point A on the bank of a straight river, 2 mi wide. To reach point B, 7 mi downstream on the opposite bank, he first rows his boat to point P on the opposite bank and then walks the remaining distance x to B, as shown in the figure. He can row at a speed of 2 mi/h and walk at a speed of 5 mi/h.
(a) Find a function that models the time needed for the trip.
(b) Where should he land so that he reaches B as soon as possible?

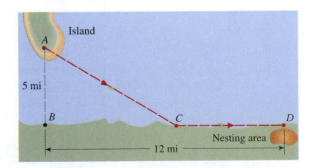

35. A bird is released from point A on an island, 5 mi from the nearest point B on a straight shoreline. The bird flies to a point C on the shoreline, and then flies along the shoreline to its nesting area D (see the figure). Suppose the bird requires 10 kcal/mi of energy to fly over land and 14 kcal/mi to fly over water (see Example 9 in Section 1.6).
(a) Find a function that models the energy expenditure of the bird.
(b) If the bird instinctively chooses a path that minimizes its energy expenditure, to what point does it fly?

36. A kite frame is to be made from six pieces of wood. The
four pieces that form its border have been cut to the lengths
indicated in the figure. Let x be as shown in the figure.
(a) Show that the area of the kite is given by the function

$$A(x) = x\left(\sqrt{25 - x^2} + \sqrt{144 - x^2}\right).$$

(b) How long should each of the two crosspieces be to
maximize the area of the kite?

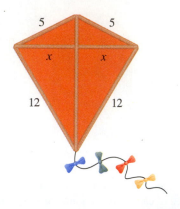

2.8   COMBINING FUNCTIONS

Two functions f and g can be combined to form new functions $f + g$, $f - g$, fg, and
f/g in a manner similar to the way we add, subtract, multiply, and divide real numbers. For example, we define the function $f + g$ by

$$(f + g)(x) = f(x) + g(x)$$

> The sum of f and g is defined by
> $$(f + g)(x) = f(x) + g(x)$$
> The name of the new function is
> "$f + g$." So this $+$ sign stands for the
> operation of addition of *functions*.
> The $+$ sign on the right side, however,
> stands for addition of the *numbers* $f(x)$
> and $g(x)$.

The new function $f + g$ is called the **sum** of the functions f and g; its value at x is
$f(x) + g(x)$. Of course, the sum on the right-hand side makes sense only if both
$f(x)$ and $g(x)$ are defined, that is, if x belongs to the domain of f and also to the
domain of g. So, if the domain of f is A and the domain of g is B, then the domain
of $f + g$ is the intersection of these domains, that is, $A \cap B$. Similarly, we can
define the **difference** $f - g$, the **product** fg, and the **quotient** f/g of the functions
f and g. Their domains are $A \cap B$, but in the case of the quotient we must remember not to divide by 0.

ALGEBRA OF FUNCTIONS

Let f and g be functions with domains A and B. Then the functions $f + g$,
$f - g$, fg, and f/g are defined as follows.

$$(f + g)(x) = f(x) + g(x) \qquad \text{Domain } A \cap B$$

$$(f - g)(x) = f(x) - g(x) \qquad \text{Domain } A \cap B$$

$$(fg)(x) = f(x)g(x) \qquad \text{Domain } A \cap B$$

$$\left(\frac{f}{g}\right)(x) = \frac{f(x)}{g(x)} \qquad \text{Domain } \{x \in A \cap B \mid g(x) \neq 0\}$$

EXAMPLE 1 ■ Combining Functions

Let $f(x) = x^3$ and $g(x) = \sqrt{x}$.

(a) Find the functions $f + g$, $f - g$, fg, and f/g and their domains.
(b) Find $(f + g)(4)$, $(f - g)(4)$, $(fg)(3)$, and $(f/g)(1)$.

SOLUTION

(a) The domain of f is $\mathbb{R}$ and the domain of g is $\{x \mid x \geqslant 0\}$. We have

$$(f + g)(x) = f(x) + g(x) = x^3 + \sqrt{x} \qquad \text{Domain } \{x \mid x \geqslant 0\}$$

$$(f - g)(x) = f(x) - g(x) = x^3 - \sqrt{x} \qquad \text{Domain } \{x \mid x \geqslant 0\}$$

$$(fg)(x) = f(x)g(x) = x^3\sqrt{x} \qquad \text{Domain } \{x \mid x \geqslant 0\}$$

$$\left(\frac{f}{g}\right)(x) = \frac{f(x)}{g(x)} = \frac{x^3}{\sqrt{x}} \qquad \text{Domain } \{x \mid x > 0\}$$

Note that in the domain of f/g we exclude 0 because $g(0) = 0$.

(b) $(f + g)(4) = f(4) + g(4) = 4^3 + \sqrt{4} = 66$

$(f - g)(4) = f(4) - g(4) = 4^3 - \sqrt{4} = 62$

$(fg)(3) = f(3)g(3) = 3^3\sqrt{3} = 27\sqrt{3}$

$\left(\dfrac{f}{g}\right)(1) = \dfrac{f(1)}{g(1)} = \dfrac{1^3}{\sqrt{1}} = 1$ ■

EXAMPLE 2 ■ Finding Domains of Combinations of Functions

If $f(x) = \sqrt{x}$ and $g(x) = \sqrt{4 - x^2}$, find the functions $f + g$, $f - g$, fg, and f/g and their domains.

SOLUTION

The domain of $f(x) = \sqrt{x}$ is $[0, \infty)$. The domain of $g(x) = \sqrt{4 - x^2}$ is the interval $[-2, 2]$ (from Example 7 in Section 2.2). The intersection of the domains of f and g is

$$[0, \infty) \cap [-2, 2] = [0, 2]$$

Thus, we have

$$(f + g)(x) = \sqrt{x} + \sqrt{4 - x^2} \qquad \text{Domain } \{x \mid 0 \leqslant x \leqslant 2\}$$

$$(f - g)(x) = \sqrt{x} - \sqrt{4 - x^2} \qquad \text{Domain } \{x \mid 0 \leqslant x \leqslant 2\}$$

$$(fg)(x) = \sqrt{x}\sqrt{4 - x^2} = \sqrt{4x - x^3} \qquad \text{Domain } \{x \mid 0 \leqslant x \leqslant 2\}$$

$$\left(\frac{f}{g}\right)(x) = \frac{\sqrt{x}}{\sqrt{4 - x^2}} = \sqrt{\frac{x}{4 - x^2}} \qquad \text{Domain } \{x \mid 0 \leqslant x < 2\}$$

Notice that the domain of f/g is the interval $[0, 2)$ because we must exclude the points where $g(x) = 0$, that is, $x = \pm 2$. ∎

The graph of the function $f + g$ can be obtained from the graphs of f and g by **graphical addition**. This means that we add corresponding y-coordinates, as illustrated in the next example.

EXAMPLE 3 ■ Using Graphical Addition

The graphs of f and g are shown in Figure 1. Use graphical addition to graph the function $f + g$.

SOLUTION

In Figure 2 we obtain the graph of $f + g$ by "graphically adding" the value of $f(x)$ to $g(x)$. This is implemented by copying the line segment PQ on top of PR to obtain the point S on the graph of $f + g$.

FIGURE 1

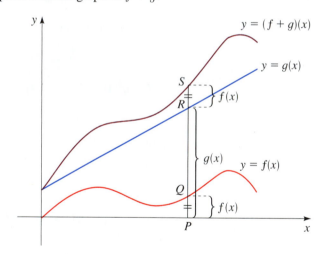

FIGURE 2
Graphical addition

∎

Composition of Functions

Now let's consider a very important way of combining two functions to get a new function. Suppose $f(x) = \sqrt{x}$ and $g(x) = x^2 + 1$. We may define a function h as

$$h(x) = f(g(x)) = f(x^2 + 1) = \sqrt{x^2 + 1}$$

The function h is made up of the functions f and g in an interesting way: Given a number x, we first apply to it the function g, then apply f to the result. In this case, f is the rule "take the square root," g is the rule "square, then add 1," and h is the rule "square, then add 1, then take the square root." In other words, we get the rule h by applying the rule g and then the rule f.

In general, given any two functions f and g, we start with a number x in the domain of g and find its image $g(x)$. If this number $g(x)$ is in the domain of f, we can then calculate the value of $f(g(x))$. The result is a new function $h(x) = f(g(x))$ obtained by substituting g into f. It is called the *composition* (or *composite*) of f and g and is denoted by $f \circ g$ ("f composed with g").

COMPOSITION OF FUNCTIONS

Given two functions f and g, the **composite function** $f \circ g$ (also called the **composition** of f and g) is defined by

$$(f \circ g)(x) = f(g(x))$$

The domain of $f \circ g$ is the set of all x in the domain of g such that $g(x)$ is in the domain of f. In other words, $(f \circ g)(x)$ is defined whenever both $g(x)$ and $f(g(x))$ are defined. We can picture $f \circ g$ using a machine diagram (Figure 3) or an arrow diagram (Figure 4).

FIGURE 3

The $f \circ g$ machine is composed of the g machine (first) and then the f machine.

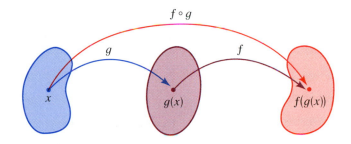

FIGURE 4

Arrow diagram for $f \circ g$

EXAMPLE 4 ■ Finding the Composition of Functions

Let $f(x) = x^2$ and $g(x) = x - 3$.

(a) Find the functions $f \circ g$ and $g \circ f$ and their domains.
(b) Find $(f \circ g)(5)$ and $(g \circ f)(7)$.

SOLUTION

(a) We have

$$(f \circ g)(x) = f(g(x)) \qquad \text{Definition of } f \circ g$$
$$= f(x - 3) \qquad \text{Definition of } g$$
$$= (x - 3)^2 \qquad \text{Definition of } f$$

In Example 4, f is the rule "square" and g is the rule "subtract 3." The function $f \circ g$ *first* subtracts 3 and *then* squares; the function $g \circ f$ *first* squares and *then* subtracts 3.

and

$$(g \circ f)(x) = g(f(x)) \qquad \text{Definition of } g \circ f$$

$$= g(x^2) \qquad \text{Definition of } f$$

$$= x^2 - 3 \qquad \text{Definition of } g$$

The domains of both $f \circ g$ and $g \circ f$ are $\mathbb{R}$.

(b) We have

$$(f \circ g)(5) = f(g(5)) = f(2) = 2^2 = 4$$

$$(g \circ f)(7) = g(f(7)) = g(49) = 49 - 3 = 46 \qquad \blacksquare$$

You can see from Example 4 that, in general, $f \circ g \neq g \circ f$. Remember that the notation $f \circ g$ means that the function g is applied first and then f is applied second.

EXAMPLE 5 ■ Finding the Composition of Functions

The graphs of f and g of Example 5, as well as $f \circ g$, $g \circ f$, $f \circ f$, and $g \circ g$, are shown below and on page 223. These graphs indicate that the operation of composition can produce functions quite different from the original functions.

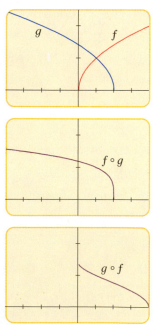

If $f(x) = \sqrt{x}$ and $g(x) = \sqrt{2 - x}$, find the following functions and their domains.

(a) $f \circ g$ (b) $g \circ f$ (c) $f \circ f$ (d) $g \circ g$

SOLUTION

(a)
$$(f \circ g)(x) = f(g(x)) \qquad \text{Definition of } f \circ g$$

$$= f(\sqrt{2 - x}) \qquad \text{Definition of } g$$

$$= \sqrt{\sqrt{2 - x}} \qquad \text{Definition of } f$$

$$= \sqrt[4]{2 - x}$$

The domain of $f \circ g$ is $\{x \mid 2 - x \geq 0\} = \{x \mid x \leq 2\} = (-\infty, 2]$.

(b)
$$(g \circ f)(x) = g(f(x)) \qquad \text{Definition of } g \circ f$$

$$= g(\sqrt{x}) \qquad \text{Definition of } f$$

$$= \sqrt{2 - \sqrt{x}} \qquad \text{Definition of } g$$

For $\sqrt{x}$ to be defined, we must have $x \geq 0$. For $\sqrt{2 - \sqrt{x}}$ to be defined, we must have $2 - \sqrt{x} \geq 0$, that is, $\sqrt{x} \leq 2$, or $x \leq 4$. Thus, we have $0 \leq x \leq 4$, so the domain of $g \circ f$ is the closed interval $[0, 4]$.

(c)
$$(f \circ f)(x) = f(f(x)) \qquad \text{Definition of } f \circ f$$

$$= f(\sqrt{x}) \qquad \text{Definition of } f$$

$$= \sqrt{\sqrt{x}} \qquad \text{Definition of } f$$

$$= \sqrt[4]{x}$$

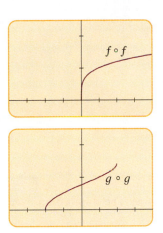

The domain of $f \circ f$ is $[0, \infty)$.

(d)
$$(g \circ g)(x) = g(g(x)) \qquad \text{Definition of } g \circ g$$

$$= g\left(\sqrt{2 - x}\right) \qquad \text{Definition of } g$$

$$= \sqrt{2 - \sqrt{2 - x}} \qquad \text{Definition of } g$$

This expression is defined when both $2 - x \geq 0$ and $2 - \sqrt{2 - x} \geq 0$. The first inequality means $x \leq 2$, and the second is equivalent to $\sqrt{2 - x} \leq 2$, or $2 - x \leq 4$, or $x \geq -2$. Thus, $-2 \leq x \leq 2$, so the domain of $g \circ g$ is $[-2, 2]$. ∎

It is possible to take the composition of three or more functions. For instance, the composite function $f \circ g \circ h$ is found by first applying h, then g, and then f as follows:

$$(f \circ g \circ h)(x) = f(g(h(x)))$$

EXAMPLE 6 ■ A Composition of Three Functions

Find $f \circ g \circ h$ if $f(x) = x/(x + 1)$, $g(x) = x^{10}$, and $h(x) = x + 3$.

SOLUTION

$$(f \circ g \circ h)(x) = f(g(h(x))) \qquad \text{Definition of } f \circ g \circ h$$

$$= f(g(x + 3)) \qquad \text{Definition of } h$$

$$= f\left((x + 3)^{10}\right) \qquad \text{Definition of } g$$

$$= \frac{(x + 3)^{10}}{(x + 3)^{10} + 1} \qquad \text{Definition of } f$$

■

So far we have used composition to build complicated functions from simpler ones. But in calculus it is useful to be able to "decompose" a complicated function into simpler ones, as shown in the following example.

EXAMPLE 7 ■ Recognizing a Composition of Functions

Given $F(x) = \sqrt[4]{x + 9}$, find functions f and g such that $F = f \circ g$.

SOLUTION

Since the formula for F says to first add 9 and then take the fourth root, we let

$$g(x) = x + 9 \qquad \text{and} \qquad f(x) = \sqrt[4]{x}$$

Then

$$\bullet (f \circ g)(x) = f(g(x))$$

$$= f(x + 9)$$

$$= \sqrt[4]{x + 9}$$

$$= F(x) \quad \blacksquare$$

EXAMPLE 8 ■ An Application of Composition of Functions

A ship is traveling at 20 mi/h parallel to a straight shoreline. The ship is 5 mi from shore. It passes a lighthouse at noon.

(a) Express the distance s between the lighthouse and the ship as a function of d, the distance the ship has traveled since noon; that is, find f so that $s = f(d)$.
(b) Express d as a function of t, the time elapsed since noon; that is, find g so that $d = g(t)$.
(c) Find $f \circ g$. What does this function represent?

SOLUTION

We first draw a diagram as in Figure 5.

(a) We can relate the distances s and d by the Pythagorean Theorem. Thus, s can be expressed as a function of d by

$$s = f(d) = \sqrt{25 + d^2}$$

(b) Since the ship is traveling at 20 mi/h, the distance d it has traveled is a function of t as follows:

$$d = g(t) = 20t$$

(c) We have

$$(f \circ g)(t) = f(g(t)) \qquad \text{Definition of } f \circ g$$

$$= f(20t) \qquad \text{Definition of } g$$

$$= \sqrt{25 + (20t)^2} \qquad \text{Definition of } f$$

The function $f \circ g$ gives the distance of the ship from the lighthouse as a function of time.   ■

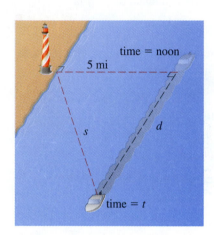

time = noon

5 mi

s

d

time = t

FIGURE 5

distance = rate × time

2.8 EXERCISES

1–6 ■ Find $f + g$, $f - g$, fg, and f/g and their domains.

1. $f(x) = x^2$, $g(x) = x + 2$

2. $f(x) = x^3 + 2x^2$, $g(x) = 3x^2 - 1$

3. $f(x) = \sqrt{1 + x^2}$, $g(x) = \sqrt{1 - x}$

4. $f(x) = \sqrt{9 - x^2}$, $g(x) = \sqrt{x^2 - 1}$

5. $f(x) = \dfrac{2}{x}$, $g(x) = -\dfrac{2}{x + 4}$

6. $f(x) = \dfrac{1}{x + 1}$, $g(x) = \dfrac{x}{x + 1}$

7–10 ■ Find the domain of the function.

7. $f(x) = \sqrt{x} + \sqrt{1 - x}$

8. $g(x) = \sqrt{x + 1} - \dfrac{1}{x}$

9. $h(x) = (x + 1)^2 (2x - 8)^{1/4}$

10. $k(x) = \dfrac{\sqrt{x + 3}}{x - 1}$

11–12 ■ Use graphical addition to sketch the graph of $f + g$.

11.

12.

13–16 ■ Draw the graphs of f, g, and $f + g$ on a common screen to illustrate graphical addition.

13. $f(x) = \sqrt{1 + x}$, $g(x) = \sqrt{1 - x}$

14. $f(x) = x^2$, $g(x) = \sqrt{x}$

15. $f(x) = x^2$, $g(x) = x^3$

16. $f(x) = \sqrt[4]{1 - x}$, $g(x) = \sqrt{1 - \dfrac{x^2}{9}}$

17–22 ■ Use $f(x) = 3x - 5$ and $g(x) = 2 - x^2$ to evaluate the expression.

17. (a) $f(g(0))$ (b) $g(f(0))$

18. (a) $f(f(4))$ (b) $g(g(3))$

19. (a) $(f \circ g)(-2)$ (b) $(g \circ f)(-2)$

20. (a) $(f \circ f)(-1)$ (b) $(g \circ g)(2)$

21. (a) $(f \circ g)(x)$ (b) $(g \circ f)(x)$

22. (a) $(f \circ f)(x)$ (b) $(g \circ g)(x)$

23–28 ■ Use the given graphs of f and g to evaluate the expression.

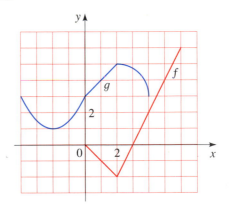

23. $f(g(2))$ **24.** $g(f(0))$

25. $(g \circ f)(4)$ **26.** $(f \circ g)(0)$

27. $(g \circ g)(-2)$

28. $(f \circ f)(4)$

29–40 ■ Find the functions $f \circ g$, $g \circ f$, $f \circ f$, and $g \circ g$ and their domains.

29. $f(x) = 2x + 3$, $\quad g(x) = 4x - 1$

30. $f(x) = 6x - 5$, $\quad g(x) = \dfrac{x}{2}$

31. $f(x) = x^2$, $\quad g(x) = x + 1$

32. $f(x) = x^3 + 2$, $\quad g(x) = \sqrt[3]{x}$

33. $f(x) = \dfrac{1}{x}$, $\quad g(x) = 2x + 4$

34. $f(x) = x^2$, $\quad g(x) = \sqrt{x - 3}$

35. $f(x) = |x|$, $\quad g(x) = 2x + 3$

36. $f(x) = x - 4$, $\quad g(x) = |x + 4|$

37. $f(x) = \dfrac{x}{x + 1}$, $\quad g(x) = 2x - 1$

38. $f(x) = \dfrac{1}{\sqrt{x}}$, $\quad g(x) = x^2 - 4x$

39. $f(x) = \sqrt[3]{x}$, $\quad g(x) = \sqrt[4]{x}$

40. $f(x) = \dfrac{2}{x}$, $\quad g(x) = \dfrac{x}{x - 2}$

41–44 ■ Find $f \circ g \circ h$.

41. $f(x) = x - 1$, $\quad g(x) = \sqrt{x}$, $\quad h(x) = x - 1$

42. $f(x) = \dfrac{1}{x}$, $\quad g(x) = x^3$, $\quad h(x) = x^2 + 2$

43. $f(x) = x^4 + 1$, $\quad g(x) = x - 5$, $\quad h(x) = \sqrt{x}$

44. $f(x) = \sqrt{x}$, $\quad g(x) = \dfrac{x}{x - 1}$, $\quad h(x) = \sqrt[3]{x}$

45–50 ■ Express the function in the form $f \circ g$.

45. $F(x) = (x - 9)^5$

46. $F(x) = \sqrt{x} + 1$

47. $G(x) = \dfrac{x^2}{x^2 + 4}$

48. $G(x) = \dfrac{1}{x + 3}$

49. $H(x) = |1 - x^3|$ $\qquad$ **50.** $H(x) = \sqrt{1 + \sqrt{x}}$

51–54 ■ Express the function in the form $f \circ g \circ h$.

51. $F(x) = \dfrac{1}{x^2 + 1}$

52. $F(x) = \sqrt[3]{\sqrt{x} - 1}$

53. $G(x) = (4 + \sqrt[3]{x})^9$

54. $G(x) = \dfrac{2}{(3 + \sqrt{x})^2}$

55. A stone is dropped in a lake, creating a circular ripple that travels outward at a speed of 60 cm/s. Express the area of this circle as a function of time t (in seconds).

56. A spherical balloon is being inflated. The radius of the balloon is increasing at the rate of 1 cm/s.
 (a) Find a function f that models the radius as a function of time.
 (b) Find a function g that models the volume as a function of the radius.
 (c) Find $g \circ f$. What does this function represent?

57. A spherical weather balloon is being inflated. The radius of the balloon is increasing at the rate of 2 cm/s. Express the surface area of the balloon as a function of time t (in seconds).

58. You have a $50 coupon from the manufacturer good for the purchase of a cell phone. The store where you are purchasing your cell phone is offering a 20% discount on all cell phones. Let x represent the regular price of the cell phone.

(a) Suppose only the 20% discount applies. Find a function f that models the purchase price of the cell phone as a function of the regular price x.

(b) Suppose only the $50 coupon applies. Find a function g that models the purchase price of the cell phone as a function of the sticker price x.

(c) If you can use the coupon and the discount, then the purchase price is either $f \circ g(x)$ or $g \circ f(x)$, depending on the order in which they are applied to the price. Find both $f \circ g(x)$ and $g \circ f(x)$. Which composition gives the lower price?

59. An appliance dealer advertises a 10% discount on all his washing machines. In addition, the manufacturer offers a $100 rebate on the purchase of a washing machine. Let x represent the sticker price of the washing machine.

(a) Suppose only the 10% discount applies. Find a function f that models the purchase price of the washer as a function of the sticker price x.

(b) Suppose only the $100 rebate applies. Find a function g that models the purchase price of the washer as a function of the sticker price x.

(c) Find $f \circ g$ and $g \circ f$. What do these functions represent? Which is the better deal?

60. An airplane is flying at a speed of 350 mi/h at an altitude of one mile. The plane passes directly above a radar station at time $t = 0$.

(a) Express the distance s (in miles) between the plane and the radar station as a function of the horizontal distance d (in miles) that the plane has flown.

(b) Express d as a function of the time t (in hours) that the plane has flown.

(c) Use composition to express s as a function of t.

61. Compound Interest   A savings account earns 5% interest compounded annually. If you invest x dollars in such an account, then the amount $A(x)$ of the investment after one year is the initial investment plus 5%; that is, $A(x) = x + 0.05x = 1.05x$. Find

$$A \circ A, \quad A \circ A \circ A \quad \text{and} \quad A \circ A \circ A \circ A$$

What do these compositions represent? Find a formula for what you get when you compose n copies of A.

62. Composing Linear Functions   The graphs of the functions

$$f(x) = m_1 x + b_1 \quad \text{and} \quad g(x) = m_2 x + b_2$$

are lines with slopes m_1 and m_2, respectively. Is the graph of $f \circ g$ a line? If so, what is its slope?

63. Solving an Equation for an Unknown Function
Suppose that

$$g(x) = 2x + 1 \quad \text{and} \quad h(x) = 4x^2 + 4x + 7$$

Find a function f such that $f \circ g = h$. (Think about what operations you would have to perform on the formula for g to end up with the formula for h.) Now suppose that

$$f(x) = 3x + 5 \quad \text{and} \quad h(x) = 3x^2 + 3x + 2$$

Use the same sort of reasoning to find a function g such that $f \circ g = h$.

64. Compositions of Odd and Even Functions   Suppose that

$$h = f \circ g$$

If g is an even function, is h necessarily even? If g is odd, is h odd? What if g is odd and f is odd? What if g is odd and f is even?

Laboratory Project

Iteration and Chaos

The **iterates** of a function f at a point x_0 are $f(x_0)$, $f(f(x_0))$, $f(f(f(x_0)))$, and so on. We write

$$x_1 = f(x_0) \qquad \text{The first iterate}$$

$$x_2 = f(f(x_0)) \qquad \text{The second iterate}$$

$$x_3 = f(f(f(x_0))) \qquad \text{The third iterate}$$

For example, if $f(x) = x^2$, then the iterates of f at 2 are $x_1 = 4$, $x_2 = 16$, $x_3 = 256$, and so on. (Check this.) Iterates can be described graphically as in Figure 1. Start with x_0 on the x-axis, move vertically to the graph of f, then horizontally to the line $y = x$, then vertically to the graph of f, and so on. The x-coordinates of the points on the graph of f are the iterates of f at x_0.

FIGURE 1

Iterates are important in studying the **logistic function**

$$f(x) = kx(1 - x)$$

which models the population of a species with limited potential for growth (such as rabbits on an island or fish in a pond). In this model the maximum population that the environment can support is 1 (that is, 100%); if we start with a fraction of that population, say 0.1 (10%), then the iterates of f at 0.1 give the population after each time interval (days, months, or years, depending on the species). The constant k depends on the rate of growth of the species being modeled; it is called the **growth constant**. For example, for $k = 2.6$ and $x_0 = 0.1$ the iterates shown in the table give the population of the species for the first 12 time intervals. The population seems to be stabilizing around 0.615 (that is, 61.5% of maximum).

In the three graphs in Figure 2 we plot the iterates of f at 0.1 for different values of the growth constant k. For $k = 2.6$ the population appears to stabilize at a value 0.615 of maximum, for $k = 3.1$ the population appears to oscillate between

n	x_n
0	0.1
1	0.234
2	0.46603
3	0.64700
4	0.59382
5	0.62712
6	0.60799
7	0.61968
8	0.61276
9	0.61694
10	0.61444
11	0.61595
12	0.61505

two values, and for $k = 3.8$ no obvious pattern emerges. This latter situation is described mathematically by the word **chaos**.

$k = 2.6$

$k = 3.1$

$k = 3.8$

FIGURE 2

1. Use the graphical procedure illustrated in Figure 1 to find the first five iterates of $f(x) = 2x(1 - x)$ at $x = 0.1$.

2. Find the iterates of $f(x) = x^2$ at $x = 1$.

3. Find the iterates of $f(x) = 1/x$ at $x = 2$.

4. Find the first six iterates of $f(x) = 1/(1 - x)$ at $x = 2$. What is 1000th iterate of f at 2?

5. Find the first 10 iterates of the logistic function at $x = 0.1$ for the given value of k. Does the population appear to stabilize, oscillate, or is it chaotic?
   (a) $k = 2.1$ (b) $k = 3.2$ (c) $k = 3.9$

6. It is easy to find iterates using a graphing calculator. The following steps show how to find the iterates of $f(x) = kx(1 - x)$ at 0.1 for $k = 3$ on a TI-85 calculator. (The procedure can be adapted for any graphing calculator.)

`y1 = K * x * (1 - x)`	Enter f as **Y1** on the graph list
`3 → K`	Store 3 in the variable **K**
`0.1 → x`	Store 0.1 in the variable **x**
`y1 → x`	Evaluate f at **x** and store result back in **x**
`0.27`	Press ENTER and obtain first iterate
`0.5913`	Keep pressing ENTER to re-execute the
`0.72499293`	command and obtain successive iterates
`0.59813454435`	

Use a graphing calculator to experiment with how the value of k affects the iterates of $f(x) = kx(1 - x)$ at 0.1. Find several different values of k that make the iterates stabilize at one value, oscillate between two values, and exhibit chaos. (Use values of k between 1 and 4). Can you find a value of k that makes the iterates oscillate between *four* values?

2.9   ONE-TO-ONE FUNCTIONS AND THEIR INVERSES

Let's compare the functions f and g whose arrow diagrams are shown in Figure 1. Note that f never takes on the same value twice (any two numbers in A have different images), whereas g does take on the same value twice (both 2 and 3 have the same image, 4). In symbols, $g(2) = g(3)$ but $f(x_1) \neq f(x_2)$ whenever $x_1 \neq x_2$. Functions that have this latter property are called *one-to-one*.

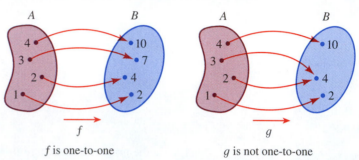

FIGURE 1

f is one-to-one g is not one-to-one

DEFINITION OF A ONE-TO-ONE FUNCTION

A function within domain A is called a **one-to-one function** if no two elements of A have the same image, that is,

$$f(x_1) \neq f(x_2) \qquad \text{whenever } x_1 \neq x_2$$

An equivalent way of writing the condition for a one-to-one function is this:

$$\text{If } f(x_1) = f(x_2), \text{ then } x_1 = x_2.$$

If a horizontal line intersects the graph of f at more than one point, then we see from Figure 2 that there are numbers $x_1 \neq x_2$ such that $f(x_1) = f(x_2)$. This means that f is not one-to-one. Therefore, we have the following geometric method for determining whether a function is one-to-one.

HORIZONTAL LINE TEST

A function is one-to-one if and only if no horizontal line intersects its graph more than once.

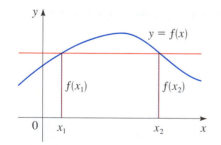

FIGURE 2
This function is not one-to-one because $f(x_1) = f(x_2)$.

EXAMPLE 1 ■ Deciding whether a Function Is One-to-One

Is the function $f(x) = x^3$ one-to-one?

SOLUTION 1

If $x_1 \neq x_2$, then $x_1^3 \neq x_2^3$ (two different numbers cannot have the same cube). Therefore, $f(x) = x^3$ is one-to-one.

FIGURE 3

$f(x) = x^3$ is one-to-one.

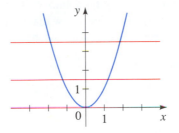

FIGURE 4

$g(x) = x^2$ is not one-to-one.

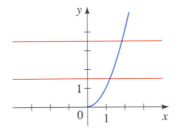

FIGURE 5

$h(x) = x^2$ ($x \geqslant 0$) is one-to-one.

SOLUTION 2

From Figure 3 we see that no horizontal line intersects the graph of $f(x) = x^3$ more than once. Therefore, by the Horizontal Line Test, f is one-to-one.   ■

Notice that the function f of Example 1 is increasing and is also one-to-one. In fact, it can be proved that every increasing function and every decreasing function is one-to-one.

EXAMPLE 2 ■ Deciding whether a Function Is One-to-One

Is the function $g(x) = x^2$ one-to-one?

SOLUTION 1

This function is not one-to-one because, for instance,

$$g(1) = 1 = g(-1)$$

and so 1 and -1 have the same image.

SOLUTION 2

From Figure 4 we see that there are horizontal lines that intersect the graph of g more than once. Therefore, by the Horizontal Line Test, g is not one-to-one.   ■

Although the function g in Example 2 is not one-to-one, it is possible to restrict its domain so that the resulting function is one-to-one. In fact, if we define

$$h(x) = x^2, \qquad x \geqslant 0$$

then h is one-to-one, as you can see from Figure 5 and the Horizontal Line Test.

EXAMPLE 3 ■ Showing That a Function Is One-to-One

Show that the function $f(x) = 3x + 4$ is one-to-one.

SOLUTION

Suppose there are numbers x_1 and x_2 such that $f(x_1) = f(x_2)$. Then

$$3x_1 + 4 = 3x_2 + 4$$

$$3x_1 = 3x_2$$

$$x_1 = x_2$$

Therefore, f is one-to-one.   ■

The Inverse of a Function

One-to-one functions are important because they are precisely the functions that possess inverse functions according to the following definition.

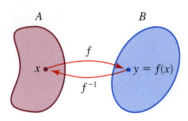

⊘ Don't mistake the -1 in f^{-1} for an exponent.

$$f^{-1} \quad does\ not\ mean \quad \frac{1}{f(x)}$$

The reciprocal $1/f(x)$ is written as $(f(x))^{-1}$.

DEFINITION OF THE INVERSE OF A FUNCTION

Let f be a one-to-one function with domain A and range B. Then its **inverse function** f^{-1} has domain B and range A and is defined by

$$f^{-1}(y) = x \quad \Leftrightarrow \quad f(x) = y$$

for any y in B.

This definition says that if f takes x into y, then f^{-1} takes y back into x. (If f were not one-to-one, then f^{-1} would not be defined uniquely.) The arrow diagram in Figure 6 indicates that f^{-1} reverses the effect of f. From the definition we have

$$\text{domain of } f^{-1} = \text{range of } f$$

$$\text{range of } f^{-1} = \text{domain of } f$$

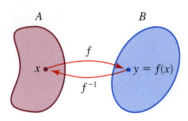

FIGURE 6

EXAMPLE 4 ■ Finding f^{-1} for Specific Values

If $f(1) = 5$, $f(3) = 7$, and $f(8) = -10$, find $f^{-1}(5)$, $f^{-1}(7)$, and $f^{-1}(-10)$.

SOLUTION

From the definition of f^{-1} we have

$$f^{-1}(5) = 1 \qquad \text{because} \qquad f(1) = 5$$

$$f^{-1}(7) = 3 \qquad \text{because} \qquad f(3) = 7$$

$$f^{-1}(-10) = 8 \qquad \text{because} \qquad f(8) = -10$$

Figure 7 shows how f^{-1} reverses the effect of f in this case.

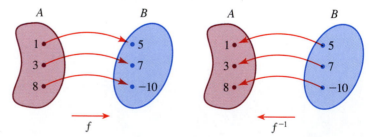

FIGURE 7

By definition the inverse function f^{-1} undoes what f does: If we start with x, apply f, and then apply f^{-1}, we arrive back at x, where we started. Similarly, f undoes what f^{-1} does. In general, any function that reverses the effect of f in this way must be the inverse of f. These observations are expressed precisely as follows.

PROPERTY OF INVERSE FUNCTIONS

Let f be a one-to-one function with domain A and range B. The inverse function f^{-1} satisfies the following cancellation properties.

$$f^{-1}(f(x)) = x \qquad \text{for every } x \text{ in } A$$

$$f(f^{-1}(x)) = x \qquad \text{for every } x \text{ in } B$$

Conversely, any function f^{-1} satisfying these equations is the inverse of f.

These properties indicate that f is the inverse function of f^{-1}, so we say that f and f^{-1} are *inverses of each other.*

EXAMPLE 5 ■ Verifying That Two Functions Are Inverses

Show that $f(x) = x^3$ and $g(x) = x^{1/3}$ are inverses of each other.

SOLUTION

Note that the domain and range of both f and g is $\mathbb{R}$. We have

$$g(f(x)) = g(x^3) = (x^3)^{1/3} = x$$

$$f(g(x)) = f(x^{1/3}) = (x^{1/3})^3 = x$$

So, by the Property of Inverse Functions, f and g are inverses of each other. These equations simply say that the cube function and the cube root function, when composed, cancel each other. ■

Now let's examine how we compute inverse functions. We first observe from the definition of f^{-1} that

$$y = f(x) \quad \Leftrightarrow \quad f^{-1}(y) = x$$

So, if $y = f(x)$ and if we are able to solve this equation for x in terms of y, then we must have $x = f^{-1}(y)$. If we then interchange x and y, we have $y = f^{-1}(x)$, which is the desired equation.

HOW TO FIND THE INVERSE OF
A ONE-TO-ONE FUNCTION

1. Write $y = f(x)$.

2. Solve this equation for x in terms of y (if possible).

3. Interchange x and y. The resulting equation is $y = f^{-1}(x)$.

Note that Steps 2 and 3 can be reversed. In other words, we can interchange x and y first and then solve for y in terms of x.

In Example 6 note how f^{-1} reverses the effect of f. The function f is the rule "multiply by 3, then subtract 2," whereas f^{-1} is the rule "add 2, then divide by 3."

EXAMPLE 6 ■ Finding the Inverse of a Function

Find the inverse of the function $f(x) = 3x - 2$.

SOLUTION

First we write $y = f(x)$.

$$y = 3x - 2$$

Then we solve this equation for x:

$$3x = y + 2 \qquad \text{Add 2}$$

$$x = \frac{y + 2}{3} \qquad \text{Divide by 3}$$

Finally, we interchange x and y:

$$y = \frac{x + 2}{3}$$

Therefore, the inverse function is $f^{-1}(x) = \dfrac{x + 2}{3}$.

CHECK YOUR ANSWER
We use the Inverse Function Property.

$$f^{-1}(f(x)) = f^{-1}(3x - 2) \qquad\qquad f(f^{-1}(x)) = f\left(\frac{x + 2}{3}\right)$$

$$= \frac{(3x - 2) + 2}{3} \qquad\qquad\qquad = 3\left(\frac{x + 2}{3}\right) - 2$$

$$= \frac{3x}{3} = x \qquad\qquad\qquad\qquad = x + 2 - 2 = x \quad ✔$$

EXAMPLE 7 ■ Finding the Inverse of a Function

In Example 7 note how f^{-1} reverses the effect of f. The function f is the rule "take the fifth power, subtract 3, then divide by 2," whereas f^{-1} is the rule "multiply by 2, add 3, then take the fifth root."

Find the inverse of the function $f(x) = \dfrac{x^5 - 3}{2}$.

SOLUTION

We first write $y = (x^5 - 3)/2$ and solve for x.

CHECK YOUR ANSWER
We use the Inverse Function Property.

$$f^{-1}(f(x)) = f^{-1}\left(\frac{x^5 - 3}{2}\right)$$

$$= \left[2\left(\frac{x^5 - 3}{2}\right) + 3\right]^{1/5}$$

$$= (x^5 - 3 + 3)^{1/5}$$

$$= (x^5)^{1/5} = x$$

$$f(f^{-1}(x)) = f((2x + 3)^{1/5})$$

$$= \frac{[(2x + 3)^{1/5}]^5 - 3}{2}$$

$$= \frac{2x + 3 - 3}{2}$$

$$= \frac{2x}{2} = x \qquad \checkmark$$

$$y = \frac{x^5 - 3}{2} \qquad \text{Equation defining function}$$

$$2y = x^5 - 3 \qquad \text{Multiply by 2}$$

$$x^5 = 2y + 3 \qquad \text{Add 3}$$

$$x = (2y + 3)^{1/5} \qquad \text{Take fifth roots}$$

Then we interchange x and y to get $y = (2x + 3)^{1/5}$. Therefore, the inverse function is $f^{-1}(x) = (2x + 3)^{1/5}$. ■

The principle of interchanging x and y to find the inverse function also gives us a method for obtaining the graph of f^{-1} from the graph of f. If $f(a) = b$, then $f^{-1}(b) = a$. Thus, the point (a, b) is on the graph of f if and only if the point (b, a) is on the graph of f^{-1}. But we get the point (b, a) from the point (a, b) by reflecting in the line $y = x$ (see Figure 8).

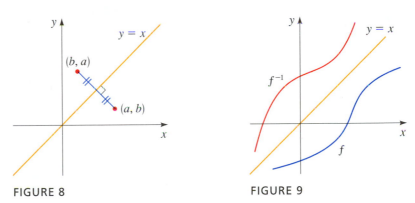

FIGURE 8   FIGURE 9

Therefore, as Figure 9 illustrates, the following is true.

The graph of f^{-1} is obtained by reflecting the graph of f in the line $y = x$.

EXAMPLE 8 ■ **Finding the Inverse of a Function**

(a) Sketch the graph of $f(x) = \sqrt{x - 2}$.
(b) Use the graph of f to sketch the graph of f^{-1}.
(c) Find an equation for f^{-1}.

SOLUTION

(a) Using the transformations from Section 2.5, we sketch the graph of $y = \sqrt{x - 2}$ by plotting the graph of the function $y = \sqrt{x}$ (Example 1(c) in Section 2.2) and moving it to the right 2 units.

(b) The graph of f^{-1} is obtained from the graph of f in part (a) by reflecting it in the line $y = x$, as shown in Figure 10.

(c) Solve $y = \sqrt{x - 2}$ for x, noting that $y \geqslant 0$.

$$\sqrt{x - 2} = y$$

$$x - 2 = y^2 \qquad \text{Square each side}$$

$$x = y^2 + 2, \quad y \geqslant 0 \qquad \text{Add 2}$$

Interchange x and y:

$$y = x^2 + 2, \qquad x \geqslant 0$$

Thus $\qquad f^{-1}(x) = x^2 + 2, \qquad x \geqslant 0$

In Example 8 note how f^{-1} reverses the effect of f. The function f is the rule "subtract 2, then take the square root"; f^{-1} is the rule "square, then add 2."

This expression shows that the graph of f^{-1} is the right half of the parabola $y = x^2 + 2$ and, from the graph shown in Figure 10, this seems reasonable.

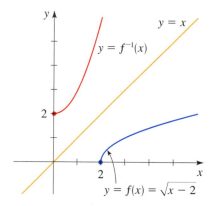

FIGURE 10

2.9 EXERCISES

1–6 ■ The graph of a function f is given. Determine whether f is one-to-one.

1.

2.

3.

4.

5.

6.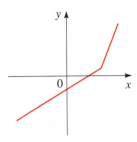

7–16 ■ Determine whether the function is one-to-one.

7. $f(x) = 7x - 3$

8. $f(x) = x^2 - 2x + 5$

9. $g(x) = \sqrt{x}$

10. $g(x) = |x|$

11. $h(x) = x^3 + 1$

12. $h(x) = \sqrt[3]{x}$

13. $f(x) = x^4 + 5$

14. $f(x) = x^4 + 5, \quad 0 \leqslant x \leqslant 2$

15. $f(x) = \dfrac{1}{x^2}$

16. $f(x) = \dfrac{1}{x}$

17–20 ■ Assume f is a one-to-one function.

17. (a) If $f(2) = 7$, find $f^{-1}(7)$.
 (b) If $f^{-1}(3) = -1$, find $f(-1)$.

18. (a) If $f(5) = 18$, find $f^{-1}(18)$.
 (b) If $f^{-1}(4) = 2$, find $f(2)$.

19. If $f(x) = 5 - 2x$, find $f^{-1}(3)$.

20. If $g(x) = x^2 + 4x$ with $x \geqslant -2$, find $g^{-1}(5)$.

21–30 ■ Use the Property of Inverse Functions to show that f and g are inverses of each other.

21. $f(x) = x + 3, \quad g(x) = x - 3$

22. $f(x) = 2x, \quad g(x) = \dfrac{x}{2}$

23. $f(x) = 2x - 5; \quad g(x) = \dfrac{x + 5}{2}$

24. $f(x) = \dfrac{3 - x}{4}; \quad g(x) = 3 - 4x$

25. $f(x) = \dfrac{1}{x}, \quad g(x) = \dfrac{1}{x}$

26. $f(x) = x^5, \quad g(x) = \sqrt[5]{x}$

27. $f(x) = x^2 - 4, \quad x \geqslant 0;$
 $g(x) = \sqrt{x + 4}, \quad x \geqslant -4$

28. $f(x) = x^3 + 1; \quad g(x) = (x - 1)^{1/3}$

29. $f(x) = \dfrac{1}{x - 1}, \quad x \neq 1;$

 $g(x) = \dfrac{1}{x} + 1, \quad x \neq 0$

30. $f(x) = \sqrt{4 - x^2}, \quad 0 \leqslant x \leqslant 2;$
 $g(x) = \sqrt{4 - x^2}, \quad 0 \leqslant x \leqslant 2$

31–50 ■ Find the inverse function of f.

31. $f(x) = 2x + 1$

32. $f(x) = 6 - x$

33. $f(x) = 4x + 7$

34. $f(x) = 3 - 5x$

35. $f(x) = \dfrac{x}{2}$

36. $f(x) = \dfrac{1}{x^2} \quad (x > 0)$

37. $f(x) = \dfrac{1}{x + 2}$

38. $f(x) = \dfrac{x - 2}{x + 2}$

39. $f(x) = \dfrac{1 + 3x}{5 - 2x}$

40. $f(x) = 5 - 4x^3$

41. $f(x) = \sqrt{2 + 5x}$

42. $f(x) = x^2 + x, \quad x \geqslant -\frac{1}{2}$

43. $f(x) = 4 - x^2, \quad x \geqslant 0$

44. $f(x) = \sqrt{2x - 1}$

45. $f(x) = 4 + \sqrt[3]{x}$

46. $f(x) = (2 - x^3)^5$

47. $f(x) = 1 + \sqrt{1 + x}$

48. $f(x) = \sqrt{9 - x^2}, \quad 0 \leqslant x \leqslant 3$

49. $f(x) = x^4, \quad x \geqslant 0$

50. $f(x) = 1 - x^3$

51–54 ■ A function f is given.
(a) Sketch the graph of f
(b) Use the graph of f to sketch the graph of f^{-1}.
(c) Find f^{-1}.

51. $f(x) = 3x - 6$

52. $f(x) = 16 - x^2, \quad x \geqslant 0$

53. $f(x) = \sqrt{x + 1}$

54. $f(x) = x^3 - 1$

55–60 ■ Draw the graph of f and use it to determine whether the function is one-to-one.

55. $f(x) = x^3 - x$

56. $f(x) = x^3 + x$

57. $f(x) = \dfrac{x + 12}{x - 6}$

58. $f(x) = \sqrt{x^3 - 4x + 1}$

59. $f(x) = |x| - |x - 6|$

60. $f(x) = x \cdot |x|$

61–64 ■ The given function is not one-to-one. Restrict its domain so that the resulting function *is* one-to-one. Find the inverse of the function with the restricted domain. (There is more than one correct answer.)

61. $f(x) = 4 - x^2$

62. $g(x) = (x - 1)^2$

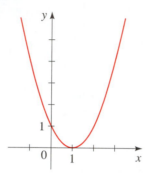

63. $h(x) = (x + 2)^2$

64. $k(x) = |x - 3|$

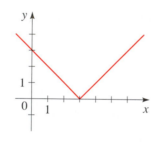

65–66 ■ Use the graph of f to sketch the graph of f^{-1}.

65.

66.

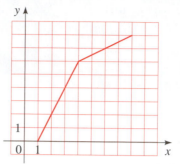

67. For his services, a private investigator requires a $500 retention fee plus $80 per hour. Let x represent the number of hours the investigator spends working on a case.
   (a) Find a function f that models the investigator's fee as a function of x.
   (b) Find f^{-1}. What does f^{-1} represent?
   (c) Find $f^{-1}(1220)$. What does your answer represent?

68. A tank holds 100 gallons of water, which drains from a leak at the bottom, causing the tank to empty in 40 minutes. Toricelli's Law gives the volume of water remaining in the tank after t minutes as

$$V(t) = 100\left(1 - \frac{t}{40}\right)^2$$

   (a) Find V^{-1}. What does V^{-1} represent?
   (b) Find $V^{-1}(15)$. What does your answer represent?

69. As blood moves through a vein or artery, its velocity v is greatest along the central axis and decreases as the distance r from the central axis increases. For an artery with radius 0.5 cm, v is given as a function of r by

$$v(r) = 18,500(0.25 - r^2)$$

   (a) Find v^{-1}. What does v^{-1} represent?
   (b) Find $v^{-1}(30)$. What does your answer represent?

70. The amount of a commodity sold is called the *demand* for the commodity (see Section 1.10). The demand D for a certain commodity is a function of the price given by

$$D(p) = -3p + 150$$

   (a) Find D^{-1}. What does D^{-1} represent?
   (b) Find $D^{-1}(30)$. What does your answer represent?

71. The relationship between the Fahrenheit (F) and Celsius (C) scales is given by

$$F(C) = \tfrac{9}{5}C + 32$$

(a) Find F^{-1}. What does F^{-1} represent?

(b) Find $F^{-1}(86)$. What does your answer represent?

72. The relative value of currencies fluctuates every day. When this problem was written, one Canadian dollar was worth 0.6478 U.S. dollars.

(a) Find a function f that gives the U.S. dollar value $f(x)$ of x Canadian dollars.

(b) Find f^{-1}. What does f^{-1} represent?

(c) How much Canadian money would $12,250 in U.S. currency be worth?

73. In a certain country, the tax on incomes less than or equal to €20,000 is 10%. For incomes more than €20,000, the tax is €2000 plus 20% of the amount over €20,000.

(a) Find a function f that gives the income tax on an income x. Express f as a piecewise defined function.

(b) Find f^{-1}. What does f^{-1} represent?

(c) How much income would require paying a tax of €10,000?

74. A car dealership advertises a 15% discount on all its new cars. In addition, the manufacturer offers a $1000 rebate on the purchase of a new car. Let x represent the sticker price of the car.

(a) Suppose only the 15% discount applies. Find a function f that models the purchase price of the car as a function of the sticker price x.

(b) Suppose only the $1000 rebate applies. Find a function g that models the purchase price of the car as a function of the sticker price x.

(c) Find a formula for $H = f \circ g$.

(d) Find H^{-1}. What does H^{-1} represent?

(e) Find $H^{-1}(13,000)$. What does your answer represent?

75. Marcello's Pizza charges a base price of $7 for a large pizza, plus $2 for each topping. Thus, if you order a large pizza with x toppings, the price of your pizza is given by the function $f(x) = 7 + 2x$. Find f^{-1}. What does the function f^{-1} represent?

 DISCOVERY • DISCUSSION

76. Determining when a Linear Function Has an Inverse
For the linear function $f(x) = mx + b$ to be one-to-one, what must be true about its slope? If it is one-to-one, find its inverse. Is the inverse linear? If so, what is its slope?

77. Finding an Inverse "In Your Head"   In the margin notes in this section we pointed out that the inverse of a function can be found by simply reversing the operations that make up the function. For instance, in Example 6 we saw that the inverse of

$$f(x) = 3x - 2 \quad \text{is} \quad f^{-1}(x) = \frac{x + 2}{3}$$

because the "reverse" of "multiply by 3 and subtract 2" is "add 2 and divide by 3." Use the same procedure to find the inverse of the following functions.

(a) $f(x) = \dfrac{2x + 1}{5}$ 　　　　(b) $f(x) = 3 - \dfrac{1}{x}$

(c) $f(x) = \sqrt{x^3 + 2}$

Now consider another function:

$$f(x) = \frac{3x - 2}{x + 7}$$

Is it possible to use the same sort of simple reversal of operations to find the inverse of this function? If so, do it. If not, explain what is different about this function that makes this task difficult.

78. The Identity Function   The function $I(x) = x$ is called the **identity function**. Show that for any function f we have $f \circ I = f$, $I \circ f = f$, and $f \circ f^{-1} = f^{-1} \circ f = I$. (This means that the identity function I behaves for functions and composition just like the number 1 behaves for real numbers and multiplication.)

79. Solving an Equation for an Unknown Function   In Exercise 63 of Section 2.8 you were asked to solve equations in which the unknowns were functions. Now that we know about inverses and the identity function (see Exercise 78), we can use algebra to solve such equations. For instance, to solve $f \circ g = h$ for the unknown function f, we perform the following steps:

$$
\begin{array}{ll}
f \circ g = h & \text{Problem: Solve for } f \\
f \circ g \circ g^{-1} = h \circ g^{-1} & \text{Compose with } g^{-1} \text{ on the right} \\
f \circ I = h \circ g^{-1} & g \circ g^{-1} = I \\
f = h \circ g^{-1} & f \circ I = f
\end{array}
$$

So the solution is $f = h \circ g^{-1}$. Use this technique to solve the equation $f \circ g = h$ for the indicated unknown function.

(a) Solve for f, where $g(x) = 2x + 1$ and $h(x) = 4x^2 + 4x + 7$

(b) Solve for g, where $f(x) = 3x + 5$ and $h(x) = 3x^2 + 3x + 2$

2 REVIEW

CONCEPT CHECK

1. Define each concept in your own words. (Check by referring to the definition in the text.)
   (a) Function
   (b) Domain and range of a function
   (c) Graph of a function
   (d) Independent and dependent variables

2. Give an example of each type of function.
   (a) Constant function
   (b) Linear function
   (c) Quadratic function

3. Sketch by hand, on the same axes, the graphs of the following functions.
   (a) $f(x) = x$ (b) $g(x) = x^2$
   (c) $h(x) = x^3$ (d) $j(x) = x^4$

4. (a) State the Vertical Line Test.
   (b) State the Horizontal Line Test.

5. Write an equation that expresses each relationship.
   (a) y is directly proportional to x.
   (b) y is inversely proportional to x.
   (c) z is jointly proportional to x and y.

6. How is the average rate of change of the function f between two points defined?

7. Define each concept in your own words.
   (a) Increasing function
   (b) Decreasing function
   (c) Constant function

8. Suppose the graph of f is given. Write an equation for each graph that is obtained from the graph of f as follows.
   (a) Shift 3 units upward.
   (b) Shift 3 units downward.
   (c) Shift 3 units to the right.
   (d) Shift 3 units to the left.
   (e) Reflect in the x-axis.
   (f) Reflect in the y-axis.
   (g) Stretch vertically by a factor of 3.
   (h) Shrink vertically by a factor of $\frac{1}{3}$.

9. (a) What is an even function? What symmetry does its graph possess? Give an example of an even function.
   (b) What is an odd function? What symmetry does its graph possess? Give an example of an odd function.

10. Write the standard form of a quadratic function.

11. What does it mean to say that $f(3)$ is a local maximum value of f?

12. Suppose that f has domain A and g has domain B.
    (a) What is the domain of $f + g$?
    (b) What is the domain of fg?
    (c) What is the domain of f/g?

13. How is the composite function $f \circ g$ defined?

14. (a) What is a one-to-one function?
    (b) How can you tell from the graph of a function whether it is one-to-one?
    (c) Suppose f is a one-to-one function with domain A and range B. How is the inverse function f^{-1} defined? What is the domain of f^{-1}? What is the range of f^{-1}?
    (d) If you are given a formula for f, how do you find a formula for f^{-1}?
    (e) If you are given the graph of f, how do you find the graph of f^{-1}?

EXERCISES

1. If $f(x) = x^3 + 2x - 4$, find $f(0)$, $f(2)$, $f(-2)$, $f(a)$, $f(-a)$, $f(x + 1)$, $f(2x)$, and $2f(x) - 2$.

2. If $f(x) = 2 - \sqrt{2x - 6}$, find $f(5)$, $f(9)$, $f(a + 3)$, $f(-x)$, $f(x^2)$, and $[f(x)]^2$.

3. The graph of a function is given (to the right).
   (a) Find $f(-2)$ and $f(2)$.
   (b) Find the domain of f.
   (c) Find the range of f.
   (d) On what intervals is f increasing? On what intervals is f decreasing?
   (e) Is f one-to-one?

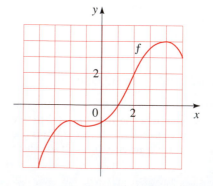

4. Which of the following figures are graphs of functions? Which of the functions are one-to-one?

(a)

(b)

(c)

(d)
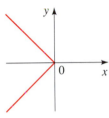

5–6 ■ Find the domain and range of the function.

5. $f(x) = \sqrt{x + 3}$

6. $F(t) = t^2 + 2t + 5$

7–14 ■ Find the domain of the function.

7. $f(x) = 7x + 15$

8. $f(x) = \dfrac{2x + 1}{2x - 1}$

9. $f(x) = \sqrt{x + 4}$

10. $f(x) = 3x - \dfrac{2}{\sqrt{x + 1}}$

11. $f(x) = \dfrac{1}{x} + \dfrac{1}{x + 1} + \dfrac{1}{x + 2}$

12. $g(x) = \dfrac{2x^2 + 5x + 3}{2x^2 - 5x - 3}$

13. $h(x) = \sqrt{4 - x} + \sqrt{x^2 - 1}$

14. $f(x) = \dfrac{\sqrt[3]{2x + 1}}{\sqrt[3]{2x + 2}}$

15–32 ■ Sketch the graph of the function.

15. $f(x) = 1 - 2x$

16. $f(x) = \frac{1}{3}(x - 5), \quad 2 \leqslant x \leqslant 8$

17. $f(t) = 1 - \frac{1}{2}t^2$

18. $g(t) = t^2 - 2t$

19. $f(x) = x^2 - 6x + 6$

20. $f(x) = 3 - 8x - 2x^2$

21. $y = 1 - \sqrt{x}$

22. $y = -|x|$

23. $y = \frac{1}{2}x^3$

24. $y = \sqrt{x + 3}$

25. $h(x) = \sqrt[3]{x}$

26. $H(x) = x^3 - 3x^2$

27. $g(x) = \dfrac{1}{x^2}$

28. $G(x) = \dfrac{1}{(x - 3)^2}$

29. $f(x) = \begin{cases} 1 - x & \text{if } x < 0 \\ 1 & \text{if } x \geqslant 0 \end{cases}$

30. $f(x) = \begin{cases} 1 - 2x & \text{if } x \leqslant 0 \\ 2x - 1 & \text{if } x > 0 \end{cases}$

31. $f(x) = \begin{cases} x + 6 & \text{if } x < -2 \\ x^2 & \text{if } x \geqslant -2 \end{cases}$

32. $f(x) = \begin{cases} -x & \text{if } x < 0 \\ x^2 & \text{if } 0 \leqslant x < 2 \\ 1 & \text{if } x \geqslant 2 \end{cases}$

33. Determine which viewing rectangle produces the most appropriate graph of the function $f(x) = 6x^3 - 15x^2 + 4x - 1$.
  (i) $[-2, 2]$ by $[-2, 2]$
  (ii) $[-8, 8]$ by $[-8, 8]$
  (iii) $[-4, 4]$ by $[-12, 12]$
  (iv) $[-100, 100]$ by $[-100, 100]$

34. Determine which viewing rectangle produces the most appropriate graph of the function $f(x) = \sqrt{100 - x^3}$.
  (i) $[-4, 4]$ by $[-4, 4]$
  (ii) $[-10, 10]$ by $[-10, 10]$
  (iii) $[-10, 10]$ by $[-10, 40]$
  (iv) $[-100, 100]$ by $[-100, 100]$

35–38 ■ Draw the graph of the function in an appropriate viewing rectangle.

35. $f(x) = x^2 + 25x + 173$

36. $f(x) = 1.1x^3 - 9.6x^2 - 1.4x + 3.2$

37. $y = \dfrac{x}{\sqrt{x^2 + 16}}$

38. $y = |x(x + 2)(x + 4)|$

39. Find, approximately, the domain of the function $f(x) = \sqrt{x^3 - 4x + 1}$.

40. Find, approximately, the range of the function $f(x) = x^4 - x^3 + x^2 + 3x - 6$.

41. Suppose that M varies directly as z, and $M = 120$ when $z = 15$. Write an equation that expresses this variation.

42. Suppose that z is inversely proportional to y, and that $z = 12$ when $y = 16$. Write an equation that expresses z in terms of y.

43. The intensity of illumination I from a light varies inversely as the square of the distance d from the light.
  (a) Write this statement as an equation.
  (b) Determine the constant of proportionality if it is known that a lamp has an intensity of 1000 candles at a distance of 8 m.
  (c) What is the intensity of this lamp at a distance of 20 m?

44. The frequency of a vibrating string under constant tension is inversely proportional to its length. If a violin string 12 inches long vibrates 440 times per second, to what length must it be shortened to vibrate 660 times per second?

45. The terminal velocity of a parachutist is directly proportional to the square root of his weight. A 160-lb parachutist attains a terminal velocity of 9 mi/h. What is the terminal velocity for a parachutist weighing 240 lb?

46. The maximum range of a projectile is directly proportional to the square of its velocity. A baseball pitcher throws a ball at 60 mi/h, with a maximum range of 242 ft. What is his maximum range if he throws the ball at 70 mi/h?

47–50 ■ Find the average rate of change of the function between the given points.

47. $f(x) = x^2 + 3x$; $x = 0, x = 2$

48. $f(x) = \dfrac{1}{x-2}$; $x = 4, x = 8$

49. $f(x) = \dfrac{1}{x}$; $x = 3, x = 3 + h$

50. $f(x) = (x+1)^2$; $x = a, x = a + h$

51–52 ■ Draw a graph of the function f, and determine the intervals on which f is increasing and on which f is decreasing.

51. $f(x) = x^3 - 4x^2$ **52.** $f(x) = |x^4 - 16|$

53. Suppose the graph of f is given. Describe how the graphs of the following functions can be obtained from the graph of f.
(a) $y = f(x) + 8$ (b) $y = f(x + 8)$
(c) $y = 1 + 2f(x)$ (d) $y = f(x - 2) - 2$
(e) $y = f(-x)$ (f) $y = -f(-x)$
(g) $y = -f(x)$ (h) $y = f^{-1}(x)$

54. The graph of f is given. Draw the graphs of the following functions.
(a) $y = f(x - 2)$ (b) $y = -f(x)$
(c) $y = 3 - f(x)$ (d) $y = \frac{1}{2}f(x) - 1$
(e) $y = f^{-1}(x)$ (f) $y = f(-x)$

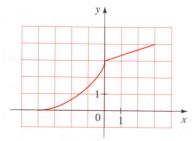

55. Determine whether f is even, odd, or neither.
(a) $f(x) = 2x^5 - 3x^2 + 2$ (b) $f(x) = x^3 - x^7$
(c) $f(x) = \dfrac{1 - x^2}{1 + x^2}$ (d) $f(x) = \dfrac{1}{x + 2}$

56. Determine whether the function in the figure is even, odd, or neither.
(a)   (b)

(c)

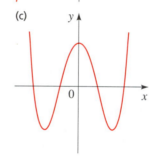

57. Express the quadratic function $f(x) = x^2 + 4x + 1$ in standard form.

58. Express the quadratic function $f(x) = -2x^2 + 12x + 12$ in standard form.

59. Find the minimum value of the function $g(x) = 2x^2 + 4x - 5$.

60. Find the maximum value of the function $f(x) = 1 - x - x^2$.

61. A stone is thrown upward from the top of a building. Its height (in feet) above the ground after t seconds is given by $h(t) = -16t^2 + 48t + 32$. What maximum height does it reach?

62. The profit P (in dollars) generated by selling x units of a certain commodity is given by

$$P(x) = -1500 + 12x - 0.0004x^2$$

What is the maximum profit, and how many units must be sold to generate it?

63–64 ■ Find the local maximum and minimum values of the function and the values of x at which they occur. State each answer correct to two decimal places.

63. $f(x) = 3.3 + 1.6x - 2.5x^3$

64. $f(x) = x^{2/3}(6 - x)^{1/3}$

65. The number of air conditioners sold by an appliance store depends on the time of year. Sketch a rough graph of the number of A/C units sold as a function of the time of year.

66. An isosceles triangle has a perimeter of 8 cm. Express the area A of the triangle as a function of the length b of the base of the triangle.

67. A rectangle is inscribed in an equilateral triangle with a perimeter of 30 cm as in the figure.
(a) Express the area A of the rectangle as a function of the length x shown in the figure.
(b) Find the dimensions of the rectangle with the largest area.

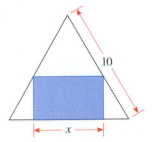

68. A piece of wire 10 m long is cut into two pieces. One piece, of length x, is bent into the shape of a square. The other piece is bent into the shape of an equilateral triangle.
(a) Express the total area enclosed as a function of x.
(b) For what value of x is this total area a minimum?

69. If $f(x) = x^2 - 3x + 2$ and $g(x) = 4 - 3x$, find the following functions.
(a) $f + g$ (b) $f - g$ (c) fg
(d) f/g (e) $f \circ g$ (f) $g \circ f$

70. If $f(x) = 1 + x^2$ and $g(x) = \sqrt{x - 1}$, find the following functions.
(a) $f \circ g$ (b) $g \circ f$ (c) $(f \circ g)(2)$
(d) $(f \circ f)(2)$ (e) $f \circ g \circ f$ (f) $g \circ f \circ g$

71–72 ■ Find the functions $f \circ g$, $g \circ f$, $f \circ f$, and $g \circ g$ and their domains.

71. $f(x) = 3x - 1$, $g(x) = 2x - x^2$

72. $f(x) = \sqrt{x}$, $g(x) = \dfrac{2}{x - 4}$

73. Find $f \circ g \circ h$, where $f(x) = \sqrt{1 - x}$, $g(x) = 1 - x^2$, and $h(x) = 1 + \sqrt{x}$.

74. If $T(x) = \dfrac{1}{\sqrt{1 + \sqrt{x}}}$, find functions f, g, and h such that

$$f \circ g \circ h = T.$$

75–80 ■ Determine whether the function is one-to-one.

75. $f(x) = 3 + x^3$

76. $g(x) = 2 - 2x + x^2$

77. $h(x) = \dfrac{1}{x^4}$

78. $r(x) = 2 + \sqrt{x + 3}$

 79. $p(x) = 3.3 + 1.6x - 2.5x^3$

 80. $q(x) = 3.3 + 1.6x + 2.5x^3$

81–84 ■ Find the inverse of the function.

81. $f(x) = 3x - 2$

82. $f(x) = \dfrac{2x + 1}{3}$

83. $f(x) = (x + 1)^3$

84. $f(x) = 1 + \sqrt[5]{x - 2}$

85. (a) Sketch the graph of the function

$$f(x) = x^2 - 4, \quad x \geqslant 0$$

(b) Use part (a) to sketch the graph of f^{-1}.
(c) Find an equation for f^{-1}.

86. (a) Show that the function $f(x) = 1 + \sqrt[4]{x}$ is one-to-one.
(b) Sketch the graph of f.
(c) Use part (b) to sketch the graph of f^{-1}.
(d) Find an equation for f^{-1}.

2 TEST

1. Which of the following are graphs of functions? If the graph is that of a function, is it one-to-one?

(a)

(b)

(c)

(d)

2. Let $f(x) = \dfrac{\sqrt{x}}{x - 1}$.

   (a) Evaluate $f(4)$, $f(6)$, and $f(a + 1)$.
   (b) Find the domain of f.

3. The maximum weight M that can be supported by a beam is jointly proportional to its width w and the square of its height h, and inversely proportional to its length L.
   (a) Write an equation that expresses this proportionality.
   (b) Determine the constant of proportionality if a beam 4 in. wide, 6 in. high, and 12 ft long can support a weight of 4800 lb.
   (c) If a 10-ft beam made of the same material is 3 in. wide and 10 in. high, what is the maximum weight it can support?

4. Determine the average rate of change for the function $f(t) = t^2 + 3t$ between $t = 2$ and $t = 2 + h$.

5. (a) Sketch the graph of the function $f(x) = x^3$.
   (b) Use part (a) to graph the function $g(x) = (x - 1)^3 - 2$.

6. (a) How is the graph of $y = f(x - 3) + 2$ obtained from the graph of f?
   (b) How is the graph of $y = f(-x)$ obtained from the graph of f?

7. (a) Write the quadratic function $f(x) = 2x^2 - 8x + 13$ in standard form.
   (b) Sketch a graph of f.
   (c) What is the minimum value of f?

8. Let $f(x) = \begin{cases} 1 - x^2 & \text{if } x \leq 0 \\ 2x + 1 & \text{if } x > 0 \end{cases}$
   (a) Evaluate $f(-2)$ and $f(1)$.
   (b) Sketch the graph of f.

9. (a) If 800 ft of fencing is available to build three adjacent pens, as shown in the diagram, express the total area of the pens as a function of x.
   (b) What value of x will maximize the total area?

10. If $f(x) = x^2 + 2x - 1$ and $g(x) = 2x - 3$, find the following functions.
   (a) $f \circ g$ (b) $g \circ f$
   (c) $f(g(2))$ (d) $g(f(2))$
   (e) $g \circ g \circ g$

11. (a) If $f(x) = \sqrt{3 - x}$, find the inverse function f^{-1}.
   (b) Sketch the graphs of f and f^{-1} on the same coordinate axes.

12. The graph of a function f is given.
   (a) Find the domain and range of f.
   (b) Sketch the graph of f^{-1}.
   (c) Find the average rate of change of f between $x = 2$ and $x = 6$.

13. Let $f(x) = 3x^4 - 14x^2 + 5x - 3$.
   (a) Draw the graph of f in an appropriate viewing rectangle.
   (b) Is f one-to-one?
   (c) Find the local maximum and minimum values of f and the values of x at which they occur. State each answer correct to two decimal places.
   (d) Use the graph to determine the range of f.
   (e) Find the intervals on which f is increasing and on which f is decreasing.

Focus on Modeling
Fitting Lines to Data

A model is a representation of an object or process. For example, a toy Ferrari is a *model* of the actual car; a road map is a model of the streets and highways in a city. A model usually represents just one aspect of the original thing. The toy Ferrari is not an actual car, but it does represent what a real Ferrari looks like; a road map does not contain the actual streets in a city, but it does represent the relationship of the streets to each other.

A **mathematical model** is a mathematical representation of an object or process. Often a mathematical model is a function that describes a certain phenomenon. In Example 11 of Section 1.10 we found that the function $T = -10h + 20$ models the atmospheric temperature T at elevation h. We then used this function to predict the temperature at a certain height. Figure 1 illustrates the process of mathematical modeling.

FIGURE 1

Mathematical models are useful because they enable us to isolate critical aspects of the thing we are studying and then to predict how it will behave. Models are used extensively in engineering, industry, and manufacturing. For example, engineers use computer models of skyscrapers to predict their strength and how they would behave in an earthquake. Aircraft manufacturers use elaborate mathematical models to predict the aerodynamic properties of a new design *before* the aircraft is actually built.

How are mathematical models developed? How are they used to predict the behavior of a process? In the next few pages and in subsequent *Focus on Modeling* sections, we explain how mathematical models can be constructed from real-world data, and we describe some of their applications.

Linear Functions as Models

Consider the following table, which gives the winning Olympic pole vaults in the 20th century.

Year	Gold medalist(s)	Height (ft)	Year	Gold medalist	Height (ft)
1900	Irving Baxter, USA	10.83	1956	Robert Richards, USA	14.96
1904	Charles Dvorak, USA	11.48	1960	Don Bragg, USA	15.42
1908	A. Gilbert, E. Cook, USA	12.17	1964	Fred Hansen, USA	16.73
1912	Harry Babcock, USA	12.96	1968	Bob Seagren, USA	17.71
1920	Frank Foss, USA	13.42	1972	W. Nordwig, E. Germany	18.04
1924	Lee Barnes, USA	12.96	1976	Tadeusz Slusarski, Poland	18.04
1928	Sabin Carr, USA	13.77	1980	W. Kozakiewicz, Poland	18.96
1932	William Miller, USA	14.15	1984	Pierre Quinon, France	18.85
1936	Earle Meadows, USA	14.27	1988	Sergei Bubka, USSR	19.77
1948	Guinn Smith, USA	14.10	1992	M. Tarassob, Unified Team	19.02
1952	Robert Richards, USA	14.92	1996	Jean Jalfione, France	19.42

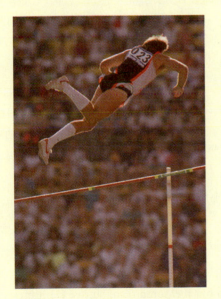

From the table we see that athletes generally pole-vault higher as time goes on. To see this trend better we make a **scatter plot**, as shown in Figure 2. How can we model this data mathematically? Since it appears that the points lie more or less along a line, we can try to fit a line visually to approximate the points in the scatter plot. In other words, we may be able to use a linear function as our model.

FIGURE 2

Scatter plot of pole-vault data

What Is the Line of Best Fit?

How do we find a line that best fits the data? It seems reasonable to choose the line that is as close as possible to all the data points in the scatter plot. So, we want the

FIGURE 3

Distances from the points to the line

```
LinReg
  y=ax+b
  a=.089119747
  b=11.08719447
  r=.9755015763
```

Output of the LinReg
function on the TI-83.

FIGURE 4

Scatter plot and regression line

line for which the sum of the distances from the data points to the line is smallest (see Figure 3).

For technical reasons it is better to find the line where the sum of the squares of these distances is smallest. The resulting line is called the **regression line** or the **least squares line**. The formula for the regression line is found using calculus.* Fortunately, this formula is programmed into most graphing calculators (such as the TI-83, TI-85, or TI-86). Using the pole-vault data and the `LinReg` command on a TI-83 calculator, we find that the regression line is

$$y = 0.0891x + 11.09 \qquad \text{Model}$$

In Figure 4 the regression line is graphed, together with the scatter plot.

What does the model predict for the Olympic pole-vault record in the year 2000? In the year 2000, $x = 100$ so the regression line gives

$$y = 0.0891(100) + 11.09 \approx 19.99 \text{ ft}$$

(Look up the actual record for 2000, and compare with this prediction.) Such predictions are reasonable for points close to our measured data, but we can't predict too far away from the measured data. Is it reasonable to use this model to predict the record 100 years from now?

---

*This formula is discussed in Exercise 55 on page 637.

Another Example

When laboratory rats are exposed to asbestos fibers, some of them develop lung tumors. The following table lists the results of several such experiments by different scientists. Using a graphing calculator we get the regression line

$$y = 0.0177x + 0.54047 \qquad \text{Model}$$

The scatter plot and the regression line are graphed in Figure 5. Is it reasonable to use a linear model for this data?

Asbestos exposure (fibers/mL)	Percent that develop lung tumors
50	2
400	6
500	5
900	10
1100	26
1600	42
1800	37
2000	28
3000	50

FIGURE 5   Scatter plot and regression line for asbestos-tumor data

How Good Is the Fit?

For any given data it is always possible to find the regression line, even if the data do not tend to lie along a line. Consider the three scatter plots in Figure 6.

FIGURE 6

The data in the first scatter plot appear to lie along a line, in the second they appear to lie along a parabola, and in the third there appears to be no discernible trend. We can easily find the regression lines for each scatter plot using a graphing calculator. But how well do these lines represent the data? The calculator gives a **correlation coefficient** r, which is a statistical measure of how well the data lie along the regression line, or how well the two variables are **correlated**. The correlation coefficient is a number between -1 and 1. A correlation coefficient r close to 1 or -1 indicates strong correlation and a coefficient close to 0 indicates very little correlation; the slope of the line determines whether the correlation coefficient is positive or negative. Also, the more data points we have, the more meaningful the correlation coefficient will be. Using a calculator we find that the correlation coefficient between asbestos fibers and lung tumors in rats is $r = 0.92$. We can reasonably conclude that the presence of asbestos and the risk of lung tumors in rats are related. Can we conclude that asbestos *causes* lung tumors in rats?

If two variables are correlated, it does not necessarily mean that a change in one variable *causes* a change in the other. For example, the mathematician John Allen Paulos points out that shoe size is strongly correlated to mathematics scores among school children. Does this mean that big feet cause high math scores? Certainly not—both shoe size and math skills increase independently as children get older. So it is important not to jump to conclusions: Correlation and causation are not the same thing. Correlation is a useful tool in bringing important cause and effect relationships to light, but to prove causation, we must explain the mechanism by which one variable affects the other. For example, the link between smoking and lung cancer was observed as a correlation long before science found the mechanism by which smoking causes lung cancer.

Problems

1. The table lists average carbon dioxide (CO_2) levels in the atmosphere, measured in parts per million (ppm) at Mauna Loa Observatory from 1980 to 1996.
   (a) Make a scatter plot of the data.
   (b) Find and graph the regression line.
   (c) Use the linear model in part (b) to estimate the CO_2 level in the atmosphere in 1998. Compare your answer with the actual CO_2 level of 366.7 measured in 1998.

2. Biologists have observed that the chirping rate of crickets of a certain species appears to be related to temperature. The table shows the chirping rates for various temperatures.
   (a) Make a scatter plot of the data.
   (b) Find and graph the regression line.
   (c) Use the linear model in part (b) to estimate the chirping rate at 100°F.

Year	CO_2 level (ppm)
1980	338.5
1982	341.0
1984	344.3
1986	347.0
1988	351.3
1990	354.0
1992	356.3
1994	358.9
1996	362.7

Temperature (°F)	Chirping rate (chirps/min)
50	20
55	46
60	79
65	91
70	113
75	140
80	173
85	198
90	211

Income	Ulcer rate
$4,000	14.1
$6,000	13.0
$8,000	13.4
$12,000	12.4
$16,000	12.0
$20,000	12.5
$30,000	10.5
$45,000	9.4
$60,000	8.2

3. The table shows (lifetime) peptic ulcer rates (per 100 population) for various family incomes as reported by the 1989 National Health Interview Survey.
  (a) Make a scatter plot of the data.
  (b) Find and graph the regression line.
  (c) Estimate the peptic ulcer rate for an income level of $25,000 according to the linear model in part (b).
  (d) Estimate the peptic ulcer rate for an income level of $80,000 according to the linear model in part (b).

4. The table lists the relative abundance of mosquitoes (as measured by the mosquito positive rate) versus the flow rate (measured as a percentage of maximum flow) of canal networks in Saga City, Japan.
  (a) Make a scatter plot of the data.
  (b) Find and graph the regression line.
  (c) Use the linear model in part (b) to estimate the mosquito positive rate if the canal flow is 70% of maximum.

Flow rate (%)	Mosquito positive rate (%)
0	22
10	16
40	12
60	11
90	6
100	2

5. The tables show the results of experiments measuring the average number of mosquito bites per 15 minutes versus temperature and wind speed.
  (a) Make a scatter plot for each set of data.
  (b) Find and graph the regression line for the temperature-bites data and for the wind-bites data.
  (c) Find the correlation coefficient for each set of data. Are linear models appropriate for these data?

Temperature (°C)	Bites/15 min
10	0.03
15	7.2
20	46.2
25	65.5
30	37.8

Wind speed (km/h)	Bites/15 min
0.5	59.3
2	35.7
4	24.8
6	21.9
8	12.0
14	4.8

Education level (years)	Ulcer rate
8	12
11	12.5
12	11
15	10
16	8
18	9

6. The table shows (lifetime) peptic ulcer rates (per 100 population) for various education levels as reported by the 1989 National Health Interview Survey.
  (a) Make a scatter plot of the data.
  (b) Find and graph the regression line.
  (c) Use the linear model in part (b) to estimate the ulcer rate for a person with 20 years of education.

7. Audiologists study the intelligibility of spoken sentences under different noise levels. Intelligibility, the MRT score, is measured as the percent of a spoken sentence that the listener can decipher at a certain noise level in decibels (dB). The table shows the results of one such test.
(a) Make a scatter plot of the data.
(b) Find and graph the regression line.
(c) Find the correlation coefficient. Is a linear model appropriate?
(d) Use the linear model in part (b) to estimate the intelligibility of a sentence at a 94-dB noise level.

Noise level (dB)	MRT score (%)
80	99
84	91
88	84
92	70
96	47
100	23
104	11

Year	IQ index
1940	85
1950	89
1960	90
1970	95
1980	97
1990	100

8. Average IQ scores in the United States have risen dramatically since the tests were first introduced, as shown in the table at the left.
(a) Make a scatter plot of the data.
(b) Find and graph the regression line.
(c) Use the linear model you found in part (b) to predict the IQ index in the year 2000.

9. The average life expectancy in the United States has been rising steadily over the past few decades, as shown in the table.
(a) Make a scatter plot of the data.
(b) Find and graph the regression line.
(c) Use the linear model you found in part (b) to predict the life expectancy in the year 2000.
(d) Search the Internet or your campus library to find the actual 2000 average life expectancy. Compare to your answer in part (c).

Year	Life expectancy
1920	54.1
1930	59.7
1940	62.9
1950	68.2
1960	69.7
1970	70.8
1980	73.7
1990	75.4

10. The table at the top of page 253 gives the heights and number of stories for 10 tall buildings.
(a) Make a scatter plot of the data.
(b) Find and graph the regression line.
(c) What is the slope of your regression line? What does its value indicate?

Building	Height (ft)	Stories
Empire State Building, New York	1250	102
One Liberty Place, Philadelphia	945	61
Canada Trust Tower, Toronto	863	51
Bank of America Tower, Seattle	943	76
Sears Tower, Chicago	1450	110
Petronas Tower I, Malaysia	1483	88
Commerzbank Tower, Germany	850	60
Palace of Culture and Science, Poland	758	42
Republic Plaza, Singapore	919	66
Transamerica Pyramid, San Francisco	853	48

11. The table below gives U.S. high school teachers' average pay and the high school graduation rate by state for the 1995–1996 academic year.

(a) Make a scatter plot of the data.

(b) Find and graph the regression line.

(c) Calculate the correlation coefficient r. With 50 data points, a correlation coefficient of 0.3 or greater indicates strong correlation. Do you think that teachers' salaries affect the graduation rate?

State	Teachers' average pay ($)	Graduation rate (%)	State	Teachers' average pay ($)	Graduation rate (%)
Alabama	32,818	57.8	Montana	30,617	82.8
Alaska	51,738	64.7	Nebraska	32,688	82.9
Arizona	33,850	58.4	Nevada	37,093	65.4
Arkansas	30,576	74.9	New Hampshire	36,640	74.9
California	43,725	65.3	New Jersey	50,442	82.9
Colorado	37,052	71.9	New Mexico	30,152	63.4
Connecticut	50,730	73.5	New York	49,034	62.0
Delaware	42,439	65.8	North Carolina	33,315	62.4
Florida	34,475	57.8	North Dakota	28,230	89.0
Georgia	37,378	55.0	Ohio	38,977	70.6
Hawaii	38,377	74.8	Oklahoma	30,606	73.0
Idaho	32,775	79.6	Oregon	42,150	66.6
Illinois	43,873	80.0	Pennsylvania	47,650	76.3
Indiana	39,682	70.1	Rhode Island	44,300	71.4
Iowa	34,040	85.3	South Carolina	33,608	54.4
Kansas	36,811	75.8	South Dakota	27,341	86.6
Kentucky	34,525	68.1	Tennessee	35,340	63.4
Louisiana	29,650	57.9	Texas	33,648	58.4
Maine	34,349	72.4	Utah	32,950	78.4
Maryland	41,739	73.9	Vermont	36,299	89.9
Massachusetts	43,930	75.8	Virginia	36,654	75.5
Michigan	49,277	69.6	Washington	38,788	72.2
Minnesota	39,106	85.3	West Virginia	33,398	76.1
Mississippi	29,547	56.8	Wisconsin	39,899	80.4
Missouri	33,975	71.2	Wyoming	32,022	77.8

12. The tables give the gold medal times in the men's and women's 100-m freestyle Olympic swimming event. The data are graphed in the scatter plot, with the women's times in red and the men's in blue.

(a) Find the regression lines for the men's data and the women's data.

(b) Sketch both regression lines on the same graph. When do these lines predict that the women will overtake the men in this event? Does this conclusion seem reasonable?

MEN

Year	Gold medalist	Time (s)
1908	C. Daniels, USA	65.6
1912	D. Kahanamoku, USA	63.4
1920	D. Kahanamoku, USA	61.4
1924	J. Weissmuller, USA	59.0
1928	J. Weissmuller, USA	58.6
1932	Y. Miyazaki, Japan	58.2
1936	F. Csik, Hungary	57.6
1948	W. Ris, USA	57.3
1952	C. Scholes, USA	57.4
1956	J. Henricks, Australia	55.4
1960	J. Devitt, Australia	55.2
1964	D. Schollander, USA	53.4
1968	M. Wenden, Australia	52.2
1972	M. Spitz, USA	51.22
1976	J. Montgomery, USA	49.99
1980	J. Woithe, E. Germany	50.40
1984	R. Gaines, USA	49.80
1988	M. Biondi, USA	48.63
1992	A. Popov, Russia	49.02
1996	A. Popov, Russia	48.74

WOMEN

Year	Gold medalist(s)	Time (s)
1912	F. Durack, Australia	82.2
1920	E. Bleibtrey, USA	73.6
1924	E. Lackie, USA	72.4
1928	A. Osipowich, USA	71.0
1932	H. Madison, USA	66.8
1936	H. Mastenbroek, Holland	65.9
1948	G. Andersen, Denmark	66.3
1952	K. Szoke, Hungary	66.8
1956	D. Fraser, Australia	62.0
1960	D. Fraser, Australia	61.2
1964	D. Fraser, Australia	59.5
1968	J. Henne, USA	60.0
1972	S. Nielson, USA	58.59
1976	K. Ender, E. Germany	55.65
1980	B. Krause, E. Germany	54.79
1984	(Tie) C. Steinseifer, USA	55.92
	N. Hogshead, USA	55.92
1988	K. Otto, E. Germany	54.93
1992	Z. Yong, China	54.64
1996	L. Jingyi, China	54.50

13. Do you think that shoe size and height are correlated? Find out by surveying the shoe sizes and heights of people in your class. (Of course, the data for men and women should be separate.) Find the correlation coefficient.

14. In this problem you will determine a linear demand equation that describes the demand for candy bars in your class. (See page 123 for a description of supply and demand equations.) Survey your classmates to determine what price they would be willing to pay for a candy bar. Your survey form might look like the sample below.

(a) Make a table of the number of respondents who answered "yes" at each price level.

(b) Make a scatter plot of your data.

(c) Find and graph the regression line $y = mp + b$, which gives the number of respondents y who would buy a candy bar if the price were p cents. This is the demand equation. Why is the slope m negative?

(d) What is the p-intercept of the demand equation? What does this intercept tell you about pricing candy bars?

Would you buy a candy bar from the vending machine in the hallway if the price is as indicated?

Price	Yes or No
30¢	
40¢	
50¢	
60¢	
70¢	
80¢	
90¢	
$1.00	
$1.10	
$1.20	

3 Polynomial and Rational Functions

Many real-life situations can be modeled by polynomial or rational functions. The volume of a silo is a polynomial function of its radius; the observed pitch of a train whistle is a rational function of the speed of the train.

In this chapter we study functions defined by polynomial expressions. Polynomials are constructed using just addition, subtraction, multiplication, and taking powers, so their values are easy to calculate. For this reason virtually all the functions used in mathematics and the sciences are evaluated numerically by using polynomial approximations. That is why polynomial functions play such an important role in mathematics. We will also study rational functions, which are quotients of polynomial functions.

3.1 POLYNOMIAL FUNCTIONS AND THEIR GRAPHS

Before we work with polynomials, we must agree on some terminology.

POLYNOMIAL FUNCTIONS

A **polynomial of degree n** is a function of the form

$$P(x) = a_n x^n + a_{n-1} x^{n-1} + \cdots + a_1 x + a_0$$

where n is a nonnegative integer and $a_n \neq 0$. The numbers $a_0, a_1, a_2, \ldots, a_n$ are called the **coefficients** of the polynomial. The number a_0 is the **constant coefficient** or **constant term**. The number a_n, the coefficient of the highest power, is the **leading coefficient**, and the term $a_n x^n$ is the **leading term**.

For example, the following are polynomial functions.

$$P(x) = 3 \qquad \text{degree 0}$$

$$Q(x) = 4x - 7 \qquad \text{degree 1}$$

$$R(x) = x^2 + x \qquad \text{degree 2}$$

$$S(x) = 2x^3 - 6x^2 - 10 \qquad \text{degree 3}$$

If a polynomial consists of just a single term, then it is called a **monomial**. For example, $P(x) = x^3$ and $Q(x) = -6x^5$ are monomials.

Graphs of Polynomials

The graphs of polynomials of degree 0 or 1 are lines (Section 1.10), and the graphs of polynomials of degree 2 are parabolas (Section 2.6). The greater the degree of the polynomial, the more complicated its graph can be. However, the graph of a polynomial function is always a smooth curve; that is, it has no breaks or corners (see Figure 1). The proof of this fact requires calculus.

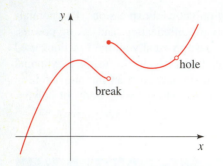

Not the graph of a polynomial function

Not the graph of a polynomial function

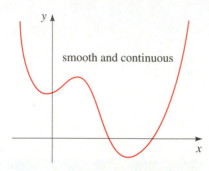

Graph of a polynomial function

FIGURE 1

The simplest polynomial functions are the monomials $P(x) = x^n$, whose graphs are shown in Figure 2. As the figure suggests, the graph of $P(x) = x^n$ has the same general shape as $y = x^2$ when n is even, and the same general shape as $y = x^3$ when n is odd. However, as the degree n becomes larger, the graphs become flatter around the origin and steeper elsewhere.

(a) $y = x$

(b) $y = x^2$

(c) $y = x^3$

(d) $y = x^4$

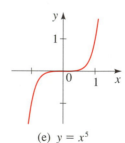

(e) $y = x^5$

FIGURE 2

EXAMPLE 1 ■ Transformations of Monomials

Sketch the graphs of the following functions.

(a) $P(x) = -x^3$ (b) $Q(x) = (x - 2)^4$ (c) $R(x) = -2x^5 + 4$

SOLUTION

We use the graphs in Figure 2 and transform them using the techniques of Section 2.5.

(a) The graph of $P(x) = -x^3$ is the reflection of the graph of $y = x^3$ in the x-axis, as shown in Figure 3(a).

(b) The graph of $Q(x) = (x - 2)^4$ is the graph of $y = x^4$ shifted to the right 2 units, as shown in Figure 3(b).

(c) We begin with the graph of $y = x^5$. The graph of $y = -2x^5$ is obtained by stretching the graph vertically and reflecting it in the x-axis (see the dashed blue graph in Figure 3(c)). Finally, the graph of $R(x) = -2x^5 + 4$ is obtained by shifting upward 4 units (see the red graph in Figure 3(c)).

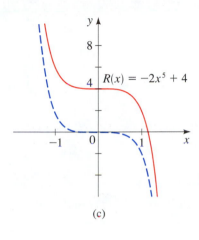

(a)   (b)   (c)

FIGURE 3

End Behavior of Polynomials

The **end behavior** of a polynomial is a description of what happens as x becomes large in the positive or negative direction. To describe end behavior, we use the following notation:

$$x \to \infty \qquad \text{means} \qquad \text{“}x \text{ becomes large in the positive direction”}$$

$$x \to -\infty \qquad \text{means} \qquad \text{“}x \text{ becomes large in the negative direction”}$$

For example, the monomial $y = x^2$ in Figure 2(b) has the following end behavior:

$$y \to \infty \quad \text{as} \quad x \to \infty \qquad \text{and} \qquad y \to \infty \quad \text{as} \quad x \to -\infty$$

The monomial $y = x^3$ in Figure 2(c) has the end behavior

$$y \to \infty \quad \text{as} \quad x \to \infty \qquad \text{and} \qquad y \to -\infty \quad \text{as} \quad x \to -\infty$$

For any polynomial, *the end behavior is determined by the term that contains the highest power of x*, because when x is large, the other terms are relatively insignificant in size. The following box shows the four possible types of end behavior, based on the highest power and the sign of its coefficient.

END BEHAVIOR OF POLYNOMIALS

The polynomial $P(x) = a_n x^n + a_{n-1} x^{n-1} + \cdots + a_1 x + a_0$ has the same end behavior as the monomial $Q(x) = a_n x^n$, so its end behavior is determined by the degree n and the sign of the leading coefficient a_n.

$y = P(x)$ has odd degree

$y \to \infty$ as $x \to \infty$

$y \to -\infty$ as $x \to -\infty$

Leading coefficient positive

$y \to \infty$ as $x \to -\infty$

$y \to -\infty$ as $x \to \infty$

Leading coefficient negative

$y = P(x)$ has even degree

$y \to \infty$ as $x \to -\infty$

$y \to \infty$ as $x \to \infty$

$y \to -\infty$ as $x \to -\infty$

Leading coefficient positive

$y \to -\infty$ as $x \to \infty$

Leading coefficient negative

(a)

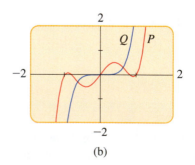

(b)

FIGURE 4
$P(x) = 3x^5 - 5x^3 + 2x$
$Q(x) = 3x^5$

 EXAMPLE 2 ■ End Behavior of a Polynomial

(a) Determine the end behavior of the polynomial $P(x) = 3x^5 - 5x^3 + 2x$.
(b) Confirm that P and its leading term $Q(x) = 3x^5$ have the same end behavior by graphing them together on a large viewing rectangle.
(c) Compare the graphs of P and Q on a small viewing rectangle.

SOLUTION

(a) Since P has odd degree and positive leading coefficient, it has the following end behavior:

$$y \to \infty \quad \text{as} \quad x \to \infty \quad \text{and} \quad y \to -\infty \quad \text{as} \quad x \to -\infty$$

(b) On the large viewing rectangle in Figure 4(a), the graphs of P and Q look almost the same. This confirms that they have the same end behavior.

(c) On the small viewing rectangle in Figure 4(b), the graphs of P and Q look quite different. For instance, P has five x-intercepts, while Q has only one. ■

To see algebraically why P and Q in Example 2 have the same end behavior, factor P as follows and compare with Q.

$$P(x) = 3x^5 \left(1 - \frac{5}{3x^2} + \frac{2}{3x^4} \right)$$

$$Q(x) = 3x^5$$

When x is large, the terms $5/3x^2$ and $2/3x^4$ are close to 0 (see Exercise 60 on page 13). So for large x, we have

$$P(x) \approx 3x^5(1 - 0 - 0)$$

$$= 3x^5 = Q(x)$$

x	$P(x)$	$Q(x)$
15	2,261,280	2,278,125
30	72,765,060	72,900,000
50	936,875,100	937,500,000

So, when x is large, P and Q have approximately the same values. We can also see this numerically by making a table like the one in the margin.

By the same reasoning we can show that the end behavior of *any* polynomial is determined by its leading term.

Using Zeros to Graph Polynomials

If $y = P(x)$ is a polynomial and if c is a number such that $P(c) = 0$, then we say that c is a **zero** of P. In other words, the zeros of P are the solutions of the polynomial equation $P(x) = 0$. When we deal with polynomial equations, we often refer to solutions as **roots**. To find the roots, we factor the polynomial and then use the Zero-Product Property, as in Section 1.5. Note that if $P(c) = 0$, then the graph of $y = P(x)$ has an x-intercept at $x = c$, so the x-intercepts of the graph are the zeros of the function.

ZEROS OF POLYNOMIALS

If P is a polynomial and if c is a number such that $P(c) = 0$, then we say that c is a **zero** of P. The following are equivalent ways of saying the same thing.

1. c is a zero of P

2. $x = c$ is a root of the equation $P(x) = 0$

3. $x - c$ is a factor of $P(x)$

For example, to find the zeros of $P(x) = x^2 + x - 6$, we factor P to get

$$P(x) = (x - 2)(x + 3)$$

From this factored form we easily see that

1. 2 is a zero of P

2. $x = 2$ is a root of the equation $x^2 + x - 6 = 0$

3. $x - 2$ is a factor of $x^2 + x - 6$

4. $x = 2$ is an x-intercept of the graph of P

The same facts are true for the other zero, -3.

The following theorem has many important consequences. (See, for instance, the Discovery Project on page 291.) Here we use it to help us graph polynomial functions.

FIGURE 5

INTERMEDIATE VALUE THEOREM FOR POLYNOMIALS

If P is a polynomial function and $P(a)$ and $P(b)$ have opposite signs, then there exists at least one value c between a and b for which $P(c) = 0$.

We will not prove this theorem, but Figure 5 shows why it is intuitively plausible.

One important consequence of this theorem is that between any two successive zeros, the values of a polynomial are either all positive or all negative. That is, between two successive zeros the graph of a polynomial lies *entirely above* or *entirely below* the x-axis. To see why, suppose c_1 and c_2 are successive zeros of P. If P has both positive and negative values between c_1 and c_2, then by the Intermediate Value Theorem P must have another zero between c_1 and c_2. But that's not possible because c_1 and c_2 are successive zeros.

So to sketch the graph of P, we first find all the zeros of P. Then we choose **test points** between (and to the right and left of) successive zeros to determine whether $P(x)$ is positive or negative on each interval determined by the zeros. For example, the polynomial

$$P(x) = (x + 2)(x - 1)(x - 2)$$

has zeros at $x = -2$, 1, and 2. These determine the intervals $(-\infty, -2)$, $(-2, 1)$, $(1, 2)$, and $(2, \infty)$. Using test points in these intervals, we get the information in the following sign diagram (see Section 1.7).

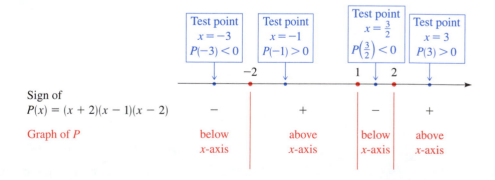

Plotting a few additional points and connecting them with a smooth curve helps us complete the graph in Figure 6.

	x	$P(x)$
Test point →	-3	-20
	-2	0
Test point →	-1	6
	0	4
	1	0
Test point →	$\frac{3}{2}$	$-\frac{7}{8}$
	2	0
Test point →	3	10

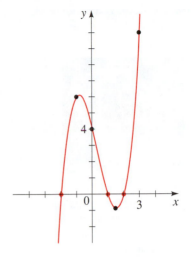

FIGURE 6

$P(x) = (x + 2)(x - 1)(x - 2)$

GUIDELINES FOR GRAPHING POLYNOMIAL FUNCTIONS

1. ZEROS.   Factor the polynomial to find all its real zeros; these are the x-intercepts of the graph.

2. TEST POINTS.   Make a table of values for the polynomial. Include test points to determine whether the graph of the polynomial lies above or below the x-axis on the intervals determined by the zeros. Include the y-intercept in the table.

3. END BEHAVIOR.   Determine the end behavior of the polynomial.

4. GRAPH.   Plot the intercepts and other points you found in the table. Sketch a smooth curve that passes through these points and exhibits the required end behavior.

EXAMPLE 3 ■ Finding Zeros and Graphing a Polynomial Function

Let $P(x) = x^3 - 5x^2 + 6x$.

(a) Find the zeros of P.   (b) Sketch the graph of P.

SOLUTION

(a) To find the zeros, we factor completely.

$$P(x) = x^3 - 5x^2 + 6x$$

$$= x(x^2 - 2x - 3) \qquad \text{Factor } x$$

$$= x(x - 3)(x + 1) \qquad \text{Factor the quadratic}$$

Thus, the zeros are $x = 0$, $x = 3$, and $x = -1$.

(b) The x-intercepts are $x = 0$, $x = 3$, and $x = -1$. The y-intercept is $P(0) = 0$. We make a table of values of $P(x)$, making sure we choose test points between (and to the right and left of) successive zeros.

Since P is of odd degree and its leading coefficient is positive, it has the following end behavior:

$$y \rightarrow \infty \quad \text{as} \quad x \rightarrow \infty \quad \text{and} \quad y \rightarrow -\infty \quad \text{as} \quad x \rightarrow -\infty$$

We plot the points in the table and connect them by a smooth curve to complete the graph, as shown in Figure 7.

	x	$P(x)$
Test point →	-2	-10
	-1	0
Test point →	$-\frac{1}{2}$	$\frac{7}{8}$
	0	0
Test point →	1	-4
	2	-6
	3	0
Test point →	4	20

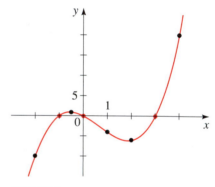

FIGURE 7
$P(x) = x^3 - 2x^2 - 3x$

EXAMPLE 4 ■ Finding Zeros and Graphing a Polynomial Function

Let $P(x) = -2x^4 - x^3 + 3x^2$.

(a) Find the zeros of P.
(b) Sketch the graph of P.

SOLUTION

(a) To find the zeros, we factor completely.

$$P(x) = -2x^4 - x^3 + 3x^2$$
$$= -x^2(2x^2 + x - 3) \qquad \text{Factor } -x^2$$
$$= -x^2(2x + 3)(x - 1) \qquad \text{Factor the quadratic}$$

Thus, the zeros are $x = 0$, $x = -\frac{3}{2}$, and $x = 1$.

(b) The x-intercepts are $x = 0$, $x = -\frac{3}{2}$, and $x = 1$. The y-intercept is $P(0) = 0$. We make a table of values of $P(x)$, making sure we choose test points between (and to the right and left of) successive zeros.

rotated, or seen from the inside. These manipulations of the car on the computer monitor translate mathematically into solving large systems of linear equations.

Tables of values are most easily calculated using a programmable calculator or a graphing calculator.

Since P is of even degree and its leading coefficient is negative, it has the following end behavior:

$$y \to -\infty \quad \text{as} \quad x \to \infty \qquad \text{and} \qquad y \to -\infty \quad \text{as} \quad x \to -\infty$$

We plot the points from the table and connect the points by a smooth curve to complete the graph in Figure 8.

x	$P(x)$
-2	-12
-1.5	0
-1	2
-0.5	0.75
0	0
0.5	0.5
1	0
1.5	-6.75

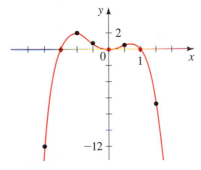

FIGURE 8
$P(x) = -2x^4 - x^3 + 3x^2$

EXAMPLE 5 ■ Finding Zeros and Graphing a Polynomial Function

Let $P(x) = x^3 - 2x^2 - 4x + 8$

(a) Find the zeros of P.   (b) Sketch the graph of P.

SOLUTION

(a) To find the zeros, we factor completely.

$$P(x) = x^3 - 2x^2 - 4x + 8$$

$$= x^2(x - 2) - 4(x - 2) \qquad \text{Group and factor}$$

$$= (x^2 - 4)(x - 2) \qquad \text{Factor } x - 2$$

$$= (x + 2)(x - 2)(x - 2) \qquad \text{Difference of squares}$$

$$= (x + 2)(x - 2)^2 \qquad \text{Simplify}$$

Thus, the zeros are $x = -2$ and $x = 2$.

(b) The x-intercepts are $x = -2$ and $x = 2$. The y-intercept is $P(0) = 8$. The table gives additional values of $P(x)$.

Since P is of odd degree and its leading coefficient is positive, it has the following end behavior:

$$y \to \infty \quad \text{as} \quad x \to \infty \qquad \text{and} \qquad y \to -\infty \quad \text{as} \quad x \to -\infty$$

We connect the points by a smooth curve to complete the graph in Figure 9.

x	$P(x)$
-3	-25
-2	0
-1	9
0	8
1	3
2	0
3	5

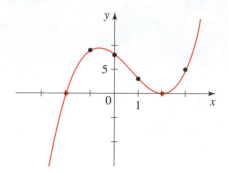

FIGURE 9
$P(x) = x^3 - 2x^2 - 4x + 8$

Although $x = 2$ is a zero of the polynomial in Example 5, the graph does not cross the x-axis at the x-intercept 2. This is because the factor $(x - 2)^2$ corresponding to that zero is raised to an even power, so it doesn't change sign as we test points on either side of 2. In the same way, the graph does not cross the x-axis at $x = 0$ in Example 4.

In general, suppose c is a zero of P and the corresponding factor $(x - c)^m$ occurs *exactly* m times in the factorization of P. Then by considering test points on either side of the x-intercept c, we conclude that the graph crosses the x-axis at c if m is odd and does not cross the x-axis if m is even. Moreover, it can be shown using calculus that near $x = c$ the graph has the same general shape as $y = A(x - c)^m$.

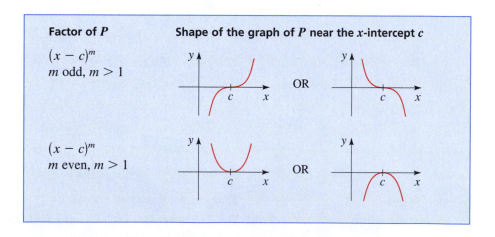

Factor of P	Shape of the graph of P near the x-intercept c
$(x - c)^m$ m odd, $m > 1$	OR
$(x - c)^m$ m even, $m > 1$	OR

EXAMPLE 6 ■ Graphing a Polynomial Function Using Its Zeros

Graph the polynomial $P(x) = x^4(x - 2)^3(x + 1)^2$.

SOLUTION

The zeros of P are -1, 0, and 2, so the x-intercepts are at $x = -1$, 0, and 2. These zeros correspond to the factors $(x + 1)^2$, x^4, and $(x - 2)^3$. So the graph crosses the x-axis at the x-intercept 2 (the corresponding factor has odd exponent 3), but does not cross at the x-intercepts 0 and -1 (the corresponding factors have even exponents 4 and 2). Near the x-intercepts -1, 0, and 2, the graph has the same general shape as $y = A(x + 1)^2$, $y = Ax^4$, and $y = A(x - 2)^3$, respectively.

Since P is a polynomial of degree 9 and has positive leading coefficient, it has the following end behavior:

$$y \to \infty \quad \text{as} \quad x \to \infty \qquad \text{and} \qquad y \to -\infty \quad \text{as} \quad x \to -\infty$$

With this information and a table of values, we sketch the graph in Figure 10.

x	$P(x)$
-1.3	-9.2
-1	0
-0.5	-3.9
0	0
1	-4
2	0
2.3	8.2

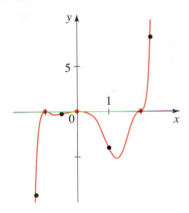

FIGURE 10
$P(x) = x^4(x - 2)^3(x + 1)^2$

Local Extrema of Polynomials

Recall from Section 2.6 that if the point $(a, f(a))$ is the highest point on the graph of f within some viewing rectangle, then $f(a)$ is a local maximum value of f, and if $(b, f(b))$ is the lowest point on the graph of f within a viewing rectangle, then $f(b)$ is a local minimum value (see Figure 11). We say that such a point $(a, f(a))$ is a **local**

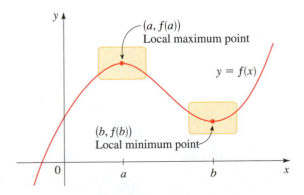

FIGURE 11

maximum point on the graph and that $(b, f(b))$ is a **local minimum point**. The set of all local maximum and minimum points on the graph of a function is called its **local extrema**.

For a polynomial function the number of local extrema must be less than the degree, as the following principle indicates. (A proof of this principle requires calculus.)

LOCAL EXTREMA OF POLYNOMIALS

If $P(x) = a_n x^n + a_{n-1} x^{n-1} + \cdots + a_1 x + a_0$ is a polynomial of degree n, then the graph of P has at most $n - 1$ local extrema.

A polynomial of degree n may in fact have less than $n - 1$ local extrema. For example, $P(x) = x^5$ (graphed in Figure 2) has *no* local extrema, even though it is of degree 5. The preceding principle tells us only that a polynomial of degree n can have no more than $n - 1$ local extrema.

EXAMPLE 7 ■ The Number of Local Extrema

Determine how many local extrema each polynomial has.

(a) $P_1(x) = x^4 + x^3 - 16x^2 - 4x + 48$

(b) $P_2(x) = x^5 + 3x^4 - 5x^3 - 15x^2 + 4x - 15$

(c) $P_3(x) = 7x^4 + 3x^2 - 10x$

SOLUTION

The graphs are shown in Figure 12.

(a) P_1 has two local minimum points and one local maximum point, for a total of three local extrema.

(b) P_2 has two local minimum points and two local maximum points, for a total of four local extrema.

(c) P_3 has just one local extremum, a local minimum.

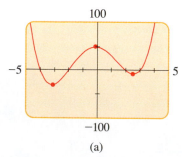

(a)

$P_1(x) = x^4 + x^3 - 16x^2 - 4x + 48$

(b)

$P_2(x) = x^5 + 3x^4 - 5x^3 - 15x^2 + 4x - 15$

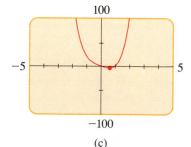

(c)

$P_3(x) = 7x^4 + 3x^2 - 10x$

FIGURE 12

With a graphing calculator we can quickly draw the graphs of many functions at once, on the same viewing screen. This allows us to see how changing a value in the definition of the functions affects the shape of its graph. In the next example we apply this principle to a family of third-degree polynomials.

EXAMPLE 8 ■ A Family of Polynomials

Sketch the family of polynomials $P(x) = x^3 - cx^2$ for $c = 0, 1, 2$, and 3. How does changing the value of c affect the graph?

SOLUTION

The polynomials

$$P_0(x) = x^3 \qquad\qquad P_1(x) = x^3 - x^2$$

$$P_2(x) = x^3 - 2x^2 \qquad\qquad P_3(x) = x^3 - 3x^2$$

are graphed in Figure 13. We see that increasing the value of c causes the graph to develop an increasingly deep "valley" to the right of the y-axis, creating a local maximum at the origin and a local minimum at a point in quadrant IV. This local minimum moves lower and farther to the right as c increases. To see why this happens, factor $P(x) = x^2(x - c)$. The polynomial P has zeros at 0 and c, and the larger c gets, the farther to the right the minimum between 0 and c will be. ■

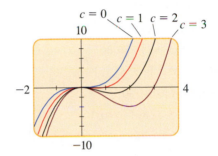

FIGURE 13
A family of polynomials
$P(x) = x^3 - cx^2$

3.1 EXERCISES

1–10 ■ Sketch the graph of the function by transforming the graph of an appropriate function of the form $y = x^n$. Indicate all x- and y-intercepts on each graph.

1. $P(x) = x^3 - 27$

2. $P(x) = -x^3 + 8$

3. $P(x) = -(x + 2)^3$

4. $P(x) = 2(x - 1)^3$

5. $P(x) = 2x^4 + 8$

6. $P(x) = 3x^4 - 27$

7. $P(x) = -(x - 1)^4 + 16$

8. $P(x) = (x + 2)^4 - 1$

9. $P(x) = 2(x + 3)^5 - 64$

10. $P(x) = -\frac{1}{2}(x - 2)^5 + 16$

11–36 ■ Sketch the graph of the function. Make sure your graph shows all intercepts and exhibits the proper end behavior.

11. $P(x) = (x - 1)(x + 2)$

12. $P(x) = (x - 1)(x + 1)(x - 2)$

13. $P(x) = x(x - 3)(x + 2)$

14. $P(x) = (2x - 1)(x + 1)(x + 3)$

15. $P(x) = (x - 3)(x + 2)(3x - 2)$

16. $P(x) = \frac{1}{5}x(x - 5)^2$

17. $P(x) = (x - 1)^2(x - 3)$

18. $P(x) = \frac{1}{4}(x + 1)^3(x - 3)$

19. $P(x) = \frac{1}{12}(x + 2)^2(x - 3)^2$

20. $P(x) = (x - 1)^2(x + 2)^3$

21. $P(x) = x^3(x + 2)(x - 3)^2$ **22.** $P(x) = (x^2 - 2x - 3)^2$

23. $P(x) = x^3 - x^2 - 6x$ **24.** $P(x) = x^3 + 2x^2 - 8x$

25. $P(x) = -x^3 + x^2 + 12x$ **26.** $P(x) = -2x^3 - x^2 + x$

27. $P(x) = x^4 - 3x^3 + 2x^2$ **28.** $P(x) = x^5 - 9x^3$

29. $P(x) = x^3 + x^2 - x - 1$

30. $P(x) = x^3 + 3x^2 - 4x - 12$

31. $P(x) = 2x^3 - x^2 - 18x + 9$

32. $P(x) = \frac{1}{8}(2x^4 + 3x^3 - 16x - 24)^2$

33. $P(x) = x^4 - 2x^3 - 8x + 16$

34. $P(x) = x^4 - 2x^3 + 8x - 16$

35. $P(x) = x^4 - 3x^2 - 4$ **36.** $P(x) = x^6 - 2x^3 + 1$

37–42 ■ Match the polynomial function with one of the graphs I–VI. Give reasons for your choice.

37. $P(x) = x(x^2 - 4)$

38. $Q(x) = -x^2(x^2 - 4)$

39. $R(x) = -x^5 + 5x^3 - 4x$

40. $S(x) = \frac{1}{2}x^6 - 2x^4$

41. $T(x) = x^4 + 2x^3$

42. $U(x) = -x^3 + 2x^2$

I

II

III

IV

V

VI

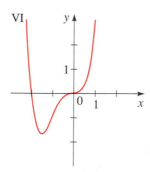

43–48 ■ Determine the end behavior of P. Compare the graphs of P and Q on a large and small viewing rectangle, as in Example 2.

43. $P(x) = 3x^3 - x^2 + 5x + 1;$ $Q(x) = 3x^3$

44. $P(x) = -\frac{1}{8}x^3 + \frac{1}{4}x^2 + 12x;$ $Q(x) = -\frac{1}{8}x^3$

45. $P(x) = x^4 - 7x^2 + 5x + 5;$ $Q(x) = x^4$

46. $P(x) = -x^5 + 2x^2 + x;$ $Q(x) = -x^5$

47. $P(x) = x^{11} - 9x^9;$ $Q(x) = x^{11}$

48. $P(x) = 2x^2 - x^{12};$ $Q(x) = -x^{12}$

 49–56 ■ Graph the polynomial in the given viewing rectangle. Find the coordinates of all local extrema. State each answer correct to two decimal places.

49. $y = -x^2 + 8x,$ $[-4, 12]$ by $[-50, 30]$

50. $y = x^3 - 3x^2,$ $[-2, 5]$ by $[-10, 10]$

51. $y = x^3 - 12x + 9,$ $[-5, 5]$ by $[-30, 30]$

52. $y = 2x^3 - 3x^2 - 12x - 32,$ $[-5, 5]$ by $[-60, 30]$

53. $y = x^4 + 4x^3,$ $[-5, 5]$ by $[-30, 30]$

54. $y = x^4 - 18x^2 + 32,$ $[-5, 5]$ by $[-100, 100]$

55. $y = 3x^5 - 5x^3 + 3,$ $[-3, 3]$ by $[-5, 10]$

56. $y = x^5 - 5x^2 + 6,$ $[-3, 3]$ by $[-5, 10]$

57–66 ■ Graph the polynomial and determine how many local maxima and minima it has.

57. $y = -2x^2 + 3x + 5$ **58.** $y = x^3 + 12x$

59. $y = x^3 - x^2 - x$ **60.** $y = 6x^3 + 3x + 1$

61. $y = x^4 - 5x^2 + 4$

62. $y = 1.2x^5 + 3.75x^4 - 7x^3 - 15x^2 + 18x$

63. $y = (x - 2)^5 + 32$

64. $y = (x^2 - 2)^3$

65. $y = x^8 - 3x^4 + x$

66. $y = \frac{1}{3}x^7 - 17x^2 + 7$

67–72 ■ Graph the family of polynomials in the same viewing rectangle, using the given values of c. Explain how changing the value of c affects the graph.

67. $P(x) = cx^3;$ $c = 1, 2, 5, \frac{1}{2}$

68. $P(x) = (x - c)^4;$ $c = -1, 0, 1, 2$

69. $P(x) = x^4 + c;$ $c = -1, 0, 1, 2$

70. $P(x) = x^3 + cx;$ $c = 2, 0, -2, -4$

71. $P(x) = x^4 - cx;$ $c = 0, 1, 8, 27$

72. $P(x) = x^c;$ $c = 1, 3, 5, 7$

73. (a) On the same coordinate axes, sketch graphs (as accurately as possible) of the functions

$$y = x^3 - 2x^2 - x + 2 \quad \text{and} \quad y = -x^2 + 5x + 2$$

(b) Based on your sketch in part (a), at how many points do the two graphs appear to intersect?

(c) Find the coordinates of all intersection points.

74. Portions of the graphs of $y = x^2$, $y = x^3$, $y = x^4$, $y = x^5$, and $y = x^6$ are plotted in the figures. Determine which function belongs to each graph.

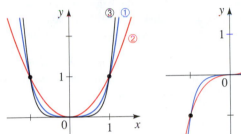

75. Recall that a function f is *odd* if $f(-x) = -f(x)$ or *even* if $f(-x) = f(x)$ for all real x.

(a) Show that if f and g are both odd, then so is the function $f + g$.

(b) Show that if f and g are both even, then so is the function $f + g$.

(c) Show that if f is odd and g is even, and neither is the constant function 0, then the function $f + g$ is neither even nor odd.

(d) Show that a polynomial $P(x)$ that contains only odd powers of x is an odd function.

(e) Show that a polynomial $P(x)$ that contains only even powers of x is an even function.

(f) Show that if a polynomial $P(x)$ contains both odd and even powers of x, then it is neither an odd nor an even function.

(g) Express the function

$$P(x) = x^5 + 6x^3 - x^2 - 2x + 5$$

as the sum of an odd function and an even function.

76. An open box is to be constructed from a piece of cardboard 20 cm by 40 cm by cutting squares of side length x from each corner and folding up the sides, as shown in the figure.

(a) Express the volume V of the box as a function of x.

(b) What is the domain of V? (Use the fact that length and volume must be positive.)

 (c) Draw a graph of the function V and use it to estimate the maximum volume for such a box.

77. A cardboard box has a square base, with each edge of the base having length x inches, as shown in the figure. The total length of all 12 edges of the box is 144 in.

(a) Show that the volume of the box is given by the function $V(x) = 2x^2(18 - x)$.

(b) What is the domain of V? (Use the fact that length and volume must be positive.)

 (c) Draw a graph of the function V and use it to estimate the maximum volume for such a box.

78. A market analyst working for a small-appliance manufacturer finds that if the firm produces and sells x blenders annually, the total profit (in dollars) is

$$P(x) = 8x + 0.3x^2 - 0.0013x^3 - 372$$

Graph the function P in an appropriate viewing rectangle and use the graph to answer the following questions.

(a) When just a few blenders are manufactured, the firm loses money (profit is negative). [For example, $P(10) = -263.3$, so the firm loses $263.30 if it produces and sells only 10 blenders.] How many blenders must the firm produce to break even?

(b) Does profit increase indefinitely as more blenders are produced and sold? If not, what is the largest possible profit the firm could have?

79. The rabbit population on a small island is observed to be given by the function

$$P(t) = 120t - 0.4t^4 + 1000$$

where t is the time (in months) since observations of the island began.

(a) When is the maximum population attained, and what is that maximum population?

(b) When does the rabbit population disappear from the island?

 80. (a) Graph the function $P(x) = (x - 1)(x - 3)(x - 4)$ and find all local extrema, correct to the nearest tenth.

(b) Graph the function

$$Q(x) = (x - 1)(x - 3)(x - 4) + 5$$

and use your answers to part (a) to find all local extrema, correct to the nearest tenth.

(c) If $a < b < c$, explain why the function

$$P(x) = (x - a)(x - b)(x - c)$$

must have two local extrema.

(d) If $a < b < c$ and d is any real number, explain why the function $Q(x) = (x - a)(x - b)(x - c) + d$ must have two local extrema.

81. (a) How many x-intercepts and how many local extrema does the polynomial $P(x) = x^3 - 4x$ have?

(b) How many x-intercepts and how many local extrema does the polynomial $Q(x) = x^3 + 4x$ have?

(c) If $a > 0$, how many x-intercepts and how many local extrema does each of the polynomials $P(x) = x^3 - ax$ and $Q(x) = x^3 + ax$ have? Explain your answer.

● **DISCOVERY · DISCUSSION**

82. Graphs of Large Powers   Graph the functions $y = x^2$, $y = x^3$, $y = x^4$, and $y = x^5$, for $-1 \le x \le 1$, on the same coordinate axes. What do you think the graph of $y = x^{100}$ would look like on this same interval? What about $y = x^{101}$? Make a table of values to confirm your answers.

83. Maximum Number of Local Extrema   What is the smallest possible degree that the polynomial whose graph is shown can have? Explain.

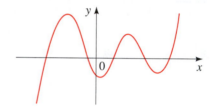

84. Possible Number of Local Extrema   Is it possible for a third-degree polynomial to have exactly one local extremum? Can a fourth-degree polynomial have exactly two local extrema? How many local extrema can polynomials of third, fourth, fifth, and sixth degree have? (Think about the end behavior of such polynomials.) Now give an example of a polynomial that has six local extrema.

85. Impossible Situation?   Is it possible for a polynomial to have two local maxima and no local minimum? Explain.

3.2 DIVIDING POLYNOMIALS

So far in this chapter we have been studying polynomial functions *graphically*. In this section we begin to study polynomials *algebraically*. Most of our work will be concerned with factoring polynomials, and to factor, we need to know how to divide polynomials.

Long Division of Polynomials

Long division for polynomials is very much like long division for numbers. For example, to divide $6x^2 - 26x + 12$ (the **dividend**) by $x - 4$ (the **divisor**), we arrange our work as follows.

$$
\begin{array}{r}
\text{quotient} \\
\downarrow \\
\end{array}
$$

$$
\begin{array}{l}
\text{divisor} \rightarrow \quad x - 4 \overline{)6x^2 - 26x + 12} \quad \leftarrow \text{dividend} \\
\qquad\qquad\quad \underline{6x^2 - 24x} \qquad\qquad\qquad \text{Multiply divisor by } 6x \\
\qquad\qquad\qquad\quad -2x + 12 \qquad\qquad \text{Subtract, then "bring down" } 12 \\
\qquad\qquad\qquad\quad \underline{-2x + \ \ 8} \qquad\qquad \text{Multiply divisor by } -2 \\
\qquad\qquad\qquad\qquad\qquad 4 \quad \leftarrow \text{remainder} \quad \text{Subtract}
\end{array}
$$

The division process ends when the last line is of lesser degree than the divisor. The last line then contains the **remainder**, and the top line contains the **quotient**. The result of the division can be interpreted in either of two ways.

$$
\frac{6x^2 - 26x + 12}{x - 4} = 6x - 2 + \frac{4}{x - 4}
$$

or

$$
6x^2 - 26x + 12 = (x - 4)(6x - 2) + 4
$$

We summarize what happens in any long division problem in the following theorem.

DIVISION ALGORITHM

If $P(x)$ and $D(x)$ are polynomials, with $D(x) \neq 0$, then there exist unique polynomials $Q(x)$ and $R(x)$ such that

$$
P(x) = D(x) \cdot Q(x) + R(x)
$$

where $R(x)$ is either 0 or of degree less than the degree of $D(x)$. The polynomials $P(x)$ and $D(x)$ are called the **dividend** and **divisor**, respectively, $Q(x)$ is the **quotient**, and $R(x)$ is the **remainder**.

EXAMPLE 1 ■ Long Division of Polynomials

Let $P(x) = 8x^4 + 6x^2 - 3x + 1$ and $D(x) = 2x^2 - x + 2$. Find polynomials $Q(x)$ and $R(x)$ such that $P(x) = D(x) \cdot Q(x) + R(x)$.

SOLUTION

We use long division after first inserting the term $0x^3$ into the dividend to ensure that the columns line up correctly.

$$
\begin{array}{r}
4x^2 + 2x \\
2x^2 - x + 2 \overline{)8x^4 + 0x^3 + 6x^2 - 3x + 1} \\
\underline{8x^4 - 4x^3 + 8x^2} \\
4x^3 - 2x^2 - 3x \\
\underline{4x^3 - 2x^2 + 4x} \\
-7x + 1
\end{array}
$$

The process is complete at this point because $-7x + 1$ is of lesser degree than the divisor $2x^2 - x + 2$. From the long division table we see that $Q(x) = 4x^2 + 2x$ and $R(x) = -7x + 1$, so

$$8x^4 + 6x^2 - 3x + 1 = (2x^2 - x + 2)(4x^2 + 2x) + (-7x + 1) \quad \blacksquare$$

■ Synthetic Division

Synthetic division is a quick method of dividing polynomials; it can be used when the divisor is of the form $x - c$. In synthetic division we write only the essential part of the long division table. Compare these long division and synthetic division tables, in which we divide $2x^3 - 7x^2 + 5$ by $x - 3$:

Long Division

$$
\begin{array}{r}
2x^2 - x - 3 \\
x - 3 \overline{)2x^3 - 7x^2 + 0x + 5} \\
\underline{2x^3 - 6x^2} \\
-x^2 + 0x \\
\underline{-x^2 + 3x} \\
-3x + 5 \\
\underline{-3x + 9} \\
-4 \leftarrow \text{remainder}
\end{array}
$$

Synthetic Division

$$
\begin{array}{c|cccc}
3 & 2 & -7 & 0 & 5 \\
  &   & 6 & -3 & -9 \\
\hline
  & 2 & -1 & -3 & -4 \\
\end{array}
$$

quotient   remainder

Note that in synthetic division we abbreviate $2x^3 - 7x^2 + 5$ by writing only the coefficients: $2 \quad -7 \quad 0 \quad 5$, and instead of $x - 3$, we simply write 3. (Writing 3 instead of -3 allows us to add instead of subtract, but this changes the sign of all the numbers that appear in the gold boxes.)

Here is how you get the synthetic division table in practice. Start by writing the divisor and dividend:

$$
\begin{array}{c|cccc}
3 & 2 & -7 & 0 & 5 \\
\end{array}
$$

Mathematical solutions to fair-division problems have recently been applied in an international treaty, the Convention on the Law of the Sea. If a country wants to develop a portion of the sea floor, it is required to divide the portion into two parts, one part to be used by itself, the other by a consortium that will preserve it for later use by a less developed country. The consortium gets first pick.

Bring down the 2, multiply $3 \cdot 2 = 6$, and write the result in the middle row. Then add:

$$
\begin{array}{r|rrrr}
3 & 2 & -7 & 0 & 5 \\
  &   & 6 &   &   \\
\hline
  & 2 & -1 & & \\
\end{array}
$$

Repeat this process of multiplying and then adding until the table is complete.

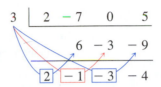

$$
\begin{array}{r|rrrr}
3 & 2 & -7 & 0 & 5 \\
  &   & 6 & -3 & -9 \\
\hline
  & 2 & -1 & -3 & -4 \\
\end{array}
$$

SYNTHETIC DIVISION

To divide $a_n x^n + a_{n-1} x^{n-1} + \cdots + a_1 x + a_0$ by $x - c$, we proceed as follows:

$$
\begin{array}{c|cccccccc}
c & a_n & a_{n-1} & a_{n-2} & a_{n-3} & \cdots & a_2 & a_1 & a_0 \\
  &   & cb_{n-1} & cb_{n-2} & cb_{n-3} & \cdots & cb_2 & cb_1 & cb_0 \\
\hline
  & b_{n-1} & b_{n-2} & b_{n-3} & b_{n-4} & \cdots & b_1 & b_0 & r \\
\end{array}
$$

Here $b_{n-1} = a_n$, and each number in the bottom row is obtained by adding the numbers above it. The remainder is r and the quotient is

$$b_{n-1} x^{n-1} + b_{n-2} x^{n-2} + \cdots + b_1 x + b_0$$

The Remainder and Factor Theorems

The next theorem shows how synthetic division can be used to evaluate polynomials easily.

REMAINDER THEOREM

If the polynomial $P(x)$ is divided by $x - c$, then the remainder is the value $P(c)$.

■ **Proof** If the divisor in the Division Algorithm is of the form $x - c$ for some real number c, then the remainder must be a constant (since the degree of the remainder is less than the degree of the divisor). If we call this constant r, then

$$P(x) = (x - c) \cdot Q(x) + r$$

Setting $x = c$ in this equation, we get $P(c) = (c - c) \cdot Q(x) + r = 0 + r = r$, that is, $P(c)$ is the remainder r.

□

EXAMPLE 2 ■ Using the Remainder Theorem to Find the Value of a Polynomial

Let $P(x) = 3x^5 + 5x^4 - 4x^3 + 7x + 3$.

(a) Find the quotient and remainder when $P(x)$ is divided by $x + 2$.
(b) Use the Remainder Theorem to find $P(-2)$.

SOLUTION

(a) Since $x + 2 = x - (-2)$, the synthetic division table for this problem takes the following form.

$$
\begin{array}{r|rrrrrr}
-2 & 3 & 5 & -4 & 0 & 7 & 3 \\
   &   & -6 & 2 & 4 & -8 & 2 \\
\hline
   & 3 & -1 & -2 & 4 & -1 & 5
\end{array}
$$

The quotient is $3x^4 - x^3 - 2x^2 + 4x - 1$ and the remainder is 5. Thus

$$3x^5 + 5x^4 - 4x^3 + 7x + 3 = (x + 2)(3x^4 - x^3 - 2x^2 + 4x - 1) + 5$$

or

$$\frac{3x^5 + 5x^4 - 4x^3 + 7x + 3}{x + 2} = 3x^4 - x^3 - 2x^2 + 4x - 1 + \frac{5}{x + 2}$$

(b) By the Remainder Theorem, $P(-2)$ is the remainder when $P(x)$ is divided by $x - (-2) = x + 2$. From part (a) the remainder is 5, so $P(-2) = 5$. ■

The next theorem says that *zeros* of polynomials correspond to *factors*; we used this fact in Section 3.1 to graph polynomials.

FACTOR THEOREM

c is a zero of P if and only if $x - c$ is a factor of $P(x)$.

■ **Proof** If $P(x)$ factors as $P(x) = (x - c) \cdot Q(x)$, then

$$P(c) = (c - c) \cdot Q(c) = 0 \cdot Q(c) = 0$$

Conversely, if $P(c) = 0$, then by the Remainder Theorem

$$P(x) = (x - c) \cdot Q(x) + 0 = (x - c) \cdot Q(x)$$

so $x - c$ is a factor of $P(x)$.

□

EXAMPLE 3 ■ Factoring a Polynomial Using the Factor Theorem

$$1 \,\big|\, 1 \quad 0 \quad -7 \quad 6$$
$$ 1 \quad 1 \quad -6$$
$$\overline{\, 1 \quad 1 \quad -6 \quad 0}$$

Let $P(x) = x^3 - 7x + 6$. Show that $P(1) = 0$, and use this fact to factor $P(x)$ completely.

SOLUTION

Substituting, we see that $P(1) = 1^3 - 7 \cdot 1 + 6 = 0$. By the Factor Theorem, this means that $x - 1$ is a factor of $P(x)$. Using synthetic or long division (shown in the margin), we see that

$$P(x) = x^3 - 7x + 6$$

$$ = (x - 1)(x^2 + x - 6) \qquad \text{See margin}$$

$$ = (x - 1)(x - 2)(x + 3) \qquad \text{Factor the quadratic } x^2 + x - 6 \qquad ■$$

$$\begin{array}{r} x^2 + x - 6 \\ x - 1\,\overline{)\,x^3 + 0x^2 - 7x + 6} \\ \underline{x^3 - x^2} \\ x^2 - 7x \\ \underline{x^2 - x} \\ -6x + 6 \\ \underline{-6x + 6} \\ 0 \end{array}$$

EXAMPLE 4 ■ Finding a Polynomial with Specified Zeros

Find a polynomial of degree 4 that has zeros -3, 0, 1, and 5.

SOLUTION

By the Factor Theorem, $x - (-3)$, $x - 0$, $x - 1$, and $x - 5$ must all be factors of the desired polynomial, so let

$$P(x) = (x + 3)(x - 0)(x - 1)(x - 5) = x^4 - 3x^3 - 13x^2 + 15x$$

Since $P(x)$ is of degree 4 it is a solution of the problem. Any other solution of the problem must be a constant multiple of $P(x)$, since only multiplication by a constant does not change the degree.   ■

The polynomial P of Example 4 is graphed in Figure 1. Note that the zeros of P correspond to the x-intercepts of the graph.

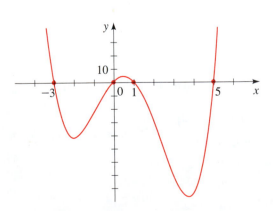

FIGURE 1
$P(x) = (x + 3)x(x - 1)(x - 5)$
has zeros -3, 0, 1, and 5

3.2 EXERCISES

1–10 ■ Find the quotient and remainder using long division.

1. $\dfrac{x^2 + 4x - 8}{x + 3}$

2. $\dfrac{x^3 - x^2 - 2x + 6}{x - 2}$

3. $\dfrac{x^3 + 6x + 5}{x - 4}$

4. $\dfrac{x^3 + 3x^2 + 4x + 3}{3x + 6}$

5. $\dfrac{x^3 + 6x + 3}{x^2 - 2x + 2}$

6. $\dfrac{3x^4 - 5x^3 - 20x - 5}{x^2 + x + 3}$

7. $\dfrac{6x^3 + 2x^2 + 22x}{2x^2 + 5}$

8. $\dfrac{9x^2 - x + 5}{3x^2 - 7x}$

9. $\dfrac{x^6 + x^4 + x^2 + 1}{x^2 + 1}$

10. $\dfrac{2x^5 - 7x^4 - 13}{4x^2 - 6x + 8}$

11–24 ■ Find the quotient and remainder using synthetic division.

11. $\dfrac{x^2 - 5x + 4}{x - 3}$

12. $\dfrac{x^2 - 5x + 4}{x - 1}$

13. $\dfrac{3x^2 + 5x}{x - 6}$

14. $\dfrac{4x^2 - 3}{x + 5}$

15. $\dfrac{x^3 + 2x^2 + 2x + 1}{x + 2}$

16. $\dfrac{3x^3 - 12x^2 - 9x + 1}{x - 5}$

17. $\dfrac{x^3 - 8x + 2}{x + 3}$

18. $\dfrac{x^4 - x^3 + x^2 - x + 2}{x - 2}$

19. $\dfrac{x^5 + 3x^3 - 6}{x - 1}$

20. $\dfrac{x^3 - 9x^2 + 27x - 27}{x - 3}$

21. $\dfrac{2x^3 + 3x^3 - 2x + 1}{x - \frac{1}{2}}$

22. $\dfrac{6x^4 + 10x^3 + 5x^2 + x + 1}{x + \frac{2}{3}}$

23. $\dfrac{x^3 - 27}{x - 3}$

24. $\dfrac{x^4 - 16}{x + 2}$

25–37 ■ Use synthetic division and the Remainder Theorem to evaluate $P(c)$.

25. $P(x) = 4x^2 + 12x + 5, \quad c = -1$

26. $P(x) = 2x^2 + 9x + 1, \quad c = \frac{1}{2}$

27. $P(x) = x^3 + 3x^2 - 7x + 6, \quad c = 2$

28. $P(x) = x^3 - x^2 + x + 5, \quad c = -1$

29. $P(x) = x^3 + 2x^2 - 7, \quad c = -2$

30. $P(x) = 2x^3 - 21x^2 + 9x - 200, \quad c = 11$

31. $P(x) = 5x^4 + 30x^3 - 40x^2 + 36x + 14, \quad c = -7$

32. $P(x) = 6x^5 + 10x^3 + x + 1, \quad c = -2$

33. $P(x) = x^7 - 3x^2 - 1, \quad c = 3$

34. $P(x) = -2x^6 + 7x^5 + 40x^4 - 7x^2 + 10x + 112,$ $c = -3$

35. $P(x) = 3x^3 + 4x^2 - 2x + 1, \quad c = \frac{2}{3}$

36. $P(x) = x^3 - x + 1, \quad c = \frac{1}{4}$

37. $P(x) = x^3 + 2x^2 - 3x - 8, \quad c = 0.1$

38. Let

$$P(x) = 6x^7 - 40x^6 + 16x^5 - 200x^4$$
$$- 60x^3 - 69x^2 + 13x - 139$$

Calculate $P(7)$ by **(a)** using synthetic division and **(b)** substituting $x = 7$ into the polynomial and evaluating directly.

39–42 ■ Use the Factor Theorem to show that $x - c$ is a factor of $P(x)$ for the given value(s) of c.

39. $P(x) = x^3 - 3x^2 + 3x - 1, \quad c = 1$

40. $P(x) = x^3 + 2x^2 - 3x - 10, \quad c = 2$

41. $P(x) = 2x^3 + 7x^2 + 6x - 5, \quad c = \frac{1}{2}$

42. $P(x) = x^4 + 3x^3 - 16x^2 - 27x + 63, \quad c = 3, -3$

43–44 ■ Show that the given value(s) of c are zeros of $P(x)$, and find all other zeros of $P(x)$.

43. $P(x) = x^3 - x^2 - 11x + 15, \quad c = 3$

44. $P(x) = 3x^4 - x^3 - 21x^2 - 11x + 6, \quad c = \frac{1}{3}, -2$

45–48 ■ Find a polynomial of the specified degree that has the given zeros.

45. Degree 3;   zeros $-1, 1, 3$

46. Degree 4;   zeros $-2, 0, 2, 4$

47. Degree 4;   zeros $-1, 1, 3, 5$

48. Degree 5;   zeros $-2, -1, 0, 1, 2$

49. Find a polynomial of degree 3 that has zeros 1, -2, and 3, and in which the coefficient of x^2 is 3.

50. Find a polynomial of degree 4 that has integer coefficients and zeros 1, -1, 2, and $\frac{1}{2}$.

DISCOVERY · DISCUSSION

51. Impossible Division?   Suppose you were asked to solve the following two problems on a test:

A. Find the remainder when $6x^{1000} - 17x^{562} + 12x + 26$ is divided by $x + 1$.

B. Is $x - 1$ a factor of $x^{567} - 3x^{400} + x^9 + 2$?

Obviously, it's impossible to solve these problems by dividing, because the polynomials are of such large degree. Use one or more of the theorems in this section to solve these problems *without* actually dividing.

52. Nested Form of a Polynomial   Expand Q to prove that the polynomials P and Q are the same.

$$P(x) = 3x^4 - 5x^3 + x^2 - 3x + 5$$

$$Q(x) = (((3x - 5)x + 1)x - 3)x + 5$$

Try to evaluate $P(2)$ and $Q(2)$ in your head, using the forms given. Which is easier? Now write the polynomial $R(x) = x^5 - 2x^4 + 3x^3 - 2x^2 + 3x + 4$ in "nested" form, like the polynomial Q. Use the nested form to find $R(3)$ in your head.

Do you see how calculating with the nested form follows the same arithmetic steps as calculating the value of a polynomial using synthetic division?

3.3   REAL ZEROS OF POLYNOMIALS

The Factor Theorem tells us that finding the zeros of a polynomial is really the same thing as factoring it into linear factors. In this section we study some algebraic methods that help us find the real zeros of a polynomial, and thereby factor the polynomial. We begin with the *rational* zeros of a polynomial.

Rational Zeros of Polynomials

To help us understand the next theorem, let's consider the polynomial

$$P(x) = (x - 2)(x - 3)(x + 4) \qquad \text{Factored form}$$

$$= x^3 - x^2 - 14x + 24 \qquad \text{Expanded form}$$

From the factored form we see that the zeros of P are 2, 3, and -4. When the polynomial is expanded, the constant 24 is obtained by multiplying $(-2) \times (-3) \times 4$. This means that the zeros of the polynomial are all factors of the constant term. The following generalizes this observation.

RATIONAL ZEROS THEOREM

If the polynomial $P(x) = a_n x^n + a_{n-1}x^{n-1} + \cdots + a_1 x + a_0$ has integer coefficients, then every rational zero of P is of the form

$$\frac{p}{q}$$

where p is a factor of the constant coefficient a_0

and q is a factor of the leading coefficient a_n.

Evariste Galois (1811–1832) is one of the very few mathematicians to have an entire theory named in his honor. Not yet 21 when he died, he completely settled the central problem in the theory of equations by describing a criterion that reveals whether a polynomial equation can be solved by algebraic operations. Galois was one of the greatest mathematicians in the world at that time, although no one knew it but him. He repeatedly sent his work to the eminent mathematicians Cauchy and Poisson, who either lost his letters or did not understand his ideas. Galois wrote in a terse style and included few details, which probably played a role in his failure to pass the entrance exams at the Ecole Polytechnique in Paris. A political radical, Galois spent several months in prison for his revolutionary activities. His brief life came to a tragic end when he was killed in a duel over a love affair. The night before his duel, fearing he would die, Galois wrote down the essence of his ideas and entrusted them to his friend Auguste Chevalier. He concluded by writing ". . . there will, I hope, be people who will find it to their advantage to decipher all this mess." The mathematician Camille Jordan did just that, 14 years later.

■ **Proof**   If p/q is a rational zero, in lowest terms, of the polynomial P, then we have

$$a_n\left(\frac{p}{q}\right)^n + a_{n-1}\left(\frac{p}{q}\right)^{n-1} + \cdots + a_1\left(\frac{p}{q}\right) + a_0 = 0$$

$$a_n p^n + a_{n-1}p^{n-1}q + \cdots + a_1 pq^{n-1} + a_0 q^n = 0 \qquad \text{Multiply by } q^n$$

$$p(a_n p^{n-1} + a_{n-1}p^{n-2}q + \cdots + a_1 q^{n-1}) = -a_0 q^n \qquad \text{Subtract } a_0 q^n \text{ and factor LHS}$$

Now p is a factor of the left side, so it must be a factor of the right as well. Since p/q is in lowest terms, p and q have no factor in common, and so p must be a factor of a_0. A similar proof shows that q is a factor of a_n. ☐

If $a_n = 1$ in the Rational Zeros Theorem, then q must be 1 or -1, so in this case, any rational zero p/q is in fact an *integer* factor of a_0. Thus, if the highest-power term in a polynomial is just x^n, all its rational zeros are integers, not fractions.

EXAMPLE 1 ■ Using the Rational Zeros Theorem to Factor a Polynomial

Factor the polynomial $P(x) = 2x^3 + x^2 - 13x + 6$.

SOLUTION

By the Rational Zeros Theorem, if p/q is a zero of $P(x)$, then p divides 6 and q divides 2, so p/q is of the form

$$\frac{\text{factor of 6}}{\text{factor of 2}}$$

The factors of 6 are ± 1, ± 2, ± 3, ± 6, and the factors of 2 are ± 1, ± 2. Thus, the possible values of p/q are

$$\pm\frac{1}{1}, \quad \pm\frac{2}{1}, \quad \pm\frac{3}{1}, \quad \pm\frac{6}{1}, \quad \pm\frac{1}{2}, \quad \pm\frac{2}{2}, \quad \pm\frac{3}{3}, \quad \pm\frac{6}{2}$$

Simplifying the fractions and eliminating duplicates, we get the following list of possible values of p/q:

$$\pm 1, \quad \pm 2, \quad \pm 3, \quad \pm 6, \quad \pm\frac{1}{2}, \quad \pm\frac{3}{2}$$

Now we check which of these *possible* zeros actually *are* zeros by substituting them, one at a time, into the polynomial P until we find one that makes $P(x) = 0$. We have

$$P(1) = 2(1)^3 + (1)^2 - 13(1) + 6 = -4 \qquad \text{1 is } not \text{ a zero of } P$$

$$P(2) = 2(2)^3 + (2)^2 - 13(2) + 6 = 0 \qquad \text{2 } is \text{ a zero of } P$$

Since $x = 2$ is a zero of P, it follows that $x - 2$ is a factor of $P(x)$. Using synthetic division (shown in the margin), we obtain the following factorization:

$$P(x) = 2x^3 + x^2 - 13x + 6$$

$$= (x - 2)(2x^2 + 5x - 3) \qquad \text{See margin}$$

$$= (x - 2)(2x - 1)(x + 3) \qquad \text{Factor } 2x^2 + 5x - 3 \qquad ■$$

$$
\begin{array}{r|rrrr}
2 & 2 & 1 & -13 & 6 \\
  &   & 4 & 10 & -6 \\
\hline
  & 2 & 5 & -3 & 0 \\
\end{array}
$$

The following box explains how we use the Rational Zeros Theorem with synthetic division to factor a polynomial.

FINDING THE RATIONAL ZEROS OF A POLYNOMIAL

1. LIST POSSIBLE ZEROS.   List all possible rational zeros using the Rational Zeros Theorem.

2. DIVIDE.   Use synthetic division to evaluate the polynomial at each of the candidates for rational zeros that you found in Step 1. When the remainder is 0, note the quotient you have obtained.

3. REPEAT.   Repeat Steps 1 and 2 for the quotient. Stop when you reach a quotient that is quadratic or factors easily, and use the quadratic formula or factor to find the remaining zeros.

EXAMPLE 2 ■ Using the Rational Zeros Theorem and the Quadratic Formula

Let $P(x) = x^4 - 5x^3 - 5x^2 + 23x + 10$.

(a) Find the zeros of P.
(b) Sketch the graph of P.

SOLUTION

(a) The leading coefficient of P is 1, so all the rational zeros are integers: They are divisors of the constant term 10. Thus, the possible candidates are

$$\pm 1, \quad \pm 2, \quad \pm 5, \quad \pm 10$$

$$
\begin{array}{r|rrrrr}
1 & 1 & -5 & -5 & 23 & 10 \\
  &   & 1 & -4 & -9 & 14 \\
\hline
  & 1 & -4 & -9 & 14 & 24 \\
\end{array}
$$

Using synthetic division (see the margin) we find that 1 and 2 are not zeros, but that 5 is a zero and that P factors as

$$
\begin{array}{r|rrrrr}
2 & 1 & -5 & -5 & 23 & 10 \\
  &   & 2 & -6 & -22 & 2 \\
\hline
  & 1 & -3 & -11 & 1 & 12 \\
\end{array}
$$

$$x^4 - 5x^3 - 5x^2 + 23x + 10 = (x - 5)(x^3 - 5x - 2)$$

$$
\begin{array}{r|rrrrr}
5 & 1 & -5 & -5 & 23 & 10 \\
  &   & 5 & 0 & -25 & -10 \\
\hline
  & 1 & 0 & -5 & -2 & 0 \\
\end{array}
$$

We now try to factor the quotient $x^3 - 5x - 2$. Its possible zeros are the divisors of -2, namely,

$$\pm 1, \quad \pm 2$$

Since we already know that 1 and 2 are not zeros of the original polynomial P, we don't need to try them again. Checking the remaining candidates -1 and -2, we see that -2 is a zero (see the margin), and P factors as

$$
\begin{array}{r|rrrr}
-2 & 1 & 0 & -5 & -2 \\
   &   & -2 & 4 & 2 \\
\hline
   & 1 & -2 & -1 & 0
\end{array}
$$

$$x^4 - 5x^3 - 5x^2 + 23x + 10 = (x - 5)(x^3 - 5x - 2)$$

$$= (x - 5)(x + 2)(x^2 - 2x - 1)$$

Now we use the quadratic formula to obtain the two remaining zeros of P:

$$x = \frac{2 \pm \sqrt{(-2)^2 - 4(1)(-1)}}{2} = 1 \pm \sqrt{2}$$

The solutions of the equation are 5, -2, $1 + \sqrt{2}$, and $1 - \sqrt{2}$.

(b) Now that we know the zeros of P, we can use the methods of Section 3.1 to sketch the graph. If we want to use a graphing calculator instead, knowing the zeros allows us to choose an appropriate viewing rectangle—one that is wide enough to contain all the x-intercepts of P. Numerical approximations to the zeros of P are

$$5, \qquad -2, \qquad 2.4, \qquad \text{and} \qquad -0.4$$

So in this case we choose the rectangle $[-3, 6]$ by $[-50, 50]$ and draw the graph shown in Figure 1.

FIGURE 1
$P(x) = x^4 - 5x^3 - 5x^2 + 23x + 10$

Descartes' Rule of Signs and Upper and Lower Bounds for Roots

In some cases, the following rule—discovered by the French philosopher and mathematician René Descartes around 1637 (see page 90)—is helpful in eliminating candidates from lengthy lists of possible rational roots. To describe this rule, we need the concept of *variation in sign*. If $P(x)$ is a polynomial with real coefficients, written with descending powers of x (and omitting powers with coefficient 0), then a **variation in sign** occurs whenever adjacent coefficients have opposite signs. For example

$$P(x) = 5x^7 - 3x^5 - x^4 + 2x^2 + x - 3$$

has three variations in sign.

Polynomial	Variations in sign
$x^2 + 4x + 1$	0
$2x^3 + x - 6$	1
$x^4 - 3x^2 - x + 4$	2

DESCARTES' RULE OF SIGNS

Let P be a polynomial with real coefficients.

1. The number of positive real zeros of $P(x)$ is either equal to the number of variations in sign in $P(x)$ or is less than that by an even whole number.

2. The number of negative real zeros of $P(x)$ is either equal to the number of variations in sign in $P(-x)$ or is less than that by an even whole number.

EXAMPLE 3 ■ Using Descartes' Rule

Use Descartes' Rule of Signs to determine the possible number of positive and negative real zeros of the polynomial

$$P(x) = 3x^6 + 4x^5 + 3x^3 - x - 3$$

SOLUTION

The polynomial has one variation in sign and so it has one positive zero. Now

$$P(-x) = 3(-x)^6 + 4(-x)^5 + 3(-x)^3 - (-x) - 3$$

$$= 3x^6 - 4x^5 - 3x^3 + x - 3$$

So, $P(-x)$ has three variations in sign. Thus, $P(x)$ has either three or one negative zero(s), making a total of either two or four real zeros.   ■

We say that a is a **lower bound** and b is an **upper bound** for the roots of a polynomial equation if every real root c of the equation satisfies $a \leq c \leq b$. The next theorem helps us find such bounds for any polynomial equation.

THE UPPER AND LOWER BOUNDS THEOREM

Let P be a polynomial with real coefficients.

1. If we divide $P(x)$ by $x - b$ (with $b > 0$) using synthetic division, and if the row that contains the quotient and remainder has no negative entry, then b is an upper bound for the real zeros of P.

2. If we divide $P(x)$ by $x - a$ (with $a < 0$) using synthetic division, and if the row that contains the quotient and remainder has entries that are alternately nonpositive and nonnegative, then a is a lower bound for the real zeros of P.

A proof of this theorem is suggested in Exercises 90 and 91. The phrase "alternately nonpositive and nonnegative" simply means that the signs of the numbers alternate, with 0 considered to be positive or negative as required.

EXAMPLE 4 ■ Upper and Lower Bounds for Zeros of a Polynomial

Show that all the real zeros of the polynomial $P(x) = x^4 - 3x^2 + 2x - 5$ lie between -3 and 2.

SOLUTION

We divide $P(x)$ by $x - 2$ and $x + 3$ using synthetic division.

$$
\begin{array}{r|rrrrr}
2 & 1 & 0 & -3 & 2 & -5 \\
  &   & 2 & 4 & 2 & 8 \\
\hline
  & 1 & 2 & 1 & 4 & 3 \quad \leftarrow \text{All entries} \\
  &   &   &   &   & \quad\text{positive}
\end{array}
\qquad
\begin{array}{r|rrrrr}
-3 & 1 & 0 & -3 & 2 & -5 \\
   &   & -3 & 9 & -18 & 48 \\
\hline
   & 1 & -3 & 6 & -16 & 43 \quad \leftarrow \begin{array}{l}\text{Entries}\\ \text{alternate}\\ \text{in sign}\end{array}
\end{array}
$$

By the Upper and Lower Bounds Theorem, -3 is a lower bound and 2 is an upper bound for the zeros. Since neither -3 nor 2 is a zero (the remainders are not 0 in the division table), all the real zeros lie between these numbers.   ■

EXAMPLE 5 ■ Factoring a Fifth-Degree Polynomial

Factor completely the polynomial

$$P(x) = 2x^5 + 5x^4 - 8x^3 - 14x^2 + 6x + 9$$

SOLUTION

The possible rational zeros of P are $\pm\frac{1}{2}$, ± 1, $\pm\frac{3}{2}$, ± 3, $\pm\frac{9}{2}$, and ± 9. We check the positive candidates first, beginning with the smallest.

$$
\begin{array}{r|rrrrrr}
\frac{1}{2} & 2 & 5 & -8 & -14 & 6 & 9 \\
 &   & 1 & 3 & -\frac{5}{2} & -\frac{33}{4} & -\frac{9}{8} \\
\hline
 & 2 & 6 & -5 & -\frac{33}{2} & -\frac{9}{4} & \frac{63}{8} \;\; \leftarrow \begin{array}{l}\frac{1}{2}\ \text{is not}\\ \text{a zero}\end{array}
\end{array}
\qquad
\begin{array}{r|rrrrrr}
1 & 2 & 5 & -8 & -14 & 6 & 9 \\
 &   & 2 & 7 & -1 & -15 & -9 \\
\hline
 & 2 & 7 & -1 & -15 & -9 & 0 \;\; \leftarrow P(1) = 0
\end{array}
$$

So 1 is a zero, and $P(x) = (x - 1)(2x^4 + 7x^3 - x^2 - 15x - 9)$. We continue by factoring the quotient.

$$
\begin{array}{r|rrrrr}
1 & 2 & 7 & -1 & -15 & -9 \\
  &   & 2 & 9 & 8 & -7 \\
\hline
  & 2 & 9 & 8 & -7 & -16 \;\; \leftarrow \begin{array}{l}1\ \text{is not}\\ \text{a zero}\end{array}
\end{array}
\qquad
\begin{array}{r|rrrrr}
\frac{3}{2} & 2 & 7 & -1 & -15 & -9 \\
  &   & 3 & 15 & 21 & 9 \\
\hline
  & 2 & 10 & 14 & 6 & 0 \;\; \leftarrow \begin{array}{l}P\left(\frac{3}{2}\right) = 0,\\ \text{all entries}\\ \text{nonnegative}\end{array}
\end{array}
$$

We see that $\frac{3}{2}$ is both a zero and an upper bound for the zeros of $P(x)$, so we don't need to check any further for positive zeros, because all the remaining candidates are greater than $\frac{3}{2}$.

$$P(x) = (x - 1)(x - \tfrac{3}{2})(2x^3 + 10x^2 + 14x + 6)$$

$$= (x - 1)(2x - 3)(x^3 + 5x^2 + 7x + 3) \qquad \text{\color{red}Factor 2 from last factor,}$$
$$\color{red}\text{multiply into second factor}$$

By Descartes' Rule of Signs, $x^3 + 5x^2 + 7x + 3$ has no positive zero, so the only possible rational zeros are -1 and -3.

$$
\begin{array}{r|rrrr}
-1 & 1 & 5 & 7 & 3 \\
   &   & -1 & -4 & -3 \\
\hline
   & 1 & 4 & 3 & 0 \quad \leftarrow P(-1) = 0
\end{array}
$$

Therefore

$$P(x) = (x - 1)(2x - 3)(x + 1)(x^2 + 4x + 3)$$

$$= (x - 1)(2x - 3)(x + 1)^2(x + 3) \qquad \text{\color{red}Factor quadratic}$$

This means that the zeros of P are 1, $\frac{3}{2}$, -1, and -3. ■

FIGURE 2
$P(x) = 2x^5 + 5x^4 - 8x^3 - 14x^2 + 6x + 9$
$\quad\quad = (x - 1)(2x - 3)(x + 1)^2(x + 3)$

The graph of the polynomial in Example 5 is shown in Figure 2.

Using Algebra and Graphing Devices to Solve Polynomial Equations

In Section 1.9 we used graphing devices to solve equations graphically. We can now use the algebraic techniques we've learned to select an appropriate viewing rectangle when solving a polynomial equation graphically.

EXAMPLE 6 ■ Solving a Fourth-Degree Equation Graphically

Find all real solutions of the following equation, correct to the nearest tenth.

$$3x^4 + 4x^3 - 7x^2 - 2x - 3 = 0$$

SOLUTION

To solve the equation graphically, we graph

$$P(x) = 3x^4 + 4x^3 - 7x^2 - 2x - 3$$

We use the Upper and Lower Bounds Theorem to see where the roots can be found.

First we use the Upper and Lower Bounds Theorem to find two numbers between which all the solutions must lie. This allows us to choose a viewing rectangle that is certain to contain all the x-intercepts of P. We use synthetic division and proceed by trial and error.

To find an upper bound, we try the whole numbers, 1, 2, 3, . . . as potential candidates. We see that 2 is an upper bound for the roots.

$$
\begin{array}{r|rrrrr}
2 & 3 & 4 & -7 & -2 & -3 \\
  &   & 6 & 20 & 26 & 48 \\
\hline
  & 3 & 10 & 13 & 24 & 45 \\
\end{array}
$$
← All positive

Now we look for a lower bound, trying the numbers -1, -2, and -3 as potential candidates. We see that -3 is a lower bound for the roots.

$$
\begin{array}{r|rrrrr}
-3 & 3 & 4 & -7 & -2 & -3 \\
   &   & -9 & 15 & -24 & 78 \\
\hline
   & 3 & -5 & 8 & -26 & 75 \\
\end{array}
$$
Entries ← alternate in sign

Thus, all the roots lie between -3 and 2. So the viewing rectangle $[-3, 2]$ by $[-20, 20]$ contains all the x-intercepts of P. The graph in Figure 3 has two x-intercepts, one between -3 and -2 and the other between 1 and 2. Zooming in, we find that the solutions of the equation, to the nearest tenth, are -2.3 and 1.3. ■

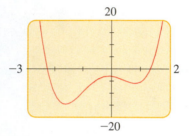

FIGURE 3
$y = 3x^4 + 4x^3 - 7x^2 - 2x - 3$

EXAMPLE 7 ■ Determining the Size of a Fuel Tank

A fuel tank consists of a cylindrical center section 4 ft long and two hemispherical end sections, as shown in Figure 4. If the tank has a volume of 100 ft³, what is the radius r shown in the figure, correct to the nearest hundredth of a foot?

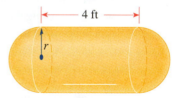

FIGURE 4

SOLUTION

Using the volume formula listed on the inside front cover of this book, we see that the volume of the cylindrical section of the tank is

$$\pi \cdot r^2 \cdot 4$$

Volume of a cylinder: $V = \pi r^2 h$

The two hemispherical parts together form a complete sphere whose volume is

$$\tfrac{4}{3}\pi r^3$$

Volume of a sphere: $V = \frac{4}{3}\pi r^3$

Because the total volume of the tank is 100 ft³, we get the following equation:

$$\tfrac{4}{3}\pi r^3 + 4\pi r^2 = 100$$

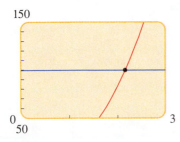

FIGURE 5
$y = \frac{4}{3}\pi x^3 + 4\pi x^2$ and $y = 100$

A negative solution for r would be meaningless in this physical situation, and by substitution we can verify that $r = 3$ leads to a tank that is over 226 ft^3 in volume, much larger than the required 100 ft^3. Thus, we know the correct radius lies somewhere between 0 and 3 ft, and so we use a viewing rectangle of $[0, 3]$ by $[50, 150]$ to graph the function $y = \frac{4}{3}\pi x^3 + 4\pi x^2$, as shown in Figure 5. Since we want the value of this function to be 100, we also graph the horizontal line $y = 100$ in the same viewing rectangle. The correct radius will be the x-coordinate of the point of intersection of the curve and the line. Using the cursor and zooming in, we see that at the point of intersection $x \approx 2.15$, correct to two decimal places. Thus, the tank has a radius of about 2.15 ft.

Note that we also could have solved the equation in Example 7 by first writing it as

$$\frac{4}{3}\pi r^3 + 4\pi r^2 - 100 = 0$$

and then finding the x-intercept of the function $y = \frac{4}{3}\pi x^3 + 4\pi x^2 - 100$.

3.3   EXERCISES

1–4 ■ List all possible rational zeros given by the Rational Zeros Theorem (but don't check to see which actually are zeros).

1. $P(x) = x^3 - 4x^2 + 3$

2. $Q(x) = x^4 - 3x^3 - 6x + 20$

3. $R(x) = 2x^5 + 3x^3 + 4x^2 - 8$

4. $S(x) = 6x^4 - x^2 + 2x + 12$

5–32 ■ Find all rational zeros of the polynomial.

5. $P(x) = x^3 + 3x^2 - 4$

6. $P(x) = x^3 - 7x^2 + 14x - 8$

7. $P(x) = x^3 - 3x - 2$

8. $P(x) = x^3 + 4x^2 - 3x - 18$

9. $P(x) = x^3 - 6x^2 + 12x - 8$

10. $P(x) = x^3 - x^2 - 8x + 12$

11. $P(x) = x^3 - 4x^2 + x + 6$

12. $P(x) = x^3 - 4x^2 - 7x + 10$

13. $P(x) = x^3 + 3x^2 + 6x + 4$

14. $P(x) = x^3 - 2x^2 - 2x - 3$

15. $P(x) = x^4 - 5x^2 + 4$

16. $P(x) = x^4 - 2x^3 - 3x^2 + 8x - 4$

17. $P(x) = x^4 + 6x^3 + 7x^2 - 6x - 8$

18. $P(x) = x^4 - x^3 - 23x^2 - 3x + 90$

19. $P(x) = 4x^4 - 25x^2 + 36$

20. $P(x) = x^4 - x^3 - 5x^2 + 3x + 6$

21. $P(x) = x^4 + 8x^3 + 24x^2 + 32x + 16$

22. $P(x) = 2x^3 + 7x^2 + 4x - 4$

23. $P(x) = 4x^3 + 4x^2 - x - 1$

24. $P(x) = 2x^3 - 3x^2 - 2x + 3$

25. $P(x) = 4x^3 - 7x + 3$

26. $P(x) = 8x^3 + 10x^2 - x - 3$

27. $P(x) = 2x^4 - 7x^3 + 3x^2 + 8x - 4$

28. $P(x) = 6x^4 - 7x^3 - 12x^2 + 3x + 2$

29. $P(x) = x^5 + 3x^4 - 9x^3 - 31x^2 + 36$

30. $P(x) = x^5 - 4x^4 - 3x^3 + 22x^2 - 4x - 24$

31. $P(x) = 3x^5 - 14x^4 - 14x^3 + 36x^2 + 43x + 10$

32. $P(x) = 2x^6 - 3x^5 - 13x^4 + 29x^3 - 27x^2 + 32x - 12$

33–42 ■ Find all the real zeros of the polynomial. Use the quadratic formula if necessary, as in Example 2.

33. $P(x) = x^3 + 4x^2 + 3x - 2$

34. $P(x) = x^3 - 5x^2 + 2x + 12$

35. $P(x) = x^4 - 6x^3 + 4x^2 + 15x + 4$

36. $P(x) = x^4 + 2x^3 - 2x^2 - 3x + 2$

37. $P(x) = x^4 - 7x^3 + 14x^2 - 3x - 9$

38. $P(x) = x^5 - 4x^4 - x^3 + 10x^2 + 2x - 4$

39. $P(x) = 4x^3 - 6x^2 + 1$

40. $P(x) = 3x^3 - 5x^2 - 8x - 2$

41. $P(x) = 2x^4 + 15x^3 + 17x^2 + 3x - 1$

42. $P(x) = 4x^5 - 18x^4 - 6x^3 + 91x^2 - 60x + 9$

43–50 ■ A polynomial P is given. (a) Find all the real zeros of P. (b) Sketch the graph of P.

43. $P(x) = x^3 - 3x^2 - 4x + 12$

44. $P(x) = -x^3 - 2x^2 + 5x + 6$

45. $P(x) = -x^3 + 2x^2 + 15x - 36$

46. $P(x) = 3x^3 + 17x^2 + 21x - 9$

47. $P(x) = x^4 - 5x^3 + 6x^2 + 4x - 8$

48. $P(x) = -x^4 + 10x^2 + 8x - 8$

49. $P(x) = x^5 - x^4 - 5x^3 + x^2 + 8x + 4$

50. $P(x) = x^5 - x^4 - 6x^3 + 14x^2 - 11x + 3$

51–56 ■ Use Descartes' Rule of Signs to determine how many positive and how many negative real zeros the polynomial can have. Then determine the possible total number of real zeros.

51. $P(x) = x^3 - x^2 - x - 3$

52. $P(x) = 2x^3 - x^2 + 4x - 7$

53. $P(x) = 2x^6 + 5x^4 - x^3 - 5x - 1$

54. $P(x) = x^4 + x^3 + x^2 + x + 12$

55. $P(x) = x^5 + 4x^3 - x^2 + 6x$

56. $P(x) = x^8 - x^5 + x^4 - x^3 + x^2 - x + 1$

57–60 ■ Show that the given values for a and b are lower and upper bounds for the real zeros of the polynomial.

57. $P(x) = 2x^3 + 5x^2 + x - 2; \quad a = -3, b = 1$

58. $P(x) = x^4 - 2x^3 - 9x^2 + 2x + 8; \quad a = -3, b = 5$

59. $P(x) = 8x^3 + 10x^2 - 39x + 9; \quad a = -3, b = 2$

60. $P(x) = 3x^4 - 17x^3 + 24x^2 - 9x + 1; \quad a = 0, b = 6$

61–64 ■ Find integers that are upper and lower bounds for the real zeros of the polynomial.

61. $P(x) = x^3 - 3x^2 + 4$

62. $P(x) = 2x^3 - 3x^2 - 8x + 12$

63. $P(x) = x^4 - 2x^3 + x^2 - 9x + 2$

64. $P(x) = x^5 - x^4 + 1$

65–70 ■ Find all rational zeros of the polynomial, and then find the irrational zeros, if any. Whenever appropriate, use the Rational Zeros Theorem, the Upper and Lower Bounds Theorem, Descartes' Rule of Signs, the quadratic formula, or other factoring techniques.

65. $P(x) = 2x^4 + 3x^3 - 4x^2 - 3x + 2$

66. $P(x) = 2x^4 + 15x^3 + 31x^2 + 20x + 4$

67. $P(x) = 4x^4 - 21x^2 + 5$

68. $P(x) = 6x^4 - 7x^3 - 8x^2 + 5x$

69. $P(x) = x^5 - 7x^4 + 9x^3 + 23x^2 - 50x + 24$

70. $P(x) = 8x^5 - 14x^4 - 22x^3 + 57x^2 - 35x + 6$

71–74 ■ Show that the polynomial does not have any rational zeros.

71. $P(x) = x^3 - x - 2$

72. $P(x) = 2x^4 - x^3 + x + 2$

73. $P(x) = 3x^3 - x^2 - 6x + 12$

74. $P(x) = x^{50} - 5x^{25} + x^2 - 1$

75–78 ■ The real solutions of the given equation are rational. List all possible rational roots using the Rational Zeros Theorem, and then graph the polynomial in the given viewing rectangle to determine which values are actually solutions. (All solutions can be seen in the given viewing rectangle.)

75. $x^3 - 3x^2 - 4x + 12 = 0; \quad [-4, 4]$ by $[-15, 15]$

76. $x^4 - 5x^2 + 4 = 0; \quad [-4, 4]$ by $[-30, 30]$

77. $2x^4 - 5x^3 - 14x^2 + 5x + 12 = 0; \quad [-2, 5]$ by $[-40, 40]$

78. $3x^3 + 8x^2 + 5x + 2 = 0; \quad [-3, 3]$ by $[-10, 10]$

79–82 ■ Use a graphing device to find all real solutions of the equation, correct to two decimal places.

79. $x^4 - x - 4 = 0$

80. $2x^3 - 8x^2 + 9x - 9 = 0$

81. $4.00x^4 + 4.00x^3 - 10.96x^2 - 5.88x + 9.09 = 0$

82. $x^5 + 2.00x^4 + 0.96x^3 + 5.00x^2 + 10.00x + 4.80 = 0$

83. A grain silo consists of a cylindrical main section and a hemispherical roof. If the total volume of the silo (including the part inside the roof section) is 15,000 ft³ and the cylindrical part is 30 ft tall, what is the radius of the silo, correct to the nearest tenth of a foot?

30 ft

84. A rectangular parcel of land has an area of 5000 ft². A diagonal between opposite corners is measured to be 10 ft longer than one side of the parcel. What are the dimensions of the land, correct to the nearest foot?

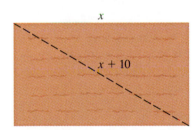

x

$x + 10$

85. Snow began falling at noon on Sunday. The amount of snow on the ground at a certain location at time t was given by the function

$$h(t) = 11.60t - 12.41t^2 + 6.20t^3 \\ - 1.58t^4 + 0.20t^5 - 0.01t^6$$

where t is measured in days from the start of the snowfall and $h(t)$ is the depth of snow in inches. Draw a graph of this function and use your graph to answer the following questions.

(a) What happened shortly after noon on Tuesday?

(b) Was there ever more than 5 in. of snow on the ground? If so, on what day(s)?

(c) On what day and at what time (to the nearest hour) did the snow disappear completely?

86. An open box with a volume of 1500 cm³ is to be constructed by taking a piece of cardboard 20 cm by 40 cm, cutting squares of side length x cm from each corner, and folding up the sides. Show that this can be done in two different ways, and find the exact dimensions of the box in each case.

40 cm

20 cm

x

x

87. A rocket consists of a right circular cylinder of height 20 m surmounted by a cone whose height and diameter are equal and whose radius is the same as that of the cylindrical section. What should this radius be (correct to two decimal places) if the total volume is to be $500\pi/3$ m³?

20 m

88. A rectangular box with a volume of $2\sqrt{2}$ ft³ has a square base. The diagonal of the box (between a pair of opposite corners) is 1 ft longer than each side of the base.

(a) If the base has sides of length x feet, show that

$$x^6 - 2x^5 - x^4 + 8 = 0$$

(b) Show that two different boxes satisfy the given conditions. Find the dimensions in each case, correct to the nearest hundredth of a foot.

x

x

89. A box with a square base has length plus girth of 108 in. (Girth is the distance "around" the box.) What is the length of the box if its volume is 2200 in^3?

90. Let $P(x)$ be a polynomial with real coefficients and let $b > 0$. Use the Division Algorithm to write

$$P(x) = (x - b) \cdot Q(x) + r$$

Suppose that $r \geq 0$ and that all the coefficients in $Q(x)$ are nonnegative. Let $z > b$.
(a) Show that $P(z) > 0$.
(b) Prove the first part of the Upper and Lower Bounds Theorem.

91. Use the first part of the Upper and Lower Bounds Theorem to prove the second part.   [*Hint:* Show that if $P(x)$ satisfies the second part of the theorem, then $P(-x)$ satisfies the first part.]

92. Show that the equation

$$x^5 - x^4 - x^3 - 5x^2 - 12x - 6 = 0$$

has exactly one rational root, and then prove that it must have either two or four irrational roots.

 DISCOVERY · DISCUSSION

93. How Many Real Zeros Can a Polynomial Have? Give examples of polynomials that have the following properties, or explain why it is impossible to find such a polynomial.

(a) A polynomial of degree 3 that has no real zeros

(b) A polynomial of degree 4 that has no real zeros

(c) A polynomial of degree 3 that has three real zeros, only one of which is rational

(d) A polynomial of degree 4 that has four real zeros, none of which is rational

What must be true about the degree of a polynomial with integer coefficients if it has no real zeros?

94. The Depressed Cubic   The most general cubic (third-degree) equation with rational coefficients can be written as $x^3 + ax^2 + bx + c = 0$.

(a) Show that if we replace x by $X - a/3$ and simplify, we end up with an equation that doesn't have an X^2 term, that is, an equation of the form $X^3 + pX + q = 0$. This is called a *depressed cubic*, because we have "depressed" the quadratic term.

(b) Use the procedure described in part (a) to depress the equation $x^3 + 6x^2 + 9x + 4 = 0$.

95. The Cubic Formula   The quadratic formula can be used to solve any quadratic (or second-degree) equation. You may have wondered if similar formulas exist for cubic (third-degree), quartic (fourth-degree), and higher-degree equations. For the depressed cubic $x^3 + px + q = 0$, Cardano (page 304) found the following formula for one solution:

$$x = \sqrt[3]{\frac{-q}{2} + \sqrt{\frac{q^2}{4} + \frac{p^3}{27}}} + \sqrt[3]{\frac{-q}{2} - \sqrt{\frac{q^2}{4} + \frac{p^3}{27}}}$$

A formula for quartic equations was discovered by the Italian mathematician Ferrari in 1540. In 1824 the Norwegian mathematician Niels Henrik Abel proved that it is impossible to write a quintic formula, that is, a formula for fifth-degree equations. Finally, Galois (page 280) gave a criterion for determining which equations can be solved by a formula involving radicals.

Use the cubic formula to find a solution for the following equations. Then solve the equations using the methods you learned in this section. Which method is easier?

(a) $x^3 - 3x + 2 = 0$

(b) $x^3 - 27x - 54 = 0$

(c) $x^3 + 3x + 4 = 0$

Discovery Project

Zeroing in on a Zero

We have seen how to find the zeros of a polynomial algebraically and graphically. Let's work through a **numerical method** for finding the zeros. With this method we can find the value of any real zero to as many decimal places as we wish.

The Intermediate Value Theorem states: If P is a polynomial and if $P(a)$ and $P(b)$ are of opposite sign, then P has a zero between a and b. (See page 262.) The Intermediate Value Theorem is an example of an **existence theorem**—it tells us that a zero exists, but doesn't tell us exactly where it is. Nevertheless, we can use the theorem to zero in on the zero.

For example, consider the polynomial $P(x) = x^3 + 8x - 30$. Notice that $P(2) < 0$ and $P(3) > 0$. By the Intermediate Value Theorem P must have a zero between 2 and 3. To "trap" the zero in a smaller interval, we evaluate P at successive tenths between 2 and 3 until we find where P changes sign, as in Table 1. From the table we see that the zero we are looking for lies between 2.2 and 2.3, as shown in Figure 1.

x	$P(x)$	
2.1	−3.94	
2.2	−1.75	}change of sign
2.3	0.57	

TABLE 1

x	$P(x)$	
2.26	−0.38	
2.27	−0.14	
2.28	0.09	}change of sign

TABLE 2

FIGURE 1

FIGURE 2

We can repeat this process by evaluating P at successive 100ths between 2.2 and 2.3, as in Table 2. By repeating this process over and over again, we can get a numerical value for the zero as accurately as we want. From Table 2 we see that the zero is between 2.27 and 2.28. To see whether it is closer to 2.27 or 2.28, we check the value of P halfway between these two numbers: $P(2.275) \approx -0.03$. Since this value is negative, the zero we are looking for lies between 2.275 and 2.28, as illustrated in Figure 2. Correct to the nearest 100th, the zero is 2.28.

1. (a) Show that $P(x) = x^2 - 2$ has a zero between 1 and 2.
   (b) Find the zero of P to the nearest tenth.
   (c) Find the zero of P to the nearest 100th.
   (d) Explain why the zero you found is an approximation to $\sqrt{2}$. Repeat the process several times to obtain $\sqrt{2}$ correct to three decimal places. Compare your results to $\sqrt{2}$ obtained by a calculator.

2. Find a polynomial that has $\sqrt[3]{5}$ as a zero. Use the process described here to zero in on $\sqrt[3]{5}$ to four decimal places.

3. Show that the polynomial has a zero between the given integers, and then zero in on that zero, correct to two decimals.
   (a) $P(x) = x^3 + x - 7$; between 1 and 2
   (b) $P(x) = x^3 - x^2 - 5$; between 2 and 3
   (c) $P(x) = 2x^4 - 4x^2 + 1$; between 1 and 2
   (d) $P(x) = 2x^4 - 4x^2 + 1$; between -1 and 0

4. Find the indicated irrational zero, correct to two decimals.
   (a) The positive zero of $P(x) = x^4 + 2x^3 + x^2 - 1$
   (b) The negative zero of $P(x) = x^4 + 2x^3 + x^2 - 1$

3.4   COMPLEX NUMBERS

In Section 1.5 we saw that if the discriminant of a quadratic equation is negative, the equation has no real solution. For example, the equation

$$x^2 + 4 = 0$$

has no real solution. If we try to solve this equation, we get $x^2 = -4$, so

$$x = \pm\sqrt{-4}$$

But this is impossible, since the square of any real number is positive. [For example, $(-2)^2 = 4$, a positive number.] Thus, negative numbers don't have real square roots.

See the note on Cardano, page 304, for an example of how complex numbers are used to find real solutions of polynomial equations.

To make it possible to solve *all* quadratic equations, mathematicians invented an expanded number system, called the *complex number system*. First they defined the new number

$$i = \sqrt{-1}$$

This means $i^2 = -1$. A complex number is then a number of the form $a + bi$, where a and b are real numbers.

> ### DEFINITION OF COMPLEX NUMBERS
>
> A **complex number** is an expression of the form
>
> $$a + bi$$
>
> where a and b are real numbers and $i^2 = -1$. The **real part** of this complex number is a and the **imaginary part** is b. Two complex numbers are **equal** if and only if their real parts are equal and their imaginary parts are equal.

Note that both the real and imaginary parts of a complex number are real numbers.

EXAMPLE 1 ■ Complex Numbers

The following are examples of complex numbers.

$3 + 4i$	Real part 3, imaginary part 4
$\frac{1}{2} - \frac{2}{3}i$	Real part $\frac{1}{2}$, imaginary part $-\frac{2}{3}$
$6i$	Real part 0, imaginary part 6
-7	Real part -7, imaginary part 0

A number such as $6i$, which has real part 0, is called a **pure imaginary number**. A real number like -7 can be thought of as a complex number with imaginary part 0.

In the complex number system every quadratic equation has solutions. The numbers $2i$ and $-2i$ are solutions of $x^2 = -4$ because

$$(2i)^2 = 2^2 i^2 = 4(-1) = -4 \qquad \text{and} \qquad (-2i)^2 = (-2)^2 i^2 = 4(-1) = -4$$

Although we use the term *imaginary* in this context, imaginary numbers should not be thought of as any less "real" (in the ordinary rather than the mathematical sense of that word) than negative numbers or irrational numbers. All numbers (except possibly the positive integers) are creations of the human mind—the numbers -1 and $\sqrt{2}$ as well as the number i. We study complex numbers because they complete, in a useful and elegant fashion, the study of the solutions of polynomial equations. In fact, imaginary numbers are useful not only in algebra and mathematics, but in the other sciences as well. To give just one example, in electrical theory the *reactance* of a circuit is a quantity whose measure is an imaginary number.

Leonhard Euler (1707–1783) was born in Basel, Switzerland, the son of a pastor. At age 13 his father sent him to the University at Basel to study theology, but Euler soon decided to devote himself to the sciences. Besides theology he studied mathematics, medicine, astronomy, physics, and oriental languages. It is said that Euler could calculate as effortlessly as "men breathe or as eagles fly." One hundred years before Euler, Fermat (see page 532) had conjectured that $2^{2^n} + 1$ is a prime number for all n. The first five of these numbers are 5, 17, 257, 65537, and 4,294,967,297. It's easy to show that the first four are prime. The fifth was also thought to be prime until Euler, with his phenomenal calculating ability, showed that it is the product $641 \times 6,700,417$ and so is not prime. Euler published more than any other mathematician in history. His collected works comprise 75 large volumes. Although he was blind for the last 17 years of his life, he continued to work and publish. In his writings he popularized the use of the symbols π, e, and i, which you will find in this textbook. One of Euler's most lasting contributions is the development of complex numbers.

■ Arithmetic Operations on Complex Numbers

Complex numbers are added, subtracted, multiplied, and divided just as we would any number of the form $a + b\sqrt{c}$. The only difference we must keep in mind is that $i^2 = -1$. Thus, the following calculations are valid.

$$(a + bi)(c + di) = ac + (ad + bc)i + bdi^2 \qquad \text{Multiply and collect like terms}$$

$$= ac + (ad + bc)i + bd(-1) \qquad i^2 = -1$$

$$= (ac - bd) + (ad + bc)i \qquad \text{Combine real and imaginary parts}$$

We therefore define the sum, difference, and product of complex numbers as follows.

ADDING, SUBTRACTING, AND MULTIPLYING COMPLEX NUMBERS

Definition	Description
Addition	
$(a + bi) + (c + di) = (a + c) + (b + d)i$	To add complex numbers, add the real parts and the imaginary parts.
Subtraction	
$(a + bi) - (c + di) = (a - c) + (b - d)i$	To subtract complex numbers, subtract the real parts and the imaginary parts.
Multiplication	
$(a + bi) \cdot (c + di) = (ac - bd) + (ad + bc)i$	Multiply complex numbers like binomials, using $i^2 = -1$.

EXAMPLE 2 ■ Adding, Subtracting, and Multiplying Complex Numbers

Express the following in the form $a + bi$.

(a) $(3 + 5i) + (4 - 2i)$ (b) $(3 + 5i) - (4 - 2i)$

(c) $(3 + 5i)(4 - 2i)$ (d) i^{23}

SOLUTION

Graphing calculators can perform arithmetic operations on complex numbers.

```
(3+5i)+(4-2i)
              7+3i
(3+5i)*(4-2i)
             22+14i
```

(a) According to the definition, we add the real parts and we add the imaginary parts.

$$(3 + 5i) + (4 - 2i) = (3 + 4) + (5 - 2)i = 7 + 3i$$

(b) $(3 + 5i) - (4 - 2i) = (3 - 4) + [5 - (-2)]i = -1 + 7i$

(c) $(3 + 5i)(4 - 2i) = [3 \cdot 4 - 5(-2)] + [3(-2) + 5 \cdot 4]i = 22 + 14i$

(d) $i^{23} = i^{20+3} = (i^2)^{10}i^3 = (-1)^{10}i^2i = (1)(-1)i = -i$ ■

Complex Conjugates

Number	Conjugate
$3 + 2i$	$3 - 2i$
$1 - i$	$1 + i$
$4i$	$-4i$
5	5

Division of complex numbers is much like rationalizing the denominator of a radical expression, which we considered in Section 1.2. For the complex number $z = a + bi$ we define its **complex conjugate** to be $\bar{z} = a - bi$. Note that

$$z \cdot \bar{z} = (a + bi)(a - bi) = a^2 + b^2$$

So the product of a complex number and its conjugate is always a nonnegative real number. We use this property to divide complex numbers.

DIVIDING COMPLEX NUMBERS

To simplify the quotient $\dfrac{a + bi}{c + di}$, multiply the numerator and the denominator by the complex conjugate of the denominator:

$$\frac{a + bi}{c + di} = \left(\frac{a + bi}{c + di}\right)\left(\frac{c - di}{c - di}\right) = \frac{(ac + bd) + (bc - ad)i}{c^2 + d^2}$$

Rather than memorize this entire formula, it's best just to remember the first step and then mutiply out the numerator and the denominator as usual.

EXAMPLE 3 ■ Dividing Complex Numbers

Express the following in the form $a + bi$.

(a) $\dfrac{3 + 5i}{1 - 2i}$

(b) $\dfrac{7 + 3i}{4i}$

SOLUTION

We multiply both numerator and denominator by the complex conjugate of the denominator to make the new denominator a real number.

(a) The complex conjugate of $1 - 2i$ is $\overline{1 - 2i} = 1 + 2i$.

$$\frac{3 + 5i}{1 - 2i} = \left(\frac{3 + 5i}{1 - 2i}\right)\left(\frac{1 + 2i}{1 + 2i}\right) = \frac{-7 + 11i}{5} = -\frac{7}{5} + \frac{11}{5}i$$

(b) The complex conjugate of $4i$ is $-4i$. Therefore

$$\frac{7 + 3i}{4i} = \left(\frac{7 + 3i}{4i}\right)\left(\frac{-4i}{-4i}\right) = \frac{12 - 28i}{16} = \frac{3}{4} - \frac{7}{4}i \qquad ■$$

Square Roots of Negative Numbers

Just as every positive real number r has two square roots ($\sqrt{r}$ and $-\sqrt{r}$), every negative number has two square roots as well. If $-r$ is a negative number, then its square roots are $\pm i\sqrt{r}$, because $(i\sqrt{r})^2 = i^2 r = -r$ and $(-i\sqrt{r})^2 = i^2 r = -r$.

SQUARE ROOTS OF NEGATIVE NUMBERS

If $-r$ is negative, then the **principal square root** of $-r$ is

$$\sqrt{-r} = i\sqrt{r}$$

The two square roots of $-r$ are $i\sqrt{r}$ and $-i\sqrt{r}$.

We usually write $i\sqrt{b}$ instead of $\sqrt{b}\,i$ to avoid confusion with $\sqrt{bi}$.

EXAMPLE 4 ■ Square Roots of Negative Numbers

(a) $\sqrt{-1} = i\sqrt{1} = i$ (b) $\sqrt{-16} = i\sqrt{16} = 4i$ (c) $\sqrt{-3} = i\sqrt{3}$ ■

Special care must be taken when performing calculations involving square roots of negative numbers. Although $\sqrt{a} \cdot \sqrt{b} = \sqrt{ab}$ when a and b are positive, this is *not* true when both are negative. For example,

$$\sqrt{-2} \cdot \sqrt{-3} = i\sqrt{2} \cdot i\sqrt{3} = i^2\sqrt{6} = -\sqrt{6}$$

but

$$\sqrt{(-2)(-3)} = \sqrt{6}$$

so

$$\sqrt{-2} \cdot \sqrt{-3} \neq \sqrt{(-2)(-3)}$$

When multiplying radicals of negative numbers, express them first in the form $i\sqrt{r}$ (where $r > 0$) to avoid possible errors of this type.

EXAMPLE 5 ■ Using Square Roots of Negative Numbers

Evaluate $(\sqrt{12} - \sqrt{-3})(3 + \sqrt{-4})$ and express in the form $a + bi$.

SOLUTION

$$
\begin{aligned}
(\sqrt{12} - \sqrt{-3})(3 + \sqrt{-4}) &= (\sqrt{12} - i\sqrt{3})(3 + i\sqrt{4}) \\
&= (2\sqrt{3} - i\sqrt{3})(3 + 2i) \\
&= (6\sqrt{3} + 2\sqrt{3}) + i(2 \cdot 2\sqrt{3} - 3\sqrt{3}) \\
&= 8\sqrt{3} + i\sqrt{3}
\end{aligned}
$$

■

Complex Roots of Quadratic Equations

We have already seen that, if $a \neq 0$, the solutions of the quadratic equation $ax^2 + bx + c = 0$ are

$$x = \frac{-b \pm \sqrt{b^2 - 4ac}}{2a}$$

If $b^2 - 4ac < 0$, then the equation has no real solution. But in the complex number system, this equation will always have solutions, because negative numbers have square roots in this expanded setting.

EXAMPLE 6 ■ Quadratic Equations with Complex Solutions

Solve each equation: (a) $x^2 + 9 = 0$ (b) $x^2 + 4x + 5 = 0$

SOLUTION

(a) The equation $x^2 + 9 = 0$ means $x^2 = -9$, so

$$x = \pm\sqrt{-9} = \pm i \sqrt{9} = \pm 3i$$

The solutions are therefore $3i$ and $-3i$.

(b) By the quadratic formula we have

$$x = \frac{-4 \pm \sqrt{4^2 - 4 \cdot 5}}{2}$$

$$= \frac{-4 \pm \sqrt{-4}}{2}$$

$$= \frac{-4 \pm 2i}{2} = \frac{2(-2 \pm i)}{2} = -2 \pm i$$

So, the solutions are $-2 + i$ and $-2 - i$.   ■

The two solutions of *any* quadratic equation that has real coefficients are complex conjugates of each other. To understand why this is true, think about the $\pm$ sign in the quadratic formula.

EXAMPLE 7 ■ Complex Conjugates as Solutions of a Quadratic

Show that the solutions of the equation

$$4x^2 - 24x + 37 = 0$$

are complex conjugates of each other.

SOLUTION

We use the quadratic formula to get

$$x = \frac{24 \pm \sqrt{(24)^2 - 4(4)(37)}}{2(4)}$$

$$= \frac{24 \pm \sqrt{-16}}{8} = \frac{24 \pm 4i}{8} = 3 \pm \frac{1}{2}i$$

So, the solutions are $3 + \frac{1}{2}i$ and $3 - \frac{1}{2}i$, and these are complex conjugates.   ■

3.4 EXERCISES

1–10 ■ Find the real and imaginary parts of the complex number.

1. $5 - 7i$

2. $-6 + 4i$

3. $\dfrac{-2 - 5i}{3}$

4. $\dfrac{4 + 7i}{2}$

5. 3

6. $-\frac{1}{2}$

7. $-\frac{2}{3}i$

8. $i\sqrt{3}$

9. $\sqrt{3} + \sqrt{-4}$

10. $2 - \sqrt{-5}$

11–52 ■ Evaluate the expression and write the result in the form $a + bi$.

11. $(2 - 5i) + (3 + 4i)$

12. $(2 + 5i) + (4 - 6i)$

13. $(-6 + 6i) + (9 - i)$

14. $(3 - 2i) + \left(-5 - \frac{1}{3}i\right)$

15. $\left(7 - \frac{1}{2}i\right) - \left(5 + \frac{3}{2}i\right)$

16. $(-4 + i) - (2 - 5i)$

17. $(-12 + 8i) - (7 + 4i)$

18. $6i - (4 - i)$

19. $4(-1 + 2i)$

20. $2i\left(\frac{1}{2} - i\right)$

21. $(7 - i)(4 + 2i)$

22. $(5 - 3i)(1 + i)$

23. $(3 - 4i)(5 - 12i)$

24. $\left(\frac{2}{3} + 12i\right)\left(\frac{1}{6} + 24i\right)$

25. $(6 + 5i)(2 - 3i)$

26. $(-2 + i)(3 - 7i)$

27. $\dfrac{1}{i}$

28. $\dfrac{1}{1 + i}$

29. $\dfrac{2 - 3i}{1 - 2i}$

30. $\dfrac{5 - i}{3 + 4i}$

31. $\dfrac{26 + 39i}{2 - 3i}$

32. $\dfrac{25}{4 - 3i}$

33. $\dfrac{10i}{1 - 2i}$

34. $(2 - 3i)^{-1}$

35. $\dfrac{4 + 6i}{3i}$

36. $\dfrac{-3 + 5i}{15i}$

37. $\dfrac{1}{1 + i} - \dfrac{1}{1 - i}$

38. $\dfrac{(1 + 2i)(3 - i)}{2 + i}$

39. i^3

40. $(2i)^4$

41. i^{100}

42. i^{1002}

43. $\sqrt{-25}$

44. $\sqrt{\dfrac{-9}{4}}$

45. $\sqrt{-3}\,\sqrt{-12}$

46. $\sqrt{\frac{1}{3}}\,\sqrt{-27}$

47. $(3 - \sqrt{-5})(1 + \sqrt{-1})$

48. $\dfrac{1 - \sqrt{-1}}{1 + \sqrt{-1}}$

49. $\dfrac{2 + \sqrt{-8}}{1 + \sqrt{-2}}$

50. $(\sqrt{3} - \sqrt{-4})(\sqrt{6} - \sqrt{-8})$

51. $\dfrac{\sqrt{-36}}{\sqrt{-2}\,\sqrt{-9}}$

52. $\dfrac{\sqrt{-7}\,\sqrt{-49}}{\sqrt{28}}$

53–66 ■ Find all solutions of the equation and express them in the form $a + bi$.

53. $x^2 + 9 = 0$

54. $9x^2 + 4 = 0$

55. $x^2 - 4x + 5 = 0$

56. $x^2 + 2x + 2 = 0$

57. $x^2 + x + 1 = 0$

58. $x^2 - 3x + 3 = 0$

59. $2x^2 - 2x + 1 = 0$

60. $2x^2 + 3 = 2x$

61. $t + 3 + \dfrac{3}{t} = 0$

62. $z + 4 + \dfrac{12}{z} = 0$

63. $6x^2 + 12x + 7 = 0$

64. $4x^2 - 16x + 19 = 0$

65. $\frac{1}{2}x^2 - x + 5 = 0$

66. $x^2 + \frac{1}{2}x + 1 = 0$

67–74 ■ Recall that the symbol $\bar{z}$ represents the complex conjugate of z. If $z = a + bi$ and $w = c + di$, prove each statement.

67. $\bar{z} + \bar{w} = \overline{z + w}$

68. $\overline{zw} = \bar{z} \cdot \bar{w}$

69. $(\bar{z})^2 = \overline{z^2}$

70. $\bar{\bar{z}} = z$

71. $z + \bar{z}$ is a real number

72. $z - \bar{z}$ is a pure imaginary number

73. $z \cdot \bar{z}$ is a real number

74. $z = \bar{z}$ if and only if z is real

◤ DISCOVERY · DISCUSSION

75. Complex Conjugate Roots   Suppose that the equation $ax^2 + bx + c = 0$ has real coefficients and complex roots. Why must the roots be complex conjugates of each other? (Think about how you would find the roots using the quadratic formula.)

76. Powers of i   Calculate the first 12 powers of i, that is, $i, i^2, i^3, \ldots, i^{12}$. Do you notice a pattern? Explain how you would calculate any whole number power of i, using the pattern you have discovered. Use this procedure to calculate i^{4446}.

3.5 COMPLEX ZEROS AND THE FUNDAMENTAL THEOREM OF ALGEBRA

We have already seen that an nth-degree polynomial can have at most n real zeros. In the complex number system an nth-degree polynomial has *exactly* n zeros, and so can be factored into exactly n linear factors. This fact is a consequence of the Fundamental Theorem of Algebra, which was proved by the German mathematician C. F. Gauss in 1799 (see page 300).

The Fundamental Theorem of Algebra and Complete Factorization

The following theorem is the basis for much of our work in factoring polynomials and solving polynomial equations.

> **FUNDAMENTAL THEOREM OF ALGEBRA**
>
> Every polynomial
>
> $$P(x) = a_n x^n + a_{n-1}x^{n-1} + \cdots + a_1 x + a_0 \qquad (n \geq 1, a_n \neq 0)$$
>
> with complex coefficients has at least one complex zero.

Because any real number is also a complex number, the theorem applies to polynomials with real coefficients as well.

The Fundamental Theorem of Algebra and the Factor Theorem together show that a polynomial can be factored completely into linear factors, as we now prove.

> **COMPLETE FACTORIZATION THEOREM**
>
> If $P(x)$ is a polynomial of degree $n > 0$, then there exist complex numbers $a, c_1, c_2, \ldots, c_n$ (with $a \neq 0$) such that
>
> $$P(x) = a(x - c_1)(x - c_2) \cdots (x - c_n)$$

■ **Proof** By the Fundamental Theorem of Algebra, P has at least one zero. Let's call it c_1. By the Factor Theorem, $P(x)$ can be factored as

$$P(x) = (x - c_1) \cdot Q_1(x)$$

where $Q_1(x)$ is of degree $n - 1$. Applying the Fundamental Theorem to the quotient $Q_1(x)$ gives us the factorization

$$P(x) = (x - c_1) \cdot (x - c_2) \cdot Q_2(x)$$

where $Q_2(x)$ is of degree $n - 2$ and c_2 is a zero of $Q_1(x)$. Continuing this process for n steps, we get a final quotient $Q_n(x)$ of degree 0, a nonzero constant that we will call a. This means that P has been factored as

$$P(x) = a(x - c_1)(x - c_2) \cdots (x - c_n)$$

To actually find the complex zeros of an nth-degree polynomial, we usually first factor as much as possible, then use the quadratic formula on parts that we can't factor further, as illustrated in the next example.

EXAMPLE 1 ■ Factoring and Using the Quadratic Formula to Find Complex Solutions

Find all solutions of the equation $x^6 - 64 = 0$.

SOLUTION

We first factor the polynomial as much as possible.

$$x^6 - 64 = (x^3)^2 - 8^2$$

$$= (x^3 - 8)(x^3 + 8) \qquad \text{Difference of squares formula}$$

$$= (x - 2)(x^2 + 2x + 4)(x + 2)(x^2 - 2x + 4) \qquad \text{Difference of cubes and sum of cubes formulas}$$

This expression will equal 0 when any factor is 0, so we find the solutions as follows:

$$x - 2 = 0 \qquad \text{means} \qquad x = 2$$

$$x^2 + 2x + 4 = 0 \qquad \text{means} \qquad x = \frac{-2 \pm \sqrt{2^2 - 4 \cdot 4}}{2} \qquad \text{Quadratic formula}$$

$$= \frac{-2 \pm \sqrt{-12}}{2} = -1 \pm i\sqrt{3}$$

$$x + 2 = 0 \qquad \text{means} \qquad x = -2$$

$$x^2 - 2x + 4 = 0 \qquad \text{means} \qquad x = \frac{2 \pm \sqrt{(-2)^2 - 4 \cdot 4}}{2} \qquad \text{Quadratic formula}$$

$$= \frac{2 \pm \sqrt{-12}}{2} = 1 \pm i\sqrt{3}$$

The roots of the equation are the zeros of the polynomial:

$$2, \quad -2, \quad -1 + i\sqrt{3}, \quad -1 - i\sqrt{3}, \quad 1 + i\sqrt{3}, \quad \text{and} \quad 1 - i\sqrt{3} \qquad ■$$

Carl Friedrich Gauss (1777–1855) is considered the greatest mathematician of modern times. His contemporaries called him the "Prince of Mathematics." He was born into a poor family; his father made a living as a mason. As a very small child, Gauss found a calculation error in his father's accounts, the first of many incidents that gave evidence of his mathematical precocity. (See also page 819.) At 19 Gauss demonstrated that the regular 17-sided polygon can be constructed with straight-edge and compass alone. This was remarkable because, since the time of Euclid, it was thought that the only regular polygons constructible in this way were the triangle and pentagon. Because of this discovery Gauss decided to pursue a career in mathematics instead of languages, his other passion. In his doctoral dissertation, written at the age of 22, Gauss proved the Fundamental Theorem of Algebra: A polynomial of degree n with complex coefficients has n roots. His other accomplishments range over every branch of mathematics, as well as physics and astronomy.

■ Zeros and Their Multiplicities

In the Complete Factorization Theorem the numbers $c_1, c_2, \ldots, c_n$ are the zeros of P. These zeros need not all be different. If the factor $x - c$ appears k times in the complete factorization of $P(x)$, then we say that c is a zero of **multiplicity k**. For example, the polynomial

$$P(x) = (x - 1)^3(x + 2)^2(x + 3)^5$$

has the following zeros:

$$1 \text{ (multiplicity 3)}, \qquad -2 \text{ (multiplicity 2)}, \qquad -3 \text{ (multiplicity 5)}$$

The polynomial P has the same number of zeros as its degree—it has degree 10 and has 10 zeros, provided we count multiplicities. This is true for all polynomials, as we prove in the following theorem.

ZEROS THEOREM

Every polynomial of degree $n \geq 1$ has exactly n zeros, provided that a zero of multiplicity k is counted k times.

■ **Proof**   Let P be a polynomial of degree n. By the Complete Factorization Theorem

$$P(x) = a(x - c_1)(x - c_2) \cdots (x - c_n)$$

Now suppose that c is a zero of P other than $c_1, c_2, \ldots, c_n$. Then

$$P(c) = a(c - c_1)(c - c_2) \cdots (c - c_n) = 0$$

Thus, by the Zero-Product Property one of the factors $c - c_i$ must be 0, so $c = c_i$ for some i. It follows that P has exactly the n zeros $c_1, c_2, \ldots, c_n$. ☐

EXAMPLE 2 ■ Factoring a Polynomial with Complex Zeros

Find the complete factorization and all five zeros of the polynomial

$$P(x) = 3x^5 + 24x^3 + 48x$$

SOLUTION

Since $3x$ is a common factor, we have

$$P(x) = 3x(x^4 + 8x^2 + 16)$$

$$= 3x(x^2 + 4)^2$$

To factor $x^2 + 4$, note that $2i$ and $-2i$ are zeros of this polynomial. Thus $x^2 + 4 = (x - 2i)(x + 2i)$, and so

$$P(x) = 3x[(x - 2i)(x + 2i)]^2$$

$$= 3x(x - 2i)(x - 2i)(x + 2i)(x + 2i)$$

The zeros of P are 0, $2i$, and $-2i$. Since the factors $x - 2i$ and $x + 2i$ each occur twice in the complete factorization of P, the zeros $2i$ and $-2i$ are of multiplicity 2 (or *double* zeros). So the total number of zeros is five.   ■

The following table gives further examples of polynomials with their complete factorizations and zeros.

Degree	Polynomial	Zero(s)	Number of zeros
1	$P(x) = x - 4$	4	1
2	$P(x) = x^2 - 10x + 25$ $= (x - 5)(x - 5)$	5 (multiplicity 2)	2
3	$P(x) = x^3 + x$ $= x(x - i)(x + i)$	$0, i, -i$	3
4	$P(x) = x^4 + 8x^2 + 16$ $= (x - 2i)(x - 2i)(x + 2i)(x + 2i)$	$2i$ (multiplicity 2), $-2i$ (multiplicity 2)	4
5	$P(x) = x^5 - 2x^4 + x^3$ $= x^3(x - 1)^2$	0 (multiplicity 3), 1 (multiplicity 2)	5

EXAMPLE 3   ■   Finding Polynomials with Specified Zeros

Find a polynomial that satisfies the given description.

(a) A polynomial $P(x)$ of degree 4, with zeros i, $-i$, 2, and -2 and with $P(3) = 25$

(b) A polynomial $Q(x)$ of degree 4, with zeros -2 and 0, where -2 is a zero of multiplicity 3

SOLUTION

(a) The required polynomial has the form

$$P(x) = a(x - i)(x - (-i))(x - 2)(x - (-2))$$

$$= a(x^2 + 1)(x^2 - 4) \qquad \text{Difference of squares}$$

$$= a(x^4 - 3x^2 - 4) \qquad \text{Multiply}$$

We know that $P(3) = a(3^4 - 3 \cdot 3^2 - 4) = 50a = 25$, so $a = \frac{1}{2}$. Thus

$$P(x) = \tfrac{1}{2}x^4 - \tfrac{3}{2}x^2 - 2$$

(b) We require

$$Q(x) = a[x - (-2)]^3(x - 0)$$

$$= a(x + 2)^3 x$$

$$= a(x^3 + 6x^2 + 12x + 8)x \qquad \text{Product Formula 4 (Section 1.3)}$$

$$= a(x^4 + 6x^3 + 12x^2 + 8x)$$

Since we are given no information about Q other than its zeros and their multiplicity, we can choose any number for a. If we use $a = 1$, we get

$$Q(x) = x^4 + 6x^3 + 12x^2 + 8x \qquad \blacksquare$$

EXAMPLE 4 ■ Finding All the Zeros of a Polynomial

Find all four zeros of $P(x) = 3x^4 - 2x^3 - x^2 - 12x - 4$.

SOLUTION

Using the Rational Zeros Theorem from Section 3.3, we obtain the following list of possible rational zeros: $\pm 1, \pm 2, \pm 4, \pm\frac{1}{3}, \pm\frac{2}{3}, \pm\frac{4}{3}$. Checking these using synthetic division, we find that 2 and $-\frac{1}{3}$ are zeros, and we get the following factorization.

$$P(x) = 3x^4 - 2x^3 - x^2 - 12x - 4$$

$$= (x - 2)(3x^3 + 4x^2 + 7x + 2) \qquad \text{Factor } x - 2$$

$$= (x - 2)\left(x + \tfrac{1}{3}\right)(3x^2 + 3x + 6) \qquad \text{Factor } x + \tfrac{1}{3}$$

$$= 3(x - 2)\left(x + \tfrac{1}{3}\right)(x^2 + x + 2) \qquad \text{Factor 3}$$

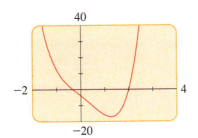

FIGURE 1

$P(x) = 3x^4 - 2x^3 - x^2 - 12x - 4$

Figure 1 shows the graph of the polynomial P in Example 4. The x-intercepts correspond to the real zeros of P. The imaginary zeros cannot be determined from the graph.

The zeros of the quadratic factor are

$$x = \frac{-1 \pm \sqrt{1 - 8}}{2} = -\frac{1}{2} \pm i\frac{\sqrt{7}}{2} \qquad \text{Quadratic formula}$$

so the zeros of $P(x)$ are

$$2, \quad -\frac{1}{3}, \quad -\frac{1}{2} + i\frac{\sqrt{7}}{2}, \quad \text{and} \quad -\frac{1}{2} - i\frac{\sqrt{7}}{2} \qquad \blacksquare$$

As you may have noticed from the examples so far, the complex zeros of polynomials with real coefficients come in pairs. Whenever $a + bi$ is a zero, its complex conjugate $a - bi$ is also a zero.

CONJUGATE ZEROS THEOREM

If the polynomial P has real coefficients, and if the complex number z is a zero of P, then its complex conjugate $\bar{z}$ is also a zero of P.

Gerolamo Cardano (1501–1576) is certainly one of the most colorful figures in the history of mathematics. He was the most renowned physician in Europe in his day, yet throughout his life he was plagued by numerous maladies, including ruptures, hemorrhoids, and an irrational fear of encountering rabid dogs. A doting father, his beloved sons broke his heart—his favorite was eventually beheaded for murdering his wife. Cardano was also a compulsive gambler; indeed, this vice may have driven him to write the *Book on Games of Chance*, the first study of probability from a mathematical point of view.

In Cardano's major mathematical work, the *Ars Magna*, he detailed the solution of the general third- and fourth-degree polynomial equations. At the time of its publication, mathematicians were uncomfortable even with negative numbers, but Cardano's formulas paved the way for the acceptance not just of negative numbers, but also of imaginary numbers, because they occurred naturally in solving polynomial equations. For example, for the cubic equation

$$x^3 - 15x - 4 = 0$$

one of his formulas gives the solution

$$x = \sqrt[3]{2 + \sqrt{-121}} + \sqrt[3]{2 - \sqrt{-121}}$$

(See page 290, Exercise 95.) This value for x actually turns out to be the *integer* 4, yet to find it Cardano had to use the imaginary number $\sqrt{-121} = 11i$.

■ **Proof**   Let

$$P(x) = a_n x^n + a_{n-1} x^{n-1} + \cdots + a_1 x + a_0$$

where each coefficient is real. Suppose that $P(z) = 0$. We must prove that $P(\bar{z}) = 0$. We use the facts that the complex conjugate of a sum of two complex numbers is the sum of the conjugates and that the conjugate of a product is the product of the conjugates (see Exercises 75 and 76 in Section 3.4).

$$
\begin{aligned}
P(\bar{z}) &= a_n (\bar{z})^n + a_{n-1} (\bar{z})^{n-1} + \cdots + a_1 \bar{z} + a_0 \\
&= \overline{a_n} \, \overline{z^n} + \overline{a_{n-1}} \, \overline{z^{n-1}} + \cdots + \overline{a_1} \overline{z} + \overline{a_0} \qquad \text{\color{red}{Because the coefficients are real}} \\
&= \overline{a_n z^n} + \overline{a_{n-1} z^{n-1}} + \cdots + \overline{a_1 z} + \overline{a_0} \\
&= \overline{a_n z^n + a_{n-1} z^{n-1} + \cdots + a_1 z + a_0} \\
&= \overline{P(z)} = \overline{0} = 0
\end{aligned}
$$

This shows that $\bar{z}$ is also a zero of $P(x)$, which proves the theorem.   □

EXAMPLE 5 ■ A Polynomial with a Specified Complex Zero

Find a polynomial $P(x)$ of degree 3 that has integer coefficients and zeros $\frac{1}{2}$ and $3 - i$.

SOLUTION

Since $3 - i$ is a zero, then so is $3 + i$ by the Conjugate Zeros Theorem. This means that $P(x)$ has the form

$$
\begin{aligned}
P(x) &= a\left(x - \tfrac{1}{2}\right)[x - (3 - i)][x - (3 + i)] \\
&= a\left(x - \tfrac{1}{2}\right)[(x - 3) + i][(x - 3) - i] \qquad \text{\color{red}{Regroup}} \\
&= a\left(x - \tfrac{1}{2}\right)[(x - 3)^2 - i^2] \qquad \text{\color{red}{Difference of squares formula}} \\
&= a\left(x - \tfrac{1}{2}\right)(x^2 - 6x + 10) \qquad \text{\color{red}{Expand}} \\
&= a\left(x^3 - \tfrac{13}{2}x^2 + 13x - 5\right) \qquad \text{\color{red}{Expand}}
\end{aligned}
$$

To make all coefficients integers, we set $a = 2$ and get

$$P(x) = 2x^3 - 13x^2 + 26x - 10$$

Any other polynomial that satisfies the given requirements must be an integer multiple of this one.   ■

EXAMPLE 6 ■ Using Descartes' Rule to Count Real and Imaginary Zeros

Without actually factoring, determine how many positive real zeros, negative real zeros, and imaginary zeros the following polynomial could have:

$$P(x) = x^4 + 6x^3 - 12x^2 - 14x - 24$$

SOLUTION

Since there is one change of sign, by Descartes' Rule of Signs, P has one positive real zero. Also, $P(-x) = x^4 - 6x^3 - 12x^2 + 14x - 24$ has three changes of sign, so there are either three or one negative real zero(s). So P has a total of either four or two real zeros. Since P is of degree 4, it has four zeros in all, which gives the following possibilities.

Positive real zeros	Negative real zeros	Imaginary zeros
1	3	0
1	1	2

Linear and Quadratic Factors

We have seen that a polynomial factors completely into linear factors if we use complex numbers. If we do not use complex numbers, then a polynomial with real coefficients can always be factored into linear and quadratic factors. We will use this property in Section 8.9 when we study partial fractions. A quadratic polynomial with no real zeros is called **irreducible** over the real numbers. Such a polynomial cannot be factored without using complex numbers.

> ### LINEAR AND QUADRATIC FACTORS THEOREM
>
> Every polynomial with real coefficients can be factored into a product of linear and irreducible quadratic factors with real coefficients.

■ **Proof**   We first observe that if $c = a + bi$ is a complex number, then

$$(x - c)(x - \bar{c}) = [x - (a + bi)][x - (a - bi)]$$

$$= [(x - a) - bi][(x - a) + bi]$$

$$= (x - a)^2 - (bi)^2$$

$$= x^2 - 2ax + (a^2 + b^2)$$

The last expression is a quadratic with *real* coefficients.

Now, if P is a polynomial with real coefficients, then by the Complete Factorization Theorem

$$P(x) = a(x - c_1)(x - c_2) \cdots (x - c_n)$$

Since the complex roots occur in conjugate pairs, we can multiply the factors corresponding to each such pair to get a quadratic factor with real coefficients. This results in P being factored into linear and irreducible quadratic factors.   □

EXAMPLE 7 ■ Factoring a Polynomial into Linear and Quadratic Factors

Let $P(x) = x^4 + 2x^2 - 8$.

(a) Factor P into linear and irreducible quadratic factors with real coefficients.
(b) Factor P completely into linear factors with complex coefficients.

SOLUTION

(a)
$$P(x) = x^4 + 2x^2 - 8$$
$$= (x^2 - 2)(x^2 + 4)$$
$$= (x - \sqrt{2})(x + \sqrt{2})(x^2 + 4)$$

The factor $x^2 + 4$ is irreducible since it has only the imaginary zeros $\pm 2i$.

(b) To get the complete factorization, we factor the remaining quadratic factor.

$$P(x) = (x - \sqrt{2})(x + \sqrt{2})(x^2 + 4)$$
$$= (x - \sqrt{2})(x + \sqrt{2})(x - 2i)(x + 2i)$$ ■

3.5 EXERCISES

1–10 ■ Find all solutions of the polynomial equation by factoring and using the quadratic formula, as in Example 1.

1. $x^3 + 8 = 0$

2. $x^3 - 8 = 0$

3. $x^3 - 2x^2 + 2x = 0$

4. $x^3 + x^2 + x = 0$

5. $x^4 + 2x^2 + 1 = 0$

6. $x^4 - x^2 - 2 = 0$

7. $x^4 - 16 = 0$

8. $x^4 + 6x^2 + 9 = 0$

9. $x^6 - 1 = 0$

10. $x^6 - 7x^3 - 8 = 0$

11–28 ■ Factor the polynomial completely and find all its zeros. State the multiplicity of each zero.

11. $P(x) = x^2 + 25$

12. $P(x) = 4x^2 + 9$

13. $Q(x) = x^2 + 2x + 2$

14. $Q(x) = x^2 - 8x + 17$

15. $P(x) = x^3 + 4x$

16. $P(x) = x^3 + x^2 + x$

17. $Q(x) = x^4 - 1$

18. $Q(x) = x^4 - 625$

19. $P(x) = 16x^4 - 81$

20. $P(x) = x^3 - 64$

21. $P(x) = x^3 + x^2 + 9x + 9$

22. $P(x) = x^6 - 729$

23. $Q(x) = x^4 + 2x^2 + 1$

24. $Q(x) = x^4 + 10x^2 + 25$

25. $P(x) = x^4 + 3x^2 - 4$

26. $P(x) = x^5 + 7x^3$

27. $P(x) = x^5 + 6x^3 + 9x$

28. $P(x) = x^6 + 16x^3 + 64$

29–38 ■ Find a polynomial with integer coefficients that satisfies the given conditions.

29. P has degree 2, and zeros $1 + i$ and $1 - i$.

30. P has degree 2, and zeros $1 + i\sqrt{2}$ and $1 - i\sqrt{2}$.

31. Q has degree 3, and zeros 3, $2i$, and $-2i$.

32. Q has degree 3, and zeros 0 and i.

33. P has degree 3, and zeros 2 and i.

34. Q has degree 3, and zeros -3 and $1 + i$.

35. R has degree 4, and zeros $1 - 2i$ and 1, with 1 a zero of multiplicity 2.

36. S has degree 4, and zeros $2i$ and $3i$.

37. T has degree 4, zeros i and $1 + i$, and constant coefficient 12.

38. U has degree 5, zeros $\frac{1}{2}$, -1, and $-i$, and leading coefficient 4; the zero -1 has multiplicity 2.

39–54 ■ Find all zeros of the polynomial.

39. $P(x) = x^3 + 2x^2 + 4x + 8$

40. $P(x) = x^3 - 7x^2 + 17x - 15$

41. $P(x) = x^3 - 2x^2 + 2x - 1$

42. $P(x) = x^3 + 7x^2 + 18x + 18$

43. $P(x) = x^3 - 3x^2 + 3x - 2$

44. $P(x) = x^3 - x - 6$

45. $P(x) = 2x^3 + 7x^2 + 12x + 9$

46. $P(x) = 2x^3 - 8x^2 + 9x - 9$

47. $P(x) = x^4 + x^3 + 7x^2 + 9x - 18$

48. $P(x) = x^4 - 2x^3 - 2x^2 - 2x - 3$

49. $P(x) = x^5 - x^4 + 7x^3 - 7x^2 + 12x - 12$

50. $P(x) = x^5 + x^3 + 8x^2 + 8$ [*Hint:* Factor by grouping.]

51. $P(x) = x^4 - 6x^3 + 13x^2 - 24x + 36$

52. $P(x) = x^4 - x^2 + 2x + 2$

53. $P(x) = x^5 - 3x^4 + 12x^3 - 28x^2 + 27x - 9$

54. $P(x) = x^5 - 2x^4 + 2x^3 - 4x^2 + x - 2$

55–60 ■ A polynomial P is given.
(a) Factor P into linear and irreducible quadratic factors with real coefficients.
(b) Factor P completely into linear factors with complex coefficients.

55. $P(x) = x^3 - 5x^2 + 4x - 20$

56. $P(x) = x^3 - 2x - 4$

57. $P(x) = x^4 + 8x^2 - 9$

58. $P(x) = x^4 + 8x^2 + 16$

59. $P(x) = x^6 - 64$

60. $P(x) = x^5 - 16x$

 61. By the Zeros Theorem, every nth-degree polynomial equation has exactly n solutions (including possibly some that are repeated). Some of these may be real and some may be

imaginary. Use a graphing device to determine how many real and imaginary solutions each equation has.
(a) $x^4 - 2x^3 - 11x^2 + 12x = 0$
(b) $x^4 - 2x^3 - 11x^2 + 12x - 5 = 0$
(c) $x^4 - 2x^3 - 11x^2 + 12x + 40 = 0$

62–64 ■ So far we have worked only with polynomials that have real coefficients. These exercises involve polynomials with real and imaginary coefficients.

62. Find all solutions of the equation.
(a) $2x + 4i = 1$ (b) $x^2 - ix = 0$
(c) $x^2 + 2ix - 1 = 0$ (d) $ix^2 - 2x + i = 0$

63. (a) Show that $2i$ and $1 - i$ are both solutions of the equation

$$x^2 - (1 + i)x + (2 + 2i) = 0$$

but that their complex conjugates $-2i$ and $1 + i$ are not.
(b) Explain why the result of part (a) does not violate the Conjugate Roots Theorem.

64. (a) Find the polynomial with *real* coefficients of the smallest possible degree for which i and $1 + i$ are zeros and in which the coefficient of the highest power is 1.
(b) Find the polynomial with *complex* coefficients of the smallest possible degree for which i and $1 + i$ are zeros and in which the coefficient of the highest power is 1.

DISCOVERY · DISCUSSION

65. Polynomials of Odd Degree   The Conjugate Zeros Theorem says that the complex zeros of a polynomial with real coefficients occur in complex conjugate pairs. Explain how this fact proves that a polynomial with real coefficients and odd degree has at least one real zero.

66. Roots of Unity   There are two square roots of 1, namely 1 and -1. These are the solutions of $x^2 = 1$. The fourth roots of 1 are the solutions of the equation $x^4 = 1$ or $x^4 - 1 = 0$. How many fourth roots of 1 are there? Find them. The cube roots of 1 are the solutions of the equation $x^3 = 1$ or $x^3 - 1 = 0$. How many cube roots of 1 are there? Find them. How would you find the sixth roots of 1? How many are there? Make a conjecture about the number of nth roots of 1.

3.6 RATIONAL FUNCTIONS

A rational function is a function of the form

$$r(x) = \frac{P(x)}{Q(x)}$$

where P and Q are polynomials. We assume that $P(x)$ and $Q(x)$ have no factor in common. Even though rational functions are constructed from polynomials, their graphs look quite different than the graphs of polynomial functions.

Rational Functions and Asymptotes

Rational functions are not defined for those values of x for which the denominator is zero. When graphing a rational function, we must pay special attention to the behavior of the graph near those x-values. We begin by graphing a very simple rational function.

EXAMPLE 1 ■ A Simple Rational Function

Sketch a graph of the rational function $f(x) = \dfrac{1}{x}$.

SOLUTION

The function f is not defined for $x = 0$. The following tables show that when x is close to zero, the value of $|f(x)|$ is large, and the closer x gets to zero, the larger $|f(x)|$ gets.

For positive real numbers,

$$\frac{1}{\text{BIG NUMBER}} = \text{small number}$$

$$\frac{1}{\text{small number}} = \text{BIG NUMBER}$$

x	$f(x)$
-0.1	-10
-0.01	-100
-0.00001	$-100{,}000$

↑ approaching 0^- ↑ approaching $-\infty$

x	$f(x)$
0.1	10
0.01	100
0.00001	$100{,}000$

↑ approaching 0^+ ↑ approaching ∞

We describe this behavior in words and in symbols as follows. The first table shows that as x approaches 0 from the left, the values of $y = f(x)$ decrease without bound. In symbols,

$$f(x) \to -\infty \quad \text{as} \quad x \to 0^- \qquad \text{"y approaches negative infinity as } x \text{ approaches 0 from the left"}$$

The second table shows that as x approaches 0 from the right, the values of $f(x)$ increase without bound. In symbols,

$$f(x) \to \infty \quad \text{as} \quad x \to 0^+ \qquad \text{"y approaches infinity as } x \text{ approaches 0 from the right"}$$

The next two tables show how $f(x)$ changes as $|x|$ becomes large.

x	$f(x)$
-10	-0.1
-100	-0.01
$-100,000$	-0.00001

x	$f(x)$
10	0.1
100	0.01
$100,000$	0.00001

↑ approaching $-\infty$ ↑ approaching 0 ↑ approaching ∞ ↑ approaching 0

These tables show that as $|x|$ becomes large, the value of $f(x)$ gets closer and closer to zero. We describe this situation in symbols by writing

$$f(x) \to 0 \quad \text{as} \quad x \to -\infty \quad \text{and} \quad f(x) \to 0 \quad \text{as} \quad x \to \infty$$

Using the information in these tables and plotting a few additional points, we obtain the graph shown in Figure 1.

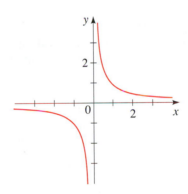

FIGURE 1

$f(x) = \dfrac{1}{x}$

In Example 1 we used the following arrow notation.

Symbol	Meaning
$x \to a^-$	x approaches a from the left
$x \to a^+$	x approaches a from the right
$x \to -\infty$	x goes to negative infinity; that is, x decreases without bound
$x \to \infty$	x goes to infinity; that is, x increases without bound

The line $x = 0$ is called a *vertical asymptote* of the graph in Figure 1, and the line $y = 0$ is a *horizontal asymptote*. Informally speaking, an asymptote of a function is a line that the graph of the function gets closer and closer to as one travels along that line in either direction.

DEFINITION OF ASYMPTOTES

1. The line $x = a$ is a **vertical asymptote** of the function $y = f(x)$ if

$$y \to \infty \quad \text{or} \quad y \to -\infty \qquad \text{as} \qquad x \to a^+ \quad \text{or} \quad x \to a^-$$

2. The line $y = b$ is a **horizontal asymptote** of the function $y = f(x)$ if

$$y \to b \qquad \text{as} \qquad x \to \infty \quad \text{or} \quad x \to -\infty$$

A rational function has vertical asymptotes where the function is undefined, that is, where the denominator is zero.

Transformations of $y = \frac{1}{x}$

A rational function of the form

$$r(x) = \frac{ax + b}{cx + d}$$

can be graphed by shifting, stretching, and/or reflecting the graph of $f(x) = \frac{1}{x}$ shown in Figure 1, using the transformations studied in Section 2.5. (Such functions are called *linear fractional transformations*.)

EXAMPLE 2 ■ **Using Transformations to Graph Simple Rational Functions**

Sketch a graph of each rational function.

(a) $r(x) = \dfrac{2}{x - 3}$

(b) $s(x) = \dfrac{3x + 5}{x + 2}$

SOLUTION

(a) Let $f(x) = \frac{1}{x}$. Then we can express r in terms of f as follows:

$$r(x) = \frac{2}{x - 3}$$

$$= 2\left(\frac{1}{x - 3}\right) \qquad \text{\color{red}Factor 2}$$

$$= 2(f(x - 3)) \qquad \text{\color{red}Since } f(x) = \frac{1}{x}$$

From this form we see that the graph of r is obtained from the graph of f by shifting 3 units to the right and stretching vertically by a factor of 2. Thus, r has vertical asymptote $x = 3$ and horizontal asymptote $y = 0$. The graph of r is shown in Figure 2.

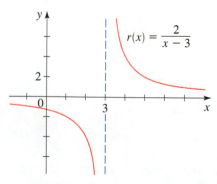

FIGURE 2

$$x + 2 \overline{\smash{)}\begin{array}{r} 3 \\ 3x + 5 \\ \underline{3x + 6} \\ -1 \end{array}}$$

(b) Using long division (see the margin), we get $s(x) = 3 - \frac{1}{x+2}$. Thus, we can express s in terms of f as follows:

$$s(x) = 3 - \frac{1}{x+2}$$

$$= -\frac{1}{x+2} + 3 \qquad \text{Rearrange terms}$$

$$= -f(x+2) + 3 \qquad \text{Since } f(x) = \frac{1}{x}$$

From this form we see that the graph of s is obtained from the graph of f by shifting 2 units to the left, reflecting in the x-axis, and shifting upward 3 units. Thus, s has vertical asymptote $x = -2$ and horizontal asymptote $y = 3$. The graph of s is shown in Figure 3.

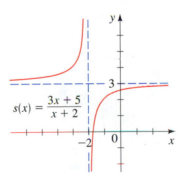

$$s(x) = \frac{3x+5}{x+2}$$

FIGURE 3

Graphing Rational Functions

The methods of Example 2 work only for simple rational functions. To graph more complicated ones, we need to take a closer look at the behavior of a rational function near its vertical and horizontal asymptotes.

EXAMPLE 3 ■ Asymptotes of a Rational Function

Sketch a graph of the rational function $r(x) = \dfrac{x-1}{x-2}$.

SOLUTION

We find the asymptotes of r and use them to sketch the graph.

VERTICAL ASYMPTOTE   The line $x = 2$ is a vertical asymptote because the denominator of r is 0 when $x = 2$.

The rational function r can also be graphed using the methods of Example 2.

BEHAVIOR NEAR VERTICAL ASYMPTOTE   We make tables of values for x-values to the left and right of 2. From the tables shown below we see that

$$y \to -\infty \quad \text{as} \quad x \to 2^- \qquad \text{and} \qquad y \to \infty \quad \text{as} \quad x \to 2^+$$

FIGURE 4

$x \to 2^-$	
x	**y**
1	0
1.5	−1
1.9	−9
1.95	−19
1.99	−99
1.999	−999

$x \to 2^+$	
x	**y**
3	2
2.5	3
2.1	11
2.05	21
2.01	101
2.001	1001

↑ approaching 2^- ↑ approaching $-\infty$ ↑ approaching 2^+ ↑ approaching ∞

The graph of $y = r(x)$ therefore has the shape near $x = 2$ shown in Figure 4.

HORIZONTAL ASYMPTOTE   The horizontal asymptote is the value that y approaches as $x \to \pm\infty$. To find this value, divide both numerator and denominator by x:

$$y = r(x) = \frac{x-1}{x-2} \cdot \frac{\dfrac{1}{x}}{\dfrac{1}{x}} = \frac{1 - \dfrac{1}{x}}{1 - \dfrac{2}{x}}$$

The fractional expressions $\frac{1}{x}$ and $\frac{2}{x}$ both approach 0 as $x \to \pm\infty$ (see Exercise 61, Section 1.1). So as $x \to \pm\infty$, we have

FIGURE 5

$$y = \frac{1 - \dfrac{1}{x}}{1 - \dfrac{2}{x}} \quad \to \quad \frac{1 - 0}{1 - 0} = 1$$

Thus, the horizontal asymptote is the line $y = 1$.

GRAPH   Since the graph must approach the horizontal asymptote, we can complete it as shown in Figure 5.

We have seen that asymptotes are important when graphing simple rational functions. In general, we use the following guidelines to graph rational functions.

SKETCHING GRAPHS OF RATIONAL FUNCTIONS

1. FACTOR.   Factor the numerator and denominator.

2. INTERCEPTS.   Find the x-intercepts by determining the zeros of the numerator, and the y-intercept from the value of the function at $x = 0$.

3. VERTICAL ASYMPTOTES.   Find the vertical asymptotes by determining the zeros of the denominator, and then see if $y \to \infty$ or $y \to -\infty$ on each side of every vertical asymptote.

4. HORIZONTAL ASYMPTOTE.   Find the horizontal asymptote (if any) by dividing both numerator and denominator by the highest power of x that appears in the denominator, and then letting $x \to \pm\infty$.

5. SKETCH THE GRAPH.   Graph the information provided by the first four steps. Then plot as many additional points as needed to fill in the rest of the graph of the function.

EXAMPLE 4 ■ Graphing a Rational Function

Sketch a graph of the rational function $r(x) = \dfrac{2x^2 + 7x - 4}{x^2 + x - 2}$.

SOLUTION

We factor the numerator and denominator, find the intercepts and asymptotes, and sketch the graph.

FACTOR: $r(x) = \dfrac{(2x - 1)(x + 4)}{(x - 1)(x + 2)}$

x-INTERCEPTS: The x-intercepts are the zeros of the numerator, $x = \frac{1}{2}$ and $x = -4$.

y-INTERCEPT: To find the y-intercept, we substitute $x = 0$ into the original form of the function:

$$r(0) = \frac{2(0)^2 + 7(0) - 4}{(0)^2 + (0) - 2} = \frac{-4}{-2} = 2$$

The y-intercept is 2.

HORIZONTAL ASYMPTOTE: The horizontal asymptote (if it exists) is the value that y approaches as $x \to \pm\infty$. To find this value, we divide both the numerator

A fraction is 0 if and only if its numerator is 0.

and the denominator by the highest power of x that appears in the denominator (in this case, x^2).

$$y = r(x) = \frac{2x^2 + 7x - 4}{x^2 + x - 2} \cdot \frac{\dfrac{1}{x^2}}{\dfrac{1}{x^2}} = \frac{2 + \dfrac{7}{x} - \dfrac{4}{x^2}}{1 + \dfrac{1}{x} - \dfrac{2}{x^2}}$$

x	$4/x^2$
10	0.04
100	0.0004
1000	0.000004
10,000	0.00000004

↑ approaching ∞   ↑ approaching 0

Any expression of the form c/x^n approaches 0 as $x \to \pm\infty$ (if $n > 0$). The table in the margin illustrates this for the term $4/x^2$. So as $x \to \pm\infty$, we have

$$y = \frac{2 + \dfrac{7}{x} - \dfrac{4}{x^2}}{1 + \dfrac{1}{x} - \dfrac{2}{x^2}} \quad \to \quad \frac{2 + 0 - 0}{1 + 0 - 0} = 2$$

Thus, the horizontal asymptote is $y = 2$.

VERTICAL ASYMPTOTES: The vertical asymptotes occur where the denominator is 0, that is, where the function is undefined. From the factored form we see that the vertical asymptotes are $x = 1$ and $x = -2$.

BEHAVIOR NEAR VERTICAL ASYMPTOTES: We need to know whether $y \to \infty$ or $y \to -\infty$ on each side of each vertical asymptote. To determine the sign of y for x-values near the vertical asymptotes, we use test values. For instance, as $x \to 1^-$, we use a test value close to and to the left of 1 ($x = 0.9$, say) to check whether y is positive or negative to the left of $x = 1$:

$$y = \frac{(2(0.9) - 1)((0.9) + 4)}{((0.9) - 1)((0.9) + 2)} \qquad \text{whose sign is} \qquad \frac{(+)(+)}{(-)(+)} \quad \text{(negative)}$$

So $y \to -\infty$ as $x \to 1^-$. On the other hand, as $x \to 1^+$, we use a test value close to and to the right of 1 ($x = 1.1$, say), to get

$$y = \frac{(2(1.1) - 1)((1.1) + 4)}{((1.1) - 1)((1.1) + 2)} \qquad \text{whose sign is} \qquad \frac{(+)(+)}{(+)(+)} \quad \text{(positive)}$$

So $y \to \infty$ as $x \to 1^+$. The other entries in the following table are calculated similarly.

As $x \to$	-2^-	-2^+	1^-	1^+
the sign of $y = \dfrac{(2x - 1)(x + 4)}{(x - 1)(x + 2)}$ is	$\dfrac{(-)(+)}{(-)(-)}$	$\dfrac{(-)(+)}{(-)(+)}$	$\dfrac{(+)(+)}{(-)(+)}$	$\dfrac{(+)(+)}{(+)(+)}$
so $y \to$	$-\infty$	∞	$-\infty$	∞

ADDITIONAL VALUES:

x	y
-6	0.93
-3	-1.75
-1	4.50
1.5	6.29
2	4.50
3	3.50

GRAPH:

FIGURE 6

The Three Forms of A Rational Function

To find the **y-intercept** of a rational function, we use the *original form* of the function.

To find the **x-intercepts** and **vertical asymptotes**, we use the *factored form* of the function.

To find the **horizontal asymptote**, we use a *compound-fraction form* of the function.

Notice that we use three forms of the equation defining the rational function when we perform this analysis. In Example 4 the *original form*

$$r(x) = \frac{2x^2 + 7x - 4}{x^2 + x - 2}$$

easily gave us the y-intercept $r(0) = (-4)/(-2) = 2$. The *factored form*

$$r(x) = \frac{(2x - 1)(x + 4)}{(x - 1)(x + 2)}$$

gave us the x-intercepts $\frac{1}{2}$ and -4, and the vertical asymptotes $x = 1$ and $x = -2$. Finally, the *compound-fraction form*

$$r(x) = \frac{2 + \dfrac{7}{x} - \dfrac{4}{x^2}}{1 + \dfrac{1}{x} - \dfrac{2}{x^2}}$$

tells us that the horizontal asymptote is $y = \frac{2}{1} = 2$.

We can determine whether a rational function $r(x) = P(x)/Q(x)$ has a horizontal asymptote by considering the degrees of the numerator and denominator. If the degrees of P and Q are the same (both n, say), then dividing numerator and denominator by x^n shows that the horizontal asymptote is

$$y = \frac{\text{leading coefficient of } P}{\text{leading coefficient of } Q}$$

as we saw in Example 4. The following box summarizes the procedure for finding asymptotes.

> ### ASYMPTOTES OF RATIONAL FUNCTIONS
>
> Let r be the rational function
>
> $$r(x) = \frac{a_n x^n + a_{n-1} x^{n-1} + \cdots + a_1 x + a_0}{b_m x^m + b_{m-1} x^{m-1} + \cdots + b_1 x + b_0}$$
>
> **1.** The vertical asymptotes of r are the lines $x = a$, where a is a zero of the denominator.
>
> **2.** (a) If $n < m$, then r has horizontal asymptote $y = 0$.
>
>   (b) If $n = m$, then r has horizontal asymptote $y = \dfrac{a_n}{b_m}$.
>
>   (c) If $n > m$, then r has no horizontal asymptote.

EXAMPLE 5 ■ **Graphing a Rational Function**

Sketch a graph of the rational function $r(x) = \dfrac{x - 2}{x^2 - 1}$.

SOLUTION

FACTOR: $\quad y = \dfrac{x - 2}{(x - 1)(x + 1)}$

x-**INTERCEPT:** 2, from $x - 2 = 0$

y-**INTERCEPT:** 2, because $r(0) = \dfrac{0 - 2}{0^2 - 1} = 2$

HORIZONTAL ASYMPTOTE: $y = 0$, since degree of numerator $<$ degree of denominator

VERTICAL ASYMPTOTES: $x = 1$ and $x = -1$, from the zeros of the denominator

BEHAVIOR NEAR VERTICAL ASYMPTOTES:

As $x \to$	1^+	1^-	-1^+	-1^-
the sign of $y = \dfrac{x-2}{(x-1)(x+1)}$ is	$\dfrac{(-)}{(+)(+)}$	$\dfrac{(-)}{(-)(+)}$	$\dfrac{(-)}{(-)(+)}$	$\dfrac{(-)}{(-)(-)}$
so $y \to$	$-\infty$	∞	∞	$-\infty$

ADDITIONAL VALUES: **GRAPH:**

x	y
-2	-1.33
-0.5	3.33
0.5	2
1.5	-0.4
3	0.125
4	0.133
5	0.125

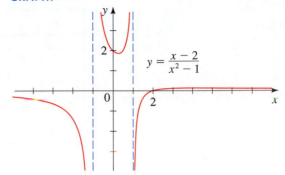

$$y = \frac{x - 2}{x^2 - 1}$$

FIGURE 7   ∎

    From the graph in Figure 7, we see that, contrary to common misconception, a graph may cross a horizontal asymptote.

EXAMPLE 6 ■ Graphing a Rational Function

Sketch a graph of the rational function $r(x) = \dfrac{x^2 - 3x - 4}{2x^2 + 4x}$.

SOLUTION

FACTOR: $y = \dfrac{(x + 1)(x - 4)}{2x(x + 2)}$

x-INTERCEPTS: -1 and 4, from $x + 1 = 0$ and $x - 4 = 0$

y-INTERCEPT: None, because $r(0)$ is undefined

HORIZONTAL ASYMPTOTE: $y = \frac{1}{2}$, because degree of numerator and denominator are the same and

$$\frac{\text{leading coefficient of numerator}}{\text{leading coefficient of denominator}} = \frac{1}{2}$$

VERTICAL ASYMPTOTES: $x = 0$ and $x = -2$, from the zeros of the denominator

BEHAVIOR NEAR VERTICAL ASYMPTOTES:

As $x \to$	-2^-	-2^+	0^-	0^+
the sign of $y = \dfrac{(x + 1)(x - 4)}{2x(x + 2)}$ is	$\dfrac{(-)(-)}{(-)(-)}$	$\dfrac{(-)(-)}{(-)(+)}$	$\dfrac{(+)(-)}{(-)(+)}$	$\dfrac{(+)(-)}{(+)(+)}$
so $y \to$	∞	$-\infty$	∞	$-\infty$

The graph of r is shown in Figure 8 on the following page.

ADDITIONAL VALUES:

x	y
-3	2.33
-2.5	3.90
-0.5	1.50
1	-1.00
3	-0.13
5	0.09

GRAPH:

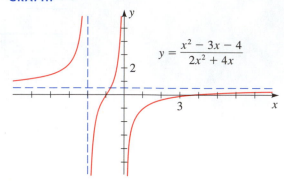

$$y = \frac{x^2 - 3x - 4}{2x^2 + 4x}$$

FIGURE 8

Slant Asymptotes and End Behavior

If $r(x) = P(x)/Q(x)$ is a rational function in which the degree of the numerator is one more than the degree of the denominator, we can use the Division Algorithm to express the function in the form

$$r(x) = ax + b + \frac{R(x)}{Q(x)}$$

where the degree of R is less than the degree of Q and $a \neq 0$. This means that as $x \to \pm\infty$, $R(x)/Q(x) \to 0$, so for large values of $|x|$, the graph of $y = r(x)$ approaches the graph of the line $y = ax + b$. In this situation we say that $y = ax + b$ is a **slant asymptote**, or an **oblique asymptote**.

EXAMPLE 7 ■ **A Rational Function with a Slant Asymptote**

Sketch a graph of the rational function $r(x) = \dfrac{x^2 - 4x - 5}{x - 3}$.

SOLUTION

FACTOR: $y = \dfrac{(x + 1)(x - 5)}{x - 3}$

x-INTERCEPTS: -1 and 5, from $x + 1 = 0$ and $x - 5 = 0$

y-INTERCEPTS: $\dfrac{5}{3}$, because $r(0) = \dfrac{0^2 - 4 \cdot 0 - 5}{0 - 3} = \dfrac{5}{3}$

HORIZONTAL ASYMPTOTE: None, because degree of numerator is greater than degree of denominator

VERTICAL ASYMPTOTE: $x = 3$, from the zero of the denominator

BEHAVIOR NEAR VERTICAL ASYMPTOTE: $y \to \infty$ as $x \to 3^-$ and $y \to -\infty$ as $x \to 3^+$

$$
\begin{array}{r}
x - 1 \\
x - 3 \overline{)\, x^2 - 4x - 5} \\
\underline{x^2 - 3x } \\
-x - 5 \\
\underline{-x + 3} \\
-8
\end{array}
$$

SLANT ASYMPTOTE: Since the degree of the numerator is one more than the degree of the denominator, the function has a slant asymptote. Dividing (see the margin), we obtain

$$r(x) = x - 1 + \frac{8}{x - 3}$$

Thus, $y = x - 1$ is the slant asymptote.

ADDITIONAL VALUES:

x	y
-2	-1.4
1	4
2	9
4	-5
6	2.33

GRAPH:

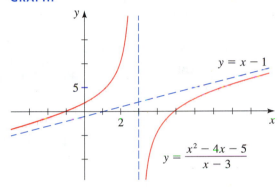

FIGURE 9

So far we have considered only horizontal and slant asymptotes as end behaviors for rational functions. In the next example we graph a function whose end behavior is like that of a parabola.

EXAMPLE 8 ■ End Behavior of a Rational Function

Graph the rational function

$$r(x) = \frac{x^3 - 2x^2 + 3}{x - 2}$$

and describe its end behavior.

SOLUTION

FACTOR: $y = \dfrac{(x + 1)(x^2 - 3x + 3)}{x - 2}$

x-INTERCEPTS: -1, from $x + 1 = 0$ (The other factor in the numerator has no real zeros.)

y-INTERCEPTS: $-\dfrac{3}{2}$, because $r(0) = \dfrac{0^3 - 2 \cdot 0^2 + 3}{0 - 2} = -\dfrac{3}{2}$

HORIZONTAL ASYMPTOTE: None, because degree of numerator is greater than degree of denominator

VERTICAL ASYMPTOTE: $x = 2$, from the zero of the denominator

BEHAVIOR NEAR VERTICAL ASYMPTOTE: $y \to -\infty$ as $x \to 2^-$ and $y \to \infty$ as $x \to 2^+$

END BEHAVIOR: Dividing (see the margin), we get

$$r(x) = x^2 + \frac{3}{x - 2}$$

This shows that the end behavior of r is like that of the parabola $y = x^2$ because $3/(x - 2)$ is small when $|x|$ is large. That is, $3/(x - 2) \to 0$ as $x \to \pm\infty$. This means that the graph of r will be close to the graph of $y = x^2$ for large $|x|$.

GRAPH: In Figure 10(a) we graph r in a small viewing rectangle; we can see the intercepts, the vertical asymptotes, and the local minimum. In Figure 10(b) we graph r in a larger viewing rectangle; here the graph looks almost like the graph of a parabola. In Figure 10(c) we graph both $y = r(x)$ and $y = x^2$; these graphs are very close to each other except near the vertical asymptote.

$$
\begin{array}{r}
x^2 \\
x - 2 \overline{)\,x^3 - 2x^2 + 0x + 3} \\
\underline{x^3 - 2x^2 } \\
3
\end{array}
$$

(a)

(b)

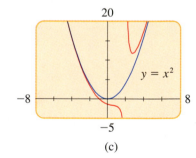

(c)

FIGURE 10

$r(x) = \dfrac{x^3 - 2x^2 + 3}{x - 2}$

Rational functions occur frequently in scientific applications of algebra. In the next example, we analyze the graph of a function from the theory of electricity.

EXAMPLE 9 ■ Electrical Resistance

When two resistors with resistances R_1 and R_2 are connected in parallel, their combined resistance R is given by the formula

$$R = \frac{R_1 R_2}{R_1 + R_2}$$

FIGURE 11

Suppose that a fixed 8-ohm resistor is connected in parallel with a variable resistor, as shown in Figure 11. If the resistance of the variable resistor is denoted by x, then the combined resistance R is a function of x. Graph R and give a physical interpretation of the graph.

SOLUTION

Substituting $R_1 = 8$ and $R_2 = x$ into the formula gives the function

$$R(x) = \frac{8x}{8 + x}$$

Since resistance cannot be negative, this function has physical meaning only when $x > 0$. The function is graphed in Figure 12(a) using the viewing rectangle $[0, 20]$ by $[0, 10]$. The function has no vertical asymptote when x is restricted to positive values. The combined resistance R increases as the variable resistance x increases. If we widen the viewing rectangle to $[0, 100]$ by $[0, 10]$, we obtain the graph in Figure 12(b). For large x, the combined resistance R levels off, getting closer and closer to the horizontal asymptote $R = 8$. No matter how large the variable resistance x, the combined resistance is never greater than 8 ohms.

FIGURE 12

$$R(x) = \frac{8x}{8 + x}$$

(a)

(b)

3.6 EXERCISES

1–6 ■ Find the x- and y-intercepts of the rational function.

1. $r(x) = \dfrac{x - 2}{x + 3}$

2. $s(x) = \dfrac{2x}{3x + 5}$

3. $t(x) = \dfrac{x^2 - x - 2}{x - 6}$

4. $r(x) = \dfrac{2}{x^2 + 3x - 4}$

5. $r(x) = \dfrac{x^2 - 9}{x^2}$

6. $r(x) = \dfrac{x^3 + 8}{x^2 + 4}$

7–8 ■ From the graph, determine the x- and y-intercepts and the vertical and horizontal asymptotes.

7.

8.

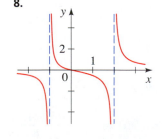

9–18 ■ Find all horizontal and vertical asymptotes (if any).

9. $r(x) = \dfrac{3}{x + 2}$ **10.** $s(x) = \dfrac{2x + 3}{x - 1}$

11. $t(x) = \dfrac{x^2}{x^2 - x - 6}$ **12.** $r(x) = \dfrac{2x - 4}{x^2 + 2x + 1}$

13. $s(x) = \dfrac{6}{x^2 + 2}$ **14.** $t(x) = \dfrac{(x - 1)(x - 2)}{(x - 3)(x - 4)}$

15. $r(x) = \dfrac{6x - 2}{x^2 + 5x - 6}$ **16.** $s(x) = \dfrac{3x^2}{x^2 + 2x + 5}$

17. $t(x) = \dfrac{x^2 + 2}{x - 1}$ **18.** $r(x) = \dfrac{x^3 + 3x^2}{x^2 - 4}$

19–24 ■ Use transformations of the graph of $y = \frac{1}{x}$ to graph the rational function, as in Example 2.

19. $r(x) = \dfrac{1}{x - 1}$ **20.** $r(x) = \dfrac{1}{x + 4}$

21. $s(x) = \dfrac{3}{x + 1}$ **22.** $s(x) = \dfrac{-2}{x - 2}$

23. $t(x) = \dfrac{2x - 3}{x - 2}$ **24.** $t(x) = \dfrac{3x - 3}{x + 2}$

25–48 ■ Find the intercepts and asymptotes, and then sketch a graph of the rational function.

25. $r(x) = \dfrac{4x - 4}{x + 2}$ **26.** $r(x) = \dfrac{2x + 6}{-6x + 3}$

27. $s(x) = \dfrac{4 - 3x}{x + 7}$ **28.** $s(x) = \dfrac{1 - 2x}{2x + 3}$

29. $r(x) = \dfrac{18}{(x - 3)^2}$ **30.** $r(x) = \dfrac{x - 2}{(x + 1)^2}$

31. $s(x) = \dfrac{4x - 8}{(x - 4)(x + 1)}$ **32.** $s(x) = \dfrac{x + 2}{(x + 3)(x - 1)}$

33. $s(x) = \dfrac{6}{x^2 - 5x - 6}$ **34.** $s(x) = \dfrac{2x - 4}{x^2 + x - 2}$

35. $t(x) = \dfrac{3x + 6}{x^2 + 2x - 8}$ **36.** $t(x) = \dfrac{x - 2}{x^2 - 4x}$

37. $r(x) = \dfrac{(x - 1)(x + 2)}{(x + 1)(x - 3)}$ **38.** $r(x) = \dfrac{2x(x + 4)}{(x - 1)(x - 2)}$

39. $r(x) = \dfrac{x^2 - 2x + 1}{x^2 + 2x + 1}$ **40.** $r(x) = \dfrac{4x^2}{x^2 - 2x - 3}$

41. $r(x) = \dfrac{2x^2 + 10x - 12}{x^2 + x - 6}$ **42.** $r(x) = \dfrac{2x^2 + 2x - 4}{x^2 + x}$

43. $r(x) = \dfrac{x^2 - x - 6}{x^2 + 3x}$ **44.** $r(x) = \dfrac{x^2 + 3x}{x^2 - x - 6}$

45. $r(x) = \dfrac{3x^2 + 6}{x^2 - 2x - 3}$ **46.** $r(x) = \dfrac{5x^2 + 5}{x^2 + 4x + 4}$

47. $s(x) = \dfrac{x^2 - 2x + 1}{x^3 - 3x^2}$ **48.** $t(x) = \dfrac{x^3 - x^2}{x^3 - 3x - 2}$

49–56 ■ Find the slant asymptote, the vertical asymptotes, and sketch a graph of the function.

49. $r(x) = \dfrac{x^2}{x - 2}$ **50.** $r(x) = \dfrac{x^2 + 2x}{x - 1}$

51. $r(x) = \dfrac{x^2 - 2x - 8}{x}$ **52.** $r(x) = \dfrac{3x - x^2}{2x - 2}$

53. $r(x) = \dfrac{x^2 + 5x + 4}{x - 3}$ **54.** $r(x) = \dfrac{x^3 + 4}{2x^2 + x - 1}$

55. $r(x) = \dfrac{x^3 + x^2}{x^2 - 4}$ **56.** $r(x) = \dfrac{2x^3 + 2x}{x^2 - 1}$

57–60 ■ Graph the rational function f and determine all vertical asymptotes from your graph. Then graph f and g in a sufficiently large viewing rectangle to show that they have the same end behavior.

57. $f(x) = \dfrac{2x^2 + 6x + 6}{x + 3}$, $g(x) = 2x$

58. $f(x) = \dfrac{-x^3 + 6x^2 - 5}{x^2 - 2x}$, $g(x) = -x + 4$

59. $f(x) = \dfrac{x^3 - 2x^2 + 16}{x - 2}$, $g(x) = x^2$

60. $f(x) = \dfrac{-x^4 + 2x^3 - 2x}{(x - 1)^2}$, $g(x) = 1 - x^2$

61–66 ■ Graph the rational function and find all vertical asymptotes, x- and y-intercepts, and local extrema, correct to the nearest decimal. Then use long division to find a polynomial that has the same end behavior as the rational function, and graph both functions in a sufficiently large viewing rectangle to verify that the end behaviors of the polynomial and the rational function are the same.

61. $y = \dfrac{2x^2 - 5x}{2x + 3}$

62. $y = \dfrac{x^4 - 3x^3 + x^2 - 3x + 3}{x^2 - 3x}$

63. $y = \dfrac{x^5}{x^3 - 1}$ **64.** $y = \dfrac{x^4}{x^2 - 2}$

65. $r(x) = \dfrac{x^4 - 3x^3 + 6}{x - 3}$ **66.** $r(x) = \dfrac{4 + x^2 - x^4}{x^2 - 1}$

67. The rabbit population on Mr. Jenkins' farm follows the formula

$$p(t) = \frac{3000t}{t + 1}$$

where $t \geq 0$ is the time (in months) since the beginning of the year.

(a) Draw a graph of the rabbit population.

(b) What eventually happens to the rabbit population?

68. After a certain drug is injected into a patient, the concentration c of the drug in the bloodstream is monitored. At time $t \geq 0$ (in minutes since the injection), the concentration (in mg/L) is given by

$$c(t) = \frac{30t}{t^2 + 2}$$

(a) Draw a graph of the drug concentration.

(b) What eventually happens to the concentration of drug in the bloodstream?

69. A drug is administered to a patient and the concentration of the drug in the bloodstream is monitored. At time $t \geq 0$ (in hours since giving the drug), the concentration (in mg/L) is given by

$$c(t) = \frac{5t}{t^2 + 1}$$

Graph the function c with a graphing device.

(a) What is the highest concentration of drug that is reached in the patient's bloodstream?

(b) What happens to the drug concentration after a long period of time?

(c) How long does it take for the concentration to drop below 0.3 mg/L?

70. Suppose a rocket is fired upward from the surface of the earth with an initial velocity v (measured in m/s). Then the maximum height h (in meters) reached by the rocket is given by the function

$$h(v) = \frac{Rv^2}{2gR - v^2}$$

where $R = 6.4 \times 10^6$ m is the radius of the earth and $g = 9.8$ m/s^2 is the acceleration due to gravity. Use a graphing device to draw a graph of the function h. (Note that h and v must both be positive, so the viewing rectangle need not contain negative values.) What does the vertical asymptote represent physically?

71. As a train moves toward an observer (see the figure), the pitch of its whistle sounds higher to the observer than it would if the train were at rest, because the crests of the sound waves are compressed closer together. This phenomenon is called the *Doppler effect*. The observed pitch P is a function of the speed v of the train and is given by

$$P(v) = P_0\left(\frac{s_0}{s_0 - v}\right)$$

where P_0 is the actual pitch of the whistle at the source and $s_0 = 332$ m/s is the speed of sound in air. Suppose that a train has a whistle pitched at $P_0 = 440$ Hz. Graph the function $y = P(v)$ using a graphing device. How can the vertical asymptote of this function be interpreted physically?

72. For a camera with a lens of fixed focal length F to focus on an object located a distance x from the lens, the film must be placed a distance y behind the lens, where F, x, and y are related by

$$\frac{1}{x} + \frac{1}{y} = \frac{1}{F}$$

(See the figure.) Suppose the camera has a 55-mm lens ($F = 55$).

(a) Express y as a function of x and graph the function.

(b) What happens to the focusing distance y as the object moves far away from the lens?

(c) What happens to the focusing distance y as the object moves close to the lens?

● **DISCOVERY · DISCUSSION**

73. Constructing a Rational Function from Its Asymptotes
Give an example of a rational function that has vertical asymptote $x = 3$. Now give an example of one that has vertical asymptote $x = 3$ *and* horizontal asymptote $y = 2$. Now give an example of a rational function with vertical asymptotes $x = 1$ and $x = -1$, horizontal asymptote $y = 0$, and x-intercept 4.

74. A Rational Function with No Asymptote   Explain how you can tell (without graphing it) that the function

$$r(x) = \frac{x^6 + 10}{x^4 + 8x^2 + 15}$$

has no x-intercept and no horizontal, vertical, or slant asymptote. What is its end behavior?'

75. Graphs with Holes   In this chapter we adopted the convention that in rational functions, the numerator and denominator don't share a common factor. In this exercise we consider the graph of a rational function that doesn't satisfy this rule.

(a) Show that the graph of

$$r(x) = \frac{3x^2 - 3x - 6}{x - 2}$$

is the line $y = 3x + 3$ with the point $(2, 9)$ removed. [*Hint:* Factor. What is the domain of r?]

(b) Graph the rational functions:

$$s(x) = \frac{x^2 + x - 20}{x + 5} \qquad t(x) = \frac{2x^2 - x - 1}{x - 1}$$

$$u(x) = \frac{x - 2}{x^2 - 2x}$$

76. Transformations of $y = 1/x^2$ In Example 2 we saw that some simple rational functions can be graphed by shifting, stretching, or reflecting the graph of $y = 1/x$. In this exercise we consider rational functions tht can be graphed by transforming the graph of $y = 1/x^2$, shown below.

(a) Graph the function

$$r(x) = \frac{1}{(x - 2)^2}$$

by transforming the graph of $y = 1/x^2$.

(b) Use long division and factoring to show that the function

$$s(x) = \frac{2x^2 + 4x + 5}{x^2 + 2x + 1}$$

can be written as

$$s(x) = 2 + \frac{3}{(x + 1)^2}$$

Then graph s by transforming the graph of $y = 1/x^2$.

(c) One of the following functions can be graphed by transforming the graph of $y = 1/x^2$; the other cannot. Use transformations to graph the one that can be, and explain why this method doesn't work for the other one.

$$p(x) = \frac{2 - 3x^2}{x^2 - 4x + 4} \qquad q(x) = \frac{12x - 3x^2}{x^2 - 4x + 4}$$

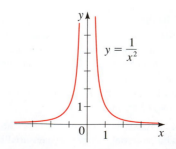

3 REVIEW

CONCEPT CHECK

1. (a) Write the defining equation for a polynomial P of degree n.
(b) What does it mean to say that c is a zero of P?

2. Sketch graphs showing the possible end behaviors of polynomials of odd degree and of even degree.

3. What steps would you follow to graph a polynomial by hand?

4. (a) What is meant by a local maximum point or local minimum point of a polynomial?
(b) How many local extrema can a polynomial of degree n have?

5. State the Division Algorithm and identify the dividend, divisor, quotient, and remainder.

6. How does synthetic division work?

7. (a) State the Remainder Theorem.
(b) State the Factor Theorem.

8. (a) State the Rational Zeros Theorem.
(b) What steps would you take to find the rational zeros of a polynomial?

9. State Descartes' Rule of Signs.

10. (a) What does it mean to say that a is a lower bound and b is an upper bound for the zeros of a polynomial?
(b) State the Upper and Lower Bounds Theorem.

11. (a) What is a complex number?
(b) What are the real and imaginary parts of a complex number?
(c) What is the complex conjugate of a complex number?
(d) How do you add, subtract, multiply, and divide complex numbers?
(e) How do you graph a complex number in the complex plane?
(f) What is the modulus of a complex number?

12. (a) State the Fundamental Theorem of Algebra.
(b) State the Complete Factorization Theorem.
(c) What does it mean to say that c is a zero of multiplicity k of a polynomial P?
(d) State the Zeros Theorem.
(e) State the Conjugate Roots Theorem.

13. (a) What is a rational function?
(b) What does it mean to say that $x = a$ is a vertical asymptote of $y = f(x)$?
(c) How do you locate a vertical asymptote?
(d) What does it mean to say that $y = b$ is a horizontal asymptote of $y = f(x)$?
(e) How do you locate a horizontal asymptote?
(f) What steps do you follow to sketch the graph of a rational function by hand?
(g) Under what circumstances does a rational function have a slant asymptote? If one exists, how do you find it?
(h) How do you determine the end behavior of a rational function?

EXERCISES

1–8 ■ Graph the polynomial. Show clearly all x- and y-intercepts.

1. $P(x) = x^3 + 27$

2. $P(x) = -2x^4 + 32$

3. $P(x) = -(x - 2)^4$

4. $P(x) = 2(x + 1)^3$

5. $P(x) = x^3 - 16x$

6. $P(x) = x^3 - 3x^2 - 4x$

7. $P(x) = x^4 + x^3 - 2x^2$

8. $P(x) = x^4 - 5x^2 + 4$

9–12 ■ Use a graphing device to graph the polynomial. Find the x- and y-intercepts and the coordinates of all local extrema, correct to the nearest decimal. Describe the end behavior of the polynomial.

9. $P(x) = x^3 - 4x + 1$

10. $P(x) = -2x^3 + 6x^2 - 2$

11. $P(x) = 3x^4 - 4x^3 - 10x - 1$

12. $P(x) = x^5 + x^4 - 7x^3 - x^2 + 6x + 3$

13. The strength S of a wooden beam of width x and depth y is given by the formula $S = 13.8xy^2$. A beam is to be cut from a log of diameter 10 in., as shown in the figure.
(a) Express the strength S of this beam as a function of x only.
(b) What is the domain of the function S?
(c) Draw a graph of S.
(d) What width will make the beam the strongest?

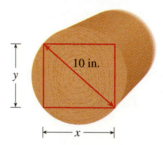

14. A small shelter for delicate plants is to be constructed of thin plastic material. It will have square ends and a rectangular top and back, with an open bottom and front, as shown in the figure. The total area of the four plastic sides is to be 1200 in^2.
(a) Express the volume V of the shelter as a function of the depth x.
(b) Draw a graph of V.
(c) What dimensions will maximize the volume of the shelter?

15–22 ■ Find the quotient and remainder.

15. $\dfrac{x^2 - 3x + 5}{x - 2}$

16. $\dfrac{x^2 + x - 12}{x - 3}$

17. $\dfrac{x^3 - x^2 + 11x + 2}{x - 4}$

18. $\dfrac{x^3 + 2x^2 - 10}{x + 3}$

19. $\dfrac{x^4 - 8x^2 + 2x + 7}{x + 5}$

20. $\dfrac{2x^4 + 3x^3 - 12}{x + 4}$

21. $\dfrac{2x^3 + x^2 - 8x + 15}{x^2 + 2x - 1}$

22. $\dfrac{x^4 - 2x^2 + 7x}{x^2 - x + 3}$

23–24 ■ Find the indicated value of the polynomial using the Remainder Theorem.

23. $P(x) = 2x^3 - 9x^2 - 7x + 13;$ find $P(5)$

24. $Q(x) = x^4 + 4x^3 + 7x^2 + 10x + 15;$ find $Q(-3)$

25. Show that $\frac{1}{2}$ is a zero of the polynomial

$$P(x) = 2x^4 + x^3 - 5x^2 + 10x - 4$$

26. Use the Factor Theorem to show that $x + 4$ is a factor of the polynomial

$$P(x) = x^5 + 4x^4 - 7x^3 - 23x^2 + 23x + 12$$

27. What is the remainder when the polynomial

$$P(x) = x^{500} + 6x^{201} - x^2 - 2x + 4$$

is divided by $x - 1$?

28. What is the remainder when $x^{101} - x^4 + 2$ is divided by $x + 1$?

29–30 ■ (a) List all possible rational roots (without testing to see if they actually are roots). (b) Determine the possible number of positive and negative real roots using Descartes' Rule of Signs.

29. $x^5 - 6x^3 - x^2 + 2x + 18 = 0$

30. $6x^4 + 3x^3 + x^2 + 3x + 4 = 0$

31–40 ■ Evaluate the expression and write in the form $a + bi$.

31. $(2 - 3i) + (1 + 4i)$

32. $(3 - 6i) - (6 - 4i)$

33. $(2 + i)(3 - 2i)$

34. $4i(2 - \frac{1}{2}i)$

35. $\dfrac{4 + 2i}{2 - i}$

36. $\dfrac{8 + 3i}{4 + 3i}$

37. i^{25}

38. $(1 + i)^3$

39. $(1 - \sqrt{-1})(1 + \sqrt{-1})$

40. $\sqrt{-10} \cdot \sqrt{-40}$

41. Find a polynomial of degree 3 with constant coefficient 12 and zeros $-\frac{1}{2}$, 2, and 3.

42. Find a polynomial of degree 4 having integer coefficients and zeros $3i$ and 4, with 4 a double zero.

43. Does there exist a polynomial of degree 4 with integer coefficients that has zeros i, $2i$, $3i$, and $4i$? If so, find it. If not, explain why.

44. Prove that the equation $3x^4 + 5x^2 + 2 = 0$ has no real root.

45–54 ■ Find all rational, irrational, and complex zeros (and state their multiplicities). Use Descartes' Rule of Signs, the Upper and Lower Bounds Theorem, the quadratic formula, or other factoring techniques to help you whenever possible.

45. $P(x) = x^3 - 3x^2 - 13x + 15$

46. $P(x) = 2x^3 + 5x^2 - 6x - 9$

47. $P(x) = x^4 + 6x^3 + 17x^2 + 28x + 20$

48. $P(x) = x^4 + 7x^3 + 9x^2 - 17x - 20$

49. $P(x) = x^5 - 3x^4 - x^3 + 11x^2 - 12x + 4$

50. $P(x) = x^4 - 81$

51. $P(x) = x^6 - 64$

52. $P(x) = 18x^3 + 3x^2 - 4x - 1$

53. $P(x) = 6x^4 - 18x^3 + 6x^2 - 30x + 36$

54. $P(x) = x^4 + 15x^2 + 54$

 55–58 ■ Use a graphing device to find all real solutions of the equation.

55. $2x^2 = 5x + 3$

56. $x^3 + x^2 - 14x - 24 = 0$

57. $x^4 - 3x^3 - 3x^2 - 9x - 2 = 0$

58. $x^5 = x + 3$

59–64 ■ Graph the rational function. Show clearly all x- and y-intercepts and asymptotes.

59. $r(x) = \dfrac{3x - 12}{x + 1}$

60. $r(x) = \dfrac{1}{(x + 2)^2}$

61. $r(x) = \dfrac{x - 2}{x^2 - 2x - 8}$

62. $r(x) = \dfrac{2x^2 - 6x - 7}{x - 4}$

63. $r(x) = \dfrac{x^2 - 9}{2x^2 + 1}$

64. $r(x) = \dfrac{x^3 + 27}{x + 4}$

 65–68 ■ Use a graphing device to analyze the graph of the rational function. Find all x- and y-intercepts, and all vertical, horizontal, and slant asymptotes. If the function has no horizontal or slant asymptote, find a polynomial that has the same end behavior as the rational function.

65. $r(x) = \dfrac{x - 3}{2x + 6}$

66. $r(x) = \dfrac{2x - 7}{x^2 + 9}$

67. $r(x) = \dfrac{x^3 + 8}{x^2 - x - 2}$

68. $r(x) = \dfrac{2x^3 - x^2}{x + 1}$

69. Find the coordinates of all points of intersection of the graphs of

$$y = x^4 + x^2 + 24x \qquad \text{and} \qquad y = 6x^3 + 20$$

3 TEST

1. Graph the polynomial $P(x) = (x - 1)^3 + 8$, showing clearly all x- and y-intercepts.

2. (a) Use synthetic division to find the quotient and remainder when $x^4 - 4x^2 + 2x + 5$ is divided by $x - 2$.
   (b) Use long division to find the quotient and remainder when $2x^5 + 4x^4 - x^3 - x^2 + 7$ is divided by $2x^2 - 1$.

3. Let $P(x) = 2x^3 - 5x^2 - 4x + 3$.
   (a) List all possible rational zeros of P.
   (b) Find the complete factorization of P.
   (c) Find the zeros of P.
   (d) Sketch the graph of P.

4. Calculate and write the result in the form $a + bi$.
   (a) $(6 - 2i) - \left(7 - \frac{1}{2}i\right)$
   (b) $(1 + i)(3 - 2i)$
   (c) $\dfrac{5 + 10i}{3 - 4i}$
   (d) i^{50}
   (e) $(2 - \sqrt{-2})(\sqrt{8} + \sqrt{-4})$

5. Find all real and complex zeros of $P(x) = x^3 - x^2 - 4x - 6$.

6. Find the complete factorization of $P(x) = x^4 - 2x^3 + 5x^2 - 8x + 4$.

7. Find a fourth-degree polynomial with integer coefficients that has zeros $3i$ and -1, with -1 a zero of multiplicity 2.

8. Let $P(x) = 2x^4 - 7x^3 + x^2 - 18x + 3$.
   (a) Use Descartes' Rule of Signs to determine how many positive and how many negative real zeros P can have.
   (b) Show that 4 is an upper bound and -1 is a lower bound for the real zeros of P.
   (c) Draw a graph of P and use it to estimate the real zeros of P, correct to two decimal places.
   (d) Find the coordinates of all local extrema of P, correct to two decimals.

9. Consider the following rational functions:

$$r(x) = \frac{2x - 1}{x^2 - x - 2} \qquad s(x) = \frac{x^3 + 27}{x^2 + 4} \qquad t(x) = \frac{x^3 - 9x}{x + 2} \qquad u(x) = \frac{x^2 + x - 6}{x^2 - 25}$$

   (a) Which of these rational functions has a horizontal asymptote?
   (b) Which of these functions has a slant asymptote?
   (c) Which of these functions has no vertical asymptote?
   (d) Graph $y = u(x)$, showing clearly any asymptotes and x- and y-intercepts the function may have.
   (e) Use long division to find a polynomial P that has the same end behavior as t. Graph both P and t on the same screen to verify that they have the same end behavior.

Focus on Problem Solving
Adding Something Extra

It is often necessary to combine different problem-solving principles. The principles of **drawing a diagram** and **introducing something extra** are used in the two problems that follow. Euler's solution of the "Königsberg bridge problem" is so brilliant that it is the basis for a whole field of mathematics called *network theory*, which has practical applications to electric circuits and economics.

■ The Königsberg Bridge Problem

The old city of Königsberg, now called Kaliningrad, is situated on both banks of the Pregel River and on two islands in the river. Seven bridges connect the parts of the city, as shown in Figure 1. After the last bridge was built, the Königsberg townspeople posed the problem of finding a route through the city that would cross all seven bridges in a continuous walk, without recrossing any bridge. No one was able to find such a route. When Euler heard of the problem, he realized that an interesting mathematical principle must be at work. To discover the principle, Euler first stripped away the nonessential parts of the problem by drawing a simpler diagram of the city in which he represented the land masses as points and the bridges as lines connecting the points (Figure 2).

FIGURE 1

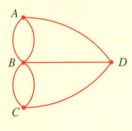

FIGURE 2

The problem now was to draw the diagram without lifting the pencil and without retracing any line. To show that this was impossible, Euler introduced something extra by labeling each point with the number of line segments that end at that point (Figure 3). He then argued that every pass through a point accounts for two line ends (one entering and one leaving). Thus for the walk to be possible, each point, except possibly the starting and ending points, must be labeled by an even number. Since more than two points are labeled by odd numbers, the walk is impossible!

The beauty of Euler's solution is that it is completely general. In modern language a diagram consisting of points and connecting lines is called a *network*, each point is a *vertex*, each line an *edge*, and the number of edges ending at a vertex is the *order* of the vertex. A network is *traversable* if it can be drawn in one stroke

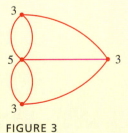

FIGURE 3

without retracing any edge. Euler's reasoning about the Königsberg bridge problem proved the following principle: *If a network has more than two vertices with odd order, then it is not traversable.*

Here is another clever use of the principle of **introducing something extra**.

■ The Impossible Museum Tour

The floor plan of a museum is in the shape of a square with six square rooms to a side. Each room is connected to each adjacent room by a door. The museum entrance and exit are at diagonally opposite corners, as shown in Figure 4.

A visitor making a tour of this museum would like to visit each room exactly once and then exit. Can you find such a path? Here are examples of attempts that failed:

Entrance

Exit

FIGURE 4

Trial and error will make it plausible to you that such a path is in fact not possible. But trial and error does not *prove* that our proposed tour of the museum is impossible. How do we give a convincing argument for this? Here is a clever idea: Imagine that the rooms are colored alternately black and white, like the checkerboard in Figure 5. So there are 18 black rooms and 18 white rooms. By coloring the rooms, we've introduced something extra that at first does not seem relevant. Indeed, the tour either can or cannot be done, regardless of the colors of the rooms. But notice how coloring the rooms in this fashion allows us to give a convincing argument that explains why this tour is impossible. We reason as follows.

FIGURE 5

Entrance

Exit

If a visitor is in a white room, he must next go to a black room, and if he is in a black room, he must next go to a white room. The entrance is in a white room, so the visitor's path, in terms of colors of rooms, is

W   B   W   B   W   B   . . .

Since the museum contains an even number of rooms and the visitor starts in a white room, he must end his tour in a black room. But the exit is in a white room, so the proposed tour is impossible!

■ Problems

1–4 ■ Determine whether each network is traversable. If it is, find a path that traverses it.

1.

2.

3.

4.

5. Is the museum tour possible if there are five rooms to a side instead of six? Find a general result about when the tour is possible for an $n \times n$ museum.

6. A Tibetan monk leaves the monastery at 7:00 A.M. and walks along his usual path to the top of the mountain, arriving at 7:00 P.M. The following morning, he starts at 7:00 A.M. at the top and takes the same path back, arriving at the monastery at 7:00 P.M. Show that there is exactly one point on the path that the monk will cross at exactly the same time of day on both trips.

7. A curve that starts and ends at the same point and does not intersect itself is called a **Jordan curve**. Such a curve always has an "inside" and an "outside." Is the point A in the figure inside or outside this Jordan curve? Find an easy method of determining whether a point is inside or outside a Jordan cuve.   [*Hint:* The point B is clearly outside the curve. Add a line segment connecting A and B.]

Jordan Curve

8. Consider the following sequence of polynomials:

$$P_1(x) = (x - a)$$
$$P_2(x) = (x - a)(x - b)$$
$$P_3(x) = (x - a)(x - b)(x - c)$$

(a) Expand the polynomials P_2 and P_3.
(b) Using the pattern you observe from part (a), write the expansions of P_4 and P_5 without actually multiplying out these polynomials. The patterns you have found are called **Viète's relations** (see page 51).

Srinivasa Ramanujan (1887–1920) was born into a poor family in the small town of Kumbakonam in India. Self-taught in mathematics, he worked in virtual isolation from other mathematicians. At the age of 25 he wrote a letter to G. H. Hardy, the leading British mathematician at the time, listing some of his discoveries. Hardy immediately recognized Ramanujan's genius and for the next six years the two worked together in London until Ramanujan fell ill and returned to his hometown in India, where he died a year later. Ramanujan was a genius with phenomenal ability to see hidden patterns in the properties of numbers. Most of his discoveries were written as complicated infinite series, the importance of which was not recognized until many years after his death. In the last year of his life he wrote 130 pages of mysterious formulas, many of which still defy proof. Hardy tells the story that when he visited Ramanujan in a hospital and arrived in a taxi, he remarked to Ramanujan that the cab's number, 1729, was uninteresting. Ramanujan replied "No, it is a very interesting number. It is the smallest number expressible as the sum of two cubes in two different ways." (See Problem 12.)

(c) For a polynomial of degree n, express the coefficient of x^{n-1} and the constant coefficient in terms of the zeros of the polynomial.

(d) Let $P(x) = x^6 - 7x^5 + 2x^3 + 17$. Without finding the zeros, find the product and the sum of the zeros of P.

9. If the equation $x^4 + ax^2 + bx + c = 0$ has roots 1, 2, and 3, find c.   [*Hint:* Use the Complete Factorization Theorem together with Problem 8(c).]

10. Find a polynomial of degree 3 with integer coefficients that has $2 - \sqrt[3]{2}$ as one of its zeros. How many other real zeros does the polynomial have?

11. You have an 8×8 grid with squares removed from two diagonally opposite corners, as shown in the figure. You also have a set of dominoes, each of which covers exactly two squares of the grid. Is it possible to cover all the remaining squares of the grid with dominoes, without overlapping any dominoes?   [*Hint:* Color the grid as a checkerboard.]

12. The number 1729 is the smallest positive integer that can be represented in two different ways as the sum of two cubes. What are the two ways?

13. If $f_0(x) = \dfrac{1}{1-x}$ and $f_{n+1} = f_0 \circ f_n$ for $n = 0, 1, 2, \ldots$, find $f_{100}(3)$.

14. Sketch a graph of the function $g(x) = |x^2 - 1| - |x^2 - 4|$.

15. Sketch a graph of the function $f(x) = x - [\![x]\!]$, where $[\![x]\!]$ is the greatest integer function defined on page 158.

16. Sketch the region in the plane defined by the equation

$$[\![x]\!]^2 + [\![y]\!]^2 = 1$$

17. (a) Show that if you multiply four consecutive integers and then add 1 to the result, you get a perfect square.

(b) Show that if you multiply three consecutive integers and then add the middle integer to the result, you get a perfect cube.

18. The equation $x^3 = 1$ has only one real solution, so it must have two imaginary solutions. Let w be one of the imaginary solutions. Show that the expression

$$(1 - w + w^2)(1 + w - w^2)$$

is a real number, and find its value.

19. A rectangle is divided into smaller rectangles in such a way that each of the smaller rectangles has at least one side of integer length. Show that one side of the original, large rectangle must have integer length.   [*Hint:* Imagine the rectangle lying in the coordinate plane, which has been tiled into a checkerboard pattern of black and white squares, each $\frac{1}{2}$ unit by $\frac{1}{2}$ unit. Observe that a rectangle has at least one integer side if and only if its area is composed of equal amounts of black and white.]

4 Exponential and Logarithmic Functions

Population growth—such as the growth of a herd of zebras in the wild—is modeled by exponential functions. These same functions describe such diverse phenomena as temperature change, compound interest, or the decay of radioisotopes such as those used in medical testing and treatment.

> *Mathematics compares the most diverse phenomena, and discovers the secret analogies which unite them.*
>
> JOSEPH FOURIER

So far we have studied relatively simple functions such as polynomial and rational functions. We now turn our attention to two of the most important functions in mathematics, the *exponential function* and its inverse function, the *logarithmic function*. We use these functions to describe exponential growth in biology and economics and radioactive decay in physics and chemistry, and other phenomena.

4.1 EXPONENTIAL FUNCTIONS

In Section 1.2 we defined a^x if $a > 0$ and x is a rational number, but we have not yet defined irrational powers. For instance, what is meant by $2^{\sqrt{3}}$ or 5^{π}? To help us answer these questions, we first look at the graph of the function $f(x) = 2^x$, where x is rational. A very crude representation of this graph is shown in Figure 1(a). The graph has a hole wherever x is irrational, and we want to fill in these holes with a smooth curve. To do this, we must define irrational powers of 2 appropriately. For example, to define $2^{\sqrt{3}}$ we use rational approximations of $\sqrt{3}$. Since

$$\sqrt{3} \approx 1.73205\ldots$$

we successively approximate $2^{\sqrt{3}}$ by the following rational powers:

$$2^{1.7}, \quad 2^{1.73}, \quad 2^{1.732}, \quad 2^{1.7320}, \quad 2^{1.73205}, \ldots$$

To demonstrate just how quickly $f(x) = 2^x$ increases, let's perform the following thought experiment. Suppose we start with a piece of paper a thousandth of an inch thick, and we fold it in half 50 times. Each time we fold the paper, the thickness of the paper stack doubles, so the thickness of the resulting stack would be $2^{50}/1000$ inches. How thick do you think that is? It works out to be more than 17 million miles!

Using advanced mathematics, it can be shown that there is exactly one number that these powers approach. We define $2^{\sqrt{3}}$ to be this number. Intuitively, these rational powers of 2 are getting closer and closer to $2^{\sqrt{3}}$. By this process we can approximate $2^{\sqrt{3}}$ to as many decimal places as we want:

$$2^{\sqrt{3}} \approx 3.321997\ldots$$

Similarly, we can define 2^x (or a^x if $a > 0$) where x is any irrational number.

Using this definition of irrational powers, we can graph $f(x) = 2^x$ on all of $\mathbb{R}$, as shown in Figure 1(b). It is an increasing function and, in fact, it increases very rapidly when $x > 0$ (see the margin note).

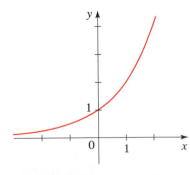

FIGURE 1   (a) $f(x) = 2^x$ for x rational   (b) $f(x) = 2^x$ for x real

It can be proved that *the Laws of Exponents are still true when the exponents are real numbers.*

EXAMPLE 1 ■ Graphing Exponential Functions by Plotting Points

Draw the graph of each function.

(a) $f(x) = 3^x$ (b) $g(x) = \left(\dfrac{1}{3}\right)^x$

SOLUTION

We calculate values of $f(x)$ and $g(x)$ and plot points to sketch the graphs in Figure 2.

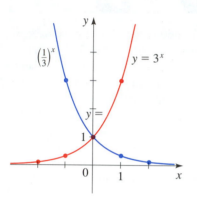

FIGURE 2

x	$f(x) = 3^x$	$g(x) = \left(\frac{1}{3}\right)^x$
-3	$\frac{1}{27}$	27
-2	$\frac{1}{9}$	9
-1	$\frac{1}{3}$	3
0	1	1
1	3	$\frac{1}{3}$
2	9	$\frac{1}{9}$
3	27	$\frac{1}{27}$

Notice that

$$g(x) = \left(\frac{1}{3}\right)^x = \frac{1}{3^x} = 3^{-x} = f(-x)$$

Reflecting graphs is explained in Section 2.5.

and so we could have obtained the graph of g from the graph of f by reflecting in the y-axis. ■

Figure 3 shows the graphs of the family of exponential functions $f(x) = a^x$ for various values of the base a. All of these graphs pass through the point $(0, 1)$

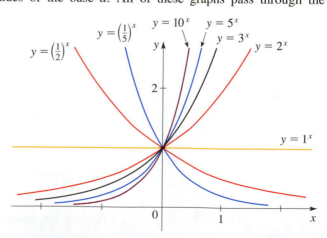

FIGURE 3

A family of exponential functions

The **Gateway Arch** in St. Louis, Missouri, is shaped in the form of the graph of a combination of exponential functions (*not* a parabola, as it might first appear). Specifically, it is a **catenary**, which is the graph of an equation of the form

$$y = a(e^{bx} + e^{-bx})$$

(see Exercise 65). This shape was chosen because it is optimal for distributing the internal structural forces of the arch. Chains and cables suspended between two points (for example, the stretches of cable between pairs of telephone poles) hang in the shape of a catenary.

because $a^0 = 1$ for $a \neq 0$. You can see from Figure 3 that there are three kinds of exponential functions $y = a^x$: If $0 < a < 1$, the exponential function decreases rapidly. If $a = 1$, it is constant. If $a > 1$, the function increases rapidly, and the larger the base, the more rapid the increase.

If $a > 1$, then the graph of $y = a^x$ approaches $y = 0$ as x decreases through negative values, and so the x-axis is a horizontal asymptote. If $0 < a < 1$, the graph approaches $y = 0$ as x increases indefinitely and, again, the x-axis is a horizontal asymptote. In either case the graph never touches the x-axis because $a^x > 0$ for all x. Thus, for $a \neq 1$, the exponential function $f(x) = a^x$ has domain $\mathbb{R}$ and range $(0, \infty)$. Let's summarize the preceding discussion.

EXPONENTIAL FUNCTIONS

For $a > 0$, the **exponential function with base a** is defined by

$$f(x) = a^x$$

For $a \neq 1$, the domain of f is $\mathbb{R}$, the range of f is $(0, \infty)$, and the graph of f has one of the following shapes:

$f(x) = a^x$ for $a > 1$

$f(x) = a^x$ for $0 < a < 1$

EXAMPLE 2 ■ Identifying Graphs of Exponential Functions

Find the exponential function $f(x) = a^x$ whose graph is given.

(a)

(b)

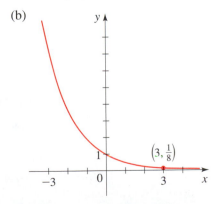

SOLUTION

(a) Since $f(2) = a^2 = 25$, we see by inspection that the base is $a = 5$. So
$f(x) = 5^x$.

(b) Since $f(3) = a^3 = \frac{1}{8}$, we see by inspection that the base is $a = \frac{1}{2}$.
So $f(x) = \left(\frac{1}{2}\right)^x$. ■

In the next two examples we see how to graph certain functions, not by plotting points, but by taking the basic graphs of the exponential functions in Figure 3 and applying the shifting and reflecting transformations of Section 2.5.

EXAMPLE 3 ■ Transformations of Exponential Functions

Use the graph $f(x) = 2^x$ to sketch the graph of each function.

(a) $g(x) = 1 + 2^x$ (b) $h(x) = -2^x$ (c) $k(x) = 2^{x-1}$

SOLUTION

Shifting and reflecting of graphs is explained in Section 2.5.

(a) To obtain the graph of $g(x) = 1 + 2^x$, we start with the graph of $f(x) = 2^x$ and shift it upward 1 unit. Notice from Figure 4(a) that the line $y = 1$ is now a horizontal asymptote.

(b) Again we start with the graph of $f(x) = 2^x$, but here we reflect in the x-axis to get the graph of $h(x) = -2^x$ shown in Figure 4(b).

(c) This time we start with the graph of $f(x) = 2^x$ and shift it to the right by 1 unit, to get the graph of $k(x) = 2^{x-1}$ shown in Figure 4(c).

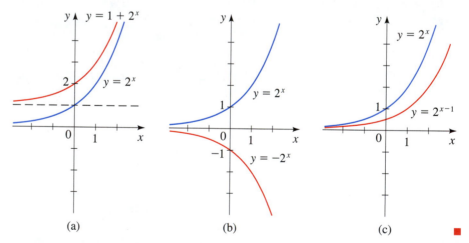

FIGURE 4   (a)   (b)   (c)   ■

EXAMPLE 4 ■ Comparing Exponential and Power Functions

Compare the rates of growth of the exponential function $f(x) = 2^x$ and the power function $g(x) = x^2$ by drawing the graphs of both functions in the following viewing rectangles.

(a) $[0, 3]$ by $[0, 8]$ (b) $[0, 6]$ by $[0, 25]$ (c) $[0, 20]$ by $[0, 1000]$

SOLUTION

(a) Figure 5(a) shows that the graph of $g(x) = x^2$ catches up with, and becomes higher than, the graph of $f(x) = 2^x$ at $x = 2$.

(b) The larger viewing rectangle in Figure 5(b) shows that the graph of $f(x) = 2^x$ overtakes that of $g(x) = x^2$ when $x = 4$.

(c) Figure 5(c) gives a more global view and shows that, when x is large, $f(x) = 2^x$ is much larger than $g(x) = x^2$.

(a)

(b)

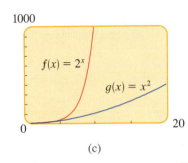

(c)

FIGURE 5

The Natural Exponential Function

Any positive number can be used as the base for an exponential function, but some bases are used more frequently than others. We will see in the remaining sections of this chapter that the bases 2 and 10 are convenient for certain applications, but the most important base is the number denoted by the letter e.

The number e is defined as the value that $(1 + 1/n)^n$ approaches as n becomes large. (In calculus this idea is made more precise through the concept of a limit. See Exercise 63.) The table in the margin shows the values of the expression $(1 + 1/n)^n$ for increasingly large values of n. It appears that, correct to five decimal places, $e \approx 2.71828$; in fact, the approximate value to 20 decimal places is

$$e \approx 2.71828182845904523536$$

It can be shown that e is an irrational number, so we cannot write its exact value.

Why use such a strange base for an exponential function? It may seem at first that a base such as 10 is easier to work with. We will see, however, that in certain applications the number e is the best possible base. In this section we study how e occurs in the description of compound interest and population growth.

n	$\left(1 + \dfrac{1}{n}\right)^n$
1	2.00000
5	2.48832
10	2.59374
100	2.70481
1000	2.71692
10,000	2.71815
100,000	2.71827
1,000,000	2.71828
10,000,000	2.71828

The notation e for the base of the natural exponential function was chosen by the Swiss mathematician Leonhard Euler (see page 293), probably because it is the first letter of the word *exponential*.

THE NATURAL EXPONENTIAL FUNCTION

The **natural exponential function** is the exponential function

$$f(x) = e^x$$

with base e. It is often referred to as *the* exponential function.

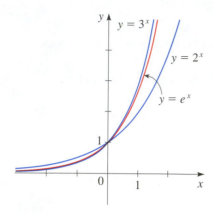

FIGURE 6

Graph of the natural exponential function

Since $2 < e < 3$, the graph of the natural exponential function lies between the graphs of $y = 2^x$ and $y = 3^x$, as shown in Figure 6.

Scientific calculators have a special key for the function $f(x) = e^x$. We use this key in the next example.

EXAMPLE 5 ■ Evaluating the Exponential Function

Evaluate each expression correct to five decimal places.

(a) e^3 (b) $2e^{-0.53}$ (c) $e^{4.8}$

SOLUTION

We use the $\boxed{e^x}$ key on a calculator to evaluate the exponential function.

(a) $e^3 \approx 20.08554$

(b) $2e^{-0.53} \approx 1.17721$

(c) $e^{4.8} \approx 121.51042$

EXAMPLE 6 ■ Transformations of the Exponential Function

Sketch the graph of each function.

(a) $f(x) = e^{-x}$ (b) $g(x) = 3e^{0.5x}$

SOLUTION

(a) We start with the graph of $y = e^x$ and reflect in the y-axis to obtain the graph of $y = e^{-x}$ as in Figure 7.

(b) We calculate several values, plot the resulting points, then connect the points with a smooth curve. The graph is shown in Figure 8.

Reflecting graphs is explained in Section 2.5.

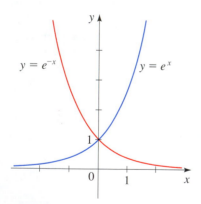

FIGURE 7

x	$f(x) = 3e^{0.5x}$
-3	0.67
-2	1.10
-1	1.82
0	3.00
1	4.95
2	8.15
3	13.45

FIGURE 8

EXAMPLE 7 ■ An Exponential Model for the Spread of a Virus

An infectious disease begins to spread in a small city of population 10,000. After t days, the number of persons who have succumbed to the virus is

modeled by the function

$$v(t) = \frac{10{,}000}{5 + 1245e^{-0.97t}}$$

(a) How many infected people are there initially (at time $t = 0$)?

(b) Determine the number of infected people after one day, two days, and five days.

(c) Graph the function v and describe its behavior.

SOLUTION

(a) Since $v(0) = 10{,}000/(5 + 1245e^0) = 10{,}000/1250 = 8$, we conclude that 8 people initially have the disease.

(b) Using a calculator, we evaluate $v(1)$, $v(2)$, and $v(5)$, and then round off to obtain the following values.

Days	Infected people
1	21
2	54
5	678

3000

$$v(t) = \frac{10{,}000}{5 + 1245e^{-0.97t}}$$

0   12

FIGURE 9

(c) From the graph in Figure 9, we see that the number of infected people first rises slowly; then between day 3 and day 8 rises quickly, and then levels off when about 2000 people are infected.   ■

The graph in Figure 9 is called a *logistic curve* or a *logistic growth model*. Curves like it occur frequently in the study of population growth. (See Exercises 47 and 48 on page 346.)

Compound Interest

Exponential functions occur in calculating compound interest. If an amount of money P, called the **principal**, is invested at an interest rate i per time period, then after one time period the interest is Pi, and the amount A of money is

$$A = P + Pi = P(1 + i)$$

If the interest is reinvested, then the new principal is $P(1 + i)$, and the amount after another time period is $A = P(1 + i)(1 + i) = P(1 + i)^2$. Similarly, after a third time period the amount is $A = P(1 + i)^3$. In general, after k periods the amount is

$$A = P(1 + i)^k$$

Notice that this is an exponential function with base $1 + i$.

If the annual interest rate is r and if interest is compounded n times per year, then in each time period the interest rate is $i = r/n$, and there are nt time periods in t years. This leads to the following formula for the amount after t years.

COMPOUND INTEREST

r is often referred to as the *nominal annual interest rate*.

Compound interest is calculated by the formula

$$A(t) = P\left(1 + \frac{r}{n}\right)^{nt}$$

where $A(t)$ = amount after t years

P = principal

r = interest rate per year

n = number of times interest is compounded per year

t = number of years

EXAMPLE 8 ■ Calculating Compound Interest

A sum of $1000 is invested at an interest rate of 12% per year. Find the amounts in the account after 3 years if interest is compounded annually, semiannually, quarterly, monthly, and daily.

SOLUTION

We use the compound interest formula with $P = \$1000$, $r = 0.12$, and $t = 3$.

Compounding	n	Amount after 3 years
Annual	1	$1000\left(1 + \dfrac{0.12}{1}\right)^{1(3)} = \$1404.93$
Semiannual	2	$1000\left(1 + \dfrac{0.12}{2}\right)^{2(3)} = \$1418.52$
Quarterly	4	$1000\left(1 + \dfrac{0.12}{4}\right)^{4(3)} = \$1425.76$
Monthly	12	$1000\left(1 + \dfrac{0.12}{12}\right)^{12(3)} = \$1430.77$
Daily	365	$1000\left(1 + \dfrac{0.12}{365}\right)^{365(3)} = \$1433.24$

We see from Example 8 that the interest paid increases as the number of compounding periods n increases. Let's see what happens as n increases indefinitely. If we let $m = n/r$, then

$$A(t) = P\left(1 + \frac{r}{n}\right)^{nt} = P\left[\left(1 + \frac{r}{n}\right)^{n/r}\right]^{rt} = P\left[\left(1 + \frac{1}{m}\right)^m\right]^{rt}$$

Recall that as m becomes large, the quantity $(1 + 1/m)^m$ approaches the number e. Thus, the amount approaches $A = Pe^{rt}$. This expression gives the amount when the interest is compounded at "every instant."

CONTINUOUSLY COMPOUNDED INTEREST

Continuously compounded interest is calculated by the formula

$$A(t) = Pe^{rt}$$

where $A(t) =$ amount after t years

$P =$ principal

$r =$ interest rate per year

$t =$ number of years

EXAMPLE 9 ■ Calculating Continuously Compounded Interest

Find the amount after 3 years if $1000 is invested at an interest rate of 12% per year, compounded continuously.

SOLUTION

We use the formula for continuously compounded interest with $P = \$1000$, $r = 0.12$, and $t = 3$ to get

$$A(3) = 1000e^{(0.12)3} = 1000e^{0.36} = \$1433.33$$

Compare this amount with the amounts in Example 8. ■

4.1 EXERCISES

1–6 ■ Sketch the graph of the function by making a table of values. Use a calculator if necessary.

1. $f(x) = 2^x$

2. $g(x) = 8^x$

3. $f(x) = \left(\frac{1}{3}\right)^x$

4. $h(x) = (1.1)^x$

5. $g(x) = 3e^x$

6. $h(x) = 2e^{-0.5x}$

7–8 ■ Graph both functions on one set of axes.

7. $y = 4^x$ and $y = 7^x$

8. $y = \left(\frac{2}{3}\right)^x$ and $y = \left(\frac{4}{3}\right)^x$

9–12 ■ Find the exponential function $f(x) = a^x$ whose graph is given.

9.

(2, 9)

10.

$\left(-1, \frac{1}{5}\right)$

11.

$\left(2, \frac{1}{16}\right)$

12.

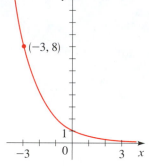

(−3, 8)

13–18 ■ Match the exponential function with one of the graphs labeled I–VI.

13. $f(x) = 5^x$

14. $f(x) = -5^x$

15. $f(x) = 5^{-x}$

16. $f(x) = 5^x + 3$

17. $f(x) = 5^{x-3}$

18. $f(x) = 5^{x+1} - 4$

I

(−1, 5)

II

(3, 1)

III

(0, 1)

IV

1

(−1, −3)

V

−3   3

(0, −1)

VI

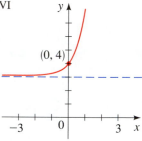

(0, 4)

19–32 ■ Graph the function, not by plotting points, but by starting from the graphs in Figures 3 and 6. State the domain, range, and asymptote.

19. $f(x) = -3^x$

20. $f(x) = 10^{-x}$

21. $g(x) = 2^x - 3$

22. $g(x) = 2^{x-3}$

23. $h(x) = 4 + \left(\frac{1}{2}\right)^x$

24. $h(x) = 6 - 3^x$

25. $f(x) = 10^{x+3}$

26. $f(x) = -\left(\frac{1}{5}\right)^x$

27. $f(x) = -e^x$

28. $y = 1 - e^x$

29. $y = e^{-x} - 1$

30. $f(x) = -e^{-x}$

31. $f(x) = e^{x-2}$

32. $y = e^{x-3} + 4$

33–34 ■ Find the function of the form $f(x) = Ca^x$ whose graph is given.

33.

34.

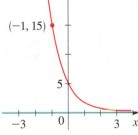

35. (a) Sketch the graphs of $f(x) = 2^x$ and $g(x) = 3(2^x)$.
(b) How are the graphs related?

36. (a) Sketch the graphs of $f(x) = 9^{x/2}$ and $g(x) = 3^x$.
(b) Use the Laws of Exponents to explain the relationship between these graphs.

37. If $f(x) = 10^x$, show that

$$\frac{f(x + h) - f(x)}{h} = 10^x \left(\frac{10^h - 1}{h} \right)$$

38. Compare the functions $f(x) = x^3$ and $g(x) = 3^x$ by evaluating both of them for $x = 0, 1, 2, 3, 4, 5, 6, 7, 8, 9, 10, 15,$ and 20. Then draw the graphs of f and g on the same set of axes.

39. (a) Compare the rates of growth of the functions $f(x) = 2^x$ and $g(x) = x^5$ by drawing the graphs of both functions in the following viewing rectangles.
  (i) $[0, 5]$ by $[0, 20]$
  (ii) $[0, 25]$ by $[0, 10^7]$
  (iii) $[0, 50]$ by $[0, 10^8]$
(b) Find the solutions of the equation $2^x = x^5$, correct to one decimal place.

40. (a) Compare the rates of growth of the functions $f(x) = 3^x$ and $g(x) = x^4$ by drawing the graphs of both functions in the following viewing rectangles:
  (i) $[-4, 4]$ by $[0, 20]$
  (ii) $[0, 10]$ by $[0, 5000]$
  (iii) $[0, 20]$ by $[0, 10^5]$
(b) Find the solutions of the equation $3^x = x^4$, correct to two decimal places.

41–42 ■ Draw graphs of the given family of functions for $c = 0.25, 0.5, 1, 2, 4$. How are the graphs related?

41. $f(x) = c2^x$ **42.** $f(x) = 2^{cx}$

43. A radioactive substance decays in such a way that the amount of mass remaining after t days is given by the function

$$m(t) = 13e^{-0.015t}$$

where $m(t)$ is measured in kilograms.
(a) Find the mass at time $t = 0$.
(b) How much of the mass remains after 45 days?

44. Radioactive iodine is used by doctors as a tracer in diagnosing certain thyroid gland disorders. This type of iodine decays in such a way that the mass remaining after t days is given by the function

$$m(t) = 6e^{-0.087t}$$

where $m(t)$ is measured in grams.
(a) Find the mass at time $t = 0$.
(b) How much of the mass remains after 20 days?

45. A sky diver jumps from a reasonable height above the ground. The air resistance she experiences is proportional to her velocity, and the constant of proportionality is 0.2. It can be shown that the downward velocity of the sky diver at time t is given by

$$v(t) = 80(1 - e^{-0.2t})$$

where t is measured in seconds and $v(t)$ is measured in feet per second (ft/s).
(a) Find the initial velocity of the sky diver.
(b) Find the velocity after 5 s and after 10 s.
(c) Draw a graph of the velocity function $v(t)$.
(d) The maximum velocity of a falling object with wind resistance is called its *terminal velocity*. From the graph in part (c) find the terminal velocity of this sky diver.

$$v(t) = 80(1 - e^{-0.2t})$$

46. A 50-gallon barrel is filled completely with pure water. Salt water with a concentration of 0.3 lb/gal is then pumped into the barrel, and the resulting mixture overflows at the same rate. The amount of salt in the barrel at time t is given by

$$Q(t) = 15(1 - e^{-0.04t})$$

where t is measured in minutes and $Q(t)$ is measured in pounds.

(a) How much salt is in the barrel after 5 min?
(b) How much salt is in the barrel after 10 min?
(c) Draw a graph of the function $Q(t)$.
(d) Use the graph in part (c) to determine the value that the amount of salt in the barrel approaches as t becomes large. Is this what you would expect?

$$Q(t) = 15(1 - e^{-0.04t})$$

47. Assume that a population of rabbits behaves according to the *logistic growth model*

$$n(t) = \frac{300}{0.05 + \left(\dfrac{300}{n_0} - 0.05\right)e^{-0.55t}}$$

where n_0 is the initial rabbit population.

(a) If the initial population is 50 rabbits, what will the population be after 12 years?
(b) Draw graphs of the function $n(t)$ for $n_0 = 50, 500, 2000, 8000,$ and $12,000$ in the viewing rectangle $[0, 15]$ by $[0, 12,000]$.
(c) From the graphs in part (b), observe that, regardless of the initial population, the rabbit population seems to approach a certain number as time goes on. What is that number? (This is the number of rabbits that the island can support.)

48. The population of a certain species of bird is limited by the type of habitat required for nesting. The population behaves according to the *logistic growth model*

$$n(t) = \frac{5600}{0.5 + 27.5e^{-0.044t}}$$

where t is measured in years.

(a) Find the initial bird population.
(b) Draw a graph of the function $n(t)$.
(c) What size does the population approach as time goes on?

49. If $10,000 is invested at an interest rate of 10% per year, compounded semiannually, find the value of the investment after the given number of years.

(a) 5 years   (b) 10 years   (c) 15 years

50. If $4000 is borrowed at a rate of 16% interest per year, compounded quarterly, find the amount due at the end of the given number of years.

(a) 4 years   (b) 6 years   (c) 8 years

51. If $3000 is invested at an interest rate of 9% per year, find the amount of the investment at the end of 5 years for the following compounding methods.

(a) Annual   (b) Semiannual
(c) Monthly   (d) Weekly
(e) Daily   (f) Hourly
(g) Continuously

52. If $4000 is invested in an account for which interest is compounded quarterly, find the amount of the investment at the end of 5 years for the following interest rates.

(a) 6%   (b) $6\frac{1}{2}$%   (c) 7%   (d) 8%

53. Which of the given interest rates and compounding periods would provide the best investment?

(i) $8\frac{1}{2}$% per year, compounded semiannually
(ii) $8\frac{1}{4}$% per year, compounded quarterly
(iii) 8% per year, compounded continuously

54. Which of the given interest rates and compounding periods would provide the better investment?

(i) $9\frac{1}{4}$% per year, compounded semiannually
(ii) 9% per year, compounded continuously

55. The **present value** of a sum of money is the amount that must be invested now, at a given rate of interest, to produce the desired sum at a later date.

(a) Find the present value of $10,000 if interest is paid at a rate of 9% per year, compounded semiannually, for 3 years.

(b) Find the present value of $100,000 if interest is paid at a rate of 8% per year, compounded monthly, for 5 years.

 56. A sum of $5000 is invested at an interest rate of 9% per year, compounded semiannually.
   (a) Find the value $A(t)$ of the investment after t years.
   (b) Draw a graph of $A(t)$.
   (c) Use the graph of $A(t)$ to determine when this investment will amount to $25,000.

57. The *hyperbolic cosine function* is defined by

$$\cosh(x) = \frac{e^x + e^{-x}}{2}$$

Sketch the graphs of the functions $y = \frac{1}{2}e^x$ and $y = \frac{1}{2}e^{-x}$ on the same axes and use graphical addition (see Section 2.8) to sketch the graph of $y = \cosh(x)$.

58. The *hyperbolic sine function* is defined by

$$\sinh(x) = \frac{e^x - e^{-x}}{2}$$

Sketch the graph of this function using graphical addition as in Exercise 57.

59–62 ■ Use the definitions in Exercises 57 and 58 to prove the identity.

59. $\cosh(-x) = \cosh(x)$

60. $\sinh(-x) = -\sinh(x)$

61. $[\cosh(x)]^2 - [\sinh(x)]^2 = 1$

62. $\sinh(x + y) = \sinh(x)\cosh(y) + \cosh(x)\sinh(y)$

 63. Illustrate the definition of the number e by graphing the curve $y = (1 + 1/x)^x$ and the line $y = e$ on the same screen using the viewing rectangle $[0, 40]$ by $[0, 4]$.

 64. Investigate the behavior of the function

$$f(x) = \left(1 - \frac{1}{x}\right)^x$$

as $x \to \infty$ by graphing f and the line $y = 1/e$ on the same screen using the viewing rectangle $[0, 20]$ by $[0, 1]$.

65. (a) Draw the graphs of the family of functions

$$f(x) = \frac{a}{2}(e^{x/a} + e^{-x/a})$$

   for $a = 0.5, 1, 1.5,$ and 2.
   (b) How does a larger value of a affect the graph?

66–67 ■ Graph the function and comment on vertical and horizontal asymptotes.

66. $y = 2^{1/x}$

67. $y = \dfrac{e^x}{x}$

68–69 ■ Find the local maximum and minimum values of the function and the value of x at which each occurs. State each answer correct to two decimal places.

68. $g(x) = x^x \quad (x > 0)$

69. $g(x) = e^x + e^{-3x}$

70–71 ■ Find, correct to two decimal places, (a) the intervals on which the function is increasing or decreasing, and (b) the range of the function.

70. $y = 10^{x-x^2}$

71. $y = xe^{-x}$

🔴 **DISCOVERY · DISCUSSION**

72. Growth of an Exponential Function Suppose you are offered a job that lasts one month, and you are to be very well paid. Which of the following methods of payment is more profitable for you?
   (a) One million dollars at the end of the month
   (b) Two cents on the first day of the month, 4 cents on the second day, 8 cents on the third day, and, in general, 2^n cents on the nth day

73. The Height of the Graph of an Exponential Function Your mathematics instructor asks you to sketch a graph of the exponential function

$$f(x) = 2^x$$

for x between 0 and 40, using a scale of 10 units to one inch. What are the dimensions of the sheet of paper you will need to sketch this graph?

Discovery Project

Exponential Explosion

To see how difficult it is to comprehend exponential growth, let's try a thought experiment.

Suppose you put a penny in your piggy bank today, two pennies tomorrow, four pennies the next day, and so on, doubling the number of pennies you add to the bank each day (see the table). How many pennies will you put in your piggy bank on day 30? The answer is 2^{30} pennies. That's simple, but can you guess how many dollars that is? 2^{30} pennies is more than 10 million dollars!

Day	Pennies
0	1
1	2
2	4
3	8
4	16
$\vdots$	$\vdots$
n	2^n
$\vdots$	$\vdots$

As you can see, the exponential function $f(x) = 2^x$ grows extremely fast. This is the principle behind atomic explosions. An atom splits releasing two neutrons which cause two atoms to split, each releasing two neutrons, causing four atoms to split, and so on. At the nth stage 2^n atoms split—an exponential explosion!

As we know, populations grow exponentially. Let's see what this means for a type of bacteria that splits every minute. Suppose that at 12:00 noon a single bacterium colonizes a discarded food can. The bacterium and his descendants are all happy, but they fear the time when the can is completely full of bacteria—doomsday.

1. How many bacteria are in the can at 12:05? at 12:10?

2. The can is completely full of bacteria at 1:00 P.M. At what time was the can only half full of bacteria?

3. When the can is exactly half full, the president of the bacteria colony reassures his constituents that doomsday is far away—after all, there is as much room left in the can as has been used in the entire previous history of the colony. Is the president correct? How much time is left before doomsday?

4. When the can is one-quarter full, how much time remains till doomsday?

5. A wise bacterium decides to start a new colony in another can and slow down splitting time to 2 minutes. How much time does this new colony have?

4.2 LOGARITHMIC FUNCTIONS

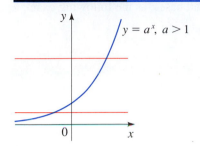

FIGURE 1

$f(x) = a^x$ is one-to-one

We read $\log_a x = y$ as "log base a of x is y."

Every exponential function $f(x) = a^x$, with $a > 0$ and $a \neq 1$, is a one-to-one function by the Horizontal Line Test (see Figure 1 for the case $a > 1$) and therefore has an inverse function. The inverse function f^{-1} is called the *logarithmic function with base a* and is denoted by $\log_a$. Recall from Section 2.9 that f^{-1} is defined by

$$f^{-1}(x) = y \iff f(y) = x$$

This leads to the following definition of the logarithmic function.

> **DEFINITION OF THE LOGARITHMIC FUNCTION**
>
> Let a be a positive number with $a \neq 1$. The **logarithmic function with base a**, denoted by $\log_a$, is defined by
>
> $$\log_a x = y \iff a^y = x$$

In words, this says that

> $\log_a x$ is the exponent to which the base a must be raised to give x.

By tradition, the name of the logarithmic function is $\log_a$, not just a single letter. Also, we usually omit the parentheses in the function notation and write

$$\log_a(x) = \log_a x$$

When we use the definition of logarithms to switch back and forth between the **logarithmic form** $\log_a x = y$ and the **exponential form** $a^y = x$, it's helpful to notice that, in both forms, the base is the same:

Logarithmic form   Exponential form

exponent   exponent
↓   ↓
$\log_a x = y$ $a^y = x$
↑   ↑
base   base

The following examples illustrate how to change an equation from one of these forms to the other.

EXAMPLE 1 ■ Logarithmic and Exponential Forms

The logarithmic and exponential forms are equivalent equations—if one is true, then so is the other. So, we can switch from one form to the other as in the following illustrations.

Logarithmic form	Exponential form
$\log_{10} 100{,}000 = 5$	$10^5 = 100{,}000$
$\log_2 8 = 3$	$2^3 = 8$
$\log_2\left(\frac{1}{8}\right) = -3$	$2^{-3} = \frac{1}{8}$
$\log_5 s = r$	$5^r = s$

■

x	$\log_{10} x$
10^4	4
10^3	3
10^2	2
10	1
1	0
10^{-1}	-1
10^{-2}	-2
10^{-3}	-3
10^{-4}	-4

It's important to understand that $\log_a x$ is an *exponent*. For example, the numbers in the right column of the table in the margin are the logarithms (base 10) of the numbers in the left column. This is the case for all bases, as the following example illustrates.

EXAMPLE 2 ■ **Evaluating Logarithms**

(a) $\log_{10} 1000 = 3$ because $10^3 = 1000$
(b) $\log_2 32 = 5$ because $2^5 = 32$
(c) $\log_{10} 0.1 = -1$ because $10^{-1} = 0.1$
(d) $\log_{16} 4 = \frac{1}{2}$ because $16^{1/2} = 4$ ■

Graphs of Logarithmic Functions

Recall that if a one-to-one function f has domain A and range B, then its inverse function f^{-1} has domain B and range A. Since the exponential function $f(x) = a^x$ with $a \neq 1$ has domain $\mathbb{R}$ and range $(0, \infty)$, we conclude that its inverse function, $f^{-1}(x) = \log_a x$, has domain $(0, \infty)$ and range $\mathbb{R}$.

The graph of $f^{-1}(x) = \log_a x$ is obtained by reflecting the graph of $f(x) = a^x$ in the line $y = x$. Figure 2 shows the case $a > 1$. The fact that $y = a^x$ (for $a > 1$) is a very rapidly increasing function for $x > 0$ implies that $y = \log_a x$ is a very slowly increasing function for $x > 1$ (see Exercise 76).

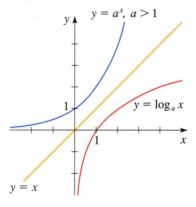

FIGURE 2

Graph of the logarithmic function
$f(x) = \log_a x$

Notice that since $a^0 = 1$, we have

$$\log_a 1 = 0$$

and so the x-intercept of the function $y = \log_a x$ is 1. Notice also that because the x-axis is a horizontal asymptote of $y = a^x$, the y-axis is a vertical asymptote of $y = \log_a x$. In fact, using the notation of Section 3.6, we can write

$$\log_a x \to -\infty \quad \text{as} \quad x \to 0^+$$

for the case $a > 1$.

EXAMPLE 3 ■ Graphing a Logarithmic Function by Plotting Points

Sketch the graph of $f(x) = \log_2 x$.

SOLUTION

To make a table of values, we choose the x-values to be powers of 2 so that we can easily find their logarithms. We plot these points and connect them with a smooth curve as in Figure 3.

x	$\log_2 x$
2^4	4
2^3	3
2^2	2
2	1
1	0
2^{-1}	-1
2^{-2}	-2
2^{-3}	-3
2^{-4}	-4

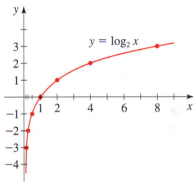

FIGURE 3

Figure 4 shows the graphs of the family of logarithmic functions with bases 2, 3, 5, and 10. These graphs are drawn by reflecting the graphs of $y = 2^x$, $y = 3^x$, $y = 5^x$, and $y = 10^x$ (see Figure 3 in Section 4.1) in the line $y = x$. We can also plot points as an aid to sketching these graphs, as illustrated in Example 3.

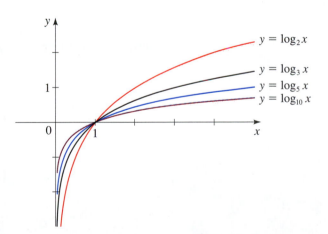

FIGURE 4

A family of logarithmic functions

In the next two examples we graph logarithmic functions by starting with the basic graphs in Figure 4 and using the transformations of Section 2.5.

EXAMPLE 4 ■ Reflecting Graphs of Logarithmic Functions

Sketch the graph of each function.

(a) $f(x) = -\log_2 x$

(b) $g(x) = \log_2(-x)$

SOLUTION

Reflecting graphs is explained in Section 2.5.

(a) We start with the graph of $y = \log_2 x$ in Figure 5(a) and reflect in the x-axis to get the graph of $f(x) = -\log_2 x$ in Figure 5(b).

(b) To obtain the graph of $g(x) = \log_2(-x)$, we reflect the graph of $y = \log_2 x$ in the y-axis. See Figure 5(c).

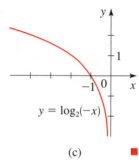

FIGURE 5   (a)   (b)   (c)   ■

EXAMPLE 5 ■ Shifting Graphs of Logarithmic Functions

Find the domain of each function, and sketch the graph.

(a) $f(x) = 2 + \log_5 x$

(b) $g(x) = \log_{10}(x - 3)$

SOLUTION

(a) The graph of f is obtained from the graph of $y = \log_5 x$ (Figure 4) by shifting upward 2 units (see Figure 6). The domain of f is $(0, \infty)$.

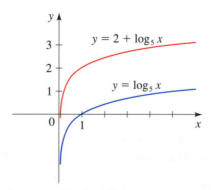

FIGURE 6

(b) The graph of g is obtained from the graph of $y = \log_{10} x$ (Figure 4) by shifting to the right 3 units (see Figure 7). The line $x = 3$ is a vertical asymptote. Since

$\log_{10} x$ is defined only when $x > 0$, the domain of $g(x) = \log_{10}(x - 3)$ is

$$\{x \mid x - 3 > 0\} = \{x \mid x > 3\} = (3, \infty)$$

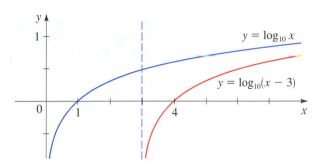

FIGURE 7

In Section 2.9 we saw that a function f and its inverse function f^{-1} satisfy the equations

$$f^{-1}(f(x)) = x \qquad \text{for } x \text{ in the domain of } f$$

$$f(f^{-1}(x)) = x \qquad \text{for } x \text{ in the domain of } f^{-1}$$

When applied to $f(x) = a^x$ and $f^{-1}(x) = \log_a x$, these equations become

$$\log_a(a^x) = x \qquad x \in \mathbb{R}$$

$$a^{\log_a x} = x \qquad x > 0$$

We list these and other properties of logarithms discussed in this section.

PROPERTIES OF LOGARITHMS

Property	Reason
1. $\log_a 1 = 0$	We must raise a to the power 0 to get 1.
2. $\log_a a = 1$	We must raise a to the power 1 to get a.
3. $\log_a a^x = x$	We must raise a to the power x to get a^x.
4. $a^{\log_a x} = x$	$\log_a x$ is the power to which a must be raised to get x.

EXAMPLE 6 ■ Applying Properties of Logarithms

We illustrate the properties of logarithms when the base is 5.

$$\log_5 1 = 0 \quad \text{Property 1} \qquad\qquad \log_5 5 = 1 \quad \text{Property 2}$$

$$\log_5 5^8 = 8 \quad \text{Property 3} \qquad\qquad 5^{\log_5 12} = 12 \quad \text{Property 4}$$

■ Common Logarithms

We now study logarithms with base 10.

> **COMMON LOGARITHM**
>
> The logarithm with base 10 is called the **common logarithm** and is denoted by omitting the base:
>
> $$\log x = \log_{10} x$$

From the definition of logarithms we can easily find that

$$\log 10 = 1 \qquad \text{and} \qquad \log 100 = 2$$

But how do we find $\log 50$? We need to find the exponent y such that $10^y = 50$. Clearly, 1 is too small and 2 is too large. So

$$1 < \log 50 < 2$$

To get a better approximation, we can experiment to find a power of 10 closer to 50. Fortunately, scientific calculators are equipped with a $\boxed{\text{LOG}}$ key that directly gives values of common logarithms.

EXAMPLE 7 ■ Evaluating Common Logarithms

Use a calculator to find appropriate values of $f(x) = \log x$ and use the values to sketch the graph.

SOLUTION

We make a table of values, using a calculator to evaluate the function at those values of x that are not powers of 10. We plot those points and connect them by a smooth curve as in Figure 8.

x	$\log x$
0.01	-2
0.1	-1
0.5	-0.301
1	0
4	0.602
5	0.699
10	1
15	1.176

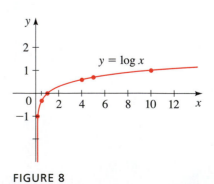

FIGURE 8

■

Natural Logarithms

Of all possible bases a for logarithms, it turns out that the most convenient choice for the purposes of calculus is the number e, which we defined in Section 4.1.

The notation ln is an abbreviation for the Latin name *logarithmus naturalis*.

NATURAL LOGARITHM

The logarithm with base e is called the **natural logarithm** and is denoted by **ln**:

$$\ln x = \log_e x$$

The natural logarithmic function $y = \ln x$ is the inverse function of the exponential function $y = e^x$. Both functions are graphed in Figure 9. By the definition of inverse functions we have

$$\ln x = y \quad \Leftrightarrow \quad e^y = x$$

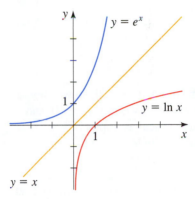

FIGURE 9

Graph of the natural logarithmic function

If we substitute $a = e$ and write "ln" for "$\log_e$" in the properties of logarithms mentioned earlier, we obtain the following properties of natural logarithms.

PROPERTIES OF NATURAL LOGARITHMS

Property	Reason
1. $\ln 1 = 0$	We must raise e to the power 0 to get 1.
2. $\ln e = 1$	We must raise e to the power 1 to get e.
3. $\ln e^x = x$	We must raise e to the power x to get e^x.
4. $e^{\ln x} = x$	$\ln x$ is the power to which e must be raised to get x.

Calculators are equipped with an $\boxed{\text{LN}}$ key that directly gives the values of natural logarithms.

EXAMPLE 8 ■ Evaluating the Natural Logarithm Function

(a) $\ln e^8 = 8$ Definition of natural logarithm

(b) $\ln\left(\dfrac{1}{e^2}\right) = \ln e^{-2} = -2$ Definition of natural logarithm

(c) $\ln 5 \approx 1.609$ Use the $\boxed{\text{LN}}$ key on a calculator   ■

EXAMPLE 9 ■ Finding the Domain of a Logarithmic Function

Find the domain of the function $f(x) = \ln(4 - x^2)$.

SOLUTION

As with any logarithmic function, $\ln x$ is defined when $x > 0$. Thus, the domain of f is

$$\{x \mid 4 - x^2 > 0\} = \{x \mid x^2 < 4\} = \{x \mid |x| < 2\}$$

$$= \{x \mid -2 < x < 2\} = (-2, 2) \qquad \blacksquare$$

EXAMPLE 10 ■ **Drawing the Graph of a Logarithmic Function**

Draw the graph of the function $y = x \ln(4 - x^2)$ and use it to find the asymptotes and local maximum and minimum values.

SOLUTION

As in Example 9 the domain of this function is the interval $(-2, 2)$, so we choose the viewing rectangle $[-3, 3]$ by $[-3, 3]$. The graph is shown in Figure 10, and from it we see that the lines $x = -2$ and $x = 2$ are vertical asymptotes.

The function has a local maximum point to the right of $x = 1$ and a local minimum point to the left of $x = -1$. By zooming in and tracing along the graph with the cursor, we find that the local maximum value is approximately 1.13 and this occurs when $x \approx 1.15$. Similarly (or by noticing that the function is odd), we find that the local minimum value is about -1.13, and it occurs when $x \approx -1.15$. ■

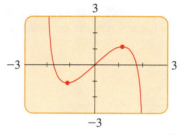

FIGURE 10
$y = x \ln(4 - x^2)$

4.2 EXERCISES

1–6 ■ Express the equation in exponential form.

1. (a) $\log_5 25 = 2$ (b) $\log_5 1 = 0$

2. (a) $\log_{10} 0.1 = -1$ (b) $\log_8 512 = 3$

3. (a) $\log_8 2 = \frac{1}{3}$ (b) $\log_2\left(\frac{1}{8}\right) = -3$

4. (a) $\log_3 81 = 4$ (b) $\log_8 4 = \frac{2}{3}$

5. (a) $\ln 5 = x$ (b) $\ln y = 5$

6. (a) $\ln(x + 1) = 2$ (b) $\ln(x - 1) = 4$

7–12 ■ Express the equation in logarithmic form.

7. (a) $5^3 = 125$ (b) $10^{-4} = 0.0001$

8. (a) $10^3 = 1000$ (b) $81^{1/2} = 9$

9. (a) $8^{-1} = \frac{1}{8}$ (b) $2^{-3} = \frac{1}{8}$

10. (a) $4^{-3/2} = 0.125$ (b) $7^3 = 343$

11. (a) $e^x = 2$ (b) $e^3 = y$

12. (a) $e^{x+1} = 0.5$ (b) $e^{0.5x} = t$

13–22 ■ Evaluate the expression.

13. (a) $\log_3 3$ (b) $\log_3 1$ (c) $\log_3 3^2$

14. (a) $\log_5 5^4$ (b) $\log_4 64$ (c) $\log_9 9$

15. (a) $\log_6 36$ (b) $\log_9 81$ (c) $\log_7 7^{10}$

16. (a) $\log_2 32$ (b) $\log_8 8^{17}$ (c) $\log_6 1$

17. (a) $\log_3\left(\frac{1}{27}\right)$ (b) $\log_{10} \sqrt{10}$ (c) $\log_5 0.2$

18. (a) $\log_5 125$ (b) $\log_{49} 7$ (c) $\log_9 \sqrt{3}$

19. (a) $2^{\log_2 37}$ (b) $3^{\log_3 8}$ (c) $e^{\ln \sqrt{5}}$

20. (a) $e^{\ln \pi}$ (b) $10^{\log 5}$ (c) $10^{\log 87}$

21. (a) $\log_8 0.25$ (b) $\ln e^4$ (c) $\ln(1/e)$

22. (a) $\log_4 \sqrt{2}$ (b) $\log_4\left(\frac{1}{2}\right)$ (c) $\log_4 8$

23–30 ■ Use the definition of the logarithmic function to find x.

23. (a) $\log_2 x = 5$ (b) $\log_2 16 = x$

24. (a) $\log_5 x = 4$ (b) $\log_{10} 0.1 = x$

25. (a) $\log_3 243 = x$ (b) $\log_3 x = 3$

26. (a) $\log_4 2 = x$ (b) $\log_4 x = 2$

27. (a) $\log_{10} x = 2$ (b) $\log_5 x = 2$

28. (a) $\log_x 1000 = 3$ (b) $\log_x 25 = 2$

29. (a) $\log_x 16 = 4$ (b) $\log_x 8 = \frac{3}{2}$

30. (a) $\log_x 6 = \frac{1}{2}$ (b) $\log_x 3 = \frac{1}{3}$

31–34 ■ Use a calculator to evaluate the expression, correct to four decimal places.

31. (a) $\log 2$ (b) $\log 35.2$ (c) $\log\left(\frac{2}{3}\right)$

32. (a) $\log 50$ (b) $\log \sqrt{2}$ (c) $\log\left(3\sqrt{2}\right)$

33. (a) $\ln 5$ (b) $\ln 25.3$ (c) $\ln\left(1 + \sqrt{3}\right)$

34. (a) $\ln 27$ (b) $\ln 7.39$ (c) $\ln 54.6$

35–38 ■ Find the function of the form $y = \log_a x$ whose graph is given.

35.

36.

37.

38.

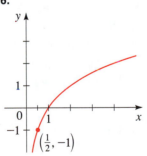

39–44 ■ Match the logarithmic function with one of the graphs labeled I–VI.

39. $f(x) = -\ln x$ **40.** $f(x) = \ln(x - 2)$

41. $f(x) = 2 + \ln x$ **42.** $f(x) = \ln(-x)$

43. $f(x) = \ln(2 - x)$ **44.** $f(x) = -\ln(-x)$

I

II

III

IV

V

VI

45. Draw the graph of $y = 4^x$, then use it to draw the graph of $y = \log_4 x$.

46. Draw the graph of $y = 3^x$, then use it to draw the graph of $y = \log_3 x$.

47–56 ■ Graph the function, not by plotting points, but by starting from the graphs in Figures 4 and 9. State the domain, range, and asymptote.

47. $f(x) = \log_2(x - 4)$ **48.** $f(x) = -\log_{10} x$

49. $g(x) = \log_5(-x)$ **50.** $g(x) = \ln(x + 2)$

51. $y = 2 + \log_3 x$

52. $y = \log_3(x - 1) - 2$

53. $y = 1 - \log_{10} x$

54. $y = 1 + \ln(-x)$

55. $y = |\ln x|$

56. $y = \ln |x|$

57–62 ■ Find the domain of the function.

57. $f(x) = \log_{10}(x + 3)$

58. $f(x) = \log_5(8 - 2x)$

59. $g(x) = \log_3(x^2 - 1)$

60. $g(x) = \ln(x - x^2)$

61. $h(x) = \ln x + \ln(2 - x)$

62. $h(x) = \sqrt{x - 2} - \log_5(10 - x)$

 63–68 ■ Draw the graph of the function in a suitable viewing rectangle and use it to find the domain, the asymptotes, and the local maximum and minimum values.

63. $y = \log_{10}(1 - x^2)$

64. $y = \ln(x^2 - x)$

65. $y = x + \ln x$

66. $y = x(\ln x)^2$

67. $y = \dfrac{\ln x}{x}$

68. $y = x \log_{10}(x + 10)$

 69. Compare the rates of growth of the functions $f(x) = \ln x$ and $g(x) = \sqrt{x}$ by drawing their graphs on a common screen using the viewing rectangle $[-1, 30]$ by $[-1, 6]$.

 70. (a) By drawing the graphs of the functions

$$f(x) = 1 + \ln(1 + x) \quad \text{and} \quad g(x) = \sqrt{x}$$

in a suitable viewing rectangle, show that even when a logarithmic function starts out higher than a root function, it is ultimately overtaken by the root function.

(b) Find, correct to two decimal places, the solutions of the equation $\sqrt{x} = 1 + \ln(1 + x)$.

71–72 ■ A family of functions is given.
(a) Draw graphs of the family for $c = 1, 2, 3,$ and 4.
(b) How are the graphs in part (a) related?

71. $f(x) = \log(cx)$

72. $f(x) = c \log x$

73–74 ■ A function $f(x)$ is given.
(a) Find the domain of the function f.
(b) Find the inverse function of f.

73. $f(x) = \log_2(\log_{10} x)$

74. $f(x) = \ln(\ln(\ln x))$

75. (a) Find the inverse of the function $f(x) = \dfrac{2^x}{1 + 2^x}$.

(b) What is the domain of the inverse function?

DISCOVERY · DISCUSSION

76. The Height of the Graph of a Logarithmic Function
Suppose that the graph of $y = 2^x$ is drawn on a coordinate plane where the unit of measurement is an inch.
(a) Show that at a distance 2 ft to the right of the origin the height of the graph is about 265 mi.
(b) If the graph of $y = \log_2 x$ is drawn on the same set of axes, how far to the right of the origin do we have to go before the height of the curve reaches 2 ft?

77. The Googolplex A **googol** is 10^{100}, and a **googolplex** is 10^{googol}. Find

$$\log(\log(\text{googol})) \quad \text{and} \quad \log(\log(\log(\text{googolplex})))$$

78. Comparing Logarithms Which is larger, $\log_4 17$ or $\log_5 24$? Explain your reasoning.

79. The Number of Digits in an Integer Compare log 1000 to the number of digits in 1000. Do the same for 10,000. How many digits does any number between 1000 and 10,000 have? Between what two values must the common logarithm of such a number lie? Use your observations to explain why the number of digits in any positive integer x is $[\![\log x]\!] + 1$. (The symbol $[\![n]\!]$ is the greatest integer function defined in Section 2.2.)

4.3 LAWS OF LOGARITHMS

Since logarithms are exponents, the Laws of Exponents give rise to the Laws of Logarithms. These properties give logarithmic functions a wide range of applications, as we will see in Section 4.5.

LAWS OF LOGARITHMS

Let a be a positive number, with $a \neq 1$. Let $A > 0$, $B > 0$, and C be any real numbers.

Law	Description
1. $\log_a(AB) = \log_a A + \log_a B$	The logarithm of a product of numbers is the sum of the logarithms of the numbers.
2. $\log_a\left(\dfrac{A}{B}\right) = \log_a A - \log_a B$	The logarithm of a quotient of numbers is the difference of the logarithms of the numbers.
3. $\log_a(A^C) = C \log_a A$	The logarithm of a power of a number is the exponent times the logarithm of the number.

■ **Proof**   We make use of the property $\log_a a^x = x$ from Section 4.2.

Law 1. Let
$$\log_a A = u \qquad \text{and} \qquad \log_a B = v$$

When written in exponential form, these equations become
$$a^u = A \qquad \text{and} \qquad a^v = B$$

Thus
$$\log_a(AB) = \log_a(a^u a^v) = \log_a(a^{u+v})$$
$$= u + v = \log_a A + \log_a B$$

Law 2. Using Law 1, we have
$$\log_a A = \log_a\left[\left(\frac{A}{B}\right)B\right] = \log_a\left(\frac{A}{B}\right) + \log_a B$$

so
$$\log_a\left(\frac{A}{B}\right) = \log_a A - \log_a B$$

Law 3. Let $\log_a A = u$. Then $a^u = A$, so
$$\log_a(A^C) = \log_a(a^u)^C = \log_a(a^{uC}) = uC = C \log_a A \qquad \square$$

John Napier (1550–1617) was a Scottish landowner for whom mathematics was a hobby. We know him today because of his key invention—logarithms, which he published in 1614 under the title *A Description of the Marvelous Rule of Logarithms*. In Napier's time, logarithms were used exclusively for simplifying complicated calculations. For example, to multiply two large numbers we would write them as powers of 10. The exponents are simply the logarithms of the numbers. For instance,

$$4532 \times 57783$$
$$\approx 10^{3.65629} \times 10^{4.76180}$$
$$= 10^{8.41809}$$
$$\approx 261{,}872{,}564$$

The idea is that multiplying powers of 10 is easy (we simply add their exponents). Napier produced extensive tables giving the logarithms (or exponents) of numbers. Since the advent of calculators and computers, logarithms are no longer used for this purpose. The logarithmic functions, however, have found many applications, some of which are described in this chapter.

Napier wrote on many topics. One of his most colorful works is a book entitled *A Plaine Discovery of the Whole Revelation of Saint John*, in which he predicted that the world would end in the year 1700.

As the following examples illustrate, these laws are used in both directions. Since the domain of any logarithmic function is the interval $(0, \infty)$, we assume that all quantities whose logarithms occur are positive.

EXAMPLE 1 ■ Using the Laws of Logarithms to Expand Expressions

Use the Laws of Logarithms to rewrite each expression.

(a) $\log_2(6x)$

(b) $\log \sqrt{5}$

(c) $\log_5(x^3 y^6)$

(d) $\ln\left(\dfrac{ab}{\sqrt[3]{c}}\right)$

SOLUTION

(a) $\log_2(6x) = \log_2 6 + \log_2 x$ Law 1

(b) $\log \sqrt{5} = \log 5^{1/2} = \frac{1}{2} \log 5$ Law 3

(c) $\log_5(x^3 y^6) = \log_5 x^3 + \log_5 y^6$ Law 1

$$= 3 \log_5 x + 6 \log_5 y \quad \text{Law 3}$$

(d) $\ln\left(\dfrac{ab}{\sqrt[3]{c}}\right) = \ln(ab) - \ln \sqrt[3]{c}$ Law 2

$$= \ln a + \ln b - \ln c^{1/3} \quad \text{Law 1}$$

$$= \ln a + \ln b - \tfrac{1}{3} \ln c \quad \text{Law 3}$$ ■

EXAMPLE 2 ■ Using the Laws of Logarithms to Evaluate Expressions

Evaluate each expression.

(a) $\log_4 2 + \log_4 32$ (b) $\log_2 80 - \log_2 5$ (c) $-\frac{1}{3} \log 8$

SOLUTION

(a) $\log_4 2 + \log_4 32 = \log_4(2 \cdot 32)$ Law 1

$$= \log_4 64 = 3 \quad \text{Because } 4^3 = 64$$

(b) $\log_2 80 - \log_2 5 = \log_2\left(\frac{80}{5}\right)$ Law 2

$$= \log_2 16 = 4 \quad \text{Because } 2^4 = 16$$

(c) $-\frac{1}{3} \log 8 = \log 8^{-1/3}$ Law 3

$$= \log\left(\tfrac{1}{2}\right) \quad \text{Property of negative exponents}$$

$$\approx -0.301 \quad \text{Use a calculator}$$ ■

EXAMPLE 3 ■ Writing an Expression as a Single Logarithm

Express $3 \log x + \frac{1}{2} \log(x + 1)$ as a single logarithm.

SOLUTION

$$3 \log x + \tfrac{1}{2} \log(x+1) = \log x^3 + \log(x+1)^{1/2} \qquad \text{Law 3}$$

$$= \log(x^3(x+1)^{1/2}) \qquad \text{Law 1} \quad \blacksquare$$

EXAMPLE 4 ■ Writing an Expression as a Single Logarithm

Express $3 \ln s + \tfrac{1}{2} \ln t - 4 \ln(t^2+1)$ as a single logarithm.

SOLUTION

$$3 \ln s + \tfrac{1}{2} \ln t - 4 \ln(t^2+1) = \ln s^3 + \ln t^{1/2} - \ln(t^2+1)^4 \qquad \text{Law 3}$$

$$= \ln(s^3 t^{1/2}) - \ln(t^2+1)^4 \qquad \text{Law 1}$$

$$= \ln\left(\frac{s^3 \sqrt{t}}{(t^2+1)^4}\right) \qquad \text{Law 2} \quad \blacksquare$$

WARNING Although the Laws of Logarithms tell us how to compute the logarithm of a product or a quotient, *there is no corresponding rule for the logarithm of a sum or a difference.* For instance,

$$\log_a(x+y) \neq \log_a x + \log_a y$$

In fact, we know that the right side is equal to $\log_a(xy)$.

Also, don't improperly simplify quotients or powers of logarithms. For instance,

$$\frac{\log 6}{\log 2} \neq \log\left(\frac{6}{2}\right)$$

$$(\log_2 x)^3 \neq 3 \log_2 x$$

Change of Base

For some purposes, we find it useful to change from logarithms in one base to logarithms in another base. Suppose we are given $\log_a x$ and want to find $\log_b x$. Let

$$y = \log_b x$$

We write this in exponential form and take the logarithm, with base a, of each side.

$$b^y = x \qquad \text{Exponential form}$$

$$\log_a(b^y) = \log_a x \qquad \text{Take } \log_a \text{ of each side}$$

$$y \log_a b = \log_a x \qquad \text{Law 3}$$

$$y = \frac{\log_a x}{\log_a b} \qquad \text{Divide by } \log_a b$$

This proves the following formula.

We may write the Change of Base Formula as

$$\log_b x = \left(\frac{1}{\log_a b}\right)\log_a x$$

So, $\log_b x$ is just a constant multiple of $\log_a x$; the constant is $\dfrac{1}{\log_a b}$.

CHANGE OF BASE FORMULA

$$\log_b x = \frac{\log_a x}{\log_a b}$$

In particular, if we put $x = a$, then $\log_a a = 1$ and this formula becomes

$$\log_b a = \frac{1}{\log_a b}$$

We can now evaluate a logarithm to *any* base by using the Change of Base Formula to express the logarithm in terms of common logarithms or natural logarithms and then using a calculator.

EXAMPLE 5 ■ **Using the Change of Base Formula to Evaluate Logarithms**

Use the Change of Base Formula and common or natural logarithms to evaluate each logarithm, correct to five decimal places.

(a) $\log_8 5$

(b) $\log_9 20$

SOLUTION

(a) We use the Change of Base Formula with $b = 8$ and $a = 10$:

$$\log_8 5 = \frac{\log_{10} 5}{\log_{10} 8} \approx 0.77398$$

(b) We use the Change of Base Formula with $b = 9$ and $a = e$:

$$\log_9 20 = \frac{\ln 20}{\ln 9} \approx 1.36342$$ ■

EXAMPLE 6 ■ **Using the Change of Base Formula to Graph a Logarithmic Function**

Use a graphing calculator to graph $f(x) = \log_6 x$.

SOLUTION

Calculators don't have a key for $\log_6$, so we use the Change of Base Formula to write

$$f(x) = \log_6 x = \frac{\ln x}{\ln 6}$$

Since calculators do have an $\boxed{\text{LN}}$ key, we can enter this new form of the function and graph it. The graph is shown in Figure 1. ■

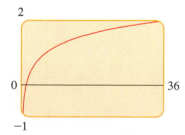

FIGURE 1

$$f(x) = \log_6 x = \frac{\ln x}{\ln 6}$$

4.3   EXERCISES

1–26 ■ Use the Laws of Logarithms to rewrite the expression in a form with no logarithm of a product, quotient, root, or power.

1. $\log_2 (2x)$

2. $\log_3 (5y)$

3. $\log_2(x(x-1))$

4. $\log_5\left(\dfrac{x}{2}\right)$

5. $\log 6^{10}$

6. $\ln\left(\sqrt{z}\right)$

7. $\log_2(AB^2)$

8. $\log_6 \sqrt[4]{17}$

9. $\log_3(x\sqrt{y})$

10. $\log_2(xy)^{10}$

11. $\log_5 \sqrt[3]{x^2 + 1}$

12. $\log_a\left(\dfrac{x^2}{yz^3}\right)$

13. $\ln\sqrt{ab}$

14. $\ln\sqrt[3]{3r^2 s}$

15. $\log\left(\dfrac{x^3 y^4}{z^6}\right)$

16. $\log\left(\dfrac{a^2}{b^4 \sqrt{c}}\right)$

17. $\log_2\left(\dfrac{x(x^2 + 1)}{\sqrt{x^2 - 1}}\right)$

18. $\log_5 \sqrt{\dfrac{x-1}{x+1}}$

19. $\ln\left(x\sqrt{\dfrac{y}{z}}\right)$

20. $\ln\dfrac{3x^2}{(x+1)^{10}}$

21. $\log \sqrt[4]{x^2 + y^2}$

22. $\log\left(\dfrac{x}{\sqrt[3]{1-x}}\right)$

23. $\log \sqrt{\dfrac{x^2 + 4}{(x^2 + 1)(x^3 - 7)^2}}$

24. $\log \sqrt{x\sqrt{y\sqrt{z}}}$

25. $\ln\left(\dfrac{x^3 \sqrt{x-1}}{3x + 4}\right)$

26. $\log\left(\dfrac{10^x}{x(x^2 + 1)(x^4 + 2)}\right)$

27–38 ■ Evaluate the expression.

27. $\log_5 \sqrt{125}$

28. $\log_2 112 - \log_2 7$

29. $\log 2 + \log 5$

30. $\log \sqrt{0.1}$

31. $\log_4 192 - \log_4 3$

32. $\log_{12} 9 + \log_{12} 16$

33. $\ln 6 - \ln 15 + \ln 20$

34. $e^{3\ln 5}$

35. $10^{2\log 4}$

36. $\log_2 8^{33}$

37. $\log(\log 1000^{10,000})$

38. $\ln(\ln(\ln e^{e^{200}}))$

39–48 ■ Rewrite the expression as a single logarithm.

39. $\log_3 5 + 5\log_3 2$

40. $\log 12 + \frac{1}{2}\log 7 - \log 2$

41. $\log_2 A + \log_2 B - 2\log_2 C$

42. $\log_5(x^2 - 1) - \log_5(x - 1)$

43. $4\log x - \frac{1}{3}\log(x^2 + 1) + 2\log(x - 1)$

44. $\ln(a + b) + \ln(a - b) - 2\ln c$

45. $\ln 5 + 2\ln x + 3\ln(x^2 + 5)$

46. $2(\log_5 x + 2\log_5 y - 3\log_5 z)$

47. $\frac{1}{3}\log(2x + 1) + \frac{1}{2}\left[\log(x - 4) - \log(x^4 - x^2 - 1)\right]$

48. $\log_a b + c\log_a d - r\log_a s$

49–56 ■ Use the Change of Base Formula and a calculator to evaluate the logarithm, correct to six decimal places. Use either natural or common logarithms.

49. $\log_2 5$

50. $\log_5 2$

51. $\log_3 16$

52. $\log_6 92$

53. $\log_7 2.61$

54. $\log_6 532$

55. $\log_4 125$

56. $\log_{12} 2.5$

 57. Use the Change of Base Formula to show that

$$\log_3 x = \frac{\ln x}{\ln 3}$$

Then use this fact to draw the graph of the function $f(x) = \log_3 x$.

58. Draw graphs of the family of functions $y = \log_a x$ for $a = 2, e, 5$, and 10 on the same screen, using the viewing rectangle $[0, 5]$ by $[-3, 3]$. How are these graphs related?

59. Use the Change of Base Formula to show that

$$\log e = \frac{1}{\ln 10}$$

60. Simplify: $(\log_2 5)(\log_5 7)$

61. Show that $-\ln\left(x - \sqrt{x^2 - 1}\right) = \ln\left(x + \sqrt{x^2 - 1}\right)$.

▲ DISCOVERY · DISCUSSION

62. Is the Equation an Identity?   Discuss each equation and determine whether it is true for all possible values of the variables. (Ignore values of the variables for which any term is undefined.)

(a) $\log\left(\dfrac{x}{y}\right) = \dfrac{\log x}{\log y}$

(b) $\log_2(x - y) = \log_2 x - \log_2 y$

(c) $\log_5\left(\dfrac{a}{b^2}\right) = \log_5 a - 2\log_5 b$

(d) $\log 2^z = z \log 2$

(e) $(\log P)(\log Q) = \log P + \log Q$

(f) $\dfrac{\log a}{\log b} = \log a - \log b$

(g) $(\log_2 7)^x = x \log_2 7$

(h) $\log_a a^a = a$

(i) $\log(x - y) = \dfrac{\log x}{\log y}$

(j) $-\ln\left(\dfrac{1}{A}\right) = \ln A$

63. Find the Error   What is wrong with the following argument?

$$\log 0.1 < 2 \log 0.1$$
$$= \log(0.1)^2$$
$$= \log 0.01$$
$$\log 0.1 < \log 0.01$$
$$0.1 < 0.01$$

64. Shifting, Shrinking, and Stretching Graphs of Functions   Let $f(x) = x^2$. Show that $f(2x) = 4f(x)$, and explain how this shows that shrinking the graph of f horizontally has the same effect as stretching it vertically. Then use the identities $e^{2+x} = e^2 e^x$ and $\ln(2x) = \ln 2 + \ln x$ to show that for $g(x) = e^x$, a horizontal shift is the same as a vertical stretch and for $h(x) = \ln x$, a horizontal shrinking is the same as a vertical shift.

4.4   EXPONENTIAL AND LOGARITHMIC EQUATIONS

In this section we solve equations that involve exponential or logarithmic functions. The techniques we develop here will be used in the next section for solving applied problems.

■ Exponential Equations

An exponential equation is one in which the variable occurs in the exponent. For example,

$$2^x = 7$$

The variable x presents a difficulty because it is in the exponent. To deal with this difficulty, we take the logarithm of each side and then use the Laws of Logarithms to "bring down x" from the exponent.

$$2^x = 7$$

$$\ln 2^x = \ln 7 \qquad \text{Take ln of each side}$$

$$x \ln 2 = \ln 7 \qquad \text{Law 3 (bring down the exponent)}$$

$$x = \frac{\ln 7}{\ln 2} \qquad \text{Solve for } x$$

$$\approx 2.807 \qquad \text{Use a calculator}$$

Recall that Law 3 of the Laws of Logarithms says that $\log_a A^C = C \log_a A$.

The method we used to solve $2^x = 7$ is typical of the methods we use to solve all exponential equations, and it can be summarized as follows.

GUIDELINES FOR SOLVING EXPONENTIAL EQUATIONS

1. Isolate the exponential expression on one side of the equation.

2. Take the logarithm of each side, then use the Laws of Logarithms to "bring down the exponent."

3. Solve for the variable.

EXAMPLE 1 ■ Solving an Exponential Equation

Find the solution of the equation $3^{x+2} = 7$, correct to six decimal places.

SOLUTION

We take the common logarithm of each side and use Law 3.

$$3^{x+2} = 7$$

$$\log(3^{x+2}) = \log 7 \qquad \text{Take log of each side}$$

$$(x + 2) \log 3 = \log 7 \qquad \text{Law 3 (bring down the exponent)}$$

$$x + 2 = \frac{\log 7}{\log 3} \qquad \text{Divide by log 3}$$

$$x = \frac{\log 7}{\log 3} - 2 \qquad \text{Subtract 2}$$

$$\approx -0.228756 \qquad \text{Use a calculator}$$

We could have used natural logarithms instead of common logarithms. In fact, using the same steps, we get

$$x = \frac{\ln 7}{\ln 3} - 2 \approx -0.228756$$

CHECK YOUR ANSWER ■ Substituting $x = -0.228756$ into the original equation and using a calculator, we get

$$3^{(-0.228756)+2} \approx 7 \quad \checkmark$$

EXAMPLE 2 ■ Solving an Exponential Equation

Solve the equation $8e^{2x} = 20$.

SOLUTION

We first divide by 8 in order to isolate the exponential term on one side of the equation.

$$8e^{2x} = 20$$

$$e^{2x} = \frac{20}{8} \qquad \text{Divide by 8}$$

$$\ln e^{2x} = \ln 2.5 \qquad \text{Take ln of each side}$$

$$2x = \ln 2.5 \qquad \text{Property of ln}$$

$$x = \frac{\ln 2.5}{2} \qquad \text{Divide by 2}$$

$$\approx 0.458 \qquad \text{Use a calculator}$$

CHECK YOUR ANSWER

Substituting $x = 0.458$ into the original equation and using a calculator, we get

$$8e^{2(0.458)} \approx 20 \quad \checkmark$$

EXAMPLE 3 ■ Solving an Exponential Equation Algebraically and Graphically

Solve the equation $e^{3-2x} = 4$ algebraically and graphically.

SOLUTION 1: ALGEBRAIC

Since the base of the exponential term is e, we use natural logarithms to solve this equation.

$$\ln(e^{3-2x}) = \ln 4 \qquad \text{Take ln of each side}$$

$$3 - 2x = \ln 4 \qquad \text{Property of ln}$$

$$2x = 3 - \ln 4$$

$$x = \tfrac{1}{2}(3 - \ln 4) \approx 0.807$$

You should check that this answer satisfies the original equation.

SOLUTION 2: GRAPHICAL

We graph the equations $y = e^{3-2x}$ and $y = 4$ in the same viewing rectangle as in Figure 1. The solutions occur where the graphs intersect. Zooming in on the point of intersection of the two graphs, we see that $x \approx 0.81$.

We can also solve the equation by graphing $y = e^{3-2x} - 4$ and finding the x-intercepts.

FIGURE 1

EXAMPLE 4 ■ **Solving an Exponential Equation**

Solve the equation $e^{2x} - e^x - 6 = 0$.

SOLUTION

To isolate the exponential term, we factor.

$$e^{2x} - e^x - 6 = 0$$

$$(e^x)^2 - e^x - 6 = 0 \qquad \text{Law of Exponents}$$

$$(e^x - 3)(e^x + 2) = 0 \qquad \text{Factor (a quadratic in } e^x)$$

$$e^x - 3 = 0 \quad \text{or} \quad e^x + 2 = 0 \qquad \text{Zero-Product Property}$$

$$e^x = 3 \qquad\qquad\quad e^x = -2$$

The equation $e^x = 3$ leads to $x = \ln 3$. But the equation $e^x = -2$ has no solution because $e^x > 0$ for all x. Thus, $x = \ln 3 \approx 1.0986$ is the only solution. You should check that this answer satisfies the original equation. ■

EXAMPLE 5 ■ **Solving an Exponential Equation**

Solve the equation $3x^2 e^x + x^3 e^x = 0$.

SOLUTION

First we factor the left side of the equation.

$$3x^2 e^x + x^3 e^x = 0$$

$$(3x^2 + x^3)e^x = 0 \qquad \text{Factor out } e^x$$

$$x^2(3 + x)e^x = 0 \qquad \text{Factor out } x^2$$

$$x^2 = 0 \quad \text{or} \quad 3 + x = 0 \quad \text{or} \quad e^x = 0 \qquad \text{Zero-Product Property}$$

$$x = 0 \qquad\qquad x = -3$$

CHECK YOUR ANSWERS

$x = 0$:

$\quad 3(0)^2 e^0 + 0^3 e^0 = 0 \qquad$ ✔

$x = -3$:

$\quad 3(-3)^2 e^{-3} + (-3)^3 e^{-3}$

$\qquad = 27e^{-3} - 27e^{-3} = 0$ ✔

The equation $e^x = 0$ has no solution because $e^x > 0$ for all x. Thus, $x = 0$ and $x = -3$ are the only solutions. ■

Logarithmic Equations

A *logarithmic equation* is one in which a logarithm of the variable occurs. For example,

$$\log_2(x + 2) = 5$$

To solve for x, we write the equation in exponential form.

$$x + 2 = 2^5 \qquad\qquad \text{Exponential form}$$

$$x = 32 - 2 = 30 \qquad \text{Solve for } x$$

Radiocarbon dating is a method archeologists use to determine the age of ancient objects. The carbon dioxide in the atmosphere always contains a fixed fraction of radioactive carbon, carbon-14 (^{14}C), with a half-life of about 5730 years. Plants absorb carbon dioxide from the atmosphere, which then makes its way to animals through the food chain. Thus, all living creatures contain the same fixed proportions of ^{14}C to nonradioactive ^{12}C as the atmosphere.

After an organism dies, it stops assimilating ^{14}C, and the amount of ^{14}C in it begins to decay exponentially. We can then determine the time elapsed since the death of the organism by measuring the amount of ^{14}C left in it.

For example, if a donkey bone contains 73% as much ^{14}C as a living donkey and it died t years ago, then by the formula for radioactive decay (Section 4.5),

$$0.73 = (1.00)e^{-(t\ln 2)/5730}$$

We solve this exponential equation to find $t \approx 2600$, so the bone is about 2600 years old.

Another way of looking at the first step is to raise the base, 2, to each side of the equation.

$$2^{\log_2(x+2)} = 2^5 \qquad \text{Raise 2 to each side}$$

$$x + 2 = 2^5 \qquad \text{Property of logarithms}$$

$$x = 32 - 2 = 30 \qquad \text{Solve for } x$$

The methods used to solve this simple problem are typical. We summarize the steps as follows.

> ### GUIDELINES FOR SOLVING LOGARITHMIC EQUATIONS
>
> 1. Isolate the logarithmic term on one side of the equation; you may need to first combine the logarithmic terms.
>
> 2. Write the equation in exponential form (or raise the base to each side of the equation).
>
> 3. Solve for the variable.

EXAMPLE 6 ■ Solving Logarithmic Equations

Solve each equation for x.

(a) $\ln x = 8$

(b) $\log_2(25 - x) = 3$

SOLUTION

(a)
$$\ln x = 8$$

$$x = e^8 \qquad \text{Exponential form}$$

Therefore, $x = e^8 \approx 2981$.

We can also solve this problem another way:

$$\ln x = 8$$

$$e^{\ln x} = e^8 \qquad \text{Raise } e \text{ to each side}$$

$$x = e^8 \qquad \text{Property of ln}$$

(b) The first step is to rewrite the equation in exponential form.

$$\log_2(25 - x) = 3$$

$$25 - x = 2^3 \qquad \text{Exponential form (or raise 2 to each side)}$$

$$25 - x = 8$$

$$x = 25 - 8 = 17$$

CHECK YOUR ANSWER

If $x = 17$, we get

$$\log_2(25 - 17) = \log_2 8 = 3 \qquad \checkmark$$

■

EXAMPLE 7 ■ **Solving a Logarithmic Equation**

Solve the equation $4 + 3 \log(2x) = 16$.

SOLUTION

We first isolate the logarithmic term. This allows us to write the equation in exponential form.

$$4 + 3 \log(2x) = 16$$

$$3 \log(2x) = 12 \qquad \text{Subtract 4}$$

$$\log(2x) = 4 \qquad \text{Divide by 3}$$

$$2x = 10^4 \qquad \text{Exponential form (or raise 10 to each side)}$$

$$x = 5000 \qquad \text{Divide by 2}$$

CHECK YOUR ANSWER

If $x = 5000$, we get

$$4 + 3 \log 2(5000) = 4 + 3 \log 10{,}000$$
$$= 4 + 3(4)$$
$$= 16 \qquad ✓$$

EXAMPLE 8 ■ **Solving a Logarithmic Equation Algebraically and Graphically**

Solve the equation $\log(x + 2) + \log(x - 1) = 1$ algebraically and graphically.

SOLUTION 1: ALGEBRAIC

We first combine the logarithmic terms using the Laws of Logarithms.

$$\log[(x + 2)(x - 1)] = 1 \qquad \text{Law 1}$$

$$(x + 2)(x - 1) = 10 \qquad \text{Exponential form (or raise 10 to each side)}$$

$$x^2 + x - 2 = 10 \qquad \text{Expand left side}$$

$$x^2 + x - 12 = 0 \qquad \text{Subtract 10}$$

$$(x + 4)(x - 3) = 0 \qquad \text{Factor}$$

$$x = -4 \qquad \text{or} \qquad x = 3$$

CHECK YOUR ANSWERS

$x = -4$:

$$\log(-4 + 2) + \log(-4 - 1)$$
$$= \log(-2) + \log(-5)$$
$$\qquad \text{undefined} \qquad ✗$$

$x = 3$:

$$\log(3 + 2) + \log(3 - 1)$$
$$= \log 5 + \log 2 = \log(5 \cdot 2)$$
$$= \log 10 = 1 \qquad ✓$$

We check these potential solutions in the original equation and find that $x = -4$ is not a solution (because logarithms of negative numbers are undefined), but $x = 3$ is a solution. (See *Check Your Answers*.)

SOLUTION 2: GRAPHICAL

We first move all terms to one side of the equation:

$$\log(x + 2) + \log(x - 1) - 1 = 0$$

Then we graph

$$y = \log(x + 2) + \log(x - 1) - 1$$

as in Figure 2. The solutions are the x-intercepts of the graph. Thus, the only solution is $x \approx 3$.

FIGURE 2

EXAMPLE 9 ■ Solving a Logarithmic Equation Graphically

Solve the equation $x^2 = 2 \ln(x + 2)$.

SOLUTION

In Example 9, it's not possible to isolate x algebraically, so we must solve the equation graphically.

We first move all terms to one side of the equation

$$x^2 - 2 \ln(x + 2) = 0$$

Then we graph

$$y = x^2 - 2 \ln(x + 2)$$

as in Figure 3. The solutions are the x-intercepts of the graph. Zooming in on the x-intercepts, we see that there are two solutions:

$$x \approx -0.71 \qquad \text{and} \qquad x \approx 1.60$$

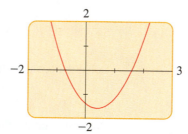

FIGURE 3

■ Compound Interest

Recall the formulas for interest that we found in Section 4.1. If a principal P is invested at an interest rate r for a period of t years, then the amount A of the investment is given by

$$A = P(1 + r) \qquad \text{Simple interest (for one year)}$$

$$A(t) = P\left(1 + \frac{r}{n}\right)^{nt} \qquad \text{Interest compounded } n \text{ times per year}$$

$$A(t) = Pe^{rt} \qquad \text{Interest compounded continuously}$$

We can use logarithms to determine the time it takes for the principal to increase to a given amount.

EXAMPLE 10 ■ Finding the Term for an Investment to Double

A sum of $5000 is invested at an interest rate of 9% per year. Find the time required for the money to double if the interest is compounded according to the following method.

(a) Semiannual

(b) Continuous

SOLUTION

(a) We use the formula for compound interest with $P = \$5000$, $A(t) = \$10,000$, $r = 0.09$, $n = 2$, and solve the resulting exponential equation for t.

$$5000\left(1 + \frac{0.09}{2}\right)^{2t} = 10,000$$

$$(1.045)^{2t} = 2 \qquad \text{Divide by 5000}$$

$$\log 1.045^{2t} = \log 2 \qquad \text{Take log of each side}$$

$$2t \log 1.045 = \log 2 \qquad \text{Law 3 (bring down the exponent)}$$

$$t = \frac{\log 2}{2 \log 1.045} \qquad \text{Divide by 2 log 1.045}$$

$$t \approx 7.9 \qquad \text{Use a calculator}$$

The money will double in 7.9 years.

(b) We use the formula for continuously compounded interest with $P = \$5000$, $A(t) = \$10,000$, $r = 0.09$, and solve the resulting exponential equation for t.

$$5000e^{0.09t} = 10,000$$

$$e^{0.09t} = 2 \qquad \text{Divide by 5000}$$

$$\ln e^{0.09t} = \ln 2 \qquad \text{Take ln of each side}$$

$$0.09t = \ln 2 \qquad \text{Property of ln}$$

$$t = \frac{\ln 2}{0.09} \qquad \text{Divide by 0.09}$$

$$t \approx 7.702 \qquad \text{Use a calculator}$$

The money will double in 7.7 years. ∎

EXAMPLE 11 ■ Time Required to Grow an Investment

A sum of $1000 is invested at an interest rate of 8% per year. Find the time required for the amount to grow to $4000 if interest is compounded continuously.

SOLUTION

We use the formula for continuously compounded interest with $P = \$1000$, $A(t) = \$4000$, $r = 0.08$, and solve the resulting exponential equation for t.

$$1000e^{0.08t} = 4000$$

$$e^{0.08t} = 4 \qquad \text{Divide by 1000}$$

$$0.08t = \ln 4 \qquad \text{Take ln of each side}$$

$$t = \frac{\ln 4}{0.08} \qquad \text{Divide by 0.08}$$

$$t \approx 17.33 \qquad \text{Use a calculator}$$

The amount will be \$4000 in about 17 years and 4 months. ∎

If an investment earns compound interest, then the **annual percentage yield** (APY) is the *simple* interest rate that yields the same amount at the end of one year.

EXAMPLE 12 ■ Calculating the Annual Percentage Yield

Find the annual percentage yield for an investment that earns interest at a rate of 6% per year, compounded daily.

SOLUTION

After one year, a principal P will grow to the amount

$$A = P\left(1 + \frac{0.06}{365}\right)^{365} = P(1.06183)$$

The formula for simple interest is

$$A = P(1 + r)$$

Comparing, we see that $r = 0.06183$, so the annual percentage yield is 6.183%. ∎

4.4   EXERCISES

1–26 ■ Find the solution of the exponential equation, correct to four decimal places.

1. $e^x = 16$

2. $10^{-x} = 2$

3. $10^{2x} = 5$

4. $e^{-2x} = \frac{1}{10}$

5. $2^{1-x} = 3$

6. $3^{2x-1} = 5$

7. $3e^x = 10$

8. $2e^{12x} = 17$

9. $e^{1-4x} = 2$

10. $4(1 + 10^{5x}) = 9$

11. $4 + 3^{5x} = 8$

12. $2^{3x} = 34$

13. $8^{0.4x} = 5$

14. $3^{x/14} = 0.1$

15. $5^{-x/100} = 2$

16. $e^{3-5x} = 16$

17. $e^{2x+1} = 200$

18. $\left(\frac{1}{4}\right)^x = 75$

19. $5^x = 4^{x+1}$

20. $10^{1-x} = 6^x$

21. $2^{3x+1} = 3^{x-2}$

22. $7^{x/2} = 5^{1-x}$

23. $\dfrac{50}{1 + e^{-x}} = 4$

24. $\dfrac{10}{1 + e^{-x}} = 2$

25. $100(1.04)^{2t} = 300$

26. $(1.00625)^{12t} = 2$

27–34 ■ Solve the equation.

27. $x^2 2^x - 2^x = 0$

28. $x^2 10^x - x10^x = 2(10^x)$

29. $4x^3 e^{-3x} - 3x^4 e^{-3x} = 0$

30. $x^2 e^x + xe^x - e^x = 0$

31. $e^{2x} - 3e^x + 2 = 0$

32. $e^{2x} - e^x - 6 = 0$

33. $e^{4x} + 4e^{2x} - 21 = 0$

34. $e^x - 12e^{-x} - 1 = 0$

35–50 ■ Solve the logarithmic equation for x.

35. $\ln x = 10$

36. $\ln(2 + x) = 1$

37. $\log x = -2$　　　　**38.** $\log(x - 4) = 3$

39. $\log(3x + 5) = 2$　　　**40.** $\log_3(2 - x) = 3$

41. $2 - \ln(3 - x) = 0$　　　**42.** $\log_2(x^2 - x - 2) = 2$

43. $\log_2 3 + \log_2 x = \log_2 5 + \log_2(x - 2)$

44. $2 \log x = \log 2 + \log(3x - 4)$

45. $\log x + \log(x - 1) = \log(4x)$

46. $\log_5 x + \log_5(x + 1) = \log_5 20$

47. $\log_5(x + 1) - \log_5(x - 1) = 2$

48. $\log x + \log(x - 3) = 1$

49. $\log_9(x - 5) + \log_9(x + 3) = 1$

50. $\ln(x - 1) + \ln(x + 2) = 1$

51. For what value of x is the following true?

$$\log(x + 3) = \log x + \log 3$$

52. For what value of x is it true that $(\log x)^3 = 3 \log x$?

53. Solve for x:　$2^{2/\log_5 x} = \frac{1}{16}$

54. Solve for x:　$\log_2(\log_3 x) = 4$

55. A man invests \$5000 in an account that pays 8.5% interest per year, compounded quarterly.
(a) Find the amount after 3 years.
(b) How long will it take for the investment to double?

56. A man invests \$6500 in an account that pays 6% interest per year, compounded continuously.
(a) What is the amount after 2 years?
(b) How long will it take for the amount to be \$8000?

57. Find the time required for an investment of \$5000 to grow to \$8000 at an interest rate of 7.5% per year, compounded quarterly.

58. Nancy wants to invest \$4000 in saving certificates that bear an interest rate of 9.75% per year, compounded semiannually. How long a time period should she choose in order to save an amount of \$5000?

59. How long will it take for an investment of \$1000 to double in value if the interest rate is 8.5% per year, compounded continuously?

60. A sum of \$1000 was invested for 4 years, and the interest was compounded semiannually. If this sum amounted to \$1435.77 in the given time, what was the interest rate?

61. Find the annual percentage yield for an investment that earns 8% per year, compounded monthly.

62. Find the annual percentage yield for an investment that earns $5\frac{1}{2}$% per year, compounded continuously.

63. A 15-g sample of radioactive iodine decays in such a way that the mass remaining after t days is given by $m(t) = 15e^{-0.087t}$ where $m(t)$ is measured in grams. After how many days is there only 5 g remaining?

64. The velocity of a sky diver t seconds after jumping is given by $v(t) = 80(1 - e^{-0.2t})$. After how many seconds is the velocity 70 ft/s?

65. A small lake is stocked with a certain species of fish. The fish population is modeled by the function

$$P = \frac{10}{1 + 4e^{-0.8t}}$$

where P is the number of fish in thousands and t is measured in years since the lake was stocked.
(a) Find the fish population after 3 years.
(b) After how many years will the fish population reach 5000 fish?

66. Environmental scientists measure the intensity of light at various depths in a lake to find the "transparency" of the water. Certain levels of transparency are required for the biodiversity of the submerged macrophyte population. In a certain lake the intensity of light at depth x is given by

$$I = 4e^{-0.6x}$$

where I is measured in lux (a measure of light intensity) and x in cm.
(a) Find the intensity at a depth of 3 cm.
(b) At what depth has the light intensity dropped to $I = 2$?

67. An electric circuit contains a battery that produces a voltage of 60 volts (V), a resistor with a resistance of 13 ohms (Ω), and an inductor with an inductance of 5 henrys (H), as shown in the figure. Using calculus, it can be shown that the current $I = I(t)$ (in amperes, A) t seconds after the switch is closed is $I = \frac{60}{13}(1 - e^{-13t/5})$.

(a) Use this equation to express the time t as a function of the current I.

(b) After how many seconds is the current 2 A?

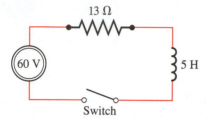

$13\ \Omega$

$60\ V$ $5\ H$

Switch

68. A *learning curve* is a graph of a function $P(t)$ that measures the performance of someone learning a skill as a function of the training time t. At first, the rate of learning is rapid. Then, as performance increases and approaches a maximal value M, the rate of learning decreases. It has been found that the function $P(t) = M - Ce^{-kt}$, where k and C are positive constants and $C < M$, is a reasonable model for learning.

(a) Express the learning time t as a function of the performance level P.

(b) For a pole-vaulter in training, the learning curve is given by $P(t) = 20 - 14e^{-0.024t}$, where $P(t)$ is the height he is able to pole-vault after t months. After how many months of training is he able to vault 12 ft?

(c) Draw a graph of the learning curve in part (b).

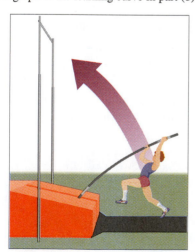

69–76 ■ Use a graphing device to find all solutions of the equation, correct to two decimal places.

69. $\ln x = 3 - x$

70. $\log x = x^2 - 2$

71. $x^3 - x = \log(x + 1)$

72. $x = \ln(4 - x^2)$

73. $e^x = -x$

74. $2^{-x} = x - 1$

75. $4^{-x} = \sqrt{x}$

76. $e^{x^2} - 2 = x^3 - x$

77–80 ■ Solve the inequality.

77. $\log(x - 2) + \log(9 - x) < 1$

78. $3 \leq \log_2 x \leq 4$

79. $2 < 10^x < 5$

80. $x^2 e^x - 2e^x < 0$

81–83 ■ Solve the equation.

81. $(x - 1)^{\log(x-1)} = 100(x - 1)$

82. $\log_2 x + \log_4 x + \log_8 x = 11$

83. $4^x - 2^{x+1} = 3$
[*Hint:* First write the equation as a quadratic equation in 2^x.]

DISCOVERY · DISCUSSION

84. **Estimating a Solution**   Without actually solving the equation, find two whole numbers between which the solution of $9^x = 20$ must lie. Do the same for $9^x = 100$. Explain how you reached your conclusions.

85. **A Surprising Equation**   Take logarithms to show that the equation

$$x^{1/\log x} = 5$$

has no solution. For what values of k does the equation

$$x^{1/\log x} = k$$

have a solution? What does this tell us about the graph of the function $f(x) = x^{1/\log x}$? Confirm your answer using a graphing device.

Many processes that occur in nature, such as population growth, radioactive decay, heat diffusion, and numerous others, can be modeled using exponential functions. Logarithmic functions are used in models for the loudness of sounds, the intensity of earthquakes, and many other phenomena. In this section we study exponential and logarithmic models.

Exponential Models of Population Growth

Biologists have observed that the population of a species doubles its size in a fixed period of time. For example, under ideal conditions a certain population of bacteria doubles in size every 3 hours. If the culture is started with 1000 bacteria, then after 3 hours there will be 2000 bacteria, after another 3 hours there will be 4000, and so on. If we let $n = n(t)$ be the number of bacteria after t hours, then

$$n(0) = 1000$$

$$n(3) = 1000 \cdot 2$$

$$n(6) = (1000 \cdot 2) \cdot 2 = 1000 \cdot 2^2$$

$$n(9) = (1000 \cdot 2^2) \cdot 2 = 1000 \cdot 2^3$$

$$n(12) = (1000 \cdot 2^3) \cdot 2 = 1000 \cdot 2^4$$

From this pattern it appears that the number of bacteria after t hours is modeled by the function

$$n(t) = 1000 \cdot 2^{t/3}$$

In general, suppose that the initial size of a population is n_0 and the doubling period is a. Then the size of the population at time t is modeled by

$$n(t) = n_0 2^{ct}$$

where $c = 1/a$. If we knew the tripling time b, then the formula would be $n(t) = n_0 3^{ct}$ where $c = 1/b$. These formulas indicate that the growth of the bacteria is modeled by an exponential function. But what base should we use? The answer is e, because then it can be shown (using calculus) that the population is modeled by

$$n(t) = n_0 e^{rt}$$

where r is the *relative rate of growth of population, expressed as a proportion of the population at any time*. For instance, if $r = 0.02$, then at any time t the growth rate is 2% of the population at time t.

Notice that the formula for population growth is the same as that for continuously compounded interest. In fact, the same principle is at work in both cases: The growth of a population (or an investment) per time period is proportional to the size of the population (or the amount of the investment). A population of 1,000,000 will increase more in one year than a population of 1000; in exactly the same way, an investment of $1,000,000 will increase more in one year than an investment of $1000.

EXPONENTIAL GROWTH MODEL

A population that experiences **exponential growth** increases according to the model

$$n(t) = n_0 e^{rt}$$

where $n(t)$ = population at time t

n_0 = initial size of the population

r = relative rate of growth (expressed as a proportion of the population)

t = time

In the following examples we assume that the populations grow exponentially.

EXAMPLE 1 ■ Predicting the Size of a Population

The initial bacterium count in a culture is 500. A biologist later makes a sample count of bacteria in the culture and finds that the relative rate of growth is 40% per hour.

(a) Find a function that models the number of bacteria after t hours.
(b) What is the estimated count after 10 hours?
(c) Sketch the graph of the function $n(t)$.

SOLUTION

(a) We use the exponential growth model with $n_0 = 500$ and $r = 0.4$ to get

$$n(t) = 500e^{0.4t}$$

where t is measured in hours.

(b) Using the function in part (a), we find that the bacterium count after 10 hours is

$$n(10) = 500e^{0.4(10)} = 500e^4 \approx 27{,}300$$

(c) The graph is shown in Figure 1.   ■

FIGURE 1

EXAMPLE 2 ■ Comparing Effects of Different Rates of Population Growth

In 2000 the population of the world was 6.1 billion and the relative rate of growth was 1.4% per year. It is claimed that a rate of 1.0% per year would make a significant difference in the total population in just a few decades. Test this claim by estimating the population of the world in the year 2050 using a relative rate of growth of (a) 1.4% per year and (b) 1.0% per year.

Graph the population functions for the next 100 years for the two relative growth rates in the same viewing rectangle.

SOLUTION

(a) By the exponential growth model, we have

$$n(t) = 6.1e^{0.014t}$$

where $n(t)$ is measured in billions and t is measured in years since 2000. Because the year 2050 is 50 years after 2000, we find

$$n(50) = 6.1e^{0.014(50)} = 6.1e^{0.7} \approx 12.3$$

Thus, the estimated population in the year 2050 is about 12.3 billion.

(b) We use the function

$$n(t) = 6.1e^{0.010t}$$

and find $n(50) = 6.1e^{0.010(50)} = 6.1e^{0.50} \approx 10.1$

So, the estimated population in the year 2050 is about 10.1 billion.

The graphs in Figure 2 show that a small change in the relative rate of growth will, over time, make a large difference in population size.   ■

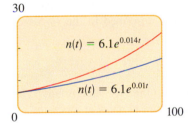

30

$n(t) = 6.1e^{0.014t}$

$n(t) = 6.1e^{0.01t}$

0   100

FIGURE 2

EXAMPLE 3 ■ Finding the Initial Population

A certain breed of rabbit was introduced onto a small island about 8 years ago. The current rabbit population on the island is estimated to be 4100, with a relative growth rate of 55% per year.

(a) What was the initial size of the rabbit population?
(b) Estimate the population 12 years from now.

SOLUTION

(a) From the exponential growth model, we have

$$n(t) = n_0 e^{0.55t}$$

and we know that the population at time $t = 8$ is $n(8) = 4100$.

We substitute what we know into the equation and solve for n_0:

$$4100 = n_0 e^{0.55(8)}$$

$$n_0 = \frac{4100}{e^{0.55(8)}} \approx \frac{4100}{81.45} \approx 50$$

Thus, we estimate that 50 rabbits were introduced onto the island.

Another way to solve part (b) is to let t be the number of years from now. In this case, $n_0 = 4100$ (the current population), and the population 12 years from now will be

$$n(12) = 4100e^{0.55(12)} \approx 3 \text{ million}$$

(b) Now that we know n_0, we can write a formula for population growth:

$$n(t) = 50e^{0.55t}$$

Twelve years from now, $t = 20$ and

$$n(20) = 50e^{0.55(20)} \approx 2{,}993{,}707$$

So, we estimate that the rabbit population on the island 12 years from now will be about 3 million. ■

Can the rabbit population in Example 3(b) actually reach such a high number? In reality, as the island becomes overpopulated with rabbits, the rabbit population growth will be slowed due to food shortage and other factors. One model that takes into account such factors is the *logistic growth model* described in the *Focus on Modeling* section, Problem 10, page 407.

EXAMPLE 4 ■ World Population Projections

The population of the world in 2000 was 6.1 billion, and the estimated relative growth rate was 1.4% per year. If the population continues to grow at this rate, when will it reach 122 billion?

SOLUTION

We use the population growth function with $n_0 = 6.1$ billion, $r = 0.014$, and $n(t) = 122$ billion. This leads to an exponential equation, which we solve for t.

$$6.1e^{0.014t} = 122$$

$$e^{0.014t} = 20 \qquad \text{Divide by 6.1}$$

$$\ln e^{0.014t} = \ln 20 \qquad \text{Take ln of each side}$$

$$0.014t = \ln 20 \qquad \text{Property of ln}$$

$$t = \frac{\ln 20}{0.014} \qquad \text{Divide by 0.014}$$

$$t \approx 213.98 \qquad \text{Use a calculator}$$

Standing Room Only

The population of the world was about 6.1 billion in 2000, and was increasing at 1.4% per year. Using the exponential model for population growth, and assuming that each person occupies an average of 4 ft² of the surface of the earth, we find that by the year 2801 there will be standing room only! (The total land surface area of the world is about 1.8×10^{15} ft².)

Thus, the population will reach 122 billion in approximately 214 years, that is, in the year $2000 + 214 = 2214$. ■

EXAMPLE 5 ■ The Number of Bacteria in a Culture

A culture starts with 10,000 bacteria, and the number doubles every 40 min.

(a) Find a function that models the number of bacteria at time t.
(b) Find the number of bacteria after one hour.
(c) After how many minutes will there be 50,000 bacteria?
 (d) Sketch a graph of the number of bacteria at time t.

SOLUTION

(a) To find the function that models this population growth, we need to find the rate r. To do this, we use the formula for population growth with $n_0 = 10{,}000$, $t = 40$, and $n(t) = 20{,}000$, and then solve for r.

$$10{,}000 \cdot e^{r(40)} = 20{,}000$$

$$e^{40r} = 2 \qquad \text{Divide by 10,000}$$

$$\ln e^{40r} = \ln 2 \qquad \text{Take ln of each side}$$

$$40r = \ln 2 \qquad \text{Property of ln}$$

$$r = \frac{\ln 2}{40} \qquad \text{Divide by 40}$$

$$r \approx 0.01733 \qquad \text{Use a calculator}$$

Now that we know $r \approx 0.01733$, we can write the function for the population growth:

$$n(t) = 10{,}000 \cdot e^{0.01733t}$$

(b) Using the function we found in part (a) with $t = 60$ min (one hour), we get

$$n(60) = 10{,}000 \cdot e^{0.01733(60)} \approx 28{,}287$$

Thus, the number of bacteria after one hour is approximately 28,000.

(c) We use the function we found in part (a) with $n(t) = 50{,}000$ and solve the resulting exponential equation for t.

$$10{,}000 \cdot e^{0.01733t} = 50{,}000$$

$$e^{0.01733t} = 5 \qquad \text{Divide by 10,000}$$

$$\ln e^{0.01733t} = \ln 5 \qquad \text{Take ln of each side}$$

$$0.01733t = \ln 5 \qquad \text{Property of ln}$$

$$t = \frac{\ln 5}{0.01733} \qquad \text{Divide by 0.01733}$$

$$t \approx 92.9 \qquad \text{Use a calculator}$$

The bacterium count will reach 50,000 in approximately 93 min.

(d) The graph of the function $n(t) = 10{,}000 \cdot e^{0.01733t}$ is shown in Figure 3. ■

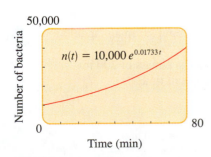

FIGURE 3

Radioactive Decay

Radioactive substances decay by spontaneously emitting radiation. The rate of decay is directly proportional to the mass of the substance. This is analogous to population growth, except that the mass of radioactive material *decreases*. It can be shown that the mass $m(t)$ remaining at time t is modeled by the function

$$m(t) = m_0 e^{-rt}$$

where r is the rate of decay expressed as a proportion of the mass and m_0 is the initial mass. Physicists express the rate of decay in terms of **half-life**, the time required for half the mass to decay. We can obtain the rate r from this as follows. If h is the half-life, then a mass of 1 unit becomes $\frac{1}{2}$ unit when $t = h$. Substituting this into the model, we get

$$\frac{1}{2} = 1 \cdot e^{-rh}$$

$$\ln\left(\tfrac{1}{2}\right) = -rh \qquad \text{Take ln of each side}$$

$$r = -\frac{1}{h}\ln(2^{-1}) \qquad \text{Solve for } r$$

$$r = \frac{\ln 2}{h} \qquad \ln 2^{-1} = -\ln 2 \text{ by Law 3}$$

This last equation allows us to find the rate r from the half-life h.

Half-lives of **radioactive elements** vary from very long to very short. Here are some examples.

Element	Half-life
Thorium-232	14.5 billion years
Uranium-235	4.5 billion years
Thorium-230	80,000 years
Plutonium-239	24,360 years
Carbon-14	5,730 years
Radium-226	1,600 years
Cesium-137	30 years
Strontium-90	28 years
Polonium-210	140 days
Thorium-234	25 days
Iodine-135	8 days
Radon-222	3.8 days
Lead-211	3.6 minutes
Krypton-91	10 seconds

RADIOACTIVE DECAY MODEL

If m_0 is the initial mass of a radioactive substance with half-life h, then the mass remaining at time t is modeled by the function

$$m(t) = m_0 e^{-rt}$$

where $r = \dfrac{\ln 2}{h}$.

EXAMPLE 6 ■ Radioactive Decay

Polonium-210 (^{210}Po) has a half-life of 140 days. Suppose a sample of this substance has a mass of 300 mg.

(a) Find a function that models the amount of the sample remaining at time t.
(b) Find the mass remaining after one year.
(c) How long will it take for the sample to decay to a mass of 200 mg?
 (d) Draw a graph of the sample mass as a function of time.

SOLUTION

(a) Using the model for radioactive decay with $m_0 = 300$ and $r = (\ln 2/140) \approx 0.00495$, we have

$$m(t) = 300e^{-0.00495t}$$

(b) We use the function we found in part (a) with $t = 365$ (one year).

$$m(365) = 300e^{-0.00495(365)} \approx 49.256$$

Thus, approximately 49 mg of ^{210}Po remains after one year.

(c) We use the function we found in part (a) with $m(t) = 200$ and solve the resulting exponential equation for t.

$$300e^{-0.00495t} = 200$$

$$e^{-0.00495t} = \tfrac{2}{3} \qquad \text{Divide by 300}$$

$$\ln e^{-0.00495t} = \ln \tfrac{2}{3} \qquad \text{Take ln of each side}$$

$$-0.00495t = \ln \tfrac{2}{3} \qquad \text{Property of ln}$$

$$t = -\frac{\ln \tfrac{2}{3}}{0.00495} \qquad \text{Divide by } -0.00495$$

$$t \approx 81.9 \qquad \text{Use a calculator}$$

The time required for the sample to decay to 200 mg is about 82 days.

(d) A graph of the function $m(t) = 300e^{-0.00495t}$ is shown in Figure 4. ∎

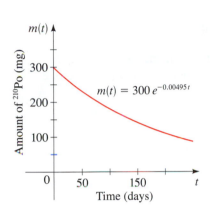

FIGURE 4

Newton's Law of Cooling

Newton's Law of Cooling states that the rate of cooling of an object is proportional to the temperature difference between the object and its surroundings, provided that the temperature difference is not too large. Using calculus, the following model can be deduced from this law.

NEWTON'S LAW OF COOLING

If D_0 is the initial temperature difference between an object and its surroundings, and if its surroundings have temperature T_s, then the temperature of the object at time t is modeled by the function

$$T(t) = T_s + D_0 e^{-kt}$$

where k is a positive constant that depends on the type of object.

EXAMPLE 7 ■ Newton's Law of Cooling

A cup of coffee has a temperature of 200°F and is placed in a room that has a temperature of 70°F. After 10 min the temperature of the coffee is 150°F.

(a) Find a function that models the temperature of the coffee at time t.
(b) Find the temperature of the coffee after 15 min.
(c) When will the coffee have cooled to 100°F?
 (d) Illustrate by drawing a graph of the temperature function.

Radioactive Waste

Harmful radioactive isotopes are produced whenever a nuclear reaction occurs, whether as the result of an atomic bomb test, a nuclear accident such as the one at Chernobyl in 1986, or the uneventful production of electricity at a nuclear power plant.

One radioactive material produced in atomic bombs is the isotope strontium-90 (^{90}Sr), with a half-life of 28 years. This is deposited like calcium in human bone tissue, where it can cause leukemia and other cancers. However, in the decades since atmospheric testing of nuclear weapons was halted, ^{90}Sr levels in the environment have fallen to a level that no longer poses a threat to health.

Nuclear power plants produce radioactive plutonium-239 (^{239}Pu), which has a half-life of 24,360 years. Because of its long half-life, ^{239}Pu could pose a threat to the environment for thousands of years. So, great care must be taken to dispose of it properly. The difficulty of ensuring the safety of the disposed radioactive waste is one reason that nuclear power plants remain controversial.

SOLUTION

(a) The temperature of the room is $T_s = 70°F$, and the initial temperature difference is

$$D_0 = 200 - 70 = 130°F$$

So, by Newton's Law of Cooling, the temperature after t minutes is modeled by the function

$$T(t) = 70 + 130e^{-kt}$$

We need to find the constant k associated with this cup of coffee. To do this, we use the fact that when $t = 10$, the temperature is $T(10) = 150$. So we have

$$70 + 130e^{-10k} = 150$$

$$130e^{-10k} = 80 \qquad \text{Subtract 70}$$

$$e^{-10k} = \tfrac{8}{13} \qquad \text{Divide by 130}$$

$$-10k = \ln \tfrac{8}{13} \qquad \text{Take ln of each side}$$

$$k = -\tfrac{1}{10} \ln \tfrac{8}{13} \qquad \text{Divide by } -10$$

$$k \approx 0.04855 \qquad \text{Use a calculator}$$

Substituting this value of k into the expression for $T(t)$, we get

$$T(t) = 70 + 130e^{-0.04855t}$$

(b) We use the function we found in part (a) with $t = 15$.

$$T(15) = 70 + 130e^{-0.04855(15)} \approx 133°F$$

(c) We use the function we found in part (a) with $T(t) = 100$ and solve the resulting exponential equation for t.

$$70 + 130e^{-0.04855t} = 100$$

$$130e^{-0.04855t} = 30 \qquad \text{Subtract 70}$$

$$e^{-0.04855t} = \tfrac{3}{13} \qquad \text{Divide by 130}$$

$$-0.04855t = \ln \tfrac{3}{13} \qquad \text{Take ln of each side}$$

$$t = \frac{\ln \tfrac{3}{13}}{-0.04855} \qquad \text{Divide by 0.04855}$$

$$t \approx 30.2 \qquad \text{Use a calculator}$$

The coffee will have cooled to 100°F after about half an hour.

(d) The graph of the temperature function is sketched in Figure 5. Notice that the line $t = 70$ is a horizontal asymptote. (Why?)

FIGURE 5

Temperature of coffee after t minutes

Logarithmic Scales

When a physical quantity varies over a very large range, it is often convenient to take its logarithm in order to have a more manageable set of numbers. We discuss three such situations: the pH scale, which measures acidity; the Richter scale, which measures the intensity of earthquakes; and the decibel scale, which measures the loudness of sounds. Other quantities that are measured on logarithmic scales are light intensity, information capacity, and radiation.

THE pH SCALE   Chemists measured the acidity of a solution by giving its hydrogen ion concentration until Sorensen, in 1909, proposed a more convenient measure. He defined

$$pH = -\log[H^+]$$

where $[H^+]$ is the concentration of hydrogen ions measured in moles per liter (M). He did this to avoid very small numbers and negative exponents. For instance,

$$\text{if}\quad [H^+] = 10^{-4}\,M,\qquad \text{then}\qquad pH = -\log_{10}(10^{-4}) = -(-4) = 4$$

Solutions with a pH of 7 are defined as *neutral*, those with pH < 7 are *acidic*, and those with pH > 7 are *basic*. Notice that when the pH increases by one unit, $[H^+]$ decreases by a factor of 10.

EXAMPLE 8 ■ pH Scale and Hydrogen Ion Concentration

(a) The hydrogen ion concentration of a sample of human blood was measured to be $[H^+] = 3.16 \times 10^{-8}$ M. Find the pH and classify the blood as acidic or basic.

(b) The most acidic rainfall ever measured occurred in Scotland in 1974; its pH was 2.4. Find the hydrogen ion concentration.

SOLUTION

(a) A calculator gives

$$pH = -\log[H^+] = -\log(3.16 \times 10^{-8}) \approx 7.5$$

Since this is greater than 7, the blood is basic.

(b) To find the hydrogen ion concentration, we need to solve for $[H^+]$ in the logarithmic equation

$$\log[H^+] = -pH$$

So, we write it in exponential form.

$$[H^+] = 10^{-pH}$$

In this case, pH = 2.4, so

$$[H^+] = 10^{-2.4} \approx 4.0 \times 10^{-3} \text{ M} \qquad \blacksquare$$

THE RICHTER SCALE   In 1935 the American geologist Charles Richter (1900–1984) defined the magnitude M of an earthquake to be

$$M = \log \frac{I}{S}$$

where I is the intensity of the earthquake (measured by the amplitude of a seismograph reading taken 100 km from the epicenter of the earthquake) and S is the intensity of a "standard" earthquake (whose amplitude is 1 micron = 10^{-4} cm). The magnitude of a standard earthquake is

$$M = \log \frac{S}{S} = \log 1 = 0$$

Richter studied many earthquakes that occurred between 1900 and 1950. The largest had magnitude 8.9 on the Richter scale, and the smallest had magnitude 0. This corresponds to a ratio of intensities of 800,000,000, so the Richter scale provides more manageable numbers to work with. For instance, an earthquake of magnitude 6 is ten times stronger than an earthquake of magnitude 5.

EXAMPLE 9 ■ Magnitude of Earthquakes

The 1906 earthquake in San Francisco had an estimated magnitude of 8.3 on the Richter scale. In the same year the strongest earthquake ever recorded occurred on changes over time. Models of facial characteristics can be programmed into a computer to give a picture of how a person's appearance changes over time. These pictures aid law enforcement agencies in locating missing persons.

the Colombia-Ecuador border and was four times as intense. What was the magnitude of the Colombia-Ecuador earthquake on the Richter scale?

SOLUTION

If I is the intensity of the San Francisco earthquake, then from the definition of magnitude we have

$$M = \log \frac{I}{S} = 8.3$$

The intensity of the Colombia-Ecuador earthquake was $4I$, so its magnitude was

$$M = \log \frac{4I}{S} = \log 4 + \log \frac{I}{S} = \log 4 + 8.3 \approx 8.9 \qquad ■$$

EXAMPLE 10 ■ Intensity of Earthquakes

The 1989 Loma Prieta earthquake that shook San Francisco had a magnitude of 7.1 on the Richter scale. How many times more intense was the 1906 earthquake (see Example 9) than the 1989 event?

SOLUTION

If I_1 and I_2 are the intensities of the 1906 and 1989 earthquakes, then we are required to find I_1/I_2. To relate this to the definition of magnitude, we divide numerator and denominator by S.

$$\log \frac{I_1}{I_2} = \log \frac{I_1/S}{I_2/S} \qquad \text{Divide numerator and denominator by } S$$

$$= \log \frac{I_1}{S} - \log \frac{I_2}{S} \qquad \text{Law 2 of Logarithms}$$

$$= 8.3 - 7.1 = 1.2 \qquad \text{Definition of earthquake magnitude}$$

Therefore

$$\frac{I_1}{I_2} = 10^{\log(I_1/I_2)} = 10^{1.2} \approx 16$$

The 1906 earthquake was about 16 times as intense as the 1989 earthquake. ■

THE DECIBEL SCALE The ear is sensitive to an extremely wide range of sound intensities. We take as a reference intensity $I_0 = 10^{-12}$ W/m² (watts per square meter) at a frequency of 1000 hertz, which measures a sound that is just barely audible (the threshold of hearing). The psychological sensation of loudness varies with the logarithm of the intensity (the Weber-Fechner Law) and so the **intensity level** β, measured in decibels (dB), is defined as

$$\beta = 10 \log \frac{I}{I_0}$$

The intensity level of the barely audible reference sound is

$$\beta = 10 \log \frac{I_0}{I_0} = 10 \log 1 = 0 \text{ dB}$$

The **intensity levels of sounds** that we can hear vary from very loud to very soft. Here are some examples of the decibel levels of commonly heard sounds.

Source of sound	β (dB)
Jet takeoff	140
Jackhammer	130
Rock concert	120
Subway	100
Heavy traffic	80
Ordinary traffic	70
Normal conversation	50
Whisper	30
Rustling leaves	10–20
Threshold of hearing	0

EXAMPLE 11 ■ Sound Intensity of a Jet Takeoff

Find the decibel intensity level of a jet engine during takeoff if the intensity was measured at 100 W/m².

SOLUTION

From the definition of intensity level we see that

$$\beta = 10 \log \frac{I}{I_0} = 10 \log \frac{10^2}{10^{-12}} = 10 \log 10^{14} = 140 \text{ dB}$$

Thus, the intensity level is 140 dB.   ■

The table in the margin lists decibel intensity levels for some common sounds ranging from the threshold of hearing to the jet takeoff of Example 11. The threshold of pain is about 120 dB.

4.5 EXERCISES

1–13 ■ These exercises use the population growth model.

1. The number of bacteria in a culture is modeled by the function

$$n(t) = 500e^{0.45t}$$

where t is measured in hours.
(a) What is the initial number of bacteria?
(b) What is the relative rate of growth of this bacterium population? Express your answer as a percentage.
(c) How many bacteria are in the culture after 3 hours?
(d) After how many hours will the number of bacteria reach 10,000?

2. The number of a certain species of fish is modeled by the function

$$n(t) = 12e^{0.012t}$$

where t is measured in years and $n(t)$ is measured in millions.
(a) What is the relative rate of growth of the fish population? Express your answer as a percentage.
(b) What will the fish population be after 5 years?
(c) After how many years will the number of fish reach 30 million?

(d) Sketch a graph of the fish population function $n(t)$.

3. The fox population in a certain region has a relative growth rate of 8% per year. It is estimated that the population in 2000 was 18,000.
(a) Find a function that models the population t years after 2000.
(b) Use the function from part (a) to estimate the fox population in the year 2008.
(c) Sketch a graph of the fox population function for the years 2000–2008.

4. The population of a country has relative growth rate of 3% per year. The government is trying to reduce the growth rate to 2%. The population in 1995 was approximately 110 million. Find the projected population for the year 2020 for the following conditions.
   (a) The relative growth rate remains at 3% per year.
   (b) The relative growth rate is reduced to 2% per year.

5. The population of a certain city was 112,000 in 1998, and the observed relative growth rate is 4% per year.
   (a) Find a function that models the population after t years.
   (b) Find the projected population in the year 2004.
   (c) In what year will the population reach 200,000?

6. The frog population in a small pond grows exponentially. The current population is 85 frogs, and the relative growth rate is 18% per year.
   (a) Find a function that models the population after t years.
   (b) Find the projected population after 3 years.
   (c) Find the number of years required for the frog population to reach 600.

7. The graph shows the deer population in a Pennsylvania county between 1996 and 2000. Assume that the population grows exponentially.
   (a) What was the deer population in 1996?
   (b) Find a function that models the deer population t years after 1996.
   (c) What is the projected deer population in 2004?
   (d) In what year will the deer population reach 100,000?

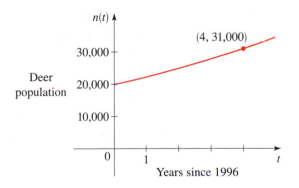

8. A culture contains 1500 bacteria initially and doubles every 30 min.
   (a) Find a function that models the number of bacteria $n(t)$ after t minutes.
   (b) Find the number of bacteria after 2 hours.
   (c) After how many minutes will the culture contain 4000 bacteria?

9. A culture starts with 8600 bacteria. After one hour the count is 10,000.
   (a) Find a function that models the number of bacteria $n(t)$ after t hours.
   (b) Find the number of bacteria after 2 hours.
   (c) After how many hours will the number of bacteria double?

10. The count in a culture of bacteria was 400 after 2 hours and 25,600 after 6 hours.
   (a) What is the relative rate of growth of the bacterium population? Express your answer as a percentage.
   (b) What was the initial size of the culture?
   (c) Find a function that models the number of bacteria $n(t)$ after t hours.
   (d) Find the number of bacteria after 4.5 hours.
   (e) When will the number of bacteria be 50,000?

11. The population of the world was 5.7 billion in 1995 and the observed relative growth rate was 2% per year.
   (a) By what year will the population have doubled?
   (b) By what year will the population have tripled?

12. The population of California was 10,586,223 in 1950 and 23,668,562 in 1980. Assume the population grows exponentially.
   (a) Find a function that models the population t years after 1950.
   (b) Find the time required for the population to double.
   (c) Use the function from part (a) to predict the population of California in the year 2000. Look up the actual California population in 2000, and compare.

13. An infectious strain of bacteria increases in number at a relative growth rate of 200% per hour. When a certain critical number of bacteria are present in the bloodstream, a person becomes ill. If a single bacterium infects a person, the critical level is reached in 24 hours. How long will it take for the critical level to be reached if the same person is infected with 10 bacteria?

14–22 ■ These exercises use the radioactive decay model.

14. The half-life of radium-226 is 1600 years. Suppose we have a 22-mg sample.
   (a) Find a function that models the mass remaining after t years.
   (b) How much of the sample will remain after 4000 years?
   (c) After how long will only 18 mg of the sample remain?

15. The half-life of cesium-137 is 30 years. Suppose we have a 10-g sample.
(a) Find a function that models the mass remaining after t years.
(b) How much of the sample will remain after 80 years?
(c) After how long will only 2 g of the sample remain?

16. The mass $m(t)$ remaining after t days from a 40-g sample of thorium-234 is given by

$$m(t) = 40e^{-0.0277t}$$

(a) How much of the sample will remain after 60 days?
(b) After how long will only 10 g of the sample remain?
(c) Find the half-life of thorium-234.

17. The half-life of strontium-90 is 28 years. How long will it take a 50-mg sample to decay to a mass of 32 mg?

18. Radium-221 has a half-life of 30 s. How long will it take for 95% of a sample to decay?

19. If 250 mg of a radioactive element decays to 200 mg in 48 hours, find the half-life of the element.

20. After 3 days a sample of radon-222 has decayed to 58% of its original amount.
(a) What is the half-life of radon-222?
(b) How long will it take the sample to decay to 20% of its original amount?

21. A wooden artifact from an ancient tomb contains 65% of the carbon-14 that is present in living trees. How long ago was the artifact made? (The half-life of carbon-14 is 5730 years.)

22. The burial cloth of an Egyptian mummy is estimated to contain 59% of the carbon-14 it contained originally. How long ago was the mummy buried? (The half-life of carbon-14 is 5730 years.)

23–26 ■ These exercises use Newton's Law of Cooling.

23. A hot bowl of soup is served at a dinner party. It starts to cool according to Newton's Law of Cooling so that its temperature at time t is given by

$$T(t) = 65 + 145e^{-0.05t}$$

where t is measured in minutes and T is measured in °F.
(a) What is the initial temperature of the soup?
(b) What is the temperature after 10 min?
(c) After how long will the temperature be 100°F?

24. Newton's Law of Cooling is used in homicide investigations to determine the time of death. The normal body temperature is 98.6°F. Immediately following death, the body begins to cool. It has been determined experimentally that the constant in Newton's Law of Cooling is approximately $k = 0.1947$, assuming time is measured in hours. Suppose that the temperature of the surroundings is 60°F.
(a) Find a function $T(t)$ that models the temperature t hours after death.
(b) If the temperature of the body is now 72°F, how long ago was the time of death?

25. A roasted turkey is taken from an oven when its temperature has reached 185°F and is placed on a table in a room where the temperature is 75°F.
(a) If the temperature of the turkey is 150°F after half an hour, what is its temperature after 45 min?
(b) When will the turkey cool to 100°F?

26. A kettle full of water is brought to a boil in a room with temperature 20°C. After 15 min the temperature of the water has decreased from 100°C to 75°C. Find the temperature after another 10 min. Illustrate by sketching a graph of the temperature function.

27–41 ■ These exercises are about logarithmic scales.

27. The hydrogen ion concentration of a sample of each substance is given. Calculate the pH of the substance.
(a) Lemon juice: $[H^+] = 5.0 \times 10^{-3}$ M
(b) Tomato juice: $[H^+] = 3.2 \times 10^{-4}$ M
(c) Seawater: $[H^+] = 5.0 \times 10^{-9}$ M

28. An unknown substance has a hydrogen ion concentration of $[H^+] = 3.1 \times 10^{-8}$ M. Find the pH and classify the substance as acidic or basic.

29. The pH reading of a sample of each substance is given. Calculate the hydrogen ion concentration of the substance.
(a) Vinegar: pH = 3.0
(b) Milk: pH = 6.5

30. The pH reading of a glass of liquid is given. Find the hydrogen ion concentration of the liquid.
(a) Beer: pH = 4.6
(b) Water: pH = 7.3

31. The hydrogen ion concentrations in cheeses range from 4.0×10^{-7} M to 1.6×10^{-5} M. Find the corresponding range of pH readings.

32. The pH readings for wines vary from 2.8 to 3.8. Find the corresponding range of hydrogen ion concentrations.

33. If one earthquake is 20 times as intense as another, how much larger is its magnitude on the Richter scale?

34. The 1906 earthquake in San Francisco had a magnitude of 8.3 on the Richter scale. At the same time in Japan an earthquake with magnitude 4.9 caused only minor damage. How many times more intense was the San Francisco earthquake than the Japanese earthquake?

35. The Alaska earthquake of 1964 had a magnitude of 8.6 on the Richter scale. How many times more intense was this than the 1906 San Francisco earthquake? (See Exercise 34.)

36. The Northridge, California, earthquake of 1994 had a magnitude of 6.8 on the Richter scale. A year later, a 7.2-magnitude earthquake struck Kobe, Japan. How many times more intense was the Kobe earthquake than the Northridge earthquake?

37. The 1985 Mexico City earthquake had a magnitude of 8.1 on the Richter scale. The 1976 earthquake in Tangshan, China, was 1.26 times as intense. What was the magnitude of the Tangshan earthquake?

38. The intensity of the sound of traffic at a busy intersection was measured at 2.0×10^{-5} W/m^2. Find the intensity level in decibels.

39. The intensity level of the sound of a subway train was measured at 98 dB. Find the intensity in W/m^2.

40. The noise from a power mower was measured at 106 dB. The noise level at a rock concert was measured at 120 dB. Find the ratio of the intensity of the rock music to that of the power mower.

41. A law of physics states that the intensity of sound is inversely proportional to the square of the distance d from the source:

$$I = \frac{k}{d^2}$$

(a) Use this model and the equation

$$\beta = 10 \log \frac{I}{I_0}$$

(described in this section) to show that the decibel levels β_1 and β_2 at distances d_1 and d_2 from a sound source are related by the equation

$$\beta_2 = \beta_1 + 20 \log \frac{d_1}{d_2}$$

(b) The intensity level at a rock concert is 120 dB at a distance 2 m from the speakers. Find the intensity level at a distance of 10 m.

4   REVIEW

CONCEPT CHECK

1. (a) Write an equation that defines the exponential function with base a.
   (b) What is the domain of this function?
   (c) If $a \neq 1$, what is the range of this function?
   (d) Sketch the general shape of the graph of the exponential function for each of the following cases.
      (i) $a > 1$ (ii) $a = 1$ (iii) $a < 1$

2. If x is large, which function grows faster, $y = 2^x$ or $y = x^2$?

3. (a) How is the number e defined?
   (b) What is the natural exponential function?

4. (a) How is the logarithmic function $y = \log_a x$ defined?
   (b) What is the domain of this function?
   (c) What is the range of this function?

(d) Sketch the general shape of the graph of the function $y = \log_a x$ if $a > 1$.

(e) What is the natural logarithm?

(f) What is the common logarithm?

5. State the three Laws of Logarithms.

6. State the Change of Base Formula.

7. (a) How do you solve an exponential equation?

(b) How do you solve a logarithmic equation?

8. Suppose an amount P is invested at an interest rate r and A is the amount after t years.

(a) Write an expression for A if the interest is compounded n times per year.

(b) Write an expression for A if the interest is compounded continuously.

9. If the initial size of a population is n_0 and the population grows exponentially with relative growth rate r, write an expression for the population $n(t)$ at time t.

10. (a) What is the half-life of a radioactive substance?

(b) If a radioactive substance has initial mass m_0 and half-life h, write an expression for the mass $m(t)$ remaining at time t.

11. What does Newton's Law of Cooling say?

12. What do the pH scale, the Richter scale, and the decibel scale have in common? What do they measure?

EXERCISES

1–12 ▪ Sketch the graph of the function. State the domain, range, and asymptote.

1. $f(x) = 3^{-x}$

2. $g(x) = 2^{x-1}$

3. $y = 5 - 10^x$

4. $y = 1 + 5^{-x}$

5. $f(x) = \log_3(x - 1)$

6. $g(x) = \log(-x)$

7. $y = 2 - \log_2 x$

8. $y = 3 + \log_5(x + 4)$

9. $F(x) = e^x - 1$

10. $G(x) = \frac{1}{2} e^{x-1}$

11. $y = 2 \ln x$

12. $y = \ln(x^2)$

13–16 ▪ Find the domain of the function.

13. $f(x) = 10^{x^2} + \log(1 - 2x)$

14. $g(x) = \ln(2 + x - x^2)$

15. $h(x) = \ln(x^2 - 4)$

16. $k(x) = \ln |x|$

17–20 ▪ Write the equation in exponential form.

17. $\log_2 1024 = 10$

18. $\log_6 37 = x$

19. $\log x = y$

20. $\ln c = 17$

21–24 ▪ Write the equation in logarithmic form.

21. $2^6 = 64$

22. $49^{-1/2} = \frac{1}{7}$

23. $10^x = 74$

24. $e^k = m$

25–40 ▪ Evaluate the expression without using a calculator.

25. $\log_2 128$

26. $\log_8 1$

27. $10^{\log 45}$

28. $\log 0.000001$

29. $\ln(e^6)$

30. $\log_4 8$

31. $\log_3\left(\frac{1}{27}\right)$

32. $2^{\log_2 13}$

33. $\log_5 \sqrt{5}$

34. $e^{2\ln 7}$

35. $\log 25 + \log 4$

36. $\log_3 \sqrt{243}$

37. $\log_2 16^{23}$

38. $\log_5 250 - \log_5 2$

39. $\log_8 6 - \log_8 3 + \log_8 2$

40. $\log \log 10^{100}$

41–46 ▪ Rewrite the expression in a form with no logarithms of products, quotients, or powers.

41. $\log(AB^2C^3)$

42. $\log_2\left(x\sqrt{x^2 + 1}\right)$

43. $\ln \sqrt{\dfrac{x^2 - 1}{x^2 + 1}}$

44. $\log\left(\dfrac{4x^3}{y^2(x - 1)^5}\right)$

45. $\log_5\left(\dfrac{x^2(1 - 5x)^{3/2}}{\sqrt{x^3 - x}}\right)$

46. $\ln\left(\dfrac{\sqrt[3]{x^4 + 12}}{(x + 16)\sqrt{x - 3}}\right)$

47–52 ▪ Rewrite the expression as a single logarithm.

47. $\log 6 + 4 \log 2$

48. $\log x + \log(x^2 y) + 3 \log y$

49. $\frac{3}{2} \log_2(x - y) - 2 \log_2(x^2 + y^2)$

50. $\log_5 2 + \log_5(x + 1) - \frac{1}{3} \log_5(3x + 7)$

51. $\log(x - 2) + \log(x + 2) - \frac{1}{2} \log(x^2 + 4)$

52. $\frac{1}{2}[\ln(x - 4) + 5 \ln(x^2 + 4x)]$

53–62 ■ Use a calculator to find the solution of the equation, correct to two decimal places.

53. $\log_2(1 - x) = 4$

54. $2^{3x-5} = 7$

55. $5^{5-3x} = 26$

56. $\ln(2x - 3) = 14$

57. $e^{3x/4} = 10$

58. $2^{1-x} = 3^{2x+5}$

59. $\log x + \log(x + 1) = \log 12$

60. $\log_8(x + 5) - \log_8(x - 2) = 1$

61. $x^2 e^{2x} + 2x e^{2x} = 8 e^{2x}$

62. $2^{3^x} = 5$

63–66 ■ Use a calculator to find the solution of the equation, correct to six decimal places.

63. $5^{-2x/3} = 0.63$

64. $2^{3x-5} = 7$

65. $5^{2x+1} = 3^{4x-1}$

66. $e^{-15k} = 10,000$

 67–70 ■ Draw a graph of the function and use it to determine the asymptotes and the local maximum and minimum values.

67. $y = e^{x/(x+2)}$

68. $y = 2x^2 - \ln x$

69. $y = \log(x^3 - x)$

70. $y = 10^x - 5^x$

71–72 ■ Find the solutions of the equation, correct to two decimal places.

71. $3 \log x = 6 - 2x$

72. $4 - x^2 = e^{-2x}$

73–74 ■ Solve the inequality graphically.

73. $\ln x > x - 2$

74. $e^x < 4x^2$

75. Use a graph of $f(x) = e^x - 3e^{-x} - 4x$ to find, approximately, the intervals on which f is increasing and on which f is decreasing.

76. Find an equation of the line shown in the figure.

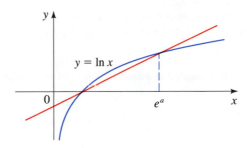

77. Evaluate $\log_4 15$, correct to six decimal places.

78. Solve the inequality: $0.2 \le \log x < 2$

79. Which is larger, $\log_4 258$ or $\log_5 620$?

80. Find the inverse of the function $f(x) = 2^{3^x}$ and state its domain and range.

81. If $12,000 is invested at an interest rate of 10% per year, find the amount of the investment at the end of 3 years for each compounding method.
(a) Semiannual   (b) Monthly
(c) Daily   (d) Continuous

82. A sum of $5000 is invested at an interest rate of $8\frac{1}{2}$% per year, compounded semiannually.
(a) Find the amount of the investment after $1\frac{1}{2}$ years.
(b) After what period of time will the investment amount to $7000?

83. The stray-cat population in a small town grows exponentially. In 1999, the town had 30 stray cats and the relative growth rate was 15% per year.
(a) Find a function that models the stray-cat population $n(t)$ after t years.
(b) Find the projected population after 4 years.
(c) Find the number of years required for the stray-cat population to reach 500.

84. A culture contains 10,000 bacteria initially. After an hour the bacteria count is 25,000.
(a) Find the doubling period.
(b) Find the number of bacteria after 3 hours.

85. Uranium-234 has a half-life of 2.7×10^5 years.
(a) Find the amount remaining from a 10-mg sample after a thousand years.
(b) How long will it take this sample to decompose until its mass is 7 mg?

86. A sample of bismuth-210 decayed to 33% of its original mass after 8 days.
  (a) Find the half-life of this element.
  (b) Find the mass remaining after 12 days.

87. The half-life of radium-226 is 1590 years.
  (a) If a sample has a mass of 150 mg, find a function that models the mass that remains after t years.
  (b) Find the mass that will remain after 1000 years.
  (c) After how many years will only 50 mg remain?

88. The half-life of palladium-100 is 4 days. After 20 days a sample has been reduced to a mass of 0.375 g.
  (a) What was the initial mass of the sample?
  (b) Find a function that models the mass remaining after t days.
  (c) What is the mass after 3 days?
  (d) After how many days will only 0.15 g remain?

89. The graph shows the population of a rare species of bird, where t represents years since 1994 and $n(t)$ is measured in thousands.
  (a) Find a function that models the bird population at time t in the form $n(t) = n_0 e^{rt}$.
  (b) What is the bird population expected to be in the year 2005?

90. A car engine runs at a temperature of 190°F. When the engine is turned off, it cools according to Newton's Law of Cooling with constant $k = 0.0341$, where the time is measured in minutes. Find the time needed for the engine to cool to 90°F if the surrounding temperature is 60°F.

91. The hydrogen ion concentration of fresh egg whites was measured as

$$[\text{H}^+] = 1.3 \times 10^{-8} \text{ M}$$

Find the pH, and classify the substance as acidic or basic.

92. The pH of lime juice is 1.9. Find the hydrogen ion concentration.

93. If one earthquake has magnitude 6.5 on the Richter scale, what is the magnitude of another quake that is 35 times as intense?

94. The drilling of a jackhammer was measured at 132 dB. The sound of whispering was measured at 28 dB. Find the ratio of the intensity of the drilling to that of the whispering.

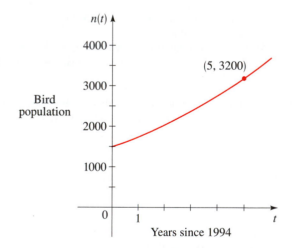

Bird population

(5, 3200)

Years since 1994

4 TEST

1. Graph the functions $y = 2^x$ and $y = \log_2 x$ on the same axes.

2. Sketch the graph of the function $f(x) = \log(x + 2)$ and state the domain, range, and asymptote.

3. Evaluate each logarithmic expression.
   (a) $\log_5 \sqrt{125}$
   (b) $\log_2 56 - \log_2 7$
   (c) $\log_8 4$
   (d) $\log_6 4 + \log_6 9$

4. Use the Laws of Logarithms to rewrite the expression without logarithms of products, quotients, powers, or roots.

   $$\log \sqrt{\frac{x + 2}{x^4(x^2 + 4)}}$$

5. Write as a single logarithm: $\ln x - 2 \ln(x^2 + 1) + \frac{1}{2} \ln(3 - x^4)$

6. Find the solution of the equation, correct to two decimal places.
   (a) $2^{x-1} = 10$
   (b) $5 \ln(3 - x) = 4$
   (c) $10^{x+3} = 6^{2x}$
   (d) $\log_2(x + 2) + \log_2(x - 1) = 2$

7. The initial size of a culture of bacteria is 1000. After one hour the bacterium count is 8000.
   (a) Find a function that models the population after t hours.
   (b) Find the population after 1.5 hours.
   (c) When will the population reach 15,000?
   (d) Sketch the graph of the population function.

8. Suppose that $12,000 is invested in a savings account paying 5.6% interest per year.
   (a) Write the formula for the amount in the account after t years if interest is compounded monthly.
   (b) Find the amount in the account after 3 years if interest is compounded daily.
   (c) How long will it take for the amount in the account to grow to $20,000 if interest is compounded semiannually?

  9. Let $f(x) = \dfrac{e^x}{x^3}$.

   (a) Graph f in an appropriate viewing rectangle.
   (b) State the asymptotes of f.
   (c) Find, correct to two decimal places, the local minimum value of f and the value of x at which it occurs.
   (d) Find the range of f.
   (e) Solve the equation $\dfrac{e^x}{x^3} = 2x + 1$. State each solution correct to two decimal places.

Focus on Modeling
Fitting Exponential and Power Functions to Data

In *Focus on Modeling* (pages 246–255), we learned how to construct linear models from data. Figure 1 shows some scatter plots of data; the first plot appears to be linear but the others are not. What do we do when the data we are studying are not linear? In this case, our model would be some other type of function that best fits the data. The type of function we choose is determined by the shape of the scatter plot or by some physical principle that underlies the data. In this *Focus* we learn how to construct exponential, power, and polynomial models. Most graphing calculators are capable of applying the method of least squares (see page 248) to obtain the exponential, power, or polynomial function that best fits the data.

We use the symbols for the various regression commands as they are given on the TI-83. On other calculators, the commands have slightly different names. Consult your calculator manual.

FIGURE 1

Modeling with Exponential Functions

To model data, we first draw a scatter plot. The shape of the scatter plot can help us decide on the type of function we need to model the data. If the scatter plot increases rapidly, we might seek an *exponential model*, that is, a function of the form

$$y = Ce^{kx}$$

where C and k are constants.

In the first example we model world population. Recall from Section 4.5 that population tends to increase exponentially.

EXAMPLE 1 ■ An Exponential Model for World Population

Table 1 gives the population of the world in the 20th century.
(a) Draw a scatter plot and note that the population grows too rapidly for a linear model to be appropriate.
(b) Find an exponential function that models population growth.
(c) Draw a graph of the function you found together with the scatter plot. How well does the model fit the data?

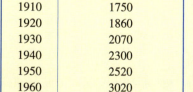

TABLE 1 World population

Year (t)	World population (P, in millions)
1900	1650
1910	1750
1920	1860
1930	2070
1940	2300
1950	2520
1960	3020
1970	3700
1980	4450
1990	5300
2000	6060

(d) Use the model you found to predict world population in the year 2020.

SOLUTION

(a) The scatter plot is shown in Figure 2. The plotted points do not appear to lie along a straight line, so a linear model is not appropriate.

FIGURE 2
Scatter plot of world population

(b) Using a graphing calculator and the **ExpReg** command, we get the exponential model

$$P(t) = (0.0082543) \cdot (1.0137186)^t$$

```
ExpReg
 y=a*b^x
 a=.0082543035
 b=1.013718645
```

This is a model of the form $y = Cb^t$. To convert this to the form $y = Ce^{kt}$, we need to find a value of k so that $b = e^k$, that is,

$$1.0137186 = e^k$$

$$k = \ln 1.0137186 \approx 0.013625 \qquad \text{Solve for } k$$

So $\qquad P(t) = 0.0082543e^{0.013625t}$

(c) From the graph in Figure 3, we see that the model appears to fit the data fairly well. The period of relatively slow population growth is explained by the depression of the 1930s and the two world wars.

FIGURE 3

Exponential model
for world population

(d) The model predicts that the world population in 2020 will be

$$P(2020) = 0.0082543e^{(0.013625)(2020)}$$

$$\approx 7{,}405{,}400{,}000 \qquad \blacksquare$$

Modeling with Power Functions

If the scatter plot of the data we are studying resembles the graph of $y = ax^2$, $y = ax^{1.32}$, or some other power function, then we seek a *power model*, that is, a function of the form

$$y = ax^n$$

where a is a positive constant and n is any real number.

In the next example we seek a power model for some astronomical data. In astronomy, distance in the solar system is often measured in astronomical units. An *astronomical unit* (AU) is the mean distance from the earth to the sun. The *period* of a planet is the time it takes the planet to make a complete revolution around the sun (measured in earth years). In this example we derive the remarkable relationship, first discovered by Johannes Kepler (see page 739), between the mean distance of a planet from the sun and its period.

TABLE 2 Distances and
periods of the planets

Planet	d	T
Mercury	0.387	0.241
Venus	0.723	0.615
Earth	1.000	1.000
Mars	1.523	1.881
Jupiter	5.203	11.861
Saturn	9.541	29.457
Uranus	19.190	84.008
Neptune	30.086	164.784
Pluto	39.507	248.350

EXAMPLE 2 ■ A Power Model for Planetary Periods

Table 2 gives the mean distance d of each planet from the sun in astronomical units and its period T in years.

(a) Sketch a scatter plot. Is a linear model appropriate?

(b) Find a power function that models the data.

(c) Draw a graph of the function you found and the scatter plot on the same graph. How well does the model fit the data?

(d) Use the model you found to find the period of an asteroid whose mean distance from the sun is 5 AU.

SOLUTION

(a) The scatter plot shown in Figure 4 indicates that the plotted points do not lie along a straight line, so a linear model is not appropriate.

FIGURE 4

(b) Using a graphing calculator and the **PwrReg** command, we get the power model

$$T = 1.000396d^{1.49966}$$

```
PwrReg
 y=a*x^b
 a=1.000396048
 b=1.499661718
```

If we round both the coefficient and the exponent to three significant figures, we can write the model as

$$T = d^{1.5}$$

This is the relationship discovered by Kepler. Sir Isaac Newton (see page 854) later used his Law of Gravity to derive this relationship theoretically, thereby providing strong scientific evidence that the Law of Gravity must be true.

(c) The graph is shown in Figure 5. The model appears to fit the data very well.

FIGURE 5

(d) In this case, $d = 5$ AU and so our model gives

$$T = 1.00039 \cdot 5^{1.49966} \approx 11.22$$

The period of the asteroid is about 11.2 years.   ■

Modeling with Polynomial Functions

Polynomial functions can also be used to model data. A *polynomial model* is a polynomial function of a given degree. For example, a polynomial model of degree 3 is a function of the form

$$y = ax^3 + bx^2 + cx + d$$

where a, b, c, and d, are constants. Most graphing calculators use the method of least squares to fit a polynomial of specified degree to a given set of data.

EXAMPLE 3 ■ A Polynomial Model

Consider the world population data in Table 1 on page 394.

(a) Find a polynomial function of degree 3 that models the data.
(b) Draw a graph of the function you found and the scatter plot on the same graph. How well does this model fit the data?
(c) Use the model you found to estimate the population in the year 1975.
(d) Would it be appropriate to use the model to predict the population in 2100?

SOLUTION

(a) Using the `CubicReg` command on a graphing calculator, we obtain the cubic model

$$P(t) = at^3 + bt^2 + ct + d$$

where

$$a = 1252.914 \qquad\qquad b = -6.817424 \times 10^6$$

$$c = 1.233675 \times 10^{10} \qquad\qquad d = -7.420959 \times 10^{12}$$

(b) Figure 6 shows that the cubic function models the world population of the 20th century very well. In fact, this model seems even better than the exponential model of Example 1.

FIGURE 6

Cubic model for world population

(c) For 1975, the population estimate provided by the model is

$$P(1975) \approx 4{,}024{,}400{,}000$$

This is very close to the official U.N. estimate for 1975.

(d) The cubic model is not likely to be accurate for dates outside the 20th century, because we know that the physical mechanism of population growth is inherently exponential. For instance, the model gives the following estimate for 1650:

$$P(1650) \approx 2{,}488{,}000{,}000$$

But historical records indicate that the world's population in 1650 was only about 550,000,000. In fact, this model gives negative populations for dates earlier than about 1550. [Try evaluating $P(1500)$.] So the model's prediction of $P(2100) \approx 24{,}600{,}000{,}000$ is extremely unlikely to be anywhere near the correct figure. ■

Linearizing Data

We have used the shape of a scatter plot to decide which type of model to use—linear, exponential, or power. This works well if the data points lie on a straight line. But it's difficult to distinguish a scatter plot that is exponential from one that requires a power model. So, to help decide which model to use, we can *linearize* the data, that is, apply a function to the data that results in "straightening" the scatter plot. The inverse of the linearizing function is then an appropriate model. We now describe how to linearize data that can be modeled by exponential or power functions.

■ Linearizing exponential data.

If we suspect that the data points (x, y) lie on an exponential curve $y = Ce^{kx}$, then the points

$$(x, \ln y)$$

should lie on a straight line. We can see this from the following calculations:

$$\ln y = \ln Ce^{kx} \qquad \text{Assume } y = Ce^{kx}$$
$$= \ln e^{kx} + \ln C \qquad \text{Property of ln}$$
$$= kx + \ln C \qquad \text{Property of ln}$$

To see that $\ln y$ is a linear function of x, let $Y = \ln y$ and $A = \ln C$; then

$$Y = kx + A$$

We apply this technique to the world population data (t, P) to obtain the points $(t, \ln P)$ in Table 3. The scatter plot in Figure 7 shows that the linearized data lie approximately on a straight line, so an exponential model should be appropriate.

TABLE 3 World population data

t	Population P (in millions)	$\ln P$
1900	1650	21.224
1910	1750	21.283
1920	1860	21.344
1930	2070	21.451
1940	2300	21.556
1950	2520	21.648
1960	3020	21.829
1970	3700	22.032
1980	4450	22.216
1990	5300	22.391
2000	6060	22.525

FIGURE 7

■ Linearizing power data.

If we suspect that the data points (x, y) lie on a power curve $y = ax^n$, then the points

$$(\ln x, \ln y)$$

should be on a straight line. We can see this from the following calculations:

$$\ln y = \ln ax^n \qquad \text{Assume } y = ax^n$$
$$= \ln a + \ln x^n \qquad \text{Property of ln}$$
$$= \ln a + n \ln x \qquad \text{Property of ln}$$

To see that $\ln y$ is a linear function of $\ln x$, let $Y = \ln y$, $X = \ln x$, and $A = \ln a$; then

$$Y = nX + A$$

We apply this technique to the planetary data (d, T) in Table 2, to obtain the points $(\ln d, \ln T)$ in Table 4. The scatter plot in Figure 8 shows that the data lie on a straight line, so a power model seems appropriate.

TABLE 4 Log-log table

ln d	ln T
-0.94933	-1.4230
-0.32435	-0.48613
0	0
0.42068	0.6318
1.6492	2.4733
2.2556	3.3829
2.9544	4.4309
3.4041	5.1046
3.6765	5.5148

FIGURE 8
Log-log plot of data in Table 4

EXAMPLE 4 ■ An Exponential or Power Model?

Data points (x, y) are shown in Table 5.

(a) Draw a scatter plot of the data.
(b) Draw scatter plots of $(x, \ln y)$ and $(\ln x, \ln y)$.
(c) Is an exponential function or a power function appropriate for modeling this data?
(d) Find an appropriate function to model the data.

SOLUTION

(a) The scatter plot of the data is shown in Figure 9.

TABLE 5

x	y
1	2
2	6
3	14
4	22
5	34
6	46
7	64
8	80
9	102
10	130

FIGURE 9

(b) We use the values from Table 6 to graph the scatter plots in Figures 10 and 11.

TABLE 6

x	$\ln x$	$\ln y$
1	0	0.7
2	0.7	1.8
3	1.1	2.6
4	1.4	3.1
5	1.6	3.5
6	1.8	3.8
7	1.9	4.2
8	2.1	4.4
9	2.2	4.6
10	2.3	4.9

FIGURE 10   FIGURE 11

(c) The scatter plot of $(x, \ln y)$ in Figure 10 does not appear linear, so an exponential model is not appropriate. On the other hand, the scatter plot of $(\ln x, \ln y)$ in Figure 11 is very nearly linear, so a power model is appropriate.

(d) Using the `PwrReg` command on a calculator, we find that the power function that best fits the data point is

$$y = 1.85x^{1.82}$$

The graph of this function and the original data points are shown in Figure 12.

FIGURE 12

Before graphing calculators and statistical software became common, exponential and power models for data were often constructed by first finding a linear model for the linearized data. Then the model for the actual data was found by taking exponentials. For instance, if it was found that $\ln y = A \ln x + B$, then by taking exponentials we get the model $y = e^B \cdot e^{A \ln x}$, or $y = Cx^A$ (where $C = e^B$).

Special graphing paper called "log paper" or "log-log paper" was used to facilitate this process.

We can also construct models from data using other types of functions. In Problems 9 and 10, for instance, we use logarithmic and logistic functions to model data.

■ Problems

1. The U.S. Constitution requires a census every 10 years. The census data for 1790–1990 is given in the table.
   (a) Make a scatter plot of the data.
   (b) Use a calculator to find an exponential model for the data.
   (c) Use your model to predict the population at the 2000 census.
   (d) Use your model to estimate the population in 1965.
   (e) Compare your answers from parts (c) and (d) to the values in the table. Do you think an exponential model is appropriate for these data?

Year	Population (in millions)	Year	Population (in millions)	Year	Population (in millions)
1790	3.9	1860	31.4	1930	123.2
1800	5.3	1870	38.6	1940	132.2
1810	7.2	1880	50.2	1950	151.3
1820	9.6	1890	63.0	1960	179.3
1830	12.9	1900	76.2	1970	203.3
1840	17.1	1910	92.2	1980	226.5
1850	23.2	1920	106.0	1990	248.7

2. In a physics experiment a lead ball is dropped from a height of 5 m. The students record the distance the ball has fallen every one-tenth of a second. (This can be done using a camera and a strobe light.)
   (a) Make a scatter plot of the data.
   (b) Use a calculator to find a power model.
   (c) Use the model you found to predict how far a dropped ball would fall in 3 s.

Time (s)	Distance (m)
0.1	0.048
0.2	0.197
0.3	0.441
0.4	0.882
0.5	1.227
0.6	1.765
0.7	2.401
0.8	3.136
0.9	3.969
1.0	4.902

3. The U.S. health-care expenditures for 1960–1993 are given in the table, and a scatter plot of the data is shown in the figure.

(a) Does the scatter plot shown suggest an exponential model?

(b) Make a table of the values $(t, \ln E)$ and a scatter plot. Does the scatter plot appear to be linear?

(c) Find the regression line for the data in part (b).

(d) Use the results of part (b) to find an exponential model for the growth of health-care expenditures.

(e) Use the model you found in part (d) to predict the total health-care expenditures in 1996.

Year	Health expenditures (in billions of dollars)
1960	27.1
1970	74.3
1980	251.1
1985	434.5
1987	506.2
1989	623.9
1990	696.6
1991	755.8
1992	820.3
1993	884.2

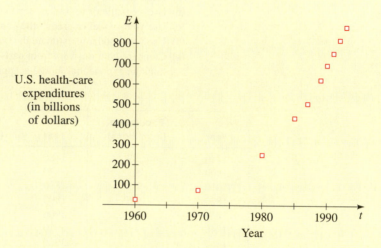

4. A student is trying to determine the half-life of radioactive iodine-131. He measures the amount of iodine-131 in a sample solution every 8 hours. His data are shown in the table.

(a) Make a scatter plot of the data.

(b) Use a calculator to find an exponential model.

(c) Use your model to find the half-life of iodine-131.

Time (h)	Amount (g)
0	4.80
8	4.66
16	4.51
24	4.39
32	4.29
40	4.14
48	4.04

Year	Lead emissions
1970	199.1
1975	143.8
1980	68.0
1985	18.3
1988	5.9
1989	5.5
1990	5.1
1991	4.5
1992	4.7

5. The table gives U.S. lead emissions into the environment in millions of metric tons for 1970–1992.
   (a) Find an exponential model for these data.
   (b) Find a fourth-degree polynomial model for these data.
   (c) Which of these curves gives a better model for the data? Use graphs of the two models to decide.
   (d) Use each model to estimate the lead emissions in 1972 and 1982.

6. A study of the U.S. Office of Science and Technology in 1972 estimated the cost of reducing automobile emissions by certain percentages. Find a model that captures the "diminishing returns" trend of these data.

Reduction in emissions (%)	Cost per car ($)
50	45
55	55
60	62
65	70
70	80
75	90
80	100
85	200
90	375
95	600

7. Data points (x, y) are shown in the table.
   (a) Draw a scatter plot of the data.
   (b) Draw scatter plots of $(x, \ln y)$ and $(\ln x, \ln y)$.
   (c) Is an exponential function or a power function appropriate for modeling this data?
   (d) Find an appropriate function to model the data.

x	y
2	0.08
4	0.12
6	0.18
8	0.25
10	0.36
12	0.52
14	0.73
16	1.06

8. Data points (x, y) are shown in the table.
   (a) Draw a scatter plot of the data.
   (b) Draw scatter plots of $(x, \ln y)$ and $(\ln x, \ln y)$.
   (c) Is an exponential function or a power function appropriate for modeling this data?
   (d) Find an appropriate function to model the data.

x	y
10	29
20	82
30	151
40	235
50	330
60	430
70	546
80	669
90	797

9. A **logarithmic model** is a function of the form

$$y = a + b \ln x$$

Many relationships between variables in the real world can be modeled by this type of function. The table and the scatter plot show the coal production (in metric tons) from a small mine in northern British Columbia.
   (a) Use the **LnReg** command on your calculator to find a logarithmic model for these production figures.
   (b) Use the model to predict coal production from this mine in 2005.

Year	Metric tons of coal
1950	882
1960	889
1970	894
1980	899
1990	905
2000	909

10. A **logistic model** is a function of the form

$$y = \frac{c}{1 + ae^{-bx}}$$

Logistic functions are appropriate for modeling population growth where the growth is limited by available resources. The table and scatter plot give the population of black flies in a closed laboratory container over an 18-day period.

(a) Use the `Logistic` command on your calculator to find a logistic model for these data.

(b) Use the model to estimate the time when there were 400 flies in the container.

Time (days)	Number of flies
0	10
2	25
4	66
6	144
8	262
10	374
12	446
16	492
18	498

Number of flies

Days

5

Trigonometric Functions of Real Numbers

Trigonometric functions are used to describe periodic phenomena such as the ebb and flow of tides, or the vibrations of a violin string.

Trigonometry contains the science of continually
undulating magnitude . . .

AUGUSTUS DE MORGAN

In this chapter and the next, we introduce two different but equivalent ways of viewing the trigonometric functions. One way is to view them as *functions of real numbers* (Chapter 5), the other as *functions of angles* (Chapter 6). The two approaches to trigonometry are independent of each other, so either Chapter 5 or Chapter 6 may be studied first.

The trigonometric functions defined in these two different ways are identical—they assign the same value to a given real number. In the first case, the real number is the length of an arc along the unit circle; in the second case, it is the measure of an angle. We study both approaches because different applications require that we view these functions differently. One approach (Chapter 5) lends itself to dynamic applications such as the modeling of harmonic motion, whereas the other approach (Chapter 6) lends itself to static applications such as the measurement of distance using triangles. The power and versatility of trigonometry stems from the fact that it can be viewed in these different ways. In modern times, trigonometry has found application in such diverse fields as signal processing, coding of music on compact discs, designing guidance systems for the space shuttle, finding distances to stars, producing CAT scans for medical imaging, and many others.

5.1 THE UNIT CIRCLE

In this section we explore some properties of the circle of radius 1 centered at the origin. These properties are used in the next section to define the trigonometric functions.

The Unit Circle

The set of points at a distance 1 from the origin is a circle of radius 1 (see Figure 1). In Section 1.8 we learned that the equation of this circle is $x^2 + y^2 = 1$.

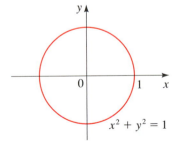

FIGURE 1
The unit circle

> **THE UNIT CIRCLE**
>
> The **unit circle** is the circle of radius 1 centered at the origin in the xy-plane. Its equation is
>
> $$x^2 + y^2 = 1$$

EXAMPLE 1 ■ A Point on the Unit Circle

Show that the point $P\left(\dfrac{\sqrt{3}}{3}, \dfrac{\sqrt{2}}{\sqrt{3}}\right)$ is on the unit circle.

SOLUTION

We need to show that this point satisfies the equation of the unit circle, that is, $x^2 + y^2 = 1$. Since

$$\left(\frac{\sqrt{3}}{3}\right)^2 + \left(\frac{\sqrt{2}}{\sqrt{3}}\right)^2 = \frac{3}{9} + \frac{2}{3} = \frac{1}{3} + \frac{2}{3} = 1$$

P is on the unit circle.   ■

EXAMPLE 2 ■ **Locating a Point on the Unit Circle**

The point $P(\sqrt{3}/2, y)$ is on the unit circle in quadrant IV. Find its y-coordinate.

SOLUTION   Since the point is on the unit circle, we have

$$\left(\frac{\sqrt{3}}{2}\right)^2 + y^2 = 1$$

$$y^2 = 1 - \frac{3}{4} = \frac{1}{4}$$

$$y = \pm\frac{1}{2}$$

Since the point is in quadrant IV, its y-coordinate must be negative, so $y = -\frac{1}{2}$.   ■

■ Terminal Points on the Unit Circle

Suppose t is a real number. Let's mark off a distance t along the unit circle, starting at the point $(1, 0)$ and moving in a counterclockwise direction if t is positive or in a clockwise direction if t is negative (Figure 2). In this way we arrive at a point $P(x, y)$ on the unit circle. The point $P(x, y)$ obtained in this way is called the **terminal point** determined by the real number t.

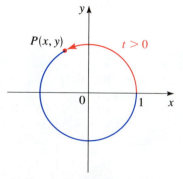

(a) Terminal point $P(x, y)$ determined
by $t > 0$

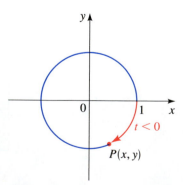

(b) Terminal point $P(x, y)$ determined
by $t < 0$

FIGURE 2

The circumference of the unit circle is $C = 2\pi(1) = 2\pi$. So, if a point starts at $(1, 0)$ and moves counterclockwise all the way around the unit circle and returns to $(1, 0)$, it travels a distance of 2π. To move halfway around the circle, it travels a distance of $\frac{1}{2}(2\pi) = \pi$. To move a quarter of the distance around the circle, it travels a distance of $\frac{1}{4}(2\pi) = \pi/2$. Where does the point end up when it travels these distances along the circle? From Figure 3 we see, for example, that when it travels a distance of π starting at $(1, 0)$, its terminal point is $(-1, 0)$.

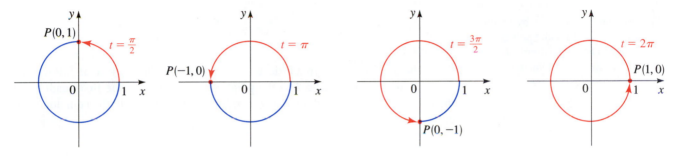

FIGURE 3
Terminal points determined by $t = \frac{\pi}{2}$, π, $\frac{3\pi}{2}$, and 2π

EXAMPLE 3 ■ Finding Terminal Points

Find the terminal point on the unit circle determined by each real number t.

(a) $t = 3\pi$ (b) $t = -\pi$ (c) $t = -\dfrac{\pi}{2}$

SOLUTION

From Figure 4 we get the following.

(a) The terminal point determined by 3π is $(-1, 0)$.

(b) The terminal point determined by $-\pi$ is $(-1, 0)$.

(c) The terminal point determined by $-\pi/2$ is $(0, -1)$.

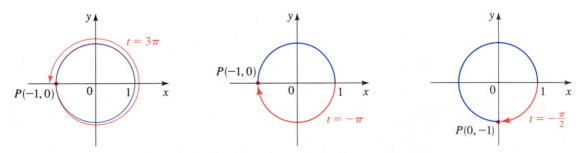

FIGURE 4

Notice that different values of t can determine the same terminal point.   ■

The terminal point $P(x, y)$ determined by $t = \pi/4$ is the same distance from $(1, 0)$ as from $(0, 1)$ along the unit circle (see Figure 5).

FIGURE 5

Since the unit circle is symmetric with respect to the line $y = x$, it follows that P lies on the line $y = x$. So P is the point of intersection (in the first quadrant) of the circle $x^2 + y^2 = 1$ and the line $y = x$. Substituting x for y in the equation of the circle, we get

$$x^2 + x^2 = 1$$

$$2x^2 = 1 \qquad \text{Combine like terms}$$

$$x^2 = \frac{1}{2} \qquad \text{Divide by 2}$$

$$x = \pm\frac{1}{\sqrt{2}} \qquad \text{Take square roots}$$

Since P is in the first quadrant, $x = 1/\sqrt{2}$ and since $y = x$, we have $y = 1/\sqrt{2}$ also. Thus, the terminal point determined by $\pi/4$ is

$$P\left(\frac{1}{\sqrt{2}}, \frac{1}{\sqrt{2}}\right) = P\left(\frac{\sqrt{2}}{2}, \frac{\sqrt{2}}{2}\right)$$

Similar methods can be used to find the terminal points determined by $t = \pi/6$ and $t = \pi/3$ (see Exercises 47 and 48). Table 1 and Figure 6 give the terminal points for some special values of t.

TABLE 1

t	Terminal point determined by t
0	$(1, 0)$
$\frac{\pi}{6}$	$\left(\frac{\sqrt{3}}{2}, \frac{1}{2}\right)$
$\frac{\pi}{4}$	$\left(\frac{\sqrt{2}}{2}, \frac{\sqrt{2}}{2}\right)$
$\frac{\pi}{3}$	$\left(\frac{1}{2}, \frac{\sqrt{3}}{2}\right)$
$\frac{\pi}{2}$	$(0, 1)$

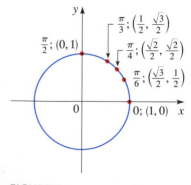

FIGURE 6

EXAMPLE 4 ■ Finding Terminal Points

Find the terminal point determined by each given real number t.

(a) $t = -\dfrac{\pi}{4}$ (b) $t = \dfrac{3\pi}{4}$ (c) $t = -\dfrac{5\pi}{6}$

SOLUTION

(a) Let P be the terminal point determined by $-\pi/4$, and let Q be the terminal point determined by $\pi/4$. From Figure 7(a) we see that the point P has the same coordinates as Q except for sign. Since P is in quadrant IV, its x-coordinate is positive and its y-coordinate is negative. Thus, the terminal point is $P(\sqrt{2}/2, -\sqrt{2}/2)$.

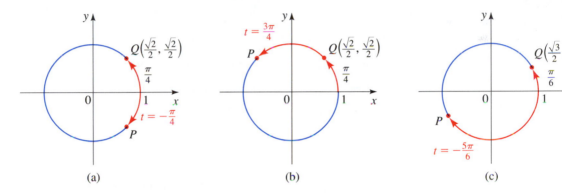

FIGURE 7   (a)   (b)   (c)

(b) Let P be the terminal point determined by $3\pi/4$, and let Q be the terminal point determined by $\pi/4$. From Figure 7(b) we see that the point P has the same coordinates as Q except for sign. Since P is in quadrant II, its x-coordinate is negative and its y-coordinate is positive. Thus, the terminal point is $P(-\sqrt{2}/2, \sqrt{2}/2)$.

(c) Let P be the terminal point determined by $-5\pi/6$, and let Q be the terminal point determined by $\pi/6$. From Figure 7(c) we see that the point P has the same coordinates as Q except for sign. Since P is in quadrant III, its coordinates are both negative. Thus, the terminal point is $P\left(-\sqrt{3}/2, -\tfrac{1}{2}\right)$. ■

The Reference Number

From Examples 3 and 4, we see that to find a terminal point in any quadrant we need only know the "corresponding" terminal point in the first quadrant. We give a procedure for finding such terminal points using the idea of the *reference number*.

REFERENCE NUMBER

Let t be a real number. The **reference number** $\bar{t}$ associated with t is the shortest distance along the unit circle between the terminal point determined by t and the x-axis.

Figure 8 shows that to find the reference number $\bar{t}$ it's helpful to know the quadrant in which the terminal point determined by t lies.

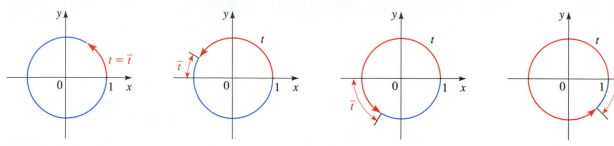

FIGURE 8

The reference number $\bar{t}$ for t

EXAMPLE 5 ■ Finding Reference Numbers

Find the reference number for each value of t.

(a) $t = \dfrac{5\pi}{6}$ (b) $t = \dfrac{7\pi}{4}$ (c) $t = -\dfrac{2\pi}{3}$ (d) $t = 5.80$

SOLUTION

From Figure 9 we find the reference numbers as follows.

(a) $\bar{t} = \pi - \dfrac{5\pi}{6} = \dfrac{\pi}{6}$

(b) $\bar{t} = 2\pi - \dfrac{7\pi}{4} = \dfrac{\pi}{4}$

(c) $\bar{t} = \pi - \dfrac{2\pi}{3} = \dfrac{\pi}{3}$

(d) $\bar{t} = 2\pi - 5.80 \approx 0.48$

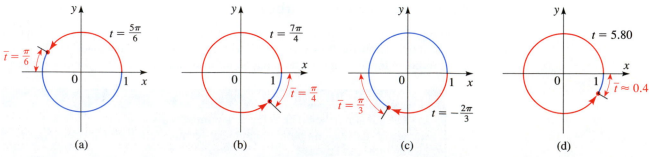

(a)   (b)   (c)   (d)

FIGURE 9

> ### USING REFERENCE NUMBERS TO FIND TERMINAL POINTS
>
> To find the terminal point P determined by any value of t, we use the following steps:
>
> 1. Find the reference number $\bar{t}$.
>
> 2. Find the terminal point $Q(a, b)$ determined by $\bar{t}$.
>
> 3. The terminal point determined by t is $P(\pm a, \pm b)$, where the signs are chosen according to the quadrant in which this terminal point lies.

EXAMPLE 6 ■ Using Reference Numbers to Find Terminal Points

Find the terminal point determined by each given real number t.

(a) $t = \dfrac{5\pi}{6}$ (b) $t = \dfrac{7\pi}{4}$ (c) $t = -\dfrac{2\pi}{3}$

SOLUTION

The reference numbers associated with these values of t were found in Example 5.

(a) The reference number is $\bar{t} = \pi/6$, which determines the terminal point $\left(\sqrt{3}/2, \frac{1}{2}\right)$ from Table 1. Since the terminal point determined by t is in quadrant II, its x-coordinate is negative and its y-coordinate is positive. Thus, the desired terminal point is

$$\left(-\frac{\sqrt{3}}{2}, \frac{1}{2}\right)$$

(b) The reference number is $\bar{t} = \pi/4$, which determines the terminal point $\left(\sqrt{2}/2, \sqrt{2}/2\right)$ from Table 1. Since the terminal point is in quadrant IV, its x-coordinate is positive and its y-coordinate is negative. Thus, the desired terminal point is

$$\left(\frac{\sqrt{2}}{2}, -\frac{\sqrt{2}}{2}\right)$$

(c) The reference number is $\bar{t} = \pi/3$, which determines the terminal point $\left(\frac{1}{2}, \sqrt{3}/2\right)$ from Table 1. Since the terminal point determined by t is in quadrant III, its coordinates are both negative. Thus, the desired terminal point is

$$\left(-\frac{1}{2}, -\frac{\sqrt{3}}{2}\right)$$

■

Since the circumference of the unit circle is 2π, the terminal point determined by t is the same as that determined by $t + 2\pi$ or $t - 2\pi$. In general, we can add or subtract 2π any number of times without changing the terminal point determined by t. We use this observation in the next example to find terminal points for large t.

EXAMPLE 7 ■ Finding the Terminal Point for Large *t*

Find the terminal point determined by $t = \dfrac{29\pi}{6}$.

SOLUTION

Since

$$t = \frac{29\pi}{6} = 4\pi + \frac{5\pi}{6}$$

we see that the terminal point of *t* is the same as that of $5\pi/6$ (that is, we subtract 4π). So by Example 6(a) the terminal point is $\left(-\sqrt{3}/2, \tfrac{1}{2}\right)$. ■

5.1 EXERCISES

1–4 ■ Show that the point is on the unit circle.

1. $\left(\tfrac{3}{5}, \tfrac{4}{5}\right)$

2. $\left(\tfrac{5}{13}, -\tfrac{12}{13}\right)$

3. $\left(-\tfrac{2}{3}, -\tfrac{\sqrt{5}}{3}\right)$

4. $\left(\tfrac{\sqrt{11}}{6}, \tfrac{5}{6}\right)$

5–10 ■ The point *P* is on the unit circle. Find $P(x, y)$ from the given information.

5. The *x*-coordinate of *P* is $\tfrac{4}{5}$ and *P* is in quadrant I.

6. The *y*-coordinate of *P* is $-\tfrac{1}{3}$ and *P* is in quadrant IV.

7. The *y*-coordinate of *P* is $\tfrac{2}{3}$ and the *x*-coordinate is negative.

8. The *x*-coordinate of *P* is positive and the *y*-coordinate of *P* is $-\sqrt{5}/5$.

9. The *x*-coordinate of *P* is $-\sqrt{2}/3$ and *P* is in quadrant III.

10. The *x*-coordinate of *P* is $-\tfrac{2}{5}$ and *P* is in quadrant II.

11–12 ■ Find *t* and the terminal point determined by *t* for each point in the figure. In Exercise 11, *t* increases in increments of $\pi/4$; in Exercise 12, *t* increases in increments of $\pi/6$.

11.

12.

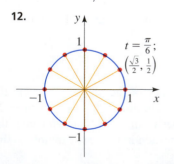

13–22 ■ Find the terminal point $P(x, y)$ on the unit circle determined by the given value of *t*.

13. $t = \dfrac{\pi}{2}$

14. $t = \dfrac{3\pi}{2}$

15. $t = \dfrac{5\pi}{6}$

16. $t = \dfrac{7\pi}{6}$

17. $t = -\dfrac{\pi}{3}$

18. $t = \dfrac{5\pi}{3}$

19. $t = \dfrac{2\pi}{3}$

20. $t = -\dfrac{\pi}{2}$

21. $t = -\dfrac{3\pi}{4}$

22. $t = \dfrac{11\pi}{6}$

23. Suppose that the terminal point determined by *t* is the point $\left(\tfrac{3}{5}, \tfrac{4}{5}\right)$ on the unit circle. Find the terminal point determined by each of the following.
  (a) $\pi - t$ (b) $-t$ (c) $\pi + t$ (d) $2\pi + t$

24. Suppose that the terminal point determined by *t* is the point $\left(\tfrac{3}{4}, \sqrt{7}/4\right)$ on the unit circle. Find the terminal point determined by each of the following.
  (a) $-t$ (b) $4\pi + t$ (c) $\pi - t$ (d) $t - \pi$

25–28 ■ Find the reference number for each value of *t*.

25. (a) $t = \dfrac{5\pi}{4}$ (b) $t = \dfrac{7\pi}{3}$

  (c) $t = -\dfrac{4\pi}{3}$ (d) $t = \dfrac{\pi}{6}$

26. (a) $t = \dfrac{5\pi}{7}$ (b) $t = -\dfrac{9\pi}{8}$

    (c) $t = 3.55$ (d) $t = -2.9$

27. (a) $t = -\dfrac{11\pi}{5}$ (b) $t = \dfrac{13\pi}{6}$

    (c) $t = \dfrac{7\pi}{3}$ (d) $t = -\dfrac{5\pi}{6}$

28. (a) $t = 3$ (b) $t = 6$
    (c) $t = -3$ (d) $t = -6$

29–42 ■ Find (a) the reference number for each value of t, and (b) the terminal point determined by t.

29. $t = \dfrac{2\pi}{3}$ 30. $t = \dfrac{4\pi}{3}$

31. $t = \dfrac{3\pi}{4}$ 32. $t = \dfrac{7\pi}{3}$

33. $t = -\dfrac{2\pi}{3}$ 34. $t = -\dfrac{7\pi}{6}$

35. $t = \dfrac{13\pi}{4}$ 36. $t = \dfrac{13\pi}{6}$

37. $t = \dfrac{7\pi}{6}$ 38. $t = \dfrac{17\pi}{4}$

39. $t = -\dfrac{11\pi}{3}$ 40. $t = \dfrac{31\pi}{6}$

41. $t = \dfrac{16\pi}{3}$ 42. $t = -\dfrac{41\pi}{4}$

43–46 ■ Use the figure to find the terminal point determined by the real number t, with coordinates correct to one decimal place.

43. $t = 1$

44. $t = 2.5$

45. $t = -1.1$

46. $t = 4.2$

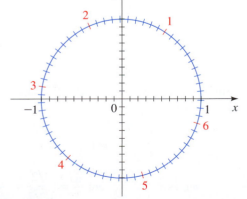

DISCOVERY · DISCUSSION

47. **Finding the Terminal Point for $\pi/6$**   Suppose the terminal point determined by $t = \pi/6$ is $P(x, y)$ and the points Q and R are as shown in the figure. Why are the distances PQ and PR the same? Use this fact, together with the Distance Formula, to show that the coordinates of P satisfy the equation $2y = \sqrt{x^2 + (y - 1)^2}$. Simplify this equation using the fact that $x^2 + y^2 = 1$. Solve the simplified equation to find $P(x, y)$.

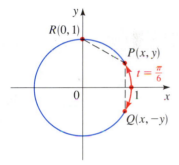

48. **Finding the Terminal Point for $\pi/3$**   Now that you know the terminal point determined by $t = \pi/6$, use symmetry to find the terminal point determined by $t = \pi/3$ (see the figure). Explain your reasoning.

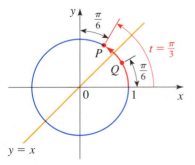

5.2 TRIGONOMETRIC FUNCTIONS OF REAL NUMBERS

A function is a rule that assigns to each real number another real number. In this section we use properties of the unit circle from the preceding section to define certain functions of real numbers, the trigonometric functions.

The Trigonometric Functions

Recall that to find the terminal point $P(x, y)$ for a given real number t, we move a distance t along the unit circle, starting at the point $(1, 0)$. We move in a counterclockwise direction if t is positive and in a clockwise direction if t is negative (see Figure 1). We now use the x- and y-coordinates of the point $P(x, y)$ to define several functions. For instance, we define the function called *sine* by assigning to each real number t the y-coordinate of the terminal point $P(x, y)$ determined by t. The functions *cosine, tangent, cosecant, secant,* and *cotangent* are also defined using the coordinates of $P(x, y)$.

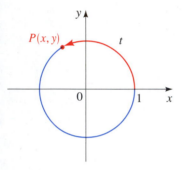

FIGURE 1

DEFINITION OF THE TRIGONOMETRIC FUNCTIONS

Let t be any real number and let $P(x, y)$ be the terminal point on the unit circle determined by t. We define

$$\sin t = y \qquad \cos t = x \qquad \tan t = \frac{y}{x} \quad (x \neq 0)$$

$$\csc t = \frac{1}{y} \quad (y \neq 0) \qquad \sec t = \frac{1}{x} \quad (x \neq 0) \qquad \cot t = \frac{x}{y} \quad (y \neq 0)$$

Because the trigonometric functions can be defined in terms of the unit circle, they are sometimes called the **circular functions**.

EXAMPLE 1 ■ Evaluating Trigonometric Functions

Find the six trigonometric functions of each given real number t.

(a) $t = \dfrac{\pi}{3}$ (b) $t = \dfrac{\pi}{2}$

SOLUTION

(a) The terminal point determined by $t = \pi/3$ is $P\left(\frac{1}{2}, \sqrt{3}/2\right)$. Since the coordinates are $x = \frac{1}{2}$ and $y = \sqrt{3}/2$, we have

$$\sin \frac{\pi}{3} = \frac{\sqrt{3}}{2} \qquad \cos \frac{\pi}{3} = \frac{1}{2} \qquad \tan \frac{\pi}{3} = \frac{\sqrt{3}/2}{1/2} = \sqrt{3}$$

$$\csc \frac{\pi}{3} = \frac{2\sqrt{3}}{3} \qquad \sec \frac{\pi}{3} = 2 \qquad \cot \frac{\pi}{3} = \frac{1/2}{\sqrt{3}/2} = \frac{\sqrt{3}}{3}$$

Relationship to the Trigonometric Functions of Angles

If you have previously studied trigonometry of right triangles (Chapter 6), you are probably wondering how the sine and cosine of an *angle* relate to those of this section. To see how, let's start with a right triangle, $\triangle OPQ$.

Right triangle OPQ

Place the triangle in the coordinate plane as shown, with angle θ in standard position.

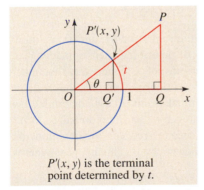

$P'(x, y)$ is the terminal point determined by t.

The point $P'(x, y)$ in the figure is the terminal point determined by the arc t. Note that triangle OPQ is similar to the small triangle $OP'Q'$ whose legs have lengths x and y.

Now, by the definition of the trigonometric functions of the *angle* θ we have

$$\sin \theta = \frac{\text{opp}}{\text{hyp}} = \frac{PQ}{OP} = \frac{P'Q'}{OP'}$$

$$= \frac{y}{1} = y$$

$$\cos \theta = \frac{\text{adj}}{\text{hyp}} = \frac{OQ}{OP} = \frac{OQ'}{OP'}$$

$$= \frac{x}{1} = x$$

By the definition of the trigonometric functions of the *real number t*, we have

$$\sin t = y$$

$$\cos t = x$$

Now, if θ is measured in radians, then $\theta = t$ (see the figure). So the trigonometric functions of the angle with radian measure θ are exactly the same as the trigonometric functions defined in terms of the terminal point determined by the real number t.

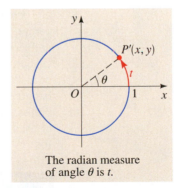

The radian measure of angle θ is t.

Why then study trigonometry in two different ways? Because different applications require that we view the trigonometric functions differently. (See *Focus on Modeling*, page 455, and Sections 6.2, 6.4, and 6.5.)

(b) The terminal point determined by $\pi/2$ is $P(0, 1)$. So

$$\sin \frac{\pi}{2} = 1 \qquad \cos \frac{\pi}{2} = 0 \qquad \csc \frac{\pi}{2} = \frac{1}{1} = 1 \qquad \cot \frac{\pi}{2} = \frac{0}{1} = 0$$

But $\tan \pi/2$ and $\sec \pi/2$ are undefined because $x = 0$ appears in the denominator in each of their definitions. ◼

Some special values of the trigonometric functions are listed in Table 1. This table is easily obtained from Table 1 of Section 5.1, together with the definitions of the trigonometric functions.

TABLE 1 Special values of the trigonometric functions

t	$\sin t$	$\cos t$	$\tan t$	$\csc t$	$\sec t$	$\cot t$
0	0	1	0	—	1	—
$\dfrac{\pi}{6}$	$\dfrac{1}{2}$	$\dfrac{\sqrt{3}}{2}$	$\dfrac{\sqrt{3}}{3}$	2	$\dfrac{2\sqrt{3}}{3}$	$\sqrt{3}$
$\dfrac{\pi}{4}$	$\dfrac{\sqrt{2}}{2}$	$\dfrac{\sqrt{2}}{2}$	1	$\sqrt{2}$	$\sqrt{2}$	1
$\dfrac{\pi}{3}$	$\dfrac{\sqrt{3}}{2}$	$\dfrac{1}{2}$	$\sqrt{3}$	$\dfrac{2\sqrt{3}}{3}$	2	$\dfrac{\sqrt{3}}{3}$
$\dfrac{\pi}{2}$	1	0	—	1	—	0

Example 1 shows that some of the trigonometric functions fail to be defined for certain real numbers. So we need to determine their domains. The functions sine and cosine are defined for all values of t. Since the functions cotangent and cosecant have y in the denominator of their definitions, they are not defined whenever the y-coordinate of the terminal point $P(x, y)$ determined by t is 0. This happens when $t = n\pi$ for any integer n, so their domains do not include these points. The functions tangent and secant have x in the denominator in their definitions, so they are not defined whenever $x = 0$. This happens when $t = (\pi/2) + n\pi$ for any integer n.

DOMAINS OF THE TRIGONOMETRIC FUNCTIONS

Function	Domain
sin, cos	All real numbers
tan, sec	All real numbers other than $\dfrac{\pi}{2} + n\pi$ for any integer n
cot, csc	All real numbers other than $n\pi$ for any integer n

Values of the Trigonometric Functions

To compute other values of the trigonometric functions, we first determine their signs. The signs of the trigonometric functions depend on the quadrant in which the terminal point of t lies. For example, if the terminal point $P(x, y)$ determined by t lies in quadrant III, then its coordinates are both negative. So $\sin t$, $\cos t$, $\csc t$, and $\sec t$ are all negative, while $\tan t$ and $\cot t$ are positive. You can check the other entries in the following box.

The following mnemonic device can be used to remember which trigonometric functions are positive in each quadrant: All of them, Sine, Tangent, or Cosine.

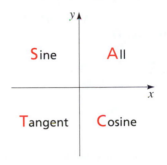

You can remember this as "All Students Take Calculus."

SIGNS OF THE TRIGONOMETRIC FUNCTIONS

Quadrant	Positive Functions	Negative functions
I	all	none
II	sin, csc	cos, sec, tan, cot
III	tan, cot	sin, csc, cos, sec
IV	cos, sec	sin, csc, tan, cot

EXAMPLE 2 ■ Determining the Sign of a Trigonometric Function

(a) $\cos \dfrac{\pi}{3} > 0$, since the terminal point of $t = \dfrac{\pi}{3}$ is in quadrant I.

(b) $\tan 4 > 0$, since the terminal point of $t = 4$ is in quadrant III.

(c) If $\cos t < 0$ and $\sin t > 0$, then the terminal point of t must be in quadrant II. ■

In Section 5.1 we used the reference number to find the terminal point determined by a real number t. Since the trigonometric functions are defined in terms of the coordinates of terminal points, we can use the reference number to find values of the trigonometric functions. Suppose that $\bar{t}$ is the reference number for t. Then the terminal point of $\bar{t}$ has the same coordinates, except possibly for sign, as the terminal point of t. So the values of the trigonometric functions at t are the same, except possibly for sign, as their values at $\bar{t}$. We illustrate the procedure in the next example.

EXAMPLE 3 ■ Evaluating Trigonometric Functions

Find each value.

(a) $\cos \dfrac{2\pi}{3}$ (b) $\tan\left(-\dfrac{\pi}{3}\right)$ (c) $\sin \dfrac{19\pi}{4}$

SOLUTION

(a) The reference number for $2\pi/3$ is $\pi/3$. Since the terminal point of $2\pi/3$ is in quadrant II, $\cos(2\pi/3)$ is negative. Thus

$$\cos \frac{2\pi}{3} = -\cos \frac{\pi}{3} = -\frac{1}{2}$$

$$\uparrow \qquad \uparrow \qquad \uparrow$$

sign   reference   from
number   Table 1

(b) The reference number for $-\pi/3$ is $\pi/3$. Since the terminal point of $-\pi/3$ is in quadrant IV, $\tan(-\pi/3)$ is negative. Thus

$$\tan\left(-\frac{\pi}{3}\right) = -\tan\frac{\pi}{3} = -\sqrt{3}$$

$$\underset{\substack{\uparrow \\ \text{sign}}}{} \quad \underset{\substack{\uparrow \\ \text{reference} \\ \text{number}}}{} \quad \underset{\substack{\uparrow \\ \text{from} \\ \text{Table 1}}}{}$$

(c) Since $(19\pi/4) - 4\pi = 3\pi/4$, the terminal points determined by $19\pi/4$ and $3\pi/4$ are the same. The reference number for $3\pi/4$ is $\pi/4$. Since the terminal point of $3\pi/4$ is in quadrant II, $\sin(3\pi/4)$ is positive. Thus

$$\sin\frac{19\pi}{4} = \sin\frac{3\pi}{4} = +\sin\frac{\pi}{4} = \frac{\sqrt{2}}{2}$$

$$\underset{\substack{\uparrow \\ \text{subtract } 4\pi}}{} \qquad \underset{\substack{\uparrow \\ \text{sign}}}{} \quad \underset{\substack{\uparrow \\ \text{reference} \\ \text{number}}}{} \quad \underset{\substack{\nwarrow \\ \text{from} \\ \text{Table 1}}}{}$$

So far we have been able to compute the values of the trigonometric functions only for certain values of t. In fact, we can compute the values of the trigonometric functions whenever t is a multiple of $\pi/6$, $\pi/4$, $\pi/3$, and $\pi/2$. How can we compute the trigonometric functions for other values of t? For example, how can we find $\sin 1.5$? One way is to carefully sketch a diagram and read the value (see Exercises 35–42); however, this method is not very accurate. Fortunately, pro-grammed directly into scientific calculators are mathematical procedures (see the margin) that find the values of *sine, cosine,* and *tangent* correct to the number of digits in the display. The calculator must be put in *radian mode* to evaluate these functions. To find values of cosecant, secant, and cotangent using a calculator, we need to use the following *reciprocal relations*:

$$\csc t = \frac{1}{\sin t} \qquad \sec t = \frac{1}{\cos t} \qquad \cot t = \frac{1}{\tan t}$$

These identities follow from the definitions of the trigonometric functions. For instance, since $\sin t = y$ and $\csc t = 1/y$, we have $\csc t = 1/y = 1/(\sin t)$. The others follow similarly.

EXAMPLE 4 ■ Using a Calculator to Evaluate Trigonometric Functions

Making sure our calculator is set to radian mode and rounding the results to six decimal places, we get

(a) $\sin 2.2 \approx 0.808496$

(b) $\cos 1.1 \approx 0.453596$

(c) $\cot 28 = \dfrac{1}{\tan 28} \approx -3.553286$

(d) $\csc 0.98 = \dfrac{1}{\sin 0.98} \approx 1.204098$ ■

Let's consider the relationship between the trigonometric functions of t and those of $-t$. From Figure 2 we see that

$$\sin(-t) = -y = -\sin t$$

$$\cos(-t) = x = \cos t$$

$$\tan(-t) = \dfrac{-y}{x} = -\dfrac{y}{x} = -\tan t$$

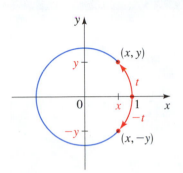

FIGURE 2

These equations show that sine and tangent are odd functions, whereas cosine is an even function. It's easy to see that the reciprocal of an even function is even and the reciprocal of an odd function is odd (see Section 2.5). This fact, together with the reciprocal relations, completes our knowledge of the even-odd properties for all the trigonometric functions.

EVEN-ODD PROPERTIES

Sine, cosecant, tangent, and cotangent are odd functions; cosine and secant are even functions.

$$\sin(-t) = -\sin t \qquad \cos(-t) = \cos t \qquad \tan(-t) = -\tan t$$

$$\csc(-t) = -\csc t \qquad \sec(-t) = \sec t \qquad \cot(-t) = -\cot t$$

EXAMPLE 5 ■ Even and Odd Trigonometric Functions

Use the even-odd properties of the trigonometric functions to determine each value.

(a) $\sin\left(-\dfrac{\pi}{6}\right)$ (b) $\cos\left(-\dfrac{\pi}{4}\right)$

SOLUTION

By the even-odd properties and Table 1, we have

(a) $\sin\left(-\dfrac{\pi}{6}\right) = -\sin\dfrac{\pi}{6} = -\dfrac{1}{2}$ Sine is odd

(b) $\cos\left(-\dfrac{\pi}{4}\right) = \cos\dfrac{\pi}{4} = \dfrac{\sqrt{2}}{2}$ Cosine is even   ■

■ Fundamental Identities

The trigonometric functions are related to each other through equations called **trigonometric identities**. We give the most important ones in the following box.*

FUNDAMENTAL IDENTITIES

Reciprocal Identities

$$\csc t = \frac{1}{\sin t} \qquad \sec t = \frac{1}{\cos t} \qquad \cot t = \frac{1}{\tan t}$$

$$\tan t = \frac{\sin t}{\cos t} \qquad \cot t = \frac{\cos t}{\sin t}$$

Pythagorean Identities

$$\sin^2 t + \cos^2 t = 1 \qquad \tan^2 t + 1 = \sec^2 t \qquad 1 + \cot^2 t = \csc^2 t$$

■ **Proof**   The reciprocal identities follow immediately from the definition on page 418. We now prove the Pythagorean identities. By definition, $\cos t = x$ and $\sin t = y$, where x and y are the coordinates of a point $P(x, y)$ on the unit circle. Since $P(x, y)$ is on the unit circle, we have $x^2 + y^2 = 1$. Thus

$$\sin^2 t + \cos^2 t = 1$$

Dividing both sides by $\cos^2 t$ (provided $\cos t \neq 0$), we get

$$\frac{\sin^2 t}{\cos^2 t} + \frac{\cos^2 t}{\cos^2 t} = \frac{1}{\cos^2 t}$$

$$\left(\frac{\sin t}{\cos t}\right)^2 + 1 = \left(\frac{1}{\cos t}\right)^2$$

$$\tan^2 t + 1 = \sec^2 t$$

We have used the reciprocal identities $\sin t/\cos t = \tan t$ and $1/\cos t = \sec t$. Similarly, dividing both sides of the first Pythagorean identity by $\sin^2 t$ (provided $\sin t \neq 0$) gives us $1 + \cot^2 t = \csc^2 t$.   □

   As their name indicates, the fundamental identities play a central role in trigonometry because we can use them to relate any trigonometric function to any

---

*We follow the usual convention of writing $\sin^2 t$ for $(\sin t)^2$. In general, we write $\sin^n t$ for $(\sin t)^n$ for all integers n except $n = -1$. The exponent $n = -1$ will be assigned another meaning in Section 7.4. Of course, the same convention applies to the other five trigonometric functions.

other. So, if we know the value of any one of the trigonometric functions at t, then we can find the values of all the others at t.

EXAMPLE 6 ■ Finding All Trigonometric Functions from the Value of One

If $\cos t = \frac{3}{5}$ and t is in quadrant IV, find the values of all the trigonometric functions at t.

SOLUTION

From the Pythagorean identities we have

$$\sin^2 t + \cos^2 t = 1$$

$$\sin^2 t + \left(\tfrac{3}{5}\right)^2 = 1 \qquad \text{Substitute } \cos t = \tfrac{3}{5}$$

$$\sin^2 t = 1 - \tfrac{9}{25} = \tfrac{16}{25} \qquad \text{Solve for } \sin^2 t$$

$$\sin t = \pm\tfrac{4}{5} \qquad \text{Take square roots}$$

Since this point is in quadrant IV, $\sin t$ is negative, so $\sin t = -\frac{4}{5}$. Now that we know both $\sin t$ and $\cos t$, we can find the values of the other trigonometric functions using the reciprocal identities:

$$\sin t = -\frac{4}{5} \qquad\qquad \cos t = \frac{3}{5} \qquad\qquad \tan t = \frac{\sin t}{\cos t} = \frac{-\frac{4}{5}}{\frac{3}{5}} = -\frac{4}{3}$$

$$\csc t = \frac{1}{\sin t} = -\frac{5}{4} \qquad \sec t = \frac{1}{\cos t} = \frac{5}{3} \qquad \cot t = \frac{1}{\tan t} = -\frac{3}{4} \qquad ■$$

EXAMPLE 7 ■ Writing One Trigonometric Function in Terms of Another

Write $\tan t$ in terms of $\cos t$, where t is in quadrant III.

SOLUTION

Since $\tan t = \sin t / \cos t$, we need to write $\sin t$ in terms of $\cos t$. By the Pythagorean identities we have

$$\sin^2 t + \cos^2 t = 1$$

$$\sin^2 t = 1 - \cos^2 t \qquad \text{Solve for } \sin^2 t$$

$$\sin t = \pm\sqrt{1 - \cos^2 t} \qquad \text{Take square roots}$$

Since $\sin t$ is negative in quadrant III, the negative sign applies here. Thus

$$\tan t = \frac{\sin t}{\cos t} = \frac{-\sqrt{1 - \cos^2 t}}{\cos t} \qquad\qquad ■$$

5.2   EXERCISES

1–2 ■ Find sin t and cos t for the values of t whose terminal points are shown on the unit circle in the figure. In Exercise 1, t increases in increments of $\pi/4$; in Exercise 2, t increases in increments of $\pi/6$. (See Exercises 11 and 12 in Section 5.1.)

1.

2.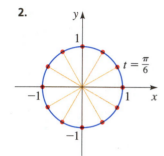

3–22 ■ Find the exact value of the trigonometric function at the given real number.

3. (a) $\sin\left(-\dfrac{\pi}{3}\right)$

  (b) $\cos\left(-\dfrac{\pi}{3}\right)$

4. (a) $\sin \pi$

  (b) $\sin(-\pi)$

5. (a) $\cos \pi$

  (b) $\cos(-\pi)$

6. (a) $\cos \dfrac{\pi}{6}$

  (b) $\cos \dfrac{5\pi}{6}$

7. (a) $\sin \dfrac{\pi}{2}$

  (b) $\sin \dfrac{3\pi}{2}$

8. (a) $\sin \dfrac{7\pi}{6}$

  (b) $\cos \dfrac{7\pi}{6}$

9. (a) $\cos \dfrac{\pi}{2}$

  (b) $\cos \dfrac{5\pi}{2}$

10. (a) $\sin \dfrac{5\pi}{6}$

  (b) $\csc \dfrac{5\pi}{6}$

11. (a) $\cos \dfrac{7\pi}{3}$

  (b) $\sec \dfrac{7\pi}{3}$

12. (a) $\sin \dfrac{3\pi}{4}$

  (b) $\cos \dfrac{3\pi}{4}$

13. (a) $\cos \dfrac{\pi}{3}$

  (b) $\cos\left(-\dfrac{\pi}{3}\right)$

14. (a) $\sin \dfrac{\pi}{6}$

  (b) $\sin\left(-\dfrac{\pi}{6}\right)$

15. (a) $\tan \dfrac{\pi}{6}$

  (b) $\tan\left(-\dfrac{\pi}{6}\right)$

16. (a) $\tan \dfrac{\pi}{3}$

  (b) $\cot \dfrac{\pi}{3}$

17. (a) $\sec \dfrac{11\pi}{3}$

  (b) $\csc \dfrac{11\pi}{3}$

18. (a) $\sec \dfrac{13\pi}{6}$

  (b) $\sec\left(-\dfrac{13\pi}{6}\right)$

19. (a) $\sin \dfrac{9\pi}{4}$

  (b) $\csc \dfrac{9\pi}{4}$

20. (a) $\sec \pi$

  (b) $\csc \dfrac{\pi}{2}$

21. (a) $\tan\left(-\dfrac{\pi}{4}\right)$

  (b) $\cot\left(-\dfrac{\pi}{4}\right)$

22. (a) $\tan \dfrac{3\pi}{4}$

  (b) $\tan \dfrac{11\pi}{4}$

23–26 ■ Find the value of each of the six trigonometric functions (if it is defined) at the given real number t. Use your answers to complete the table.

23. $t = 0$ **24.** $t = \dfrac{\pi}{2}$ **25.** $t = \pi$ **26.** $t = \dfrac{3\pi}{2}$

t	$\sin t$	$\cos t$	$\tan t$	$\csc t$	$\sec t$	$\cot t$
0	0	1		undefined		
$\dfrac{\pi}{2}$						
π			0			undefined
$\dfrac{3\pi}{2}$						

27–34 ■ The terminal point $P(x, y)$ determined by t is given. Find sin t, cos t, and tan t.

27. $\left(\dfrac{3}{5}, \dfrac{4}{5}\right)$

28. $\left(-\dfrac{3}{5}, \dfrac{4}{5}\right)$

29. $\left(\dfrac{\sqrt{5}}{4}, -\dfrac{\sqrt{11}}{4}\right)$

30. $\left(-\dfrac{1}{3}, -\dfrac{2\sqrt{2}}{3}\right)$

31. $\left(\dfrac{40}{41}, \dfrac{9}{41}\right)$

32. $\left(-\dfrac{6}{7}, \dfrac{\sqrt{13}}{7}\right)$

33. $\left(-\dfrac{5}{13}, -\dfrac{12}{13}\right)$

34. $\left(\dfrac{\sqrt{5}}{5}, \dfrac{2\sqrt{5}}{5}\right)$

35–42 ■ Find the approximate value of the given trigonometric function by using (a) the figure and (b) a calculator. Compare the two values.

35. $\sin 1$

36. $\cos 0.8$

37. $\sin 1.2$

38. $\cos 5$

39. $\tan 0.8$

40. $\tan(-1.3)$

41. $\cos 4.1$

42. $\sin(-5.2)$

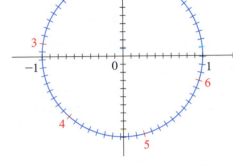

43–46 ■ Find the sign of the expression if the terminal point determined by t is in the given quadrant.

43. $\sin t \cos t$, quadrant II

44. $\tan t \sec t$, quadrant IV

45. $\dfrac{\tan t \sin t}{\cot t}$, quadrant III

46. $\cos t \sec t$, any quadrant

47–50 ■ From the information given, find the quadrant in which the terminal point determined by t lies.

47. $\sin t > 0$ and $\cos t < 0$

48. $\tan t > 0$ and $\sin t < 0$

49. $\csc t > 0$ and $\sec t < 0$

50. $\cos t < 0$ and $\cot t < 0$

51–60 ■ Write the first expression in terms of the second if the terminal point determined by t is in the given quadrant.

51. $\sin t$, $\cos t$; quadrant II

52. $\cos t$, $\sin t$; quadrant IV

53. $\tan t$, $\sin t$; quadrant IV

54. $\tan t$, $\cos t$; quadrant III

55. $\sec t$, $\tan t$; quadrant II

56. $\csc t$, $\cot t$; quadrant III

57. $\tan t$, $\sec t$; quadrant III

58. $\sin t$, $\sec t$; quadrant IV

59. $\tan^2 t$, $\sin t$; any quadrant

60. $\sec^2 t \sin^2 t$, $\cos t$; any quadrant

61–68 ■ Find the values of the trigonometric functions of t from the given information.

61. $\sin t = \frac{3}{5}$, terminal point of t is in quadrant II

62. $\cos t = -\frac{4}{5}$, terminal point of t is in quadrant III

63. $\sec t = 3$, terminal point of t is in quadrant IV

64. $\tan t = \frac{1}{4}$, terminal point of t is in quadrant III

65. $\tan t = -\frac{3}{4}$, $\cos t > 0$

66. $\sec t = 2$, $\sin t < 0$

67. $\sin t = -\frac{1}{4}$, $\sec t < 0$

68. $\tan t = -4$, $\csc t > 0$

69–76 ■ Determine whether the function is even, odd, or neither.

69. $f(x) = x^2 \sin x$

70. $f(x) = x^2 \cos 2x$

71. $f(x) = \sin x \cos x$

72. $f(x) = e^x \sin x$

73. $f(x) = |x| \cos x$

74. $f(x) = x \sin^3 x$

75. $f(x) = x^3 + \cos x$

76. $f(x) = \cos(\sin x)$

DISCOVERY · DISCUSSION

77. Reduction Formulas   Explain how the figure shows that the following "reduction formulas" are valid:

$$\sin(t + \pi) = -\sin t \qquad \cos(t + \pi) = -\cos t$$

$$\tan(t + \pi) = \tan t$$

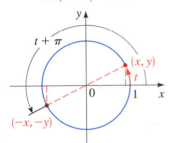

78. More Reduction Formulas   By the "Angle-Side-Angle" theorem from elementary geometry, triangles CDO and AOB in the figure are congruent. Explain how this proves that if B has coordinates (x, y), then D has coordinates $(-y, x)$. Then explain how the figure shows that the following "reduction formulas" are valid:

$$\sin\left(t + \frac{\pi}{2}\right) = \cos t$$

$$\cos\left(t + \frac{\pi}{2}\right) = -\sin t$$

$$\tan\left(t + \frac{\pi}{2}\right) = -\cot t$$

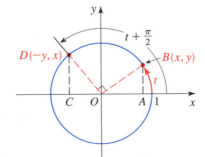

5.3   TRIGONOMETRIC GRAPHS

The graph of a function gives us a better idea of its behavior. So, in this section we graph the sine and cosine functions and certain transformations of these functions. The other trigonometric functions are graphed in the next section.

Graphs of the Sine and Cosine Functions

To help us graph the sine and cosine functions, we first observe that these functions repeat their values in a regular fashion. To see exactly how this happens, recall that the circumference of the unit circle is 2π. It follows that the terminal point $P(x, y)$ determined by the real number t is the same as that determined by $t + 2\pi$. Since the sine and cosine functions are defined in terms of the coordinates of $P(x, y)$, it follows that their values are unchanged by the addition of any integer multiple of 2π. In other words,

$$\sin(t + 2n\pi) = \sin t \qquad \text{for any integer } n$$

$$\cos(t + 2n\pi) = \cos t \qquad \text{for any integer } n$$

Thus, the sine and cosine functions are *periodic* according to the following definition: A function f is **periodic** if there is a positive number p such that $f(t + p) = f(t)$ for every t. The least such positive number (if it exists) is the **period** of f. If f has period p, then the graph of f on any interval of length p is called **one complete period** of f.

PERIODIC PROPERTIES OF SINE AND COSINE

The function sine has period 2π: $\sin(t + 2\pi) = \sin t$

The function cosine has period 2π: $\cos(t + 2\pi) = \cos t$

TABLE 1

t	$\sin t$	$\cos t$
$0 \to \dfrac{\pi}{2}$	$0 \to 1$	$1 \to 0$
$\dfrac{\pi}{2} \to \pi$	$1 \to 0$	$0 \to -1$
$\pi \to \dfrac{3\pi}{2}$	$0 \to -1$	$-1 \to 0$
$\dfrac{3\pi}{2} \to 2\pi$	$-1 \to 0$	$0 \to 1$

So the sine and cosine functions repeat their values in any interval of length 2π. To sketch their graphs, we first graph one period. To sketch the graphs on the interval $0 \le t \le 2\pi$, we could try to make a table of values and use those points to draw the graph. Since no such table can be complete, let's look more closely at the definitions of these functions.

Recall that $\sin t$ is the y-coordinate of the terminal point $P(x, y)$ on the unit circle determined by the real number t. How does the y-coordinate of this point vary as t increases? It's easy to see that the y-coordinate of $P(x, y)$ increases to 1, then decreases to -1 repeatedly as the point $P(x, y)$ travels around the unit circle. (See Figure 1.) In fact, as t increases from 0 to $\pi/2$, $y = \sin t$ increases from 0 to 1. As t increases from $\pi/2$ to π, the value of $y = \sin t$ decreases from 1 to 0. Table 1 shows the variation of the sine and cosine functions for t between 0 and 2π.

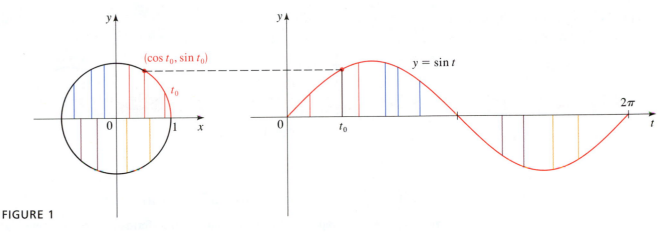

FIGURE 1

To draw the graphs more accurately, we find a few other values of $\sin t$ and $\cos t$ in Table 2. We could find still other values with the aid of a calculator.

TABLE 2

t	0	$\dfrac{\pi}{6}$	$\dfrac{\pi}{3}$	$\dfrac{\pi}{2}$	$\dfrac{2\pi}{3}$	$\dfrac{5\pi}{6}$	π	$\dfrac{7\pi}{6}$	$\dfrac{4\pi}{3}$	$\dfrac{3\pi}{2}$	$\dfrac{5\pi}{3}$	$\dfrac{11\pi}{6}$	2π
$\sin t$	0	$\dfrac{1}{2}$	$\dfrac{\sqrt{3}}{2}$	1	$\dfrac{\sqrt{3}}{2}$	$\dfrac{1}{2}$	0	$-\dfrac{1}{2}$	$-\dfrac{\sqrt{3}}{2}$	-1	$-\dfrac{\sqrt{3}}{2}$	$-\dfrac{1}{2}$	0
$\cos t$	1	$\dfrac{\sqrt{3}}{2}$	$\dfrac{1}{2}$	0	$-\dfrac{1}{2}$	$-\dfrac{\sqrt{3}}{2}$	-1	$-\dfrac{\sqrt{3}}{2}$	$-\dfrac{1}{2}$	0	$\dfrac{1}{2}$	$\dfrac{\sqrt{3}}{2}$	1

Now we use this information to graph the functions $\sin t$ and $\cos t$ for t between 0 and 2π in Figures 2 and 3. These are the graphs of one period. Using the fact that these functions are periodic with period 2π, we get their complete graphs by continuing the same pattern to the left and to the right in every successive interval of length 2π.

The graph of the sine function is symmetric with respect to the origin. This is as expected, since sine is an odd function. Since the cosine function is an even function, its graph is symmetric with respect to the y-axis.

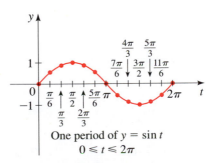

One period of $y = \sin t$
$0 \le t \le 2\pi$

FIGURE 2 Graph of $\sin t$

FIGURE 3 Graph of cos t

Graphs of Transformations of Sine and Cosine

We now consider graphs of functions that are transformations of the sine and cosine functions. Thus, the graphing techniques of Section 2.5 are very useful here. The graphs we obtain are important for understanding applications to physical situations such as harmonic motion (see *Focus on Modeling*, page 455), but some of them are beautiful graphs that are interesting in their own right.

It's traditional to use the letter x to denote the variable in the domain of a function. So, from here on we use the letter x and write $y = \sin x$, $y = \cos x$, $y = \tan x$, and so on to denote these functions.

EXAMPLE 1 ■ Cosine Curves

Sketch the graph of each function.

(a) $f(x) = 2 + \cos x$ (b) $g(x) = -\cos x$

SOLUTION

(a) The graph of $y = 2 + \cos x$ is the same as the graph of $y = \cos x$, but shifted up 2 units [see Figure 4(a)].

(b) The graph of $y = -\cos x$ in Figure 4(b) is the reflection of the graph of $y = \cos x$ in the x-axis.

(a)

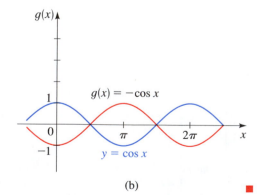

(b)

FIGURE 4

Vertical stretching and shrinking of graphs is discussed in Section 2.5.

Let's graph $y = 2 \sin x$. We start with the graph of $y = \sin x$ and multiply the y-coordinate of each point by 2. This has the effect of stretching the graph vertically by a factor of 2. To graph $y = \frac{1}{2} \sin x$, we start with the graph of $y = \sin x$ and multiply the y-coordinate of each point by $\frac{1}{2}$. This has the effect of shrinking the graph vertically by a factor of $\frac{1}{2}$ (see Figure 5).

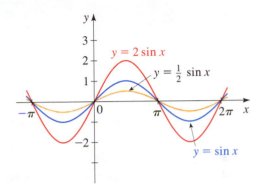

FIGURE 5

In general, for the functions

$$y = a \sin x \qquad \text{and} \qquad y = a \cos x$$

the number $|a|$ is called the **amplitude** and is the largest value these functions attain. Graphs of $y = a \sin x$ for several values of a are shown in Figure 6.

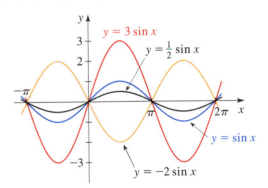

FIGURE 6

EXAMPLE 2 ■ Stretching a Cosine Curve

Find the amplitude of $y = -3 \cos x$ and sketch its graph.

SOLUTION

The amplitude is $|-3| = 3$, so the largest value the graph attains is 3 and the smallest value is -3. To sketch the graph, we begin with the graph of $y = \cos x$, stretch the graph vertically by a factor of 3, and reflect in the x-axis, arriving at the graph in Figure 7.

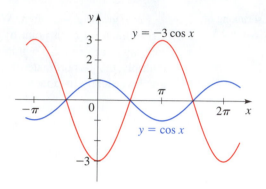

FIGURE 7

Since the sine and cosine functions have period 2π, the functions

$$y = a \sin kx \qquad \text{and} \qquad y = a \cos kx \qquad (k > 0)$$

complete one period as kx varies from 0 to 2π, that is, for $0 \leq kx \leq 2\pi$ or for $0 \leq x \leq 2\pi/k$. So these functions complete one period as x varies between 0 and $2\pi/k$ and thus have period $2\pi/k$. The graphs of these functions are called **sine curves** and **cosine curves**, respectively. (Collectively, sine and cosine curves are often referred to as **sinusoidal** curves.)

SINE AND COSINE CURVES

The sine and cosine curves

$$y = a \sin kx \qquad \text{and} \qquad y = a \cos kx \qquad (k > 0)$$

have amplitude $|a|$ and period $2\pi/k$.

An appropriate interval on which to graph one complete period is $[0, 2\pi/k]$.

Horizontal stretching and shrinking of graphs is discussed in Section 2.5.

To see how the value of k affects the graph of $y = \sin kx$, let's graph the sine curve $y = \sin 2x$. Since the period is $2\pi/2 = \pi$, the graph completes one period in the interval $0 \leq x \leq \pi$ [see Figure 8(a)]. For the sine curve $y = \sin \frac{1}{2}x$, the period is $2\pi \div \frac{1}{2} = 4\pi$, and so the graph completes one period in the interval $0 \leq x \leq 4\pi$ [see Figure 8(b)]. We see that the effect is to shrink the graph if $k > 1$ or to stretch the graph if $k < 1$.

FIGURE 8   (a)   (b)

For comparison, in Figure 9 we show the graphs of one period of the sine curve $y = a \sin kx$ for several values of k.

FIGURE 9

EXAMPLE 3 ■ Amplitude and Period

Find the amplitude and period of each function, and sketch its graph.

(a) $y = 4 \cos 3x$ (b) $y = -2 \sin \frac{1}{2} x$

SOLUTION

(a) For $y = 4 \cos 3x$,

$$\text{amplitude} = |a| = 4$$

$$\text{period} = \frac{2\pi}{3}$$

The graph is shown in Figure 10.

(b) For $y = -2 \sin \frac{1}{2} x$,

$$\text{amplitude} = |a| = |-2| = 2$$

$$\text{period} = \frac{2\pi}{\frac{1}{2}} = 4\pi$$

The graph is shown in Figure 11. ■

The graphs of functions of the form $y = a \sin k(x - b)$ and $y = a \cos k(x - b)$ are simply sine and cosine curves shifted horizontally by an amount $|b|$. They are shifted to the right if $b > 0$ or to the left if $b < 0$. The number b is the **phase shift**. We summarize the properties of these functions in the following box.

FIGURE 10

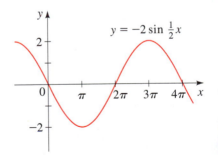

FIGURE 11

SHIFTED SINE AND COSINE CURVES

The sine and cosine curves

$$y = a \sin k(x - b) \qquad \text{and} \qquad y = a \cos k(x - b) \qquad (k > 0)$$

have amplitude $|a|$, period $2\pi/k$, and phase shift b.

An appropriate interval on which to graph one complete period is $[b, b + (2\pi/k)]$.

The graphs of $y = \sin\left(x - \dfrac{\pi}{3}\right)$ and $y = \sin\left(x + \dfrac{\pi}{6}\right)$ are shown in Figure 12.

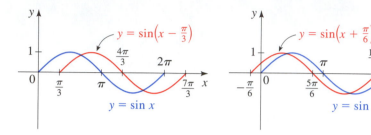

FIGURE 12

EXAMPLE 4 ■ A Shifted Sine Curve

Find the amplitude, period, and phase shift of $y = 3 \sin 2\left(x - \dfrac{\pi}{4}\right)$, and graph one complete period.

SOLUTION

We have

$$\text{amplitude} = |a| = 3$$

$$\text{period} = \frac{2\pi}{2} = \pi$$

$$\text{phase shift} = \frac{\pi}{4} \qquad \text{Shift } \tfrac{\pi}{4} \text{ to the } right$$

Since the phase shift is $\pi/4$ and the period is π, one complete period occurs on the interval

$$\left[\frac{\pi}{4}, \frac{\pi}{4} + \pi\right] = \left[\frac{\pi}{4}, \frac{5\pi}{4}\right]$$

As an aid in sketching the graph, we divide this interval into four equal parts, then graph a sine curve with amplitude 3 as in Figure 13.

Here is another way to find an appropriate interval on which to graph one complete period. Since the period of $y = \sin x$ is 2π, the function $y = 3 \sin 2(x - \frac{\pi}{4})$ will go through one complete period as $2(x - \frac{\pi}{4})$ varies from 0 to 2π.

Start of period:	End of period:
$2\left(x - \frac{\pi}{4}\right) = 0$	$2\left(x - \frac{\pi}{4}\right) = 2\pi$
$x - \frac{\pi}{4} = 0$	$x - \frac{\pi}{4} = \pi$
$x = \frac{\pi}{4}$	$x = \frac{5\pi}{4}$

So we graph one period on the interval $\left[\frac{\pi}{4}, \frac{5\pi}{4}\right]$.

FIGURE 13

EXAMPLE 5 ■ A Shifted Cosine Curve

Find the amplitude, period, and phase shift of

$$y = \frac{3}{4}\cos\left(2x + \frac{2\pi}{3}\right)$$

and graph one complete period.

SOLUTION

We first write this function in the form $y = a\cos k(x - b)$. To do this, we factor 2 from the expression $2x + \frac{2\pi}{3}$ to get

$$y = \frac{3}{4}\cos 2\left[x - \left(-\frac{\pi}{3}\right)\right]$$

Thus, we have

$$\text{amplitude} = |a| = \frac{3}{4}$$

$$\text{period} = \frac{2\pi}{k} = \frac{2\pi}{2} = \pi$$

$$\text{phase shift} = b = -\frac{\pi}{3} \qquad \text{Shift } \tfrac{\pi}{3} \text{ to the } \textit{left}$$

We can also find one complete period as follows:

Start of period:

$2x + \frac{2\pi}{3} = 0$

$2x = -\frac{2\pi}{3}$

$x = -\frac{\pi}{3}$

End of period:

$2x + \frac{2\pi}{3} = 2\pi$

$2x = \frac{4\pi}{3}$

$x = \frac{2\pi}{3}$

So we graph one period on the interval $\left[-\frac{\pi}{3}, \frac{2\pi}{3}\right]$.

From this information it follows that one period of this cosine curve begins at $-\pi/3$ and ends at $(-\pi/3) + \pi = 2\pi/3$. To sketch the graph over the interval $[-\pi/3, 2\pi/3]$, we divide this interval into four equal parts and graph a cosine curve with amplitude $\frac{3}{4}$ as shown in Figure 14.

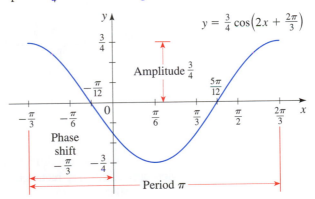

FIGURE 14

Using Graphing Devices to Graph Trigonometric Functions

In Section 1.9 we discussed the use of graphing calculators and computers, and we saw that it's important to choose a viewing rectangle carefully in order to produce a reasonable graph of a function. This is especially true for trigonometric functions; Example 6 shows that, if care is not taken, it's easy to produce a very misleading graph of a trigonometric function.

EXAMPLE 6 ■ Choosing the Viewing Rectangle

Graph the function $f(x) = \sin 50x$ in an appropriate viewing rectangle.

SOLUTION

Figure 15(a) shows the graph of f produced by a graphing calculator using the viewing rectangle $[-12, 12]$ by $[-1.5, 1.5]$. At first glance the graph appears to be reasonable. But if we change the viewing rectangle to the ones shown in Figure 15, the graphs look very different. Something strange is happening.

The appearance of the graphs in Figure 15 depends on the machine used. The graphs you get with your own graphing device might not look like these figures, but they will also be quite inaccurate.

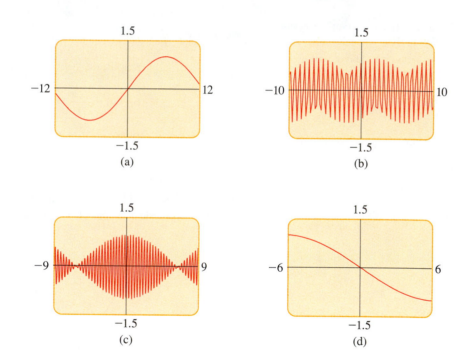

FIGURE 15

Graphs of $f(x) = \sin 50x$ in different viewing rectangles

To explain the big differences in appearance of these graphs and to find an appropriate viewing rectangle, we need to find the period of the function $y = \sin 50x$:

$$\text{period} = \frac{2\pi}{50} = \frac{\pi}{25} \approx 0.126$$

This suggests that we should deal only with small values of x in order to show just a few oscillations of the graph. If we choose the viewing rectangle $[-0.25, 0.25]$ by $[-1.5, 1.5]$, we get the graph shown in Figure 16.

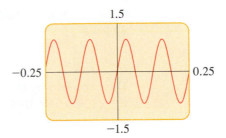

FIGURE 16

$f(x) = \sin 50x$

The function h in Example 7 is **periodic** with period 2π. In general, functions that are sums of functions from the following list

$1, \cos kx, \cos 2kx, \cos 3kx, \ldots$

$\sin kx, \sin 2kx, \sin 3kx, \ldots$

are periodic. Although these functions appear to be special, they are actually fundamental to describing all periodic functions that arise in practice. The French mathematician J. B. J. Fourier (see page 551) discovered that nearly every periodic function can be written as a sum (usually an infinite sum) of these functions (see Section 10.3). This is remarkable because it means that any situation in which periodic variation occurs can be described mathematically using the functions sine and cosine. A modern application of Fourier's discovery is the digital encoding of sound on compact discs.

Now we see what went wrong in Figure 15. The oscillations of $y = \sin 50x$ are so rapid that when the calculator plots points and joins them, it misses most of the maximum and minimum points and therefore gives a very misleading impression of the graph.

EXAMPLE 7 ■ A Sum of Sine and Cosine Curves

Graph $f(x) = 2 \cos x$, $g(x) = \sin 2x$, and $h(x) = 2 \cos x + \sin 2x$ on a common screen to illustrate the method of graphical addition.

SOLUTION

Notice that $h = f + g$, so its graph is obtained by adding the corresponding y-coordinates of the graphs of f and g. The graphs of f, g, and h are shown in Figure 17.

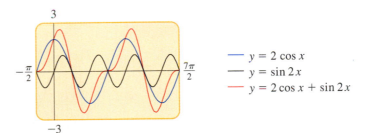

FIGURE 17

$\quad\quad y = 2 \cos x$
$\quad\quad y = \sin 2x$
$\quad\quad y = 2 \cos x + \sin 2x$

EXAMPLE 8 ■ A Cosine Curve with Exponential Amplitude

Graph the functions $y = e^{-x}$, $y = -e^{-x}$, and $y = e^{-x} \cos 6\pi x$ on a common screen. Comment on and explain the relationship among the graphs.

SOLUTION

Figure 18 shows all three graphs in the viewing rectangle $[-1, 2]$ by $[-3, 3]$. It appears that the graph of $y = e^{-x} \cos 6\pi x$ lies between the graphs of the exponential functions $y = e^{-x}$ and $y = -e^{-x}$.

To understand this, recall that the values of $\cos 6\pi x$ lie between -1 and 1, that is,

$$-1 \leq \cos 6\pi x \leq 1$$

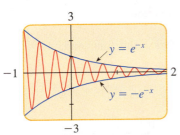

FIGURE 18

$y = e^{-x} \cos 6\pi x$

for all values of x. Multiplying the inequalities by e^{-x}, and noting that $e^{-x} \geq 0$, we get

$$-e^{-x} \leq e^{-x} \cos 6\pi x \leq e^{-x}$$

This explains why the exponential functions form a boundary for the graph of $y = e^{-x} \cos 6\pi x$. (Note that the graphs touch when $\cos 6\pi x = \pm 1$.)   ■

Example 8 shows that the function $y = e^{-x}$ controls the amplitude of the graph of $y = e^{-x} \cos 6\pi x$. In general, if $f(x) = a(x) \sin kx$ or $a(x) \cos kx$, the function a determines how the amplitude of f varies, and the graph of f lies between the graphs of $y = -a(x)$ and $y = a(x)$. Here is another example.

EXAMPLE 9 ■ A Cosine Curve with Variable Amplitude

Graph the function $f(x) = \cos 2\pi x \cos 16\pi x$.

SOLUTION

The graph is shown in Figure 19. Although it was drawn by a computer, we could have drawn it by hand, by first sketching the boundary curves $y = \cos 2\pi x$ and $y = -\cos 2\pi x$. The graph of f is a cosine curve that lies between the graphs of these two functions.

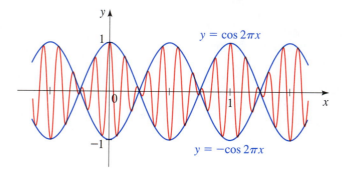

FIGURE 19

$f(x) = \cos 2\pi x \cos 16\pi x$

■

EXAMPLE 10 ■ A Sine Curve with Decaying Amplitude

The function $f(x) = \dfrac{\sin x}{x}$ is important in calculus. Graph this function and comment on its behavior when x is close to 0.

SOLUTION

The viewing rectangle $[-15, 15]$ by $[-0.5, 1.5]$ shown in Figure 20(a) gives a good global view of the graph of f. The viewing rectangle $[-1, 1]$ by $[-0.5, 1.5]$ in Figure 20(b) focuses on the behavior of f when $x \approx 0$. Notice that although $f(x)$

is not defined when $x = 0$ (in other words, 0 is not in the domain of f), the values of f seem to approach 1 when x gets close to 0. This fact is crucial in calculus.

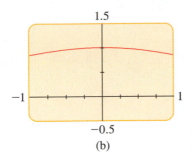

FIGURE 20

$$f(x) = \frac{\sin x}{x}$$

(a)   (b)

The function in Example 10 can be written as

$$f(x) = \left(\frac{1}{x}\right) \cdot \sin x$$

and may thus be viewed as a sine function whose amplitude is controlled by the function $a(x) = 1/x$.

5.3 EXERCISES

1–10 ■ Graph the function.

1. $y = 1 + \sin x$

2. $y = -\sin x$

3. $y = 1 - \cos x$

4. $y = -2 + \cos x$

5. $y = -2 \sin x$

6. $y = 3 \cos x$

7. $y = 4 - 2 \cos x$

8. $y = 3 + 3 \sin x$

9. $y = |\cos x|$

10. $y = |\sin x|$

11–20 ■ Find the amplitude and period of the function, and sketch its graph.

11. $y = \cos 4x$

12. $y = -\sin 2x$

13. $y = 3 \sin 3x$

14. $y = -2 \sin 2\pi x$

15. $y = 10 \sin \frac{1}{2}x$

16. $y = \cos 10\pi x$

17. $y = -\cos \frac{1}{3}x$

18. $y = \sin(-2x)$

19. $y = 3 \cos 3\pi x$

20. $y = 5 - 2 \sin 2x$

21–34 ■ Find the amplitude, period, and phase shift of the function, and graph one complete period.

21. $y = \cos\left(x - \frac{\pi}{2}\right)$

22. $y = 2 \sin\left(x - \frac{\pi}{3}\right)$

23. $y = -2 \sin\left(x - \frac{\pi}{6}\right)$

24. $y = 3 \cos\left(x + \frac{\pi}{4}\right)$

25. $y = 5 \cos\left(3x - \frac{\pi}{4}\right)$

26. $y = -4 \sin 2\left(x + \frac{\pi}{2}\right)$

27. $y = 2 \sin\left(\frac{2}{3}x - \frac{\pi}{6}\right)$

28. $y = \sin \frac{1}{2}\left(x + \frac{\pi}{4}\right)$

29. $y = 3 \cos \pi\left(x + \frac{1}{2}\right)$

30. $y = 1 + \cos\left(3x + \frac{\pi}{2}\right)$

31. $y = -\frac{1}{2} \cos\left(2x - \frac{\pi}{3}\right)$

32. $y = 3 + 2 \sin 3(x + 1)$

33. $y = \sin(3x + \pi)$

34. $y = \cos\left(\frac{\pi}{2} - x\right)$

35–40 ■ The graph of one complete period of a sine or cosine curve is given.

(a) Find the amplitude, period, and phase shift.

(b) Write an equation that represents the curve in the form

$$y = a \sin k(x - b) \qquad \text{or} \qquad y = a \cos k(x - b)$$

35.

36.

37.

38.

39.

40.

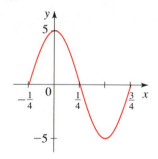

41–48 ■ Determine an appropriate viewing rectangle for each function, and use it to draw the graph.

41. $f(x) = \cos 100x$

42. $f(x) = 3 \sin 120x$

43. $f(x) = \sin(x/40)$

44. $f(x) = \cos(x/80)$

45. $y = \tan 25x$

46. $y = \csc 40x$

47. $y = e^{\sin 20x}$

48. $y = \sqrt{\tan 10\pi x}$

49. As a wave passes by an offshore piling, the height of the water is modeled by the function

$$h(t) = 3 \cos\left(\frac{\pi}{10}t\right)$$

where $h(t)$ is the height in feet above mean sea level at time t seconds.

(a) Find the period of the wave.
(b) Find the wave height, that is, the vertical distance between the trough and the crest of the wave.

crest

trough

50. A tuning fork is struck, producing a pure tone as its tines vibrate. The vibrations are modeled by the function

$$v(t) = 0.7 \sin(880\pi t)$$

where $v(t)$ is the displacement of the tines in millimeters at time t seconds.

(a) Find the period of the vibration.
(b) Find the frequency of the vibration, that is, the number of times the fork vibrates per second.
(c) Graph the function v.

51. Each time your heart beats, your blood pressure first increases and then decreases as the heart rests between beats. The maximum and minimum blood pressures are called the *systolic* and *diastolic* pressures, respectively. Your *blood pressure reading* is written as systolic/diastolic. A reading of 120/80 is considered normal.

A certain person's blood pressure is modeled by the function

$$p(t) = 115 + 25 \sin(160\pi t)$$

where $p(t)$ is the pressure in mmHg, at time t measured in minutes.

(a) Find the period of p.
(b) Find the number of heartbeats per minute.
(c) Graph the function p.
(d) Find the blood pressure reading. How does this compare to normal blood pressure?

52. Variable stars are ones whose brightness varies periodically. One of the most visible is R Leonis; its brightness is modeled by the function

$$b(t) = 7.9 - 2.1 \cos\left(\frac{\pi}{156}t\right)$$

where t is measured in days.

(a) Find the period of R Leonis.

(b) Find the maximum and minimum brightness.
(c) Graph the function b.

53–54 ■ Graph f, g, and $f + g$ on a common screen to illustrate graphical addition.

53. $f(x) = x$, $g(x) = \sin x$

54. $f(x) = \sin x$, $g(x) = \sin 2x$

55–60 ■ Graph the three functions on a common screen. How are the graphs related?

55. $y = x^2$, $y = -x^2$, $y = x^2 \sin x$

56. $y = x$, $y = -x$, $y = x \cos x$

57. $y = e^x$, $y = -e^x$, $y = e^x \sin 5\pi x$

58. $y = \dfrac{1}{1+x^2}$, $y = -\dfrac{1}{1+x^2}$, $y = \dfrac{\cos 2\pi x}{1+x^2}$

59. $y = \cos 3\pi x$, $y = -\cos 3\pi x$, $y = \cos 3\pi x \cos 21\pi x$

60. $y = \sin 2\pi x$, $y = -\sin 2\pi x$, $y = \sin 2\pi x \sin 10\pi x$

61–64 ■ Find the maximum and minimum values of the function.

61. $y = \sin x + \sin 2x$

62. $y = x - 2 \sin x$, $0 \le x \le 2\pi$

63. $y = 2 \sin x + \sin^2 x$

64. $y = \dfrac{\cos x}{2 + \sin x}$

65–68 ■ Find all solutions of the equation that lie in the interval $[0, \pi]$. State each answer correct to two decimal places.

65. $\cos x = 0.4$

66. $\tan x = 2$

67. $\csc x = 3$

68. $\cos x = x$

69. Let $f(x) = \dfrac{1 - \cos x}{x}$.

(a) Is the function f even, odd, or neither?
(b) Find the x-intercepts of the graph of f.
(c) Graph f in an appropriate viewing rectangle.
(d) Describe the behavior of the function as $x \to \infty$ and as $x \to -\infty$.
(e) Notice that $f(x)$ is not defined when $x = 0$. What happens as $x \to 0$?

▲ **DISCOVERY · DISCUSSION**

70. Compositions Involving Trigonometric Functions
This exercise explores the effect of the inner function g on a composite function $y = f(g(x))$.
(a) Graph the function $y = \sin(\sqrt{x})$ using the viewing rectangle $[0, 400]$ by $[-1.5, 1.5]$. In what ways does this graph differ from the graph of the sine function?
(b) Graph the function $y = \sin(x^2)$ using the viewing rectangle $[-5, 5]$ by $[-1.5, 1.5]$. In what ways does this graph differ from the graph of the sine function?

71. Periodic Functions I Recall that a function f is *periodic* if there is a positive number p such that $f(t + p) = f(t)$ for every t, and the least such p (if it exists) is the *period* of f. The graph of a function of period p looks the same on each interval of length p, so we can easily determine the period from the graph. Determine whether the function whose graph is shown is periodic; if it is periodic, find the period.

(a)

(b)

(c)

(d)

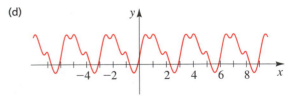

72. Periodic Functions II Use a graphing device to graph the following functions. From the graph, determine whether the function is periodic; if it is periodic find the period.
(a) $y = |\sin x|$
(b) $y = \sin|x|$
(c) $y = 2^{\cos x}$
(d) $y = x - [\![x]\!]$

73. Sinusoidal Curves The graph of $y = \sin x$ is the same as the graph of $y = \cos x$ shifted to the right $\pi/2$ units. So the sine curve $y = \sin x$ is also at the same time a cosine curve: $y = \cos(x - \pi/2)$. In fact, any sine curve is also a cosine curve with a different phase shift, and any cosine curve is also a sine curve. Sine and cosine curves are collectively referred to as *sinusoidal*. Find all possible ways of expressing the curve whose graph is shown as a sine curve $y = a \sin(x - b)$ and as a cosine curve $y = a \cos(x - b)$. Explain why you think you have found all possible choices for a and b in each case.

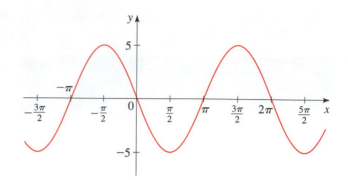

Discovery Project

Predator/Prey Models

Sine and cosine functions are used primarily in physics and engineering to model oscillatory behavior, such as the motion of a pendulum or the current in an AC electrical circuit. (See *Focus on Modeling*, pages 455–471.) But these functions also arise in the other sciences. In this project, we consider an application to biology—we use sine functions to model the population of a predator and its prey.

An isolated island is inhabited by two species of mammals: lynx and hares. The lynx are *predators* who feed on the hares, their *prey*. The lynx and hare populations change cyclically, as graphed in Figure 1. In part A of the graph, hares are abundant, so the lynx have plenty to eat and their population increases. By the time portrayed in part B, so many lynx are feeding on the hares that the hare population declines. In part C, the hare population has declined so much that there is not enough food for the lynx, so the lynx population starts to decrease. In part D, so many lynx have died that the hares have few enemies and their population increases again. This takes us back where we started, and the cycle repeats over and over again.

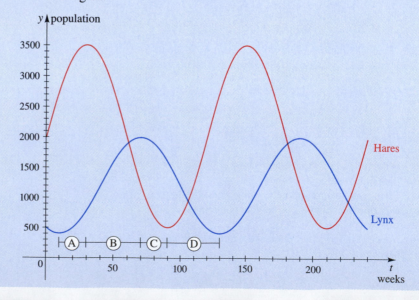

FIGURE 1

The graphs in Figure 1 are sine curves that have been shifted upward, so they are graphs of functions of the form

$$y = a \sin k(t - b) + c$$

Here c is the amount by which the sine curve has been shifted vertically (see Section 2.5). Note that c is the average value of the function, halfway between the highest and lowest values on the graph. The amplitude $|a|$ is the amount by which the graph varies above and below the average value (see Figure 2).

FIGURE 2
$y = a \sin k(t - b) + c$

1. Find functions of the form $y = a \sin k(t - b) + c$ that model the lynx and hare populations graphed in Figure 1. Graph both functions on your calculator and compare to Figure 1 to verify that your functions are the right ones.

2. Add the lynx and hare population functions to get a new function that models the total *mammal* population on this island. Graph this function on your calculator, and find its average value, amplitude, period, and phase shift. How are the average value and period of the mammal population function related to the average value and period of the lynx and hare population functions?

3. A small lake on the island contains two species of fish: hake and redfish. The hake are predators that eat the redfish. The fish population in the lake varies periodically with period 180 days. The number of hake varies between 500 and 1500, and the number of redfish varies between 1000 and 3000. The hake reach their maximum population 30 days after the redfish have reached *their* maximum population in the cycle.

   (a) Sketch a graph (like the one in Figure 1) that shows two complete periods of the population cycle for these species of fish. Assume that $t = 0$ corresponds to a time when the redfish population is at a maximum.

   (b) Find cosine functions of the form $y = a \cos k(t - b) + c$ that model the hake and redfish populations in the lake.

4. In real life, most predator/prey populations do not behave as simply as the examples we have described here. In most cases, the populations of predator and prey oscillate, but the amplitude of the oscillations gets smaller and smaller, so that eventually both populations stabilize near a constant value.

   (a) Sketch a graph that illustrates how the populations of predator and prey might behave in this case.

   (b) What form of function could we use to model the populations in this case?

5.4 MORE TRIGONOMETRIC GRAPHS

In this section we graph the tangent, cotangent, secant, and cosecant functions, and transformations of these functions.

Graphs of the Tangent, Cotangent, Secant, and Cosecant Function

We begin by stating the periodic properties of these functions. Recall that sine and cosine have period 2π. Since cosecant and secant are the reciprocals of sine and cosine, respectively, they also have period 2π (see Exercise 47). Tangent and cotangent, however, have period π (see Exercise 77 of Section 5.2).

> **PERIODIC PROPERTIES**
>
> The functions tangent and cotangent have period π:
>
> $$\tan(x + \pi) = \tan x \qquad \cot(x + \pi) = \cot x$$
>
> The functions cosecant and secant have period 2π:
>
> $$\csc(x + 2\pi) = \csc x \qquad \sec(x + 2\pi) = \sec x$$

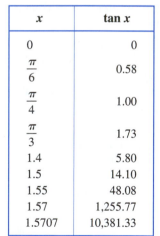

x	$\tan x$
0	0
$\dfrac{\pi}{6}$	0.58
$\dfrac{\pi}{4}$	1.00
$\dfrac{\pi}{3}$	1.73
1.4	5.80
1.5	14.10
1.55	48.08
1.57	1,255.77
1.5707	10,381.33

We first sketch the graph of tangent. Since it has period π, we need only sketch the graph on any interval of length π and then repeat the pattern to the left and to the right. We sketch the graph on the interval $(-\pi/2, \pi/2)$. Since $\tan \pi/2$ and $\tan(-\pi/2)$ aren't defined, we need to be careful in sketching the graph at points near $\pi/2$ and $-\pi/2$. As x gets near $\pi/2$ through values less than $\pi/2$, the value of $\tan x$ becomes large. To see this, notice that as x gets close to $\pi/2$, $\cos x$ approaches 0 and $\sin x$ approaches 1 and so $\tan x = \sin x/\cos x$ is large. A table of values of $\tan x$ for x close to $\pi/2$ (≈ 1.570796) is shown in the margin.

Thus, by choosing x close enough to $\pi/2$ through values less than $\pi/2$, we can make the value of $\tan x$ larger than any given positive number. In a similar way, by choosing x close to $-\pi/2$ through values greater than $-\pi/2$, we can make $\tan x$ smaller than any given negative number. In the notation of Section 3.6 we have

$$\tan x \to \infty \qquad \text{as} \qquad x \to \frac{\pi^-}{2}$$

$$\tan x \to -\infty \qquad \text{as} \qquad x \to -\frac{\pi^+}{2}$$

Thus, $x = \pi/2$ and $x = -\pi/2$ are vertical asymptotes (see Section 3.6). With the information we have so far, we sketch the graph of $\tan x$ for $-\pi/2 < x < \pi/2$ in

Figure 1. The complete graph of tangent [see Figure 5(a) on page 446] is now obtained using the fact that tangent is periodic with period π.

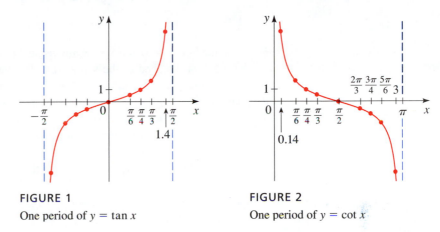

FIGURE 1

One period of $y = \tan x$

FIGURE 2

One period of $y = \cot x$

The function $y = \cot x$ is graphed on the interval $(0, \pi)$ by a similar analysis (see Figure 2). Since $\cot x$ is undefined for $x = n\pi$ with n an integer, its complete graph [in Figure 5(b)] has vertical asymptotes at these values.

To graph the cosecant and secant functions, we use the reciprocal identities

$$\csc x = \frac{1}{\sin x} \qquad \text{and} \qquad \sec x = \frac{1}{\cos x}$$

So, to graph $y = \csc x$, we take the reciprocals of the y-coordinates of the points of the graph of $y = \sin x$. (See Figure 3.) Similarly, to graph $y = \sec x$, we take the reciprocals of the y-coordinates of the points of the graph of $y = \cos x$. (See Figure 4.)

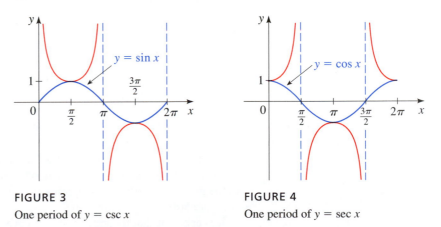

FIGURE 3

One period of $y = \csc x$

FIGURE 4

One period of $y = \sec x$

Let's consider more closely the graph of the function $y = \csc x$ on the interval $0 < x < \pi$. We need to examine the values of the function near 0 and π since at these values $\sin x = 0$, and $\csc x$ is thus undefined. We see that

$$\csc x \to \infty \qquad \text{as} \qquad x \to 0^{+}$$

$$\csc x \to \infty \qquad \text{as} \qquad x \to \pi^{-}$$

Thus, the lines $x = 0$ and $x = \pi$ are vertical asymptotes. In the interval $\pi < x < 2\pi$ the graph is sketched in the same way. The values of $\csc x$ in that interval are the same as those in the interval $0 < x < \pi$ except for sign (see Figure 3). The complete graph in Figure 5(c) is now obtained from the fact that the function cosecant is periodic with period 2π. Note that the graph has vertical asymptotes at the points where $\sin x = 0$, that is, at $x = n\pi$, for n an integer.

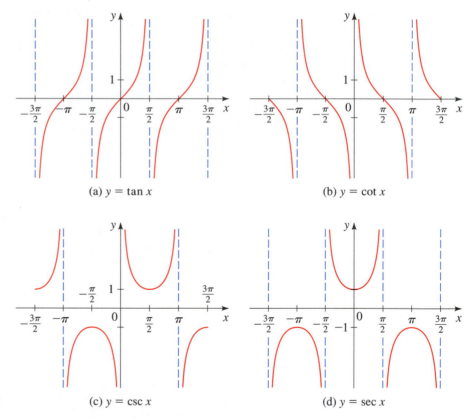

(a) $y = \tan x$

(b) $y = \cot x$

(c) $y = \csc x$

(d) $y = \sec x$

FIGURE 5

The graph of $y = \sec x$ is sketched in a similar manner. Observe that the domain of $\sec x$ is the set of all real numbers other than $x = (\pi/2) + n\pi$, for n an integer, so the graph has vertical asymptotes at those points. The complete graph is shown in Figure 5(d).

It is apparent that the graphs of $y = \tan x$, $y = \cot x$, and $y = \csc x$ are symmetric about the origin, whereas that of $y = \sec x$ is symmetric about the y-axis. This is because tangent, cotangent, and cosecant are odd functions, whereas secant is an even function.

■ Graphs Involving Tangent and Cotangent Functions

We now consider graphs of transformations of the tangent and cotangent functions.

EXAMPLE 1 ■ **Graphing Tangent Curves**

Graph each function.

(a) $y = 2 \tan x$ (b) $y = -\tan x$

SOLUTION

We first graph $y = \tan x$ and then transform it as required.

(a) To graph $y = 2 \tan x$, we multiply the y-coordinate of each point on the graph of $y = \tan x$ by 2. The resulting graph is shown in Figure 6(a).

(b) The graph of $y = -\tan x$ in Figure 6(b) is obtained from that of $y = \tan x$ by reflecting in the x-axis.

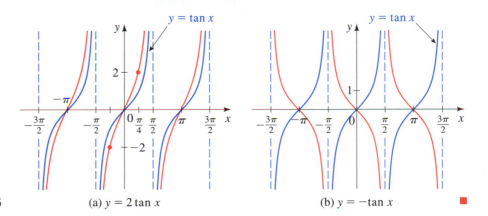

FIGURE 6   (a) $y = 2 \tan x$ (b) $y = -\tan x$ ■

Since the tangent and cotangent functions have period π, the functions

$$y = a \tan kx \qquad \text{and} \qquad y = a \cot kx \qquad (k > 0)$$

complete one period as kx varies from 0 to π, that is, for $0 \le kx \le \pi$. Solving this inequality, we get $0 \le x \le \pi/k$. So they each have period π/k.

> ### TANGENT AND COTANGENT CURVES
>
> The functions
>
> $$y = a \tan kx \qquad \text{and} \qquad y = a \cot kx \qquad (k > 0)$$
>
> have period π/k.

Thus, one complete period of the graphs of these functions occurs on any interval of length π/k. To sketch a complete period of these graphs, it's convenient to select an interval between vertical asymptotes:

To graph one period of $y = a \tan kx$, an appropriate interval is $\left(-\dfrac{\pi}{2k}, \dfrac{\pi}{2k}\right)$.

To graph one period of $y = a \cot kx$, an appropriate interval is $\left(0, \dfrac{\pi}{k}\right)$.

EXAMPLE 2 ■ Graphing Tangent Curves

Graph each function.

(a) $y = \tan 2x$ (b) $y = \tan 2\left(x - \dfrac{\pi}{4}\right)$

SOLUTION

(a) The period is $\pi/2$ and an appropriate interval is $(-\pi/4, \pi/4)$. The endpoints $x = -\pi/4$ and $x = \pi/4$ are vertical asymptotes. Thus, we graph one complete period of the function on $(-\pi/4, \pi/4)$. The graph has the same shape as that of the tangent function, but is shrunk horizontally by a factor of $\frac{1}{2}$. We then repeat that portion of the graph to the left and to the right. See Figure 7(a).

(b) The graph is the same as that in part (a), but it is shifted to the right $\pi/4$, as shown in Figure 7(b).

Since $y = \tan x$ completes one period between $x = -\frac{\pi}{2}$ and $x = \frac{\pi}{2}$, the function $y = \tan 2\left(x - \frac{\pi}{4}\right)$ completes one period as $2\left(x - \frac{\pi}{4}\right)$ varies from $-\frac{\pi}{2}$ to $\frac{\pi}{2}$.

Start of period: End of period:

$2\left(x - \frac{\pi}{4}\right) = -\frac{\pi}{2}$ $2\left(x - \frac{\pi}{4}\right) = \frac{\pi}{2}$

$x - \frac{\pi}{4} = -\frac{\pi}{4}$ $x - \frac{\pi}{4} = \frac{\pi}{4}$

$x = 0$ $x = \frac{\pi}{2}$

So we graph one period on the interval $\left(0, \frac{\pi}{2}\right)$.

FIGURE 7

(a) $y = \tan 2x$

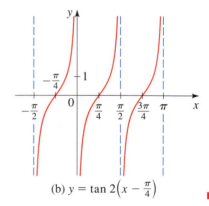

(b) $y = \tan 2\left(x - \frac{\pi}{4}\right)$

EXAMPLE 3 ■ A Shifted Cotangent Curve

Graph $y = 2 \cot\left(3x - \dfrac{\pi}{2}\right)$.

SOLUTION

We first put this in the form $y = a \cot k(x - b)$ by factoring 3 from the expression $3x - \dfrac{\pi}{2}$:

$$y = 2 \cot\left(3x - \frac{\pi}{2}\right) = 2 \cot 3\left(x - \frac{\pi}{6}\right)$$

Since $y = \cot x$ completes one period between $x = 0$ and $x = \pi$, the function $y = 2 \cot\left(3x - \frac{\pi}{2}\right)$ completes one period as $3x - \frac{\pi}{2}$ varies from 0 to π.

Start of period: End of period:

$3x - \frac{\pi}{2} = 0$ $3x - \frac{\pi}{2} = \pi$

$3x = \frac{\pi}{2}$ $3x = \frac{3\pi}{2}$

$x = \frac{\pi}{6}$ $x = \frac{\pi}{2}$

So we graph one period on the interval $\left(\frac{\pi}{6}, \frac{\pi}{2}\right)$.

Thus, the graph is the same as that of $y = 2 \cot 3x$, but is shifted to the right $\pi/6$. The period of $y = 2 \cot 3x$ is $\pi/3$, and an appropriate interval is $(0, \pi/3)$. To get the corresponding interval for the desired graph, we shift this interval to the right $\pi/6$. This gives

$$\left(0 + \frac{\pi}{6}, \frac{\pi}{3} + \frac{\pi}{6}\right) = \left(\frac{\pi}{6}, \frac{\pi}{2}\right)$$

Finally, we graph one period in the shape of cotangent on the interval $(\pi/6, \pi/2)$ and repeat that portion of the graph to the left and to the right. See Figure 8.

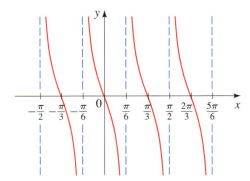

FIGURE 8
$y = 2 \cot\left(3x - \frac{\pi}{2}\right)$

Graphs Involving the Cosecant and Secant Functions

We have already observed that the cosecant and secant functions are the reciprocals of the sine and cosine functions. Thus, the following result is the counterpart of the result for sine and cosine curves in Section 5.3.

COSECANT AND SECANT CURVES

The functions

$$y = a \csc kx \qquad \text{and} \qquad y = a \sec kx \qquad (k > 0)$$

have period $2\pi/k$.

An appropriate interval on which to graph one complete period is $[0, 2\pi/k]$.

EXAMPLE 4 ■ Graphing Cosecant Curves

Graph each function.

(a) $y = \dfrac{1}{2} \csc 2x$

(b) $y = \dfrac{1}{2} \csc\left(2x + \dfrac{\pi}{2}\right)$

SOLUTION

(a) The period is $2\pi/2 = \pi$. An appropriate interval is $[0, \pi]$, and the asymptotes occur in this interval whenever $\sin 2x = 0$. So the asymptotes in this interval are $x = 0$, $x = \pi/2$, and $x = \pi$. With this information we sketch on the interval $[0, \pi]$ a graph with the same general shape as that of one period of the cosecant function. The complete graph in Figure 9(a) is obtained by repeating this portion of the graph to the left and to the right.

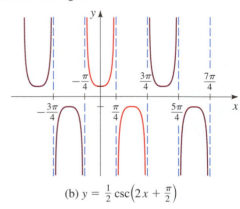

FIGURE 9

(a) $y = \frac{1}{2} \csc 2x$

(b) $y = \frac{1}{2} \csc\left(2x + \frac{\pi}{2}\right)$

Since $y = \csc x$ completes one period between $x = 0$ and $x = 2\pi$, the function $y = \frac{1}{2} \csc\left(2x + \frac{\pi}{2}\right)$ completes one period as $2x + \frac{\pi}{2}$ varies from 0 to 2π.

Start of period: End of period:

$2x + \frac{\pi}{2} = 0$ $2x + \frac{\pi}{2} = 2\pi$

$2x = -\frac{\pi}{2}$ $2x = \frac{3\pi}{2}$

$x = -\frac{\pi}{4}$ $x = \frac{3\pi}{4}$

So we graph one period on the interval $\left(-\frac{\pi}{4}, \frac{3\pi}{4}\right)$.

(b) We first write

$$y = \frac{1}{2} \csc\left(2x + \frac{\pi}{2}\right) = \frac{1}{2} \csc 2\left(x + \frac{\pi}{4}\right)$$

From this we see that the graph is the same as that in part (a), but shifted to the left $\pi/4$. The graph is shown in Figure 9(b). ■

EXAMPLE 5 ■ Graphing a Secant Curve

Graph $y = 3 \sec \frac{1}{2} x$.

SOLUTION

The period is $2\pi \div \frac{1}{2} = 4\pi$. An appropriate interval is $[0, 4\pi]$, and the asymptotes occur in this interval wherever $\cos \frac{1}{2} x = 0$. Thus, the asymptotes in this interval are $x = \pi$, $x = 3\pi$. With this information we sketch on the interval $[0, 4\pi]$ a graph with the same general shape as that of one period of the secant function. The complete graph in Figure 10 is obtained by repeating this portion of the graph to the left and to the right. ■

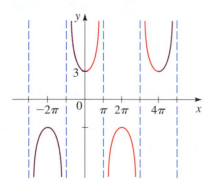

FIGURE 10
$y = 3 \sec \frac{1}{2} x$

5.4   EXERCISES

1–46 ■ Find the period and graph the function.

1. $y = 4 \tan x$

2. $y = -4 \tan x$

3. $y = -\frac{1}{2} \tan x$

4. $y = \frac{1}{2} \tan x$

5. $y = -\cot x$

6. $y = 2 \cot x$

7. $y = 2 \csc x$

8. $y = \frac{1}{2} \csc x$

9. $y = 3 \sec x$

10. $y = -3 \sec x$

11. $y = \tan\left(x + \frac{\pi}{2}\right)$

12. $y = \tan\left(x - \frac{\pi}{4}\right)$

13. $y = \csc\left(x - \frac{\pi}{2}\right)$

14. $y = \sec\left(x + \frac{\pi}{4}\right)$

15. $y = \cot\left(x + \frac{\pi}{4}\right)$

16. $y = 2 \csc\left(x - \frac{\pi}{3}\right)$

17. $y = \frac{1}{2} \sec\left(x - \frac{\pi}{6}\right)$

18. $y = 3 \csc\left(x + \frac{\pi}{2}\right)$

19. $y = \tan 2x$

20. $y = \tan \frac{1}{2}x$

21. $y = \tan \pi x$

22. $y = \cot \frac{\pi}{2}x$

23. $y = \sec 2x$

24. $y = 5 \csc 3x$

25. $y = \csc 2x$

26. $y = \csc \frac{1}{2}x$

27. $y = 2 \tan 3x$

28. $y = 2 \tan \frac{\pi}{2}x$

29. $y = 5 \csc 3x$

30. $y = 5 \sec 2\pi x$

31. $y = \tan 2\left(x + \frac{\pi}{2}\right)$

32. $y = \csc 2\left(x + \frac{\pi}{2}\right)$

33. $y = \tan 2(x - \pi)$

34. $y = \sec 2\left(x - \frac{\pi}{2}\right)$

35. $y = \cot\left(2x - \frac{\pi}{2}\right)$

36. $y = \frac{1}{2} \tan(\pi x - \pi)$

37. $y = 2 \csc\left(\pi x - \frac{\pi}{3}\right)$

38. $y = 2 \sec\left(\frac{1}{2}x - \frac{\pi}{3}\right)$

39. $y = 5 \sec\left(3x - \frac{\pi}{2}\right)$

40. $y = \frac{1}{2} \sec(2\pi x - \pi)$

41. $y = \tan\left(\frac{2}{3}x - \frac{\pi}{6}\right)$

42. $y = \tan \frac{1}{2}\left(x + \frac{\pi}{4}\right)$

43. $y = 3 \sec \pi\left(x + \frac{1}{2}\right)$

44. $y = \sec\left(3x + \frac{\pi}{2}\right)$

45. $y = -2 \tan\left(2x - \frac{\pi}{3}\right)$

46. $y = 2 \csc(3x + 3)$

47. (a) Prove that if f is periodic with period p, then $1/f$ is also periodic with period p.

(b) Prove that cosecant and secant each have period 2π.

 DISCOVERY • DISCUSSION

48. Reduction Formulas   Use the graphs in Figure 5 to explain why the following formulas are true.

$$\tan\left(x - \frac{\pi}{2}\right) = -\cot x$$

$$\sec\left(x - \frac{\pi}{2}\right) = \csc x$$

5   REVIEW

CONCEPT CHECK

1. (a) What is the unit circle?

(b) Use a diagram to explain what is meant by the terminal point determined by a real number t.

(c) What is the reference number $\bar{t}$ associated with t?

(d) If t is a real number and $P(x, y)$ is the terminal point determined by t, write equations that define $\sin t$, $\cos t$, $\tan t$, $\cot t$, $\sec t$, and $\csc t$.

(e) What are the domains of the six functions that you defined in part (d)?

(f) Which trigonometric functions are positive in quadrants I, II, III, and IV?

2. (a) What is an even function?

(b) Which trigonometric functions are even?

(c) What is an odd function?

(d) Which trigonometric functions are odd?

3. (a) State the reciprocal identities.

(b) State the Pythagorean identities.

4. (a) What is a periodic function?
(b) What are the periods of the six trigonometric functions?

5. Graph the sine and cosine functions. How is the graph of cosine related to the graph of sine?

6. Write expressions for the amplitude, period, and phase shift of the sine curve $y = a \sin k(x - b)$ and the cosine curve $y = a \cos k(x - b)$.

7. (a) Graph the tangent and cotangent functions.
(b) State the periods of the tangent curve $y = a \tan kx$ and the cotangent curve $y = a \cot kx$.

8. (a) Graph the secant and cosecant functions.
(b) State the periods of the secant curve $y = a \sec kx$ and the cosecant curve $y = a \csc kx$.

EXERCISES

1–2 ■ A point $P(x, y)$ is given.
(a) Show that P is on the unit circle.
(b) Suppose that P is the terminal point determined by t. Find $\sin t$, $\cos t$, and $\tan t$.

1. $P\left(-\dfrac{\sqrt{3}}{2}, \dfrac{1}{2}\right)$ **2.** $P\left(\dfrac{3}{5}, -\dfrac{4}{5}\right)$

3–6 ■ A real number t is given.
(a) Find the reference number for t.
(b) Find the terminal point $P(x, y)$ on the unit circle determined by t.
(c) Find the six trigonometric functions of t.

3. $t = \dfrac{2\pi}{3}$ **4.** $t = \dfrac{5\pi}{3}$

5. $t = -\dfrac{11\pi}{4}$ **6.** $t = -\dfrac{7\pi}{6}$

7–16 ■ Find the value of the trigonometric function. If possible, give the exact value; otherwise, use a calculator to find an approximate value correct to five decimal places.

7. (a) $\sin \dfrac{3\pi}{4}$ (b) $\cos \dfrac{3\pi}{4}$

8. (a) $\tan \dfrac{\pi}{3}$ (b) $\tan\left(-\dfrac{\pi}{3}\right)$

9. (a) $\sin 1.1$ (b) $\cos 1.1$

10. (a) $\cos \dfrac{\pi}{5}$ (b) $\cos\left(-\dfrac{\pi}{5}\right)$

11. (a) $\cos \dfrac{9\pi}{2}$ (b) $\sec \dfrac{9\pi}{2}$

12. (a) $\sin \dfrac{\pi}{7}$ (b) $\csc \dfrac{\pi}{7}$

13. (a) $\tan \dfrac{5\pi}{2}$ (b) $\cot \dfrac{5\pi}{2}$

14. (a) $\sin 2\pi$ (b) $\csc 2\pi$

15. (a) $\tan \dfrac{5\pi}{6}$ (b) $\cot \dfrac{5\pi}{6}$

16. (a) $\cos \dfrac{\pi}{3}$ (b) $\sin \dfrac{\pi}{6}$

17–20 ■ Use the fundamental identities to write the first expression in terms of the second.

17. $\dfrac{\tan t}{\cos t}$, $\sin t$ **18.** $\tan^2 t \sec t$, $\cos t$

19. $\tan t$, $\sin t$; t in quadrant IV

20. $\sec t$, $\sin t$; t in quadrant II

21–24 ■ Find the values of the remaining trigonometric functions at t from the given information.

21. $\sin t = \frac{5}{13}$, $\cos t = -\frac{12}{13}$

22. $\sin t = -\frac{1}{2}$, $\cos t > 0$

23. $\cot t = -\frac{1}{2}$, $\csc t = \sqrt{5}/2$

24. $\cos t = -\frac{3}{5}$, $\tan t < 0$

25. If $\tan t = \frac{1}{4}$ and the terminal point for t is in quadrant III, find $\sec t + \cot t$.

26. If $\sin t = -\frac{8}{17}$ and the terminal point for t is in quadrant IV, find $\csc t + \sec t$.

27. If $\cos t = \frac{3}{5}$ and the terminal point for t is in quadrant I, find $\tan t + \sec t$.

28. If $\sec t = -5$ and the terminal point for t is in quadrant II, find $\sin^2 t + \cos^2 t$.

29–36 ■ (a) Find the amplitude, period, and phase shift of the function, and (b) sketch the graph.

29. $y = 10 \cos \frac{1}{2} x$ **30.** $y = 4 \sin 2\pi x$

31. $y = -\sin \frac{1}{2}x$

32. $y = 2 \sin\left(x - \frac{\pi}{4}\right)$

33. $y = 3 \sin(2x - 2)$

34. $y = \cos 2\left(x - \frac{\pi}{2}\right)$

35. $y = -\cos\left(\frac{\pi}{2}x + \frac{\pi}{6}\right)$

36. $y = 10 \sin\left(2x - \frac{\pi}{2}\right)$

37–40 ■ The graph of one period of a function of the form $y = a \sin k(x - b)$ or $y = a \cos k(x - b)$ is shown. Determine the function.

37.

38.

(1, 2)

39.

40.

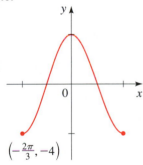

$\left(-\frac{2\pi}{3}, -4\right)$

41–48 ■ Find the period, and sketch the graph.

41. $y = 3 \tan x$

42. $y = \tan \pi x$

43. $y = 2 \cot\left(x - \frac{\pi}{2}\right)$

44. $y = \sec\left(\frac{1}{2}x - \frac{\pi}{2}\right)$

45. $y = 4 \csc(2x + \pi)$

46. $y = \tan\left(x + \frac{\pi}{6}\right)$

47. $y = \tan\left(\frac{1}{2}x - \frac{\pi}{8}\right)$

48. $y = -4 \sec 4\pi x$

 49–54 ■ A function is given.
(a) Use a graphing device to graph the function.
(b) Determine from the graph whether the function is periodic and, if so, determine the period.
(c) Determine from the graph whether the function is odd, even, or neither.

49. $y = |\cos x|$

50. $y = \sin(\cos x)$

51. $y = \cos(2^{0.1x})$

52. $y = 1 + 2^{\cos x}$

53. $y = e^{-|x|} \cos 3x$

54. $y = \ln x \sin 3x \quad (x > 0)$

 55–58 ■ Graph the three functions on a common screen. How are the graphs related?

55. $y = x, \quad y = -x, \quad y = x \sin x$

56. $y = 2^{-x}, \quad y = -2^{-x}, \quad y = 2^{-x} \cos 4\pi x$

57. $y = x, \quad y = \sin 4x, \quad y = x + \sin 4x$

58. $y = \sin^2 x, \quad y = \cos^2 x, \quad y = \sin^2 x + \cos^2 x$

 59–60 ■ Find the maximum and minimum values of the function.

59. $y = \cos x + \sin 2x$

60. $y = \cos x + \sin^2 x$

61. Find the solutions of $\sin x = 0.3$ in the interval $[0, 2\pi]$.

62. Find the solutions of $\cos 3x = x$ in the interval $[0, \pi]$.

63. Let $f(x) = \dfrac{\sin^2 x}{x}$.
(a) Is the function f even, odd, or neither?
(b) Find the x-intercepts of the graph of f.
(c) Graph f in an appropriate viewing rectangle.
(d) Describe the behavior of the function as $x \to \infty$ and as $x \to -\infty$.
(e) Notice that $f(x)$ is not defined when $x = 0$. What happens as $x \to 0$?

 64. Let $y_1 = \cos(\sin x)$ and $y_2 = \sin(\cos x)$.
(a) Graph y_1 and y_2 in the same viewing rectangle.
(b) Determine the period of each of these functions from its graph.
(c) Find an inequality between $\sin(\cos x)$ and $\cos(\sin x)$ that is valid for all x.

5 TEST

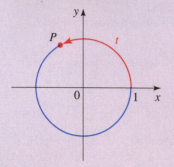

1. The point $P(x, y)$ is on the unit circle in quadrant IV. If $x = \sqrt{11}/6$, find y.

2. The point P in the figure at the left has y-coordinate $\frac{4}{5}$. Find each of the following.
   (a) $\sin t$ (b) $\cos t$
   (c) $\tan t$ (d) $\sec t$

3. Find the exact value of each of the following.
   (a) $\sin \dfrac{7\pi}{6}$ (b) $\cos \dfrac{13\pi}{4}$
   (c) $\tan \left(-\dfrac{5\pi}{3}\right)$ (d) $\csc \dfrac{3\pi}{2}$

4. Express $\tan t$ in terms of $\sin t$, if the terminal point determined by t is in quadrant II.

5. If $\cos t = -\frac{8}{17}$ and if the terminal point determined by t is in quadrant III, find $\tan t \cot t + \csc t$.

6–7 ■ A trigonometric function is given.
(a) Find the amplitude, period, and phase shift of the function.
(b) Sketch the graph.

6. $y = -5 \cos 4x$ 7. $y = 2 \sin\left(\dfrac{1}{2}x - \dfrac{\pi}{6}\right)$

8–9 ■ Find the period, and graph the function.

8. $y = -\csc 2x$ 9. $y = \tan\left(2x - \dfrac{\pi}{2}\right)$

10. The graph shown is one period of a function of the form $y = a \sin k(x - b)$. Determine the function.

 11. (a) Use a graphing device to graph the function in an appropriate viewing rectangle.
   (b) Determine from the graph if the function is even, odd, or neither.
   (c) Find the minimum and maximum values of the function.

$$y = \dfrac{\cos x}{1 + x^2}$$

Focus on Modeling

Harmonic Motion

Periodic behavior—behavior that repeats over and over again—is common in nature. Perhaps the most familiar example is the daily rising and setting of the sun, which results in the repetitive pattern of day, night, day, night, ... Another example is the daily variation of tide levels at the beach, which results in the repetitive pattern of high tide, low tide, high tide, low tide, ... Certain animal populations increase and decrease in a predictable periodic pattern: A large population exhausts the food supply, which causes the population to dwindle; this in turn results in a more plentiful food supply, which makes it possible for the population to increase; and the pattern then repeats over and over (see pages 442–443).

Other common examples of periodic behavior involve motion that is caused by vibration or oscillation. A mass suspended from a spring that has been compressed and then allowed to vibrate vertically is a simple example. This same "back and forth" motion also occurs in such diverse phenomena as sound waves, light waves, alternating electrical current, and pulsating stars, to name a few. In this section we consider the problem of modeling periodic behavior.

■ Modeling Periodic Behavior

The trigonometric functions are ideally suited for modeling periodic behavior. A glance at the graphs of the sine and cosine functions, for instance, shows that these functions themselves exhibit periodic behavior. Figure 1 shows the graph of $y = \sin t$. If we think of t as time, we see that as time goes on, $y = \sin t$ increases and decreases over and over again. Figure 2 shows that the motion of a vibrating mass on a spring is modeled very accurately by $y = \sin t$.

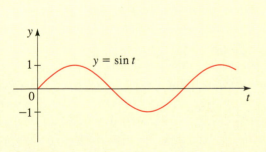

FIGURE 1

$y = \sin t$

FIGURE 2

Motion of a vibrating spring is modeled by $y = \sin t$.

Notice that the mass returns to its original position over and over again. A **cycle** is one complete vibration of an object, so the mass in Figure 2 completes one cycle of its motion between O and P. Our observations about how the sine and cosine functions model periodic behavior are summarized in the following box.

The main difference between the two equations describing simple harmonic motion is the starting point. At $t = 0$, we get

$$y = a \sin \omega \cdot 0 = 0$$

or

$$y = a \cos \omega \cdot 0 = a$$

In other words, in the first case the motion "starts" with zero displacement, while in the second case the motion "starts" with the displacement at maximum (at the amplitude a).

SIMPLE HARMONIC MOTION

If the equation describing the displacement y of an object at time t is

$$y = a \sin \omega t \qquad \text{or} \qquad y = a \cos \omega t$$

then the object is in **simple harmonic motion**. In this case,

amplitude $= |a|$ Maximum displacement of the object

period $= \dfrac{2\pi}{\omega}$ Time required to complete one cycle

frequency $= \dfrac{\omega}{2\pi}$ Number of cycles per unit of time

---

Notice that the functions

$$y = a \sin 2\pi \nu t \qquad \text{and} \qquad y = a \cos 2\pi \nu t$$

have frequency ν, because $2\pi\nu/(2\pi) = \nu$. Since we can immediately read the frequency from these equations, we often write equations of simple harmonic motion in this form.

EXAMPLE 1 ■ A Vibrating Spring

The displacement of a mass suspended by a spring is modeled by the function

$$y = 10 \sin 4\pi t$$

where y is measured in inches and t in seconds (see Figure 3).

(a) Find the amplitude, period, and frequency of the motion of the mass.
(b) Sketch the graph of the displacement of the mass.

SOLUTION

(a) From the formulas for amplitude, period, and frequency, we get

$$\text{amplitude} = |a| = 10 \text{ in.}$$

$$\text{period} = \frac{2\pi}{\omega} = \frac{2\pi}{4\pi} = \frac{1}{2} \text{ s}$$

$$\text{frequency} = \frac{\omega}{2\pi} = \frac{4\pi}{2\pi} = 2 \text{ Hz}$$

(b) The graph of the displacement of the mass at time t is shown in Figure 4.   ■

An important situation where simple harmonic motion occurs is in the production of sound. Sound is produced by a regular variation in air pressure from the nor-

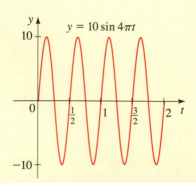

Rest position

FIGURE 3

$y = 10 \sin 4\pi t$

FIGURE 4

mal pressure. If the pressure varies in simple harmonic motion, then a pure sound is produced. The tone of the sound depends on the frequency and the loudness depends on the amplitude.

EXAMPLE 2 ■ Vibrations of a Musical Note

A tuba player plays the note E and sustains the sound for some time. For a pure E the variation in pressure from normal air pressure is given by

$$V(t) = 0.2 \sin 80\pi t$$

where V is measured in pounds per square inch and t in seconds.

(a) Find the amplitude, period, and frequency of V.
(b) Sketch a graph of V.
(c) If the tuba player increases the loudness of the note, how does the equation for V change?
(d) If the player is playing the note incorrectly and it is a little flat, how does the equation for V change?

SOLUTION

(a) From the formulas for amplitude, period, and frequency, we get

$$\text{amplitude} = |0.2| = 0.2$$

$$\text{period} = \frac{2\pi}{80\pi} = \frac{1}{40}$$

$$\text{frequency} = \frac{80\pi}{2\pi} = 40$$

(b) The graph of V is shown in Figure 5.

(c) If the player increases the loudness the amplitude increases. So the number 0.2 is replaced by a larger number.

(d) If the note is flat, then the frequency is decreased. Thus, the coefficient of t is less than 80π. ■

$y = 0.2 \sin 80\pi t$

FIGURE 5

EXAMPLE 3 ■ Modeling a Vibrating Spring

A mass is suspended from a spring. The spring is compressed a distance of 4 cm and then released. It is observed that the mass returns to the compressed position after $\frac{1}{3}$ s.

(a) Find a function that models the displacement of the mass.
(b) Sketch the graph of the displacement of the mass.

SOLUTION

(a) The motion of the mass is given by one of the equations for simple harmonic motion. The amplitude of the motion is 4 cm. Since this amplitude is reached

4 cm

Rest
position

at time $t = 0$, an appropriate function that models the displacement is of the form

$$y = a \cos \omega t$$

Since the period is $p = \frac{1}{3}$, we can find ω from the following equation:

$$\text{period} = \frac{2\pi}{\omega}$$

$$\frac{1}{3} = \frac{2\pi}{\omega} \qquad \text{Period} = \frac{1}{3}$$

$$\omega = 6\pi \qquad \text{Solve for } \omega$$

So, the motion of the mass is modeled by the function

$$y = 4 \cos 6\pi t$$

where y is the displacement from the rest position at time t. Notice that when $t = 0$, the displacement is $y = 4$, as we expect.

(b) The graph of the displacement of the mass at time t is shown in Figure 6. ■

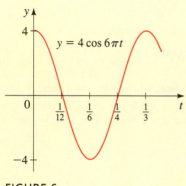

FIGURE 6

In general, the sine or cosine functions representing harmonic motion may be shifted horizontally or vertically. In this case, the equations take the form

$$y = a \sin(\omega(t - c)) + b \qquad \text{or} \qquad y = a \cos(\omega(t - c)) + b$$

The vertical shift b indicates that the variation occurs around an average value b. The horizontal shift c indicates the position of the object at $t = 0$. (See Figure 7.)

FIGURE 7   (a)   (b)

EXAMPLE 4 ■ Modeling the Brightness of a Variable Star

A variable star is one whose brightness alternately increases and decreases. For the variable star Delta Cephei, the time between periods of maximum brightness is 5.4

days. The average brightness (or magnitude) of the star is 4.0, and its brightness varies by ±0.35 magnitude.

(a) Find a function that models the brightness of Delta Cephei as a function of time.

(b) Sketch a graph of the brightness of Delta Cephei as a function of time.

SOLUTION

(a) Let's find a function in the form

$$y = a \cos(\omega(t - c)) + b$$

The amplitude is the maximum variation from average brightness, so the amplitude is $a = 0.35$ magnitude. We are given that the period is 5.4 days, so

$$\omega = \frac{2\pi}{5.4} \approx 1.164$$

Since the brightness varies from an average value of 4.0 magnitudes, the graph is shifted upward by $b = 4.0$. If we take $t = 0$ to be a time when the star is at maximum brightness, there is no horizontal shift, so $c = 0$ (this is because a cosine curve achieves its maximum at $t = 0$). Thus, the function we want is

$$y = 0.35 \cos(1.16t) + 4.0$$

where t is the number of days from a time when the star is at maximum brightness.

(b) The graph is sketched in Figure 8.

FIGURE 8

The number of hours of daylight varies throughout the course of a year. In the Northern Hemisphere, the longest day is June 21, and the shortest is December 21. The average length of daylight is 12 h, and the variation from this average depends on the latitude. (For example, Fairbanks, Alaska, experiences more than 20 h of daylight on the longest day and less than 4 h on the shortest day!) The graph in Figure 9 shows the number of hours of daylight at different times of the year for various latitudes. It's apparent from the graph that the variation in hours of daylight is simple harmonic.

FIGURE 9

Graph of the length of daylight from March 21 through December 21 at various latitudes

EXAMPLE 5 ■ Modeling the Number of Hours of Daylight

In Philadelphia (40° N latitude), the longest day of the year has 14 h 50 min of daylight and the shortest day has 9 h 10 min of daylight.

(a) Find a function L that models the length of daylight as a function of t, the number of days from January 1.

(b) An astronomer needs at least 11 hours of darkness for a long exposure astronomical photograph. During which days of the year can he do this?

SOLUTION

(a) We need to find a function in the form

$$y = a \sin(\omega(t - c)) + b$$

whose graph is the 40° N latitude curve in Figure 9. From the information given, we see that the amplitude is

$$a = \tfrac{1}{2}\left(14\tfrac{5}{6} - 9\tfrac{1}{6}\right) \approx 2.83 \text{ h}$$

Since there are 365 days in a year, the period is 365, so

$$\omega = \frac{2\pi}{365} \approx 0.0172$$

Since the average length of daylight is 12 h, the graph is shifted upward by 12, so $b = 12$. Since the curve attains the average value (12) on March 21, the

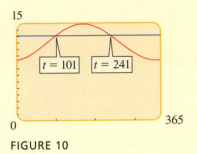

FIGURE 10

80th day of the year, the curve is shifted 80 units to the right. Thus, $c = 80$. So a function that models the number of hours of daylight is

$$y = 2.83 \sin(0.0172(t - 80)) + 12$$

where t is the number of days from January 1.

(b) Since a day has 24 h, 11 h of night correspond to 13 h of daylight. So we need to solve the inequality $y \le 13$. To solve this inequality graphically, we graph $y = 2.83 \sin 0.0172(t - 80) + 12$ and $y = 13$ on the same graph. From the graph in Figure 10 we see that there are fewer than 13 h of daylight between day 1 (January 1) and day 101 (April 11) and from day 241 (August 29) to day 365 (December 31). ∎

Modeling Periodic Behavior from Data

To determine what type of function to use in modeling data, we first draw a scatter plot of the data. The shape of the graph often suggests the appropriate function (see page 394). If the scatter plot has the shape of a sine or cosine curve, then we might try to model the data with a sine or cosine function. The next example illustrates the process.

EXAMPLE 6 ■ Modeling the Height of a Tide

The water depth in a narrow channel varies with the tides. Table 1 shows the water depth over a 12-hour period.

(a) Make a scatter plot of the data in Table 1.
(b) Find a function that models the water depth with respect to time.
(c) Graph the function you found and compare with the data.
(d) If a boat needs at least 11 ft of water to cross the channel, during which times can it safely do so?

SOLUTION

(a) The scatter plot is shown in Figure 11. The scatter plot helps us visualize the data; we see that the data appear to lie on a sine or cosine curve.

TABLE 1

Time	Depth (ft)
12:00 A.M.	9.8
1:00 A.M.	11.4
2:00 A.M.	11.6
3:00 A.M.	11.2
4:00 A.M.	9.6
5:00 A.M.	8.5
6:00 A.M.	6.5
7:00 A.M.	5.7
8:00 A.M.	5.4
9:00 A.M.	6.0
10:00 A.M.	7.0
11:00 A.M.	8.6
12:00 P.M.	10.0

FIGURE 11

(b) We find a function that models the data in the form

$$y = a \cos \omega(t - c) + b$$

First, we need to find the amplitude a. We have

$$\text{amplitude} = \tfrac{1}{2}(\text{maximum value} - \text{minimum value}) = \tfrac{1}{2}(11.6 - 5.4) = 3.1$$

To find the period, we note that the time between consecutive maximum and minimum values represents half of one period. Thus

$$\text{period} = 2(\text{time of maximum value} - \text{time of minimum value}) = 2(8 - 2) = 12$$

Thus, $\omega = 2\pi/12 = 0.52$. The vertical shift is the average of the maximum and minimum values

$$\text{vertical shift} = \tfrac{1}{2}(\text{maximum value} + \text{minimum value}) = \tfrac{1}{2}(11.6 + 5.4) = 8.5$$

From the graph we see that the cosine curve is shifted 2 h to the right (because the maximum value occurs at approximately $t = 2.0$), so $c = 2.0$. Thus, a function that models the tides over the given time period is given by

$$y = 3.1 \cos 0.52(t - 2.0) + 8.5$$

(c) A graph of the function and the scatter plot are shown in Figure 12. It appears that the model we found is a good approximation to the data.

FIGURE 12

(d) We need to solve the inequality $y \geqslant 11$. We solve this inequality graphically by graphing $y = 3.1 \cos 0.52(t - 2.0) + 8.5$ and $y = 11$ on the same graph. From the graph in Figure 13 we see the water depth is higher than 11 ft between $t \approx 0.8$ and $t \approx 3.2$. This corresponds to the times 12:48 A.M. to 3:12 A.M.

FIGURE 13

For the TI-83 and TI-86 the command **SinReg** (for sine regression) finds the sine curve that best fits the given data.

In Example 6 we used the scatter plot to guide us in finding a cosine curve that gives an approximate model of the data. Some graphing calculators are capable of finding a sine or cosine curve that best fits a given set of data points. The method the calculator uses is similar to the method of finding a line of best fit, as explained on pages 247–248.

EXAMPLE 7 ■ Fitting a Sine Curve to Data

(a) Use a graphing device to find the sine curve that best fits the depth of water data in Table 1 on page 461.

(b) Compare the result to the model found in Example 6.

SOLUTION

(a) Using the data in Table 1 and the **SinReg** command on the TI-83 calculator, we get a function of the form

$$y = a \sin(bt + c) + d$$

where

$$a = 3.1 \qquad\qquad b = 0.53$$
$$c = 0.55 \qquad\qquad d = 8.42$$

So, the sine function that best fits the data is

$$y = 3.1 \sin(0.53t + 0.55) + 8.42$$

(b) To compare this with the function in Example 6, we change the sine function to a cosine function by using the reduction formula $\sin u = \cos\left(u - \frac{\pi}{2}\right)$.

$$y = 3.1 \sin(0.53t + 0.55) + 8.42$$

$$= 3.1 \cos\left(0.53t + 0.55 - \frac{\pi}{2}\right) + 8.42 \qquad \text{Reduction formula}$$

$$= 3.1 \cos(0.53t - 1.02) + 8.42$$

$$= 3.1 \cos 0.53(t - 1.92) + 8.42 \qquad \text{Factor 0.53}$$

```
SinReg
 y=a*sin(bx+c)+d
 a=3.097877596
 b=.5268322697
 c=.5493035195
 d=8.424021899
```

Output of the **SinReg** function on the TI-83.

Comparing this with the function we obtained in Example 6, we see that there are small differences in the coefficients. In Figure 14 we graph a scatter plot of the data together with the sine function of best fit.

FIGURE 14

In Example 6 we estimated the values of the amplitude, period, and shifts from the data. In Example 7 the calculator computed the sine curve that best fits the data (that is, the curve that deviates least from the data as explained on page 248). The different ways of obtaining the model account for the differences in the functions.

Damped Harmonic Motion

The spring in Figure 2 is assumed to oscillate in a frictionless environment. In this hypothetical case, the amplitude of the oscillation will not change. In the presence of friction, however, the motion of the spring eventually "dies down"; that is, the amplitude of the motion decreases with time. Motion of this type is called *damped harmonic motion*.

(a) Harmonic motion: $y = \sin 8\pi t$

(b) Damped harmonic motion:
$y = e^{-t} \sin 8\pi t$

FIGURE 15

DAMPED HARMONIC MOTION

If the equation describing the displacement y of an object at time t is

$$y = ke^{-ct} \sin \omega t \qquad \text{or} \qquad y = ke^{-ct} \cos \omega t \qquad (c > 0)$$

then the object is in **damped harmonic motion**. The constant c is the **damping constant**.

Damped harmonic motion is simply harmonic motion for which the amplitude is governed by the function $a(t) = ke^{-ct}$. Figure 15 shows the difference between harmonic motion and damped harmonic motion.

The larger the damping constant c, the quicker the oscillation dies down. When a guitar string is plucked and then allowed to vibrate freely, a point on that string undergoes damped harmonic motion. We can hear the damping of the motion as the sound produced by the vibration of the string fades. How fast the damping of the string occurs (as measured by the size of the constant c) is a property of the size of the string and the material it is made of. Another example of damped harmonic motion is the motion that a shock absorber on a car undergoes when the car hits a bump in the road. In this case, the shock absorber is engineered to damp the motion as quickly as possible (large c) and to have the frequency as small as possible (small ω). On the other hand, the sound produced by a tuba player playing a note is undamped as long as the player can maintain the loudness of the note. The electromagnetic waves that produce light move in simple harmonic motion that is not damped.

EXAMPLE 8 ■ A Vibrating Violin String

The G-string on a violin is pulled a distance of 0.5 cm above its rest position, then released and allowed to vibrate. The damping constant c for this string is determined to be 1.4. Suppose that the note produced is a pure G (frequency = 200 cycles per second). Find an equation that describes the motion of the point at which the string was plucked.

SOLUTION

Let P be the point at which the string was plucked. We will find a function $f(t)$ that gives the distance at time t of the point P from its original rest position. Since the maximum displacement occurs at $t = 0$, we find an equation in the form

$$y = ke^{-ct} \cos \omega t$$

From this equation, we see that $f(0) = k$. But we know that the original displacement of the string is 0.5 cm. Thus, $k = 0.5$. Since the frequency of the vibration is 200, we have $\omega/(2\pi) = 200$ or $\omega = (200)(2\pi)$. Finally, since we know that the damping constant is 1.4, we get

$$f(t) = 0.5e^{-1.4t} \cos 400\pi t \qquad ■$$

EXAMPLE 9 ■ Ripples on a Pond

A stone is dropped in a calm lake, causing waves to form. The up-and-down motion of a point on the surface of the water is modeled by damped harmonic motion. At some time the amplitude of the wave is measured, and 20 s later it is found that the amplitude has dropped to $\frac{1}{10}$ of this value. Find the damping constant c.

SOLUTION

The amplitude is governed by the coefficient ke^{-ct} in the equations for damped harmonic motion. Thus, the amplitude at time t is ke^{-ct}, and 20 s later, it is $ke^{-c(t+20)}$. So, from the given information

$$\frac{ke^{-ct}}{ke^{-c(t+20)}} = 10$$

$$ke^{-c(t+20)} = \frac{1}{10}ke^{-ct}$$

We now solve this equation for c. Canceling k and using the Laws of Exponents, we get

$$e^{-ct} \cdot e^{-20c} = \tfrac{1}{10}e^{-ct}$$

$$e^{-20c} = \tfrac{1}{10} \qquad \text{Cancel } e^{-ct}$$

$$e^{20c} = 10 \qquad \text{Take reciprocals}$$

Taking the natural logarithm of each side gives

$$20c = \ln(10)$$

$$c = \tfrac{1}{20}\ln(10) \approx \tfrac{1}{20}(2.30) \approx 0.12$$

Thus, the damping constant is $c \approx 0.12$. ∎

▪ Problems

1–4 ▪ A mass suspended from a spring vibrates in simple harmonic motion. The given function models the displacement of the mass from its rest position.
(a) Find the amplitude, period, and frequency of the motion of the mass.
(b) Sketch a graph of the displacement of the mass over one complete period.

1. $y = 2 \sin 3t$

2. $y = 3 \cos \tfrac{1}{2}t$

3. $y = -0.25 \cos\left(1.5t - \tfrac{\pi}{3}\right)$

4. $y = -\tfrac{3}{2}\sin(0.2t + 1.4)$

5. A cork floating in a lake is bobbing in simple harmonic motion. Its displacement above the bottom of the lake is modeled by

$$y = 0.2 \cos 20\pi t + 8$$

where y is measured in meters and t is measured in minutes.
(a) Find the frequency of the motion of the cork.
(b) Sketch a graph of y.
(c) Find the maximum displacement of the cork above the lake bottom.

6. The carrier wave for an FM radio signal is modeled by the function

$$y = a \sin(2\pi(9.15 \times 10^7)t)$$

where t is measured in seconds. Find the period and frequency of the carrier wave.

7. In a predator/prey model (see page 442), the predator population is modeled by the function

$$y = 900 \cos 2t + 8000$$

where t is measured in years.
(a) What is the maximum population?
(b) Find the length of time between successive periods of maximum population.

8. Each time your heart beats, your blood pressure increases, then decreases as the heart rests between beats. A certain person's blood pressure is modeled by the function

$$p(t) = 115 + 25 \sin(160\pi t)$$

where $p(t)$ is the pressure in mmHg, at time t measured in minutes.
(a) Find the amplitude, period, and frequency of p.
(b) Sketch a graph of p.
(c) If a person is exercising, his heart beats faster. How does this affect the period and frequency of p?

Rest
position

2 ft

9–22 ■ These problems require you to find a function that models a real-life case of simple harmonic motion.

9. A mass attached to a spring is moving up and down in simple harmonic motion. The graph at the left gives its displacement $d(t)$ from equilibrium at time t. Express the function d in the form $d(t) = a \sin \omega t$.

10. The graph shows the variation of the water level relative to mean sea level in Commencement Bay at Tacoma, Washington, for a particular 24-hour period. Assuming that this variation is modeled by simple harmonic motion, find an equation of the form $y = a \sin \omega t$ that describes the variation in water level as a function of the number of hours after midnight.

11. The Bay of Fundy in Nova Scotia has the highest tides in the world. In one 12-hour period, the water starts at mean sea level, rises to 21 ft above, drops to 21 ft below, then returns to mean sea level. Assuming that the motion of the tides is simple harmonic, find an equation that describes the height of the tide in the Bay of Fundy above mean sea level. Sketch a graph that shows the level of the tides over a 12-hour period.

12. A mass suspended from a spring is pulled down a distance of 2 ft from its rest position, as shown in the figure at the left. The mass is released at time $t = 0$ and allowed to oscillate. If the mass returns to this position after 1 s, find an equation that describes its motion.

13. A mass is suspended on a spring. The spring is compressed so that the mass is located 5 cm above its rest position. The mass is released at time $t = 0$ and allowed to oscillate. It is observed that the mass reaches its lowest point $\frac{1}{2}$ s after it is released. Find an equation that describes the motion of the mass.

14. The frequency of oscillation of an object suspended on a spring depends on the stiffness k of the spring (called the *spring constant*) and the mass m of the object. If the spring is compressed a distance a and then allowed to oscillate, its displacement is given by

$$f(t) = a \cos \sqrt{k/m}\, t$$

(a) A 10-g mass is suspended from a spring with stiffness $k = 3$. If the spring is compressed a distance 5 cm and then released, find the equation that describes the oscillation of the spring.
(b) Find a general formula for the frequency (in terms of k and m).
(c) How is the frequency affected if the mass is increased? Is the oscillation faster or slower?
(d) How is the frequency affected if a stiffer spring is used (larger k)? Is the oscillation faster or slower?

15. A ferris wheel has a radius of 10 m, and the bottom of the wheel passes 1 m above the ground. If the ferris wheel makes one complete revolution every 20 s, find an equation that gives the height above the ground of a person on the ferris wheel as a function of time.

16. The pendulum in a grandfather clock makes one complete swing every 2 s. The maximum angle that the pendulum makes with respect to its rest position is 10°. We know from physical principles that the angle θ between the pendulum and its rest position changes in simple harmonic fashion. Find an equation that describes the size of the angle θ as a function of time. (Take $t = 0$ to be a time when the pendulum is vertical.)

17. The variable star Zeta Gemini has a period of 10 days. The average brightness of the star is 3.8 magnitudes, and the maximum variation from the average is 0.2 magnitude. Assuming that the variation in brightness is simple harmonic, find an equation that gives the brightness of the star as a function of time.

18. Astronomers believe that the radius of a variable star increases and decreases with the brightness of the star. The variable star Delta Cephei (Example 4) has an average radius of 20 million miles and changes by a maximum of 1.5 million miles from this average during a single pulsation. Find an equation that describes the radius of this star as a function of time.

19. An *alternator*, or alternating current (AC) generator, produces an electrical current by rotating an *armature* (or coil) in a magnetic field. The figure represents a simple version of such a generator. As the wire passes through the magnetic field, a voltage E is generated in the wire. It can be shown that the voltage generated is given by

$$E(t) = E_0 \cos \omega t$$

where E_0 is the maximum voltage produced (which depends on the strength of the magnetic field) and $\omega/(2\pi)$ is the number of revolutions per second of the armature (the frequency). Ordinary 110-V household alternating current varies from $+155$ to -155 V with a frequency of 60 Hz (cycles per second). Find an equation that describes this variation in voltage.

Magnets

N   S

Wire

20. The armature in an electric generator is rotating at the rate of 100 revolutions per second (rps). If the maximum voltage produced is 310 V, find an equation that describes this variation in voltage. What is the rms voltage (see the marginal note)?

21. The graph shows an oscilloscope reading of the variation in voltage of an AC current produced by a simple generator.
(a) Find the maximum voltage produced.
(b) Find the frequency (cycles per second) of the generator.

Why do we say that household current is 110 V when the maximum voltage produced is 155 V? From the symmetry of the cosine function, we see that the average voltage produced is zero. This average value would be the same for all AC generators and so gives no information about the voltage generated. To obtain a more informative measure of voltage, engineers use the **root-mean-square** (rms) method. It can be shown that the rms voltage is $1/\sqrt{2}$ times the maximum voltage. So, for household current the rms voltage is

$$155 \times \frac{1}{\sqrt{2}} \approx 110 \text{ V}$$

(c) How many revolutions per second does the armature in the generator make?

(d) Find a formula that describes the variation in voltage as a function of time.

22. A train whistle produces a sharp, pure tone. While the train is moving toward us, the pitch of the whistle increases; it then decreases as the train moves away from us. This phenomenon is known as the *Doppler effect*. Suppose that the maximum change in pressure caused by the whistle is 10 N/m^2 and the frequency is 20 cycles per second.

(a) Find an equation that describes the variation in pressure as a function of time, and sketch the graph.

(b) How does the graph in part (a) change as the train moves toward us? Away from us?

23–26 ■ These problems require you to find a function that models simple harmonic motion from data. See Examples 6 and 7.

23. The table gives the average monthly temperature in Montgomery County, Maryland.

(a) Make a scatter plot of the data.

(b) Find a cosine curve that models the data (as in Example 6).

(c) Graph the function you found in part (b) together with the scatter plot.

(d) Use a graphing calculator to find the sine curve that best fits the data (as in Example 7).

Month	Average temperature (°F)
January	40.0
February	43.1
March	54.6
April	64.2
May	73.8
June	81.8
July	85.8
August	83.9
September	76.9
October	66.8
November	55.5
December	44.5

Time	Body temperature (°C)
0	36.8
2	36.7
4	36.6
6	36.7
8	36.8
10	37.0
12	37.2
14	37.3
16	37.4
18	37.3
20	37.2
22	37.0
24	36.8

 24. Circadian rhythm (from the Latin *circa*—about, and *diem*—day) is the daily biological pattern by which body temperature, blood pressure, and other physiological variables change. The data show typical changes in human body temperature over a 24-hour period ($t = 0$ corresponds to midnight).
(a) Make a scatter plot of the data.
(b) Find a cosine curve that models the data (as in Example 6).
(c) Graph the function you found in part (b) together with the scatter plot.
(d) Use a graphing calculator to find the sine curve that best fits the data (as in Example 7).

 25. When two species interact in a predator/prey relationship (see page 442), the populations of both species tend to vary in a sinusoidal fashion. In a certain midwestern county, the main food source for barn owls consists of field mice and other small mammals. The table gives the population of barn owls in this county every July 1 over a 12-year period.
(a) Make a scatter plot of the data.
(b) Find a sine curve that models the data (as in Example 6).
(c) Graph the function you found in part (b) together with the scatter plot.
(d) Use a graphing calculator to find the sine curve that best fits the data (as in Example 7). Compare to your answer from part (b).

Year	Owl population
0	50
1	62
2	73
3	80
4	71
5	60
6	51
7	43
8	29
9	20
10	28
11	41
12	49

 26. For reasons not yet fully understood, the number of fingerling salmon that survive the trip from their riverbed spawning grounds to the open ocean varies approximately sinusoidally from year to year. The table shows the number of salmon that hatch in a certain British Columbia creek and then make their way to the Strait of Georgia. The data is given in thousands of fingerlings, over a period of 16 years.
(a) Make a scatter plot of the data.
(b) Find a sine curve that models the data (as in Example 6).
(c) Graph the function you found in part (b) together with the scatter plot.

(d) Use a graphing calculator to find the sine curve that best fits the data (as in Example 7). Compare to your answer from part (b).

Year	Salmon (× 1000)	Year	Salmon (× 1000)
1985	43	1993	56
1986	36	1994	63
1987	27	1995	57
1988	23	1996	50
1989	26	1997	44
1990	33	1998	38
1991	43	1999	30
1992	50	2000	22

27–30 ■ These problems deal with modeling damped harmonic motion.

27. A strong gust of wind strikes a tall building, causing it to sway back and forth in damped harmonic motion. The frequency of the oscillation is 0.5 cycle per second and the damping constant is $c = 0.9$. Find an equation that describes the motion of the building. (Assume $k = 1$ and take $t = 0$ to be the instant when the gust of wind strikes the building.)

28. When a car hits a certain bump on the road, a shock absorber on the car is compressed a distance of 6 in., then released (see the figure). The shock absorber vibrates in damped harmonic motion with a frequency of 2 cycles per second. The damping constant for this particular shock absorber is 2.8.
  (a) Find an equation that describes the displacement of the shock absorber from its rest position as a function of time. Take $t = 0$ to be the instant that the shock absorber is released.
  (b) How long does it take for the amplitude of the vibration to decrease to 0.5 in?

29. A tuning fork is struck and oscillates in damped harmonic motion. The amplitude of the motion is measured, and 3 s later it is found that the amplitude has dropped to $\frac{1}{4}$ of this value. Find the damping constant c for this tuning fork.

30. A guitar string is pulled at point P a distance of 3 cm above its rest position. It is then released and vibrates in damped harmonic motion with a frequency of 165 cycles per second. After 2 s, it is observed that the amplitude of the vibration at point P is 0.6 cm.
  (a) Find the damping constant c.
  (b) Find an equation that describes the position of point P above its rest position as a function of time. Take $t = 0$ to be the instant that the string is released.

6

Trigonometric Functions of Angles

Trigonometric functions are important in surveying, navigation, and astronomy. They are used, for instance, to find heights of mountains or distances to nearby stars.

The trigonometric functions can be viewed in two different but equivalent ways. One way is to view them as *functions of angles* (Chapter 6), the other as *functions of real numbers* (Chapter 5). The two approaches to trigonometry are independent of each other, so either Chapter 5 or Chapter 6 may be studied first.

The trigonometric functions defined in these two different ways are identical— they assign the same value to a given real number. In the first case, the real number is the measure of an angle; in the second, it is the length of an arc along the unit circle. We study both approaches because different applications require that we view these functions differently. One approach (Chapter 6) lends itself to static applications such as the measurement of distance using triangles. The other approach (Chapter 5) lends itself to dynamic applications such as modeling harmonic motion. The power and versatility of trigonometry stem from the fact that it can be viewed in these different ways. Today, trigonometry is an indispensable tool in physics, engineering, computer science, biology, and in practically all the sciences.

6.1 ANGLE MEASURE

An **angle** AOB consists of two rays R_1 and R_2 with a common vertex O (see Figure 1). We often interpret an angle as a rotation of the ray R_1 onto R_2. In this case, R_1 is called the **initial side**, and R_2 is called the **terminal side** of the angle. If the rotation is counterclockwise, the angle is considered **positive**, and if the rotation is clockwise, the angle is considered **negative**.

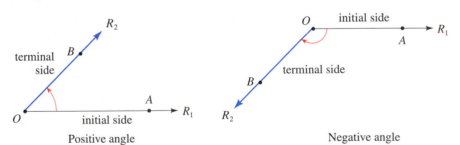

FIGURE 1   Positive angle   Negative angle

■ Angle Measure

The **measure** of an angle is the amount of rotation about the vertex required to move R_1 onto R_2. Intuitively, this is how much the angle "opens." One unit of measurement for angles is the **degree**. An angle of measure 1 degree is formed by rotating the initial side $\frac{1}{360}$ of a complete revolution. In calculus and other branches of mathematics, a more natural method of measuring angles is used—*radian measure*. The amount an angle opens is measured along the arc of a circle of radius 1 with its center at the vertex of the angle.

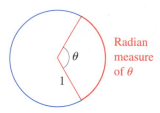

FIGURE 2

DEFINITION OF RADIAN MEASURE

If a circle of radius 1 is drawn with the vertex of an angle at its center, then the measure of this angle in **radians** (abbreviated **rad**) is the length of the arc that subtends the angle (see Figure 2).

The circumference of the circle of radius 1 is 2π and so a complete revolution has measure 2π rad, a straight angle has measure π rad, and a right angle has measure $\pi/2$ rad. An angle that is subtended by an arc of length 2 along the unit circle has radian measure 2 (see Figure 3).

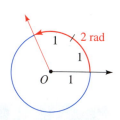

FIGURE 3
Radian measure

Since a complete revolution measured in degrees is 360° and measured in radians is 2π rad, we get the following simple relationship between these two methods of angle measurement.

RELATIONSHIP BETWEEN DEGREES AND RADIANS

$$180° = \pi \text{ rad} \qquad 1 \text{ rad} = \left(\frac{180}{\pi}\right)° \qquad 1° = \frac{\pi}{180} \text{ rad}$$

1. To convert degrees to radians, multiply by $\dfrac{\pi}{180}$.

2. To convert radians to degrees, multiply by $\dfrac{180}{\pi}$.

To get some idea of the size of a radian, notice that

$$1 \text{ rad} \approx 57.296° \qquad \text{and} \qquad 1° \approx 0.01745 \text{ rad}$$

Measure of $\theta = 1$ rad
Measure of $\theta \approx 57.296°$

FIGURE 4

An angle θ of measure 1 rad is shown in Figure 4.

EXAMPLE 1 ■ Converting between Radians and Degrees

(a) Express 60° in radians.

(b) Express $\dfrac{\pi}{6}$ rad in degrees.

SOLUTION

The relationship between degrees and radians gives

(a) $60° = 60\left(\dfrac{\pi}{180}\right)$ rad $= \dfrac{\pi}{3}$ rad

(b) $\dfrac{\pi}{6}$ rad $= \left(\dfrac{\pi}{6}\right)\left(\dfrac{180}{\pi}\right) = 30°$ ∎

A note on terminology: We often use a phrase such as "a 30° angle" to mean *an angle whose measure is* 30°. Also, for an angle θ, we write θ = 30° or θ = π/6 to mean *the measure of θ is 30° or π/6 rad.* When no unit is given, the angle is assumed to be measured in radians.

◼ Angles in Standard Position

An angle is in **standard position** if it is drawn in the *xy*-plane with its vertex at the origin and its initial side on the positive *x*-axis. Figure 5 gives examples of angles in standard position.

(a)

(b)

(c)

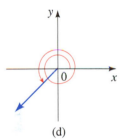

(d)

FIGURE 5
Angles in standard position

Two angles in standard position are **coterminal** if their sides coincide. In Figure 5 the angles in (a) and (c) are coterminal.

EXAMPLE 2 ◼ **Coterminal Angles**

(a) Find angles that are coterminal with the angle θ = 30° in standard position.

(b) Find angles that are coterminal with the angle $\theta = \dfrac{\pi}{3}$ in standard position.

SOLUTION

(a) To find positive angles that are coterminal with θ, we add any multiple of 360°. Thus

$$30° + 360° = 390° \quad \text{and} \quad 30° + 720° = 750°$$

are coterminal with $\theta = 30°$. To find negative angles that are coterminal with θ, we subtract any multiple of 360°. Thus

$$30° - 360° = -330° \qquad \text{and} \qquad 30° - 720° = -690°$$

are coterminal with θ. See Figure 6.

FIGURE 6

(b) To find positive angles that are coterminal with θ, we add any multiple of 2π. Thus

$$\frac{\pi}{3} + 2\pi = \frac{7\pi}{3} \qquad \text{and} \qquad \frac{\pi}{3} + 4\pi = \frac{13\pi}{3}$$

are coterminal with $\theta = \pi/3$. To find negative angles that are coterminal with θ, we subtract any multiple of 2π. Thus

$$\frac{\pi}{3} - 2\pi = -\frac{5\pi}{3} \qquad \text{and} \qquad \frac{\pi}{3} - 4\pi = -\frac{11\pi}{3}$$

are coterminal with θ. See Figure 7.

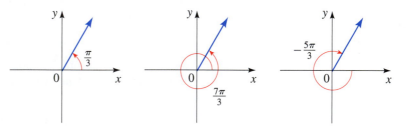

FIGURE 7

EXAMPLE 3 ■ Coterminal Angles

Find an angle with measure between 0° and 360° that is coterminal with the angle of measure 1290° in standard position.

SOLUTION

We can subtract 360° as many times as we wish from 1290°, and the resulting angle will be coterminal with 1290°. Thus, $1290° - 360° = 930°$ is coterminal with 1290°, and so is the angle $1290° - 2(360)° = 570°$.

To find the angle we want between 0° and 360°, we subtract 360° from 1290° as many times as necessary. An efficient way to do this is to determine how many times 360° goes into 1290°, that is, divide 1290 by 360, and the remainder will be the angle we are looking for. We see that 360 goes into 1290 three times with a remainder of 210. Thus, 210° is the desired angle (see Figure 8).

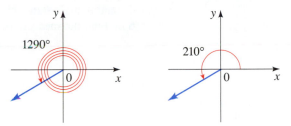

FIGURE 8

Length of a Circular Arc

(Figure 9, left margin)

FIGURE 9
$s = \theta r$

An angle whose radian measure is θ is subtended by an arc that is the fraction $\theta/(2\pi)$ of the circumference of a circle. Thus, in a circle of radius r, the length s of an arc that subtends the angle θ (see Figure 9) is

$$s = \frac{\theta}{2\pi} \times \text{circumference of circle}$$

$$= \frac{\theta}{2\pi}(2\pi r) = \theta r$$

> **LENGTH OF A CIRCULAR ARC**
>
> In a circle of radius r, the length s of an arc that subtends a central angle of θ radians is
>
> $$s = r\theta$$

Solving for θ, we get the important formula

$$\theta = \frac{s}{r}$$

This formula allows us to define radian measure using a circle of any radius r: The radian measure of an angle θ is s/r, where s is the length of the circular arc that subtends θ in a circle of radius r (see Figure 10).

FIGURE 10

The radian measure of θ is the number of "radiuses" that can fit in the arc that subtends θ; hence the term *radian*.

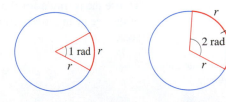

EXAMPLE 4 ■ Arc Length and Angle Measure

(a) Find the length of an arc of a circle with radius 10 m that subtends a central angle of 30°.

(b) A central angle θ in a circle of radius 4 m is subtended by an arc of length 6 m. Find the measure of θ in radians.

SOLUTION

(a) From Example 1(b) we see that $30° = \pi/6$ rad. So the length of the arc is

⊘ The formula $s = r\theta$ is true only when θ is measured in radians.

$$s = r\theta = (10)\frac{\pi}{6} = \frac{5\pi}{3} \text{ m}$$

(b) By the formula $\theta = s/r$, we have

$$\theta = \frac{s}{r} = \frac{6}{4} = \frac{3}{2} \text{ rad}$$

■

Area of a Circular Sector

The area of a circle of radius r is $A = \pi r^2$. A sector of this circle with central angle θ has an area that is the fraction $\theta/(2\pi)$ of the area of the entire circle (see Figure 11). So the area of this sector is

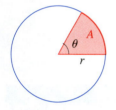

FIGURE 11
$A = \frac{1}{2}r^2\theta$

$$A = \frac{\theta}{2\pi} \times \text{area of circle}$$

$$= \frac{\theta}{2\pi}(\pi r^2) = \frac{1}{2}r^2\theta$$

AREA OF A CIRCULAR SECTOR

In a circle of radius r, the area A of a sector with a central angle of θ radians is

$$A = \frac{1}{2}r^2\theta$$

EXAMPLE 5 ■ Area of a Sector

Find the area of a sector of a circle with central angle 60° if the radius of the circle is 3 m.

SOLUTION

To use the formula for the area of a circular sector, we must find the central angle of the sector in radians: $60° = 60(\pi/180)$ rad $= \pi/3$ rad. Thus, the area of the sector is

⊘ The formula $A = \frac{1}{2}r^2\theta$ is true only when θ is measured in radians.

$$A = \frac{1}{2}r^2\theta = \frac{1}{2}(3)^2\left(\frac{\pi}{3}\right) = \frac{3\pi}{2} \text{ m}^2$$

■

Circular Motion

Suppose a point moves along a circle as shown in Figure 12. There are two ways to describe the motion of the point—linear speed and angular speed. **Linear speed** is the rate at which the distance traveled is changing, so linear speed is the distance traveled divided by the time elapsed. **Angular speed** is the rate at which the central angle θ is changing, so angular speed is the number of radians this angle changes divided by the time elapsed.

FIGURE 12

> ### LINEAR SPEED AND ANGULAR SPEED
>
> Suppose a point moves along a circle of radius r and the ray from the center of the circle to the point traverses θ radians in time t. Let $s = r\theta$ be the distance the point travels in time t. Then the speed of the object is given by
>
> $$\text{Angular speed} \qquad \omega = \frac{\theta}{t}$$
>
> $$\text{Linear speed} \qquad v = \frac{s}{t}$$

EXAMPLE 6 ■ Finding Linear and Angular Speed

A boy rotates a stone in a 3-ft-long sling at the rate of 15 revolutions every 10 seconds. Find the angular and linear velocities of the stone.

SOLUTION

In 10 s, the angle θ changes by $15 \cdot 2\pi = 30\pi$ radians. So the *angular speed* of the stone is

$$\omega = \frac{\theta}{t} = \frac{30\pi \text{ rad}}{10 \text{ s}} = 3\pi \text{ rad/s}$$

The distance traveled by the stone in 10 s is $s = 15 \cdot 2\pi r = 15 \cdot 2\pi \cdot 3 = 90\pi$ ft. So the *linear speed* of the stone is

$$v = \frac{s}{t} = \frac{90\pi \text{ ft}}{10 \text{ s}} = 9\pi \text{ ft/s} \qquad ■$$

Notice that angular speed does *not* depend on the radius of the circle, but only on the angle θ. However, if we know the angular speed ω and the radius r, we can find linear speed as follows: $v = s/t = r\theta/t = r(\theta/t) = r\omega$.

> ### RELATIONSHIP BETWEEN LINEAR AND ANGULAR SPEED
>
> If a point moves along a circle of radius r with angular speed ω, then its linear speed v is given by
>
> $$v = r\omega$$

EXAMPLE 7 ■ Finding Linear Speed from Angular Speed

A woman is riding a bicycle whose wheels are 26 inches in diameter. If the wheels rotate at 125 revolutions per minute (rpm), find the speed at which she is traveling, in mi/h.

SOLUTION

The angular speed of the wheels is $2\pi \cdot 125 = 250\pi$ rad/min. Since the wheels have radius 13 in. (half the diameter), the linear speed is

$$v = r\omega = 13 \cdot 250\pi \approx 10{,}210.2 \text{ in./min}$$

Since there are 12 inches per foot, 5280 feet per mile, and 60 minutes per hour, her speed in miles per hour is

$$\frac{10{,}210.2 \text{ in./min} \times 60 \text{ min/h}}{12 \text{ in./ft} \times 5280 \text{ ft/mi}} \approx 9.7 \text{ mi/h}$$ ■

6.1 EXERCISES

1–8 ■ Find the radian measure of the angle with the given degree measure.

1. 36°   **2.** 200°   **3.** −480°

4. −72°   **5.** 60°   **6.** 45°

7. −135°   **8.** 150°

9–16 ■ Find the degree measure of the angle with the given radian measure.

9. $\dfrac{3\pi}{4}$ **10.** $-\dfrac{7\pi}{2}$ **11.** $\dfrac{5\pi}{6}$

12. 2   **13.** −1.5   **14.** $\dfrac{2\pi}{9}$

15. $-\dfrac{\pi}{12}$ **16.** $\dfrac{\pi}{18}$

17–22 ■ The measure of an angle in standard position is given. Find two positive angles and two negative angles that are coterminal with the given angle.

17. 50°   **18.** 135°   **19.** $\dfrac{3\pi}{4}$

20. $\dfrac{11\pi}{6}$ **21.** $-\dfrac{\pi}{4}$ **22.** −45°

23–28 ■ The measures of two angles in standard position are given. Determine whether the angles are coterminal.

23. 70°, 430°   **24.** −30°, 330°

25. $\dfrac{5\pi}{6}$, $\dfrac{17\pi}{6}$ **26.** $\dfrac{32\pi}{3}$, $\dfrac{11\pi}{3}$

27. 155°, 875°   **28.** 50°, 340°

29–34 ■ Find an angle between 0° and 360° that is coterminal with the given angle.

29. 733°   **30.** 361°

31. 1110°   **32.** −100°

33. −800°   **34.** 1270°

35–40 ■ Find an angle between 0 and 2π that is coterminal with the given angle.

35. $\dfrac{17\pi}{6}$ **36.** $-\dfrac{7\pi}{3}$ **37.** 87π

38. 10   **39.** $\dfrac{17\pi}{4}$ **40.** $\dfrac{51\pi}{2}$

41. Find the length of the arc s in the figure.

42. Find the angle θ in the figure.

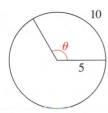

43. Find the radius r of the circle in the figure.

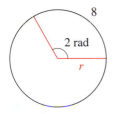

44. Find the length of an arc that subtends a central angle of 45° in a circle of radius 10 m.

45. Find the length of an arc that subtends a central angle of 2 rad in a circle of radius 2 mi.

46. A central angle θ in a circle of radius 5 m is subtended by an arc of length 6 m. Find the measure of θ in degrees and in radians.

47. An arc of length 100 m subtends a central angle θ in a circle of radius 50 m. Find the measure of θ in degrees and in radians.

48. A circular arc of length 3 ft subtends a central angle of 25°. Find the radius of the circle.

49. Find the radius of the circle if an arc of length 6 m on the circle subtends a central angle of $\pi/6$ rad.

50. How many revolutions will a car wheel of diameter 30 in. make as the car travels a distance of one mile?

51. Pittsburgh, Pennsylvania, and Miami, Florida, lie approximately on the same meridian. Pittsburgh has a latitude of 40.5° N and Miami, 25.5° N. Find the distance between these two cities. (The radius of the earth is 3960 mi.)

52. Memphis, Tennessee, and New Orleans, Louisiana, lie approximately on the same meridian. Memphis has latitude 35° N and New Orleans, 30° N. Find the distance between these two cities. (The radius of the earth is 3960 mi.)

53. Find the distance that the earth travels in one day in its path around the sun. Assume that a year has 365 days and that the path of the earth around the sun is a circle of radius 93 million miles. [The path of the earth around the sun is actually an *ellipse* with the sun at one focus (see Section 9.2). This ellipse, however, has very small eccentricity, so it is nearly circular.]

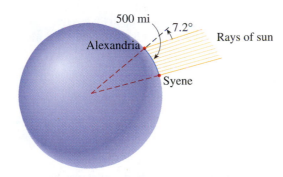

54. The Greek mathematician Eratosthenes (ca. 276–195 B.C.) measured the circumference of the earth from the following observations. He noticed that on a certain day the sun shone directly down a deep well in Syene (modern Aswan). At the same time in Alexandria, 500 miles north (on the same meridian), the rays of the sun shone at an angle of 7.2° to the zenith. Use this information and the figure to find the radius and circumference of the earth. (The data used in this problem are more accurate than those available to Eratosthenes.)

55. Find the distance along an arc on the surface of the earth that subtends a central angle of 1 minute (1 minute = $\frac{1}{60}$ degree). This distance is called a *nautical mile*. (The radius of the earth is 3960 mi.)

56. Find the area of the sector shown in each figure.

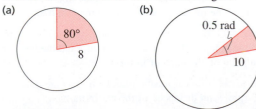

57. Find the area of a sector with central angle 1 rad in a circle of radius 10 m.

58. A sector of a circle has a central angle of 60°. Find the area of the sector if the radius of the circle is 3 mi.

59. The area of a sector of a circle with a central angle of 2 rad is 16 m². Find the radius of the circle.

60. A sector of a circle of radius 24 mi has an area of 288 mi². Find the central angle of the sector.

61. The area of a circle is 72 cm². Find the area of a sector of this circle that subtends a central angle of $\pi/6$ rad.

62. Three circles with radii 1, 2, and 3 ft are externally tangent to one another, as shown in the figure. Find the area of the sector of the circle of radius 1 that is cut off by the line segments joining the center of that circle to the centers of the other two circles.

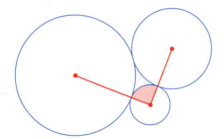

63. A winch of radius 2 ft is used to lift heavy loads. If the winch makes 8 revolutions every 15 s, find the speed at which the load is rising.

64. A ceiling fan with 16-in. blades rotates at 45 rpm.
  (a) Find the angular speed of the fan in rad/min.
  (b) Find the linear speed of the tips of the blades in in/min.

65. A radial saw has a blade with a 6-in. radius. Suppose that the blade spins at 1000 rpm.
  (a) Find the angular speed of the blade in rad/min.
  (b) Find the linear speed of the sawteeth in ft/s.

66. The earth rotates about its axis once every 23 h 56 min 4 s, and the radius of the earth is 3960 mi. Find the linear speed of a point on the equator in mi/h.

67. The wheels of a car have radius 11 in. and are rotating at 600 rpm. Find the speed of the car in mi/h.

68. A truck with 48-in. diameter wheels is traveling at 50 mi/h.
  (a) Find the angular speed of the wheels in rad/min.
  (b) How many revolutions per minute do the wheels make?

69. To measure the speed of a current, scientists place a paddle wheel in the stream and observe the rate at which it rotates. If the paddle wheel has radius 0.20 m and rotates at 100 rpm, find the speed of the current in m/s.

70. The sprockets and chain of a bicycle are shown in the figure. The pedal sprocket has a radius of 4 in, the wheel sprocket a radius of 2 in, and the wheel a radius of 13 in. The cyclist pedals at 40 rpm.
  (a) Find the angular speed of the wheel sprocket.
  (b) Find the speed of the bicycle. (Assume that the wheel turns at the same rate as the wheel sprocket.)

13 in.

2 in.

4 in.

 DISCOVERY · DISCUSSION

71. Different Ways of Measuring Angles   The custom of measuring angles using degrees, with 360° in a circle, dates back to the ancient Babylonians, who used a number system based on groups of 60. Another system of measuring angles divides the circle into 400 units, called *grads*. In this system a right angle is 100 grad, so this fits in with our base 10 number system.

Write a short essay comparing the advantages and disadvantages of these two systems and the radian system of measuring angles. Which system do you prefer?

72. Clocks and Angles   In one hour, the minute hand on a clock moves through a complete circle, and the hour hand moves through $\frac{1}{12}$ of a circle. Through how many radians do the minute and the hour hand move between 1:00 P.M. and 6:45 P.M. (on the same day)?

6.2   TRIGONOMETRY OF RIGHT TRIANGLES

In this section we study certain ratios of the sides of right triangles, called trigonometric ratios, and give several applications.

■ Trigonometric Ratios

Consider a right triangle with θ as one of its acute angles. The trigonometric ratios are defined as follows (see Figure 1).

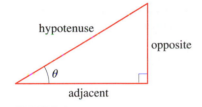

FIGURE 1

THE TRIGONOMETRIC RATIOS		
$\sin \theta = \dfrac{\text{opposite}}{\text{hypotenuse}}$	$\cos \theta = \dfrac{\text{adjacent}}{\text{hypotenuse}}$	$\tan \theta = \dfrac{\text{opposite}}{\text{adjacent}}$
$\csc \theta = \dfrac{\text{hypotenuse}}{\text{opposite}}$	$\sec \theta = \dfrac{\text{hypotenuse}}{\text{adjacent}}$	$\cot \theta = \dfrac{\text{adjacent}}{\text{opposite}}$

The symbols we use for these ratios are abbreviations for their full names: **sine, cosine, tangent, cosecant, secant, cotangent**. Since any two right triangles with angle θ are similar, these ratios are the same, regardless of the size of the triangle; they depend only on the angle θ (see Figure 2).

Hipparchus (circa 140 B.C.) is considered the founder of trigonometry. He constructed tables for a function closely related to the modern sine function and evaluated for angles at half-degree intervals. These are considered the first trigonometric tables. He used his tables mainly to calculate the paths of the planets through the heavens.

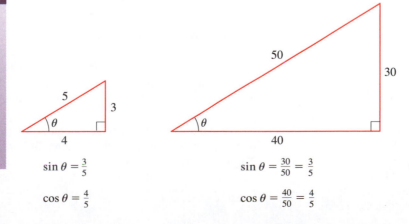

FIGURE 2

$\sin \theta = \frac{3}{5}$ $\sin \theta = \frac{30}{50} = \frac{3}{5}$

$\cos \theta = \frac{4}{5}$ $\cos \theta = \frac{40}{50} = \frac{4}{5}$

FIGURE 3

EXAMPLE 1 ■ Finding Trigonometric Ratios

Find the six trigonometric ratios of the angle θ in Figure 3.

SOLUTION

$$\sin \theta = \frac{2}{3} \qquad \cos \theta = \frac{\sqrt{5}}{3} \qquad \tan \theta = \frac{2}{\sqrt{5}}$$

$$\csc \theta = \frac{3}{2} \qquad \sec \theta = \frac{3}{\sqrt{5}} \qquad \cot \theta = \frac{\sqrt{5}}{2}$$

■

EXAMPLE 2 ■ Finding Trigonometric Ratios

If $\cos \alpha = \frac{3}{4}$, sketch a right triangle with acute angle α, and find the other five trigonometric ratios of α.

SOLUTION

Since $\cos \alpha$ is defined as the ratio of the adjacent side to the hypotenuse, we sketch a triangle with hypotenuse of length 4 and a side of length 3 adjacent to α. If the opposite side is x, then by the Pythagorean Theorem, $3^2 + x^2 = 4^2$ or $x^2 = 7$, so $x = \sqrt{7}$. We then use the triangle in Figure 4 to find the ratios.

$$\sin \alpha = \frac{\sqrt{7}}{4} \qquad \cos \alpha = \frac{3}{4} \qquad \tan \alpha = \frac{\sqrt{7}}{3}$$

$$\csc \alpha = \frac{4}{\sqrt{7}} \qquad \sec \alpha = \frac{4}{3} \qquad \cot \alpha = \frac{3}{\sqrt{7}}$$

■

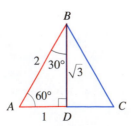

FIGURE 4

FIGURE 5

FIGURE 6

■ Special Triangles

Certain right triangles have ratios that can be calculated easily from the Pythagorean Theorem. Since they are used frequently, we mention them here.

The first triangle is obtained by drawing a diagonal in a square of side 1 (see Figure 5). By the Pythagorean Theorem this diagonal has length $\sqrt{2}$. The resulting triangle has angles 45°, 45°, and 90° (or $\pi/4$, $\pi/4$, and $\pi/2$). To get the second triangle, we start with an equilateral triangle ABC of side 2 and draw the perpendicular bisector DB of the base, as in Figure 6. By the Pythagorean Theorem the length of DB is $\sqrt{3}$. Since DB bisects angle ABC, we obtain a triangle with angles 30°, 60°, and 90° (or $\pi/6$, $\pi/3$, and $\pi/2$).

We can now use the special triangles in Figures 5 and 6 to calculate the trigonometric ratios for angles with measures 30°, 45°, and 60° (or $\pi/6$, $\pi/4$, and $\pi/3$). These are listed in Table 1.

TABLE 1 Values of the trigonometric ratios for special angles

θ in degrees	θ in radians	sin θ	cos θ	tan θ	csc θ	sec θ	cot θ
30°	$\dfrac{\pi}{6}$	$\dfrac{1}{2}$	$\dfrac{\sqrt{3}}{2}$	$\dfrac{\sqrt{3}}{3}$	2	$\dfrac{2\sqrt{3}}{3}$	$\sqrt{3}$
45°	$\dfrac{\pi}{4}$	$\dfrac{\sqrt{2}}{2}$	$\dfrac{\sqrt{2}}{2}$	1	$\sqrt{2}$	$\sqrt{2}$	1
60°	$\dfrac{\pi}{3}$	$\dfrac{\sqrt{3}}{2}$	$\dfrac{1}{2}$	$\sqrt{3}$	$\dfrac{2\sqrt{3}}{3}$	2	$\dfrac{\sqrt{3}}{3}$

It's useful to remember these special trigonometric ratios because they occur often. Of course, they can be recalled easily if we remember the triangles from which they are obtained.

To find the values of the trigonometric ratios for other angles, we use a calculator. Mathematical methods (called *numerical methods*) used in finding the trigonometric ratios are programmed directly into scientific calculators. For instance, when the $\boxed{\text{SIN}}$ key is pressed, the calculator computes an approximation to the value of the sine of the given angle. Calculators give the values of sine, cosine, and tangent; the other ratios can be easily calculated from these using the following *reciprocal relations:*

$$\csc t = \frac{1}{\sin t} \qquad \sec t = \frac{1}{\cos t} \qquad \cot t = \frac{1}{\tan t}$$

For an explanation of numerical methods, see the marginal note on page 422.

You should check that these relations follow immediately from the definitions of the trigonometric ratios.

We follow the convention that when we write sin *t, we mean the sine of the angle whose radian measure is t.* For instance, sin 1 means the sine of the angle whose radian measure is 1. When using a calculator to find an approximate value for this number, set your calculator to radian mode; you will find that

$$\sin 1 \approx 0.841471$$

If you want to find the sine of the angle whose measure is 1°, set your calculator to degree mode; you will find that

$$\sin 1° \approx 0.0174524$$

Aristarchus of Samos (310–230 B.C.) was a famous Greek scientist, musician, astronomer, and geometer. In his book *On the Sizes and Distances of the Sun and the Moon,* he estimated the distance to the sun by observing that when the moon is exactly half full, the triangle formed by the sun, moon, and the earth has a right angle at the moon. His method was similar to the one described in Exercise 51 in this section. Aristarchus was the first to advance the theory that the earth and planets move around the sun, an idea that did not gain full acceptance until after the time of Copernicus, 1800 years later. For this reason he is often called the "Copernicus of antiquity."

EXAMPLE 3 ■ Using a Calculator to Find Trigonometric Ratios

With our calculator in degree mode, and writing the results correct to five decimal places, we find

$$\sin 17° \approx 0.29237 \qquad \sec 88° = \frac{1}{\cos 88°} \approx 28.65371$$

With our calculator in radian mode, and writing the results correct to five decimal places, we find

$$\cos 1.2 \approx 0.36236 \qquad \cot 1.54 = \frac{1}{\tan 1.54} \approx 0.03081 \qquad \blacksquare$$

■ Applications of Trigonometry of Right Triangles

A triangle has six parts: three angles and three sides. To **solve a triangle** means to determine all of its parts from the information known about the triangle, that is, to determine the lengths of the three sides and the measures of the three angles.

EXAMPLE 4 ■ Solving a Right Triangle

Solve triangle ABC, shown in Figure 7.

SOLUTION

It's clear that $\angle B = 60°$. To find a, we look for an equation that relates a to the lengths and angles we already know. In this case, we have $\sin 30° = a/12$, so

$$a = 12 \sin 30° = 12\left(\tfrac{1}{2}\right) = 6$$

Similarly, $\cos 30° = b/12$, so

$$b = 12 \cos 30° = 12\left(\frac{\sqrt{3}}{2}\right) = 6\sqrt{3} \qquad \blacksquare$$

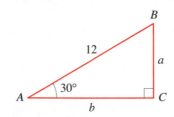

FIGURE 7

It's very useful to know that, using the information given in Figure 8, the lengths of the legs of a right triangle are

$$a = r \sin \theta \qquad \text{and} \qquad b = r \cos \theta$$

The ability to solve right triangles using the trigonometric ratios is fundamental to many problems in navigation, surveying, astronomy, and the measurement of distances. The applications we consider in this section always involve right triangles but, as we will see in the next three sections, trigonometry is also useful in solving triangles that are not right triangles.

To discuss the next examples, we need some terminology. If an observer is looking at an object, then the line from the observer's eye to the object is called the **line of sight** (Figure 9). If the object being observed is above the horizontal, then the angle between the line of sight and the horizontal is called the **angle of elevation**. If the object is below the horizontal, then the angle between the line of sight and the horizontal is called the **angle of depression**. In many of the examples and exercises in this chapter, angles of elevation and depression will be given for a hypothetical observer at ground level. If the line of sight follows a physical object, such as an inclined plane or a hillside, we use the term **angle of inclination**.

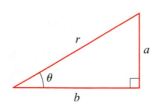

FIGURE 8
$a = r \sin \theta$
$b = r \cos \theta$

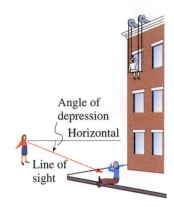

FIGURE 9

Thales of Miletus (circa 625–547 B.C.) is the legendary founder of Greek geometry. It is said that he calculated the height of a Greek column by comparing the length of the shadow of his staff with that of the column. Using properties of similar triangles, he argued that the ratio of the height h of the column to the height h' of his staff was equal to the ratio of the length s of the column's shadow to the length s' of the staff's shadow:

$$\frac{h}{h'} = \frac{s}{s'}$$

Since three of these quantities are known, Thales was able to calculate the height of the column.

According to legend, Thales used a similar method to find the height of the Great Pyramid in Egypt, a feat that impressed Egypt's king. Plutarch wrote that "although he [the king of Egypt] admired you [Thales] for other things, yet he particularly liked the manner by which you measured the height of the pyramid without any trouble or instrument." The principle Thales used, the fact that ratios of corresponding sides of similar triangles are equal, is the foundation of the subject of trigonometry.

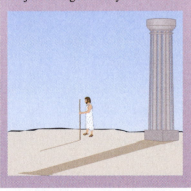

The next example gives an important application of trigonometry to the problem of measurement: We measure the height of a tall tree without having to climb it! Although the example is simple, the result is fundamental to the method of applying the trigonometric ratios to such problems.

EXAMPLE 5 ■ Finding the Height of a Tree

A giant redwood tree casts a shadow 532 ft long. Find the height of the tree if the angle of elevation of the sun is 25.7°.

SOLUTION

Let the height of the tree be h. From Figure 10 we see that

$$\frac{h}{532} = \tan 25.7° \qquad \text{Definition of tan}$$

$$h = 532 \tan 25.7° \qquad \text{Multiply by 532}$$

$$\approx 532(0.48127) \approx 256 \qquad \text{Use a calculator}$$

Therefore, the height of the tree is about 256 ft.

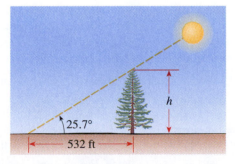

25.7°

532 ft

h

FIGURE 10

EXAMPLE 6 ■ A Problem Involving Right Triangles

From a point on the ground 500 ft from the base of a building, it is observed that the angle of elevation to the top of the building is 24° and the angle of elevation to the top of a flagpole atop the building is 27°. Find the height of the building and the length of the flagpole.

SOLUTION

Figure 11 illustrates the situation. The height of the building is found in the same way that we found the height of the tree in Example 5.

$$\frac{h}{500} = \tan 24° \qquad \text{Definition of tan}$$

$$h = 500 \tan 24° \qquad \text{Multiply by 500}$$

$$\approx 500(0.4452) \approx 223 \qquad \text{Use a calculator}$$

FIGURE 11

The height of the building is approximately 223 ft. To find the length of the flagpole, let's first find the height from the ground to the top of the pole:

$$\frac{k}{500} = \tan 27°$$

$$k = 500 \tan 27° \approx 500(0.5095) \approx 255$$

To find the length of the flagpole, we subtract h from k. So the length of the pole is approximately $255 - 223 = 32$ ft. ■

The key labels $\boxed{\text{SIN}^{-1}}$ or $\boxed{\text{INV}}$ $\boxed{\text{SIN}}$ stand for "inverse sine." The inverse trigonometric functions are studied in Section 7.4.

In some problems we need to find an angle in a right triangle whose sides are given. To do this, we use Table 1 (page 485) "backward"; that is, we find the *angle* with the specified trigonometric ratio. For example, if $\sin \theta = \frac{1}{2}$, what is the angle θ? From Table 1 we can tell that $\theta = 30°$. To find an angle whose sine is not given in the table, we use the $\boxed{\text{SIN}^{-1}}$ or $\boxed{\text{INV}}$ $\boxed{\text{SIN}}$ or $\boxed{\text{ARCSIN}}$ keys on a calculator. For example, if $\sin \theta = 0.8$, we apply the $\boxed{\text{SIN}^{-1}}$ key to get $\theta = 53.13°$ or 0.927 rad. The calculator also gives angles whose cosine or tangent are known, using the $\boxed{\text{COS}^{-1}}$ or $\boxed{\text{TAN}^{-1}}$ key.

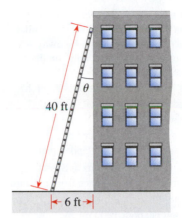

FIGURE 12

EXAMPLE 7 ■ Solving for an Angle in a Right Triangle

A 40-ft ladder leans against a building. If the base of the ladder is 6 ft from the base of the building, what is the angle formed by the ladder and the building?

SOLUTION

First we sketch a diagram as in Figure 12. If θ is the angle between the ladder and the building, then

$$\sin \theta = \frac{6}{40} = 0.15$$

So θ is the angle whose sine is 0.15. To find the angle θ, we use a calculator and use the $\boxed{\text{SIN}^{-1}}$ key. With our calculator in degree mode, we get

$$\theta \approx 8.6°$$ ■

6.2 EXERCISES

1–6 ■ Find the exact values of the six trigonometric ratios of the angle θ in the triangle.

1.

2.

3.

4.
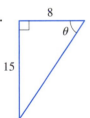

5.

6.

7–8 ■ Find (a) $\sin \alpha$ and $\cos \beta$, (b) $\tan \alpha$ and $\cot \beta$, and (c) $\sec \alpha$ and $\csc \beta$.

7.

8.

9–14 ■ Find the side labeled x. In Exercises 13 and 14 state your answer correct to five decimal places.

9.

10.

11.

12.

13.

14.

15–16 ■ Express x and y in terms of trigonometric ratios of θ.

15.

16.

17–22 ■ Sketch a triangle that has acute angle θ, and find the other five trigonometric ratios of θ.

17. $\sin \theta = \frac{3}{5}$

18. $\cos \theta = \frac{9}{40}$

19. $\cot \theta = 1$

20. $\tan \theta = \sqrt{3}$

21. $\sec \theta = \frac{7}{2}$

22. $\csc \theta = \frac{13}{12}$

23–28 ■ Evaluate the expression without using a calculator.

23. $\sin \dfrac{\pi}{6} + \cos \dfrac{\pi}{6}$

24. $\sin 30° \csc 30°$

25. $\sin 30° \cos 60° + \sin 60° \cos 30°$

26. $(\sin 60°)^2 + (\cos 60°)^2$

27. $(\cos 30°)^2 - (\sin 30°)^2$

28. $\left(\sin \dfrac{\pi}{3} \cos \dfrac{\pi}{4} - \sin \dfrac{\pi}{4} \cos \dfrac{\pi}{3}\right)^2$

29–32 ■ Solve the right triangle.

29.

30.

31.

32.

33. Use a ruler to carefully measure the sides of the triangle, and then use your measurements to estimate the six trigonometric ratios of θ.

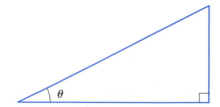

34. Using a protractor, sketch a right triangle that has the acute angle 40°. Measure the sides carefully and use your results to estimate the six trigonometric ratios of 40°.

35. The angle of elevation to the top of the Empire State Building in New York is found to be 11° from the ground at a distance of 1 mi from the base of the building. Using this information, find the height of the Empire State Building.

36. A plane is flying within sight of the Gateway Arch in St. Louis, Missouri, at an elevation of 35,000 ft. The pilot would like to estimate her distance from the Gateway Arch. She finds that the angle of depression to a point on the ground below the arch is 22°.
(a) What is the distance between the plane and the arch?
(b) What is the distance between a point on the ground directly below the plane and the arch?

37. A laser beam is to be directed toward the center of the moon, but the beam strays 0.5° from its intended path.
(a) How far has the beam diverged from its assigned target when it reaches the moon? (The distance from the earth to the moon is 240,000 mi.)
(b) The radius of the moon is about 1000 mi. Will the beam strike the moon?

38. From the top of a 200-ft lighthouse, the angle of depression to a ship in the ocean is 23°. How far is the ship from the base of the lighthouse?

39. A 20-ft ladder leans against a building so that the angle between the ground and the ladder is 72°. How high does the ladder reach on the building?

40. A 20-ft ladder is leaning against a building. If the base of the ladder is 6 ft from the base of the building, what is the angle of elevation of the ladder? How high does the ladder reach on the building?

41. A 96-ft tree casts a shadow that is 120 ft long. What is the angle of elevation of the sun?

42. A 600-ft guy wire is attached to the top of a communication tower. If the wire makes an angle of 65° with the ground, how tall is the communication tower?

43. A man is lying on the beach, flying a kite. He holds the end of the kite string at ground level, and estimates the angle of elevation of the kite to be 50°. If the string is 450 ft long, how high is the kite above the ground?

44. The height of a steep cliff is to be measured from a point on the opposite side of the river. Find the height of the cliff from the information given in the figure.

45. A water tower is located 325 ft from a building (see the figure). From a window in the building it is observed that the angle of elevation to the top of the tower is 39° and the angle of depression to the bottom of the tower is 25°. How tall is the tower? How high is the window?

46. An airplane is flying at an elevation of 5150 ft, directly above a straight highway. Two motorists are driving cars on the highway on opposite sides of the plane, and the angle of depression to one car is 35° and to the other is 52°. How far apart are the cars?

47. If both cars in Exercise 46 are on one side of the plane and if the angle of depression to one car is 38° and to the other car is 52°, how far apart are the cars?

48. A hot-air balloon is floating above a straight road. To estimate their height above the ground, the balloonists simultaneously measure the angle of depression to two consecutive mileposts on the road on the same side of the balloon. The angles of depression are found to be 20° and 22°. How high is the balloon?

49. To estimate the height of a mountain above a level plain, the angle of elevation to the top of the mountain is measured to be 32°. One thousand feet closer to the mountain along the plain, it is found that the angle of elevation is 35°. Estimate the height of the mountain.

50. (a) Show that the height h of the mountain in the figure is given by
$$h = d \frac{\tan \beta \tan \alpha}{\tan \beta - \tan \alpha} = \frac{d}{\cot \alpha - \cot \beta}$$
(b) Use the formula in part (a) to find the height h of the mountain if $\alpha \approx 25°$, $\beta \approx 29°$, and $d \approx 800$ ft.

51. When the moon is exactly half full, the earth, moon, and sun form a right angle (see the figure). At that time the angle formed by the sun, earth, and moon is measured to be 89.85°. If the distance from the earth to the moon is 240,000 mi, estimate the distance from the earth to the sun.

52. To find the distance to the sun as in Exercise 51, we needed to know the distance to the moon. Here is a way to estimate that distance: When the moon is seen at its zenith at a point A on the earth, it is observed to be at the horizon from point B (see the figure). Points A and B are 6155 mi apart, and the radius of the earth is 3960 mi.
(a) Find the angle θ in degrees.
(b) Estimate the distance from point A to the moon.

53. In Exercise 54 of Section 6.1 a method was given for finding the radius of the earth. Here is a more modern method: From a satellite 600 mi above the earth, it is observed that the angle formed by the vertical and the line of sight to the horizon is 60.276°. Use this information to find the radius of the earth.

54. To find the distance to nearby stars, the method of parallax is used. The idea is to find a triangle with the star at one vertex and with a base as large as possible. To do this, the star is observed at two different times exactly 6 months apart, and its apparent change in position is recorded. From these two observations, $\angle E_1 S E_2$ can be calculated. (The times are chosen so that $\angle E_1 S E_2$ is as large as possible, which guarantees that $\angle E_1 O S$ is 90°.) The angle $E_1 S O$ is called the *parallax* of the star. Alpha Centauri, the star nearest the earth, has a parallax of 0.000211°. Estimate the distance to this star. (Take the distance from the earth to the sun to be 9.3×10^7 mi.)

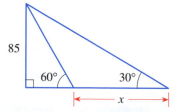

55–58 ■ Find *x* correct to one decimal place.

55.

56.

57.

58.

59. Express the lengths *a*, *b*, *c*, and *d* in the figure in terms of the trigonometric ratios of θ.

     DISCOVERY · DISCUSSION

60. Similar Triangles   If two triangles are similar, what properties do they share? Explain how these properties make it possible to define the trigonometric ratios without regard to the size of the triangle.

In the preceding section we defined the trigonometric ratios for acute angles. Here we extend the trigonometric ratios to all angles by defining the trigonometric functions of angles. With these functions we can solve practical problems that involve angles which are not necessarily acute.

■ Trigonometric Functions of Angles

Let POQ be a right triangle with acute angle θ as shown in Figure 1(a). Place θ in standard position as shown in Figure 1(b).

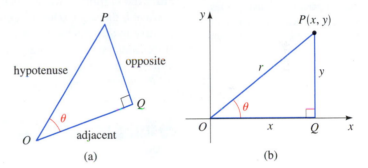

FIGURE 1

(a)   (b)

Then $P = P(x, y)$ is a point on the terminal side of θ. In triangle POQ, the opposite side has length y and the adjacent side has length x. Using the Pythagorean Theorem, we see that the hypotenuse has length $r = \sqrt{x^2 + y^2}$. So

$$\sin\theta = \frac{y}{r}, \qquad \cos\theta = \frac{x}{r}, \qquad \tan\theta = \frac{y}{x}$$

The other trigonometric ratios can be found in the same way.

These observations allow us to extend the trigonometric ratios to any angle. We define the trigonometric functions of angles as follows (see Figure 2).

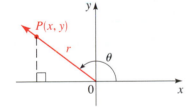

FIGURE 2

DEFINITION OF THE TRIGONOMETRIC FUNCTIONS

Let θ be an angle in standard position and let $P(x, y)$ be a point on the terminal side. If $r = \sqrt{x^2 + y^2}$ is the distance from the origin to the point $P(x, y)$, then

$$\sin\theta = \frac{y}{r} \qquad\qquad \cos\theta = \frac{x}{r} \qquad\qquad \tan\theta = \frac{y}{x} \ \ (x \neq 0)$$

$$\csc\theta = \frac{r}{y} \ \ (y \neq 0) \qquad \sec\theta = \frac{r}{x} \ \ (x \neq 0) \qquad \cot\theta = \frac{x}{y} \ \ (y \neq 0)$$

FIGURE 3

Since division by 0 is an undefined operation, certain trigonometric functions are not defined for certain angles. For example, $\tan 90° = y/x$ is undefined because $x = 0$. The angles for which the trigonometric functions may be undefined are the angles for which either the x- or y-coordinate of a point on the terminal side of the angle is 0. These are **quadrantal angles**—angles that are coterminal with the coordinate axes.

It is a crucial fact that the values of the trigonometric functions do *not* depend on the choice of the point $P(x, y)$. This is because if $P'(x', y')$ is any other point on the terminal side, as in Figure 3, then triangles POQ and $P'OQ'$ are similar.

■ Evaluating Trigonometric Functions at Any Angle

The following mnemonic device can be used to remember which trigonometric functions are positive in each quadrant: All of them, Sine, Tangent, or Cosine.

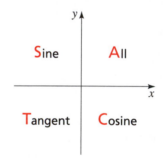

You can remember this as "All Students Take Calculus."

From the definition we see that the values of the trigonometric functions are all positive if the angle θ has its terminal side in quadrant I. This is because x and y are positive in this quadrant. [Of course, r is always positive, since it is simply the distance from the origin to the point $P(x, y)$.] If the terminal side of θ is in quadrant II, however, then x is negative and y is positive. Thus, in quadrant II the functions $\sin \theta$ and $\csc \theta$ are positive, and all the other trigonometric functions have negative values. You can check the other entries in the following table.

SIGNS OF THE TRIGONOMETRIC FUNCTIONS		
Quadrant	Positive functions	Negative functions
I	all	none
II	sin, csc	cos, sec, tan, cot
III	tan, cot	sin, csc, cos, sec
IV	cos, sec	sin, csc, tan, cot

We now turn our attention to finding the values of the trigonometric functions for angles that are not acute.

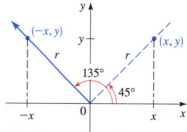

FIGURE 4

EXAMPLE 1   ■   Finding Trigonometric Functions of Angles

Find (a) $\cos 135°$ and (b) $\tan 390°$.

SOLUTION

(a) From Figure 4 we see that $\cos 135° = -x/r$. But $\cos 45° = x/r$, and since $\cos 45° = \sqrt{2}/2$, we have

$$\cos 135° = -\frac{\sqrt{2}}{2}$$

(b) The angles 390° and 30° are coterminal. From Figure 5 it's clear that $\tan 390° = \tan 30°$ and, since $\tan 30° = \sqrt{3}/3$, we have

$$\tan 390° = \frac{\sqrt{3}}{3}$$

FIGURE 5

■

Relationship to the Trigonometric Functions of Real Numbers

You may have already studied the trigonometric functions defined using the unit circle (Chapter 5). To see how they relate to the trigonometric functions of an *angle*, let's start with the unit circle in the coordinate plan.

$P(x, y)$ is the terminal point determined by t

Let $P(x, y)$ be the terminal point determined by an arc of length t on the unit circle. Then t subtends an angle θ at the center of the circle. If we drop a perpendicular from P onto the point Q on the x-axis, then triangle $\triangle OPQ$ is a right triangle with legs of length x and y, as shown in the figure.

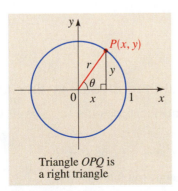

Triangle OPQ is a right triangle

Now, by the definition of the trigonometric functions of the *real number t*, we have

$$\sin t = y$$

$$\cos t = x$$

By the definition of the trigonometric functions of the *angle* θ, we have

$$\sin \theta = \frac{\text{opp}}{\text{hyp}} = \frac{y}{1} = y$$

$$\cos \theta = \frac{\text{adj}}{\text{hyp}} = \frac{x}{1} = x$$

If θ is measured in radians, then $\theta = t$. (See the figure below.) Comparing the two ways of defining the trigonometric functions, we see that they are identical. In other words, as functions, they assign identical values to a given real number (the real number is the radian measure of θ in one case or the length t of an arc in the other).

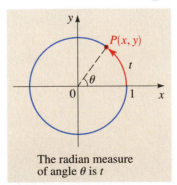

The radian measure of angle θ is t

Why then do we study trigonometry in two different ways? Because different applications require that we view the trigonometric functions differently. (See *Focus on Modeling*, page 455, and Sections 6.2, 6.4, and 6.5.)

From Example 1 we see that the trigonometric functions for angles that aren't acute have the same value, except possibly for sign, as the corresponding trigonometric functions of an acute angle. That acute angle will be called the *reference angle.*

REFERENCE ANGLE

Let θ be an angle in standard position. The **reference angle** $\bar{\theta}$ associated with θ is the acute angle formed by the terminal side of θ and the x-axis.

Figure 6 shows that to find a reference angle it's useful to know the quadrant in which the terminal side of the angle lies.

   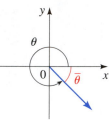

FIGURE 6
The reference angle $\bar{\theta}$ for an angle θ

FIGURE 7

FIGURE 8

EXAMPLE 2 ■ Finding Reference Angles

Find the reference angle for (a) $\theta = \dfrac{5\pi}{3}$ and (b) $\theta = 870°$.

SOLUTION

(a) The reference angle is the acute angle formed by the terminal side of the angle $5\pi/3$ and the x-axis (see Figure 7). Since the terminal side of this angle is in quadrant IV, the reference angle is

$$\bar{\theta} = 2\pi - \frac{5\pi}{3} = \frac{\pi}{3}$$

(b) The angles 870° and 150° are coterminal [because $870 - 2(360) = 150$]. Thus, the terminal side of this angle is in quadrant II (see Figure 8). So the reference angle is

$$\bar{\theta} = 180° - 150° = 30°$$ ■

EVALUATING TRIGONOMETRIC FUNCTIONS AT ANY ANGLE

To find the values of the trigonometric functions for any angle θ, we carry out the following steps.

1. Find the reference angle $\bar{\theta}$ associated with the angle θ.

2. Determine the sign of the trigonometric function of θ.

3. The value of the trigonometric function of θ is the same, except possibly for sign, as the value of the trigonometric function of $\bar{\theta}$.

FIGURE 9

$\frac{S|A}{T|C}$ sin 240° *is negative.*

FIGURE 10

$\frac{S|A}{T|C}$ tan 495° *is negative,*
so cot 495° *is negative.*

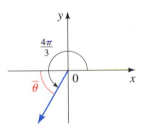

FIGURE 11

$\frac{S|A}{T|C}$ sin $\frac{16\pi}{3}$ *is negative.*

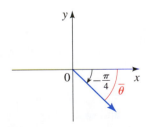

FIGURE 12

$\frac{S|A}{T|C}$ cos$\left(-\frac{\pi}{4}\right)$ *is positive,*

so sec$\left(-\frac{\pi}{4}\right)$ *is positive.*

EXAMPLE 3 ■ Using the Reference Angle to Evaluate
Trigonometric Functions

Find (a) sin 240° and (b) cot 495°.

SOLUTION

(a) This angle has its terminal side in quadrant III, as shown in Figure 9. The reference angle is therefore 240° − 180° = 60°, and the value of sin 240° is negative. Thus

$$\sin 240° = -\sin 60° = -\frac{\sqrt{3}}{2}$$

↑ sign   ↑ reference angle

(b) The angle 495° is coterminal with the angle 135°, and the terminal side of this angle is in quadrant II, as shown in Figure 10. So the reference angle is 180° − 135° = 45°, and the value of cot 495° is negative. We have

$$\cot 495° = \cot 135° = -\cot 45° = -1$$

↑ coterminal angles   ↑ sign   ↑ reference angle   ■

EXAMPLE 4 ■ Using the Reference Angle to Evaluate
Trigonometric Functions

Find (a) sin $\dfrac{16\pi}{3}$ and (b) sec$\left(-\dfrac{\pi}{4}\right)$.

SOLUTION

(a) The angle 16π/3 is coterminal with 4π/3, and these angles are in quadrant III (see Figure 11). Thus, the reference angle is $(4\pi/3) - \pi = \pi/3$. Since the value of sine is negative in quadrant III, we have

$$\sin \frac{16\pi}{3} = \sin \frac{4\pi}{3} = -\sin \frac{\pi}{3} = -\frac{\sqrt{3}}{2}$$

↑ coterminal angles   ↑ sign   ↑ reference angle

(b) The angle −π/4 is in quadrant IV, and its reference angle is π/4 (see Figure 12). Since secant is positive in this quadrant, we get

$$\sec\left(-\frac{\pi}{4}\right) = +\sec \frac{\pi}{4} = \frac{\sqrt{2}}{2}$$

↑ sign   ↑ reference angle   ■

■ Trigonometric Identities

The trigonometric functions of angles are related to each other through several important equations called **trigonometric identities**. We've already encountered the reciprocal identities. These identities continue to hold for any angle θ, provided both sides of the equation are defined. The Pythagorean identities are a consequence of the Pythagorean Theorem.*

FUNDAMENTAL IDENTITIES

Reciprocal Identities

$$\csc \theta = \frac{1}{\sin \theta} \qquad \sec \theta = \frac{1}{\cos \theta} \qquad \cot \theta = \frac{1}{\tan \theta}$$

$$\tan \theta = \frac{\sin \theta}{\cos \theta} \qquad \cot \theta = \frac{\cos \theta}{\sin \theta}$$

Pythagorean Identities

$$\sin^2\theta + \cos^2\theta = 1 \qquad \tan^2\theta + 1 = \sec^2\theta \qquad 1 + \cot^2\theta = \csc^2\theta$$

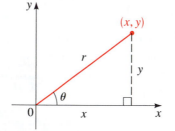

FIGURE 13

■ **Proof** Let's prove the first Pythagorean identity. Using $x^2 + y^2 = r$ (the Pythagorean Theorem) in Figure 13, we have

$$\sin^2\theta + \cos^2\theta = \left(\frac{y}{r}\right)^2 + \left(\frac{x}{r}\right)^2 = \frac{x^2 + y^2}{r^2} = \frac{r^2}{r^2} = 1$$

Thus, $\sin^2\theta + \cos^2\theta = 1$. (Although the figure indicates an acute angle, you should check that the proof holds for all angles θ.)   □

See Exercises 59 and 60 for the proofs of the other two Pythagorean identities.

EXAMPLE 5 ■ **Expressing One Trigonometric Function in Terms of Another**

(a) Express $\sin \theta$ in terms of $\cos \theta$.
(b) Express $\tan \theta$ in terms of $\sin \theta$, where θ is in quadrant II.

SOLUTION

(a) From the first Pythagorean identity we get

$$\sin \theta = \pm\sqrt{1 - \cos^2\theta}$$

where the sign depends on the quadrant. If θ is in quadrant I or II, then $\sin \theta$ is positive, and hence

$$\sin \theta = \sqrt{1 - \cos^2\theta}$$

whereas if θ is in quadrant III or IV, $\sin \theta$ is negative and so

$$\sin \theta = -\sqrt{1 - \cos^2\theta}$$

---

*We follow the usual convention of writing $\sin^2\theta$ for $(\sin \theta)^2$. In general, we write $\sin^n\theta$ for $(\sin \theta)^n$ for all integers n except $n = -1$. The exponent $n = -1$ will be assigned another meaning in Section 7.4. Of course, the same convention applies to the other five trigonometric functions.

(b) Since $\tan \theta = \sin \theta / \cos \theta$, we need to write $\cos \theta$ in terms of $\sin \theta$. By part (a)

$$\cos \theta = \pm\sqrt{1 - \sin^2\theta}$$

and since $\cos \theta$ is negative in quadrant II, the negative sign applies here. Thus

$$\tan \theta = \frac{\sin \theta}{\cos \theta} = \frac{\sin \theta}{-\sqrt{1 - \sin^2\theta}} \qquad ■$$

EXAMPLE 6 ■ Evaluating a Trigonometric Function

If $\tan \theta = \frac{2}{3}$ and θ is in quadrant III, find $\cos \theta$.

SOLUTION 1

We need to write $\cos \theta$ in terms of $\tan \theta$. From the identity $\tan^2\theta + 1 = \sec^2\theta$, we get $\sec \theta = \pm\sqrt{\tan^2\theta + 1}$. In quadrant III, $\sec \theta$ is negative, so

$$\sec \theta = -\sqrt{\tan^2\theta + 1}$$

Thus

$$\cos \theta = \frac{1}{\sec \theta} = \frac{1}{-\sqrt{\tan^2\theta + 1}}$$

$$= \frac{1}{-\sqrt{\left(\frac{2}{3}\right)^2 + 1}} = \frac{1}{-\sqrt{\frac{13}{9}}} = -\frac{3}{\sqrt{13}}$$

SOLUTION 2

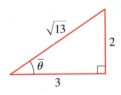

FIGURE 14

This problem can be solved more easily using the method of Example 2 of Section 6.2. Recall that, except for sign, the values of the trigonometric functions of any angle are the same as those of an acute angle (the reference angle). So, ignoring the sign for the moment, let's sketch a right triangle with an acute angle $\bar{\theta}$ satisfying $\tan \bar{\theta} = \frac{2}{3}$ (see Figure 14). By the Pythagorean Theorem the hypotenuse of this triangle has length $\sqrt{13}$. From the triangle in Figure 14 we immediately see that $\cos \bar{\theta} = 3/\sqrt{13}$. Since θ is in quadrant III, $\cos \theta$ is negative and so

$$\cos \theta = -\frac{3}{\sqrt{13}} \qquad ■$$

EXAMPLE 7 ■ Evaluating Trigonometric Functions

If $\sec \theta = 2$ and θ is in quadrant IV, find the other five trigonometric functions of θ.

SOLUTION

FIGURE 15

We sketch a triangle as in Figure 15 so that $\sec \bar{\theta} = 2$. Taking into account the fact

that θ is in quadrant IV, we get

$$\sin \theta = -\frac{\sqrt{3}}{2} \qquad \cos \theta = \frac{1}{2} \qquad \tan \theta = -\sqrt{3}$$

$$\csc \theta = -\frac{2}{\sqrt{3}} \qquad \sec \theta = 2 \qquad \cot \theta = -\frac{1}{\sqrt{3}} \qquad ■$$

■ Areas of Triangles

(a)

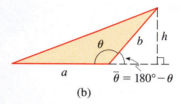

(b)

FIGURE 16

We end this section by giving an application of the trigonometric functions that involves angles that are not necessarily acute. More extensive applications appear in the next two sections.

The area of a triangle is $\mathcal{A} = \frac{1}{2} \times$ base $\times$ height. If we know two sides and the included angle of a triangle, then we can find the height using the trigonometric functions, and from this we can find the area.

If θ is an acute angle, then the height of the triangle in Figure 16(a) is given by $h = b \sin \theta$. Thus, the area is

$$\mathcal{A} = \tfrac{1}{2} \times \text{base} \times \text{height} = \tfrac{1}{2} ab \sin \theta$$

If the angle θ is not acute, then from Figure 16(b) we see that the height of the triangle is

$$h = b \sin(180° - \theta) = b \sin \theta$$

This is so because the reference angle of θ is the angle $180° - \theta$. Thus, in this case also, the area of the triangle is

$$\mathcal{A} = \tfrac{1}{2} \times \text{base} \times \text{height} = \tfrac{1}{2} ab \sin \theta$$

AREA OF A TRIANGLE

The area $\mathcal{A}$ of a triangle with sides of lengths a and b and with included angle θ is

$$\mathcal{A} = \tfrac{1}{2} ab \sin \theta$$

EXAMPLE 8 ■ Finding the Area of a Triangle

Find the area of triangle ABC shown in Figure 17.

SOLUTION

FIGURE 17

The triangle has sides of length 10 cm and 3 cm, with included angle 120°. Therefore

$$\mathcal{A} = \tfrac{1}{2} ab \sin \theta$$

$$= \tfrac{1}{2} (10)(3) \sin 120°$$

$$= 15 \sin 60° \qquad \text{Reference angle}$$

$$= 15 \frac{\sqrt{3}}{2} \approx 13 \text{ cm}^2 \qquad ■$$

6.3 EXERCISES

1–6 ■ Find the reference angle for the given angle.

1. (a) $210°$ (b) $300°$ (c) $-120°$

2. (a) $225°$ (b) $450°$ (c) $-200°$

3. (a) $\dfrac{9\pi}{4}$ (b) $\dfrac{23\pi}{6}$ (c) $-\dfrac{\pi}{3}$

4. (a) $\dfrac{17\pi}{6}$ (b) $\dfrac{31\pi}{4}$ (c) $-\dfrac{4\pi}{3}$

5. (a) $\dfrac{11\pi}{5}$ (b) $\dfrac{11}{5}$ (c) $-\dfrac{4\pi}{7}$

6. (a) 2.3 (b) 2.3π (c) $-\pi$

7–30 ■ Find the exact value of the trigonometric function.

7. $\sin 150°$ **8.** $\sin 225°$ **9.** $\cos 135°$

10. $\cos(-60°)$ **11.** $\tan(-60°)$ **12.** $\sec 300°$

13. $\csc(-630°)$ **14.** $\cot 210°$ **15.** $\cos 570°$

16. $\sec 120°$ **17.** $\tan 750°$ **18.** $\cos 660°$

19. $\sin \dfrac{2\pi}{3}$ **20.** $\sin \dfrac{5\pi}{3}$ **21.** $\sin \dfrac{3\pi}{2}$

22. $\cos \dfrac{7\pi}{3}$ **23.** $\cos\left(-\dfrac{7\pi}{3}\right)$ **24.** $\tan \dfrac{5\pi}{6}$

25. $\sec \dfrac{17\pi}{3}$ **26.** $\csc \dfrac{5\pi}{4}$ **27.** $\cot\left(-\dfrac{\pi}{4}\right)$

28. $\cos \dfrac{7\pi}{4}$ **29.** $\tan \dfrac{5\pi}{2}$ **30.** $\sin \dfrac{11\pi}{6}$

31–34 ■ Find the quadrant in which θ lies from the information given.

31. $\sin \theta < 0$ and $\cos \theta < 0$

32. $\tan \theta < 0$ and $\sin \theta < 0$

33. $\sec \theta > 0$ and $\tan \theta < 0$

34. $\csc \theta > 0$ and $\cos \theta < 0$

35–40 ■ Write the first trigonometric function in terms of the second for θ in the given quadrant.

35. $\tan \theta$, $\cos \theta$; θ in quadrant III

36. $\cot \theta$, $\sin \theta$; θ in quadrant II

37. $\cos \theta$, $\sin \theta$; θ in quadrant IV

38. $\sec \theta$, $\sin \theta$; θ in quadrant I

39. $\sec \theta$, $\tan \theta$; θ in quadrant II

40. $\csc \theta$, $\cot \theta$; θ in quadrant III

41–48 ■ Find the values of the trigonometric functions of θ from the information given.

41. $\sin \theta = \frac{3}{5}$, θ in quadrant II

42. $\cos \theta = -\frac{7}{12}$, θ in quadrant III

43. $\tan \theta = -\frac{3}{4}$, $\cos \theta > 0$

44. $\sec \theta = 5$, $\sin \theta < 0$

45. $\csc \theta = 2$, θ in quadrant I

46. $\cot \theta = \frac{1}{4}$, $\sin \theta < 0$

47. $\cos \theta = -\frac{2}{7}$, $\tan \theta < 0$

48. $\tan \theta = -4$, $\sin \theta > 0$

49. If $\theta = \pi/3$, find the value of each expression.

(a) $\sin 2\theta$, $2 \sin \theta$ (b) $\sin \frac{1}{2}\theta$, $\dfrac{\sin \theta}{2}$

(c) $\sin^2\theta$, $\sin(\theta^2)$

50. Find the area of a triangle with sides of length 7 and 9 and included angle $72°$.

51. Find the area of a triangle with sides of length 10 and 22 and included angle $10°$.

52. Find the area of an equilateral triangle with side of length 10.

53. A triangle has an area of 16 in^2, and two of the sides of the triangle have lengths 5 in. and 7 in. Find the angle included by these two sides.

54. Find the area of the shaded region in the figure.

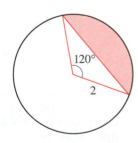

55–58 ■ In the following exercises you are asked to use the trigonometric functions to model real-life situations. See the guidelines in Section 2.7 on modeling with functions.

55. A steel pipe is being carried down a hallway 9 ft wide. At the end of the hall there is a right-angled turn into a narrower hallway 6 ft wide.

(a) Show that the length of the pipe in the figure is modeled by the function

$$L(\theta) = 9 \csc \theta + 6 \sec \theta$$

(b) Graph the function L for $0 < \theta < \pi/2$.

(c) Find the minimum value of the function L.

(d) Explain why the value of L you found in part (c) is the length of the longest pipe that can be carried around the corner.

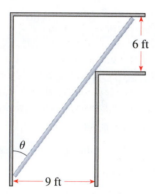

56. A rain gutter is to be constructed from a metal sheet of width 30 cm by bending up one-third of the sheet on each side through an angle θ.

(a) Show that the cross-sectional area of the gutter is modeled by the function

$$A(\theta) = 100 \sin \theta + 100 \sin \theta \cos \theta$$

(b) Graph the function A for $0 \le \theta \le \pi/2$.

(c) For what angle θ is the largest cross-sectional area achieved?

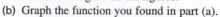

57. A rectangular beam is to be cut from a cylindrical log of diameter 20 cm, as shown in the figure.

(a) Express the cross-sectional area of the beam as a function of the angle θ in the figures.

(b) Graph the function you found in part (a).

(c) Find the dimensions of the beam with largest cross-sectional area.

58. The strength of a beam is proportional to the width and the square of the depth. A beam is cut from a log as in Exercise 57. Express the strength of the beam as a function of the angle θ in the figures.

59. Use the first Pythagorean identity to prove the second. [*Hint:* Divide by $\cos^2\theta$.]

60. Use the first Pythagorean identity to prove the third.

▲ DISCOVERY · DISCUSSION

61. Using a Calculator   To solve a certain problem, you need to find the sine of 4 rad. Your study partner uses his calculator and tells you that sin 4 is 0.0697564737. On your calculator you get −0.7568024953. What is wrong? What mistake did your partner make?

62. Viète's Trigonometric Diagram   In the 16th century, the French mathematician François Viète (see page 51) published the following remarkable diagram. Each of the six trigonometric functions of θ is equal to the length of a line segment in the figure. For instance, $\sin \theta = |PR|$, since from $\triangle OPR$ we see that

$$\sin \theta = \frac{\text{opp}}{\text{hyp}} = \frac{|PR|}{|OR|} = \frac{|PR|}{1}$$

For each of the five other trigonometric functions, find a line segment in the figure whose length equals the value of the function at θ. (Note that the radius of the circle is 1, the center is O, segment QS is tangent to the circle at R, and $\angle SOQ$ is a right angle.)

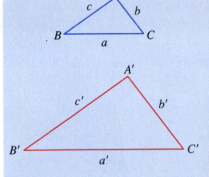

Thales used similar triangles to find the height of a tall column. See page 487.

Similarity

In geometry you learned that two triangles are similar if they have the same angles. In this case, the ratios of corresponding sides are equal. Triangles ABC and $A'B'C'$ in the margin are similar, so

$$\frac{a'}{a} = \frac{b'}{b} = \frac{c'}{c}$$

Similarity is the crucial idea underlying trigonometry. We can define $\sin\theta$ as the ratio of the opposite side to the hypotenuse in *any* right triangle with an angle θ, because all such right triangles are similar. So the ratio represented by $\sin\theta$ does not depend on the size of the right triangle but only on the angle θ. This is a powerful idea because angles are often easier to measure than distances. For example, the angle formed by the sun, earth, and moon can be measured from the earth. The secret to finding the distance to the sun is that the trigonometric ratios are the same for the huge triangle formed by the sun, earth, and moon as for any other similar triangle (see Exercise 51 in Section 6.2).

In general, two objects are **similar** if they have the same shape even though they may not be the same size.* For example, we recognize the following as representations of the letter A because they are all similar.

If two figures are similar, then the distances between corresponding points in the figures are proportional. The blue and red A's above are similar—the ratio of distances between corresponding points is $\frac{3}{2}$. We say that the **similarity ratio** is $s = \frac{3}{2}$. To obtain the distance d' between any two points in the blue A, we multiply the corresponding distance d in the red A by $\frac{3}{2}$. So

$$d' = sd \qquad \text{or} \qquad d' = \tfrac{3}{2}d$$

Likewise, the similarity ratio between the first and last letters is $s = 5$, so $x' = 5x$.

1. Write a short paragraph explaining how the concept of similarity is used to define the trigonometric ratios.

---

*If they have the same shape *and* size, they are congruent, which is a special case of similarity.

2. How is similarity used in map making? How are distances on a city road map related to actual distances?

3. How is your yearbook photograph similar to you? Compare distances between different points on your face (such as distance between ears, length of nose, distance between eyes, and so on) to the corresponding distances in a photograph. What is the similarity ratio?

4. The figure illustrates a method for drawing an apple twice the size of a given apple. Use the method to draw a tie 3 times the size (similarity ratio 3) of the blue tie.

5. Give conditions under which two rectangles are similar to each other. Do the same for two isosceles triangles.

6. Suppose that two similar triangles have similarity ratio s.
   (a) How are the perimeters of the triangles related?
   (b) How are the areas of the triangles related?

1   2

If the side of a square is doubled, its area is multiplied by 2^2.

7. (a) If two squares have similarity ratio s, show that their areas A_1 and A_2 have the property that $A_2 = s^2 A_1$.
   (b) If the side of a square is tripled, its area is multiplied by what factor?
   (c) A plane figure can be approximated by squares (as shown). Explain how we can conclude that for any two plane figures with similarity ratio s, their areas satisfy $A_2 = s^2 A_1$. [Use part (a).]

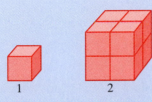

If the side of a cube is doubled, its volume is multiplied by 2^3.

8. (a) If two cubes have similarity ratio s, show that their volumes V_1 and V_2 have the property that $V_2 = s^3 V_1$.

(b) If the side of a cube is multiplied by 10, by what factor is the volume multiplied?

(c) How can we use the fact that a solid object can be "filled" by little cubes to show that for any two solids with similarity ratio s, the volumes satisfy $V_2 = s^3 V_1$?

9. King Kong is 10 times as tall as Joe, a normal-sized 300-lb gorilla. Assuming that King Kong and Joe are similar, use the results from Problems 7 and 8 to answer the following questions.

(a) How much does King Kong weigh?

(b) If Joe's hand is 13 in. long, how long is King Kong's hand?

(c) If it takes 2 square yards of material to make a shirt for Joe, how much material would a shirt for King Kong require?

6.4   THE LAW OF SINES

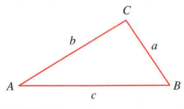

FIGURE 1

In Section 6.2 we used the trigonometric ratios to solve right triangles. The trigonometric functions can also be used to solve *oblique triangles*, that is, triangles with no right angles. To do this, we first study the Law of Sines here and then the Law of Cosines in the next section. To state these laws (or formulas) more easily, we follow the convention of labeling the angles of a triangle as A, B, C, and the lengths of the corresponding opposite sides as a, b, c, as in Figure 1.

To solve a triangle, we need to know certain information about its sides and angles. To decide whether we have enough information, it's often helpful to make a sketch. For instance, if we are given two angles and the included side, then it's clear that one and only one triangle can be formed [see Figure 2(a)]. Similarly, if two sides and the included angle are known, then a unique triangle is determined [Figure 2(c)]. But if we know all three angles and no sides, we cannot uniquely determine the triangle because many triangles can have the same three angles. (All these triangles would be similar, of course.) So we won't consider this last case.

(a) ASA or SAA

(b) SSA

(c) SAS

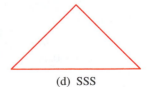

(d) SSS

FIGURE 2

In general, a triangle is determined by three of its six parts (angles and sides) as long as at least one of these three parts is a side. So, the possibilities, illustrated in Figure 2, are as follows:

Case 1	One side and two angles (ASA or SAA)
Case 2	Two sides and the angle opposite one of those sides (SSA)
Case 3	Two sides and the included angle (SAS)
Case 4	Three sides (SSS)

The first two cases are solved using the Law of Sines; Cases 3 and 4 require the Law of Cosines.

■ The Law of Sines

The **Law of Sines** says that in any triangle the lengths of the sides are proportional to the sines of the corresponding opposite angles.

THE LAW OF SINES

In triangle ABC we have

$$\frac{\sin A}{a} = \frac{\sin B}{b} = \frac{\sin C}{c}$$

■ **Proof**   To see why the Law of Sines is true, refer to Figure 3. By the formula in Section 6.3 the area of triangle ABC is $\frac{1}{2} ab \sin C$. By the same formula the area of this triangle is also $\frac{1}{2} ac \sin B$ and $\frac{1}{2} bc \sin A$. Thus

$$\tfrac{1}{2} bc \sin A = \tfrac{1}{2} ac \sin B = \tfrac{1}{2} ab \sin C$$

Multiplying by $2/(abc)$ gives the Law of Sines.   □

FIGURE 3

EXAMPLE 1 ■ Tracking a Satellite (ASA)

A satellite orbiting the earth passes directly overhead at observation stations in Phoenix and Los Angeles, 340 mi apart. At an instant when the satellite is between these two stations, its angle of elevation is simultaneously observed to be 60° at Phoenix and 75° at Los Angeles. How far is the satellite from Los Angeles? In other words, find the distance AC in Figure 4.

SOLUTION

Whenever two angles in a triangle are known, the third angle can be determined immediately because the sum of the angles of a triangle is 180°. In this case, $\angle C = 180° - (75° + 60°) = 45°$ (see Figure 3), so we have

$$\frac{\sin B}{b} = \frac{\sin C}{c} \qquad \text{Law of Sines}$$

$$\frac{\sin 60°}{b} = \frac{\sin 45°}{340} \qquad \text{Substitute}$$

$$b = \frac{340 \sin 60°}{\sin 45°} \approx 416 \qquad \text{Solve for } b$$

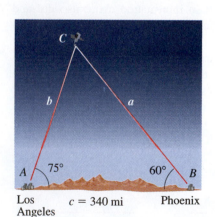

FIGURE 4

The distance of the satellite from Los Angeles is approximately 416 mi.   ■

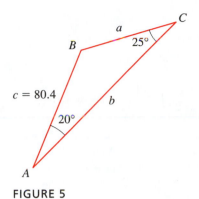

FIGURE 5

EXAMPLE 2 ■ Solving a Triangle (SAA)

Solve the triangle in Figure 5.

SOLUTION

First, $\angle B = 180° - (20° + 25°) = 135°$. Since side c is known, to find side a we use the relation

$$\frac{\sin A}{a} = \frac{\sin C}{c} \qquad \text{Law of Sines}$$

$$a = \frac{c \sin A}{\sin C} = \frac{80.4 \sin 20°}{\sin 25°} \approx 65.1 \qquad \text{Solve for } a$$

Similarly, to find b we use

$$\frac{\sin B}{b} = \frac{\sin C}{c} \qquad \text{Law of Sines}$$

$$b = \frac{c \sin B}{\sin C} = \frac{80.4 \sin 135°}{\sin 25°} \approx 134.5 \qquad \text{Solve for } b \qquad ■$$

■ The Ambiguous Case

In Examples 1 and 2 a unique triangle was determined by the information given. This is always true of Case 1 (ASA or SAA). But in Case 2 (SSA) there may be two triangles, one triangle, or no triangle with the given properties. For this reason, Case 2 is sometimes called the **ambiguous case**. To see why this is so, we show in Figure 6 the possibilities when angle A and sides a and b are given. In part (a) no solution is possible, since side a is too short to complete the triangle. In part (b) the solution is a right triangle. In part (c) two solutions are possible, and in part (d) there is a unique triangle with the given properties. We illustrate the possibilities of Case 2 in the following examples.

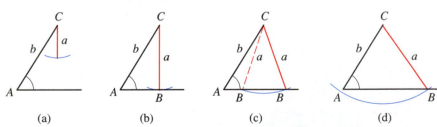

FIGURE 6
The ambiguous case

(a)   (b)   (c)   (d)

EXAMPLE 3 ■ SSA, the One-Solution Case

Solve triangle ABC, where $\angle A = 45°$, $a = 7\sqrt{2}$, and $b = 7$.

SOLUTION

We first sketch the triangle with the information we have (see Figure 7). Our sketch is necessarily tentative, since we don't yet know the other angles. Nevertheless, we can now see the possibilities.

FIGURE 7

We first find $\angle B$.

$$\frac{\sin A}{a} = \frac{\sin B}{b} \qquad \text{Law of Sines}$$

$$\sin B = \frac{b \sin A}{a} = \frac{7}{7\sqrt{2}} \sin 45° = \left(\frac{1}{\sqrt{2}}\right)\left(\frac{\sqrt{2}}{2}\right) = \frac{1}{2} \qquad \text{Solve for } \sin B$$

Which angles B have $\sin B = \frac{1}{2}$? From the preceding section we know that there are two such angles smaller than 180° (they are 30° and 150°). Which of these angles is compatible with what we know about triangle ABC? Since $\angle A = 45°$, we cannot have $\angle B = 150°$, because $45° + 150° > 180°$. So $\angle B = 30°$, and the remaining angle is $\angle C = 180° - (30° + 45°) = 105°$.

Now we can find side c.

$$\frac{\sin B}{b} = \frac{\sin C}{c} \qquad \text{Law of Sines}$$

$$c = \frac{b \sin C}{\sin B} = \frac{7 \sin 105°}{\sin 30°} = \frac{7 \sin 105°}{\frac{1}{2}} \approx 13.5 \qquad \text{Solve for } c$$

■

In Example 3 there were two possibilities for angle B, and one of these was not compatible with the rest of the information. In general, if $\sin A < 1$, we must check the angle and its supplement as possibilities, because any angle smaller than 180° can be in the triangle. To decide whether either possibility works, we check to see whether the resulting sum of the angles exceeds 180°. It can happen, as in Figure 6(c), that both possibilities are compatible with the given information. In that case, two different triangles are solutions to the problem.

EXAMPLE 4 ■ SSA, the Two-Solution Case

Solve triangle ABC if $\angle A = 43.1°$, $a = 186.2$, and $b = 248.6$.

SOLUTION

From the given information we sketch the triangle shown in Figure 8. Note that side a may be drawn in two possible positions to complete the triangle. From the Law of Sines

$$\sin B = \frac{b \sin A}{a} = \frac{248.6 \sin 43.1°}{186.2} \approx 0.91225$$

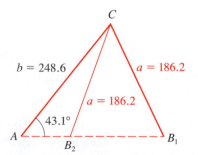

FIGURE 8

We consider only angles smaller than 180°, since no triangle can contain an angle of 180° or larger.

Surveying is a method of land measurement used for mapmaking. Surveyors use a process called *triangulation* in which a network of thousands of interlocking triangles is created on the area to be mapped. The process is started by measuring the length of a *baseline* between two surveying stations. Then, using an instrument called a *theodolite*, the angles between these two stations and a third station are measured. The Law of Sines is then used to calculate the two other sides of the triangle formed by the three stations. The calculated sides are used as baselines, and the process is repeated over and over to create a network of triangles. In this

(continued)

method, the only distance measured is the initial baseline; all other distances are calculated from the Law of Sines. This method is practical because it is much easier to measure angles than distances.

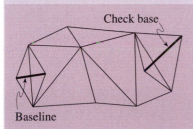

Check base

Baseline

One of the most ambitious map-making efforts of all time was the Great Trigonometric Survey of India, which required several expeditions and took over a century to complete. The famous expedition of 1823, led by **Sir George Everest**, lasted 20 years. Ranging over treacherous terrain and encountering the dreaded malaria-carrying mosquitoes, this expedition reached the foothills of the Himalayas. A later expedition, using triangulation, calculated the height of the highest peak of the Himalayas to be 29,002 ft. The peak was named in honor of Sir George Everest. Today, using satellites, the height of Mt. Everest is estimated to be 29,028 ft. The very close agreement of these two estimates shows the great accuracy of the trigonometric method.

There are two possible angles B between $0°$ and $180°$ such that $\sin B = 0.91225$. Using the $\boxed{\text{SIN}^{-1}}$ key on a calculator (or $\boxed{\text{INV}}$ $\boxed{\text{SIN}}$ or $\boxed{\text{ARCSIN}}$), we find that one of these angles is approximately $65.8°$. The other is approximately $180° - 65.8° = 114.2°$. We denote these two angles by B_1 and B_2 so that

$$\angle B_1 \approx 65.8° \qquad \text{and} \qquad \angle B_2 \approx 114.2°$$

Thus, two triangles satisfy the given conditions: triangle $A_1B_1C_1$ and triangle $A_2B_2C_2$.

Solve triangle $A_1B_1C_1$:

$$\angle C_1 \approx 180° - (43.1° + 65.8°) = 71.1° \qquad \text{Find } \angle C_1$$

Thus $\qquad c_1 = \dfrac{a_1 \sin C_1}{\sin A_1} \approx \dfrac{186.2 \sin 71.1°}{\sin 43.1°} \approx 257.8 \qquad$ Law of Sines

Solve triangle $A_2B_2C_2$:

$$\angle C_2 \approx 180° - (43.1° + 114.2°) = 22.7° \qquad \text{Find } \angle C_2$$

Thus $\qquad c_2 = \dfrac{a_2 \sin C_2}{\sin A_2} \approx \dfrac{186.2 \sin 22.7°}{\sin 43.1°} \approx 105.2 \qquad$ Law of Sines

Triangles $A_1B_1C_1$ and $A_2B_2C_2$ are shown in Figure 9.

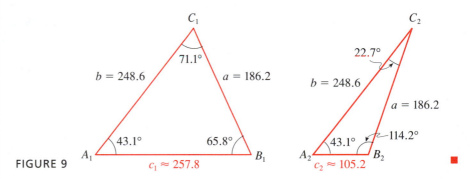

FIGURE 9

The next example presents a situation for which no triangle is compatible with the given data.

EXAMPLE 5 ■ SSA, the No-Solution Case

Solve triangle ABC, where $\angle A = 42°$, $a = 70$, and $b = 122$.

SOLUTION

First, let's try to find $\angle B$. We have

$$\frac{\sin A}{a} = \frac{\sin B}{b}$$ Law of Sines

$$\sin B = \frac{b \sin A}{a} = \frac{122 \sin 42°}{70} \approx 1.17$$ Solve for sin B

Since the sine of an angle is never greater than 1, we conclude that no triangle satisfies the conditions given in this problem.   ■

6.4 EXERCISES

1–6 ■ Use the Law of Sines to find the indicated side x or angle θ.

1.

2.

3.

4.

5.

6.

7–8 ■ Solve the triangle using the Law of Sines.

7.

8.

9–14 ■ Sketch each triangle and then solve the triangle using the Law of Sines.

9. $\angle A = 50°$, $\angle B = 68°$, $c = 230$

10. $\angle A = 23°$, $\angle B = 110°$, $c = 50$

11. $\angle A = 30°$, $\angle C = 65°$, $b = 10$

12. $\angle A = 22°$, $\angle B = 95°$, $a = 420$

13. $\angle B = 29°$, $\angle C = 51°$, $b = 44$

14. $\angle B = 10°$, $\angle C = 100°$, $c = 115$

15–22 ■ Use the Law of Sines to solve for all possible triangles that satisfy the given conditions.

15. $a = 28$, $b = 15$, $\angle A = 110°$

16. $a = 30$, $c = 40$, $\angle A = 37°$

17. $a = 20$, $c = 45$, $\angle A = 125°$

18. $b = 45$, $c = 42$, $\angle C = 38°$

19. $b = 25$, $c = 30$, $\angle B = 25°$

20. $a = 75$, $b = 100$, $\angle A = 30°$

21. $a = 50$, $b = 100$, $\angle A = 50°$

22. $a = 100$, $b = 80$, $\angle A = 135°$

23. To find the distance across a river, a surveyor chooses points A and B, which are 200 ft apart on one side of the river (see the figure). She then chooses a reference point C on the opposite side of the river and finds that

$\angle BAC \approx 82°$ and $\angle ABC \approx 52°$. Approximate the distance from A to C.

24. A pilot is flying over a straight highway. He determines the angles of depression to two mileposts, 5 mi apart, to be 32° and 48°, as shown in the figure.
  (a) Find the distance of the plane from point A.
  (b) Find the elevation of the plane.

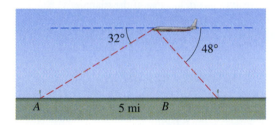

25. The path of a satellite orbiting the earth causes it to pass directly over two tracking stations A and B, which are 50 mi apart. When the satellite is on one side of the two stations, the angles of elevation at A and B are measured to be 87.0° and 84.2°, respectively.
  (a) How far is the satellite from station A?
  (b) How high is the satellite above the ground?

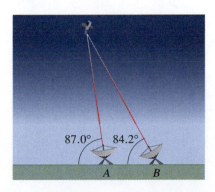

26. A tree on a hillside casts a shadow 215 ft down the hill. If the angle of inclination of the hillside is 22° to the horizontal and the angle of elevation of the sun is 52°, find the height of the tree.

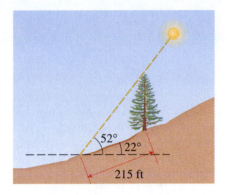

27. A communication tower is located at the top of a steep hill, as shown. The angle of inclination of the hill is 58°. A guy wire is to be attached to the top of the tower and to the ground, 100 m downhill from the base of the tower. The angle α in the figure is determined to be 12°. Find the length of cable required for the guy wire.

28. Points A and B are separated by a lake. To find the distance between them, a surveyor locates a point C on land such that $\angle CAB = 48.6°$. He also measures CA as 312 ft and CB as 527 ft. Find the distance between A and B.

29. To calculate the height of a mountain, angles α, β, and distance d are determined, as shown in the figure on page 512.
  (a) Find the length of BC in terms of α, β, and d.
  (b) Show that the height h of the mountain is given by the formula

$$h = d \frac{\sin \alpha \sin \beta}{\sin(\beta - \alpha)}$$

(c) Use the formula in part (b) to find the height of a mountain if $\alpha \approx 25°$, $\beta \approx 29°$, and $d \approx 800$ ft.

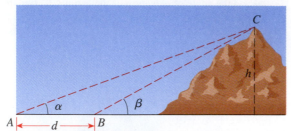

30. Observers at P and Q are located on the side of a hill that is inclined 32° to the horizontal, as shown. The observer at P determines the angle of elevation to a hot-air balloon to be 62°. At the same instant, the observer at Q measures the angle of elevation to the balloon to be 71°. If P is 60 m down the hill from Q, find the distance from Q to the balloon.

31. A water tower 30 m tall is located at the top of a hill. From a distance of 120 m down the hill, it is observed that the angle formed between the top and base of the tower is 8°. Find the angle of inclination of the hill.

32. For the triangle shown, find (a) $\angle BCD$ and (b) $\angle DCA$.

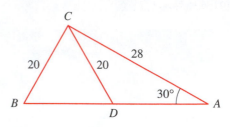

33. In triangle ABC, $\angle A = 40°$, $a = 15$, and $b = 20$.
(a) Show that there are two triangles, ABC and $A'B'C'$, that satisfy these conditions.
(b) Show that the areas of the triangles in part (a) are proportional to the sines of the angles C and C', that is,

$$\frac{\text{area of } \triangle ABC}{\text{area of } \triangle A'B'C'} = \frac{\sin C}{\sin C'}$$

34. Show that, given the three angles A, B, C of a triangle and one side, say a, the area of the triangle is

$$\text{area} = \frac{a^2 \sin B \sin C}{2 \sin A}$$

 DISCOVERY · DISCUSSION

35. Number of Solutions in the Ambiguous Case   We have seen that when using the Law of Sines to solve a triangle in the SSA case, there may be two, one, or no solution(s). Sketch triangles like those in Figure 6 to verify the criteria in the table for the number of solutions if you are given $\angle A$ and sides a and b.

Criterion	Number of solutions
$a \geq b$	1
$b > a > b \sin A$	2
$a = b \sin A$	1
$a < b \sin A$	0

If $\angle A = 30°$ and $b = 100$, use these crieteria to find the range of values of a for which the triangle ABC has two solutions, one solution, or no solution.

6.5   THE LAW OF COSINES

The Law of Sines cannot be used directly to solve triangles if we know two sides and the angle between them or if we know all three sides (these are Cases 3 and 4 of the preceding section). In these two cases, the **Law of Cosines** applies.

THE LAW OF COSINES

In any triangle ABC, we have

$$a^2 = b^2 + c^2 - 2bc \cos A$$

$$b^2 = a^2 + c^2 - 2ac \cos B$$

$$c^2 = a^2 + b^2 - 2ab \cos C$$

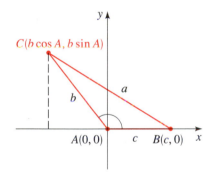

FIGURE 1

■ **Proof**   To prove the Law of Cosines, place triangle ABC so that $\angle A$ is at the origin, as shown in Figure 1. The coordinates of the vertices B and C are $(c, 0)$ and $(b \cos A, b \sin A)$, respectively. (You should check that the coordinates of these points will be the same if we draw angle A as an acute angle.) Using the Distance Formula, we get

$$a^2 = (b \cos A - c)^2 + (b \sin A - 0)^2$$

$$= b^2 \cos^2 A - 2bc \cos A + c^2 + b^2 \sin^2 A$$

$$= b^2(\cos^2 A + \sin^2 A) - 2bc \cos A + c^2$$

$$= b^2 + c^2 - 2bc \cos A \qquad \text{Because } \sin^2 A + \cos^2 A = 1$$

This proves the first formula. The other two formulas are obtained in the same way by placing each of the other vertices of the triangle at the origin and repeating the preceding argument.   □

In words, the Law of Cosines says that the square of any side of a triangle is equal to the sum of the squares of the other two sides, minus twice the product of those two sides times the cosine of the included angle.

If one of the angles of a triangle, say $\angle C$, is a right angle, then $\cos C = 0$ and the Law of Cosines reduces to the Pythagorean Theorem, $c^2 = a^2 + b^2$. Thus, the Pythagorean Theorem is a special case of the Law of Cosines.

EXAMPLE 1 ■ Length of a Tunnel

A tunnel is to be built through a mountain. To estimate the length of the tunnel, a surveyor makes the measurements shown in Figure 2. Use the surveyor's data to approximate the length of the tunnel.

FIGURE 2

SOLUTION

To approximate the length c of the tunnel, we use the Law of Cosines:

$$c^2 = a^2 + b^2 - 2ab \cos C \qquad \text{Law of Cosines}$$

$$= 388^2 + 212^2 - 2(388)(212) \cos 82.4° \qquad \text{Substitute}$$

$$\approx 173730.2367 \qquad \text{Use a calculator}$$

$$c \approx \sqrt{173730.2367} \approx 416.8 \qquad \text{Take square roots}$$

Thus, the tunnel will be approximately 417 ft long.   ■

EXAMPLE 2 ■ SSS, the Law of Cosines

The sides of a triangle are $a = 5$, $b = 8$, and $c = 12$ (see Figure 3). Find the angles of the triangle.

SOLUTION

We first find $\angle A$. From the Law of Cosines, we have $a^2 = b^2 + c^2 - 2bc \cos A$. Solving for $\cos A$, we get

$$\cos A = \frac{b^2 + c^2 - a^2}{2bc} = \frac{8^2 + 12^2 - 5^2}{2(8)(12)} = \frac{183}{192} = 0.953125$$

Using a calculator, we find that $\angle A \approx 18°$. In the same way the equations

$$\cos B = \frac{a^2 + c^2 - b^2}{2ac} = \frac{5^2 + 12^2 - 8^2}{2(5)(12)} = 0.875$$

$$\cos C = \frac{a^2 + b^2 - c^2}{2ab} = \frac{5^2 + 8^2 - 12^2}{2(5)(8)} = -0.6875$$

give $\angle B \approx 29°$ and $\angle C \approx 133°$. Of course, once two angles are calculated, the third can more easily be found from the fact that the sum of the angles of a triangle is 180°. However, it's a good idea to calculate all three angles using the Law of Cosines and add the three angles as a check on your computations.   ■

EXAMPLE 3 ■ SAS, the Law of Cosines and the Law of Sines

Solve triangle ABC, where $\angle A = 46.5°$, $b = 10.5$, and $c = 18.0$.

SOLUTION

We can find a using the Law of Cosines.

$$a^2 = b^2 + c^2 - 2bc \cos A$$

$$= (10.5)^2 + (18.0)^2 - 2(10.5)(18.0)(\cos 46.5°) \approx 174.05$$

Thus, $a \approx \sqrt{174.05} \approx 13.2$. The two remaining angles can now be found using the

FIGURE 3

Law of Sines. We have

$$\sin B = \frac{b \sin A}{a} \approx \frac{10.5 \sin 46.5°}{13.2} \approx 0.577$$

So B is the angle whose sine is 0.577. For this, we use our calculator to get $\angle B \approx 35.2°$. Since angle B can have measure between 0° and 180°, another possibility for angle B is $\angle B = 180° - 35.2° = 144.8°$. It's a simple matter to choose between these two possibilities, since the largest angle in a triangle must be opposite the longest side. So the correct choice is $\angle B \approx 35.2°$. In this case, $\angle C \approx 180° - (46.5° + 35.2°) = 98.3°$, and indeed the largest angle, $\angle C$, is opposite the longest side, $c = 18.0$.

To summarize: $\angle B \approx 35.2°$, $\angle C \approx 98.3°$, and $a \approx 13.2$. (See Figure 4.) ■

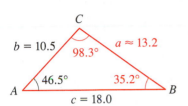

FIGURE 4

We see from Example 3 that, when using both the Law of Cosines and the Law of Sines to solve a triangle in the SAS case, we must be careful to choose the correct measure for the remaining angle. In any triangle, the longest side is opposite the largest angle, and the shortest side is opposite the smallest angle. Thus, when using the Law of Sines, we must choose the angle so that this condition is satisfied. That's why it's a good idea to sketch the triangle so you can check your final answer.

In navigation a direction is often given as a **bearing**, that is, as an acute angle measured from due north or due south. The bearing N 30° E, for example, indicates a direction that points 30° to the east of due north (see Figure 5).

FIGURE 5   N 30° E   N 60° W   S 70° W   S 50° E

EXAMPLE 4 ■ Navigation

A pilot sets out from an airport and heads in the direction N 20° E, flying at 200 mi/h. After one hour, he makes a course correction and heads in the direction N 40° E. Half an hour after that, engine trouble forces him to make an emergency landing.

(a) Find the distance between the airport and his final landing point.
(b) Find the bearing from the airport to his final landing point.

SOLUTION

(a) In one hour the plane travels 200 mi, and in half an hour it travels 100 mi, so we can plot the pilot's course as in Figure 6. When he makes his course correction, he turns 20° to the right, so the angle between the two legs of his trip

FIGURE 6

is $180° - 20° = 160°$. So by the Law of Cosines we have

$$b^2 = 200^2 + 100^2 - 2 \cdot 200 \cdot 100 \cos 160°$$

$$\approx 87,587.70$$

Thus, $b \approx 295.95$. The pilot lands about 296 mi from his starting point.

(b) We first use the Law of Sines to find $\angle A$.

$$\frac{\sin A}{100} = \frac{\sin 160°}{295.95}$$

$$\sin A = 100 \cdot \frac{\sin 160°}{295.95} \approx 0.11557$$

Using the $\boxed{\text{SIN}^{-1}}$ key on a calculator, we find that $\angle A \approx 6.636°$. From Figure 6 we see that the line from the airport to the final landing site points in the direction $20° + 6.636° = 26.636°$ east of due north. Thus, the bearing is about N 26.6° E. ∎

> Another angle with sine 0.11557 is $180° - 6.636° = 173.364°$. But this is clearly too large to be $\angle A$ in $\angle ABC$.

The Area of a Triangle

An interesting application of the Law of Cosines involves a formula for finding the area of a triangle from the lengths of its three sides (see Figure 7).

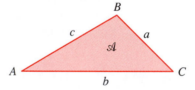

FIGURE 7

HERON'S FORMULA

The area $\mathcal{A}$ of triangle ABC is given by

$$\mathcal{A} = \sqrt{s(s - a)(s - b)(s - c)}$$

where $s = \frac{1}{2}(a + b + c)$ is the **semiperimeter** of the triangle; that is, s is half the perimeter.

■ **Proof** We start with the formula $\mathcal{A} = \frac{1}{2}ab \sin C$ from Section 6.3. Thus

$$\mathcal{A}^2 = \tfrac{1}{4}a^2b^2\sin^2 C$$

$$= \tfrac{1}{4}a^2b^2(1 - \cos^2 C) \qquad \text{\color{red}Pythagorean identity}$$

$$= \tfrac{1}{4}a^2b^2(1 - \cos C)(1 + \cos C) \qquad \text{\color{red}Factor}$$

Next, we write the expressions $1 - \cos C$ and $1 + \cos C$ in terms of a, b and c.

By the Law of Cosines we have

$$\cos C = \frac{a^2 + b^2 - c^2}{2ab}$$ Law of Cosines

$$1 + \cos C = 1 + \frac{a^2 + b^2 - c^2}{2ab}$$ Add 1

$$= \frac{2ab + a^2 + b^2 - c^2}{2ab}$$ Common denominator

$$= \frac{(a + b)^2 - c^2}{2ab}$$ Factor

$$= \frac{(a + b + c)(a + b - c)}{2ab}$$ Difference of squares

Similarly

$$1 - \cos C = \frac{(c + a - b)(c - a + b)}{2ab}$$

Substituting these expressions in the formula we obtained for $\mathcal{A}^2$ gives

$$\mathcal{A}^2 = \tfrac{1}{4} a^2 b^2 \, \frac{(a + b + c)(a + b - c)}{2ab} \, \frac{(c + a - b)(c - a + b)}{2ab}$$

$$= \frac{(a + b + c)}{2} \, \frac{(a + b - c)}{2} \, \frac{(c + a - b)}{2} \, \frac{(c - a + b)}{2}$$

$$= s(s - c)(s - b)(s - a)$$

Showing that each factor in the last expression equals the corresponding factor in the preceding expression is left as an exercise. Heron's Formula now follows by taking the square root of each side.   ☐

EXAMPLE 5 ■ Area of a Lot

A businessman wishes to buy a triangular lot in a busy downtown location (see Figure 8). The lot frontages on the three adjacent streets are 125, 280, and 315 ft. Find the area of the lot.

SOLUTION

The semiperimeter of the lot is

$$s = \frac{125 + 280 + 315}{2} = 360$$

By Heron's Formula the area is

$$\mathcal{A} = \sqrt{360(360 - 125)(360 - 280)(360 - 315)} \approx 17{,}451.6$$

Thus, the area is approximately 17,452 ft^2.   ■

FIGURE 8

6.5 EXERCISES

1–8 ■ Use the Law of Cosines to determine the indicated side x or angle θ.

1.

2.

3.

4.

5.

6.

7.

8.

9–18 ■ Solve triangle ABC.

9.

10.

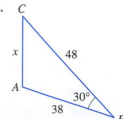

11. $a = 3.0,\quad b = 4.0,\quad \angle C = 53°$

12. $b = 60,\quad c = 30,\quad \angle A = 70°$

13. $a = 20,\quad b = 25,\quad c = 22$

14. $a = 10,\quad b = 12,\quad c = 16$

15. $b = 125,\quad c = 162,\quad \angle B = 40°$

16. $a = 65,\quad c = 50,\quad \angle C = 52°$

17. $a = 50,\quad b = 65,\quad \angle A = 55°$

18. $a = 73.5,\quad \angle B = 61°,\quad \angle C = 83°$

19–26 ■ Find the indicated side x or angle θ. (Use either the Law of Sines or the Law of Cosines, as appropriate.)

19.

20.

21.

22.

23.

24.

25.

26.

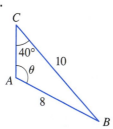

27. To find the distance across a small lake, a surveyor has taken the measurements shown. Find the distance across the lake using this information.

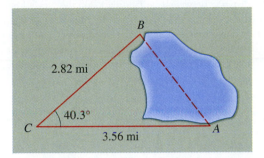

28. A parallelogram has sides of lengths 3 and 5, and one angle is 50°. Find the lengths of the diagonals.

29. Two straight roads diverge at an angle of 65°. Two cars leave the intersection at 2:00 P.M., one traveling at 50 mi/h and the other at 30 mi/h. How far apart are the cars at 2:30 P.M.?

30. A car travels along a straight road, heading east for 1 h, then traveling for 30 min on another road that leads northeast. If the car has maintained a constant speed of 40 mi/h, how far is it from its starting position?

31. A pilot flies in a straight path for 1 h 30 min. She then makes a course correction, heading 10° to the right of her original course, and flies 2 h in the new direction. If she maintains a constant speed of 625 mi/h, how far is she from her starting position?

32. Two boats leave the same port at the same time. One travels at a speed of 30 mi/h in the direction N 50° E and the other travels at a speed of 26 mi/h in a direction S 70° E (see the figure). How far apart are the two boats after one hour?

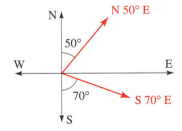

33. A fisherman leaves his home port and heads in the direction N 70° W. He travels 30 mi and reaches Egg Island. The next day he sails N 10° E for 50 mi, reaching Forrest Island.
(a) Find the distance between the fisherman's home part and Forrest Island.

(b) Find the bearing from Forrest Island back to his home port.

34. Airport B is 300 mi from airport A at a bearing N 50° E (see the figure). A pilot wishing to fly from A to B mistakenly flies due east at 200 mi/h for 30 minutes, when he notices his error.
(a) How far is the pilot from his destination at the time he notices the error?
(b) What bearing should he head his plane in order to arrive at airport B?

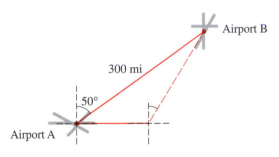

35. A triangular field has sides of lengths 22, 36, and 44 yd. Find the largest angle.

36. Two tugboats that are 120 ft apart pull a barge, as shown. If the length of one cable is 212 ft and the length of the other is 230 ft, find the angle formed by the two cables.

37. A boy is flying two kites at the same time. He has 380 ft of line out to one kite and 420 ft to the other. He estimates the angle between the two lines to be 30°. Approximate the distance between the kites.

38. A 125-ft tower is located on the side of a mountain that is inclined 32° to the horizontal. A guy wire is to be attached to the top of the tower and anchored at a point 55 ft down-hill from the base of the tower. Find the shortest length of wire needed.

39. A steep mountain is inclined 74° to the horizontal and rises 3400 ft above the surrounding plain. A cable car is to be installed from a point 800 ft from the base to the top of the mountain, as shown. Find the shortest length of cable needed.

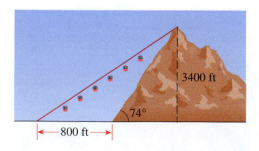

40. The CN Tower in Toronto, Canada, is the tallest free-standing structure in the world. A woman on the observation deck, 1150 ft above the ground, wants to determine the distance between two landmarks on the ground below. She observes that the angle formed by the lines of sight to these two landmarks is 43°. She also observes that the angle between the vertical and the line of sight to one of the landmarks is 62° and to the other landmark is 54°. Find the distance between the two landmarks.

41. Three circles of radii 4, 5, and 6 cm are mutually tangent. Find the shaded area enclosed between the circles.

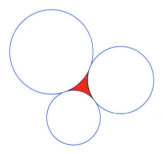

42. Prove that in triangle ABC

$$a = b \cos C + c \cos B$$
$$b = c \cos A + a \cos C$$
$$c = a \cos B + b \cos A$$

These are called the *Projection Laws*. [*Hint:* To get the first equation, add together the second and third equations in the Law of Cosines and solve for a.]

43. A surveyor wishes to find the distance between two points A and B on the opposite side of the river. On her side of the river, she chooses two points C and D that are 20 m apart and measures the angles shown. Find the distance between *A* and *B*.

44. Find the area of a triangle with sides of lengths 12, 18, and 24 m.

45. Land in downtown Columbia is valued at $20 a square foot. What is the value of a triangular lot with sides of lengths 112, 148, and 190 ft?

46. Find the area of the quadrilateral in the figure, correct to two decimal places.

● **DISCOVERY · DISCUSSION**

47. Solving for the Angles in a Triangle   When we solved for ∠*B* in Example 3 using the Law of Sines, we had to be careful because there were two possibilities, 35.2° and 144.8°. That is because these are the *two* angles between 0° and 180° whose sine is 0.577. What if we use the Law of Cosines to find ∠*B*? Would the problem of two possible measures for ∠*B* still arise? Which method do you prefer here: the Law of Sines or the Law of Cosines?

---

6   REVIEW

CONCEPT CHECK

1. (a) Explain the difference between a positive angle and a negative angle.
  (b) How is an angle of measure 1 degree formed?
  (c) How is an angle of measure 1 radian formed?
  (d) How is the radian measure of an angle θ defined?
  (e) How do you convert from degrees to radians?
  (f) How do you convert from radians to degrees?

2. (a) When is an angle in standard position?
  (b) When are two angles coterminal?

3. (a) What is the length *s* of an arc of a circle with radius *r* that subtends a central angle of θ radians?
  (b) What is the area *A* of a sector of a circle with radius *r* and central angle θ radians?

4. If θ is an acute angle in a right triangle, define the six trigonometric ratios in terms of the adjacent and opposite sides and the hypotenuse.

5. What does it mean to solve a triangle?

6. If θ is an angle in standard position, $P(x, y)$ is a point on the terminal side, and *r* is the distance from the origin to *P*, write expressions for the six trigonometric functions of θ.

7. Which trigonometric functions are positive in quadrants I, II, III, and IV?

8. If θ is an angle in standard position, what is its reference angle $\bar{\theta}$?

9. (a) State the reciprocal identities.
  (b) State the Pythagorean identities.

10. (a) What is the area of a triangle with sides of length *a* and *b* and with included angle θ?
  (b) What is the area of a triangle with sides of length *a*, *b* and *c*?

11. (a) State the Law of Sines.
  (b) State the Law of Cosines.

12. Explain the ambiguous case in the Law of Sines.

EXERCISES

1–2 ■ Find the radian measure that corresponds to the given degree measure.

1. (a) 60°   (b) 330°   (c) −135°   (d) −90°

2. (a) 24°   (b) −330°   (c) 750°   (d) 5°

3–4 ■ Find the degree measure that corresponds to the given radian measure.

3. (a) $\dfrac{5\pi}{2}$ (b) $-\dfrac{\pi}{6}$ (c) $\dfrac{9\pi}{4}$ (d) 3.1

4. (a) 8   (b) $-\dfrac{5}{2}$ (c) $\dfrac{11\pi}{6}$ (d) $\dfrac{3\pi}{5}$

5. Find the length of an arc of a circle of radius 8 m if the arc subtends a central angle of 1 rad.

6. Find the measure of a central angle θ in a circle of radius 5 ft if the angle is subtended by an arc of length 7 ft.

7. A circular arc of length 100 ft subtends a central angle of 70°. Find the radius of the circle.

8. How many revolutions will a car wheel of diameter 28 in. make over a period of half an hour if the car is traveling at 60 mi/h?

9. New York and Los Angeles are 2450 mi apart. Find the angle that the arc between these two cities subtends at the center of the earth. (The radius of the earth is 3960 mi.)

10. Find the area of a sector with central angle 2 rad in a circle of radius 5 m.

11. Find the area of a sector with central angle 52° in a circle of radius 200 ft.

12. A sector in a circle of radius 25 ft has an area of 125 ft². Find the central angle of the sector.

13. A potter's wheel with radius 8 in. spins at 150 rpm. Find the angular and linear speeds of a point on the rim of the wheel.

14. In an automobile transmission a *gear ratio g* is the ratio

$$g = \frac{\text{angular speed of engine}}{\text{angular speed of wheels}}$$

The angular speed of the engine is shown on the tachometer (in rpm).

A certain sports car has wheels with radius 11 in. Its gear ratios are shown in the table. Suppose the car is in fourth gear and the tachometer reads 3500 rpm.

(a) Find the angular speed of the engine.

(b) Find the angular speed of the wheels.

(c) How fast (in mi/h) is the car traveling?

Gear	Ratio
1st	4.1
2nd	3.0
3rd	1.6
4th	0.9
5th	0.7

15–16 ■ Find the values of the six trigonometric ratios of θ.

15.

16.

17–20 ■ Find the sides labeled x and y, correct to two decimal places.

17.

18.

19.

20.

21–22 ■ Solve the triangle.

21.

3   20°

22.

60°

20

23. Express the lengths a and b in the figure in terms of the trigonometric ratios of θ.

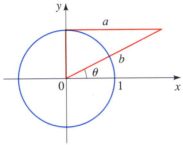

24. The highest tower in the world is the CN Tower in Toronto, Canada. From a distance of 1 km from its base, the angle of elevation to the top of the tower is 28.81°. Find the height of the tower.

25. Find the perimeter of a regular hexagon that is inscribed in a circle of radius 8 m.

26. The pistons in a car engine move up and down repeatedly to turn the crankshaft, as shown. Find the height of the point P above the center O of the crankshaft in terms of the angle θ.

8 in.

27. As viewed from the earth, the angle subtended by the full moon is 0.518°. Use this information and the fact that the distance AB from the earth to the moon is 236,900 mi to find the radius of the moon.

0.518°

A B

28. A pilot measures the angles of depression to two ships to be 40° and 52° (see the figure). If the pilot is flying at an elevation of 35,000 ft, find the distance between the two ships.

40°

52°

29–40 ■ Find the exact value.

29. $\sin 315°$

30. $\csc \dfrac{9\pi}{4}$

31. $\tan(-135°)$

32. $\cos \dfrac{5\pi}{6}$

33. $\cot\left(-\dfrac{22\pi}{3}\right)$

34. $\sin 405°$

35. $\cos 585°$

36. $\sec \dfrac{22\pi}{3}$

37. $\csc \dfrac{8\pi}{3}$

38. $\sec \dfrac{13\pi}{6}$

39. $\cot(-390°)$

40. $\tan \dfrac{23\pi}{4}$

41. Find the values of the six trigonometric ratios of the angle θ in standard position if the point $(-5, 12)$ is on the terminal side of θ.

42. Find $\sin \theta$ if θ is in standard position and its terminal side intersects the circle of radius 1 centered at the origin at the point $\left(-\sqrt{3}/2, \frac{1}{2}\right)$.

43. Find the acute angle that is formed by the line $y - \sqrt{3}\,x + 1 = 0$ and the x-axis.

44. Find the six trigonometric ratios of the angle θ in standard position if its terminal side is in quadrant III and is parallel to the line $4y - 2x - 1 = 0$.

45–48 ■ Write the first expression in terms of the second, for θ in the given quadrant.

45. $\tan \theta$, $\cos \theta$; θ in quadrant II

46. $\sec \theta$, $\sin \theta$; θ in quadrant III

47. $\tan^2\theta$, $\sin \theta$; θ in any quadrant

48. $\csc^2\theta \cos^2\theta$, $\sin \theta$; θ in any quadrant

49–52 ■ Find the values of the six trigonometric functions of θ from the information given.

49. $\tan \theta = \sqrt{7}/3$, $\sec \theta = \frac{4}{3}$ **50.** $\sec \theta = \frac{41}{40}$, $\csc \theta = -\frac{41}{9}$

51. $\sin \theta = \frac{3}{5}$, $\cos \theta < 0$ **52.** $\sec \theta = -\frac{13}{5}$, $\tan \theta > 0$

53. If $\tan \theta = -\frac{1}{2}$ for θ in quadrant II, find $\sin \theta + \cos \theta$.

54. If $\sin \theta = \frac{1}{2}$ for θ in quadrant I, find $\tan \theta + \sec \theta$.

55. If $\tan \theta = -1$, find $\sin^2\theta + \cos^2\theta$.

56. If $\cos \theta = -\sqrt{3}/2$ and $\pi/2 < \theta < \pi$, find $\sin 2\theta$.

57–62 ■ Find the side labeled x.

57.

58.

59.

60.

61.

62.

63. Two ships leave a port at the same time. One travels at 20 mi/h in a direction N 32° E, and the other travels at 28 mi/h in a direction S 42° E (see the figure). How far apart are the two ships after 2 h?

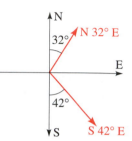

64. From a point A on the ground, the angle of elevation to the top of a tall building is 24.1°. From a point B, which is 600 ft closer to the building, the angle of elevation is measured to be 30.2°. Find the height of the building.

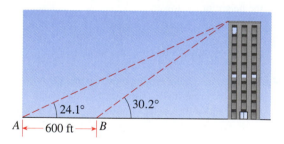

65. Find the distance between points A and B on opposite sides of a lake from the information shown.

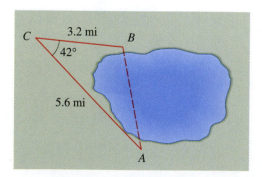

66. A boat is cruising the ocean off a straight shoreline. Points A and B are 120 mi apart on the shore, as shown. It is found that $\angle A = 42.3°$ and $\angle B = 68.9°$. Find the shortest distance from the boat to the shore.

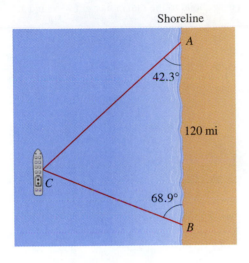

Shoreline

42.3°

120 mi

68.9°

67. To measure the height of an inaccessible cliff on the opposite side of a river, a surveyor makes the measurements shown. Find the height of the cliff.

33.1°

69.4°

200 m

51.6°

68. Find the area of a triangle with sides of length 8 and 14 and included angle 35°.

69. Find the area of a triangle with sides of length 5, 6, and 8.

6 TEST

1. Find the radian measures that correspond to the degree measures $300°$ and $-18°$.

2. Find the degree measures that correspond to the radian measures $\dfrac{5\pi}{6}$ and 2.4.

3. The rotor blades of a helicopter are 25 ft long and are rotating at 200 rpm.
   (a) Find the angular speed of the rotor.
   (b) Find the linear speed of a point on the tip of a blade.

4. Find the exact value of each of the following.
   (a) $\sin 405°$ (b) $\tan(-150°)$
   (c) $\sec \dfrac{5\pi}{3}$ (d) $\csc \dfrac{5\pi}{2}$

5. Find $\tan \theta + \sin \theta$ for the angle θ shown.

6. Find the lengths a and b shown in the figure in terms of θ.

7. If $\cos \theta = -\frac{1}{3}$ and θ is in quadrant III, find $\tan \theta \cot \theta + \csc \theta$.

8. If $\sin \theta = \frac{5}{13}$ and $\tan \theta = -\frac{5}{12}$, find $\sec \theta$.

9. Express $\tan \theta$ in terms of $\sec \theta$ for θ in quadrant II.

10. The base of the ladder in the figure is 6 ft from the building, and the angle formed by the ladder and the ground is $73°$. How high up the building does the ladder touch?

11–14 ■ Find the side labeled x.

11.

12.

13.

14.

15. Refer to the figure at the left.
  (a) Find the area of the shaded region.
  (b) Find the perimeter of the shaded region.

16. Refer to the figure at the right.
  (a) Find the angle opposite the longest side.
  (b) Find the area of the triangle.

17. Two wires tether a balloon to the ground, as shown. How high is the balloon above the ground?

Focus on Problem Solving
Taking Cases

One important problem-solving technique is the strategy of **taking cases**. A classic use of this strategy is in the classification of the regular polyhedra. These are called *Platonic solids* because they were mentioned in the writings of Plato.

A *regular polygon* is one in which all sides and all angles are equal. There are infinitely many regular polygons, as indicated in Figure 1.

triangle   square   pentagon   hexagon   heptagon   octagon

For three-dimensional shapes, the analogous concept is that of regular polyhedra. A *regular polyhedron* is a solid in which all faces are congruent regular polygons, and the same number of polygons meet at each corner, or *vertex*. We would like to find all possible regular polyhedra. It might seem at first that there should be infinitely many, just as for regular polygons. But we will see that there are just finitely many polyhedra.

■ Classifying the Regular Polyhedra

We now prove that there are exactly five regular polyhedra (see Figure 2). To show this, we consider all possible cases.

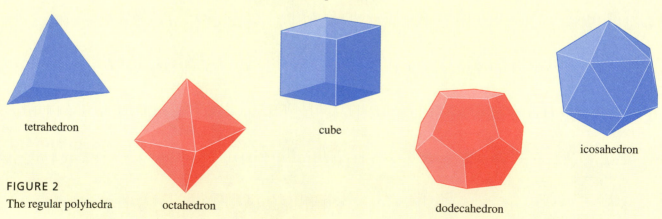

tetrahedron

cube

icosahedron

FIGURE 2

The regular polyhedra

octahedron

dodecahedron

■ **CASE 1** Suppose the faces of a regular polyhedron are equilateral triangles. How many such triangles can meet at a corner? To make a corner, there must be at least three triangles. We can also have four or five. But six equilateral triangles can-

Three, four, and five equilateral triangles can be folded up to make a corner, but six such triangles lie flat.

not meet at a point to make a corner. (Why?) If three triangles meet at each vertex, we can complete the polyhedron by adding one more triangle to make a *tetrahedron*. If four triangles meet at each vertex, we have an *octahedron*. If five triangles meet at each vertex, the resulting regular polyhedron is an *icosahedron*. Thus, we have found all the regular polyhedra with triangular faces.

■ **CASE 2** Suppose the faces of a regular polygon are squares. If three squares meet at each point, then the polyhedron is a *cube*. It's impossible for four or more squares to meet at a point to make a corner. (Why?) Thus, the only regular polyhedron with square faces is the cube.

■ **CASE 3** Suppose the faces of a regular polygon are pentagons. If three pentagons meet at each vertex, the resulting polyhedron is a *dodecahedron*. Since the angles of a regular pentagon are 108°, it's impossible for more than three regular pentagons to meet at a vertex. Thus, the only regular polyhedron with pentagonal faces is the dodecahedron.

■ **CASE 4** Is it possible for the faces of a regular polygon to be regular hexagons? Since the angle of a regular hexagon is 120°, when three such hexagons meet at a point they do not form a corner, so it's impossible for a regular polyhedron to have hexagonal faces. The same reasoning shows that no other regular polygon can be the face of a regular polyhedron.

Since these four cases account for all the possibilities, we have shown that there are exactly five regular polyhedra.

Euler's Formula

How many faces, edges, and vertices does a regular polyhedron have? In the 18th century, Euler observed that

$$F - E + V = 2 \qquad \text{Euler's Formula}$$

where F is the number of faces, E is the number of edges, and V is the number of vertices. We can use Euler's Formula to answer the question. For example, the icosahedron is assembled from F equilateral triangles:

In these triangles the total number of sides is $3F$, and the total number of angles is also $3F$. In an icosahedron, five angles of these triangles meet to form a vertex, so the total number of vertices must be

$$V = \frac{3F}{5}$$

Since two sides of adjacent triangles meet to form one edge of the polyhedron, the number of edges must be

$$E = \frac{3F}{2}$$

It's interesting to note that there are infinitely many 2-dimensional regular polygons but only five 3-dimensional regular polyhedra. Although it's impossible to draw the 4-dimensional regular polyhedra, mathematicians have shown that there are six of them. Surprisingly, in all higher dimensions there are exactly three regular polyhedra—the n-dimensional cube, tetrahedron, and octahedron.

Substituting into Euler's Formula gives

$$F - \frac{3F}{2} + \frac{3F}{5} = 2$$

Solving gives $F = 20$. Substituting this value of F into the formulas for edges and faces gives $E = 30$ and $V = 12$. Thus, for the icosahedron, $F = 20$, $E = 30$, and $V = 12$. Using similar reasoning we can find the number of faces, edges, and vertices for the other regular polyhedra.

■ Problems

1. Find the number of faces, edges, and vertices for each of the regular polyhedra, not by counting them, but by using Euler's Formula.

2. Describe the polyhedron whose edges are the line segments joining the centers of the faces of an octahedron, as shown in the figure. Do the same for the other Platonic solids.

3. As the following figures indicate, it's possible to *tile* the plane (that is, completely cover it) with equilateral triangles and with squares. Find all other regular polygons that tile the plane. Prove your answer.

4. (a) Find a formula for the area A_n of the regular polygon with n sides inscribed in a circle of radius 1 (see the figure). Express your answer as a trigonometric function of n.

A_3 A_4 A_5 ... A_n

The Four-Color Problem

A famous problem that was solved by taking cases is the *four-color problem*: Prove that every map can be colored using no more than four colors. The requirement is that any two regions that share a common border should have different colors. (If two regions meet only at a single point, as do Utah and New Mexico, for example, then they may have the same color.) The four-color problem was first stated as a conjecture in 1852, but it wasn't until 1976 that Kenneth Appel and Wolfgang Haken were able to prove it. In their proof they considered thousands of cases corresponding to different types of maps, and most of those cases were so complicated that they had to use a high-speed computer for the verification.

(b) Find A_3, A_4, A_{100}, and A_{1000}, correct to six decimal places. Notice that the values get closer and closer to π. Why?

5. A group of n pulleys, all of radius 1, is fixed so that their centers form a convex n-gon of perimeter P. (The figure shows the case $n = 4$.) Find the length of the belt that fits around the pulleys. [*Hint:* Try fitting together the sectors of the pulleys that touch the belt.]

6. A pair of pulleys is connected by a belt, as shown in each figure. Find the length of the belt.

(a)

(b)

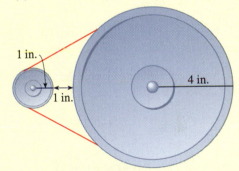

7. Two circles of radius 1 are placed so that their centers are one unit apart. Find the area of the region common to both circles.

8. Suppose that the lengths of the sides of a triangle are rational numbers. Prove that the cosine of each angle is a rational number.

9. Find the exact value of

$$\sin \frac{\pi}{100} + \sin \frac{2\pi}{100} + \sin \frac{3\pi}{100} + \sin \frac{4\pi}{100} + \cdots + \sin \frac{199\pi}{100}$$

where the sum contains the sines of all numbers of the form $k\pi/100$, with k varying from 1 to 199.

10. In order to draw a map of the spherical earth on a flat sheet of paper, several ingenious methods have been developed. One of these is the *cylindrical projection.** In this

---

*This type of map is often erroneously referred to as the Mercator "projection." The Mercator map actually involves procedures much more complicated than simple projection—in fact, constructing an accurate Mercator map requires knowledge of calculus.

Pierre de Fermat (1601–1665) was a French lawyer who became interested in mathematics at the age of 30. Because of his job as a magistrate, Fermat had little time to write complete proofs of his discoveries and often wrote them in the margin of whatever book he was reading at the time. After his death, his copy of Diophantus' *Arithmetica* (see page 37) was found to contain a particularly tantalizing comment. Where Diophantus discusses the solutions of $x^2 + y^2 = z^2$ (for example, $x = 3$, $y = 4$, $z = 5$), Fermat states in the margin that for $n \geq 3$ there are no natural number solutions to the equation $x^n + y^n = z^n$. In other words, it's impossible for a cube to equal the sum of two cubes, a fourth power to equal the sum of two fourth powers, and so on. Fermat writes "I have discovered a truly wonderful proof for this but the margin is too small to contain it." All the other margin comments in Fermat's copy of *Arithmetica* have been proved. This one, however, remained unproved, and it came to be known as "Fermat's Last Theorem."

In 1994, Andrew Wiles of Princeton University announced a proof of Fermat's Last Theorem, an astounding 350 years after it was conjectured. His proof is one of the most widely reported mathematical results in the popular press.

method, illustrated in the figure, each point on the spherical earth is projected onto a circumscribed cylinder (tangent to the sphere on the equator) by a line through the center of the earth. By what factor are the following distances distorted using this method of projection? That is, what is the ratio of the projected distance on the cylinder to the actual distance on the sphere?

(a) The distance between 20° and 21° N latitude along a meridian

(b) The distance between 40° and 41° N latitude along a meridian

(c) The distance between 80° and 81° N latitude along a meridian

(d) The distance between two points that are 1° apart on the 20th parallel

(e) The distance between two points that are 1° apart on the 40th parallel

(f) The distance between two points that are 1° apart on the 80th parallel

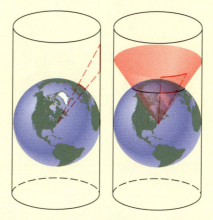

11. A bug is sitting at point A in one corner of a room and wants to crawl to point B in the corner diagonally opposite (see the figure). Find the shortest path for the bug.

12. Prove that every prime number is the leg of exactly one right triangle with integer sides. (This problem was first stated by Fermat.)

13. Show that the equation $x^2 + y^2 = 4z + 3$ has no solution in integers. [*Hint:* Recall that an even number is of the form $2n$ and an odd number is of the form $2n + 1$. Consider all possible cases for x and y even or odd.]

14. (a) Find all prime numbers p such that $2p + 1$ is a perfect square.
    [*Hint:* Write the equation $2p + 1 = n^2$ as $2p = n^2 - 1$ and factor.
    Then consider cases.]
    (b) Find all prime numbers p such that $2p + 1$ is a perfect cube.

15. (a) Write 13 as the sum of two squares. Then do the same for 41.
    (b) Verify that $(a^2 + b^2)(c^2 + d^2) = (ac + bd)^2 + (ad - bc)^2$.
    (c) Express 533 as the sum of two squares in two different ways.
        [*Hint:* Factor 533 and use parts (a) and (b).]

16. Find the area of the region between the two concentric circles shown in the figure.

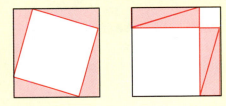

Bhaskara (born 1114) was an Indian mathematician, astronomer, and astrologer. Among his many accomplishments was an ingenious proof of the Pythagorean Theorem (see Problem 17). His important mathematical book *Lilavati* [*The Beautiful*] consists of algebra problems posed in the form of stories to his daughter Lilavati. Many of the problems begin "Oh beautiful maiden, suppose . . ." The story is told that using astrology, Bhaskara had determined that great misfortune would befall his daughter if she married at any time other than at a certain hour of a certain day. On her wedding day, as she was anxiously watching the water clock, a pearl fell unnoticed from her headdress. It stopped the flow of water in the clock, causing her to miss the opportune moment for marriage. Bhaskara's *Lilavati* was written to console her.

17. The Indian mathematician Bhaskara sketched the two figures shown here and wrote below them, "Behold!" Explain how his sketches prove the Pythagorean Theorem.

18. If the lengths of the sides of a right triangle, in increasing order, are a, b, and c, show that $a^3 + b^3 < c^3$.

7

Analytic Trigonometry

Trigonometry is used to analyze directed quantities, or vectors. The path of a jet aircraft is determined by resolving the vector forces of thrust, wind, and shear that act on the plane; the heading of a sailboat is determined by resolving the forces of the wind on the sails and the water currents on the hull of the boat.

A mathematician must often use a rule, and, naturally, he begins by proving the rule.

HENRI POINCARÉ

In Chapters 5 and 6 we studied the graphical and geometric properties of trigonometric functions. In this chapter we study the algebraic aspects of trigonometry, that is, simplifying and factoring expressions and solving equations that involve trigonometric functions. The basic tools in the algebra of trigonometry are trigonometric identities. We will find identities for trigonometric functions of sums and differences of real numbers, multiple-angle formulas, and other related identities. These identities are used in the study of complex numbers, vectors, and analytic geometry.

7.1 TRIGONOMETRIC IDENTITIES

We begin this section by reviewing some of the basic trigonometric identities that we studied in Chapters 5 and 6.

FUNDAMENTAL TRIGONOMETRIC IDENTITIES

Reciprocal Identities

$$\csc x = \frac{1}{\sin x} \qquad \sec x = \frac{1}{\cos x} \qquad \cot x = \frac{1}{\tan x}$$

$$\tan x = \frac{\sin x}{\cos x} \qquad \cot x = \frac{\cos x}{\sin x}$$

Pythagorean Identities

$$\sin^2 x + \cos^2 x = 1 \qquad \tan^2 x + 1 = \sec^2 x \qquad 1 + \cot^2 x = \csc^2 x$$

Even-Odd Identities

$$\sin(-x) = -\sin x \qquad \cos(-x) = \cos x \qquad \tan(-x) = -\tan x$$

Cofunction Identities

$$\sin\left(\frac{\pi}{2} - u\right) = \cos u \qquad \tan\left(\frac{\pi}{2} - u\right) = \cot u \qquad \sec\left(\frac{\pi}{2} - u\right) = \csc u$$

$$\cos\left(\frac{\pi}{2} - u\right) = \sin u \qquad \cot\left(\frac{\pi}{2} - u\right) = \tan u \qquad \csc\left(\frac{\pi}{2} - u\right) = \sec u$$

■ Simplifying Trigonometric Expressions

Identities enable us to write the same expression in different ways. It is often possible to rewrite a complicated looking expression as a much simpler one. To simplify algebraic expressions, we used factoring, common denominators, and the Special Product Formulas. To simplify trigonometric expressions, we use these same techniques together with the fundamental trigonometric identities.

EXAMPLE 1 ■ Simplifying a Trigonometric Expression

Simplify the expression $\cos t + \tan t \sin t$.

SOLUTION

We start by rewriting the expression in terms of sine and cosine.

$$\cos t + \tan t \sin t = \cos t + \left(\frac{\sin t}{\cos t}\right) \sin t \qquad \text{Reciprocal identity}$$

$$= \frac{\cos^2 t + \sin^2 t}{\cos t} \qquad \text{Common denominator}$$

$$= \frac{1}{\cos t} \qquad \text{Pythagorean identity}$$

$$= \sec t \qquad \text{Reciprocal identity} \qquad ■$$

EXAMPLE 2 ■ Simplifying by Combining Fractions

Simplify the expression $\dfrac{\sin \theta}{\cos \theta} + \dfrac{\cos \theta}{1 + \sin \theta}$.

SOLUTION

We combine the fractions by using a common denominator.

$$\frac{\sin \theta}{\cos \theta} + \frac{\cos \theta}{1 + \sin \theta} = \frac{\sin \theta \,(1 + \sin \theta) + \cos^2 \theta}{\cos \theta \,(1 + \sin \theta)} \qquad \text{Common denominator}$$

$$= \frac{\sin \theta + \sin^2 \theta + \cos^2 \theta}{\cos \theta \,(1 + \sin \theta)} \qquad \text{Expand}$$

$$= \frac{\sin \theta + 1}{\cos \theta \,(1 + \sin \theta)} \qquad \text{Pythagorean identity}$$

$$= \frac{1}{\cos \theta} = \sec \theta \qquad \text{Cancel and use reciprocal identity} \qquad ■$$

■ Proving Trigonometric Identities

Many identities follow from the fundamental identities. In the examples that follow, we learn how to prove that a given trigonometric equation is an identity, and in the process we will see how to discover new identities.

First, it's easy to decide when a given equation is *not* an identity. All we need to do is show that the equation does not hold for some value of the variable (or variables). Thus, the equation

$$\sin x + \cos x = 1$$

is not an identity, because when $x = \pi/4$, we have

$$\sin \frac{\pi}{4} + \cos \frac{\pi}{4} = \frac{\sqrt{2}}{2} + \frac{\sqrt{2}}{2} = \sqrt{2} \neq 1$$

To verify that a trigonometric equation is an identity, we transform one side of the equation into the other side by a series of steps, each of which is itself an identity.

GUIDELINES FOR PROVING TRIGONOMETRIC IDENTITIES

1. **START WITH ONE SIDE.**   Pick one side of the equation and write it down. Your goal is to transform it into the other side. It's usually easier to start with the more complicated side.

2. **USE KNOWN IDENTITIES.**   Use algebra and the identities you know to change the side you started with. Bring fractional expressions to a common denominator, factor, and use the fundamental identities to simplify expressions.

3. **CONVERT TO SINES AND COSINES.**   If you are stuck, you may find it helpful to rewrite all functions in terms of sines and cosines.

EXAMPLE 3 ■ **Proving an Identity by Rewriting in Terms of Sine and Cosine**

Verify the identity $\cos \theta \, (\sec \theta - \cos \theta) = \sin^2\theta$.

SOLUTION

The left-hand side looks more complicated, so we start with it and try to transform it into the right-hand side.

$$\text{LHS} = \cos \theta \, (\sec \theta - \cos \theta)$$

$$= \cos \theta \left(\frac{1}{\cos \theta} - \cos \theta \right) \qquad \text{\color{red}{Reciprocal identity}}$$

$$= 1 - \cos^2\theta \qquad\qquad\qquad \text{\color{red}{Expand}}$$

$$= \sin^2\theta = \text{RHS} \qquad\qquad \text{\color{red}{Pythagorean identity}} \qquad ■$$

In Example 3 it isn't easy to see how to change the right-hand side into the left-hand side, but it's definitely possible. Simply notice that each step is reversible. In other words, if we start with the last expression in the proof and work backward through the steps, the right side is transformed into the left side. You will probably agree, however, that it's more difficult to prove the identity this way. That's why it's often better to change the more complicated side of the identity into the simpler side.

EXAMPLE 4 ■ Proving an Identity by Combining Fractions

Verify the identity

$$2 \tan x \sec x = \frac{1}{1 - \sin x} - \frac{1}{1 + \sin x}$$

SOLUTION

Finding a common denominator and combining the fractions on the right-hand side of this equation, we get

$$\text{RHS} = \frac{1}{1 - \sin x} - \frac{1}{1 + \sin x}$$

$$= \frac{(1 + \sin x) - (1 - \sin x)}{(1 - \sin x)(1 + \sin x)} \qquad \text{Common denominator}$$

$$= \frac{2 \sin x}{1 - \sin^2 x} \qquad \text{Simplify}$$

$$= \frac{2 \sin x}{\cos^2 x} \qquad \text{Pythagorean identity}$$

$$= 2 \frac{\sin x}{\cos x} \left(\frac{1}{\cos x} \right) \qquad \text{Factor}$$

$$= 2 \tan x \sec x = \text{LHS} \qquad \text{Reciprocal identities} \qquad ■$$

See *Focus on Problem Solving,* pages 134–137.

In Example 5 we introduce "something extra" to the problem by multiplying the numerator and the denominator by a trigonometric expression, chosen so that we can simplify the result.

EXAMPLE 5 ■ Proving an Identity by Introducing Something Extra

Verify the identity $\dfrac{\cos u}{1 - \sin u} = \sec u + \tan u$.

SOLUTION

We start with the left-hand side and multiply numerator and denominator by $1 + \sin u$.

$$\text{LHS} = \frac{\cos u}{1 - \sin u}$$

We multiply by $1 + \sin u$ because we know by the difference of squares formula that $(1 - \sin u)(1 + \sin u) = 1 - \sin^2 u$, and this is just $\cos^2 u$, a simpler expression.

$$= \frac{\cos u}{1 - \sin u} \cdot \frac{1 + \sin u}{1 + \sin u} \qquad \text{Multiply numerator and denominator by } 1 + \sin u$$

$$= \frac{\cos u(1 + \sin u)}{1 - \sin^2 u} \qquad \text{Expand denominator}$$

$$= \frac{\cos u(1 + \sin u)}{\cos^2 u} \qquad \text{Pythagorean identity}$$

$$= \frac{1 + \sin u}{\cos u} \qquad \text{Cancel common factor}$$

$$= \frac{1}{\cos u} + \frac{\sin u}{\cos u} \qquad \text{Separate into two fractions}$$

$$= \sec u + \tan u \qquad \text{Reciprocal identities} \qquad \blacksquare$$

Here is another method for proving that an equation is an identity. If we can transform each side of the equation separately, by way of identities, to arrive at the same result, then the equation is an identity. Example 6 illustrates this procedure.

EXAMPLE 6 ■ Proving an Identity by Working with Both Sides

Verify the identity $\dfrac{1 + \cos \theta}{\cos \theta} = \dfrac{\tan^2\theta}{\sec \theta - 1}$.

SOLUTION

We prove the identity by changing each side separately into the same expression. Supply the reasons for each step.

$$\text{LHS} = \frac{1 + \cos \theta}{\cos \theta} = \frac{1}{\cos \theta} + \frac{\cos \theta}{\cos \theta} = \sec \theta + 1$$

$$\text{RHS} = \frac{\tan^2\theta}{\sec \theta - 1} = \frac{\sec^2\theta - 1}{\sec \theta - 1} = \frac{(\sec \theta - 1)(\sec \theta + 1)}{\sec \theta - 1} = \sec \theta + 1$$

It follows that $\text{LHS} = \text{RHS}$, so the equation is an identity. $\qquad \blacksquare$

 Warning: To prove an identity, we do *not* just perform the same operations on both sides of the equation. For example, if we start with an equation that is not an identity, such as

(1) $$\sin x = -\sin x$$

and square both sides, we get the equation

(2) $$\sin^2 x = \sin^2 x$$

which is clearly an identity. Does this mean that the original equation is an identity? Of course not. The problem here is that the operation of squaring is not **reversible** in the sense that we cannot arrive back at (1) from (2) by taking square roots (reversing the procedure). Only operations that are reversible will necessarily transform an identity into an identity.

We end this section by describing the technique of *trigonometric substitution*, which we use to convert algebraic expressions to trigonometric ones. This is often useful in calculus, for instance, in finding the area of a circle or an ellipse.

EXAMPLE 7 ■ Trigonometric Substitution

Substitute $\sin \theta$ for x in the expression $\sqrt{1 - x^2}$ and simplify. Assume that $0 \le \theta \le \pi/2$.

SOLUTION

Setting $x = \sin \theta$, we have

$$\sqrt{1 - x^2} = \sqrt{1 - \sin^2\theta} \qquad \text{Substitute } x = \sin \theta$$

$$= \sqrt{\cos^2\theta} \qquad \text{Pythagorean identity}$$

$$= \cos \theta \qquad \text{Take square root}$$

The last equality is true because $\cos \theta \ge 0$ for the values of θ in question. ■

7.1 EXERCISES

1–8 ■ Write the trigonometric expression in terms of sine and cosine, and then simplify.

1. $\sin t \cot t$

2. $\sin t \sec t$

3. $\tan x \csc x$

4. $\sin x \cos x \sec x$

5. $\tan^2 x - \sec^2 x$

6. $\dfrac{\sec x}{\csc x}$

7. $\sin u + \cot u \cos u$

8. $\cos^2\theta(1 + \tan^2\theta)$

9–22 ■ Simplify the trigonometric expression.

9. $\dfrac{\sin x \sec x}{\tan x}$

10. $\cos^3 x + \sin^2 x \cos x$

11. $\dfrac{1 + \cos y}{1 + \sec y}$

12. $\dfrac{\tan x}{\sec(-x)}$

13. $\dfrac{\sec^2 x - 1}{\sec^2 x}$

14. $\dfrac{\sec x - \cos x}{\tan x}$

15. $\dfrac{1 + \csc x}{\cos x + \cot x}$

16. $\dfrac{\sin x}{\csc x} + \dfrac{\cos x}{\sec x}$

17. $\dfrac{1 + \sin u}{\cos u} + \dfrac{\cos u}{1 + \sin u}$

18. $\tan x \cos x \csc x$

19. $\dfrac{2 + \tan^2 x}{\sec^2 x} - 1$

20. $\dfrac{1 + \cot A}{\csc A}$

21. $\tan \theta + \cos(-\theta) + \tan(-\theta)$

22. $\dfrac{\cos x}{\sec x + \tan x}$

23–84 ■ Verify the identity.

23. $\dfrac{\sin \theta}{\tan \theta} = \cos \theta$

24. $\dfrac{\tan x}{\sec x} = \sin x$

25. $\dfrac{\cos u \sec u}{\tan u} = \cot u$

26. $\dfrac{\cot x \sec x}{\csc x} = 1$

27. $\dfrac{\tan y}{\csc y} = \sec y - \cos y$

28. $\dfrac{\cos v}{\sec v \sin v} = \csc v - \sin v$

29. $\sin B + \cos B \cot B = \csc B$

30. $\cos(-x) - \sin(-x) = \cos x + \sin x$

31. $\cot(-\alpha) \cos(-\alpha) + \sin(-\alpha) = -\csc \alpha$

32. $\csc x \left[\csc x + \sin(-x)\right] = \cot^2 x$

33. $\tan \theta + \cot \theta = \sec \theta \csc \theta$

34. $(\sin x + \cos x)^2 = 1 + 2 \sin x \cos x$

35. $(1 - \cos \beta)(1 + \cos \beta) = \dfrac{1}{\csc^2 \beta}$

36. $\dfrac{\cos x}{\sec x} + \dfrac{\sin x}{\csc x} = 1$

37. $\dfrac{(\sin x + \cos x)^2}{\sin^2 x - \cos^2 x} = \dfrac{\sin^2 x - \cos^2 x}{(\sin x - \cos x)^2}$

38. $(\sin x + \cos x)^4 = (1 + 2 \sin x \cos x)^2$

39. $\dfrac{\sec t - \cos t}{\sec t} = \sin^2 t$

40. $\dfrac{1 - \sin x}{1 + \sin x} = (\sec x - \tan x)^2$

41. $\dfrac{1}{1 - \sin^2 y} = 1 + \tan^2 y$

42. $\csc x - \sin x = \cos x \cot x$

43. $(\cot x - \csc x)(\cos x + 1) = -\sin x$

44. $\sin^4 \theta - \cos^4 \theta = \sin^2 \theta - \cos^2 \theta$

45. $(1 - \cos^2 x)(1 + \cot^2 x) = 1$

46. $\cos^2 x - \sin^2 x = 2 \cos^2 x - 1$

47. $2 \cos^2 x - 1 = 1 - 2 \sin^2 x$

48. $(\tan y + \cot y)\sin y \cos y = 1$

49. $\dfrac{1 - \cos \alpha}{\sin \alpha} = \dfrac{\sin \alpha}{1 + \cos \alpha}$

50. $\sin^2 \alpha + \cos^2 \alpha + \tan^2 \alpha = \sec^2 \alpha$

51. $\dfrac{\sin x - 1}{\sin x + 1} = \dfrac{-\cos^2 x}{(\sin x + 1)^2}$

52. $\dfrac{\sin w}{\sin w + \cos w} = \dfrac{\tan w}{1 + \tan w}$

53. $\dfrac{(\sin t + \cos t)^2}{\sin t \cos t} = 2 + \sec t \csc t$

54. $\sec t \csc t \, (\tan t + \cot t) = \sec^2 t + \csc^2 t$

55. $\dfrac{1 + \tan^2 u}{1 - \tan^2 u} = \dfrac{1}{\cos^2 u - \sin^2 u}$

56. $\dfrac{1 + \sec^2 x}{1 + \tan^2 x} = 1 + \cos^2 x$

57. $\dfrac{\sec x}{\sec x - \tan x} = \sec x \, (\sec x + \tan x)$

58. $\dfrac{\sec x + \csc x}{\tan x + \cot x} = \sin x + \cos x$

59. $\sec v - \tan v = \dfrac{1}{\sec v + \tan v}$

60. $\dfrac{\sin A}{1 - \cos A} - \cot A = \csc A$

61. $\dfrac{\sin x + \cos x}{\sec x + \csc x} = \sin x \cos x$

62. $\dfrac{1 - \cos x}{\sin x} + \dfrac{\sin x}{1 - \cos x} = 2 \csc x$

63. $\dfrac{\csc x - \cot x}{\sec x - 1} = \cot x$

64. $\dfrac{\csc^2 x - \cot^2 x}{\sec^2 x} = \cos^2 x$

65. $\tan^2 u - \sin^2 u = \tan^2 u \sin^2 u$

66. $\dfrac{\tan v \sin v}{\tan v + \sin v} = \dfrac{\tan v - \sin v}{\tan v \sin v}$

67. $\sec^4 x - \tan^4 x = \sec^2 x + \tan^2 x$

68. $\dfrac{\cos \theta}{1 - \sin \theta} = \sec \theta + \tan \theta$

69. $\dfrac{\cos \theta}{1 - \sin \theta} = \dfrac{\sin \theta - \csc \theta}{\cos \theta - \cot \theta}$

70. $\dfrac{1 + \tan x}{1 - \tan x} = \dfrac{\cos x + \sin x}{\cos x - \sin x}$

71. $\dfrac{\cos^2 t + \tan^2 t - 1}{\sin^2 t} = \tan^2 t$

72. $\dfrac{1}{1 - \sin x} - \dfrac{1}{1 + \sin x} = 2 \sec x \tan x$

73. $\dfrac{1}{\sec x + \tan x} + \dfrac{1}{\sec x - \tan x} = 2 \sec x$

74. $\dfrac{1 + \sin x}{1 - \sin x} - \dfrac{1 - \sin x}{1 + \sin x} = 4 \tan x \sec x$

75. $(\tan x + \cot x)^2 = \sec^2 x + \csc^2 x$

76. $\tan^2 x - \cot^2 x = \sec^2 x - \csc^2 x$

77. $\dfrac{\sec u - 1}{\sec u + 1} = \dfrac{1 - \cos u}{1 + \cos u}$ **78.** $\dfrac{\cot x + 1}{\cot x - 1} = \dfrac{1 + \tan x}{1 - \tan x}$

79. $\dfrac{\sin^3 x + \cos^3 x}{\sin x + \cos x} = 1 - \sin x \cos x$

80. $\dfrac{\tan v - \cot v}{\tan^2 v - \cot^2 v} = \sin v \cos v$

81. $\dfrac{1 + \sin x}{1 - \sin x} = (\tan x + \sec x)^2$

82. $\dfrac{\tan x + \tan y}{\cot x + \cot y} = \tan x \tan y$

83. $(\tan x + \cot x)^4 = \csc^4 x \sec^4 x$

84. $(\sin \alpha - \tan \alpha)(\cos \alpha - \cot \alpha) = (\cos \alpha - 1)(\sin \alpha - 1)$

85–90 ■ Make the indicated trigonometric substitution in the given algebraic expression and simplify (see Example 7). Assume $0 \le \theta < \pi/2$.

85. $\dfrac{x}{\sqrt{1 - x^2}}, \quad x = \sin \theta$

86. $\sqrt{1 + x^2}, \quad x = \tan \theta$

87. $\sqrt{x^2 - 1}, \quad x = \sec \theta$

88. $\dfrac{1}{x^2 \sqrt{4 + x^2}}, \quad x = 2 \tan \theta$

89. $\sqrt{9 - x^2}, \quad x = 3 \sin \theta$

90. $\dfrac{\sqrt{x^2 - 25}}{x}, \quad x = 5 \sec \theta$

91–94 ■ Graph f and g in the same viewing rectangle. Do the graphs suggest that the equation $f(x) = g(x)$ is an identity? Prove your answer.

91. $f(x) = \cos^2 x - \sin^2 x, \quad g(x) = 1 - 2 \sin^2 x$

92. $f(x) = \tan x\,(1 + \sin x), \quad g(x) = \dfrac{\sin x \cos x}{1 + \sin x}$

93. $f(x) = (\sin x + \cos x)^2, \quad g(x) = 1$

94. $f(x) = \cos^4 x - \sin^4 x, \quad g(x) = 2 \cos^2 x - 1$

95. Show that the equation is not an identity.

(a) $\sin 2x = 2 \sin x$

(b) $\sin(x + y) = \sin x + \sin y$

(c) $\sec^2 x + \csc^2 x = 1$

(d) $\dfrac{1}{\sin x + \cos x} = \csc x + \sec x$

● DISCOVERY · DISCUSSION

96. Cofunction Identities In the right triangle shown, explain why

$$v = \frac{\pi}{2} - u$$

Explain how you can obtain all six cofunction identities from this triangle, for $0 < u < \pi/2$.

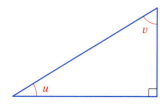

97. Graphs and Identities Suppose you graph two functions, f and g, on a graphing device, and their graphs appear identical in the viewing rectangle. Does this prove that the equation $f(x) = g(x)$ is an identity? Explain.

98. Making Up Your Own Identity If you start with a trigonometric expression and rewrite it or simplify it, then setting the original expression equal to the rewritten expression yields a trigonometric identity. For instance, from Example 1, we get the identity

$$\cos t + \tan t \sin t = \sec t$$

Use this technique to make up your own identity, then give it to a classmate to verify.

7.2 ADDITION AND SUBTRACTION FORMULAS

We now derive identities for trigonometric functions of sums and differences.

ADDITION AND SUBTRACTION FORMULAS

Formulas for sine

$$\sin(s + t) = \sin s \cos t + \cos s \sin t$$

$$\sin(s - t) = \sin s \cos t - \cos s \sin t$$

Formulas for cosine

$$\cos(s + t) = \cos s \cos t - \sin s \sin t$$

$$\cos(s - t) = \cos s \cos t + \sin s \sin t$$

Formulas for tangent

$$\tan(s + t) = \frac{\tan s + \tan t}{1 - \tan s \tan t}$$

$$\tan(s - t) = \frac{\tan s - \tan t}{1 + \tan s \tan t}$$

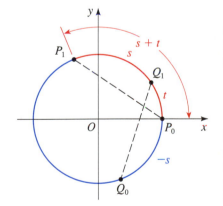

FIGURE 1

■ **Proof of Addition Formula for Cosine** To prove the formula $\cos(s + t) = \cos s \cos t - \sin s \sin t$, we use Figure 1. In the figure, the distances t, $s + t$, and $-s$ have been marked on the unit circle, starting at $P_0(1, 0)$ and terminating at Q_1, P_1, and Q_0, respectively. The coordinates of these points are

$$P_0(1, 0) \qquad\qquad Q_0(\cos(-s), \sin(-s))$$

$$P_1(\cos(s + t), \sin(s + t)) \qquad Q_1(\cos t, \sin t)$$

Since $\cos(-s) = \cos s$ and $\sin(-s) = -\sin s$, it follows that the point Q_0 has the coordinates $Q_0(\cos s, -\sin s)$. Notice that the distances between P_0 and P_1 and between Q_0 and Q_1 measured along the arc of the circle are equal. Since equal arcs are subtended by equal chords, it follows that $d(P_0, P_1) = d(Q_0, Q_1)$. Using the Distance Formula, we get

$$\sqrt{[\cos(s + t) - 1]^2 + [\sin(s + t) - 0]^2} = \sqrt{(\cos t - \cos s)^2 + (\sin t + \sin s)^2}$$

Squaring both sides and expanding, we have

these add to 1

$$\cos^2(s + t) - 2\cos(s + t) + 1 + \sin^2(s + t)$$

$$= \cos^2 t - 2\cos s \cos t + \cos^2 s + \sin^2 t + 2\sin s \sin t + \sin^2 s$$

these add to 1

these add to 1

Using the Pythagorean identity $\sin^2 z + \cos^2 z = 1$ three times gives

$$2 - 2\cos(s + t) = 2 - 2\cos s \cos t + 2 \sin s \sin t$$

Finally, subtracting 2 from each side and dividing both sides by -2, we get

$$\cos(s + t) = \cos s \cos t - \sin s \sin t$$

which proves the addition formula for cosine.

■ **Proof of Subtraction Formula for Cosine** Replacing t with $-t$ in the addition formula for cosine, we get

$$\cos(s - t) = \cos(s + (-t))$$

$$= \cos s \cos(-t) - \sin s \sin(-t) \qquad \text{Addition formula for cosine}$$

$$= \cos s \cos t + \sin s \sin t \qquad \text{Even-odd identities}$$

This proves the subtraction formula for cosine.   □

See Exercises 52 and 53 for proofs of the other addition formulas.

EXAMPLE 1 ■ **Using the Addition and Subtraction Formulas**

Find the exact value of each expression: (a) $\cos 75°$ (b) $\cos \dfrac{\pi}{12}$

SOLUTION

(a) Notice that $75° = 45° + 30°$. Since we know the exact values of sine and cosine at $45°$ and $30°$, we use the addition formula for cosine to get

$$\cos 75° = \cos(45° + 30°)$$

$$= \cos 45° \cos 30° - \sin 45° \sin 30°$$

$$= \frac{\sqrt{2}}{2} \frac{\sqrt{3}}{2} - \frac{\sqrt{2}}{2} \frac{1}{2} = \frac{\sqrt{2}\sqrt{3} - \sqrt{2}}{4} = \frac{\sqrt{6} - \sqrt{2}}{4}$$

(b) Since $\dfrac{\pi}{12} = \dfrac{\pi}{4} - \dfrac{\pi}{6}$, the subtraction formula for cosine gives

$$\cos \frac{\pi}{12} = \cos\left(\frac{\pi}{4} - \frac{\pi}{6}\right)$$

$$= \cos \frac{\pi}{4} \cos \frac{\pi}{6} + \sin \frac{\pi}{4} \sin \frac{\pi}{6}$$

$$= \frac{\sqrt{2}}{2} \frac{\sqrt{3}}{2} + \frac{\sqrt{2}}{2} \frac{1}{2} = \frac{\sqrt{6} + \sqrt{2}}{4} \qquad ■$$

EXAMPLE 2 ■ Using the Addition Formula for Sine

Find the exact value of the expression: $\sin 20° \cos 40° + \cos 20° \sin 40°$

SOLUTION

We recognize the expression as the right-hand side of the addition formula for sine with $s = 20°$ and $t = 40°$. So we have

$$\sin 20° \cos 40° + \cos 20° \sin 40° = \sin(20° + 40°) = \sin 60° = \frac{\sqrt{3}}{2}$$ ■

EXAMPLE 3 ■ Proving an Identity

Prove the cofunction identity: $\cos\left(\frac{\pi}{2} - u\right) = \sin u$

SOLUTION

By the subtraction formula for cosine,

$$\cos\left(\frac{\pi}{2} - u\right) = \cos\frac{\pi}{2}\cos u + \sin\frac{\pi}{2}\sin u$$

$$= 0 \cdot \cos u + 1 \cdot \sin u = \sin u$$ ■

EXAMPLE 4 ■ Proving an Identity

Verify the identity: $\frac{1 + \tan x}{1 - \tan x} = \tan\left(\frac{\pi}{4} + x\right)$

SOLUTION

Starting with the right-hand side and using the addition formula for tangent, we get

$$\text{RHS} = \tan\left(\frac{\pi}{4} + x\right) = \frac{\tan\frac{\pi}{4} + \tan x}{1 - \tan\frac{\pi}{4}\tan x}$$

$$= \frac{1 + \tan x}{1 - \tan x} = \text{LHS}$$ ■

The next example is a typical use of the addition and subtraction formulas in calculus.

EXAMPLE 5 ■ An Identity from Calculus

If $f(x) = \sin x$, show that

$$\frac{f(x + h) - f(x)}{h} = \sin x\left(\frac{\cos h - 1}{h}\right) + \cos x\left(\frac{\sin h}{h}\right)$$

SOLUTION

$$\frac{f(x + h) - f(x)}{h} = \frac{\sin(x + h) - \sin x}{h} \qquad \text{Definition of } f$$

$$= \frac{\sin x \cos h + \cos x \sin h - \sin x}{h} \qquad \text{Addition formula for sine}$$

$$= \frac{\sin x \,(\cos h - 1) + \cos x \sin h}{h} \qquad \text{Factor}$$

$$= \sin x \left(\frac{\cos h - 1}{h}\right) + \cos x \left(\frac{\sin h}{h}\right) \qquad \text{Separate the fraction} \qquad ■$$

Expressions of the Form $A \sin x + B \cos x$

We can write expressions of the form $A \sin x + B \cos x$ in terms of a single trigonometric function using the addition formula for sine. For example, consider the expression

$$\frac{1}{2} \sin x + \frac{\sqrt{3}}{2} \cos x$$

If we set $\phi = \pi/3$, then $\cos \phi = \frac{1}{2}$ and $\sin \phi = \sqrt{3}/2$, and we can write

$$\frac{1}{2} \sin x + \frac{\sqrt{3}}{2} \cos x = \cos \phi \sin x + \sin \phi \cos x$$

$$= \sin(x + \phi) = \sin\left(x + \frac{\pi}{3}\right)$$

We are able to do this because the coefficients $\frac{1}{2}$ and $\sqrt{3}/2$ are precisely the cosine and sine of a particular number, in this case, $\pi/3$. We can use this same idea in general to write $A \sin x + B \cos x$ in the form $k \sin(x + \phi)$. We start by multiplying the numerator and denominator by $\sqrt{A^2 + B^2}$ to get

$$A \sin x + B \cos x = \sqrt{A^2 + B^2} \left(\frac{A}{\sqrt{A^2 + B^2}} \sin x + \frac{B}{\sqrt{A^2 + B^2}} \cos x\right)$$

We need a number ϕ with the property that

$$\cos \phi = \frac{A}{\sqrt{A^2 + B^2}} \qquad \text{and} \qquad \sin \phi = \frac{B}{\sqrt{A^2 + B^2}}$$

Figure 2 shows that the point (A, B) in the plane determines a number ϕ with precisely this property. With this ϕ, we have

$$A \sin x + B \cos x = \sqrt{A^2 + B^2} \,(\cos \phi \sin x + \sin \phi \cos x)$$

$$= \sqrt{A^2 + B^2} \,\sin(x + \phi)$$

FIGURE 2

We have proved the following theorem.

> ### SUMS OF SINES AND COSINES
>
> If A and B are real numbers, then
>
> $$A \sin x + B \cos x = k \sin(x + \phi)$$
>
> where $k = \sqrt{A^2 + B^2}$ and ϕ satisfies
>
> $$\cos \phi = \frac{A}{\sqrt{A^2 + B^2}} \quad \text{and} \quad \sin \phi = \frac{B}{\sqrt{A^2 + B^2}}$$

EXAMPLE 6 ■ A Sum of Sine and Cosine Terms

Express $3 \sin x + 4 \cos x$ in the form $k \sin(x + \phi)$.

SOLUTION

By the preceding theorem, $k = \sqrt{A^2 + B^2} = \sqrt{3^2 + 4^2} = 5$. The angle ϕ has the property that $\sin \phi = \frac{4}{5}$ and $\cos \phi = \frac{3}{5}$. Using a calculator, we find $\phi \approx 53.1°$. Thus

$$3 \sin x + 4 \cos x \approx 5 \sin(x + 53.1°)$$ ■

EXAMPLE 7 ■ Graphing a Trigonometric Function

Write the function $f(x) = -\sin 2x + \sqrt{3} \cos 2x$ in the form $k \sin(2x + \phi)$ and use the new form to graph the function.

SOLUTION

Since $A = -1$ and $B = \sqrt{3}$, we have $k = \sqrt{A^2 + B^2} = \sqrt{1 + 3} = 2$. The angle ϕ satisfies $\cos \phi = -\frac{1}{2}$ and $\sin \phi = \sqrt{3}/2$. From the signs of these quantities we conclude that ϕ is in quadrant II. Thus, $\phi = 2\pi/3$. By the preceding theorem we can write

$$f(x) = -\sin 2x + \sqrt{3} \cos 2x = 2 \sin\left(2x + \frac{2\pi}{3}\right)$$

Using the form

$$f(x) = 2 \sin 2\left(x + \frac{\pi}{3}\right)$$

we see that the graph is a sine curve with amplitude 2, period $2\pi/2 = \pi$, and phase shift $-\pi/3$. The graph is shown in Figure 3. ■

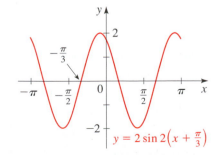

$y = 2 \sin 2\left(x + \frac{\pi}{3}\right)$

FIGURE 3

7.2   EXERCISES

1–10 ■ Use an addition or subtraction formula to find the exact value of the expression, as demonstrated in Example 1.

1. $\cos 15°$

2. $\sin 105°$

3. $\sin 165°$

4. $\tan 195°$

5. $\tan 75°$

6. $\cos \dfrac{11\pi}{12}$

7. $\sin \dfrac{19\pi}{12}$

8. $\tan \dfrac{17\pi}{12}$

9. $\cos\left(-\dfrac{\pi}{12}\right)$

10. $\sin\left(-\dfrac{5\pi}{12}\right)$

11–16 ■ Use an addition or subtraction formula to write the expression as a trigonometric function of one number, and find its exact value.

11. $\sin 18° \cos 27° + \cos 18° \sin 27°$

12. $\cos 10° \cos 80° - \sin 10° \sin 80°$

13. $\cos \dfrac{3\pi}{7} \cos \dfrac{2\pi}{21} + \sin \dfrac{3\pi}{7} \sin \dfrac{2\pi}{21}$

14. $\dfrac{\tan \dfrac{\pi}{18} + \tan \dfrac{\pi}{9}}{1 - \tan \dfrac{\pi}{18} \tan \dfrac{\pi}{9}}$

15. $\dfrac{\tan 73° - \tan 13°}{1 + \tan 73° \tan 13°}$

16. $\cos \dfrac{13\pi}{15} \cos\left(-\dfrac{\pi}{5}\right) - \sin \dfrac{13\pi}{15} \sin\left(-\dfrac{\pi}{5}\right)$

17–20 ■ Prove the cofunction identity using the addition and subtraction formulas.

17. $\tan\left(\dfrac{\pi}{2} - u\right) = \cot u$

18. $\cot\left(\dfrac{\pi}{2} - u\right) = \tan u$

19. $\sec\left(\dfrac{\pi}{2} - u\right) = \csc u$

20. $\csc\left(\dfrac{\pi}{2} - u\right) = \sec u$

21–38 ■ Prove the identity.

21. $\sin\left(x - \dfrac{\pi}{2}\right) = -\cos x$

22. $\cos\left(x - \dfrac{\pi}{2}\right) = \sin x$

23. $\sin(x - \pi) = -\sin x$

24. $\cos(x - \pi) = -\cos x$

25. $\tan(x - \pi) = \tan x$

26. $\sin\left(\dfrac{\pi}{2} - x\right) = \sin\left(\dfrac{\pi}{2} + x\right)$

27. $\cos\left(x + \dfrac{\pi}{6}\right) + \sin\left(x - \dfrac{\pi}{3}\right) = 0$

28. $\tan\left(x - \dfrac{\pi}{4}\right) = \dfrac{\tan x - 1}{\tan x + 1}$

29. $\sin(x + y) - \sin(x - y) = 2 \cos x \sin y$

30. $\cos(x + y) + \cos(x - y) = 2 \cos x \cos y$

31. $\cot(x - y) = \dfrac{\cot x \cot y + 1}{\cot y - \cot x}$

32. $\cot(x + y) = \dfrac{\cot x \cot y - 1}{\cot x + \cot y}$

33. $\tan x - \tan y = \dfrac{\sin(x - y)}{\cos x \cos y}$

34. $1 - \tan x \tan y = \dfrac{\cos(x + y)}{\cos x \cos y}$

35. $\dfrac{\sin(x + y) - \sin(x - y)}{\cos(x + y) + \cos(x - y)} = \tan y$

36. $\cos(x + y) \cos(x - y) = \cos^2 x - \sin^2 y$

37. $\sin(x + y + z) = \sin x \cos y \cos z + \cos x \sin y \cos z$
$\qquad\qquad + \cos x \cos y \sin z - \sin x \sin y \sin z$

38. $\tan(x - y) + \tan(y - z) + \tan(z - x)$
$\qquad\qquad = \tan(x - y) \tan(y - z) \tan(z - x)$

39–42 ■ Write the expression in terms of sine only.

39. $-\sqrt{3} \sin x + \cos x$

40. $\sin x + \cos x$

41. $5(\sin 2x - \cos 2x)$

42. $3 \sin \pi x + 3\sqrt{3} \cos \pi x$

43–44 ■ (a) Express the function in terms of sine only.
(b) Graph the function.

43. $f(x) = \sin x + \cos x$

44. $g(x) = \cos 2x + \sqrt{3} \sin 2x$

45. Show that if $\beta - \alpha = \pi/2$, then

$$\sin(x + \alpha) + \cos(x + \beta) = 0$$

46. Let $g(x) = \cos x$. Show that

$$\frac{g(x + h) - g(x)}{h} = -\cos x\left(\frac{1 - \cos h}{h}\right) - \sin x\left(\frac{\sin h}{h}\right)$$

47. Refer to the figure. Show that $\alpha + \beta = \gamma$, and find $\tan \gamma$.

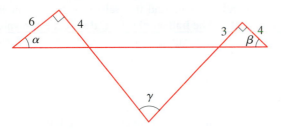

48. (a) If L is a line in the plane and θ is the angle formed by the line and the x-axis as shown in the figure, show that the slope m of the line is given by

$$m = \tan \theta$$

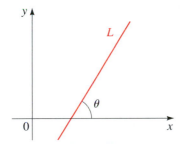

(b) Let L_1 and L_2 be two nonparallel lines in the plane with slopes m_1 and m_2, respectively. Let ψ be the acute angle formed by the two lines (see the figure). Show that

$$\tan \psi = \frac{m_2 - m_1}{1 + m_1 m_2}$$

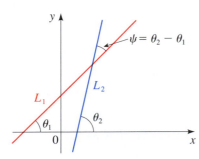

(c) Find the acute angle formed by the two lines

$$y = \tfrac{1}{3} x + 1 \qquad \text{and} \qquad y = -\tfrac{1}{2} x - 3$$

(d) Show that if two lines are perpendicular, then the slope of one is the negative reciprocal of the slope of the other. [*Hint:* First find an expression for $\cot \psi$.]

49–50 ■ (a) Graph the function and make a conjecture, then (b) prove that your conjecture is true.

49. $y = \sin^2\left(x + \dfrac{\pi}{4}\right) + \sin^2\left(x - \dfrac{\pi}{4}\right)$

50. $y = -\tfrac{1}{2}[\cos(x + \pi) + \cos(x - \pi)]$

51. A digital delay-device echoes an input signal by repeating it a fixed length of time after it is received. If such a device receives the pure note $f_1(t) = 5 \sin t$ and echoes the pure note $f_2(t) = 5 \cos t$, then the combined sound is $f(t) = f_1(t) + f_2(t)$.

(a) Graph $y = f(t)$ and observe that the graph has the form of a sine curve $y = k \sin(t + \phi)$.

(b) Find k and ϕ.

◼ DISCOVERY · DISCUSSION

52. Addition Formula for Sine   In the text we proved only the addition and subtraction formulas for cosine. Use these formulas and the cofunction identities

$$\sin x = \cos\left(\frac{\pi}{2} - x\right)$$

$$\cos x = \sin\left(\frac{\pi}{2} - x\right)$$

to prove the addition formula for sine. [*Hint:* To get started, use the first cofunction identity to write

$$\sin(s + t) = \cos\left(\frac{\pi}{2} - (s + t)\right)$$

$$= \cos\left(\left(\frac{\pi}{2} - s\right) - t\right)$$

and use the subtraction formula for cosine.]

53. Addition Formula for Tangent   Use the addition formulas for cosine and sine to prove the addition formula for tangent. [*Hint:* Use

$$\tan(s + t) = \frac{\sin(s + t)}{\cos(s + t)}$$

and divide the numerator and denominator by $\cos s \cos t$.]

7.3 DOUBLE-ANGLE, HALF-ANGLE, AND PRODUCT-SUM FORMULAS

The identities we consider in this section are consequences of the addition formulas. The **double-angle formulas** allow us to find the values of the trigonometric functions at $2x$ from their values at x. The **half-angle formulas** relate the values of the trigonometric functions at $\frac{1}{2}x$ to their values at x. The **product-sum formulas** relate products of sines and cosines to sums of sines and cosines.

Double-Angle Formulas

The formulas in the following box are consequences of the addition formulas, which we proved in the preceding section.

DOUBLE-ANGLE FORMULAS

Formula for sine

$$\sin 2x = 2 \sin x \cos x$$

Formulas for cosine

$$\cos 2x = \cos^2 x - \sin^2 x$$

$$= 1 - 2 \sin^2 x$$

$$= 2 \cos^2 x - 1$$

Formula for tangent

$$\tan 2x = \frac{2 \tan x}{1 - \tan^2 x}$$

■ **Proof of Double-Angle Formula for Sine**   We use the addition formula for sine.

$$\sin 2x = \sin(x + x)$$

$$= \sin x \cos x + \sin x \cos x$$

$$= 2 \sin x \cos x$$

■ **Proof of Double-Angle Formulas for Cosine**

$$\cos 2x = \cos(x + x)$$

$$= \cos x \cos x - \sin x \sin x$$

$$= \cos^2 x - \sin^2 x$$

The second and third formulas for $\cos 2x$ are obtained from the formula we just proved and the Pythagorean identity. Substituting $\cos^2 x = 1 - \sin^2 x$ gives

$$\cos 2x = \cos^2 x - \sin^2 x$$

$$= (1 - \sin^2 x) - \sin^2 x$$

$$= 1 - 2\sin^2 x$$

The third formula is obtained in the same way, by substituting $\sin^2 x = 1 - \cos^2 x$.

■ **Proof of Double-Angle Formula for Tangent**   We use the addition formula for tangent.

$$\tan 2x = \tan(x + x)$$

$$= \frac{\tan x + \tan x}{1 - \tan x \tan x}$$

$$= \frac{2\tan x}{1 - \tan^2 x} \qquad \square$$

EXAMPLE 1 ■ Using the Double-Angle Formulas

If $\cos x = -\frac{2}{3}$ and x is in quadrant II, find $\cos 2x$ and $\sin 2x$.

SOLUTION

Using one of the double-angle formulas for cosine, we get

$$\cos 2x = 2\cos^2 x - 1$$

$$= 2\left(-\frac{2}{3}\right)^2 - 1 = \frac{8}{9} - 1 = -\frac{1}{9}$$

To use the formula $\sin 2x = 2\sin x \cos x$, we need to find $\sin x$ first. We have

$$\sin x = \sqrt{1 - \cos^2 x} = \sqrt{1 - \left(-\frac{2}{3}\right)^2} = \frac{\sqrt{5}}{3}$$

where we have used the positive square root because $\sin x$ is positive in quadrant II. Thus

$$\sin 2x = 2\sin x \cos x$$

$$= 2\left(\frac{\sqrt{5}}{3}\right)\left(-\frac{2}{3}\right) = -\frac{4\sqrt{5}}{9}$$

EXAMPLE 2 ■ A Triple-Angle Formula

Write $\cos 3x$ in terms of $\cos x$.

SOLUTION

$$\cos 3x = \cos(2x + x)$$

$$= \cos 2x \cos x - \sin 2x \sin x \qquad \text{Addition formula}$$

$$= (2 \cos^2 x - 1) \cos x - (2 \sin x \cos x) \sin x \qquad \text{Double-angle formulas}$$

$$= 2 \cos^3 x - \cos x - 2 \sin^2 x \cos x \qquad \text{Expand}$$

$$= 2 \cos^3 x - \cos x - 2 \cos x (1 - \cos^2 x) \qquad \text{Pythagorean identity}$$

$$= 2 \cos^3 x - \cos x - 2 \cos x + 2 \cos^3 x \qquad \text{Expand}$$

$$= 4 \cos^3 x - 3 \cos x \qquad \text{Simplify} \qquad ■$$

Example 2 shows that $\cos 3x$ can be written as a polynomial of degree 3 in $\cos x$. The identity $\cos 2x = 2 \cos^2 x - 1$ shows that $\cos 2x$ is a polynomial of degree 2 in $\cos x$. In fact, for any natural number n, we can write $\cos nx$ as a polynomial in $\cos x$ of degree n (see Exercise 83). The analogous result for $\sin nx$ is not true in general.

EXAMPLE 3 ■ Proving an Identity

Prove the identity: $\dfrac{\sin 3x}{\sin x \cos x} = 4 \cos x - \sec x$

SOLUTION

We start with the left-hand side.

$$\frac{\sin 3x}{\sin x \cos x} = \frac{\sin(x + 2x)}{\sin x \cos x}$$

$$= \frac{\sin x \cos 2x + \cos x \sin 2x}{\sin x \cos x} \qquad \text{Addition formula}$$

$$= \frac{\sin x (2 \cos^2 x - 1) + \cos x (2 \sin x \cos x)}{\sin x \cos x} \qquad \text{Double-angle formulas}$$

$$= \frac{\sin x (2 \cos^2 x - 1)}{\sin x \cos x} + \frac{\cos x (2 \sin x \cos x)}{\sin x \cos x} \qquad \text{Separate fraction}$$

$$= \frac{2 \cos^2 x - 1}{\cos x} + 2 \cos x \qquad \text{Cancel}$$

$$= 2 \cos x - \frac{1}{\cos x} + 2 \cos x \qquad \text{Separate fraction}$$

$$= 4 \cos x - \sec x \qquad \text{Reciprocal identity} \qquad ■$$

▮ Half-Angle Formulas

The following formulas allow us to write any trigonometric expression involving even powers of sine and cosine in terms of the first power of cosine only. This technique is important in calculus. The half-angle formulas are immediate consequences of these formulas.

FORMULAS FOR LOWERING POWERS

$$\sin^2 x = \frac{1 - \cos 2x}{2} \qquad \cos^2 x = \frac{1 + \cos 2x}{2}$$

$$\tan^2 x = \frac{1 - \cos 2x}{1 + \cos 2x}$$

■ **Proof**   The first formula is obtained by solving for $\sin^2 x$ in the double-angle formula $\cos 2x = 1 - 2\sin^2 x$. Similarly, the second formula is obtained by solving for $\cos^2 x$ in the double-angle formula $\cos 2x = 2\cos^2 x - 1$.

The last formula follows from the first two and the reciprocal identities as follows:

$$\tan^2 x = \frac{\sin^2 x}{\cos^2 x} = \frac{\dfrac{1 - \cos 2x}{2}}{\dfrac{1 + \cos 2x}{2}} = \frac{1 - \cos 2x}{1 + \cos 2x}$$

$\square$

EXAMPLE 4 ■ Lowering Powers in a Trigonometric Expression

Express $\sin^2 x \cos^2 x$ in terms of the first power of cosine.

SOLUTION

We use the formulas for lowering powers repeatedly.

$$\sin^2 x \cos^2 x = \left(\frac{1 - \cos 2x}{2}\right)\left(\frac{1 + \cos 2x}{2}\right)$$

$$= \frac{1 - \cos^2 2x}{4} = \frac{1}{4} - \frac{1}{4}\cos^2 2x$$

$$= \frac{1}{4} - \frac{1}{4}\left(\frac{1 + \cos 4x}{2}\right) = \frac{1}{4} - \frac{1}{8} - \frac{\cos 4x}{8}$$

$$= \tfrac{1}{8} - \tfrac{1}{8}\cos 4x = \tfrac{1}{8}(1 - \cos 4x)$$

Another way to obtain this identity is to use the double-angle formula for sine in the form $\sin x \cos x = \frac{1}{2}\sin 2x$. Thus

$$\sin^2 x \cos^2 x = \frac{1}{4}\sin^2 2x = \frac{1}{4}\left(\frac{1 - \cos 4x}{2}\right) = \frac{1}{8}(1 - \cos 4x)$$

■

HALF-ANGLE FORMULAS

$$\sin \frac{u}{2} = \pm \sqrt{\frac{1 - \cos u}{2}} \qquad \cos \frac{u}{2} = \pm \sqrt{\frac{1 + \cos u}{2}}$$

$$\tan \frac{u}{2} = \frac{1 - \cos u}{\sin u} = \frac{\sin u}{1 + \cos u}$$

The choice of the $+$ or $-$ sign depends on the quadrant in which $u/2$ lies.

■ **Proof** We substitute $x = u/2$ in the formulas for lowering powers and take the square root of each side. This yields the first two half-angle formulas. In the case of the half-angle formula for tangent, we get

$$\tan \frac{u}{2} = \pm \sqrt{\frac{1 - \cos u}{1 + \cos u}}$$

$$= \pm \sqrt{\left(\frac{1 - \cos u}{1 + \cos u}\right)\left(\frac{1 - \cos u}{1 - \cos u}\right)} \qquad \text{Multiply numerator and denominator by } 1 - \cos u$$

$$= \pm \sqrt{\frac{(1 - \cos u)^2}{1 - \cos^2 u}} \qquad \text{Simplify}$$

$$= \pm \frac{|1 - \cos u|}{|\sin u|} \qquad \sqrt{A^2} = |A|$$

Now, $1 - \cos u$ is nonnegative for all values of u. It is also true that $\sin u$ and $\tan(u/2)$ always have the same sign. (Verify this.) It follows that

$$\tan \frac{u}{2} = \frac{1 - \cos u}{\sin u}$$

The other half-angle formula for tangent is derived from this by multiplying the numerator and denominator by $1 + \cos u$. □

EXAMPLE 5 ■ Using a Half-Angle Formula

Find the exact value of $\sin 22.5°$.

SOLUTION

Since $22.5°$ is half of $45°$, we use the half-angle formula for sine with $u = 45°$. We choose the $+$ sign because $22.5°$ is in the first quadrant.

$$\sin \frac{45°}{2} = \sqrt{\frac{1 - \cos 45°}{2}} \qquad \text{Half-angle formula}$$

$$= \sqrt{\frac{1 - \sqrt{2}/2}{2}} \qquad \cos 45° = \sqrt{2}/2$$

$$= \sqrt{\frac{2 - \sqrt{2}}{4}} \qquad \text{Common denominator}$$

$$= \tfrac{1}{2}\sqrt{2 - \sqrt{2}} \qquad \text{Simplify} \qquad ■$$

EXAMPLE 6 ■ Using a Half-Angle Formula

Find $\tan(u/2)$ if $\sin u = \frac{2}{5}$ and u is in quadrant II.

SOLUTION

To use the half-angle formulas for tangent, we first need to find $\cos u$. Since cosine is negative in quadrant II, we have

$$\cos u = -\sqrt{1 - \sin^2 u}$$

$$= -\sqrt{1 - \left(\tfrac{2}{5}\right)^2} = -\frac{\sqrt{21}}{5}$$

Thus

$$\tan \frac{u}{2} = \frac{1 - \cos u}{\sin u}$$

$$= \frac{1 + \sqrt{21}/5}{\tfrac{2}{5}} = \frac{5 + \sqrt{21}}{2} \qquad ■$$

Product-Sum Formulas

It is possible to write the product $\sin u \cos v$ as a sum of trigonometric functions. To see this, consider the addition and subtraction formulas for the sine function:

$$\sin(u + v) = \sin u \cos v + \cos u \sin v$$

$$\sin(u - v) = \sin u \cos v - \cos u \sin v$$

Adding the left- and right-hand sides of these formulas gives

$$\sin(u + v) + \sin(u - v) = 2 \sin u \cos v$$

Dividing by 2 yields the formula

$$\sin u \cos v = \tfrac{1}{2}[\sin(u + v) + \sin(u - v)]$$

The other three **product-to-sum formulas** follow from the addition formulas in a similar way.

PRODUCT-TO-SUM FORMULAS

$$\sin u \cos v = \tfrac{1}{2}[\sin(u + v) + \sin(u - v)]$$

$$\cos u \sin v = \tfrac{1}{2}[\sin(u + v) - \sin(u - v)]$$

$$\cos u \cos v = \tfrac{1}{2}[\cos(u + v) + \cos(u - v)]$$

$$\sin u \sin v = \tfrac{1}{2}[\cos(u - v) - \cos(u + v)]$$

EXAMPLE 7 ■ **Expressing a Trigonometric Product as a Sum**

Express $\sin 3x \sin 5x$ as a sum of trigonometric functions.

SOLUTION

Using the fourth product-to-sum formula with $u = 3x$ and $v = 5x$ and the fact that cosine is an even function, we get

$$\sin 3x \sin 5x = \tfrac{1}{2}\left[\cos(3x - 5x) - \cos(3x + 5x)\right]$$

$$= \tfrac{1}{2}\cos(-2x) - \tfrac{1}{2}\cos 8x$$

$$= \tfrac{1}{2}\cos 2x - \tfrac{1}{2}\cos 8x \qquad\blacksquare$$

The product-to-sum formulas can also be used as sum-to-product formulas. This is possible because the right-hand side of each product-to-sum formula is a sum and the left side is a product. For example, if we let

$$u = \frac{x + y}{2} \qquad \text{and} \qquad v = \frac{x - y}{2}$$

in the first product-to-sum formula, we get

$$\sin\frac{x + y}{2}\cos\frac{x - y}{2} = \tfrac{1}{2}\left[\sin x + \sin y\right]$$

so

$$\sin x + \sin y = 2\sin\frac{x + y}{2}\cos\frac{x - y}{2}$$

The remaining three of the following **sum-to-product formulas** are obtained in a similar manner.

SUM-TO-PRODUCT FORMULAS

$$\sin x + \sin y = 2\sin\frac{x + y}{2}\cos\frac{x - y}{2}$$

$$\sin x - \sin y = 2\cos\frac{x + y}{2}\sin\frac{x - y}{2}$$

$$\cos x + \cos y = 2\cos\frac{x + y}{2}\cos\frac{x - y}{2}$$

$$\cos x - \cos y = -2\sin\frac{x + y}{2}\sin\frac{x - y}{2}$$

EXAMPLE 8 ■ Expressing a Trigonometric Sum as a Product

Write $\sin 7x + \sin 3x$ as a product.

SOLUTION

The first sum-to-product formula gives

$$\sin 7x + \sin 3x = 2 \sin \frac{7x + 3x}{2} \cos \frac{7x - 3x}{2}$$

$$= 2 \sin 5x \cos 2x$$ ■

EXAMPLE 9 ■ Proving an Identity

Verify the identity: $\dfrac{\sin 3x - \sin x}{\cos 3x + \cos x} = \tan x$

SOLUTION

We apply the second sum-to-product formula to the numerator and the third formula to the denominator.

$$\text{LHS} = \frac{\sin 3x - \sin x}{\cos 3x + \cos x} = \frac{2 \cos \dfrac{3x + x}{2} \sin \dfrac{3x - x}{2}}{2 \cos \dfrac{3x + x}{2} \cos \dfrac{3x - x}{2}}$$ Sum-to-product formulas

$$= \frac{2 \cos 2x \sin x}{2 \cos 2x \cos x}$$ Simplify

$$= \frac{\sin x}{\cos x} = \tan x = \text{RHS}$$ Cancel   ■

7.3 EXERCISES

1–8 ■ Find $\sin 2x$, $\cos 2x$, and $\tan 2x$ from the given information.

1. $\sin x = \frac{5}{13}$, x in quadrant I

2. $\tan x = -\frac{4}{3}$, x in quadrant II

3. $\cos x = \frac{4}{5}$, $\csc x < 0$ **4.** $\csc x = 4$, $\tan x < 0$

5. $\sin x = -\frac{3}{5}$, x in quadrant III

6. $\sec x = 2$, x in quadrant IV

7. $\tan x = -\frac{1}{3}$, $\cos x > 0$

8. $\cot x = \frac{2}{3}$, $\sin x > 0$

9–14 ■ Use the formulas for lowering powers to rewrite the expression in terms of the first power of cosine, as in Example 4.

9. $\sin^4 x$ **10.** $\cos^4 x$

11. $\cos^4 x \sin^4 x$ **12.** $\cos^4 x \sin^2 x$

13. $\cos^2 x \sin^4 x$ **14.** $\cos^6 x$

15–22 ■ Use an appropriate half-angle formula to find the exact value of the expression.

15. $\sin 15°$ **16.** $\tan 15°$

17. $\cos 22.5°$ **18.** $\sin 75°$

19. $\tan \dfrac{\pi}{8}$

20. $\cos \dfrac{3\pi}{8}$

21. $\sin \dfrac{\pi}{12}$

22. $\cos \dfrac{5\pi}{12}$

23–28 ■ Simplify the expression by using a double-angle formula or a half-angle formula.

23. (a) $2 \sin 18° \cos 18°$ (b) $2 \sin 3\theta \cos 3\theta$

24. (a) $\dfrac{2 \tan 7°}{1 - \tan^2 7°}$ (b) $\dfrac{2 \tan 7\theta}{1 - \tan^2 7\theta}$

25. (a) $\cos^2 34° - \sin^2 34°$ (b) $\cos^2 5\theta - \sin^2 5\theta$

26. (a) $\cos^2 \dfrac{\theta}{2} - \sin^2 \dfrac{\theta}{2}$ (b) $2 \sin \dfrac{\theta}{2} \cos \dfrac{\theta}{2}$

27. (a) $\dfrac{\sin 8°}{1 + \cos 8°}$ (b) $\dfrac{1 - \cos 4\theta}{\sin 4\theta}$

28. (a) $\sqrt{\dfrac{1 - \cos 30°}{2}}$ (b) $\sqrt{\dfrac{1 - \cos 8\theta}{2}}$

29–34 ■ Find $\sin \dfrac{x}{2}$, $\cos \dfrac{x}{2}$, and $\tan \dfrac{x}{2}$ from the given information.

29. $\sin x = \frac{3}{5}$, $0° < x < 90°$

30. $\cos x = -\frac{4}{5}$, $180° < x < 270°$

31. $\csc x = 3$, $90° < x < 180°$

32. $\tan x = 1$, $0° < x < 90°$

33. $\sec x = \frac{3}{2}$, $270° < x < 360°$

34. $\cot x = 5$, $180° < x < 270°$

35–40 ■ Write the product as a sum.

35. $\sin 2x \cos 3x$

36. $\sin x \sin 5x$

37. $\cos x \sin 4x$

38. $\cos 5x \cos 3x$

39. $3 \cos 4x \cos 7x$

40. $11 \sin \dfrac{x}{2} \cos \dfrac{x}{4}$

41–46 ■ Write the sum as a product.

41. $\sin 5x + \sin 3x$

42. $\sin x - \sin 4x$

43. $\cos 4x - \cos 6x$

44. $\cos 9x + \cos 2x$

45. $\sin 2x - \sin 7x$

46. $\sin 3x + \sin 4x$

47–52 ■ Find the value of the product or sum.

47. $2 \sin 52.5° \sin 97.5°$

48. $3 \cos 37.5° \cos 7.5°$

49. $\cos 37.5° \sin 7.5°$

50. $\sin 75° + \sin 15°$

51. $\cos 255° - \cos 195°$

52. $\cos \dfrac{\pi}{12} + \cos \dfrac{5\pi}{12}$

53–70 ■ Prove the identity.

53. $\cos^2 5x - \sin^2 5x = \cos 10x$

54. $\sin 8x = 2 \sin 4x \cos 4x$

55. $(\sin x + \cos x)^2 = 1 + \sin 2x$

56. $\dfrac{2 \tan x}{1 + \tan^2 x} = \sin 2x$

57. $\dfrac{\sin 4x}{\sin x} = 4 \cos x \cos 2x$

58. $\dfrac{1 + \sin 2x}{\sin 2x} = 1 + \frac{1}{2} \sec x \csc x$

59. $\dfrac{2(\tan x - \cot x)}{\tan^2 x - \cot^2 x} = \sin 2x$

60. $\cot 2x = \dfrac{1 - \tan^2 x}{2 \tan x}$

61. $\tan 3x = \dfrac{3 \tan x - \tan^3 x}{1 - 3 \tan^2 x}$

62. $4(\sin^6 x + \cos^6 x) = 4 - 3 \sin^2 2x$

63. $\cos^4 x - \sin^4 x = \cos 2x$

64. $\tan^2\left(\dfrac{x}{2} + \dfrac{\pi}{4}\right) = \dfrac{1 + \sin x}{1 - \sin x}$

65. $\dfrac{\sin x + \sin 5x}{\cos x + \cos 5x} = \tan 3x$

66. $\dfrac{\sin 3x + \sin 7x}{\cos 3x - \cos 7x} = \cot 2x$

67. $\dfrac{\sin 10x}{\sin 9x + \sin x} = \dfrac{\cos 5x}{\cos 4x}$

68. $\dfrac{\sin x + \sin 3x + \sin 5x}{\cos x + \cos 3x + \cos 5x} = \tan 3x$

69. $\dfrac{\sin x + \sin y}{\cos x + \cos y} = \tan\left(\dfrac{x + y}{2}\right)$

70. $\tan y = \dfrac{\sin(x + y) - \sin(x - y)}{\cos(x + y) + \cos (x - y)}$

71. Show that $\sin 45° + \sin 15° = \sin 75°$.

72. Show that $\cos 87° + \cos 33° = \sin 63°$.

73. Prove the identity

$$\dfrac{\sin x + \sin 2x + \sin 3x + \sin 4x + \sin 5x}{\cos x + \cos 2x + \cos 3x + \cos 4x + \cos 5x} = \tan 3x$$

74. Use the identity

$$\sin 2x = 2 \sin x \cos x$$

n times to show that

$$\sin(2^n x) = 2^n \sin x \cos x \cos 2x \cos 4x \cdots \cos 2^{n-1} x$$

75. (a) Graph $f(x) = \dfrac{\sin 3x}{\sin x} - \dfrac{\cos 3x}{\cos x}$ and make a conjecture.
(b) Prove the conjecture you made in part (a).

76. (a) Graph $f(x) = \cos 2x + 2 \sin^2 x$ and make a conjecture.
    (b) Prove the conjecture you made in part (a).

77. Let $f(x) = \sin 6x + \sin 7x$.
    (a) Graph $y = f(x)$.
    (b) Verify that $f(x) = 2 \cos \frac{1}{2}x \sin \frac{13}{2}x$.
    (c) Graph $y = 2 \cos \frac{1}{2}x$ and $y = -2 \cos \frac{1}{2}x$, together with the graph in part (a), in the same viewing rectangle. How are these graphs related to the graph of f?

78. When two pure notes that are close in frequency are played together, their sounds interfere to produce *beats*; that is, the loudness (or amplitude) of the sound alternately increases and decreases. If the two notes are given by

$$f_1(t) = \cos 11t \qquad \text{and} \qquad f_2(t) = \cos 13t$$

the resulting sound is $f(t) = f_1(t) + f_2(t)$.
    (a) Graph the function $y = f(t)$.
    (b) Verify that $f(t) = 2 \cos t \cos 12t$.
    (c) Graph $y = 2 \cos t$ and $y = -2 \cos t$, together with the graph in part (a), in the same viewing rectangle. How do these graphs describe the variation in the loudness of the sound?

79–81 ■ In these exercises you are asked to use the trigonometric functions to model real-life situations. See the guidelines on modeling with functions in Section 2.7.

79. A rectangular beam is to be cut from a cylindrical log of diameter 20 in.
    (a) Show that the cross-sectional area of the beam is modeled by the function

$$A(\theta) = 200 \sin 2\theta$$

    where θ is as shown in the figure.

    (b) Show that the maximum cross-sectional area of such a beam is 200 in². [*Hint:* Use the fact that $\sin u$ achieves its maximum value at $u = \pi/2$.]

80. A rectangle is to be inscribed in a semicircle of radius 5 cm as shown in the figure.
    (a) Show that the area of the rectangle is modeled by the function

$$A(\theta) = 25 \sin 2\theta$$

    (b) Find the largest possible area for such an inscribed rectangle.
    (c) Find the dimensions of the inscribed rectangle with the largest possible area.

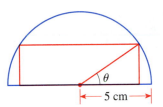

81. The lower right-hand corner of a long piece of paper 6 in. wide is folded over to the left-hand edge as shown. The length L of the fold depends on the angle θ. Show that

$$L = \frac{3}{\sin \theta \cos^2 \theta}$$

82. Let $3x = \pi/3$ and let $y = \cos x$. Use the result of Example 2 to show that y satisfies the equation

$$8y^3 - 6y - 1 = 0$$

    NOTE This equation has roots of a certain kind that are used to show that the angle $\pi/3$ cannot be trisected using a ruler and compass only.

83. (a) Show that there is a polynomial $P(t)$ of degree 4 such that $\cos 4x = P(\cos x)$ (see Example 2).
    (b) Show that there is a polynomial $Q(t)$ of degree 5 such that $\cos 5x = Q(\cos x)$.

    NOTE In general, there is a polynomial $P_n(t)$ of degree n such that $\cos nx = P_n(\cos x)$. These polynomials are called Tchebycheff polynomials, after the Russian mathematician P. L. Tchebycheff (1821–1894).

84. In triangle *ABC* (see the figure) the line segment *s* bisects angle *C*. Show that the length of *s* is given by

$$s = \frac{2ab \cos x}{a + b}$$

Hint: Find the area of triangle *ABC* in two different ways. You will need the following facts from geometry:

An angle inscribed in a semicircle is a right angle, so $\angle ACB$ is a right angle.

 DISCOVERY · DISCUSSION

85. Geometric Proof of a Double-Angle Formula Use the figure to prove that $\sin 2\theta = 2 \sin \theta \cos \theta$.

The central angle subtended by the chord of a circle is twice the angle subtended by the chord on the circle, so $\angle BOC$ is 2θ.

7.4 INVERSE TRIGONOMETRIC FUNCTIONS

If *f* is a one-to-one function with domain *A* and range *B*, then its inverse f^{-1} is the function with domain *B* and range *A* defined by

$$f^{-1}(x) = y \iff f(y) = x$$

(See Section 2.9.) In other words, f^{-1} is the rule that reverses the action of *f*. Figure 1 represents the actions of *f* and f^{-1} graphically.

For a function to have an inverse, it must be one-to-one. Since the trigonometric functions are not one-to-one, they do not have inverses. It is possible, however, to restrict the domains of the trigonometric functions in such a way that the resulting functions are one-to-one.

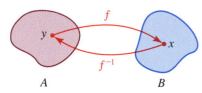

FIGURE 1

$f^{-1}(x) = y \Leftrightarrow f(y) = x$

The Inverse Sine Function

Let's first consider the sine function. There are many ways to restrict the domain of sine so that the new function is one-to-one. A natural way to do this is to restrict the domain to the interval $[-\pi/2, \pi/2]$. The reason for this choice is that sine attains each of its values exactly once on this interval. We write Sin *x* (with a capital S) for the new function, which has the domain $[-\pi/2, \pi/2]$ and the same values as sin *x* on this interval. The graphs of sin *x* and Sin *x* are shown in Figure 2. The function Sin *x* is one-to-one (by the Horizontal Line Test), and so has an inverse.

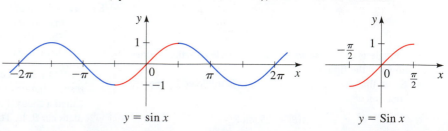

FIGURE 2

$y = \sin x$

$y = \text{Sin } x$

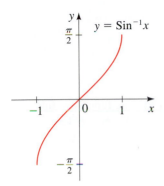

FIGURE 3

The inverse of the function Sin is the function Sin^{-1} defined by

$$\text{Sin}^{-1}x = y \iff \text{Sin } y = x$$

for $-1 \leq x \leq 1$ and $-\pi/2 \leq y \leq \pi/2$. The graph of $y = \text{Sin}^{-1}x$ is shown in Figure 3; it is obtained by reflecting the graph of $y = \text{Sin } x$ in the line $y = x$. It is customary to write $\text{Sin}^{-1}x$ simply as $\sin^{-1}x$.

DEFINITION OF THE INVERSE SINE FUNCTION

The **inverse sine function** is the function $\sin^{-1}$ with domain $[-1, 1]$ and range $[-\pi/2, \pi/2]$ defined by

$$\sin^{-1}x = y \iff \sin y = x$$

The inverse sine function is also called **arcsine** and is denoted by **arcsin**.

Thus, $\sin^{-1}x$ *is the number in the interval* $[-\pi/2, \pi/2]$ *whose sine is x.* In other words, $\sin(\sin^{-1}x) = x$. In fact, from the general properties of inverse functions studied in Section 2.9, we have the following relations.

$$\sin(\sin^{-1}x) = x \qquad \text{for } -1 \leq x \leq 1$$

$$\sin^{-1}(\sin x) = x \qquad \text{for } -\frac{\pi}{2} \leq x \leq \frac{\pi}{2}$$

EXAMPLE 1 ■ Evaluating the Inverse Sine Function

Find (a) $\sin^{-1}\frac{1}{2}$, (b) $\sin^{-1}\left(-\frac{1}{2}\right)$, and (c) $\sin^{-1}\frac{3}{2}$.

SOLUTION

(a) The number in the interval $[-\pi/2, \pi/2]$ whose sine is $\frac{1}{2}$ is $\pi/6$. Thus, $\sin^{-1}\frac{1}{2} = \pi/6$.

(b) The number in the interval $[-\pi/2, \pi/2]$ whose sine is $-\frac{1}{2}$ is $-\pi/6$. Thus, $\sin^{-1}\left(-\frac{1}{2}\right) = -\pi/6$.

(c) Since $\frac{3}{2} > 1$, it is not in the domain of $\sin^{-1}x$, so $\sin^{-1}\frac{3}{2}$ is not defined. ■

EXAMPLE 2 ■ Using a Calculator to Evaluate Inverse Sine

Find approximate values for (a) $\sin^{-1}(0.82)$ and (b) $\sin^{-1}\frac{1}{3}$.

SOLUTION

Since no rational multiple of π has a sine of 0.82 or $\frac{1}{3}$, we use a calculator to approximate these values. Using the $\boxed{\text{INV}}$ $\boxed{\text{SIN}}$, or $\boxed{\text{SIN}^{-1}}$, or $\boxed{\text{ARCSIN}}$ key(s) on the calculator (with the calculator in radian mode), we get

(a) $\sin^{-1}(0.82) \approx 0.96141$ (b) $\sin^{-1}\frac{1}{3} \approx 0.33984$ ■

EXAMPLE 3 ■ Composing Trigonometric Functions and Their Inverses

Find $\cos\left(\sin^{-1}\frac{3}{5}\right)$.

SOLUTION 1

It's easy to find $\sin\left(\sin^{-1}\frac{3}{5}\right)$. In fact, by the properties of inverse functions, this value is exactly $\frac{3}{5}$. To find $\cos\left(\sin^{-1}\frac{3}{5}\right)$, we reduce this to the easier problem by writing the cosine function in terms of the sine function. Let $u = \sin^{-1}\frac{3}{5}$. Since $-\pi/2 \le u \le \pi/2$, $\cos u$ is positive and we can write

$$\cos u = +\sqrt{1 - \sin^2 u}$$

Thus
$$\cos\left(\sin^{-1}\tfrac{3}{5}\right) = \sqrt{1 - \sin^2\left(\sin^{-1}\tfrac{3}{5}\right)}$$

$$= \sqrt{1 - \left(\tfrac{3}{5}\right)^2} = \sqrt{1 - \tfrac{9}{25}} = \sqrt{\tfrac{16}{25}} = \tfrac{4}{5}$$

SOLUTION 2

Let $\theta = \sin^{-1}\frac{3}{5}$. Then θ is the number in the interval $[-\pi/2, \pi/2]$ whose sine is $\frac{3}{5}$. Let's interpret θ as an angle and draw a right triangle with θ as one of its acute angles, with opposite side 3 and hypotenuse 5 (see Figure 4). The remaining leg of the triangle is found by the Pythagorean Theorem to be 4. From the figure we get

$$\cos\left(\sin^{-1}\tfrac{3}{5}\right) = \cos\theta = \tfrac{4}{5}$$ ■

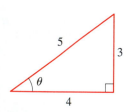

FIGURE 4

From Solution 2 of Example 3 we can immediately find the values of the other trigonometric functions of $\theta = \sin^{-1}\frac{3}{5}$ from the triangle. Thus

$$\tan\left(\sin^{-1}\tfrac{3}{5}\right) = \tfrac{3}{4} \qquad \sec\left(\sin^{-1}\tfrac{3}{5}\right) = \tfrac{5}{4} \qquad \csc\left(\sin^{-1}\tfrac{3}{5}\right) = \tfrac{5}{3}$$

The Inverse Cosine Function

If the domain of the cosine function is restricted to the interval $[0, \pi]$, the resulting function is one-to-one and so has an inverse. We choose this interval because on it, cosine attains each of its values exactly once (see Figure 5).

FIGURE 5

$y = \cos x$

$y = \cos x,\ 0 \le x \le \pi$

DEFINITION OF THE INVERSE COSINE FUNCTION

The **inverse cosine function** is the function $\cos^{-1}$ with domain $[-1, 1]$ and range $[0, \pi]$ defined by

$$\cos^{-1}x = y \iff \cos y = x$$

The inverse cosine function is also called **arccosine** and is denoted by **arccos**.

Thus, $y = \cos^{-1}x$ *is the number in the interval* $[0, \pi]$ *whose cosine is* x. The following relations follow from the inverse function properties.

$$\cos(\cos^{-1}x) = x \qquad \text{for } -1 \leq x \leq 1$$

$$\cos^{-1}(\cos x) = x \qquad \text{for } 0 \leq x \leq \pi$$

The graph of $y = \cos^{-1}x$ is shown in Figure 6; it is obtained by reflecting the graph of $y = \cos x, 0 \leq x \leq \pi$, in the line $y = x$.

$y = \cos^{-1}x$

FIGURE 6

EXAMPLE 4 ■ Evaluating the Inverse Cosine Function

Find (a) $\cos^{-1}\left(\sqrt{3}/2\right)$, (b) $\cos^{-1}0$, and (c) $\cos^{-1}\frac{5}{7}$.

SOLUTION

(a) The number in the interval $[0, \pi]$ whose cosine is $\sqrt{3}/2$ is $\pi/6$. Thus, $\cos^{-1}\left(\sqrt{3}/2\right) = \pi/6$.

(b) The number in the interval $[0, \pi]$ whose cosine is 0 is $\pi/2$. Thus, $\cos^{-1}0 = \pi/2$.

(c) Since no rational multiple of π has cosine $\frac{5}{7}$, we use a calculator to find this value approximately: $\cos^{-1}\frac{5}{7} \approx 0.77519$. ■

EXAMPLE 5 ■ Composing Trigonometric Functions and Their Inverses

Write $\sin(\cos^{-1}x)$ and $\tan(\cos^{-1}x)$ as algebraic expressions in x for $-1 \leq x \leq 1$.

SOLUTION 1

Let $u = \cos^{-1}x$. We need to find $\sin u$ and $\tan u$ in terms of x. As in Example 3 the idea here is to write sine and tangent in terms of cosine. We have

$$\sin u = \pm\sqrt{1 - \cos^2u} \qquad \text{and} \qquad \tan u = \frac{\sin u}{\cos u} = \frac{\pm\sqrt{1 - \cos^2u}}{\cos u}$$

To choose the proper signs, note that u lies in the interval $[0, \pi]$ because $u = \cos^{-1}x$. Since $\sin u$ is positive on this interval, the $+$ sign is the correct choice. Substituting $u = \cos^{-1}x$ in the displayed equations and using the relation $\cos(\cos^{-1}x) = x$ gives

$$\sin(\cos^{-1}x) = \sqrt{1 - x^2} \qquad \text{and} \qquad \tan(\cos^{-1}x) = \frac{\sqrt{1 - x^2}}{x}$$

SOLUTION 2

Let $\theta = \cos^{-1}x$, so $\cos\theta = x$. In Figure 7 we draw a right triangle with an acute angle θ, adjacent side x, and hypotenuse 1. By the Pythagorean Theorem, the remaining leg is $\sqrt{1 - x^2}$. From the figure,

$$\sin(\cos^{-1}x) = \sin\theta = \sqrt{1 - x^2} \qquad \text{and} \qquad \tan(\cos^{-1}x) = \tan\theta = \frac{\sqrt{1 - x^2}}{x} \qquad ■$$

FIGURE 7

$\cos\theta = \dfrac{x}{1} = x$

NOTE In Solution 2 of Example 5 it may seem that because we are sketching a triangle, the angle $\theta = \cos^{-1}x$ must be acute. But it turns out that the triangle method works for any θ and for any x. The domains and ranges of all six inverse trigonometric functions have been chosen in such a way that we can always use a triangle to find $S(T^{-1}(x))$, where S and T are any trigonometric functions.

EXAMPLE 6 ■ **Composing a Trigonometric Function and an Inverse**

Write $\sin(2\cos^{-1}x)$ as an algebraic expression in x for $-1 \le x \le 1$.

SOLUTION

Let $\theta = \cos^{-1}x$ and sketch a triangle as shown in Figure 8. We need to find $\sin 2\theta$, but from the triangle we can find trigonometric functions only of θ, not of 2θ. The double-angle identity for sine is useful here. We have

$$\sin(2\cos^{-1}x) = \sin 2\theta$$

$$= 2\sin\theta\cos\theta \qquad \text{Double-angle formula}$$

$$= 2\sqrt{1 - x^2}\,x \qquad \text{From triangle}$$

$$= 2x\sqrt{1 - x^2} \qquad\qquad\qquad ■$$

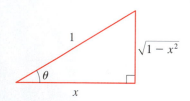

FIGURE 8

$\cos\theta = \dfrac{x}{1} = x$

■ The Inverse Tangent Function

We restrict the domain of the tangent function to the interval $(-\pi/2, \pi/2)$ in order to obtain a one-to-one function.

DEFINITION OF THE INVERSE TANGENT FUNCTION

The **inverse tangent function** is the function $\tan^{-1}$ with domain $\mathbb{R}$ and range $(-\pi/2, \pi/2)$ defined by

$$\tan^{-1}x = y \iff \tan y = x$$

The inverse tangent function is also called **arctangent** and is denoted by **arctan**.

Thus, $\tan^{-1}x$ *is the number in the interval* $(-\pi/2, \pi/2)$ *whose tangent is x.* The following relations follow from the inverse function properties.

$$\tan(\tan^{-1}x) = x \qquad \text{for } x \in \mathbb{R}$$

$$\tan^{-1}(\tan x) = x \qquad \text{for } -\frac{\pi}{2} < x < \frac{\pi}{2}$$

Figure 9 shows the graph of $y = \tan x$ on the interval $(-\pi/2, \pi/2)$ and the graph of its inverse function, $y = \tan^{-1}x$.

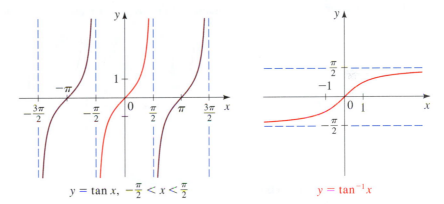

FIGURE 9 $y = \tan x, \ -\frac{\pi}{2} < x < \frac{\pi}{2}$ $y = \tan^{-1}x$

EXAMPLE 7 ■ Evaluating the Inverse Tangent Function

Find (a) $\tan^{-1}1$, (b) $\tan^{-1}\sqrt{3}$, and (c) $\tan^{-1}(-20)$.

SOLUTION

(a) The number in the interval $(-\pi/2, \pi/2)$ with tangent 1 is $\pi/4$. Thus, $\tan^{-1}1 = \pi/4$.

(b) The number in the interval $(-\pi/2, \pi/2)$ with tangent $\sqrt{3}$ is $\pi/3$. Thus, $\tan^{-1}\sqrt{3} = \pi/3$.

(c) We use a calculator to find that $\tan^{-1}(-20) \approx -1.52084$. ■

EXAMPLE 8 ■ The Angle of a Beam of Light

A lighthouse is located on an island that is 2 mi off a straight shoreline (see Figure 10). Express the angle formed by the beam of light and the shoreline in terms of the distance d in the figure.

SOLUTION

From the figure we see that $\tan \theta = 2/d$. Taking the inverse tangent of both sides, we get

$$\tan^{-1}(\tan \theta) = \tan^{-1}\left(\frac{2}{d}\right)$$

$$\theta = \tan^{-1}\left(\frac{2}{d}\right) \qquad \text{Cancellation property}$$ ■

Shoreline

Lighthouse

2 mi

FIGURE 10

The Inverse Secant, Cosecant, and Cotangent Functions

To define the inverse functions of the secant, cosecant, and cotangent functions, we restrict the domain of each function to a set on which it is one-to-one and on which it attains all its values. Although any interval satisfying these criteria is appropriate, we choose to restrict the domains in a way that simplifies the choice of sign in computations involving inverse trigonometric functions. The choices we make are also appropriate for calculus. This explains the seemingly strange restriction for the domains of the secant and cosecant functions. We end this section by displaying the graphs of the secant, cosecant, and cotangent functions with their restricted domains and the graphs of their inverse functions (Figures 11–13).

See Exercise 58 for a way of finding the values of these inverse trigonometric functions on a calculator.

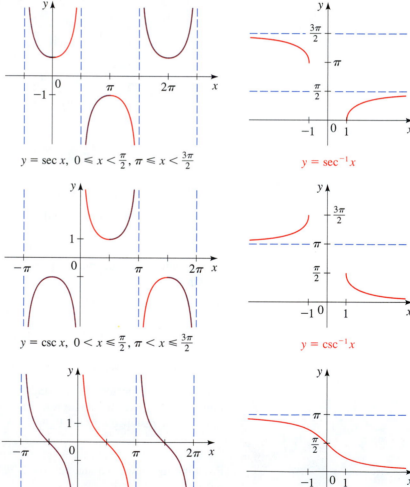

FIGURE 11

The inverse secant function

$$y = \sec x, \ 0 \leq x < \frac{\pi}{2}, \ \pi \leq x < \frac{3\pi}{2}$$

$$y = \sec^{-1}x$$

FIGURE 12

The inverse cosecant function

$$y = \csc x, \ 0 < x \leq \frac{\pi}{2}, \ \pi < x \leq \frac{3\pi}{2}$$

$$y = \csc^{-1}x$$

FIGURE 13

The inverse cotangent function

$$y = \cot x, \ 0 < x < \pi$$

$$y = \cot^{-1}x$$

7.4   EXERCISES

1–8 ■ Find the exact value of each expression, if it is defined.

1. (a) $\sin^{-1}\frac{1}{2}$ (b) $\cos^{-1}\frac{1}{2}$ (c) $\cos^{-1}2$

2. (a) $\sin^{-1}\frac{\sqrt{3}}{2}$ (b) $\cos^{-1}\frac{\sqrt{3}}{2}$ (c) $\cos^{-1}\left(-\frac{\sqrt{3}}{2}\right)$

3. (a) $\sin^{-1}\frac{\sqrt{2}}{2}$ (b) $\cos^{-1}\frac{\sqrt{2}}{2}$ (c) $\sin^{-1}\left(-\frac{\sqrt{2}}{2}\right)$

4. (a) $\tan^{-1}\sqrt{3}$ (b) $\tan^{-1}\left(-\sqrt{3}\right)$ (c) $\sin^{-1}\sqrt{3}$

5. (a) $\sin^{-1}1$ (b) $\cos^{-1}1$ (c) $\cos^{-1}(-1)$

6. (a) $\tan^{-1}1$ (b) $\tan^{-1}(-1)$ (c) $\tan^{-1}0$

7. (a) $\tan^{-1}\frac{\sqrt{3}}{3}$ (b) $\tan^{-1}\left(-\frac{\sqrt{3}}{3}\right)$ (c) $\sin^{-1}(-2)$

8. (a) $\sin^{-1}0$ (b) $\cos^{-1}0$ (c) $\cos^{-1}\left(-\frac{1}{2}\right)$

9–12 ■ Use a calculator to find an approximate value of each expression correct to five decimal places, if it is defined.

9. (a) $\sin^{-1}(0.13844)$ (b) $\cos^{-1}(-0.92761)$

10. (a) $\cos^{-1}(0.31187)$ (b) $\tan^{-1}(26.23110)$

11. (a) $\tan^{-1}(1.23456)$ (b) $\sin^{-1}(1.23456)$

12. (a) $\cos^{-1}(-0.25713)$ (b) $\tan^{-1}(-0.25713)$

13–28 ■ Find the exact value of the expression, if it is defined.

13. $\sin\left(\sin^{-1}\frac{1}{4}\right)$ **14.** $\cos\left(\cos^{-1}\frac{2}{3}\right)$

15. $\tan(\tan^{-1}5)$ **16.** $\sin(\sin^{-1}5)$

17. $\cos^{-1}\left(\cos\frac{\pi}{3}\right)$ **18.** $\tan^{-1}\left(\tan\frac{\pi}{6}\right)$

19. $\sin^{-1}\left[\sin\left(-\frac{\pi}{6}\right)\right]$ **20.** $\sin^{-1}\left(\sin\frac{5\pi}{6}\right)$

21. $\tan^{-1}\left(\tan\frac{2\pi}{3}\right)$ **22.** $\cos^{-1}\left[\cos\left(-\frac{\pi}{4}\right)\right]$

23. $\tan\left(\sin^{-1}\frac{1}{2}\right)$ **24.** $\sin(\sin^{-1}0)$

25. $\cos\left(\sin^{-1}\frac{\sqrt{3}}{2}\right)$ **26.** $\tan\left(\sin^{-1}\frac{\sqrt{2}}{2}\right)$

27. $\tan^{-1}\left(2\sin\frac{\pi}{3}\right)$ **28.** $\cos^{-1}\left(\sqrt{3}\sin\frac{\pi}{6}\right)$

29–40 ■ Evaluate the expression by sketching a triangle, as in Solution 2 of Example 3.

29. $\sin\left(\cos^{-1}\frac{3}{5}\right)$ **30.** $\tan\left(\sin^{-1}\frac{4}{5}\right)$

31. $\sin\left(\tan^{-1}\frac{12}{5}\right)$ **32.** $\cos(\tan^{-1}5)$

33. $\sec\left(\sin^{-1}\frac{12}{13}\right)$ **34.** $\csc\left(\cos^{-1}\frac{7}{25}\right)$

35. $\cos(\tan^{-1}2)$ **36.** $\cot\left(\sin^{-1}\frac{2}{3}\right)$

37. $\sin\left(2\cos^{-1}\frac{3}{5}\right)$ **38.** $\tan\left(2\tan^{-1}\frac{5}{13}\right)$

39. $\sin\left(\sin^{-1}\frac{1}{2}+\cos^{-1}\frac{1}{2}\right)$ **40.** $\cos\left(\sin^{-1}\frac{3}{5}-\cos^{-1}\frac{3}{5}\right)$

41–48 ■ Rewrite the expression as an algebraic expression in x.

41. $\cos(\sin^{-1}x)$ **42.** $\sin(\tan^{-1}x)$

43. $\tan(\sin^{-1}x)$ **44.** $\cos(\tan^{-1}x)$

45. $\cos(2\tan^{-1}x)$ **46.** $\sin(2\sin^{-1}x)$

47. $\cos(\cos^{-1}x+\sin^{-1}x)$ **48.** $\sin(\tan^{-1}x-\sin^{-1}x)$

49–52 ■ In these exercises you are asked to use the inverse trigonometric functions to model real-life situations. See the guidelines on modeling with functions in Section 2.7.

49. An observer views the space shuttle from a distance of 2 miles from the launch pad.
   (a) Express the height of the space shuttle as a function of the angle of elevation θ.
   (b) Express the angle of elevation θ as a function of the height h of the space shuttle.

50. A 50-ft pole casts a shadow as shown in the figure.
   (a) Express the angle of elevation θ of the sun as a function of the length s of the shadow.

(b) Find the angle θ of elevation of the sun when the shadow is 20 ft long.

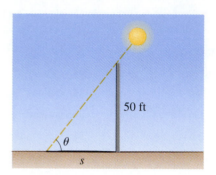

51. A 680-ft rope anchors a hot-air balloon as shown in the figure.

(a) Express the angle θ as a function of the height h of the balloon.

(b) Find the angle θ if the balloon is 500 ft high.

52. The figures indicate that the higher the orbit of a satellite, the more of the earth the satellite can "see." Let θ, s, and h be as in the figure, and assume the earth is a sphere of radius 3960 mi.

(a) Express the angle θ as a function of h.

(b) Express the distance s as a function of θ.

(c) Express the distance s as a function of h. [Find the composition of the functions in parts (a) and (b).]

(d) If the satellite is 100 mi above the earth, what is the distance s that it can see?

(e) How high does the satellite have to be in order to see both Los Angeles and New York, 2450 mi apart?

 53–54 ■ (a) Graph the function and make a conjecture, and (b) prove that your conjecture is true.

53. $y = \sin^{-1}x + \cos^{-1}x$ **54.** $y = \tan^{-1}x + \tan^{-1}\dfrac{1}{x}$

 55–56 ■ (a) Use a graphing device to find all solutions of the equation, correct to two decimal places, and (b) find the exact solution.

55. $\tan^{-1}x + \tan^{-1}2x = \dfrac{\pi}{4}$ **56.** $\sin^{-1}x - \cos^{-1}x = 0$

● DISCOVERY · DISCUSSION

57. Two Different Compositions The functions

$$f(x) = \sin(\sin^{-1}x) \qquad \text{and} \qquad g(x) = \sin^{-1}(\sin x)$$

both simplify to just x for suitable values of x. But these functions are not the same for all x. Graph both f and g to show how the functions differ. (Think carefully about the domain and range of $\sin^{-1}$.)

58. Inverse Trigonometric Functions on a Calculator Most calculators do not have keys for $\sec^{-1}$, $\csc^{-1}$, or $\cot^{-1}$. Prove the following identities, then use these identities and a calculator to find $\sec^{-1}2$, $\csc^{-1}3$, and $\cot^{-1}4$.

$$\sec^{-1}x = \cos^{-1}\left(\frac{1}{x}\right), \quad x \geq 1$$

$$\csc^{-1}x = \sin^{-1}\left(\frac{1}{x}\right), \quad x \geq 1$$

$$\cot^{-1}x = \tan^{-1}\left(\frac{1}{x}\right), \quad x > 0$$

Discovery Project

Where to Sit at the Movies

Everyone knows that the apparent size of an object depends on its distance from the viewer. The farther away an object, the smaller its apparent size. The apparent size is determined by the angle the object subtends at the eye of the viewer.

   If you are looking at a painting hanging on a wall, how far away should you stand to get the maximum view? If the painting is hung above eye level, then the following figures show that the angle subtended at the eye is small if you are too close or too far away. The same situation occurs when choosing where to sit in a movie theatre.

Small θ Large θ Small θ

1. The screen in a theatre is 22 ft high and is positioned 10 ft above the floor, which is flat. The first row of seats is 7 ft from the screen and the rows are 3 ft apart. You decide to sit in the row where you get the maximum view, that is, where the angle θ subtended by the screen at your eyes is a maximum. Suppose your eyes are 4 ft above the floor, as in the figure, and you sit at a distance x from the screen.

   (a) Show that $\theta = \tan^{-1}\left(\dfrac{28}{x}\right) - \tan^{-1}\left(\dfrac{6}{x}\right)$.

   (b) Use the subtraction formula for tangent to show that

   $$\theta = \tan^{-1}\left(\frac{22x}{x^2 + 168}\right)$$

   (c) Use a graphing device to graph θ as a function of x. What value of x maximizes θ? In which row should you sit? What is the viewing angle in this row?

2. Now suppose that, starting with the first row of seats, the floor of the seating area is inclined at an angle of $\alpha = 25°$ above the horizontal, and the distance that you sit up the incline is x, as shown in the figure.

   (a) Use the Law of Cosines to show that

   $$\theta = \cos^{-1}\left(\frac{a^2 + b^2 - 484}{2ab}\right)$$

where

$$a^2 = (7 + x \cos \alpha)^2 + (28 - x \sin \alpha)^2$$

and

$$b^2 = (7 + x \cos \alpha)^2 + (x \sin \alpha - 6)^2$$

(b) Use a graphing device to graph θ as a function of x, and estimate the value of x that maximizes θ. In which row should you sit? What is the viewing angle θ in this row?

7.5 TRIGONOMETRIC EQUATIONS

An equation that contains trigonometric functions is called a **trigonometric equation**. For example, the following are trigonometric equations:

$$\sin^2 x + \cos^2 x = 1 \qquad 2 \sin x - 1 = 0 \qquad \tan^2 2x - 1 = 0$$

The first equation is an *identity*—that is, it is true for every value of the variable x. The other two equations are true only for certain values of x. To solve a trigonometric equation, we find all the values of the variable that make the equation true.

■ Solving Trigonometric Equations

To solve a trigonometric equation, we use the rules of algebra to isolate the trigonometric function on one side of the equal sign. Then we use our knowledge of the values of the trigonometric functions to solve for the variable.

EXAMPLE 1 ■ Solving a Trigonometric Equation

Solve the equation $2 \sin x - 1 = 0$.

SOLUTION

We start by isolating $\sin x$.

$$2 \sin x - 1 = 0 \qquad \text{Given equation}$$

$$2 \sin x = 1 \qquad \text{Add 1}$$

$$\sin x = \frac{1}{2} \qquad \text{Divide by 2}$$

Because sine has period 2π, we first find the solutions in the interval $[0, 2\pi)$. These are $x = \pi/6$ and $x = 5\pi/6$. To get all other solutions, we add any integer multiple of 2π to these solutions. Thus, the solutions are

$$x = \frac{\pi}{6} + 2k\pi, \qquad x = \frac{5\pi}{6} + 2k\pi$$

where k is any integer. Figure 1 gives a graphical representation of the solutions.

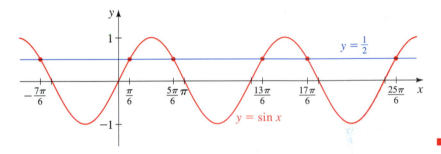

FIGURE 1

EXAMPLE 2 ■ Solving a Trigonometric Equation

Solve the equation $\tan^2 x - 3 = 0$.

SOLUTION

We start by isolating $\tan x$.

$$\tan^2 x - 3 = 0 \qquad \text{Given equation}$$

$$\tan^2 x = 3 \qquad \text{Add 3}$$

$$\tan x = \pm\sqrt{3} \qquad \text{Take square roots}$$

Because tangent has period π, we first find the solutions in the interval $(-\pi/2, \pi/2)$. These are $x = -\pi/3$ and $x = \pi/3$. To get all other solutions, we add any integer multiple of π to these solutions. Thus, the solutions are

$$x = -\frac{\pi}{3} + k\pi, \qquad x = \frac{\pi}{3} + k\pi$$

where k is any integer.

EXAMPLE 3 ■ Finding Intersection Points

Find the values of x for which the graphs of $f(x) = \sin x$ and $g(x) = \cos x$ intersect.

SOLUTION 1: GRAPHICAL

The graphs intersect where $f(x) = g(x)$. In Figure 2 we graph $y_1 = \sin x$ and $y_2 = \cos x$ on the same screen, for x between 0 and 2π. Using TRACE or the

Intersect command on the graphing calculator, we see that the two points of intersection in this interval occur where $x \approx 0.785$ and $x \approx 3.927$. Since sin and cos are periodic with period 2π, the intersection points occur where

$$x \approx 0.785 + 2k\pi \qquad \text{and} \qquad x \approx 3.927 + 2k\pi$$

where k is any integer.

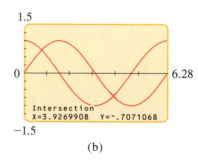

FIGURE 2   (a)   (b)

SOLUTION 2: ALGEBRAIC

To find the exact solution, we set $f(x) = g(x)$ and solve the resulting equation algebraically.

$$\sin x = \cos x \qquad \text{Equate functions}$$

Since the numbers x for which $\cos x = 0$ are not solutions of the equation, we can divide both sides by $\cos x$.

$$\frac{\sin x}{\cos x} = 1 \qquad \text{Divide by } \cos x$$

$$\tan x = 1 \qquad \text{Reciprocal identity}$$

Because tangent has period π, we first find the solutions in the interval $(-\pi/2, \pi/2)$. The only solution in this interval is $x = \pi/4$. To get all solutions, we add any integer multiple of π to this solution. Thus, the solutions are

$$x = \frac{\pi}{4} + k\pi$$

where k is any integer. The graphs intersect for these values of x. You should use your calculator to check that, correct to three decimals, these are the same values as we obtained in Solution 1. ∎

Solving Trigonometric Equations by Factoring

Factoring is one of the most useful techniques for solving equations, including trigonometric equations. The idea is to move all terms to one side of the equation, factor, then use the Zero-Product Property (Section 1.5).

Zero-Product Property

If $AB = 0$, then $A = 0$ or $B = 0$.

EXAMPLE 4 ■ An Equation of Quadratic Type

Solve the equation $2\cos^2 x - 7\cos x + 3 = 0$.

SOLUTION

We factor the left-hand side of the equation.

Equation of Quadratic Type

$2C^2 - 7C + 3 = 0$

$(2C - 1)(C - 3) = 0$

$$2\cos^2 x - 7\cos x + 3 = 0 \qquad \text{Given equation}$$

$$(2\cos x - 1)(\cos x - 3) = 0 \qquad \text{Factor}$$

$$2\cos x - 1 = 0 \quad \text{or} \quad \cos x - 3 = 0 \qquad \text{Set each factor equal to 0}$$

$$\cos x = \tfrac{1}{2} \quad \text{or} \quad \cos x = 3 \qquad \text{Solve for } \cos x$$

Because cosine has period 2π, we first find the solutions in the interval $[0, 2\pi)$. For the first equation these are $x = \pi/3$ and $x = 5\pi/3$. The second equation has no solutions because $\cos x$ is never greater than 1. Thus, the solutions are

$$x = \frac{\pi}{3} + 2k\pi, \qquad x = \frac{5\pi}{3} + 2k\pi$$

where k is any integer.   ■

EXAMPLE 5 ■ Using a Trigonometric Identity

Solve the equation $1 + \sin x = 2\cos^2 x$.

SOLUTION

We use a trigonometric identity to rewrite the equation in terms of a single trigonometric function.

$$1 + \sin x = 2\cos^2 x \qquad \text{Given equation}$$

$$1 + \sin x = 2(1 - \sin^2 x) \qquad \text{Pythagorean identity}$$

Equation of Quadratic Type

$2S^2 + S - 1 = 0$

$(2S - 1)(S + 1) = 0$

$$2\sin^2 x + \sin x - 1 = 0 \qquad \begin{array}{l}\text{Put all terms on one}\\\text{side of the equation}\end{array}$$

$$(2\sin x - 1)(\sin x + 1) = 0 \qquad \text{Factor}$$

$$2\sin x - 1 = 0 \quad \text{or} \quad \sin x + 1 = 0 \qquad \begin{array}{l}\text{Set each factor}\\\text{equal to 0}\end{array}$$

$$\sin x = \frac{1}{2} \quad \text{or} \quad \sin x = -1 \qquad \text{Solve for } \sin x$$

$$x = \frac{\pi}{6}, \frac{5\pi}{6} \quad \text{or} \quad x = \frac{3\pi}{2} \qquad \begin{array}{l}\text{Solve for } x \text{ in the}\\\text{interval } [0, 2\pi)\end{array}$$

Because sine has period 2π, we get all the solutions of the equation by adding any integer multiple of 2π to these solutions. Thus, the solutions are

$$x = \frac{\pi}{6} + 2k\pi, \qquad x = \frac{5\pi}{6} + 2k\pi, \qquad x = \frac{3\pi}{2} + 2k\pi$$

where k is any integer.   ∎

EXAMPLE 6 ■ Using a Trigonometric Identity

Solve the equation $\sin 2x - \cos x = 0$.

SOLUTION

The first term is a function of $2x$ and the second is a function of x, so we begin by using a trigonometric identity to rewrite the first term as a function of x only.

$\sin 2x - \cos x = 0$		Given equation
$2 \sin x \cos x - \cos x = 0$		Double-angle formula
$\cos x\,(2 \sin x - 1) = 0$		Factor
$\cos x = 0 \qquad$ or $\qquad 2 \sin x - 1 = 0$		Set each factor equal to 0
$\sin x = \dfrac{1}{2}$		Solve for $\sin x$
$x = \dfrac{\pi}{2}, \dfrac{3\pi}{2} \qquad$ or $\qquad x = \dfrac{\pi}{6}, \dfrac{5\pi}{6}$		Solve for x in the interval $[0, 2\pi)$

Both sine and cosine have period 2π, so we get all the solutions of the equation by adding any integer multiple of 2π to these solutions. Thus, the solutions are

$$x = \frac{\pi}{2} + 2k\pi, \qquad x = \frac{3\pi}{2} + 2k\pi, \qquad x = \frac{\pi}{6} + 2k\pi, \qquad x = \frac{5\pi}{6} + 2k\pi$$

where k is any integer.   ∎

EXAMPLE 7 ■ Squaring and Using an Identify

Solve the equation $\cos x + 1 = \sin x$ in the interval $[0, 2\pi)$.

SOLUTION

To get an equation that involves either sine only or cosine only, we square both sides and use a Pythagorean identity.

$$\cos x + 1 = \sin x \qquad \text{Given equation}$$

$$\cos^2 x + 2 \cos x + 1 = \sin^2 x \qquad \text{Square both sides}$$

$$\cos^2 x + 2 \cos x + 1 = 1 - \cos^2 x \qquad \text{Pythagorean identity}$$

$$2 \cos^2 x + 2 \cos x = 0 \qquad \text{Simplify}$$

$$2 \cos x \,(\cos x + 1) = 0 \qquad \text{Factor}$$

$2 \cos x = 0$	or	$\cos x + 1 = 0$	Set each factor equal to 0
$\cos x = 0$	or	$\cos x = -1$	Solve for $\cos x$
$x = \dfrac{\pi}{2}, \dfrac{3\pi}{2}$	or	$x = \pi$	Solve for x in the interval $[0, 2\pi)$

Because we squared both sides, we need to check for extraneous solutions. From *Check Your Answers*, we see that the solutions of the given equation are $\pi/2$ and π.

CHECK YOUR ANSWERS

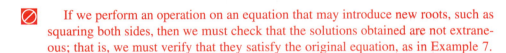

$$x = \frac{\pi}{2}: \qquad\qquad x = \frac{3\pi}{2}: \qquad\qquad x = \pi:$$

$$\cos \frac{\pi}{2} + 1 \stackrel{?}{=} \sin \frac{\pi}{2} \qquad \cos \frac{3\pi}{2} + 1 \stackrel{?}{=} \sin \frac{3\pi}{2} \qquad \cos \pi + 1 \stackrel{?}{=} \sin \pi$$

$$0 + 1 = 1 \;\checkmark \qquad\qquad 0 + 1 \stackrel{?}{=} -1 \;\times \qquad\qquad -1 + 1 = 0 \;\checkmark \qquad\blacksquare$$

🚫 If we perform an operation on an equation that may introduce new roots, such as squaring both sides, then we must check that the solutions obtained are not extraneous; that is, we must verify that they satisfy the original equation, as in Example 7.

Equations with Trigonometric Functions of Multiple Angles

When solving trigonometric equations that involve functions of multiples of angles, we first solve for the multiple of the angle, then divide to solve for the angle.

EXAMPLE 8 ■ Trigonometric Functions of Multiple Angles

Consider the equation $2 \sin 3x - 1 = 0$.

(a) Find all solutions of the equation.
(b) Find the solutions in the interval $[0, 2\pi)$.

SOLUTION

(a) We start by isolating sin $3x$, and then solve for the multiple angle $3x$.

$$2 \sin 3x - 1 = 0 \qquad \text{Given equation}$$

$$2 \sin 3x = 1 \qquad \text{Add 1}$$

$$\sin 3x = \frac{1}{2} \qquad \text{Divide by 2}$$

$$3x = \frac{\pi}{6}, \frac{5\pi}{6} \qquad \text{Solve for } 3x \text{ in the interval } [0, 2\pi)$$

To get all solutions, we add any integer multiple of 2π to these solutions. Thus, the solutions are of the form

$$3x = \frac{\pi}{6} + 2k\pi, \qquad 3x = \frac{5\pi}{6} + 2k\pi$$

To solve for x, we divide by 3 to get the solutions

$$x = \frac{\pi}{18} + \frac{2k\pi}{3}, \qquad x = \frac{5\pi}{18} + \frac{2k\pi}{3}$$

where k is any integer.

(b) The solutions from part (a) that are in the interval $[0, 2\pi)$ correspond to $k = 0$, 1, and 2. For all other values of k, the corresponding values of x lie outside this interval. Thus, the solutions in the interval $[0, 2\pi)$ are

$$x = \frac{\pi}{18}, \frac{5\pi}{18}, \frac{13\pi}{18}, \frac{17\pi}{18}, \frac{25\pi}{18}, \frac{29\pi}{18} \qquad \blacksquare$$

EXAMPLE 9 ■ **Trigonometric Functions of Multiple Angles**

Consider the equation $\sqrt{3} \tan \dfrac{x}{2} - 1 = 0$.

(a) Find all solutions of the equation.

(b) Find the solutions in the interval $[0, 4\pi)$.

SOLUTION

(a) We start by isolating $\tan(x/2)$.

$$\sqrt{3} \tan \frac{x}{2} - 1 = 0 \qquad \text{Given equation}$$

$$\sqrt{3} \tan \frac{x}{2} = 1 \qquad \text{Add 1}$$

$$\tan \frac{x}{2} = \frac{1}{\sqrt{3}} \qquad \text{Divide by } \sqrt{3}$$

$$\frac{x}{2} = \frac{\pi}{3} \qquad \text{Solve for } \frac{x}{2} \text{ in the interval } \left(-\frac{\pi}{2}, \frac{\pi}{2}\right)$$

Since tan has period π, to get all solutions we add any integer multiple of π to these solutions. Thus, the solutions are of the form

$$\frac{x}{2} = \frac{\pi}{3} + k\pi$$

Multiplying by 2, we get the solutions

$$x = \frac{2\pi}{3} + 2k\pi$$

where k is any integer.

(b) The solutions from part (a) that are in the interval $[0, 4\pi)$ correspond to $k = 0$ and $k = 1$. For all other values of k, the corresponding values of x lie outside this interval. Thus, the solutions in the interval $[0, 4\pi)$ are

$$x = \frac{2\pi}{3}, \frac{8\pi}{3} \qquad \blacksquare$$

Using Inverse Trigonometric Functions to Solve Trigonometric Equations

So far, all the equations we've solved have had solutions like $\pi/4$, $\pi/3$, $5\pi/6$, and so on. We were able to find these solutions from the special values of the trigonometric functions that we've memorized. We now consider equations whose solution requires us to use the inverse trigonometric functions.

EXAMPLE 10 ■ **Using Inverse Trigonometric Functions**

Solve the equation $\tan^2 x - \tan x - 2 = 0$.

SOLUTION

We start by factoring the left-hand side.

Equation of Quadratic Type

$T^2 - T - 2 = 0$

$(T - 2)(T + 1) = 0$

$\tan^2 x - \tan x - 2 = 0$	Given equation
$(\tan x - 2)(\tan x + 1) = 0$	Factor
$\tan x - 2 = 0 \quad$ or $\quad \tan x + 1 = 0$	Set each factor equal to 0
$\tan x = 2 \qquad$ or $\qquad \tan x = -1$	Solve for $\tan x$
$x = \tan^{-1} 2$ or $\qquad x = -\dfrac{\pi}{4}$	Solve for x in the interval $(-\pi/2, \pi/2)$

Because tangent has period π, we get all solutions by adding integer multiples of π to these solutions. Thus, all the solutions are

$$x = \tan^{-1}2 + k\pi, \qquad x = -\frac{\pi}{4} + k\pi$$

where k is any integer.   ∎

If we are using inverse trigonometric functions to solve an equation, we must keep in mind that $\sin^{-1}$ and $\tan^{-1}$ give values in quadrants I and IV, and $\cos^{-1}$ gives values in quadrants I and II. To find other solutions, we must look at the quadrant where the trigonometric function in the equation can take on the value we need.

EXAMPLE 11 ■ Using Inverse Trigonometric Functions

(a) Solve the equation $3 \sin \theta - 2 = 0$.

(b) Use a calculator to approximate the solutions in the interval $[0, 2\pi)$, correct to 5 decimals.

SOLUTION

(a) We start by isolating $\sin \theta$.

$$3 \sin \theta - 2 = 0 \qquad \text{\color{red}Given equation}$$

$$3 \sin \theta = 2 \qquad \text{\color{red}Add 2}$$

$$\sin \theta = \frac{2}{3} \qquad \text{\color{red}Divide by 3}$$

From Figure 3 we see that $\sin \theta$ equals $\frac{2}{3}$ in quadrants I and II. The solution in quadrant I is $\theta = \sin^{-1}\frac{2}{3}$. The solution in quadrant II is $\theta = \pi - \sin^{-1}\frac{2}{3}$. Since these are the solutions in the interval $[0, 2\pi)$, we get all other solutions by adding integer multiples of 2π to these. Thus, all the solutions of the equation are

$$\theta = \left(\sin^{-1}\tfrac{2}{3}\right) + 2k\pi, \qquad \theta = \left(\pi - \sin^{-1}\tfrac{2}{3}\right) + 2k\pi$$

where k is any integer.

(b) Using a calculator set in radian mode, we see that $\sin^{-1}\frac{2}{3} \approx 0.72973$ and $\pi - \sin^{-1}\frac{2}{3} \approx 2.41186$, so the solutions in the interval $[0, 2\pi)$ are

$$\theta \approx 0.72973, \qquad \theta \approx 2.41186$$ ∎

FIGURE 3

7.5 EXERCISES

1–38 ■ Find all solutions of the equation.

1. $\sin x - 1 = 0$ **2.** $\cos x + 1 = 0$

3. $2 \cos x - 1 = 0$ **4.** $\sqrt{2} \sin x - 1 = 0$

5. $2 \sin x + \sqrt{3} = 0$ **6.** $\tan x + 1 = 0$

7. $4 \cos^2 x - 1 = 0$ **8.** $2 \cos^2 x - 1 = 0$

9. $\sec^2 x - 2 = 0$

10. $\csc^2 x - 4 = 0$

11. $\cos x \,(2 \sin x + 1) = 0$

12. $\sec x \,(2 \cos x - \sqrt{2}) = 0$

13. $(\tan x + \sqrt{3})(\cos x + 2) = 0$

14. $(2 \cos x + \sqrt{3})(2 \sin x - 1) = 0$

15. $\cos x \sin x - 2 \cos x = 0$

16. $\tan x \sin x + \sin x = 0$

17. $4 \cos^2 x - 4 \cos x + 1 = 0$

18. $2 \sin^2 x - \sin x - 1 = 0$

19. $\sin^2 x = 2 \sin x + 3$ **20.** $3 \tan^3 x = \tan x$

21. $\sin^2 x = 4 - 2 \cos^2 x$ **22.** $2 \cos^2 x + \sin x = 1$

23. $2 \sin 3x + 1 = 0$ **24.** $2 \cos 2x + 1 = 0$

25. $\sec 4x - 2 = 0$ **26.** $\sqrt{3} \tan 3x + 1 = 0$

27. $\sqrt{3} \sin 2x = \cos 2x$ **28.** $\cos 3x = \sin 3x$

29. $\cos \dfrac{x}{2} - 1 = 0$ **30.** $2 \sin \dfrac{x}{3} + \sqrt{3} = 0$

31. $\tan \dfrac{x}{4} + \sqrt{3} = 0$

32. $\sec \dfrac{x}{2} = \cos \dfrac{x}{2}$

33. $\tan^5 x - 9 \tan x = 0$

34. $3 \tan^3 x - 3 \tan^2 x - \tan x + 1 = 0$

35. $4 \sin x \cos x + 2 \sin x - 2 \cos x - 1 = 0$

36. $\sin 2x = 2 \tan 2x$

37. $\cos^2 2x - \sin^2 2x = 0$

38. $\sec x - \tan x = \cos x$

39–46 ■ Find all solutions of the equation in the interval $[0, 2\pi)$.

39. $2 \cos 3x = 1$ **40.** $3 \csc^2 x = 4$

41. $2 \sin x \tan x - \tan x = 1 - 2 \sin x$

42. $\sec x \tan x - \cos x \cot x = \sin x$

43. $\tan x - 3 \cot x = 0$ **44.** $2 \sin^2 x - \cos x = 1$

45. $\tan 3x + 1 = \sec 3x$ **46.** $3 \sec^2 x + 4 \cos^2 x = 7$

47–54 ■ (a) Find all solutions of the equation. (b) Use a calculator to solve the equation in the interval $[0, 2\pi)$, correct to five decimal places.

47. $\cos x = 0.4$ **48.** $2 \tan x = 13$

49. $\sec x - 5 = 0$ **50.** $3 \sin x = 7 \cos x$

51. $5 \sin^2 x - 1 = 0$

52. $2 \sin 2x - \cos x = 0$

53. $3 \sin^2 x - 7 \sin x + 2 = 0$

54. $\tan^4 x - 13 \tan^2 x + 36 = 0$

55–58 ■ Graph f and g on the same axes, and find their points of intersection.

55. $f(x) = 3 \cos x + 1, \quad g(x) = \cos x - 1$

56. $f(x) = \sin 2x, \quad g(x) = 2 \sin 2x + 1$

57. $f(x) = \tan x, \quad g(x) = \sqrt{3}$

58. $f(x) = \sin x - 1, \quad g(x) = \cos x$

59. If a projectile is fired with velocity v_0 at an angle θ, then its *range*, the horizontal distance it travels (in feet), is modeled by the function

$$R(\theta) = \frac{v_0^2 \sin 2\theta}{32}$$

(See page 804.) If $v_0 = 2200$ ft/s, what angle should be chosen in order for the projectile to hit a target on the ground 5000 ft away?

60. In Philadelphia the number of hours of daylight on day t (where t is the number of days after January 1) is modeled by the function

$$L(t) = 12 + 2.83 \sin\left(\frac{2\pi}{365}(t - 80)\right)$$

(a) Which days of the year have about 10 hours of daylight?

(b) How many days of the year have more than 10 hours of daylight?

61. It has been observed since ancient times that light refracts or "bends" as it travels from one medium to another (from air to water, for example). If v_1 is the speed of light in one medium and v_2 its speed in another medium, then according to **Snell's Law,**

$$\frac{\sin\theta_1}{\sin\theta_2} = \frac{v_1}{v_2}$$

where θ_1 is the *angle of incidence* and θ_2 is the *angle of refraction* (see the figure). The number v_1/v_2 is called the *index of refraction*. The index of refraction from air to water is 1.33. If a ray of light passes through the surface of a lake at an angle of incidence of 70°, find the angle of refraction.

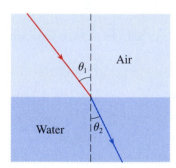

62. The displacement of a spring vibrating in damped harmonic motion is given by

$$y = 4e^{3t} \sin 2\pi t$$

Find the times when the spring is at its equilibrium position $(y = 0)$.

63–66 ■ Use an addition or subtraction formula to simplify the equation. Then find all solutions in the interval $[0, 2\pi)$.

63. $\cos x \cos 3x - \sin x \sin 3x = 0$

64. $\cos x \cos 2x + \sin x \sin 2x = \frac{1}{2}$

65. $\sin 2x \cos x + \cos 2x \sin x = \sqrt{3}/2$

66. $\sin 3x \cos x - \cos 3x \sin x = 0$

67–70 ■ Use a double- or half-angle formula to solve the equation in the interval $[0, 2\pi)$.

67. $\sin 2x + \cos x = 0$

68. $\tan \dfrac{x}{2} - \sin x = 0$

69. $\cos 2x + \cos x = 2$

70. $\tan x + \cot x = 4 \sin 2x$

71–74 ■ Solve the equation by first using a sum-to-product formula.

71. $\sin x + \sin 3x = 0$

72. $\cos 5x - \cos 7x = 0$

73. $\cos 4x + \cos 2x = \cos x$

74. $\sin 5x - \sin 3x = \cos 4x$

75–80 ■ Use a graphing device to find the solutions of the equation, correct to two decimal places.

75. $\sin 2x = x$

76. $\cos x = \dfrac{x}{3}$

77. $2^{\sin x} = x$

78. $\sin x = x^3$

79. $\dfrac{\cos x}{1 + x^2} = x^2$

80. $\cos x = \frac{1}{2}(e^x + e^{-x})$

● **DISCOVERY · DISCUSSION**

81. Equations and Identities Which of the following statements is true?

A. Every identity is an equation.

B. Every equation is an identity.

Give examples to illustrate your answer. Write a short paragraph to explain the difference between an equation and an identity.

82. A Special Trigonometric Equation What makes the equation $\sin(\cos x) = 0$ different from all the other equations we've looked at in this section? Find all solutions of this equation.

7.6   TRIGONOMETRIC FORM OF COMPLEX NUMBERS; DeMOIVRE'S THEOREM

Complex numbers were introduced in Chapter 3 in order to solve certain algebraic equations. The applications of complex numbers go far beyond this initial use, however. Complex numbers are now used routinely in physics, electrical engineering, aerospace engineering, and many other fields. In this section we represent complex numbers in trigonometric form, using the functions sine and cosine. This will enable us to find the nth roots of complex numbers. To describe the trigonometric form of complex numbers, we must first learn to work with complex numbers graphically.

Graphing Complex Numbers

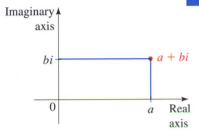

FIGURE 1

To graph real numbers or sets of real numbers, we have been using the number line, which has just one dimension. Complex numbers, however have two components: the real part and the imaginary part. This suggests that we need two axes to graph complex numbers: one for the real part and one for the imaginary part. We call these the **real axis** and the **imaginary axis**, respectively. The plane determined by these two axes is called the **complex plane**. To graph the complex number $a + bi$, we plot the ordered pair of numbers (a, b) in this plane, as indicated in Figure 1.

EXAMPLE 1 ■ Graphing Complex Numbers

Graph the complex numbers $z_1 = 2 + 3i$, $z_2 = 3 - 2i$, and $z_1 + z_2$.

SOLUTION

We have $z_1 + z_2 = (2 + 3i) + (3 - 2i) = 5 + i$. The graph is shown in Figure 2.

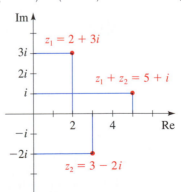

FIGURE 2

EXAMPLE 2 ■ Graphing Sets of Complex Numbers

Graph each of the following sets of complex numbers.

(a) $S = \{a + bi \mid a \geq 0\}$ (b) $T = \{a + bi \mid a < 1, b \geq 0\}$

SOLUTION

(a) S is the set of complex numbers whose real part is nonnegative. The graph is shown in Figure 3(a).

(b) T is the set of complex numbers for which the real part is less than 1 and the imaginary part is nonnegative. The graph is shown in Figure 3(b).

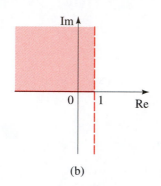

FIGURE 3   (a)   (b)   ■

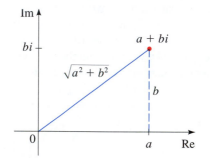

FIGURE 4

The plural of modulus is moduli.

Recall from Section 1.1 that the absolute value of a real number can be thought of as its distance from the origin on the real number line. We define absolute value for complex numbers in a similar fashion. From Figure 4 we can see, using the Pythagorean Theorem, that the distance between $a + bi$ and the origin in the complex plane is $\sqrt{a^2 + b^2}$. This leads to the following definition.

> The **modulus** (or **absolute value**) of the complex number $z = a + bi$ is
> $$|z| = \sqrt{a^2 + b^2}$$

EXAMPLE 3 ■ Calculating the Modulus

Find the moduli of the complex numbers $3 + 4i$ and $8 - 5i$.

SOLUTION

$$|3 + 4i| = \sqrt{3^2 + 4^2} = \sqrt{25} = 5$$
$$|8 - 5i| = \sqrt{8^2 + (-5)^2} = \sqrt{89}$$

■

EXAMPLE 4 ■ Absolute Value of Complex Numbers

Graph each of the following sets of complex numbers.

(a) $C = \{z \mid |z| = 1\}$ (b) $D = \{z \mid |z| \leq 1\}$

SOLUTION

(a) C is the set of complex numbers whose distance from the origin is 1. Thus, C is a circle of radius 1 with center at the origin.

(b) D is the set of complex numbers whose distance from the origin is less than or equal to 1. Thus, D is the disk that consists of all complex numbers on and inside the circle C of part (a).

The graphs of C and D are shown in Figure 5.

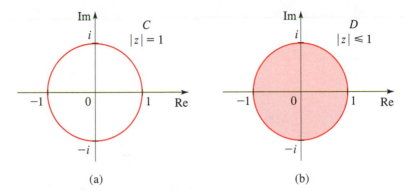

FIGURE 5   (a)   (b)

Trigonometric Form of Complex Numbers

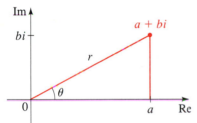

FIGURE 6

Let $z = a + bi$ be a complex number, and in the complex plane let's draw the line segment joining the origin to the point $a + bi$ (see Figure 6). The length of this line segment is denoted by $r = |z| = \sqrt{a^2 + b^2}$. If θ is an angle in standard position whose terminal side coincides with this line segment, then by the definitions of sine and cosine (see Section 4.2)

$$a = r \cos \theta \quad \text{and} \quad b = r \sin \theta$$

so $z = r \cos \theta + ir \sin \theta = r(\cos \theta + i \sin \theta)$. We have shown the following.

TRIGONOMETRIC FORM OF COMPLEX NUMBERS

A complex number $z = a + bi$ has the **trigonometric form**

$$z = r(\cos \theta + i \sin \theta)$$

where $r = |z| = \sqrt{a^2 + b^2}$ and $\tan \theta = b/a$. The number r is the **modulus** of z, and θ is an **argument** of z.

The argument of z is not unique, but any two arguments of z differ by a multiple of 2π.

EXAMPLE 5 ■ **Writing Complex Numbers in Trigonometric Form**

Write each of the following complex numbers in trigonometric form.

(a) $1 + i$ (b) $-1 + \sqrt{3}\, i$ (c) $-4\sqrt{3} - 4i$ (d) $3 + 4i$

SOLUTION

These complex numbers are graphed in Figure 7, which helps us find their arguments.

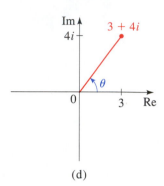

(a)   (b)   (c)   (d)

FIGURE 7

$\tan \theta = \frac{1}{1} = 1$

$\theta = \frac{\pi}{4}$

$\tan \theta = \frac{\sqrt{3}}{-1} = -\sqrt{3}$

$\theta = \frac{2\pi}{3}$

$\tan \theta = \frac{-4}{-4\sqrt{3}} = \frac{1}{\sqrt{3}}$

$\theta = \frac{7\pi}{6}$

$\tan \theta = \frac{4}{3}$

$\theta = \tan^{-1} \frac{4}{3}$

(a) An argument is $\theta = \pi/4$ and $r = \sqrt{1+1} = \sqrt{2}$. Thus

$$1 + i = \sqrt{2}\left(\cos \frac{\pi}{4} + i \sin \frac{\pi}{4}\right)$$

(b) An argument is $\theta = 2\pi/3$ and $r = \sqrt{1+3} = 2$. Thus

$$-1 + \sqrt{3}\, i = 2\left(\cos \frac{2\pi}{3} + i \sin \frac{2\pi}{3}\right)$$

(c) An argument is $\theta = 7\pi/6$ (or we could use $\theta = -5\pi/6$), and $r = \sqrt{48 + 16} = 8$. Thus

$$-4\sqrt{3} - 4i = 8\left(\cos \frac{7\pi}{6} + i \sin \frac{7\pi}{6}\right)$$

(d) An argument is $\theta = \tan^{-1} \frac{4}{3}$ and $r = \sqrt{3^2 + 4^2} = 5$. So

$$3 + 4i = 5\left[\cos\left(\tan^{-1} \tfrac{4}{3}\right) + i \sin\left(\tan^{-1} \tfrac{4}{3}\right)\right]$$

■

The addition formulas for sine and cosine discussed in Section 7.2 greatly simplify the multiplication and division of complex numbers in trigonometric form. The following theorem shows how.

MULTIPLICATION AND DIVISION OF COMPLEX NUMBERS

If the two complex numbers z_1 and z_2 have the trigonometric forms

$$z_1 = r_1(\cos \theta_1 + i \sin \theta_1) \qquad \text{and} \qquad z_2 = r_2(\cos \theta_2 + i \sin \theta_2)$$

then

$$z_1 z_2 = r_1 r_2 [\cos(\theta_1 + \theta_2) + i \sin(\theta_1 + \theta_2)] \qquad \text{Multiplication}$$

$$\frac{z_1}{z_2} = \frac{r_1}{r_2}[\cos(\theta_1 - \theta_2) + i \sin(\theta_1 - \theta_2)] \qquad (z_2 \neq 0) \qquad \text{Division}$$

This theorem says:

To multiply two complex numbers, multiply the moduli and add the arguments.

To divide two complex numbers, divide the moduli and subtract the arguments.

■ **Proof** To prove the multiplication formula, we simply multiply the two complex numbers.

$$z_1 z_2 = r_1 r_2 (\cos \theta_1 + i \sin \theta_1)(\cos \theta_2 + i \sin \theta_2)$$

$$= r_1 r_2 [\cos \theta_1 \cos \theta_2 - \sin \theta_1 \sin \theta_2 + i(\sin \theta_1 \cos \theta_2 + \cos \theta_1 \sin \theta_2)]$$

$$= r_1 r_2 [\cos(\theta_1 + \theta_2) + i \sin(\theta_1 + \theta_2)]$$

In the last step we used the addition formulas for sine and cosine. □

The proof of the division formula is left as an exercise.

EXAMPLE 6 ■ Multiplying and Dividing Complex Numbers

Let

$$z_1 = 2\left(\cos \frac{\pi}{4} + i \sin \frac{\pi}{4}\right) \qquad \text{and} \qquad z_2 = 5\left(\cos \frac{\pi}{3} + i \sin \frac{\pi}{3}\right)$$

Find (a) $z_1 z_2$ and (b) z_1/z_2.

SOLUTION

(a) By the multiplication formula

$$z_1 z_2 = (2)(5)\left[\cos\left(\frac{\pi}{4} + \frac{\pi}{3}\right) + i \sin\left(\frac{\pi}{4} + \frac{\pi}{3}\right)\right]$$

$$= 10\left(\cos \frac{7\pi}{12} + i \sin \frac{7\pi}{12}\right)$$

To approximate the answer, we use a calculator in radian mode and get

$$z_1 z_2 \approx 10(-0.2588 + 0.9659i) = -2.588 + 9.659i$$

(b) By the division formula

$$\frac{z_1}{z_2} = \frac{2}{5}\left[\cos\left(\frac{\pi}{4} - \frac{\pi}{3}\right) + i \sin\left(\frac{\pi}{4} - \frac{\pi}{3}\right)\right]$$

$$= \frac{2}{5}\left[\cos\left(-\frac{\pi}{12}\right) + i \sin\left(-\frac{\pi}{12}\right)\right]$$

$$= \frac{2}{5}\left(\cos \frac{\pi}{12} - i \sin \frac{\pi}{12}\right)$$

Using a calculator in radian mode, we get the approximate answer:

$$\frac{z_1}{z_2} \approx \frac{2}{5}(0.9659 - 0.2588i) = 0.3864 - 0.1035i$$

■

DeMoivre's Theorem

Repeated use of the multiplication formula gives the following useful formula for raising a complex number to a power n for any positive integer n.

> ### DeMOIVRE'S THEOREM
>
> If $z = r(\cos \theta + i \sin \theta)$, then for any integer n
>
> $$z^n = r^n(\cos n\theta + i \sin n\theta)$$

This theorem says: *To take the nth power of a complex number, we take the nth power of the modulus and multiply the argument by n.*

■ **Proof**   By the multiplication formula

$$z^2 = zz = r^2[\cos(\theta + \theta) + i \sin(\theta + \theta)]$$

$$= r^2(\cos 2\theta + i \sin 2\theta)$$

Now we multiply z^2 by z to get

$$z^3 = z^2z = r^3[\cos(2\theta + \theta) + i \sin(2\theta + \theta)]$$

$$= r^3(\cos 3\theta + i \sin 3\theta)$$

Repeating this argument, we see that for any positive integer n

$$z^n = r^n(\cos n\theta + i \sin n\theta)$$

A similar argument using the division formula shows that this also holds for negative integers. ☐

EXAMPLE 7 ■ Finding a Power Using DeMoivre's Theorem

Find $\left(\frac{1}{2} + \frac{1}{2}i\right)^{10}$.

SOLUTION

Since $\frac{1}{2} + \frac{1}{2}i = \frac{1}{2}(1 + i)$, it follows from Example 5(a) that

$$\frac{1}{2} + \frac{1}{2}i = \frac{\sqrt{2}}{2}\left(\cos \frac{\pi}{4} + i \sin \frac{\pi}{4}\right)$$

So by DeMoivre's Theorem,

$$\left(\frac{1}{2} + \frac{1}{2}i\right)^{10} = \left(\frac{\sqrt{2}}{2}\right)^{10}\left(\cos\frac{10\pi}{4} + i\sin\frac{10\pi}{4}\right)$$

$$= \frac{2^5}{2^{10}}\left(\cos\frac{5\pi}{2} + i\sin\frac{5\pi}{2}\right) = \frac{1}{32}i \qquad \blacksquare$$

*n*th Roots of Complex Numbers

DeMoivre's Theorem gives us a method for calculating the *n*th roots of any complex number.

> ### *n*TH ROOTS OF COMPLEX NUMBERS
>
> If $z = r(\cos\theta + i\sin\theta)$ and n is a positive integer, then z has the n distinct *n*th roots
>
> $$w_k = r^{1/n}\left[\cos\left(\frac{\theta + 2k\pi}{n}\right) + i\sin\left(\frac{\theta + 2k\pi}{n}\right)\right]$$
>
> for $k = 0, 1, 2, \ldots, n - 1$.

■ **Proof** To find the *n*th roots of z, we need to find a complex number w such that

$$w^n = z$$

Let's write z in trigonometric form:

$$z = r(\cos\theta + i\sin\theta)$$

One *n*th root of z is

$$w = r^{1/n}\left(\cos\frac{\theta}{n} + i\sin\frac{\theta}{n}\right)$$

since by DeMoivre's Theorem, $w^n = z$. But the argument θ of z can be replaced by $\theta + 2k\pi$ for any integer k. Since this expression gives a different value of w for $k = 0, 1, 2, \ldots, n - 1$, we have proved the formula in the theorem. □

The following observations help us use the preceding formula.

> 1. The modulus of each *n*th root is $r^{1/n}$.
>
> 2. The argument of the first root is θ/n.
>
> 3. We repeatedly add $2\pi/n$ to get the argument of each successive root.

These observations show that, when graphed, the *n*th roots of z are spaced equally on the circle of radius $r^{1/n}$.

EXAMPLE 8 ■ Finding Roots of a Complex Number

Find the six sixth roots of $z = -64$, and graph these roots in the complex plane.

SOLUTION

In trigonometric form, $z = 64(\cos \pi + i \sin \pi)$. Applying the formula for nth roots with $n = 6$, we get

$$w_k = 64^{1/6}\left[\cos\left(\frac{\pi + 2k\pi}{6}\right) + i \sin\left(\frac{\pi + 2k\pi}{6}\right)\right]$$

for $k = 0, 1, 2, 3, 4, 5$. Thus, the six sixth roots of -64 are

We add $2\pi/6 = \pi/3$ to each argument to get the argument of the next root.

$$w_0 = 64^{1/6}\left(\cos\frac{\pi}{6} + i \sin\frac{\pi}{6}\right) = \sqrt{3} + i$$

$$w_1 = 64^{1/6}\left(\cos\frac{\pi}{2} + i \sin\frac{\pi}{2}\right) = 2i$$

$$w_2 = 64^{1/6}\left(\cos\frac{5\pi}{6} + i \sin\frac{5\pi}{6}\right) = -\sqrt{3} + i$$

$$w_3 = 64^{1/6}\left(\cos\frac{7\pi}{6} + i \sin\frac{7\pi}{6}\right) = -\sqrt{3} - i$$

$$w_4 = 64^{1/6}\left(\cos\frac{3\pi}{2} + i \sin\frac{3\pi}{2}\right) = -2i$$

$$w_5 = 64^{1/6}\left(\cos\frac{11\pi}{6} + i \sin\frac{11\pi}{6}\right) = \sqrt{3} - i$$

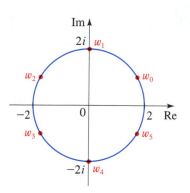

FIGURE 8
The six sixth roots of $z = -64$

All these points lie on the circle of radius 2, as shown in Figure 8. ■

When finding roots of complex numbers, we sometimes write the argument θ of the complex number in degrees. In this case, the nth roots are obtained from the formula

$$w_k = r^{1/n}\left[\cos\left(\frac{\theta + 360°k}{n}\right) + i \sin\left(\frac{\theta + 360°k}{n}\right)\right]$$

for $k = 0, 1, 2, \ldots, n - 1$.

EXAMPLE 9 ■ Finding Cube Roots of a Complex Number

Find the three cube roots of $z = 2 + 2i$, and graph these roots in the complex plane.

SOLUTION

First we write z in trigonometric form using degrees. We have $r = \sqrt{2^2 + 2^2} = 2\sqrt{2}$ and $\theta = 45°$. Thus

$$z = 2\sqrt{2}(\cos 45° + i\sin 45°)$$

Applying the formula for nth roots (in degrees) with $n = 3$, we find the cube roots of z are of the form

$(2\sqrt{2})^{1/3} = (2^{3/2})^{1/3} = 2^{1/2} = \sqrt{2}$

$$w_k = (2\sqrt{2})^{1/3}\left[\cos\left(\frac{45° + 360°k}{3}\right) + i\sin\left(\frac{45° + 360°k}{3}\right)\right]$$

We add $360°/3 = 120°$ to each argument to get the argument of the next root.

where $k = 0, 1, 2$. Thus, the three cube roots are

$$w_0 = \sqrt{2}(\cos 15° + i\sin 15°) \approx 1.366 + 0.366i$$

$$w_1 = \sqrt{2}(\cos 135° + i\sin 135°) = -1 + i$$

$$w_2 = \sqrt{2}(\cos 255° + i\sin 255°) \approx -0.366 + 1.366i$$

The three cube roots of z are graphed in Figure 9. These roots are spaced equally on the circle of radius $\sqrt{2}$.

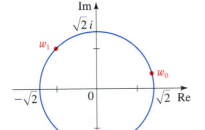

EXAMPLE 10 ■ Solving an Equation Using the nth Roots Formula

Solve the equation $z^6 + 64 = 0$.

SOLUTION

This equation can be written as $z^6 = -64$. Thus, the solutions are the sixth roots of -64, which we found in Example 8.

FIGURE 9
The three cube roots of $z = 2 + 2i$

7.6   EXERCISES

1–8 ■ Graph the complex number and find its modulus.

1. $4i$

2. -2

3. $5 + 2i$

4. $7 - 3i$

5. $\sqrt{3} + i$

6. $-1 - \dfrac{\sqrt{3}}{3}i$

7. $\dfrac{3 + 4i}{5}$

8. $\dfrac{-\sqrt{2} + i\sqrt{2}}{2}$

9–10 ■ Sketch the complex number z, and also sketch $2z$, $-z$, and $\frac{1}{2}z$ on the same complex plane.

9. $z = 1 + i$

10. $z = -1 + i\sqrt{3}$

11–12 ■ Sketch the complex number z and its complex conjugate $\bar{z}$ on the same complex plane.

11. $z = 8 + 2i$

12. $z = -5 + 6i$

13–14 ■ Sketch z_1, z_2, $z_1 + z_2$, and z_1z_2 on the same complex plane.

13. $z_1 = 2 - i$, $z_2 = 2 + i$

14. $z_1 = -1 + i$, $z_2 = 2 - 3i$

15–22 ■ Sketch the set in the complex plane.

15. $\{z = a + bi \mid a \le 0, b \ge 0\}$

16. $\{z = a + bi \mid a > 1, b > 1\}$

17. $\{z \mid |z| = 3\}$ **18.** $\{z \mid |z| \geq 1\}$

19. $\{z \mid |z| < 2\}$ **20.** $\{z \mid 2 \leq |z| \leq 5\}$

21. $\{z = a + bi \mid a + b < 2\}$

22. $\{z = a + bi \mid a \geq b\}$

23–46 ■ Write the complex number in trigonometric form with argument θ between 0 and 2π.

23. $1 + i$ **24.** $1 + \sqrt{3}\,i$ **25.** $\sqrt{2} - \sqrt{2}\,i$

26. $1 - i$ **27.** $2\sqrt{3} - 2i$ **28.** $-1 + i$

29. $-3i$ **30.** $-3 - 3\sqrt{3}\,i$ **31.** $5 + 5i$

32. 4 **33.** $4\sqrt{3} - 4i$ **34.** $8i$

35. -20 **36.** $\sqrt{3} + i$ **37.** $3 + 4i$

38. $i(2 - 2i)$ **39.** $3i(1 + i)$ **40.** $2(1 - i)$

41. $4(\sqrt{3} + i)$ **42.** $-3 - 3i$ **43.** $2 + i$

44. $3 + \sqrt{3}\,i$ **45.** $\sqrt{2} + \sqrt{2}\,i$ **46.** $-\pi i$

47–54 ■ Find the product $z_1 z_2$ and the quotient z_1/z_2. Express your answer in trigonometric form.

47. $z_1 = \cos \pi + i \sin \pi$, $z_2 = \cos \dfrac{\pi}{3} + i \sin \dfrac{\pi}{3}$

48. $z_1 = \cos \dfrac{\pi}{4} + i \sin \dfrac{\pi}{4}$, $z_2 = \cos \dfrac{3\pi}{4} + i \sin \dfrac{3\pi}{4}$

49. $z_1 = 3\left(\cos \dfrac{\pi}{6} + i \sin \dfrac{\pi}{6}\right)$, $z_2 = 5\left(\cos \dfrac{4\pi}{3} + i \sin \dfrac{4\pi}{3}\right)$

50. $z_1 = 7\left(\cos \dfrac{9\pi}{8} + i \sin \dfrac{9\pi}{8}\right)$, $z_2 = 2\left(\cos \dfrac{\pi}{8} + i \sin \dfrac{\pi}{8}\right)$

51. $z_1 = 4(\cos 120° + i \sin 120°)$,
  $z_2 = 2(\cos 30° + i \sin 30°)$

52. $z_1 = \sqrt{2}\,(\cos 75° + i \sin 75°)$,
  $z_2 = 3\sqrt{2}\,(\cos 60° + i \sin 60°)$

53. $z_1 = 4(\cos 200° + i \sin 200°)$,
  $z_2 = 25(\cos 150° + i \sin 150°)$

54. $z_1 = \tfrac{4}{5}(\cos 25° + i \sin 25°)$,
  $z_2 = \tfrac{1}{5}(\cos 155° + i \sin 155°)$

55–62 ■ Write z_1 and z_2 in trigonometric form, and then find the product $z_1 z_2$ and the quotients z_1/z_2 and $1/z_1$.

55. $z_1 = \sqrt{3} + i$, $z_2 = 1 + \sqrt{3}\,i$

56. $z_1 = \sqrt{2} - \sqrt{2}\,i$, $z_2 = 1 - i$

57. $z_1 = 2\sqrt{3} - 2i$, $z_2 = -1 + i$

58. $z_1 = -\sqrt{2}\,i$, $z_2 = -3 - 3\sqrt{3}\,i$

59. $z_1 = 5 + 5i$, $z_2 = 4$ **60.** $z_1 = 4\sqrt{3} - 4i$, $z_2 = 8i$

61. $z_1 = -20$, $z_2 = \sqrt{3} + i$ **62.** $z_1 = 3 + 4i$, $z_2 = 2 - 2i$

63–74 ■ Find the indicated power using DeMoivre's Theorem.

63. $(1 + i)^{20}$ **64.** $\left(1 - \sqrt{3}\,i\right)^5$

65. $\left(2\sqrt{3} + 2i\right)^5$ **66.** $(1 - i)^8$

67. $\left(\dfrac{\sqrt{2}}{2} + \dfrac{\sqrt{2}}{2}i\right)^{12}$ **68.** $\left(\sqrt{3} - i\right)^{-10}$

69. $(2 - 2i)^8$ **70.** $\left(-\dfrac{1}{2} - \dfrac{\sqrt{3}}{2}i\right)^{15}$

71. $(-1 - i)^7$ **72.** $\left(3 + \sqrt{3}\,i\right)^4$

73. $\left(2\sqrt{3} + 2i\right)^{-5}$ **74.** $(1 - i)^{-8}$

75–84 ■ Find the indicated roots, and graph the roots in the complex plane.

75. The square roots of $4\sqrt{3} + 4i$

76. The cube roots of $4\sqrt{3} + 4i$

77. The fourth roots of $-81i$ **78.** The fifth roots of 32

79. The eighth roots of 1 **80.** The cube roots of $1 + i$

81. The cube roots of i **82.** The fifth roots of i

83. The fourth roots of -1

84. The fifth roots of $-16 - 16\sqrt{3}\,i$

85–90 ■ Solve the equation.

85. $z^4 + 1 = 0$ **86.** $z^8 - i = 0$

87. $z^3 - 4\sqrt{3} - 4i = 0$ **88.** $z^6 - 1 = 0$

89. $z^3 + 1 = -i$ **90.** $z^3 - 1 = 0$

91. (a) Let $w = \cos \dfrac{2\pi}{n} + i \sin \dfrac{2\pi}{n}$ where n is a positive integer. Show that $1, w, w^2, w^3, \ldots, w^{n-1}$ are the n distinct nth roots of 1.

(b) If $z \neq 0$ is any complex number and $s^n = z$, show that the n distinct nth roots of z are

$$s, sw, sw^2, sw^3, \ldots, sw^{n-1}$$

 DISCOVERY · DISCUSSION

92. Sums of Roots of Unity Find the exact values of all three cube roots of 1 (see Exercise 91) and then add them.

Do the same for the fourth, fifth, sixth, and eighth roots of 1. What do you think is the sum of the nth roots of 1, for any n?

93. Products of Roots of Unity   Find the product of the three cube roots of 1 (see Exercise 91). Do the same for the fourth, fifth, sixth, and eighth roots of 1. What do you think is the product of the nth roots of 1, for any n?

94. Complex Coefficients and the Quadratic Formula
The quadratic formula works whether the coefficients of the equation are real or complex. Solve these equations using the quadratic formula, and, if necessary, DeMoivre's Theorem.

(a) $z^2 + (1 + i)z + i = 0$

(b) $z^2 - iz + 1 = 0$

(c) $z^2 - (2 - i)z - \frac{1}{4}i = 0$

7.7   VECTORS

In applications of mathematics, certain quantities are determined completely by their magnitude—for example, length, mass, area, temperature, and energy. We speak of a length of 5 m or a mass of 3 kg; only one number is needed to describe each of these quantities. Such a quantity is called a **scalar**.

On the other hand, to describe the displacement of an object, two numbers are required: the *magnitude* and the *direction* of the displacement. To describe the velocity of a moving object, we must specify both the *speed* and the *direction* of travel. Quantities such as displacement, velocity, acceleration, and force that involve magnitude as well as direction are called *directed quantities*. One way to represent such quantities mathematically is through the use of *vectors*.

Geometric Description of Vectors

A **vector** in the plane is a line segment with an assigned direction. We sketch a vector as shown in Figure 1 with an arrow to specify the direction. We denote this vector by $\overrightarrow{AB}$. Point A is the **initial point**, and B is the **terminal point** of the vector $\overrightarrow{AB}$. The length of the line segment AB is called the **magnitude** or **length** of the vector and is denoted by $|\overrightarrow{AB}|$. We use boldface letters to denote vectors. Thus, we write $\mathbf{u} = \overrightarrow{AB}$.

Two vectors are considered **equal** if they have equal magnitude and the same direction. Thus, all the vectors in Figure 2 are equal. This definition of equality makes sense if we think of a vector as representing a displacement. Two such displacements are the same if they have equal magnitudes and the same direction. So the vectors in Figure 2 can be thought of as the *same* displacement applied to objects in different locations in the plane.

If the displacement $\mathbf{u} = \overrightarrow{AB}$ is followed by the displacement $\mathbf{v} = \overrightarrow{BC}$, then the resulting displacement is $\overrightarrow{AC}$ as shown in Figure 3. In other words, the single displacement represented by the vector $\overrightarrow{AC}$ has the same effect as the other two displacements together. We call the vector $\overrightarrow{AC}$ the **sum** of the vectors $\overrightarrow{AB}$ and $\overrightarrow{BC}$ and we write $\overrightarrow{AC} = \overrightarrow{AB} + \overrightarrow{BC}$. (The **zero vector**, denoted by $\mathbf{0}$, represents no displacement). Thus, to find the sum of any two vectors $\mathbf{u}$ and $\mathbf{v}$, we sketch vectors equal to $\mathbf{u}$ and $\mathbf{v}$ with the initial point of one at the terminal point of the other

$\mathbf{u} = \overrightarrow{AB}$

FIGURE 1

FIGURE 2

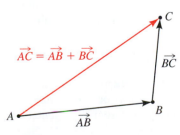

$\overrightarrow{AC} = \overrightarrow{AB} + \overrightarrow{BC}$

FIGURE 3

[see Figure 4(a)]. If we draw **u** and **v** starting at the same point, then **u** + **v** is the vector that is the diagonal of the parallelogram formed by **u** and **v**, as shown in Figure 4(b).

(a)   (b)

FIGURE 4
Addition of vectors

If a is a real number and **v** is a vector, we define a new vector a**v** as follows: The vector a**v** has magnitude $|a||\mathbf{v}|$ and has the same direction as **v** if $a > 0$, or the opposite direction if $a < 0$. If $a = 0$, then $a\mathbf{v} = \mathbf{0}$, the zero vector. This process is called **multiplication of a vector by a scalar**. Multiplying a vector by a scalar has the effect of stretching or shrinking the vector. Figure 5 shows graphs of the vector a**v** for different values of a. We write the vector $(-1)\mathbf{v}$ as $-\mathbf{v}$. Thus, $-\mathbf{v}$ is the vector with the same length as **v** but with the opposite direction.

FIGURE 5
Multiplication of a vector by a scalar

The **difference** of two vectors **u** and **v** is defined by $\mathbf{u} - \mathbf{v} = \mathbf{u} + (-\mathbf{v})$. Figure 6 shows that the vector $\mathbf{u} - \mathbf{v}$ is the other diagonal of the parallelogram formed by **u** and **v**.

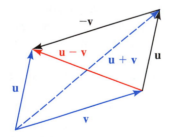

FIGURE 6
Subtraction of vectors

■ Analytic Description of Vectors

So far we've discussed vectors geometrically. By placing a vector in a coordinate plane, we can describe it analytically (that is, by using components). In Figure 7(a), to go from the initial point of the vector **v** to the terminal point, we move a units to the right and b units upward. We represent **v** as an ordered pair of real numbers.

Note the distinction between the *vector* $\langle a, b \rangle$ and the *point* (a, b).

$$\mathbf{v} = \langle a, b \rangle$$

where a is the **horizontal component** of **v** and b is the **vertical component** of **v**. Remember that a vector represents a magnitude and a direction, not a particular arrow in the plane. Thus, the vector $\langle a, b \rangle$ has many different representations, depending on its initial point [see Figure 7(b)].

FIGURE 7

(a)   (b)

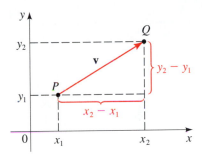

FIGURE 8

Using Figure 8, the relationship between a geometric representation of a vector and the analytic one can be stated as follows.

> **COMPONENT FORM OF A VECTOR**
>
> If a vector **v** is represented in the plane with initial point $P(x_1, y_1)$ and terminal point $Q(x_2, y_2)$, then
>
> $$\mathbf{v} = \langle x_2 - x_1, y_2 - y_1 \rangle$$

EXAMPLE 1 ■ Describing Vectors in Component Form

(a) Find the component form of the vector **u** with initial point $(-2, 5)$ and terminal point $(3, 7)$.

(b) If the vector $\mathbf{v} = \langle 3, 7 \rangle$ is sketched with initial point $(2, 4)$, what is its terminal point?

(c) Sketch representations of the vector $\mathbf{w} = \langle 2, 3 \rangle$ with initial points at $(0, 0)$, $(2, 2)$, $(-2, -1)$, and $(1, 4)$.

SOLUTION

(a) The desired vector is

$$\mathbf{u} = \langle 3 - (-2), 7 - 5 \rangle = \langle 5, 2 \rangle$$

(b) Let the terminal point of **v** be (x, y). Then

$$\langle x - 2, y - 4 \rangle = \langle 3, 7 \rangle$$

So $x - 2 = 3$ and $y - 4 = 7$, or $x = 5$ and $y = 11$. The terminal point is $(5, 11)$.

(c) Representations of the vector **w** are sketched in Figure 9.   ■

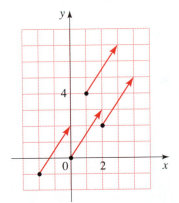

FIGURE 9

We now give analytic definitions of the various operations on vectors that we have described geometrically. Let's start with equality of vectors. We've said that two vectors are equal if they have equal magnitude and the same direction. For the

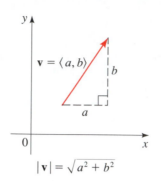

$$|\mathbf{v}| = \sqrt{a^2 + b^2}$$

FIGURE 10

vectors $\mathbf{u} = \langle a_1, b_1 \rangle$ and $\mathbf{v} = \langle a_2, b_2 \rangle$, this means that $a_1 = a_2$ and $b_1 = b_2$. In other words, two vectors are **equal** if and only if their corresponding components are equal. Thus, all the arrows in Figure 7(b) represent the same vector, as do all the arrows in Figure 9.

Applying the Pythagorean Theorem to the triangle in Figure 10, we obtain the following formula for the magnitude of a vector.

MAGNITUDE OF A VECTOR

The **magnitude** or **length** of a vector $\mathbf{v} = \langle a, b \rangle$ is

$$|\mathbf{v}| = \sqrt{a^2 + b^2}$$

EXAMPLE 2 ■ Magnitudes of Vectors

Find the magnitude of each vector.

(a) $\mathbf{u} = \langle 2, -3 \rangle$ (b) $\mathbf{v} = \langle 5, 0 \rangle$ (c) $\mathbf{w} = \left\langle \frac{3}{5}, \frac{4}{5} \right\rangle$

SOLUTION

(a) $|\mathbf{u}| = \sqrt{2^2 + (-3)^2} = \sqrt{13}$

(b) $|\mathbf{v}| = \sqrt{5^2 + 0^2} = \sqrt{25} = 5$

(c) $|\mathbf{w}| = \sqrt{\left(\frac{3}{5}\right)^2 + \left(\frac{4}{5}\right)^2} = \sqrt{\frac{9}{25} + \frac{16}{25}} = 1$ ■

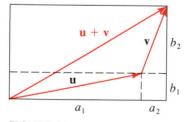

FIGURE 11

The following definitions of addition, subtraction, and scalar multiplication of vectors correspond to the geometric descriptions given earlier. Figure 11 shows how the analytic definition of addition corresponds to the geometric one.

ALGEBRAIC OPERATIONS ON VECTORS

If $\mathbf{u} = \langle a_1, b_1 \rangle$ and $\mathbf{v} = \langle a_2, b_2 \rangle$, then

$$\mathbf{u} + \mathbf{v} = \langle a_1 + a_2, b_1 + b_2 \rangle$$

$$\mathbf{u} - \mathbf{v} = \langle a_1 - a_2, b_1 - b_2 \rangle$$

$$c\mathbf{u} = \langle ca_1, cb_1 \rangle, \qquad c \in \mathbb{R}$$

EXAMPLE 3 ■ Operations with Vectors

If $\mathbf{u} = \langle 2, -3 \rangle$ and $\mathbf{v} = \langle -1, 2 \rangle$, find $\mathbf{u} + \mathbf{v}$, $\mathbf{u} - \mathbf{v}$, $2\mathbf{u}$, $-3\mathbf{v}$, and $2\mathbf{u} + 3\mathbf{v}$.

SOLUTION

By the definitions of the vector operations, we have

$$\mathbf{u} + \mathbf{v} = \langle 2, -3 \rangle + \langle -1, 2 \rangle = \langle 1, -1 \rangle$$

$$\mathbf{u} - \mathbf{v} = \langle 2, -3 \rangle - \langle -1, 2 \rangle = \langle 3, -5 \rangle$$

$$2\mathbf{u} = 2\langle 2, -3 \rangle = \langle 4, -6 \rangle$$

$$-3\mathbf{v} = -3\langle -1, 2 \rangle = \langle 3, -6 \rangle$$

$$2\mathbf{u} + 3\mathbf{v} = 2\langle 2, -3 \rangle + 3\langle -1, 2 \rangle = \langle 4, -6 \rangle + \langle -3, 6 \rangle = \langle 1, 0 \rangle \qquad \blacksquare$$

The following properties for vector operations can be easily proved from the definitions. The **zero vector** is the vector $\mathbf{0} = \langle 0, 0 \rangle$. It plays the same role for addition of vectors as the number 0 does for addition of real numbers.

PROPERTIES OF VECTORS

Vector addition	Multiplication by a scalar
$\mathbf{u} + \mathbf{v} = \mathbf{v} + \mathbf{u}$	$c(\mathbf{u} + \mathbf{v}) = c\mathbf{u} + c\mathbf{v}$
$\mathbf{u} + (\mathbf{v} + \mathbf{w}) = (\mathbf{u} + \mathbf{v}) + \mathbf{w}$	$(c + d)\mathbf{u} = c\mathbf{u} + d\mathbf{u}$
$\mathbf{u} + \mathbf{0} = \mathbf{u}$	$(cd)\mathbf{u} = c(d\mathbf{u}) = d(c\mathbf{u})$
$\mathbf{u} + (-\mathbf{u}) = \mathbf{0}$	$1\mathbf{u} = \mathbf{u}$
Length of a vector	$0\mathbf{u} = \mathbf{0}$
$|c\mathbf{u}| = |c| \, |\mathbf{u}|$	$c\mathbf{0} = \mathbf{0}$

A vector of length 1 is called a **unit vector**. For instance, in Example 2(c), the vector $\mathbf{w} = \langle \frac{3}{5}, \frac{4}{5} \rangle$ is a unit vector. Two useful unit vectors are $\mathbf{i}$ and $\mathbf{j}$, defined by

$$\mathbf{i} = \langle 1, 0 \rangle \qquad \mathbf{j} = \langle 0, 1 \rangle$$

These vectors are special because any vector can be expressed in terms of them.

VECTORS IN TERMS OF i AND j

The vector $\mathbf{v} = \langle a, b \rangle$ can be expressed in terms of $\mathbf{i}$ and $\mathbf{j}$ by

$$\mathbf{v} = \langle a, b \rangle = a\mathbf{i} + b\mathbf{j}$$

EXAMPLE 4 ■ Vectors in Terms of i and j

(a) Write the vector $\mathbf{u} = \langle 5, -8 \rangle$ in terms of $\mathbf{i}$ and $\mathbf{j}$.

(b) If $\mathbf{u} = 3\mathbf{i} + 2\mathbf{j}$ and $\mathbf{v} = -\mathbf{i} + 6\mathbf{j}$, write $2\mathbf{u} + 5\mathbf{v}$ in terms of $\mathbf{i}$ and $\mathbf{j}$.

SOLUTION

(a) $\mathbf{u} = 5\mathbf{i} + (-8)\mathbf{j} = 5\mathbf{i} - 8\mathbf{j}$

(b) The properties of addition and scalar multiplication of vectors show that we can manipulate vectors in the same way as algebraic expressions. Thus

$$2\mathbf{u} + 5\mathbf{v} = 2(3\mathbf{i} + 2\mathbf{j}) + 5(-\mathbf{i} + 6\mathbf{j})$$

$$= (6\mathbf{i} + 4\mathbf{j}) + (-5\mathbf{i} + 30\mathbf{j})$$

$$= \mathbf{i} + 34\mathbf{j} \qquad \blacksquare$$

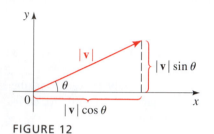

FIGURE 12

Let $\mathbf{v}$ be a vector in the plane with its initial point at the origin. The **direction** of $\mathbf{v}$ is θ, the smallest positive angle in standard position formed by the positive x-axis and $\mathbf{v}$ (see Figure 12). If we know the magnitude and direction of a vector, then Figure 12 shows that we can find the horizontal and vertical components of the vector.

HORIZONTAL AND VERTICAL COMPONENTS OF A VECTOR

Let $\mathbf{v}$ be a vector with magnitude $|\mathbf{v}|$ and direction θ.
Then $\mathbf{v} = \langle a, b \rangle = a\mathbf{i} + b\mathbf{j}$, where

$$a = |\mathbf{v}| \cos \theta \qquad \text{and} \qquad b = |\mathbf{v}| \sin \theta$$

EXAMPLE 5 ■ Components and Direction of a Vector

(a) A vector $\mathbf{v}$ has length 8 and direction $\pi/3$. Find the horizontal and vertical components, and write $\mathbf{v}$ in terms of $\mathbf{i}$ and $\mathbf{j}$.

(b) Find the direction of the vector $\mathbf{u} = -\sqrt{3}\,\mathbf{i} + \mathbf{j}$.

SOLUTION

(a) We have $\mathbf{v} = \langle a, b \rangle$, where the components are given by

$$a = 8 \cos \frac{\pi}{3} = 4 \qquad \text{and} \qquad b = 8 \sin \frac{\pi}{3} = 4\sqrt{3}$$

Thus, $\mathbf{v} = \langle 4, 4\sqrt{3} \rangle = 4\mathbf{i} + 4\sqrt{3}\,\mathbf{j}$.

(b) From Figure 13 we see that the direction θ has the property that

FIGURE 13

$$\tan \theta = \frac{1}{-\sqrt{3}} = -\frac{\sqrt{3}}{3}$$

Thus, the reference angle for θ is $\pi/6$. Since the terminal point of the vector $\mathbf{u}$ is in quadrant II, it follows that $\theta = 5\pi/6$. $\qquad \blacksquare$

Using Vectors to Model Velocity and Force

The use of bearings (such as N 30° E) to describe directions is explained on page 515 in Section 6.5.

The **velocity** of a moving object is modeled by a vector whose direction is the direction of motion and whose magnitude is the speed. Figure 14 shows some vectors **u**, representing the velocity of wind flowing in the direction N 30° E, and a vector **v**, representing the velocity of an airplane flying through this wind at the point P. It's obvious from our experience that wind affects both the speed and the direction of an airplane. Figure 15 indicates that the true velocity of the plane (relative to the ground) is given by the vector **w** = **u** + **v**.

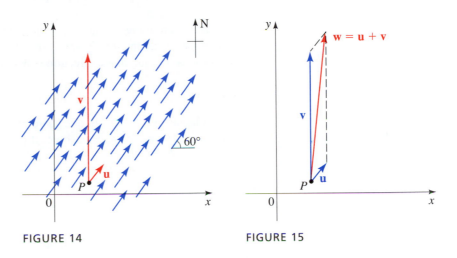

FIGURE 14   FIGURE 15

EXAMPLE 6 ■ The True Speed and Direction of an Airplane

An airplane heads due north at 300 mi/h. It experiences a 40 mi/h crosswind flowing in the direction N 30° E, as shown in Figure 14.
(a) Express the velocity **v** of the airplane relative to the air, and the velocity **u** of the wind, in component form.
(b) Find the true velocity of the airplane as a vector.
(c) Find the true speed and direction of the airplane.

SOLUTION

(a) The velocity of the airplane relative to the air is **v** = 0**i** + 300**j** = 300**j**.
   By the formulas for the components of a vector, we find that the velocity of the wind is

$$\mathbf{u} = (40 \cos 60°)\mathbf{i} + (40 \sin 60°)\mathbf{j}$$

$$= 20\mathbf{i} + 20\sqrt{3}\,\mathbf{j}$$

$$\approx 20\mathbf{i} + 34.64\,\mathbf{j}$$

(b) The true velocity of the airplane is given by the vector $\mathbf{w} = \mathbf{u} + \mathbf{v}$.

$$\mathbf{w} = \mathbf{u} + \mathbf{v} = (20\mathbf{i} + 20\sqrt{3}\,\mathbf{j}) + (300\,\mathbf{j})$$

$$= 20\mathbf{i} + (20\sqrt{3} + 300)\mathbf{j}$$

$$\approx 20\mathbf{i} + 334.64\,\mathbf{j}$$

(c) The true speed of the airplane is given by the magnitude of $\mathbf{w}$.

$$|\mathbf{w}| \approx \sqrt{(20)^2 + (334.64)^2} \approx 335.2 \text{ mi/h}$$

The direction of the airplane is the direction θ of the vector $\mathbf{w}$. The angle θ has the property that $\tan\theta \approx 334.64/20 = 16.732$ and so $\theta \approx 86.6°$. Thus, the airplane is heading in the direction N $3.4°$ E. ■

EXAMPLE 7 ■ Calculating a Heading

A woman launches a boat from one shore of a straight river and wants to land at the point directly on the opposite shore. If the speed of the boat (relative to the water) is 10 mi/h and the river is flowing east at the rate of 5 mi/h, in what direction should she head the boat in order to arrive at the desired landing point?

SOLUTION

We choose a coordinate system with the origin at the initial position of the boat as shown in Figure 16. Let $\mathbf{u}$ and $\mathbf{v}$ represent the velocities of the river and the boat, respectively. Clearly, $\mathbf{u} = 5\mathbf{i}$ and, since the speed of the boat is 10 mi/h, we have $|\mathbf{v}| = 10$, so

$$\mathbf{v} = (10\cos\theta)\mathbf{i} + (10\sin\theta)\mathbf{j}$$

where the angle θ is as shown in Figure 16. The true course of the boat is given by the vector $\mathbf{w} = \mathbf{u} + \mathbf{v}$. We have

$$\mathbf{w} = \mathbf{u} + \mathbf{v} = 5\mathbf{i} + (10\cos\theta)\mathbf{i} + (10\sin\theta)\mathbf{j}$$

$$= (5 + 10\cos\theta)\mathbf{i} + (10\sin\theta)\mathbf{j}$$

Since the woman wants to land at a point directly across the river, her direction should have horizontal component 0. In other words, she should choose θ in such a way that

$$5 + 10\cos\theta = 0$$

$$\cos\theta = -\tfrac{1}{2}$$

$$\theta = 120°$$

Thus, she should head the boat in the direction $\theta = 120°$ (or N $30°$ W). ■

FIGURE 16

Force is also represented by a vector. Intuitively, we can think of force as describing a push or a pull on an object, for example, a horizontal push of a book across a table or the downward pull of the earth's gravity on a ball. Force is measured in pounds (or in newtons, in the metric system). For instance, a man weighing 200 lb exerts a force of 200 lb downward on the ground. If several forces are acting on an object, the **resultant force** experienced by the object is the vector sum of these forces.

EXAMPLE 8 ■ Resultant Force

Two forces $\mathbf{F}_1$ and $\mathbf{F}_2$ with magnitudes 10 and 20 lb, respectively, act on an object at a point P as shown in Figure 17. Find the resultant force acting at P.

SOLUTION

We write $\mathbf{F}_1$ and $\mathbf{F}_2$ in component form:

$$\mathbf{F}_1 = (10 \cos 45°)\mathbf{i} + (10 \sin 45°)\mathbf{j} = 10\,\frac{\sqrt{2}}{2}\,\mathbf{i} + 10\,\frac{\sqrt{2}}{2}\,\mathbf{j} = 5\sqrt{2}\,\mathbf{i} + 5\sqrt{2}\,\mathbf{j}$$

$$\mathbf{F}_2 = (20 \cos 150°)\mathbf{i} + (20 \sin 150°)\mathbf{j} = -20\,\frac{\sqrt{3}}{2}\,\mathbf{i} + 20\left(\frac{1}{2}\right)\mathbf{j}$$

$$= -10\sqrt{3}\,\mathbf{i} + 10\,\mathbf{j}$$

So the resultant force $\mathbf{F}$ is

$$\mathbf{F} = \mathbf{F}_1 + \mathbf{F}_2$$
$$= \left(5\sqrt{2}\,\mathbf{i} + 5\sqrt{2}\,\mathbf{j}\right) + \left(-10\sqrt{3}\,\mathbf{i} + 10\,\mathbf{j}\right)$$
$$= \left(5\sqrt{2} - 10\sqrt{3}\right)\mathbf{i} + \left(5\sqrt{2} + 10\right)\mathbf{j}$$
$$\approx -10\,\mathbf{i} + 17\,\mathbf{j}$$

The resultant force $\mathbf{F}$ is shown in Figure 18. ■

FIGURE 17

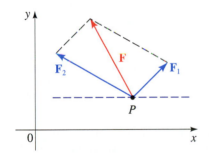

FIGURE 18

7.7   EXERCISES

1–6 ■ Sketch the vector indicated. (The vectors **u** and **v** are shown in the figure.)

1. $2\mathbf{u}$

2. $-\mathbf{v}$

3. $\mathbf{u} + \mathbf{v}$

4. $\mathbf{u} - \mathbf{v}$

5. $\mathbf{v} - 2\mathbf{u}$

6. $2\mathbf{u} + \mathbf{v}$

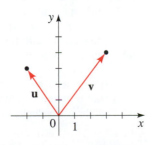

7–16 ■ Express the vector with initial point P and terminal point Q in component form.

7.

8.

9.

10.

11. $P(3, 2)$, $Q(8, 9)$

12. $P(1, 1)$, $Q(9, 9)$

13. $P(5, 3)$, $Q(1, 0)$

14. $P(-1, 3)$, $Q(-6, -1)$

15. $P(-1, -1)$, $Q(-1, 1)$

16. $P(-8, -6)$, $Q(-1, -1)$

17–22 ■ Find $2\mathbf{u}$, $-3\mathbf{v}$, $\mathbf{u} + \mathbf{v}$, and $3\mathbf{u} - 4\mathbf{v}$ for the given vectors $\mathbf{u}$ and $\mathbf{v}$.

17. $\mathbf{u} = \langle 2, 7 \rangle$, $\mathbf{v} = \langle 3, 1 \rangle$

18. $\mathbf{u} = \langle -2, 5 \rangle$, $\mathbf{v} = \langle 2, -8 \rangle$

19. $\mathbf{u} = \langle 0, -1 \rangle$, $\mathbf{v} = \langle -2, 0 \rangle$

20. $\mathbf{u} = \mathbf{i}$, $\mathbf{v} = -2\mathbf{j}$

21. $\mathbf{u} = 2\mathbf{i}$, $\mathbf{v} = 3\mathbf{i} - 2\mathbf{j}$

22. $\mathbf{u} = \mathbf{i} + \mathbf{j}$, $\mathbf{v} = \mathbf{i} - \mathbf{j}$

23–26 ■ Find $|\mathbf{u}|$, $|\mathbf{v}|$, $|2\mathbf{u}|$, $\left|\frac{1}{2}\mathbf{v}\right|$, $|\mathbf{u} + \mathbf{v}|$, $|\mathbf{u} - \mathbf{v}|$, and $|\mathbf{u}| - |\mathbf{v}|$.

23. $\mathbf{u} = 2\mathbf{i} + \mathbf{j}$, $\mathbf{v} = 3\mathbf{i} - 2\mathbf{j}$

24. $\mathbf{u} = -2\mathbf{i} + 3\mathbf{j}$, $\mathbf{v} = \mathbf{i} - 2\mathbf{j}$

25. $\mathbf{u} = \langle 10, -1 \rangle$, $\mathbf{v} = \langle -2, -2 \rangle$

26. $\mathbf{u} = \langle -6, 6 \rangle$, $\mathbf{v} = \langle -2, -1 \rangle$

27–32 ■ Find the horizontal and vertical components of the vector with given length and direction, and write the vector in terms of the vectors $\mathbf{i}$ and $\mathbf{j}$.

27. $|\mathbf{v}| = 40$, $\theta = 30°$

28. $|\mathbf{v}| = 50$, $\theta = 120°$

29. $|\mathbf{v}| = 1$, $\theta = 225°$

30. $|\mathbf{v}| = 800$, $\theta = 125°$

31. $|\mathbf{v}| = 4$, $\theta = 10°$

32. $|\mathbf{v}| = \sqrt{3}$, $\theta = 300°$

33. A man pushes a lawn mower with a force of 30 lb exerted at an angle of 30° to the ground. Find the horizontal and vertical components of the force.

34. A jet is flying in a direction N 20° E with a speed of 500 mi/h. Find the north and east components of the velocity.

35–40 ■ Find the magnitude and direction (in degrees) of the vector.

35. $\mathbf{v} = \langle 3, 4 \rangle$

36. $\mathbf{v} = \left\langle -\dfrac{\sqrt{2}}{2}, -\dfrac{\sqrt{2}}{2} \right\rangle$

37. $\mathbf{v} = \langle -12, 5 \rangle$

38. $\mathbf{v} = \langle 40, 9 \rangle$

39. $\mathbf{v} = \mathbf{i} + \sqrt{3}\,\mathbf{j}$

40. $\mathbf{v} = \mathbf{i} + \mathbf{j}$

41. A river flows due south at 3 mi/h. A swimmer attempting to cross the river heads due east swimming at 2 mi/h relative to the water. Find the true velocity of the swimmer.

42. A migrating salmon heads in the direction N 45° E, swimming at 5 mi/h relative to the water. The prevailing ocean currents flow due east at 3 mi/h. Find the true velocity of the fish.

43. A pilot heads his jet due east. The jet has a speed of 425 mi/h relative to the air. The wind is blowing due north with a speed of 40 mi/h.
(a) Express the velocity of the wind as a vector in component form.
(b) Express the velocity of the jet relative to the air as a vector in component form.
(c) Find the true velocity of the jet.
(d) Find the true speed and direction of the jet.

44. A jet is flying through a wind that is blowing with a speed of 55 mi/h in the direction N 30° E (see the figure). The jet has a speed of 765 mi/h relative to the air, and the pilot heads the jet in the direction N 45° E.
(a) Express the velocity of the wind as a vector in component form.
(b) Express the velocity of the jet relative to the air as a vector in component form.
(c) Find the true velocity of the jet.

(d) Find the true speed and direction of the jet.

45. Find the true speed and direction of the jet in Exercise 44 if the pilot heads the plane in the direction N 30° W.

46. In what direction should the pilot in Exercise 44 head the plane for the true course to be due north?

47. A straight river flows east at a speed of 10 mi/h. A boater starts at the south shore of the river and heads in a direction 60° from the shore (see the figure). The motorboat has a speed of 20 mi/h relative to the water.
  (a) Express the velocity of the river as a vector in component form.
  (b) Express the velocity of the motorboat relative to the water as a vector in component form.
  (c) Find the true velocity of the motorboat.
  (d) Find the true speed and direction of the motorboat.

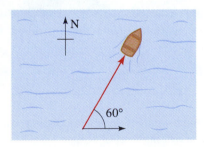

48. The boater in Exercise 47 wants to arrive at a point on the north shore of the river directly opposite the starting point. In what direction should the boat be headed?

49. A boat heads in the direction N 72° E. The speed of the boat relative to the water is 24 mi/h. The water is flowing directly south. It is observed that the true direction of the boat is directly east.
  (a) Express the velocity of the boat relative to the water as a vector in component form.
  (b) Find the speed of the water and the true speed of the boat.

50. A woman walks due west on the deck of an ocean liner at 2 mi/h. The ocean liner is moving due north at a speed of 25 mi/h. Find the speed and direction of the woman relative to the surface of the water.

51–56 ■ The forces $\mathbf{F}_1, \mathbf{F}_2, \ldots, \mathbf{F}_n$ acting at the same point P are said to be in *equilibrium* if the resultant force is zero, that is, if $\mathbf{F}_1 + \mathbf{F}_2 + \cdots + \mathbf{F}_n = \mathbf{0}$. Find (a) the resultant forces acting at P, and (b) the additional force required (if any) for the forces to be in equilibrium.

51. $\mathbf{F}_1 = \langle 2, 5 \rangle, \quad \mathbf{F}_2 = \langle 3, -8 \rangle$

52. $\mathbf{F}_1 = \langle 3, -7 \rangle, \quad \mathbf{F}_2 = \langle 4, -2 \rangle, \quad \mathbf{F}_3 = \langle -7, 9 \rangle$

53. $\mathbf{F}_1 = 4\mathbf{i} - \mathbf{j}, \quad \mathbf{F}_2 = 3\mathbf{i} - 7\mathbf{j},$
$\mathbf{F}_3 = -8\mathbf{i} + 3\mathbf{j}, \quad \mathbf{F}_4 = \mathbf{i} + \mathbf{j}$

54. $\mathbf{F}_1 = \mathbf{i} - \mathbf{j}, \quad \mathbf{F}_2 = \mathbf{i} + \mathbf{j}, \quad \mathbf{F}_3 = -2\mathbf{i} + \mathbf{j}$

55.

56.

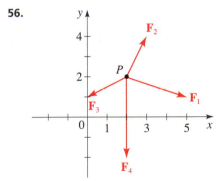

57. A 100-lb weight hangs from a string as shown in the figure. Find the tensions $\mathbf{T}_1$ and $\mathbf{T}_2$ in the string.

58. The cranes in the figure are lifting an object that weighs 18,278 lb. Find the tensions T_1 and T_2.

  DISCOVERY · DISCUSSION

59. Vectors That Form a Polygon Suppose that n vectors can be placed head to tail in the plane so that they form a polygon. (The figure shows the case of a hexagon.) Explain why the sum of these vectors is **0**.

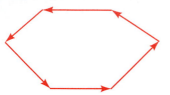

7.8 THE DOT PRODUCT

In this section we define an operation on vectors called the dot product. This concept is very useful in calculus and in applications of vectors to physics and engineering.

■ The Dot Product of Vectors

We begin by defining the dot product of two vectors.

> **DEFINITION OF THE DOT PRODUCT**
>
> If $\mathbf{u} = \langle a_1, b_1 \rangle$ and $\mathbf{v} = \langle a_2, b_2 \rangle$ are vectors, then their **dot product**, denoted by $\mathbf{u} \cdot \mathbf{v}$, is defined by
>
> $$\mathbf{u} \cdot \mathbf{v} = a_1 a_2 + b_1 b_2$$

Thus, to find the dot product of $\mathbf{u}$ and $\mathbf{v}$ we multiply corresponding components and add. The result is *not* a vector; it is a real number, or scalar.

EXAMPLE 1 ■ Calculating Dot Products

(a) If $\mathbf{u} = \langle 3, -2 \rangle$ and $\mathbf{v} = \langle 4, 5 \rangle$, then

$$\mathbf{u} \cdot \mathbf{v} = (3)(4) + (-2)(5) = 2$$

(b) If $\mathbf{u} = 2\mathbf{i} + \mathbf{j}$ and $\mathbf{v} = 5\mathbf{i} - 6\mathbf{j}$, then

$$\mathbf{u} \cdot \mathbf{v} = (2)(5) + (1)(-6) = 4$$ ■

The proofs of the following properties of the dot product follow easily from the definition.

PROPERTIES OF THE DOT PRODUCT

$$\mathbf{u} \cdot \mathbf{v} = \mathbf{v} \cdot \mathbf{u}$$

$$(a\mathbf{u}) \cdot \mathbf{v} = a(\mathbf{u} \cdot \mathbf{v}) = \mathbf{u} \cdot (a\mathbf{v})$$

$$(\mathbf{u} + \mathbf{v}) \cdot \mathbf{w} = \mathbf{u} \cdot \mathbf{w} + \mathbf{v} \cdot \mathbf{w}$$

$$|\mathbf{u}|^2 = \mathbf{u} \cdot \mathbf{u}$$

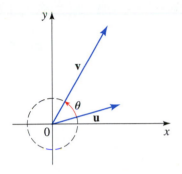

FIGURE 1

■   **Proof**   We prove only the last property. The proofs of the others are left as exercises. Let $\mathbf{u} = \langle a\ b \rangle$. Then

$$\mathbf{u} \cdot \mathbf{u} = \langle a, b \rangle \cdot \langle a, b \rangle = a^2 + b^2 = |\mathbf{u}|^2 \qquad \square$$

Let $\mathbf{u}$ and $\mathbf{v}$ be vectors and sketch them with initial points at the origin. We define the **angle θ between u and v** to be the smaller of the angles formed by these representations of $\mathbf{u}$ and $\mathbf{v}$ (see Figure 1). Thus, $0 \le \theta \le \pi$. The next theorem relates the angle between two vectors to their dot product.

THE DOT PRODUCT THEOREM

If θ is the angle between two nonzero vectors $\mathbf{u}$ and $\mathbf{v}$, then

$$\mathbf{u} \cdot \mathbf{v} = |\mathbf{u}|\,|\mathbf{v}| \cos \theta$$

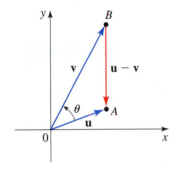

FIGURE 2

■   **Proof**   The proof is a nice application of the Law of Cosines. Applying the Law of Cosines to triangle AOB in Figure 2 gives

$$|\mathbf{u} - \mathbf{v}|^2 = |\mathbf{u}|^2 + |\mathbf{v}|^2 - 2|\mathbf{u}|\,|\mathbf{v}| \cos \theta$$

Using the properties of the dot product, we write the left-hand side as follows:

$$|\mathbf{u} - \mathbf{v}|^2 = (\mathbf{u} - \mathbf{v}) \cdot (\mathbf{u} - \mathbf{v})$$

$$= \mathbf{u} \cdot \mathbf{u} - \mathbf{u} \cdot \mathbf{v} - \mathbf{v} \cdot \mathbf{u} + \mathbf{v} \cdot \mathbf{v}$$

$$= |\mathbf{u}|^2 - 2(\mathbf{u} \cdot \mathbf{v}) + |\mathbf{v}|^2$$

Equating the right-hand sides of the displayed equations, we get

$$|\mathbf{u}|^2 - 2(\mathbf{u} \cdot \mathbf{v}) + |\mathbf{v}|^2 = |\mathbf{u}|^2 + |\mathbf{v}|^2 - 2|\mathbf{u}|\,|\mathbf{v}| \cos \theta$$

$$-2(\mathbf{u} \cdot \mathbf{v}) = -2|\mathbf{u}|\,|\mathbf{v}| \cos \theta$$

$$\mathbf{u} \cdot \mathbf{v} = |\mathbf{u}|\,|\mathbf{v}| \cos \theta$$

This proves the theorem. $\qquad \square$

The Dot Product Theorem is useful because it allows us to find the angle between two vectors if we know the components of the vectors. The angle is obtained simply by solving the equation in the Dot Product Theorem for $\cos \theta$. We state this important result explicitly.

ANGLE BETWEEN TWO VECTORS

If θ is the angle between two nonzero vectors $\mathbf{u}$ and $\mathbf{v}$, then

$$\cos \theta = \frac{\mathbf{u} \cdot \mathbf{v}}{|\mathbf{u}||\mathbf{v}|}$$

EXAMPLE 2 ■ Finding the Angle between Two Vectors

Find the angle between the vectors $\mathbf{u} = \langle 2, 5 \rangle$ and $\mathbf{v} = \langle 4, -3 \rangle$.

SOLUTION

By the formula for the angle between two vectors, we have

$$\cos \theta = \frac{\mathbf{u} \cdot \mathbf{v}}{|\mathbf{u}||\mathbf{v}|} = \frac{(2)(4) + (5)(-3)}{\sqrt{4 + 25}\,\sqrt{16 + 9}} = \frac{-7}{5\sqrt{29}}$$

Thus, the angle between $\mathbf{u}$ and $\mathbf{v}$ is

$$\theta = \cos^{-1}\left(\frac{-7}{5\sqrt{29}}\right) \approx 105.1°$$ ■

Two nonzero vectors $\mathbf{u}$ and $\mathbf{v}$ are called **perpendicular**, or **orthogonal**, if the angle between them is $\pi/2$. The following theorem shows that we can determine if two vectors are perpendicular by finding their dot product.

ORTHOGONAL VECTORS

Two nonzero vectors $\mathbf{u}$ and $\mathbf{v}$ are perpendicular if and only if $\mathbf{u} \cdot \mathbf{v} = 0$.

■ **Proof**   If $\mathbf{u}$ and $\mathbf{v}$ are perpendicular, then the angle between them is $\pi/2$ and so

$$\mathbf{u} \cdot \mathbf{v} = |\mathbf{u}||\mathbf{v}| \cos \frac{\pi}{2} = 0$$

Conversely, if $\mathbf{u} \cdot \mathbf{v} = 0$, then

$$|\mathbf{u}||\mathbf{v}| \cos \theta = 0$$

Since $\mathbf{u}$ and $\mathbf{v}$ are nonzero vectors, we conclude that $\cos \theta = 0$, and so $\theta = \pi/2$. Thus, $\mathbf{u}$ and $\mathbf{v}$ are orthogonal.   □

EXAMPLE 3 ■ Checking Vectors for Perpendicularity

Determine whether the vectors in each pair are perpendicular.

(a) $\mathbf{u} = \langle 3, 5 \rangle$ and $\mathbf{v} = \langle 2, -8 \rangle$ (b) $\mathbf{u} = \langle 2, 1 \rangle$ and $\mathbf{v} = \langle -1, 2 \rangle$

SOLUTION

(a) $\mathbf{u} \cdot \mathbf{v} = (3)(2) + (5)(-8) = -34 \neq 0$, so $\mathbf{u}$ and $\mathbf{v}$ are not perpendicular.

(b) $\mathbf{u} \cdot \mathbf{v} = (2)(-1) + (1)(2) = 0$, so $\mathbf{u}$ and $\mathbf{v}$ are perpendicular. ∎

■ The Component of u Along v

The **component of u along v** (or the **component of u in the direction of v**) is defined to be

$$|\mathbf{u}| \cos \theta$$

where θ is the angle between $\mathbf{u}$ and $\mathbf{v}$. Figure 3 gives a geometric interpretation of this concept. Intuitively, the component of $\mathbf{u}$ along $\mathbf{v}$ is the magnitude of the portion of $\mathbf{u}$ that points in the direction of $\mathbf{v}$. Notice that the component of $\mathbf{u}$ along $\mathbf{v}$ is negative if $\pi/2 < \theta \leq \pi$.

Note that the component of $\mathbf{u}$ along $\mathbf{v}$ is a scalar, not a vector.

FIGURE 3

 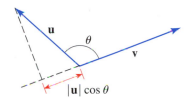

When analyzing forces in physics and engineering, it's often helpful to express a vector as a sum of two vectors lying in perpendicular directions. For example, suppose a car is parked on an inclined driveway as in Figure 4. The weight of the car is a vector $\mathbf{w}$ that points directly downward. We can write

$$\mathbf{w} = \mathbf{u} + \mathbf{v}$$

where $\mathbf{u}$ is parallel to the driveway and $\mathbf{v}$ is perpendicular to the driveway. The vector $\mathbf{u}$ is the force that tends to roll the car down the driveway, and $\mathbf{v}$ is the force experienced by the surface of the driveway. The magnitudes of these forces are the components of $\mathbf{w}$ along $\mathbf{u}$ and $\mathbf{v}$, respectively.

FIGURE 4

FIGURE 5

EXAMPLE 4 ■ Resolving a Force into Components

A car weighing 3000 lb is parked on a driveway that is inclined 15° to the horizontal, as shown in Figure 5.

(a) Find the magnitude of the force required to prevent the car from rolling down the driveway.

(b) Find the magnitude of the force experienced by the driveway due to the weight of the car.

SOLUTION

The car exerts a force **w** of 3000 lb directly downward. We resolve **w** into the sum of two vectors **u** and **v**, one parallel to the surface of the driveway and the other perpendicular to it, as shown in Figure 5.

(a) The magnitude of the part of the force **w** that causes the car to roll down the driveway is

$$|\mathbf{u}| = \text{component of } \mathbf{w} \text{ along } \mathbf{u} = 3000 \cos 75° \approx 776$$

Thus, the force needed to prevent the car from rolling down the driveway is about 776 lb.

(b) The magnitude of the force exerted by the car on the driveway is

$$|\mathbf{v}| = \text{component of } \mathbf{w} \text{ along } \mathbf{v} = 3000 \cos 15° \approx 2898$$

The force experienced by the driveway is about 2898 lb.   ■

The component of **u** along **v** can be computed using dot products:

$$|\mathbf{u}| \cos \theta = \frac{|\mathbf{v}| \, |\mathbf{u}| \cos \theta}{|\mathbf{v}|} = \frac{\mathbf{u} \cdot \mathbf{v}}{|\mathbf{v}|}$$

We have shown the following.

CALCULATING COMPONENTS

The component of **u** along **v** is $\dfrac{\mathbf{u} \cdot \mathbf{v}}{|\mathbf{v}|}$

EXAMPLE 5 ■ Finding Components

Let $\mathbf{u} = \langle 1, 4 \rangle$ and $\mathbf{v} = \langle -2, 1 \rangle$. Find the component of **u** along **v**.

SOLUTION

We have

$$\text{component of } \mathbf{u} \text{ along } \mathbf{v} = \frac{\mathbf{u} \cdot \mathbf{v}}{|\mathbf{v}|} = \frac{(1)(-2) + (4)(1)}{\sqrt{4 + 1}} = \frac{2}{\sqrt{5}}$$ ■

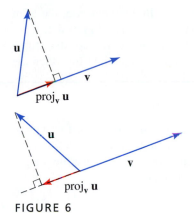

FIGURE 6

■ The Projection of u onto v

Figure 6 shows representations of the vectors **u** and **v**. The projection of **u** onto **v**, denoted by $\text{proj}_{\mathbf{v}}\, \mathbf{u}$, is the vector whose *direction* is the same as **v** and whose *length* is the component of **u** along **v**. To find an expression for $\text{proj}_{\mathbf{v}}\, \mathbf{u}$, we first find a unit vector in the direction of **v** and then multiply it by the component of **u** along **v**.

$$\text{proj}_{\mathbf{v}}\, \mathbf{u} = (\text{component of } \mathbf{u} \text{ along } \mathbf{v})(\text{unit vector in direction of } \mathbf{v})$$

$$= \left(\frac{\mathbf{u} \cdot \mathbf{v}}{|\mathbf{v}|}\right)\frac{\mathbf{v}}{|\mathbf{v}|} = \left(\frac{\mathbf{u} \cdot \mathbf{v}}{|\mathbf{v}|^2}\right)\mathbf{v}$$

We often need to **resolve** a vector **u** into the sum of two vectors, one parallel to **v** and one orthogonal to **v**. That is, we want to write $\mathbf{u} = \mathbf{u}_1 + \mathbf{u}_2$ where $\mathbf{u}_1$ is parallel to **v** and $\mathbf{u}_2$ is orthogonal to **v**. In this case, $\mathbf{u}_1 = \text{proj}_{\mathbf{v}}\, \mathbf{u}$ and $\mathbf{u}_2 = \mathbf{u} - \text{proj}_{\mathbf{v}}\, \mathbf{u}$ (see Exercise 45).

CALCULATING PROJECTIONS

The **projection of u onto v** is the vector $\text{proj}_{\mathbf{v}}\, \mathbf{u}$ given by

$$\text{proj}_{\mathbf{v}}\, \mathbf{u} = \left(\frac{\mathbf{u} \cdot \mathbf{v}}{|\mathbf{v}|^2}\right)\mathbf{v}$$

If the vector **u** is **resolved** into $\mathbf{u}_1$ and $\mathbf{u}_2$, where $\mathbf{u}_1$ is parallel to **v** and $\mathbf{u}_2$ is orthogonal to **v**, then

$$\mathbf{u}_1 = \text{proj}_{\mathbf{v}}\, \mathbf{u} \qquad \text{and} \qquad \mathbf{u}_2 = \mathbf{u} - \text{proj}_{\mathbf{v}}\, \mathbf{u}$$

EXAMPLE 6 ■ Resolving a Vector into Orthogonal Vectors

Let $\mathbf{u} = \langle -2, 9 \rangle$ and $\mathbf{v} = \langle -1, 2 \rangle$.

(a) Find $\text{proj}_{\mathbf{v}}\, \mathbf{u}$.

(b) Resolve **u** into $\mathbf{u}_1$ and $\mathbf{u}_2$, where $\mathbf{u}_1$ is parallel to **v** and $\mathbf{u}_2$ is orthogonal to **v**.

SOLUTION

(a) By the formula for the projection of one vector onto another we have

$$\text{proj}_{\mathbf{v}}\, \mathbf{u} = \left(\frac{\mathbf{u} \cdot \mathbf{v}}{|\mathbf{v}|^2}\right)\mathbf{v} \qquad \textcolor{red}{\text{Formula for projection}}$$

$$= \left(\frac{\langle -2, 9 \rangle \cdot \langle -1, 2 \rangle}{(-1)^2 + 2^2}\right)\langle -1, 2 \rangle \qquad \textcolor{red}{\text{Definition of } \mathbf{u} \text{ and } \mathbf{v}}$$

$$= 4\langle -1, 2 \rangle = \langle -4, 8 \rangle$$

(b) By the formula in the preceding box we have $\mathbf{u} = \mathbf{u}_1 + \mathbf{u}_2$, where

$$\mathbf{u}_1 = \text{proj}_{\mathbf{v}}\, \mathbf{u} = \langle -4, 8 \rangle \qquad \textcolor{red}{\text{From part (a)}}$$

$$\mathbf{u}_2 = \mathbf{u} - \text{proj}_{\mathbf{v}}\, \mathbf{u} = \langle -2, 9 \rangle - \langle -4, 8 \rangle = \langle 2, 1 \rangle$$

■

■ Work

One use of the dot product occurs in calculating work. In everyday use, the term *work* means the total amount of effort required to perform a task. In physics, *work* has a technical meaning that conforms to this intuitive meaning. If a constant force of magnitude F moves an object through a distance d along a straight line, then the **work** done is

$$W = Fd \qquad \text{or} \qquad \text{work} = \text{force} \times \text{distance}$$

If F is measured in pounds and d in feet, then the unit of work is a foot-pound (ft-lb). For example, how much work is done in lifting a 20-lb weight 6 ft off the ground? Since a force of 20 lb is required to lift this weight and since the weight moves through a distance of 6 ft, the amount of work done is

$$W = Fd = (20)(6) = 120 \text{ ft-lb}$$

This formula applies only when the force is directed along the direction of motion. In the general case, if the force $\mathbf{F}$ moves an object from P to Q, as in Figure 7, then only the component of the force in the direction of $\mathbf{D} = \overrightarrow{PQ}$ affects the object. Thus, the effective magnitude of the force on the object is

$$\text{component of } \mathbf{F} \text{ along } \mathbf{D} = |\mathbf{F}| \cos \theta$$

FIGURE 7

So, the work done is

$$W = \text{force} \times \text{distance} = (|\mathbf{F}| \cos \theta) |\mathbf{D}| = |\mathbf{F}| |\mathbf{D}| \cos \theta = \mathbf{F} \cdot \mathbf{D}$$

We have derived the following simple formula for calculating work.

WORK

The **work** W done by a force $\mathbf{F}$ in moving along a vector $\mathbf{D}$ is

$$W = \mathbf{F} \cdot \mathbf{D}$$

EXAMPLE 7 ■ Calculating Work

A force is given by the vector $\mathbf{F} = \langle 2, 3 \rangle$ and moves an object from the point $(1, 3)$ to the point $(5, 9)$. Find the work done.

SOLUTION

The displacement vector is

$$\mathbf{D} = \langle 5 - 1, 9 - 3 \rangle = \langle 4, 6 \rangle$$

So the work done is

$$W = \mathbf{F} \cdot \mathbf{D} = \langle 2, 3 \rangle \cdot \langle 4, 6 \rangle = 26$$

If the unit of force is pounds and the distance is measured in feet, then the work done is 26 ft-lb. ■

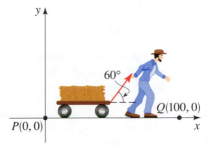

FIGURE 8

EXAMPLE 8 ■ Calculating Work

A man pulls a wagon horizontally by exerting a force of 20 lb on the handle. If the handle makes an angle of 60° with the horizontal, find the work done in moving the wagon 100 ft.

SOLUTION

We choose a coordinate system with the origin at the initial position of the wagon (see Figure 8). That is, the wagon moves from the point $P(0, 0)$ to the point $Q(100, 0)$. The vector that represents this displacement is

$$\mathbf{D} = 100\,\mathbf{i}$$

The force on the handle can be written in terms of components (see Section 7.7) as

$$\mathbf{F} = (20 \cos 60°)\,\mathbf{i} + (20 \sin 60°)\,\mathbf{j} = 10\,\mathbf{i} + 10\,\sqrt{3}\,\mathbf{j}$$

Thus, the work done is

$$W = \mathbf{F} \cdot \mathbf{D} = (10\,\mathbf{i} + 10\,\sqrt{3}\,\mathbf{j}) \cdot (100\,\mathbf{i}) = 1000 \text{ ft-lb} \qquad ■$$

7.8   EXERCISES

1–8 ■ Find (a) $\mathbf{u} \cdot \mathbf{v}$ and (b) the angle between $\mathbf{u}$ and $\mathbf{v}$ to the nearest degree.

1. $\mathbf{u} = \langle 2, 0 \rangle$, $\mathbf{v} = \langle 1, 1 \rangle$

2. $\mathbf{u} = \mathbf{i} + \sqrt{3}\,\mathbf{j}$, $\mathbf{v} = -\sqrt{3}\,\mathbf{i} + \mathbf{j}$

3. $\mathbf{u} = \langle 2, 7 \rangle$, $\mathbf{v} = \langle 3, 1 \rangle$

4. $\mathbf{u} = \langle -6, 6 \rangle$, $\mathbf{v} = \langle 1, -1 \rangle$

5. $\mathbf{u} = \langle 3, -2 \rangle$, $\mathbf{v} = \langle 1, 2 \rangle$

6. $\mathbf{u} = 2\,\mathbf{i} + \mathbf{j}$, $\mathbf{v} = 3\,\mathbf{i} - 2\,\mathbf{j}$

7. $\mathbf{u} = -5\,\mathbf{j}$, $\mathbf{v} = -\mathbf{i} - \sqrt{3}\,\mathbf{j}$

8. $\mathbf{u} = \mathbf{i} + \mathbf{j}$, $\mathbf{v} = \mathbf{i} - \mathbf{j}$

9–14 ■ Determine whether the given vectors are orthogonal.

9. $\mathbf{u} = \langle 6, 4 \rangle$, $\mathbf{v} = \langle -2, 3 \rangle$

10. $\mathbf{u} = \langle 0, -5 \rangle$, $\mathbf{v} = \langle 4, 0 \rangle$

11. $\mathbf{u} = \langle -2, 6 \rangle$, $\mathbf{v} = \langle 4, 2 \rangle$

12. $\mathbf{u} = 2\,\mathbf{i}$, $\mathbf{v} = -7\,\mathbf{j}$

13. $\mathbf{u} = 2\,\mathbf{i} - 8\,\mathbf{j}$, $\mathbf{v} = -12\,\mathbf{i} - 3\,\mathbf{j}$

14. $\mathbf{u} = 4\,\mathbf{i}$, $\mathbf{v} = -\mathbf{i} + 3\,\mathbf{j}$

15–18 ■ Find the indicated quantity, assuming $\mathbf{u} = 2\,\mathbf{i} + \mathbf{j}$, $\mathbf{v} = \mathbf{i} - 3\,\mathbf{j}$, and $\mathbf{w} = 3\,\mathbf{i} + 4\,\mathbf{j}$.

15. $\mathbf{u} \cdot \mathbf{v} + \mathbf{u} \cdot \mathbf{w}$

16. $\mathbf{u} \cdot (\mathbf{v} + \mathbf{w})$

17. $(\mathbf{u} + \mathbf{v}) \cdot (\mathbf{u} - \mathbf{v})$

18. $(\mathbf{u} \cdot \mathbf{v})(\mathbf{u} \cdot \mathbf{w})$

19–22 ■ Find the component of $\mathbf{u}$ along $\mathbf{v}$.

19. $\mathbf{u} = \langle 4, 6 \rangle$, $\mathbf{v} = \langle 3, -4 \rangle$

20. $\mathbf{u} = \langle -3, 5 \rangle$, $\mathbf{v} = \langle 1/\sqrt{2}, 1/\sqrt{2} \rangle$

21. $\mathbf{u} = 7\,\mathbf{i} - 24\,\mathbf{j}$, $\mathbf{v} = \mathbf{j}$

22. $\mathbf{u} = 7\,\mathbf{i}$, $\mathbf{v} = 8\,\mathbf{i} + 6\,\mathbf{j}$

23–28 ■ (a) Calculate $\text{proj}_{\mathbf{v}}\,\mathbf{u}$. (b) Resolve $\mathbf{u}$ into $\mathbf{u}_1$ and $\mathbf{u}_2$, where $\mathbf{u}_1$ is parallel to $\mathbf{v}$ and $\mathbf{u}_2$ is orthogonal to $\mathbf{v}$.

23. $\mathbf{u} = \langle -2, 4 \rangle$, $\mathbf{v} = \langle 1, 1 \rangle$

24. $\mathbf{u} = \langle 7, -4 \rangle$, $\mathbf{v} = \langle 2, 1 \rangle$

25. $\mathbf{u} = \langle 1, 2 \rangle$, $\mathbf{v} = \langle 1, -3 \rangle$

26. $\mathbf{u} = \langle 11, 3 \rangle$, $\mathbf{v} = \langle -3, -2 \rangle$

27. $\mathbf{u} = \langle 2, 9 \rangle$, $\mathbf{v} = \langle -3, 4 \rangle$

28. $\mathbf{u} = \langle 1, 1 \rangle$, $\mathbf{v} = \langle 2, -1 \rangle$

29–32 ■ Find the work done by the force **F** in moving an object from P to Q.

29. $\mathbf{F} = 4\mathbf{i} - 5\mathbf{j}$; $P(0, 0), Q(3, 8)$

30. $\mathbf{F} = 400\mathbf{i} + 50\mathbf{j}$; $P(-1, 1), Q(200, 1)$

31. $\mathbf{F} = 10\mathbf{i} + 3\mathbf{j}$; $P(2, 3), Q(6, -2)$

32. $\mathbf{F} = -4\mathbf{i} + 20\mathbf{j}$; $P(0, 10), Q(5, 25)$

33. The force $\mathbf{F} = 4\mathbf{i} - 7\mathbf{j}$ moves an object 4 ft along the x-axis in the positive direction. Find the work done if the unit of force is the pound.

34. A constant force $\mathbf{F} = \langle 2, 8 \rangle$ moves an object along a straight line from the point $(2, 5)$ to the point $(11, 13)$. Find the work done if the distance is measured in feet and the force is measured in pounds.

35. A lawn mower is pushed a distance of 200 ft along a horizontal path by a constant force of 50 lb. The handle of the lawn mower is held at an angle of 30° from the horizontal (see the figure). Find the work done.

36. A car drives 500 ft on a road that is inclined 12° to the horizontal, as shown in the figure. The car weighs 2500 lb. Thus, gravity acts straight down on the car with a constant force $\mathbf{F} = -2500\mathbf{j}$. Find the work done by the car in overcoming gravity.

37. A car is on a driveway that is inclined 25° to the horizontal. If the car weighs 2755 lb, find the force required to keep it from rolling down the driveway.

38. A car is on a driveway that is inclined 10° to the horizontal. A force of 490 lb is required to keep the car from rolling down the driveway.

(a) Find the weight of the car.
(b) Find the force the car exerts against the driveway.

39. A package that weighs 200 lb is placed on an inclined plane. If a force of 80 lb is just sufficient to keep the package from sliding, find the angle of inclination of the plane. (Ignore the effects of friction.)

40. A sailboat has its sail inclined in the direction N 20° E. The wind is blowing into the sail in the direction S 45° W with a force of 220 lb (see the figure).

(a) Find the effective force of the wind on the sail. [*Hint:* Find the components of the wind parallel to the sail and perpendicular to the sail. The component of the wind parallel to the sail slips by and does not push on the sail.]

(b) If the keel of the ship is aligned due north, find the effective force of the wind that drives the boat forward. [*Hint:* Only the component of the force found in part (a) that is parallel to the keel drives the boat forward.]

41–44 ■ Let **u**, **v**, and **w** be vectors and let a be a scalar. Prove the given property.

41. $\mathbf{u} \cdot \mathbf{v} = \mathbf{v} \cdot \mathbf{u}$

42. $(a\mathbf{u}) \cdot \mathbf{v} = a(\mathbf{u} \cdot \mathbf{v}) = \mathbf{u} \cdot (a\mathbf{v})$

43. $(\mathbf{u} + \mathbf{v}) \cdot \mathbf{w} = \mathbf{u} \cdot \mathbf{w} + \mathbf{v} \cdot \mathbf{w}$

44. $(\mathbf{u} - \mathbf{v}) \cdot (\mathbf{u} + \mathbf{v}) = |\mathbf{u}|^2 - |\mathbf{v}|^2$

45. Show that the vectors $\text{proj}_\mathbf{v}\, \mathbf{u}$ and $\mathbf{u} - \text{proj}_\mathbf{v}\, \mathbf{u}$ are orthogonal.

DISCOVERY · DISCUSSION

46. Distance from a Point to a Line   Let L be the line $2x + 4y = 8$ and let P be the point $(3, 4)$.

(a) Show that the points $Q(0, 2)$ and $R(2, 1)$ lie on L.

(b) Let $\mathbf{u} = \overrightarrow{QP}$ and $\mathbf{v} = \overrightarrow{QR}$, as shown in the figure. Find $\mathbf{w} = \text{proj}_{\mathbf{v}}\, \mathbf{u}$.

(c) Sketch a graph that explains why $|\mathbf{u} - \mathbf{w}|$ is the distance from P to L. Find this distance.

(d) Write a short paragraph describing the steps you would take to find the distance from a given point to a given line.

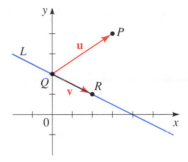

7   REVIEW

CONCEPT CHECK

1. (a) State the reciprocal identities.
   (b) State the Pythagorean identities.
   (c) State the even-odd identities.
   (d) State the cofunction identities.

2. Explain the difference between an equation and an identity.

3. How do you prove a trigonometric identity?

4. (a) State the addition formulas for sine, cosine, and tangent.
   (b) State the subtraction formulas for sine, cosine, and tangent.

5. (a) State the double-angle formulas for sine, cosine, and tangent.
   (b) State the formulas for lowering powers.
   (c) State the half-angle formulas.

6. (a) State the product-to-sum formulas.
   (b) State the sum-to-product formulas.

7. (a) Define the inverse sine function $\sin^{-1}$. What are its domain and range?
   (b) For what values of x is the equation $\sin(\sin^{-1}x) = x$ true?
   (c) For what values of x is the equation $\sin^{-1}(\sin x) = x$ true?

8. (a) Define the inverse cosine function $\cos^{-1}$. What are its domain and range?
   (b) For what values of x is the equation $\cos(\cos^{-1}x) = x$ true?
   (c) For what values of x is the equation $\cos^{-1}(\cos x) = x$ true?

9. (a) Define the inverse tangent function $\tan^{-1}$. What are its domain and range?
   (b) For what values of x is the equation $\tan(\tan^{-1}x) = x$ true?
   (c) For what values of x is the equation $\tan^{-1}(\tan x) = x$ true?

10. How do you graph a complex number z? What is the trigonometric form of a complex number z? What is the modulus of z? What is the argument of z?

11. (a) How do you multiply two complex numbers if they are given in trigonometric form?
   (b) How do you divide two such numbers?

12. (a) State DeMoivre's Theorem.
   (b) How do you find the nth roots of a complex number?

13. (a) What is the difference between a scalar and a vector?
   (b) Draw a diagram to show how to add two vectors.
   (c) Draw a diagram to show how to subtract two vectors.
   (d) Draw a diagram to show how to multiply a vector by the scalars 2, $\frac{1}{2}$, -2, and $-\frac{1}{2}$.

14. If $\mathbf{u} = \langle a_1, b_1 \rangle$, $\mathbf{v} = \langle a_2, b_2 \rangle$ and c is a scalar, write expressions for $\mathbf{u} + \mathbf{v}$, $\mathbf{u} - \mathbf{v}$, $c\mathbf{u}$, and $|\mathbf{u}|$.

15. (a) If $\mathbf{v} = \langle a, b \rangle$, write $\mathbf{v}$ in terms of $\mathbf{i}$ and $\mathbf{j}$.
   (b) Write the components of $\mathbf{v}$ in terms of the magnitude and direction of $\mathbf{v}$.

16. If $\mathbf{u} = \langle a_1, b_1 \rangle$ and $\mathbf{v} = \langle a_2, b_2 \rangle$, what is the dot product $\mathbf{u} \cdot \mathbf{v}$?

17. (a) How do you use the dot product to find the angle between two vectors?

(b) How do you use the dot product to determine whether two vectors are perpendicular?

18. What is the component of **u** along **v**, and how do you calculate it?

19. What is the projection of **u** onto **v**, and how do you calculate it?

20. How much work is done by the force **F** in moving an object along a displacement **D**?

EXERCISES

1–24 ■ Verify the identity.

1. $\sin\theta(\cot\theta + \tan\theta) = \sec\theta$

2. $(\sec\theta - 1)(\sec\theta + 1) = \tan^2\theta$

3. $\cos^2x\csc x - \csc x = -\sin x$

4. $\dfrac{1}{1 - \sin^2x} = 1 + \tan^2x$

5. $\dfrac{\cos^2x - \tan^2x}{\sin^2x} = \cot^2x - \sec^2x$

6. $\dfrac{1 + \sec x}{\sec x} = \dfrac{\sin^2x}{1 - \cos x}$ **7.** $\dfrac{\cos^2x}{1 - \sin x} = \dfrac{\cos x}{\sec x - \tan x}$

8. $(1 - \tan x)(1 - \cot x) = 2 - \sec x\csc x$

9. $\sin^2x\cot^2x + \cos^2x\tan^2x = 1$

10. $(\tan x + \cot x)^2 = \csc^2x\sec^2x$

11. $\dfrac{\sin 2x}{1 + \cos 2x} = \tan x$

12. $\dfrac{\cos(x + y)}{\cos x\sin y} = \cot y - \tan x$

13. $\tan\dfrac{x}{2} = \csc x - \cot x$

14. $\dfrac{\sin(x + y) + \sin(x - y)}{\cos(x + y) + \cos(x - y)} = \tan x$

15. $\sin(x + y)\sin(x - y) = \sin^2x - \sin^2y$

16. $\csc x - \tan\dfrac{x}{2} = \cot x$ **17.** $1 + \tan x\tan\dfrac{x}{2} = \sec x$

18. $\dfrac{\sin 3x + \cos 3x}{\cos x - \sin x} = 1 + 2\sin 2x$

19. $\left(\cos\dfrac{x}{2} - \sin\dfrac{x}{2}\right)^2 = 1 - \sin x$

20. $\dfrac{\cos 3x - \cos 7x}{\sin 3x + \sin 7x} = \tan 2x$ **21.** $\dfrac{\sin 2x}{\sin x} - \dfrac{\cos 2x}{\cos x} = \sec x$

22. $(\cos x + \cos y)^2 + (\sin x - \sin y)^2 = 2 + 2\cos(x + y)$

23. $\tan\left(x + \dfrac{\pi}{4}\right) = \dfrac{1 + \tan x}{1 - \tan x}$ **24.** $\dfrac{\sec x - 1}{\sin x\sec x} = \tan\dfrac{x}{2}$

25–28 ■ (a) Graph f and g. (b) Do the graphs suggest that the equation $f(x) = g(x)$ is an identity? Prove your answer.

25. $f(x) = 1 - \left(\cos\dfrac{x}{2} - \sin\dfrac{x}{2}\right)^2$, $g(x) = \sin x$

26. $f(x) = \sin x + \cos x$, $g(x) = \sqrt{\sin^2x + \cos^2x}$

27. $f(x) = \tan x\tan\dfrac{x}{2}$, $g(x) = \dfrac{1}{\cos x}$

28. $f(x) = 1 - 8\sin^2x + 8\sin^4x$, $g(x) = \cos 4x$

29–30 ■ (a) Graph the function(s) and make a conjecture, and (b) prove your conjecture.

29. $f(x) = 2\sin^2 3x + \cos 6x$

30. $f(x) = \sin x\cot\dfrac{x}{2}$, $g(x) = \cos x$

31–46 ■ Solve the equation in the interval $[0, 2\pi)$.

31. $\cos x\sin x - \sin x = 0$ **32.** $\sin x - 2\sin^2x = 0$

33. $2\sin^2x - 5\sin x + 2 = 0$

34. $\sin x - \cos x - \tan x = -1$

35. $2\cos^2x - 7\cos x + 3 = 0$ **36.** $4\sin^2x + 2\cos^2x = 3$

37. $\dfrac{1 - \cos x}{1 + \cos x} = 3$ **38.** $\sin x = \cos 2x$

39. $\tan^3x + \tan^2x - 3\tan x - 3 = 0$

40. $\cos 2x\csc^2x = 2\cos 2x$ **41.** $\tan\tfrac{1}{2}x + 2\sin 2x = \csc x$

42. $\cos 3x + \cos 2x + \cos x = 0$

43. $\tan x + \sec x = \sqrt{3}$ **44.** $2\cos x - 3\tan x = 0$

45. $\cos x = x^2 - 1$ **46.** $e^{\sin x} = x$

47. If a projectile is fired with velocity v_0 at an angle θ, then the maximum height it reaches (in feet) is modeled by the function

$$M(\theta) = \dfrac{v_0^2\sin^2\theta}{64}$$

Suppose $v_0 = 400$ ft/s.
(a) At what angle θ should the projectile be fired so that the maximum height it reaches is 2000 ft?

(b) Is it possible for the projectile to reach a height of 3000 ft?

(c) Find the angle θ for which the projectile will travel highest.

48. The displacement of an automobile shock absorber is modeled by the function

$$f(t) = e^{-0.2t} \sin 4\pi t$$

Find the times when the shock absorber is at its equilibrium position (that is, when $f(t) = 0$).

49–58 ■ Find the exact value of the expression.

49. $\cos 15°$

50. $\sin \dfrac{5\pi}{12}$

51. $\tan \dfrac{\pi}{8}$

52. $2 \sin \dfrac{\pi}{12} \cos \dfrac{\pi}{12}$

53. $\sin 5° \cos 40° + \cos 5° \sin 40°$

54. $\dfrac{\tan 66° - \tan 6°}{1 + \tan 66° \tan 6°}$

55. $\cos^2 \dfrac{\pi}{8} - \sin^2 \dfrac{\pi}{8}$

56. $\dfrac{1}{2} \cos \dfrac{\pi}{12} + \dfrac{\sqrt{3}}{2} \sin \dfrac{\pi}{12}$

57. $\cos 37.5° \cos 7.5°$

58. $\cos 67.5° + \cos 22.5°$

59–64 ■ Find the exact value of the expression given that $\sec x = \frac{3}{2}$, $\csc y = 3$, and x and y are in quadrant I.

59. $\sin(x + y)$

60. $\cos(x - y)$

61. $\tan(x + y)$

62. $\sin 2x$

63. $\cos \dfrac{y}{2}$

64. $\tan \dfrac{y}{2}$

65–72 ■ Find the exact value of the expression.

65. $\sin^{-1}(\sqrt{3}/2)$

66. $\tan^{-1}(\sqrt{3}/3)$

67. $\cos(\tan^{-1} \sqrt{3})$

68. $\sin(\cos^{-1}(\sqrt{3}/2))$

69. $\tan(\sin^{-1} \frac{2}{5})$

70. $\sin(\cos^{-1} \frac{3}{8})$

71. $\cos(2 \sin^{-1} \frac{1}{3})$

72. $\cos(\sin^{-1} \frac{5}{13} - \cos^{-1} \frac{4}{5})$

73–74 ■ Rewrite the expression as an algebraic function of x.

73. $\sin(\tan^{-1} x)$

74. $\sec(\sin^{-1} x)$

75–76 ■ Express θ in terms of x.

75.

76.

77. A 10-ft-wide highway sign is adjacent to a roadway, as shown in the figure. As a driver approaches the sign, the viewing angle θ changes.

(a) Express viewing angle θ as a function of the distance x between the driver and the sign.

(b) For what distance x is the viewing angle largest?

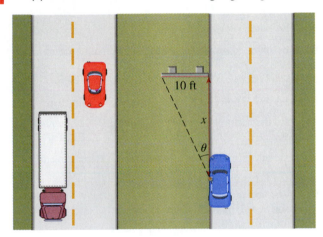

78. A 380-ft-tall building supports a 40-ft communications tower (see the figure). As a driver approaches the building, the viewing angle θ of the tower changes.

(a) Express the viewing angle θ as a function of the distance x between the driver and the building.

(b) At what distance from the building is the viewing angle θ as large as possible?

79–84 ■ A complex number is given.
(a) Graph the complex number in the complex plane.
(b) Find the modulus and argument.
(c) Write the number in trigonometric form.

79. $4 + 4i$

80. $-10i$

81. $5 + 3i$

82. $1 + \sqrt{3}\, i$

83. $-1 + i$

84. -20

85–88 ■ Find the indicated power.

85. $(1 - \sqrt{3}\, i)^4$

86. $(1 + i)^8$

87. $(\sqrt{3} + i)^{-4}$

88. $\left(\dfrac{1}{2} + \dfrac{\sqrt{3}}{2} i\right)^{20}$

89–92 ■ Find the indicated roots.

89. The square roots of $-16i$

90. The cube roots of $4 + 4\sqrt{3}\, i$

91. The sixth roots of 1   **92.** The eighth roots of i

93–94 ■ Find $|\mathbf{u}|$, $\mathbf{u} + \mathbf{v}$, $\mathbf{u} - \mathbf{v}$, $2\mathbf{u}$, and $3\mathbf{u} - 2\mathbf{v}$.

93. $\mathbf{u} = \langle -2, 3 \rangle$, $\mathbf{v} = \langle 8, 1 \rangle$ **94.** $\mathbf{u} = 2\mathbf{i} + \mathbf{j}$, $\mathbf{v} = \mathbf{i} - 2\mathbf{j}$

95. Find the vector $\mathbf{u}$ with initial point $P(0, 3)$ and terminal point $Q(3, -1)$.

96. Find the vector $\mathbf{u}$ having length $|\mathbf{u}| = 20$ and direction $\theta = 60°$.

97. If the vector $5\mathbf{i} - 8\mathbf{j}$ is placed in the plane with its initial point at $P(5, 6)$, find its terminal point.

98. Find the direction of the vector $2\mathbf{i} - 5\mathbf{j}$.

99. Two tugboats are pulling a barge, as shown. One pulls with a force of 2.0×10^4 lb in the direction N 50° E and the other with a force of 3.4×10^4 lb in the direction S 75° E.
(a) Find the resultant force on the barge as a vector.
(b) Find the magnitude and direction of the resultant force.

100. An airplane heads N 60° E at a speed of 600 mi/h relative to the air. A wind begins to blow in the direction N 30° W at 50 mi/h.
(a) Find the velocity of the airplane as a vector.
(b) Find the true speed and direction of the airplane.

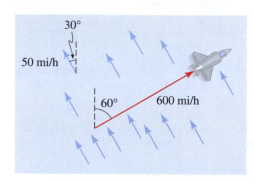

101–104 ■ Find $|\mathbf{u}|$, $\mathbf{u} \cdot \mathbf{u}$, and $\mathbf{u} \cdot \mathbf{v}$.

101. $\mathbf{u} = \langle 4, -3 \rangle$, $\mathbf{v} = \langle 9, -8 \rangle$

102. $\mathbf{u} = \langle 5, 12 \rangle$, $\mathbf{v} = \langle 10, -4 \rangle$

103. $\mathbf{u} = -2\mathbf{i} + 2\mathbf{j}$, $\mathbf{v} = \mathbf{i} + \mathbf{j}$

104. $\mathbf{u} = 10\mathbf{j}$, $\mathbf{v} = 5\mathbf{i} - 3\mathbf{j}$

105–108 ■ Are $\mathbf{u}$ and $\mathbf{v}$ orthogonal? If not, find the angle between them.

105. $\mathbf{u} = \langle -4, 2 \rangle$, $\mathbf{v} = \langle 3, 6 \rangle$

106. $\mathbf{u} = \langle 5, 3 \rangle$, $\mathbf{v} = \langle -2, 6 \rangle$

107. $\mathbf{u} = 2\mathbf{i} + \mathbf{j}$, $\mathbf{v} = \mathbf{i} + 3\mathbf{j}$

108. $\mathbf{u} = \mathbf{i} - \mathbf{j}$, $\mathbf{v} = \mathbf{i} + \mathbf{j}$

109–110 ■ The vectors $\mathbf{u}$ and $\mathbf{v}$ are given.
(a) Find the component of $\mathbf{u}$ along $\mathbf{v}$.
(b) Find $\text{proj}_{\mathbf{v}}\, \mathbf{u}$.
(c) Resolve $\mathbf{u}$ into the vectors $\mathbf{u}_1$ and $\mathbf{u}_2$, where $\mathbf{u}_1$ is parallel to $\mathbf{v}$ and $\mathbf{u}_2$ is perpendicular to $\mathbf{v}$.

109. $\mathbf{u} = \langle 3, 1 \rangle$, $\mathbf{v} = \langle 6, -1 \rangle$

110. $\mathbf{u} = \langle -8, 6 \rangle$, $\mathbf{v} = \langle 20, 20 \rangle$

111. Find the work done by the force $\mathbf{F} = 2\mathbf{i} + 9\mathbf{j}$ in moving an object from the point $(1, 1)$ to the point $(7, -1)$.

112. A force $\mathbf{F}$ with magnitude 250 lb moves an object in the direction of the vector $\mathbf{D}$ a distance of 20 ft. If the work done is 3800 ft-lb, find the angle between $\mathbf{F}$ and $\mathbf{D}$.

7 TEST

1. Verify each identity.

   (a) $\tan \theta \sin \theta + \cos \theta = \sec \theta$

   (b) $\dfrac{\tan x}{1 - \cos x} = \csc x (1 + \sec x)$

   (c) $\dfrac{2 \tan x}{1 + \tan^2 x} = \sin 2x$

2. Let $x = 2 \sin \theta$, $-\pi/2 < \theta < \pi/2$. Simplify the expression

$$\frac{x}{\sqrt{4 - x^2}}$$

3. Find the exact value of each expression.

   (a) $\sin 8° \cos 22° + \cos 8° \sin 22°$ (b) $\sin 75°$ (c) $\sin \dfrac{\pi}{12}$

4. For the angles α and β in the figures, find $\cos(\alpha + \beta)$.

5. (a) Write $\sin 3x \cos 5x$ as a sum of trigonometric functions.
   (b) Write $\sin 2x - \sin 5x$ as a product of trigonometric functions.

6. If $\sin \theta = -\frac{4}{5}$ and θ is in quadrant III, find $\tan(\theta/2)$.

7. Graph $y = \sin x$ and $y = \sin^{-1}x$, and specify the domain of each function.

8. Express θ in each figure in terms of x.

   (a)

   (b)

9. Solve each trigonometric equation in the interval $[0, 2\pi)$.
   (a) $2 \cos^2 x + 5 \cos x + 2 = 0$ (b) $\sin 2x - \cos x = 0$

10. Find all solutions in the interval $[0, 2\pi)$, correct to five decimal places:

$$5 \cos 2x = 2$$

11. Find the exact value of $\cos\left(\tan^{-1}\frac{9}{40}\right)$.

12. Let $z = 1 + \sqrt{3}\,i$
   (a) Graph z in the complex plane.
   (b) Write z in trigonometric form.
   (c) Find the complex number z^9.

13. Let

$$z_1 = 4\left(\cos\frac{7\pi}{12} + i\sin\frac{7\pi}{12}\right)$$

$$z_2 = 2\left(\cos\frac{5\pi}{12} + i\sin\frac{5\pi}{12}\right)$$

Find $z_1 z_2$ and $\dfrac{z_1}{z_2}$.

14. Find the cube roots of $27i$, and sketch these roots in the complex plane.

15. Let $\mathbf{u}$ be the vector with initial point $P(3, -1)$ and terminal point $Q(-3, 9)$.
   (a) Express $\mathbf{u}$ in terms of $\mathbf{i}$ and $\mathbf{j}$.
   (b) Find the length of $\mathbf{u}$.

16. Let $\mathbf{u} = \langle 1, 3 \rangle$ and $\mathbf{v} = \langle -6, 2 \rangle$.
   (a) Find $\mathbf{u} - 3\mathbf{v}$.
   (b) Find $|\mathbf{u} + \mathbf{v}|$.
   (c) Find $\mathbf{u} \cdot \mathbf{v}$.
   (d) Are $\mathbf{u}$ and $\mathbf{v}$ perpendicular?

17. A river is flowing due east at 8 mi/h. A man heads his motorboat in a direction N 30° E in the river. The speed of the motorboat relative to the water is 12 mi/h.
   (a) Express the true velocity of the motorboat as a vector.
   (b) Find the true speed and direction of the motorboat.

18. Let $\mathbf{u} = 3\mathbf{i} + 2\mathbf{j}$ and $\mathbf{v} = 5\mathbf{i} - \mathbf{j}$.
   (a) Find the angle between $\mathbf{u}$ and $\mathbf{v}$.
   (b) Find the component of $\mathbf{u}$ along $\mathbf{v}$.
   (c) Find $\text{proj}_\mathbf{v}\,\mathbf{u}$.

19. Find the work done by the force $\mathbf{F} = 3\mathbf{i} - 5\mathbf{j}$ in moving an object from the point $(2, 2)$ to the point $(7, -13)$.

Focus on Problem Solving
Using Analogy

Often a difficult problem can be successfully tackled by **using analogy**. We first solve a similar but simpler problem. The simpler problem can reveal insights that give us a new way to attack the original problem. This problem-solving principle is particularly useful when we've hit a dead end in a problem or if we're having trouble getting started. Mathematicians frequently use this principle to guide them in the uncharted waters of modern mathematical research.

■ A Trigonometric Equation

How many solutions does the following equation have?

$$\sin x = \frac{x}{100}$$

■ Thinking about the Problem

There is no obvious algebraic technique for solving this equation, so let's try to solve it graphically, by graphing $y = \sin x$ and $y = x/100$ on the same axes. We should be able to count the solutions easily from the graph.

In Figure 1 we graph these functions over three complete periods of the sine function, that is, for $-3\pi \le x \le 3\pi$. The line $y = x/100$ has such a shallow slope that it is virtually indistinguishable from the x-axis in this graph, and it's obvious that if we extend the graph to the left and right, we will find many more solutions of the equation.

FIGURE 1

FIGURE 2

We'll use a graphing calculator to get a wider, more appropriate graph. Figure 2 shows the result when we graph $y = \sin x$ and $y = x/100$ in the viewing rectangle $[-110, 110]$ by $[-1.2, 1.2]$. The graph still doesn't tell us much, because it is such a mess—the graphing calculator doesn't have enough resolution to draw this many periods of the sine function accurately.

■ Solving a Simpler Problem

The graphs in Figures 1 and 2 didn't help us because the equation has too many solutions to count easily. If we change 100 to 10, then the line will be much less shallow. So, let's see if we can count the number of solutions of the equation

$$\sin x = \frac{x}{10}$$

The graph in Figure 3 shows that this equation has seven solutions. The figure captures all the solutions, because to the left and right of the portion shown the graph of $y = x/10$ lies below or above the graph of $y = \sin x$.

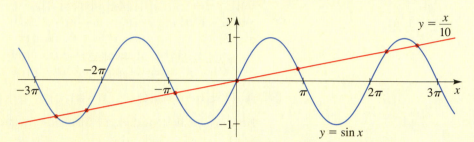

FIGURE 3

Examining the solutions more closely, we see that the graphs of $y = x/10$ and $y = \sin x$ intersect twice in each period of $y = \sin x$. This means that there are *two* solutions in each period. So to count the number of solutions we need only count the number of periods for which the two graphs intersect. From Figure 3 we see that there are *four* such periods: two for which $x \geq 0$ and two for which $x \leq 0$. Since the solution $x = 0$ occurs in both cases, the total number of solutions of the equation is

$$2 \cdot 4 - 1 = 7$$

■ **Solving the Original Problem**

We now use the insight we gained in solving the simpler equation to solve the original equation. When the slope is $1/100$, there will also be two solutions in each period. All we need to find is the number of periods in which the graphs of $y = x/100$ and $y = \sin x$ intersect. Since $-1 \leq \sin x \leq 1$, the graphs can intersect only when $-1 \leq x/100 \leq 1$ as well. Solving this inequality, we see that the graphs intersect only in the interval $-100 \leq x \leq 100$. How many complete periods of $y = \sin x$ occur in this interval? Since $100/2\pi \approx 16$, there are 16 periods between 0 and 100 and 16 periods between -100 and 0, for a total of 32 periods. So, the number of solutions of the original equation is

$$2 \cdot 32 - 1 = 63$$

As in the simpler case, we have subtracted 1 to avoid counting $x = 0$ twice.   ■

■ **Problems**

1. Find the number of solutions of the equation $\sin x = \dfrac{x}{1{,}000{,}000}$.

2. Find the number of solutions of the equation $\cos x = \dfrac{x^2}{400}$.

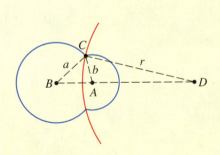

Lattice points in the plane

3. Three tangent circles of radius 10 cm are drawn as shown. All centers lie on the line AB. The tangent AC to the right-hand circle is drawn, intersecting the middle circle at D and E. Find the length of the segment DE.

4. A ball of radius a is set inside a cone so that the surface of the cone is tangent to the ball. A larger ball of radius b fits inside the cone in such a way that it is tangent to both the ball of radius a and the sides of the cone. Express b in terms of a and the angle θ shown in the figure to the left.

5. Points (m, n) in the coordinate plane, both of whose coordinates are integers, are called *lattice points*. Show that it's impossible for any equilateral triangle to have each of its vertices at a lattice point. [*Hint:* See Section 7.2, Exercise 48(b).]

6. Find $\angle A + \angle B + \angle C$ in the figure.

7. If $0 < \theta < \pi/2$ and $\sin 2\theta = a$, find $\sin \theta + \cos \theta$.

8. Show that $\pi/4 = \tan^{-1}\frac{1}{2} + \tan^{-1}\frac{1}{5} + \tan^{-1}\frac{1}{8}$.

   N O T E This identity was used by Zacharias Dase in 1844 to calculate the decimal expansion of π to 200 places.

9. When two bubbles cling together in midair, their common surface is part of a sphere whose center D lies on the line passing through the centers of the bubbles. Also, the angles ACB and ACD both have measure $60°$ (see the figure).
   (a) Show that the radius r of the common face is given by
   $$r = \frac{ab}{a - b}$$
   (b) If the radii of the two bubbles are 4 cm and 3 cm, respectively, find the radius of the common face.
   (c) What shape does the common face take if the two bubbles have radii of the same length?

10. Solve the inequality $|\tan x| \leqslant 1$.

11. Solve the equation $|\sin x| = \sin x + 2 \cos x$, $0 \leqslant x \leqslant 2\pi$.

12. Prove that $\log_2 5$ is irrational.

13. Evaluate $(\log_2 3)(\log_3 4)(\log_4 5) \cdots (\log_{31} 32)$.

14. Show that if $x > 0$ and $x \neq 1$, then
$$\frac{1}{\log_2 x} + \frac{1}{\log_3 x} + \frac{1}{\log_5 x} = \frac{1}{\log_{30} x}$$

Systems of Equations and Inequalities

Systems of equations and inequalities are essential in modeling situations that involve many varying quantities. For instance, systems of equations are used in manipulating three-dimensional images on a computer screen; systems of inequalities are used to find the most efficient routing for a telephone call or for a data packet sent over the Internet.

Many of the problems to which we can apply the techniques of algebra give rise to sets of equations with several variables, rather than to just a single equation in a single variable. A set of equations with common variables is called a *system of equations*. In this chapter we develop techniques for finding simultaneous solutions of such systems. We first consider the simplest case: pairs of equations with two unknowns. To help us solve *linear* equations in an arbitrary number of variables, we study the algebra of matrices and determinants. We also study systems of inequalities, and in the *Focus on Modeling* section we examine linear programming, an optimization technique used widely in business and the social sciences.

8.1 SYSTEMS OF EQUATIONS

In this section we learn how to solve systems of two equations in two unknowns. Here is an example of how such systems arise.

EXAMPLE 1 ■ Gasoline Sales

A gas station in a remote area sells two types of gasoline: regular at $2.20 per gallon and premium at $3.00 per gallon. At the end of a business day the cashier finds that receipts totaled $680 and that 280 gallons of gasoline were sold. Suppose that x gallons of regular and y gallons of premium were sold.
(a) Find a system of equations that relates x and y.
(b) Show that $x = 200$ and $y = 80$ is a solution of the system.

SOLUTION

(a) A total of 280 gallons were sold, so $x + y = 280$. Also, since the total revenue was $680, we have

$$\boxed{\text{revenue from regular}} + \boxed{\text{revenue from premium}} = \boxed{\$680}$$

$$2.20x + 3.00y = 680$$

Thus, we have the system

$$\begin{cases} x + y = 280 \\ 2.20x + 3.00y = 680 \end{cases}$$

(b) A solution of the system is a pair of values for x and y that makes both equations true. Substituting $x = 200$ and $y = 80$ into the system, we get

$$\begin{cases} 200 + 80 = 280 \\ 2.20(200) + 3.00(80) = 680 \end{cases}$$

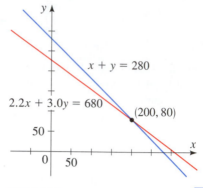

FIGURE 1

Both equations are true, so the ordered pair (200, 80) is a solution of the system. This means that 200 gallons of regular and 80 gallons of premium were sold. Graphically, the solution of the system is the point where the graphs of the two equations intersect (see Figure 1). ■

In general, a **system of equations** is a set of equations that involve the same variables. A **solution** of a system is an assignment of values for the variables that makes *each* equation in the system true. To **solve** a system means to find all solutions of the system. In this section we learn three methods for solving systems of equations: by substitution, by elimination, and graphically.

■ Substitution Method

In the **substitution method** we start with one equation in the system and solve for one variable in terms of the other variable. The following box describes the procedure.

SUBSTITUTION METHOD

1. SOLVE FOR ONE VARIABLE.   Choose one equation and solve for one variable in terms of the other variable.

2. SUBSTITUTE.   Substitute the expression you found in Step 1 into the other equation to get an equation in one variable, then solve for that variable.

3. BACK-SUBSTITUTE.   Substitute the value you found in Step 2 back into the expression found in Step 1 to solve for the remaining variable.

EXAMPLE 2 ■ Substitution Method

Find all solutions of the system.

$$\begin{cases} 2x + y = 1 \\ 3x + 4y = 14 \end{cases}$$

SOLUTION

We solve for y in the first equation.

$$y = 1 - 2x \qquad \text{Solve for } y \text{ in the first equation}$$

Now we substitute for y in the second equation and solve for x:

$$3x + 4(1 - 2x) = 14 \qquad \text{Substitute } y = 1 - 2x \text{ into the second equation}$$

$$3x + 4 - 8x = 14 \qquad \text{Expand}$$

$$-5x + 4 = 14 \qquad \text{Simplify}$$

$$-5x = 10 \qquad \text{Subtract 4}$$

$$x = -2 \qquad \text{Solve for } x$$

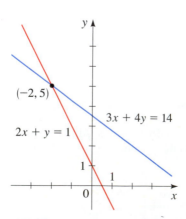

FIGURE 2

Next we back-substitute $x = -2$ into the equation $y = 1 - 2x$:

$$y = 1 - 2(-2) = 5 \qquad \text{Back-substitute}$$

Thus, $x = -2$ and $y = 5$, so the solution is the ordered pair $(-2, 5)$. Figure 2 shows that the graphs of the two equations intersect at the point $(-2, 5)$. ∎

EXAMPLE 3 ■ Substitution Method

Find all solutions of the system.

$$\begin{cases} x^2 + y^2 = 100 \\ 3x - y = 10 \end{cases}$$

SOLUTION

We start by solving for y in the second equation.

$$y = 3x - 10 \qquad \text{Solve for } y \text{ in the second equation}$$

Next we substitute for y in the first equation and solve for x:

$$x^2 + (3x - 10)^2 = 100 \qquad \begin{array}{l}\text{Substitute } y = 3x - 10 \\ \text{into the first equation}\end{array}$$
$$x^2 + (9x^2 - 60x + 100) = 100 \qquad \text{Expand}$$
$$10x^2 - 60x = 0 \qquad \text{Simplify}$$
$$10x(x - 6) = 0 \qquad \text{Factor}$$
$$x = 0 \quad \text{or} \quad x = 6 \qquad \text{Solve for } x$$

Now we back-substitute these values of x into the equation $y = 3x - 10$.

For $x = 0$: $\quad y = 3(0) - 10 = -10 \qquad \text{Back-substitute}$

For $x = 6$: $\quad y = 3(6) - 10 = 8 \qquad \text{Back-substitute}$

So we have two solutions: $(0, -10)$ and $(6, 8)$.

The graph of the first equation is a circle, and the graph of the second equation is a line; Figure 3 shows that the graphs intersect at the two points $(0, -10)$ and $(6, 8)$.

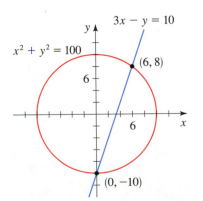

FIGURE 3 ∎

■ Elimination Method

To solve a system using the **elimination method**, we try to combine the equations using sums or differences so as to eliminate one of the variables.

ELIMINATION METHOD

1. ADJUST THE COEFFICIENTS.   Multiply one or more of the equations by appropriate numbers so that the coefficient of one variable in one equation is the negative of its coefficient in the other equation.

2. ADD THE EQUATIONS.   Add the two equations to eliminate one variable, then solve for the remaining variable.

3. BACK-SUBSTITUTE.   Substitute the value you found in Step 2 back into one of the original equations, and solve for the remaining variable.

EXAMPLE 4 ■ Elimination Method

Find all solutions of the system.

$$\begin{cases} 3x + 2y = 14 \\ x - 2y = 2 \end{cases}$$

SOLUTION

Since the coefficients of the y-terms are negatives of each other, we can add the equations to eliminate y.

$$\begin{cases} 3x + 2y = 14 \\ x - 2y = 2 \end{cases} \qquad \text{System}$$

$$4x = 16 \qquad \text{Add}$$

$$x = 4 \qquad \text{Solve for } x$$

Now we back-substitute $x = 4$ into one of the original equations and solve for y. Let's choose the second equation because it looks simpler.

$$x - 2y = 2 \qquad \text{Second equation}$$

$$4 - 2y = 2 \qquad \text{Back-substitute } x = 4 \text{ into second equation}$$

$$-2y = -2 \qquad \text{Subtract 4}$$

$$y = 1 \qquad \text{Solve for } y$$

The solution is $(4, 1)$. Figure 4 shows that the graphs of the equations in the system intersect at the point $(4, 1)$. ■

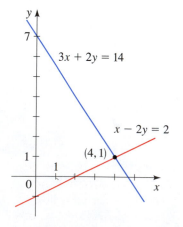

$3x + 2y = 14$

$x - 2y = 2$

$(4, 1)$

FIGURE 4

EXAMPLE 5 ■ Elimination Method

Find all solutions of the system.

$$\begin{cases} 3x^2 + 2y = 26 \\ 5x^2 + 7y = 3 \end{cases}$$

SOLUTION

We choose to eliminate the x-term, so we multiply the first equation by 5 and the second equation by -3. Then we add the two equations and solve for y.

$$\begin{cases} 15x^2 + 10y = 130 & \text{First equation times 5} \\ -15x^2 - 21y = -9 & \text{Second equation times } -3 \end{cases}$$

$$-11y = 121 \qquad \text{Add}$$

$$y = -11 \qquad \text{Solve for } y$$

Now we back-substitute $y = -11$ into one of the original equations, say $3x^2 + 2y = 26$, and solve for x:

$$3x^2 + 2(-11) = 26 \qquad \text{Back-substitute } y = -11 \text{ into the first equation}$$

$$3x^2 = 48 \qquad \text{Add 22}$$

$$x^2 = 16 \qquad \text{Divide by 3}$$

$$x = -4 \qquad \text{or} \qquad x = 4 \qquad \text{Solve for } x$$

So we have two solutions: $(-4, -11)$ and $(4, -11)$.

The graphs of both equations are parabolas; Figure 5 shows that the graphs intersect at the two points $(-4, -11)$ and $(4, -11)$.

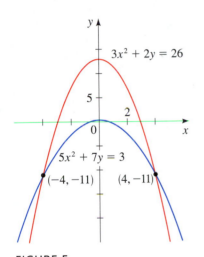

FIGURE 5

The graphs of quadratic functions $y = ax^2 + bx + c$ are called *parabolas*; see Section 2.6.

CHECK YOUR ANSWERS

$x = -4, y = -11$:
$$\begin{cases} 3(-4)^2 + 2(-11) = 26 \\ 5(-4)^2 + 7(-11) = 3 \end{cases} ✔$$

$x = 4, y = -11$:
$$\begin{cases} 3(4)^2 + 2(-11) = 26 \\ 5(4)^2 + 7(-11) = 3 \end{cases} ✔$$

Graphical Method

In the **graphical method** we use a graphing device to solve the system of equations. Note that with many graphing devices, any equation must first be expressed in terms of one or more functions of the form $y = f(x)$ before we can use the calculator to graph it. Not all equations can be readily expressed in this way, so not all systems can be solved by this method.

Global Positioning System (GPS)
On a cold foggy day in 1707, a British naval fleet was sailing at a fast clip directly home. Although the fleet was only a few yards from the rocky shores of England, the captain didn't know it. In the ensuing disaster the fleet was totally destroyed. This tragedy could have been avoided had the captain known his position accurately. In those days latitude was determined by the position of the north star (and this could only be done at night in good weather) and longitude by the position of the sun relative to where it would be in England *at that same time*. So navigation required an accurate method of telling time on ships. (The invention of the spring-loaded clock brought about the eventual solution.)

Since then, several different methods have been developed to determine position, and all rely heavily on mathematics (see LORAN, page 750). The latest method, called the Global Positioning System, uses triangulation. In

(continued)

GRAPHICAL METHOD

1. **GRAPH EACH EQUATION.** To express each equation in a form suitable for the graphing calculator, solve for *y* as a function of *x*. Graph the equations on the same screen.

2. **FIND THE INTERSECTION POINTS.** The solutions are the *x*- and *y*-coordinates of the points of intersection.

It may be more convenient to solve for *x* in terms of *y* in the equations. In that case, in Step 1 graph *x* as a function of *y* instead.

EXAMPLE 6 ■ Graphical Method

Find all solutions of the system.

$$\begin{cases} x^2 - y = 2 \\ 2x - y = -1 \end{cases}$$

SOLUTION

Solving for *y* in terms of *x*, we get the equivalent system

$$\begin{cases} y = x^2 - 2 \\ y = 2x + 1 \end{cases}$$

Figure 6 shows that the graphs of these equations intersect at two points. Zooming in, we see that the solutions are

$$(-1, -1) \qquad \text{and} \qquad (3, 7)$$

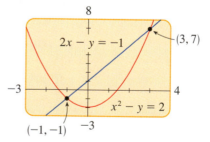

FIGURE 6

CHECK YOUR ANSWERS

$x = -1, y = -1$:
$$\begin{cases} (-1)^2 - (-1) = 2 \\ 2(-1) - (-1) = -1 \end{cases} ✔$$

$x = 3, y = 7$:
$$\begin{cases} 3^2 - 7 = 2 \\ 2(3) - 7 = -1 \end{cases} ✔$$

this system 24 satellites are strategically located above the surface of the earth. A hand-held GPS device measures distance from a satellite using the travel time of radio signals emitted from the satellite. Knowing the distance to three different satellites tells us that we are at the point of intersection of three different spheres. This uniquely determines our position (see Exercise 53, page 629).

EXAMPLE 7 ■ **Solving a System of Equations Graphically**

Find all solutions of the system, correct to one decimal place.

$$\begin{cases} x^2 + y^2 = 12 \\ y = 2x^2 - 5x \end{cases}$$

SOLUTION

The graph of the first equation is a circle and the second a parabola. To graph the circle on a graphing calculator, we must first solve for y in terms of x (see Section 1.9).

$$x^2 + y^2 = 12$$

$$y^2 = 12 - x^2 \qquad \text{Isolate } y^2 \text{ on LHS}$$

$$y = \pm\sqrt{12 - x^2} \qquad \text{Take square roots}$$

To graph the circle, we must graph both functions:

$$y = \sqrt{12 - x^2} \qquad \text{and} \qquad y = -\sqrt{12 - x^2}$$

In Figure 7 the graph of the circle is shown in red and the parabola in blue. The graphs intersect in quadrants I and II. Zooming in, or using the **Intersect** command, we see that the intersection points are $(-0.559, 3.419)$ and $(2.847, 1.974)$. There also appears to be an intersection point in quadrant IV. However, when we zoom in, we see that the curves come close to each other but don't intersect (see Figure 8). Thus, the system has two solutions; correct to the nearest tenth, they are

$$(-0.6, 3.2) \qquad \text{and} \qquad (2.9, 2.0)$$

(a)

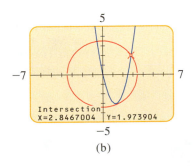

(b)

FIGURE 7
$x^2 + y^2 = 12, \quad y = 2x^2 - 5x$

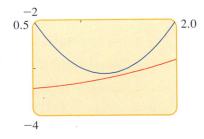

FIGURE 8
Zooming in ■

8.1 EXERCISES

1–8 ■ Use the substitution method to find all solutions of the system of equations.

1. $\begin{cases} x - y = 2 \\ 2x + 3y = 9 \end{cases}$

2. $\begin{cases} 2x + y = 7 \\ x + 2y = 2 \end{cases}$

3. $\begin{cases} y = x^2 \\ y = x + 12 \end{cases}$

4. $\begin{cases} x^2 + y^2 = 25 \\ y = 2x \end{cases}$

5. $\begin{cases} x^2 + y^2 = 8 \\ x + y = 0 \end{cases}$

6. $\begin{cases} x^2 + y = 9 \\ x - y + 3 = 0 \end{cases}$

7. $\begin{cases} x + y^2 = 0 \\ 2x + 5y^2 = 75 \end{cases}$

8. $\begin{cases} x^2 - y = 1 \\ 2x^2 + 3y = 17 \end{cases}$

9–16 ■ Use the elimination method to find all solutions of the system of equations.

9. $\begin{cases} x + 2y = 5 \\ 2x + 3y = 8 \end{cases}$

10. $\begin{cases} 4x - 3y = 11 \\ 8x + 4y = 12 \end{cases}$

11. $\begin{cases} x^2 - 2y = 1 \\ x^2 + 5y = 29 \end{cases}$

12. $\begin{cases} 3x^2 + 4y = 17 \\ 2x^2 + 5y = 2 \end{cases}$

13. $\begin{cases} 3x^2 - y^2 = 11 \\ x^2 + 4y^2 = 8 \end{cases}$

14. $\begin{cases} 2x^2 + 4y = 13 \\ x^2 - y^2 = \frac{7}{2} \end{cases}$

15. $\begin{cases} x - y^2 + 3 = 0 \\ 2x^2 + y^2 - 4 = 0 \end{cases}$

16. $\begin{cases} x^2 - y^2 = 1 \\ 2x^2 - y^2 = x + 3 \end{cases}$

17–30 ■ Find all solutions of the system of equations.

17. $\begin{cases} y + x^2 = 4x \\ y + 4x = 16 \end{cases}$

18. $\begin{cases} x - y^2 = 0 \\ y - x^2 = 0 \end{cases}$

19. $\begin{cases} x - 2y = 2 \\ y^2 - x^2 = 2x + 4 \end{cases}$

20. $\begin{cases} y = 4 - x^2 \\ y = x^2 - 4 \end{cases}$

21. $\begin{cases} x - y = 4 \\ xy = 12 \end{cases}$

22. $\begin{cases} xy = 24 \\ 2x^2 - y^2 + 4 = 0 \end{cases}$

23. $\begin{cases} x^2 y = 16 \\ x^2 + 4y + 16 = 0 \end{cases}$

24. $\begin{cases} x + \sqrt{y} = 0 \\ y^2 - 4x^2 = 12 \end{cases}$

25. $\begin{cases} x^2 + y^2 = 9 \\ x^2 - y^2 = 1 \end{cases}$

26. $\begin{cases} x^2 + 2y^2 = 2 \\ 2x^2 - 3y = 15 \end{cases}$

27. $\begin{cases} 2x^2 - 8y^3 = 19 \\ 4x^2 + 16y^3 = 34 \end{cases}$

28. $\begin{cases} x^4 - y^3 = 17 \\ 3x^4 + 5y^3 = 53 \end{cases}$

29. $\begin{cases} \dfrac{2}{x} - \dfrac{3}{y} = 1 \\ -\dfrac{4}{x} + \dfrac{7}{y} = 1 \end{cases}$

30. $\begin{cases} \dfrac{4}{x^2} + \dfrac{6}{y^4} = \dfrac{7}{2} \\ \dfrac{1}{x^2} - \dfrac{2}{y^4} = 0 \end{cases}$

31–40 ■ Use the graphical method to find all solutions of the system of equations, correct to two decimal places.

31. $\begin{cases} y = 2x + 6 \\ y = -x + 5 \end{cases}$

32. $\begin{cases} y = -2x + 12 \\ y = x + 3 \end{cases}$

33. $\begin{cases} y = x^2 + 8x \\ y = 2x + 16 \end{cases}$

34. $\begin{cases} y = x^2 - 4x \\ 2x - y = 2 \end{cases}$

35. $\begin{cases} x^2 + y^2 = 25 \\ x + 3y = 2 \end{cases}$

36. $\begin{cases} x^2 + y^2 = 17 \\ x^2 - 2x + y^2 = 13 \end{cases}$

37. $\begin{cases} \dfrac{x^2}{9} + \dfrac{y^2}{18} = 1 \\ y = -x^2 + 6x - 2 \end{cases}$

38. $\begin{cases} x^2 - y^2 = 3 \\ y = x^2 - 2x - 8 \end{cases}$

39. $\begin{cases} x^4 + 16y^4 = 32 \\ x^2 + 2x + y = 0 \end{cases}$

40. $\begin{cases} y = e^x + e^{-x} \\ y = 5 - x^2 \end{cases}$

41. A rectangle has an area of 180 cm² and a perimeter of 54 cm. What are its dimensions?

42. A right triangle has an area of 84 ft² and a hypotenuse 25 ft long. What are the lengths of its other two sides?

43. The perimeter of a rectangle is 70 and its diagonal is 25. Find its length and width.

44. A circular piece of sheet metal has a diameter of 20 in. The edges are to be cut off to form a rectangle of area 160 in² (see the figure). What are the dimensions of the rectangle?

45. A hill is inclined so that its "slope" is $\frac{1}{2}$, as shown in the figure. We introduce a coordinate system with the origin at the base of the hill and with the scales on the axes measured in meters. A rocket is fired from the base of the hill in such a way that its trajectory is the parabola $y = -x^2 + 401x$. At what point does the rocket strike the hillside? How far is this point from the base of the hill (to the nearest cm)?

46. A rectangular piece of sheet metal with an area of 1200 in^2 is to be bent into a cylindrical length of stovepipe having a volume of 600 in^3. What are the dimensions of the sheet metal?

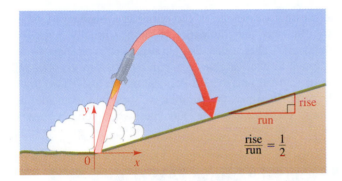

47. Find an equation for the line that passes through the points of intersection of the circles $x^2 + y^2 = 25$ and $x^2 - 3x + y^2 + y = 30$. [*Hint:* Try to eliminate the x^2- and y^2-terms.]

48–51 ■ Find all solutions of the system.

48. $\begin{cases} x - y = 3 \\ x^3 - y^3 = 387 \end{cases}$ [*Hint:* Factor the left side of the second equation.]

49. $\begin{cases} x^2 + xy = 1 \\ xy + y^2 = 3 \end{cases}$ [*Hint:* Add the equations and factor the result.]

50. $\begin{cases} 2^x + 2^y = 10 \\ 4^x + 4^y = 68 \end{cases}$ [*Hint:* Note that $4^x = 2^{2x} = (2^x)^2$.]

51. $\begin{cases} \log x + \log y = \frac{3}{2} \\ 2\log x - \log y = 0 \end{cases}$

● DISCOVERY · DISCUSSION

52. Intersection of a Parabola and a Line   On a sheet of graph paper, or using a graphing calculator, draw the parabola $y = x^2$. Then draw the graphs of the linear equation $y = x + k$ on the same coordinate plane for various values of k. Try to choose values of k so that the line and the parabola intersect at two points for some of your k's, and not for others. For what value of k is there exactly one intersection point? Use the results of your experiment to make a conjecture about the values of k for which the following system has two solutions, one solution, and no solution. Prove your conjecture.

$$\begin{cases} y = x^2 \\ y = x + k \end{cases}$$

53. Global Positioning System (GPS)   The Global Positioning System determines the location of an object from its distances to satellites in orbit around the earth. In the simplified, two-dimensional situation shown in the figure, determine the coordinates of P from the fact that P is 26 units from satellite A and 20 units from satellite B.

How many satellites do you think are needed to determine position in the real world (which is three-dimensional)?

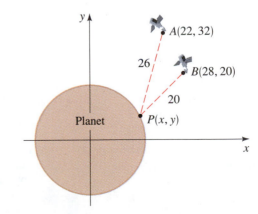

8.2 SYSTEMS OF LINEAR EQUATIONS IN TWO VARIABLES

An equation of the form $Ax + By = C$ is called **linear** since its graph is a line (see Section 1.10). In this section we study systems of linear equations in two variables:

$$\begin{cases} a_1 x + b_1 y = c_1 \\ a_2 x + b_2 y = c_2 \end{cases}$$

We can use either the substitution method or the elimination method to solve such systems algebraically. But since the elimination method is usually easier for linear systems, we use elimination rather than substitution in our examples.

The graph of a linear system in two variables is a pair of lines, so to solve the system graphically, we must find the intersection point(s) of the lines. Two lines may intersect in a single point, they may be parallel, or they may coincide, as shown in Figure 1. So there are three possible outcomes when solving such a system.

NUMBER OF SOLUTIONS OF A LINEAR SYSTEM IN TWO VARIABLES

For a system of linear equations in two variables, exactly one of the following is true. (See Figure 1.)

1. The system has exactly one solution.

2. The system has no solution.

3. The system has infinitely many solutions.

A system that has no solution is said to be **inconsistent**. A system with infinitely many solutions is called **dependent**.

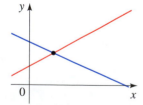

(a) Linear system with one solution. Lines intersect at a single point.

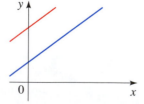

(b) Linear system with no solution. Lines are parallel—they do not intersect.

(c) Linear system with infinitely many solutions. Lines coincide—equations are for the same line.

FIGURE 1

EXAMPLE 1 ■ A Linear System with One Solution

Solve the system and graph the lines.

$$\begin{cases} 3x - y = 0 \\ 5x + 2y = 22 \end{cases}$$

SOLUTION

We eliminate y from the equations and solve for x.

$$\begin{cases} 6x - 2y = 0 & \text{First equation times 2} \\ \underline{5x + 2y = 22} & \end{cases}$$

$$11x \qquad = 22 \qquad \text{Add}$$

$$x = \tfrac{22}{11} = 2 \qquad \text{Solve for } x$$

Now we back-substitute into the first equation and solve for y:

$$6(2) - 2y = 0 \qquad \text{Back-substitute } x = 2$$

$$-2y = -12 \qquad \text{Subtract } 6 \cdot 2 = 12$$

$$y = 6 \qquad \text{Solve for } y$$

The solution of the system is the ordered pair $(2, 6)$, that is,

$$x = 2, \qquad y = 6$$

The graph in Figure 2 shows that the lines in the system intersect at the point $(2, 6)$.

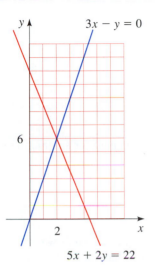

$3x - y = 0$

$5x + 2y = 22$

FIGURE 2

CHECK YOUR ANSWER

$x = 2, y = 6$:

$$\begin{cases} 3(2) - (6) = 0 \\ 5(2) + 2(6) = 22 \end{cases} \quad \checkmark$$

EXAMPLE 2 ■ A Linear System with No Solution

Solve the system.

$$\begin{cases} 8x - 2y = 5 \\ -12x + 3y = 7 \end{cases}$$

SOLUTION

This time we try to find a suitable combination of the two equations to eliminate the variable y. Multiplying the first equation by 3 and the second by 2 gives

$$\begin{cases} 24x - 6y = 15 & \text{First equation times 3} \\ \underline{-24x + 6y = 14} & \text{Second equation times 2} \end{cases}$$

$$0 = 29 \qquad \text{Add}$$

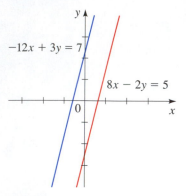

$-12x + 3y = 7$

$8x - 2y = 5$

FIGURE 3

Adding the two equations eliminates *both x and y* in this case, and we end up with $0 = 29$, which is obviously false. No matter what values we assign to x and y, we cannot make this statement true, so the system has *no solution*. Figure 3 shows that the lines in the system are parallel and do not intersect. The system is inconsistent.   ■

EXAMPLE 3 ■ **A Linear System with Infinitely Many Solutions**

Solve the system.

$$\begin{cases} 3x - 6y = 12 \\ 4x - 8y = 16 \end{cases}$$

SOLUTION

We multiply the first equation by 4 and the second by 3 to prepare for subtracting the equations to eliminate x. The new equations are

$$\begin{cases} 12x - 24y = 48 & \text{First equation times 4} \\ 12x - 24y = 48 & \text{Second equation times 3} \end{cases}$$

We see that the two equations in the original system are simply different ways of expressing the equation of one single line. The coordinates of any point on this line give a solution of the system. Writing the equation in slope-intercept form, we have $y = \frac{1}{2}x - 2$. So if we let t represent any real number, we can write the solution as

$$x = t$$

$$y = \tfrac{1}{2}t - 2$$

We can also write the solution in ordered-pair form as

$$\left(t, \tfrac{1}{2}t - 2 \right)$$

where t is any real number. The system has infinitely many solutions.   ■

In Example 3, to get specific solutions we have to assign values to t. For instance, if $t = 1$, we get the solution $\left(1, -\frac{3}{2}\right)$. If $t = 4$, we get the solution $(4, 0)$. For every value of t we get a different solution.

■ Modeling with Linear Systems

Frequently, when we use equations to solve problems in the sciences or in other areas, we obtain systems like the ones we've been considering. When modeling with systems of equations, we use the following guidelines, similar to those in Section 1.6.

GUIDELINES FOR MODELING WITH SYSTEMS OF EQUATIONS

1. IDENTIFY THE VARIABLES. Identify the quantities the problem asks you to find. These are usually determined by a careful reading of the question posed at the end of the problem. Introduce notation for the variables (call them x and y or some other letters).

2. EXPRESS ALL UNKNOWN QUANTITIES IN TERMS OF THE VARIABLES. Read the problem again and express all the quantities mentioned in the problem in terms of the variables you defined in Step 1.

3. SET UP A SYSTEM OF EQUATIONS. Find the crucial facts in the problem that give the relationships between the expressions you found in Step 2. Set up a system of equations (or a model) that expresses these relationships.

4. SOLVE THE SYSTEM AND INTERPRET THE RESULTS. Solve the system you found in Step 3, check your solutions, and state your final answer as a sentence that answers the question posed in the problem.

The next two examples illustrate how to model with systems of equations.

EXAMPLE 4 ■ A Distance-Speed-Time Problem

A woman rows a boat upstream from one point on a river to another point 4 mi away in $1\frac{1}{2}$ hours. The return trip, traveling with the current, takes only 45 min. How fast does she row relative to the water, and at what speed is the current flowing?

SOLUTION

We are asked to find the rowing speed and the speed of the current, so we let

$$x = \text{rowing speed (mi/h)}$$

$$y = \text{current speed (mi/h)}$$

The woman's speed when she rows upstream is her rowing speed minus the speed of the current; her speed downstream is her rowing speed plus the speed of the current. Now we translate this information into the language of algebra.

In Words	In Algebra
Rowing speed	x
Current speed	y
Speed upstream	$x - y$
Speed downstream	$x + y$

Identify the variables

Express unknown quantities in terms of the variable

The distance upstream and downstream is 4 mi, so using the fact that speed × time = distance for both legs of the trip, we get

$$\boxed{\text{speed upstream}} \times \boxed{\text{time upstream}} = \boxed{\text{distance traveled}}$$

$$\boxed{\text{speed downstream}} \times \boxed{\text{time downstream}} = \boxed{\text{distance traveled}}$$

In algebraic notation this translates into the following equations.

Set up a system of equations

$$(x - y)\tfrac{3}{2} = 4$$

$$(x + y)\tfrac{3}{4} = 4$$

(The times have been converted to hours, since we are expressing the speeds in miles per *hour*.) If we multiply the equations by 2 and 4, respectively, to clear the denominators, we get the system

Solve the system

$$\begin{cases} 3x - 3y = \ \ 8 & \text{\color{red}First equation times 2} \\ 3x + 3y = 16 & \text{\color{red}Second equation times 4} \end{cases}$$

$$\begin{aligned} 6x \ \ \ \ \ \ &= 24 & &\text{\color{red}Add} \\ x \ \ \ \ \ \ &= 4 & &\text{\color{red}Solve for } x \end{aligned}$$

Back-substituting this value of x into the first equation (the second works just as well) and solving for y gives

$$\begin{aligned} 3(4) - 3y &= 8 & &\text{\color{red}Back-substitute } x = 4 \\ -3y &= 8 - 12 & &\text{\color{red}Subtract 12} \\ y &= \tfrac{4}{3} & &\text{\color{red}Solve for } y \end{aligned}$$

The woman rows at 4 mi/h and the current flows at $1\tfrac{1}{3}$ mi/h. ∎

CHECK YOUR ANSWER

Speed upstream is

$$\frac{\text{distance}}{\text{time}} = \frac{4 \text{ mi}}{1\tfrac{1}{2} \text{ h}} = 2\tfrac{2}{3} \text{ mi/h}$$

and this should equal

rowing speed − current flow

$$= 4 \text{ mi/h} - \tfrac{4}{3} \text{ mi/h} = 2\tfrac{2}{3} \text{ mi/h}$$

Speed downstream is

$$\frac{\text{distance}}{\text{time}} = \frac{4 \text{ mi}}{\tfrac{3}{4} \text{ h}} = 5\tfrac{1}{3} \text{ mi/h}$$

and this should equal

rowing speed + current flow

$$= 4 \text{ mi/h} + \tfrac{4}{3} \text{ mi/h} = 5\tfrac{1}{3} \text{ mi/h} ✓$$

EXAMPLE 5 ■ **A Mixture Problem**

A vintner fortifies wine that contains 10% alcohol by adding 70% alcohol solution to it. The resulting mixture has an alcoholic strength of 16% and fills 1000 one-liter bottles. How many liters (L) of the wine and of the alcohol solution does he use?

SOLUTION

Since we are asked for the amounts of wine and alcohol, we let

Identify the variables

$$x = \text{amount of wine used (L)}$$

$$y = \text{amount of alcohol used (L)}$$

From the fact that the wine contains 10% alcohol and the solution 70% alcohol, we get the following.

In Words	In Algebra
Amount of wine used (L)	x
Amount of alcohol used (L)	y
Amount of alcohol in wine (L)	$0.10x$
Amount of alcohol in solution (L)	$0.70y$

Express unknown quantities in terms of the variable

The volume of the mixture must be the total of the two volumes the vintner is adding together, so

$$x + y = 1000$$

Also, the amount of alcohol in the mixture must be the total of the alcohol contributed by the wine and by the alcohol solution, that is

$$0.10x + 0.70y = (0.16)1000$$

$$0.10x + 0.70y = 160 \qquad \text{Simplify}$$

$$x + 7y = 1600 \qquad \text{Multiply by 10 to clear decimals}$$

Thus, we get the system

Set up a system of equations

$$\begin{cases} x + y = 1000 \\ x + 7y = 1600 \end{cases}$$

Subtracting the first equation from the second eliminates the variable x, and we get

Solve the system

$$6y = 600 \qquad \text{Subtract first equation from second}$$

$$y = 100 \qquad \text{Solve for } y$$

We now back-substitute $y = 100$ into the first equation and solve for x:

$$x + 100 = 1000 \qquad \text{Back-substitute } y = 100$$

$$x = 900 \qquad \text{Solve for } x$$

The vintner uses 900 L of wine and 100 L of the alcohol solution.   ■

8.2   EXERCISES

1–6 ■ Graph each linear system, either by hand or using a graphing device. Use the graph to determine if the system has one solution, no solution, or infinitely many solutions. If there is exactly one solution, use the graph to find it.

1. $\begin{cases} x + y = 4 \\ 2x - y = 2 \end{cases}$

2. $\begin{cases} 2x + y = 11 \\ x - 2y = 4 \end{cases}$

3. $\begin{cases} 2x - 3y = 12 \\ -x + \frac{3}{2}y = 4 \end{cases}$

4. $\begin{cases} 2x + 6y = 0 \\ -3x - 9y = 18 \end{cases}$

5. $\begin{cases} -x + \frac{1}{2}y = -5 \\ 2x - \ y = 10 \end{cases}$ **6.** $\begin{cases} 12x + 15y = -18 \\ 2x + \frac{5}{2}y = \ -3 \end{cases}$

7–30 ■ Solve the system, or show that it has no solution. If the system has infinitely many solutions, express them in the ordered-pair form given in Example 3.

7. $\begin{cases} x + y = 4 \\ -x + y = 0 \end{cases}$ **8.** $\begin{cases} x - \ y = 3 \\ x + 3y = 7 \end{cases}$

9. $\begin{cases} 2x - 3y = 9 \\ 4x + 3y = 9 \end{cases}$ **10.** $\begin{cases} 3x + 2y = 0 \\ -x - 2y = 8 \end{cases}$

11. $\begin{cases} x + 3y = 5 \\ 2x - \ y = 3 \end{cases}$ **12.** $\begin{cases} x + \ y = \ 7 \\ 2x - 3y = -1 \end{cases}$

13. $\begin{cases} -x + \ y = \ 2 \\ 4x - 3y = -3 \end{cases}$ **14.** $\begin{cases} 4x - 3y = \ 28 \\ 9x - \ y = -6 \end{cases}$

15. $\begin{cases} x + 2y = 7 \\ 5x - \ y = 2 \end{cases}$ **16.** $\begin{cases} -4x + 12y = \ \ 0 \\ 12x + \ 4y = 160 \end{cases}$

17. $\begin{cases} \frac{1}{2}x + \frac{1}{3}y = 2 \\ \frac{1}{5}x - \frac{2}{3}y = 8 \end{cases}$ **18.** $\begin{cases} 0.2x - 0.2y = -1.8 \\ -0.3x + 0.5y = \ \ 3.3 \end{cases}$

19. $\begin{cases} 3x + 2y = 8 \\ x - 2y = 0 \end{cases}$ **20.** $\begin{cases} 4x + 2y = 16 \\ x - 5y = 70 \end{cases}$

21. $\begin{cases} x + \ 4y = 8 \\ 3x + 12y = 2 \end{cases}$ **22.** $\begin{cases} -3x + \ 5y = 2 \\ 9x - 15y = 6 \end{cases}$

23. $\begin{cases} 2x - 6y = \ \ 10 \\ -3x + 9y = -15 \end{cases}$ **24.** $\begin{cases} 2x - \ 3y = -8 \\ 14x - 21y = \ \ 3 \end{cases}$

25. $\begin{cases} 6x + 4y = 12 \\ 9x + 6y = 18 \end{cases}$ **26.** $\begin{cases} 25x - 75y = \ 100 \\ -10x + 30y = -40 \end{cases}$

27. $\begin{cases} 8s - 3t = -3 \\ 5s - 2t = -1 \end{cases}$ **28.** $\begin{cases} u - 30v = -5 \\ -3u + 80v = \ \ 5 \end{cases}$

29. $\begin{cases} \frac{1}{2}x + \frac{3}{5}y = \ 3 \\ \frac{5}{3}x + 2y = 10 \end{cases}$ **30.** $\begin{cases} \frac{3}{2}x - \frac{1}{3}y = \ \frac{1}{2} \\ 2x - \frac{1}{2}y = -\frac{1}{2} \end{cases}$

31–34 ■ Find x and y in terms of a and b.

31. $\begin{cases} x + \ y = 0 \\ x + ay = 1 \end{cases}$ $(a \neq 1)$

32. $\begin{cases} ax + by = 0 \\ x + \ y = 1 \end{cases}$ $(a \neq b)$

33. $\begin{cases} ax + by = 1 \\ bx + ay = 1 \end{cases}$ $(a^2 - b^2 \neq 0)$

34. $\begin{cases} ax + \ by = 0 \\ a^2x + b^2y = 1 \end{cases}$ $(a \neq 0, b \neq 0, a \neq b)$

35. Find two numbers whose sum is 34 and whose difference is 10.

36. The sum of two numbers is twice their difference. The larger number is 6 more than twice the smaller. Find the numbers.

37. A man has 14 coins in his pocket, all of which are dimes and quarters. If the total value of his change is $2.75, how many dimes and how many quarters does he have?

38. The admission fee at an amusement park is $1.50 for children and $4.00 for adults. On a certain day, 2200 people entered the park, and the admission fees collected totaled $5050. How many children and how many adults were admitted?

39. A man flies a small airplane from Fargo to Bismarck, North Dakota—a distance of 180 mi. Because he is flying into a head wind, the trip takes him 2 hours. On the way back, the wind is still blowing at the same speed, so the return trip takes only 1 h 12 min. What is his speed in still air, and how fast is the wind blowing?

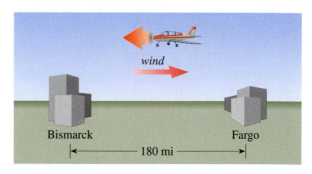

40. A boat on a river travels downstream between two points, 20 mi apart, in one hour. The return trip against the current takes $2\frac{1}{2}$ hours. What is the boat's speed, and how fast does the current in the river flow?

41. A woman keeps fit by bicycling and running every day. On Monday she spends $\frac{1}{2}$ hour at each activity, covering a total of $12\frac{1}{2}$ mi. On Tuesday, she runs for 12 min and cycles for 45 min, covering a total of 16 mi. Assuming her running and cycling speeds don't change from day to day, find these speeds.

42. A biologist has two brine solutions, one containing 5% salt and another containing 20% salt. How many milliliters of each solution should he mix to obtain 1 L of a solution that contains 14% salt?

43. A researcher performs an experiment to test a hypothesis that involves the nutrients niacin and retinol. She feeds one group of laboratory rats a daily diet of precisely 32 units of niacin and 22,000 units of retinol. She uses two types of commercial pellet foods. Food A contains 0.12 unit of niacin and 100 units of retinol per gram. Food B contains 0.20 unit of niacin and 50 units of retinol per gram. How many grams of each food does she feed this group of rats each day?

44. A customer in a coffee shop purchases a blend of two coffees: Kenyan, costing $3.50 a pound, and Sri Lankan, costing $5.60 a pound. He buys 3 lb of the blend, which costs him $11.55. How many pounds of each kind went into the mixture?

45. A chemist has two large containers of sulfuric acid solution, with different concentrations of acid in each container. Blending 300 mL of the first solution and 600 mL of the second gives a mixture that is 15% acid, whereas 100 mL of the first mixed with 500 mL of the second gives a $12\frac{1}{2}\%$ acid mixture. What are the concentrations of sulfuric acid in the original containers?

46. A woman invests a total of $20,000 in two accounts, one paying 5% and the other paying 8% simple interest per year. Her annual interest is $1180. How much did she invest at each rate?

47. A man invests his savings in two accounts, one paying 6% and the other paying 10% simple interest per year. He puts twice as much in the lower-yielding account because it is less risky. His annual interest is $3520. How much did he invest at each rate?

48. John and Mary leave their house at the same time and drive in opposite directions. John drives at 60 mi/h and travels 35 mi farther than Mary, who drives at 40 mi/h. Mary's trip takes 15 min longer than John's. For what length of time does each of them drive?

49. The sum of the digits of a two-digit number is 7. When the digits are reversed, the number is increased by 27. Find the number.

50. Find the area of the triangle that lies in the first quadrant (with its base on the x-axis) and that is bounded by the lines $y = 2x - 4$ and $y = -4x + 20$.

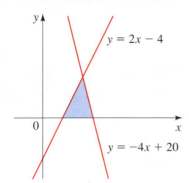

51–54 ■ Use a graphing device to graph both lines in the same viewing rectangle. (Note that you must solve for y in terms of x before graphing if you are using a graphing calculator.) Solve the system correct to two decimal places, either by zooming in and using TRACE or by using Intersect.

51. $\begin{cases} 0.21x + 3.17y = 9.51 \\ 2.35x - 1.17y = 5.89 \end{cases}$

52. $\begin{cases} 18.72x - 14.91y = 12.33 \\ 6.21x - 12.92y = 17.82 \end{cases}$

53. $\begin{cases} 2371x - 6552y = 13{,}591 \\ 9815x + 992y = 618{,}555 \end{cases}$

54. $\begin{cases} -435x + 912y = 0 \\ 132x + 455y = 994 \end{cases}$

DISCOVERY • DISCUSSION

55. The Least Squares Line   The *least squares* line or *regression* line is the line that best fits a set of points in the plane. We studied this line in *Focus on Modeling* (see page 248). Using calculus, it can be shown that the line that best fits the n data points $(x_1, y_1), (x_2, y_2), \ldots, (x_n, y_n)$ is the line $y = ax + b$, where the coefficients a and b satisfy the following pair of linear equations. [The notation $\Sigma_{k=1}^{n} x_k$

stands for the sum of all the x's. See Section 10.1 for a complete description of sigma (Σ) notation.]

$$\left(\sum_{k=1}^{n} x_k\right)a + nb = \sum_{k=1}^{n} y_k$$

$$\left(\sum_{k=1}^{n} x_k^2\right)a + \left(\sum_{k=1}^{n} x_k\right)b = \sum_{k=1}^{n} x_k y_k$$

Use these equations to find the least squares line for the following data points.

$$(1, 3), \quad (2, 5), \quad (3, 6), \quad (5, 6), \quad (7, 9)$$

Sketch the points and your line to confirm that the line fits these points well. If your calculator computes regression lines, see whether it gives you the same line as the formulas.

8.3 SYSTEMS OF LINEAR EQUATIONS IN SEVERAL VARIABLES

Linear equations

$6x_1 - 3x_2 + \sqrt{5}\,x_3 = 1000$

$x + y + z = 2w - \frac{1}{2}$

Nonlinear equations

$x^2 + 3y - \sqrt{z} = 5$

$x_1 x_2 + 6x_3 = -6$

A **linear equation in n variables** is an equation that can be put in the form

$$a_1 x_1 + a_2 x_2 + \cdots + a_n x_n = c$$

where $a_1, a_2, \ldots, a_n$ and c are real numbers, and $x_1, x_2, \ldots, x_n$ are the variables. If we have only three or four variables, we generally use x, y, z, and w instead of x_1, x_2, x_3, and x_4. Such equations are called *linear* because if we have just two variables the equation is $a_1 x + a_2 y = c$, which is the equation of a line.

In this section we study systems of linear equations in three or more variables.

Solving a Linear System

The following are two examples of systems of linear equations in three variables. The second system is in **triangular form**; that is, the variable x doesn't appear in the second equation and the variables x and y do not appear in the third equation.

A system of linear equations

$$\begin{cases} x - 2y - z = 1 \\ -x + 3y + 3z = 4 \\ 2x - 3y + z = 10 \end{cases}$$

A system in triangular form

$$\begin{cases} x - 2y - z = 1 \\ y + 2z = 5 \\ z = 3 \end{cases}$$

It is easy to solve a system that is in triangular form using back-substitution. So, our goal in this section is to start with a system of linear equations and change it to a system in triangular form that has the same solutions as the original system. We start by showing how to use back-substitution to solve a system that is already in triangular form.

EXAMPLE 1 ■ Solving a Triangular System Using Back-Substitution

Solve the system using back-substitution:

$$\begin{cases} x - 2y - z = 1 \\ y + 2z = 5 \\ z = 3 \end{cases}$$

SOLUTION

From the last equation we know that $z = 3$. We back-substitute this into the second equation and solve for y.

$$y + 2(3) = 5 \qquad \text{Back-substitute } z = 3 \text{ into second equation}$$

$$y = -1 \qquad \text{Solve for } y$$

Then we back-substitute $y = -1$ and $z = 3$ into the first equation and solve for x.

$$x - 2(-1) - (3) = 1 \qquad \text{Back-substitute } y = -1 \text{ and } z = 3 \text{ into first equation}$$

$$x = 2 \qquad \text{Solve for } x$$

The solution of the system is $x = 2$, $y = -1$, $z = 3$. We can also write the solution as the ordered triple $(2, -1, 3)$. ■

To change a system of linear equations to an **equivalent system** (that is, a system with the same solutions as the original system), we use the elimination method. This means we can use the following operations.

OPERATIONS THAT YIELD AN EQUIVALENT SYSTEM

1. Add a nonzero multiple of one equation to another.

2. Multiply an equation by a nonzero constant.

3. Interchange the positions of two equations.

To solve a linear system, we use these operations to change the system to an equivalent triangular system. Then we use back-substitution as in Example 1. This process is called **Gaussian elimination**.

EXAMPLE 2 ■ Solving a System of Three Equations in Three Variables

Solve the system using Gaussian elimination.

$$\begin{cases} x - 2y + 3z = 1 \\ x + 2y - z = 13 \\ 3x + 2y - 5z = 3 \end{cases}$$

SOLUTION

We need to change this to a triangular system, so we begin by eliminating the x-term from the second equation.

$$\begin{array}{ll} x + 2y - z = 13 & \text{Second equation} \\ \underline{x - 2y + 3z = 1} & \text{First equation} \\ \quad\; 4y - 4z = 12 & \text{Second equation} + (-1) \times \text{first equation} = \text{new second equation} \end{array}$$

This gives us a new, equivalent system that is one step closer to triangular form:

$$\begin{cases} x - 2y + 3z = 1 \\ \qquad 4y - 4z = 12 \\ 3x + 2y - 5z = 3 \end{cases}$$

Now we eliminate the x-term from the third equation.

$$\begin{aligned} 3x + 2y - 5z &= 3 \\ -3x + 6y - 9z &= -3 \\ \hline 8y - 14z &= 0 \end{aligned}$$

$$\begin{cases} x - 2y + 3z = 1 \\ \qquad 4y - 4z = 12 \\ \qquad 8y - 14z = 0 \qquad \text{\color{red}{Third equation} + (-3) \times first equation = new third equation} \end{cases}$$

Then we eliminate the y-term from the third equation.

$$\begin{aligned} 8y - 14z &= 0 \\ -8y + 8z &= -24 \\ \hline -6z &= -24 \end{aligned}$$

$$\begin{cases} x - 2y + 3z = 1 \\ \qquad 4y - 4z = 12 \\ \qquad\quad -6z = -24 \qquad \text{\color{red}{Third equation} + (-2) \times second equation = new third equation} \end{cases}$$

The system is now in triangular form, but it will be easier to work with if we divide the second and third equations by the common factors of each term.

$$x - 2y + 3z = 1$$
$$y - z = 3 \qquad \text{\color{red}{$\frac{1}{4}$ \times second equation = new second equation}}$$
$$z = 4 \qquad \text{\color{red}{$-\frac{1}{6}$ \times third equation = new third equation}}$$

Now we use back-substitution to solve the system. From the third equation we get $z = 4$. We back-substitute this into the second equation and solve for y.

$$y - (4) = 3 \qquad \text{\color{red}{Back-substitute $z = 3$ into second equation}}$$
$$y = 7 \qquad \text{\color{red}{Solve for y}}$$

Now we back-substitute $y = 7$ and $z = 4$ into the first equation and solve for x.

$$x - 2(7) + 3(4) = 1 \qquad \text{\color{red}{Back-substitute $y = 7$ and $z = 4$ into first equation}}$$
$$x = 3 \qquad \text{\color{red}{Solve for x}}$$

The solution of the system is $x = 3$, $y = 7$, $z = 4$, which we can write as the ordered triple $(3, 7, 4)$.

CHECK YOUR ANSWER

$x = 3, y = 7, z = 4$:

$$(3) - 2(7) + 3(4) = 1$$
$$(3) + 2(7) - (4) = 13$$
$$3(3) + 2(7) - 5(4) = 3 \quad \checkmark$$

The Number of Solutions of a Linear System

When you study calculus or linear algebra, you will learn that the graph of a linear equation in three variables is a *plane* in a three-dimensional coordinate system. For a system of three equations in three variables, the following situations arise:

1. The three planes intersect in a single point.
   The system has a unique solution.

2. The three planes intersect in more than one point.
   The system has infinitely many solutions.

3. The three planes have no point in common.
   The system has no solution.

Just as in the case of two variables, a system of equations in several variables may have one solution, no solution, or infinitely many solutions. The graphical interpretation of the solutions of a linear system is analogous to that for systems of equations in two variables (see the margin note).

> ### NUMBER OF SOLUTIONS OF A LINEAR SYSTEM
>
> For a system of linear equations, exactly one of the following is true.
>
> 1. The system has exactly one solution.
> 2. The system has no solution.
> 3. The system has infinitely many solutions.

A system with no solutions is said to be **inconsistent**, and a system with infinitely many solutions is said to be **dependent**.

EXAMPLE 3 ■ A System with No Solution

Solve the following system.

$$\begin{cases} x + 2y - 2z = 1 \\ 2x + 2y - \ \ z = 6 \\ 3x + 4y - 3z = 5 \end{cases}$$

SOLUTION

To put this in triangular form, we begin by eliminating the x-terms from the second equation and the third equation.

$$\begin{cases} x + 2y - 2z = 1 \\ -2y + 3z = 4 \\ 3x + 4y + 3z = 5 \end{cases} \qquad \text{Second equation} + (-2) \times \text{first equation} = \text{new second equation}$$

$$\begin{cases} x + 2y - 2z = 1 \\ -2y + 3z = 4 \\ -2y + 3z = 2 \end{cases} \qquad \text{Third equation} + (-3) \times \text{first equation} = \text{new third equation}$$

Now we eliminate the y-term from the third equation.

$$\begin{cases} x + 2y - 2z = 1 \\ -2y + 3z = 4 \\ 0 = 2 \end{cases} \qquad \text{Third equation} + (-1) \times \text{second equation} = \text{new third equation}$$

The system is now in triangular form, but the third equation says $0 = 2$, which is false. No matter what values we assign x, y, and z, the third equation will never be true. This means the system has *no solution*.

■

A linear system has no solution if we end up with a *false equation* after applying Gaussian elimination to the system.

EXAMPLE 4 ■ A System with Infinitely Many Solutions

Solve the following system.

$$\begin{cases} x - y + 5z = -2 \\ 2x + y + 4z = 2 \\ 2x + 4y - 2z = 8 \end{cases}$$

SOLUTION

To put this in triangular form, we begin by eliminating the *x*-terms from the second equation and the third equation.

$$\begin{cases} x - y + 5z = -2 \\ 3y - 6z = 6 \\ 2x + 4y - 2z = 8 \end{cases}$$ Second equation + (−2) × first equation = new second equation

$$\begin{cases} x - y + 5z = -2 \\ 3y - 6z = 6 \\ 6y - 12z = 12 \end{cases}$$ Third equation + (−2) × first equation = new third equation

Now we eliminate the *y*-term from the third equation.

$$\begin{cases} x - y + 5z = -2 \\ 3y - 6z = 6 \\ 0 = 0 \end{cases}$$ Third equation + (−2) × second equation = new third equation

The new third equation is true, but it gives us no new information, so we can drop it from the system. Only two equations are left. We can use them to solve for *x* and *y* in terms of *z*, but *z* can take on any value, so there are infinitely many solutions.

To find the complete solution of the system, we begin by using the new second equation to solve for *y* in terms of *z*.

$$3y - 6z = 6$$ Second equation

$$y - 2z = 2$$ Multiply by $\frac{1}{3}$

$$y = 2z + 2$$ Solve for y

Then we substitute $y = 2z + 2$ into the first equation and solve for *x* in terms of *z*.

$$x - (2z + 2) + 5z = -2$$ Substitute $y = 2z + 2$ into first equation

$$x + 3z - 2 = -2$$ Simplify

$$x = -3z$$ Solve for x

To describe the complete solution, we let t represent any real number. The solution is

$$x = -3t$$

$$y = 2t + 2$$

$$z = t$$

We can also write this as the ordered triple $(-3t, 2t + 2, t)$.   ■

In the solution of Example 4 the variable t is called a **parameter**. To get a specific solution, we give a specific value to the parameter t. For instance, if we set $t = 2$, we get

$$x = -3(2) = -6$$

$$y = 2(2) + 2 = 6$$

$$z = 2$$

Thus, $(-6, 6, 2)$ is a solution of the system. If we set $t = -1$, we get a different solution, because then we have

$$x = -3(-1) = 3$$

$$y = 2(-1) + 2 = 0$$

$$z = -1$$

So $(3, 0, -1)$ is also a solution. (You should check that both of these points satisfy the original equations.) Since there are infinitely many choices for the parameter t, the system has infinitely many solutions.

Recall that if we are given several points in the plane then we can find the line of *best* fit, or the quadratic polynomial of *best* fit (see pages 247 and 398). Not all the given points will necessarily lie on the line or polynomial of best fit. But given just two points we can find the line of "*exact*" fit, that is, the line that actually passes through both points. Similarly, given three points (which are not on a line), we can find the quadratic polynomial of *exact* fit. That's what we are doing in Example 5. In Exercise 47, Section 8.4, we explore exact curve fitting for higher-degree polynomials.

■ Curve Fitting and Modeling Using Linear Systems

We know that given *two* points in the coordinate plane, there is a unique line $y = ax + b$ that passes through both points. It is also true that given three points that don't lie on a line, there is a unique quadratic (or second-degree) polynomial $y = ax^2 + bx + c$ whose graph passes through the three points.*

EXAMPLE 5 ■ Fitting a Quadratic Polynomial to Three Points

Find the quadratic polynomial that passes through the following points:

$$(-1, 6), \quad (1, 2), \quad (2, 3)$$

---

*Of course, all the points have to have different x-coordinates; otherwise there would be no function of any kind whose graph passes through the given points.

SOLUTION

We need to find the function of the form $y = ax^2 + bx + c$ that contains each of these points. Substituting the x- and y-coordinates of each point into the equation, we get the following three equations.

$$6 = a(-1)^2 + b(-1) + c \qquad \text{Substitute } x = -1 \text{ and } y = 6$$

$$2 = a(1)^2 + b(1) + c \qquad \text{Substitute } x = 1 \text{ and } y = 2$$

$$3 = a(2)^2 + b(2) + c \qquad \text{Subsitute } x = 2 \text{ and } y = 3$$

This simplifies into the following system in the variables a, b, and c, which we solve using Gaussian elimination.

$$\begin{cases} a - b + c = 6 \\ a + b + c = 2 \\ 4a + 2b + c = 3 \end{cases}$$

$$\begin{cases} a - b + c = 6 \\ 2b = -4 \qquad \text{Second equation} + (-1) \times \text{first equation} = \text{new second equation} \\ 6b - 3c = -21 \qquad \text{Third equation} + (-4) \times \text{first equation} = \text{new third equation} \end{cases}$$

$$\begin{cases} a - b + c = 6 \\ 2b = -4 \\ -3c = -9 \qquad \text{Third equation} + (-3) \times \text{second equation} = \text{new third equation} \end{cases}$$

CHECK YOUR ANSWER

$6 = (-1)^2 - 2(-1) + 3$
$2 = (2)^2 - 2(1) + 3$
$3 = (2)^2 - 2(2) + 3$ ✔

Now that the system is in triangular form, we use back-substitution to find that $a = 1$, $b = -2$, and $c = 3$. Thus, the required quadratic polynomial is

$$y = x^2 - 2x + 3$$

From Figure 1 we see that the graph of the polynomial passes through the given points.

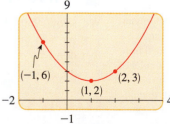

FIGURE 1

EXAMPLE 6 ■ Modeling a Financial Problem Using a Linear System

John receives an inheritance of $50,000. His financial advisor suggests that he invest this in three mutual funds: a money-market fund, a blue-chip stock fund, and a high-tech stock fund. The advisor estimates that the money-market fund will return 5% over the next year, the blue-chip fund 9%, and the high-tech fund 16%. John wants a total first-year return of $4000. To avoid excessive risk, he decides to invest three times as much in the money-market fund as in the high-tech stock fund. How much should he invest in each fund?

SOLUTION

Let x = amount invested in the money-market fund

y = amount invested in the blue-chip stock fund

z = amount invested in the high-tech stock fund

We convert each fact given in the problem into an equation.

$$x + y + z = 50{,}000 \qquad \text{Total amount invested is \$50,000}$$

$$0.05x + 0.09y + 0.16z = 4000 \qquad \text{Total investment return is \$4000}$$

$$x = 3z \qquad \text{Money-market amount is 3 times high-tech amount}$$

Multiplying the second equation by 100 and rewriting the third gives the following system, which we solve using Gaussian elimination.

$$\begin{cases} x + y + z = 50{,}000 \\ 5x + 9y + 16z = 400{,}000 \quad \text{100 × second equation} \\ x \quad\quad - 3z = \quad\quad 0 \quad \text{Subtract } 3z \end{cases}$$

$$\begin{cases} x + y + z = 50{,}000 \\ 4y + 11z = 150{,}000 \quad \text{Second equation} + (-5) \times \text{first equation} = \text{new second equation} \\ -y - 4z = -50{,}000 \quad \text{Third equation} + (-1) \times \text{first equation} = \text{new third equation} \end{cases}$$

$$\begin{cases} x + y + z = 50{,}000 \\ -5z = -50{,}000 \quad \text{Second equation} + 4 \times \text{third equation} = \text{new second equation} \\ -y - 4z = -50{,}000 \end{cases}$$

$$\begin{cases} x + y + z = 50{,}000 \\ z = 10{,}000 \quad \left(-\frac{1}{5}\right) \times \text{second equation} \\ y + 4z = 50{,}000 \quad (-1) \times \text{third equation} \end{cases}$$

$$\begin{cases} x + y + z = 50{,}000 \\ y + 4z = 50{,}000 \quad \text{Interchange second and third equations} \\ z = 10{,}000 \end{cases}$$

Now that the system is in triangular form, we use back-substitution to find that $x = 30{,}000$, $y = 10{,}000$, and $z = 10{,}000$. This means that he should invest

$30,000 in the money market fund

$10,000 in the blue-chip stock fund

$10,000 in the high-tech stock fund ■

8.3 EXERCISES

1–4 ■ State whether the equation or system of equations is linear.

1. $6x - \sqrt{3y} + \frac{1}{2}z = 0$

2. $x^2 + y^2 + z^2 = 4$

3. $\begin{cases} xy - 3y + z = 5 \\ x - y^2 + 5z = 0 \\ 2x + yz = 3 \end{cases}$

4. $\begin{cases} x - 2y + 3z = 10 \\ 2x + 5y = 2 \\ y + 2z = 4 \end{cases}$

5–10 ■ Use back-substitution to solve the triangular system.

5. $\begin{cases} x - 2y + 4z = 3 \\ y + 2z = 7 \\ z = 2 \end{cases}$

6. $\begin{cases} x + y - 3z = 8 \\ y - 3z = 5 \\ z = -1 \end{cases}$

7. $\begin{cases} x + 2y + z = 7 \\ -y + 3z = 9 \\ 2z = 6 \end{cases}$

8. $\begin{cases} x - 2y + 3z = 10 \\ 2y - z = 2 \\ 3z = 12 \end{cases}$

9. $\begin{cases} 2x - y + 6z = 5 \\ y + 4z = 0 \\ -2z = 1 \end{cases}$

10. $\begin{cases} 4x + 3z = 10 \\ 2y - z = -6 \\ \frac{1}{2}z = 4 \end{cases}$

11–14 ■ Perform an operation on the given system that eliminates the indicated variable. Write the new equivalent system.

11. $\begin{cases} x - 2y - z = 4 \\ x - y + 3z = 0 \\ 2x + y + z = 0 \end{cases}$

Eliminate the x-term from the second equation.

12. $\begin{cases} x + y - 3z = 3 \\ -2x + 3y + z = 2 \\ x - y + 2z = 0 \end{cases}$

Eliminate the x-term from the second equation.

13. $\begin{cases} 2x - y + 3z = 2 \\ x + 2y - z = 4 \\ -4x + 5y + z = 10 \end{cases}$

Eliminate the x-term from the third equation.

14. $\begin{cases} x - 4y + z = 3 \\ y - 3z = 10 \\ 3y - 8z = 24 \end{cases}$

Eliminate the y-term from the third equation.

15–32 ■ Find the complete solution of the linear system, or show that it is inconsistent.

15. $\begin{cases} x + y + z = 4 \\ x + 3y + 3z = 10 \\ 2x + y - z = 3 \end{cases}$

16. $\begin{cases} x + y + z = 0 \\ -x + 2y + 5z = 3 \\ 3x - y = 6 \end{cases}$

17. $\begin{cases} x - 4z = 1 \\ 2x - y - 6z = 4 \\ 2x + 3y - 2z = 8 \end{cases}$

18. $\begin{cases} x - y + 2z = 2 \\ 3x + y + 5z = 8 \\ 2x - y - 2z = -7 \end{cases}$

19. $\begin{cases} 2x + 4y - z = 2 \\ x + 2y - 3z = -4 \\ 3x - y + z = 1 \end{cases}$

20. $\begin{cases} 2x + y - z = -8 \\ -x + y + z = 3 \\ -2x + 4z = 18 \end{cases}$

21. $\begin{cases} y - 2z = 0 \\ 2x + 3y = 2 \\ -x - 2y + z = -1 \end{cases}$

22. $\begin{cases} 2y + z = 3 \\ 5x + 4y + 3z = -1 \\ x - 3y = -2 \end{cases}$

23. $\begin{cases} x + 2y - z = 1 \\ 2x + 3y - 4z = -3 \\ 3x + 6y - 3z = 4 \end{cases}$

24. $\begin{cases} -x + 2y + 5z = 4 \\ x - 2z = 0 \\ 4x - 2y - 11z = 2 \end{cases}$

25. $\begin{cases} 2x + 3y - z = 1 \\ x + 2y = 3 \\ x + 3y + z = 4 \end{cases}$

26. $\begin{cases} x - 2y - 3z = 5 \\ 2x + y - z = 5 \\ 4x - 3y - 7z = 5 \end{cases}$

27. $\begin{cases} x + y - z = 0 \\ x + 2y - 3z = -3 \\ 2x + 3y - 4z = -3 \end{cases}$

28. $\begin{cases} x - 2y + z = 3 \\ 2x - 5y + 6z = 7 \\ 2x - 3y - 2z = 5 \end{cases}$

29. $\begin{cases} x + 3y - 2z = 0 \\ 2x + 4z = 4 \\ 4x + 6y = 4 \end{cases}$

30. $\begin{cases} 2x + 4y - z = 3 \\ x + 2y + 4z = 6 \\ x + 2y - 2z = 0 \end{cases}$

31.
$$\begin{cases} x + & z + 2w = 6 \\ y - 2z & = -3 \\ x + 2y - z & = -2 \\ 2x + y + 3z - 2w = 0 \end{cases}$$

32.
$$\begin{cases} x + y + z + w = 0 \\ x + y + 2z + 2w = 0 \\ 2x + 2y + 3z + 4w = 1 \\ 2x + 3y + 4z + 5w = 2 \end{cases}$$

33–36 ■ Find the quadratic polynomial $y = ax^2 + bx + c$ whose graph passes through the given points.

33. $(-2, 3),\ (-1, 1),\ (1, 9)$

34. $(-1, -3)\ (2, 0),\ (3, -3)$

35. $(-1, -10),\ (1, 0),\ (2, 15)$

36. $(-2, 9),\ (2, 5),\ (3, 19)$

37–38 ■ An investor has $100,000 to invest in three types of bonds: short-term, intermediate-term, and long-term. How much should she invest in each type to satisfy the given conditions?

37. Short-term bonds pay 4% annually, intermediate-term bonds pay 5%, and long-term bonds pay 6%. The investor wishes to realize a total annual income of 5.1%, with equal amounts invested in short- and intermediate-term bonds.

38. Short-term bonds pay 4% annually, intermediate-term bonds pay 6%, and long-term bonds pay 8%. The investor wishes to have a total annual return of $6700 on her investment, with equal amounts invested in intermediate- and long-term bonds.

39. A biologist is performing an experiment on the effects of various combinations of vitamins. She wishes to feed each of her laboratory rabbits a diet that contains exactly 9 mg of niacin, 14 mg of thiamin, and 32 mg of riboflavin. She has available three different types of commercial rabbit pellets whose content of the relevant vitamins per ounce is given in the following table. How many ounces of each type of food should each rabbit be given daily to satisfy the experiment requirements?

	Type A	Type B	Type C
Niacin (mg)	2	3	1
Thiamin (mg)	3	1	3
Riboflavin (mg)	8	5	7

40. Using Kirchhoff's Laws, it can be shown that the currents I_1, I_2, and I_3 that pass through the three branches of the circuit in the figure satisfy the given linear system. Solve the system to find I_1, I_2, and I_3.

$$\begin{cases} I_1 + I_2 - I_3 = 0 \\ 16I_1 - 8I_2 = 4 \\ 8I_2 + 4I_3 = 5 \end{cases}$$

DISCOVERY · DISCUSSION

41. Exact Fit Versus Best Fit

(a) Find the quadratic function $y = ax^2 + bx + c$ whose graph passes through the given points. (This is the quadratic curve of *exact* fit.) Graph the points and the quadratic curve that you found.

$$(-2, 10),\ (1, -5)\ (2, -6),\ (4, -2)$$

(b) Now use the **QuadReg** command on your calculator to find the quadratic curve that *best* fits the points in part (a). How does this compare to the function you found in part (a)?

(c) Show that no quadratic function passes through the points

$$(-2, 11),\ (1, -6),\ (2, -5),\ (4, -1)$$

(d) Use the **QuadReg** command on your calculator to find the quadratic curve that best fits the points in part (b). Graph the points and the quadratic curve that you found.

(e) Explain how the curve of exact fit differs from the curve of best fit.

42. Can a Linear System Have Exactly Two Solutions?
(a) Suppose that (x_0, y_0, z_0) and (x_1, y_1, z_1) are solutions of the system

$$\begin{cases} a_1x + b_1y + c_1z = d_1 \\ a_2x + b_2y + c_2z = d_2 \\ a_3x + b_3y + c_3z = d_3 \end{cases}$$

Show that $\left(\dfrac{x_0 + x_1}{2}, \dfrac{y_0 + y_1}{2}, \dfrac{z_0 + z_1}{2}\right)$ is also a solution.

(b) Use the result of part (a) to prove that if the system has two different solutions, then it has infinitely many solutions.

8.4 SYSTEMS OF LINEAR EQUATIONS: MATRICES

In this section we express a system of equations as a rectangular array of numbers called a *matrix*. Matrices* provide us with efficient methods for solving systems of linear equations.

> ### DEFINITION OF MATRIX
>
> An $m \times n$ **matrix** is a rectangular array of numbers with m **rows** and n **columns**.
>
> $$\begin{bmatrix} a_{11} & a_{12} & a_{13} & \cdots & a_{1n} \\ a_{21} & a_{22} & a_{23} & \cdots & a_{2n} \\ a_{31} & a_{32} & a_{33} & \cdots & a_{3n} \\ \vdots & \vdots & \vdots & \ddots & \vdots \\ a_{m1} & a_{m2} & a_{m3} & \cdots & a_{mn} \end{bmatrix} \begin{matrix} \leftarrow \\ \leftarrow \\ \leftarrow \\ \\ \leftarrow \end{matrix} \left. \right\} m \text{ rows}$$
>
> $\underbrace{\qquad\qquad\qquad\qquad}_{n \text{ columns}}$
>
> The numbers a_{ij} are the **entries** of the matrix. The subscript on the entry a_{ij} indicates that it is in the ith row and the jth column.

We write a system of linear equations in matrix form by writing only the coefficients and constants that appear in the equations. The resulting matrix is called the **augmented matrix** of the system. For example, the linear system on the left below corresponds to the augmented matrix on the right.

Linear system

$$\begin{cases} 3x - 2y + z = 5 \\ x + 3y - z = 0 \\ -x + \quad 4z = 11 \end{cases}$$

Augmented matrix

$$\begin{bmatrix} 3 & -2 & 1 & 5 \\ 1 & 3 & -1 & 0 \\ -1 & 0 & 4 & 11 \end{bmatrix}$$

Notice that a missing variable in an equation corresponds to a 0 entry in the augmented matrix. Writing linear systems carefully with the variables lined up in columns makes it easy to find the augmented matrix of the system.

*The plural of *matrix* is *matrices*.

EXAMPLE 1 ■ Finding the Augmented Matrix of a Linear System

Write the augmented matrix of the system of equations.

$$\begin{cases} 6x - 2y - z = 4 \\ x + 3z = 1 \\ 7y + z = 5 \end{cases}$$

SOLUTION

First we write the linear system with the variables lined up in columns.

$$\begin{cases} 6x - 2y - z = 4 \\ x \qquad + 3z = 1 \\ \qquad 7y + z = 5 \end{cases}$$

The augmented matrix is the matrix whose entries are the coefficients and the constants in this system.

$$\begin{bmatrix} 6 & -2 & -1 & 4 \\ 1 & 0 & 3 & 1 \\ 0 & 7 & 1 & 5 \end{bmatrix}$$ ■

Elementary Row Operations

The operations that we used in Section 8.3 to solve linear systems correspond to operations on the rows of the augmented matrix of the system. For example, adding a multiple of one equation to another corresponds to adding a multiple of one row to another.

ELEMENTARY ROW OPERATIONS

1. Add a multiple of one row to another.

2. Multiply a row by a nonzero constant.

3. Interchange two rows.

Note that performing any of these operations on the matrix form of a system does not change its solution. We use the following notation to describe the elementary row operations:

Symbol	Description
$R_i + kR_j \rightarrow R_i$	Change the ith row by adding k times row j to it, and then put the result back in row i.
kR_i	Multiply the ith row by k.
$R_i \leftrightarrow R_j$	Interchange the ith and jth rows.

In the next example we compare the two ways of writing systems of linear equations.

EXAMPLE 2 ■ Using Elementary Row Operations to Solve a Linear System

Solve the system of linear equations.

$$\begin{cases} x - y + 3z = 4 \\ x + 2y - 2z = 10 \\ 3x - y + 5z = 14 \end{cases}$$

SOLUTION

Our goal is to eliminate the x-term from the second equation and the x- and y-terms from the third equation. For comparison, we write both the system of equations and its augmented matrix form.

	System	Matrix form

$$\begin{cases} x - y + 3z = 4 \\ x + 2y - 2z = 10 \\ 3x - y + 5z = 14 \end{cases} \qquad \begin{bmatrix} 1 & -1 & 3 & 4 \\ 1 & 2 & -2 & 10 \\ 3 & -1 & 5 & 14 \end{bmatrix}$$

Add -1 times the first equation to the second.

Add -3 times the first to the third.

$$\begin{cases} x - y + 3z = 4 \\ 3y - 5z = 6 \\ 2y - 4z = 2 \end{cases} \qquad \xrightarrow[\,R_3 - 3R_1 \to R_3\,]{R_2 - R_1 \to R_2} \qquad \begin{bmatrix} 1 & -1 & 3 & 4 \\ 0 & 3 & -5 & 6 \\ 0 & 2 & -4 & 2 \end{bmatrix}$$

Multiply the third equation by $\frac{1}{2}$.

$$\begin{cases} x - y + 3z = 4 \\ 3y - 5z = 6 \\ y - 2z = 1 \end{cases} \qquad \xrightarrow{\frac{1}{2}R_3} \qquad \begin{bmatrix} 1 & -1 & 3 & 4 \\ 0 & 3 & -5 & 6 \\ 0 & 1 & -2 & 1 \end{bmatrix}$$

Add -3 times the third equation to the second (to eliminate y from the second equation).

$$\begin{cases} x - y + 3z = 4 \\ z = 3 \\ y - 2z = 1 \end{cases} \qquad \xrightarrow{R_2 - 3R_3 \to R_2} \qquad \begin{bmatrix} 1 & -1 & 3 & 4 \\ 0 & 0 & 1 & 3 \\ 0 & 1 & -2 & 1 \end{bmatrix}$$

Interchange the second and third equations.

$$\begin{cases} x - y + 3z = 4 \\ y - 2z = 1 \\ z = 3 \end{cases} \qquad \xrightarrow{R_2 \leftrightarrow R_3} \qquad \begin{bmatrix} 1 & -1 & 3 & 4 \\ 0 & 1 & -2 & 1 \\ 0 & 0 & 1 & 3 \end{bmatrix}$$

Now we use back-substitution to find that $x = 2$, $y = 7$, and $z = 3$. The solution is $(2, 7, 3)$. ■

Row-Echelon Form and Gaussian Elimination

In general, to solve a system of linear equations using its augmented matrix, we use elementary row operations to arrive at a matrix in a certain form. This form is described in the following box.

> ## ROW-ECHELON FORM AND REDUCED ROW-ECHELON FORM OF A MATRIX
>
> A matrix is in **row-echelon form** if it satisfies the following conditions.
>
> 1. The first nonzero number in each row (reading from left to right) is 1. This is called the **leading entry**.
>
> 2. The leading entry in each row is to the right of the leading entry in the row immediately above it.
>
> 3. All rows consisting entirely of zeros are at the bottom of the matrix.
>
> A matrix in row-echelon form is in **reduced row-echelon form** if it also satisfies the following condition.
>
> 4. Every number above and below each leading entry is a 0.

In the following matrices the first matrix is in reduced row-echelon form, but the second one is just in row-echelon form. The third matrix is not in row-echelon form. The entries in red are the leading entries.

Reduced row-echelon form   Row-echelon form   Not in row-echelon form

$$
\begin{bmatrix} 1 & 3 & 0 & 0 & 0 \\ 0 & 0 & 1 & 0 & -3 \\ 0 & 0 & 0 & 1 & \frac{1}{2} \\ 0 & 0 & 0 & 0 & 0 \end{bmatrix}
\qquad
\begin{bmatrix} 1 & 3 & -6 & 10 & 0 \\ 0 & 0 & 1 & 4 & -3 \\ 0 & 0 & 0 & 1 & \frac{1}{2} \\ 0 & 0 & 0 & 0 & 0 \end{bmatrix}
\qquad
\begin{bmatrix} 0 & 1 & -\frac{1}{2} & 0 & 7 \\ 1 & 0 & 3 & 4 & -5 \\ 0 & 0 & 0 & 1 & 0.4 \\ 0 & 1 & 1 & 0 & 0 \end{bmatrix}
$$

A systematic way of using the elementary row operations to put a matrix in row-echelon form is as follows:

- Start by obtaining 1 in the top left corner. Then obtain zeros below that 1 by adding appropriate multiples of the first row to the rows below it.

- Next, obtain a leading 1 in the next row, and then obtain zeros below that 1.

- At each stage make sure that every leading entry is to the right of the leading entry in the row above it—rearrange the rows if necessary

- Continue this process until you arrive at a matrix in row-echelon form.

The figure below illustrates how the process might work for a 3×4 matrix.

$$
\begin{bmatrix} 1 & \blacksquare & \blacksquare & \blacksquare \\ 0 & \blacksquare & \blacksquare & \blacksquare \\ 0 & \blacksquare & \blacksquare & \blacksquare \end{bmatrix}
\longrightarrow
\begin{bmatrix} 1 & \blacksquare & \blacksquare & \blacksquare \\ 0 & 1 & \blacksquare & \blacksquare \\ 0 & 0 & \blacksquare & \blacksquare \end{bmatrix}
\longrightarrow
\begin{bmatrix} 1 & \blacksquare & \blacksquare & \blacksquare \\ 0 & 1 & \blacksquare & \blacksquare \\ 0 & 0 & 1 & \blacksquare \end{bmatrix}
$$

Once an augmented matrix is in row-echelon form, we can solve the corresponding linear system by using back-substitution. This technique is called **Gaussian elimination**, in honor of its inventor, the German mathematician C. F. Gauss (see page 300).

SOLVING A SYSTEM USING GAUSSIAN ELIMINATION

1. AUGMENTED MATRIX.   Write the augmented matrix of the system.

2. ROW-ECHELON FORM.   Use elementary row operations to change the augmented matrix to row-echelon form.

3. BACK-SUBSTITUTION.   Write the new system of equations that corresponds to the row-echelon form of the augmented matrix and solve by back-substitution.

EXAMPLE 3 ■ Solving a System Using Row-Echelon Form

Solve the system of linear equations using Gaussian elimination.

$$\begin{cases} 4x + 8y - 4z = 4 \\ 3x + 6y + 5z = -13 \\ -2x + y + 12z = -17 \end{cases}$$

SOLUTION

We first write the augmented matrix of the system, and then use elementary row operations to put it in row-echelon form.

$$\begin{bmatrix} 4 & 8 & -4 & 4 \\ 3 & 6 & 5 & -13 \\ -2 & 1 & 12 & -17 \end{bmatrix} \xrightarrow{\frac{1}{4}R_1} \begin{bmatrix} 1 & 2 & -1 & 1 \\ 3 & 6 & 5 & -13 \\ -2 & 1 & 12 & -17 \end{bmatrix} \xrightarrow[R_3 + 2R_1 \to R_3]{R_2 - 3R_1 \to R_2} \begin{bmatrix} 1 & 2 & -1 & 1 \\ 0 & 0 & 8 & -16 \\ 0 & 5 & 10 & -15 \end{bmatrix}$$

$$\xrightarrow{R_2 \leftrightarrow R_3} \begin{bmatrix} 1 & 2 & -1 & 1 \\ 0 & 5 & 10 & -15 \\ 0 & 0 & 8 & -16 \end{bmatrix} \xrightarrow[\frac{1}{8}R_3]{\frac{1}{5}R_2} \begin{bmatrix} 1 & 2 & -1 & 1 \\ 0 & 1 & 2 & -3 \\ 0 & 0 & 1 & -2 \end{bmatrix}$$

We now have an equivalent matrix in row-echelon form, and the corresponding system of equations is

$$\begin{cases} x + 2y - z = 1 \\ y + 2z = -3 \\ z = -2 \end{cases}$$

We use back-substitution to solve the system.

$$y + 2(-2) = -3 \qquad \text{Back-substitute } z = -2 \text{ into the second equation}$$

$$y = 1 \qquad \text{Solve for } y$$

$$x + 2(1) - (-2) = 1 \qquad \text{Back-substitute } y = 1 \text{ and } z = -2 \text{ into the first equation}$$

$$x = -3 \qquad \text{Solve for } x$$

So the solution of the system is $(-3, 1, -2)$. ■

```
ref([A])
   [[1 2 -1  1 ]
    [0  1  2  -3]
    [0  0  1  -2]]
```

We can use the "row-echelon form" matrix command on a graphing calculator to put a matrix in row-echelon form. (On the TI-83 the command is **ref**.) For the augmented matrix in Example 3, the **ref** command gives the output shown in Figure 1.

If we put the augmented matrix of a linear system in *reduced* row-echelon form then we don't need to use back-substitution to solve the system. To put a matrix in reduced row-echelon form, we use the following steps.

- Use the elementary row operations to put the matrix in row-echelon form.

- Obtain zeros above each leading entry by adding multiples of the row containing that entry to the rows above it. It is most efficient to begin with the last leading entry and work up.

Here is how the process might work for a 3×4 matrix:

$$\begin{bmatrix} 1 & \blacksquare & \blacksquare & \blacksquare \\ 0 & 1 & \blacksquare & \blacksquare \\ 0 & 0 & 1 & \blacksquare \end{bmatrix} \rightarrow \begin{bmatrix} 1 & \blacksquare & 0 & \blacksquare \\ 0 & 1 & 0 & \blacksquare \\ 0 & 0 & 1 & \blacksquare \end{bmatrix} \rightarrow \begin{bmatrix} 1 & 0 & 0 & \blacksquare \\ 0 & 1 & 0 & \blacksquare \\ 0 & 0 & 1 & \blacksquare \end{bmatrix}$$

Solving a system using the reduced row-echelon form is called **Gauss-Jordan elimination**. The process is illustrated in the next example.

EXAMPLE 4 ■ Solving a System Using Reduced Row-Echelon Form

Solve the system of linear equations, using Gauss-Jordan elimination.

$$\begin{cases} 4x + 8y - 4z = 4 \\ 3x + 6y + 5z = -13 \\ -2x + y + 12z = -17 \end{cases}$$

SOLUTION

In Example 3 we used Gaussian elimination on the augmented matrix of this system to arrive at an equivalent matrix in row-echelon form. We continue using elementary row operations on the last matrix in Example 3 to arrive at an equivalent matrix in reduced row-echelon form.

$$\begin{bmatrix} 1 & 2 & -1 & 1 \\ 0 & 1 & 2 & -3 \\ 0 & 0 & 1 & -2 \end{bmatrix} \xrightarrow[R_1 + R_3 \to R_1]{R_2 - 2R_3 \to R_2} \begin{bmatrix} 1 & 2 & 0 & -1 \\ 0 & 1 & 0 & 1 \\ 0 & 0 & 1 & -2 \end{bmatrix} \xrightarrow{R_1 - 2R_2 \to R_1} \begin{bmatrix} 1 & 0 & 0 & -3 \\ 0 & 1 & 0 & 1 \\ 0 & 0 & 1 & -2 \end{bmatrix}$$

We now have an equivalent matrix in reduced row-echelon form, and the corresponding system of equations is

Since the system is in reduced row-echelon form, back-substitution is not required to get the solution.

$$\begin{cases} x = -3 \\ y = 1 \\ z = -2 \end{cases}$$

Hence we immediately arrive at the solution $(-3, 1, -2)$. ■

FIGURE 2

We can use the "reduced row-echelon form" matrix command on a graphing calculator to put a matrix in reduced row-echelon form. (On the TI-83 the command is `rref`.) For the augmented matrix in Example 4, the `rref` command gives the output shown in Figure 2.

▪ Inconsistent and Dependent Systems

The systems of linear equations that we considered in Examples 1–4 had exactly one solution. But as we know from Section 8.3 a linear system may have one solution, no solution, or infinitely many solutions. Fortunately, the row-echelon form of a system allows us to determine which of these cases applies, as described in the following box.

First we need some terminology. A **leading variable** in a linear system is one that corresponds to a leading entry in the row-echelon form of the augmented matrix of the system.

THE SOLUTIONS OF A LINEAR SYSTEM IN ROW-ECHELON FORM

Suppose the augmented matrix of a system of linear equations has been transformed by Gaussian elimination into row-echelon form. Then exactly one of the following is true.

1. **No solution.** If the row-echelon form contains a row that represents the equation $0 = c$ where c is not zero, then the system has no solution. A system with no solution is called **inconsistent**.

2. **One solution.** If each variable in the row-echelon form is a leading variable then the system has exactly one solution, which we find using back-substitution or Gauss-Jordan elimination.

3. **Infinitely many solutions.** If the variables in the row-echelon form are not all leading variables, and if the system is not inconsistent, then it has infinitely many solutions. In this case, the system is called **dependent**. We solve the system by putting the matrix in reduced row-echelon form and then expressing the leading variables in terms of the nonleading variables. The nonleading variables may take on any real numbers as their values.

The matrices below, all in row-echelon form, illustrate the three cases described in the box.

<table>
<tr><td align="center">No solution</td><td align="center">One solution</td><td align="center">Infinitely many solutions</td></tr>
</table>

$$\begin{bmatrix} 1 & 2 & 5 & 7 \\ 0 & 1 & 3 & 4 \\ 0 & 0 & 0 & 1 \end{bmatrix} \quad \begin{bmatrix} 1 & 6 & -1 & 3 \\ 0 & 1 & 2 & -2 \\ 0 & 0 & 1 & 8 \end{bmatrix} \quad \begin{bmatrix} 1 & 2 & -3 & 1 \\ 0 & 1 & 5 & -2 \\ 0 & 0 & 0 & 0 \end{bmatrix}$$

EXAMPLE 5 ■ A System with No Solution

Solve the system.

$$\begin{cases} x - 3y + 2z = 12 \\ 2x - 5y + 5z = 14 \\ x - 2y + 3z = 20 \end{cases}$$

SOLUTION

We transform the system into row-echelon form.

$$\begin{bmatrix} 1 & -3 & 2 & 12 \\ 2 & -5 & 5 & 14 \\ 1 & -2 & 3 & 20 \end{bmatrix} \xrightarrow[\begin{array}{c} R_2 - 2R_1 \to R_2 \\ R_3 - R_1 \to R_3 \end{array}]{} \begin{bmatrix} 1 & -3 & 2 & 12 \\ 0 & 1 & 1 & -10 \\ 0 & 1 & 1 & 8 \end{bmatrix}$$

$$\xrightarrow{R_3 - R_2 \to R_3} \begin{bmatrix} 1 & -3 & 2 & 12 \\ 0 & 1 & 1 & -10 \\ 0 & 0 & 0 & 18 \end{bmatrix} \xrightarrow{\frac{1}{18}R_3} \begin{bmatrix} 1 & -3 & 2 & 12 \\ 0 & 1 & 1 & -10 \\ 0 & 0 & 0 & 1 \end{bmatrix}$$

This last matrix is in row-echelon form, so we can stop the Gaussian elimination process. Now if we translate the last row back into equation form, we get $0x + 0y + 0z = 1$, or $0 = 1$, which is false. No matter what values we pick for x, y, and z, the last equation will never be a true statement. This means the system *has no solution*. ■

```
ref([A])
[[1 -2.5 2.5 7   ]
 [0 1    1   -10]
 [0 0    0   1  ]]
```

FIGURE 3

Figure 3 shows the row-echelon form produced by a TI-83 calculator for the augmented matrix in Example 5. This not the same as the row-echelon form that we obtained in Example 5. In fact, every matrix has infinitely many different row-echelon forms, but each leads to the same solution of the original system. Note, however, that the *reduced* row-echelon form of a matrix *is* unique.

EXAMPLE 6 ■ A System with Infinitely Many Solutions

Find the complete solution of the system.

$$\begin{cases} -3x - 5y + 36z = 10 \\ -x \qquad + 7z = 5 \\ x + y - 10z = -4 \end{cases}$$

SOLUTION

We transform the system into reduced row-echelon form.

$$
\begin{bmatrix} -3 & -5 & 36 & 10 \\ -1 & 0 & 7 & 5 \\ 1 & 1 & -10 & -4 \end{bmatrix}
\xrightarrow{R_1 \leftrightarrow R_3}
\begin{bmatrix} 1 & 1 & -10 & -4 \\ -1 & 0 & 7 & 5 \\ -3 & -5 & 36 & 10 \end{bmatrix}
$$

$$
\xrightarrow[R_3 + 3R_1 \to R_3]{R_2 + R_1 \to R_2}
\begin{bmatrix} 1 & 1 & -10 & -4 \\ 0 & 1 & -3 & 1 \\ 0 & -2 & 6 & -2 \end{bmatrix}
\xrightarrow{R_3 + 2R_2 \to R_3}
\begin{bmatrix} 1 & 1 & -10 & -4 \\ 0 & 1 & -3 & 1 \\ 0 & 0 & 0 & 0 \end{bmatrix}
$$

$$
\xrightarrow{R_1 - R_2 \to R_1}
\begin{bmatrix} 1 & 0 & -7 & -5 \\ 0 & 1 & -3 & 1 \\ 0 & 0 & 0 & 0 \end{bmatrix}
$$

The third row corresponds to the equation $0 = 0$. This equation is always true, no matter what values are used for x, y, and z. Since the equation adds no new information about the variables, we can drop it from the system. So the last matrix corresponds to the system

Leading variables
$$
\begin{cases} x & -7z = -5 \\ y - 3z = 1 \end{cases}
$$

We solve for the leading variables x and y in terms of the nonleading variable z:

$$x = 7z - 5 \qquad \text{Solve for } x \text{ in the first equation}$$

$$y = 3z + 1 \qquad \text{Solve for } y \text{ in the second equation}$$

To obtain the complete solution, we let t represent any real number, and we express x, y, and z in terms of t:

$$x = 7t - 5$$

$$y = 3t + 1$$

$$z = t$$

We can also write the solution as the ordered triple $(7t - 5, \, 3t + 1, \, t)$, where t is any real number. ∎

In Example 6, to get specific solutions we give a specific value to t. For example, if $t = 1$, then

$$x = 7(1) - 5 = 2$$

$$y = 3(1) + 1 = 4$$

$$z = 1$$

Reduced row-echelon form on the TI-83 calculator:

```
rref([A])
  [[1 0 -7 -5]
   [0 1 -3 1 ]
   [0 0 0  0 ]]
```

Thus, $(2, 4, 1)$ is a solution to the system. We would get a different solution if we let $t = 2$, because we would then have

$$x = 7(2) - 5 = 9$$

$$y = 3(2) + 1 = 7$$

$$z = 2$$

So $(9, 7, 2)$ is also a solution. There are infinitely many choices for t, so there are infinitely many solutions to the system.

EXAMPLE 7 ■ A System with Infinitely Many Solutions

Find the complete solution of the system.

$$\begin{cases} x + 2y - 3z - 4w = 10 \\ x + 3y - 3z - 4w = 15 \\ 2x + 2y - 6z - 8w = 10 \end{cases}$$

SOLUTION

We transform the system into reduced row-echelon form.

$$\begin{bmatrix} 1 & 2 & -3 & -4 & 10 \\ 1 & 3 & -3 & -4 & 15 \\ 2 & 2 & -6 & -8 & 10 \end{bmatrix} \xrightarrow[R_3 - 2R_1 \rightarrow R_3]{R_2 - R_1 \rightarrow R_2} \begin{bmatrix} 1 & 2 & -3 & -4 & 10 \\ 0 & 1 & 0 & 0 & 5 \\ 0 & -2 & 0 & 0 & -10 \end{bmatrix}$$

$$\xrightarrow{R_3 + 2R_2 \rightarrow R_3} \begin{bmatrix} 1 & 2 & -3 & -4 & 10 \\ 0 & 1 & 0 & 0 & 5 \\ 0 & 0 & 0 & 0 & 0 \end{bmatrix} \xrightarrow{R_1 - 2R_2 \rightarrow R_1} \begin{bmatrix} 1 & 0 & -3 & -4 & 0 \\ 0 & 1 & 0 & 0 & 5 \\ 0 & 0 & 0 & 0 & 0 \end{bmatrix}$$

This is in reduced row-echelon form. Since the last row represents the equation $0 = 0$, we may discard it. So the last matrix corresponds to the system

Leading variables $\begin{cases} x \quad\quad -3z - 4w = 0 \\ \quad\quad y \quad\quad\quad\quad = 5 \end{cases}$

To obtain the complete solution, we solve for the leading variables x and y in terms of the nonleading variables z and w, and we let z and w be any real numbers. Thus, the complete solution is

$$x = 3s + 4t$$

$$y = 5$$

$$z = s$$

$$w = t$$

where s and t are any real numbers.

Note that s and t do *not* have to be the *same* real number in the solution for Example 7. We can choose arbitrary values for each if we wish to construct a specific solution to the system. For example, if we let $s = 1$ and $t = 2$, then we get the solution $(11, 5, 1, 2)$. You should check that this does indeed satisfy all three of the original equations in Example 7.

Examples 6 and 7 illustrate this general fact: If a system in row-echelon form has n nonzero equations in m variables $(m > n)$, then the complete solution will have $m - n$ nonleading variables. For instance, in Example 6 we arrived at two nonzero equations in the three variables x, y, and z, with $3 - 2 = 1$ nonleading variable.

■ Modeling with Linear Systems

Linear equations, often containing hundreds or even thousands of variables, occur frequently in the applications of algebra to the sciences and to other fields. For now, let's consider an example that involves only three variables.

EXAMPLE 8 ■ Nutritional Analysis Using a System of Linear Equations

A nutritionist is performing an experiment on student volunteers. He wishes to feed one of his subjects a daily diet that consists of a combination of three commercial diet foods: MiniCal, LiquiFast, and SlimQuick. For the experiment it's important that the subject consume exactly 500 mg of potassium, 75 g of protein, and 1150 units of vitamin D every day. The amounts of these nutrients in one ounce of each food are given in the table. How many ounces of each food should the subject eat every day to satisfy the nutrient requirements exactly?

	MiniCal	LiquiFast	SlimQuick
Potassium (mg)	50	75	10
Protein (g)	5	10	3
Vitamin D (units)	90	100	50

SOLUTION

Let x, y, and z represent the number of ounces of MiniCal, LiquiFast, and SlimQuick, respectively, that the subject should eat every day. This means that he will get $50x$ mg of potassium from MiniCal, $75y$ mg from LiquiFast, and $10z$ mg from SlimQuick, for a total of $50x + 75y + 10z$ mg potassium in all. Since the potassium requirement is 500 mg, we get the first equation below. Similar reasoning for the protein and vitamin D requirements leads to the system

$$\begin{cases} 50x + 75y + 10z = 500 & \text{Potassium} \\ 5x + 10y + 3z = 75 & \text{Protein} \\ 90x + 100y + 50z = 1150 & \text{Vitamin D} \end{cases}$$

```
rref([A])
    [[1 0 0 5 ]
     [0 1 0 2 ]
     [0 0 1 10]]
```

FIGURE 4

CHECK YOUR ANSWER

$x = 5, y = 2, z = 10$:

$$\begin{cases} 10(5) + 15(2) + 2(10) = 100 \\ 5(5) + 10(2) + 3(10) = 75 \\ 9(5) + 10(2) + 5(10) = 115 \end{cases} \checkmark$$

Dividing the first equation by 5 and the third one by 10 gives the system

$$\begin{cases} 10x + 15y + 2z = 100 \\ 5x + 10y + 3z = 75 \\ 9x + 10y + 5z = 115 \end{cases}$$

We can solve this system using Gaussian elimination, or we can use a graphing calculator to find the reduced row-echelon form of the augmented matrix of the system. Using the **rref** command on the TI-83, we get the output in Figure 4. From the reduced row-echelon form we see that $x = 5$, $y = 2$, $z = 10$. The subject should be fed 5 oz of MiniCal, 2 oz of LiquiFast, and 10 oz of SlimQuick every day.   ∎

A more practical application might involve dozens of foods and nutrients rather than just three. Such problems lead to systems with large numbers of variables and equations. Computers or graphing calculators are essential for solving such large systems.

8.4 EXERCISES

1–8 ■ (a) Determine whether the matrix is in row-echelon form. (b) Determine whether the matrix is in reduced row-echelon form. (c) Write the system of equations for which the given matrix is the augmented matrix.

1. $\begin{bmatrix} 1 & 0 & -3 \\ 0 & 1 & 5 \end{bmatrix}$

2. $\begin{bmatrix} 1 & 3 & -3 \\ 0 & 1 & 5 \end{bmatrix}$

3. $\begin{bmatrix} 1 & 2 & 8 & 0 \\ 0 & 1 & 3 & 2 \\ 0 & 0 & 0 & 0 \end{bmatrix}$

4. $\begin{bmatrix} 1 & 0 & -7 & 0 \\ 0 & 1 & 3 & 0 \\ 0 & 0 & 0 & 1 \end{bmatrix}$

5. $\begin{bmatrix} 1 & 0 & 0 & 0 \\ 0 & 0 & 0 & 0 \\ 0 & 1 & 5 & 1 \end{bmatrix}$

6. $\begin{bmatrix} 1 & 0 & 0 & 1 \\ 0 & 1 & 0 & 2 \\ 0 & 0 & 1 & 3 \end{bmatrix}$

7. $\begin{bmatrix} 1 & 3 & 0 & -1 & 0 \\ 0 & 0 & 1 & 2 & 0 \\ 0 & 0 & 0 & 0 & 1 \\ 0 & 0 & 0 & 0 & 0 \end{bmatrix}$

8. $\begin{bmatrix} 1 & 3 & 0 & 1 & 0 & 0 \\ 0 & 1 & 0 & 4 & 0 & 0 \\ 0 & 0 & 0 & 1 & 1 & 2 \\ 0 & 0 & 0 & 1 & 0 & 0 \end{bmatrix}$

9–18 ■ The system of linear equations has a unique solution. Find the solution using Gaussian elimination or Gauss-Jordan elimination.

9. $\begin{cases} x - 2y + z = 1 \\ y + 2z = 5 \\ x + y + 3z = 8 \end{cases}$

10. $\begin{cases} x + y + 6z = 3 \\ x + y + 3z = 3 \\ x + 2y + 4z = 7 \end{cases}$

11. $\begin{cases} x + y + z = 2 \\ 2x - 3y + 2z = 4 \\ 4x + y - 3z = 1 \end{cases}$

12. $\begin{cases} x + y + z = 4 \\ -x + 2y + 3z = 17 \\ 2x - y = -7 \end{cases}$

13. $\begin{cases} x + 2y - z = -2 \\ x + z = 0 \\ 2x - y - z = -3 \end{cases}$

14. $\begin{cases} 2y + z = 4 \\ x + y = 4 \\ 3x + 3y - z = 10 \end{cases}$

15. $\begin{cases} x_1 + 2x_2 - x_3 = 9 \\ 2x_1 - x_3 = -2 \\ 3x_1 + 5x_2 + 2x_3 = 22 \end{cases}$

16. $\begin{cases} 2x_1 + x_2 = 7 \\ 2x_1 - x_2 + x_3 = 6 \\ 3x_1 - 2x_2 + 4x_3 = 11 \end{cases}$

17. $\begin{cases} 2x - 3y - z = 13 \\ -x + 2y - 5z = 6 \\ 5x - y - z = 49 \end{cases}$

18. $\begin{cases} 10x + 10y - 20z = 60 \\ 15x + 20y + 30z = -25 \\ -5x + 30y - 10z = 45 \end{cases}$

19–28 ■ Determine whether the system of linear equations is inconsistent or dependent. If it is dependent, find the complete solution.

19. $\begin{cases} x + y + z = 2 \\ y - 3z = 1 \\ 2x + y + 5z = 0 \end{cases}$

20. $\begin{cases} x \quad\;\; + 3z = 3 \\ 2x + y - 2z = 5 \\ \quad\;\; - y + 8z = 8 \end{cases}$

21. $\begin{cases} 2x - 3y - 9z = -5 \\ x \quad\quad + 3z = 2 \\ -3x + y - 4z = -3 \end{cases}$

22. $\begin{cases} x - 2y + 5z = 3 \\ -2x + 6y - 11z = 1 \\ 3x - 16y + 20z = -26 \end{cases}$

23. $\begin{cases} x - y + 3z = 3 \\ 4x - 8y + 32z = 24 \\ 2x - 3y + 11z = 4 \end{cases}$

24. $\begin{cases} -2x + 6y - 2z = -12 \\ x - 3y + 2z = 10 \\ -x + 3y + 2z = 6 \end{cases}$

25. $\begin{cases} x + 4y - 2z = -3 \\ 2x - y + 5z = 12 \\ 8x + 5y + 11z = 30 \end{cases}$

26. $\begin{cases} 3r + 2s - 3t = 10 \\ r - s - t = -5 \\ r + 4s - t = 20 \end{cases}$

27. $\begin{cases} 2x + y - 2z = 12 \\ -x - \frac{1}{2}y + z = -6 \\ 3x + \frac{3}{2}y - 3z = 18 \end{cases}$

28. $\begin{cases} y - 5z = 7 \\ 3x + 2y = 12 \\ 3x + 10z = 80 \end{cases}$

29–40 ■ Solve the system of linear equations.

29. $\begin{cases} 4x - 3y + z = -8 \\ -2x + y - 3z = -4 \\ x - y + 2z = 3 \end{cases}$

30. $\begin{cases} 2x - 3y + 5z = 14 \\ 4x - y - 2z = -17 \\ -x - y + z = 3 \end{cases}$

31. $\begin{cases} x + 2y - 3z = -5 \\ -2x - 4y - 6z = 10 \\ 3x + 7y - 2z = -13 \end{cases}$

32. $\begin{cases} 3x - y + 2z = -1 \\ 4x - 2y + z = -7 \\ -x + 3y - 2z = -1 \end{cases}$

33. $\begin{cases} -x + 2y + z - 3w = 3 \\ 3x - 4y + z + w = 9 \\ -x - y + z + w = 0 \\ 2x + y + 4z - 2w = 3 \end{cases}$

34. $\begin{cases} x + y - z - w = 6 \\ 2x + z - 3w = 8 \\ x - y + 4w = -10 \\ 3x + 5y - z - w = 20 \end{cases}$

35. $\begin{cases} x + y + 2z - w = -2 \\ 3y + z + 2w = 2 \\ x + y + 3w = 2 \\ -3x + z + 2w = 5 \end{cases}$

36. $\begin{cases} x - 3y + 2z + w = -2 \\ x - 2y - 2w = -10 \\ z + 5w = 15 \\ 3x + 2z + w = -3 \end{cases}$

37. $\begin{cases} x + z + w = 4 \\ y - z = -4 \\ x - 2y + 3z + w = 12 \\ 2x - 2z + 5w = -1 \end{cases}$

38. $\begin{cases} y - z + 2w = 0 \\ 3x + 2y + w = 0 \\ 2x + 4w = 12 \\ -2x - 2z + 5w = 6 \end{cases}$

39. $\begin{cases} x - y + w = 0 \\ 3x - z + 2w = 0 \\ x - 4y + z + 2w = 0 \end{cases}$

40. $\begin{cases} 2x - y + 2z + w = 5 \\ -x + y + 4z - w = 3 \\ 3x - 2y - z = 0 \end{cases}$

41. A doctor recommends that a patient take 50 mg each of niacin, riboflavin, and thiamin daily to alleviate a vitamin deficiency. In his medicine chest at home, the patient finds three brands of vitamin pills. The amounts of the relevant vitamins per pill are given in the table. How many pills of each type should he take every day to get 50 mg of each vitamin?

	VitaMax	Vitron	VitaPlus
Niacin (mg)	5	10	15
Riboflavin (mg)	15	20	0
Thiamin (mg)	10	10	10

42. A chemist has three acid solutions at various concentrations. The first is 10% acid, the second is 20%, and the third is 40%. How many milliliters of each should he use to make 100 mL of 18% solution, if he has to use four times as much of the 10% solution as the 40% solution?

43. Amanda, Bryce, and Corey enter a race in which they have to run, swim, and cycle over a marked course. Their average speeds are given in the table. Corey finishes first with a total time of 1 h 45 min. Amanda comes in second with a time of 2 h 30 min. Bryce finishes last with a time of 3 h. Find the distance (in miles) for each part of the race.

	Average Speed (mi/h)		
	Running	Swimming	Cycling
Amanda	10	4	20
Bryce	$7\frac{1}{2}$	6	15
Corey	15	3	40

44. A small school has 100 students who occupy three classrooms: A, B, and C. After the first period of the school day, half the students in room A move to room B, one-fifth of the students in room B move to room C, and one-third of the students in room C move to room A. Nevertheless, the total number of students in each room is the same for both periods. How many students occupy each room?

45. A furniture factory makes wooden tables, chairs, and armoires. Each piece of furniture requires three operations: cutting the wood, assembling, and finishing. Each operation requires the number of hours (h) given in the table. The workers in the factory can provide 300 hours of cutting, 400 hours of assembling, and 590 hours of finishing each work week. How many tables, chairs, and armoires should be produced so that all available labor-hours are used? Or is this impossible?

	Table	Chair	Armoire
Cutting (h)	$\frac{1}{2}$	1	1
Assembling (h)	$\frac{1}{2}$	$1\frac{1}{2}$	1
Finishing (h)	1	$1\frac{1}{2}$	2

46. A diagram of a section of a city's street network is shown in the figure. The arrows indicate one-way streets, and the numbers show how many cars enter or leave this section of the city via the indicated street in a certain one-hour period. The variables x, y, z, and w represent the number of cars that travel along the portions of First, Second, Avocado,

and Birch Streets during this period. Find x, y, z, and w, assuming that none of the cars stop or park on any of the streets shown.

DISCOVERY · DISCUSSION

47. Polynomials Determined by a Set of Points   We all know that two points uniquely determine a line $y = ax + b$ in the coordinate plane. Similarly, three points uniquely determine a quadratic (second-degree) polynomial

$$y = ax^2 + bx + c$$

four points uniquely determine a cubic (third-degree) polynomial

$$y = ax^3 + bx^2 + cx + d$$

and so on. (Some exceptions to this rule are if the three points actually lie on a line, or the four points lie on a quadratic or line, and so on.) For the following set of five points, find the line that contains the first two points, the quadratic that contains the first three points, the cubic that contains the first four points, and the fourth-degree polynomial that contains all five points.

$$(0, 0), \quad (1, 12), \quad (2, 40), \quad (3, 6), \quad (-1, -14)$$

Graph the points and functions in the same viewing rectangle using a graphing device.

8.5   THE ALGEBRA OF MATRICES

Thus far we've used matrices simply for notational convenience when solving linear systems. Matrices have many other uses in mathematics and the sciences, and for most of these applications a knowledge of matrix algebra is essential. Like numbers, matrices can be added, subtracted, multiplied, and divided. In this section we learn how to perform these algebraic operations on matrices.

Recall that an $m \times n$ matrix A is a rectangular array of numbers with m rows and n columns. We say that A has **dimension** $m \times n$. The individual numbers that make up the matrix are called its **entries**. The entry that appears in the ith row and jth column is denoted a_{ij}. We sometimes write the matrix A as

$$A = [a_{ij}]$$

For example, the matrix

$$A = \begin{bmatrix} 2 & 7 & -4 & 0 \\ 1 & \frac{1}{2} & 12 & 10 \\ -6 & -1 & 5 & 8 \end{bmatrix}$$

has dimension 3×4 because it has 3 horizontal rows and 4 vertical columns. Here are some entries in this matrix: $a_{12} = 7$, $a_{24} = 10$, $a_{32} = -1$, and $a_{34} = 8$.

Equal matrices

$$\begin{bmatrix} \sqrt{4} & 2^2 & e^0 \\ 0.5 & 1 & 1-1 \end{bmatrix} = \begin{bmatrix} 2 & 4 & 1 \\ \frac{1}{2} & \frac{2}{2} & 0 \end{bmatrix}$$

Unequal matrices

$$\begin{bmatrix} 1 & 2 \\ 3 & 4 \\ 5 & 6 \end{bmatrix} \neq \begin{bmatrix} 1 & 3 & 5 \\ 2 & 4 & 6 \end{bmatrix}$$

> ## EQUALITY OF MATRICES
>
> The matrices $A = [a_{ij}]$ and $B = [b_{ij}]$ are **equal** if and only if they have the same dimension $m \times n$, and corresponding entries are equal, that is,
>
> $$a_{ij} = b_{ij}$$
>
> for $i = 1, 2, \ldots, m$ and $j = 1, 2, \ldots, n$.

■ Algebraic Operations on Matrices

Two matrices can be added or subtracted if they have the same dimension. (Otherwise, their sum or difference is undefined.) We add or subtract the matrices by adding or subtracting corresponding entries. To multiply a matrix by a number, we multiply every element of the matrix by that number. This is called the *scalar product*.

> ## SUM, DIFFERENCE, AND SCALAR PRODUCT OF MATRICES
>
> Let $A = [a_{ij}]$ and $B = [b_{ij}]$ be matrices of the same dimension $m \times n$, and let c be any real number.
>
> **1.** The **sum** $A + B$ is the $m \times n$ matrix obtained by adding corresponding entries of a and B. That is
>
> $$A + B = [a_{ij} + b_{ij}]$$
>
> **2.** The **difference** $A - B$ is the $m \times n$ matrix obtained by subtracting corresponding entries of A and B. That is
>
> $$A - B = [a_{ij} - b_{ij}]$$
>
> **3.** The **scalar product** cA is the $m \times n$ matrix obtained by multiplying each entry of A by c. That is
>
> $$cA = [ca_{ij}]$$

EXAMPLE 1 ■ Performing Algebraic Operations on Matrices

Let

$$A = \begin{bmatrix} 2 & -3 \\ 0 & 5 \\ 7 & -\frac{1}{2} \end{bmatrix} \qquad B = \begin{bmatrix} 1 & 0 \\ -3 & 1 \\ 2 & 2 \end{bmatrix}$$

$$C = \begin{bmatrix} 7 & -3 & 0 \\ 0 & 1 & 5 \end{bmatrix} \qquad D = \begin{bmatrix} 6 & 0 & -6 \\ 8 & 1 & 9 \end{bmatrix}$$

Carry out each indicated operation, or explain why it cannot be performed.

(a) $A + B$ (b) $C - D$ (c) $C + A$ (d) $5A$

SOLUTION

(a) $A + B = \begin{bmatrix} 2 & -3 \\ 0 & 5 \\ 7 & -\frac{1}{2} \end{bmatrix} + \begin{bmatrix} 1 & 0 \\ -3 & 0 \\ 2 & 2 \end{bmatrix} = \begin{bmatrix} 3 & -3 \\ -3 & 5 \\ 9 & \frac{3}{2} \end{bmatrix}$

(b) $C - D = \begin{bmatrix} 7 & -3 & 0 \\ 0 & 1 & 5 \end{bmatrix} - \begin{bmatrix} 6 & 0 & -6 \\ 8 & 1 & 9 \end{bmatrix} = \begin{bmatrix} 1 & -3 & 6 \\ -8 & 0 & -4 \end{bmatrix}$

(c) $C + A$ is undefined because we can't add matrices of different dimensions.

(d) $5A = 5 \begin{bmatrix} 2 & -3 \\ 0 & 5 \\ 7 & -\frac{1}{2} \end{bmatrix} = \begin{bmatrix} 10 & -15 \\ 0 & 25 \\ 35 & -\frac{5}{2} \end{bmatrix}$ ■

■ Matrix Multiplication

Multiplication of two matrices is more difficult to describe than other matrix operations. In later examples we will see why taking the matrix product involves a rather complex procedure, which we now describe.

First, the product AB (or $A \cdot B$) of two matrices A and B is defined only when the number of columns in A is equal to the number of rows in B. This means that if we write their dimensions side by side, the two inner numbers must match:

Matrices	A	B
Dimensions	$m \times n$	$n \times k$
	↑	↑
	columns in A	rows in B

If the dimensions of A and B match in this fashion, then the product AB will have dimension $m \times k$. Before describing the procedure for obtaining the elements of AB, we define the *inner product* of a row of A and a column of B.

Julia Robinson (1919–1985) was born in St. Louis, Missouri, and grew up at Point Loma, California. Due to an illness, Robinson missed two years of school but later, with the aid of a tutor, she completed fifth, sixth, seventh, and eighth grades, all in one year. At San Diego State University, she found mathematics especially interesting after reading biographies of mathematicians in the book *Men of Mathematics* by E. T. Bell. She said, "I cannot overemphasize the importance of such books . . . in the intellectual life of a student." Robinson is famous for her work on Hilbert's tenth problem (page 689), which asks for a general procedure for determining whether an equation has integer solutions. Her ideas led to a complete answer to the problem. Interestingly, the answer involved certain properties of the Fibonacci numbers (page 811) discovered by the then 22-year-old Russian mathematician Yuri Matijasevič. As a result of her brilliant work on Hilbert's tenth problem, Robinson was offered a professorship at the University of California, Berkeley, and became the first woman mathematician elected to the National Academy of Sciences. She also served as president of the American Mathematical Society.

If $[a_1 \quad a_2 \quad \cdots \quad a_n]$ is a row of A, and if $\begin{bmatrix} b_1 \\ b_2 \\ \cdot \\ \cdot \\ \cdot \\ b_n \end{bmatrix}$ is a column of B, then

their **inner product** is the number $a_1b_1 + a_2b_2 + \cdots + a_nb_n$.

For example,

$$[2 \quad -1 \quad 0 \quad 4] \cdot \begin{bmatrix} 5 \\ 4 \\ -3 \\ \frac{1}{2} \end{bmatrix} = 2 \cdot 5 + (-1) \cdot 4 + 0 \cdot (-3) + 4 \cdot \tfrac{1}{2} = 8$$

We now define the **product** AB of two matrices.

THE PRODUCT OF TWO MATRICES

If $A = [a_{ij}]$ is an $m \times n$ matrix and $B = [b_{ij}]$ an $n \times k$ matrix, then their product is the $m \times k$ matrix

$$C = [c_{ij}]$$

where c_{ij} is the inner product of the ith row of A and the jth column of B. We write the product as

$$C = AB$$

EXAMPLE 2 ■ Multiplying Matrices

Let

$$A = \begin{bmatrix} 1 & 3 \\ -1 & 0 \end{bmatrix} \quad \text{and} \quad B = \begin{bmatrix} -1 & 5 & 2 \\ 0 & 4 & 7 \end{bmatrix}$$

Calculate, if possible, the products AB and BA.

SOLUTION

Since A has dimension 2×2 and B has dimension 2×3, the product AB is defined and has dimension 2×3. We can thus write

$$AB = \begin{bmatrix} 1 & 3 \\ -1 & 0 \end{bmatrix}\begin{bmatrix} -1 & 5 & 2 \\ 0 & 4 & 7 \end{bmatrix} = \begin{bmatrix} ? & ? & ? \\ ? & ? & ? \end{bmatrix}$$

Inner numbers match, so product is defined.
Outer numbers give dimension of product: 2×3

where the question marks must be filled in using the rule defining the entries of a matrix product. If we define $C = AB = [c_{ij}]$, then the entry c_{11} is the inner product of the first row of A and the first column of B:

$$\begin{bmatrix} 1 & 3 \\ -1 & 0 \end{bmatrix} \begin{bmatrix} -1 & 5 & 2 \\ 0 & 4 & 7 \end{bmatrix} \qquad 1 \cdot (-1) + 3 \cdot 0 = -1$$

Similarly, we calculate the remaining entries of the product as follows.

Entry	Inner product of:	Value	Product matrix
c_{12}	$\begin{bmatrix} 1 & 3 \\ -1 & 0 \end{bmatrix} \begin{bmatrix} -1 & 5 & 2 \\ 0 & 4 & 7 \end{bmatrix}$	$1 \cdot 5 + 3 \cdot 4 = 17$	$\begin{bmatrix} -1 & 17 & \\ & & \end{bmatrix}$
c_{13}	$\begin{bmatrix} 1 & 3 \\ -1 & 0 \end{bmatrix} \begin{bmatrix} -1 & 5 & 2 \\ 0 & 4 & 7 \end{bmatrix}$	$1 \cdot 2 + 3 \cdot 7 = 23$	$\begin{bmatrix} -1 & 17 & 23 \\ & & \end{bmatrix}$
c_{21}	$\begin{bmatrix} 1 & 3 \\ -1 & 0 \end{bmatrix} \begin{bmatrix} -1 & 5 & 2 \\ 0 & 4 & 7 \end{bmatrix}$	$(-1) \cdot (-1) + 0 \cdot 0 = 1$	$\begin{bmatrix} -1 & 17 & 23 \\ 1 & & \end{bmatrix}$
c_{22}	$\begin{bmatrix} 1 & 3 \\ -1 & 0 \end{bmatrix} \begin{bmatrix} -1 & 5 & 2 \\ 0 & 4 & 7 \end{bmatrix}$	$(-1) \cdot 5 + 0 \cdot 4 = -5$	$\begin{bmatrix} -1 & 17 & 23 \\ 1 & -5 & \end{bmatrix}$
c_{23}	$\begin{bmatrix} 1 & 3 \\ -1 & 0 \end{bmatrix} \begin{bmatrix} -1 & 5 & 2 \\ 0 & 4 & 7 \end{bmatrix}$	$(-1) \cdot 2 + 0 \cdot 7 = -2$	$\begin{bmatrix} -1 & 17 & 23 \\ 1 & -5 & -2 \end{bmatrix}$

Thus, we have

$$AB = \begin{bmatrix} -1 & 17 & 23 \\ 1 & -5 & -2 \end{bmatrix}$$

The product BA is not defined, however, because the dimensions are

$$2 \times 3 \qquad \text{and} \qquad 2 \times 2$$

The inner two numbers are not the same, so the rows and columns won't match up when we try to calculate the product.   ■

Graphing calculators and computers are capable of performing matrix algebra. For instance, if we enter the matrices in Example 2 into the matrix variables $[A]$ and $[B]$ on a TI-83 calculator, then the calculator finds their product as shown in Figure 1.

The next example shows that even when both AB and BA are defined, they aren't necessarily equal. This result proves that matrix multiplication is *not* commutative.

FIGURE 1

EXAMPLE 3 ■ Matrix Multiplication Is Not Commutative

Let

$$A = \begin{bmatrix} 5 & 7 \\ -3 & 0 \end{bmatrix} \qquad \text{and} \qquad B = \begin{bmatrix} 1 & 2 \\ 9 & -1 \end{bmatrix}$$

Calculate the products AB and BA.

SOLUTION

Since both matrices A and B have dimension 2×2, both products AB and BA are defined, and each product is also a 2×2 matrix.

$$AB = \begin{bmatrix} 5 & 7 \\ -3 & 0 \end{bmatrix} \begin{bmatrix} 1 & 2 \\ 9 & -1 \end{bmatrix} = \begin{bmatrix} 5 \cdot 1 + 7 \cdot 9 & 5 \cdot 2 + 7 \cdot (-1) \\ (-3) \cdot 1 + 0 \cdot 9 & (-3) \cdot 2 + 0 \cdot (-1) \end{bmatrix} = \begin{bmatrix} 68 & 3 \\ -3 & -6 \end{bmatrix}$$

$$BA = \begin{bmatrix} 1 & 2 \\ 9 & -1 \end{bmatrix} \begin{bmatrix} 5 & 7 \\ -3 & 0 \end{bmatrix} = \begin{bmatrix} 1 \cdot 5 + 2 \cdot (-3) & 1 \cdot 7 + 2 \cdot 0 \\ 9 \cdot 5 + (-1) \cdot (-3) & 9 \cdot 7 + (-1) \cdot 0 \end{bmatrix} = \begin{bmatrix} -1 & 7 \\ 48 & 63 \end{bmatrix}$$

This shows that, in general, $AB \neq BA$. In fact, in this example AB and BA don't even have an entry in common. ∎

Although matrix multiplication is not commutative, it does obey the Associative and Distributive Properties.

PROPERTIES OF MATRIX MULTIPLICATION

Let A, B, and C be matrices for which the following products are defined. Then

$$A(BC) = (AB)C \qquad \text{Associative Property}$$

$$A(B + C) = AB + AC$$
$$\qquad\qquad\qquad\qquad \text{Distributive Property}$$
$$(B + C)A = BA + CA$$

The next two examples give some indication of why mathematicians chose to define the matrix product in such an apparently bizarre fashion.

EXAMPLE 4 ■ Writing a Linear System as a Matrix Equation

Matrix equations are described in more detail on page 679.

Show that the following matrix equation is equivalent to the system of equations in Example 2 of Section 8.3.

$$\begin{bmatrix} 1 & -2 & 3 \\ 1 & 2 & -1 \\ 3 & 2 & -5 \end{bmatrix} \begin{bmatrix} x \\ y \\ z \end{bmatrix} = \begin{bmatrix} 1 \\ 13 \\ 3 \end{bmatrix}$$

SOLUTION

If we perform the matrix multiplication on the left side of the equation, we get

$$\begin{bmatrix} x - 2y + 3z \\ x + 2y - z \\ 3x + 2y - 5z \end{bmatrix} = \begin{bmatrix} 1 \\ 13 \\ 3 \end{bmatrix}$$

Because two matrices are equal only if their corresponding entries are equal, we equate entries to get

$$\begin{cases} x - 2y + 3z = 1 \\ x + 2y - z = 13 \\ 3x + 2y - 5z = 3 \end{cases}$$

This is exactly the system of equations in Example 2 of Section 8.3. ■

The preceding example shows that our definition of matrix product allows us to express a system of linear equations as a single matrix equation.

EXAMPLE 5 ■ Representing Demographic Data in Terms of Matrices

In a certain city the proportion of voters in each age group who are registered as Democrats, Republicans, or Independents is given by the following matrix.

	18–30	31–50	Over 50
Democrat	0.30	0.60	0.50
Republican	0.50	0.35	0.25
Independent	0.20	0.05	0.25

$$A = \begin{bmatrix} 0.30 & 0.60 & 0.50 \\ 0.50 & 0.35 & 0.25 \\ 0.20 & 0.05 & 0.25 \end{bmatrix}$$

(Age column heading across 18–30, 31–50, Over 50)

The next matrix gives the distribution, by age and sex, of the voting population of this city.

		Male	Female
Age	18–30	5,000	6,000
	31–50	10,000	12,000
	Over 50	12,000	15,000

$$= B$$

For this problem, let's make the (highly unrealistic) assumption that within each age group, political preference is not related to gender. That is, the percentage of Democrat males in the 18–30 group, for example, is the same as the percentage of Democrat females in this group.

(a) Calculate the product AB.
(b) How many males are registered as Democrats in this city?
(c) How many females are registered as Republicans?

SOLUTION

(a) $$AB = \begin{bmatrix} 0.30 & 0.60 & 0.50 \\ 0.50 & 0.35 & 0.25 \\ 0.20 & 0.05 & 0.25 \end{bmatrix} \begin{bmatrix} 5,000 & 6,000 \\ 10,000 & 12,000 \\ 12,000 & 15,000 \end{bmatrix} = \begin{bmatrix} 13,500 & 16,500 \\ 9,000 & 10,950 \\ 4,500 & 5,550 \end{bmatrix}$$

Olga Taussky-Todd (1906–1995) was instrumental in developing applications of Matrix Theory. Described as "in love with anything matrices can do," she successfully applied matrices to aerodynamics, a field used in the design of airplanes and rockets. Taussky-Todd was also famous for her work in Number Theory, which deals with prime numbers and divisibility. Although Number Theory has often been called the least applicable branch of mathematics, it is now used in significant ways throughout the computer industry. Taussky-Todd studied mathematics at a time when young women rarely aspired to be mathematicians. She said, "When I entered university I had no idea what it meant to study mathematics." One of the most respected mathematicians of her day, she was for many years a professor of mathematics at Caltech in Pasadena.

(b) When we take the inner product of a row in A with a column in B, we are adding the number of people in each age group who belong to the category in question. For example, the entry c_{21} of $C = AB$ (the 9000) is obtained by taking the inner product of the Republican row in A with the Male column in B. This number is therefore the total number of male Republicans in this city. We can label the rows and columns of AB as follows.

$$\begin{array}{c} \\ \text{Democrat} \\ \text{Republican} \\ \text{Independent} \end{array} \begin{array}{cc} \text{Male} & \text{Female} \\ \begin{bmatrix} 13{,}500 & 16{,}500 \\ 9{,}000 & 10{,}950 \\ 4{,}500 & 5{,}550 \end{bmatrix} = AB \end{array}$$

Thus, 13,500 males are registered as Democrats in this city.

(c) There are 10,950 females registered as Republicans. ■

If we add the entries in the columns of matrix A in Example 5, we see that in each case the sum is 1. (Can you see why this has to be true, given what the matrix describes?) A matrix with this property is called **stochastic**. Stochastic matrices are used extensively in statistics, where they arise frequently in situations like the one described here.

■ Computer Graphics

One important use of matrices is in the digital representation of images. A digital camera or a scanner converts an image into a matrix by dividing the image into a rectangular array of elements called pixels. Each pixel is assigned a value that represents the color, brightness, or some other feature of that location. For example, in a 256-level gray-scale image each pixel is assigned a value between 0 and 255, where 0 represents white, 255 black, and the numbers in between increasing gradations of gray. The gradations of a much simpler 8-level gray scale are shown in Figure 2. We use this 8-level gray scale to illustrate the process.

To digitize the black and white image in Figure 3(a), we place a grid over the picture as shown in Figure 3(b). Each cell in the grid is compared to the gray scale, and then assigned a value between 0 and 7 depending on which gray square in the

FIGURE 2

(a) Original image

(b) 10 × 10 grid

$$\begin{bmatrix} 1 & 1 & 1 & 1 & 1 & 1 & 1 & 2 & 2 & 1 \\ 1 & 1 & 1 & 1 & 1 & 1 & 4 & 6 & 5 & 2 \\ 1 & 1 & 1 & 1 & 2 & 3 & 3 & 5 & 5 & 3 \\ 1 & 1 & 1 & 1 & 3 & 5 & 4 & 6 & 3 & 2 \\ 1 & 1 & 1 & 1 & 1 & 2 & 3 & 2 & 2 & 1 \\ 1 & 1 & 1 & 1 & 1 & 3 & 3 & 2 & 1 & 1 \\ 1 & 1 & 1 & 1 & 1 & 1 & 4 & 1 & 1 & 1 \\ 1 & 1 & 1 & 1 & 2 & 2 & 4 & 2 & 2 & 2 \\ 2 & 2 & 3 & 5 & 5 & 2 & 2 & 3 & 4 & 4 \\ 3 & 3 & 3 & 4 & 3 & 2 & 3 & 3 & 3 & 4 \end{bmatrix}$$

(c) Matrix representation

(d) Digital image

FIGURE 3

scale most closely matches the "darkness" of the cell. (If the cell is not uniformly gray, an average value is assigned.) The values are stored in the matrix shown in Figure 3(c). The digital image corresponding to this matrix is shown in Figure 3(d). Obviously the grid that we have used is far too coarse to provide good image resolution. In practice, currently available high-resolution digital cameras use matrices of dimension as large as 2048×2048.

Once the image is stored as a matrix, it can be manipulated using matrix operations. For example, to darken the image, we add a constant to each entry in the matrix; to lighten the image, we subtract. To increase the contrast, we darken the darker areas and lighten the lighter areas, so we could add 1 to each entry that is 4, 5, or 6 and subtract 1 from each entry that is 1, 2, or 3. (Note that we cannot darken an entry of 7 or lighten a 0.) Applying this process to the matrix in Figure 3(c) produces the new matrix in Figure 4(a). This generates the high-contrast image shown in Figure 4(b).

$$\begin{bmatrix} 0 & 0 & 0 & 0 & 0 & 0 & 0 & 1 & 1 & 0 \\ 0 & 0 & 0 & 0 & 0 & 0 & 5 & 7 & 6 & 1 \\ 0 & 0 & 0 & 0 & 1 & 2 & 2 & 6 & 6 & 2 \\ 0 & 0 & 0 & 0 & 2 & 6 & 5 & 7 & 2 & 1 \\ 0 & 0 & 0 & 0 & 0 & 1 & 2 & 1 & 1 & 0 \\ 0 & 0 & 0 & 0 & 0 & 2 & 2 & 1 & 0 & 0 \\ 0 & 0 & 0 & 0 & 0 & 0 & 5 & 0 & 0 & 0 \\ 0 & 0 & 0 & 0 & 1 & 1 & 5 & 1 & 1 & 1 \\ 1 & 1 & 2 & 6 & 6 & 1 & 1 & 2 & 5 & 5 \\ 2 & 2 & 2 & 5 & 2 & 1 & 2 & 2 & 2 & 5 \end{bmatrix}$$

(a) Matrix modified to increase contrast

(b) High contrast image

FIGURE 4

Other ways of representing and manipulating images using matrices are discussed on pages 672 and 684.

8.5 EXERCISES

1–2 ■ Determine whether the matrices A and B are equal.

1. $A = \begin{bmatrix} 1 & -2 & 0 \\ \frac{1}{2} & 6 & 0 \end{bmatrix}$, $\quad B = \begin{bmatrix} 1 & -2 \\ \frac{1}{2} & 6 \end{bmatrix}$

2. $A = \begin{bmatrix} \frac{1}{4} & \ln 1 \\ 2 & 3 \end{bmatrix}$, $\quad B = \begin{bmatrix} 0.25 & 0 \\ \sqrt{4} & \frac{6}{2} \end{bmatrix}$

3–10 ■ Perform the matrix operation, or if it is impossible, explain why.

3. $\begin{bmatrix} 2 & 6 \\ -5 & 3 \end{bmatrix} + \begin{bmatrix} -1 & -3 \\ 6 & 2 \end{bmatrix}$

4. $\begin{bmatrix} 0 & 1 & 1 \\ 1 & 1 & 0 \end{bmatrix} - \begin{bmatrix} 2 & 1 & -1 \\ 1 & 3 & -2 \end{bmatrix}$

5. $3 \begin{bmatrix} 1 & 2 \\ 4 & -1 \\ 1 & 0 \end{bmatrix}$

6. $2 \begin{bmatrix} 1 & 1 & 0 \\ 1 & 0 & 1 \\ 0 & 1 & 1 \end{bmatrix} + \begin{bmatrix} 1 & 1 \\ 2 & 1 \\ 3 & 1 \end{bmatrix}$

7. $\begin{bmatrix} 2 & 6 \\ 1 & 3 \\ 2 & 4 \end{bmatrix} \begin{bmatrix} 1 & -2 \\ 3 & 6 \\ -2 & 0 \end{bmatrix}$

8. $\begin{bmatrix} 2 & 1 & 2 \\ 6 & 3 & 4 \end{bmatrix} \begin{bmatrix} 1 & -2 \\ 3 & 6 \\ -2 & 0 \end{bmatrix}$

9. $\begin{bmatrix} 1 & 2 \\ -1 & 4 \end{bmatrix} \begin{bmatrix} 1 & -2 & 3 \\ 2 & 2 & -1 \end{bmatrix}$

10. $\begin{bmatrix} 2 & -3 \\ 0 & 1 \\ 1 & 2 \end{bmatrix} \begin{bmatrix} 5 \\ 1 \end{bmatrix}$

11–31 ■ The matrices A, B, C, D, E, F, and G are defined as follows.

$$A = \begin{bmatrix} 2 & -5 \\ 0 & 7 \end{bmatrix} \qquad B = \begin{bmatrix} 3 & \frac{1}{2} & 5 \\ 1 & -1 & 3 \end{bmatrix} \qquad C = \begin{bmatrix} 2 & -\frac{5}{2} & 0 \\ 0 & 2 & -3 \end{bmatrix}$$

$$D = \begin{bmatrix} 7 & 3 \end{bmatrix} \qquad E = \begin{bmatrix} 1 \\ 2 \\ 0 \end{bmatrix} \qquad F = \begin{bmatrix} 1 & 0 & 0 \\ 0 & 1 & 0 \\ 0 & 0 & 1 \end{bmatrix}$$

$$G = \begin{bmatrix} 5 & -3 & 10 \\ 6 & 1 & 0 \\ -5 & 2 & 2 \end{bmatrix}$$

Carry out the indicated algebraic operation, or explain why it cannot be performed.

11. $B + C$

12. $B + F$

13. $C - B$

14. $5A$

15. $3B + 2C$

16. $C - 5A$

17. $2C - 6B$

18. DA

19. AD

20. BC

21. BF

22. GF

23. $(DA)B$

24. $D(AB)$

25. GE

26. A^2

27. A^3

28. $DB + DC$

29. B^2

30. F^2

31. $BF + FE$

32. What must be true about the dimensions of the matrices A and B if both products AB and BA are defined?

33–36 ■ Write the system of equations as a matrix equation (see Example 4).

33. $\begin{cases} 2x - 5y = 7 \\ 3x + 2y = 4 \end{cases}$

34. $\begin{cases} 6x - y + z = 12 \\ 2x \quad\; + z = 7 \\ \quad\;\; y - 2z = 4 \end{cases}$

35. $\begin{cases} 3x_1 + 2x_2 - x_3 + x_4 = 0 \\ x_1 \qquad - x_3 \qquad = 5 \\ \quad\; 3x_2 + x_3 - x_4 = 4 \end{cases}$

36. $\begin{cases} x - y + z = 2 \\ 4x - 2y - z = 2 \\ x + y + 5z = 2 \\ -x - y - z = 2 \end{cases}$

37–40 ■ Solve for x and y.

37. $\begin{bmatrix} x & 2y \\ 4 & 6 \end{bmatrix} = \begin{bmatrix} 2 & -2 \\ 2x & -6y \end{bmatrix}$

38. $3\begin{bmatrix} x & y \\ y & x \end{bmatrix} = \begin{bmatrix} 6 & -9 \\ -9 & 6 \end{bmatrix}$

39. $\begin{bmatrix} 6 & x \\ 1 & 0 \end{bmatrix}\begin{bmatrix} y & 2 \\ -1 & 2 \end{bmatrix} = \begin{bmatrix} 4 & 16 \\ 1 & 2 \end{bmatrix}$

40. $\begin{bmatrix} 2 & 5 \\ 1 & 3 \end{bmatrix}\begin{bmatrix} x \\ y \end{bmatrix} = \begin{bmatrix} -2 \\ -3 \end{bmatrix}$

41–44 ■ Solve the matrix equation for the unknown matrix X, or explain why no solution exists.

$$A = \begin{bmatrix} 4 & 6 \\ 1 & 3 \end{bmatrix} \qquad B = \begin{bmatrix} 2 & 5 \\ 3 & 7 \end{bmatrix}$$

$$C = \begin{bmatrix} 2 & 3 \\ 1 & 0 \\ 0 & 2 \end{bmatrix} \qquad D = \begin{bmatrix} 10 & 20 \\ 30 & 20 \\ 10 & 0 \end{bmatrix}$$

41. $2X - A = B$

42. $5(X - C) = D$

43. $3X + B = C$

44. $A + D = 3X$

45. A small fast-food chain with restaurants in Santa Monica, Long Beach, and Anaheim sells only hamburgers, hot dogs, and milk shakes. On a certain day, sales were distributed according to the following matrix.

	Number of items sold		
	Santa Monica	Long Beach	Anaheim
Hamburgers	4000	1000	3500
Hot dogs	400	300	200
Milk shakes	700	500	9000

$= A$

The price of each item is given by the following matrix.

Hamburger	Hot dog	Milk shake
[$0.90	$0.80	$1.10]

$= B$

(a) Calculate the product BA.

(b) Interpret the entries in the product matrix BA.

46. A specialty-car manufacturer has plants in Auburn, Biloxi, and Chattanooga. Three models are produced, with daily production given in the following matrix.

	Cars produced each day		
	Model K	Model R	Model W
Auburn	12	10	0
Biloxi	4	4	20
Chattanooga	8	9	12

$= A$

Because of a wage increase, February profits are less than January profits. The profit per car is tabulated by model in the following matrix.

predators eat the prey too fast they will be left without food and ensure their own extinction.

Since Lotka and Volterra's time, more detailed mathematical models of animal populations have been developed. For many species the population is divided into several stages—immature, juvenile, adult, and so on. The proportion of each stage that survives or reproduces in a given time period is entered into a matrix (called a transition matrix); matrix multiplication is then used to predict the population in succeeding time periods. (See the *Laboratory Project*, page 672.) As you can see, the power of mathematics to model and predict is an invaluable tool in the ongoing debate over the environment.

on the *entire* new matrix:

$$\xrightarrow[R_3 + 3R_1 \to R_3]{R_2 - 2R_1 \to R_2} \left[\begin{array}{ccc|ccc} 1 & -2 & -4 & 1 & 0 & 0 \\ 0 & 1 & 2 & -2 & 1 & 0 \\ 0 & 0 & 3 & 3 & 0 & 1 \end{array}\right]$$

$$\xrightarrow{\frac{1}{3}R_3} \left[\begin{array}{ccc|ccc} 1 & -2 & -4 & 1 & 0 & 0 \\ 0 & 1 & 2 & -2 & 1 & 0 \\ 0 & 0 & 1 & 1 & 0 & \frac{1}{3} \end{array}\right]$$

$$\xrightarrow{R_1 + 2R_2 \to R_1} \left[\begin{array}{ccc|ccc} 1 & 0 & 0 & -3 & 2 & 0 \\ 0 & 1 & 2 & -2 & 1 & 0 \\ 0 & 0 & 1 & 1 & 0 & \frac{1}{3} \end{array}\right]$$

$$\xrightarrow{R_2 - 2R_3 \to R_2} \left[\begin{array}{ccc|ccc} 1 & 0 & 0 & -3 & 2 & 0 \\ 0 & 1 & 0 & -4 & 1 & -\frac{2}{3} \\ 0 & 0 & 1 & 1 & 0 & \frac{1}{3} \end{array}\right]$$

We have now transformed the left half of this matrix into an identity matrix. (This means we've put the entire matrix in reduced row-echelon form.) Note that to do this in as systematic a fashion as possible, we first changed the elements below the main diagonal to zeros, just as we would if we were using Gaussian elimination. We then changed each main diagonal element to a 1 by multiplying by the appropriate constant(s). Finally, we completed the process by changing the remaining entries on the left side to zeros. The right half is now A^{-1}.

$$A^{-1} = \left[\begin{array}{ccc} -3 & 2 & 0 \\ -4 & 1 & -\frac{2}{3} \\ 1 & 0 & \frac{1}{3} \end{array}\right]$$

(b) We calculate AA^{-1} and $A^{-1}A$, and verify that both products give the identity matrix I_3.

$$AA^{-1} = \left[\begin{array}{ccc} 1 & -2 & -4 \\ 2 & -3 & -6 \\ -3 & 6 & 15 \end{array}\right]\left[\begin{array}{ccc} -3 & 2 & 0 \\ -4 & 1 & -\frac{2}{3} \\ 1 & 0 & \frac{1}{3} \end{array}\right] = \left[\begin{array}{ccc} 1 & 0 & 0 \\ 0 & 1 & 0 \\ 0 & 0 & 1 \end{array}\right]$$

$$A^{-1}A = \left[\begin{array}{ccc} -3 & 2 & 0 \\ -4 & 1 & -\frac{2}{3} \\ 1 & 0 & \frac{1}{3} \end{array}\right]\left[\begin{array}{ccc} 1 & -2 & -4 \\ 2 & -3 & -6 \\ -3 & 6 & 15 \end{array}\right] = \left[\begin{array}{ccc} 1 & 0 & 0 \\ 0 & 1 & 0 \\ 0 & 0 & 1 \end{array}\right]$$

Graphing calculators are capable of calculating matrix inverses. On the TI-82 and TI-83 calculators, matrices are stored in memory using names such as [A],

```
[A]⁻¹▶Frac
  [[-3  2   0   ]
   [-4  1  -2/3]
   [1   0   1/3 ]]
```

FIGURE 1

[B], [C], ... To find the inverse of [A], we key in

$$[A] \boxed{x^{-1}} \boxed{ENTER}$$

For the matrix of Example 2, this results in the output shown in Figure 1 (where we have also used the ▶Frac command to display the output in fraction form rather than in decimal form).

The next example shows that not every square matrix has an inverse.

EXAMPLE 3 ■ A Matrix That Does Not Have an Inverse

Find the inverse of the matrix.

$$\begin{bmatrix} 2 & -3 & -7 \\ 1 & 2 & 7 \\ 1 & 1 & 4 \end{bmatrix}$$

SOLUTION

We proceed as follows.

$$\begin{bmatrix} 2 & -3 & -7 & | & 1 & 0 & 0 \\ 1 & 2 & 7 & | & 0 & 1 & 0 \\ 1 & 1 & 4 & | & 0 & 0 & 1 \end{bmatrix} \xrightarrow{R_1 \leftrightarrow R_2} \begin{bmatrix} 1 & 2 & 7 & | & 0 & 1 & 0 \\ 2 & -3 & -7 & | & 1 & 0 & 0 \\ 1 & 1 & 4 & | & 0 & 0 & 1 \end{bmatrix}$$

$$\xrightarrow[R_3 - R_1 \to R_3]{R_2 - 2R_1 \to R_2} \begin{bmatrix} 1 & 2 & 7 & | & 0 & 1 & 0 \\ 0 & -7 & -21 & | & 1 & -2 & 0 \\ 0 & -1 & -3 & | & 0 & -1 & 1 \end{bmatrix}$$

$$\xrightarrow{-\frac{1}{7}R_2} \begin{bmatrix} 1 & 2 & 7 & | & 0 & 1 & 0 \\ 0 & 1 & 3 & | & -\frac{1}{7} & \frac{2}{7} & 0 \\ 0 & -1 & -3 & | & 0 & -1 & 1 \end{bmatrix}$$

$$\xrightarrow[R_1 - 2R_2 \to R_1]{R_3 + R_2 \to R_3} \begin{bmatrix} 1 & 0 & 1 & | & \frac{2}{7} & \frac{3}{7} & 0 \\ 0 & 1 & 3 & | & -\frac{1}{7} & \frac{2}{7} & 0 \\ 0 & 0 & 0 & | & -\frac{1}{7} & -\frac{5}{7} & 1 \end{bmatrix}$$

At this point, we would like to change the 0 in the (3, 3) position of this matrix to a 1, without changing the zeros in the (3, 1) and (3, 2) positions. But there is no way to accomplish this, because no matter what multiple of rows 1 and/or 2 we add to row 3, we can't change the third zero in row 3 without changing the first or second as well. Thus, we cannot change the left half to the identity matrix. The original matrix doesn't have an inverse. ■

```
ERR:SINGULAR MAT
1:Quit
2:Goto
```

FIGURE 2

 If we encounter a row of zeros on the left when trying to find an inverse, as in Example 3, then the original matrix does not have an inverse. If we try to calculate the inverse of the matrix from Example 3 on a TI-83 calculator, we get the error message shown in Figure 2.

Matrix Equations

We saw in Section 8.5 that a system of linear equations can be written as a single matrix equation. For example, the system

$$\begin{cases} x - 2y - 4z = 7 \\ 2x - 3y - 6z = 5 \\ -3x + 6y + 15z = 0 \end{cases}$$

is equivalent to the matrix equation

$$\begin{bmatrix} 1 & -2 & -4 \\ 2 & -3 & -6 \\ -3 & 6 & 15 \end{bmatrix} \begin{bmatrix} x \\ y \\ z \end{bmatrix} = \begin{bmatrix} 7 \\ 5 \\ 0 \end{bmatrix}$$

If we let

$$A = \begin{bmatrix} 1 & -2 & -4 \\ 2 & -3 & -6 \\ -3 & 6 & 15 \end{bmatrix} \qquad X = \begin{bmatrix} x \\ y \\ z \end{bmatrix} \qquad B = \begin{bmatrix} 7 \\ 5 \\ 0 \end{bmatrix}$$

then this matrix equation can be written as

$$AX = B$$

The matrix A is called the **coefficient matrix**.

We solve this matrix equation by multiplying each side by the inverse of A (provided this inverse exists):

Solving the matrix equation $AX = B$ is very similar to solving the simple real-number equation

$$3x = 12$$

which we solve by multiplying each side by the reciprocal (or inverse) of 3:

$$\tfrac{1}{3}(3x) = \tfrac{1}{3}(12)$$

$$x = 4$$

$$AX = B$$

$$A^{-1}(AX) = A^{-1}B \qquad \text{Multiply on left by } A^{-1}$$

$$(A^{-1}A)X = A^{-1}B \qquad \text{Associative property}$$

$$I_3X = A^{-1}B \qquad \text{Property of inverses}$$

$$X = A^{-1}B \qquad \text{Property of identity matrix}$$

In Example 2 we showed that

$$A^{-1} = \begin{bmatrix} -3 & 2 & 0 \\ -4 & 1 & -\frac{2}{3} \\ 1 & 0 & \frac{1}{3} \end{bmatrix}$$

So, from $X = A^{-1}B$ we have

$$\begin{bmatrix} x \\ y \\ z \end{bmatrix} = \begin{bmatrix} -3 & 2 & 0 \\ -4 & 1 & -\frac{2}{3} \\ 1 & 0 & \frac{1}{3} \end{bmatrix} \begin{bmatrix} 7 \\ 5 \\ 0 \end{bmatrix} = \begin{bmatrix} -11 \\ -23 \\ 7 \end{bmatrix}$$

Thus, $x = -11$, $y = -23$, $z = 7$ is the solution of the original system.

SOLVING A MATRIX EQUATION

If A is a square $n \times n$ matrix that has an inverse A^{-1}, and if X is a variable matrix and B a known matrix, both with n rows, then the solution of the matrix equation

$$AX = B$$

is given by

$$X = A^{-1}B$$

EXAMPLE 4 ■ Solving a System Using the Matrix Inverse

Solve the system of equations.

$$\begin{cases} 2x - 5y = 15 \\ 3x - 6y = 36 \end{cases}$$

SOLUTION

We first convert this to a matrix equation of the form $AX = B$:

$AX = B$

$$\begin{bmatrix} 2 & -5 \\ 3 & -6 \end{bmatrix} \begin{bmatrix} x \\ y \end{bmatrix} = \begin{bmatrix} 15 \\ 36 \end{bmatrix}$$

Using the rule for calculating the inverse of a 2×2 matrix, we get

Find A^{-1}.

$$\begin{bmatrix} 2 & -5 \\ 3 & -6 \end{bmatrix}^{-1} = \frac{1}{2(-6) - (-5)3} \begin{bmatrix} -6 & -(-5) \\ -3 & 2 \end{bmatrix} = \frac{1}{3} \begin{bmatrix} -6 & 5 \\ -3 & 2 \end{bmatrix}$$

Multiplying each side of the matrix equation by this inverse matrix, we get

$X = A^{-1}B$

$$\begin{bmatrix} x \\ y \end{bmatrix} = \frac{1}{3} \begin{bmatrix} -6 & 5 \\ -3 & 2 \end{bmatrix} \begin{bmatrix} 15 \\ 36 \end{bmatrix} = \begin{bmatrix} 30 \\ 9 \end{bmatrix}$$

So $x = 30$ and $y = 9$. ■

EXAMPLE 5 ■ Modeling Nutritional Requirements Using Matrix Equations

A pet-store owner feeds his hamsters and gerbils different mixtures of three types of rodent food: KayDee Food, Pet Pellets, and Rodent Chow. He wishes to feed his animals the correct amount of each brand to satisfy their daily requirements for protein, fat, and carbohydrates exactly. Suppose that hamsters require 340 mg of protein, 280 mg of fat, and 440 mg of carbohydrates, and gerbils need 480 mg of protein, 360 mg of fat, and 680 mg of carbohydrates each day. The amount of each nutrient in one gram of each brand is given in the following table. How many grams of each food should the storekeeper feed his hamsters and gerbils daily to satisfy their nutrient requirements?

	KayDee Food	Pet Pellets	Rodent Chow
Protein (mg)	10	0	20
Fat (mg)	10	20	10
Carbohydrates (mg)	5	10	30

SOLUTION

If we let x_1, x_2, and x_3 be the grams of KayDee Food, Pet Pellets, and Rodent Chow, respectively, that the hamsters should eat, and if we let y_1, y_2, and y_3 be the corresponding amounts for the gerbils, then we want to solve the matrix equations

$$\begin{bmatrix} 10 & 0 & 20 \\ 10 & 20 & 10 \\ 5 & 10 & 30 \end{bmatrix} \begin{bmatrix} x_1 \\ x_2 \\ x_3 \end{bmatrix} = \begin{bmatrix} 340 \\ 280 \\ 440 \end{bmatrix} \quad \text{Hamster equation}$$

$$\begin{bmatrix} 10 & 0 & 20 \\ 10 & 20 & 10 \\ 5 & 10 & 30 \end{bmatrix} \begin{bmatrix} y_1 \\ y_2 \\ y_3 \end{bmatrix} = \begin{bmatrix} 480 \\ 360 \\ 680 \end{bmatrix} \quad \text{Gerbil equation}$$

Let

$$A = \begin{bmatrix} 10 & 0 & 20 \\ 10 & 20 & 10 \\ 5 & 10 & 30 \end{bmatrix}, B = \begin{bmatrix} 340 \\ 280 \\ 440 \end{bmatrix}, C = \begin{bmatrix} 480 \\ 360 \\ 680 \end{bmatrix}, X = \begin{bmatrix} x_2 \\ x_2 \\ x_3 \end{bmatrix}, Y = \begin{bmatrix} y_1 \\ y_2 \\ y_3 \end{bmatrix}$$

Then we can write these matrix equations as

$$AX = B \qquad \text{Hamster equation}$$

$$AY = C \qquad \text{Gerbil equation}$$

We want to solve for X and Y, so we multiply both sides of each equation by A^{-1}, the inverse of the coefficient matrix. We can find A^{-1} by hand, but it is more convenient to use a graphing calculator as shown in Figure 3.

FIGURE 3   (a)   (b)

So

$$X = A^{-1}B = \begin{bmatrix} 10 \\ 3 \\ 12 \end{bmatrix}, \qquad Y = A^{-1}C = \begin{bmatrix} 8 \\ 4 \\ 20 \end{bmatrix}$$

Thus, each hamster should be fed 10 g of KayDee Food, 3 g of Pet Pellets, and 12 g of Rodent Chow, and each gerbil should be fed 8 g of KayDee Food, 4 g of Pet Pellets, and 20 g of Rodent Chow daily. ∎

Since finding the inverse of a 3 × 3 or larger matrix usually involves a lot of work, the method used in Example 5 is really useful only when we are solving several systems of equations with the same coefficient matrix, or if we are using a graphing calculator or computer.

8.6 EXERCISES

1–4 ■ Calculate the products AB and BA to verify that B is the inverse of A.

1. $A = \begin{bmatrix} 4 & 1 \\ 7 & 2 \end{bmatrix}$, $B = \begin{bmatrix} 2 & -1 \\ -7 & 4 \end{bmatrix}$

2. $A = \begin{bmatrix} 2 & -3 \\ 4 & -7 \end{bmatrix}$, $B = \begin{bmatrix} \frac{7}{2} & -\frac{3}{2} \\ 2 & -1 \end{bmatrix}$

3. $A = \begin{bmatrix} 1 & 3 & -1 \\ 1 & 4 & 0 \\ -1 & -3 & 2 \end{bmatrix}$, $B = \begin{bmatrix} 8 & -3 & 4 \\ -2 & 1 & -1 \\ 1 & 0 & 1 \end{bmatrix}$

4. $A = \begin{bmatrix} 3 & 2 & 4 \\ 1 & 1 & -6 \\ 2 & 1 & 12 \end{bmatrix}$, $B = \begin{bmatrix} 9 & -10 & -8 \\ -12 & 14 & 11 \\ -\frac{1}{2} & \frac{1}{2} & \frac{1}{2} \end{bmatrix}$

5–6 ■ Find the inverse of the matrix and verify that $A^{-1}A = AA^{-1} = I_2$ and $B^{-1}B = BB^{-1} = I_3$.

5. $A = \begin{bmatrix} 7 & 4 \\ 3 & 2 \end{bmatrix}$

6. $B = \begin{bmatrix} 1 & 3 & 2 \\ 0 & 2 & 2 \\ -2 & -1 & 0 \end{bmatrix}$

7–22 ■ Find the inverse of the matrix if it exists.

7. $\begin{bmatrix} 5 & 3 \\ 3 & 2 \end{bmatrix}$

8. $\begin{bmatrix} 3 & 4 \\ 7 & 9 \end{bmatrix}$

9. $\begin{bmatrix} 2 & 5 \\ -5 & -13 \end{bmatrix}$

10. $\begin{bmatrix} -7 & 4 \\ 8 & -5 \end{bmatrix}$

11. $\begin{bmatrix} 6 & -3 \\ -8 & 4 \end{bmatrix}$

12. $\begin{bmatrix} \frac{1}{2} & \frac{1}{3} \\ 5 & 4 \end{bmatrix}$

13. $\begin{bmatrix} 0.4 & -1.2 \\ 0.3 & 0.6 \end{bmatrix}$

14. $\begin{bmatrix} 4 & 2 & 3 \\ 3 & 3 & 2 \\ 1 & 0 & 1 \end{bmatrix}$

15. $\begin{bmatrix} 2 & 4 & 1 \\ -1 & 1 & -1 \\ 1 & 4 & 0 \end{bmatrix}$

16. $\begin{bmatrix} 5 & 7 & 4 \\ 3 & -1 & 3 \\ 6 & 7 & 5 \end{bmatrix}$

17. $\begin{bmatrix} 1 & 2 & 3 \\ 4 & 5 & -1 \\ 1 & -1 & -10 \end{bmatrix}$

18. $\begin{bmatrix} 2 & 1 & 0 \\ 1 & 1 & 4 \\ 2 & 1 & 2 \end{bmatrix}$

19. $\begin{bmatrix} 0 & -2 & 2 \\ 3 & 1 & 3 \\ 1 & -2 & 3 \end{bmatrix}$

20. $\begin{bmatrix} 3 & -2 & 0 \\ 5 & 1 & 1 \\ 2 & -2 & 0 \end{bmatrix}$

21. $\begin{bmatrix} 1 & 2 & 0 & 3 \\ 0 & 1 & 1 & 1 \\ 0 & 1 & 0 & 1 \\ 1 & 2 & 0 & 2 \end{bmatrix}$

22. $\begin{bmatrix} 1 & 0 & 1 & 0 \\ 0 & 1 & 0 & 1 \\ 1 & 1 & 1 & 0 \\ 1 & 1 & 1 & 1 \end{bmatrix}$

23–30 ■ Solve the system of equations by converting to a matrix equation and using the inverse of the coefficient matrix, as in Example 4. Use the inverses from Exercises 7–10, 15, 16, 19, and 21.

23. $\begin{cases} 5x + 3y = 4 \\ 3x + 2y = 0 \end{cases}$

24. $\begin{cases} 3x + 4y = 10 \\ 7x + 9y = 20 \end{cases}$

25. $\begin{cases} 2x + 5y = 2 \\ -5x - 13y = 20 \end{cases}$

26. $\begin{cases} -7x + 4y = 0 \\ 8x - 5y = 100 \end{cases}$

27. $\begin{cases} 2x + 4y + z = 7 \\ -x + y - z = 0 \\ x + 4y = -2 \end{cases}$

28. $\begin{cases} 5x + 7y + 4z = 1 \\ 3x - y + 3z = 1 \\ 6x + 7y + 5z = 1 \end{cases}$

29. $\begin{cases} -2y + 2z = 12 \\ 3x + y + 3z = -2 \\ x - 2y + 3z = 8 \end{cases}$

30. $\begin{cases} x + 2y + 3w = 0 \\ y + z + w = 1 \\ y + w = 2 \\ x + 2y + 2w = 3 \end{cases}$

31–36 ■ Use a calculator that can perform matrix operations to solve the system, as in Example 5.

31. $\begin{cases} x + y - 2z = 3 \\ 2x + 5z = 11 \\ 2x + 3y = 12 \end{cases}$ **32.** $\begin{cases} 3x + 4y - z = 2 \\ 2x - 3y + z = -5 \\ 5x - 2y + 2z = -3 \end{cases}$

33. $\begin{cases} 12x + \frac{1}{2}y - 7z = 21 \\ 11x - 2y + 3z = 43 \\ 13x + y - 4z = 29 \end{cases}$ **34.** $\begin{cases} x + \frac{1}{2}y - \frac{1}{3}z = 4 \\ x - \frac{1}{4}y + \frac{1}{6}z = 7 \\ x + y - z = -6 \end{cases}$

35. $\begin{cases} x + y - 3w = 0 \\ x - 2z = 8 \\ 2y - z + w = 5 \\ 2x + 3y - 2w = 13 \end{cases}$

36. $\begin{cases} x + y + z + w = 15 \\ x - y + z - w = 5 \\ x + 2y + 3z + 4w = 26 \\ x - 2y + 3z - 4w = 2 \end{cases}$

37–38 ■ Solve the matrix equation by multiplying each side by the appropriate inverse matrix.

37. $\begin{bmatrix} 3 & -2 \\ -4 & 3 \end{bmatrix} \begin{bmatrix} x & y & z \\ u & v & w \end{bmatrix} = \begin{bmatrix} 1 & 0 & -1 \\ 2 & 1 & 3 \end{bmatrix}$

38. $\begin{bmatrix} 0 & -2 & 2 \\ 3 & 1 & 3 \\ 1 & -2 & 3 \end{bmatrix} \begin{bmatrix} x & u \\ y & v \\ z & w \end{bmatrix} = \begin{bmatrix} 3 & 6 \\ 6 & 12 \\ 0 & 0 \end{bmatrix}$

39. A nutritionist is studying the effects of the nutrients folic acid, choline, and inositol. He has three types of food available, and each type contains the following amounts of these nutrients per ounce:

	Type A	Type B	Type C
Folic acid (mg)	3	1	3
Choline (mg)	4	2	4
Inositol (mg)	3	2	4

(a) Find the inverse of the matrix

$$\begin{bmatrix} 3 & 1 & 3 \\ 4 & 2 & 4 \\ 3 & 2 & 4 \end{bmatrix}$$

and use it to solve the remaining parts of this problem.

(b) How many ounces of each food should the nutritionist feed his laboratory rats if he wants their daily diet to contain 10 mg of folic acid, 14 mg of choline, and 13 mg of inositol?

(c) How much of each food is needed to supply 9 mg of folic acid, 12 mg of choline, and 10 mg of inositol?

(d) Will any combination of these foods supply 2 mg of folic acid, 4 mg of choline, and 11 mg of inositol?

40. Refer to Exercise 39. Suppose food type C has been improperly labeled, and it actually contains 4 mg of folic acid, 6 mg of choline, and 5 mg of inositol per ounce. Would it still be possible to use matrix inversion to solve parts (b), (c), and (d) of Exercise 39? Why or why not?

41. An encyclopedia saleswoman works for a company that offers three different grades of bindings for its encyclopedias: standard, deluxe, and leather. For each set she sells, she earns a commission that is based on the set's binding grade. One week she sells one standard, one deluxe, and two leather sets and makes $675 in commission. The next week she sells two standard, one deluxe, and one leather set for a $600 commission. The third week she sells one standard, two deluxe, and one leather set, earning $625 in commission.

(a) Let x, y, and z represent the commission she earns on standard, deluxe, and leather sets, respectively. Translate the given information into a system of equations in x, y, and z.

(b) Express the system of equations you found in part (a) as a matrix equation of the form $AX = B$.

(c) Find the inverse of the coefficient matrix A and use it to solve the matrix equation in part (b). How much commission does the saleswoman earn on a set of encyclopedias in each grade of binding?

42–45 ■ Find the inverse of the matrix. For what value(s) of x, if any, does the matrix have no inverse?

42. $\begin{bmatrix} x & 1 \\ -1 & 1/x \end{bmatrix}$ **43.** $\begin{bmatrix} e^x & -e^{2x} \\ e^{2x} & e^{3x} \end{bmatrix}$

44. $\begin{bmatrix} 1 & e^x & 0 \\ e^x & -e^{2x} & 0 \\ 0 & 0 & 2 \end{bmatrix}$ **45.** $\begin{bmatrix} \cos\theta & \sin\theta \\ -\sin\theta & \cos\theta \end{bmatrix}$

46. Find the inverse of the following matrix, where $abcd \neq 0$:

$$\begin{bmatrix} a & 0 & 0 & 0 \\ 0 & b & 0 & 0 \\ 0 & 0 & c & 0 \\ 0 & 0 & 0 & d \end{bmatrix}$$

DISCOVERY · DISCUSSION

47. No Zero-Product Property for Matrices

We have used the Zero-Product Property to solve algebraic equations. Matrices do *not* have this property. Let O represent the

2×2 zero matrix:

$$O = \begin{bmatrix} 0 & 0 \\ 0 & 0 \end{bmatrix}$$

Find 2×2 matrices $A \neq O$ and $B \neq O$ such that $AB = O$. Can you find a matrix $A \neq O$ such that $A^2 = O$?

 Discovery Project

Computer Graphics I

Matrix algebra is the basic tool used in computer graphics to manipulate images on a computer screen. To see how this works, let's represent the point (x, y) by a 2×1 matrix:

$$(x, y) \quad \leftrightarrow \quad \begin{bmatrix} x \\ y \end{bmatrix}$$

For example, the point $(3, 2)$ is represented by the matrix

$$P = \begin{bmatrix} 3 \\ 2 \end{bmatrix}$$

Multiplying by a 2×2 matrix *moves* the point in the plane. For example, if

$$T = \begin{bmatrix} 1 & 0 \\ 0 & -1 \end{bmatrix}$$

then multiplying P by T we get

$$TP = \begin{bmatrix} 1 & 0 \\ 0 & -1 \end{bmatrix} \begin{bmatrix} 3 \\ 2 \end{bmatrix} = \begin{bmatrix} 3 \\ -2 \end{bmatrix}$$

We see that the point $(3, 2)$ has been moved to the point $(3, -2)$. In general, multiplication by this matrix T reflects points in the x-axis. If every point in an image is multiplied by this matrix, then the entire image will be flipped upside down about the x-axis. Matrix multiplication "transforms" a point to a new point in the plane. For this reason, a matrix used in this way is called a **transformation**.

Table 1 gives some standard transformations and their effects on the points in the square in Figure 1.

FIGURE 1

TABLE 1

Transformation matrix	Effect
$T = \begin{bmatrix} 1 & 0 \\ 0 & -1 \end{bmatrix}$ Reflection in x-axis	
$T = \begin{bmatrix} c & 0 \\ 0 & 1 \end{bmatrix}$ Expansion (or contraction) in the x-direction	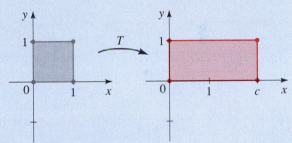
$T = \begin{bmatrix} 1 & c \\ 0 & 1 \end{bmatrix}$ Shear in x-direction	

FIGURE 2

Simple line drawings such as the house in Figure 2 consist of a collection of vertex points and connecting line segments.

The entire image in Figure 2 can be represented in a computer by the 2×11 **data matrix**

$$D = \begin{bmatrix} 2 & 0 & 0 & 2 & 4 & 4 & 3 & 3 & 2 & 2 & 3 \\ 0 & 0 & 3 & 5 & 3 & 0 & 0 & 2 & 2 & 0 & 0 \end{bmatrix}$$

The columns of D represent the vertex points of the image. To draw the house, we connect successive points (columns) in D by line segments. Now we can transform the whole house by multiplying D by an appropriate transformation matrix. For example, if we apply the shear transformation $T = \begin{bmatrix} 1 & 0.5 \\ 0 & 1 \end{bmatrix}$, we get the following matrix.

$$TD = \begin{bmatrix} 1 & 0.5 \\ 0 & 1 \end{bmatrix} \begin{bmatrix} 2 & 0 & 0 & 2 & 4 & 4 & 3 & 3 & 2 & 2 & 3 \\ 0 & 0 & 3 & 5 & 3 & 0 & 0 & 2 & 2 & 0 & 0 \end{bmatrix}$$

$$= \begin{bmatrix} 2 & 0 & 1.5 & 4.5 & 5.5 & 4 & 3 & 4 & 3 & 2 & 3 \\ 0 & 0 & 3 & 5 & 3 & 0 & 0 & 2 & 2 & 0 & 0 \end{bmatrix}$$

To draw the image represented by TD, we start with the point $\begin{bmatrix} 2 \\ 0 \end{bmatrix}$, connect it by a line segment to the point $\begin{bmatrix} 0 \\ 0 \end{bmatrix}$, then follow that by a line segment to $\begin{bmatrix} 1.5 \\ 3 \end{bmatrix}$, and so on. The resulting tilted house is shown in Figure 3.

A convenient way to draw an image corresponding to a given data matrix is to use a graphing calculator. The TI-83 program in the margin converts a data matrix stored in `[A]` into the corresponding image, as shown in Figure 4. (To use this program for a data matrix with m columns, store the matrix in `[A]` and change the "10" in the `For` command to $m - 1$.)

We will revisit computer graphics in the *Discovery Project* on page 770, where we will find special matrices that rotate an image by any specified angle.

1. The square in Figure 1 has the following vertices.

$$\begin{bmatrix} 0 \\ 0 \end{bmatrix}, \quad \begin{bmatrix} 1 \\ 0 \end{bmatrix}, \quad \begin{bmatrix} 1 \\ 1 \end{bmatrix}, \quad \begin{bmatrix} 0 \\ 1 \end{bmatrix}$$

Apply each of the three transformations given in Table 1 to these vertices and sketch the result, to verify that each transformation has the indicated effect. Use $c = 2$ in the expansion matrix and $c = 1$ in the shear matrix.

2. Verify that multiplication by the given matrix has the indicated effect when applied to the square in Figure 1. Use $c = 3$ in the expansion matrix and $c = 1$ in the shear matrix.

$$T_1 = \begin{bmatrix} -1 & 0 \\ 0 & 1 \end{bmatrix} \qquad T_2 = \begin{bmatrix} 1 & 0 \\ 0 & c \end{bmatrix} \qquad T_3 = \begin{bmatrix} 1 & 0 \\ c & 1 \end{bmatrix}$$

Reflection in y-axis   Expansion (or contraction) in y-direction   Shear in y-direction

3. Let $T = \begin{bmatrix} 1 & 1.5 \\ 0 & 1 \end{bmatrix}$,

(a) What effect does T have on the square in Figure 1?

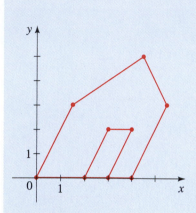

FIGURE 3

```
PROGRAM:IMAGE
:For(N,1,10)
:Line([A](1,N),
[A](2,N),[A](1,N+1),
[A](2,N+1))
:End
```

(a)

(b)

FIGURE 4

(b) Find T^{-1}.

(c) What effect does T^{-1} have on the square in Figure 1?

(d) What happens to the square if we first apply T, then T^{-1}?

4. (a) Let $T = \begin{bmatrix} 3 & 0 \\ 0 & 1 \end{bmatrix}$. What effect does T have on the square in Figure 1?

(b) Let $S = \begin{bmatrix} 1 & 0 \\ 0 & 2 \end{bmatrix}$. What effect does S have on the square in Figure 1?

(c) Apply S to the vertices of the square, and then apply T to the result. What is the effect of the combined transformation?

(d) Find the product matrix $W = TS$.

(e) Apply the transformation W to the square. Compare to your final result in part (c). What do you notice?

5. The figure shows three outline versions of the letter **F**. The second one is obtained from the first by shrinking horizontally by a factor of 0.75, and the third is obtained from the first by shearing horizontally by a factor of 0.25.

(a) Find a data matrix D for the first letter **F**.

(b) Find the transformation matrix T that transforms the first **F** into the second. Calculate TD and verify that this is a data matrix for the second **F**.

(c) Find the transformation matrix S that transforms the first **F** into the third. Calculate SD and verify that this is a data matrix for the third **F**.

6. Here is a data matrix for a line drawing.

$$D = \begin{bmatrix} 0 & 1 & 2 & 1 & 0 & 0 \\ 0 & 0 & 2 & 4 & 4 & 0 \end{bmatrix}$$

(a) Draw the image represented by D.

(b) Let $T = \begin{bmatrix} 1 & 1 \\ 0 & -1 \end{bmatrix}$. Calculate the matrix product TD and draw the image represented by this product. What is the effect of the transformation T?

(c) Express T as a product of a shear matrix and a reflection matrix.

If a matrix is **square** (that is, if it has the same number of rows as columns), then we can assign to it a number called its **determinant**. Determinants can be used to solve systems of linear equations, as we will see later in this section. They are also useful in determining whether a matrix has an inverse.

We denote the determinant of a square matrix A by the symbol $\det(A)$ or $|A|$. We first define $\det(A)$ for the simplest cases. If $A = [a]$ is a 1×1 matrix, then $\det(A) = a$. The following box gives the definition of a 2×2 determinant.

We will use both notations, $\det(A)$ and $|A|$, for the determinant of A. Although the symbol $|A|$ looks like the absolute value symbol, it will be clear from the context which meaning is intended.

DETERMINANT OF A 2 × 2 MATRIX

The determinant of the 2×2 matrix $A = \begin{bmatrix} a & b \\ c & d \end{bmatrix}$ is

$$\det(A) = |A| = \begin{vmatrix} a & b \\ c & d \end{vmatrix} = ad - bc$$

EXAMPLE 1 ■ Determinant of a 2 × 2 Matrix

Evaluate $|A|$ for $A = \begin{bmatrix} 6 & -3 \\ 2 & 3 \end{bmatrix}$.

SOLUTION

$$\begin{vmatrix} 6 & -3 \\ 2 & 3 \end{vmatrix} = 6 \cdot 3 - (-3)2 = 18 - (-6) = 24$$

To evaluate a 2×2 determinant, we take the product of the diagonal from top left to bottom right, and subtract the product from top right to bottom left, as indicated by the arrows.

To define the concept of determinant for an arbitrary $n \times n$ matrix, we need the following terminology.

Let A be an $n \times n$ matrix.

1. The **minor** M_{ij} of the element a_{ij} is the determinant of the matrix obtained by deleting the ith row and jth column of A.

2. The **cofactor** A_{ij} of the element a_{ij} is

$$A_{ij} = (-1)^{i+j} M_{ij}$$

For example, if A is the matrix

$$\begin{bmatrix} 2 & 3 & -1 \\ 0 & 2 & 4 \\ -2 & 5 & 6 \end{bmatrix}$$

David Hilbert (1862–1943) was born in Königsberg, Germany, and became a professor at Göttingen University. He is considered by many to be the greatest mathematician of the 20th century. At the International Congress of Mathematicians held in Paris in 1900, Hilbert set the direction of mathematics for the about-to-dawn 20th century by posing 23 problems he believed to be of crucial importance. He said that "these are problems whose solutions we expect from the future." Most of Hilbert's problems have now been solved (see Julia Robinson, page 663 and Alan Turing, page 106), and their solutions have led to important new areas of mathematical research. Yet as we enter the new millennium, some of his problems remain unsolved. In his work, Hilbert emphasized structure, logic, and the foundations of mathematics. Part of his genius lay in his ability to see the most general possible statement of a problem. For instance, Euler proved that every whole number is the sum of four squares; Hilbert proved a similar statement for all powers of positive integers.

then the minor M_{12} is the determinant of the matrix obtained by deleting the first row and second column from A. Thus

$$M_{12} = \begin{vmatrix} 2 & 3 & -1 \\ 0 & 2 & 4 \\ -2 & 5 & 6 \end{vmatrix} = \begin{vmatrix} 0 & 4 \\ -2 & 6 \end{vmatrix} = 0(6) - 4(-2) = 8$$

So, the cofactor $A_{12} = (-1)^{1+2} M_{12} = -8$. Similarly

$$M_{33} = \begin{vmatrix} 2 & 3 & -1 \\ 0 & 2 & 4 \\ -2 & 5 & 6 \end{vmatrix} = \begin{vmatrix} 2 & 3 \\ 0 & 2 \end{vmatrix} = 2 \cdot 2 - 3 \cdot 0 = 4$$

So, $A_{33} = (-1)^{3+3} M_{33} = 4$.

Note that the cofactor of a_{ij} is simply the minor of a_{ij} multiplied by either 1 or -1, depending on whether $i + j$ is even or odd. Thus, in a 3×3 matrix we obtain the cofactor of any element by prefixing its minor with the sign obtained from the following checkerboard pattern:

$$\begin{bmatrix} + & - & + \\ - & + & - \\ + & - & + \end{bmatrix}$$

We are now ready to define the determinant of any square matrix.

THE DETERMINANT OF A SQUARE MATRIX

If A is an $n \times n$ matrix, then the **determinant** of A is obtained by multiplying each element of the first row by its cofactor, and then adding the results. In symbols,

$$\det(A) = |A| = \begin{vmatrix} a_{11} & a_{12} & \cdots & a_{1n} \\ a_{21} & a_{22} & \cdots & a_{2n} \\ \cdot & \cdot & \cdot & \cdot \\ \cdot & \cdot & \cdot & \cdot \\ \cdot & \cdot & \cdot & \cdot \\ a_{n1} & a_{n2} & \cdots & a_{nn} \end{vmatrix} = a_{11}A_{11} + a_{12}A_{12} + \cdots + a_{1n}A_{1n}$$

EXAMPLE 2 ■ **Determinant of a 3 × 3 Matrix**

Evaluate the determinant of the matrix.

$$A = \begin{bmatrix} 2 & 3 & -1 \\ 0 & 2 & 4 \\ -2 & 5 & 6 \end{bmatrix}$$

SOLUTION

$$\det(A) = \begin{vmatrix} 2 & 3 & -1 \\ 0 & 2 & 4 \\ -2 & 5 & 6 \end{vmatrix} = 2\begin{vmatrix} 2 & 4 \\ 5 & 6 \end{vmatrix} - 3\begin{vmatrix} 0 & 4 \\ -2 & 6 \end{vmatrix} + (-1)\begin{vmatrix} 0 & 2 \\ -2 & 5 \end{vmatrix}$$

$$= 2(2 \cdot 6 - 4 \cdot 5) - 3[0 \cdot 6 - 4(-2)] - [0 \cdot 5 - 2(-2)]$$

$$= -16 - 24 - 4$$

$$= -44 \qquad\blacksquare$$

In our definition of the determinant we used the cofactors of elements in the first row only. This is called **expanding the determinant by the first row**. In fact, *we can expand the determinant by any row or column in the same way, and obtain the same result in each case* (although we won't prove this). The next example illustrates this principle.

EXAMPLE 3 ■ **Expanding a Determinant about a Row and a Column**

Expand the determinant of the matrix A in Example 2 by the second row and by the third column, and show that the value obtained is the same in each case.

SOLUTION

Expanding by the second row, we get

$$\det(A) = \begin{vmatrix} 2 & 3 & -1 \\ 0 & 2 & 4 \\ -2 & 5 & 6 \end{vmatrix} = -0\begin{vmatrix} 3 & -1 \\ 5 & 6 \end{vmatrix} + 2\begin{vmatrix} 2 & -1 \\ -2 & 6 \end{vmatrix} - 4\begin{vmatrix} 2 & 3 \\ -2 & 5 \end{vmatrix}$$

$$= 0 + 2[2 \cdot 6 - (-1)(-2)] - 4[2 \cdot 5 - 3(-2)]$$

$$= 0 + 20 - 64 = -44$$

Expanding by the third column gives

$$\det(A) = \begin{vmatrix} 2 & 3 & -1 \\ 0 & 2 & 4 \\ -2 & 5 & 6 \end{vmatrix}$$

$$= -1\begin{vmatrix} 0 & 2 \\ -2 & 5 \end{vmatrix} - 4\begin{vmatrix} 2 & 3 \\ -2 & 5 \end{vmatrix} + 6\begin{vmatrix} 2 & 3 \\ 0 & 2 \end{vmatrix}$$

$$= -[0 \cdot 5 - 2(-2)] - 4[2 \cdot 5 - 3(-2)] + 6(2 \cdot 2 - 3 \cdot 0)$$

$$= -4 - 64 + 24 = -44$$

In both cases, we obtain the same value for the determinant as when we expanded by the first row in Example 2. ■

Graphing calculators are capable of computing determinants. The figure shows the output when the TI-83 is used to calculate the determinant in Example 3.

```
[A]
        [[2   3  -1]
         [0   2   4]
         [-2  5   6]]
det([A])
                  -44
```

The following principle allows us to determine whether a square matrix has an inverse without actually calculating the inverse. This is one of the most important uses of the determinant in matrix algebra, and it is the reason for the name *determinant*.

INVERTIBILITY CRITERION

If A is a square matrix, then A has an inverse if and only if $\det(A) \neq 0$.

We will not prove this fact, but from the formula for the inverse of a 2×2 matrix (page 675), you can see why it is true in the 2×2 case.

EXAMPLE 4 ■ Using the Determinant to Show That a Matrix Is Not Invertible

Show that the matrix A has no inverse.

$$A = \begin{bmatrix} 1 & 2 & 0 & 4 \\ 0 & 0 & 0 & 3 \\ 5 & 6 & 2 & 6 \\ 2 & 4 & 0 & 9 \end{bmatrix}$$

SOLUTION

We begin by calculating the determinant of A. Since all but one of the elements of the second row is zero, we expand the determinant by the second row. If we do this, we see from the following equation that only the cofactor A_{24} will have to be calculated.

$$\det(A) = -0 \cdot A_{21} + 0 \cdot A_{22} - 0 \cdot A_{23} + 3 \cdot A_{24} = 3A_{24}$$

$$= 3 \begin{vmatrix} 1 & 2 & 0 \\ 5 & 6 & 2 \\ 2 & 4 & 0 \end{vmatrix} \qquad \text{Expand this by the third column}$$

$$= 3(-2) \begin{vmatrix} 1 & 2 \\ 2 & 4 \end{vmatrix}$$

$$= 3(-2)(1 \cdot 4 - 2 \cdot 2) = 0$$

Since the determinant of A is zero, A cannot have an inverse, by the Invertibility Criterion.   ■

Row and Column Transformations

The preceding example shows that if we expand a determinant about a row or column that contains many zeros, our work is reduced considerably because we don't

Emmy Noether (1882–1935) was one of the foremost mathematicians of the early 20th century. Her groundbreaking work in abstract algebra provided much of the foundation for this field, and her work in Invariant Theory was essential in the development of Einstein's theory of general relativity. Although women weren't allowed to study at German universities at that time, she audited courses unofficially and went on to receive a doctorate at Erlangen *summa cum laude*, despite the opposition of the academic senate, which declared that women students would "overthrow all academic order." She subsequently taught mathematics at Göttingen, Moscow, and Frankfurt. In 1933 she left Germany to escape Nazi persecution, accepting a position at Bryn Mawr College in suburban Philadelphia. She lectured there and at the Institute for Advanced Study in Princeton, New Jersey, until her untimely death in 1935.

have to evaluate the cofactors of the elements that are zero. The following principle often enables us to simplify the process of finding a determinant by introducing zeros into it without changing its value.

> ### ROW AND COLUMN TRANSFORMATIONS OF A DETERMINANT
>
> If A is a square matrix, and if the matrix B is obtained from A by adding a multiple of one row to another, or a multiple of one column to another, then $\det(A) = \det(B)$.

EXAMPLE 5 ■ Using Row and Column Transformations to Calculate a Determinant

Find the determinant of the matrix A. Does it have an inverse?

$$A = \begin{bmatrix} 8 & 2 & -1 & -4 \\ 3 & 5 & -3 & 11 \\ 24 & 6 & 1 & -12 \\ 2 & 2 & 7 & -1 \end{bmatrix}$$

SOLUTION

If we add -3 times row 1 to row 3, we change all but one element of row 3 to zeros:

$$\begin{bmatrix} 8 & 2 & -1 & -4 \\ 3 & 5 & -3 & 11 \\ 0 & 0 & 4 & 0 \\ 2 & 2 & 7 & -1 \end{bmatrix}$$

This new matrix has the same determinant as A, and if we expand its determinant by the third row, we get

$$\det(A) = 4 \begin{vmatrix} 8 & 2 & -4 \\ 3 & 5 & 11 \\ 2 & 2 & -1 \end{vmatrix}$$

Now, adding 2 times column 3 to column 1 in this determinant gives us

$$\det(A) = 4 \begin{vmatrix} 0 & 2 & -4 \\ 25 & 5 & 11 \\ 0 & 2 & -1 \end{vmatrix} \qquad \text{Expand this by the first column}$$

$$= 4(-25) \begin{vmatrix} 2 & -4 \\ 2 & -1 \end{vmatrix}$$

$$= 4(-25)[2(-1) - (-4)2] = -600$$

Since the determinant of A is not zero, A does have an inverse.   ■

Cramer's Rule

The solutions of linear equations can sometimes be expressed using determinants. To illustrate, let's solve the following pair of linear equations for the variable x.

$$\begin{cases} ax + by = r \\ cx + dy = s \end{cases}$$

To eliminate the variable y, we multiply the first equation by d and the second by b, and subtract.

$$\begin{aligned} adx + bdy &= rd \\ bcx + bdy &= bs \\ \hline adx - bcx &= rd - bs \end{aligned}$$

Factoring the left-hand side, we get $(ad - bc)x = rd - bs$. Assuming that $ad - bc \neq 0$, we can now solve this equation for x, obtaining

$$x = \frac{rd - bs}{ad - bc}$$

The numerator and denominator of this fraction look like the determinants of 2×2 matrices. In fact, we can write the solution of the system as

$$x = \frac{\begin{vmatrix} r & b \\ s & d \end{vmatrix}}{\begin{vmatrix} a & b \\ c & d \end{vmatrix}} \qquad y = \frac{\begin{vmatrix} a & r \\ c & s \end{vmatrix}}{\begin{vmatrix} a & b \\ c & d \end{vmatrix}}$$

where the solution for y was obtained using the same sort of technique as the solution for x. Notice that the denominator in each case is the determinant of the coefficient matrix, which we will call D. The numerator in the solution for x is the determinant of the matrix obtained from D by replacing the first column, the coefficients of x, by r and s, respectively. Similarly, in the solution for y the numerator is the determinant of the matrix obtained from D by replacing the second column, the coefficients of y, by r and s. Thus, if we define

$$D = \begin{bmatrix} a & b \\ c & d \end{bmatrix} \qquad D_x = \begin{bmatrix} r & b \\ s & d \end{bmatrix} \qquad D_y = \begin{bmatrix} a & r \\ c & s \end{bmatrix}$$

then we can write the solution of the system as

$$x = \frac{|D_x|}{|D|} \qquad \text{and} \qquad y = \frac{|D_y|}{|D|}$$

This pair of formulas is known as **Cramer's Rule**, and it can be used to solve any pair of linear equations in two unknowns in which the determinant of the coefficient matrix is not zero.

EXAMPLE 6 ■ Using Cramer's Rule to Solve a System with Two Variables

Use Cramer's Rule to solve the system.

$$\begin{cases} 2x + 6y = -1 \\ x + 8y = 2 \end{cases}$$

SOLUTION

For this system we have

$$|D| = \begin{vmatrix} 2 & 6 \\ 1 & 8 \end{vmatrix} = 2 \cdot 8 - 6 \cdot 1 = 10$$

$$|D_x| = \begin{vmatrix} -1 & 6 \\ 2 & 8 \end{vmatrix} = (-1)8 - 6 \cdot 2 = -20$$

$$|D_y| = \begin{vmatrix} 2 & -1 \\ 1 & 2 \end{vmatrix} = 2 \cdot 2 - (-1)1 = 5$$

The solution is

$$x = \frac{|D_x|}{|D|} = \frac{-20}{10} = -2$$

$$y = \frac{|D_y|}{|D|} = \frac{5}{10} = \frac{1}{2}$$

■

Cramer's Rule can be extended to apply to any system of n linear equations in n variables in which the determinant of the coefficient matrix is not zero. As we saw in the preceding section, any such system can be written in matrix form as

$$\begin{bmatrix} a_{11} & a_{12} & \cdots & a_{1n} \\ a_{21} & a_{22} & \cdots & a_{2n} \\ \vdots & \vdots & & \vdots \\ a_{n1} & a_{n2} & \cdots & a_{nn} \end{bmatrix} \begin{bmatrix} x_1 \\ x_2 \\ \vdots \\ x_n \end{bmatrix} = \begin{bmatrix} b_1 \\ b_2 \\ \vdots \\ b_n \end{bmatrix}$$

By analogy with our derivation of Cramer's Rule in the case of two equations in two unknowns, we let D be the coefficient matrix in this system, and D_{x_i} be the matrix obtained by replacing the ith column of D by the numbers $b_1, b_2, \ldots, b_n$ that appear to the right of the equal sign. The solution of the system is then given by the following rule.

CRAMER'S RULE

If a system of n linear equations in the n variables $x_1, x_2, \ldots, x_n$ is equivalent to the matrix equation $DX = B$, and if $|D| \neq 0$, then its solutions are

$$x_1 = \frac{|D_{x_1}|}{|D|}, \quad x_2 = \frac{|D_{x_2}|}{|D|}, \quad \ldots, \quad x_n = \frac{|D_{x_n}|}{|D|}$$

where D_{x_i} is the matrix obtained by replacing the ith column of D by the $n \times 1$ matrix B.

EXAMPLE 7 ■ **Using Cramer's Rule to Solve a System with Three Variables**

Use Cramer's Rule to solve the system.

$$\begin{cases} 2x - 3y + 4z = 1 \\ x \qquad + 6z = 0 \\ 3x - 2y \qquad = 5 \end{cases}$$

SOLUTION

First, we evaluate the determinants that appear in Cramer's Rule.

$$|D| = \begin{vmatrix} 2 & -3 & 4 \\ 1 & 0 & 6 \\ 3 & -2 & 0 \end{vmatrix} = -38 \qquad |D_x| = \begin{vmatrix} 1 & -3 & 4 \\ 0 & 0 & 6 \\ 5 & -2 & 0 \end{vmatrix} = -78$$

$$|D_y| = \begin{vmatrix} 2 & 1 & 4 \\ 1 & 0 & 6 \\ 3 & 5 & 0 \end{vmatrix} = -22 \qquad |D_z| = \begin{vmatrix} 2 & -3 & 1 \\ 1 & 0 & 0 \\ 3 & -2 & 5 \end{vmatrix} = 13$$

Now we use Cramer's Rule to get the solution:

$$x = \frac{|D_x|}{|D|} = \frac{-78}{-38} = \frac{39}{19}$$

$$y = \frac{|D_y|}{|D|} = \frac{-22}{-38} = \frac{11}{19}$$

$$z = \frac{|D_z|}{|D|} = \frac{13}{-38} = -\frac{13}{38} \qquad ■$$

Solving the system in Example 7 using Gaussian elimination would involve matrices whose elements are fractions with fairly large denominators. Thus, in cases like Examples 6 and 7, Cramer's Rule provides an efficient method for solving systems of linear equations. But in systems with more than three equations, evaluating the various determinants involved is usually a long and tedious task (unless you are using a graphing calculator). Moreover, the rule doesn't apply if $|D| = 0$ or if D is not a square matrix. So, Cramer's Rule is a useful alternative to Gaussian elimination, but only in some situations.

■ Areas of Triangles Using Determinants

Determinants provide a simple way of calculating the area of a triangle in the coordinate plane.

AREA OF A TRIANGLE

If a triangle in the coordinate plane has vertices (a_1, b_1), (a_2, b_2), and (a_3, b_3), then its area is

$$\text{area} = \pm\tfrac{1}{2} \begin{vmatrix} a_1 & b_1 & 1 \\ a_2 & b_2 & 1 \\ a_3 & b_3 & 1 \end{vmatrix}$$

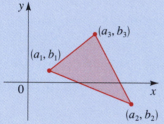

where the sign is chosen to make the area positive.

You are asked to prove this formula in Exercise 56.

EXAMPLE 8 ■ Area of a Triangle

Find the area of the triangle shown in Figure 1.

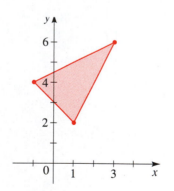

FIGURE 1

We can calculate the determinant by hand or by using a graphing calculator.

```
[A]
        [[-1  4  1]
         [3   6  1]
         [1   2  1]]
det([A])
            -12
```

SOLUTION

The vertices are $(1, 2)$, $(3, 6)$, and $(-1, 4)$. Using the formula in the preceding box, we get:

$$\text{area} = \pm\tfrac{1}{2} \begin{vmatrix} -1 & 4 & 1 \\ 3 & 6 & 1 \\ 1 & 2 & 1 \end{vmatrix} = \pm\tfrac{1}{2}(-12)$$

To make the area positive, we choose the negative sign in the formula. Thus, the area of the triangle is

$$\text{area} = -\tfrac{1}{2}(-12) = 6$$

■

8.7 EXERCISES

1–8 ■ Find the determinant of the matrix, if it exists.

1. $\begin{bmatrix} 2 & 0 \\ 0 & 3 \end{bmatrix}$

2. $\begin{bmatrix} 0 & -1 \\ 2 & 0 \end{bmatrix}$

3. $\begin{bmatrix} 4 & 5 \\ 0 & -1 \end{bmatrix}$

4. $\begin{bmatrix} -2 & 1 \\ 3 & -2 \end{bmatrix}$

5. $\begin{bmatrix} 2 & 5 \end{bmatrix}$

6. $\begin{bmatrix} 3 \\ 0 \end{bmatrix}$

7. $\begin{bmatrix} \frac{1}{2} & \frac{1}{8} \\ 1 & \frac{1}{2} \end{bmatrix}$

8. $\begin{bmatrix} 2.2 & -1.4 \\ 0.5 & 1.0 \end{bmatrix}$

9–14 ■ Evaluate the minor and cofactor using the matrix A.

$$A = \begin{bmatrix} 1 & 0 & \frac{1}{2} \\ -3 & 5 & 2 \\ 0 & 0 & 4 \end{bmatrix}$$

9. M_{11}, A_{11}

10. M_{33}, A_{33}

11. M_{12}, A_{12}

12. M_{13}, A_{13}

13. M_{23}, A_{23}

14. M_{32}, A_{32}

15–22 ■ Find the determinant of the matrix. Determine whether the matrix has an inverse, but don't calculate the inverse.

15. $\begin{bmatrix} 2 & 1 & 0 \\ 0 & -2 & 4 \\ 0 & 1 & -3 \end{bmatrix}$

16. $\begin{bmatrix} 0 & -1 & 0 \\ 2 & 6 & 4 \\ 1 & 0 & 3 \end{bmatrix}$

17. $\begin{bmatrix} 1 & 3 & 7 \\ 2 & 0 & -1 \\ 0 & 2 & 6 \end{bmatrix}$

18. $\begin{bmatrix} -2 & -\frac{3}{2} & \frac{1}{2} \\ 2 & 4 & 0 \\ \frac{1}{2} & 2 & 1 \end{bmatrix}$

19. $\begin{bmatrix} 30 & 0 & 20 \\ 0 & -10 & -20 \\ 40 & 0 & 10 \end{bmatrix}$

20. $\begin{bmatrix} 1 & 2 & 5 \\ -2 & -3 & 2 \\ 3 & 5 & 3 \end{bmatrix}$

21. $\begin{bmatrix} 1 & 3 & 3 & 0 \\ 0 & 2 & 0 & 1 \\ -1 & 0 & 0 & 2 \\ 1 & 6 & 4 & 1 \end{bmatrix}$

22. $\begin{bmatrix} 1 & 2 & 0 & 2 \\ 3 & -4 & 0 & 4 \\ 0 & 1 & 6 & 0 \\ 1 & 0 & 2 & 0 \end{bmatrix}$

23–26 ■ Evaluate the determinant, using row or column operations whenever possible to simplify your work.

23. $\begin{vmatrix} 0 & 0 & 4 & 6 \\ 2 & 1 & 1 & 3 \\ 2 & 1 & 2 & 3 \\ 3 & 0 & 1 & 7 \end{vmatrix}$

24. $\begin{vmatrix} -2 & 3 & -1 & 7 \\ 4 & 6 & -2 & 3 \\ 7 & 7 & 0 & 5 \\ 3 & -12 & 4 & 0 \end{vmatrix}$

25. $\begin{vmatrix} 1 & 2 & 3 & 4 & 5 \\ 0 & 2 & 4 & 6 & 8 \\ 0 & 0 & 3 & 6 & 9 \\ 0 & 0 & 0 & 4 & 8 \\ 0 & 0 & 0 & 0 & 5 \end{vmatrix}$

26. $\begin{vmatrix} 2 & -1 & 6 & 4 \\ 7 & 2 & -2 & 5 \\ 4 & -2 & 10 & 8 \\ 6 & 1 & 1 & 4 \end{vmatrix}$

27. Let

$$B = \begin{bmatrix} 4 & 1 & 0 \\ -2 & -1 & 1 \\ 4 & 0 & 3 \end{bmatrix}$$

(a) Evaluate $\det(B)$ by expanding by the second row.
(b) Evaluate $\det(B)$ by expanding by the third column.
(c) Do your results in parts (a) and (b) agree?

28. Consider the system

$$\begin{cases} x + 2y + 6z = 5 \\ -3x - 6y + 5z = 8 \\ 2x + 6y + 9z = 7 \end{cases}$$

(a) Verify that $x = -1, y = 0, z = 1$ is a solution of the system.
(b) Find the determinant of the coefficient matrix.
(c) Without solving the system, determine whether there are any other solutions.
(d) Can Cramer's Rule be used to solve this system? Why or why not?

29–44 ■ Use Cramer's Rule to solve the system.

29. $\begin{cases} 2x - y = -9 \\ x + 2y = 8 \end{cases}$

30. $\begin{cases} 6x + 12y = 33 \\ 4x + 7y = 20 \end{cases}$

31. $\begin{cases} x - 6y = 3 \\ 3x + 2y = 1 \end{cases}$

32. $\begin{cases} \frac{1}{2}x + \frac{1}{3}y = 1 \\ \frac{1}{4}x - \frac{1}{6}y = -\frac{3}{2} \end{cases}$

33. $\begin{cases} 0.4x + 1.2y = 0.4 \\ 1.2x + 1.6y = 3.2 \end{cases}$

34. $\begin{cases} 10x - 17y = 21 \\ 20x - 31y = 39 \end{cases}$

35. $\begin{cases} x - y + 2z = 0 \\ 3x \quad\quad + z = 11 \\ -x + 2y \quad\quad = 0 \end{cases}$ **36.** $\begin{cases} 5x - 3y + z = 6 \\ \quad\quad 4y - 6z = 22 \\ 7x + 10y \quad\quad = -13 \end{cases}$

37. $\begin{cases} 2x_1 + 3x_2 - 5x_3 = 1 \\ x_1 + x_2 - x_3 = 2 \\ \quad\quad 2x_2 + x_3 = 8 \end{cases}$ **38.** $\begin{cases} -2a \quad\quad + c = 2 \\ a + 2b - c = 9 \\ 3a + 5b + 2c = 22 \end{cases}$

39. $\begin{cases} \frac{1}{3}x - \frac{1}{5}y + \frac{1}{2}z = \frac{7}{10} \\ -\frac{2}{3}x + \frac{2}{5}y + \frac{3}{2}z = \frac{11}{10} \\ x - \frac{4}{5}y + z = \frac{9}{5} \end{cases}$ **40.** $\begin{cases} 2x - y = 5 \\ 5x \quad\quad + 3z = 19 \\ 4y + 7z = 17 \end{cases}$

41. $\begin{cases} 3y + 5z = 4 \\ 2x \quad\quad - z = 10 \\ 4x + 7y = 0 \end{cases}$ **42.** $\begin{cases} 2x - 5y = 4 \\ x + y - z = 8 \\ 3x \quad\quad + 5z = 0 \end{cases}$

43. $\begin{cases} x + y + z + w = 0 \\ 2x \quad\quad\quad\quad + w = 0 \\ \quad y - z \quad\quad = 0 \\ x \quad\quad + 2z \quad\quad = 1 \end{cases}$ **44.** $\begin{cases} x + y = 1 \\ y + z = 2 \\ z + w = 3 \\ w - x = 4 \end{cases}$

45–46 ■ Evaluate the determinants.

45. $\begin{vmatrix} a & 0 & 0 & 0 & 0 \\ 0 & b & 0 & 0 & 0 \\ 0 & 0 & c & 0 & 0 \\ 0 & 0 & 0 & d & 0 \\ 0 & 0 & 0 & 0 & e \end{vmatrix}$ **46.** $\begin{vmatrix} a & a & a & a & a \\ 0 & a & a & a & a \\ 0 & 0 & a & a & a \\ 0 & 0 & 0 & a & a \\ 0 & 0 & 0 & 0 & a \end{vmatrix}$

47–50 ■ Solve for x.

47. $\begin{vmatrix} x & 12 & 13 \\ 0 & x-1 & 23 \\ 0 & 0 & x-2 \end{vmatrix} = 0$ **48.** $\begin{vmatrix} x & 1 & 1 \\ 1 & 1 & x \\ x & 1 & x \end{vmatrix} = 0$

49. $\begin{vmatrix} 1 & 0 & x \\ x^2 & 1 & 0 \\ x & 0 & 1 \end{vmatrix} = 0$ **50.** $\begin{vmatrix} a & b & x-a \\ x & x+b & x \\ 0 & 1 & 1 \end{vmatrix} = 0$

51–54 ■ Sketch the triangle with the given vertices and use a determinant to find its area.

51. $(0, 0)$, $(6, 2)$, $(3, 8)$ **52.** $(1, 0)$, $(3, 5)$, $(-2, 2)$

53. $(-1, 3)$, $(2, 9)$, $(5, -6)$ **54.** $(-2, 5)$, $(7, 2)$, $(3, -4)$

55. Show that

$$\begin{vmatrix} 1 & x & x^2 \\ 1 & y & y^2 \\ 1 & z & z^2 \end{vmatrix} = (x - y)(y - z)(z - x)$$

◆ **DISCOVERY · DISCUSSION**

56. Determinant Formula for the Area of a Triangle   The figure shows a triangle in the plane with vertices (a_1, b_1), (a_2, b_2), and (a_3, b_3).

(a) Find the coordinates of the vertices of the surrounding rectangle and find its area.

(b) Find the area of the red triangle by subtracting the areas of the three blue triangles from the area of the rectangle.

(c) Use your answer to part (b) to show that the area of the red triangle is given by

$$\text{area} = \pm\frac{1}{2} \begin{vmatrix} a_1 & b_1 & 1 \\ a_2 & b_2 & 1 \\ a_3 & b_3 & 1 \end{vmatrix}$$

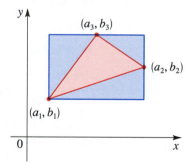

57. Collinear Points and Determinates

(a) If three points lie on a line, what is the area of the "triangle" that they determine? Use the answer to this question, together with the determinant formula for the area of a triangle, to explain why the points (a_1, b_1), (a_2, b_2), and (a_3, b_3) are collinear if and only if

$$\begin{vmatrix} a_1 & b_1 & 1 \\ a_2 & b_2 & 1 \\ a_3 & b_3 & 1 \end{vmatrix} = 0$$

(b) Use a determinant to check whether each set of points is collinear. Graph them to verify your answer.

(i) $(-6, 4)$, $(2, 10)$, $(6, 13)$

(ii) $(-5, 10)$, $(2, 6)$, $(15, -2)$

58. Determinant Form for the Equation of a Line

(a) Use the result of Exercise 57(a) to show that the equation of the line containing the points (x_1, y_1) and (x_2, y_2) is

$$\begin{vmatrix} x & y & 1 \\ x_1 & y_1 & 1 \\ x_2 & y_2 & 1 \end{vmatrix} = 0$$

(b) Use the result of part (a) to find an equation for the line containing the points $(20, 50)$ and $(-10, 25)$.

59. Matrices with Determinant Zero   Use the definition of the determinant and the elementary row and column operations to explain why matrices of the following types have determinant 0.

(a) A matrix with a row or column consisting entirely of zeros

(b) A matrix with two rows the same or two columns the same

(c) A matrix in which one row is a multiple of another row, or one column is a multiple of another column

60. Solving Linear Systems   Suppose you had to solve a linear system with five equations and five variables without the assistance of a calculator or computer. Which method would you prefer: Cramer's Rule or Gaussian elimination? Write a short paragraph explaining the reasons for your answer.

8.8   SYSTEMS OF INEQUALITIES

In this section we study systems of inequalities in two variables from a graphical point of view. First we consider the graph of a single inequality. We already know that the graph of $y = x^2$, for example, is the *parabola* in Figure 1. If we replace the equal sign by the symbol $\geqslant$, we obtain the *inequality*

$$y \geqslant x^2$$

Its graph consists of not just the parabola in Figure 1, but also every point whose y-coordinate is *larger* than x^2. We indicate the solution in Figure 2(a) by shading the points *above* the parabola.

Similarly, the graph of $y \leqslant x^2$ in Figure 2(b) consists of all points on and *below* the parabola, whereas the graphs of $y > x^2$ and $y < x^2$ don't include the points on the parabola itself, as indicated by the dashed curves in Figures 2(c) and 2(d).

FIGURE 1

(a) $y \geqslant x^2$

(b) $y \leqslant x^2$

(c) $y > x^2$

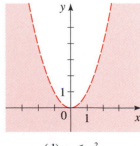

(d) $y < x^2$

FIGURE 2

The graph of an inequality, in general, consists of a region in the plane whose boundary is the graph of the equation obtained by replacing the inequality sign

($\geqslant$, $\leqslant$, $>$, or $<$) with an equal sign. To determine which side of the graph gives the solution set of the inequality, we need only check **test points**.

GRAPHING INEQUALITIES

To graph an inequality, we carry out the following steps.

1. GRAPH EQUATION. Graph the equation corresponding to the inequality. Use a dashed curve for $>$ or $<$, or a solid curve for $\leqslant$ or $\geqslant$.

2. TEST POINTS. Test one point in each of the regions formed by the graph in Step 1. If the point satisfies the inequality, then all the points in that region satisfy the inequality. (In that case, shade the region to indicate it is part of the graph.) If the test point does not satisfy the inequality, then the region isn't part of the graph.

EXAMPLE 1 ■ Graphs of Nonlinear Inequalities

Graph each inequality.

(a) $x^2 + y^2 < 25$ (b) $x + 2y \geqslant 5$

SOLUTION

(a) The graph of $x^2 + y^2 = 25$ is a circle of radius 5 centered at the origin. The points on the circle itself do not satisfy the inequality because it is of the form $<$, so we graph the circle with a dashed curve, as shown in Figure 3.

To determine whether the inside or the outside of the circle satisfies the inequality, we use the test points $(0, 0)$ on the inside and $(6, 0)$ on the outside. To do this, we substitute the coordinates of each point into the inequality and determine whether the result satisfies the inequality. (Note that *any* point inside or outside the circle can serve as a test point. We have chosen these points for simplicity.)

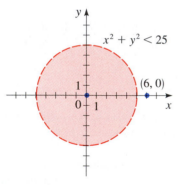

FIGURE 3

Test point	$x^2 + y^2 < 25$	Conclusion
$(0, 0)$	$0^2 + 0^2 = 0 < 25$	Part of graph
$(6, 0)$	$6^2 + 0^2 = 36 \not< 25$	Not part of graph

Thus, the graph of $x^2 + y^2 < 25$ is the set of all points *inside* the circle (see Figure 3).

(b) The graph of $x + 2y = 5$ is the line shown in Figure 4. We use the test points $(0, 0)$ and $(5, 5)$ on opposite sides of the line.

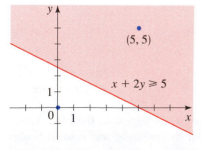

FIGURE 4

Test point	$x + 2y \geqslant 5$	Conclusion
$(0, 0)$	$0 + 2(0) = 0 \not\geqslant 5$	Not part of graph
$(5, 5)$	$5 + 2(5) = 15 \geqslant 5$	Part of graph

Our check shows that the points *above* the line satisfy the inequality.

Alternatively, we could put the inequality into slope-intercept form and graph it directly:

$$x + 2y \geqslant 5$$

$$2y \geqslant -x + 5$$

$$y \geqslant -\tfrac{1}{2}x + \tfrac{5}{2}$$

From this form we see that the graph includes all points whose y-coordinates are *greater* than those on the line $y = -\tfrac{1}{2}x + \tfrac{5}{2}$; that is, the graph consists of the points *on or above* this line, as shown in Figure 4. ■

EXAMPLE 2 ■ A System of Two Inequalities

Graph the solution set of the system of inequalities.

$$\begin{cases} x^2 + y^2 < 25 \\ x + 2y \geqslant 5 \end{cases}$$

SOLUTION

These are the two inequalities of Example 1. In this example we wish to graph only those points that simultaneously satisfy both inequalities. The solution consists of the intersection of the graphs in Example 1. In Figure 5(a) we show the two regions on the same coordinate plane (in different colors), and in Figure 5(b) we show their intersection.

Vertices: The points $(-3, 4)$ and $(5, 0)$ in Figure 5(b) are the **vertices** of the solution set. They are obtained by solving the system of *equations*

$$\begin{cases} x^2 + y^2 = 25 \\ x + 2y = 5 \end{cases}$$

We solve this system of equations by substitution. Solving for x in the second equation gives $x = 5 - 2y$, and substituting this into the first equation gives

$$(5 - 2y)^2 + y^2 = 25 \qquad \text{Substitute } x = 5 - 2y$$

$$(25 - 20y + 4y^2) + y^2 = 25 \qquad \text{Expand}$$

$$-20y + 5y^2 = 0 \qquad \text{Simplify}$$

$$-5y(4 - y) = 0 \qquad \text{Factor}$$

Thus, $y = 0$ or $y = 4$. When $y = 0$, we have $x = 5 - 2(0) = 5$, and when $y = 4$, we have $x = 5 - 2(4) = -3$. So the points of intersection of these curves are $(5, 0)$ and $(-3, 4)$.

Note that in this case the vertices are not part of the solution set, since they don't satisfy the inequality $x^2 + y^2 < 25$ (and so they are graphed as open circles in the figure). They simply show where the "corners" of the solution set lie. ■

(a)

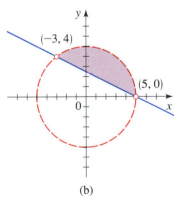

(b)

FIGURE 5

$$\begin{cases} x^2 + y^2 < 25 \\ x + 2y \geqslant 5 \end{cases}$$

An inequality is **linear** if it can be put into one of the following forms:

$$ax + by \geq c \qquad ax + by \leq c \qquad ax + by > c \qquad ax + by < c$$

In the next example we graph the solution set of a system of linear inequalities.

EXAMPLE 3 ■ A System of Four Linear Inequalities

Graph the solution set of the system, and label its vertices.

$$\begin{cases} x + 3y \leq 12 \\ x + y \leq 8 \\ x \geq 0 \\ y \geq 0 \end{cases}$$

SOLUTION

In Figure 6 we first graph the lines given by the equations that correspond to each inequality. To determine the graphs of the linear inequalities, we only need to check one test point. For simplicity let's use the point $(0, 0)$.

Inequality	Test point $(0, 0)$	Conclusion
$x + 3y \leq 12$	$0 + 3(0) = 0 \leq 12$	Satisfies inequality
$x + y \leq 8$	$0 + 0 = 0 \leq 8$	Satisfies inequality

Since $(0, 0)$ is below the line $x + 3y = 12$, our check shows that the region on or below the line must satisfy the inequality. Likewise, since $(0, 0)$ is below the line $x + y = 8$, our check shows that the region on or below this line must satisfy the inequality. The inequalities $x \geq 0$ and $y \geq 0$ say that x and y are nonnegative. These regions are sketched in Figure 6(a), and the intersection—the solution set—is sketched in Figure 6(b).

Vertices: The coordinates of each vertex are obtained by simultaneously solving the equations of the lines that intersect at that vertex. From the system

$$\begin{cases} x + 3y = 12 \\ x + y = 8 \end{cases}$$

we get the vertex $(6, 2)$. The other two vertices are at the x- and y-intercepts of the corresponding lines: $(8, 0)$ and $(0, 4)$. In this case, all the vertices *are* part of the solution set. ■

(a)

(b)

FIGURE 6

EXAMPLE 4 ■ A System of Linear Inequalities

Graph the solution set of the system.

$$\begin{cases} x + 2y \geq 8 \\ -x + 2y \leq 4 \\ 3x - 2y \leq 8 \end{cases}$$

FIGURE 7

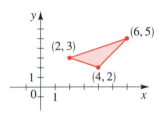

FIGURE 8

SOLUTION

We must graph the lines that correspond to these inequalities and then shade the appropriate regions, as in Example 3. We will use a graphing calculator, so we must first isolate y on the left-hand side of each inequality.

$$\begin{cases} y \geq -\frac{1}{2}x + 4 \\ y \leq \frac{1}{2}x + 2 \\ y \geq \frac{3}{2}x - 4 \end{cases}$$

Using the shading feature of the calculator, we obtain the graph in Figure 7. The solution set is the triangular region that is shaded in all three patterns. We then use $\boxed{\text{TRACE}}$ or the `Intersect` command to find the vertices of the region. The solution set is graphed in Figure 8. ∎

When a region in the plane can be covered by a (sufficiently large) circle, it's said to be **bounded**. A region that is not bounded is called **unbounded**. For example, the regions graphed in Figure 3, 5(b), 6(b), and 8 are bounded, whereas those in Figures 2 and 4 are unbounded. An unbounded region cannot be "fenced in"—it extends infinitely far in at least one direction.

8.8 EXERCISES

1–12 ■ Graph the inequality.

1. $x \leq 2$

2. $y > -3$

3. $y > x$

4. $y < x + 2$

5. $y \leq 2x + 2$

6. $y < -x + 5$

7. $2x - y \leq 8$

8. $3x + 4y + 12 > 0$

9. $4x + 5y < 25$

10. $-x^2 + y \geq 10$

11. $y > x^2 + 1$

12. $x^2 + y^2 \geq 5$

13–34 ■ Graph the solution of the system of inequalities. Find the coordinates of all vertices, and determine whether the solution set is bounded.

13. $\begin{cases} x + y \leq 4 \\ y \geq x \end{cases}$

14. $\begin{cases} 2x + 3y > 12 \\ 3x - y < 21 \end{cases}$

15. $\begin{cases} y < \frac{1}{4}x + 2 \\ y \geq 2x - 5 \end{cases}$

16. $\begin{cases} x - y > 0 \\ 4 + y \leq 2x \end{cases}$

17. $\begin{cases} x \geq 0 \\ y \geq 0 \\ 3x + 5y \leq 15 \\ 3x + 2y \leq 9 \end{cases}$

18. $\begin{cases} x > 2 \\ y < 12 \\ 2x - 4y > 8 \end{cases}$

19. $\begin{cases} y < 9 - x^2 \\ y \geq x + 3 \end{cases}$

20. $\begin{cases} y \geq x^2 \\ x + y \geq 6 \end{cases}$

21. $\begin{cases} x^2 + y^2 \leq 4 \\ x - y > 0 \end{cases}$

22. $\begin{cases} x > 0 \\ y > 0 \\ x + y < 10 \\ x^2 + y^2 > 9 \end{cases}$

23. $\begin{cases} x^2 - y \leq 0 \\ 2x^2 + y \leq 12 \end{cases}$

24. $\begin{cases} x^2 + y^2 < 9 \\ 2x + y^2 \geq 1 \end{cases}$

25. $\begin{cases} x + 2y \leq 14 \\ 3x - y \geq 0 \\ x - y \geq 2 \end{cases}$

26. $\begin{cases} y < x + 6 \\ 3x + 2y \geq 12 \\ x - 2y \leq 2 \end{cases}$

27. $\begin{cases} x \geq 0 \\ y \geq 0 \\ x \leq 5 \\ x + y \leq 7 \end{cases}$

28. $\begin{cases} x \geq 0 \\ y \geq 0 \\ y \leq 4 \\ 2x + y \leq \end{cases}$

29. $\begin{cases} y > x + 1 \\ x + 2y \leq 12 \\ x + 1 > 0 \end{cases}$

30. $\begin{cases} x + \\ 3y \end{cases}$

31. $\begin{cases} x^2 + y^2 \le 8 \\ \quad x \ge 2 \\ \quad y \ge 0 \end{cases}$ **32.** $\begin{cases} x^2 - y \ge 0 \\ x + y < 6 \\ x - y < 6 \end{cases}$

33. $\begin{cases} x^2 + y^2 < 9 \\ \quad x + y > 0 \\ \quad x \le 0 \end{cases}$ **34.** $\begin{cases} \quad y \ge x^3 \\ \quad y \le 2x + 4 \\ x + y \ge 0 \end{cases}$

 35–38 ■ Use a graphing calculator to graph the solution of the system of inequalities. Find the coordinates of all vertices, correct to one decimal place.

35. $\begin{cases} y \ge x - 3 \\ y \ge -2x + 6 \\ y \le 8 \end{cases}$ **36.** $\begin{cases} \quad x + y \ge 12 \\ 2x + y \le 24 \\ \quad x - y \ge -6 \end{cases}$

37. $\begin{cases} y \le 6x - x^2 \\ x + y \ge 4 \end{cases}$ **38.** $\begin{cases} \quad y \ge x^3 \\ 2x + y \ge 0 \\ \quad y \le 2x + 6 \end{cases}$

39. A publishing company publishes a total of no more than 100 books every year. At least 20 of these are nonfiction, but the company always publishes at least as much fiction as nonfiction. Find a system of inequalities that describes the possible numbers of fiction and nonfiction books that the company can produce each year consistent with these policies. Graph the solution set.

40. A man and his daughter manufacture unfinished tables and chairs. Each table requires 3 hours of sawing and 1 hour of assembly. Each chair requires 2 hours of sawing and 2 hours of assembly. The two of them can put in up to 12 hours of sawing and 8 hours of assembly work each day. Find a system of inequalities that describes all possible combinations of tables and chairs that they can make daily. Graph the solution set.

 DISCOVERY · DISCUSSION

41. Shading Unwanted Regions   To graph the solution of a system of inequalities, we have shaded the solution of each inequality in a different color; the solution of the system is the region where all the shaded parts overlap. Here is a different method: for each inequality, shade the region that does *not* satisfy the inequality. Explain why the part of the plane that is left unshaded is the solution of the system. Solve the following system by both methods. Which do you prefer?

$$\begin{cases} \quad x + 2y > 4 \\ -x + y < 1 \\ \quad x + 3y < 9 \\ \quad x < 3 \end{cases}$$

8.9   PARTIAL FRACTIONS

To write a sum or difference of fractional expressions as a single fraction, we bring them to a common denominator. For example,

$$\frac{1}{x - 1} + \frac{1}{2x + 1} = \frac{(2x + 1) + (x - 1)}{(x - 1)(2x + 1)} = \frac{3x}{2x^2 - x - 1}$$

But for some applications of algebra to calculus, we must reverse this process—that is, we must express a fraction such as $3x/(2x^2 - x - 1)$ as the sum of the simpler fractions $1/(x - 1)$ and $1/(2x + 1)$. These simpler fractions are called *partial fractions*; we learn how to find them in this section.

Let r be the rational function

$$r(x) = \frac{P(x)}{Q(x)}$$

where the degree of P is less than the degree of Q. By the Linear and Quadratic Factors Theorem in Section 3.5, every polynomial with real coefficients can be factored completely into linear and irreducible quadratic factors, that is, factors of the form $ax + b$ and $ax^2 + bx + c$, where a, b, and c are real numbers. For instance,

$$x^4 - 1 = (x^2 - 1)(x^2 + 1) = (x - 1)(x + 1)(x^2 + 1)$$

After we have completely factored the denominator Q of r, we can express $r(x)$ as a sum of **partial fractions** of the form

$$\frac{A}{(ax + b)^i} \quad \text{and} \quad \frac{Ax + B}{(ax^2 + bx + c)^j}$$

This sum is called the **partial fraction decomposition** of r. Let's examine the details of the four possible cases.

CASE 1: THE DENOMINATOR IS A PRODUCT OF DISTINCT LINEAR FACTORS

Suppose that we can factor $Q(x)$ as

$$Q(x) = (a_1 x + b_1)(a_2 x + b_2) \cdots (a_n x + b_n)$$

with no factor repeated. In this case, the partial fraction decomposition of $P(x)/Q(x)$ takes the form

$$\frac{P(x)}{Q(x)} = \frac{A_1}{a_1 x + b_1} + \frac{A_2}{a_2 x + b_2} + \cdots + \frac{A_n}{a_n x + b_n}$$

The constants $A_1, A_2, \ldots, A_n$ are determined as in the following example.

EXAMPLE 1 ■ Distinct Linear Factors

Find the partial fraction decomposition of $\dfrac{5x + 7}{x^3 + 2x^2 - x - 2}$.

SOLUTION

The denominator factors as follows:

$$x^3 + 2x^2 - x - 2 = x^2(x + 2) - (x + 2) = (x^2 - 1)(x + 2)$$
$$= (x - 1)(x + 1)(x + 2)$$

This gives us the partial fraction decomposition

$$\frac{5x + 7}{x^3 + 2x^2 - x - 2} = \frac{A}{x - 1} + \frac{B}{x + 1} + \frac{C}{x + 2}$$

Multiplying each side by the common denominator, $(x - 1)(x + 1)(x + 2)$, we get

$$5x + 7 = A(x + 1)(x + 2) + B(x - 1)(x + 2) + C(x - 1)(x + 1)$$
$$= A(x^2 + 3x + 2) + B(x^2 + x - 2) + C(x^2 - 1) \qquad \text{Expand}$$
$$= (A + B + C)x^2 + (3A + B)x + (2A - 2B - C) \qquad \text{Combine like terms}$$

If two polynomials are equal, then their coefficients are equal. Thus, since $5x + 7$ has no x^2-term, we have $A + B + C = 0$. Similarly, by comparing the coefficients of x, we see that $3A + B = 5$, and by comparing constant terms, we get $2A - 2B - C = 7$. This leads to the following system of linear equations for A, B, and C.

$$\begin{cases} A + B + C = 0 \\ 3A + B = 5 \\ 2A - 2B - C = 7 \end{cases}$$

We use the methods developed in Section 8.4 to solve this system.

$$\begin{bmatrix} 1 & 1 & 1 & 0 \\ 3 & 1 & 0 & 5 \\ 2 & -2 & -1 & 7 \end{bmatrix} \xrightarrow[\substack{R_3 - 2R_1 \to R_3}]{R_2 - 3R_1 \to R_2} \begin{bmatrix} 1 & 1 & 1 & 0 \\ 0 & -2 & -3 & 5 \\ 0 & -4 & -3 & 7 \end{bmatrix} \xrightarrow[-R_2]{R_3 - 2R_2 \to R_3}$$

$$\begin{bmatrix} 1 & 1 & 1 & 0 \\ 0 & 2 & 3 & -5 \\ 0 & 0 & 3 & -3 \end{bmatrix} \xrightarrow[\frac{1}{3}R_3]{R_2 - R_3 \to R_2} \begin{bmatrix} 1 & 1 & 1 & 0 \\ 0 & 2 & 0 & -2 \\ 0 & 0 & 1 & -1 \end{bmatrix}$$

Thus, $C = -1$, $B = -1$, and $A = 2$, so the required partial fraction decomposition is

$$\frac{5x + 7}{x^3 + 2x^2 - x - 2} = \frac{2}{x - 1} + \frac{-1}{x + 1} + \frac{-1}{x + 2}$$

◼

The same method of attack works in the remaining cases. We set up the partial fraction decomposition with the unknown constants, A, B, C, ... Then we multiply each side of the resulting equation by the common denominator, simplify the right-hand side of the equation, and equate coefficients. This gives a set of linear equations that will always have a unique solution (provided that the partial fraction decomposition has been set up correctly).

CASE 2: THE DENOMINATOR IS A PRODUCT OF LINEAR FACTORS, SOME OF WHICH ARE REPEATED

Suppose the complete factorization of $Q(x)$ contains the linear factor $ax + b$ repeated k times; that is, $(ax + b)^k$ is a factor of $Q(x)$. Then, corresponding to each such factor, the partial fraction decomposition for $P(x)/Q(x)$ contains

$$\frac{A_1}{ax + b} + \frac{A_2}{(ax + b)^2} + \cdots + \frac{A_k}{(ax + b)^k}$$

EXAMPLE 2 ◼ Repeated Linear Factors

Find the partial fraction decomposition of $\dfrac{x^2 + 1}{x(x - 1)^3}$.

SOLUTION

Because the factor $x - 1$ is repeated three times in the denominator, the partial fraction decomposition has the form

$$\frac{x^2 + 1}{x(x - 1)^3} = \frac{A}{x} + \frac{B}{x - 1} + \frac{C}{(x - 1)^2} + \frac{D}{(x - 1)^3}$$

Multiplying each side by the common denominator, $x(x - 1)^3$, gives

$$x^2 + 1 = A(x - 1)^3 + Bx(x - 1)^2 + Cx(x - 1) + Dx$$
$$= A(x^3 - 3x^2 + 3x - 1) + B(x^3 - 2x^2 + x) + C(x^2 - x) + Dx \qquad \text{Expand}$$
$$= (A + B)x^3 + (-3A - 2B + C)x^2 + (3A + B - C + D)x - A \qquad \text{Combine like terms}$$

Equating coefficients, we get the equations

$$\begin{cases} A + B & = 0 \\ -3A - 2B + C & = 1 \\ 3A + B - C + D & = 0 \\ -A & = 1 \end{cases}$$

If we rearrange these equations by putting the last one in the first position, we can easily see (using substitution) that the solution to the system is $A = -1$, $B = 1$, $C = 0$, $D = 2$, and so the partial fraction decomposition is

$$\frac{x^2 + 1}{x(x - 1)^3} = \frac{-1}{x} + \frac{1}{x - 1} + \frac{2}{(x - 1)^3} \qquad \blacksquare$$

CASE 3: THE DENOMINATOR HAS IRREDUCIBLE QUADRATIC FACTORS, NONE OF WHICH IS REPEATED

Suppose the complete factorization of $Q(x)$ contains the quadratic factor $ax^2 + bx + c$ (which can't be factored further). Then, corresponding to this, the partial fraction decomposition of $P(x)/Q(x)$ will have a term of the form

$$\frac{Ax + B}{ax^2 + bx + c}$$

EXAMPLE 3 ■ Distinct Quadratic Factors

Find the partial fraction decomposition of $\dfrac{2x^2 - x + 4}{x^3 + 4x}$.

SOLUTION

Since $x^3 + 4x = x(x^2 + 4)$, which can't be factored further, we write

$$\frac{2x^2 - x + 4}{x^3 + 4x} = \frac{A}{x} + \frac{Bx + C}{x^2 + 4}$$

Multiplying by $x(x^2 + 4)$, we get

$$2x^2 - x + 4 = A(x^2 + 4) + (Bx + C)x$$
$$= (A + B)x^2 + Cx + 4A$$

Equating coefficients gives us the equations

$$\begin{cases} A + B = 2 \\ \quad\quad C = -1 \\ 4A = 4 \end{cases}$$

and so $A = 1$, $B = 1$, and $C = -1$. The required partial fraction decomposition is

$$\frac{2x^2 - x + 4}{x^3 + 4x} = \frac{1}{x} + \frac{x - 1}{x^2 + 4}$$

■

CASE 4: THE DENOMINATOR HAS A REPEATED IRREDUCIBLE QUADRATIC FACTOR

Suppose the complete factorization of $Q(x)$ contains the factor $(ax^2 + bx + c)^k$, where $ax^2 + bx + c$ can't be factored further. Then the partial fraction decomposition of $P(x)/Q(x)$ will have the terms

$$\frac{A_1x + B_1}{ax^2 + bx + c} + \frac{A_2x + B_2}{(ax^2 + bx + c)^2} + \cdots + \frac{A_kx + B_k}{(ax^2 + bx + c)^k}$$

EXAMPLE 4 ■ Repeated Quadratic Factors

Write the form of the partial fraction decomposition of

$$\frac{x^5 - 3x^2 + 12x - 1}{x^3(x^2 + x + 1)(x^2 + 2)^3}$$

SOLUTION

$$\frac{x^5 - 3x^2 + 12x - 1}{x^3(x^2 + x + 1)(x^2 + 2)^3}$$

$$= \frac{A}{x} + \frac{B}{x^2} + \frac{C}{x^3} + \frac{Dx + E}{x^2 + x + 1} + \frac{Fx + G}{x^2 + 2} + \frac{Hx + I}{(x^2 + 2)^2} + \frac{Jx + K}{(x^2 + 2)^3}$$

■

To find the values of A, B, C, D, E, F, G, H, I, J, and K in Example 4, we would have to solve a system of 11 linear equations. Although possible, this would certainly involve a great deal of work!

The techniques we have described in this section apply only to rational functions $P(x)/Q(x)$ in which the degree of P is less than the degree of Q. If this isn't the case, we must first use long division to divide Q into P.

EXAMPLE 5 ■ **Using Long Division to Prepare for Partial Fractions**

Find the partial fraction decomposition of

$$\frac{2x^4 + 4x^3 - 2x^2 + x + 7}{x^3 + 2x^2 - x - 2}$$

SOLUTION

Since the degree of the numerator is larger than the degree of the denominator, we use long division to obtain

$$
\begin{array}{r}
2x \\
x^3 + 2x^2 - x - 2 \overline{)\, 2x^4 + 4x^3 - 2x^2 + x + 7} \\
\underline{2x^4 + 4x^3 - 2x^2 - 4x} \\
5x + 7
\end{array}
$$

$$\frac{2x^4 + 4x^3 - 2x^2 + x + 7}{x^3 + 2x^2 - x - 2} = 2x + \frac{5x + 7}{x^3 + 2x^2 - x - 2}$$

The remainder term now satisfies the requirement that the degree of the numerator is less than the degree of the denominator. At this point we proceed as in Example 1 to obtain the decomposition

$$\frac{2x^4 + 4x^3 - 2x^2 + x + 7}{x^3 + 2x^2 - x - 2} = 2x + \frac{2}{x - 1} + \frac{-1}{x + 1} + \frac{-1}{x + 2}$$ ■

8.9 EXERCISES

1–10 ■ Write the form of the partial fraction decomposition of the function (as in Example 4). Do not determine the numerical values of the coefficients.

1. $\dfrac{1}{(x - 1)(x + 2)}$

2. $\dfrac{x}{x^2 + 3x - 4}$

3. $\dfrac{x^2 - 3x + 5}{(x - 2)^2(x + 4)}$

4. $\dfrac{1}{x^4 - x^3}$

5. $\dfrac{x^2}{(x - 3)(x^2 + 4)}$

6. $\dfrac{1}{x^4 - 1}$

7. $\dfrac{x^3 - 4x^2 + 2}{(x^2 + 1)(x^2 + 2)}$

8. $\dfrac{x^4 + x^2 + 1}{x^2(x^2 + 4)^2}$

9. $\dfrac{x^3 + x + 1}{x(2x - 5)^3(x^2 + 2x + 5)^2}$

10. $\dfrac{1}{(x^6 - 1)(x^4 - 1)}$

11–42 ■ Find the partial fraction decomposition of the rational function.

11. $\dfrac{2}{(x - 1)(x + 1)}$

12. $\dfrac{2x}{(x - 1)(x + 1)}$

13. $\dfrac{5}{(x - 1)(x + 4)}$

14. $\dfrac{x + 6}{x(x + 3)}$

15. $\dfrac{12}{x^2 - 9}$

16. $\dfrac{x - 12}{x^2 - 4x}$

17. $\dfrac{4}{x^2 - 4}$

18. $\dfrac{2x + 1}{x^2 + x - 2}$

19. $\dfrac{x + 14}{x^2 - 2x - 8}$

20. $\dfrac{8x - 3}{2x^2 - x}$

21. $\dfrac{x}{8x^2 - 10x + 3}$

22. $\dfrac{7x - 3}{x^3 + 2x^2 - 3x}$

23. $\dfrac{9x^2 - 9x + 6}{2x^3 - x^2 - 8x + 4}$

24. $\dfrac{-3x^2 - 3x + 27}{(x + 2)(2x^2 + 3x - 9)}$

25. $\dfrac{x^2 + 1}{x^3 + x^2}$

26. $\dfrac{3x^2 + 5x - 13}{(3x + 2)(x^2 - 4x + 4)}$

27. $\dfrac{2x}{4x^2 + 12x + 9}$

28. $\dfrac{x - 4}{(2x - 5)^2}$

29. $\dfrac{4x^2 - x - 2}{x^4 + 2x^3}$

30. $\dfrac{x^3 - 2x^2 - 4x + 3}{x^4}$

31. $\dfrac{-10x^2 + 27x - 14}{(x - 1)^3(x + 2)}$

32. $\dfrac{-2x^2 + 5x - 1}{x^4 - 2x^3 + 2x - 1}$

33. $\dfrac{3x^3 + 22x^2 + 53x + 41}{(x + 2)^2(x + 3)^2}$

34. $\dfrac{3x^2 + 12x - 20}{x^4 - 8x^2 + 16}$

35. $\dfrac{x - 3}{x^3 + 3x}$

36. $\dfrac{3x^2 - 2x + 8}{x^3 - x^2 + 2x - 2}$

37. $\dfrac{2x^3 + 7x + 5}{(x^2 + x + 2)(x^2 + 1)}$

38. $\dfrac{x^2 + x + 1}{2x^4 + 3x^2 + 1}$

39. $\dfrac{x^4 + x^3 + x^2 - x + 1}{x(x^2 + 1)^2}$

40. $\dfrac{2x^2 - x + 8}{(x^2 + 4)^2}$

41. $\dfrac{x^5 - 2x^4 + x^3 + x + 5}{x^3 - 2x^2 + x - 2}$

42. $\dfrac{x^5 - 3x^4 + 3x^3 - 4x^2 + 4x + 12}{(x - 2)^2(x^2 + 2)}$

43. Determine A and B in terms of a and b:

$$\frac{ax + b}{x^2 - 1} = \frac{A}{x - 1} + \frac{B}{x + 1}$$

44. Determine A, B, C, and D in terms of a and b:

$$\frac{ax^3 + bx^2}{(x^2 + 1)^2} = \frac{Ax + B}{x^2 + 1} + \frac{Cx + D}{(x^2 + 1)^2}$$

◼ DISCOVERY · DISCUSSION

45. Recognizing Partial Fraction Decompositions   For each expression, determine whether it is already a partial fraction decomposition, or whether it can be decomposed further.

(a) $\dfrac{x}{x^2 + 1} + \dfrac{1}{x + 1}$

(b) $\dfrac{x}{(x + 1)^2}$

(c) $\dfrac{1}{x + 1} + \dfrac{2}{(x + 1)^2}$

(d) $\dfrac{x + 2}{(x^2 + 1)^2}$

46. Assembling and Disassembling Partial Fractions   The following expression is a partial fraction decomposition:

$$\frac{2}{x - 1} + \frac{1}{(x - 1)^2} + \frac{1}{x + 1}$$

Use a common denominator to combine the terms into one fraction. Then use the techniques of this section to find its partial fraction decomposition. Did you get back the original expression?

8   REVIEW

CONCEPT CHECK

1. Suppose you are asked to solve a system of two equations (not necessarily linear) in two variables. Explain how you would solve them
   (a) by the substitution method
   (b) by the elimination method
   (c) graphically

2. Suppose you are asked to solve a system of two *linear* equations in two variables.
   (a) Would you prefer to use substitution or elimination?
   (b) How many solutions are possible? Draw diagrams to illustrate the possibilities.

3. Explain how Gaussian elimination works. Your explanation should include a discussion of matrices, elementary row operations, row-echelon form, back-substitution, and leading variables.

4. (a) What is meant by an inconsistent system?
   (b) What is meant by a dependent system? What is a leading variable?

5. Suppose you have used Gaussian elimination to transform a matrix that represents a system of linear equations into row-echelon form. How can you tell if the system has
   (a) no solution?
   (b) exactly one solution?
   (c) infinitely many solutions?

6. How can you tell if a matrix is in reduced row-echelon form? What is the advantage of using a matrix in reduced row-echelon form?

7. (a) What does it mean to say that A is a matrix with dimension $m \times n$?
   (b) If A and B have the same dimension and k is a real number, how do you find $A + B$, $A - B$, and kA?

8. (a) Under what conditions is it possible to find the product AB of two matrices?
   (b) If those conditions are satisfied, how do you calculate the product AB?

9. (a) What is the identity matrix I_n?
   (b) If A is a square $n \times n$ matrix, what is its inverse matrix?
   (c) Write a formula for the inverse of a 2×2 matrix.
   (d) Explain how you would find the inverse of a 3×3 matrix.

10. Suppose A is an $n \times n$ matrix.
    (a) What is the minor M_{ij} of the element a_{ij}?
    (b) What is the cofactor A_{ij}?

(c) How do you find the determinant of A?
(d) How can you tell if A has an inverse?

11. State Cramer's Rule for solving a system of linear equations in terms of determinants. Do you prefer to use Cramer's Rule or Gaussian elimination? Explain.

12. Explain the procedure for solving a system of inequalities.

13. Explain how to find the partial fraction decomposition of a fractional expression. Your explanation should discuss each of the four cases that arise.

EXERCISES

1–6 ■ Solve the system of equations and graph the lines.

1. $\begin{cases} 3x - y = 5 \\ 2x + y = 5 \end{cases}$

2. $\begin{cases} y = 2x + 6 \\ y = -x + 3 \end{cases}$

3. $\begin{cases} 2x - 7y = 28 \\ y = \frac{2}{7}x - 4 \end{cases}$

4. $\begin{cases} 6x - 8y = 15 \\ -\frac{3}{2}x + 2y = -4 \end{cases}$

5. $\begin{cases} 2x - y = 1 \\ x + 3y = 10 \\ 3x + 4y = 15 \end{cases}$

6. $\begin{cases} 2x + 5y = 9 \\ -x + 3y = 1 \\ 7x - 2y = 14 \end{cases}$

7–10 ■ Solve the system of equations.

7. $\begin{cases} y = x^2 + 2x \\ y = 6 + x \end{cases}$

8. $\begin{cases} x^2 + y^2 = 8 \\ y = x + 2 \end{cases}$

9. $\begin{cases} 3x + \dfrac{4}{y} = 6 \\ x - \dfrac{8}{y} = 4 \end{cases}$

10. $\begin{cases} x^2 + y^2 = 10 \\ x^2 + 2y^2 - 7y = 0 \end{cases}$

11–14 ■ Use a graphing device to solve the system, correct to the nearest hundredth.

11. $\begin{cases} 0.32x + 0.43y = 0 \\ 7x - 12y = 341 \end{cases}$

12. $\begin{cases} \sqrt{12}\,x - 3\sqrt{2}\,y = 660 \\ 7137x + 3931y = 20{,}000 \end{cases}$

13. $\begin{cases} x - y^2 = 10 \\ x = \frac{1}{22}y + 12 \end{cases}$

14. $\begin{cases} y = 5^x + x \\ y = x^5 + 5 \end{cases}$

15–22 ■ Find the complete solution of the system, or show that the system has no solution.

15. $\begin{cases} x + y + 2z = 6 \\ 2x + 5z = 12 \\ x + 2y + 3z = 9 \end{cases}$

16. $\begin{cases} x - 2y + 3z = 1 \\ x - 3y - z = 0 \\ 2x - 6z = 6 \end{cases}$

17. $\begin{cases} x - 2y + 3z = 1 \\ 2x - y + z = 3 \\ 2x - 7y + 11z = 2 \end{cases}$

18. $\begin{cases} x + y + z + w = 2 \\ 2x - 3z = 5 \\ x - 2y + 4w = 9 \\ x + y + 2z + 3w = 5 \end{cases}$

19. $\begin{cases} x - 3y + z = 4 \\ 4x - y + 15z = 5 \end{cases}$

20. $\begin{cases} 2x - 3y + 4z = 3 \\ 4x - 5y + 9z = 13 \\ 2x + 7z = 0 \end{cases}$

21. $\begin{cases} -x + 4y + z = 8 \\ 2x - 6y + z = -9 \\ x - 6y - 4z = -15 \end{cases}$

22. $\begin{cases} x - z + w = 2 \\ 2x + y - 2w = 12 \\ 3y + z + w = 4 \\ x + y - z = 10 \end{cases}$

23. A man invests his savings in two accounts, one paying 6% interest per year and the other paying 7%. He has twice as much invested in the 7% account as in the 6% account, and his annual interest income is $600. How much is invested in each account?

24. Find the values of a, b, and c if the parabola

$$y = ax^2 + bx + c$$

passes through the points $(-1, -4)$, $(1, 0)$, and $(2, 11)$.

25–36 ◾ Let

$$A = [2 \quad 0 \quad -1] \qquad B = \begin{bmatrix} 1 & 2 & 4 \\ -2 & 1 & 0 \end{bmatrix}$$

$$C = \begin{bmatrix} \frac{1}{2} & 3 \\ 2 & \frac{3}{2} \\ -2 & 1 \end{bmatrix} \qquad D = \begin{bmatrix} 1 & 4 \\ 0 & -1 \\ 2 & 0 \end{bmatrix}$$

$$E = \begin{bmatrix} 2 & -1 \\ -\frac{1}{2} & 1 \end{bmatrix} \qquad F = \begin{bmatrix} 4 & 0 & 2 \\ -1 & 1 & 0 \\ 7 & 5 & 0 \end{bmatrix}$$

$$G = [5]$$

Carry out the indicated operation, or explain why it cannot be performed.

25. $A + B$ **26.** $C - D$ **27.** $2C + 3D$

28. $5B - 2C$ **29.** GA **30.** AG

31. BC **32.** CB **33.** BF

34. FC **35.** $(C + D)E$ **36.** $F(2C - D)$

37–42 ◾ Find the determinant and, if possible, the inverse of the matrix.

37. $\begin{bmatrix} 1 & 4 \\ 2 & 9 \end{bmatrix}$ **38.** $\begin{bmatrix} 2 & 2 \\ 1 & -3 \end{bmatrix}$

39. $\begin{bmatrix} 4 & -12 \\ -2 & 6 \end{bmatrix}$ **40.** $\begin{bmatrix} 2 & 4 & 0 \\ -1 & 1 & 2 \\ 0 & 3 & 2 \end{bmatrix}$

41. $\begin{bmatrix} 3 & 0 & 1 \\ 2 & -3 & 0 \\ 4 & -2 & 1 \end{bmatrix}$ **42.** $\begin{bmatrix} 1 & 0 & 0 & 1 \\ 0 & 2 & 0 & 2 \\ 0 & 0 & 3 & 3 \\ 0 & 0 & 0 & 4 \end{bmatrix}$

43–44 ◾ Express the system of linear equations as a matrix equation. Then solve the matrix equation by multiplying each side by the inverse of the coefficient matrix.

43. $\begin{cases} 12x - 5y = 10 \\ 5x - 2y = 17 \end{cases}$ **44.** $\begin{cases} 2x + y + 5z = \frac{1}{3} \\ x + 2y + 2z = \frac{1}{4} \\ x \qquad\quad + 3z = \frac{1}{6} \end{cases}$

45–48 ◾ Solve the system using Cramer's Rule.

45. $\begin{cases} 2x + 7y = 13 \\ 6x + 16y = 30 \end{cases}$ **46.** $\begin{cases} 12x - 11y = 140 \\ 7x + 9y = 20 \end{cases}$

47. $\begin{cases} 2x - y + 5z = 0 \\ -x + 7y \qquad = 9 \\ 5x + 4y + 3z = -9 \end{cases}$ **48.** $\begin{cases} 3x + 4y - z = 10 \\ x \qquad\quad - 4z = 20 \\ 2x + y + 5z = 30 \end{cases}$

49–52 ◾ Graph the solution set of the system of inequalities. Find the coordinates of all vertices, and determine whether the solution set is bounded or unbounded.

49. $\begin{cases} x^2 + y^2 < 9 \\ x + y < 0 \end{cases}$ **50.** $\begin{cases} y - x^2 \geqslant 4 \\ y < 20 \end{cases}$

51. $\begin{cases} x \geqslant 0, \quad y \geqslant 0 \\ x + 2y \leqslant 12 \\ y \leqslant x + 4 \end{cases}$ **52.** $\begin{cases} x \geqslant 4 \\ x + y \geqslant 24 \\ x \leqslant 2y + 12 \end{cases}$

53–54 ◾ Solve for x, y, and z in terms of a, b, and c.

53. $\begin{cases} -x + y + z = a \\ x - y + z = b \\ x + y - z = c \end{cases}$

54. $\begin{cases} ax + by + cz = a - b + c \\ bx + by + cz = c \\ cx + cy + cz = c \end{cases}$ $(a \neq b, b \neq c, c \neq 0)$

55. For what values of k do the following three lines have a common point of intersection?

$$x + y = 12$$
$$kx - y = 0$$
$$y - x = 2k$$

56. For what value of k does the following system have infinitely many solutions?

$$\begin{cases} kx + y + z = 0 \\ x + 2y + kz = 0 \\ -x + 2y + 3z = 0 \end{cases}$$

57–60 ◾ Find the partial fraction decomposition of the rational function.

57. $\dfrac{3x + 1}{x^2 - 2x - 15}$ **58.** $\dfrac{8}{x^3 - 4x}$

59. $\dfrac{2x - 4}{x(x - 1)^2}$ **60.** $\dfrac{x + 6}{x^3 - 2x^2 + 4x - 8}$

8 TEST

1. In $2\frac{1}{2}$ hours an airplane travels 600 km against the wind. It takes 50 min to travel 300 km with the wind. Find the speed of the wind and the speed of the airplane in still air.

2–5 ■ Find all solutions of the system. Determine whether the system is linear or non-linear. If it is linear, state whether it is inconsistent, dependent, or neither.

2. $\begin{cases} 3x - y = 10 \\ 2x + 5y = 1 \end{cases}$

3. $\begin{cases} x - y + 9z = -8 \\ x \quad - 4z = 7 \\ 3x - y + z = 5 \end{cases}$

4. $\begin{cases} 2x - y + z = 0 \\ 3x + 2y - 3z = 1 \\ x - 4y + 5z = -1 \end{cases}$

5. $\begin{cases} 2x^2 + y^2 = 6 \\ 3x^2 - 4y = 11 \end{cases}$

6. Three friends stop at roadside fruitstand and buy cherries, peaches, and pears, purchasing the amounts shown in the table. How much does each variety of fruit cost per pound?

	Peaches	Cherries	Pears	Total Cost
Mary	4 lb	3 lb	1 lb	$9
Jill	2 lb	1 lb	3 lb	$6
Fred	2 lb	2 lb	2 lb	$7

7–14 ■ Let

$$A = \begin{bmatrix} 2 & 3 \\ 2 & 4 \end{bmatrix} \qquad B = \begin{bmatrix} 2 & 4 \\ -1 & 1 \\ 3 & 0 \end{bmatrix} \qquad C = \begin{bmatrix} 1 & 0 & 4 \\ -1 & 1 & 2 \\ 0 & 1 & 3 \end{bmatrix}$$

Carry out the indicated operation, or explain why it cannot be performed.

7. $A + B$ 8. AB 9. $BA - 3B$ 10. CBA

11. A^{-1} 12. B^{-1} 13. $\det(B)$ 14. $\det(C)$

15. Write a matrix equation equivalent to the following system.

$$\begin{cases} 4x - 3y = 10 \\ 3x - 2y = 30 \end{cases}$$

Find the inverse of the coefficient matrix, and use it to solve the system.

16. Solve using Cramer's Rule:

$$\begin{cases} 2x \quad\;\; - \;\; z = 14 \\ 3x - \;\; y + 5z = \;\; 0 \\ 4x + 2y + 3z = -2 \end{cases}$$

17. Only one of the following matrices has an inverse. Find the determinant of each matrix, and use the determinants to identify the one that has an inverse. Then find the inverse.

$$A = \begin{bmatrix} 1 & 4 & 1 \\ 0 & 2 & 0 \\ 1 & 0 & 1 \end{bmatrix} \qquad B = \begin{bmatrix} 1 & 4 & 0 \\ 0 & 2 & 0 \\ -3 & 0 & 1 \end{bmatrix}$$

18. Graph the solution set of each system of inequalities. Label the vertices with their coordinates.

(a) $\begin{cases} 2x + \;\; y \leq \;\; 8 \\ \quad x - \;\; y \geq -2 \\ \quad x + 2y \geq \;\; 4 \end{cases}$
(b) $\begin{cases} x^2 - 5 + y \leq 0 \\ y \leq 5 + 2x \end{cases}$

19. Find the partial fraction decomposition of the following rational function.

$$\frac{4x - 1}{(x - 1)^2(x + 2)}$$

20. Using a graphing device to find all solutions of the following system, correct to two decimal places.

$$\begin{cases} 2x^2 + y^2 = 16 \\ y = x^4 - 4x^3 + 6x^2 - 4x \end{cases}$$

Focus on Modeling

Linear Programming

Linear programming is a modeling technique used to determine the optimal allocation of resources in business, the military, and other areas of human endeavor. For example, a manufacturer who makes several different products from the same raw materials can use linear programming to determine how much of each product should be produced to maximize the profit. This modeling technique is probably the most important practical application of systems of linear inequalities. In 1975 Leonid Kantorovich and T. C. Koopmans won the Nobel Prize in economics for their work in the development of this technique.

Although linear programming can be applied to very complex problems with hundreds or even thousands of variables, we consider only a few simple examples to which the graphical methods of Section 8.8 can be applied. (For large numbers of variables, a linear programming method based on matrices is used.) Let's examine a typical problem.

EXAMPLE 1 ■ Manufacturing for Maximum Profit

A small shoe manufacturer makes two styles of shoes: oxfords and loafers. Two machines are used in the process: a cutting machine and a sewing machine. Each type of shoe requires 15 min per pair on the cutting machine. Oxfords require 10 min of sewing per pair, and loafers require 20 min of sewing per pair. Because the manufacturer can hire only one operator for each machine, each process is available for just 8 hours per day. If the profit is $15 on each pair of oxfords and $20 on each pair of loafers, how many pairs of each type should be produced per day for maximum profit?

SOLUTION

First we organize the given information into a table. To be consistent, let's convert all times to hours.

Because loafers produce more profit, it would seem best to manufacture only loafers. Surprisingly, this does not turn out to be the most profitable solution.

	Oxfords	Loafers	Time available
Time on cutting machine (h)	$\frac{1}{4}$	$\frac{1}{4}$	8
Time on sewing machine (h)	$\frac{1}{6}$	$\frac{1}{3}$	8
Profit	$15	$20	

We describe the model and solve the problem in four steps.

Choosing the Variables   To make a mathematical model, we first give names to the variable quantities. For this problem we let

$$x = \text{number of pairs of oxfords made daily}$$

$$y = \text{number of pairs of loafers made daily}$$

Linear programming helps the telephone industry determine the most efficient way to route telephone calls. The computerized routing decisions must be made very rapidly so callers are not kept waiting for connections. Since the database of customers and routes is huge, an extremely fast method for solving linear programming problems is essential. In 1984 the 28-year-old mathematician **Narendra Karmarkar**, working at Bell Labs in Murray Hill, New Jersey, discovered just such a method. His idea is so ingenious and his method so fast that the discovery caused a sensation in the mathematical world. Although mathematical discoveries rarely make the news, this one was reported in *Time*, on December 3, 1984. Today airlines routinely use Karmarkar's technique to minimize costs in scheduling passengers, flight personnel, fuel, baggage, and maintenance workers.

FIGURE 1

Finding the Objective Function   The objective of this problem is to determine which values for x and y give maximum profit. Since each pair of oxfords provides \$15 profit and each pair of loafers \$20, the total profit is given by

$$P = 15x + 20y$$

This function is called the *objective function*.

Graphing the Feasible Region   The larger x and y are, the greater the profit. But we cannot choose arbitrarily large values for these variables, because of the restrictions, or *constraints*, in the problem. Each restriction is an inequality in the variables.

In this problem the total number of cutting hours needed is $\frac{1}{4}x + \frac{1}{4}y$. Since only 8 hours are available on the cutting machine, we have

$$\frac{1}{4}x + \frac{1}{4}y \leq 8$$

Similarly, by considering the amount of time needed and available on the sewing machine, we get

$$\frac{1}{6}x + \frac{1}{3}y \leq 8$$

We cannot produce a negative number of shoes, so we also have

$$x \geq 0 \qquad \text{and} \qquad y \geq 0$$

Thus, x and y must satisfy the constraints

$$\begin{cases} \frac{1}{4}x + \frac{1}{4}y \leq 8 \\ \frac{1}{6}x + \frac{1}{3}y \leq 8 \\ \qquad\qquad x \geq 0 \\ \qquad\qquad y \geq 0 \end{cases}$$

If we multiply the first inequality by 4 and the second by 6, we obtain the simplified system

$$\begin{cases} x + y \leq 32 \\ x + 2y \leq 48 \\ \qquad x \geq 0 \\ \qquad y \geq 0 \end{cases}$$

The solution of this system (with vertices labeled) is sketched in Figure 1. The only values that satisfy the restrictions of the problem are the ones that correspond to points of the shaded region in Figure 1. This is called the *feasible region* for the problem.

Finding Maximum Profit   As x or y increases, profit will increase as well. Thus, it seems reasonable that the maximum profit will occur at a point on one of the outside edges of the feasible region, where it's impossible to increase x or y without going outside the region. In fact, it can be shown that the maximum value occurs at a vertex. This means that we need to check the profit only at the vertices.

Vertex	$P = 15x + 20y$
(0, 0)	0
(0, 24)	$15(0) + 20(24) = \$480$
(16, 16)	$15(16) + 20(16) = \$560$ ←
(32, 0)	$15(32) + 20(0) = \$480$

maximum profit ⌐

The largest value of P occurs at the point $(16, 16)$, where $P = \$560$. Thus, the manufacturer should make 16 pairs of oxfords and 16 pairs of loafers, for a maximum daily profit of $560. ■

The linear programming problems that we consider all follow the pattern of Example 1. Each problem involves two variables. The problem describes restrictions, called **constraints**, that lead to a system of linear inequalities whose solution is called the **feasible region**. The function we wish to maximize or minimize is called the **objective function**. This function always attains its largest and smallest values at the **vertices** of the feasible region. This modeling technique involves four steps, summarized in the following box.

GUIDELINES FOR LINEAR PROGRAMMING

1. CHOOSE THE VARIABLES.   Decide what variable quantities in the problem should be named x and y.

2. FIND THE OBJECTIVE FUNCTION.   Write an expression for the function we want to maximize or minimize.

3. GRAPH THE FEASIBLE REGION.   Express the constraints as a system of inequalities and graph the solution of this system (the feasible region).

4. FIND THE MAXIMUM OR MINIMUM.   Evaluate the objective function at the vertices of the feasible region to determine its maximum or minimum value.

EXAMPLE 2 ■ A Shipping Problem

A car dealer has warehouses in Millville and Trenton and dealerships in Camden and Atlantic City. Every car sold at the dealerships must be delivered from one of the warehouses. On a certain day the Camden dealers sell 10 cars, and the Atlantic City dealers sell 12. The Millville warehouse has 15 cars available, and the Trenton warehouse has 10. The cost of shipping one car is $50 from Millville to Camden, $40 from Millville to Atlantic City, $60 from Trenton to Camden, and $55 from Trenton to Atlantic City. How many cars should be moved from each warehouse to each dealership to fill the orders at minimum cost?

SOLUTION

Our first step is to organize the given information. Rather than construct a table, we draw a diagram to show the flow of cars from the warehouses to the dealerships (see Figure 2 on page 718). The diagram shows the number of cars available at each warehouse or required at each dealership and the cost of shipping between these locations.

Choosing the Variables   The arrows in Figure 2 indicate four possible routes, so the problem seems to involve four variables. But we let

x = number of cars to be shipped from Millville to Camden

y = number of cars to be shipped from Millville to Atlantic City

FIGURE 2

To fill the orders, we must have

$$10 - x = \text{number of cars shipped from Trenton to Camden}$$

$$12 - y = \text{number of cars shipped from Trenton to Atlantic City}$$

So the only variables in the problem are x and y.

Finding the Objective Function   The objective of this problem is to minimize cost. From Figure 2 we see that the total cost of shipping the cars is

$$
\begin{aligned}
C &= 50x + 40y + 60(10 - x) + 55(12 - y) \\
&= 50x + 40y + 600 - 60x + 660 - 55y \\
&= 1260 - 10x - 15y
\end{aligned}
$$

This is the objective function.

Graphing the Feasible Region   Now we derive the constraint inequalities that define the feasible region. First, the number of cars shipped on each route can't be negative, so we have

$$x \geqslant 0 \qquad\qquad y \geqslant 0$$

$$10 - x \geqslant 0 \qquad\qquad 12 - y \geqslant 0$$

Second, the total number of cars shipped from each warehouse can't exceed the number of cars available there, so

$$x + y \leqslant 15$$

$$(10 - x) + (12 - y) \leqslant 10$$

Simplifying the latter inequality, we get

$$22 - x - y \leqslant 10$$

$$-x - y \leqslant -12$$

$$x + y \geqslant 12$$

The inequalities $10 - x \geqslant 0$ and $12 - y \geqslant 0$ can be rewritten as $x \leqslant 10$ and

FIGURE 3

$y \leq 12$, respectively. Thus, the feasible region is described by the constraints

$$\begin{cases} x + y \leq 15 \\ x + y \geq 12 \\ 0 \leq x \leq 10 \\ 0 \leq y \leq 12 \end{cases}$$

The feasible region is graphed in Figure 3.

Finding Minimum Cost   We check the value of the objective function at each vertex of the feasible region.

Vertex	$C = 1260 - 10x - 15y$
(0, 12)	$1260 - 10(0) - 15(12) = \$1080$
(3, 12)	$1260 - 10(3) - 15(12) = \$1050$ ← minimum cost
(10, 5)	$1260 - 10(10) - 15(5) = \$1085$
(10, 2)	$1260 - 10(10) - 15(2) = \$1130$

The lowest cost is incurred at the point (3, 12). Thus, the dealer should ship

3 cars from Millville to Camden
12 cars from Millville to Atlantic City
7 cars from Trenton to Camden
0 cars from Trenton to Atlantic City   ■

In the 1940s mathematicians developed matrix methods for solving linear programming problems that involve more than two variables. These methods were first used by the Allies in World War II to solve supply problems similar to (but, of course, much more complicated than) Example 2. Improving these matrix methods is an active and exciting area of current mathematical research.

■ Problems

1–4 ■ Find the maximum and minimum values of the given objective function on the indicated feasible region.

1. $M = 200 - x - y$

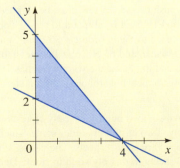

2. $N = \frac{1}{2}x + \frac{1}{4}y + 40$

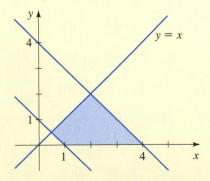

3. $P = 140 - x + 3y$

$$\begin{cases} x \geqslant 0, \quad y \geqslant 0 \\ \quad 2x + y \leqslant 10 \\ \quad 2x + 4y \leqslant 28 \end{cases}$$

4. $Q = 70x + 82y$

$$\begin{cases} x \geqslant 0, \quad y \geqslant 0 \\ x \leqslant 10, \quad y \leqslant 20 \\ \quad x + y \geqslant 5 \\ \quad x + 2y \leqslant 18 \end{cases}$$

5. A furniture manufacturer makes wooden tables and chairs. The production process involves two basic types of labor: carpentry and finishing. A table requires 2 hours of carpentry and 1 hour of finishing, whereas a chair requires 3 hours of carpentry and $\frac{1}{2}$ hour of finishing. The profit is $35 per table and $20 per chair. The manufacturer's employees can supply a maximum of 108 hours of carpentry work and 20 hours of finishing work per day. How many tables and chairs should be made each day to maximize profit?

6. A housing contractor has subdivided a farm into 100 building lots. He has designed two types of homes for these lots: colonial and ranch style. A colonial requires $30,000 of capital and produces a profit of $4000 when sold. A ranch-style house requires $40,000 of capital and provides an $8000 profit. If he has $3.6 million of capital on hand, how many houses of each type should he build for maximum profit? Will any of the lots be left vacant?

7. A trucker hauls citrus fruit from Florida to Montreal. Each crate of oranges is 4 ft^3 in volume and weighs 80 lb. Each crate of grapefruit has a volume of 6 ft^3 and weighs 100 lb. Her truck has a maximum capacity of 300 ft^3 and can carry no more than 5600 lb. Moreover, she is not permitted to carry more crates of grapefruit than crates of oranges. If she makes a profit of $2.50 on each crate of oranges and $4 on each crate of grapefruit, how many crates of each fruit should she carry for maximum profit?

8. A manufacturer of calculators produces two models: standard and scientific. Long-term demand for the two models mandates that the company manufacture at least 100 standard and 80 scientific calculators each day. However, because of limitations on production capacity, no more than 200 standard and 170 scientific calculators can be made daily. To satisfy a shipping contract, a total of at least 200 calculators must be shipped every day.
(a) If the production cost is $5 for a standard calculator and $7 for a scientific one, how many of each model should be produced daily to minimize this cost?
(b) If each standard calculator results in a $2 loss but each scientific one produces a $5 profit, how many of each model should be made daily to maximize profit?

9. An electronics discount chain has a sale on a certain brand of stereo. The chain has stores in Santa Monica and El Toro and warehouses in Long Beach and Pasadena. To satisfy rush orders, 15 sets must be shipped from the warehouses to the Santa Monica store, and 19 must be shipped to the El Toro store. The cost of shipping a set is $5 from Long Beach to Santa Monica, $6 from Long Beach to El Toro, $4 from Pasadena to Santa Monica, and $5.50 from Pasadena to El Toro. If the Long Beach warehouse has 24 sets and the Pasadena warehouse has 18 sets in stock, how many sets should be shipped from each warehouse to each store to fill the orders at a minimum shipping cost?

10. A man owns two building supply stores, one on the east side and one on the west side of a city. Two customers order some $\frac{1}{2}$-inch plywood. Customer A needs 50 sheets and customer B needs 70 sheets. The east-side store has 80 sheets and the west-side store has 45 sheets of this plywood in stock. Delivery cost per sheet is $0.50 from the east-side store to customer A, $0.60 from the east-side store to customer B, $0.40 from the west-side store to A, and $0.55 from the west-side store to B. How many sheets should be shipped from each store to each customer to minimize delivery costs?

11. A confectioner sells two types of nut mixtures. The standard-mixture package contains 100 g of cashews and 200 g of peanuts and sells for $1.95. The deluxe-mixture package contains 150 g of cashews and 50 g of peanuts and sells for $2.25. The confectioner has 15 kg of cashews and 20 kg of peanuts available. Based on past sales, he needs to have at least as many standard as deluxe packages available. How many bags of each mixture should he package to maximize his revenue?

12. A biologist wishes to feed laboratory rabbits a mixture of two types of foods. Type I contains 8 g of fat, 12 g of carbohydrate, and 2 g of protein per ounce, whereas type II contains 12 g of fat, 12 g of carbohydrate, and 1 g of protein per ounce. Type I costs $0.20 per ounce and type II cost $0.30 per ounce. The rabbits each receive a daily minimum of 24 g of fat, 36 g of carbohydrate, and 4 g of protein, but get no more than 5 oz of food per day. How many ounces of each food type should be fed to each rabbit daily to satisfy the dietary requirements at minimum cost?

13. A woman wishes to invest $12,000 in three types of bonds: municipal bonds paying 7% interest per year, bank investment certificates paying 8%, and high-risk bonds paying 12%. For tax reasons, she wants the amount invested in municipal bonds to be at least three times the amount invested in bank certificates. To keep her level of risk manageable, she will invest no more than $2000 in high-risk bonds. How much should she invest in each type of bond to maximize her annual interest yield?   [*Hint:* Let x = amount in municipal bonds and y = amount in bank certificates. Then the amount in high-risk bonds will be $12,000 - x - y$.]

14. Refer to Problem 13. Suppose the investor decides to increase to $3000 the maximum amount she will allow to be invested in high-risk bonds but leaves the other conditions unchanged. By how much will her maximum possible interest yield increase?

15. A small software company publishes computer games and educational and utility software. Their policy is to market a total of 36 new programs each year, with at least four of these being games. The number of utility programs published is never more than twice the number of educational programs. On average, the company makes an annual profit of $5000 on each computer game, $8000 on each educational program, and $6000 on each utility program. How many of each type of software should they publish annually for maximum profit?

16. All parts of this problem refer to the following feasible region and objective function.

$$\begin{cases} x \geq 0 \\ x \geq y \\ x + 2y \leq 12 \\ x + y \leq 10 \end{cases}$$

$$P = x + 4y$$

(a) Graph the feasible region.
(b) On your graph from part (a), sketch the graphs of the linear equations obtained by setting P equal to 40, 36, 32, and 28.
(c) If we continue to decrease the value of P, at which vertex of the feasible region will these lines first touch the feasible region?
(d) Verify that the maximum value of P on the feasible region occurs at the vertex you chose in part (c).

9 Topics in Analytic Geometry

The path of an object moving under the gravitational pull of another is a conic section. For instance, the trajectory of a basketball is a parabola; the orbit of a planet around the sun is an ellipse.

In this chapter we study the geometry of **conic sections** (or simply **conics**). Conic sections are the curves formed by the intersection of a plane with a pair of circular cones. These curves have four basic shapes, called **circles**, **ellipses**, **parabolas**, and **hyperbolas**, as illustrated in the figure.

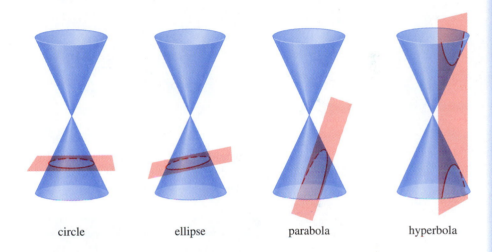

| circle | ellipse | parabola | hyperbola |

The ancient Greeks studied these curves because they considered the geometry of conic sections to be very beautiful. The mathematician Apollonius (262–190 B.C.) wrote a definitive eight-volume work on the subject. In more modern times, conics have been found to be useful as well as beautiful. Galileo discovered in 1590 that the path of a missile shot upward at an angle is a parabola. In 1609 Kepler found that the planets move in elliptical orbits around the sun. In 1668 Newton built the first reflecting telescope, whose principle is based on the properties of parabolas and hyperbolas. In this century, many further applications of conic sections have been developed. One important application is the LORAN radio navigation system, which uses the intersection points of hyperbolas to pinpoint the location of ships and aircraft. Another application is the medical procedure of lithotripsy, a method of removing kidney stones without surgery that uses a property of ellipses. These and other applications of conic sections will be considered in this chapter.

In addition to studying conics, we will examine two other ways to describe points and curves in the Cartesian plane: polar coordinates and parametric equations. Both of these topics require a thorough understanding of trigonometry.

9.1 PARABOLAS

We saw in Section 2.6 that the graph of the equation $y = ax^2 + bx + c$ is a U-shaped curve called a *parabola* that opens either upward or downward, depending on whether the sign of a is positive or negative (see Figure 1). The lowest or highest point of the parabola is called the **vertex**, and the parabola is symmetric about its **axis** (or **axis of symmetry**).

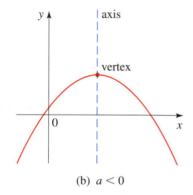

(a) $a > 0$

(b) $a < 0$

FIGURE 1
$y = ax^2 + bx + c$

In this section we study parabolas from a geometric rather than an algebraic point of view. We begin with the geometric definition of a parabola and show how this leads to the algebraic formula that we are already familiar with.

GEOMETRIC DEFINITION OF A PARABOLA

A **parabola** is the set of points in the plane equidistant from a fixed point F (called the **focus**) and a fixed line l (called the **directrix**).

This definition is illustrated in Figure 2. Note that the vertex V of the parabola lies halfway between the focus and the directrix and that the axis of symmetry is the line that runs through the focus perpendicular to the directrix.

In this section we restrict our attention to parabolas that are situated with the vertex at the origin and that have a vertical or horizontal axis of symmetry. (Parabolas in more general positions will be considered in Section 9.4.) If the focus of such a parabola is the point $F(0, p)$, then the axis of symmetry must be vertical and the directrix has the equation $y = -p$. Figure 3 illustrates the case $p > 0$.

If $P(x, y)$ is any point on the parabola, then the distance from P to the focus F (using the Distance Formula) is

$$\sqrt{x^2 + (y - p)^2}$$

The distance from P to the directrix is

$$|y - (-p)| = |y + p|$$

FIGURE 2

FIGURE 3

By the definition of a parabola, these two distances must be equal:

$$\sqrt{x^2 + (y - p)^2} = |y + p|$$

$$x^2 + (y - p)^2 = |y + p|^2 = (y + p)^2 \qquad \text{Square both sides}$$

$$x^2 + y^2 - 2py + p^2 = y^2 + 2py + p^2 \qquad \text{Expand}$$

$$x^2 - 2py = 2py \qquad \text{Simplify}$$

$$x^2 = 4py$$

If $p > 0$, then the parabola opens upward, but if $p < 0$, it opens downward. When x is replaced by $-x$, the equation remains unchanged, so the graph is symmetric about the y-axis. We summarize what we have proved in the following box.

PARABOLA WITH VERTICAL AXIS

The graph of the equation

$$x^2 = 4py$$

is a parabola with the following properties.

VERTEX	$V(0, 0)$
FOCUS	$F(0, p)$
DIRECTRIX	$y = -p$

The parabola opens upward if $p > 0$ or downward if $p < 0$.

(a) $x^2 = 4py$ with $p > 0$

(b) $x^2 = 4py$ with $p < 0$

EXAMPLE 1 ■ Finding the Equation of a Parabola

Find the equation of the parabola with vertex $V(0, 0)$ and focus $F(0, 2)$, and sketch its graph.

SOLUTION

Since the focus is $F(0, 2)$, we conclude that $p = 2$ (and so the directrix is $y = -2$).

(a)

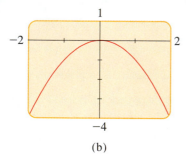

(b)

FIGURE 5

FIGURE 4

Thus, the equation of the parabola is

$$x^2 = 4(2)y \qquad \text{\textcolor{red}{$x^2 = 4py$ with $p = 2$}}$$

$$x^2 = 8y$$

Since $p = 2 > 0$, the parabola opens upward. See Figure 4.

EXAMPLE 2 ■ Finding the Focus and Directrix of a Parabola from Its Equation

Find the focus and directrix of the parabola $y = -x^2$, and sketch the graph.

SOLUTION

To find the focus and directrix, we put the given equation in the standard form $x^2 = -y$. Comparing this to the general equation $x^2 = 4py$, we see that $4p = -1$, so $p = -\frac{1}{4}$. Thus, the focus is $F\left(0, -\frac{1}{4}\right)$ and the directrix is $y = \frac{1}{4}$. The graph of the parabola, together with the focus and the directrix, is shown in Figure 5(a). We can also draw the graph using a graphing calculator as shown in Figure 5(b). ■

Reflecting the graph in Figure 3 about the diagonal line $y = x$ has the effect of interchanging the roles of x and y. This results in a parabola with horizontal axis. By the same method as before, we can prove the following properties.

PARABOLA WITH HORIZONTAL AXIS

The graph of the equation

$$y^2 = 4px$$

is a parabola with the following properties.

VERTEX	$V(0, 0)$
FOCUS	$F(p, 0)$
DIRECTRIX	$x = -p$

The parabola opens to the right if $p > 0$ or to the left if $p < 0$.

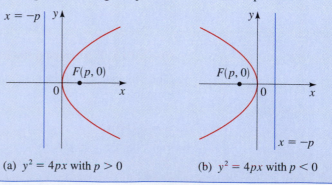

(a) $y^2 = 4px$ with $p > 0$ (b) $y^2 = 4px$ with $p < 0$

EXAMPLE 3 ■ **A Parabola with Horizontal Axis**

A parabola has the equation $6x + y^2 = 0$.

(a) Find the focus and directrix of the parabola, and sketch the graph.

(b) Use a graphing calculator to draw the graph.

SOLUTION

(a) To find the focus and directrix, we put the given equation in the standard form $y^2 = -6x$. Comparing this to the general equation $y^2 = 4px$, we see that $4p = -6$, so $p = -\frac{3}{2}$. Thus, the focus is $F\left(-\frac{3}{2}, 0\right)$ and the directrix is $x = \frac{3}{2}$. Since $p < 0$, the parabola opens to the left. The graph of the parabola, together with the focus and the directrix, is shown in Figure 6(a).

(b) To draw the graph using a graphing calculator, we need to solve for y.

$$6x + y^2 = 0$$

$$y^2 = -6x \qquad \text{Subtract } 6x$$

$$y = \pm\sqrt{-6x} \qquad \text{Take square roots}$$

To obtain the graph of the parabola, we graph both functions

$$y = \sqrt{-6x} \qquad \text{and} \qquad y = -\sqrt{-6x}$$

as shown in Figure 6(b).

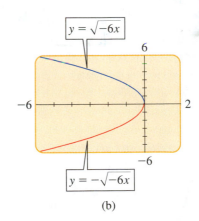

FIGURE 6   (a)   (b)   ■

The equation $y^2 = 4px$ does not define y as a function of x (see page 162). So, to use a graphing calculator to graph a parabola with horizontal axis, we must first solve for y. This leads to two functions, $y = \sqrt{4px}$ and $y = \sqrt{-4px}$. We must graph both functions to get the complete graph of the parabola. For example, in Figure 6(b) we had to graph both $y = \sqrt{-6x}$ and $y = -\sqrt{-6x}$ to graph the parabola $y^2 = -6x$.

We can use the coordinates of the focus to estimate the "width" of a parabola when sketching its graph. The line segment that runs through the focus perpendicular to the axis, with endpoints on the parabola, is called the **latus rectum**, and its length is the **focal diameter** of the parabola. From Figure 7 we can see that the distance from an endpoint Q of the latus rectum to the directrix is $|2p|$. Thus, the distance from Q to the focus must be $|2p|$ as well (by the definition of a parabola), and so the focal diameter is $|4p|$. In the next example we use the focal diameter to determine the "width" of a parabola when graphing it.

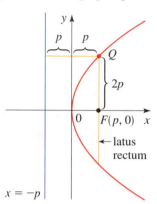

FIGURE 7

EXAMPLE 4 ■ Using the Focal Diameter to Sketch a Parabola

Find the focus, directrix, and focal diameter of the parabola $y = \frac{1}{2}x^2$, and sketch its graph.

SOLUTION

We first put the equation in the form $x^2 = 4py$.

$$y = \frac{1}{2}x^2$$

$$x^2 = 2y \qquad \text{Multiply each side by 2}$$

From this equation we see that $4p = 2$, so the focal diameter is 2. Solving for p gives $p = \frac{1}{2}$, so the focus is $\left(0, \frac{1}{2}\right)$ and the directrix is $y = -\frac{1}{2}$. Since the focal diameter is 2, the latus rectum extends 1 unit to the left and 1 unit to the right of the focus. This enables us to sketch the graph in Figure 8.

FIGURE 8

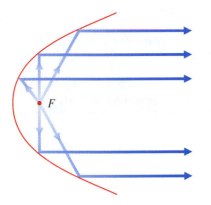

FIGURE 9

Parabolic reflector

Parabolas have an important property that makes them useful as reflectors for lamps and telescopes. Light from a source placed at the focus of a surface with parabolic cross section will be reflected in such a way that it travels parallel to the axis of the parabola (see Figure 9). Thus, a parabolic mirror reflects the light into a beam of parallel rays. Conversely, light approaching the reflector in rays parallel to its axis of symmetry is concentrated to the focus. This *reflection property*, which can be proved using calculus, is used in the construction of reflecting telescopes.

EXAMPLE 5 ■ Finding the Focal Point of a Searchlight Reflector

A searchlight has a parabolic reflector that forms a "bowl," which is 12 in. wide from rim to rim and 8 in. deep, as shown in Figure 10. If the filament of the light bulb is located at the focus, how far from the vertex of the reflector is it?

FIGURE 10

A parabolic reflector

SOLUTION

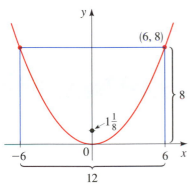

FIGURE 11

We introduce a coordinate system and place a parabolic cross section of the reflector so that its vertex is at the origin and its axis is vertical (see Figure 11). Then the equation of this parabola has the form $x^2 = 4py$. From Figure 11 we see that the point $(6, 8)$ lies on the parabola. We use this to find p.

$$6^2 = 4p(8) \qquad \text{The point } (6, 8) \text{ satisfies the equation } x^2 = 4py$$

$$36 = 32p$$

$$p = \tfrac{9}{8}$$

The focus is $F\left(0, \tfrac{9}{8}\right)$, so the distance between the vertex and the focus is $\tfrac{9}{8} = 1\tfrac{1}{8}$ in. Because the filament is positioned at the focus, it is located $1\tfrac{1}{8}$ in. from the vertex of the reflector. ■

In the next example we graph a family of parabolas, to show how changing the distance between the focus and the vertex affects the "width" of a parabola.

EXAMPLE 6 ■ A Family of Parabolas

(a) Find equations for the parabolas with vertex at the origin and foci
$F_1\left(0, \tfrac{1}{8}\right)$, $F_2\left(0, \tfrac{1}{2}\right)$, $F_3(0, 1)$, and $F_4(0, 4)$.

(b) Draw the graphs of the parabolas in part (a). What do you conclude?

SOLUTION

(a) Since the foci are on the positive y-axis, the parabolas open upward and have equations of the form $x^2 = 4py$. This leads to the following equations.

Focus	p	Equation $x^2 = 4py$	Form of the equation for graphing calculator
$F_1\left(0, \frac{1}{8}\right)$	$p = \frac{1}{8}$	$x^2 = \frac{1}{2}y$	$y = 2x^2$
$F_2\left(0, \frac{1}{2}\right)$	$p = \frac{1}{2}$	$x^2 = 2y$	$y = 0.5x^2$
$F_3(0, 1)$	$p = 1$	$x^2 = 4y$	$y = 0.25x^2$
$F_4(0, 4)$	$p = 4$	$x^2 = 16y$	$y = 0.0625x^2$

(b) The graphs are drawn in Figure 12. We see that the closer the focus to the vertex, the narrower the parabola.

$y = 2x^2$

$y = 0.5x^2$

$y = 0.25x^2$

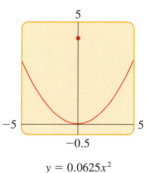

$y = 0.0625x^2$

FIGURE 12 A family of parabolas

9.1 EXERCISES

1–6 ■ Match the equation with the graphs labeled I–VI. Give reasons for your answers.

1. $y^2 = 2x$

2. $y^2 = -\frac{1}{4}x$

3. $x^2 = -6y$

4. $2x^2 = y$

5. $y^2 - 8x = 0$

6. $12y + x^2 = 0$

I

II

III

IV

V

VI

7–18 ■ Find the focus, directrix, and focal diameter of the parabola, and sketch its graph.

7. $y^2 = 4x$ **8.** $x^2 = y$

9. $x^2 = 9y$ **10.** $y^2 = 3x$

11. $y = 5x^2$ **12.** $y = -2x^2$

13. $x = -8y^2$ **14.** $x = \frac{1}{2}y^2$

15. $x^2 + 6y = 0$ **16.** $x - 7y^2 = 0$

17. $5x + 3y^2 = 0$ **18.** $8x^2 + 12y = 0$

 19–24 ■ Use a graphing device to graph the parabola.

19. $x^2 = 16y$ **20.** $x^2 = -8y$

21. $y^2 = -\frac{1}{3}x$ **22.** $8y^2 = x$

23. $4x + y^2 = 0$ **24.** $x - 2y^2 = 0$

25–36 ■ Find an equation for the parabola that has its vertex at the origin and satisfies the given condition(s).

25. Focus $F(0, 2)$ **26.** Focus $F\left(0, -\frac{1}{2}\right)$

27. Focus $F(-8, 0)$ **28.** Focus $F(5, 0)$

29. Directrix $x = 2$ **30.** Directrix $y = 6$

31. Directrix $y = -10$ **32.** Directrix $x = -\frac{1}{8}$

33. Focus on the positive x-axis, 2 units away from the directrix

34. Directrix has y-intercept 6

35. Opens upward with focus 5 units from the vertex

36. Focal diameter 8 and focus on the negative y-axis

37–46 ■ Find an equation of the parabola whose graph is shown.

37.

38.

39.

40.

41.

42.

43.

44.

45.

46.

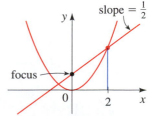

47. A lamp with a parabolic reflector is shown in the figure on the following page. The bulb is placed at the focus and the focal diameter is 12 cm.

(a) Find an equation of the parabola.
(b) Find the diameter $d(C, D)$ of the opening, 20 cm from the vertex.

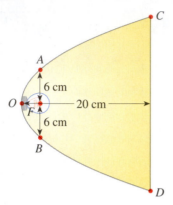

48. A reflector for a satellite dish is parabolic in cross section, with the receiver at the focus F. The reflector is 1 ft deep and 20 ft wide from rim to rim (see the figure). How far is the receiver from the vertex of the parabolic reflector?

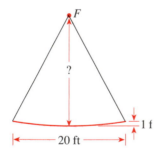

49. In a suspension bridge the shape of the suspension cables is parabolic. The bridge shown in the figure has towers that are 600 m apart, and the lowest point of the suspension cables is 150 m below the top of the towers. Find the equation of the parabolic part of the cables, placing the origin of the coordinate system at the vertex.

    NOTE This equation is used to find the length of cable needed in the construction of the bridge.

50. The Hale telescope at the Mount Palomar Observatory has a 200-in. mirror, as shown. The mirror is constructed in a parabolic shape that collects light from the stars and focuses it at the **prime focus**, that is, the focus of the parabola. The mirror is 3.79 in. deep at its center. Find the **focal length** of this parabolic mirror, that is, the distance from the vertex to the focus.

51. (a) Find equations for the family of parabolas with vertex at the origin and with directrixes $y = \frac{1}{2}$, $y = 1$, $y = 4$, and $y = 8$.
    (b) Draw the graphs. What do you conclude?

DISCOVERY · DISCUSSION

52. **Parabolas in the Real World**   Several examples of the uses of parabolas are given in the text. Find other situations in real life where parabolas occur. Consult a scientific encyclopedia in the reference section of your library, or search the Internet.

53. **Light Cone from a Flashlight**   A flashlight is held to form a lighted area on the ground, as shown in the figure. Is it possible to angle the flashlight in such a way that the boundary of the lighted area is a parabola? Explain your answer.

Laboratory Project

Rolling Down a Ramp

Galileo was the first to show that the motion of a falling ball can be modeled by a parabola (see page 803). To accomplish this, he rolled balls down a ramp and used his acute sense of timing to study the motion of the ball. In this project you will perform a similar experiment, using modern equipment. You will need the following:

■ A wide board, about 6–8 ft long, to be used as a ramp

■ A large ball, the size of a soccer or volleyball

■ A calculator-based motion detector, such as the Texas Instruments CBR system (Calculator Based Ranger)

Use books to prop up the ramp at an angle of about 15°, clamp the motion detector to the top of the ramp, and connect it to the calculator, as shown in the figure. (Read the manual carefully to make sure you have set up the calculator and motion detector correctly.)

1. Mark a spot on the ramp about 2 ft from the top. Place the ball on the mark and release it to see how many seconds it takes to roll to the bottom. Adjust the motion detector to record the ball's position every 0.05 s over this length of time. Now let the ball roll down the ramp again, this time with the motion detector running. The calculator should record at least 50 data points that indicate the distance between the ball and the motion detector every 0.05 s.

2. Make a scatter plot of your data, plotting time on the *x*-axis and distance on the *y*-axis. Does the plot look like half of a parabola?

3. Use the quadratic regression command on your calculator (called `QuadReg` on the TI-83) to find the parabola equation $y = ax^2 + bx + c$ that best fits the data. Graph this equation on your scatter plot to see how well it fits. Do the data points really form part of a parabola?

4. Repeat the experiment with the ramp inclined at a shallower angle. How does reducing the angle of the ramp affect the shape of the parabola?

5. Try rolling the ball up the ramp from the bottom to the spot you marked in Step 1, so that it rolls up and then back down again. If you perform the experiment this way instead of just letting the ball roll down, how does your graph in Step 3 change?

9.2   ELLIPSES

FIGURE 1

FIGURE 2

An ellipse is an oval curve that looks like an elongated circle. More precisely, we have the following definition.

GEOMETRIC DEFINITION OF AN ELLIPSE

An **ellipse** is the set of all points in the plane the sum of whose distances from two fixed points F_1 and F_2 is a constant. (See Figure 1.) These two fixed points are the **foci** (plural of **focus**) of the ellipse.

The geometric definition suggests a simple method for drawing an ellipse. Place a sheet of paper on a drawing board and insert thumbtacks at the two points that are to be the foci of the ellipse. Attach the ends of a string to the tacks, as shown in Figure 2. With the point of a pencil, hold the string taut. Then carefully move the pencil around the foci, keeping the string taut at all times. The pencil will trace out an ellipse, because the sum of the distances from the point of the pencil to the foci will always equal the length of the string, which is constant.

If the string is only slightly longer than the distance between the foci, then the ellipse traced out will be elongated in shape as in Figure 3(a), but if the foci are close together relative to the length of the string, the ellipse will be almost circular, as shown in Figure 3(b).

FIGURE 3   (a)   (b)

To obtain the simplest equation for an ellipse, we place the foci on the x-axis at $F_1(-c, 0)$ and $F_2(c, 0)$, so that the origin is halfway between them (see Figure 4).

For later convenience we let the sum of the distances from a point on the ellipse to the foci be $2a$. Then if $P(x, y)$ is any point on the ellipse, we have

$$d(P, F_1) + d(P, F_2) = 2a$$

So, from the Distance Formula

$$\sqrt{(x + c)^2 + y^2} + \sqrt{(x - c)^2 + y^2} = 2a$$

or $$\sqrt{(x - c)^2 + y^2} = 2a - \sqrt{(x + c)^2 + y^2}$$

FIGURE 4

Squaring each side and expanding, we get

$$x^2 - 2cx + c^2 + y^2 = 4a^2 - 4a\sqrt{(x + c)^2 + y^2} + (x^2 + 2cx + c^2 + y^2)$$

which simplifies to

$$4a\sqrt{(x + c)^2 + y^2} = 4a^2 + 4cx$$

Dividing each side by 4 and squaring again, we get

$$a^2[(x + c)^2 + y^2] = (a^2 + cx)^2$$

$$a^2x^2 + 2a^2cx + a^2c^2 + a^2y^2 = a^4 + 2a^2cx + c^2x^2$$

$$(a^2 - c^2)x^2 + a^2y^2 = a^2(a^2 - c^2)$$

Since the sum of the distances from P to the foci must be larger than the distance between the foci, we have that $2a > 2c$, or $a > c$. Thus, $a^2 - c^2 > 0$, and we can divide each side of the preceding equation by $a^2(a^2 - c^2)$ to get

$$\frac{x^2}{a^2} + \frac{y^2}{a^2 - c^2} = 1$$

For convenience let $b^2 = a^2 - c^2$ (with $b > 0$). Since $b^2 < a^2$, it follows that $b < a$. The preceding equation then becomes

$$\frac{x^2}{a^2} + \frac{y^2}{b^2} = 1 \qquad \text{with } a > b$$

This is the equation of the ellipse. To graph it, we need to know the x- and y-intercepts. Setting $y = 0$, we get

$$\frac{x^2}{a^2} = 1$$

so $x^2 = a^2$, or $x = \pm a$. Thus, the ellipse crosses the x-axis at $(a, 0)$ and $(-a, 0)$. These points are called the **vertices** of the ellipse, and the segment that joins them is called the **major axis**. Its length is $2a$.

Similarly, if we set $x = 0$, we get $y = \pm b$, so the ellipse crosses the y-axis at $(0, b)$ and $(0, -b)$. The segment that joins these points is called the **minor axis**, and it has length $2b$. Note that $2a > 2b$, so the major axis is longer than the minor axis.

In Section 1.8 we studied several tests that detect symmetry in a graph. If we replace x by $-x$ or y by $-y$ in the ellipse equation, it remains unchanged. Thus, the ellipse is symmetric about both the x- and y-axes, and hence about the origin as well. For this reason the origin is called the **center** of the ellipse. The complete graph is shown in Figure 5.

If the foci of the ellipse are placed on the y-axis at $(0, \pm c)$ rather than on the x-axis, then the roles of x and y are reversed in the preceding discussion, and we get a vertical ellipse. Thus, we have the following description of ellipses.

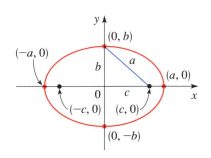

FIGURE 5

$\dfrac{x^2}{a^2} + \dfrac{y^2}{b^2} = 1$ with $a > b$

ELLIPSE WITH CENTER AT THE ORIGIN

The graph of each of the following equations is an ellipse with center at the origin and having the given properties.

EQUATION	$\dfrac{x^2}{a^2} + \dfrac{y^2}{b^2} = 1$	$\dfrac{x^2}{b^2} + \dfrac{y^2}{a^2} = 1$
	$a > b > 0$	$a > b > 0$
VERTICES	$(\pm a, 0)$	$(0, \pm a)$
MAJOR AXIS	Horizontal, length $2a$	Vertical, length $2a$
MINOR AXIS	Vertical, length $2b$	Horizontal, length $2b$
FOCI	$(\pm c, 0), \quad c^2 = a^2 - b^2$	$(0, \pm c), \quad c^2 = a^2 - b^2$
GRAPH		

In the standard equation for an ellipse, a^2 is the *larger* denominator and b^2 is the *smaller*. To find c^2, we subtract: larger denominator minus smaller denominator.

EXAMPLE 1 ■ Sketching an Ellipse

An ellipse has the equation

$$\frac{x^2}{9} + \frac{y^2}{4} = 1$$

(a) Find the foci, vertices, and the lengths of the major and minor axes, and sketch the graph.

(b) Draw the graph using a graphing calculator.

SOLUTION

(a) Since the denominator of x^2 is larger, the ellipse has horizontal major axis. This gives $a^2 = 9$ and $b^2 = 4$, so $c^2 = a^2 - b^2 = 9 - 4 = 5$. Thus, $a = 3$, $b = 2$, and $c = \sqrt{5}$.

FOCI	$(\pm \sqrt{5}, 0)$
VERTICES	$(\pm 3, 0)$
LENGTH OF MAJOR AXIS	6
LENGTH OF MINOR AXIS	4

The graph is shown in Figure 6(a).

(b) To draw the graph using a graphing calculator, we need to solve for y.

$$\frac{x^2}{9} + \frac{y^2}{4} = 1$$

$$\frac{y^2}{4} = 1 - \frac{x^2}{9} \qquad \text{Subtract } x^2/9$$

$$y^2 = 4\left(1 - \frac{x^2}{9}\right) \qquad \text{Multiply by 4}$$

$$y = \pm 2 \sqrt{1 - \frac{x^2}{9}} \qquad \text{Take square roots}$$

To obtain the graph of the ellipse, we graph both functions

$$y = 2\sqrt{1 - x^2/9} \qquad \text{and} \qquad y = -2\sqrt{1 - x^2/9}$$

as shown in Figure 6(b).

Note that the equation of an ellipse does not define y as a function of x (see page 162). That's why we need to graph two functions to graph an ellipse.

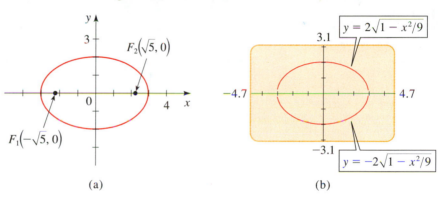

FIGURE 6
$$\frac{x^2}{9} + \frac{y^2}{4} = 1$$

(a)

(b)

EXAMPLE 2 ■ **Finding the Equation of an Ellipse**

The vertices of an ellipse are $(\pm 4, 0)$ and the foci are $(\pm 2, 0)$. Find its equation and sketch the graph.

SOLUTION

Since the vertices are $(\pm 4, 0)$, we have $a = 4$. The foci are $(\pm 2, 0)$, so $c = 2$. To write the equation, we need to find b. Since $c^2 = a^2 - b^2$, we have

$$2^2 = 4^2 - b^2$$

$$b^2 = 16 - 4 = 12$$

Thus, the equation of the ellipse is

$$\frac{x^2}{16} + \frac{y^2}{12} = 1$$

The graph is shown in Figure 7.

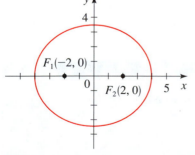

FIGURE 7
$$\frac{x^2}{16} + \frac{y^2}{12} = 1$$

EXAMPLE 3 ■ Finding the Foci of an Ellipse

Find the foci of the ellipse $16x^2 + 9y^2 = 144$, and sketch its graph.

SOLUTION

First we put the equation in standard form. Dividing by 144, we get

$$\frac{x^2}{9} + \frac{y^2}{16} = 1$$

Since $16 > 9$, this is an ellipse with its foci on the y-axis, and with $a = 4$ and $b = 3$. We have

$$c^2 = a^2 - b^2 = 16 - 9 = 7$$
$$c = \sqrt{7}$$

Thus, the foci are $(0, \pm\sqrt{7})$. The graph is shown in Figure 8(a).

We can also draw the graph using a graphing calculator as shown in Figure 8(b).

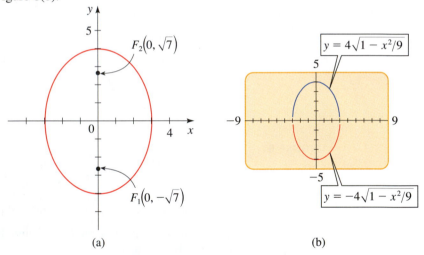

(a)   (b)

FIGURE 8
$16x^2 + 9y^2 = 144$

We saw earlier in this section (Figure 3) that if $2a$ is only slightly greater than $2c$, the ellipse is long and thin, whereas if $2a$ is much greater than $2c$, the ellipse is almost circular. We measure the deviation of an ellipse from being circular by the ratio of a and c.

DEFINITION OF ECCENTRICITY

For the ellipse $\dfrac{x^2}{a^2} + \dfrac{y^2}{b^2} = 1$ or $\dfrac{x^2}{b^2} + \dfrac{y^2}{a^2} = 1$ (with $a > b > 0$), the **eccentricity** e is the number

$$e = \frac{c}{a}$$

where $c = \sqrt{a^2 - b^2}$. The eccentricity of every ellipse satisfies $0 < e < 1$.

Johannes Kepler (1571–1630) was the first to give a correct description of the motion of the planets. The cosmology of his time postulated complicated systems of circles moving on circles to describe these motions. Kepler sought a simpler and more harmonious description. As official astronomer at the imperial court in Prague, he studied the astronomical observations of the Danish astronomer Tycho Brahe, whose data were the most accurate available at the time. After numerous attempts to find a theory, Kepler made the momentous discovery that the orbits of the planets are elliptical. His three great laws of planetary motion are

1. The orbit of each planet is an ellipse with the sun at one focus.
2. The line segment that joins the sun to a planet sweeps out equal areas in equal time (see the figure).
3. The square of the period of revolution of a planet is proportional to the cube of the length of the major axis of its orbit.

His formulation of these laws is perhaps the most impressive deduction from empirical data in the history of science.

Thus, if e is close to 1, then c is almost equal to a, and the ellipse is elongated in shape, but if e is close to 0, then the ellipse is close to a circle in shape. The eccentricity is a measure of how "stretched" the ellipse is.

In Figure 9 we show a number of ellipses to demonstrate the effect of varying the eccentricity e.

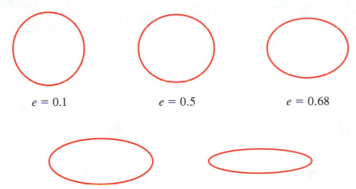

$e = 0.1$ $e = 0.5$ $e = 0.68$

$e = 0.86$ $e = 0.96$

FIGURE 9
Ellipses with various eccentricities

EXAMPLE 4 ■ Finding the Equation of an Ellipse from Its Eccentricity and Foci

Find the equation of the ellipse with foci $(0, \pm 8)$ and eccentricity $e = \frac{4}{5}$, and sketch its graph.

SOLUTION

We are given $e = \frac{4}{5}$ and $c = 8$. Thus

$$\frac{4}{5} = \frac{8}{a} \qquad \text{Eccentricity } e = \frac{c}{a}$$

$$4a = 40 \qquad \text{Cross multiply}$$

$$a = 10$$

To find b, we use the fact that $c^2 = a^2 - b^2$.

$$8^2 = 10^2 - b^2$$

$$b^2 = 10^2 - 8^2 = 36$$

$$b = 6$$

Thus, the equation of the ellipse is

$$\frac{x^2}{36} + \frac{y^2}{100} = 1$$

Because the foci are on the y-axis, the ellipse is oriented vertically. To sketch the ellipse, we find the intercepts: The x-intercepts are ± 6 and the y-intercepts are ± 10. The graph is sketched in Figure 10.

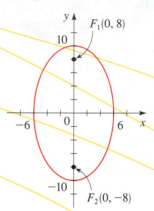

FIGURE 10

$$\frac{x^2}{36} + \frac{y^2}{100} = 1$$

Gravitational attraction causes the planets to move in elliptical orbits around the sun with the sun at one focus. This remarkable property was first observed by Johannes Kepler and was later deduced by Isaac Newton from his inverse square law of gravity, using calculus. The orbits of the planets have different eccentricities, but most are nearly circular (see the margin note on page 738).

Ellipses, like parabolas, have an interesting *reflection property* that leads to a number of practical applications. If a light source is placed at one focus of a reflecting surface with elliptical cross sections, then all the light will be reflected off the surface to the other focus, as shown in Figure 11. This principle, which works for sound waves as well as for light, is used in *lithotripsy*, a treatment for kidney stones. The patient is placed in a tub of water with elliptical cross sections in such a way that the kidney stone is accurately located at one focus. High-intensity sound waves generated at the other focus are reflected to the stone and destroy it with minimal damage to surrounding tissue. The patient is spared the trauma of surgery and recovers within days instead of weeks.

The reflection property of ellipses is also used in the construction of *whispering galleries*. Sound coming from one focus bounces off the walls and ceiling of an elliptical room and passes through the other focus. In these rooms even quiet whispers spoken at one focus can be heard clearly at the other. Famous whispering galleries include the National Statuary Gallery of the U.S. Capitol in Washington, D.C., and the Mormon Tabernacle in Salt Lake City, Utah.

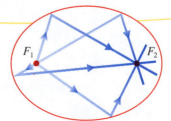

FIGURE 11

9.2 EXERCISES

1–4 ■ Match the equation with the graphs labeled I–IV. Give reasons for your answers.

1. $\dfrac{x^2}{16} + \dfrac{y^2}{4} = 1$

2. $x^2 + \dfrac{y^2}{9} = 1$

3. $4x^2 + y^2 = 4$

4. $16x^2 + 25y^2 = 400$

I

II

III

IV

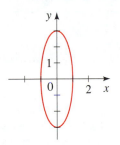

5–18 ■ Find the vertices, foci, and eccentricity of the ellipse. Determine the lengths of the major and minor axes, and sketch the graph.

5. $\dfrac{x^2}{25} + \dfrac{y^2}{9} = 1$

6. $\dfrac{x^2}{16} + \dfrac{y^2}{25} = 1$

7. $9x^2 + 4y^2 = 36$

8. $4x^2 + 25y^2 = 100$

9. $x^2 + 4y^2 = 16$

10. $4x^2 + y^2 = 16$

11. $2x^2 + y^2 = 3$

12. $5x^2 + 6y^2 = 30$

13. $x^2 + 4y^2 = 1$

14. $9x^2 + 4y^2 = 1$

15. $\tfrac{1}{2}x^2 + \tfrac{1}{8}y^2 = \tfrac{1}{4}$

16. $x^2 = 4 - 2y^2$

17. $y^2 = 1 - 2x^2$

18. $20x^2 + 4y^2 = 5$

19–24 ■ Find an equation for the ellipse whose graph is shown.

19.

20.

21.

F(0, 2)

22.

F(0, 3)

23.

(8, 6)

24.

(−1, 2)

25–28 ■ Use a graphing device to graph the ellipse.

25. $\dfrac{x^2}{25} + \dfrac{y^2}{20} = 1$

26. $x^2 + \dfrac{y^2}{12} = 1$

27. $6x^2 + y^2 = 36$

28. $x^2 + 2y^2 = 8$

29–40 ■ Find an equation for the ellipse that satisfies the given conditions.

29. Foci $(\pm 4, 0)$, vertices $(\pm 5, 0)$

30. Foci $(0, \pm 3)$, vertices $(0, \pm 5)$

31. Length of major axis 4, length of minor axis 2, foci on y-axis

32. Length of major axis 6, length of minor axis 4, foci on x-axis

33. Foci $(0, \pm 2)$, length of minor axis 6

34. Foci $(\pm 5, 0)$, length of major axis 12

35. Endpoints of major axis $(\pm 10, 0)$, distance between foci 6

36. Endpoints of minor axis $(0, \pm 3)$, distance between foci 8

37. Length of major axis 10, foci on x-axis, ellipse passes through the point $\left(\sqrt{5}, 2\right)$

38. Eccentricity $\frac{1}{9}$, foci $(0, \pm 2)$

39. Eccentricity 0.8, foci $(\pm 1.5, 0)$

40. Eccentricity $\sqrt{3}/2$, foci on y-axis, length of major axis 4

41–42 ■ Find the intersection points of the pair of ellipses. Sketch the graphs of each pair of equations on the same coordinate axes and label the points of intersection.

41. $\begin{cases} 4x^2 + y^2 = 4 \\ 4x^2 + 9y^2 = 36 \end{cases}$

42. $\begin{cases} \dfrac{x^2}{16} + \dfrac{y^2}{9} = 1 \\ \dfrac{x^2}{9} + \dfrac{y^2}{16} = 1 \end{cases}$

43. The planets move around the sun in elliptical orbits with the sun at one focus. The point in the orbit at which the planet is closest to the sun is called **perihelion**, and the point at which it is farthest is called **aphelion**. These points are the vertices of the orbit. The earth's distance from the sun is 147,000,000 km at perihelion and 153,000,000 km at aphelion. Find an equation for the earth's orbit. (Place the origin at the center of the orbit with the sun on the x-axis.)

44. With an eccentricity of 0.25, Pluto's orbit is the most eccentric in the solar system. The length of the minor axis of its orbit is approximately 10,000,000,000 km. Find the distance between Pluto and the sun at perihelion and at aphelion. (See Exercise 43.)

45. For an object in an elliptical orbit around the moon, the points in the orbit that are closest to and farthest from the center of the moon are called **perilune** and **apolune**, respectively. These are the vertices of the orbit. The center of the moon is at one focus of the orbit. The *Apollo 11* spacecraft was placed in a lunar orbit with perilune at 68 mi and apolune at 195 mi above the surface of the moon. Assuming the moon is a sphere of radius 1075 mi, find an equation for the orbit of *Apollo 11*. (Place the coordinate axes so that the origin is at the center of the orbit and the foci are located on the x-axis.)

46. A carpenter wishes to construct an elliptical table top from a sheet of plywood, 4 ft by 8 ft. He will trace out the ellipse using the "thumbtack and string" method illustrated in Figures 2 and 3. What length of string should he use, and how far apart should the tacks be located, if the ellipse is to be the largest possible that can be cut out of the plywood sheet?

47. A "sunburst" window above a doorway is constructed in the shape of the top half of an ellipse, as shown in the

figure. The window is 20 in. tall at its highest point and 80 in. wide at the bottom. Find the height of the window 25 in. from the center of the base.

48. The **ancillary circle** of an ellipse is the circle with radius equal to half the length of the minor axis and center the same as the ellipse (see the figure). The ancillary circle is thus the largest circle that can fit within an ellipse.
(a) Find an equation for the ancillary circle of the ellipse $x^2 + 4y^2 = 16$.
(b) For the ellipse and ancillary circle of part (a), show that if (s, t) is a point on the ancillary circle, then $(2s, t)$ is a point on the ellipse.

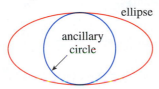

49. A **latus rectum** for an ellipse is a line segment perpendicular to the major axis at a focus, with endpoints on the ellipse, as shown. Show that the length of a latus rectum is $2b^2/a$ for the ellipse

$$\frac{x^2}{a^2} + \frac{y^2}{b^2} = 1 \qquad \text{with } a > b$$

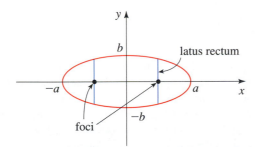

50. If $k > 0$, the following equation represents an ellipse:

$$\frac{x^2}{k} + \frac{y^2}{4 + k} = 1$$

Show that all the ellipses represented by this equation have the same foci, no matter what the value of k.

51. (a) Use a graphing device to sketch the top half (the portion in the first and second quadrants) of the family of ellipses $x^2 + ky^2 = 100$ for $k = 4, 10, 25,$ and 50.
(b) What do the members of this family of ellipses have in common? How do they differ?

DISCOVERY · DISCUSSION

52. Drawing an Ellipse on a Blackboard   Try drawing an ellipse as accurately as possible on a blackboard. How would a piece of string and two friends help this process?

53. Light Cone from a Flashlight   A flashlight shines on a wall, as shown in the figure. What is the shape of the boundary of the lighted area? Explain your answer.

54. Is It an Ellipse?   A piece of paper is wrapped around a cylindrical bottle, and then a compass is used to draw a circle on the paper, as shown in the figure. When the paper is laid flat, is the shape drawn on the paper an ellipse? (You don't need to prove your answer, but you may want to do the experiment and see what you get.)

9.3 HYPERBOLAS

Although ellipses and hyperbolas have completely different shapes, their definitions and equations are similar. Instead of using the *sum* of distances from two fixed foci, as in the case of an ellipse, we use the *difference* to define a hyperbola.

GEOMETRIC DEFINITION OF A HYPERBOLA

A **hyperbola** is the set of all points in the plane, the difference of whose distances from two fixed points F_1 and F_2 is a constant. (See Figure 1.) These two fixed points are the **foci** of the hyperbola.

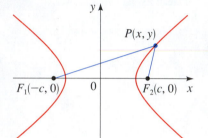

FIGURE 1

P is on the hyperbola if
$|d(P, F_1) - d(P, F_2)| = 2a$

As in the case of the ellipse, we get the simplest equation for the hyperbola by placing the foci on the x-axis at $(\pm c, 0)$, as shown in Figure 1. By definition, if $P(x, y)$ lies on the hyperbola, then either $d(P, F_1) - d(P, F_2)$ or $d(P, F_2) - d(P, F_1)$ must equal some positive constant, which we call $2a$. Thus, we have

$$d(P, F_1) - d(P, F_2) = \pm 2a$$

or

$$\sqrt{(x + c)^2 + y^2} - \sqrt{(x - c)^2 + y^2} = \pm 2a$$

Proceeding as we did in the case of the ellipse (Section 9.2), we simplify this to

$$(c^2 - a^2)x^2 - a^2 y^2 = a^2(c^2 - a^2)$$

From triangle PF_1F_2 in Figure 1 we see that $|d(P, F_1) - d(P, F_2)| < 2c$. It follows that $2a < 2c$, or $a < c$. Thus, $c^2 - a^2 > 0$, so we can set $b^2 = c^2 - a^2$. We then simplify the last displayed equation to get

$$\frac{x^2}{a^2} - \frac{y^2}{b^2} = 1$$

This is the *equation of the hyperbola*. If we replace x by $-x$ or y by $-y$ in this equation, it remains unchanged, so the hyperbola is symmetric about both the x- and y-axes and about the origin. The x-intercepts are $\pm a$, and the points $(a, 0)$ and $(-a, 0)$ are the **vertices** of the hyperbola. There is no y-intercept, because setting $x = 0$ in the equation of the hyperbola leads to $-y^2 = b^2$, which has no real solution. Furthermore, the equation of the hyperbola implies that

$$\frac{x^2}{a^2} = \frac{y^2}{b^2} + 1 \geq 1$$

so $x^2/a^2 \geq 1$; thus, $x^2 \geq a^2$, and hence $x \geq a$ or $x \leq -a$. This means that the hyperbola consists of two parts, called its **branches**. The segment joining the two vertices on the separate branches is the **transverse axis** of the hyperbola, and the origin is called its **center**.

If we place the foci of the hyperbola on the y-axis rather than on the x-axis, then this has the effect of reversing the roles of x and y in the derivation of the equation of the hyperbola. This leads to a hyperbola with a vertical transverse axis.

The main properties of hyperbolas are listed in the following box.

HYPERBOLA WITH CENTER AT THE ORIGIN

The graph of each of the following equations is a hyperbola with center at the origin and having the given properties.

EQUATION	$\dfrac{x^2}{a^2} - \dfrac{y^2}{b^2} = 1 \quad (a > 0, b > 0)$	$\dfrac{y^2}{a^2} - \dfrac{x^2}{b^2} = 1 \quad (a > 0, b > 0)$
VERTICES	$(\pm a, 0)$	$(0, \pm a)$
TRANSVERSE AXIS	Horizontal, length $2a$	Vertical, length $2a$
ASYMPTOTES	$y = \pm \dfrac{b}{a} x$	$y = \pm \dfrac{a}{b} x$
FOCI	$(\pm c, 0), \quad c^2 = a^2 + b^2$	$(0, \pm c), \quad c^2 = a^2 + b^2$
GRAPH		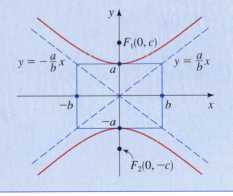

The *asymptotes* mentioned in this box are lines that the hyperbola approaches for large values of x and y. To find the asymptotes in the first case in the box, we solve the equation for y to get

Asymptotes of rational functions are discussed in Section 3.6.

$$y = \pm \frac{b}{a} \sqrt{x^2 - a^2}$$

$$= \pm \frac{b}{a} x \sqrt{1 - \frac{a^2}{x^2}}$$

As x gets large, a^2/x^2 gets closer to zero. In other words, as $x \to \infty$ we have $a^2/x^2 \to 0$. So, for large x the value of y can be approximated as $y = \pm (b/a)x$. This shows that these lines are asymptotes of the hyperbola.

Asymptotes are an essential aid for graphing a hyperbola; they help us determine its shape. A convenient way to find the asymptotes, for a parabola with horizontal transverse axis, is to first plot the points $(a, 0)$, $(-a, 0)$, $(0, b)$, and $(0, -b)$. Then sketch horizontal and vertical segments through these points to construct a rectangle, as shown in Figure 2(a). We call this rectangle the **central box** of the hyperbola. The slopes of the diagonals of the central box are $\pm b/a$, so by extending them we obtain the asymptotes $y = \pm(b/a)x$, as sketched in part (b) of the figure. Finally, we plot the vertices and use the asymptotes as a guide in sketching the hyperbola shown in part (c). (A similar procedure applies to graphing a hyperbola that has a vertical transverse axis.)

(a) Central box

(b) Asymptotes

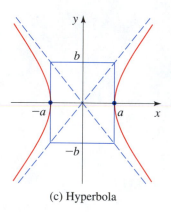

(c) Hyperbola

FIGURE 2

Steps in graphing the hyperbola $\dfrac{x^2}{a^2} - \dfrac{y^2}{b^2} = 1$

HOW TO SKETCH A HYPERBOLA

1. **SKETCH THE CENTRAL BOX.**   This is the rectangle centered at the origin, with sides parallel to the axes, that crosses one axis at $\pm a$, the other at $\pm b$.

2. **SKETCH THE ASYMPTOTES.**   These are the lines obtained by extending the diagonals of the central box.

3. **PLOT THE VERTICES.**   These are the two x-intercepts or the two y-intercepts.

4. **SKETCH THE HYPERBOLA.**   Start at a vertex and sketch a branch of the hyperbola, approaching the asymptotes. Sketch the other branch in the same way.

EXAMPLE 1 ■ A Hyperbola with Horizontal Transverse Axis

A hyperbola has the equation

$$9x^2 - 16y^2 = 144$$

(a) Find the vertices, foci, and asymptotes, and sketch the graph.

(b) Draw the graph using a graphing calculator.

SOLUTION

(a) First we divide both sides of the equation by 144 to put it into standard form:

$$\frac{x^2}{16} - \frac{y^2}{9} = 1$$

Because the x^2-term is positive, the hyperbola has a horizontal transverse axis; its vertices and foci are on the x-axis. Since $a^2 = 16$ and $b^2 = 9$, we get $a = 4$, $b = 3$, and $c = \sqrt{16 + 9} = 5$. Thus, we have

VERTICES	$(\pm 4, 0)$
FOCI	$(\pm 5, 0)$
ASYMPTOTES	$y = \pm\frac{3}{4}x$

After sketching the central box and asymptotes, we complete the sketch of the hyperbola as in Figure 3(a).

(b) To draw the graph using a graphing calculator, we need to solve for y.

$$9x^2 - 16y^2 = 144$$

$$-16y^2 = -9x^2 + 144 \qquad \text{Subtract } 9x^2$$

$$y^2 = 9\left(\frac{x^2}{16} - 1\right) \qquad \text{Divide by } -16 \text{ and factor } 9$$

$$y = \pm 3\sqrt{\frac{x^2}{16} - 1} \qquad \text{Take square roots}$$

Note that the equation of a hyperbola does not define y as a function of x (see page 162). That's why we need to graph two functions to graph a hyperbola.

To obtain the graph of the hyperbola, we graph the functions

$$y = 3\sqrt{(x^2/16) - 1} \qquad \text{and} \qquad y = -3\sqrt{(x^2/16) - 1}$$

as shown in Figure 3(b).

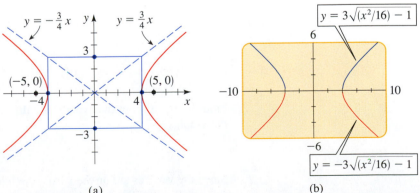

FIGURE 3

$9x^2 - 16y^2 = 144$

(a)

(b)

Paths of Comets

The path of a comet is an ellipse, a parabola, or a hyperbola with the sun at a focus. This fact can be proved using calculus and Newton's laws of motion.* If the path is a parabola or a hyperbola, the comet will never return. If the path is an ellipse, it can be determined precisely when and where the comet can be seen again. Halley's comet has an elliptical path and returns every 75 years; it was last seen in 1987. The brightest comet of the 20th century was comet Hale-Bopp, seen in 1997. Its orbit is also elliptical.

*James Stewart, *Calculus,* 4th ed. (Pacific Grove, CA: Brooks/Cole, 1999), pp. 897–899.

EXAMPLE 2 ■ A Hyperbola with Vertical Transverse Axis

Find the vertices, foci, and asymptotes of the hyperbola, and sketch its graph.

$$x^2 - 9y^2 + 9 = 0$$

SOLUTION

We begin by writing the equation in the standard form for a hyperbola.

$$x^2 - 9y^2 = -9$$

$$y^2 - \frac{x^2}{9} = 1 \qquad \text{Divide by } -9$$

Because the y^2-term is positive, the hyperbola has a vertical transverse axis; its foci and vertices are on the y-axis. Since $a^2 = 1$ and $b^2 = 9$, we get $a = 1$, $b = 3$, and $c = \sqrt{1 + 9} = \sqrt{10}$. Thus, we have

VERTICES	$(0, \pm 1)$
FOCI	$(0, \pm \sqrt{10})$
ASYMPTOTES	$y = \pm \frac{1}{3} x$

We sketch the central box and asymptotes, then complete the graph, as shown in Figure 4(a).

We can also draw the graph using a graphing calculator, as shown in Figure 4(b).

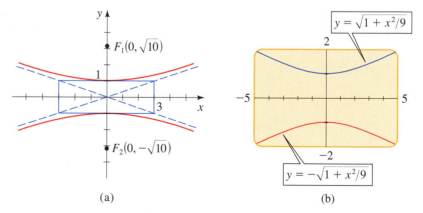

FIGURE 4
$x^2 - 9y^2 + 9 = 0$

(a)

(b)

EXAMPLE 3 ■ Finding the Equation of a Hyperbola from Its Vertices and Foci

Find the equation of the hyperbola with vertices $(\pm 3, 0)$ and foci $(\pm 4, 0)$. Sketch the graph.

SOLUTION

Since the vertices are on the x-axis, the hyperbola has a horizontal transverse axis.

Its equation is of the form

$$\frac{x^2}{3^2} - \frac{y^2}{b^2} = 1$$

We have $a = 3$ and $c = 4$. To find b, we use the relation $a^2 + b^2 = c^2$:

$$3^2 + b^2 = 4^2$$

$$b^2 = 4^2 - 3^2 = 7$$

$$b = \sqrt{7}$$

Thus, the equation of the hyperbola is

$$\frac{x^2}{9} - \frac{y^2}{7} = 1$$

The graph is shown in Figure 5.   ■

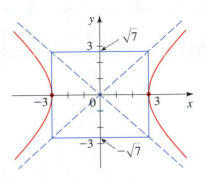

FIGURE 5
$$\frac{x^2}{9} - \frac{y^2}{7} = 1$$

EXAMPLE 4 ■ Finding the Equation of a Hyperbola from Its Vertices and Asymptotes

Find the equation and the foci of the hyperbola with vertices $(0, \pm 2)$ and asymptotes $y = \pm 2x$. Sketch the graph.

SOLUTION

Since the vertices are on the y-axis, the hyperbola has a vertical transverse axis with $a = 2$. From the asymptote equation we see that

$$\frac{a}{b} = 2$$

$$\frac{2}{b} = 2 \qquad \text{Since } a = 2$$

$$b = \frac{2}{2} = 1$$

Thus, the equation of the hyperbola is

$$\frac{y^2}{4} - x^2 = 1$$

To find the foci, we calculate $c^2 = a^2 + b^2 = 2^2 + 1^2 = 5$, so $c = \sqrt{5}$. Thus, the foci are $(0, \pm\sqrt{5})$. The graph is shown in Figure 6.   ■

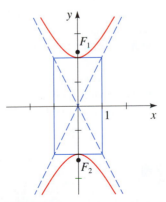

FIGURE 6
$$\frac{y^2}{4} - x^2 = 1$$

Like parabolas and ellipses, hyperbolas have an interesting *reflection property*. Light aimed at one focus of a hyperbolic mirror is reflected toward the other focus, as shown in Figure 7. This property is used in the construction of Cassegrain-type telescopes. A hyperbolic mirror is placed in the telescope tube so that light reflected from the primary parabolic reflector is aimed at one focus of the hyperbolic mirror. The light is then refocused at a more accessible point below the primary reflector (Figure 8).

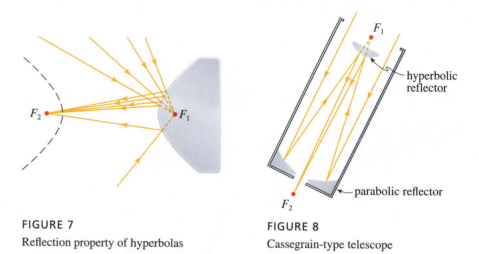

FIGURE 7

Reflection property of hyperbolas

FIGURE 8

Cassegrain-type telescope

The LORAN (LOng RAnge Navigation) system was used until the early 1990s; it has now been superseded by the GPS system (see page 626). In the LORAN system, hyperbolas are used onboard a ship to determine its location. In Figure 9 radio stations at A and B transmit signals simultaneously for reception by the ship at P. The onboard computer converts the time difference in reception of these signals into a distance difference $d(P, A) - d(P, B)$. From the definition of a hyperbola this locates the ship on one branch of a hyperbola with foci at A and B (sketched in black in the figure). The same procedure is carried out with two other radio stations at C and D, and this locates the ship on a second hyperbola (shown in red in the figure). (In practice, only three stations are needed because one station can be used as a focus for both hyperbolas.) The coordinates of the intersection point of these two hyperbolas, which can be calculated precisely by the computer, give the location of P.

FIGURE 9

LORAN system for finding the location of a ship

9.3   EXERCISES

1–4 ■ Match the equation with the graphs labeled I–IV. Give reasons for your answers.

1. $\dfrac{x^2}{4} - y^2 = 1$

2. $y^2 - \dfrac{x^2}{9} = 1$

3. $16y^2 - x^2 = 144$

4. $9x^2 - 25y^2 = 225$

I

II

III

IV

5–16 ■ Find the vertices, foci, and asymptotes of the hyperbola, and sketch its graph.

5. $\dfrac{x^2}{4} - \dfrac{y^2}{16} = 1$

6. $\dfrac{y^2}{9} - \dfrac{x^2}{16} = 1$

7. $y^2 - \dfrac{x^2}{25} = 1$

8. $\dfrac{x^2}{2} - y^2 = 1$

9. $x^2 - y^2 = 1$

10. $9x^2 - 4y^2 = 36$

11. $25y^2 - 9x^2 = 225$

12. $x^2 - y^2 + 4 = 0$

13. $x^2 - 4y^2 - 8 = 0$

14. $x^2 - 2y^2 = 3$

15. $4y^2 - x^2 = 1$

16. $9x^2 - 16y^2 = 1$

17–22 ■ Find the equation for the hyperbola whose graph is shown.

17.

18.

19.

20.

21.

22.

23–26 ■ Use a graphing device to graph the hyperbola.

23. $x^2 - 2y^2 = 8$

24. $3y^2 - 4x^2 = 24$

25. $\dfrac{y^2}{2} - \dfrac{x^2}{6} = 1$

26. $\dfrac{x^2}{100} - \dfrac{y^2}{64} = 1$

27–38 ■ Find an equation for the hyperbola that satisfies the given conditions.

27. Foci $(\pm 5, 0)$, vertices $(\pm 3, 0)$

28. Foci $(0, \pm 10)$, vertices $(0, \pm 8)$

29. Foci $(0, \pm 2)$, vertices $(0, \pm 1)$

30. Foci $(\pm 6, 0)$, vertices $(\pm 2, 0)$

31. Vertices $(\pm 1, 0)$, asymptotes $y = \pm 5x$

32. Vertices $(0, \pm 6)$, asymptotes $y = \pm \frac{1}{3}x$

33. Foci $(0, \pm 8)$, asymptotes $y = \pm \frac{1}{2}x$

34. Vertices $(0, \pm 6)$, hyperbola passes through $(-5, 9)$

35. Asymptotes $y = \pm x$, hyperbola passes through $(5, 3)$

36. Foci $(\pm 3, 0)$, hyperbola passes through $(4, 1)$

37. Foci $(\pm 5, 0)$, length of transverse axis 6

38. Foci $(0, \pm 1)$, length of transverse axis 1

39. (a) Show that the asymptotes of the hyperbola $x^2 - y^2 = 5$ are perpendicular to each other.
   (b) Find an equation for the hyperbola with foci $(\pm c, 0)$ and with asymptotes perpendicular to each other.

40. The hyperbolas

$$\frac{x^2}{a^2} - \frac{y^2}{b^2} = 1 \qquad \text{and} \qquad \frac{x^2}{a^2} - \frac{y^2}{b^2} = -1$$

are said to be **conjugate** to each other.
   (a) Show that the hyperbolas

$$x^2 - 4y^2 + 16 = 0 \qquad \text{and} \qquad 4y^2 - x^2 + 16 = 0$$

are conjugate to each other, and sketch their graphs on the same coordinate axes.
   (b) What do the hyperbolas of part (a) have in common?
   (c) Show that any pair of conjugate hyperbolas have the relationship you discovered in part (b).

41. In the derivation of the equation of the hyperbola at the beginning of this section, we said that the equation

$$\sqrt{(x + c)^2 + y^2} - \sqrt{(x - c)^2 + y^2} = \pm 2a$$

simplifies to

$$(c^2 - a^2)x^2 - a^2 y^2 = a^2(c^2 - a^2)$$

Supply the steps needed to show this.

42. (a) For the hyperbola

$$\frac{x^2}{9} - \frac{y^2}{16} = 1$$

determine the values of a, b, and c, and find the coordinates of the foci F_1 and F_2.
   (b) Show that the point $P\left(5, \frac{16}{3}\right)$ lies on this hyperbola.

(c) Find $d(P, F_1)$ and $d(P, F_2)$.
(d) Verify that the difference between $d(P, F_1)$ and $d(P, F_2)$ is $2a$.

43. Refer to Figure 9 in the text. Suppose that the radio stations at A and B are 500 mi apart, and that the ship at P receives Station A's signal 2640 microseconds (μs) before it receives the signal from B.
   (a) Assuming that radio signals travel at 980 ft/μs, find $d(P, A) - d(P, B)$.
   (b) Find an equation for the branch of the hyperbola indicated in black in the figure. (Place A and B on the y-axis with the origin halfway between them. Use miles as the unit of distance.)
   (c) If A is due north of B, and if P is due east of A, how far is P from A?

44. Some comets, such as Halley's comet, are a permanent part of the solar system, traveling in elliptical orbits around the sun. Others pass through the solar system only once, following a hyperbolic path with the sun at a focus. The figure shows the path of such a comet. Find an equation for the path, assuming that the closest the comet comes to the sun is 2×10^9 mi and that the path the comet was taking before it neared the solar system is at a right angle to the path it continues on after leaving the solar system.

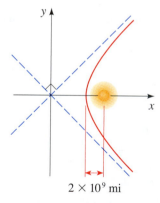

2×10^9 mi

45. Hyperbolas are called **confocal** if they have the same foci.
   (a) Show that the hyperbolas

$$\frac{y^2}{k} - \frac{x^2}{16 - k} = 1 \qquad \text{with } 0 < k < 16$$

are confocal.
   (b) Use a graphing device to draw the top branches of the family of hyperbolas in part (a) for $k = 1$, 4, 8, and 12. How does the shape of the graph change as k increases?

DISCOVERY · DISCUSSION

46. Hyperbolas in the Real World   Several examples of the uses of hyperbolas are given in the text. Find other situations in real life where hyperbolas occur. Consult a scientific encyclopedia in the reference section of your library, or search the Internet.

47. Light from a Lamp   The light from a lamp forms a lighted area on a wall, as shown in the figure. Why is the boundary of this lighted area a hyperbola? How can one hold a flashlight so that its beam forms a hyperbola on the ground?

9.4   SHIFTED CONICS

In the preceding sections we studied parabolas with vertices at the origin and ellipses and hyperbolas with centers at the origin. We restricted ourselves to these cases because these equations have the simplest form. In this section we consider conics whose vertices and centers are not necessarily at the origin, and we determine how this affects their equations.

In Section 2.5 we studied transformations of functions that have the effect of shifting their graphs. In general, for any equation in x and y, if we replace x by $x - h$ or by $x + h$, the graph of the new equation is simply the old graph shifted horizontally; if y is replaced by $y - k$ or by $y + k$, the graph is shifted vertically. The following box gives the details.

SHIFTING GRAPHS OF EQUATIONS

If h and k are positive real numbers, then replacing x by $x - h$ or by $x + h$ and replacing y by $y - k$ or by $y + k$ has the following effect(s) on the graph of any equation in x and y.

Replacement	How the graph is shifted
1. x replaced by $x - h$	Right h units
2. x replaced by $x + h$	Left h units
3. y replaced by $y - k$	Upward k units
4. y replaced by $y + k$	Downward k units

For example, consider the ellipse with equation

$$\frac{x^2}{a^2} + \frac{y^2}{b^2} = 1$$

which is shown in Figure 1. If we shift it so that its center is at the point (h, k)
instead of at the origin, then its equation becomes

$$\frac{(x - h)^2}{a^2} + \frac{(y - k)^2}{b^2} = 1$$

FIGURE 1
Shifted ellipse

EXAMPLE 1 ■ **Sketching the Graph of a Shifted Ellipse**

Sketch the graph of the ellipse

$$\frac{(x + 1)^2}{4} + \frac{(y - 2)^2}{9} = 1$$

and determine the coordinates of the foci.

SOLUTION

The ellipse

$$\frac{(x + 1)^2}{4} + \frac{(y - 2)^2}{9} = 1 \qquad \text{Shifted ellipse}$$

is shifted so that its center is at $(-1, 2)$. It is obtained from the ellipse

$$\frac{x^2}{4} + \frac{y^2}{9} = 1 \qquad \text{Ellipse with center at origin}$$

by shifting it left 1 unit and upward 2 units. The endpoints of the minor and major
axes of the unshifted ellipse are $(2, 0)$, $(-2, 0)$, $(0, 3)$, $(0, -3)$. We apply the
required shifts to these points to obtain the corresponding points on the shifted
ellipse:

$$(2, 0) \quad \rightarrow \quad (2 - 1, 0 + 2) = (1, 2)$$

$$(-2, 0) \quad \rightarrow \quad (-2 - 1, 0 + 2) = (-3, 2)$$

$$(0, 3) \quad \rightarrow \quad (0 - 1, 3 + 2) = (-1, 5)$$

$$(0, -3) \quad \rightarrow \quad (0 - 1, -3 + 2) = (-1, -1)$$

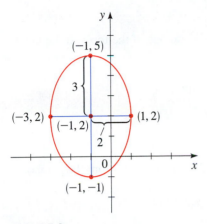

FIGURE 2

$$\frac{(x+1)^2}{4} + \frac{(y-2)^2}{9} = 1$$

This helps us sketch the graph in Figure 2.

To find the foci of the shifted ellipse, we first find the foci of the ellipse with center at the origin. Since $a^2 = 9$ and $b^2 = 4$, we have $c^2 = 9 - 4 = 5$, so $c = \sqrt{5}$. So the foci are $(0, \pm\sqrt{5})$. Shifting left 1 unit and upward 2 units, we get

$$(0, \sqrt{5}) \;\to\; (0-1, \sqrt{5}+2) = (-1, 2+\sqrt{5})$$

$$(0, -\sqrt{5}) \;\to\; (0-1, -\sqrt{5}+2) = (-1, 2-\sqrt{5})$$

Thus, the foci of the shifted ellipse are

$$(-1, 2+\sqrt{5}) \quad \text{and} \quad (-1, 2-\sqrt{5}) \qquad \blacksquare$$

Applying shifts to parabolas and hyperbolas leads to the equations and graphs shown in Figures 3 and 4.

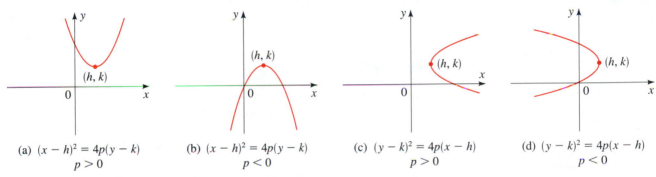

(a) $(x-h)^2 = 4p(y-k)$ $p>0$ (b) $(x-h)^2 = 4p(y-k)$ $p<0$ (c) $(y-k)^2 = 4p(x-h)$ $p>0$ (d) $(y-k)^2 = 4p(x-h)$ $p<0$

FIGURE 3
Shifted parabolas

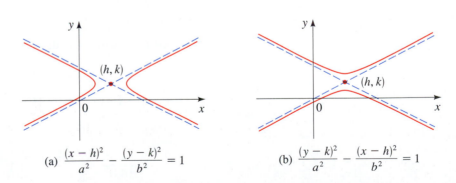

FIGURE 4
Shifted hyperbolas

(a) $\dfrac{(x-h)^2}{a^2} - \dfrac{(y-k)^2}{b^2} = 1$ (b) $\dfrac{(y-k)^2}{a^2} - \dfrac{(x-h)^2}{b^2} = 1$

EXAMPLE 2 ■ Graphing a Shifted Parabola

Determine the vertex, focus, and directrix and sketch the graph of the parabola.

$$x^2 - 4x = 8y - 28$$

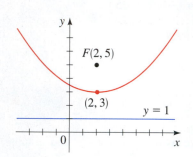

FIGURE 5
$x^2 - 4x = 8y - 28$

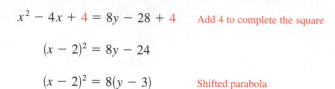

SOLUTION

We complete the square in x to put this equation into one of the forms in Figure 3.

$$x^2 - 4x + 4 = 8y - 28 + 4 \qquad \text{Add 4 to complete the square}$$

$$(x - 2)^2 = 8y - 24$$

$$(x - 2)^2 = 8(y - 3) \qquad \text{Shifted parabola}$$

This parabola opens upward with vertex at $(2, 3)$. It is obtained from the parabola

$$x^2 = 8y \qquad \text{Parabola with vertex at origin}$$

by shifting right 2 units and upward 3 units. Since $4p = 8$, we have $p = 2$, so the focus is 2 units above the vertex and the directrix is 2 units below the vertex. Thus, the focus is $(2, 5)$ and the directrix is $y = 1$. The graph is shown in Figure 5. ■

EXAMPLE 3 ■ **Graphing a Shifted Hyperbola**

A shifted conic has the equation

$$9x^2 - 72x - 16y^2 - 32y = 16$$

(a) Complete the square in x and y to show that the equation represents a hyperbola.
(b) Find the center, vertices, foci, and asymptotes of the hyperbola and sketch its graph.
 (c) Draw the graph using a graphing calculator.

SOLUTION

(a) We complete the squares in both x and y:

$$9(x^2 - 8x \qquad) - 16(y^2 + 2y \qquad) = 16$$

$$9(x^2 - 8x + 16) - 16(y^2 + 2y + 1) = 16 + 9 \cdot 16 - 16 \cdot 1 \qquad \text{Complete the squares}$$

$$9(x - 4)^2 - 16(y + 1)^2 = 144 \qquad \text{Divide this by 144}$$

$$\frac{(x - 4)^2}{16} - \frac{(y + 1)^2}{9} = 1 \qquad \text{Shifted hyperbola}$$

Comparing this to Figure 4(a), we see that this is the equation of a shifted hyperbola.

(b) The shifted hyperbola has center $(4, -1)$ and a horizontal transverse axis.

$$\text{CENTER} \quad (4, -1)$$

Its graph will have the same shape as the unshifted hyperbola

$$\frac{x^2}{16} - \frac{y^2}{9} = 1 \qquad \text{Hyperbola with center at origin}$$

Since $a^2 = 16$ and $b^2 = 9$, we have $a = 4$, $b = 3$, and $c = \sqrt{a^2 + b^2} = \sqrt{16 + 9} = 5$. Thus, the foci lie 5 units to the left and to the right of the center, and the vertices lie 4 units to either side of the center.

$$\text{FOCI} \quad (-1, -1) \text{ and } (9, -1)$$

$$\text{VERTICES} \quad (0, -1) \text{ and } (8, -1)$$

The asymptotes of the unshifted hyperbola are $y = \pm\frac{3}{4}x$, so the asymptotes of the shifted parabola are found as follows.

$$\text{ASYMPTOTES} \quad y + 1 = \pm\tfrac{3}{4}(x - 4)$$

$$y + 1 = \pm\tfrac{3}{4}x \mp 3$$

$$y = \tfrac{3}{4}x - 4 \qquad \text{and} \qquad y = -\tfrac{3}{4}x + 2$$

To help us sketch the hyperbola, we draw the central box; it extends 4 units left and right from the center and 3 units upward and downward from the center. We then draw the asymptotes and complete the graph of the shifted hyperbola as shown in Figure 6(a) on the following page.

 (c) To draw the graph using a graphing calculator, we need to solve for y. The given equation is a quadratic equation in y, so we use the quadratic formula to solve for y. Writing the equation in the form

$$16y^2 + 32y - 9x^2 + 72x + 16 = 0$$

we get

$$y = \frac{-32 \pm \sqrt{32^2 - 4(16)(-9x^2 + 72x + 16)}}{2(16)} \qquad \text{Quadratic formula}$$

$$= \frac{-32 \pm \sqrt{576x^2 - 4608x}}{32} \qquad \text{Expand}$$

$$= \frac{-32 \pm 24\sqrt{x^2 - 8x}}{32} \qquad \text{Factor 576 from under the radical}$$

$$= -1 \pm \tfrac{3}{4}\sqrt{x^2 - 8x} \qquad \text{Simplify}$$

To obtain the graph of the hyperbola, we graph the functions

$$y = -1 + 0.75\sqrt{x^2 - 8x} \qquad \text{and} \qquad y = -1 - 0.75\sqrt{x^2 - 8x}$$

as shown in Figure 6(b).

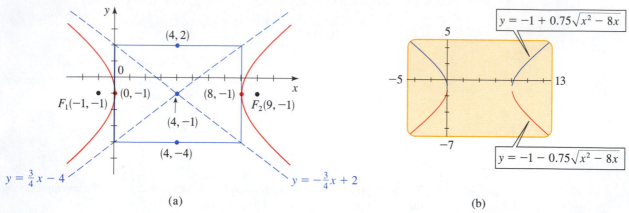

(a)   (b)

FIGURE 6
$9x^2 - 72x - 16y^2 - 32y = 16$

■

If we expand and simplify the equations of any of the shifted conics illustrated in Figures 1, 3, and 4, then we will always obtain an equation of the form

$$Ax^2 + Cy^2 + Dx + Ey + F = 0$$

where A and C are not both 0. Conversely, if we begin with an equation of this form, then we can complete the square in x and y to see which type of conic section the equation represents. In some cases, the graph of the equation turns out to be just a pair of lines, a single point, or there may be no graph at all. These cases are called **degenerate conics**. The next example illustrates such a case.

EXAMPLE 4 ■ An Equation That Leads to a Degenerate Conic

Sketch the graph of the equation

$$9x^2 - y^2 + 18x + 6y = 0$$

SOLUTION

Because the coefficients of x^2 and y^2 are of opposite sign, this equation looks as if it should represent a hyperbola (like the equation of Example 3). To see whether this is in fact the case, we complete the squares:

$$9(x^2 + 2x \quad) - (y^2 - 6y \quad) = 0$$

$$9(x^2 + 2x + 1) - (y^2 - 6y + 9) = 0 + 9 - 9$$

$$9(x + 1)^2 - (y - 3)^2 = 0$$

$$(x + 1)^2 - \frac{(y - 3)^2}{9} = 0 \qquad \text{Divide by 9}$$

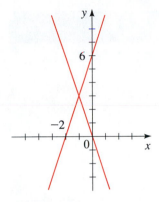

FIGURE 7
$9x^2 - y^2 + 18x + 6y = 0$

For this to fit the form of the equation of a hyperbola, we would need a nonzero constant to the right of the equal sign. In fact, further analysis shows that this is the equation of a pair of intersecting lines:

$$(y - 3)^2 = 9(x + 1)^2$$

$$y - 3 = \pm 3(x + 1) \qquad \text{Take square roots}$$

$$y = 3(x + 1) + 3 \qquad \text{or} \qquad y = -3(x + 1) + 3$$

$$y = 3x + 6 \qquad\qquad\qquad y = -3x$$

These lines are graphed in Figure 7. ■

Because the equation in Example 4 looked at first glance like the equation of a hyperbola but, in fact, turned out to represent simply a pair of lines, we refer to its graph as a **degenerate hyperbola**. Degenerate ellipses and parabolas can also arise when we complete the square(s) in an equation that seems to represent a conic. For example, the equation

$$4x^2 + y^2 - 8x + 2y + 6 = 0$$

looks as if it should represent an ellipse, because the coefficients of x^2 and y^2 have the same sign. But completing the squares leads to

$$(x - 1)^2 + \frac{(y + 1)^2}{4} = -\frac{1}{4}$$

which has no solution at all (since the sum of two squares cannot be negative). This equation is therefore degenerate.

To summarize, we have the following theorem.

GENERAL EQUATION OF A SHIFTED CONIC

The graph of the equation

$$Ax^2 + Cy^2 + Dx + Ey + F = 0$$

where A and C are not both 0, is a conic or a degenerate conic. In the nondegenerate cases, the graph is

1. a parabola if A or C is 0.

2. an ellipse if A and C have the same sign (or a circle if $A = C$).

3. a hyperbola if A and C have opposite signs.

9.4 EXERCISES

1–4 ■ Find the center, foci, and vertices of the ellipse, and determine the lengths of the major and minor axes. Then sketch the graph.

1. $\dfrac{(x-2)^2}{9} + \dfrac{(y-1)^2}{4} = 1$ **2.** $\dfrac{(x-3)^2}{16} + (y+3)^2 = 1$

3. $\dfrac{x^2}{9} + \dfrac{(y+5)^2}{25} = 1$ **4.** $\dfrac{(x+2)^2}{4} + y^2 = 1$

5–8 ■ Find the vertex, focus, and directrix of the parabola, and sketch the graph.

5. $(x-3)^2 = 8(y+1)$ **6.** $(y+5)^2 = -6x + 12$

7. $-4\left(x + \tfrac{1}{2}\right)^2 = y$ **8.** $y^2 = 16x - 8$

9–12 ■ Find the center, foci, vertices, and asymptotes of the hyperbola. Then sketch the graph.

9. $\dfrac{(x+1)^2}{9} - \dfrac{(y-3)^2}{16} = 1$ **10.** $(x-8)^2 - (y+6)^2 = 1$

11. $y^2 - \dfrac{(x+1)^2}{4} = 1$ **12.** $\dfrac{(y-1)^2}{25} - (x+3)^2 = 1$

13–18 ■ Find an equation for the conic whose graph is shown.

13.

14.

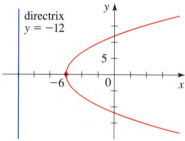

directrix
$y = -12$

15.

$F(8, 0)$

16.

17.

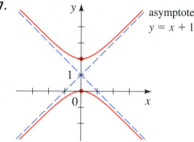

asymptote
$y = x + 1$

18.

19–30 ■ Complete the square to determine whether the equation represents an ellipse, a parabola, a hyperbola, or a degenerate conic. If the graph is an ellipse, find the center, foci, vertices, and lengths of the major and minor axes. If it is a parabola, find the vertex, focus, and directrix. If it is a hyperbola, find the center, foci, vertices, and asymptotes. Then sketch the graph of the equation. If the equation has no graph, explain why.

19. $9x^2 - 36x + 4y^2 = 0$

20. $y^2 = 4(x + 2y)$

21. $x^2 - 4y^2 - 2x + 16y = 20$

22. $x^2 + 6x + 12y + 9 = 0$

23. $4x^2 + 25y^2 - 24x + 250y + 561 = 0$

24. $2x^2 + y^2 = 2y + 1$

25. $16x^2 - 9y^2 - 96x + 288 = 0$

26. $4x^2 - 4x - 8y + 9 = 0$

27. $x^2 + 16 = 4(y^2 + 2x)$

28. $x^2 - y^2 = 10(x - y) + 1$

29. $3x^2 + 4y^2 - 6x - 24y + 39 = 0$

30. $x^2 + 4y^2 + 20x - 40y + 300 = 0$

 31–34 ■ Use a graphing device to graph the conic.

31. $2x^2 - 4x + y + 5 = 0$

32. $4x^2 + 9y^2 - 36y = 0$

33. $9x^2 + 36 = y^2 + 36x + 6y$

34. $x^2 - 4y^2 + 4x + 8y = 0$

35. Determine what the value of F must be if the graph of the equation

$$4x^2 + y^2 + 4(x - 2y) + F = 0$$

is (a) an ellipse, (b) a single point, or (c) the empty set.

36. Find an equation for the ellipse that shares a vertex and a focus with the parabola $x^2 + y = 100$ and has its other focus at the origin.

 37. This exercise deals with **confocal parabolas**, that is, families of parabolas that have the same focus.
(a) Draw graphs of the family of parabolas

$$x^2 = 4p(y + p)$$

for $p = -2, -\frac{3}{2}, -1, -\frac{1}{2}, \frac{1}{2}, 1, \frac{3}{2}, 2$.

(b) Show that each parabola in this family has its focus at the origin.

(c) Describe the effect on the graph of moving the vertex closer to the origin.

◆ DISCOVERY · DISCUSSION

38. A Family of Confocal Conics   Conics that share a focus are called **confocal**. Consider the family of conics that have a focus at $(0, 1)$ and a vertex at the origin (see the figure).
(a) Find equations of two different ellipses that have these properties.
(b) Find equations of two different hyperbolas that have these properties.
(c) Explain why only one parabola satisfies these properties. Find its equation.
(d) Sketch the conics you found in parts (a), (b), and (c) on the same coordinate axes (for the hyperbolas, sketch the top branches only).
(e) How are the ellipses and hyperbolas related to the parabola?

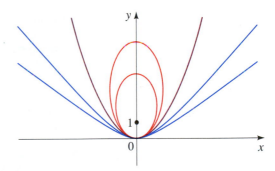

9.5   ROTATION OF AXES

In Section 9.4 we studied conics with equations of the form

$$Ax^2 + Cy^2 + Dx + Ey + F = 0$$

We saw that the graph is always an ellipse, parabola, or hyperbola with horizontal or vertical axes (except in the degenerate cases). In this section we study the most general second-degree equation

$$Ax^2 + Bxy + Cy^2 + Dx + Ey + F = 0$$

We will see that the graph of an equation of this form is also a conic. In fact, by rotating the coordinate axes through an appropriate angle, we can eliminate the term Bxy and then use our knowledge of conic sections to analyze the graph.

FIGURE 1

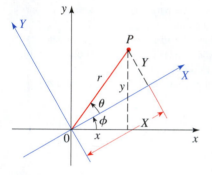

FIGURE 2

In Figure 1 the x- and y-axes have been rotated through an acute angle ϕ about the origin to produce a new pair of axes, which we call the X- and Y-axes. A point P that has coordinates (x, y) in the old system has coordinates (X, Y) in the new system. If we let r denote the distance of P from the origin and let θ be the angle that the segment OP makes with the new X-axis, then we can see from Figure 2 (by considering the two right triangles in the figure) that

$$X = r \cos \theta \qquad\qquad Y = r \sin \theta$$
$$x = r \cos(\theta + \phi) \qquad y = r \sin(\theta + \phi)$$

Using the addition formula for cosine, we see that

$$
\begin{aligned}
x &= r \cos(\theta + \phi) \\
&= r(\cos \theta \cos \phi - \sin \theta \sin \phi) \\
&= (r \cos \theta) \cos \phi - (r \sin \theta) \sin \phi \\
&= X \cos \phi - Y \sin \phi
\end{aligned}
$$

Similarly, we can apply the addition formula for sine to the expression for y to obtain $y = X \sin \phi + Y \cos \phi$. By treating these equations for x and y as a system of linear equations in the variables X and Y (see Exercise 33), we obtain expressions for X and Y in terms of x and y, as detailed in the following box.

ROTATION OF AXES FORMULAS

Suppose the x- and y-axes in a coordinate plane are rotated through the acute angle ϕ to produce the X- and Y-axes, as shown in Figure 1. Then the coordinates (x, y) and (X, Y) of a point in the xy- and the XY-planes are related as follows:

$$x = X \cos \phi - Y \sin \phi \qquad X = x \cos \phi + y \sin \phi$$
$$y = X \sin \phi + Y \cos \phi \qquad Y = -x \sin \phi + y \cos \phi$$

EXAMPLE 1 ■ Rotation of Axes

If the coordinate axes are rotated through 30°, find the XY-coordinates of the point with xy-coordinates $(2, -4)$.

SOLUTION

Using the Rotation of Axes Formulas with $x = 2$, $y = -4$, and $\phi = 30°$, we get

$$X = 2 \cos 30° + (-4) \sin 30° = 2\left(\frac{\sqrt{3}}{2}\right) - 4\left(\frac{1}{2}\right) = \sqrt{3} - 2$$

$$Y = -2 \sin 30° + (-4) \cos 30° = -2\left(\frac{1}{2}\right) - 4\left(\frac{\sqrt{3}}{2}\right) = -1 - 2\sqrt{3}$$

The XY-coordinates are $(-2 + \sqrt{3}, -1 - 2\sqrt{3})$.

EXAMPLE 2 ■ Rotating a Hyperbola

Rotate the coordinate axes through 45° to show that the graph of the equation $xy = 2$ is a hyperbola.

SOLUTION

We use the Rotation of Axes Formulas with $\phi = 45°$ to obtain

$$x = X \cos 45° - Y \sin 45° = \frac{X}{\sqrt{2}} - \frac{Y}{\sqrt{2}}$$

$$y = X \sin 45° + Y \cos 45° = \frac{X}{\sqrt{2}} + \frac{Y}{\sqrt{2}}$$

Substituting these expressions into the original equation gives

$$\left(\frac{X}{\sqrt{2}} - \frac{Y}{\sqrt{2}} \right) \left(\frac{X}{\sqrt{2}} + \frac{Y}{\sqrt{2}} \right) = 2$$

$$\frac{X^2}{2} - \frac{Y^2}{2} = 2$$

$$\frac{X^2}{4} - \frac{Y^2}{4} = 1$$

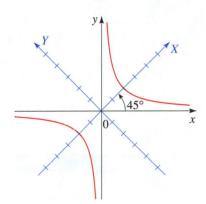

FIGURE 3
$xy = 2$

We recognize this as a hyperbola with vertices $(\pm 2, 0)$ in the XY-coordinate system. Its asymptotes are $Y = \pm X$, which correspond to the coordinate axes in the xy-system (see Figure 3). ■

The method of Example 2 can be used to transform any equation of the form

$$Ax^2 + Bxy + Cy^2 + Dx + Ey + F = 0$$

into an equation in X and Y that doesn't contain an XY-term by choosing an appropriate angle of rotation. To find the angle that works, we rotate the axes through an angle ϕ and substitute for x and y using the Rotation of Axes Formulas:

$$A(X \cos \phi - Y \sin \phi)^2 + B(X \cos \phi - Y \sin \phi)(X \sin \phi + Y \cos \phi)$$

$$+ C(X \sin \phi + Y \cos \phi)^2 + D(X \cos \phi - Y \sin \phi)$$

$$+ E(X \sin \phi + Y \cos \phi) + F = 0$$

If we expand this and collect like terms, we obtain an equation of the form

$$A'X^2 + B'XY + C'Y^2 + D'X + E'Y + F' = 0$$

where

$$A' = A \cos^2\phi + B \sin\phi \cos\phi + C \sin^2\phi$$

$$B' = 2(C - A) \sin\phi \cos\phi + B(\cos^2\phi - \sin^2\phi)$$

$$C' = A \sin^2\phi - B \sin\phi \cos\phi + C \cos^2\phi$$

$$D' = D \cos\phi + E \sin\phi$$

$$E' = -D \sin\phi + E \cos\phi$$

$$F' = F$$

To eliminate the XY-term, we would like to choose ϕ so that $B' = 0$, that is,

$$2(C - A) \sin\phi \cos\phi + B(\cos^2\phi - \sin^2\phi) = 0$$

$$(C - A) \sin 2\phi + B \cos 2\phi = 0 \qquad \text{Double-angle formulas for sine and cosine}$$

$$B \cos 2\phi = (A - C) \sin 2\phi$$

$$\cot 2\phi = \frac{A - C}{B} \qquad \text{Divide by } B \sin 2\phi$$

The preceding calculation proves the following theorem.

SIMPLIFYING THE GENERAL CONIC EQUATION

To eliminate the xy-term in the general conic equation

$$Ax^2 + Bxy + Cy^2 + Dx + Ey + F = 0$$

rotate the axes through the acute angle ϕ that satisfies

$$\cot 2\phi = \frac{A - C}{B}$$

EXAMPLE 3 ■ Eliminating the *xy*-Term

Use a rotation of axes to eliminate the xy-term in the equation

$$6 \sqrt{3}\, x^2 + 6xy + 4 \sqrt{3}\, y^2 = 21 \sqrt{3}$$

Identify and sketch the curve.

SOLUTION

To eliminate the xy-term, we rotate the axes through an angle ϕ that satisfies

$$\cot 2\phi = \frac{A - C}{B} = \frac{6\sqrt{3} - 4\sqrt{3}}{6} = \frac{\sqrt{3}}{3}$$

Thus, $2\phi = 60°$ and hence $\phi = 30°$. With this value of ϕ, we get

$$x = X\left(\frac{\sqrt{3}}{2}\right) - Y\left(\frac{1}{2}\right) \qquad \text{Rotation of Axes Formulas}$$

$$y = X\left(\frac{1}{2}\right) + Y\left(\frac{\sqrt{3}}{2}\right) \qquad \cos\phi = \frac{\sqrt{3}}{2}, \sin\phi = \frac{1}{2}$$

Substituting these values for x and y into the given equation leads to

$$6\sqrt{3}\left(\frac{X\sqrt{3}}{2} - \frac{Y}{2}\right)^2 + 6\left(\frac{X\sqrt{3}}{2} - \frac{Y}{2}\right)\left(\frac{X}{2} + \frac{Y\sqrt{3}}{2}\right) + 4\sqrt{3}\left(\frac{X}{2} + \frac{Y\sqrt{3}}{2}\right)^2 = 21\sqrt{3}$$

Expanding and collecting like terms, we get

$$7\sqrt{3}\,X^2 + 3\sqrt{3}\,Y^2 = 21\sqrt{3}$$

$$\frac{X^2}{3} + \frac{Y^2}{7} = 1 \qquad \text{Divide by } 21\sqrt{3}$$

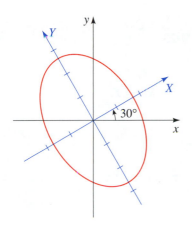

FIGURE 4
$6\sqrt{3}\,x^2 + 6xy + 4\sqrt{3}\,y^2 = 21\sqrt{3}$

This is the equation of an ellipse in the XY-coordinate system. The foci lie on the Y-axis. Because $a^2 = 7$ and $b^2 = 3$, the length of the major axis is $2\sqrt{7}$, and the length of the minor axis is $2\sqrt{3}$. The ellipse is sketched in Figure 4. ∎

In the preceding example we were able to determine ϕ without difficulty, since we remembered that $\cot 60° = \sqrt{3}/3$. In general, finding ϕ is not quite so easy. The next example illustrates how the following half-angle formulas, which are valid for $0 < \phi < \pi/2$, are useful in determining ϕ (see Section 7.3):

$$\cos\phi = \sqrt{\frac{1 + \cos 2\phi}{2}} \qquad \sin\phi = \sqrt{\frac{1 - \cos 2\phi}{2}}$$

EXAMPLE 4 ■ Graphing a Rotated Conic

A conic has the equation

$$64x^2 + 96xy + 36y^2 - 15x + 20y - 25 = 0$$

(a) Use a rotation of axes to eliminate the xy-term.
(b) Identify and sketch the graph.
(c) Draw the graph using a graphing calculator.

SOLUTION

(a) To eliminate the xy-term, we rotate the axes through an angle ϕ that satisfies

$$\cot 2\phi = \frac{A - C}{B} = \frac{64 - 36}{96} = \frac{7}{24}$$

In Figure 5 we sketch a triangle with $\cot 2\phi = \frac{7}{24}$. We see that

$$\cos 2\phi = \tfrac{7}{25}$$

FIGURE 5

so, using the half-angle formulas, we get

$$\cos\phi = \sqrt{\frac{1 + \frac{7}{25}}{2}} = \sqrt{\frac{16}{25}} = \frac{4}{5}$$

$$\sin\phi = \sqrt{\frac{1 + \frac{7}{25}}{2}} = \sqrt{\frac{9}{25}} = \frac{3}{5}$$

The Rotation of Axes Formulas then give

$$x = \tfrac{4}{5}X - \tfrac{3}{5}Y \qquad \text{and} \qquad y = \tfrac{3}{5}X + \tfrac{4}{5}Y$$

Substituting into the given equation, we have

$$64\left(\tfrac{4}{5}X - \tfrac{3}{5}Y\right)^2 + 96\left(\tfrac{4}{5}X - \tfrac{3}{5}Y\right)\left(\tfrac{3}{5}X + \tfrac{4}{5}Y\right)$$
$$+ 36\left(\tfrac{3}{5}X + \tfrac{4}{5}Y\right)^2 - 15\left(\tfrac{4}{5}X - \tfrac{3}{5}Y\right) + 20\left(\tfrac{3}{5}X + \tfrac{4}{5}Y\right) - 25 = 0$$

Expanding and collecting like terms, we get

$$100X^2 + 25Y - 25 = 0$$

$$-4X^2 = Y - 1 \qquad \text{Simplify}$$

$$X^2 = -\tfrac{1}{4}(Y - 1) \qquad \text{Divide by 4}$$

(b) We recognize this as the equation of a parabola that opens along the negative Y-axis and has vertex $(0, 1)$ in XY-coordinates. Since $4p = -\tfrac{1}{4}$, we have $p = -\tfrac{1}{16}$, so the focus is $\left(0, \tfrac{15}{16}\right)$ and the directrix is $Y = \tfrac{17}{16}$. Using

$$\phi = \cos^{-1}\tfrac{4}{5} \approx 37°$$

we sketch the graph in Figure 6(a).

(c) To draw the graph using a graphing calculator, we need to solve for y. The given equation is a quadratic equation in y, so we can use the quadratic formula to solve for y. Writing the equation in the form

$$36y^2 + (96x + 20)y + (64x^2 - 15x - 25) = 0$$

we get

$$y = \frac{-(96x + 20) \pm \sqrt{(96x + 20)^2 - 4(36)(64x^2 - 15x - 25)}}{2(36)} \qquad \text{Quadratic formula}$$

$$= \frac{-(96x + 20) \pm \sqrt{6000x + 4000}}{72} \qquad \text{Expand}$$

$$= \frac{-96x - 20 \pm 20\sqrt{15x + 10}}{72} \qquad \text{Simplify}$$

$$= \frac{-24x - 5 \pm 5\sqrt{15x + 10}}{18} \qquad \text{Simplify}$$

To obtain the graph of the parabola, we graph the functions

$$y = (-24x - 5 + 5\sqrt{15x + 10})/18 \qquad \text{and} \qquad y = (-24x - 5 - 5\sqrt{15x + 10})/18$$

as shown in Figure 6(b).

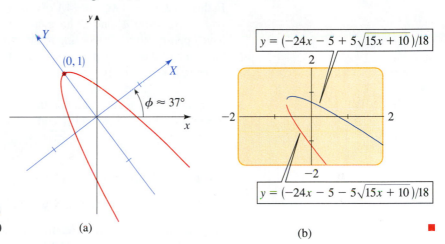

FIGURE 6

$64x^2 + 96xy + 36y^2 - 15x + 20y - 25 = 0$

(a)

(b)

In Examples 3 and 4 we were able to identify the type of conic by rotating the axes. The next theorem gives rules for identifying the type of conic directly from the equation, without rotating axes.

IDENTIFYING CONICS BY THE DISCRIMINANT

The graph of the equation

$$Ax^2 + Bxy + Cy^2 + Dx + Ey + F = 0$$

is either a conic or a degenerate conic. In the nondegenerate cases, the graph is

1. a parabola if $B^2 - 4AC = 0$.

2. an ellipse if $B^2 - 4AC < 0$.

3. a hyperbola if $B^2 - 4AC > 0$.

The quantity $B^2 - 4AC$ is called the **discriminant** of the equation.

■ **Proof**   If we rotate the axes through an angle ϕ, we get an equation of the form

$$A'X^2 + B'XY + C'Y^2 + D'X + E'Y + F' = 0$$

where $A', B', C', \ldots$ are given by the formulas on page 764. A straightforward calculation shows that

$$(B')^2 - 4A'C' = B^2 - 4AC$$

Thus, the expression $B^2 - 4AC$ remains unchanged for any rotation. In particular, if we choose a rotation that eliminates the xy-term $(B' = 0)$, we get

$$A'X^2 + C'Y^2 + D'X + E'Y + F' = 0$$

In this case, $B^2 - 4AC = -4A'C'$. So $B^2 - 4AC = 0$ if either A' or C' is zero; $B^2 - 4AC < 0$ if A' and C' have the same sign; and $B^2 - 4AC > 0$ if A' and C' have opposite signs. According to the box on page 767, these cases correspond to the graph of the last displayed equation being a parabola, an ellipse, or a hyperbola, respectively. ☐

In the proof we indicated that the discriminant is unchanged by any rotation; for this reason, the discriminant is said to be **invariant** under rotation.

EXAMPLE 5 ■ Identifying a Conic by the Discriminant

A conic has the equation

$$3x^2 + 5xy - 2y^2 + x - y + 4 = 0$$

(a) Use the discriminant to identify the conic.
(b) Confirm your answer to part (a) by graphing the conic with a graphing calculator.

SOLUTION

(a) Since $A = 3$, $B = 5$, and $C = -2$, the discriminant is

$$B^2 - 4AC = 5^2 - 4(3)(-2) = 49 > 0$$

So the conic is a hyperbola.

(b) Using the quadratic formula, we solve for y to get

$$y = \frac{5x - 1 \pm \sqrt{49x^2 - 2x + 33}}{4}$$

We graph these functions in Figure 7. The graph confirms that this is a hyperbola. ■

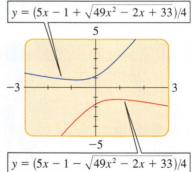

$$y = (5x - 1 + \sqrt{49x^2 - 2x + 33})/4$$

$$y = (5x - 1 - \sqrt{49x^2 - 2x + 33})/4$$

FIGURE 7

9.5 EXERCISES

1–6 ■ Determine the XY-coordinates of the given point if the coordinate axes are rotated through the indicated angle.

1. $(1, 1)$, $\phi = 45°$

2. $(-2, 1)$, $\phi = 30°$

3. $(3, -\sqrt{3})$, $\phi = 60°$

4. $(2, 0)$, $\phi = 15°$

5. $(0, 2)$, $\phi = 55°$

6. $(\sqrt{2}, 4\sqrt{2})$, $\phi = 45°$

7–12 ■ Determine the equation of the given conic in XY-coordinates when the coordinate axes are rotated through the indicated angle.

7. $x^2 - 3y^2 = 4$, $\phi = 60°$

8. $y = (x - 1)^2$, $\phi = 45°$

9. $x^2 - y^2 = 2y$, $\phi = \cos^{-1}\frac{3}{5}$

10. $x^2 + 2y^2 = 16$, $\phi = \sin^{-1}\frac{3}{5}$

11. $x^2 + 2\sqrt{3}\,xy - y^2 = 4$, $\phi = 30°$

12. $xy = x + y$, $\phi = \pi/4$

13–26 ■ (a) Use the discriminant to determine whether the graph of the equation is a parabola, an ellipse, or a hyperbola. (b) Use a rotation of axes to eliminate the xy-term. (c) Sketch the graph.

13. $xy = 8$

14. $xy + 4 = 0$

15. $x^2 + 2xy + y^2 + x - y = 0$

16. $13x^2 + 6\sqrt{3}\,xy + 7y^2 = 16$

17. $x^2 + 2\sqrt{3}\,xy - y^2 + 2 = 0$

18. $21x^2 + 10\sqrt{3}\,xy + 31y^2 = 144$

19. $11x^2 - 24xy + 4y^2 + 20 = 0$

20. $25x^2 - 120xy + 144y^2 - 156x - 65y = 0$

21. $\sqrt{3}\,x^2 + 3xy = 3$

22. $153x^2 + 192xy + 97y^2 = 225$

23. $2\sqrt{3}\,x^2 - 6xy + \sqrt{3}\,x + 3y = 0$

24. $9x^2 - 24xy + 16y^2 = 100(x - y - 1)$

25. $52x^2 + 72xy + 73y^2 = 40x - 30y + 75$

26. $(7x + 24y)^2 = 600x - 175y + 25$

27–30 ■ (a) Use the discriminant to identify the conic. (b) Confirm your answer by graphing the conic using a graphing device.

27. $2x^2 - 4xy + 2y^2 - 5x - 5 = 0$

28. $x^2 - 2xy + 3y^2 = 8$

29. $6x^2 + 10xy + 3y^2 - 6y = 36$

30. $9x^2 - 6xy + y^2 + 6x - 2y = 0$

31. (a) Use rotation of axes to show that the following equation represents a hyperbola:

$$7x^2 + 48xy - 7y^2 - 200x - 150y + 600 = 0$$

   (b) Find the XY- and xy-coordinates of the center, vertices, and foci.
   (c) Find the equations of the asymptotes in XY- and xy-coordinates.

32. (a) Use rotation of axes to show that the following equation represents a parabola:

$$2\sqrt{2}\,(x + y)^2 = 7x + 9y$$

   (b) Find the XY- and xy-coordinates of the vertex and focus.
   (c) Find the equation of the directrix in XY- and xy-coordinates.

33. Solve the equations:

$$x = X \cos \phi - Y \sin \phi$$
$$y = X \sin \phi + Y \cos \phi$$

for X and Y in terms of x and y. [*Hint:* To begin, multiply the first equation by $\cos \phi$ and the second by $\sin \phi$, and then add the two equations to solve for X.]

34. Show that the graph of the equation

$$\sqrt{x} + \sqrt{y} = 1$$

is part of a parabola by rotating the axes through an angle of 45°. [*Hint:* First convert the equation to one that does not involve radicals.]

35. Let Z, Z', and R be the matrices

$$Z = \begin{bmatrix} x \\ y \end{bmatrix} \qquad Z' = \begin{bmatrix} X \\ Y \end{bmatrix}$$

$$R = \begin{bmatrix} \cos \phi & -\sin \phi \\ \sin \phi & \cos \phi \end{bmatrix}$$

Show that the Rotation of Axes Formulas can be written as

$$Z = RZ' \quad \text{and} \quad Z' = R^{-1}Z$$

 DISCOVERY · DISCUSSION

36. Algebraic Invariants A quantity is invariant under rotation if it does not change when the axes are rotated. It was stated in the text that for the general equation of a conic, the quantity $B^2 - 4AC$ is invariant under rotation.
   (a) Use the formulas for A', B', and C' on page 764 to prove that the quantity $B^2 - 4AC$ is invariant under rotation; that is, show that

$$B^2 - 4AC = B'^2 - 4A'C'$$

   (b) Prove that $A + C$ is invariant under rotation.
   (c) Is the quantity F invariant under rotation?

37. Geometric Invariants Do you expect that the distance between two points is invariant under rotation? Prove your answer by comparing the distance $d(P, Q)$ and $d(P', Q')$ where P' and Q' are the images of P and Q under a rotation of axes.

Discovery Project

Computer Graphics II

In the Discovery Project on page 684 we saw how matrix multiplication is used in computer graphics. We found matrices that reflect, expand, or shear an image. We now consider matrices that rotate an image. Recall that a point (x, y) in the plane is represented by the 2×1 matrix $\begin{bmatrix} x \\ y \end{bmatrix}$. The matrix that rotates this point about the origin through an angle ϕ is

$$R = \begin{bmatrix} \cos \phi & -\sin \phi \\ \sin \phi & \cos \phi \end{bmatrix} \qquad \text{Rotation matrix}$$

Compare this matrix with the rotation of axes matrix in Exercise 35, Section 9.5. Note that rotating a point counterclockwise corresponds to rotating the axes clockwise.

When the point $P = \begin{bmatrix} x \\ y \end{bmatrix}$ is rotated clockwise about the origin through an angle ϕ, it moves to a new location $P' = \begin{bmatrix} x' \\ y' \end{bmatrix}$ given by the matrix product $P' = RP$, as shown in Figure 1.

$$P' = RP = \begin{bmatrix} \cos \phi & -\sin \phi \\ \sin \phi & \cos \phi \end{bmatrix}\begin{bmatrix} x \\ y \end{bmatrix} = \begin{bmatrix} x \cos \phi - y \sin \phi \\ x \sin \phi + y \cos \phi \end{bmatrix}$$

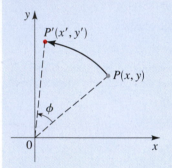

FIGURE 1

For example, if $\phi = 90°$, the rotation matrix is

$$R = \begin{bmatrix} \cos 90° & -\sin 90° \\ \sin 90° & \cos 90° \end{bmatrix} = \begin{bmatrix} 0 & -1 \\ 1 & 0 \end{bmatrix} \qquad \text{Rotation matrix } (\phi = 90°)$$

Applying a 90° rotation to the point $P = \begin{bmatrix} 1 \\ 2 \end{bmatrix}$ moves it to the point

$$P' = RP = \begin{bmatrix} 0 & -1 \\ 1 & 0 \end{bmatrix}\begin{bmatrix} 1 \\ 2 \end{bmatrix} = \begin{bmatrix} -2 \\ 1 \end{bmatrix}$$

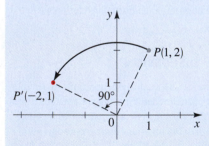

FIGURE 2

See Figure 2.

If the rotation matrix is applied to every point in an image, then the entire image is rotated. To rotate the house in Figure 3(a) through a 30° angle about the origin, we multiply its data matrix (described on page 685) by the rotation matrix that has $\phi = 30°$.

$$RD = \begin{bmatrix} \frac{\sqrt{3}}{2} & -\frac{1}{2} \\ \frac{1}{2} & \frac{\sqrt{3}}{2} \end{bmatrix}\begin{bmatrix} 2 & 0 & 0 & 2 & 4 & 4 & 3 & 3 & 2 & 2 & 3 \\ 0 & 0 & 3 & 5 & 3 & 0 & 0 & 2 & 2 & 0 & 0 \end{bmatrix}$$

$$\approx \begin{bmatrix} 1.73 & 0 & -1.50 & -0.77 & 1.96 & 3.46 & 2.60 & 1.60 & 0.73 & 1.73 & 2.60 \\ 1 & 0 & 2.60 & 5.33 & 4.60 & 2 & 1.50 & 3.23 & 2.73 & 1 & 1.50 \end{bmatrix}$$

The new data matrix RD represents the rotated house in Figure 3(b).

FIGURE 3   (a)   (b)

The Discovery Project on page 684 describes a TI-83 program that draws the image corresponding to a given data matrix. You may find it convenient to use this program in some of the following activities.

1. Use a rotation matrix to find the new coordinates of the given point when it is rotated through the given angle.

   (a) $(1, 4)$, $\phi = 90°$ **(b)** $(-2, 1)$, $\phi = 60°$

   (c) $(-2, -2)$, $\phi = 135°$ **(d)** $(7, 3)$, $\phi = -60°$

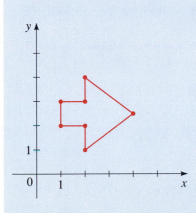

2. Find a data matrix for the line drawing in the figure shown in the margin. Multiply the data matrix by a suitable rotation matrix to rotate the image about the origin by $\phi = 120°$. Sketch the rotated image given by the new data matrix.

3. Sketch the image represented by the data matrix D.

$$D = \begin{bmatrix} 2 & 3 & 3 & 4 & 4 & 1 & 1 & 2 & 2 \\ 1 & 1 & 3 & 3 & 4 & 4 & 3 & 3 & 1 \end{bmatrix}$$

   Find the rotation matrix R that corresponds to a 45° rotation, and the transformation matrix T that corresponds to an expansion by a factor of 2 in the x-direction (see page 683). How does multiplying the data matrix by RT change the image? How about multiplying by TR? Calculate the products RTD and TRD, and sketch the corresponding images to confirm your answers.

4. Let R be the rotation matrix for the angle ϕ. Show that R^{-1} is the rotation matrix for the angle $-\phi$.

5. To **translate** an image by (h, k), we add h to each x-coordinate and k to each y-coordinate of each point in the image (see the figure on the following page). This can be done by adding an approximate matrix M to D, but the dimension of M would change depending on the dimension of D. In practice, translation is

accomplished by matrix multiplication. To see how this is done, we introduce **homogeneous coordinates**; that is, we represent the point (x, y) by a 3×1 matrix:

$$(x, y) \quad \leftrightarrow \quad \begin{bmatrix} x \\ y \\ 1 \end{bmatrix}$$

(a) Let T be the matrix

$$T = \begin{bmatrix} 1 & 0 & h \\ 0 & 1 & k \\ 0 & 0 & 1 \end{bmatrix}$$

Show that T translates the point (x, y) to the point $(x + h, y + h)$ by verifying the following matrix multiplication

$$\begin{bmatrix} 1 & 0 & h \\ 0 & 1 & k \\ 0 & 0 & 1 \end{bmatrix} \begin{bmatrix} x \\ y \\ 1 \end{bmatrix} = \begin{bmatrix} x + h \\ y + k \\ 1 \end{bmatrix}$$

(b) Find T^{-1} and describe how T^{-1} translates points.

(c) Verify that multiplying by the following matrices has the indicated effects on a point (x, y) represented by its homogeneous coordinates $\begin{bmatrix} x \\ y \\ 1 \end{bmatrix}$.

$$\begin{bmatrix} 1 & 0 & 0 \\ 0 & -1 & 0 \\ 0 & 0 & 1 \end{bmatrix} \quad \begin{bmatrix} c & 0 & 0 \\ 0 & 1 & 0 \\ 0 & 0 & 1 \end{bmatrix} \quad \begin{bmatrix} 1 & c & 0 \\ 0 & 1 & 0 \\ 0 & 0 & 1 \end{bmatrix} \quad \begin{bmatrix} \cos \phi & -\sin \phi & 0 \\ \sin \phi & \cos \phi & 0 \\ 0 & 0 & 1 \end{bmatrix}$$

| Reflection in x-axis | Expansion (or contraction) in x-direction | Shear in x-direction | Rotation about the origin by the angle ϕ |

(d) Sketch the image represented (in homogeneous coordinates) by the data matrix.

$$D = \begin{bmatrix} 3 & 5 & 5 & 7 & 7 & 9 & 9 & 7 & 7 & 5 & 5 & 3 & 3 \\ 7 & 7 & 5 & 5 & 7 & 7 & 9 & 9 & 11 & 11 & 9 & 9 & 7 \\ 1 & 1 & 1 & 1 & 1 & 1 & 1 & 1 & 1 & 1 & 1 & 1 & 1 \end{bmatrix}$$

Find a matrix T that translates the image by $(-6, -8)$, and a matrix R that rotates the image by $45°$. Sketch the images represented by the data matrices TD, RTD, and $T^{-1}RTD$. Describe how an image is changed when its data matrix is multiplied by T, by RT, and by $T^{-1}RT$.

9.6 POLAR COORDINATES

A coordinate system is a method for specifying the location of a point in the plane. Thus far we have been dealing with the rectangular (or Cartesian) coordinate system, which describes locations using a rectangular grid. Using rectangular coordinates is like describing a location in a city by saying that, for example, it's at the corner of 2nd Street and 4th Avenue. But we might also describe this same location by saying that it's 3 miles northeast of City Hall. Instead of specifying the location with respect to a grid of streets and avenues, we can describe it by giving its distance and direction from a fixed reference point.

Polar Coordinates

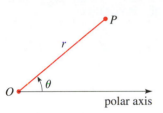

FIGURE 1

The *polar coordinate system* uses distances and directions to specify the locations of points in the plane. To set up this system, we first choose a fixed point O called the **origin** (or **pole**). We then draw a ray (half-line) starting at O, called the **polar axis**. This axis is usually drawn horizontally to the right of O and coincides with the positive x-axis in rectangular coordinates. Now let P be any point in the plane. Let r be the distance from P to the origin, and let θ be the angle between the polar axis and the segment OP, as shown in Figure 1. Then the ordered pair (r, θ) uniquely specifies the location of P. We write $P(r, \theta)$ and refer to r and θ as the **polar coordinates** of P. We use the convention that θ is positive if measured in a counterclockwise direction from the polar axis, or negative if measured in a clockwise direction. It's customary to use radian measure for θ. If $r = 0$, then $P = O$ no matter what value θ has, so $(0, \theta)$ represents the pole for any value of θ.

Because the angles $\theta + 2n\pi$ (for $n = \pm 1, \pm 2, \pm 3, \ldots$) all have the same terminal side as the angle θ, each point has infinitely many representations in polar coordinates. For example, $(2, \pi/3)$, $(2, 7\pi/3)$, and $(2, -5\pi/3)$ all represent the same point P, as shown in Figure 2.

FIGURE 2

Moreover, we also allow r to take on negative values, with the understanding that if $-r$ is negative, then the point $P(-r, \theta)$ lies r units away from the origin in the direction *opposite* to that given by θ. Thus, the point P graphed in Figure 2 can also be described by the coordinates $(-2, 4\pi/3)$ or $(-2, -2\pi/3)$. By this convention, the coordinates (r, θ) and $(-r, \theta + \pi)$ represent the same point (see Figure 3 on the following page). In fact, any point $P(r, \theta)$ can also be represented by

$$P(r, \theta + 2n\pi) \qquad \text{and} \qquad P(-r, \theta + (2n + 1)\pi)$$

for any integer n.

FIGURE 3

EXAMPLE 1 ■ Plotting Points in Polar Coordinates

Plot the points whose polar coordinates are given.

(a) $(1, 3\pi/4)$ (b) $(3, -\pi/6)$ (c) $(3, 3\pi)$ (d) $(-4, \pi/4)$

SOLUTION

The points are plotted in Figure 4. Note that the point in part (d) lies 4 units from the origin along the angle $5\pi/4$, because the given value of r is negative.

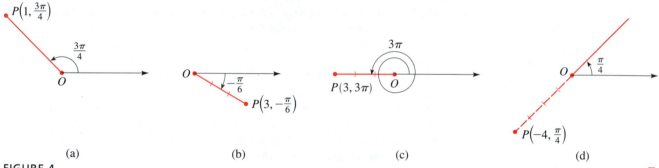

(a)   (b)   (c)   (d)

FIGURE 4   ■

Situations often arise in which we need to consider polar and rectangular coordinates simultaneously. The connection between the two systems is illustrated in Figure 5, where the polar axis coincides with the positive x-axis. The formulas in the following box are obtained from the figure using the definitions of the trigonometric functions and the Pythagorean Theorem. (Although we have pictured the case where $r > 0$ and θ is acute, the formulas hold for any angle θ and for any value of r.)

FIGURE 5

RELATIONSHIP BETWEEN POLAR AND RECTANGULAR COORDINATES

1. To change from polar to rectangular coordinates, use the formulas

$$x = r \cos \theta \quad \text{and} \quad y = r \sin \theta$$

2. To change from rectangular to polar coordinates, use the formulas

$$r^2 = x^2 + y^2 \quad \text{and} \quad \tan \theta = \frac{y}{x} \quad (x \neq 0)$$

EXAMPLE 2 ■ Converting Polar Coordinates to Rectangular Coordinates

Find rectangular coordinates for the point that has polar coordinates $(4, 2\pi/3)$.

SOLUTION

Since $r = 4$ and $\theta = 2\pi/3$, we have

$$x = r \cos \theta = 4 \cos \frac{2\pi}{3} = 4 \cdot \left(-\frac{1}{2}\right) = -2$$

$$y = r \sin \theta = 4 \sin \frac{2\pi}{3} = 4 \cdot \frac{\sqrt{3}}{2} = 2\sqrt{3}$$

Thus, the point has rectangular coordinates $(-2, 2\sqrt{3})$.   ■

EXAMPLE 3 ■ Converting Rectangular Coordinates to Polar Coordinates

Find polar coordinates for the point that has rectangular coordinates $(2, -2)$.

SOLUTION

Using $x = 2$, $y = -2$, we get

$$r^2 = x^2 + y^2 = 2^2 + (-2)^2 = 8$$

so $r = 2\sqrt{2}$ or $-2\sqrt{2}$. Also

$$\tan \theta = \frac{y}{x} = \frac{-2}{2} = -1$$

so $\theta = 3\pi/4$ or $-\pi/4$. Since the point $(2, -2)$ lies in quadrant IV (see Figure 6), we can represent it in polar coordinates as $(2\sqrt{2}, -\pi/4)$ or $(-2\sqrt{2}, 3\pi/4)$.   ■

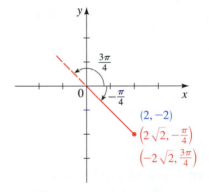

FIGURE 6

⊘   Note that the equations relating polar and rectangular coordinates do not uniquely determine r or θ. When we use these equations to find the polar coordinates of a point, we must be careful that the values we choose for r and θ give us a point in the correct quadrant, as we saw in Example 3.

Graphs of Polar Equations

The **graph of a polar equation** $r = f(\theta)$ consists of all points P that have at least one polar representation (r, θ) whose coordinates satisfy the equation. In the next two examples we see that circles centered at the origin and lines that pass through the origin have particularly simple equations in polar coordinates.

EXAMPLE 4 ■ Sketching the Graph of a Polar Equation

Sketch the graph of the equation $r = 3$.

FIGURE 7

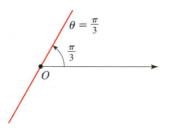

FIGURE 8

SOLUTION

The graph consists of all points whose r-coordinate is 3, that is, all points that are 3 units away from the origin. Therefore, this is the circle of radius 3 centered at the origin, as shown in Figure 7.   ■

In general, the graph of the equation $r = a$ is a circle of radius $|a|$ centered at the origin. Squaring both sides of this equation, we get $r^2 = a^2$, and since $r^2 = x^2 + y^2$, the equivalent equation in rectangular coordinates is $x^2 + y^2 = a^2$.

EXAMPLE 5 ■ Sketching the Graph of a Polar Equation

Sketch the graph of the equation $\theta = \pi/3$ and express the equation in rectangular coordinates.

SOLUTION

The graph consists of all points whose θ-coordinate is $\pi/3$. This is the straight line that passes through the origin and makes an angle of $\pi/3$ with the polar axis (see Figure 8). Note that the points $(r, \pi/3)$ on the line with $r > 0$ lie in quadrant I, whereas those with $r < 0$ lie in quadrant III. If the point (x, y) lies on this line, then

$$\frac{y}{x} = \tan \theta = \tan \frac{\pi}{3} = \sqrt{3}$$

Thus, the rectangular equation of this line is $y = \sqrt{3}\, x$.   ■

To sketch a polar curve whose graph isn't as obvious as the ones in the preceding examples, we rely on two techniques. One technique is to plot points calculated for sufficiently many values of θ and then join them in a continuous curve. This is what we did when we first learned to graph functions in rectangular coordinates. The other technique is to convert the polar equation into rectangular coordinates in the hope that the resulting equation is one that we recognize from our previous work. Both methods are used in the next example.

EXAMPLE 6 ■ Sketching the Graph of a Polar Equation

(a) Sketch the graph of the polar equation $r = 2 \sin \theta$.
(b) Convert the equation of part (a) to rectangular coordinates.

SOLUTION

(a) We first use the equation to determine the polar coordinates of several points on the curve. The results are shown in the following table.

θ	0	$\pi/6$	$\pi/4$	$\pi/3$	$\pi/2$	$2\pi/3$	$3\pi/4$	$5\pi/6$	π
$r = 2 \sin \theta$	0	1	$\sqrt{2}$	$\sqrt{3}$	2	$\sqrt{3}$	$\sqrt{2}$	1	0

We plot these points in Figure 9 and then join them to sketch the curve. The graph appears to be a circle. We have used values of θ only between 0 and π, since the same points (this time expressed with negative r-coordinates) would be obtained if we allowed θ to range from π to 2π.

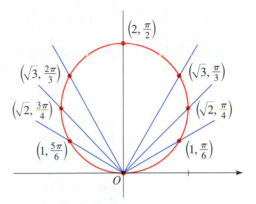

FIGURE 9
$r = 2 \sin \theta$

(b) We multiply both sides of the equation by r to get

$$r^2 = 2r \sin \theta$$

Using the formulas that relate polar and rectangular coordinates, we replace r^2 by $x^2 + y^2$ and $r \sin \theta$ by y to obtain the rectangular equation

$$x^2 + y^2 = 2y$$

$$x^2 + y^2 - 2y = 0 \qquad \text{Subtract } 2y$$

$$x^2 + (y - 1)^2 = 1 \qquad \text{Complete the square in } y$$

This is the equation of a circle of radius 1 centered at the point $(0, 1)$. ■

In general, the graphs of equations of the form

$$r = 2a \sin \theta \qquad \text{and} \qquad r = 2a \cos \theta$$

are circles with radius $|a|$ centered at the points with polar coordinates $(a, \pi/2)$ and $(a, 0)$, respectively.

EXAMPLE 7 ■ Sketching the Graph of a Polar Equation

(a) Sketch the graph of $r = 2 + 2 \cos \theta$.
(b) Convert the equation of part (a) to rectangular coordinates.

SOLUTION

(a) Instead of plotting points as in Example 6, we first sketch the graph of $r = 2 + 2 \cos \theta$ in *rectangular* coordinates in Figure 10. We can think of this graph as a table of values that enables us to read at a glance the values of r that correspond to increasing values of θ. For instance, we see that as θ increases from 0 to $\pi/2$, r (the distance from O) decreases from 4 to 2, so we sketch the

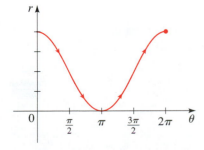

FIGURE 10
$r = 2 + 2 \cos \theta$

FIGURE 11   Steps in sketching $r = 2 + 2\cos\theta$

corresponding part of the polar graph in Figure 11(a). As θ increases from $\pi/2$ to π, Figure 10 shows that r decreases from 2 to 0, so we sketch the next part of the graph as in Figure 11(b). As θ increases from π to $3\pi/2$, r increases from 0 to 2, as shown in part (c). Finally, as θ increases from $3\pi/2$ to 2π, r increases from 2 to 4, as shown in part (d). If we let θ increase beyond 2π or decrease beyond 0, we would simply retrace our path. Combining the portions of the graph from parts (a) through (d) of Figure 11, we sketch the complete graph in part (e).

(b)   We first multiply both sides of the equation by r:

$$r^2 = 2r + 2r\cos\theta$$

Using $r^2 = x^2 + y^2$ and $y = r\cos\theta$, we can convert two of the three terms in the equation into rectangular coordinates, but eliminating the remaining r requires more work:

$$x^2 + y^2 = 2r + 2y$$
$$x^2 + y^2 - 2y = 2r$$
$$(x^2 + y^2 - 2y)^2 = 4r^2 \qquad \text{Square both sides}$$
$$(x^2 + y^2 - 2y)^2 = 4(x^2 + y^2) \qquad \text{Since } r^2 = x^2 + y^2$$

In this case, the rectangular equation is much more complicated than the polar equation, so the polar form is the more useful one.   ∎

The curve in Figure 11 is called a **cardioid** because it is heart-shaped. In general, the graph of any equation of the form

$$r = a(1 \pm \cos\theta) \qquad \text{or} \qquad r = a(1 \pm \sin\theta)$$

is a cardioid.

EXAMPLE 8 ■ Sketching the Graph of a Polar Equation

Sketch the curve $r = \cos 2\theta$.

SOLUTION

As in Example 7, we first sketch the graph of $r = \cos 2\theta$ in *rectangular* coordinates, as shown in Figure 12. As θ increases from 0 to $\pi/4$, Figure 12 shows that r decreases from 1 to 0, and so we draw the corresponding portion of the polar curve in Figure 13 (indicated by ①). As θ increases from $\pi/4$ to $\pi/2$, the value of r goes from 0 to -1. This means that the distance from the origin increases from 0 to 1, but instead of being in quadrant I, this portion of the polar curve (indicated by ②) lies on the opposite side of the origin in quadrant III. The remainder of the curve is drawn in a similar fashion, with the arrows and numbers indicating the order in which the portions are traced out. The resulting curve has four petals and is called a **four-leaved rose**.

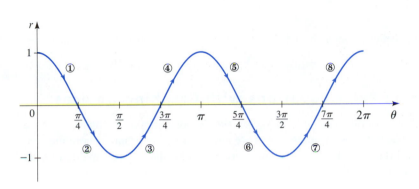

FIGURE 12

Graph of $r = \cos 2\theta$ sketched in rectangular coordinates

FIGURE 13

Four-leaved rose $r = \cos 2\theta$ sketched in polar coordinates

In general, the graph of an equation of the form

$$r = a \cos n\theta \quad \text{or} \quad r = a \sin n\theta$$

is an **n-leaved rose** if n is odd or a $2n$-leaved rose if n is even (as in Example 8).

When graphing a polar equation, it's often helpful to take advantage of symmetry. We list three tests for symmetry; Figure 14 on the following page shows why these tests work.

TESTS FOR SYMMETRY

1. If a polar equation is unchanged when we replace θ by $-\theta$, then the graph is symmetric about the polar axis [Figure 14(a)].

2. If the equation is unchanged when we replace r by $-r$, then the graph is symmetric about the pole [Figure 14(b)].

3. If the equation is unchanged when we replace θ by $\pi - \theta$, the graph is symmetric about the vertical line $\theta = \pi/2$ (the y-axis) [Figure 14(c)].

The graphs in Figures 7, 11(e), and 13 are symmetric about the polar axis. The graph in Figure 13 is also symmetric about the pole. Figures 9 and 13 show graphs that are symmetric about $\theta = \pi/2$. Note that the four-leaved rose in Figure 13 meets all three tests for symmetry.

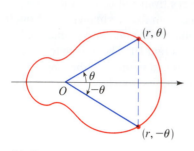

(a) Symmetry about the polar axis

(b) Symmetry about the pole

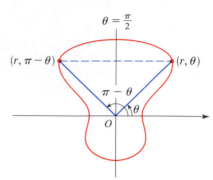

(c) Symmetry about the line $\theta = \frac{\pi}{2}$

FIGURE 14

FIGURE 15
$r = \sin\theta + \sin^3(5\theta/2)$

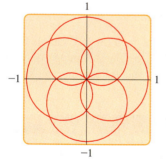

FIGURE 16
$r = \cos(2\theta/3)$

Graphing Polar Equations with Graphing Devices

Although it's useful to be able to sketch simple polar graphs by hand, we need a graphing calculator or computer when we are faced with a graph as complicated as the one in Figure 15. Fortunately, most graphing calculators are capable of graphing polar equations directly.

EXAMPLE 9 ■ Drawing the Graph of a Polar Equation

Graph the equation $r = \cos(2\theta/3)$.

SOLUTION

We need to determine the domain for θ. So we ask ourselves: How many complete rotations are required before the graph starts to repeat itself? The graph repeats itself when the same value of r is obtained at θ and $\theta + 2n\pi$. Thus, we need to find an integer n, so that

$$\cos\frac{2(\theta + 2n\pi)}{3} = \cos\frac{2\theta}{3}$$

For this equality to hold, $4n\pi/3$ must be a multiple of 2π, and this first happens when $n = 3$. Therefore, we obtain the entire graph if we choose values of θ between $\theta = 0$ and $\theta = 0 + 2(3)\pi = 6\pi$. The graph is shown in Figure 16. ■

EXAMPLE 10 ■ A Family of Polar Equations

Graph the family of polar equations $r = 1 + c\sin\theta$ for $c = 3, 2.5, 2, 1.5, 1$. How does the shape of the graph change as c changes? (These curves are called **limaçons**, after the French word for snail, because of the shape of the curve for certain values of c.)

SOLUTION

Figure 17 shows computer-drawn graphs for the given values of c. For $c > 1$, the graph has an inner loop; the loop decreases in size as c decreases. When $c = 1$, the loop disappears and the graph becomes a cardioid (see Example 7).

| $c = 3.0$ | $c = 2.5$ | $c = 2.0$ | $c = 1.5$ | $c = 1.0$ |

FIGURE 17 A family of limaçons $r = 1 + c \sin \theta$ in the viewing rectangle $[-2.5, 2.5]$ by $[-0.5, 4.5]$ ■

The following box gives a summary of some of the basic polar graphs used in calculus.

SOME COMMON POLAR CURVES

Circles and Spiral

$r = a$
circle

$r = a \sin \theta$
circle

$r = a \cos \theta$
circle

$r = a\theta$
spiral

Limaçons

$r = a \pm b \sin \theta$

$r = a \pm b \cos \theta$

$(a > 0, b > 0)$

Orientation depends on the trigonometric function (sine or cosine) and the sign of b.

$a < b$
limaçon with inner loop

$a = b$
cardioid

$a > b$
dimpled limaçon

$a \geq 2b$
convex limaçon

Roses

$r = a \sin n\theta$

$r = a \cos n\theta$

n-leaved if n is odd
$2n$-leaved if n is even

$r = a \cos 2\theta$
4-leaved rose

$r = a \cos 3\theta$
3-leaved rose

$r = a \cos 4\theta$
8-leaved rose

$r = a \cos 5\theta$
5-leaved rose

Lemniscates

Figure-eight-shaped curves

$r^2 = a^2 \sin 2\theta$
lemniscate

$r^2 = a^2 \cos 2\theta$
lemniscate

9.6 EXERCISES

1–6 ■ Plot the point that has the given polar coordinates. Then give two other polar coordinate representations of the point, one with $r < 0$ and the other with $r > 0$.

1. $(3, \pi/2)$ **2.** $(2, 3\pi/4)$ **3.** $(-1, 7\pi/6)$

4. $(-2, -\pi/3)$ **5.** $(-5, 0)$ **6.** $(3, 1)$

7–12 ■ Find the rectangular coordinates for the point whose polar coordinates are given.

7. $(4, \pi/6)$ **8.** $(6, 2\pi/3)$

9. $(\sqrt{2}, -\pi/4)$ **10.** $(-1, 5\pi/2)$

11. $(5, 5\pi)$ **12.** $(0, 13\pi)$

13–18 ■ Convert the rectangular coordinates to polar coordinates with $r > 0$ and $0 \le \theta < 2\pi$.

13. $(-1, 1)$ **14.** $(3\sqrt{3}, -3)$

15. $(\sqrt{8}, \sqrt{8})$ **16.** $(-\sqrt{6}, -\sqrt{2})$

17. $(3, 4)$ **18.** $(1, -2)$

19–24 ■ Convert the equation to polar form.

19. $x = y$ **20.** $x^2 + y^2 = 9$

21. $y = x^2$ **22.** $y = 5$

23. $x = 4$ **24.** $x^2 - y^2 = 1$

25–38 ■ Convert the polar equation to rectangular coordinates.

25. $r = 7$ **26.** $\theta = \pi$

27. $r \cos \theta = 6$ **28.** $r = 6 \cos \theta$

29. $r^2 = \tan \theta$ **30.** $r^2 = \sin 2\theta$

31. $r = \dfrac{1}{\sin \theta - \cos \theta}$ **32.** $r = \dfrac{1}{1 + \sin \theta}$

33. $r = 1 + \cos \theta$ **34.** $r = \dfrac{4}{1 + 2 \sin \theta}$

35. $r = 2 \sec \theta$ **36.** $r = 2 - \cos \theta$

37. $\sec \theta = 2$ **38.** $\cos 2\theta = 1$

39–44 ■ Match the polar equation with the graphs labeled I–VI. Use the table on page 781 to help you.

39. $r = 3 \cos \theta$ **40.** $r = 3$

41. $r = 2 + 2 \sin \theta$ **42.** $r = 1 + 2 \cos \theta$

43. $r = \sin 3\theta$ **44.** $r = \sin 4\theta$

I

II

III

IV

V

VI
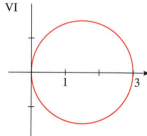

45–66 ■ Sketch the graph of the polar equation.

45. $r = 2$ **46.** $r = -1$

47. $\theta = -\pi/2$ **48.** $\theta = 5\pi/6$

49. $r = 6 \sin \theta$ **50.** $r = \cos \theta$

51. $r = -2 \cos \theta$ **52.** $r = 2 \sin \theta + 2 \cos \theta$

53. $r = 2 - 2 \cos \theta$ **54.** $r = 1 + \sin \theta$

55. $r = -3(1 + \sin \theta)$ **56.** $r = \cos \theta - 1$

57. $r = \theta, \quad \theta \ge 0$ (spiral)

58. $r\theta = 1, \quad \theta > 0$ (reciprocal spiral)

59. $r = \sin 2\theta$ (four-leaved rose)

60. $r = 2 \cos 3\theta$ (three-leaved rose)

61. $r^2 = \cos 2\theta$ (lemniscate)

62. $r^2 = 4 \sin 2\theta$ (lemniscate)

63. $r = 2 + \sin \theta$ (limaçon)

64. $r = 1 - 2 \cos \theta$ (limaçon)

65. $r = 2 + \sec \theta$ (conchoid)

66. $r = \sin \theta \tan \theta$ (cissoid)

67–70 ■ Use a graphing device to graph the polar equation. Choose the domain of θ to make sure you produce the entire graph.

67. $r = \cos(\theta/2)$ **68.** $r = \sin(8\theta/5)$

69. $r = 1 + 2 \sin(\theta/2)$ (nephroid)

70. $r = \sqrt{1 - 0.8 \sin^2\theta}$ (hippopede)

71. Graph the family of polar equations $r = 1 + \sin n\theta$ for $n = 1, 2, 3, 4,$ and 5. How is the number of loops related to n?

72. Graph the family of polar equations $r = 1 + c \sin 2\theta$ for $c = 0.3, 0.6, 1, 1.5,$ and 2. How does the graph change as c increases?

73–76 ■ Match the polar equation with the graphs labeled I–IV. Give reasons for your answers.

73. $r = \sin(\theta/2)$ **74.** $r = 1/\sqrt{\theta}$

75. $r = \theta \sin \theta$ **76.** $r = 1 + 3 \cos(3\theta)$

I

II

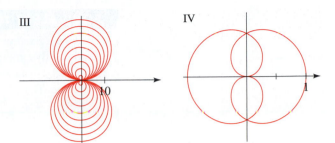

III   IV

77. (a) Show that the distance between the points whose polar coordinates are (r_1, θ_1) and (r_2, θ_2) is

$$\sqrt{r_1^2 + r_2^2 - 2r_1 r_2 \cos(\theta_2 - \theta_1)}$$

[*Hint:* Use the Law of Cosines.]

   (b) Find the distance between the points whose polar coordinates are $(3, 3\pi/4)$ and $(-1, 7\pi/6)$.

78. Show that the graph of $r = a \cos \theta + b \sin \theta$ is a circle, and find its center and radius.

DISCOVERY • DISCUSSION

79. A Transformation of Polar Graphs   How are the graphs of $r = 1 + \sin(\theta - \pi/6)$ and $r = 1 + \sin(\theta - \pi/3)$ related to the graph of $r = 1 + \sin \theta$? In general, how is the graph of $r = f(\theta - \alpha)$ related to the graph of $r = f(\theta)$?

80. Choosing a Convenient Coordinate System   Compare the polar equation of the circle $r = 2$ with its equation in rectangular coordinates. In which coordinate system is the equation simpler? Do the same for the equation of the four-leaved rose $r = \sin 2\theta$. Which coordinate system would you choose to use to study these curves?

81. Choosing a Convenient Coordinate System   Compare the rectangular equation of the line $y = 2$ with its polar equation. In which coordinate system is the equation simpler? Which coordinate system would you choose to study lines?

9.7   POLAR EQUATIONS OF CONICS

Earlier in this chapter we defined a parabola in terms of a focus and directrix, but we defined the ellipse and hyperbola in terms of two foci. In this section we give a more unified treatment of all three types of conics in terms of a focus and directrix. If we place the focus at the origin, then a conic section has a simple polar equation. Moreover, in polar form, rotation of conics becomes a simple matter. Polar equations of ellipses are crucial in the derivation of Kepler's laws of motion (see page 739).

EQUIVALENT DESCRIPTION OF CONICS

Let F be a fixed point (the **focus**), ℓ a fixed line (the **directrix**), and e a fixed positive number (the **eccentricity**). The set of all points P such that the ratio of the distance from P to F to the distance from P to ℓ is the constant e is a conic. That is, the set of all points P such that

$$\frac{d(P, F)}{d(P, \ell)} = e$$

is a conic. The conic is a parabola if $e = 1$, an ellipse if $e < 1$, or a hyperbola if $e > 1$.

■ **Proof**   If $e = 1$, then $d(P, F) = d(P, \ell)$, and so the given condition becomes the definition of a parabola as given in Section 9.1.

Now, suppose $e \neq 1$. Let's place the focus F at the origin and the directrix parallel to the y-axis and d units to the right. Thus, the directrix has equation $x = d$ and is perpendicular to the polar axis. If the point P has polar coordinates (r, θ), we see from Figure 1 that $d(P, F) = r$ and $d(P, \ell) = d - r \cos \theta$. Thus, the condition $d(P, F)/d(P, \ell) = e$, or $d(P, F) = e \cdot d(P, \ell)$, becomes

$$r = e(d - r \cos \theta)$$

If we square both sides of this polar equation and convert to rectangular coordinates, we get

$$x^2 + y^2 = e^2(d - x)^2$$

$$(1 - e^2)x^2 + 2de^2x + y^2 = e^2d^2 \qquad \text{\textcolor{red}{Expand and simplify}}$$

$$\left(x + \frac{e^2d}{1 - e^2}\right)^2 + \frac{y^2}{1 - e^2} = \frac{e^2d^2}{(1 - e^2)^2} \qquad \text{\textcolor{red}{Divide by $1 - e^2$ and complete the square}}$$

If $e < 1$, then dividing both sides of this equation by $e^2d^2/(1 - e^2)^2$ gives an equation of the form

$$\frac{(x - h)^2}{a^2} + \frac{y^2}{b^2} = 1$$

where

$$h = \frac{-e^2d}{1 - e^2} \qquad a^2 = \frac{e^2d^2}{(1 - e^2)^2} \qquad b^2 = \frac{e^2d^2}{1 - e^2}$$

This is the equation of an ellipse with center $(h, 0)$. In Section 9.2 we found that the foci of an ellipse are a distance c from the center, where $c^2 = a^2 - b^2$. In our case

$$c^2 = a^2 - b^2 = \frac{e^4d^2}{(1 - e^2)^2}$$

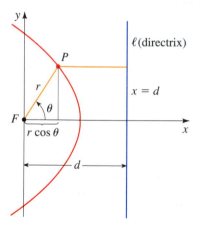

FIGURE 1

Thus, $c = e^2d/(1 - e^2) = -h$, which confirms that the focus defined in the theorem is the same as the focus defined in Section 9.2. It also follows that $e = c/a$.

If $e > 1$, a similar proof shows that the conic is a hyperbola with $e = c/a$, where $c^2 = a^2 + b^2$. $\square$

In the proof we saw that the polar equation of the conic in Figure 1 is $r = e(d - r\cos\theta)$. Solving for r, we get

$$r = \frac{ed}{1 + e\cos\theta}$$

If the directrix is chosen to be to the *left* of the focus $(x = -d)$, then we get the equation $r = ed/(1 - e\cos\theta)$. If the directrix is *parallel* to the polar axis $(y = d$ or $y = -d)$, then we get $\sin\theta$ instead of $\cos\theta$ in the equation. These observations are summarized in the following box and in Figure 2.

POLAR EQUATIONS OF CONICS

A polar equation of the form

$$r = \frac{ed}{1 \pm e\cos\theta} \qquad \text{or} \qquad r = \frac{ed}{1 \pm e\sin\theta}$$

represents a conic with one focus at the origin and with eccentricity e. The conic is

1. a parabola if $e = 1$.

2. an ellipse if $0 < e < 1$.

3. a hyperbola if $e > 1$.

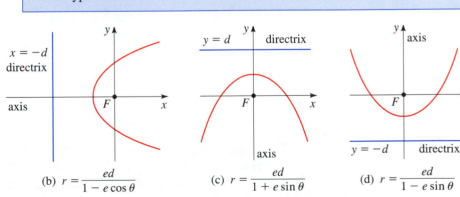

(a) $r = \dfrac{ed}{1 + e\cos\theta}$ 　　 (b) $r = \dfrac{ed}{1 - e\cos\theta}$ 　　 (c) $r = \dfrac{ed}{1 + e\sin\theta}$ 　　 (d) $r = \dfrac{ed}{1 - e\sin\theta}$

FIGURE 2

The form of the polar equation of a conic indicates the location of the directrix

To graph the polar equation of a conic, we first determine the location of the directrix from the form of the equation. The four cases that arise are shown in Figure 2. (The figure shows only the parts of the graphs that are close to the focus at the origin. The shape of the rest of the graph depends on whether the equation represents a parabola, an ellipse, or a hyperbola.) The axis of a conic is perpendicular to the directrix—specifically we have the following:

1. For a parabola, the axis of symmetry is perpendicular to the directrix.

2. For an ellipse, the major axis is perpendicular to the directrix.

3. For a hyperbola, the transverse axis is perpendicular to the directrix.

EXAMPLE 1 ■ Finding a Polar Equation for a Conic

Find a polar equation for the parabola that has its focus at the origin and whose directrix is the line $y = -6$.

SOLUTION

Using $e = 1$ and $d = 6$, and using part (d) of Figure 2, we see that the polar equation of the parabola is

$$ r = \frac{6}{1 - \sin\theta} $$

EXAMPLE 2 ■ Identifying and Sketching a Conic

A conic is given by the polar equation

$$ r = \frac{10}{3 - 2\cos\theta} $$

Identify the conic and sketch its graph.

SOLUTION

Dividing the numerator and denominator by 3, we write the equation as

$$ r = \frac{\frac{10}{3}}{1 - \frac{2}{3}\cos\theta} $$

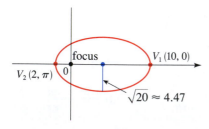

FIGURE 3
$$ r = \frac{10}{3 - 2\cos\theta} $$

We see that $e = \frac{2}{3} < 1$, so the equation represents an ellipse. Comparing the equation to those in Figure 2, we see that the major axis is parallel to the polar axis. To find the endpoints of the major axis, we let $\theta = 0$ and $\theta = \pi$ in the equation. This gives the points $V_1(10, 0)$ and $V_2(2, \pi)$. The distance between these two points is 12; thus, $2a = 12$ and so $a = 6$. The center of the ellipse is at $C(4, 0)$, the midpoint of V_1V_2. To sketch the graph, we need to find b. We have $c = ae = 6\left(\frac{2}{3}\right) = 4$, so

$$ b^2 = a^2 - c^2 = 6^2 - 4^2 = 20 $$

and $b = \sqrt{20} \approx 4.47$. With this information, we can sketch the graph in Figure 3. ■

To graph a polar conic, it is helpful to plot the points on the graph for which $\theta = 0$, $\pi/2$, π, and $3\pi/2$. These points give us a rough idea of the location and shape of the graph.

EXAMPLE 3 ■ Identifying and Sketching a Conic

A conic is given by the polar equation

$$r = \frac{12}{2 + 4\sin\theta}$$

Identify the conic and sketch its graph.

SOLUTION

Dividing numerator and denominator by 2, we write the equation as

$$r = \frac{6}{1 + 2\sin\theta}$$

We see that $e = 2 > 1$, so the equation represents a hyperbola. Comparing the equation to those in Figure 2, we see that the transverse axis is perpendicular to the polar axis. The vertices occur when $\theta = \pi/2$ and $\theta = 3\pi/2$, so they are $V_1(2, \pi/2)$ and $V_2(-6, 3\pi/2) = V_2(6, \pi/2)$. The x-intercepts occur when $\theta = 0$ and $\theta = \pi$; they are $(6, 0)$ and $(6, \pi)$. The x-intercepts help us sketch the lower branch of the hyperbola (Figure 4).

The distance between the two vertices is 4; thus, $2a = 4$ and so $a = 2$. To find b, we first find $c = ae = 2 \cdot 2 = 4$, so

$$b^2 = c^2 - a^2 = 4^2 - 2^2 = 12$$

and $b = \sqrt{12} \approx 3.46$. Knowing a and b helps us sketch the asymptotes and complete the graph in Figure 4. ■

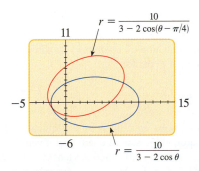

FIGURE 4
$r = \dfrac{12}{2 + 4\sin\theta}$

When we rotate conic sections, it is much more convenient to use polar equations than Cartesian equations. We use the fact that the graph of $r = f(\theta - \alpha)$ is the graph of $r = f(\theta)$ rotated counterclockwise about the origin through an angle α (see Exercise 79 in Section 9.6).

EXAMPLE 4 ■ Rotating an Ellipse

Suppose the ellipse of Example 2 is rotated through an angle $\pi/4$ about the origin. Find a polar equation for the resulting ellipse, and draw its graph.

SOLUTION

We get the equation of the rotated ellipse by replacing θ with $\theta - \pi/4$ in the equation given in Example 2. So the new equation is

$$r = \frac{10}{3 - 2\cos(\theta - \pi/4)}$$

We use this equation to graph the rotated ellipse in Figure 5. Notice that the ellipse has been rotated about the focus at the origin. ■

FIGURE 5

In Figure 6 we use a computer to sketch a number of conics to demonstrate the effect of varying the eccentricity e. Notice that when e is close to 0, the ellipse is nearly circular and becomes more elongated as e increases. When $e = 1$, of course, the conic is a parabola. As e increases beyond 1, the conic is an ever steeper hyperbola.

$e = 0.5$ $e = 0.86$ $e = 1$ $e = 1.4$ $e = 4$

FIGURE 6

9.7 EXERCISES

1–8 ■ Write a polar equation of a conic that has its focus at the origin and satisfies the given conditions.

1. Ellipse, eccentricity $\frac{2}{3}$, directrix $x = 3$

2. Hyperbola, eccentricity $\frac{4}{3}$, directrix $x = -3$

3. Parabola, directrix $y = 2$

4. Ellipse, eccentricity $\frac{1}{2}$, directrix $y = -4$

5. Hyperbola, eccentricity 4, directrix $r = 5 \sec \theta$

6. Ellipse, eccentricity 0.6, directrix $r = 2 \csc \theta$

7. Parabola, vertex at $(5, \pi/2)$

8. Ellipse, eccentricity 0.4, vertex at $(2, 0)$

9–16 ■ (a) Find the eccentricity and identify the conic. (b) Sketch the conic and label the vertices.

9. $r = \dfrac{4}{1 + 3 \cos \theta}$ **10.** $r = \dfrac{8}{3 + 3 \cos \theta}$

11. $r = \dfrac{2}{1 - \cos \theta}$ **12.** $r = \dfrac{10}{3 - 2 \sin \theta}$

13. $r = \dfrac{6}{2 + \sin \theta}$ **14.** $r = \dfrac{5}{2 - 3 \sin \theta}$

15. $r = \dfrac{7}{2 - 5 \sin \theta}$ **16.** $r = \dfrac{8}{3 + \cos \theta}$

17. (a) Find the eccentricity and directrix of the conic $r = 1/(4 - 3 \cos \theta)$ and graph the conic and its directrix.

(b) If this conic is rotated about the origin through an angle $\pi/3$, write the resulting equation and draw its graph.

18. Graph the parabola $r = 5/(2 + 2 \sin \theta)$ and its directrix. Also graph the curve obtained by rotating this parabola about its focus through an angle $\pi/6$.

19. Graph the conics $r = e/(1 - e \cos \theta)$ with $e = 0.4, 0.6, 0.8$, and 1.0 on a common screen. How does the value of e affect the shape of the curve?

20. (a) Graph the conics $r = ed/(1 + e \sin \theta)$ for $e = 1$ and various values of d. How does the value of d affect the shape of the conic?

(b) Graph these conics for $d = 1$ and various values of e. How does the value of e affect the shape of the conic?

21. (a) Show that the polar equation of an ellipse with directrix $x = -d$ can be written in the form

$$r = \frac{a(1 - e^2)}{1 - e \cos \theta}$$

(b) Find an approximate polar equation for the elliptical orbit of the earth around the sun (at one focus) given that the eccentricity is about 0.017 and the length of the major axis is about 2.99×10^8 km.

22. (a) The planets move around the sun in elliptical orbits with the sun at one focus. The positions of a planet that are closest to, and farthest from, the sun are called its **perihelion** and **aphelion**, respectively. Use Exercise 21(a) to show that the perihelion distance from a planet to the sun is $a(1 - e)$ and the aphelion distance is $a(1 + e)$.

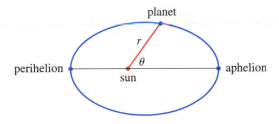

23. The distance from the planet Pluto to the sun is 4.43×10^9 km at perihelion and 7.37×10^9 km at aphelion. Use Exercise 22 to find the eccentricity of Pluto's orbit.

 DISCOVERY · DISCUSSION

24. Distance to a Focus   When we found polar equations for the conics, we placed one focus at the pole. It's easy to find the distance from that focus to any point on the conic. Explain how the polar equation gives us this distance.

25. Polar Equations of Orbits   When a satellite orbits the earth, its path is an ellipse with one focus at the center of the earth. Why do scientists use polar (rather than rectangular) coordinates to track the position of satellites? Your answer to Exercise 24 is relevant here.

(b) Use the data of Exercise 21(b) to find the distances from the earth to the sun at perihelion and at aphelion.

9.8   PARAMETRIC EQUATIONS

So far we've described a curve by giving an equation that the coordinates of all points on the curve must satisfy. For example, we know that the equation $y = x^2$ represents a parabola in rectangular coordinates and that $r = \sin \theta$ represents a circle in polar coordinates. We now study another method for describing a curve in the plane, which in many situations turns out to be more useful and natural than either rectangular or polar equations. In this method, the x- and y-coordinates of points on the curve are given separately as functions of an additional variable t, called the **parameter**:

$$x = f(t) \qquad y = g(t)$$

These are called **parametric equations** for the curve. Substituting a value of t into each equation determines the coordinates of a point (x, y). As t varies, the point $(x, y) = (f(t), g(t))$ varies and traces out the curve. If we think of t as representing time, then as t increases, we can imagine a particle at $(x, y) = (f(t), g(t))$ moving along the curve.

EXAMPLE 1 ■ Sketching a Parametric Curve

Sketch the curve defined by the parametric equations

$$x = t^2 - 3t \qquad y = t - 1$$

Eliminate the parameter t to obtain a single equation for the curve in the variables x and y.

SOLUTION

For every value of t, we get a point on the curve. For example, if $t = 0$, then $x = 0$ and $y = -1$, so the corresponding point is $(0, -1)$. In Figure 1 we plot the points (x, y) determined by the values of t shown in the following table.

t	x	y
-2	10	-3
-1	4	-2
0	0	-1
1	-2	0
2	-2	1
3	0	2
4	4	3
5	10	4

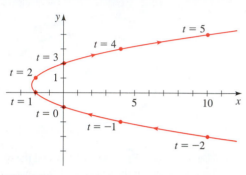

FIGURE 1

As t increases, a particle whose position is given by the parametric equations moves along the curve in the direction of the arrows. The curve seems to be a parabola. We can confirm this by eliminating the parameter t from the parametric equations and reducing them to a single equation as follows. First we solve for t in the second equation to get $t = y + 1$. Substituting this into the first equation, we get

$$x = (y + 1)^2 - 3(y + 1) = y^2 - y - 2$$

The curve is the parabola $x = y^2 - y - 2$.   ■

Notice that we would obtain the same graph as in Example 1 from the parametrization

$$x = t^2 - t - 2 \qquad y = t$$

because the points on this curve also satisfy the equation $x = y^2 - y - 2$. But the same value of t produces different points on the curve in these two parametrizations. For example, when $t = 0$, the particle that traces out the curve in Figure 1 is at $(0, -1)$, whereas in the parametrization of the preceding equations, the particle is already at $(-2, 0)$ when $t = 0$. *Thus, a parametrization contains more information than just the shape of the curve; it also indicates* how *the curve is being traced out.*

EXAMPLE 2 ■ Eliminating the Parameter

Describe and graph the curve represented by the parametric equations

$$x = \cos t \qquad y = \sin t \qquad 0 \leqslant t \leqslant 2\pi$$

SOLUTION

To identify the curve, we eliminate the parameter. Since $\cos^2 t + \sin^2 t = 1$ and since $x = \cos t$ and $y = \sin t$ for every point (x, y) on the curve, we have

$$x^2 + y^2 = (\cos t)^2 + (\sin t)^2 = 1$$

FIGURE 2

FIGURE 3

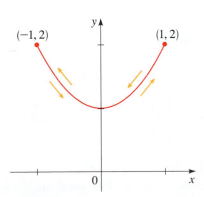

FIGURE 4

This means that all points on the curve satisfy the equation $x^2 + y^2 = 1$, so the graph is a circle of radius 1 centered at the origin. As t increases from 0 to 2π, the point given by the parametric equations starts at $(1, 0)$ and moves counterclockwise once around the circle, as shown in Figure 2. Notice that the parameter t can be interpreted as the angle shown in the figure.   ■

EXAMPLE 3 ■ Finding Parametric Equations for a Graph

Find parametric equations for the line of slope 3 that passes through the point $(2, 6)$.

SOLUTION

Let's start at the point $(2, 6)$ and move up and to the right along this line. Because the line has slope 3, for every 1 unit we move to the right, we must move up 3 units. In other words, if we increase the x-coordinate by t units, we must correspondingly increase the y-coordinate by $3t$ units. This leads to the parametric equations

$$x = 2 + t \qquad y = 6 + 3t$$

To confirm that these equations give the desired line, we eliminate the parameter. We solve for t in the first equation and substitute into the second to get

$$y = 6 + 3(x - 2) = 3x$$

Thus, the slope-intercept form of the equation of this line is $y = 3x$, which is a line of slope 3 that does pass through $(2, 6)$ as required. The graph is shown in Figure 3.   ■

EXAMPLE 4 ■ Sketching a Parametric Curve

Sketch the curve with parametric equations

$$x = \sin t \qquad y = 2 - \cos^2 t$$

SOLUTION

To eliminate the parameter, we first use the trigonometric identity $\cos^2 t = 1 - \sin^2 t$ to change the second equation:

$$y = 2 - \cos^2 t = 2 - (1 - \sin^2 t) = 1 + \sin^2 t$$

Now we can substitute $\sin t = x$ from the first equation to get

$$y = 1 + x^2$$

and so the point (x, y) moves along the parabola $y = 1 + x^2$. However, since $-1 \le \sin t \le 1$, we have $-1 \le x \le 1$, so the parametric equations represent only the part of the parabola between $x = -1$ and $x = 1$. Since $\sin t$ is periodic, the point $(x, y) = (\sin t, 2 - \cos^2 t)$ moves back and forth infinitely often along the parabola between the points $(-1, 2)$ and $(1, 2)$ as shown in Figure 4.   ■

FIGURE 6

cycloid

FIGURE 7

FIGURE 8

FIGURE 9
$x = t + 2 \sin 2t$, $y = t + 2 \cos 5t$

EXAMPLE 5 ■ Parametric Equations for the Cycloid

As a circle rolls along a straight line, the curve traced out by a fixed point P on the circumference of the circle is called a **cycloid** (see Figure 5). If the circle has radius a and rolls along the x-axis, with one position of the point P being at the origin, find parametric equations for the cycloid.

FIGURE 5

SOLUTION

Figure 6 shows the circle and the point P after the circle has rolled through an angle θ (in radians). The distance $d(O, T)$ that the circle has rolled must be the same as the length of the arc PT, which, by the arc length formula, is $a\theta$ (see Section 6.1). This means that the center of the circle is $C(a\theta, a)$.

Let the coordinates of P be (x, y). Then from Figure 6 (which illustrates the case $0 < \theta < \pi/2$), we see that

$$x = d(O, T) - d(P, Q) = a\theta - a \sin \theta = a(\theta - \sin \theta)$$
$$y = d(T, C) - d(Q, C) = a - a \cos \theta = a(1 - \cos \theta)$$

so parametric equations for the cycloid are

$$x = a(\theta - \sin \theta) \qquad y = a(1 - \cos \theta) \qquad ■$$

The cycloid has a number of interesting physical properties. It is the "curve of quickest descent" in the following sense. Let's choose two points P and Q that are not directly above each other, and join them with a wire. Suppose we allow a bead to slide down the wire under the influence of gravity (ignoring friction). Of all possible shapes that the wire can be bent into, the bead will slide from P to Q the fastest when the shape is half of an arch of an inverted cycloid (see Figure 7). The cycloid is also the "curve of equal descent" in the sense that no matter where we place a bead B on a cycloid-shaped wire, it takes the same time to slide to the bottom (see Figure 8). These rather surprising properties of the cycloid were proved (using calculus) in the 17th century by several mathematicians and physicists, including Johann Bernoulli, Blaise Pascal, and Christiaan Huygens.

🖩 Using Graphing Devices to Graph Parametric Curves

Most graphing calculators and computer graphing programs can be used to graph parametric equations. Such devices are particularly useful when sketching complicated curves like the one shown in Figure 9.

EXAMPLE 6 ■ Graphing Parametric Curves

Use a graphing device to draw the following parametric curves. Discuss their similarities and differences.

(a) $x = \sin 2t$
 $y = 2 \cos t$

(b) $x = \sin 3t$
 $y = 2 \cos t$

SOLUTION

In both parts (a) and (b), the graph will lie inside the rectangle given by $-1 \le x \le 1$, $-2 \le y \le 2$, since both the sine and the cosine of any number will be between -1 and 1. Thus, we may use the viewing rectangle $[-1.5, 1.5]$ by $[-2.5, 2.5]$.

(a) Since $2 \cos t$ is periodic with period 2π (see Section 5.3), and since $\sin 2t$ has period π, letting t vary over the interval $0 \le t \le 2\pi$ gives us the complete graph, which is shown in Figure 10(a).

(b) Again, letting t take on values between 0 and 2π gives the complete graph shown in Figure 10(b).

Both graphs are *closed curves*, which means they form loops with the same starting and ending point; also, both graphs cross over themselves. However, the graph in Figure 10(a) has two loops, like a figure eight, whereas the graph in Figure 10(b) has three loops. ∎

The curves graphed in Example 6 are called Lissajous figures. A **Lissajous figure** is the graph of a pair of parametric equations of the form

$$x = A \sin \omega_1 t \qquad y = B \cos \omega_2 t$$

where A, B, ω_1, and ω_2 are real constants. Since $\sin \omega_1 t$ and $\cos \omega_2 t$ are both between -1 and 1, a Lissajous figure will lie inside the rectangle determined by $-A \le x \le A$, $-B \le y \le B$. This fact can be used to choose a viewing rectangle when graphing a Lissajous figure, as in Example 6.

Recall from Section 9.6 that rectangular coordinates (x, y) and polar coordinates (r, θ) are related by the equations $x = r \cos \theta$, $y = r \sin \theta$. Thus, we can graph the polar equation $r = f(\theta)$ by changing it to parametric form as follows:

$$x = r \cos \theta = f(\theta) \cos \theta \qquad \text{Since } r = f(\theta)$$

$$y = r \sin \theta = f(\theta) \sin \theta$$

Replacing θ by the standard parametric variable t, we have the following result.

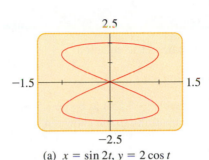

2.5

−1.5 | 1.5

−2.5

(a) $x = \sin 2t, y = 2 \cos t$

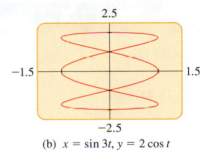

2.5

−1.5 | 1.5

−2.5

(b) $x = \sin 3t, y = 2 \cos t$

FIGURE 10

POLAR EQUATIONS IN PARAMETRIC FORM

The graph of the polar equation $r = f(\theta)$ is the same as the graph of the parametric equations

$$x = f(t) \cos t \qquad y = f(t) \sin t$$

EXAMPLE 7 ■ **Parametric Form of a Polar Equation**

Consider the polar equation $r = \ln \theta, \ 1 \leq \theta \leq 10\pi$.

(a) Express the equation in parametric form.
(b) Draw a graph of the parametric equations from part (a).

SOLUTION

(a) The given polar equation is equivalent to the parametric equation

$$x = \ln t \cos t \qquad y = \ln t \sin t$$

(b) Since $\ln 10\pi \approx 3.45$, we use the viewing rectangle $[-3.5, 3.5]$ by $[-3.5, 3.5]$, and we let t vary from 1 to $10\pi \approx 31.42$. The resulting graph shown in Figure 11 is a *logarithmic spiral*. ■

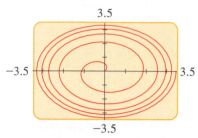

FIGURE 11
$x = \ln t \cos t, \ y = \ln t \sin t$

9.8 EXERCISES

1–22 ■ (a) Sketch the curve represented by the parametric equations. (b) Find a rectangular-coordinate equation for the curve by eliminating the parameter.

1. $x = 2t, \quad y = t + 6$

2. $x = 6t - 4, \quad y = 3t, \quad t \geq 0$

3. $x = t^2, \quad y = t - 2, \quad 2 \leq t \leq 4$

4. $x = 2t + 1, \quad y = \left(t + \frac{1}{2}\right)^2$

5. $x = \sqrt{t}, \quad y = 1 - t$

6. $x = t^2, \quad y = t^4 + 1$

7. $x = \dfrac{1}{t}, \quad y = t + 1$

8. $x = t + 1, \quad y = \dfrac{t}{t + 1}$

9. $x = 4t^2, \quad y = 8t^3$

10. $x = |t|, \quad y = |1 - |t||$

11. $x = 2 \sin t, \quad y = 2 \cos t, \quad 0 \leq t \leq \pi$

12. $x = 2 \cos t, \quad y = 3 \sin t, \quad 0 \leq t \leq 2\pi$

13. $x = \sin^2 t, \quad y = \sin^4 t$

14. $x = \sin^2 t, \quad y = \cos t$

15. $x = \cos t, \quad y = \cos 2t$

16. $x = \cos 2t, \quad y = \sin 2t$

17. $x = \sec t, \quad y = \tan t, \quad 0 \leq t < \pi/2$

18. $x = \cot t, \quad y = \csc t, \quad 0 < t < \pi$

19. $x = e^t, \quad y = e^{-t}$

20. $x = e^{2t}, \quad y = e^t, \quad t \geq 0$

21. $x = \cos^2 t, \quad y = \sin^2 t$

22. $x = \cos^3 t, \quad y = \sin^3 t, \quad 0 \leq t \leq 2\pi$

23–26 ■ Find parametric equations for the line with the given properties.

23. Slope $\frac{1}{2}$, passing through $(4, -1)$

24. Slope -2, passing through $(-10, -20)$

25. Passing through $(6, 7)$ and $(7, 8)$

26. Passing through $(12, 7)$ and the origin

27. Find parametric equations for the circle $x^2 + y^2 = a^2$.

28. Find parametric equations for the ellipse

$$\frac{x^2}{a^2} + \frac{y^2}{b^2} = 1$$

29. Show by eliminating the parameter θ that the following parametric equations represent a hyperbola:

$$x = a \tan \theta \qquad y = b \sec \theta$$

30. Show that the following parametric equations represent a part of the hyperbola of Exercise 29:

$$x = a \sqrt{t} \qquad y = b \sqrt{t + 1}$$

31–34 ■ Stretch the curve given by the parametric equations.

31. $x = t \cos t, \quad y = t \sin t, \quad t \geq 0$

32. $x = \sin t, \quad y = \sin 2t$

33. $x = \dfrac{3t}{1 + t^3}, \quad y = \dfrac{3t^2}{1 + t^3}$

34. $x = \cot t, \quad y = 2 \sin^2 t, \quad 0 < t < \pi$

35. If a projectile is fired with an initial speed of v_0 ft/s at an angle α above the horizontal, then its position after t seconds is given by the parametric equations

$$x = (v_0 \cos \alpha)t \qquad y = (v_0 \sin \alpha)t - 16t^2$$

(where x and y are measured in feet). Show that the path of the projectile is a parabola by eliminating the parameter t.

36. Referring to Exercise 35, suppose a gun fires a bullet into the air with an initial speed of 2048 ft/s at an angle of 30° to the horizontal.
  (a) After how many seconds will the bullet hit the ground?
  (b) How far from the gun will the bullet hit the ground?
  (c) What is the maximum height attained by the bullet?

37–42 ■ Use a graphing device to draw the curve represented by the parametric equations.

37. $x = \sin t$, $y = 2 \cos 3t$ **38.** $x = 2 \sin t$, $y = \cos 4t$

39. $x = 3 \sin 5t$, $y = 5 \cos 3t$

40. $x = \sin 4t$, $y = \cos 3t$

41. $x = \sin(\cos t)$, $y = \cos(t^{3/2})$, $0 \le t \le 2\pi$

42. $x = 2 \cos t + \cos 2t$, $y = 2 \sin t - \sin 2t$

43–46 ■ (a) Express the polar equation in parametric form. (b) Use a graphing device to graph the parametric equations you found in part (a).

43. $r = e^{\theta/12}$, $0 \le \theta \le 4\pi$ **44.** $r = \sin \theta + 2 \cos \theta$

45. $r = \dfrac{4}{2 - \cos \theta}$ **46.** $r = 2^{\sin \theta}$

47–50 ■ Match the parametric equations with the graphs labeled I–IV. Give reasons for your answers.

47. $x = t^3 - 2t$, $y = t^2 - t$ **48.** $x = \sin 3t$, $y = \sin 4t$

49. $x = t + \sin 2t$, $y = t + \sin 3t$

50. $x = \sin(t + \sin t)$, $y = \cos(t + \cos t)$

I

II

III

IV

51. In Example 5 suppose the point P that traces out the curve lies not on the edge of the circle, but rather at a fixed point inside the rim, at a distance b from the center (with $b < a$). The curve traced out by P is called a **curtate cycloid** (or **trochoid**). Show that parametric equations for the curtate cycloid are

$$x = a\theta - b \sin \theta \qquad y = a - b \cos \theta$$

Sketch the graph.

52. In Exercise 51 if the point P lies *outside* the circle at a distance b from the center (with $b > a$), then the curve traced out by P is called a **prolate cycloid**. Show that parametric equations for the prolate cycloid are the same as the equations for the curtate cycloid, and sketch the graph for the case where $a = 1$ and $b = 2$.

53. A circle C of radius b rolls on the inside of a larger circle of radius a centered at the origin. Let P be a fixed point on the smaller circle, with initial position at the point $(a, 0)$, as shown in the figure. The curve traced out by P is called a **hypocycloid**.

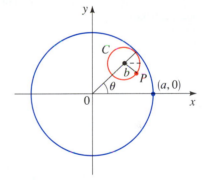

(a) Show that parametric equations for the hypocycloid are

$$x = (a - b) \cos \theta + b \cos\left(\frac{a - b}{b} \theta\right)$$

$$y = (a - b) \sin \theta - b \sin\left(\frac{a - b}{b} \theta\right)$$

(b) If $a = 4b$, the hypocycloid is called an **astroid**. Show that in this case the parametric equations can be reduced to

$$x = a \cos^3\theta \qquad y = a \sin^3\theta$$

Sketch the curve and eliminate the parameter to obtain an equation for the astroid in rectangular coordinates.

54. If the circle C of Exercise 53 rolls on the *outside* of the larger circle, the curve traced out by P is called an **epicycloid**. Find parametric equations for the epicycloid.

55. In the figure, the circle of radius a is stationary and, for every θ, the point P is the midpoint of the segment QR. The curve traced out by P for $0 < \theta < \pi$ is called the **longbow curve**. Find parametric equations for this curve.

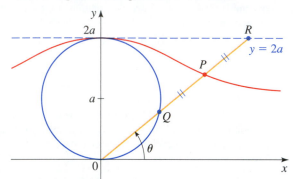

56. A string is wound around a circle and then unwound while being held taut. The curve traced out by the point P at the end of the string is called the **involute** of the circle, as shown in the figure. If the circle has radius a and is centered at the origin, and if the initial position of P is $(a, 0)$, show that parametric equations for the involute in terms of the parameter θ are

$$x = a(\cos \theta + \theta \sin \theta) \qquad y = a(\sin \theta - \theta \cos \theta)$$

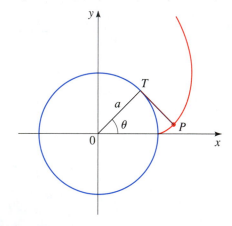

57. Eliminate the parameter θ in the parametric equations for the cycloid (Example 5) to obtain a rectangular coordinate equation for the section of the curve given by $0 \le \theta \le \pi$.

◆ DISCOVERY · DISCUSSION

58. More Information in Parametric Equations   In this section we stated that parametric equations contain more information than just the shape of a curve. Write a short paragraph explaining this statement. Use the following example and your answers to questions (a) and (b) in your explanation.

The position of a particle is given by the parametric equations

$$x = \sin t \qquad y = \cos t$$

where t represents time. We know that the shape of the path of the particle is a circle.
(a) How long does it take the particle to go once around the circle? Find parametric equations if the particle moves twice as fast around the circle.
(b) Does the particle travel clockwise or counterclockwise around the circle? Find parametric equations if the particle moves in the opposite direction around the circle.

59. Different Ways of Tracing Out a Curve   The curves C, D, E, and F are defined parametrically as follows, where the parameter t takes on all real values unless otherwise stated:

$$C: \quad x = t, \quad y = t^2$$
$$D: \quad x = \sqrt{t}, \quad y = t, \quad t \ge 0$$
$$E: \quad x = \sin t, \quad y = \sin^2 t$$
$$F: \quad x = e^t, \quad y = e^{2t}$$

(a) Show that the points on all four of these curves satisfy the same rectangular coordinate equation.
(b) Draw the graph of each curve and explain how the curves differ from one another.

9 REVIEW

CONCEPT CHECK

1. (a) Give the geometric definition of a parabola. What are the focus and directrix of the parabola?

(b) Sketch the parabola $x^2 = 4py$ for the case $p > 0$. Identify on your diagram the vertex, focus, and directrix. What happens if $p < 0$?

(c) Sketch the parabola $y^2 = 4px$, together with its vertex, focus, and directrix, for the case $p > 0$. What happens if $p < 0$?

2. (a) Give the geometric definition of an ellipse. What are the foci of the ellipse?

(b) For the ellipse with equation

$$\frac{x^2}{a^2} + \frac{y^2}{b^2} = 1$$

where $a > b > 0$, what are the coordinates of the vertices and the foci? What are the major and minor axes? Illustrate with a graph.

(c) Give an expression for the eccentricity of the ellipse in part (b).

(d) State the equation of an ellipse with foci on the y-axis.

3. (a) Give the geometric definition of a hyperbola. What are the foci of the hyperbola?

(b) For the hyperbola with equation

$$\frac{x^2}{a^2} - \frac{y^2}{b^2} = 1$$

what are the coordinates of the vertices and foci? What are the equations of the asymptotes? What is the transverse axis? Illustrate with a graph.

(c) State the equation of a hyperbola with foci on the y-axis.

(d) What steps would you take to sketch a hyperbola with a given equation?

4. Suppose h and k are positive numbers. What is the effect on the graph of an equation in x and y if

(a) x is replaced by $x - h$? By $x + h$?

(b) y is replaced by $y - k$? By $y + k$?

5. How can you tell whether the following nondegenerate conic is a parabola, an ellipse, or a hyperbola?

$$Ax^2 + Cy^2 + Dx + Ey + F = 0$$

6. Suppose the x- and y-axes are rotated through an acute angle ϕ to produce the X- and Y-axes. Write equations that relate the coordinates (x, y) and (X, Y) of a point in the xy-plane and XY-plane, respectively.

7. (a) How do you eliminate the xy-term in this equation?

$$Ax^2 + Bxy + Cy^2 + Dx + Ey + F = 0$$

(b) What is the discriminant of the conic in part (a)? How can you use the discriminant to determine whether the conic is a parabola, an ellipse, or a hyperbola?

8. (a) Describe how polar coordinates represent the position of a point in the plane.

(b) What equations do you use to change from polar to rectangular coordinates?

(c) What equations do you use to change from rectangular to polar coordinates?

9. (a) Write polar equations that represent a conic with eccentricity e.

(b) For what values of e is the conic an ellipse? A hyperbola? A parabola?

10. How do you sketch a curve with parametric equations $x = f(t),\ y = g(t)$?

EXERCISES

1–4 ■ Find the vertex, focus, and directrix of the parabola, and sketch the graph.

1. $x^2 + 8y = 0$

2. $2x - y^2 = 0$

3. $x - y^2 + 4y - 2 = 0$

4. $2x^2 + 6x + 5y + 10 = 0$

5–8 ■ Find the center, vertices, foci, and the lengths of the major and minor axes of the ellipse, and sketch the graph.

5. $x^2 + 4y^2 = 16$

6. $9x^2 + 4y^2 = 1$

7. $4x^2 + 9y^2 = 36y$

8. $2x^2 + y^2 = 2 + 4(x - y)$

9–12 ■ Find the center, vertices, foci, and asymptotes of the hyperbola, and sketch the graph.

9. $x^2 - 2y^2 = 16$

10. $x^2 - 4y^2 + 16 = 0$

11. $9y^2 + 18y = x^2 + 6x + 18$

12. $y^2 = x^2 + 6y$

13–18 ■ Find an equation for the conic whose graph is shown.

13.

14.

15.

16.

17.

18.
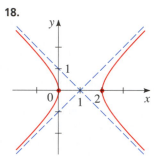

19–30 ■ Determine the type of curve represented by the equation. Find the foci and vertices (if any), and sketch the graph.

19. $\dfrac{x^2}{12} + y = 1$

20. $\dfrac{x^2}{12} + \dfrac{y^2}{144} = \dfrac{y}{12}$

21. $x^2 - y^2 + 144 = 0$

22. $x^2 + 6x = 9y^2$

23. $4x^2 + y^2 = 8(x + y)$

24. $3x^2 - 6(x + y) = 10$

25. $x = y^2 - 16y$

26. $2x^2 + 4 = 4x + y^2$

27. $2x^2 - 12x + y^2 + 6y + 26 = 0$

28. $36x^2 - 4y^2 - 36x - 8y = 31$

29. $9x^2 + 8y^2 - 15x + 8y + 27 = 0$

30. $x^2 + 4y^2 = 4x + 8$

31–38 ■ Find an equation for the conic section with the given properties.

31. The parabola with focus $F(0, 1)$ and directrix $y = -1$

32. The ellipse with center $C(0, 4)$, foci $F_1(0, 0)$ and $F_2(0, 8)$, and major axis of length 10

33. The hyperbola with vertices $V(0, \pm 2)$ and asymptotes $y = \pm\frac{1}{2}x$

34. The hyperbola with center $C(2, 4)$, foci $F_1(2, 1)$ and $F_2(2, 7)$, and vertices $V_1(2, 6)$ and $V_2(2, 2)$

35. The ellipse with foci $F_1(1, 1)$ and $F_2(1, 3)$, and with one vertex on the x-axis

36. The parabola with vertex $V(5, 5)$ and directrix the y-axis

37. The ellipse with vertices $V_1(7, 12)$ and $V_2(7, -8)$, and passing through the point $P(1, 8)$

38. The parabola with vertex $V(-1, 0)$ and horizontal axis of symmetry, and crossing the y-axis at $y = 2$

39. A cannon fires a cannonball as shown in the figure. The path of the cannonball is a parabola with vertex at the highest point of the path. If the cannonball lands 1600 ft from the cannon and the highest point it reaches is 3200 ft above the ground, find an equation for the path of the cannonball. Place the origin at the location of the cannon.

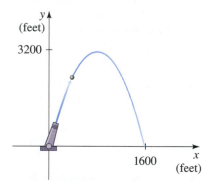

40. A satellite is in an elliptical orbit around the earth with the center of the earth at one focus. The height of the satellite above the earth varies between 140 mi and 440 mi. Assume the earth is a sphere with radius 3960 mi. Find an equation for the path of the satellite with the origin at the center of the earth.

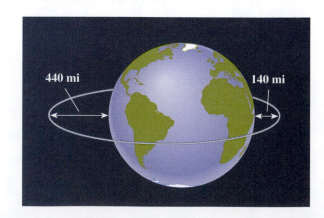

41. The path of the earth around the sun is an ellipse with the sun at one focus. The ellipse has major axis 186,000,000 mi and eccentricity 0.017. Find the distance between the earth and the sun when the earth is (a) closest to the sun and (b) farthest from the sun.

42. A ship is located 40 mi from a straight shoreline. LORAN stations A and B are located on the shoreline, 300 mi apart. From the LORAN signals, the captain determines that his ship is 80 mi closer to A than to B. Find the location of the ship. (Place A and B on the y-axis with the x-axis halfway between them. Find the x- and y-coordinates of the ship.)

 43. (a) Draw graphs of the following family of ellipses for $k = 1, 2, 4,$ and 8.

$$\frac{x^2}{16 + k^2} + \frac{y^2}{k^2} = 1$$

    (b) Prove that all the ellipses in part (a) have the same foci.

 44. (a) Draw graphs of the following family of parabolas for $k = \frac{1}{2}, 1, 2,$ and 4.

$$y = kx^2$$

    (b) Find the foci of the parabolas in part (a).
    (c) How does the location of the focus change as k increases?

45–48 ■ (a) Use the discriminant to determine whether the graph of the equation is a parabola, an ellipse, or a hyperbola. (b) Use a rotation of axes to eliminate the xy-term. (c) Sketch the graph.

45. $x^2 + 4xy + y^2 = 1$

46. $5x^2 - 6xy + 5y^2 - 8x + 8y - 8 = 0$

47. $7x^2 - 6\sqrt{3}\,xy + 13y^2 - 4\sqrt{3}\,x - 4y = 0$

48. $9x^2 + 24xy + 16y^2 = 25$

 49–52 ■ Use a graphing device to graph the conic. Identify the type of conic from the graph.

49. $5x^2 + 3y^2 = 60$ **50.** $9x^2 - 12y^2 + 36 = 0$

51. $6x + y^2 - 12y = 30$ **52.** $52x^2 - 72xy + 73y^2 = 100$

53–60 ■ (a) Sketch the graph of the polar equation. (b) Express the equation in rectangular coordinates.

53. $r = 3 + 3\cos\theta$ **54.** $r = 3\sin\theta$

55. $r = 2\sin 2\theta$ **56.** $r = 4\cos 3\theta$

57. $r^2 = \sec 2\theta$ **58.** $r^2 = 4\sin 2\theta$

59. $r = \sin\theta + \cos\theta$ **60.** $r = \dfrac{4}{2 + \cos\theta}$

 61–64 ■ Use a graphing device to graph the polar equation. Choose the domain of θ to make sure you produce the entire graph.

61. $r = \cos(\theta/3)$ **62.** $r = \sin(9\theta/4)$

63. $r = 1 + 4\cos(\theta/3)$ **64.** $r = \theta\sin\theta$

65–68 ■ (a) Find the eccentricity and identify the conic. (b) Sketch the conic and label the vertices.

65. $r = \dfrac{1}{1 - \cos\theta}$ **66.** $r = \dfrac{2}{1 + 2\sin\theta}$

67. $r = \dfrac{4}{1 + 2\sin\theta}$ **68.** $r = \dfrac{12}{1 - 4\cos\theta}$

69–72 ■ Sketch the parametric curve, and eliminate the parameter to find an equation in rectangular coordinates that all points on the curve satisfy.

69. $x = 1 - t^2,\ \ y = 1 + t$ **70.** $x = t^2 - 1,\ \ y = t^2 + 1$

71. $x = 1 + \cos t,\ \ y = 1 - \sin t,\ \ 0 \le t \le \pi/2$

72. $x = \dfrac{1}{t} + 2,\ \ y = \dfrac{2}{t^2},\ \ 0 < t \le 2$

 73–74 ■ Use a graphing device to draw the parametric curve.

73. $x = \cos 2t,\ \ y = \sin 3t$

74. $x = \sin(t + \cos 2t),\ \ y = \cos(t + \sin 3t)$

75. In the figure the point P is the midpoint of the segment QR and $0 \le \theta < \pi/2$. Using θ as the parameter, find a parametric representation for the curve traced out by P.

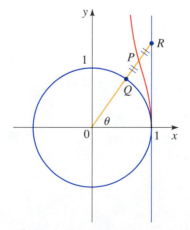

9   TEST

1. Find the focus and directrix of the parabola $x^2 = -12y$, and sketch its graph.

2. Find the vertices, foci, and the lengths of the major and minor axes for the ellipse $\dfrac{x^2}{16} + \dfrac{y^2}{4} = 1$. Then sketch its graph.

3. Find the vertices, foci, and asymptotes of the hyperbola $\dfrac{y^2}{9} - \dfrac{x^2}{16} = 1$. Then sketch its graph.

4–6 ■ Find an equation for the conic whose graph is shown.

4.

5.

6.

7–9 ■ Sketch the graph of the equation.

7. $16x^2 + 36y^2 - 96x + 36y + 9 = 0$

8. $9x^2 - 8y^2 + 36x + 64y = 164$

9. $2x + y^2 + 8y + 8 = 0$

10. Find an equation for the hyperbola with foci $(0, \pm 5)$ and with asymptotes $y = \pm \frac{3}{4} x$.

11. Find an equation for the parabola with focus $(2, 4)$ and directrix the x-axis.

12. A parabolic reflector for a car headlight forms a bowl shape that is 6 in. wide at its opening and 3 in. deep, as shown in the figure at the left. How far from the vertex should the filament of the bulb be placed if it is to be located at the focus?

13. (a) Use the discriminant to determine whether the graph of this equation is a parabola, an ellipse, or a hyperbola:

$$5x^2 + 4xy + 2y^2 = 18$$

(b) Use rotation of axes to eliminate the xy-term in the equation.
(c) Sketch the graph of the equation.
(d) Find the coordinates of the vertices of this conic (in the xy-coordinate system).

14. Graph the polar equation $r = 2 + \cos \theta$.

15. Convert the polar equation $r = 2 \cos \theta - 4 \sin \theta$ to rectangular coordinates, and identify the graph.

16. (a) Find the polar equation of the conic that has a focus at the origin, eccentricity $e = \frac{1}{2}$, and directrix $x = 2$. Sketch the graph.
(b) What type of conic is represented by the following equation? Sketch its graph.

$$r = \frac{3}{2 - \sin \theta}$$

17. (a) Sketch the graph of the parametric curve

$$x = 3 \sin \theta + 3 \qquad y = 2 \cos \theta \qquad 0 \leqslant \theta \leqslant \pi$$

(b) Eliminate the parameter θ in part (a) to obtain an equation for this curve in rectangular coordinates.

Focus on Modeling

The Path of a Projectile

Modeling motion is one of the most important ideas in both classical and modern physics. Much of Isaac Newton's work dealt with creating a mathematical model for how objects move and interact—this was the main reason for his invention of calculus (see page 921). Albert Einstein developed his Special Theory of Relativity in the early 1900s to refine Newton's laws of motion.

In this section we use coordinate geometry to model the motion of a projectile, such as a ball thrown upward into the air, a bullet fired from a gun, or any other sort of missile. A similar model was created by Galileo (see the margin), but we have the advantage of using our modern mathematical notation to make describing the model much easier than it was for Galileo!

Suppose that we fire a projectile into the air from ground level, with an initial speed v_0 and at an angle θ upward from the ground. The initial *velocity* of the projectile is a vector (see Section 7.7) with horizontal component $v_0 \cos \theta$ and vertical component $v_0 \sin \theta$, as shown in Figure 1.

FIGURE 1

If there were no gravity (and no air resistance), the projectile would just keep moving indefinitely at the same speed and in the same direction. Since distance = speed × time, the projectile's position at time t would therefore be given by the following parametric equations (assuming the origin of our coordinate system is placed at the initial location of the projectile):

$$x = (v_0 \cos \theta)t \qquad y = (v_0 \sin \theta)t \qquad \text{No gravity}$$

But, of course, we know that gravity will pull the projectile back to ground level. Using calculus, it can be shown that the effect of gravity can be accounted for by subtracting $\frac{1}{2}gt^2$ from the vertical position of the projectile. In this expression, g is the gravitational acceleration: $g \approx 32$ ft/s$^2 \approx 9.8$ m/s^2.

Thus, we have the following parametric equations for the path of the projectile:

$$x = (v_0 \cos \theta)t \qquad y = (v_0 \sin \theta)t - \tfrac{1}{2}gt^2 \qquad \text{With gravity}$$

EXAMPLE ■ The Path of a Cannonball

Find parametric equations that model the path of a cannonball fired into the air with an initial speed of 150.0 m/s at a 30° angle of elevation. Sketch the path of the cannonball. How far from the cannon does the ball hit the ground, and for how long is it in the air?

SOLUTION

Substituting the given initial speed and angle into the general parametric equations of the path of a projectile, we get

$$x = (150.0 \cos 30°)t \qquad y = (150.0 \sin 30°)t - \tfrac{1}{2}(9.8)t^2 \qquad \text{Substitute } v_0 = 150.0,\ \theta = 30°$$

$$x = 129.9t \qquad y = 75.0t - 4.9t^2 \qquad \text{Simplify}$$

This path is graphed in Figure 2.

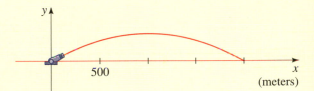

FIGURE 2
Path of a cannonball

How can we tell where and when the cannonball hits the ground? Since ground level corresponds to $y = 0$, we substitute this value for y and solve for t.

$$0 = 75.0t - 4.9t^2 \qquad \text{Set } y = 0$$

$$0 = t(75.0 - 4.9t) \qquad \text{Factor}$$

$$t = 0 \quad \text{or} \quad t = \frac{75.0}{4.9} \approx 15.3 \qquad \text{Solve for } t$$

The first solution, $t = 0$, is the time when the cannon was fired; the second solution means that the cannonball hits the ground after 15.3 s of flight. To see *where* this happens, we substitute this value into the equation for x, the horizontal location of the cannonball.

$$x = 129.9(15.3) \approx 1987.5 \text{ m}$$

The cannonball travels almost 2 km before hitting the ground.   ■

Galileo Galilei (1564–1642) was born in Pisa, Italy. He studied medicine, but later abandoned this in favor of science and mathematics. At the age of 25 he demonstrated that light objects fall at the same rate as heavier ones, by dropping cannonballs of various sizes from the Leaning Tower of Pisa. This contradicted the then-accepted view of Aristotle that heavier objects fall more quickly. He also showed that the distance an object falls is proportional to the square of the time it has been falling, and from this was able to prove that the path of a projectile is a parabola.

Galileo constructed the first telescope, and using it, discovered the moons of Jupiter. His advocacy of the Copernican view that the earth revolves around the sun (rather than being stationary) led to his being called before the Inquisition. By then an old man, he was forced to recant his views, but he is said to have muttered under his breath "the earth nevertheless does move." Galileo revolutionized science by expressing scientific principles in the language of mathematics. He said, "The great book of nature is written in mathematical symbols."

Figure 3 shows the paths of several projectiles, all fired with the same initial speed but at different angles. From the graphs we see that if the firing angle is too high or too low, the projectile doesn't travel very far.

FIGURE 3

Paths of projectiles

Let's try to find the optimal firing angle—the angle that shoots the projectile as far as possible. We'll go through the same steps as we did in the preceding example, but we'll use the general parametric equations instead. First, we solve for the time when the projectile hits the ground by substituting $y = 0$.

$$0 = (v_0 \sin \theta)t - \tfrac{1}{2} gt^2 \qquad \text{Substitute } y = 0$$

$$0 = t(v_0 \sin \theta - \tfrac{1}{2} gt) \qquad \text{Factor}$$

$$0 = v_0 \sin \theta - \tfrac{1}{2} gt \qquad \text{Set second factor equal to 0}$$

$$t = \frac{2v_0 \sin \theta}{g} \qquad \text{Solve for } t$$

Now we substitute this into the equation for x to see how far the projectile has traveled horizontally when it hits the ground.

$$x = (v_0 \cos \theta)t \qquad \text{Parametric equation for } x$$

$$= (v_0 \cos \theta)\left(\frac{2v_0 \sin \theta}{g}\right) \qquad \text{Substitute } t = (2v_0 \sin \theta)/g$$

$$= \frac{2v_0^2 \sin \theta \cos \theta}{g} \qquad \text{Simplify}$$

$$= \frac{v_0^2 \sin 2\theta}{g} \qquad \text{Use identity } \sin 2\theta = 2 \sin \theta \cos \theta$$

We want to choose θ so that x is as large as possible. Because the largest value that the sine of any angle can have is 1, the sine of 90°. Thus, we want $2\theta = 90°$, or $\theta = 45°$. So to send the projectile as far as possible, it should be shot up at an angle of 45°. From the last equation in the preceding display, we can see that it will then travel a distance $x = v_0^2/g$.

■ Problems

1. From the graphs in Figure 3 the paths of projectiles appear to be parabolas that open downward. Eliminate the parameter t from the general parametric equations to verify that these are indeed parabolas.

2. Suppose a baseball is thrown at 30 ft/s at a 60° angle to the horizontal, from a height of 4 ft above the ground. Find parametric equations for the path of the baseball, and sketch its graph. How far does the baseball travel, and when does it hit the ground?

3. Suppose that a rocket is fired at an angle of 5° from the vertical, with an initial speed of 1000 ft/s.
   (a) Find the length of time the rocket is in the air.
   (b) Find the greatest height it reaches.
   (c) Find the horizontal distance it has traveled when it hits the ground.
   (d) Graph the rocket's path.

4. The initial speed of a missile is 330 m/s. At what angle should it be fired so that it hits a target 10 km away? You should find that there are two possible angles. Graph the missile paths for both angles. For which angle is the target hit sooner?

(a)

5. Find the maximum height reached by a projectile as a function of its initial speed v_0 and its firing angle θ.

6. Suppose that a projectile is fired into a headwind that pushes it back so as to reduce its horizontal speed by a constant amount w. Find parametric equations for the path of the projectile.

(b)

7. Using the parametric equations you derived in Problem 6, draw graphs of the path of a projectile with initial speed $v_0 = 32$ ft/s, fired into a headwind of $w = 24$ ft/s, for the angles $\theta = 5°, 15°, 30°, 40°, 45°, 55°, 60°$, and 75°. Is it still true that the greatest range is attained when firing at 45°? Draw some more graphs for different angles, and use these graphs to estimate the optimal firing angle.

8. The path of a projectile can be simulated on a graphing calculator. On the TI-83 use the "Path" graph style to graph the general parametric equations for the path of a projectile and watch as the circular cursor moves, simulating the motion of the projectile. Selecting the size of the **Tstep** determines the speed of the "projectile."
   (a) Simulate the path of a projectile. Experiment with various values of θ. Use $v_0 = 10$ ft/s and **Tstep** = 0.02. Part (a) of the figure shows one such path.
   (b) Simulate the path of two projectiles, fired simultaneously, one at $\theta = 30°$ and the other at $\theta = 60°$. This can be done on the TI-83 using **Simul** mode ("simultaneous" mode). Use $v_0 = 10$ ft/s and **Tstep** = 0.02. See part (b) of the figure. Where do the projectiles land? Which lands first?
   (c) Simulate the path of a ball thrown straight up ($\theta = 90°$). Experiment with values of v_0 between 5 and 20 ft/s. Use the "Animate" graph style and **Tstep** = 0.02. Simulate the path of two balls thrown simultaneously at different speeds. To better distinguish the two balls, place them at different x-coordinates (for example, $x = 1$ and $x = 2$). See part (c) of the figure. How does doubling v_0 change the maximum height the ball reaches?

(c)

10

Sequences and Series

Sequences and series abound in nature; the Fibonacci sequence, for example, is hidden in the intricate structure of the nautilus. Sequences are also used in everyday tasks, such as calculating the monthly mortgage payment on your dream home.

A mathematician, like a painter or poet, is a maker of patterns.

G. H. HARDY

In this chapter we study sequences and series of numbers. Roughly speaking, a *sequence* is a list of numbers written in a specific order, and a *series* is what one gets by adding the numbers in a sequence. Sequences and series have many theoretical and practical uses. Among other applications, we consider how series are used to calculate the value of an annuity.

Section 10.5 introduces a special kind of proof called *mathematical induction*. In Section 10.6 we use mathematical induction to prove a formula for expanding $(a + b)^n$ for any natural number n.

10.1 SEQUENCES AND SUMMATION NOTATION

A *sequence* is a set of numbers written in a specific order:

$$a_1, a_2, a_3, a_4, \ldots, a_n, \ldots$$

The number a_1 is called the *first term*, a_2 is the *second term*, and in general a_n is the *nth term*. Since for every natural number n there is a corresponding number a_n, we can define a sequence as a function.

DEFINITION OF A SEQUENCE

A **sequence** is a function f whose domain is the set of natural numbers. The values $f(1), f(2), f(3), \ldots$ are called the **terms** of the sequence.

We usually write a_n instead of the function notation $f(n)$ for the value of the function at the number n.

Here is a simple example of a sequence:

$$2, 4, 6, 8, 10, \ldots$$

The dots indicate that the sequence continues indefinitely. We can write a sequence in this way when it's clear what the subsequent terms of the sequence are. This sequence consists of even numbers. To be more accurate, however, we need to specify a procedure for finding *all* the terms of the sequence. This can be done by giving a formula for the *nth* term a_n of the sequence. In this case,

$$a_n = 2n$$

and the sequence can be written as

Another way to write this sequence is to use function notation:

$$a(n) = 2n$$

so $a(1) = 2, a(2) = 4, a(3) = 6, \ldots$

$$2, \quad 4, \quad 6, \quad 8, \quad \ldots, \quad 2n, \quad \ldots$$

2,	4,	6,	8,	...,	2n,	...
↑	↑	↑	↑		↑	
1st term	2nd term	3rd term	4th term		nth term	

Notice how the formula $a_n = 2n$ gives all the terms of the sequence. For instance, substituting 1, 2, 3, and 4 for n gives the first four terms:

$$a_1 = 2 \cdot 1 = 2 \qquad a_2 = 2 \cdot 2 = 4$$
$$a_3 = 2 \cdot 3 = 6 \qquad a_4 = 2 \cdot 4 = 8$$

To find the 103rd term of this sequence, we use $n = 103$ to get

$$a_{103} = 2 \cdot 103 = 206$$

EXAMPLE 1 ■ Finding the Terms of a Sequence

Find the first five terms and the 100th term of the sequence defined by each formula.

(a) $a_n = 2n - 1$ (b) $c_n = n^2 - 1$

(c) $t_n = \dfrac{n}{n + 1}$ (d) $r_n = \dfrac{(-1)^n}{2^n}$

SOLUTION

To find the first five terms, we substitute $n = 1, 2, 3, 4,$ and 5 in the formula for the nth term. To find the 100th term, we substitute $n = 100$. This gives the following.

nth term	First five terms	100th term
(a) $2n - 1$	$1, 3, 5, 7, 9$	199
(b) $n^2 - 1$	$0, 3, 8, 15, 24$	9999
(c) $\dfrac{n}{n+1}$	$\dfrac{1}{2}, \dfrac{2}{3}, \dfrac{3}{4}, \dfrac{4}{5}, \dfrac{5}{6}$	$\dfrac{100}{101}$
(d) $\dfrac{(-1)^n}{2^n}$	$-\dfrac{1}{2}, \dfrac{1}{4}, -\dfrac{1}{8}, \dfrac{1}{16}, -\dfrac{1}{32}$	$\dfrac{1}{2^{100}}$

In Example 1(d) the presence of $(-1)^n$ in the sequence has the effect of making successive terms alternately negative and positive.

It is often useful to picture a sequence by sketching its graph. Since a sequence is a function whose domain is the natural numbers, we can draw its graph in the Cartesian plane. For instance, the graph of the sequence

$$1, \frac{1}{2}, \frac{1}{3}, \frac{1}{4}, \frac{1}{5}, \frac{1}{6}, \ldots, \frac{1}{n}, \ldots$$

is shown in Figure 1. Compare this to the graph of

$$1, -\frac{1}{2}, \frac{1}{3}, -\frac{1}{4}, \frac{1}{5}, -\frac{1}{6}, \ldots, \frac{(-1)^{n+1}}{n}, \ldots$$

shown in Figure 2. The graph of every sequence consists of isolated points that are *not* connected.

FIGURE 1

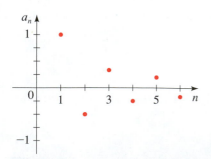

FIGURE 2

$c)\ t_n = \dfrac{n}{n+1}$

$= \dfrac{100}{100+1} = \dfrac{100}{101}\ \checkmark$

$d)\ r_n = \dfrac{(-1)^n}{2n}$

$= \dfrac{(-1)^{100}}{2(100)} = \dfrac{1}{2(100)}$

Graphing calculators are useful in analyzing sequences. To work with sequences on a TI-83, we put the calculator in **Seq** mode ("sequence" mode). If we enter the sequence $u(n) = n/(n + 1)$ of Example 1(c), we can display the terms using the ┃TABLE┃ command as shown in Figure 3(a). We can also graph the sequence as shown in Figure 3(b).

FIGURE 3

$u(n) = n/(n + 1)$

(a)   (b)

Finding patterns is an important part of mathematics. Consider a sequence that begins

$$1, 4, 9, 16, \ldots$$

Not all sequences can be defined by a formula. For example, there is no known formula for the sequence of prime numbers:

$$2, 3, 5, 7, 11, 13, 17, 19, 23, \ldots$$

Can you detect a pattern in these numbers? In other words, can you define a sequence whose first four terms are these numbers? The answer to this question seems easy; these numbers are the squares of the numbers 1, 2, 3, 4. Thus, the sequence we are looking for is defined by $a_n = n^2$. However, this is not the *only* sequence whose first four terms are 1, 4, 9, 16. In other words, the answer to our problem is not unique (see Exercise 69). In the next example we are interested in finding an *obvious* sequence whose first few terms agree with the given ones.

EXAMPLE 2 ■ Finding the *n*th Term of a Sequence

Find the *n*th term of a sequence whose first several terms are given.

(a) $\frac{1}{2}, \frac{3}{4}, \frac{5}{6}, \frac{7}{8}, \ldots$ (b) $-2, 4, -8, 16, -32, \ldots$

SOLUTION

(a) We notice that the numerators of these fractions are the odd numbers and the denominators are the even numbers. Even numbers are of the form $2n$, and odd numbers are of the form $2n - 1$ (an odd number differs from an even number by 1). So, a sequence that has these numbers for its first four terms is given by

$$a_n = \frac{2n - 1}{2n}$$

(b) These numbers are powers of 2 and they alternate in sign, so a sequence that agrees with these terms is given by

$$a_n = (-1)^n 2^n$$

You should check that these formulas do indeed generate the given terms. ■

Recursively Defined Sequences

Some sequences do not have simple defining formulas like those of the preceding example. The nth term of a sequence may depend on some or all of the terms preceding it. A sequence defined in this way is called **recursive**. Here are two examples.

EXAMPLE 3 ■ Finding the Terms of a Recursively Defined Sequence

Find the first five terms of the sequence defined recursively by $a_1 = 1$ and

$$a_n = 3(a_{n-1} + 2)$$

SOLUTION

The defining formula for this sequence is recursive. It allows us to find the nth term a_n if we know the preceding term a_{n-1}. Thus, we can find the second term from the first term, the third term from the second term, the fourth term from the third term, and so on. Since we are given the first term $a_1 = 1$, we can proceed as follows.

$$a_2 = 3(a_1 + 2) = 3(1 + 2) = 9$$

$$a_3 = 3(a_2 + 2) = 3(9 + 2) = 33$$

$$a_4 = 3(a_3 + 2) = 3(33 + 2) = 105$$

$$a_5 = 3(a_4 + 2) = 3(105 + 2) = 321$$

Thus, the first five terms of this sequence are

$$1, 9, 33, 105, 321, \ldots$$

■

Note that in order to find the 20th term of the sequence in Example 3, we must first find all 19 preceding terms. This is most easily done using a graphing calculator. Figure 4(a) shows how to enter this sequence on the TI-83 calculator. From Figure 4(b) we see that the 20th term of the sequence is $a_{20} = 4{,}649{,}045{,}865$.

FIGURE 4
$u(n) = 3(u(n - 1) + 2),\ u(1) = 1$

(a)

(b)

EXAMPLE 4 ■ The Fibonacci Sequence

Find the first 11 terms of the sequence defined recursively by $F_1 = 1$, $F_2 = 1$ and

$$F_n = F_{n-1} + F_{n-2}$$

SOLUTION

To find F_n, we need to find the two preceding terms F_{n-1} and F_{n-2}. Since we are given F_1 and F_2, we proceed as follows.

$$F_3 = F_2 + F_1 = 1 + 1 = 2$$

$$F_4 = F_3 + F_2 = 2 + 1 = 3$$

$$F_5 = F_4 + F_3 = 3 + 2 = 5$$

It's clear what is happening here. Each term is simply the sum of the two terms that precede it, so we can easily write down as many terms as we please. Here are the first 11 terms:

$$1, 1, 2, 3, 5, 8, 13, 21, 34, 55, 89, \ldots$$

■

The sequence in Example 4 is called the **Fibonacci sequence**, named after the 13th-century Italian mathematician who used it to solve a problem about the breeding of rabbits (see Exercise 68). The sequence also occurs in numerous other applications in nature. (See Figures 5 and 6.) In fact, so many phenomena behave like the Fibonacci sequence that one mathematical journal, the *Fibonacci Quarterly*, is devoted entirely to its properties.

FIGURE 5

The Fibonacci sequence in the branching of a tree

FIGURE 6

Fibonacci spiral

Nautilus shell

■ **The Partial Sums of a Sequence**

In calculus we are often interested in adding the terms of a sequence. This leads to the following definition.

> ### THE PARTIAL SUMS OF A SEQUENCE
>
> For the sequence
>
> $$a_1, \quad a_2, \quad a_3, \quad a_4, \quad \ldots, \quad a_n, \quad \ldots$$
>
> the **partial sums** are
>
> $$S_1 = a_1$$
> $$S_2 = a_1 + a_2$$
> $$S_3 = a_1 + a_2 + a_3$$
> $$S_4 = a_1 + a_2 + a_3 + a_4$$
> $$\vdots$$
> $$S_n = a_1 + a_2 + a_3 + \cdots + a_n$$
> $$\vdots$$
>
> S_1 is called the **first partial sum**, S_2 is the **second partial sum**, and so on. S_n is called the **nth partial sum**. The sequence $S_1, S_2, S_3, \ldots, S_n, \ldots$ is called the **sequence of partial sums**.

EXAMPLE 5 ■ **Finding the Partial Sums of a Sequence**

Find the first four partial sums and the nth partial sum of the sequence given by $a_n = 1/2^n$.

SOLUTION

The terms of the sequence are

$$\frac{1}{2}, \frac{1}{4}, \frac{1}{8}, \ldots$$

The first four partial sums are

$$S_1 = \frac{1}{2} \qquad\qquad = \frac{1}{2}$$

$$S_2 = \frac{1}{2} + \frac{1}{4} \qquad\qquad = \frac{3}{4}$$

$$S_3 = \frac{1}{2} + \frac{1}{4} + \frac{1}{8} \qquad = \frac{7}{8}$$

$$S_4 = \frac{1}{2} + \frac{1}{4} + \frac{1}{8} + \frac{1}{16} = \frac{15}{16}$$

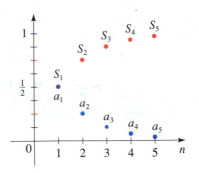

FIGURE 7

Graph of the sequence a_n and the
sequence of partial sums S_n

Notice that in the value of each partial sum the denominator is a power of 2 and
the numerator is one less than the denominator. In general, the nth partial sum is

$$S_n = \frac{2^n - 1}{2^n} = 1 - \frac{1}{2^n}$$

The first five terms of a_n and S_n are graphed in Figure 7. ■

EXAMPLE 6 ■ Finding the Partial Sums of a Sequence

Find the first four partial sums and the nth partial sum of the sequence given by

$$a_n = \frac{1}{n} - \frac{1}{n + 1}$$

SOLUTION

The first four partial sums are

$$S_1 = \left(1 - \frac{1}{2}\right) \qquad\qquad\qquad\qquad = 1 - \frac{1}{2}$$

$$S_2 = \left(1 - \frac{1}{2}\right) + \left(\frac{1}{2} - \frac{1}{3}\right) \qquad\qquad = 1 - \frac{1}{3}$$

$$S_3 = \left(1 - \frac{1}{2}\right) + \left(\frac{1}{2} - \frac{1}{3}\right) + \left(\frac{1}{3} - \frac{1}{4}\right) \qquad = 1 - \frac{1}{4}$$

$$S_4 = \left(1 - \frac{1}{2}\right) + \left(\frac{1}{2} - \frac{1}{3}\right) + \left(\frac{1}{3} - \frac{1}{4}\right) + \left(\frac{1}{4} - \frac{1}{5}\right) = 1 - \frac{1}{5}$$

Do you detect a pattern here? Of course. The nth partial sum is

$$S_n = 1 - \frac{1}{n + 1}$$ ■

Sigma Notation

Given a sequence

$$a_1, a_2, a_3, a_4, \ldots$$

we can write the sum of the first n terms using **summation notation**, or **sigma
notation**. This notation derives its name from the Greek letter Σ (capital sigma,
corresponding to our S for "sum"). Sigma notation is used as follows:

$$\sum_{k=1}^{n} a_k = a_1 + a_2 + a_3 + a_4 + \cdots + a_n$$

This tells us to
end with $k = n$.

This tells us to add.

This tells us to
start with $k = 1$.

The left side of this expression is read "The sum of a_k from $k = 1$ to $k = n$." The
letter k is called the **index of summation**, or the **summation variable**, and the idea

The ancient Greeks considered a line segment to be divided into the **golden ratio** if the ratio of the shorter part to the longer part is the same as the ratio of the longer part to the whole segment.

Thus, the segment shown is divided into the golden ratio if

$$\frac{1}{x} = \frac{x}{1 + x}$$

This leads to a quadratic equation whose positive solution is

$$x = \frac{1 + \sqrt{5}}{2} \approx 1.618$$

This ratio occurs naturally in many places. For instance, psychological experiments show that the most pleasing shape of rectangle is one whose sides are in golden ratio. The ancient Greeks agreed with this and built their temples in this ratio.

The golden ratio is related to the Fibonacci numbers (see page 811). In fact, it can be shown using calculus* that the ratio of two successive Fibonacci numbers

$$\frac{F_{n+1}}{F_n}$$

gets closer to the golden ratio the larger the value of n. Try finding this ratio for $n = 10$.

*James Stewart, *Calculus*, 4th ed. (Pacific Grove, CA: Brooks/Cole, 1999) p. 737.

is to replace k in the expression after the sigma by the integers $1, 2, 3, \ldots, n$, and add the resulting expressions, arriving at the right side of the equation.

EXAMPLE 7 ■ Sigma Notation

Find each sum.

(a) $\displaystyle\sum_{k=1}^{5} k^2$ (b) $\displaystyle\sum_{j=3}^{5} \frac{1}{j}$ (c) $\displaystyle\sum_{i=5}^{10} i$ (d) $\displaystyle\sum_{i=1}^{6} 2$

SOLUTION

(a) $\displaystyle\sum_{k=1}^{5} k^2 = 1^2 + 2^2 + 3^2 + 4^2 + 5^2 = 55$

(b) $\displaystyle\sum_{j=3}^{5} \frac{1}{j} = \frac{1}{3} + \frac{1}{4} + \frac{1}{5} = \frac{47}{60}$

(c) $\displaystyle\sum_{i=5}^{10} i = 5 + 6 + 7 + 8 + 9 + 10 = 45$

(d) $\displaystyle\sum_{i=1}^{6} 2 = 2 + 2 + 2 + 2 + 2 + 2 = 12$ ■

We can use a graphing calculator to evaluate sums. For instance, Figure 8 shows how the TI-83 can be used to evaluate the sums in parts (a) and (b) of Example 7.

```
sum(seq(K²,K,1,5,1))
                    55
sum(seq(1/J,J,3,5,
1))▶Frac
                 47/60
```

FIGURE 8

EXAMPLE 8 ■ Writing Sums in Sigma Notation

Write each sum using sigma notation.

(a) $1^3 + 2^3 + 3^3 + 4^3 + 5^3 + 6^3 + 7^3$

(b) $\sqrt{3} + \sqrt{4} + \sqrt{5} + \cdots + \sqrt{77}$

SOLUTION

(a) We can write

$$1^3 + 2^3 + 3^3 + 4^3 + 5^3 + 6^3 + 7^3 = \sum_{k=1}^{7} k^3$$

(b) A natural way to write this sum is

$$\sqrt{3} + \sqrt{4} + \sqrt{5} + \cdots + \sqrt{77} = \sum_{k=3}^{77} \sqrt{k}$$

However, there is no unique way of writing a sum in sigma notation. We could also write this sum as

$$\sqrt{3} + \sqrt{4} + \sqrt{5} + \cdots + \sqrt{77} = \sum_{k=0}^{74} \sqrt{k+3}$$

or

$$\sqrt{3} + \sqrt{4} + \sqrt{5} + \cdots + \sqrt{77} = \sum_{k=1}^{75} \sqrt{k+2} \qquad \blacksquare$$

The following properties of sums are natural consequences of properties of the real numbers.

PROPERTIES OF SUMS

Let $a_1, a_2, a_3, a_4, \ldots$ and $b_1, b_2, b_3, b_4, \ldots$ be sequences. Then for every positive integer n and any real number c, the following properties hold.

1. $\displaystyle\sum_{k=1}^{n} (a_k + b_k) = \sum_{k=1}^{n} a_k + \sum_{k=1}^{n} b_k$

2. $\displaystyle\sum_{k=1}^{n} (a_k - b_k) = \sum_{k=1}^{n} a_k - \sum_{k=1}^{n} b_k$

3. $\displaystyle\sum_{k=1}^{n} ca_k = c\left(\sum_{k=1}^{n} a_k \right)$

■ **Proof**   To prove Property 1, we write out the left side of the equation to get

$$\sum_{k=1}^{n} (a_k + b_k) = (a_1 + b_1) + (a_2 + b_2) + (a_3 + b_3) + \cdots + (a_n + b_n)$$

Because addition is commutative and associative, we can rearrange the terms on the right side to read

$$\sum_{k=1}^{n} (a_k + b_k) = (a_1 + a_2 + a_3 + \cdots + a_n) + (b_1 + b_2 + b_3 + \cdots + b_n)$$

Rewriting the right side using sigma notation gives Property 1. Property 2 is proved in a similar manner. To prove Property 3, we use the Distributive Property:

$$\sum_{k=1}^{n} ca_k = ca_1 + ca_2 + ca_3 + \cdots + ca_n$$

$$= c(a_1 + a_2 + a_3 + \cdots + a_n) = c\left(\sum_{k=1}^{n} a_k \right) \qquad \square$$

10.1 EXERCISES

1–10 ■ Find the first four terms and the 100th term of the sequence.

1. $a_n = n + 1$

2. $a_n = 2n + 3$

3. $a_n = \dfrac{1}{n + 1}$

4. $a_n = n^2 + 1$

5. $a_n = \dfrac{(-1)^n}{n^2}$

6. $a_n = \dfrac{1}{n^2}$

7. $a_n = 1 + (-1)^n$

8. $a_n = (-1)^{n+1} \dfrac{n}{n + 1}$

9. $a_n = n^n$

10. $a_n = 3$

11–16 ■ Find the first five terms of the given recursively defined sequence.

11. $a_n = 2(a_{n-1} - 2)$ and $a_1 = 3$

12. $a_n = \dfrac{a_{n-1}}{2}$ and $a_1 = -8$

13. $a_n = 2a_{n-1} + 1$ and $a_1 = 1$

14. $a_n = \dfrac{1}{1 + a_{n-1}}$ and $a_1 = 1$

15. $a_n = a_{n-1} + a_{n-2}$ and $a_1 = 1, a_2 = 2$

16. $a_n = a_{n-1} + a_{n-2} + a_{n-3}$ and $a_1 = a_2 = a_3 = 1$

 17–22 ■ Use a graphing calculator to do the following.
(a) Find the first 10 terms of the sequence.
(b) Graph the first 10 terms of the sequence.

17. $a_n = 4n + 3$

18. $a_n = n^2 + n$

19. $a_n = \dfrac{12}{n}$

20. $a_n = 4 - 2(-1)^n$

21. $a_n = \dfrac{1}{a_{n-1}}$ and $a_1 = 2$

22. $a_n = a_{n-1} - a_{n-2}$ and $a_1 = 1, a_2 = 3$

23–30 ■ Find the nth term of a sequence whose first several terms are given.

23. $2, 4, 8, 16, \ldots$

24. $-\frac{1}{3}, \frac{1}{9}, -\frac{1}{27}, \frac{1}{81}, \ldots$

25. $1, 4, 7, 10, \ldots$

26. $5, -25, 125, -625, \ldots$

27. $1, \frac{3}{4}, \frac{5}{9}, \frac{7}{16}, \frac{9}{25}, \ldots$

28. $\frac{3}{4}, \frac{4}{5}, \frac{5}{6}, \frac{6}{7}, \ldots$

29. $0, 2, 0, 2, 0, 2, \ldots$

30. $1, \frac{1}{2}, 3, \frac{1}{4}, 5, \frac{1}{6}, \ldots$

31–34 ■ Find the first six partial sums $S_1, S_2, S_3, S_4, S_5, S_6$ of the sequence.

31. $1, 3, 5, 7, \ldots$

32. $1^2, 2^2, 3^2, 4^2, \ldots$

33. $\dfrac{1}{3}, \dfrac{1}{3^2}, \dfrac{1}{3^3}, \dfrac{1}{3^4}, \ldots$

34. $-1, 1, -1, 1, \ldots$

35–38 ■ Find the first four partial sums and the nth partial sum of the sequence a_n.

35. $a_n = \dfrac{2}{3^n}$

36. $a_n = \dfrac{1}{n + 1} - \dfrac{1}{n + 2}$

37. $a_n = \sqrt{n} - \sqrt{n + 1}$

38. $a_n = \log\left(\dfrac{n}{n + 1}\right)$ [*Hint:* Use a property of logarithms to write the nth term as a difference.]

39–46 ■ Find the sum.

39. $\displaystyle\sum_{k=1}^{4} k$

40. $\displaystyle\sum_{k=1}^{4} k^2$

41. $\displaystyle\sum_{k=1}^{3} \dfrac{1}{k}$

42. $\displaystyle\sum_{j=1}^{100} (-1)^j$

43. $\displaystyle\sum_{i=1}^{8} [1 + (-1)^i]$

44. $\displaystyle\sum_{i=4}^{12} 10$

45. $\displaystyle\sum_{k=1}^{5} 2^{k-1}$

46. $\displaystyle\sum_{i=1}^{3} i 2^i$

 47–52 ■ Use a graphing calculator to evaluate the sum.

47. $\displaystyle\sum_{k=1}^{10} k^2$

48. $\displaystyle\sum_{k=1}^{100} (3k + 4)$

49. $\displaystyle\sum_{j=7}^{20} j^2(1 + j)$

50. $\displaystyle\sum_{j=5}^{15} \dfrac{1}{j^2 + 1}$

51. $\displaystyle\sum_{n=0}^{22} (-1)^n 2n$

52. $\displaystyle\sum_{n=1}^{100} \dfrac{(-1)^n}{n}$

53–58 ■ Write the sum without using sigma notation.

53. $\displaystyle\sum_{k=1}^{5} \sqrt{k}$

54. $\displaystyle\sum_{i=0}^{4} \dfrac{2i - 1}{2i + 1}$

55. $\displaystyle\sum_{k=0}^{6} \sqrt{k + 4}$

56. $\displaystyle\sum_{k=6}^{9} k(k + 3)$

57. $\displaystyle\sum_{k=3}^{100} x^k$

58. $\displaystyle\sum_{j=1}^{n} (-1)^{j+1} x^j$

59–66 ■ Write the sum using sigma notation.

59. $1 + 2 + 3 + 4 + \cdots + 100$

60. $2 + 4 + 6 + \cdots + 20$ **61.** $1^2 + 2^2 + 3^2 + \cdots + 10^2$

62. $\dfrac{1}{2 \ln 2} - \dfrac{1}{3 \ln 3} + \dfrac{1}{4 \ln 4} - \dfrac{1}{5 \ln 5} + \cdots + \dfrac{1}{100 \ln 100}$

63. $\dfrac{1}{1 \cdot 2} + \dfrac{1}{2 \cdot 3} + \dfrac{1}{3 \cdot 4} + \cdots + \dfrac{1}{999 \cdot 1000}$

64. $\dfrac{\sqrt{1}}{1^2} + \dfrac{\sqrt{2}}{2^2} + \dfrac{\sqrt{3}}{3^2} + \cdots + \dfrac{\sqrt{n}}{n^2}$

65. $1 + x + x^2 + x^3 + \cdots + x^{100}$

66. $1 - 2x + 3x^2 - 4x^3 + 5x^4 + \cdots - 100x^{99}$

67. Find a formula for the nth term of the sequence

$$\sqrt{2}, \quad \sqrt{2\sqrt{2}}, \quad \sqrt{2\sqrt{2\sqrt{2}}}, \quad \sqrt{2\sqrt{2\sqrt{2\sqrt{2}}}}, \quad \cdots$$

[*Hint:* Write each term as a power of 2.]

68. Fibonacci posed the following problem: Suppose that rabbits live forever and that every month each pair produces a new pair that becomes productive at age 2 months. If we start with one newborn pair, how many pairs of rabbits will we have in the nth month? Show that the answer is F_n, where F_n is the nth term of the Fibonacci sequence.

 DISCOVERY · DISCUSSION

69. Different Sequences That Start the Same

(a) Show that the first four terms of the sequence $a_n = n^2$ are

$$1, 4, 9, 16, \ldots$$

(b) Show that the first four terms of the sequence $a_n = n^2 + (n-1)(n-2)(n-3)(n-4)$ are also

$$1, 4, 9, 16, \ldots$$

(c) Find a sequence whose first six terms are the same as those of $a_n = n^2$ but whose succeeding terms differ from this sequence.

(d) Find two different sequences that begin

$$2, 4, 8, 16, \ldots$$

70. A Recursively Defined Sequence   Find the first 40 terms of the sequence defined by

$$a_{n+1} = \begin{cases} \dfrac{a_n}{2} & \text{if } a_n \text{ is an even number} \\ 3a_n + 1 & \text{if } a_n \text{ is an odd number} \end{cases}$$

and $a_1 = 11$. Do the same if $a_1 = 25$. Make a conjecture about this type of sequence. Try several other values for a_1, to test your conjecture.

71. A Different Type of Recursion   Find the first 10 terms of the sequence defined by

$$a_n = a_{n-a_{n-1}} + a_{n-a_{n-2}}$$

with

$$a_1 = 1 \text{ and } a_2 = 1$$

How is this recursive sequence different from the others in this section?

10.2 ARITHMETIC SEQUENCES

Perhaps the simplest way to generate a sequence is to start with a number a and add to it a fixed constant d, over and over again.

DEFINITION OF AN ARITHMETIC SEQUENCE

An **arithmetic sequence** is a sequence of the form

$$a, a + d, a + 2d, a + 3d, a + 4d, \ldots$$

The number a is the **first term**, and d is the **common difference** of the sequence. The **nth term** of an arithmetic sequence is given by

$$a_n = a + (n - 1)d$$

Fractals

Many of the things we model in this book have regular predictable shapes. But recent advances in mathematics have made it possible to mathematically model such seemingly random or even chaotic shapes as those of a cloud, a flickering flame, a mountain, or a jagged coastline. The basic tools in this type of modeling are the fractals invented by the mathematician Benoit Mandelbrot. A *fractal* is a geometric shape built up from a simple basic shape by scaling and repeating the shape indefinitely according to a given rule (an example is the snowflake curve on page 865). Fractals have infinite detail; this means the closer you look, the more you see. They are also *self-similar*; that is, zooming in on a portion of the fractal yields the same detail as the original shape. Because of their beautiful shapes, fractals are used by movie makers to create fictional landscapes or exotic backgrounds.

   Although a fractal is a complex shape, it is produced according to very simple rules. This property of fractals is exploited in a process of storing pictures on a computer called *fractal image compression*. In this process a picture is stored as a simple basic shape and a rule;

(continued)

The number d is called the common difference because any two consecutive terms of an arithmetic sequence differ by d.

EXAMPLE 1 ■ Arithmetic Sequences

(a) If $a = 2$ and $d = 3$, then we have the arithmetic sequence

$$2, 2 + 3, 2 + 6, 2 + 9, \ldots$$

or

$$2, 5, 8, 11, \ldots$$

Any two consecutive terms of this sequence differ by $d = 3$. The nth term is $a_n = 2 + 3(n - 1)$.

(b) Consider the arithmetic sequence

$$9, 4, -1, -6, -11, \ldots$$

Here the common difference is $d = -5$. The terms of an arithmetic sequence decrease if the common difference is negative. The nth term is $a_n = 9 - 5(n - 1)$.

(c) The graph of the arithmetic sequence $a_n = 1 + 2(n - 1)$ is shown in Figure 1. Notice that the points in the graph lie on a straight line with slope $d = 2$.

FIGURE 1

   An arithmetic sequence is determined completely by the first term a and the common difference d. Thus, if we know the first two terms of an arithmetic sequence, then we can find a formula for the nth term, as the next example shows.

EXAMPLE 2 ■ Finding Terms of an Arithmetic Sequence

Find the first six terms and the 300th term of the arithmetic sequence

$$13, 7, \ldots$$

SOLUTION

Since the first term is 13, we have $a = 13$. The common difference is $d = 7 - 13 = -6$. Thus, the nth term of this sequence is

$$a_n = 13 - 6(n - 1)$$

From this we find the first six terms:

$$13, 7, 1, -5, -11, -17, \ldots$$

The 300th term is $a_{300} = 13 - 6(299) = -1781$. ■

The next example shows that an arithmetic sequence is determined completely by *any* two of its terms.

EXAMPLE 3 ■ Finding Terms of an Arithmetic Sequence

The 11th term of an arithmetic sequence is 52, and the 19th term is 92. Find the 1000th term.

SOLUTION

To find the *n*th term of this sequence, we need to find a and d in the formula

$$a_n = a + (n - 1)d$$

From this formula we get

$$a_{11} = a + (11 - 1)d = a + 10d$$
$$a_{19} = a + (19 - 1)d = a + 18d$$

Since $a_{11} = 52$ and $a_{19} = 92$, we get the two equations:

$$\begin{cases} 52 = a + 10d \\ 92 = a + 18d \end{cases}$$

Solving this system for a and d, we get $a = 2$ and $d = 5$. (Verify this.) Thus, the *n*th term of this sequence is

$$a_n = 2 + 5(n - 1)$$

The 1000th term is $a_{1000} = 2 + 5(999) = 4997$. ■

Partial Sums of Arithmetic Sequences

Suppose we want to find the sum of the numbers 1, 2, 3, 4, . . . , 100, that is,

$$\sum_{k=1}^{100} k$$

When the famous mathematician C. F. Gauss was a schoolboy, his teacher posed this problem to the class and expected that it would keep the students busy for a long time. But Gauss answered the question almost immediately. His idea was this: Since we are adding numbers produced according to a fixed pattern, there must also be a pattern (or formula) for finding the sum. He started by writing the numbers

---

from 1 to 100 and below them the same numbers in reverse order. Writing S for the sum and adding corresponding terms gives

$$S = \quad 1 + \quad 2 + \quad 3 + \cdots + \quad 98 + \quad 99 + 100$$

$$S = 100 + \quad 99 + \quad 98 + \cdots + \quad 3 + \quad 2 + \quad 1$$

$$2S = 101 + 101 + 101 + \cdots + 101 + 101 + 101$$

It follows that $2S = 100(101) = 10{,}100$ and so $S = 5050$.

Of course, the sequence of natural numbers $1, 2, 3, \ldots$ is an arithmetic sequence (with $a = 1$ and $d = 1$), and the method for summing the first 100 terms of this sequence can be used to find a formula for the nth partial sum of any arithmetic sequence. We want to find the sum of the first n terms of the arithmetic sequence whose terms are $a_k = a + (k - 1)d$; that is, we want to find

$$S_n = \sum_{k=1}^{n} [a + (k - 1)d]$$

$$= a + (a + d) + (a + 2d) + (a + 3d) + \cdots + [a + (n - 1)d]$$

Using Gauss's method, we write

$$S_n = \qquad a \qquad + \quad (a + d) \quad + \cdots + \ [a + (n - 2)d] + [a + (n - 1)d]$$

$$S_n = [a + (n - 1)d] + [a + (n - 2)d] + \cdots + \qquad (a + d) \qquad + \qquad a$$

$$2S_n = [2a + (n - 1)d] + [2a + (n - 1)d] + \cdots + [2a + (n - 1)d] + [2a + (n - 1)d]$$

There are n identical terms on the right side of this equation, so

$$2S_n = n[2a + (n - 1)d]$$

$$S_n = \frac{n}{2}[2a + (n - 1)d]$$

Notice that $a_n = a + (n - 1)d$ is the nth term of this sequence. So, we can write

$$S_n = \frac{n}{2}[a + a + (n - 1)d] = n\left(\frac{a + a_n}{2}\right)$$

This last formula says that the sum of the first n terms of an arithmetic sequence is the average of the first and nth terms multiplied by n, the number of terms in the sum. We now summarize this result.

PARTIAL SUMS OF AN ARITHMETIC SEQUENCE

For the arithmetic sequence $a_n = a + (n - 1)d$, the **nth partial sum**

$$S_n = a + (a + d) + (a + 2d) + (a + 3d) + \cdots + [a + (n - 1)d]$$

is given by either of the following formulas.

1. $S_n = \dfrac{n}{2}[2a + (n - 1)d]$ \qquad\qquad **2.** $S_n = n\left(\dfrac{a + a_n}{2}\right)$

EXAMPLE 4 ■ Finding a Partial Sum of an Arithmetic Sequence

Find the sum of the first 40 terms of the arithmetic sequence

$$3, 7, 11, 15, \ldots$$

SOLUTION

For this arithmetic sequence, $a = 3$ and $d = 4$. Using Formula 1 for the partial sum of an arithmetic sequence, we get

$$S_{40} = \tfrac{40}{2}[2(3) + (40 - 1)4] = 20(6 + 156) = 3240$$ ■

EXAMPLE 5 ■ Finding a Partial Sum of an Arithmetic Sequence

Find the sum of the first 50 odd numbers.

SOLUTION

The odd numbers form an arithmetic sequence with $a = 1$ and $d = 2$. The nth term is $a_n = 1 + 2(n - 1) = 2n - 1$, so the 50th odd number is $a_{50} = 2(50) - 1 = 99$. Substituting in Formula 2 for the partial sum of an arithmetic sequence, we get

$$S_{50} = 50\left(\frac{a + a_{50}}{2}\right) = 50\left(\frac{1 + 99}{2}\right) = 50 \cdot 50 = 2500$$ ■

EXAMPLE 6 ■ Finding the Seating Capacity of an Amphitheater

An amphitheater has 50 rows of seats with 30 seats in the first row, 32 in the second, 34 in the third, and so on. Find the total number of seats.

SOLUTION

The numbers of seats in the rows form an arithmetic sequence with $a = 30$ and $d = 2$. Since there are 50 rows, the total number of seats is the sum

$$S_{50} = \tfrac{50}{2}[2(30) + 49(2)] \qquad \text{Formula 1}$$

$$= 3950$$

Thus, the amphitheater has 3950 seats.   ■

EXAMPLE 7 ■ Finding the Number of Terms in a Partial Sum

How many terms of the arithmetic sequences $5, 7, 9, \ldots$ must be added to get 572?

SOLUTION

We are asked to find n when $S_n = 572$. Substituting $a = 5$, $d = 2$, and $S_n = 572$ in Formula 1 for the partial sum of an arithmetic sequence, we get

$$572 = \frac{n}{2}[2 \cdot 5 + (n - 1)2] \qquad \text{Formula 1}$$

$$572 = 5n + n(n - 1)$$

$$0 = n^2 + 4n - 572$$

$$0 = (n - 22)(n + 26)$$

This gives $n = 22$ or $n = -26$. But since n is the *number* of terms in this partial sum, we must have $n = 22$. ∎

10.2 EXERCISES

1–4 ■ (a) Find the first five terms of the sequence. (b) What is the common difference d? (c) Graph the terms you found in (a).

1. $a_n = 5 + 2(n - 1)$ **2.** $a_n = 3 - 4(n - 1)$

3. $a_n = \frac{5}{2} - (n - 1)$ **4.** $a_n = \frac{1}{2}(n - 1)$

5–8 ■ Find the nth term of the arithmetic sequence with given first term a and common difference d. What is the 10th term?

5. $a = 3$, $d = 5$ **6.** $a = -6$, $d = 3$

7. $a = \frac{5}{2}$, $d = -\frac{1}{2}$ **8.** $a = \sqrt{3}$, $d = \sqrt{3}$

9–14 ■ Determine whether the sequence is arithmetic. If it is arithmetic, find the common difference.

9. 5, 8, 11, 14, . . .   **10.** 3, 6, 9, 13, . . .

11. 2, 4, 8, 16, . . .   **12.** 2, 4, 6, 8, . . .

13. $3, \frac{3}{2}, 0, -\frac{3}{2}, \ldots$ **14.** ln 2, ln 4, ln 8, ln 16, . . .

15. 2.6, 4.3, 6.0, 7.7, . . .   **16.** $\frac{1}{2}, \frac{1}{3}, \frac{1}{4}, \frac{1}{5}, \ldots$

17–22 ■ Find the first five terms of the sequence and determine if it is arithmetic. If it is arithmetic, find the common difference and express the nth term of the sequence in the standard form $a_n = a + (n - 1)d$.

17. $a_n = 4 + 7n$ **18.** $a_n = 4 + 2^n$

19. $a_n = \dfrac{1}{1 + 2n}$ **20.** $a_n = 1 + \dfrac{n}{2}$

21. $a_n = 6n - 10$ **22.** $a_n = 3 + (-1)^n n$

23–32 ■ Determine the common difference, the fifth term, the nth term, and the 100th term of the arithmetic sequence.

23. 2, 5, 8, 11, . . .   **24.** 1, 5, 9, 13, . . .

25. 4, 9, 14, 19, . . .   **26.** 11, 8, 5, 2, . . .

27. $-12, -8, -4, 0, \ldots$ **28.** $\frac{7}{6}, \frac{5}{3}, \frac{13}{6}, \frac{8}{3}, \ldots$

29. 25, 26.5, 28, 29.5, . . .   **30.** 15, 12.3, 9.6, 6.9, . . .

31. $2, 2 + s, 2 + 2s, 2 + 3s, \ldots$

32. $-t, -t + 3, -t + 6, -t + 9, \ldots$

33. The tenth term of an arithmetic sequence is $\frac{55}{2}$, and the second term is $\frac{7}{2}$. Find the first term.

34. The 12th term of an arithmetic sequence is 32, and the fifth term is 18. Find the 20th term.

35. The 100th term of an arithmetic sequence is 98, and the common difference is 2. Find the first three terms.

36. The 20th term of an arithmetic sequence is 101, and the common difference is 3. Find a formula for the nth term.

37. Which term of the arithmetic sequence 1, 4, 7, . . . is 88?

38. The first term of an arithmetic sequence is 1, and the common difference is 4. Is 11,937 a term of this sequence? If so, which term is it?

39–44 ■ Find the partial sum S_n of the arithmetic sequence that satisfies the given conditions.

39. $a = 1$, $d = 2$, $n = 10$ **40.** $a = 3$, $d = 2$, $n = 12$

41. $a = 4, d = 2, n = 20$ **42.** $a = 100, d = -5, n = 8$

43. $a_1 = 55, d = 12, n = 10$ **44.** $a_2 = 8, a_5 = 9.5, n = 15$

45–50 ■ A partial sum of an arithmetic sequence is given. Find the sum.

45. $1 + 5 + 9 + \cdots + 401$

46. $-3 + \left(-\frac{3}{2}\right) + 0 + \frac{3}{2} + 3 + \cdots + 30$

47. $0.7 + 2.7 + 4.7 + \cdots + 56.7$

48. $-10 - 9.9 - 9.8 - \cdots - 0.1$

49. $\displaystyle\sum_{k=0}^{10} (3 + 0.25k)$ **50.** $\displaystyle\sum_{n=0}^{20} (1 - 2n)$

51. The purchase value of an office computer is \$12,500. Its annual depreciation is \$1875. Find the value of the computer after 6 years.

52. Telephone poles are stored in a pile with 25 poles in the first layer, 24 in the second, and so on. If there are 12 layers, how many telephone poles does the pile contain?

53. A man gets a job with a salary of \$30,000 a year. He is promised a \$2300 raise each subsequent year. Find his total earnings for a 10-year period.

54. A drive-in theater has spaces for 20 cars in the first parking row, 22 in the second, 24 in the third, and so on. If there are 21 rows in the theater, find the number of cars that can be parked.

55. An architect designs a theater with 15 seats in the first row, 18 in the second, 21 in the third, and so on. If the theater is to have a seating capacity of 870, how many rows must the architect use in his design?

56. An arithmetic sequence has first term $a = 5$ and common difference $d = 2$. How many terms of this sequence must be added to get 2700?

57. When an object is allowed to fall freely near the surface of the earth, the gravitational pull is such that the object falls 16 ft in the first second, 48 ft in the next second, 80 ft in the next second, and so on.
(a) Find the total distance a ball falls in 6 s.

(b) Find a formula for the total distance a ball falls in n seconds.

58. In the well-known song "The Twelve Days of Christmas," a person gives his sweetheart k gifts on the kth day for each of the 12 days of Christmas. The person also repeats each gift identically on each subsequent day. Thus, on the 12th day the sweetheart receives a gift for the first day, 2 gifts for the second, 3 gifts for the third, and so on. Show that the number of gifts received on the 12th day is a partial sum of an arithmetic sequence. Find this sum.

59. Show that a right triangle whose sides are in arithmetic progression is similar to a 3–4–5 triangle.

60. Find the product of the numbers

$$10^{1/10}, 10^{2/10}, 10^{3/10}, 10^{4/10}, \ldots, 10^{19/10}$$

61. A sequence is **harmonic** if the reciprocals of the terms of the sequence form an arithmetic sequence. Determine whether the following sequence is harmonic:

$$1, \tfrac{3}{5}, \tfrac{3}{7}, \tfrac{1}{3}, \ldots$$

62. The **harmonic mean** of two numbers is the reciprocal of the average of the reciprocals of the two numbers. Find the harmonic mean of 3 and 5.

■ DISCOVERY · DISCUSSION

63. Arithmetic Means   The **arithmetic mean** (or average) of two numbers a and b is

$$m = \frac{a + b}{2}$$

Note that m is the same distance from a as from b, so a, m, b is an arithmetic sequence. In general, if $m_1, m_2, \ldots, m_k$ are equally spaced between a and b so that

$$a, m_1, m_2, \ldots, m_k, b$$

is an arithmetic sequence, then $m_1, m_2, \ldots, m_k$ are called k arithmetic means between a and b.
(a) Insert two arithmetic means between 10 and 18.
(b) Insert three arithmetic means between 10 and 18.
(c) Suppose a doctor needs to increase a patient's dosage of a certain medicine from 100 mg to 300 mg per day in five equal steps. How many arithmetic means must be inserted between 100 and 300 to give the progression of daily doses, and what are these means?

10.3 GEOMETRIC SEQUENCES

Another simple way of generating a sequence is to start with a number a and repeatedly multiply by a fixed nonzero constant r.

DEFINITION OF A GEOMETRIC SEQUENCE

A **geometric sequence** is a sequence of the form

$$a, ar, ar^2, ar^3, ar^4, \ldots$$

The number a is the **first term**, and r is the **common ratio** of the sequence. The ***n*th term** of a geometric sequence is given by

$$a_n = ar^{n-1}$$

The number r is called the common ratio because the ratio of any two consecutive terms of the sequence is r.

EXAMPLE 1 ■ Geometric Sequences

(a) If $a = 3$ and $r = 2$, then we have the geometric sequence

$$3,\quad 3 \cdot 2,\quad 3 \cdot 2^2,\quad 3 \cdot 2^3,\quad 3 \cdot 2^4,\quad \ldots$$

or

$$3, 6, 12, 24, 48, \ldots$$

Notice that the ratio of any two consecutive terms is $r = 2$. The nth term is $a_n = 3(2)^{n-1}$.

(b) The sequence

$$2, -10, 50, -250, 1250, \ldots$$

is a geometric sequence with $a = 2$ and $r = -5$. When r is negative, the terms of the sequence alternate in sign. The nth term is $a_n = 2(-5)^{n-1}$.

(c) The sequence

$$1, \frac{1}{3}, \frac{1}{9}, \frac{1}{27}, \frac{1}{81}, \ldots$$

is a geometric sequence with $a = 1$ and $r = \frac{1}{3}$. The nth term is $a_n = 1\left(\frac{1}{3}\right)^{n-1}$.

(d) The graph of the geometric sequence $a_n = \frac{1}{5} \cdot 2^{n-1}$ is shown in Figure 1. Notice that the points in the graph lie on the graph of the exponential function $y = \frac{1}{5} \cdot 2^{x-1}$.

FIGURE 1

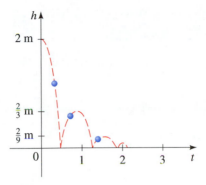

FIGURE 2

If $0 < r < 1$, then the terms of the geometric sequence ar^{n-1} decrease, but if $r > 1$, then the terms increase. (What happens if $r = 1$?) ∎

Geometric sequences occur naturally. Here is a simple example. Suppose a ball has elasticity such that when it is dropped it bounces up one-third of the distance it has fallen. If this ball is dropped from a height of 2 m, then it bounces up to a height of $2\left(\frac{1}{3}\right) = \frac{2}{3}$ m. On its second bounce, it returns to a height of $\left(\frac{2}{3}\right)\left(\frac{1}{3}\right) = \frac{2}{9}$ m, and so on (see Figure 2). Thus, the height h_n that the ball reaches on its nth bounce is given by the geometric sequence

$$h_n = \tfrac{2}{3}\left(\tfrac{1}{3}\right)^{n-1} = 2\left(\tfrac{1}{3}\right)^n$$

We can find the nth term of a geometric sequence if we know any two terms, as the following examples show.

EXAMPLE 2 ■ **Finding Terms of a Geometric Sequence**

Find the eighth term of the geometric sequence $5, 15, 45, \ldots$.

SOLUTION

To find a formula for the nth term of this sequence, we need to find a and r. Clearly, $a = 5$. To find r, we find the ratio of any two consecutive terms. For instance, $r = \frac{45}{15} = 3$. Thus

$$a_n = 5(3)^{n-1}$$

The eighth term is $a_8 = 5(3)^{8-1} = 5(3)^7 = 10{,}935$. ∎

EXAMPLE 3 ■ **Finding Terms of a Geometric Sequence**

The third term of a geometric series is $\frac{63}{4}$, and the sixth term is $\frac{1701}{32}$. Find the fifth term.

SOLUTION

Since this series is geometric, its nth term is given by the formula $a_n = ar^{n-1}$. Thus

$$a_3 = ar^{3-1} = ar^2$$

$$a_6 = ar^{6-1} = ar^5$$

From the values we are given for these two terms, we get the following system of equations:

$$\begin{cases} \frac{63}{4} = ar^2 \\ \frac{1701}{32} = ar^5 \end{cases}$$

We can solve this system by dividing.

$$\frac{ar^5}{ar^2} = \frac{\frac{1701}{32}}{\frac{63}{4}}$$

$$r^3 = \frac{27}{8} \qquad \text{\color{red}Simplify}$$

$$r = \frac{3}{2} \qquad \text{\color{red}Take the cube root of each side}$$

Substituting for r in the first equation, $\frac{63}{4} = ar^2$, gives

$$\frac{63}{4} = a\left(\frac{3}{2}\right)^2$$

$$a = 7 \qquad \text{\color{red}Solve for } a$$

It follows that the nth term of this sequence is

$$a_n = 7\left(\frac{3}{2}\right)^{n-1}$$

Thus, the fifth term is

$$a_5 = 7\left(\frac{3}{2}\right)^{5-1} = 7\left(\frac{3}{2}\right)^4 = \frac{567}{16} \qquad \blacksquare$$

Partial Sums of Geometric Sequences

For the geometric sequence $a, ar, ar^2, ar^3, ar^4, \ldots, ar^{n-1}, \ldots$, the nth partial sum is

$$S_n = \sum_{k=1}^{n} ar^{k-1} = a + ar + ar^2 + ar^3 + ar^4 + \cdots + ar^{n-1}$$

To find a formula for S_n, we multiply S_n by r and subtract from S_n to get

$$S_n = a + ar + ar^2 + ar^3 + ar^4 + \cdots + ar^{n-1}$$

$$rS_n = \quad\ \ ar + ar^2 + ar^3 + ar^4 + \cdots + ar^{n-1} + ar^n$$

$$\overline{\rule{0pt}{0pt}\quad\quad\quad\quad\quad\quad\quad\quad\quad\quad\quad\quad\quad\quad\quad\quad\quad\quad\quad}$$

$$S_n - rS_n = a - ar^n$$

So, $\qquad\qquad\qquad\qquad S_n(1 - r) = a(1 - r^n)$

$$S_n = \frac{a(1 - r^n)}{1 - r} \qquad (r \neq 1)$$

We summarize this result.

PARTIAL SUMS OF A GEOMETRIC SEQUENCE

For the geometric sequence $a_n = ar^{n-1}$, the **nth partial sum**

$$S_n = a + ar + ar^2 + ar^3 + ar^4 + \cdots + ar^{n-1} \qquad (r \neq 1)$$

is given by

$$S_n = a\,\frac{1 - r^n}{1 - r}$$

EXAMPLE 4 ■ Finding a Partial Sum of a Geometric Sequence

Find the sum of the first five terms of the geometric sequence

$$1, 0.7, 0.49, 0.343, \ldots$$

SOLUTION

The required sum is the sum of the first five terms of a geometric sequence with $a = 1$ and $r = 0.7$. Using the formula for S_n with $n = 5$, we get

$$S_5 = 1 \cdot \frac{1 - (0.7)^5}{1 - 0.7} = 2.7731$$

Thus, the sum of the first five terms of this sequence is 2.7731.

EXAMPLE 5 ■ Finding a Partial Sum of a Geometric Sequence

Find the sum $\displaystyle\sum_{k=1}^{5} 7\left(-\tfrac{2}{3}\right)^k$.

SOLUTION

The given sum is the fifth partial sum of a geometric sequence with first term $a = 7\left(-\tfrac{2}{3}\right) = -\tfrac{14}{3}$ and common ratio $r = -\tfrac{2}{3}$. Thus, by the formula for S_n, we have

$$S_5 = -\frac{14}{3} \cdot \frac{1 - \left(-\tfrac{2}{3}\right)^5}{1 - \left(-\tfrac{2}{3}\right)} = -\frac{14}{3} \cdot \frac{1 + \tfrac{32}{243}}{\tfrac{5}{3}} = -\frac{770}{243}$$

What Is an Infinite Series?

An expression of the form

$$a_1 + a_2 + a_3 + a_4 + \cdots$$

is called an **infinite series**. The dots mean that we are to continue the addition indefinitely. What meaning can we attach to the sum of infinitely many numbers? It seems at first that it is not possible to add infinitely many numbers and arrive at a finite number. But consider the following problem. You have a cake and you want to eat it by first eating half the cake, then eating half of what remains, then again eating half of what remains. This process can continue indefinitely because at each stage some of the cake remains. (See Figure 3.)

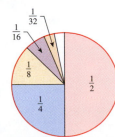

FIGURE 3

Does this mean that it's impossible to eat all of the cake? Of course not. Let's write down what you have eaten from this cake:

$$\frac{1}{2} + \frac{1}{4} + \frac{1}{8} + \frac{1}{16} + \cdots + \frac{1}{2^n} + \cdots$$

This is an infinite series, and we note two things about it: First, from Figure 3 it's clear that no matter how many terms of this series we add, the total will never exceed 1. Second, the more terms of this series we add, the closer the sum is to 1 (see Figure 3). This suggests that the number 1 can be written as the sum of infinitely many smaller numbers:

$$1 = \frac{1}{2} + \frac{1}{4} + \frac{1}{8} + \frac{1}{16} + \cdots + \frac{1}{2^n} + \cdots$$

To make this more precise, let's look at the partial sums of this series:

$$S_1 = \frac{1}{2} \qquad\qquad = \frac{1}{2}$$

$$S_2 = \frac{1}{2} + \frac{1}{4} \qquad\qquad = \frac{3}{4}$$

$$S_3 = \frac{1}{2} + \frac{1}{4} + \frac{1}{8} \qquad = \frac{7}{8}$$

$$S_4 = \frac{1}{2} + \frac{1}{4} + \frac{1}{8} + \frac{1}{16} = \frac{15}{16}$$

and, in general (see Example 5 of Section 10.1),

$$S_n = 1 - \frac{1}{2^n}$$

As n gets larger and larger, we are adding more and more of the terms of this series. Intuitively, as n gets larger, S_n gets closer to the sum of the series. Now notice that as n gets large, $1/2^n$ gets closer and closer to 0. Thus, S_n gets close to $1 - 0 = 1$. Using the notation of Section 3.6, we can write

$$S_n \to 1 \qquad \text{as} \qquad n \to \infty$$

In general, if S_n gets close to a finite number S as n gets large, we say that S is the **sum of the infinite series**.

Infinite Geometric Series

An **infinite geometric series** is a series of the form

$$a + ar + ar^2 + ar^3 + ar^4 + \cdots + ar^{n-1} + \cdots$$

Here is another way to arrive at the formula for the sum of an infinite geometric series:

$$S = a + ar + ar^2 + ar^3 + \cdots$$
$$= a + r(a + ar + ar^2 + \cdots)$$
$$= a + rS$$

Solve the equation $S = a + rS$ for S to get

$$S - rS = a$$
$$(1 - r)S = a$$
$$S = \frac{a}{1 - r}$$

We can apply the reasoning used earlier to find the sum of an infinite geometric series. The nth partial sum of such a series is given by the formula

$$S_n = a\,\frac{1 - r^n}{1 - r} \qquad (r \neq 1)$$

It can be shown that if $|r| < 1$, then r^n gets close to 0 as n gets large (you can easily convince yourself of this using a calculator). It follows that S_n gets close to $a/(1 - r)$ as n gets large, or

$$S_n \to \frac{a}{1 - r} \qquad \text{as} \qquad n \to \infty$$

Thus, the sum of this infinite geometric series is $a/(1 - r)$.

SUM OF AN INFINITE GEOMETRIC SERIES

If $|r| < 1$, then the infinite geometric series

$$a + ar + ar^2 + ar^3 + ar^4 + \cdots + ar^{n-1} + \cdots$$

has the sum

$$S = \frac{a}{1 - r}$$

EXAMPLE 6 ■ Finding the Sum of an Infinite Geometric Series

Find the sum of the infinite geometric series

$$2 + \frac{2}{5} + \frac{2}{25} + \frac{2}{125} + \cdots + \frac{2}{5^n} + \cdots$$

SOLUTION

We use the formula for the sum of an infinite geometric series. In this case, $a = 2$ and $r = \frac{1}{5}$. Thus, the sum of this infinite series is

$$S = \frac{2}{1 - \frac{1}{5}} = \frac{5}{2}$$

■

EXAMPLE 7 ■ Writing a Repeated Decimal as a Fraction

Find the fraction that represents the rational number $2.3\overline{51}$.

SOLUTION

This repeating decimal can be written as a series:

$$\frac{23}{10} + \frac{51}{1000} + \frac{51}{100,000} + \frac{51}{10,000,000} + \frac{51}{1,000,000,000} + \cdots$$

After the first term, the terms of this series form an infinite geometric series with

$$a = \frac{51}{1000} \quad \text{and} \quad r = \frac{1}{100}$$

Thus, the sum of this part of the series is

$$S = \frac{\frac{51}{1000}}{1 - \frac{1}{100}} = \frac{\frac{51}{1000}}{\frac{99}{100}} = \frac{51}{1000} \cdot \frac{100}{99} = \frac{51}{990}$$

So,

$$2.3\overline{51} = \frac{23}{10} + \frac{51}{990} = \frac{2328}{990} = \frac{388}{165}$$

■

10.3 EXERCISES

1–4 ■ The nth term of a sequence is given.
(a) Find the first five terms of the sequence.
(b) What is the common ratio r?
(c) Graph the terms you found in (a).

1. $a_n = 5(2)^{n-1}$ **2.** $a_n = 3(-4)^{n-1}$

3. $a_n = \frac{5}{2}(-\frac{1}{2})^{n-1}$ **4.** $a_n = 3^{n-1}$

5–8 ■ Find the nth term of the geometric sequence with given first term a and common ratio r. What is the fourth term?

5. $a = 3, r = 5$ **6.** $a = -6, r = 3$

7. $a = \frac{5}{2}, r = -\frac{1}{2}$ **8.** $a = \sqrt{3}, r = \sqrt{3}$

9–16 ■ Determine whether the sequence is geometric. If it is geometric, find the common ratio.

9. $2, 4, 8, 16, \ldots$ **10.** $2, 6, 18, 36, \ldots$

11. $3, \frac{3}{2}, \frac{3}{4}, \frac{3}{8}, \ldots$ **12.** $27, -9, 3, -1, \ldots$

13. $\frac{1}{2}, \frac{1}{3}, \frac{1}{4}, \frac{1}{5}, \ldots$ **14.** $e^2, e^4, e^6, e^8, \ldots$

15. $1.0, 1.1, 1.21, 1.331, \ldots$

16. $\frac{1}{2}, \frac{1}{4}, \frac{1}{6}, \frac{1}{8}, \ldots$

17–22 ■ Find the first five terms of the sequence and determine if it is geometric. If it is geometric, find the common ratio and express the nth term of the sequence in the standard form $a_n = ar^{n-1}$.

17. $a_n = 2(3)^n$ **18.** $a_n = 4 + 3^n$

19. $a_n = \frac{1}{4^n}$ **20.** $a_n = (-1)^n 2^n$

21. $a_n = \ln(5^{n-1})$

22. $a_n = n^n$

23–32 ■ Determine the common ratio, the fifth term, and the nth term of the geometric sequence.

23. $2, 6, 18, 54, \ldots$ **24.** $7, \frac{14}{3}, \frac{28}{9}, \frac{56}{27}, \ldots$

25. $0.3, -0.09, 0.027, -0.0081, \ldots$

26. $1, \sqrt{2}, 2, 2\sqrt{2}, \ldots$

27. $144, -12, 1, -\frac{1}{12}, \ldots$ **28.** $-8, -2, -\frac{1}{2}, -\frac{1}{8}, \ldots$

29. $3, 3^{5/3}, 3^{7/3}, 27, \ldots$ **30.** $t, \frac{t^2}{2}, \frac{t^3}{4}, \frac{t^4}{8}, \ldots$

31. $1, s^{2/7}, s^{4/7}, s^{6/7}, \ldots$ **32.** $5, 5^{c+1}, 5^{2c+1}, 5^{3c+1}, \ldots$

33. The first term of a geometric sequence is 8, and the second term is 4. Find the fifth term.

34. The first term of a geometric sequence is 3, and the third term is $\frac{4}{3}$. Find the fifth term.

35. The common ratio in a geometric sequence is $\frac{2}{5}$, and the fourth term is $\frac{5}{2}$. Find the third term.

36. The common ratio in a geometric sequence is $\frac{3}{2}$, and the fifth term is 1. Find the first three terms.

37. Which term of the geometric sequence $2, 6, 18, \ldots$ is 118,098?

38. The second and the fifth terms of a geometric sequence are 10 and 1250, respectively. Is 31,250 a term of this sequence? If so, which term is it?

39–42 ■ Find the partial sum S_n of the geometric sequence that satisfies the given conditions.

39. $a = 5$, $r = 2$, $n = 6$ **40.** $a = \frac{2}{3}$, $r = \frac{1}{3}$, $n = 4$

41. $a_3 = 28$, $a_6 = 224$, $n = 6$

42. $a_2 = 0.12$, $a_5 = 0.00096$, $n = 4$

43–46 ■ Find the sum.

43. $1 + 3 + 9 + \cdots + 2187$

44. $1 - \frac{1}{2} + \frac{1}{4} - \frac{1}{8} + \cdots - \frac{1}{512}$

45. $\displaystyle\sum_{k=0}^{10} 3\left(\frac{1}{2}\right)^k$ **46.** $\displaystyle\sum_{j=0}^{5} 7\left(\frac{3}{2}\right)^j$

47. A ball is dropped from a height of 80 ft. The elasticity of this ball is such that it rebounds three-fourths of the distance it has fallen. How high does the ball rebound on the fifth bounce? Find a formula for how high the ball rebounds on the nth bounce.

48. A culture initially has 5000 bacteria, and its size increases by 8% every hour. How many bacteria are present at the end of 5 hours? Find a formula for the number of bacteria present after n hours.

49. A truck radiator holds 5 gal and is filled with water. A gallon of water is removed from the radiator and replaced with a gallon of antifreeze; then, a gallon of the mixture is removed from the radiator and again replaced by a gallon of antifreeze. This process is repeated indefinitely. How much water remains in the tank after this process is repeated 3 times? 5 times? n times?

50. A very patient woman wishes to become a billionaire. She decides to follow a simple scheme: She puts aside 1 cent the first day, 2 cents the second day, 4 cents the third day, and so on, doubling the number of cents each day. How much money will she have at the end of 30 days? How many days will it take this woman to realize her wish?

51. A ball is dropped from a height of 9 ft. The elasticity of the ball is such that it always bounces up one-third the distance it has fallen.
(a) Find the total distance the ball has traveled at the instant it hits the ground the fifth time.
(b) Find a formula for the total distance the ball has traveled at the instant it hits the ground the nth time.

52. The following is a well-known children's rhyme:

> As I was going to St. Ives
> I met a man with seven wives;
> Every wife had seven sacks;
> Every sack had seven cats;
> Every cat had seven kits;
> Kits, cats, sacks, and wives,
> How many were going to St. Ives?

Assuming that the entire group is actually going to St. Ives, show that the answer to the question in the rhyme is a partial sum of a geometric sequence, and find the sum.

53–60 ■ Find the sum of the infinite geometric series.

53. $1 + \dfrac{1}{3} + \dfrac{1}{9} + \dfrac{1}{27} + \cdots$

54. $1 - \dfrac{1}{2} + \dfrac{1}{4} - \dfrac{1}{8} + \cdots$

55. $1 - \dfrac{1}{3} + \dfrac{1}{9} - \dfrac{1}{27} + \cdots$

56. $\dfrac{2}{5} + \dfrac{4}{25} + \dfrac{8}{125} + \cdots$

57. $\dfrac{1}{3^6} + \dfrac{1}{3^8} + \dfrac{1}{3^{10}} + \dfrac{1}{3^{12}} + \cdots$

58. $3 - \dfrac{3}{2} + \dfrac{3}{4} - \dfrac{3}{8} + \cdots$

59. $-\dfrac{100}{9} + \dfrac{10}{3} - 1 + \dfrac{3}{10} - \cdots$

60. $\dfrac{1}{\sqrt{2}} + \dfrac{1}{2} + \dfrac{1}{2\sqrt{2}} + \dfrac{1}{4} + \cdots$

61–66 ■ Express the repeating decimal as a fraction.

61. $0.777\ldots$ **62.** $0.25\overline{3}$

63. $0.030303\ldots$ **64.** $2.11\overline{25}$

65. $0.\overline{112}$ **66.** $0.123123123\ldots$

67. A certain ball rebounds to half the height from which it is dropped. Use an infinite geometric series to approximate the total distance the ball travels, after being dropped from 1 m above the ground, until it comes to rest.

68. If the ball in Exercise 67 is dropped from a height of 8 ft, then 1 s is required for its first complete bounce—from the instant it first touches the ground until it next touches the ground. Each subsequent complete bounce requires $1/\sqrt{2}$ as long as the preceding complete bounce. Use an infinite geometric series to estimate the time period from the instant the ball first touches the ground until it stops bouncing.

69. The midpoints of the sides of a square of side 1 are joined to form a new square. This procedure is repeated for each new square. (See the figure.)
   (a) Find the sum of the areas of all the squares.
   (b) Find the sum of the perimeters of all the squares.

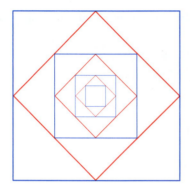

70. A circular disk of radius R is cut out of paper, as shown in figure (a). Two disks of radius $\frac{1}{2}R$ are cut out of paper and placed on top of the first disk, as in figure (b), and then four disks of radius $\frac{1}{4}R$ are placed on these two disks [figure (c)]. Assuming that this process can be repeated indefinitely, find the total area of all the disks.

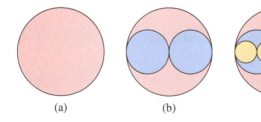

(a)   (b)   (c)

71. A yellow square of side 1 is divided into nine smaller squares, and the middle square is colored blue as shown in the figure. Each of the smaller yellow squares is in turn divided into nine squares, and each middle square is col-

ored blue. If this process is continued indefinitely, what is the total area colored blue?

72. If the numbers $a_1, a_2, \ldots, a_n$ form a geometric sequence, then $a_2, a_3, \ldots, a_{n-1}$ are **geometric means** between a_1 and a_n. Insert three geometric means between 5 and 80.

73. Find the sum of the first ten terms of the sequence

$$a + b, a^2 + 2b, a^3 + 3b, a^4 + 4b, \ldots$$

 DISCOVERY · DISCUSSION

74. Arithmetic or Geometric? The first four terms of a sequence are given. Determine whether these terms can be the terms of an arithmetic sequence, a geometric sequence, or neither. Find the next term if the sequence is arithmetic or geometric.
   (a) $5, -3, 5, -3, \ldots$ **(b)** $\frac{1}{3}, 1, \frac{5}{3}, \frac{7}{3}, \ldots$
   (c) $\sqrt{3}, 3, 3\sqrt{3}, 9, \ldots$ **(d)** $1, -1, 1, -1, \ldots$
   (e) $2, -1, \frac{1}{2}, 2, \ldots$ **(f)** $x - 1, x, x + 1, x + 2, \ldots$
   (g) $-3, -\frac{3}{2}, 0, \frac{3}{2}, \ldots$ **(h)** $\sqrt{5}, \sqrt[3]{5}, \sqrt[6]{5}, 1, \ldots$

75. Reciprocals of a Geometric Sequence If $a_1, a_2, a_3, \ldots$ is a geometric sequence with common ratio r, show that the sequence

$$\frac{1}{a_1}, \frac{1}{a_2}, \frac{1}{a_3}, \ldots$$

is also a geometric sequence, and find the common ratio.

76. Logarithms of a Geometric Sequence If $a_1, a_2, a_3, \ldots$ is a geometric sequence with a common ratio $r > 0$ and $a_1 > 0$, show that the sequence

$$\log a_1, \log a_2, \log a_3, \ldots$$

is an arithmetic sequence, and find the common difference.

77. Exponentials of an Arithmetic Sequence If $a_1, a_2, a_3, \ldots$ is an arithmetic sequence with common difference d, show that the sequence

$$10^{a_1}, 10^{a_2}, 10^{a_3}, \ldots$$

is a geometric sequence, and find the common ratio.

Discovery Project

Finding Patterns

The ancient Greeks studied triangular numbers, square numbers, pentagonal numbers, and other **polygonal numbers**, like those shown in the figure.

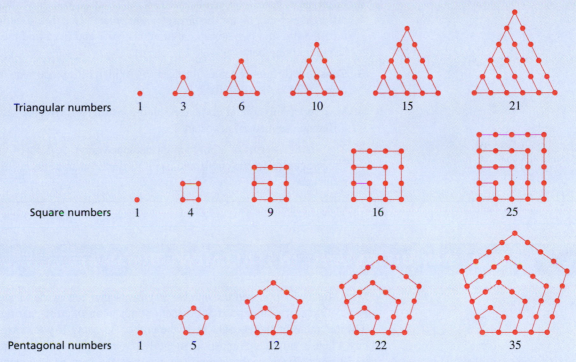

Triangular numbers   1   3   6   10   15   21

Square numbers   1   4   9   16   25

Pentagonal numbers   1   5   12   22   35

To find a pattern for such numbers, we construct a **first difference sequence** by taking differences of successive terms; we repeat the process to get a **second difference sequence**, **third difference sequence**, and so on. For the sequence of triangular numbers T_n we get the following **difference table**:

Triangular numbers	1	3	6	10	15	21
First differences		2	3	4	5	6
Second differences			1	1	1	1

We stop at the second difference sequence because it's a constant sequence. Assuming that this sequence will continue to have constant value 1, we can work backward from the bottom row to find more terms of the first difference sequence, and from these, more triangular numbers.

If a sequence is given by a polynomial function and if we calculate the first differences, the second differences, the third differences, and so on, then eventually we get a constant sequence. For example, the triangular numbers are given by the

The formula for the *n*th triangular number can be found using the formula for the sum of the first *n* whole numbers (Example 2, Section 10.5). From the definition of T_n we have

$$T_n = 1 + 2 + \cdots + n$$
$$= \frac{n(n + 1)}{2}$$
$$= \frac{1}{2}n^2 + \frac{1}{2}n$$

polynomial $T_n = \frac{1}{2}n^2 + \frac{1}{2}n$ (see the margin note); the second difference sequence is the constant sequence 1, 1, 1, ...

1. Construct a difference table for the square numbers and the pentagonal numbers. Use the table to find the tenth pentagonal number.

2. From the patterns you've observed so far, what do you think the second difference would be for the *hexagonal numbers*? Use this, together with the fact that the first two hexagonal numbers are 1 and 6, to find the first eight hexagonal numbers.

3. Construct difference tables for $C_n = n^3$. Which difference sequence is constant? Do the same for $F_n = n^4$.

4. Make up a polynomial of degree 5 and construct a difference table. Which difference sequence is constant?

5. The first few terms of a polynomial sequence are 1, 2, 4, 8, 16, 31, 57, ... Construct a difference table and use it to find four more terms of this sequence.

10.4 ANNUITIES AND INSTALLMENT BUYING

Many financial transactions involve payments that are made at regular intervals. For example, if you deposit $100 each month in an interest-bearing account, what will the value of the account be at the end of 5 years? If you borrow $100,000 to buy a house, how much must the monthly payments be in order to pay off the loan in 30 years? Each of these questions involves the sum of a sequence of numbers; we use the results of the preceding section to answer them here.

The Amount of an Annuity

An **annuity** is a sum of money that is paid in regular equal payments. Although the word *annuity* suggests annual (or yearly) payments, they can be made semi-annually, quarterly, monthly, or at some other regular interval. Payments are usually made at the end of the payment interval. The **amount of an annuity** is the sum of all the individual payments from the time of the first payment until the last payment is made, together with all the interest. We denote this sum by A_f (the subscript *f* here is used to denote *final* amount).

EXAMPLE 1 ■ Calculating the Amount of an Annuity

An investor deposits $400 every December 15 and June 15 for 10 years in an account that earns interest at the rate of 8% per year, compounded semiannually. How much will be in the account immediately after the last payment?

SOLUTION

We need to find the amount of an annuity consisting of 20 semiannual payments of \$400 each. Since the interest rate is 8% per year, compounded semiannually, the interest rate per time period is $i = 0.08/2 = 0.04$. The first payment is in the account for 19 time periods, the second for 18 time periods, and so on. The last payment receives no interest. The situation can be illustrated by the time line in Figure 1.

When using interest rates in calculators, remember to convert percentages to decimals. For example, 8% is 0.08.

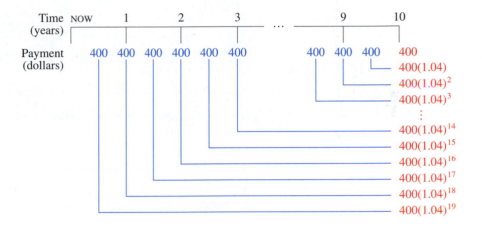

FIGURE 1

The amount A_f of the annuity is the sum of these 20 amounts. Thus

$$A_f = 400 + 400(1.04) + 400(1.04)^2 + \cdots + 400(1.04)^{19}$$

But this is a geometric series with $a = 400$, $r = 1.04$, and $n = 20$, so

$$A_f = 400 \, \frac{1 - (1.04)^{20}}{1 - 1.04} \approx 11{,}911.23$$

Thus, the amount in the account after the last payment is \$11,911.23. ■

In general, the regular annuity payment is called the **periodic rent** and is denoted by R. We also let i denote the interest rate per time period and n the number of payments. *We always assume that the time period in which interest is compounded is equal to the time between payments.* By the same reasoning as in Example 1, we see that the amount A_f of an annuity is

$$A_f = R + R(1 + i) + R(1 + i)^2 + \cdots + R(1 + i)^{n-1}$$

Since this is the nth partial sum of a geometric sequence with $a = R$ and $r = 1 + i$, the formula for the partial sum gives

$$A_f = R\,\frac{1 - (1 + i)^n}{1 - (1 + i)} = R\,\frac{1 - (1 + i)^n}{-i} = R\,\frac{(1 + i)^n - 1}{i}$$

AMOUNT OF AN ANNUITY

The amount A_f of an annuity consisting of n regular equal payments of size R with interest rate i per time period is given by

$$A_f = R\,\frac{(1 + i)^n - 1}{i}$$

EXAMPLE 2 ■ Calculating the Amount of an Annuity

How much money should be invested every month at 12% per year, compounded monthly, in order to have $4000 in 18 months?

SOLUTION

In this problem $i = 0.12/12 = 0.01$, $A_f = 4000$, and $n = 18$. We need to find the amount R of each payment. By the formula for the amount of an annuity.

$$4000 = R\,\frac{(1 + 0.01)^{18} - 1}{0.01}$$

Solving for R, we get

$$R = \frac{4000(0.01)}{(1 + 0.01)^{18} - 1} \approx 203.928$$

Thus, the monthly investment should be $203.93.　　■

The Present Value of an Annuity

If you were to receive $10,000 five years from now, it would be worth much less than getting $10,000 right now. This is because of the interest you could accumulate during the next five years if you invested the money now. What smaller amount would you be willing to accept *now* instead of receiving $10,000 in five years? This is the amount of money that, together with interest, would be worth $10,000 in five years. The amount we are looking for here is called the *discounted value* or *present value*. If the interest rate is 8% per year, compounded quarterly, then the interest per time period is $i = 0.08/4 = 0.02$, and there are $4 \times 5 = 20$ time periods. If we let PV denote the present value, then by the formula for compound interest (Section 4.1) we have

$$10,000 = PV(1 + i)^n = PV(1 + 0.02)^{20}$$

so

$$PV = 10,000(1 + 0.02)^{-20} \approx 6729.713$$

Thus, in this situation, the present value of $10,000 is $6729.71. This reasoning leads to a general formula for present value:

$$PV = A(1 + i)^{-n}$$

Similarly, the **present value of an annuity** is the amount A_p that must be invested now at the interest rate i per time period in order to provide n payments, each of amount R. Clearly, A_p is the sum of the present values of each individual payment (see Exercise 22). Another way of finding A_p is to note that A_p is the present value of A_f:

$$A_p = A_f(1 + i)^{-n} = R\,\frac{(1 + i)^n - 1}{i}\,(1 + i)^{-n} = R\,\frac{1 - (1 + i)^{-n}}{i}$$

THE PRESENT VALUE OF AN ANNUITY

The **present value** A_p of an annuity consisting of n regular equal payments of size R and interest rate i per time period is given by

$$A_p = R\,\frac{1 - (1 + i)^{-n}}{i}$$

EXAMPLE 3 ■ Calculating the Present Value of an Annuity

A person wins $10,000,000 in the California lottery, and the amount is paid in yearly installments of half a million dollars each for 20 years. What is the present value of his winnings? Assume that he can earn 10% interest, compounded annually.

SOLUTION

Since the amount won is paid as an annuity, we need to find its present value. Here $i = 0.1$, $R = \$500{,}000$, and $n = 20$. Thus

$$A_p = 500{,}000\,\frac{1 - (1 + 0.1)^{-20}}{0.1} \approx 4{,}256{,}781.859$$

This means that the winner really won only $4,256,781.86 if it were paid immediately. ∎

■ Installment Buying

When you buy a house or a car by installment, the payments you make are an annuity whose present value is the amount of the loan.

EXAMPLE 4 ■ The Amount of a Loan

A student wishes to buy a car. He can afford to pay $200 per month but has no money for a down payment. If he can make these payments for four years and the interest rate is 12%, what purchase price can he afford?

SOLUTION

The payments the student makes constitute an annuity whose present value is the price of the car (which is also the amount of the loan, in this case). Here we have $i = 0.12/12 = 0.01$, $R = 200$, $n = 12 \times 4 = 48$, so

$$A_p = R\,\frac{1 - (1 + i)^{-n}}{i} = 200\,\frac{1 - (1 + 0.01)^{-48}}{0.01} \approx 7594.792$$

Thus, the student can buy a car priced at \$7594.79. ∎

When a bank makes a loan that is to be repaid with regular equal payments R, then the payments form an annuity whose present value A_p is the amount of the loan. So, to find the size of the payments, we solve for R in the formula for the amount of an annuity. This gives the following formula for R.

INSTALLMENT BUYING

If a loan A_p is to be repaid in n regular equal payments with interest rate i per time period, then the size R of each payment is given by

$$R = \frac{iA_p}{1 - (1 + i)^{-n}}$$

EXAMPLE 5 ■ Calculating Monthly Mortgage Payments

A couple borrows \$100,000 at 9% interest as a mortage loan on a house. They expect to make monthly payments for 30 years to repay the loan. What is the size of each payment?

SOLUTION

The mortgage payments form an annuity whose present value is $A_p = \$100{,}000$. Also, $i = 0.09/12 = 0.0075$, and $n = 12 \times 30 = 360$. We are looking for the amount R of each payment. From the formula for installment buying, we get

$$R = \frac{iA_p}{1 - (1 + i)^{-n}}$$

$$= \frac{(0.0075)(100{,}000)}{1 - (1 + 0.0075)^{-360}} \approx 804.623$$

Thus, the monthly payments are \$804.62. ∎

We now illustrate the use of graphing devices in solving problems related to installment buying.

 ### EXAMPLE 6 ■ Calculating the Interest Rate from the Size of Monthly Payments

A car dealer sells a new car for \$18,000. He offers the buyer payments of \$405 per month for 5 years. What interest rate is this car dealer charging?

SOLUTION

The payments form an annuity with present value $A_p = \$18,000$, $R = 405$, and $n = 12 \times 5 = 60$. To find the interest rate, we must solve for i in the equation

$$R = \frac{iA_p}{1 - (1 + i)^{-n}}$$

A little experimentation will convince you that it's not possible to algebraically solve this equation for i. So, to find i we use a graphing device to graph R as a function of the interest rate x, and we then use the graph to find the interest rate corresponding to the value of R we want ($\$405$ in this case). Since $i = x/12$, we graph the function

$$R(x) = \frac{\dfrac{x}{12}(18,000)}{1 - \left(1 + \dfrac{x}{12}\right)^{-60}}$$

FIGURE 2

in the viewing rectangle $[0.06, 0.16] \times [350, 450]$, as shown in Figure 2. We also graph the horizontal line $R(x) = 405$ in the same viewing rectangle. Then, by moving the cursor to the point of intersection of the two graphs, we find that the corresponding x-value is approximately 0.125. Thus, the interest rate is about $12\frac{1}{2}\%$. ∎

10.4   EXERCISES

1. Find the amount of an annuity that consists of 10 annual payments of $1000 each into an account that pays 6% interest per year.

2. Find the amount of an annuity that consists of 24 monthly payments of $500 each into an account that pays 8% interest per year, compounded monthly.

3. Find the amount of an annuity that consists of 20 annual payments of $5000 each into an account that pays interest of 12% per year.

4. Find the amount of an annuity that consists of 20 semi-annual payments of $500 each into an account that pays 6% interest per year, compounded semiannually.

5. Find the amount of an annuity that consists of 16 quarterly payments of $300 each into an account that pays 8% interest per year, compounded quarterly.

6. How much money should be invested every quarter at 10% per year, compounded quarterly, in order to have $5000 in 2 years?

7. How much money should be invested monthly at 6% per year, compounded monthly, in order to have $2000 in 8 months?

8. What is the present value of an annuity that consists of 20 semiannual payments of $1000 at the interest rate of 9% per year, compounded semiannually?

9. How much money must be invested now at 9% per year, compounded semiannually, to fund an annuity of 20 payments of $200 each, paid every 6 months, the first payment being 6 months from now?

10. A 55-year-old man deposits $50,000 to fund an annuity with an insurance company. The money will be invested at 8% per year, compounded semiannually. He is to draw semiannual payments until he reaches age 65. What is the amount of each payment?

11. A woman wants to borrow $12,000 in order to buy a car. She wants to repay the loan by monthly installments for 4 years. If the interest rate on this loan is $10\frac{1}{2}\%$ per year, compounded monthly, what is the amount of each payment?

12. What is the monthly payment on a 30-year mortgage of $80,000 at 9% interest? What is the monthly payment on this same mortgage if it is to be repaid over a 15-year period?

13. What is the monthly payment on a 30-year mortgage of $100,000 at 8% interest per year, compounded monthly? What is the total amount paid on this loan over the 30-year period?

14. A couple can afford to make a monthly mortgage payment of $650. If the mortgage rate is 9% and the couple intends to secure a 30-year mortgage, how much can they borrow?

15. A couple secures a 30-year loan of $100,000 at $9\frac{3}{4}$% per year, compounded monthly, to buy a house.
   (a) What is the amount of their monthly payment?
   (b) What total amount will they pay over the 30-year period?
   (c) If, instead of taking the loan, the couple deposits the monthly payments in an account that pays $9\frac{3}{4}$% interest per year, compounded monthly, how much will be in the account at the end of the 30-year period?

16. Jane agrees to buy a car for a down payment of $2000 and payments of $220 per month for 3 years. If the interest rate is 8% per year, compounded monthly, what is the actual purchase price of her car?

17. Mike buys a ring for his fiancee by paying $30 per month for one year. If the interest rate is 10% per year, compounded monthly, what is the price of the ring?

18. Janet's payments on her $12,500 car are $420 a month for 3 years. Assuming that interest is compounded monthly, what interest rate is she paying on the car loan?

19. John buys a stereo system for $640. He agrees to pay $32 per month for 2 years. Assuming that interest is compounded monthly, what interest rate is he paying?

20. A man purchases a $2000 diamond ring for a down payment of $200 and monthly installments of $88 for 2 years. Assuming that interest is compounded monthly, what interest rate is he paying?

21. An item at a department store is priced at $189.99 and can be bought by making 20 payments of $10.50. Find the interest rate, assuming that interest is compounded monthly.

22. (a) Draw a time line as in Example 1 to show that the present value of an annuity is the sum of the present values of each payment, that is,

$$A_p = \frac{R}{1+i} + \frac{R}{(1+i)^2} + \frac{R}{(1+i)^3} + \cdots + \frac{R}{(1+i)^n}$$

   (b) Use part (a) to derive the formula for A_p given in the text.

◆ DISCOVERY · DISCUSSION

23. An Annuity That Lasts Forever   An **annuity in perpetuity** is one that continues forever. Such annuities are useful in setting up scholarship funds to ensure that the award continues.

   (a) Draw a time line (as in Example 1) to show that to set up an annuity in perpetuity of amount R per time period, the amount that must be invested now is

$$A_p = \frac{R}{1+i} + \frac{R}{(1+i)^2} + \frac{R}{(1+i)^3} + \cdots + \frac{R}{(1+i)^n} + \cdots$$

   where i is the interest rate per time period.

   (b) Find the sum of the infinite series in part (a) to show that

$$A_p = \frac{R}{i}$$

   (c) How much money must be invested now at 10% per year, compounded annually, to provide an annuity in perpetuity of $5000 per year? The first payment is due in one year.

   (d) How much money must be invested now at 8% per year, compounded quarterly, to provide an annuity in perpetuity of $3000 per year? The first payment is due in one year.

24. Amortizing a Mortgage   When they bought their house, John and Mary took out a $90,000 mortgage at 9% interest, repayable monthly over 30 years. Their payment is $724.17 per month (check this using the formula in the text). The bank gave them an **amortization schedule**, which is a table showing how much of each payment is interest, how much goes toward the principal, and the remaining principal after each payment. The table below shows the first few entries in the amortization schedule.

Payment number	Total payment	Interest payment	Principal payment	Remaining principal
1	724.17	675.00	49.17	89,950.83
2	724.17	674.63	49.54	89,901.29
3	724.17	674.26	49.91	89,851.38
4	724.17	673.89	50.28	89,801.10

After 10 years they have made 120 payments and are wondering how much they still owe, but they have lost the amortization schedule.

   (a) How much do John and Mary still owe on their mortgage? [*Hint:* The remaining balance is the present value of the 240 remaining payments.]

   (b) How much of their next payment is interest and how much goes toward the principal? [*Hint:* Since 9% ÷ 12 = 0.75%, they must pay 0.75% of the remaining principal in interest each month.]

10.5 MATHEMATICAL INDUCTION

There are two aspects to mathematics—discovery and proof—and both are of equal importance. We must discover something before we can attempt to prove it, and we can only be certain of its truth once it has been proved. In this section we examine the relationship between these two key components of mathematics more closely.

■ Conjecture and Proof

Let's try a simple experiment. We add more and more of the odd numbers as follows:

$$1 = 1$$

$$1 + 3 = 4$$

$$1 + 3 + 5 = 9$$

$$1 + 3 + 5 + 7 = 16$$

$$1 + 3 + 5 + 7 + 9 = 25$$

What do you notice about the numbers on the right side of these equations? They are in fact all perfect squares. These equations say the following:

The sum of the first 1 odd number is 1^2.

The sum of the first 2 odd numbers is 2^2.

The sum of the first 3 odd numbers is 3^2.

The sum of the first 4 odd numbers is 4^2.

The sum of the first 5 odd numbers is 5^2.

This leads naturally to the following question: Is it true that for every natural number n, the sum of the first n odd numbers is n^2? Could this remarkable property be true? We could try a few more numbers and find that the pattern persists for the first 6, 7, 8, 9, and 10 odd numbers. At this point, we feel quite sure that this is always true, so we make a *conjecture*:

The sum of the first n odd numbers is n^2.

Since we know that the nth odd number is $2n - 1$, we can write this statement more precisely as

$$1 + 3 + 5 + \cdots + (2n - 1) = n^2$$

It's important to realize that this is still a conjecture. We cannot conclude by checking a finite number of cases that a property is true for all numbers (there are infinitely many). To see this more clearly, suppose someone tells us he has added up

Consider the polynomial
$$p(n) = n^2 - n + 41$$
Here are some values of $p(n)$:

$p(1) = 41 \qquad p(2) = 43$

$p(3) = 47 \qquad p(4) = 53$

$p(5) = 61 \qquad p(6) = 71$

$p(7) = 83 \qquad p(8) = 97$

All the values so far are prime numbers. In fact, if you keep going, you will find that $p(n)$ is prime for all natural numbers up to $n = 40$. It may seem reasonable at this point to conjecture that $p(n)$ is prime for *every* natural number n. But our conjecture would be too hasty, because it is easily seen that $p(41)$ is *not* prime. This illustrates that we cannot be certain of the truth of a statement no matter how many special cases we check. We need a convincing argument—a *proof*—to determine the truth of a statement.

the first trillion odd numbers and found that they do *not* add up to 1 trillion squared. What would you tell this person? It would be silly to say that you're sure it's true because you've already checked the first five cases. You could, however, take out paper and pencil and start checking it yourself, but this task would probably take the rest of your life. The tragedy would be that after completing this task you would still not be sure of the truth of the conjecture. Do you see why?

Herein lies the power of mathematical proof. A **proof** is a clear argument that demonstrates the truth of a statement beyond doubt.

■ Mathematical Induction

Let's consider a special kind of proof called **mathematical induction**. Here is how it works: Suppose we have a statement that says something about all natural numbers n. Let's call this statement P. For example, we could consider the statement

> P: For every natural number n, the sum of the first n odd numbers is n^2.

Since this statement is about *all* natural numbers, it contains infinitely many statements; we will call them $P(1)$, $P(2)$, ...

$$P(1): \quad \text{The sum of the first 1 odd number is } 1^2.$$
$$P(2): \quad \text{The sum of the first 2 odd numbers is } 2^2.$$
$$P(3): \quad \text{The sum of the first 3 odd numbers is } 3^2.$$
$$\vdots \qquad\qquad \vdots$$

How can we prove all of these statements at once? Mathematical induction is a clever way of doing just that.

The crux of the idea is this: Suppose we can prove that whenever one of these statements is true, then the one following it in the list is also true. In other words,

> For every k, if $P(k)$ is true, then $P(k + 1)$ is true.

This is called the **induction step** because it leads us from the truth of one statement to the next. Now, suppose that we can also prove that

$$P(1) \text{ is true.}$$

The induction step now leads us through the following chain of statements:

$$P(1) \text{ is true, so } P(2) \text{ is true.}$$
$$P(2) \text{ is true, so } P(3) \text{ is true.}$$
$$P(3) \text{ is true, so } P(4) \text{ is true.}$$
$$\vdots \qquad\qquad \vdots$$

So we see that if both the induction step and $P(1)$ are proved, then statement P is proved for all n. Here is a summary of this important method of proof.

PRINCIPLE OF MATHEMATICAL INDUCTION

For each natural number n, let $P(n)$ be a statement depending on n. Suppose that the following two conditions are satisfied.

1. $P(1)$ is true.

2. For every natural number k, if $P(k)$ is true, then $P(k + 1)$ is true.

Then $P(n)$ is true for all natural numbers n.

To apply this principle, there are two steps:

Step 1 Prove that $P(1)$ is true.

Step 2 Assume that $P(k)$ is true and use this assumption to prove that $P(k + 1)$ is true.

Notice that in Step 2 we do not prove that $P(k)$ is true. We only show that *if $P(k)$ is true, then $P(k + 1)$ is also true*. The assumption that $P(k)$ is true is called the **induction hypothesis**.

We now use mathematical induction to prove that the conjecture we made at the beginning of this section is true.

EXAMPLE 1 ■ A Proof by Mathematical Induction

Prove that for all natural numbers n,

$$1 + 3 + 5 + \cdots + (2n - 1) = n^2$$

SOLUTION

Let $P(n)$ denote the statement $1 + 3 + 5 + \cdots + (2n - 1) = n^2$.

Step 1 We need to show that $P(1)$ is true. But $P(1)$ is simply the statement that $1 = 1^2$, which is of course true.

Step 2 We assume that $P(k)$ is true. Thus, our induction hypothesis is

$$1 + 3 + 5 + \cdots + (2k - 1) = k^2$$

We want to use this to show that $P(k + 1)$ is true, that is,

$$1 + 3 + 5 + \cdots + (2k - 1) + [2(k + 1) - 1] = (k + 1)^2$$

[Note that we get $P(k + 1)$ by substituting $k + 1$ for each n in the statement $P(n)$.] We start with the left side and use the induction hypothesis to obtain the right side of the equation:

$$1 + 3 + 5 + \cdots + (2k - 1) + [2(k + 1) - 1]$$

$$= [1 + 3 + 5 + \cdots + (2k - 1)] + [2(k + 1) - 1] \qquad \text{Group the first } k \text{ terms}$$

$$= k^2 + [2(k + 1) - 1] \qquad \text{Induction hypothesis}$$

$$= k^2 + [2k + 2 - 1] \qquad \text{Distributive Property}$$

$$= k^2 + 2k + 1 \qquad \text{Simplify}$$

$$= (k + 1)^2 \qquad \text{Factor}$$

Thus, $P(k + 1)$ follows from $P(k)$ and this completes the induction step.

Having proved Steps 1 and 2, we conclude by the Principle of Mathematical Induction that $P(n)$ is true for all natural numbers n. ■

EXAMPLE 2 ■ A Proof by Mathematical Induction

Prove that for every natural number n,

$$1 + 2 + 3 + \cdots + n = \frac{n(n + 1)}{2}$$

SOLUTION

Let $P(n)$ be the statement $1 + 2 + 3 + \cdots + n = n(n + 1)/2$. We want to show that $P(n)$ is true for all natural numbers n.

Step 1 We need to show that $P(1)$ is true. But $P(1)$ says that

$$1 = \frac{1(1 + 1)}{2}$$

and this statement is clearly true.

Blaise Pascal (1623–1662) is considered one of the most versatile minds in modern history. He was a writer and philosopher as well as a gifted mathematician and physicist. Among his contributions that appear in this book are the theory of probability, Pascal's triangle, and the Principle of Mathematical Induction.

Pascal's father, himself a mathematician, believed that his son should not study mathematics until he was 15 or 16. But at age 12, Blaise insisted on learning geometry, and proved most of its elementary theorems himself. At 19, he invented the first mechanical adding machine. In 1647, after writing a major treatise on the conic sections, he abruptly abandoned mathematics because he felt his intense studies were contributing to his ill health. He devoted himself instead to frivolous recreations such as gambling, but this only served to pique his interest in probability. In 1654 he miraculously survived a carriage accident in which his horses ran off a bridge. Taking this to be a sign from God, he entered a monastery, where he pursued theology and philosophy, writing his famous *Pensées*. He also continued his mathematical research. He valued faith and intuition more than reason as the source of truth, declaring that "the heart has its own reasons, which reason cannot know."

Step 2 Assume that $P(k)$ is true. Thus, our induction hypothesis is

$$1 + 2 + 3 + \cdots + k = \frac{k(k + 1)}{2}$$

We want to use this to show that $P(k + 1)$ is true, that is,

$$1 + 2 + 3 + \cdots + k + (k + 1) = \frac{(k + 1)[(k + 1) + 1]}{2}$$

So, we start with the left side and use the induction hypothesis to obtain the right side:

$1 + 2 + 3 + \cdots + k + (k + 1)$

$= [1 + 2 + 3 + \cdots + k] + (k + 1)$ Group the first k terms

$= \dfrac{k(k + 1)}{2} + (k + 1)$ Induction hypothesis

$= (k + 1)\left(\dfrac{k}{2} + 1\right)$ Factor $k + 1$

$= (k + 1)\left(\dfrac{k + 2}{2}\right)$ Common denominator

$= \dfrac{(k + 1)[(k + 1) + 1]}{2}$ Write $k + 2$ as $k + 1 + 1$

Thus, $P(k + 1)$ follows from $P(k)$ and this completes the induction step.

Having proved Steps 1 and 2, we conclude by the Principle of Mathematical Induction that $P(n)$ is true for all natural numbers n. ■

Formulas for the sums of powers of the first n natural numbers are important in calculus. Formula 1 in the following box is proved in Example 2. The other formulas are also proved using mathematical induction (see Exercises 4 and 7).

SUMS OF POWERS

0. $\displaystyle\sum_{k=1}^{n} 1 = n$ **1.** $\displaystyle\sum_{k=1}^{n} k = \frac{n(n + 1)}{2}$

2. $\displaystyle\sum_{k=1}^{n} k^2 = \frac{n(n + 1)(2n + 1)}{6}$ **3.** $\displaystyle\sum_{k=1}^{n} k^3 = \frac{n^2(n + 1)^2}{4}$

It might happen that a statement $P(n)$ is false for the first few natural numbers, but true from some number on. For example, we may want to prove that $P(n)$ is true for $n \geq 5$. Notice that if we prove that $P(5)$ is true, then this fact, together with the induction step, would imply the truth of $P(5)$, $P(6)$, $P(7)$, ... The next example illustrates this point.

EXAMPLE 3 ■ Proving an Inequality by Mathematical Induction

Prove that $4n < 2^n$ for all $n \geq 5$.

SOLUTION

Let $P(n)$ denote the statement $4n < 2^n$.

Step 1 $P(5)$ is the statement that $4 \cdot 5 < 2^5$, or $20 < 32$, which is true.

Step 2 Assume that $P(k)$ is true. Thus, our induction hypothesis is

$$4k < 2^k$$

We want to use this to show that $P(k + 1)$ is true, that is,

We get $P(k + 1)$ by replacing k by $k + 1$ in the statement $P(k)$.

$$4(k + 1) < 2^{k+1}$$

So, we start with the left side of the inequality and use the induction hypothesis to show that it is less than the right side. For $k \geq 5$, we have

$$4(k + 1) = 4k + 4$$

$$< 2^k + 4 \qquad \text{Induction hypothesis}$$

$$< 2^k + 4k \qquad \text{Because } 4 < 4k$$

$$< 2^k + 2^k \qquad \text{Induction hypothesis}$$

$$= 2 \cdot 2^k$$

$$= 2^{k+1} \qquad \text{Property of exponents}$$

Thus, $P(k + 1)$ follows from $P(k)$ and this completes the induction step.

Having proved Steps 1 and 2, we conclude by the Principle of Mathematical Induction that $P(n)$ is true for all natural numbers $n \geq 5$.   ■

10.5 EXERCISES

1–12 ■ Use mathematical induction to prove that the formula is true for all natural numbers n.

1. $2 + 4 + 6 + \cdots + 2n = n(n + 1)$

2. $1 + 4 + 7 + \cdots + (3n - 2) = \dfrac{n(3n - 1)}{2}$

3. $5 + 8 + 11 + \cdots + (3n + 2) = \dfrac{n(3n + 7)}{2}$

4. $1^2 + 2^2 + 3^2 + \cdots + n^2 = \dfrac{n(n + 1)(2n + 1)}{6}$

5. $1 \cdot 2 + 2 \cdot 3 + 3 \cdot 4 + \cdots + n(n + 1)$

$$= \dfrac{n(n + 1)(n + 2)}{3}$$

6. $1 \cdot 3 + 2 \cdot 4 + 3 \cdot 5 + \cdots + n(n + 2)$

$$= \dfrac{n(n + 1)(2n + 7)}{6}$$

7. $1^3 + 2^3 + 3^3 + \cdots + n^3 = \dfrac{n^2(n + 1)^2}{4}$

8. $1^3 + 3^3 + 5^3 + \cdots + (2n - 1)^3 = n^2(2n^2 - 1)$

9. $2^3 + 4^3 + 6^3 + \cdots + (2n)^3 = 2n^2(n + 1)^2$

10. $\dfrac{1}{1 \cdot 2} + \dfrac{1}{2 \cdot 3} + \dfrac{1}{3 \cdot 4} + \cdots + \dfrac{1}{n(n + 1)} = \dfrac{n}{(n + 1)}$

11. $1 \cdot 2 + 2 \cdot 2^2 + 3 \cdot 2^3 + 4 \cdot 2^4 + \cdots + n \cdot 2^n$
$$= 2[1 + (n - 1)2^n]$$

12. $1 + 2 + 2^2 + \cdots + 2^{n-1} = 2^n - 1$

13. Show that $n^2 + n$ is divisible by 2 for all natural numbers n.

14. Show that $5^n - 1$ is divisible by 4 for all natural numbers n.

15. Show that $n^2 - n + 41$ is odd for all natural numbers n.

16. Show that $n^3 - n + 3$ is divisible by 3 for all natural numbers n.

17. Show that $8^n - 3^n$ is divisible by 5 for all natural numbers n.

18. Show that $3^{2n} - 1$ is divisible by 8 for all natural numbers n.

19. Prove that $n < 2^n$ for all natural numbers n.

20. Prove that $(n + 1)^2 < 2n^2$ for all natural numbers $n \geqslant 3$.

21. Prove that if $x > -1$, then $(1 + x)^n \geqslant 1 + nx$ for all natural numbers n.

22. Show that $100n \leqslant n^2$ for all $n \geqslant 100$.

23. Let $a_{n+1} = 3a_n$ and $a_1 = 5$. Show that $a_n = 5 \cdot 3^{n-1}$ for all natural numbers n.

24. A sequence is defined recursively by $a_{n+1} = 3a_n - 8$ and $a_1 = 4$. Find an explicit formula for a_n and then use mathematical induction to prove that the formula you found is true.

25. Show that $x - y$ is a factor of $x^n - y^n$ for all natural numbers n.
[*Hint:* $x^{k+1} - y^{k+1} = x^k(x - y) + (x^k - y^k)y$]

26. Show that $x + y$ is a factor of $x^{2n-1} + y^{2n-1}$ for all natural numbers n.

27–31 ◼ F_n denotes the nth term of the Fibonacci sequence discussed in Section 10.1. Use mathematical induction to prove the statement.

27. F_{3n} is even for all natural numbers n.

28. $F_1 + F_2 + F_3 + \cdots + F_n = F_{n+2} - 1$

29. $F_1^2 + F_2^2 + F_3^2 + \cdots + F_n^2 = F_n F_{n+1}$

30. If $a_{n+2} = a_{n+1} \cdot a_n$ and $a_1 = a_2 = 2$, then $a_n = 2^{F_n}$ for all natural numbers n.

31. For all $n \geqslant 2$,
$$\begin{bmatrix} 1 & 1 \\ 1 & 0 \end{bmatrix}^n = \begin{bmatrix} F_{n+1} & F_n \\ F_n & F_{n-1} \end{bmatrix}$$

32. Let a_n be the nth term of the sequence defined recursively by
$$a_{n+1} = \frac{1}{1 + a_n}$$
and $a_1 = 1$. Find a formula for a_n in terms of the Fibonacci numbers F_n. Prove that the formula you found is valid for all natural numbers n.

33. Let F_n be the nth term of the Fibonacci sequence. Find and prove an inequality relating n and F_n for natural numbers n.

34. Find and prove an inequality relating $100n$ and n^3.

🔲 DISCOVERY · DISCUSSION

35. True or False? Determine whether each statement is true or false. If you think the statement is true, prove it. If you think it is false, give an example where it fails.
(a) $p(n) = n^2 - n + 11$ is prime for all n.
(b) $n^2 > n$ for all $n \geqslant 2$.
(c) $2^{2n+1} + 1$ is divisible by 3 for all $n \geqslant 1$.
(d) $n^3 \geqslant (n + 1)^2$ for all $n \geqslant 2$.
(e) $n^3 - n$ is divisible by 3 for all $n \geqslant 2$.
(f) $n^3 - 6n^2 + 11n$ is divisible by 6 for all $n \geqslant 1$.

36. All Cats Are Black? What is wrong with the following "proof" by mathematical induction that all cats are black? Let $P(n)$ denote the statement: In any group of n cats, if one is black, then they are all black.

Step 1 The statement is clearly true for $n = 1$.

Step 2 Suppose that $P(k)$ is true. We show that $P(k + 1)$ is true. Suppose we have a group of $k + 1$ cats, one of whom is black; call this cat "Midnight." Remove some other cat (call it "Sparky") from the group. We are left with k cats, one of whom (Midnight) is black, so by the induction hypothesis, all k of these are black. Now put Sparky back in the group and take out Midnight. We again have a group of k cats, all of whom—except possibly Sparky—are black. Then by the induction hyothesis, Sparky must be black, too. So all $k + 1$ cats in the original group are black.

Midnight   Sparky

Thus, by induction $P(n)$ is true for all n. Since everyone has seen at least one black cat, it follows that all cats are black.

10.6 THE BINOMIAL THEOREM

An expression of the form $a + b$ is called a **binomial**. Although in principle it's easy to raise $a + b$ to any power, raising it to a very high power would be a tedious task. In this section we find a formula that gives the expansion of $(a + b)^n$ for any natural number n and then prove it using mathematical induction. We discover this formula by finding a pattern for the successive powers of $a + b$. Let's first look at some special cases:

$$(a + b)^1 = a + b$$
$$(a + b)^2 = a^2 + 2ab + b^2$$
$$(a + b)^3 = a^3 + 3a^2b + 3ab^2 + b^3$$
$$(a + b)^4 = a^4 + 4a^3b + 6a^2b^2 + 4ab^3 + b^4$$
$$(a + b)^5 = a^5 + 5a^4b + 10a^3b^2 + 10a^2b^3 + 5ab^4 + b^5$$
$$\vdots$$

The following simple patterns emerge for the expansion of $(a + b)^n$:

1. There are $n + 1$ terms, the first being a^n and the last b^n.

2. The exponents of a decrease by 1 from term to term while the exponents of b increase by 1.

3. The sum of the exponents of a and b in each term is n.

For instance, notice how the exponents of a and b behave in the expansion of $(a + b)^5$.

The exponents of a decrease:

$$(a + b)^5 = a^{\boxed{5}} + 5a^{\boxed{4}}b^1 + 10a^{\boxed{3}}b^2 + 10a^{\boxed{2}}b^3 + 5a^{\boxed{1}}b^4 + b^5$$

The exponents of b increase:

$$(a + b)^5 = a^5 + 5a^4b^{\boxed{1}} + 10a^3b^{\boxed{2}} + 10a^2b^{\boxed{3}} + 5a^1b^{\boxed{4}} + b^{\boxed{5}}$$

With these observations we can write the form of the expansion of $(a + b)^n$ for any natural number n. For example, writing a question mark for the missing coefficients, we have

$$(a + b)^8 = a^8 + ?a^7b + ?a^6b^2 + ?a^5b^3 + ?a^4b^4 + ?a^3b^5 + ?a^2b^6 + ?ab^7 + b^8$$

To complete the expansion, we need to determine these coefficients. To find a pattern, let's write the coefficients in the expansion of $(a + b)^n$ for the first few values of n in a triangular array as shown in the following array, which is called **Pascal's triangle**.

$$
\begin{array}{cccccccccccc}
(a + b)^0 & & & & & & 1 & & & & & \\
(a + b)^1 & & & & & 1 & & 1 & & & & \\
(a + b)^2 & & & & 1 & & 2 & & 1 & & & \\
(a + b)^3 & & & 1 & & 3 & & 3 & & 1 & & \\
(a + b)^4 & & 1 & & 4 & & 6 & & 4 & & 1 & \\
(a + b)^5 & 1 & & 5 & & 10 & & 10 & & 5 & & 1 \\
\end{array}
$$

The row corresponding to $(a + b)^0$ is called the zeroth row and is included to show the symmetry of the array. The key observation about Pascal's triangle is the following property.

> ### KEY PROPERTY OF PASCAL'S TRIANGLE
>
> Every entry (other than a 1) is the sum of the two entries diagonally above it.

From this property it's easy to find any row of Pascal's triangle from the row above it. For instance, we find the sixth and seventh rows, starting with the fifth row:

$$
\begin{array}{ccccccccccccc}
(a + b)^5 & & 1 & & 5 & & 10 & & 10 & & 5 & & 1 \\
(a + b)^6 & 1 & & 6 & & 15 & & 20 & & 15 & & 6 & & 1 \\
(a + b)^7 & 1 & & 7 & & 21 & & 35 & & 35 & & 21 & & 7 & & 1 \\
\end{array}
$$

To see why this property holds, let's consider the following expansions:

$$(a + b)^5 = a^5 + 5a^4b + 10a^3b^2 + \boxed{10a^2b^3} + \boxed{5ab^4} + b^5$$

$$(a + b)^6 = a^6 + 6a^5b + 15a^4b^2 + 20a^3b^3 + \boxed{15a^2b^4} + 6ab^5 + b^6$$

We arrive at the expansion of $(a + b)^6$ by multiplying $(a + b)^5$ by $(a + b)$. Notice, for instance, that the circled term in the expansion of $(a + b)^6$ is obtained via this multiplication from the two circled terms above it. We get this term when the two terms above it are multiplied by b and a, respectively. Thus, its coefficient is the sum of the coefficients of these two terms. We will use this observation at the end of this section when we prove the Binomial Theorem.

Having found these patterns, we can now easily obtain the expansion of any binomial, at least to relatively small powers.

EXAMPLE 1 ■ Expanding a Binomial Using Pascal's Triangle

Find the expansion of $(a + b)^7$ using Pascal's triangle.

SOLUTION

The first term in the expansion is a^7, and the last term is b^7. Using the fact that the exponent of a decreases by 1 from term to term and that of b increases by 1 from term to term, we have

$$(a + b)^7 = a^7 + ?a^6b + ?a^5b^2 + ?a^4b^3 + ?a^3b^4 + ?a^2b^5 + ?ab^6 + b^7$$

The appropriate coefficients appear in the seventh row of Pascal's triangle. Thus

$$(a + b)^7 = a^7 + 7a^6b + 21a^5b^2 + 35a^4b^3 + 35a^3b^4 + 21a^2b^5 + 7ab^6 + b^7 \quad ■$$

EXAMPLE 2 ■ Expanding a Binomial Using Pascal's Triangle

Use Pascal's triangle to expand $(2 - 3x)^5$.

SOLUTION

We find the expansion of $(a + b)^5$ and then substitute 2 for a and $-3x$ for b. Using Pascal's triangle for the coefficients, we get

$$(a + b)^5 = a^5 + 5a^4b + 10a^3b^2 + 10a^2b^3 + 5ab^4 + b^5$$

Substituting $a = 2$ and $b = -3x$ gives

$$(2 - 3x)^5 = (2)^5 + 5(2)^4(-3x) + 10(2)^3(-3x)^2 + 10(2)^2(-3x)^3 + 5(2)(-3x)^4 + (-3x)^5$$

$$= 32 - 240x + 720x^2 - 1080x^3 + 810x^4 - 243x^5 \quad ■$$

■ The Binomial Coefficients and Pascal's Triangle

Although Pascal's triangle is useful in finding the binomial expansion for reasonably small values of n, it isn't practical for finding $(a + b)^n$ for large values of n. The reason is that the method we use for finding the successive rows of Pascal's triangle is recursive. Thus, to find the 100th row of this triangle, we must first find the preceding 99 rows.

We need to examine the pattern in the coefficients more carefully to develop a formula that allows us to calculate directly any coefficient in the binomial expansion. Such a formula exists, and the rest of this section is devoted to finding and proving it. However, to state this formula we need some notation.

The product of the first n natural numbers is denoted by $n!$ and is called **n factorial**:

$4! = 1 \cdot 2 \cdot 3 \cdot 4 = 24$

$7! = 1 \cdot 2 \cdot 3 \cdot 4 \cdot 5 \cdot 6 \cdot 7 = 5040$

$10! = 1 \cdot 2 \cdot 3 \cdot 4 \cdot 5 \cdot 6 \cdot 7 \cdot 8 \cdot 9 \cdot 10$
$\quad = 3{,}628{,}800$

$$n! = 1 \cdot 2 \cdot 3 \cdot \cdots \cdot (n-1) \cdot n$$

We also define 0! as follows:

$$0! = 1$$

This definition of 0! makes many formulas involving factorials shorter and easier to write.

THE BINOMIAL COEFFICIENT

Let n and r be nonnegative integers with $r \leq n$. The **binomial coefficient** is denoted by $\binom{n}{r}$ and is defined by

$$\binom{n}{r} = \frac{n!}{r!(n-r)!}$$

EXAMPLE 3 ■ Calculating Binomial Coefficients

(a) $\displaystyle \binom{9}{4} = \frac{9!}{4!(9-4)!} = \frac{9!}{4!5!} = \frac{1 \cdot 2 \cdot 3 \cdot 4 \cdot 5 \cdot 6 \cdot 7 \cdot 8 \cdot 9}{(1 \cdot 2 \cdot 3 \cdot 4)(1 \cdot 2 \cdot 3 \cdot 4 \cdot 5)}$

$$= \frac{6 \cdot 7 \cdot 8 \cdot 9}{1 \cdot 2 \cdot 3 \cdot 4} = 126$$

(b) $\displaystyle \binom{100}{3} = \frac{100!}{3!(100-3)!} = \frac{1 \cdot 2 \cdot 3 \cdots 97 \cdot 98 \cdot 99 \cdot 100}{(1 \cdot 2 \cdot 3)(1 \cdot 2 \cdot 3 \cdots 97)}$

$$= \frac{98 \cdot 99 \cdot 100}{1 \cdot 2 \cdot 3} = 161{,}700$$

(c) $\displaystyle \binom{100}{97} = \frac{100!}{97!(100-97)!} = \frac{1 \cdot 2 \cdot 3 \cdots 97 \cdot 98 \cdot 99 \cdot 100}{(1 \cdot 2 \cdot 3 \cdots 97)(1 \cdot 2 \cdot 3)}$

$$= \frac{98 \cdot 99 \cdot 100}{1 \cdot 2 \cdot 3} = 161{,}700 \qquad ■$$

Although the binomial coefficient $\binom{n}{r}$ is defined in terms of a fraction, all the results of Example 3 are natural numbers. In fact, $\binom{n}{r}$ is always a natural number (see Exercise 50). Notice that the binomial coefficients in parts (b) and (c) of Example 3 are equal. This is a special case of the following relation, which you are asked to prove in Exercise 48.

$$\binom{n}{r} = \binom{n}{n-r}$$

To see the connection between the binomial coefficients and the binomial expansion of $(a + b)^n$, let's calculate the following binomial coefficients:

$$\binom{5}{2} = \frac{5!}{2!(5-2)!} = 10$$

$$\binom{5}{0} = 1 \qquad \binom{5}{1} = 5 \qquad \binom{5}{2} = 10 \qquad \binom{5}{3} = 10 \qquad \binom{5}{4} = 5 \qquad \binom{5}{5} = 1$$

These are precisely the entries in the fifth row of Pascal's triangle. In fact, we can write Pascal's triangle as follows.

$$\binom{0}{0}$$

$$\binom{1}{0} \qquad \binom{1}{1}$$

$$\binom{2}{0} \qquad \binom{2}{1} \qquad \binom{2}{2}$$

$$\binom{3}{0} \qquad \binom{3}{1} \qquad \binom{3}{2} \qquad \binom{3}{3}$$

$$\binom{4}{0} \qquad \binom{4}{1} \qquad \binom{4}{2} \qquad \binom{4}{3} \qquad \binom{4}{4}$$

$$\binom{5}{0} \qquad \binom{5}{1} \qquad \binom{5}{2} \qquad \binom{5}{3} \qquad \binom{5}{4} \qquad \binom{5}{5}$$

$$\cdots$$

$$\binom{n}{0} \qquad \binom{n}{1} \qquad \binom{n}{2} \qquad \cdots \qquad \binom{n}{n-1} \qquad \binom{n}{n}$$

To demonstrate that this pattern holds, we need to show that any entry in this version of Pascal's triangle is the sum of the two entries diagonally above it. In other words, we must show that each entry satisfies the key property of Pascal's triangle. We now state this property in terms of the binomial coefficients.

KEY PROPERTY OF THE BINOMIAL COEFFICIENTS

For any nonnegative integers r and k with $r \le k$,

$$\binom{k}{r-1} + \binom{k}{r} = \binom{k+1}{r}$$

Notice that the two terms on the left side of this equation are adjacent entries in the kth row of Pascal's triangle and the term on the right side is the entry diagonally below them, in the $(k + 1)$st row. Thus, this equation is a restatement of the key property of Pascal's triangle in terms of the binomial coefficients. A proof of this formula is outlined in Exercise 49.

The Binomial Theorem

We are now ready to state the Binomial Theorem.

THE BINOMIAL THEOREM

$$(a + b)^n = \binom{n}{0}a^n + \binom{n}{1}a^{n-1}b + \binom{n}{2}a^{n-2}b^2 + \cdots + \binom{n}{n-1}ab^{n-1} + \binom{n}{n}b^n$$

We prove this theorem at the end of this section. First, let's look at some of its applications.

EXAMPLE 4 ■ Expanding a Binomial Using the Binomial Theorem

Use the Binomial Theorem to expand $(x + y)^4$.

SOLUTION

By the Binomial Theorem,

$$(x + y)^4 = \binom{4}{0}x^4 + \binom{4}{1}x^3y + \binom{4}{2}x^2y^2 + \binom{4}{3}xy^3 + \binom{4}{4}y^4$$

Verify that

$$\binom{4}{0} = 1 \qquad \binom{4}{1} = 4 \qquad \binom{4}{2} = 6 \qquad \binom{4}{3} = 4 \qquad \binom{4}{4} = 1$$

It follows that

$$(x + y)^4 = x^4 + 4x^3y + 6x^2y^2 + 4xy^3 + y^4$$ ■

EXAMPLE 5 ■ Expanding a Binomial Using the Binomial Theorem

Use the Binomial Theorem to expand $(\sqrt{x} - 1)^8$.

SOLUTION

We first find the expansion of $(a + b)^8$ and then substitute $\sqrt{x}$ for a and -1 for b. Using the Binomial Theorem, we have

$$(a + b)^8 = \binom{8}{0}a^8 + \binom{8}{1}a^7b + \binom{8}{2}a^6b^2 + \binom{8}{3}a^5b^3 + \binom{8}{4}a^4b^4$$

$$+ \binom{8}{5}a^3b^5 + \binom{8}{6}a^2b^6 + \binom{8}{7}ab^7 + \binom{8}{8}b^8$$

Sir Isaac Newton (1642–1727) is universally regarded as one of the giants of physics and mathematics. He is well known for discovering the laws of motion and gravity and for inventing the calculus, but he also proved the Binomial Theorem and the laws of optics, and developed methods for solving polynomial equations to any desired accuracy. He was born on Christmas Day, a few months after the death of his father. After an unhappy childhood, he entered Cambridge University, where he learned mathematics by studying the writings of Euclid and Descartes.

During the plague years of 1665 and 1666, when the university was closed, Newton thought and wrote about ideas that, once published, instantly revolutionized the sciences. Imbued with a pathological fear of criticism, he published these writings only after many years of encouragement from Edmund Halley (who discovered the now-famous comet) and other colleagues. Newton's works brought him enormous fame and prestige. Even poets were moved to praise; *(continued)*

Verify that

$$\binom{8}{0} = 1 \qquad \binom{8}{1} = 8 \qquad \binom{8}{2} = 28 \qquad \binom{8}{3} = 56 \qquad \binom{8}{4} = 70$$

$$\binom{8}{5} = 56 \qquad \binom{8}{6} = 28 \qquad \binom{8}{7} = 8 \qquad \binom{8}{8} = 1$$

So

$$(a + b)^8 = a^8 + 8a^7b + 28a^6b^2 + 56a^5b^3 + 70a^4b^4 + 56a^3b^5$$
$$+ 28a^2b^6 + 8ab^7 + b^8$$

Performing the substitutions $a = x^{1/2}$ and $b = -1$ gives

$$(\sqrt{x} - 1)^8 = (x^{1/2})^8 + 8(x^{1/2})^7(-1) + 28(x^{1/2})^6(-1)^2 + 56(x^{1/2})^5(-1)^3$$
$$+ 70(x^{1/2})^4(-1)^4 + 56(x^{1/2})^3(-1)^5 + 28(x^{1/2})^2(-1)^6$$
$$+ 8(x^{1/2})(-1)^7 + (-1)^8$$

This simplifies to

$$(\sqrt{x} - 1)^8 = x^4 - 8x^{7/2} + 28x^3 - 56x^{5/2} + 70x^2 - 56x^{3/2} + 28x - 8x^{1/2} + 1$$

The Binomial Theorem can be used to find a particular term of a binomial expansion without having to find the entire expansion.

GENERAL TERM OF THE BINOMIAL EXPANSION

The term that contains a^r in the expansion of $(a + b)^n$ is

$$\binom{n}{n - r}a^r b^{n-r}$$

EXAMPLE 6 ■ Finding a Particular Term in a Binomial Expansion

Find the term that contains x^5 in the expansion of $(2x + y)^{20}$.

SOLUTION

The term that contains x^5 is given by the formula for the general term with $a = 2x$, $b = y$, $n = 20$, and $r = 5$. So, this term is

$$\binom{20}{15}a^5b^{15} = \frac{20!}{15!(20 - 15)!}\,(2x)^5y^{15} = \frac{20!}{15!5!}\,32x^5y^{15} = 496{,}128x^5y^{15} \quad ■$$

EXAMPLE 7 ■ Finding a Particular Term in a Binomial Expansion

Find the coefficient of x^8 in the expansion of $\left(x^2 + \dfrac{1}{x}\right)^{10}$.

SOLUTION

Both x^2 and $1/x$ are powers of x, so the power of x in each term of the expansion is determined by both terms of the binomial. To find the required coefficient, we first find the general term in the expansion. By the formula we have $a = x^2$, $b = 1/x$, and $n = 10$, so the general term is

$$\binom{10}{10-r}(x^2)^r\left(\frac{1}{x}\right)^{10-r} = \binom{10}{10-r}x^{2r}(x^{-1})^{10-r} = \binom{10}{10-r}x^{3r-10}$$

Thus, the term that contains x^8 is the term in which

$$3r - 10 = 8$$

$$r = 6$$

So the required coefficient is

$$\binom{10}{10-6} = \binom{10}{4} = 210 \qquad\qquad ■$$

Proof of the Binomial Theorem

We now give a proof of the Binomial Theorem using mathematical induction.

■ **Proof** Let $P(n)$ denote the statement

$$(a + b)^n = \binom{n}{0}a^n + \binom{n}{1}a^{n-1}b + \binom{n}{2}a^{n-2}b^2 + \cdots + \binom{n}{n-1}ab^{n-1} + \binom{n}{n}b^n$$

Step 1 We show that $P(1)$ is true. But $P(1)$ is just the statement

$$(a + b)^1 = \binom{1}{0}a^1 + \binom{1}{1}b^1 = 1a + 1b = a + b$$

which is certainly true.

Step 2 We assume that $P(k)$ is true. Thus, our induction hypothesis is

$$(a + b)^k = \binom{k}{0}a^k + \binom{k}{1}a^{k-1}b + \binom{k}{2}a^{k-2}b^2 + \cdots + \binom{k}{k-1}ab^{k-1} + \binom{k}{k}b^k$$

We use this to show that $P(k + 1)$ is true.

$$(a + b)^{k+1} = (a + b)\left[(a + b)^k\right]$$

$$= (a + b)\left[\binom{k}{0}a^k + \binom{k}{1}a^{k-1}b + \binom{k}{2}a^{k-2}b^2 + \cdots + \binom{k}{k-1}ab^{k-1} + \binom{k}{k}b^k\right] \quad \text{Induction hypothesis}$$

$$= a\left[\binom{k}{0}a^k + \binom{k}{1}a^{k-1}b + \binom{k}{2}a^{k-2}b^2 + \cdots + \binom{k}{k-1}ab^{k-1} + \binom{k}{k}b^k\right]$$

$$+ b\left[\binom{k}{0}a^k + \binom{k}{1}a^{k-1}b + \binom{k}{2}a^{k-2}b^2 + \cdots + \binom{k}{k-1}ab^{k-1} + \binom{k}{k}b^k\right] \quad \text{Distributive property}$$

$$= \binom{k}{0}a^{k+1} + \binom{k}{1}a^k b + \binom{k}{2}a^{k-1}b^2 + \cdots + \binom{k}{k-1}a^2 b^{k-1} + \binom{k}{k}ab^k$$

$$+ \binom{k}{0}a^k b + \binom{k}{1}a^{k-1}b^2 + \binom{k}{2}a^{k-2}b^3 + \cdots + \binom{k}{k-1}ab^k + \binom{k}{k}b^{k+1} \quad \text{Distributive property}$$

$$= \binom{k}{0}a^{k+1} + \left[\binom{k}{0} + \binom{k}{1}\right]a^k b + \left[\binom{k}{1} + \binom{k}{2}\right]a^{k-1}b^2 + \cdots + \left[\binom{k}{k-1} + \binom{k}{k}\right]ab^k + \binom{k}{k}b^{k+1} \quad \text{Group like terms}$$

Using the key property of the binomial coefficients, we can write each of the expressions in square brackets as a single binomial coefficient. Also, writing the first and last coefficients as $\binom{k+1}{0}$ and $\binom{k+1}{k+1}$ (these are equal to 1 by Exercise 46) gives

$$(a + b)^{k+1} = \binom{k+1}{0}a^{k+1} + \binom{k+1}{1}a^k b + \binom{k+1}{2}a^{k-1}b^2 + \cdots + \binom{k+1}{k}ab^k + \binom{k+1}{k+1}b^{k+1}$$

But this last equation is precisely $P(k + 1)$, and this completes the induction step.

Having proved Steps 1 and 2, we conclude by the Principle of Mathematical Induction that the theorem is true for all natural numbers n. ☐

10.6 EXERCISES

1–12 ■ Use Pascal's triangle to expand the expression.

1. $(x + y)^6$

2. $(2x + 1)^4$

3. $\left(x + \dfrac{1}{x}\right)^4$

4. $(x - y)^5$

5. $(x - 1)^5$

6. $(\sqrt{a} + \sqrt{b})^6$

7. $(x^2 y - 1)^5$

8. $(1 + \sqrt{2})^6$

9. $(2x - 3y)^3$

10. $(1 + x^3)^3$

11. $\left(\dfrac{1}{x} - \sqrt{x}\right)^5$

12. $\left(2 + \dfrac{x}{2}\right)^5$

13–20 ■ Evaluate the expression.

13. $\binom{6}{4}$

14. $\binom{8}{3}$

15. $\binom{100}{98}$

16. $\binom{10}{5}$

17. $\binom{3}{1}\binom{4}{2}$

18. $\binom{5}{2}\binom{5}{3}$

19. $\binom{5}{0} + \binom{5}{1} + \binom{5}{2} + \binom{5}{3} + \binom{5}{4} + \binom{5}{5}$

20. $\binom{5}{0} - \binom{5}{1} + \binom{5}{2} - \binom{5}{3} + \binom{5}{4} - \binom{5}{5}$

21–24 ■ Use the Binomial Theorem to expand the expression.

21. $(x + 2y)^4$

22. $(1 - x)^5$

23. $\left(1 + \dfrac{1}{x}\right)^6$

24. $(2A + B^2)^4$

25. Find the first three terms in the expansion of $(x + 2y)^{20}$.

26. Find the first four terms in the expansion of $(x^{1/2} + 1)^{30}$.

27. Find the last two terms in the expansion of $(a^{2/3} + a^{1/3})^{25}$.

28. Find the first three terms in the expansion of

$$\left(x + \dfrac{1}{x}\right)^{40}$$

29. Find the middle term in the expansion of $(x^2 + 1)^{18}$.

30. Find the fifth term in the expansion of $(ab - 1)^{20}$.

31. Find the 24th term in the expansion of $(a + b)^{25}$.

32. Find the 28th term in the expansion of $(A - B)^{30}$.

33. Find the 100th term in the expansion of $(1 + y)^{100}$.

34. Find the second term in the expansion of

$$\left(x^2 - \dfrac{1}{x}\right)^{25}$$

35. Find the term containing x^4 in the expansion of $(x + 2y)^{10}$.

36. Find the term containing y^3 in the expansion of $(\sqrt{2} + y)^{12}$.

37. Find the term containing b^8 in the expansion of $(a + b^2)^{12}$.

38. Find the term that does not contain x in the expansion of

$$\left(8x + \dfrac{1}{2x}\right)^8$$

39–42 ■ Factor using the Binomial Theorem.

39. $x^4 + 4x^3y + 6x^2y^2 + 4xy^3 + y^4$

40. $(x - 1)^5 + 5(x - 1)^4 + 10(x - 1)^3 + 10(x - 1)^2 + 5(x - 1) + 1$

41. $8a^3 + 12a^2b + 6ab^2 + b^3$

42. $x^8 + 4x^6y + 6x^4y^2 + 4x^2y^3 + y^4$

43–44 ■ Simplify using the Binomial Theorem.

43. $\dfrac{(x + h)^3 - x^3}{h}$

44. $\dfrac{(x + h)^4 - x^4}{h}$

45. Show that $(1.01)^{100} > 2$.
[*Hint:* Note that $(1.01)^{100} = (1 + 0.01)^{100}$ and use the Binomial Theorem to show that the sum of the first two terms of the expansion is greater than 2.]

46. Show that $\dbinom{n}{0} = 1$ and $\dbinom{n}{n} = 1$.

47. Show that $\dbinom{n}{1} = \dbinom{n}{n - 1} = n$.

48. Show that $\dbinom{n}{r} = \dbinom{n}{n - r}$ for $0 \leq r \leq n$.

49. In this exercise we prove the identity

$$\binom{n}{r - 1} + \binom{n}{r} = \binom{n + 1}{r}$$

(a) Write the left side of this equation as the sum of two fractions.

(b) Show that a common denominator of the expression you found in part (a) is $r!\,(n - r + 1)!$.

(c) Add the two fractions using the common denominator in part (b), simplify the numerator, and note that the resulting expression is equal to the right side of the equation.

50. Prove that $\dbinom{n}{r}$ is an integer for all n and for $0 \leq r \leq n$. [*Suggestion:* Use induction to show that the statement is true for all n, and use Exercise 49 for the induction step.]

● DISCOVERY · DISCUSSION

51. Powers of Factorials Which is larger, $(100!)^{101}$ or $(101!)^{100}$? [*Hint:* Try factoring the expressions. Do they have any common factors?]

52. Sums of Binomial Coefficients Add each of the first five rows of Pascal's triangle, as indicated. Do you see a pattern?

$$1 + 1 = ?$$
$$1 + 2 + 1 = ?$$
$$1 + 3 + 3 + 1 = ?$$
$$1 + 4 + 6 + 4 + 1 = ?$$
$$1 + 5 + 10 + 10 + 5 + 1 = ?$$

Based on the pattern you have found, find the sum of the nth row:

$$\binom{n}{0} + \binom{n}{1} + \binom{n}{2} + \cdots + \binom{n}{n}$$

Prove your result by expanding $(1 + 1)^n$ using the Binomial Theorem.

53. Alternating Sums of Binomial Coefficients Find the sum

$$\binom{n}{0} - \binom{n}{1} + \binom{n}{2} - \cdots + (-1)^n \binom{n}{n}$$

by finding a pattern as in Exercise 52. Prove your result by expanding $(1 - 1)^n$ using the Binomial Theorem.

10 REVIEW

CONCEPT CHECK

1. (a) What is a sequence?
   (b) What is an arithmetic sequence? Write an expression for the nth term of an arithmetic sequence.
   (c) What is a geometric sequence? Write an expression for the nth term of a geometric sequence.

2. (a) What is a recursively defined sequence?
   (b) What is the Fibonacci sequence?

3. (a) What is meant by the partial sums of a sequence?
   (b) If an arithmetic sequence has first term a and common difference d, write an expression for the sum of its first n terms.
   (c) If a geometric sequence has first term a and common ratio r, write an expression for the sum of its first n terms.
   (d) Write an expression for the sum of an infinite geometric series with first term a and common ratio r. For what values of r is your formula valid?

4. (a) Write the sum $\sum_{k=1}^{n} a_k$ without using Σ-notation.

   (b) Write $b_1 + b_2 + b_3 + \cdots + b_n$ using Σ-notation.

5. Write an expression for the amount A_f of an annuity consisting of n regular equal payments of size R with interest rate i per time period.

6. State the Principle of Mathematical Induction.

7. Write the first five rows of Pascal's Triangle. How are the entries related to each other?

8. (a) What does the symbol $n!$ mean?
   (b) Write an expression for the binomial coefficient $\binom{n}{r}$.
   (c) State the Binomial Theorem.
   (d) Write the term that contains a^r in the expansion of $(a + b)^n$.

EXERCISES

1–6 ■ Find the first four terms as well as the tenth term of the sequence with the given nth term.

1. $a_n = \dfrac{n^2}{n + 1}$

2. $a_n = (-1)^n \dfrac{2^n}{n}$

3. $a_n = \dfrac{(-1)^n + 1}{n^3}$

4. $a_n = \dfrac{n(n + 1)}{2}$

5. $a_n = \dfrac{(2n)!}{2^n n!}$

6. $a_n = \binom{n + 1}{2}$

7–10 ■ A sequence is defined recursively. Find the first seven terms of the sequence.

7. $a_n = a_{n-1} + 2n - 1, \quad a_1 = 1$

8. $a_n = \dfrac{a_{n-1}}{n}, \quad a_1 = 1$

9. $a_n = a_{n-1} + 2a_{n-2}, \quad a_1 = 1, a_2 = 3$

10. $a_n = \sqrt{3a_{n-1}}, \quad a_1 = \sqrt{3}$

11–14 ■ The nth term of a sequence is given.
(a) Find the first five terms of the sequence.
(b) Graph the terms you found in part (a).

(c) Determine if the series is arithmetic or geometric. Find the common difference or the common ratio.

11. $a_n = 2n + 5$

12. $a_n = \dfrac{5}{2^n}$

13. $a_n = \dfrac{3^n}{2^{n+1}}$

14. $a_n = 4 - \dfrac{n}{2}$

15–22 ■ The first four terms of a sequence are given. Determine whether they can be the terms of an arithmetic sequence, a geometric sequence, or neither. If the sequence is arithmetic or geometric, find the fifth term.

15. $5, 5.5, 6, 6.5, \ldots$

16. $1, -\frac{3}{2}, 2, -\frac{5}{2}, \ldots$

17. $\sqrt{2}, 2\sqrt{2}, 3\sqrt{2}, 4\sqrt{2}, \ldots$

18. $\sqrt{2}, 2, 2\sqrt{2}, 4, \ldots$

19. $t - 3, t - 2, t - 1, t, \ldots$

20. $t^3, t^2, t, 1, \ldots$

21. $\dfrac{3}{4}, \dfrac{1}{2}, \dfrac{1}{3}, \dfrac{2}{9}, \ldots$

22. $a, 1, \dfrac{1}{a}, \dfrac{1}{a^2}, \ldots$

23. Show that $3, 6i, -12, -24i, \ldots$ is a geometric sequence, and find the common ratio. (Here $i = \sqrt{-1}$.)

24. Find the nth term of the geometric sequence $2, 2 + 2i, 4i, -4 + 4i, -8, \ldots$ (Here $i = \sqrt{-1}$.)

25. The sixth term of an arithmetic sequence is 17, and the fourth term is 11. Find the second term.

26. The 20th term of an arithmetic sequence is 96, and the common difference is 5. Find the nth term.

27. The third term of a geometric sequence is 9, and the common ratio is $\frac{3}{2}$. Find the fifth term.

28. The second term of a geometric sequence is 10, and the fifth term is $\frac{1250}{27}$. Find the nth term.

29. The frequencies of musical notes (measured in cycles per second) form a geometric sequence. Middle C has a frequency of 256, and the C that is an octave higher has a frequency of 512. Find the frequency of C two octaves below middle C.

30. A person has two parents, four grandparents, eight great-grandparents, and so on. How many ancestors does a person have 15 generations back?

31. A certain type of bacteria divides every 5 s. If three of these bacteria are put into a petri dish, how many bacteria are in the dish at the end of 1 min?

32. If $a_1, a_2, a_3, \ldots$ and $b_1, b_2, b_3, \ldots$ are arithmetic sequences, show that $a_1 + b_1, a_2 + b_2, a_3 + b_3, \ldots$ is also an arithmetic sequence.

33. If $a_1, a_2, a_3, \ldots$ and $b_1, b_2, b_3, \ldots$ are geometric sequences, show that $a_1b_1, a_2b_2, a_3b_3, \ldots$ is also a geometric sequence.

34. (a) If $a_1, a_2, a_3, \ldots$ is an arithmetic sequence, is the sequence $a_1 + 2, a_2 + 2, a_3 + 2, \ldots$ arithmetic?
   (b) If $a_1, a_2, a_3, \ldots$ is a geometric sequence, is the sequence $5a_1, 5a_2, 5a_3, \ldots$ geometric?

35. Find the values of x for which the sequence $6, x, 12, \ldots$ is
   (a) arithmetic   (b) geometric

36. Find the values of x and y for which the sequence $2, x, y, 17, \ldots$ is
   (a) arithmetic   (b) geometric

37–40 ■ Find the sum.

37. $\displaystyle\sum_{k=3}^{6} (k + 1)^2$

38. $\displaystyle\sum_{i=1}^{4} \dfrac{2i}{2i - 1}$

39. $\displaystyle\sum_{k=1}^{6} (k + 1)2^{k-1}$

40. $\displaystyle\sum_{m=1}^{5} 3^{m-2}$

41–44 ■ Write the sum without using sigma notation. Do not evaluate.

41. $\displaystyle\sum_{k=1}^{10} (k - 1)^2$

42. $\displaystyle\sum_{j=2}^{100} \dfrac{1}{j - 1}$

43. $\displaystyle\sum_{k=1}^{50} \dfrac{3^k}{2^{k+1}}$

44. $\displaystyle\sum_{n=1}^{10} n^2 2^n$

45–48 ■ Write the sum using sigma notation. Do not evaluate.

45. $3 + 6 + 9 + 12 + \cdots + 99$

46. $1^2 + 2^2 + 3^2 + \cdots + 100^2$

47. $1 \cdot 2^3 + 2 \cdot 2^4 + 3 \cdot 2^5 + 4 \cdot 2^6 + \cdots + 100 \cdot 2^{102}$

48. $\dfrac{1}{1 \cdot 2} + \dfrac{1}{2 \cdot 3} + \dfrac{1}{3 \cdot 4} + \cdots + \dfrac{1}{999 \cdot 1000}$

49–54 ■ Determine whether the expression is a partial sum of an arithmetic or geometric sequence. Then find the sum.

49. $1 + 0.9 + (0.9)^2 + \cdots + (0.9)^5$

50. $3 + 3.7 + 4.4 + \cdots + 10$

51. $\sqrt{5} + 2\sqrt{5} + 3\sqrt{5} + \cdots + 100\sqrt{5}$

52. $\frac{1}{3} + \frac{2}{3} + 1 + \frac{4}{3} + \cdots + 33$

53. $\displaystyle\sum_{n=0}^{6} 3(-4)^n$ **54.** $\displaystyle\sum_{k=0}^{8} 7(5)^{k/2}$

55. The first term of an arithmetic sequence is $a = 7$, and the common difference is $d = 3$. How many terms of this sequence must be added to obtain 325?

56. The sum of the first three terms of a geometric series is 52, and the common ratio is $r = 3$. Find the first term.

57. Refer to Exercise 30. What is the total number of a person's ancestors in 15 generations?

58. Find the amount of an annuity consisting of 16 annual payments of $1000 each into an account that pays 8% interest per year, compounded annually.

59. How much money should be invested every quarter at 12% per year, compounded quarterly, in order to have $10,000 in one year?

60. What are the monthly payments on a mortgage of $60,000 at 9% interest if the loan is to be repaid in
(a) 30 years?   (b) 15 years?

61–64 ■ Find the sum of the infinite geometric series.

61. $1 - \frac{2}{5} + \frac{4}{25} - \frac{8}{125} + \cdots$

62. $0.1 + 0.01 + 0.001 + 0.0001 + \cdots$

63. $1 + \dfrac{1}{3^{1/2}} + \dfrac{1}{3} + \dfrac{1}{3^{3/2}} + \cdots$

64. $a + ab^2 + ab^4 + ab^6 + \cdots$

65–67 ■ Use mathematical induction to prove that the formula is true for all natural numbers n.

65. $1 + 4 + 7 + \cdots + (3n - 2) = \dfrac{n(3n - 1)}{2}$

66. $\dfrac{1}{1\cdot 3} + \dfrac{1}{3\cdot 5} + \dfrac{1}{5\cdot 7} + \cdots + \dfrac{1}{(2n-1)(2n+1)}$

$= \dfrac{n}{2n + 1}$

67. $\left(1 + \dfrac{1}{1}\right)\left(1 + \dfrac{1}{2}\right)\left(1 + \dfrac{1}{3}\right)\cdots\left(1 + \dfrac{1}{n}\right) = n + 1$

68. Show that $7^n - 1$ is divisible by 6 for all natural numbers n.

69. Let $a_{n+1} = 3a_n + 4$ and $a_1 = 4$. Show that $a_n = 2 \cdot 3^n - 2$ for all natural numbers n.

70. Prove that the Fibonacci number F_{4n} is divisible by 3 for all natural numbers n.

71. Find and prove an inequality that relates 2^n and $n!$.

72–75 ■ Evaluate the expression.

72. $\dbinom{5}{2}\dbinom{5}{3}$

73. $\dbinom{10}{2} + \dbinom{10}{6}$

74. $\displaystyle\sum_{k=0}^{5} \dbinom{5}{k}$

75. $\displaystyle\sum_{k=0}^{8} \dbinom{8}{k}\dbinom{8}{8-k}$

76–77 ■ Expand the expression.

76. $(1 - x^2)^6$ **77.** $(2x + y)^4$

78. Find the 20th term in the expansion of $(a + b)^{22}$.

79. Find the first three terms in the expansion of $(b^{-2/3} + b^{1/3})^{20}$.

80. Find the term containing A^6 in the expansion of $(A + 3B)^{10}$.

10 TEST

1. Find the first four terms and the tenth term of the sequence whose nth term is

$$a_n = n^2 - 1$$

2. A sequence is defined recursively by

$$a_{n+2} = (a_n)^2 - a_{n+1}$$

   If $a_1 = 1$ and $a_2 = 1$, find a_5.

3. Find the common difference, the fourth term, and the nth term in the arithmetic sequence $80, 76, 72, \ldots$.

4. The first term of a geometric sequence is 25, and the fourth term is $\frac{1}{5}$. Find the common ratio and the fifth term.

5. Determine whether each statement is true or false. If it is true, prove it. If it is false, give an example where it fails.
   (a) If $a_1, a_2, a_3, \ldots$ is an arithmetic sequence, then the sequence $a_1^2, a_2^2, a_3^2, \ldots$ is also arithmetic.
   (b) If $a_1, a_2, a_3, \ldots$ is a geometric sequence, then the sequence $a_1^2, a_2^2, a_3^2, \ldots$ is also geometric.

6. (a) Write the formula for the nth partial sum of an arithmetic sequence.
   (b) The first term of an arithmetic sequence is 10, and the tenth term is 2. Find the partial sum of the first ten terms.
   (c) Find the common difference and the 100th term of the sequence in part (b).

7. (a) Write the formula for the nth partial sum of a geometric sequence.
   (b) Find the sum

$$\frac{1}{3} + \frac{2}{3^2} + \frac{2^2}{3^3} + \frac{2^3}{3^4} + \cdots + \frac{2^9}{3^{10}}$$

8. Find the sum of the infinite geometric series $1 + \dfrac{1}{2^{1/2}} + \dfrac{1}{2} + \dfrac{1}{2^{3/2}} + \cdots$.

9. Use mathematical induction to prove that, for all natural numbers n,

$$1^2 + 2^2 + 3^2 + \cdots + n^2 = \frac{n(n+1)(2n+1)}{6}$$

10. Write the expression without using sigma notation, and then find the sum.

   (a) $\displaystyle\sum_{n=1}^{5} (1 - n^2)$ (b) $\displaystyle\sum_{n=3}^{6} (-1)^n 2^{n-2}$

11. Expand $(2x + y^2)^5$.

12. Find the term containing x^3 in the binomial expansion of $(3x - 2)^{10}$.

Focus on Problem Solving
Finding Patterns

The solutions to many problems in mathematics involve **finding patterns**. The algebraic formulas we've found in this book are compact ways of describing patterns. For example, the familiar equation $(a + b)^2 = a^2 + 2ab + b^2$ gives the pattern for squaring the sum of two numbers. Another pattern we have encountered is the pattern for the sum of the first n natural numbers:

$$1 + 2 + 3 + \cdots + n = \frac{n(n + 1)}{2}$$

How do we discover patterns? In many cases, a good way to start is to experiment with the problem. To prove that a pattern always holds, we can often use mathematical induction, but other proofs are possible. A geometrical method is used in the problem we give here. It is attributed to the 11th-century mathematician Abu Bekr Mohammed ibn Al Husain Al Karchi.

The Gnomons of Al Karchi

We prove the beautiful formula

$$1^3 + 2^3 + 3^3 + \cdots + n^3 = (1 + 2 + 3 + \cdots + n)^2$$

But first, let's see how this formula was discovered. The sum of the first n natural numbers is called a *triangular number*. The name "triangular" comes from the following figures:

| 1 | $1 + 2 = 3$ | $1 + 2 + 3 = 6$ | $1 + 2 + 3 + 4 = 10$ | $1 + 2 + 3 + 4 + 5 = 15$ |

The first few triangular numbers are

$$1, 3, 6, 10, 15, \ldots$$

Now, let's look at the sums of the cubes:

$$1^3 = 1$$

$$1^3 + 2^3 = 9$$

$$1^3 + 2^3 + 3^3 = 36$$

$$1^3 + 2^3 + 3^3 + 4^3 = 100$$

$$1^3 + 2^3 + 3^3 + 4^3 + 5^3 = 225$$

We get the sequence

$$1, 9, 36, 100, 225, \ldots$$

It doesn't take long to notice that these are the squares of the triangular numbers. It appears that the sum of the first n cubes equals the square of the sum of the first n numbers.

To show that this pattern always holds, Al Karchi sketches the following diagram:

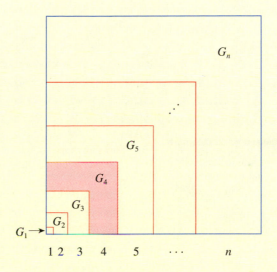

Each region G_n in the shape of an inverted L is called a *gnomon*. The area of gnomon G_4 is

$$\text{area} = 4^2 + 2[4 \times (1 + 2 + 3)] = 64$$

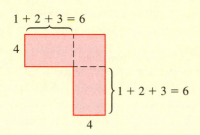

Similarly, the areas of gnomons $G_1, G_2, G_3, G_4, G_5, \ldots$ are 1, 8, 27, 64, 125, ... These are the cubes of the natural numbers. The pattern persists here also, because the area of the nth gnomon G_n is n^3, as the following calculation shows:

$$n^2 + 2(n \times [1 + 2 + \cdots + (n - 1)]) = n^2 + 2n \, \frac{(n - 1)n}{2}$$

$$= n^2 + n^3 - n^2 = n^3$$

Now comes Al Karchi's punchline: The first n gnomons form a square of side $1 + 2 + 3 + \cdots + n$, and so the sum of the areas of these gnomons equals the area of the square, that is, $1^3 + 2^3 + \cdots + n^3 = (1 + 2 + \cdots + n)^2$.

Problems

1. Use the diagram in the margin to find and prove a formula for the sum of the first n odd numbers.

2. Use the figure to find a simple formula for $F_1^2 + F_2^2 + \cdots + F_n^2$, where F_k is the kth Fibonacci number.

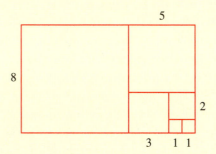

3. Show that every perfect cube is a difference of two squares.   [*Hint:* Use the formula for the sum of cubes that we have just proved.]

4. (a) Find the product $\left(1 - \dfrac{1}{2}\right)\left(1 - \dfrac{1}{3}\right)\left(1 - \dfrac{1}{4}\right) \cdots \left(1 - \dfrac{1}{200}\right)$.

   (b) Find the product $\left(1 - \dfrac{1}{4}\right)\left(1 - \dfrac{1}{9}\right)\left(1 - \dfrac{1}{16}\right) \cdots \left(1 - \dfrac{1}{n^2}\right)$.

5. An $n \times n$ *magic square* consists of the numbers from 1 to n^2 arranged in a square in such a way that the sum of the numbers in any row, column, or diagonal is the same. Let's call this constant sum S. Find a formula for S for an $n \times n$ magic square.

6. Choose n different points on the circumference of a circle, and then connect them with line segments. We are interested in the number of regions into which these segments divide the circle. Let's see what happens for $n = 1, 2, 3,$ and 4.
   (a) How many regions result from five points on the circle?
   (b) Based on the pattern you have observed, how many regions do you think would arise from n different points on the circle?
   (c) Draw a diagram of the regions that result from six points on the circle. (Do this very carefully.) Does this result fit with the pattern you conjectured in part (b)?
   (d) Take the sequence of numbers of regions for $n = 1, 2, 3, 4, 5, 6$, and construct the

1   1   1 $\cdots$ 1

The earliest known **magic square** is a 3×3 magic square from ancient China. According to Chinese legend, it appeared on the back of a turtle emerging from the river Lo.

4	9	2
3	5	7
8	1	6

Here is an example of a 4×4 magic square:

16	3	2	13
5	10	11	8
9	6	7	12
4	15	14	1

difference table (see page 833). Use the difference table to determine the number of regions for $n = 7, 8, 9$, and 10.

| 1 point | 2 points | 3 points | 4 points |
| 1 region | 2 regions | 4 regions | 8 regions |

7. Starting with an equilateral triangle of side 1, we successively construct new figures with more and more sides as shown. The *snowflake curve* is the result of repeating this process indefinitely.

    (a) Find the area enclosed by the snowflake curve.   [*Hint:* This area is the sum of an infinite geometric series.]

    (b) Show that the length of the snowflake curve is infinite.

Parts (a) and (b) show that if your lawn has the shape of a snowflake curve, you could easily mow it, because it has finite area, but you could never put a fence along its boundary.

8. Prove that the number of people who have shaken hands an odd number of times is an even number.

9. Consider a 6×6 grid as shown in the figure in the margin.

    (a) Find the number of squares of all sizes in this grid. Generalize your result to an $n \times n$ grid.

    (b) Find the number of rectangles of all sizes in this grid. Generalize your result to an $n \times n$ grid.

10. Let p be a prime number.

    (a) Show that $\binom{p}{k}$ is divisible by p for $1 \leq k \leq p$.

    (b) Use mathematical induction to prove that $n^p - n$ is divisible by p for all n.
    [*Hint:* For the induction step, use the Binomial Theorem to expand $(n + 1)^p$, and then use part (a).] This fact is known as Fermat's Little Theorem. It is used in constructing public-key codes (see page 544).

11 Counting and Probability

Probability is the mathematical study of chance and random processes. The laws of probability are essential for understanding genetics, opinion polls, pricing stock options, setting odds in horseracing and games of chance, and many other fields.

When it is not in our power to determine what is true, we ought to follow what is most probable.

RENÉ DESCARTES

Many questions in mathematics involve *counting*. For example, in how many ways can a committee of two men and three women be chosen from a group of 35 men and 40 women? How many different license plates can be made using three letters followed by three digits? How many different poker hands are possible?

Closely related to the problem of counting is that of *probability*. We consider questions such as these: If a committee of five people is chosen randomly from a group of 35 men and 40 women, what are the chances that no women will be chosen for the committee? What is the likelihood of getting a straight flush in a poker game? In studying probability, we give precise mathematical meaning to phrases such as "what are the chances . . . ?" and "what is the likelihood . . . ?"

11.1 COUNTING PRINCIPLES

Suppose that three towns, Ashbury, Brampton, and Carmichael, are located in such a way that two roads connect Ashbury to Brampton and three roads connect Brampton to Carmichael. How many different routes can one take to travel from Ashbury to Carmichael via Brampton? The key idea in answering this question is to consider the problem in stages. At the first stage—from Ashbury to Brampton—there are two choices. For each of these choices, there are three choices to make at the second stage—from Brampton to Carmichael. Thus, the number of different routes is $2 \times 3 = 6$. These routes are conveniently enumerated by a *tree diagram* as in Figure 1.

The method used to solve this problem leads to the following principle.

FUNDAMENTAL COUNTING PRINCIPLE

Suppose that two events occur in order. If the first can occur in m ways and the second in n ways (after the first has occurred), then the two events can occur in order in $m \times n$ ways.

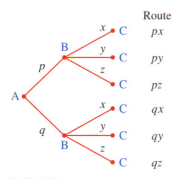

FIGURE 1
Tree diagram

There is an immediate consequence of this principle for any number of events: If $E_1, E_2, \ldots, E_k$ are events that occur in order and if E_1 can occur in n_1 ways, E_2 in n_2 ways, and so on, then the events can occur in order in $n_1 \times n_2 \times \cdots \times n_k$ ways.

EXAMPLE 1 ■ Using the Fundamental Counting Principle

An ice-cream store offers three types of cones and 31 flavors. How many different single-scoop ice-cream cones is it possible to buy at this store?

Persi Diaconis (b. 1945) is currently professor of statistics and mathematics at Stanford University in California. He was born in New York City into a musical family and studied violin until the age of 14. At that time he left home to become a magician. He was a magician (apprentice and master) for ten years. Magic is still his passion, and if there were a professorship for magic, he would certainly qualify for such a post! His interest in card tricks led him to a study of probability and statistics. He is now one of the leading statisticians in the world. With his background he approaches mathematics with an undeniable flair. He says "Statistics is the physics of numbers. Numbers seem to arise in the world in an orderly fashion. When we examine the world, the same regularities seem to appear again and again." Among his many original contributions to mathematics is a probabilistic study of the perfect card shuffle.

SOLUTION

There are two choices: type of cone and flavor of ice cream. At the first stage we choose a type of cone, and at the second stage we choose a flavor. We can think of the different stages as boxes:

stage 1   stage 2
type of cone   flavor

The first box can be filled in three ways and the second in 31 ways:

stage 1   stage 2

Thus, by the Fundamental Counting Principle, there are $3 \times 31 = 93$ ways of choosing a single-scoop ice-cream cone at this store.   ■

EXAMPLE 2 ■ Using the Fundamental Counting Principle

In a certain state, automobile license plates display three letters followed by three digits. How many such plates are possible if repetition of the letters

(a) is allowed?   (b) is not allowed?

SOLUTION

(a) There are six choices, one for each letter or digit on the license plate. As in the preceding example, we sketch a box for each stage:

26	26	26	10	10	10

letters   digits

At the first stage, we choose a letter (from 26 possible choices); at the second stage, another letter (again from 26 choices); at the third stage, another letter (26 choices); at the fourth stage, a digit (from 10 possible choices); at the fifth stage, a digit (again from 10 choices); and at the sixth stage, another digit (10 choices). By the Fundamental Counting Principle, the number of possible license plates is

$$26 \times 26 \times 26 \times 10 \times 10 \times 10 = 17,576,000$$

(b) If repetition of letters is not allowed, then we can arrange the choices as follows:

26	25	24	10	10	10

letters   digits

At the first stage, we have 26 letters to choose from, but once the first letter is chosen, there are only 25 letters to choose from at the second stage. Once the first two letters are chosen, 24 letters are left to choose from for the third stage. The digits are chosen as before. Thus, the number of possible license plates in this case is

$$26 \times 25 \times 24 \times 10 \times 10 \times 10 = 15{,}600{,}000$$

EXAMPLE 3 ■ Using Factorial Notation

In how many different ways can a race with six runners be completed? Assume there is no tie.

SOLUTION

There are six possible choices for first place, five choices for second place (since only five runners are left after first place has been decided), four choices for third place, and so on. So, by the Fundamental Counting Principle, the number of different ways this race can be completed is

Factorial notation is explained on page 851.

$$6 \times 5 \times 4 \times 3 \times 2 \times 1 = 6! = 720$$

11.1 EXERCISES

1. A vendor sells ice cream from a cart on the boardwalk. He offers vanilla, chocolate, strawberry, and pistachio ice cream, served on either a waffle, sugar, or plain cone. How many different single-scoop ice-cream cones can you buy from this vendor?

2. How many three-letter "words" (strings of letters) can be formed using the 26 letters of the alphabet if repetition of letters
  (a) is allowed?   (b) is not allowed?

3. How many three-letter "words" (strings of letters) can be formed using the letters *WXYZ* if repetition of letters
  (a) is allowed?   (b) is not allowed?

4. Eight horses are entered in a race.
  (a) How many different orders are possible for completing the race?
  (b) In how many different ways can first, second, and third places be decided? (Assume there is no tie.)

5. A multiple-choice test has five questions with four choices for each question. In how many different ways can the test be completed?

6. Telephone numbers consist of seven digits; the first digit cannot be 0 or 1. How many telephone numbers are possible?

7. In how many different ways can a race with five runners be completed? (Assume there is no tie.)

8. In how many ways can five people be seated in a row of five seats?

9. A restaurant offers six different main courses, eight types of drinks, and three kinds of desserts. How many different meals consisting of a main course, a drink, and a dessert does the restaurant offer?

10. In how many ways can five different mathematics books be placed next to each other on a shelf?

11. Towns A, B, C, and D are located in such a way that there are four roads from A to B, five roads from B to C, and six roads from C to D. How many routes are there from town A to town D via towns B and C?

12. In a family of four children, how many different boy-girl birth-order combinations are possible? (The birth orders *BBBG* and *BBGB* are different.)

13. A coin is flipped five times, and the resulting sequence of heads and tails is recorded. How many such sequences are possible?

14. A red die and a white die are rolled, and the numbers showing are recorded. How many different outcomes are possible? (The singular form of the word *dice* is *die*.)

15. A red die, a blue die, and a white die are rolled, and the numbers showing are recorded. How many different outcomes are possible?

16. Two cards are chosen in order from a deck. In how many ways can this be done if
  (a) the first card must be a spade and the second must be a heart?
  (b) both cards must be spades?

17. A girl has 5 skirts, 8 blouses, and 12 pairs of shoes. How many different skirt-blouse-shoe outfits can she wear? (Assume that each item matches all the others, so she is willing to wear any combination.)

18. A company's employee ID number system consists of one letter followed by three digits. How many different ID numbers are possible with this system?

19. A company has 2844 employees. Each employee is to be given an ID number that consists of one letter followed by two digits. Is it possible to give each employee a different ID number using this scheme? Explain.

20. An all-star baseball team has a roster of seven pitchers and three catchers. How many pitcher-catcher pairs can the manager select from this roster?

21. Standard automobile license plates in California display a nonzero digit, followed by three letters, followed by three digits. How many different standard plates are possible in this system?

22. A combination lock has 60 different positions. To open the lock, the dial is turned to a certain number in the clockwise direction, then to a number in the counterclockwise direction, and finally to a third number in the clockwise direction. If successive numbers in the combination cannot be the same, how many different combinations are possible?

23. A true-false test contains ten questions. In how many different ways can this test be completed?

24. An automobile dealer offers five models. Each model comes in a choice of four colors, three types of stereo equipment, with or without air conditioning, and with or without a sunroof. In how many different ways can a customer order an auto from this dealer?

25. The registrar at a certain university classifies students according to a major, minor, year (1, 2, 3, 4), and sex (M, F). Each student must choose one major and either one or no minor from the 32 fields taught at this university. How many different student classifications are possible?

26. How many monograms consisting of three initials are possible?

27. A state has registered 8 million automobiles. To simplify the license plate system, a state employee suggests that each plate display only two letters followed by three digits. Will this system create enough different license plates for all the vehicles registered?

28. A state license plate design has six places. Each plate begins with a fixed number of letters, and the remaining places are filled with digits. (For example, one letter followed by five digits, two letters followed by four digits, and so on.) The state has 17 million registered vehicles.
  (a) The state decides to change to a system consisting of one letter followed by five digits. Will this design allow for enough different plates to accommodate all the vehicles registered?

(b) Find a system that will be sufficient if the smallest possible number of letters is to be used.

29. In how many ways can a president, vice president, and secretary be chosen from a class of 30 students?

30. In how many ways can a president, vice president, and secretary be chosen from a class of 20 females and 30 males if the president must be a female and the vice president a male?

31. A senate subcommittee consists of ten Democrats and seven Republicans. In how many ways can a chairman, vice chairman, and secretary be chosen if the chairman must be a Democrat and the vice chairman must be a Republican?

32. Social Security numbers consist of nine digits, with the first digit between 0 and 6, inclusive. How many Social Security numbers are possible?

33. Five-letter "words" are formed using the letters *A, B, C, D, E, F, G*. How many such words are possible for each of the following conditions?
(a) No condition is imposed.
(b) No letter can be repeated in a word.
(c) Each word must begin with the letter *A*.
(d) The letter *C* must be in the middle.
(e) The middle letter must be a vowel.

34. How many five-letter palindromes are possible? (A *palindrome* is a string of letters that reads the same backward and forward, such as the string *XCZCX*.)

35. A certain computer programming language allows names of variables to consist of two characters, the first being any letter and the second any letter or digit. How many names of variables are possible?

36. How many different three-character code words consisting of letters or digits are possible for the following code designs?
(a) The first entry must be a letter.
(b) The first entry cannot be zero.

37. In how many ways can four men and four women be seated in a row of eight seats for the following situations?
(a) The women are to be seated together, and the men are to be seated together.
(b) They are to be seated alternately by gender.

38. In how many ways can five different mathematics books be placed on a shelf if the two algebra books are to be placed next to each other?

39. Eight mathematics books and three chemistry books are to be placed on a shelf. In how many ways can this be done if the mathematics books are next to each other and the chemistry books are next to each other?

40. Three-digit numbers are formed using the digits 2, 4, 5, and 7, with repetition of digits allowed. How many such numbers can be formed if
(a) the numbers are less than 700?
(b) the numbers are even?
(c) the numbers are divisible by 5?

41. How many three-digit odd numbers can be formed using the digits 1, 2, 4, and 6 if repetition of digits is not allowed?

● DISCOVERY · DISCUSSION

42. Pairs of Initials   Explain why in any group of 677 people, at least two people must have the same pair of initials.

43. Area Codes   Until recently, telephone area codes in the United States, Canada, and the Caribbean islands were chosen according to the following rules: (i) The first digit *cannot* be 0 or a 1, and (ii) the second digit *must* be a 0 or a 1. But in 1995, the second rule was abandoned when the area code 360 was introduced in parts of western Washington State. Since then, many other new area codes that violate Rule (ii) have come into use, although Rule (i) still remains in effect.
(a) How many area code + telephone number combinations were possible under the old rules? (See Exercise 6 for a description of local telephone numbers.)
(b) How many area code + telephone number combinations are now possible under the new rules?
(c) Why do you think it was necessary to make this change?
(d) How many area codes that violate Rule (ii) are you personally familiar with?

11.2 PERMUTATIONS AND COMBINATIONS

In this section we single out two important special cases of the Fundamental Counting Principle—permutations and combinations.

■ Permutations

A **permutation** of a set of distinct objects is an ordering of these objects. For example, some permutations of the letters *ABCDWXYZ* are

XAYBZWCD	*ZAYBCDWX*	*DBWAZXYC*	*YDXAWCZB*

Permutations of
three colored squares

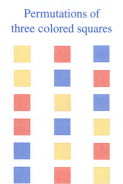

How many such permutations are possible? Since there are eight choices for the first position, seven for the second (after the first has been chosen), six for the third (after the first two have been chosen), and so on, the Fundamental Counting Principle tells us that the number of possible permutations is

$$8 \times 7 \times 6 \times 5 \times 4 \times 3 \times 2 \times 1 = 40{,}320$$

This same reasoning with 8 replaced by n leads to the following observation.

> The number of permutations of n objects is $n!$.

How many permutations consisting of five letters can be made from these same eight letters? Some of these permutations are

XYZWC	*AZDWX*	*AZXYB*	*WDXZB*

Again, there are eight choices for the first position, seven for the second, six for the third, five for the fourth, and four for the fifth. By the Fundamental Counting Principle, the number of such permutations is

$$8 \times 7 \times 6 \times 5 \times 4 = 6720$$

In general, if a set has n elements, then the number of ways of ordering r elements from the set is denoted by $P(n, r)$ and is called **the number of permutations of n objects taken r at a time**.

We have just shown that $P(8, 5) = 6720$. The same reasoning used to find $P(8, 5)$ will help us find a general formula for $P(n, r)$. Indeed, there are n objects and r positions to place them in. Thus, there are n choices for the first position, $n - 1$ choices for the second, $n - 2$ choices for the third, and so on. The last position can be filled in $n - r + 1$ ways. By the Fundamental Counting Principle,

$$P(n, r) = n(n - 1)(n - 2) \cdots (n - r + 1)$$

This formula can be written more compactly using factorial notation:

$$P(n, r) = n(n - 1)(n - 2) \cdots (n - r + 1)$$

$$= \frac{n(n - 1)(n - 2) \cdots (n - r + 1)(n - r) \cdots 3 \cdot 2 \cdot 1}{(n - r) \cdots 3 \cdot 2 \cdot 1} = \frac{n!}{(n - r)!}$$

PERMUTATIONS OF *n* OBJECTS TAKEN *r* AT A TIME

The number of permutations of *n* objects taken *r* at a time is

$$P(n, r) = \frac{n!}{(n - r)!}$$

EXAMPLE 1 ■ Finding the Number of Permutations

A club has nine members. In how many ways can a president, vice president, and secretary be chosen from the members of this club?

SOLUTION

We need the number of ways of selecting three members, *in order*, for the positions of president, vice president, and secretary from the nine club members. This number is

$$P(9, 3) = \frac{9!}{(9 - 3)!} = \frac{9!}{6!} = 9 \times 8 \times 7 = 504$$

■

EXAMPLE 2 ■ Finding the Number of Permutations

From 20 raffle tickets in a hat, four tickets are to be selected in order. The holder of the first ticket wins a car, the second a motorcycle, the third a bicycle, and the fourth a skateboard. In how many different ways can these prizes be awarded?

SOLUTION

The order in which the tickets are chosen determines who wins each prize. So, we need to find the number of ways of selecting four objects, *in order*, from 20 objects (the tickets). This number is

$$P(20, 4) = \frac{20!}{(20 - 4)!} = \frac{20!}{16!} = 20 \times 19 \times 18 \times 17 = 116{,}280$$

■

Distinguishable Permutations

If we have a collection of ten balls, each a different color, then the number of permutations of these balls is $P(10, 10) = 10!$. If all ten balls are red, then we have just one distinguishable permutation because all the ways of ordering these balls look exactly the same. In general, when considering a set of objects, some of which are of the same kind, then two permutations are **distinguishable** if one cannot be obtained from the other by interchanging the positions of elements of the same kind. For example, if we have ten balls, of which six are red and the other four are each a different color, then how many distinguishable permutations are possible? The key point here is that balls of the same color are not distinguishable. So each rearrangement of the red balls, keeping all the other balls fixed, gives essentially the same permutation. Since there are 6! rearrangements of the red balls for each fixed position of the other balls, the total number of distinguishable permutations is 10!/6!. The same type of reasoning gives the following general rule.

DISTINGUISHABLE PERMUTATIONS

If a set of n objects consists of k different kinds of objects with n_1 objects of the first kind, n_2 objects of the second kind, n_3 objects of the third kind, and so on, where $n_1 + n_2 + \cdots + n_k = n$, then the number of distinguishable permutations of these objects is

$$\frac{n!}{n_1!\, n_2!\, n_3! \cdots n_k!}$$

EXAMPLE 3 ■ **Finding the Number of Distinguishable Permutations**

Find the number of different ways of placing 15 balls in a row given that 4 are red, 3 are yellow, 6 are black, and 2 are blue.

SOLUTION

We want to find the number of distinguishable permutations of these balls. By the formula, this number is

$$\frac{15!}{4!\, 3!\, 6!\, 2!} = 6{,}306{,}300$$

    Suppose we have 15 wooden balls in a row and four colors of paint: red, yellow, black, and blue. In how many different ways can the 15 balls be painted in such a way that we have 4 red, 3 yellow, 6 black, and 2 blue balls? A little thought will show that this number is exactly the same as that calculated in Example 3. This way of looking at the problem is somewhat different, however. Here we think of the number of ways to **partition** the balls into four groups, each containing 4, 3, 6, and 2 balls to be painted red, yellow, black, and blue, respectively. The next example shows how this reasoning is used.

EXAMPLE 4 ■ **Finding the Number of Partitions**

Fourteen construction workers are to be assigned to three different tasks. Seven workers are needed for mixing cement, five for laying bricks, and two for carrying the bricks to the brick layers. In how many different ways can the workers be assigned to these tasks?

SOLUTION

We need to partition the workers into three groups containing 7, 5, and 2 workers, respectively. This number is

$$\frac{14!}{7!\,5!\,2!} = 72{,}072$$

Combinations

When finding permutations, we are interested in the number of ways of ordering elements of a set. In many counting problems, however, order is *not* important. For example, a poker hand is the same hand, regardless of how it is ordered. A poker player interested in the number of possible hands wants to know the number of ways of drawing five cards from 52 cards, without regard to the order in which the cards of a given hand are dealt. In this section we develop a formula for counting in situations such as this, where order doesn't matter.

A **combination** of r elements of a set is any subset of r elements from the set (without regard to order). If the set has n elements, then the number of combinations of r elements is denoted by $C(n, r)$ and is called the **number of combinations of n elements taken r at a time**.

For example, consider a set with the four elements, A, B, C, and D. The combinations of these four elements taken three at a time are

ABC	*ABD*	*ACD*	*BCD*

The permutations of these elements taken three at a time are

ABC	*ABD*	*ACD*	*BCD*
ACB	*ADB*	*ADC*	*BDC*
BAC	*BAD*	*CAD*	*CBD*
BCA	*BDA*	*CDA*	*CDB*
CAB	*DAB*	*DAC*	*DBC*
CBA	*DBA*	*DCA*	*DCB*

We notice that the number of combinations is a lot fewer than the number of permutations. In fact, each combination of three elements generates 3! permutations. So

$$C(4, 3) = \frac{P(4, 3)}{3!} = \frac{4!}{3!\,(4 - 3)!} = 4$$

In general, each combination of r objects gives rise to $r!$ permutations of these objects.

Thus

$$C(n, r) = \frac{P(n, r)}{r!} = \frac{n!}{r!\,(n - r)!}$$

In Section 10.6 we denoted $C(n, r)$ by $\binom{n}{r}$, but it is customary to use the notation $C(n, r)$ in the context of counting. For an explanation of why these are the same, see Exercise 76.

> ### COMBINATIONS OF n OBJECTS TAKEN r AT A TIME
>
> The number of combinations of n objects taken r at a time is
>
> $$C(n, r) = \frac{n!}{r!\,(n - r)!}$$

The key difference between permutations and combinations is *order*. If we are interested in ordered arrangements, then we are counting permutations; but if we are concerned with subsets without regard to order, then we are counting combinations. Compare Examples 5 and 6 below (where order doesn't matter) with Examples 1 and 2 (where order does matter).

EXAMPLE 5 ■ Finding the Number of Combinations

A club has nine members. In how many ways can a committee of three be chosen from the members of this club?

SOLUTION

We need the number of ways of choosing three of the nine members. Order is not important here, because the committee is the same no matter how its members are ordered. So, we want the number of combinations of nine objects (the club members) taken three at a time. This number is

$$C(9, 3) = \frac{9!}{3!(9 - 3)!} = \frac{9!}{3!\,6!} = \frac{9 \times 8 \times 7}{3 \times 2 \times 1} = 84$$

EXAMPLE 6 ■ Finding the Number of Combinations

From 20 raffle tickets in a hat, four tickets are to be chosen at random. The holders of the winning tickets are to be awarded free trips to the Bahamas. In how many ways can the four winners be chosen?

SOLUTION

We need to find the number of ways of choosing four winners from 20 entries. The order in which the tickets are chosen doesn't matter, because the same prize is awarded to each of the four winners. So, we want the number of combinations of 20 objects (the tickets) taken four at a time. This number is

$$C(20, 4) = \frac{20!}{4!\,(20 - 4)!} = \frac{20!}{4!\,16!} = \frac{20 \times 19 \times 18 \times 17}{4 \times 3 \times 2 \times 1} = 4845$$

If a set S has n elements, then $C(n, k)$ is the number of ways of choosing k elements from S, that is, the number of k-element subsets of S. Thus, the number of subsets of S of all possible sizes is given by the sum

$$C(n, 0) + C(n, 1) + C(n, 2) + \cdots + C(n, n) = 2^n$$

(See Section 10.6, Exercise 52, where this sum is discussed.)

> A set with n elements has 2^n subsets.

EXAMPLE 7 ■ Finding the Number of Subsets of a Set

A pizza parlor offers the basic cheese pizza and a choice of 16 toppings. How many different kinds of pizza can be ordered at this pizza parlor?

SOLUTION

We need the number of possible subsets of the 16 toppings (including the empty set, which corresponds to a plain cheese pizza). Thus, $2^{16} = 65{,}536$ different pizzas can be ordered. ■

■ Problem Solving with Permutations and Combinations

The crucial step in solving counting problems is deciding whether to use permutations, combinations, or the Fundamental Counting Principle. In some cases, the solution of a problem may require using more than one of these principles. Here are some general guidelines to help us decide how to apply these principles.

> ### GUIDELINES FOR SOLVING COUNTING PROBLEMS
>
> 1. FUNDAMENTAL COUNTING PRINCIPLE.   When consecutive choices are being made, use the Fundamental Counting Principal.
>
> 2. DOES THE ORDER MATTER?   When we want to find the number of ways of picking r objects from n objects, we need to ask ourselves: Does the order in which we pick the objects matter?
>
>   If the order matters, we use permutations.
>
>   If the order doesn't matter, we use combinations.

EXAMPLE 8 ■ A Problem Involving Combinations

A group of 25 campers contains 15 women and 10 men. In how many ways can a scouting party of 5 be chosen if it must consist of 3 women and 2 men?

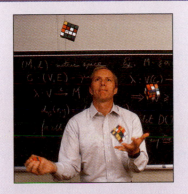

Ronald Graham, born in Taft, California, in 1935, is considered the world's leading mathematician in the field of combinatorics, the branch of mathematics that deals with counting. For many years Graham headed the Mathematical Studies Center at Bell Laboratories in Murray Hill, New Jersey, where he solved key problems for the telephone industry. During the *Apollo* program, NASA needed to evaluate mission schedules so that the three astronauts aboard the spacecraft could find the time to perform all the necessary tasks. The number of ways to allot these tasks were astronomical—too vast for even a computer to sort out. Graham, using his knowledge of combinatorics, was able to reassure NASA that there were easy ways of solving their problem that were not too far from the theoretically best possible solution. Besides being a prolific mathematician, Graham is an accomplished juggler (he has been on stage with the Cirque du Soleil and is a past president of the International Jugglers Association). Several of his research papers address the mathematical aspects of juggling. He is also fluent in Mandarin Chinese and Japanese, and once spoke with President Jiang in his native language.

SOLUTION

Three women can be chosen from the 15 women in the group in $C(15, 3)$ ways, and two men can be chosen from the 10 men in the group in $C(10, 2)$ ways. Thus, by the Fundamental Counting Principle the number of ways of choosing the scouting party is

$$C(15, 3) \times C(10, 2) = 455 \times 45 = 20{,}475$$ ■

EXAMPLE 9 ■ A Problem Involving Permutations and Combinations

A committee of seven—consisting of a chairman, a vice chairman, a secretary, and four other members—is to be chosen from a class of 20 students. In how many ways can this committee be chosen?

SOLUTION

In choosing the three officers, order is important. So, the number of ways of choosing them is

$$P(20, 3) = 6840$$

Next, we need to choose four other students from the 17 remaining. Since order doesn't matter in this case, the number of ways of doing this is

$$C(17, 4) = 2380$$

We could have first chosen the four unordered members of the committee—in $C(20, 4)$ ways—and then the three officers from the remaining 16 members, in $P(16, 3)$ ways. Check that this gives the same answer.

Thus, by the Fundamental Counting Principle the number of ways of choosing this committee is

$$P(20, 3) \times C(17, 4) = 6840 \times 2380 = 16{,}279{,}200$$ ■

EXAMPLE 10 ■ A Group Photograph

Twelve employees at a company picnic are to stand in a row for a group photograph. In how many ways can this be done if

(a) Jane and John insist on standing next to each other?
(b) Jane and John refuse to stand next to each other?

Jane John

SOLUTION

(a) Since the order in which the people stand is important, we use permutations. But we can't use the formula for permutations directly. Since Jane and John insist on standing together, let's think of them as one object. Thus, we have 11 objects to arrange in a row and there are $P(11, 11)$ ways of doing this. For each of these arrangements, there are two ways of having Jane and John stand together—Jane-John or John-Jane. Thus, by the Fundamental Counting Principle the total number of arrangements is

$$2 \times P(11, 11) = 2 \times 11! = 79{,}833{,}600$$

(b) There are $P(12, 12)$ ways of arranging the 12 people. Of these, $2 \times P(11, 11)$ have Jane and John standing together [by part (a)]. All the rest have Jane and John standing apart. So the number of arrangements with Jane and John apart is

$$P(12, 12) - 2 \times P(11, 11) = 12! - 2 \times 11! = 399{,}168{,}000$$ ■

11.2 EXERCISES

1–6 ■ Evaluate the expression.

1. $P(8, 3)$ **2.** $P(9, 2)$ **3.** $P(11, 4)$

4. $P(10, 5)$ **5.** $P(100, 1)$ **6.** $P(99, 3)$

7. In how many different ways can a president, vice president, and secretary be chosen from a class of 15 students?

8. In how many different ways can first, second, and third prizes be awarded in a game with eight contestants?

9. In how many different ways can six of ten people be seated in a row of six chairs?

10. In how many different ways can six people be seated in a row of six chairs?

11. How many three-letter "words" can be made from the letters *FGHIJK*? (Letters may not be repeated.)

12. How many permutations are possible from the letters of the word *LOVE*?

13. How many different three-digit whole numbers can be formed using the digits 1, 3, 5, and 7 if no repetition of digits is allowed?

14. A pianist plans to play eight pieces at a recital. In how many ways can she arrange these pieces in the program?

15. In how many different ways can a race with nine runners be completed, assuming there is no tie?

16. A ship carries five signal flags of different colors. How many different signals can be sent by hoisting exactly three of the five flags on the ship's flagpole in different orders?

17. In how many ways can first, second, and third prizes be awarded in a contest with 1000 contestants?

18. In how many ways can a president, vice president, secretary, and treasurer be chosen from a class of 30 students?

19. In how many ways can five students be seated in a row of five chairs if Jack insists on sitting in the first chair?

Jack

20. In how many ways can the students in Exercise 19 be seated if Jack insists on sitting in the middle chair?

21–24 ■ Find the number of distinguishable permutations of the given letters.

21. *AAABBC*   **22.** *AAABBBCCC*

23. *AABCD*   **24.** *ABCDDDEE*

25. In how many ways can two blue marbles and four red marbles be arranged in a row?

26. In how many different ways can five red balls, two white balls, and seven blue balls be arranged in a row?

27. In how many different ways can four pennies, three nickels, two dimes, and three quarters be arranged in a row?

28. In how many different ways can the letters of the word *ELEEMOSYNARY* be arranged?

29. A man bought three vanilla ice-cream cones, two chocolate cones, four strawberry cones, and five butterscotch cones for his 14 chidren. In how many ways can he distribute the cones among his children?

30. When seven students take a trip, they find a hotel with three rooms available—a room for one person, a room for two people, and a room for three people. In how many different ways can the students be assigned to these rooms? (One student has to sleep in the car.)

31. Eight workers are cleaning a large house. Five are needed to clean windows, two to clean the carpets, and one to clean the rest of the house. In how many different ways can these tasks be assigned to the eight workers?

32. A jogger jogs every morning to his health club, which is eight blocks east and five blocks north of his home. He always takes a route that is as short as possible, but he likes to vary it (see the figure). How many different routes can he take?   [*Hint:* The route shown can be thought of as *ENNEEENENEENE*, where *E* is East and *N* is North.]

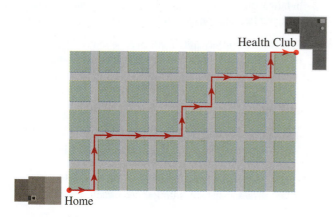

33–38 ■ Evaluate the expression.

33. $C(8, 3)$ **34.** $C(9, 2)$ **35.** $C(11, 4)$

36. $C(10, 5)$ **37.** $C(100, 1)$ **38.** $C(99, 3)$

39. In how many ways can three books be chosen from a group of six?

40. In how many ways can three pizza toppings be chosen from 12 available toppings?

41. In how many ways can six people be chosen from a group of ten?

42. In how many ways can a committee of three members be chosen from a club of 25 members?

43. How many different five-card hands can be dealt from a deck of 52 cards?

44. How many different seven-card hands can be picked from a deck of 52 cards?

45. A student must answer seven of the ten questions on an exam. In how many ways can she choose the seven questions?

46. A pizza parlor offers a choice of 16 different toppings. How many three-topping pizzas are possible?

47. A violinist has practiced 12 pieces. In how many ways can he choose eight of these pieces for a recital?

48. If a woman has eight skirts, in how many ways can she choose five of these to take on a weekend trip?

49. In how many ways can seven students from a class of 30 be chosen for a field trip?

50. In how many ways can the seven students in Exercise 49 be chosen if Jack must go on the field trip?

51. In how many ways can the seven students in Exercise 49 be chosen if Jack is not allowed to go on the field trip?

52. In the 6/49 lottery game, a player picks six numbers from 1 to 49. How many different choices does the player have?

53. In the California Lotto game, a player chooses six numbers from 1 to 53. It costs $1 to play this game. How much would it cost to buy every possible combination of six numbers to ensure picking the winning six numbers?

YOU CAN WIN $1,000,000
●●●LOTTO●●●

1	2	3	4	5	6	7	8	9	10
11	12	13	14	15	16	17	18	19	20
21	22	23	24	25	26	27	28	29	30
31	32	33	34	35	36	37	38	39	40
41	42	43	44	45	46	47	48	49	50
51	52	53							

54. A class has 20 students, of which 12 are females and 8 are males. In how many ways can a committee of five students be picked from this class under each condition?
   (a) No restriction is placed on the number of males or females on the committee.
   (b) No males are to be included on the committee.
   (c) The committee must have three females and two males.

55. A set has eight elements.
   (a) How many subsets containing five elements does this set have?
   (b) How many subsets does this set have?

56. A travel agency has limited numbers of eight different free brochures about Australia. The agent tells you to take any that you like, but no more than one of any kind. How many different ways can you choose brochures (including not choosing any)?

57. A hamburger chain gives their customers a choice of ten different hamburger toppings. In how many different ways can a customer order a hamburger?

58. Each of 20 shoppers in a shopping mall chooses to enter or not to enter the Dressfastic clothing store. How many different outcomes of their decisions are possible?

59. From a group of ten male and ten female tennis players, two men and two women are to face each other in a men-versus-women doubles match. In how many different ways can this match be arranged?

60. A school dance committee is to consist of two freshmen, three sophomores, four juniors, and five seniors. If six freshmen, eight sophomores, twelve juniors, and ten seniors are eligible to be on the committee, in how many ways can the committee be chosen?

61. A group of 22 aspiring thespians contains ten men and twelve women. For the next play the director wants to choose a leading man, a leading lady, a supporting male role, a supporting female role, and eight extras—three women and five men. In how many ways can the cast be chosen?

62. A hockey team has 20 players of which twelve play forward, six play defense, and two are goalies. In how many ways can the coach pick a starting lineup consisting of three forwards, two defense players, and one goalie?

63. A pizza parlor offers four sizes of pizza (small, medium, large, and colossus), two types of crust (thick and thin), and 14 different toppings. How many different pizzas can be made with these choices?

64. Sixteen boys and nine girls go on a camping trip. In how many ways can a group of six be selected to gather firewood, given the following conditions?

   (a) The group consists of two girls and four boys.
   (b) The group contains at least two girls.

65. In how many ways can ten students be arranged in a row for a class picture if John and Jane want to stand next to each other, and Mike and Molly also insist on standing next to each other?

66. In how many ways can the ten students in Exercise 65 be arranged if Mike and Molly insist on standing together but John and Jane refuse to stand next to each other?

67–68 ■ In how many ways can four men and four women be seated in a row of eight seats for each of the following arrangements?

67. (a) The first seat is to be occupied by a man.
   (b) The first and last seats are to be occupied by women.

68. (a) The women are to be seated together.
   (b) The men and women are to be seated alternately by gender.

69. From a group of 30 contestants, six are to be chosen as semifinalists, then two of those are chosen as finalists, and then the top prize is awarded to one of the finalists. In how many ways can these choices be made in sequence?

70. Three delegates are to be chosen from a group of four lawyers, a priest, and three professors. In how many ways can the delegation be chosen if it must include at least one professor?

71. In how many ways can a committee of four be chosen from a group of ten if two people refuse to serve together on the same committee?

72. Twelve dots are drawn on a page in such a way that no three are collinear. How many straight lines can be formed by joining the dots?

73. A five-person committee consisting of students and teachers is being formed to study the issue of student parking privileges. Of those who have expressed an interest in serving on the committee, 12 are teachers and 14 are students. In how many ways can the committee be formed if at least one student and one teacher must be included?

DISCOVERY · DISCUSSION

74. Complementary Combinations Without performing any calculations, explain in words why the number of ways of choosing two objects from ten objects is the same as the number of ways of choosing eight objects from ten objects. In general, explain why $C(n, r) = C(n, n - r)$.

75. An Identity Involving Combinations Kevin has ten different marbles, and he wants to give three of them to Luke and two to Mark. How many ways can he choose to do this? There are two ways of analyzing this problem: He could first pick three for Luke and then two for Mark, or he could first pick two for Mark and then three for Luke. Explain how these two viewpoints show that $C(10, 3) \cdot C(7, 2) = C(10, 2) \cdot C(8, 3)$. In general, explain why

$$C(n, r) \cdot C(n - r, k) = C(n, k) \cdot C(n - k, r)$$

76. Why Is $\binom{n}{r}$ the Same as $C(n, r)$? This exercise explains why the binomial coefficients $\binom{n}{r}$ that appear in the expansion of $(x + y)^n$ are the same as $C(n, r)$, the number of ways of choosing r objects from n objects. First, note that expanding a binomial using only the Distributive Property

gives

$$(x + y)^2 = (x + y)(x + y)$$
$$= (x + y)x + (x + y)y$$
$$= xx + xy + yx + yy$$
$$(x + y)^3 = (x + y)(xx + xy + yx + yy)$$
$$= xxx + xxy + xyx + xyy + yxx$$
$$+ yxy + yyx + yyy$$

(a) Expand $(x + y)^5$ using only the Distributive Property.
(b) Write all the terms that represent $x^2 y^3$ together. These are all the terms that contain two x's and three y's.
(c) Note that the two x's appear in all possible positions. Conclude that the number of terms that represent $x^2 y^3$ is $C(5, 2)$.
(d) In general, explain why $\binom{n}{r}$ in the Binomial Theorem is the same as $C(n, r)$.

11.3 PROBABILITY

If you roll a pair of dice, what are the chances of rolling a double six? What is the likelihood of winning a state lottery? The subject of probability was invented to give precise answers to questions like these. It is now an indispensable tool for making decisions in such diverse areas as business, manufacturing, psychology, genetics, and in many of the sciences. Probability is used to determine the effectiveness of new medicines, assess a fair price for an insurance policy, decide on the likelihood of a candidate winning an election, determine the opinion of many people on a certain topic (without interviewing everyone), and answer many other questions that involve a measure of uncertainty.

To discuss probability, let's begin by defining some terms. An **experiment** is a process, such as tossing a coin or rolling a die, that gives definite results, called the **outcomes** of the experiment. For tossing a coin, the possible outcomes are "heads" and "tails"; for rolling a die, the outcomes are 1, 2, 3, 4, 5, and 6. The **sample space** of an experiment is the set of all possible outcomes. If we let H stand for heads and T for tails, then the sample space of the coin-tossing experiment is

$$S = \{H, T\}$$

The sample space for rolling a die is

$$S = \{1, 2, 3, 4, 5, 6\}$$

We will be concerned only with experiments for which all the outcomes are "equally likely." We already have an intuitive feeling for what this means. When tossing a perfectly balanced coin, heads and tails are equally likely outcomes in the sense that if this experiment is repeated many times, we expect that about half the results will be heads and half will be tails.

The mathematical theory of probability was first discussed in 1654 in a series of letters between Pascal (see page 845) and Fermat (see page 532). Their correspondence was prompted by a question raised by the experienced gambler the Chevalier de Méré. The Chevalier was interested in the equitable distribution of the stakes of an interrupted gambling game (see Problem 3, page 904).

In any given experiment we are often concerned with a particular set of outcomes. We might be interested in a die showing an even number or in picking an ace from a deck of cards. Any particular set of outcomes is a subset of the sample space. This leads to the following definition.

DEFINITION OF AN EVENT

If S is the sample space of an experiment, then an **event** is any subset of the sample space.

EXAMPLE 1 ■ Events in a Sample Space

If an experiment consists of tossing a coin three times and recording the results in order, the sample space is

$$S = \{HHH, HHT, HTH, THH, TTH, THT, HTT, TTT\}$$

The event E of showing "exactly two heads" is the subset of S that consists of all outcomes with two heads. Thus

$$E = \{HHT, HTH, THH\}$$

The event F of showing "at least two heads" is

$$F = \{HHH, HHT, HTH, THH\}$$

and the event of showing "no heads" is $G = \{TTT\}$.   ■

We are now ready to define the notion of probability. Intuitively, we know that rolling a die may result in any of six equally likely outcomes, so the chance of any particular outcome occurring is $\frac{1}{6}$. What is the chance of showing an even number? Of the six equally likely outcomes possible, three are even numbers. So, it's reasonable to say that the chance of showing an even number is $\frac{3}{6} = \frac{1}{2}$. This reasoning is the intuitive basis for the following definition of probability.

DEFINITION OF PROBABILITY

Let S be the sample space of an experiment in which all outcomes are equally likely, and let E be an event. The probability of E, written $P(E)$, is

$$P(E) = \frac{n(E)}{n(S)} = \frac{\text{number of elements in } E}{\text{number of elements in } S}$$

Notice that $0 \le n(E) \le n(S)$, so the probability $P(E)$ of an event is a number between 0 and 1, that is,

$$0 \le P(E) \le 1$$

The closer the probability of an event is to 1, the more likely the event is to happen; the closer to 0, the less likely. If $P(E) = 1$, then E is called the **certain event** and if $P(E) = 0$, then E is called the **impossible event**.

EXAMPLE 2 ■ Finding the Probability of an Event

A coin is tossed three times and the results are recorded. What is the probability of getting exactly two heads? At least two heads? No heads?

SOLUTION

By the results of Example 1, the sample space S of this experiment contains eight outcomes and the event E of getting "exactly two heads" contains three outcomes, $\{HHT\ HTH,\ THH\}$, so by the definition of probability,

$$P(E) = \frac{n(E)}{n(S)} = \frac{3}{8}$$

Similarly, the event F of getting "at least two heads" has four outcomes, $\{HHH, HHT, HTH, THH\}$, and so

$$P(F) = \frac{n(F)}{n(S)} = \frac{4}{8} = \frac{1}{2}$$

The event G of getting "no heads" has one element, so

$$P(G) = \frac{n(G)}{n(S)} = \frac{1}{8}$$ ■

  To find the probability of an event, we do not need to list all the elements in the sample space and the event. What we do need is the *number* of elements in these sets. The counting techniques we've learned in the preceding sections will be very useful here.

EXAMPLE 3 ■ Finding the Probability of an Event

A five-card poker hand is drawn from a standard deck of 52 cards. What is the probability that all five cards are spades?

SOLUTION

The experiment here consists of choosing five cards from the deck, and the sample space S consists of all possible five-card hands. Thus, the number of elements in the sample space is

$$n(S) = C(52, 5) = \frac{52!}{5!\,(52-5)!} = 2{,}598{,}960$$

The event E we are interested in consists of choosing five spades. Since the deck contains only 13 spades, the number of ways of choosing five spades is

$$n(E) = C(13, 5) = \frac{13!}{5! \, (13 - 5)!} = 1287$$

Thus, the probability of drawing five spades is

$$P(E) = \frac{n(E)}{n(S)} = \frac{1287}{2,598,960} \approx 0.0005 \qquad\blacksquare$$

What does the answer to Example 3 tell us? Since $0.0005 = \frac{1}{2000}$, this means that if you play poker many, many times, on average you will be dealt a hand consisting of only spades about once in every 2000 hands.

EXAMPLE 4 ■ Finding the Probability of an Event

A bag contains 20 tennis balls, of which four are defective. If two balls are selected at random from the bag, what is the probability that both are defective?

SOLUTION

The experiment consists of choosing two balls from 20, so the number of elements in the sample space S is $C(20, 2)$. Since there are four defective balls, the number of ways of picking two defective balls is $C(4, 2)$. Thus, the probability of the event E of picking two defective balls is

$$P(E) = \frac{n(E)}{n(S)} = \frac{C(4, 2)}{C(20, 2)} = \frac{6}{190} \approx 0.032 \qquad\blacksquare$$

The **complement** of an event E is the set of outcomes in the sample space that is not in E. We denote the complement of an event E by E'. We can calculate the probability of E' using the definition and the fact that $n(E') = n(S) - n(E)$:

$$P(E') = \frac{n(E')}{n(S)} = \frac{n(S) - n(E)}{n(S)} = \frac{n(S)}{n(S)} - \frac{n(E)}{n(S)} = 1 - P(E)$$

PROBABILITY OF THE COMPLEMENT OF AN EVENT

Let S be the sample space of an experiment and E an event. Then

$$P(E') = 1 - P(E)$$

This is an extremely useful result, since it is often difficult to calculate the probability of an event E but easy to find the probability of E', from which $P(E)$ can be calculated immediately using this formula.

EXAMPLE 5 ■ Finding the Probability of the Complement of an Event

An urn contains 10 red balls and 15 blue balls. Six balls are drawn at random from the urn. What is the probability that at least one ball is red?

SOLUTION

Let E be the event that at least one red ball is drawn. It is tedious to count all the possible ways in which one or more of the balls drawn are red. So, let's consider E', the complement of this event—namely, that none of the balls chosen is red. The number of ways of choosing 6 blue balls from the 15 blue balls is $C(15, 6)$; the number of ways of choosing 6 balls from the 25 balls is $C(25, 6)$. Thus

Since

$$P(E') = 1 - P(E)$$

we have

$$P(E) = 1 - P(E')$$

$$P(E') = \frac{n(E')}{n(S)} = \frac{C(15, 6)}{C(25, 6)} = \frac{5005}{177{,}100} = \frac{13}{460}$$

By the formula for the complement of an event, we have

$$P(E) = 1 - P(E') = 1 - \frac{13}{460} = \frac{447}{460} \approx 0.97$$ ■

Mutually Exclusive Events

Two events that have no outcome in common are said to be **mutually exclusive** (see Figure 1). For example, in drawing a card from a deck, the events

E: The card is an ace

F: The card is a queen

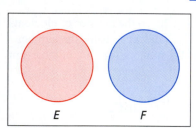

FIGURE 1

are mutually exclusive, because a card cannot be both an ace and a queen.

If E and F are mutually exclusive events, what is the probability that E or F occurs? The word *or* indicates that we want the probability of the *union* of these events, that is, $E \cup F$. Since E and F have no element in common,

$$n(E \cup F) = n(E) + n(F)$$

Thus

$$P(E \cup F) = \frac{n(E \cup F)}{n(S)} = \frac{n(E) + n(F)}{n(S)} = \frac{n(E)}{n(S)} + \frac{n(F)}{n(S)} = P(E) + P(F)$$

We have proved the following formula.

PROBABILITY OF THE UNION OF MUTUALLY EXCLUSIVE EVENTS

If E and F are mutually exclusive events in a sample space S, then the probability of E *or* F is

$$P(E \cup F) = P(E) + P(F)$$

There is a natural extension of this formula for any number of mutually exclusive events: If $E_1, E_2, \ldots, E_n$ are pairwise mutually exclusive, then

$$P(E_1 \cup E_2 \cup \cdots \cup E_n) = P(E_1) + P(E_2) + \cdots + P(E_n)$$

EXAMPLE 6 ■ The Probability of Mutually Exclusive Events

A card is drawn at random from a standard deck of 52 cards. What is the probability that the card is either a seven or a face card?

SOLUTION

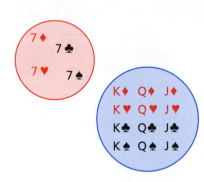

Let E and F denote the following events.

E: The card is a seven

F: The card is a face card

Since a card cannot be both a seven and a face card, the events are mutually exclusive. We want the probability of E or F; in other words, the probability of $E \cup F$. By the formula,

$$P(E \cup F) = P(E) + P(F) = \frac{4}{52} + \frac{12}{52} = \frac{4}{13}$$

■

The Probability of the Union of Two Events

If two events E and F are not mutually exclusive, then they share outcomes in common. The situation is described graphically in Figure 2. The overlap of the two sets is their intersection, that is, $E \cap F$. Again, we are interested in the event E or F, so we must count the elements in $E \cup F$. If we simply added the number of elements in E to the number of elements in F, then we would be counting the elements in the overlap twice—once in E and once in F. So, to get the correct total, we must subtract the number of elements in $E \cap F$. Thus

$$n(E \cup F) = n(E) + n(F) - n(E \cap F)$$

Using the formula for probability, we get

$$P(E \cup F) = \frac{n(E \cup F)}{n(S)} = \frac{n(E) + n(F) - n(E \cap F)}{n(S)}$$

$$= \frac{n(E)}{n(S)} + \frac{n(F)}{n(S)} - \frac{n(E \cap F)}{n(S)}$$

$$= P(E) + P(F) - P(E \cap F)$$

FIGURE 2

$E \cap F$

E F

We have proved the following.

> ## PROBABILITY OF THE UNION OF TWO EVENTS
>
> If E and F are events in a sample space S, then the probability of E *or* F is
>
> $$P(E \cup F) = P(E) + P(F) - P(E \cap F)$$

EXAMPLE 7 ■ The Probability of the Union of Events

What is the probability that a card drawn at random from a standard 52-card deck is either a face card or a spade?

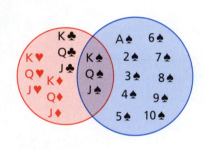

SOLUTION

We let E and F denote the following events:

$$E: \quad \text{The card is a face card}$$

$$F: \quad \text{The card is a spade}$$

There are 12 face cards and 13 spades in a 52-card deck, so

$$P(E) = \frac{12}{52} \qquad \text{and} \qquad P(F) = \frac{13}{52}$$

Since 3 cards are simultaneously face cards and spades, we have

$$P(E \cap F) = \frac{3}{52}$$

Thus, by the formula for the probability of the union of two events, we have

$$P(E \cup F) = P(E) + P(F) - P(E \cap F)$$

$$= \frac{12}{52} + \frac{13}{52} - \frac{3}{52} = \frac{11}{26} \qquad ■$$

The Intersection of Independent Events

We have considered the probability of events joined by the word *or*, that is, the union of events. Now we study the probability of events joined by the word *and*—in other words, the intersection of events.

When the occurrence of one event does not affect the probability of another event, we say that the events are **independent**. For instance, if a balanced coin is tossed, the probability of showing heads on the second toss is $\frac{1}{2}$, regardless of the outcome of the first toss. So, any two tosses of a coin are independent.

> ### PROBABILITY OF THE INTERSECTION OF INDEPENDENT EVENTS
>
> If E and F are independent events in a sample space S, then the probability of E *and* F is
>
> $$P(E \cap F) = P(E)P(F)$$

EXAMPLE 8 ■ The Probability of Independent Events

A jar contains five red balls and four black balls. A ball is drawn at random from the jar and then replaced; then another ball is picked. What is the probability that both balls are red?

SOLUTION

The events are independent. The probability that the first ball is red is $\frac{5}{9}$. The probability that the second is red is also $\frac{5}{9}$. Thus, the probability that both balls are red is

$$\frac{5}{9} \times \frac{5}{9} = \frac{25}{81} \approx 0.31$$

EXAMPLE 9 ■ The Birthday Problem

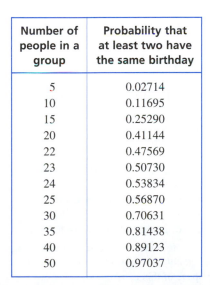

What is the probability that in a class of 35 students, at least two have the same birthday?

SOLUTION

It's reasonable to assume that the 35 birthdays are independent and that each day of the 365 days in a year is equally likely as a date of birth. (We ignore February 29.)

Let E be the event that two of the students have the same birthday. It is tedious to list all the possible ways in which at least two of the students have matching birthdays. So, we consider the complementary event E', that is, that *no* two students have the same birthday. To find this probability, we consider the students one at a time. The probability that the first student has a birthday is 1, the probability that the second has a birthday different from the first is $\frac{364}{365}$, the probability that the third has a birthday different from the first two is $\frac{363}{365}$, and so on. Thus

$$P(E') = 1 \cdot \tfrac{364}{365} \cdot \tfrac{363}{365} \cdot \tfrac{362}{365} \cdot \cdots \cdot \tfrac{331}{365} \approx 0.186$$

So

$$P(E) = 1 - P(E') \approx 1 - 0.186 = 0.814$$

Most people are surprised that the probability in Example 9 is so high. For this reason, this problem is sometimes called the "birthday paradox." The table in the margin gives the probability that two people in a group will share the same birthday for groups of various sizes.

Number of people in a group	Probability that at least two have the same birthday
5	0.02714
10	0.11695
15	0.25290
20	0.41144
22	0.47569
23	0.50730
24	0.53834
25	0.56870
30	0.70631
35	0.81438
40	0.89123
50	0.97037

11.3 EXERCISES

1. An experiment consists of tossing a coin twice.
   (a) Find the sample space.
   (b) Find the probability of getting heads exactly two times.
   (c) Find the probability of getting heads at least one time.
   (d) Find the probability of getting heads exactly one time.

2. An experiment consists of tossing a coin and rolling a die.
   (a) Find the sample space.
   (b) Find the probability of getting heads and an even number.
   (c) Find the probability of getting heads and a number greater than 4.
   (d) Find the probability of getting tails and an odd number.

3–4 ■ A die is rolled. Find the probability of the given event.

3. (a) The number showing is a six.
   (b) The number showing is an even number.
   (c) The number showing is greater than 5.

4. (a) The number showing is a two or a three.
   (b) The number showing is an odd number.
   (c) The number showing is a number divisible by 3.

5–6 ■ A card is drawn randomly from a standard 52-card deck. Find the probability of the given event.

5. (a) The card drawn is a king.
   (b) The card drawn is a face card.
   (c) The card drawn is not a face card.

6. (a) The card drawn is a heart.
   (b) The card drawn is either a heart or a spade.
   (c) The card drawn is a heart, a diamond, or a spade.

7–8 ■ A ball is drawn randomly from a jar that contains five red balls, two white balls, and one yellow ball. Find the probability of the given event.

7. (a) A red ball is drawn.
   (b) The ball drawn is not yellow.
   (c) A black ball is drawn.

8. (a) Neither a white nor yellow ball is drawn.
   (b) A red, white, or yellow ball is drawn.
   (c) The ball drawn is not white.

9. A drawer contains an unorganized collection of 18 socks—three pairs are red, two pairs are white, and four pairs are black.
   (a) If one sock is drawn at random from the drawer, what is the probability that it is red?
   (b) Once a sock is drawn and discovered to be red, what is the probability of drawing another red sock to make a matching pair?

10. A child's game has a spinner as shown in the figure. Find the probability of the given event.
   (a) The spinner stops on an even number.
   (b) The spinner stops on an odd number or a number greater than 3.

11. A letter is chosen at random from the word *EXTRATERRESTRIAL*. Find the probability of the given event.
   (a) The letter *T* is chosen.
   (b) The letter chosen is a vowel.
   (c) The letter chosen is a consonant.

12–15 ■ A poker hand, consisting of five cards, is dealt from a standard deck of 52 cards. Find the probability that the hand contains the cards described.

12. Five hearts

13. Five cards of the same suit

14. Five face cards

15. An ace, king, queen, jack, and 10 of the same suit (royal flush)

16. A pair of dice is rolled, and the numbers showing are observed.
   (a) List the sample space of this experiment.
   (b) Find the probability of getting a sum of 7.
   (c) Find the probability of getting a sum of 9.
   (d) Find the probability that the two dice show doubles (the same number).
   (e) Find the probability that the two dice show different numbers.
   (f) Find the probability of getting a sum of 9 or higher.

17. A couple intends to have four children. Assume that having a boy or a girl is an equally likely event.
   (a) List the sample space of this experiment.
   (b) Find the probability that the couple has only boys.

(c) Find the probability that the couple has two boys and two girls.

(d) Find the probability that the couple has four children of the same sex.

(e) Find the probability that the couple has at least two girls.

18. What is the probability that a 13-card bridge hand consists of all cards from the same suit?

19. An American roulette wheel has 38 slots; two slots are numbered 0 and 00, and the remaining slots are numbered from 1 to 36. Find the probability that the ball lands in an odd-numbered slot.

20. A toddler has wooden blocks showing the letters C, E, F, H, N, and R. Find the probability that the child arranges the letters in the indicated order.
(a) In the order *FRENCH*
(b) In alphabetical order

21. In the 6/49 lottery game, a player selects six numbers from 1 to 49. What is the probability of picking the six winning numbers?

22. The president of a large company selects six employees to receive a special bonus. He claims that the six employees are chosen randomly from among the 30 employees, of which 19 are women and 11 are men. What is the probability that no woman is chosen?

23. An exam has ten true-false questions. A student who has not studied answers all ten questions by just guessing. Find the probability that the student correctly answers the given number of questions.
(a) All ten questions
(b) Exactly seven questions

24. To control the quality of their product, the Bright-Light Company inspects three light bulbs out of each batch of ten bulbs manufactured. If a defective bulb is found, the batch is discarded. Suppose a batch contains two defective bulbs. What is the probability that the batch will be discarded?

25. An often-quoted example of an event of extremely low probability is that a monkey types Shakespeare's entire play *Hamlet* by randomly striking keys on a typewriter. Assume that the typewriter has 48 keys (including the space bar) and that the monkey is equally likely to hit any key.
(a) Find the probability that such a monkey will actually correctly type just the title of the play as his first word.
(b) What is the probability that the monkey will type the phrase "To be or not to be" as his first words?

26. A monkey is trained to arrange wooden blocks in a straight line. He is then given six blocks showing the letters A, E, H, L, M, T. What is the probability that he will arrange them to spell the word *HAMLET*?

27. A monkey is trained to arrange wooden blocks in a straight line. He is then given 11 blocks showing the letters A, B, B, I, I, L, O, P, R, T, Y. What is the probability that the monkey will arrange the blocks to spell the word *PROBABILITY*?

28. Eight horses are entered in a race. You randomly predict a particular order for the horses to complete the race. What is the probability that your prediction is correct?

29. Many genetic traits are controlled by two genes, one dominant and one recessive. In Gregor Mendel's original experiments with peas, the genes controlling the height of the plant are denoted by T (tall) and t (short). The gene T is dominant, so a plant with the genotype (genetic makeup) TT or Tt is tall, whereas one with genotype tt is short. By a statistical analysis of the offspring in his experiments, Mendel concluded that offspring inherit one gene from each parent, and each possible combination of the two genes is equally likely. If each parent has the genotype Tt, then the following chart gives the possible genotypes of the offspring:

		Parent 2	
		T	t
Parent 1	T	TT	Tt
	t	Tt	tt

Find the probability that a given offspring of these parents will be (a) tall or (b) short.

30. Refer to Exercise 29. Make a chart of the possible genotypes of the offspring if one parent has genotype Tt and the other tt. Find the probability that a given offspring will be (a) tall or (b) short.

31–32 ■ Determine whether the events E and F in the given experiment are mutually exclusive.

31. The experiment consists of selecting a person at random.
   (a) E: The person is male
       F: The person is female
   (b) E: The person is tall
       F: The person is blond

32. The experiment consists of choosing at random a student from your class.
   (a) E: The student is female
       F: The student wears glasses
   (b) E: The student has long hair
       F: The student is male

33–34 ■ A die is rolled and the number showing is observed. Determine whether the events E and F are mutually exclusive. Then find the probability of the event $E \cup F$.

33. (a) E: The number is even
         F: The number is odd
     (b) E: The number is even
         F: The number is greater than 4

34. (a) E: The number is greater than 3
         F: The number is less than 5
     (b) E: The number is divisible by 3
         F: The number is less than 3

35–36 ■ A card is drawn at random from a standard 52-card deck. Determine whether the events E and F are mutually exclusive. Then find the probability of the event $E \cup F$.

35. (a) E: The card is a face card
         F: The card is a spade
     (b) E: The card is a heart
         F: The card is a spade

36. (a) E: The card is a club
         F: The card is a king
     (b) E: The card is an ace
         F: The card is a spade

37–38 ■ Refer to the spinner shown in the figure. Find the probability of the given event.

37. (a) The spinner stops on red.
     (b) The spinner stops on an even number.
     (c) The spinner stops on red or an even number.

38. (a) The spinner stops on blue.
     (b) The spinner stops on an odd number.
     (c) The spinner stops on blue or an odd number.

39. An American roulette wheel has 38 slots: two of the slots are numbered 0 and 00, and the rest are numbered from 1 to 36. Find the probability that the ball lands in an odd-numbered slot or in a slot with a number higher than 31.

40. A toddler has eight wooden blocks showing the letters A, E, I, G, L, N, T, and R. What is the probability that the child will arrange the letters to spell one of the words *TRIANGLE* or *INTEGRAL*?

41. A committee of five is chosen randomly from a group of six males and eight females. What is the probability that the committee includes either all males or all females?

42. In the 6/49 lottery game a player selects six numbers from 1 to 49. What is the probability of selecting at least five of the six winning numbers?

43. A jar contains six red marbles numbered 1 to 6 and ten blue marbles numbered 1 to 10. A marble is drawn at random from the jar. Find the probability that the given event occurs.
   (a) The marble is red.
   (b) The marble is odd-numbered.
   (c) The marble is red or odd-numbered.
   (d) The marble is blue or even-numbered.

44. A coin is tossed twice. Let E be the event "the first toss shows heads" and F the event "the second toss shows heads."
   (a) Are the events E and F independent?
   (b) Find the probability of showing heads on both tosses.

45. A die is rolled twice. Let E be the event "the first roll shows a six" and F the event "the second roll shows a six."
   (a) Are the events E and F independent?
   (b) Find the probability of showing a six on both rolls.

46–47 ■ Spinners A and B shown in the figure are spun at the same time.

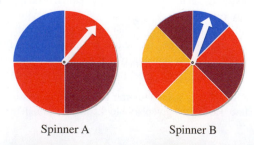

Spinner A   Spinner B

46. (a) Are the events "spinner A stops on red" and "spinner B stops on yellow" independent?
(b) Find the probability that spinner A stops on red and spinner B stops on yellow.

47. (a) Find the probability that both spinners stop on purple.
(b) Find the probability that both spinners stop on blue.

48. A die is rolled twice. What is the probability of showing a one on both rolls?

49. A die is rolled twice. What is the probability of showing a one on the first roll and an even number on the second roll?

50. A card is drawn from a deck and replaced, and then a second card is drawn.
(a) What is the probability that both cards are aces?
(b) What is the probability that the first is an ace and the second a spade?

51. A roulette wheel has 38 slots: Two slots are numbered 0 and 00, and the rest are numbered 1 to 36. A player places a bet on a number between 1 and 36 and wins if a ball thrown into the spinning roulette wheel lands in the slot with the same number. Find the probability of winning on two consecutive spins of the roulette wheel.

52. A researcher claims that she has taught a monkey to spell the word *MONKEY* using the five wooden letters *E, O, K, M, N, Y*. If the monkey has not actually learned anything and is merely arranging the blocks randomly, what is the probability that he will spell the word correctly three consecutive times?

53. What is the probability of rolling "snake eyes" (double ones) three times in a row with a pair of dice?

54. In the 6/49 lottery game, a player selects six numbers from 1 to 49 and wins if he selects the winning six numbers. What is the probability of winning the lottery two times in a row?

55. Jar A contains three red balls and four white balls. Jar B contains five red balls and two white balls. Which one of the following ways of randomly selecting balls gives the greatest probability of drawing two red balls?
(i) Draw two balls from jar B.
(ii) Draw one ball from each jar.
(iii) Put all the balls in one jar, and then draw two balls.

56. A slot machine has three wheels: Each wheel has 11 positions—a bar and the digits 0, 1, 2, . . . , 9. When the handle is pulled, the three wheels spin independently before coming to rest. Find the probability that the wheels stop on the following positions.
(a) Three bars
(b) The same number on each wheel
(c) At least one bar

57. Find the probability that in a group of eight students at least two people have the same birthday.

58. What is the probability that in a group of six students at least two have birthdays in the same month?

59. A student has locked her locker with a combination lock, showing numbers from 1 to 40, but she has forgotten the three-number combination that opens the lock. In order to open the lock, she decides to try all possible combinations. If she can try ten different combinations every minute, what is the probability that she will open the lock within one hour?

60. A mathematics department consists of ten men and eight women. Six mathematics faculty members are to be selected at random for the curriculum committee.
(a) What is the probability that two women and four men are selected?
(b) What is the probability that two or fewer women are selected?
(c) What is the probability that more than two women are selected?

61. Twenty students are arranged randomly in a row for a class picture. Paul wants to stand next to Phyllis. Find the probability that he gets his wish.

62. Eight boys and 12 girls are arranged in a row. What is the probability that all the boys will be standing at one end of the row and all the girls at the other end?

 DISCOVERY · DISCUSSION

63. The "Second Son" Paradox Mrs. Smith says, "I have two children—the older one is named William." Mrs. Jones replies, "One of my two children is also named William." For each woman, list the sample space for the genders of her children, and calculate the probability that her other child is also a son. Explain why these two probabilities are different.

64. The "Oldest Son or Daughter" Phenomenon Poll your class to determine how many of your male classmates are the oldest sons in their families, and how many of your female classmates are the oldest daughters in their families. You will most likely find that they form a majority of the class. Explain why a randomly selected individual has a high probability of being the oldest son or daughter in his or her family.

Laboratory Project

Small Samples, Big Results

A national poll finds that voter preference for presidential candidates is as follows:

Candidate A   57%

Candidate B   43%

In the poll 1600 adults were surveyed. Since over 100 million voters participate in a national election, it hardly seems possible that surveying only 1600 adults would be of any value. But it is, and it can be proved mathematically that if the sample of 1600 adults is selected *at random*, then the results are accurate to within ±3% more than 95% of the time. Scientists use these methods to determine properties of a big population by testing a small sample. For example, a small sample of fish from a lake is tested to determine the proportion that is diseased, or a small sample of a manufactured product is tested to determine the proportion that is defective.

We can get a feeling for how this works through a simple experiment. Put 1000 white beans and 1000 black beans in a bag, and mix them thoroughly. (It takes a lot of time to count out 1000 beans, so get several friends to count out 100 or so each.) Take a small cup and scoop up a small sample from the beans.

1. Record the proportion of black (or white) beans in the sample. How closely does the proportion in the sample compare with the actual proportion in the bag?

2. Take several samples and record the proportion of black (or white) beans in each sample.
   (a) Graph your results.
   (b) Average your results. How close is your average to 0.5?

3. Try the experiment again but with 500 black beans and 1500 white beans. What proportion of black (or white) beans would you expect in this sample?

4. Have some friends mix black and white beans without telling you the number of each. Estimate the proportion of black (or white) beans by taking a few small samples.

5. This bean experiment can be simulated on your graphing calculator. Let's designate the numbers in the interval $[0, 0.25]$ as "black beans" and those in $(0.25, 1]$ as "white beans." Now use the random number generator in your calculator to randomly pick a sample of numbers between 0 and 1. Try samples of size 100, 400, 1600, and larger. Determine the proportion of these that are "black beans." What would you expect this proportion to be? Do your results improve when you use a larger sample? (The TI-83 program in the margin takes a sample of 100 random numbers.)

6. Biologists use sampling techniques to estimate fish populations. For example, to estimate the number of trout in a lake, some of the trout are collected, tagged, and released. Later, a sample is taken and the number of tagged trout in the sample is recorded. The total number of trout in the lake can be estimated from the following proportionality:

$$\frac{\text{number of tagged trout in lake}}{\text{number of trout in lake}} \approx \frac{\text{number of tagged trout in sample}}{\text{number of trout in sample}}$$

Model this process using beans as follows. Start with a bag containing an unknown number of white beans. Remove a handful of the beans, count them, then "tag" them by marking them with a felt tip pen. Return the tagged beans to the bag and mix thoroughly. Now take a sample from the bag and use the number of tagged beans in the sample to estimate the total number of beans in the bag.

```
PROGRAM: SAMPLE
:100 → N
:0 → C
:For (J, 1, N)
:rand → B
:(B ≤ 0.25) + C → C
:End
:Disp "PROPORTION="
:Disp C/N
```

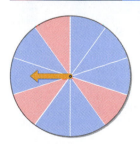

11.4   EXPECTED VALUE

In the game shown in Figure 1, you pay $1 to spin the arrow. If the arrow stops in a red region, you get $3 (the dollar you paid plus $2); otherwise, you lose the dollar you paid. If you play this game many times, how much would you expect to win? Or lose? To answer these questions, let's consider the probabilities of winning and losing. Since three of the regions are red, the probability of winning is $\frac{3}{10} = 0.3$ and that of losing is $\frac{7}{10} = 0.7$. Remember, this means that if you play this game many times, you expect to win "on average" three out of ten times. So, suppose you play the game 1000 times. Then you would expect to win 300 times and lose 700 times. Since we win $2 or lose $1 in each game, our expected payoff in 1000 games is

$$2(300) + (-1)(700) = -100$$

FIGURE 1

So the average expected return per game is $\frac{-100}{1000} = -0.1$. In other words, we expect to lose, on average, 10 cents per game. Another way to view this average is to divide each side of the preceding equation by 1000. Writing E for the result, we get

$$E = \frac{2(300) + (-1)(700)}{1000}$$

$$= 2\left(\frac{300}{1000}\right) + (-1)\frac{700}{1000}$$

$$= 2(0.3) + (-1)(0.7)$$

Thus, the expected return, or *expected value*, per game is

$$E = a_1 p_1 + a_2 p_2$$

where a_1 is the payoff that occurs with probability p_1 and a_2 is the payoff that occurs with probability p_2. This example leads us to the following definition of expected value.

DEFINITION OF EXPECTED VALUE

A game gives payoffs $a_1, a_2, \ldots, a_n$ with probabilities $p_1, p_2, \ldots, p_n$. The **expected value** (or **expectation**) E of this game is

$$E = a_1 p_1 + a_2 p_2 + \cdots + a_n p_n$$

The expected value is an average expectation per game if the game is played many times. In general, E need not be one of the possible payoffs. In the preceding example the expected value is -10 cents, but it's impossible to lose exactly 10 cents in any given trial of the game.

EXAMPLE 1 ■ Finding an Expected Value

A die is rolled, and you receive $1 for each point that shows. What is your expectation?

SOLUTION

Each face of the die has probability $\frac{1}{6}$ of showing. So you get $1 with probability $\frac{1}{6}$, $2 with probability $\frac{1}{6}$, $3 with probability $\frac{1}{6}$, and so on. Thus, the expected value is

$$E = 1\left(\frac{1}{6}\right) + 2\left(\frac{1}{6}\right) + 3\left(\frac{1}{6}\right) + 4\left(\frac{1}{6}\right) + 5\left(\frac{1}{6}\right) + 6\left(\frac{1}{6}\right) = \frac{21}{6} = 3.5$$

This means that if you play this game many times, you will make, on average, $3.50 per game.   ■

EXAMPLE 2 ■ Finding an Expected Value

In Monte Carlo, the game of roulette is played on a wheel with slots numbered 0, 1, 2, . . . , 36. The wheel is spun, and a ball dropped in the wheel is equally likely to end up in any one of the slots. To play the game, you bet $1 on any number other than zero. (For example, you may bet $1 on number 23.) If the ball stops in your slot, you get $36 (the $1 you bet plus $35). Find the expected value of this game.

SOLUTION

You gain $35 with probability $\frac{1}{37}$, and you lose $1 with probability $\frac{36}{37}$. Thus

$$E = (35)\,\frac{1}{37} + (-1)\,\frac{36}{37} \approx -0.027$$

In other words, if you play this game many times, you would expect to lose 2.7 cents on every dollar you bet (on average). Consequently, the house expects to gain 2.7 cents on every dollar that is bet. This expected value is what makes gambling very profitable for the gaming house and very unprofitable for the gambler. ■

11.4   EXERCISES

1–10 ■ Find the expected value (or expectation) of the games described.

1. Mike wins $2 if a coin toss shows heads and $1 if it shows tails.

2. Jane wins $10 if a die roll shows a six, and she loses $1 otherwise.

3. The game consists of drawing a card from a deck. You win $100 if you draw the ace of spades or lose $1 if you draw any other card.

4. Tim wins $3 if a coin toss shows heads or $2 if it shows tails.

5. Carol wins $3 if a die roll shows a six, and she wins $0.50 otherwise.

6. A coin is tossed twice. Albert wins $2 for each heads and must pay $1 for each tails.

7. A die is rolled. Tom wins $2 if the die shows an even number and he pays $2 otherwise.

8. A card is drawn from a deck. You win $104 if the card is an ace, $26 if it is a face card, and $13 if it is the 8 of clubs.

9. A bag contains two silver dollars and eight slugs. You pay 50 cents to reach into the bag and take a coin, which you get to keep.

10. A bag contains eight white balls and two black balls. John picks two balls at random from the bag, and he wins $5 if he does not pick a black ball.

11. In the game of roulette as played in Las Vegas, the wheel has 38 slots: Two slots are numbered 0 and 00, and the rest are numbered 1 to 36. A $1 bet on any number other than 0 or 00 wins $36 ($35 plus the $1 bet). Find the expected value of this game.

12. A sweepstakes offers a first prize of $1,000,000, second prize of $100,000, and third prize of $10,000. Suppose that two million people enter the contest and three names are drawn randomly for the three prizes.
    (a) Find the expected winnings for a person participating in this contest.
    (b) Is it worth paying a dollar to enter this sweepstakes?

13. A box contains 100 envelopes. Ten envelopes contain $10 each, ten contain $5 each, two are "unlucky," and the rest are empty. A player draws an envelope from the box and keeps whatever is in it. If a person draws an unlucky envelope, however, he must pay $100. What is the expectation of a person playing this game?

14. A safe containing $1,000,000 is locked with a combination lock. You pay $1 for one guess at the six-digit combination. If you open the lock, you get to keep the million dollars. What is your expectation?

15. An investor buys 1000 shares of a risky stock for $5 a share. She estimates that the probability the stock will rise in value to $20 a share is 0.1 and the probability that it will fall to $1 a share is 0.9. If the only criterion for her decision to buy this stock was the expected value of her profit, did she make a wise investment?

16. A slot machine has three wheels, and each wheel has 11 positions—the digits 0, 1, 2, . . . , 9 and the picture of a watermelon. When a quarter is placed in the machine and the handle is pulled, the three wheels spin independently and come to rest. When three watermelons show, the payout is $5; otherwise, nothing is paid. What is the expected value of this game?

17. In a 6/49 lottery game, a player pays $1 and selects six numbers from 1 to 49. Any player who has chosen the six winning numbers wins $1,000,000. Assuming this is the only way to win, what is the expected value of this game?

18. A bag contains two silver dollars and six slugs. A game consists of reaching into the bag and drawing a coin, which

you get to keep. Determine the "fair price" of playing this game, that is, the price at which the player can be expected to break even if he plays the game many times (in other words, the price at which his expectation is zero).

19. A game consists of drawing a card from a deck. You win $13 if you draw an ace. What is a "fair price" to pay to play this game? (See Exercise 18.)

 DISCOVERY · DISCUSSION

20. The Expected Value of a Sweepstakes Contest A magazine clearinghouse holds a sweepstakes contest to sell subscriptions. If you return the winning number, you win $1,000,000. You have a 1-in-20-million chance of winning, but your only cost to enter the contest is a first-class stamp to mail the entry. Use the current price of a first-class stamp to calculate your expected net winnings if you enter this contest. Is it worth entering the sweepstakes?

11   REVIEW

CONCEPT CHECK

1. What does the Fundamental Counting Principle say?

2. (a) What is a permutation of a set of distinct objects?
   (b) How many permutations are there of n objects?
   (c) How many permutations are there of n objects taken r at a time?
   (d) What is the number of distinguishable permutations of n objects if there are k different kinds of objects with n_1 objects of the first kind, n_2 objects of the second kind, and so on?

3. (a) What is a combination of r elements of a set?
   (b) How many combinations are there of n elements taken r at a time?
   (c) How many subsets does a set with n elements have?

4. In solving a problem involving picking r objects from n objects, how do you know whether to use permutations or combinations?

5. (a) What is meant by the sample space of an experiment?
   (b) What is an event?
   (c) Define the probability of an event E in a sample space S.
   (d) What is the probability of the complement of E?

6. (a) What are mutually exclusive events?
   (b) If E and F are mutually exclusive events, what is the probability of the union of E and F? What if E and F are not mutually exclusive?

7. (a) What are independent events?
   (b) If E and F are independent events, what is the probability of the intersection of E and F?

8. Suppose that a game gives payoffs $a_1, a_2, \ldots, a_n$ with probabilities $p_1, p_2, \ldots, p_n$. What is the expected value of the game? What is the significance of the expected value?

EXERCISES

1. A coin is tossed, a die is rolled, and a card is drawn from a deck. How many possible outcomes does this experiment have?

2. How many three-digit numbers can be formed using the digits 1, 2, 3, 4, 5, 6 if repetition of digits
   (a) is allowed?
   (b) is not allowed?

3. (a) How many different two-element subsets does the set {*A, E, I, O, U*} have?

(b) How many different two-letter "words" can be made using the letters from the set in part (a)?

4. An airline company overbooks a particular flight and seven passengers are "bumped" from the flight. If 120 passengers are booked on this flight, in how many ways can the airline choose the seven passengers to be bumped?

5. A quiz has ten true-false questions. How many different ways is it possible to earn a score of exactly 70% on this quiz?

6. A test has ten true-false questions and five multiple-choice questions with four choices for each. In how many ways can this test be completed?

7. If you must answer only eight of ten questions on a test, how many ways do you have of choosing the questions you will omit?

8. An ice-cream store offers 15 flavors of ice cream. The specialty is a banana split with four scoops of ice cream. If each scoop must be a different flavor, how many different banana splits may be ordered?

9. A company uses a different three-letter security code for each of its employees. What is the maximum number of codes this security system can generate?

10. A group of students determines that they can stand in a row for their class picture in 120 different ways. How many students are in this class?

11. A coin is tossed ten times. In how many different ways can the result be three heads and seven tails?

12. The Yukon Territory in Canada uses a license-plate system for automobiles that consists of two letters followed by three numbers. Explain how we can know that fewer than 700,000 autos are licensed in the Yukon.

13. A group of friends have reserved a tennis court. They find that there are ten different ways in which two of them can play a singles game on this court. How many friends are in this group?

14. A pizza parlor advertises that they prepare 2048 different types of pizza. How many toppings does this parlor offer?

15. In Morse code, each letter is represented by a sequence of dots and dashes, with repetition allowed. How many letters can be represented using Morse code if three or fewer symbols are used?

16. The genetic code is based on the four nucleotides adenine (A), cytosine (C), guanine (G), and thymine (T). These are connected in long strings to form DNA molecules. For example, a sequence in the DNA may look like CAGTGGTACC . . . . The code uses "words," all the same length, that are composed of the nucleotides A, C, G, and T. It is known that at least 20 different words exist. What is the minimum word length necessary to generate 20 words?

17. Given 16 subjects from which to choose, in how many ways can a student select fields of study as follows?

(a) A major and a minor

(b) A major, a first minor, and a second minor

(c) A major and two minors

18. (a) How many three-digit numbers can be formed using the digits 0, 1, . . . , 9? (Remember, a three-digit number cannot have 0 as the leftmost digit.)

(b) If a number is chosen randomly from the set {0, 1, 2, . . . , 1000}, what is the probability that the number chosen is a three-digit number?

19–20 ■ An **anagram** of a word is a permutation of the letters of that word. For example, anagrams of the word *triangle* include *griantle*, *integral*, and *tenalgir*.

19. How many anagrams of the word *TRIANGLE* are possible?

20. How many anagrams are possible from the word *MISSISSIPPI*?

21. A committee of seven is to be chosen from a group of ten men and eight women. In how many ways can the committee be chosen using each of the following selection requirements?

(a) No restriction is placed on the number of men and women on the committee.

(b) The committee must have exactly four men and three women.

(c) Susie refuses to serve on the committee.

(d) At least five women must serve on the committee.

(e) At most two men can serve on the committee.

(f) The committee is to have a chairman, a vice chairman, a secretary, and four other members.

22. A jar contains ten red balls labeled 0, 1, 2, . . . , 9 and five white balls labeled 0, 1, 2, 3, 4. If a ball is drawn from the jar, find the probability of the given event.

(a) The ball is red.

(b) The ball is even-numbered.

(c) The ball is white and odd-numbered.

(d) The ball is red or odd-numbered.

23. If two balls are drawn from the jar in Exercise 22, find the probability of the given event.

(a) Both balls are red.

(b) One ball is white and the other is red.

(c) At least one ball is red.

(d) Both balls are red and even-numbered.

(e) Both balls are white and odd-numbered.

24. A coin is tossed three times in a row, and the outcomes of each toss are observed.
(a) Find the sample space for this experiment.
(b) Find the probability of getting three heads.
(c) Find the probability of getting two or more heads.
(d) Find the probability of getting tails on the first toss.

25. A shelf has ten books: two mysteries, four romance novels, and four mathematics textbooks. If you select a book at random to take to the beach, what is the probability that it turns out to be a mathematics text?

26. A die is rolled and a card is selected from a standard 52-card deck. What is the probability that both the die and the card show a six?

27. Find the probability that the indicated card is drawn at random from a 52-card deck.
(a) An ace
(b) An ace or a jack
(c) An ace or a spade
(d) A red ace

28. A card is drawn from a 52-card deck, a die is rolled, and a coin is tossed. Find the probability of each outcome.
(a) The ace of spades, a six, and heads
(b) A spade, a six, and heads
(c) A face card, a number greater than 3, and heads

29. Two dice are rolled. Find the probability of each outcome.
(a) The dice show the same number.
(b) The dice show different numbers.

30. Four cards are dealt from a standard 52-card deck. Find the probability that the cards are
(a) all kings   (b) all spades   (c) all the same color

31. In the "numbers game" lottery, a player picks a three-digit number (from 000 to 999), and if the number is selected in the drawing, the player wins $500. If another number with the same digits (in any order) is drawn, the player wins $50. John plays the number 159.
(a) What is the probability that he will win $500?
(b) What is the probability that he will win $50?

32. In a television game show, a contestant is given five cards with a different digit on each and is asked to arrange them to match the price of a brand-new car. If she gets the price right, she wins the car. What is the probability that she wins, assuming that she knows the first digit but must guess the remaining four?

33. Two dice are rolled. John gets $5 if they show the same number, or he pays $1 if they show different numbers. What is the expected value of this game?

34. Three dice are rolled. John gets $5 if they all show the same number; he pays $1 otherwise. What is the expected value of this game?

35. Mary will win $1,000,000 if she can name the 13 original states in the order in which they ratified the U.S. Constitution. Mary has no knowledge of this order, so she makes a guess. What is her expectation?

36. A pizza parlor offers 12 different toppings, one of which is anchovies. If a pizza is ordered at random, what is the probability that anchovies is one of the toppings?

37. A drawer contains an unorganized collection of 50 socks—20 are red and 30 are blue. Suppose the lights go out so Kathy can't distinguish the color of the socks.
(a) What is the minimum number of socks Kathy must take out of the drawer to be sure of getting a matching pair?
(b) If two socks are taken at random from the drawer, what is the probability that they make a matching pair?

38. A volleyball team has nine players. In how many ways can a starting lineup be chosen if it consists of two forward players and three defense players?

39. Zip codes consist of five digits.
(a) How many different zip codes are possible?
(b) How many different zip codes can be read when the envelope is turned upside down? (An upside down 9 is a 6; and 0, 1, and 8 are the same when read upside down.)
(c) What is the probability that a randomly chosen zip code can be read upside down?
(d) How many zip codes read the same upside down as right side up?

40. In the Zip+4 postal code system, zip codes consist of nine digits.
(a) How many different Zip+4 codes are possible?
(b) How many different Zip+4 codes are palindromes? (A *palindrome* is a number that reads the same from left to right as right to left.)
(c) What is the probability that a randomly chosen Zip+4 code is a palindrome?

41. Let $N = 3,600,000$. (Note that $N = 2^7 3^2 5^5$.)
(a) How many divisors does N have?
(b) How many even divisors does N have?
(c) How many divisors of N are multiples of 6?
(d) What is the probability that a randomly chosen divisor of N is even?

42. The U.S. Senate has two senators from each of the 50 states. In how many ways can a committee of five senators be chosen if no state is to have two members on the committee?

11 TEST

1. How many five-letter "words" can be made from the letters $A, B, C, D, E, F, G, H, I, J$ if repetition (a) is allowed or (b) is not allowed?

2. A restaurant offers five main courses, three types of desserts, and four kinds of drinks. In how many ways can a customer order a meal consisting of one choice from each category?

3. Over the past year, John has purchased 30 books.
   (a) In how many ways can he pick four of these books and arrange them, in order, on his nightstand bookshelf?
   (b) In how many ways can he choose four of these books to take with him on his vacation at the shore?

4. A commuter must travel from Ajax to Barrie and back every day. Four roads join the two cities. The commuter likes to vary the trip as much as possible, so he always leaves and returns by different roads. In how many different ways can he make the round-trip?

5. A pizza parlor offers four sizes of pizza and 14 different toppings. A customer may choose any number of toppings (or no topping at all). How many different pizzas does this parlor offer?

6. An *anagram* of a word is a rearrangement of the letters of the word.
   (a) How many anagrams of the word *LOVE* are possible?
   (b) How many different anagrams of the word *KISSES* are possible?

7. A board of directors consisting of eight members is to be chosen from a pool of 30 candidates. The board is to have a chairman, a treasurer, a secretary, and five other members. In how many ways can the board of directors be chosen?

8. One card is drawn from a deck. Find the probability of the given event.
   (a) The card is red.
   (b) The card is a king.
   (c) The card is a red king.

9. A jar contains five red balls, numbered 1 to 5, and eight white balls, numbered 1 to 8. A ball is chosen at random from the jar. Find the probability of the given event.
   (a) The ball is red.
   (b) The ball is even-numbered.
   (c) The ball is red or even-numbered.

10. Three people are chosen at random from a group of five men and ten women. What is the probability that all three are men?

11. Two dice are rolled. What is the probability of getting doubles?

12. In a group of four students, what is the probability that at least two have the same astrological sign?

13. You are to draw one card from a deck. If it is an ace, you win $10; if it is a face card, you win $1; otherwise, you lose $0.50. What is the expected value of this game?

Focus on Modeling
The Monte Carlo Method

A good way to familiarize ourselves with a fact is to experiment with it. For instance, to convince ourselves that the earth is a sphere (which was considered a major paradox at one time), we could go up in a space shuttle to see that it is so; to see whether a given equation is an identity, we might try some special cases to make sure there are no obvious counterexamples. In problems involving probability, we can perform an experiment many times and use the results to estimate the probability in question. In fact, we often model the experiment on a computer, thereby making it feasible to perform the experiment a large number of times. This technique is called the **Monte Carlo method**, named after the famous gambling casino in Monaco.

EXAMPLE 1 ■ The Contestant's Dilemma

In a TV game show, a contestant chooses one of three doors. Behind one of them is a valuable prize—the other two doors have nothing behind them. After the contestant has made her choice, the host opens one of the other two doors, one that he knows does not conceal a prize, and then gives her the opportunity to change her choice.

Should the contestant switch, stay, or does it matter? In other words, by switching doors, does she increase, decrease, or leave unchanged her probability of winning? At first, it may seem that switching doors doesn't make any difference. After all, two doors are left—one with the prize and one without—so it seems reasonable that the contestant has an equal chance of winning or losing. But if you play this game many times, you will find that by switching doors you actually win about $\frac{2}{3}$ of the time.

The authors modeled this game on a computer and found that in one million games the simulated contestant (who always switches) won 667,049 times—very close to $\frac{2}{3}$ of the time. Thus, it seems that switching doors does make a difference: Switching increases the contestant's chances of winning. This experiment forces us to reexamine our reasoning. Here is why switching doors is the correct strategy:

1. When the contestant first made her choice, she had a $\frac{1}{3}$ chance of winning. If she doesn't switch, no matter what the host does, her probability of winning remains $\frac{1}{3}$.

2. If the contestant decides to switch, she will switch to the winning door if she had initially chosen a losing one, or to a losing door if she had initially chosen the winning one. Since the probability of having initially selected a losing door is $\frac{2}{3}$, by switching the probability of winning then becomes $\frac{2}{3}$.

We conclude that the contestant should switch, because her probability of winning is $\frac{2}{3}$ if she switches and $\frac{1}{3}$ if she doesn't. Put simply, there is a much greater chance that she initially chose a losing door (since there are more of these), so she should switch. ■

Contestant: "I choose door number 2."

Contestant: "Oh no, what should I do?"

The "contestant's dilemma" problem discussed here is an example of how subtle probability can be. This problem was posed in a nationally syndicated column in *Parade* magazine in 1990. The correct solution was presented in the column, but it generated considerable controversy, with thousands of letters arguing that the solution was wrong. This shows how problems in probability can be quite tricky. Without a lot of experience in probabilistic thinking, it's easy to make a mistake. Even great mathematicians like D'Alembert and Roberval (see Problems 2 and 3) made mistakes in probability. Professor David Burton writes in his book *The History of Mathematics*, "Probability theory abounds in paradoxes that wrench the common sense and trip the unwary."

An experiment can be modeled using any computer language or programmable calculator that has a random-number generator. This is a command or function (usually called **Rnd** or **Rand**) that returns a randomly chosen number x with $0 \le x < 1$. In the next example we see how to use this to model a simple experiment.

EXAMPLE 2 ■ Monte Carlo Model of a Coin Toss

When a balanced coin is tossed, each outcome—"heads" or "tails"—has probability $\frac{1}{2}$. This doesn't mean that if we toss a coin several times, we will necessarily get exactly half heads and half tails. We would expect, however, the proportion of heads and of tails to get closer and closer to $\frac{1}{2}$ as the number of tosses increases. To test this hypothesis, we could toss a coin a very large number of times and keep track of the results. But this is a very tedious process, so we will use the Monte Carlo method to model this process.

To model a coin toss with a calculator or computer, we use the random-number generator to get a random number x such that $0 \le x < 1$. Because the number is chosen randomly, the probability that it lies in the first half of this interval $\left(0 \le x < \frac{1}{2}\right)$ is the same as the probability that it lies in the second half $\left(\frac{1}{2} \le x < 1\right)$. Thus, we could model the outcome "heads" by the event that $0 \le x < \frac{1}{2}$ and the outcome "tails" by the event that $\frac{1}{2} \le x < 1$.

An easier way to keep track of heads and tails is to note that if $0 \le x < 1$, then $0 \le 2x < 2$, and so $[\![2x]\!]$, the integer part of $2x$, is either 0 or 1, each with probability $\frac{1}{2}$. (On most programmable calculators, the function **Int** gives the integer part of a number.) Thus, we could model "heads" with the outcome "0" and "tails" with the outcome "1" when we take the integer part of $2x$. The program in the margin models 100 tosses of a coin on the TI-83 calculator. The graph in Figure 1 shows what proportion p of the tosses have come up "heads" after n tosses. As you can see, this proportion settles down near 0.5 as the number n of tosses increases—just as we hypothesized. ■

```
PROGRAM: HEADTAIL
:0→J:0→K
:For(N, 1, 100)
:rand→X
:int(2X)→Y
:J+(1−Y)→J
:K+Y→K
:END
:Disp "HEADS=",J
:Disp "TAILS=",K
```

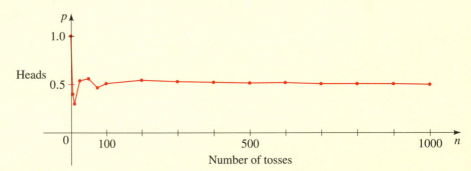

FIGURE 1

Relative frequency of "heads"

In general, if a process has n equally likely outcomes, then we can model the process using a random-number generator as follows: If our program or calculator produces the random number x, with $0 \le x < 1$, then the integer part of nx will be a random choice from the n integers $0, 1, 2, \ldots, n - 1$. Thus, we can use the outcomes $0, 1, 2, \ldots, n - 1$ as models for the outcomes of the actual experiment.

■ Problems

1. In a game show like the one described in Example 1, a prize is concealed behind one of ten doors. After the contestant chooses a door, the host opens eight losing doors and then gives the contestant the opportunity to switch to the other unopened door.

   (a) Play this game with a friend 30 or more times, using the strategy of switching doors each time. Count the number of times you win, and estimate the probability of winning with this strategy.

   (b) Calculate the probability of winning with the "switching" strategy using reasoning similar to that in Example 1. Compare with your result from part (a).

2. A couple intend to have two children. What is the probability that they will have one child of each sex? The French mathematician D'Alembert analyzed this problem (incorrectly) by reasoning that three outcomes are possible: two boys, or two girls, or one child of each sex. He concluded that the probability of having one of each sex is $\frac{1}{3}$, mistakenly assuming that the three outcomes are "equally likely."

   (a) Model this problem with a pair of coins (using "heads" for boys and "tails" for girls), or write a program to model the problem. Perform the experiment 40 or more times, counting the number of boy-girl combinations. Estimate the probability of having one child of each sex.

   (b) Calculate the correct probability of having one child of each sex, and compare this with your result from part (a).

3. A game between two players consists of tossing a coin. Player A gets a point if the coin shows heads, and player B gets a point if it shows tails. The first player to get six points wins an $8000 jackpot. As it happens, the police raid the place when player A has five points and B has three points. After everyone has calmed down, how should the jackpot be divided between the two players? In other words, what is the probability of A winning (and that of B winning) if the game were to continue?

   The French mathematicians Pascal and Fermat corresponded about this problem, and both came to the same correct conclusion (though by very different reasonings). Their friend Roberval disagreed with both of them. He argued that player A has probability $\frac{3}{4}$ of winning, because the game can end in the four ways H, TH, TTH, TTT, and in three of these, A wins. Roberval's reasoning was wrong.

   (a) Continue the game from the point at which it was interrupted, using either a coin or a modeling program. Perform this experiment 80 or more times, and estimate the probability that player A wins.

   (b) Calculate the probability that player A wins. Compare with your estimate from part (a).

4. In the World Series, the top teams in the National League and the American League play a best-of-seven series; that is, they play until one team has won four games. (No tie is allowed, so this results in a maximum of seven games.) Suppose the teams are evenly matched, so that the probability that either team wins a given game is $\frac{1}{2}$.

   (a) Use a coin or a modeling program to model a World Series, where "heads" represents a win by Team A and "tails" a win by Team B. Perform this experiment at least 80 times, keeping track of how many games are needed to decide each series. Estimate the probability that an evenly matched series will end in four games. Do the same for five, six, and seven games.

   (b) What is the probability that the series will end in four games? Five games? Six games? Seven games? Compare with your estimates from part (a).

   (c) Find the expected value for the number of games until the series ends. [*Hint:* This will be $P(\text{four games}) \times 4 + P(\text{five}) \times 5 + P(\text{six}) \times 6 + P(\text{seven}) \times 7.$]

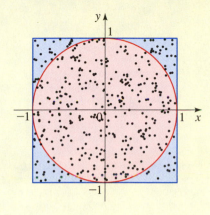

PROGRAM: PI
:0→P
:For(N, 1, 1000)
:rand→X:rand→Y
:P+((X²+Y²)<1)→P
:End
:Disp "PI IS APPROX",4∗P/N

5. In this problem we use the Monte Carlo method to estimate the value of π. The circle in the figure has radius 1, so its area is π and the square has area 4. If we choose a point at random from the square, the probability that it lies inside the circle will be

$$\frac{\text{area of circle}}{\text{area of square}} = \frac{\pi}{4}$$

The Monte Carlo method involves choosing many points inside the square. Then we have

$$\frac{\text{number of hits inside circle}}{\text{number of hits inside square}} \approx \frac{\pi}{4}$$

Thus, 4 times this ratio will give us an approximation for π.

To implement this method, we use a random-number generator to obtain the coordinates (x, y) of a random point in the square, and then check to see if it lies inside the circle (that is, we check if $x^2 + y^2 < 1$). Note that we only need to use points in the first quadrant, since the ratio of areas is the same in each quadrant. The program in the margin shows a way of doing this on the TI-83 calculator for 1000 randomly selected points.

Carry out this Monte Carlo simulation for as many points as you can. How do your results compare with the actual value of π? Do you think this is a reasonable way to get a good approximation for π?

6. The Monte Carlo method can be used to estimate the area under the graph of a function. The figure shows the region under the graph of $f(x) = x^2$, above the x-axis, between $x = 0$ and $x = 1$. If we choose a point in the square at random, the probability that it lies under the graph of $f(x) = x^2$ is the area under the graph divided by the area of the square. So if we randomly select a large number of points in the square, we have

$$\frac{\text{number of hits under graph}}{\text{number of hits in square}} \approx \frac{\text{area under graph}}{\text{area of square}}$$

Modify the program from Problem 5 to carry out this Monte Carlo simulation and approximate the required area. [*Note:* The exact value of this area will be found in Chapter 12.]

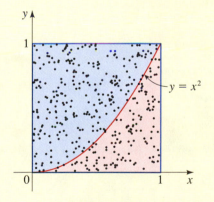

7. Choose two numbers at random from the interval $[0, 1)$. What is the probability that the sum of the two numbers is less than 1?
(a) Use a Monte Carlo model to estimate the probability.
(b) Calculate the exact value of the probability. [*Hint:* Call the numbers x and y. Choosing these numbers is the same as choosing an ordered pair (x, y) in the unit square $\{(x, y) \mid 0 \le x < 1, 0 \le y < 1\}$. What proportion of the points in this square corresponds to $x + y$ being less than 1?]

12

Limits: A Preview of Calculus

Limits are used to calculate quantities as diverse as the total water pressure on the face of a dam or the speed of a racing car.

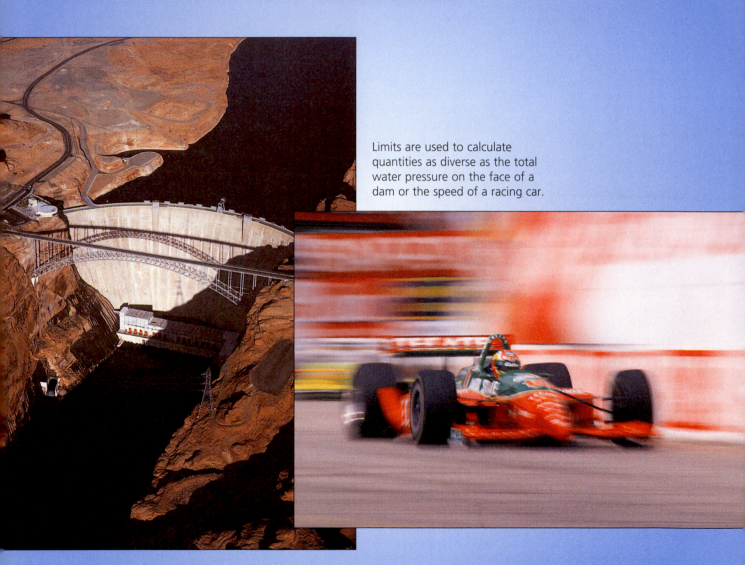

Our minds are finite, but we are surrounded by possibilities that are infinite, and the purpose of human life is to grasp as much as we can out of that infinitude.

ALFRED NORTH WHITEHEAD

The concept of a limit is the basic idea in calculus. After learning how to calculate limits of functions numerically, graphically, and algebraically in the first two sections of this chapter, we use them to find tangent lines and rates of change. Then we examine limits of sequences and employ them to calculate areas of regions with curved sides.

12.1 FINDING LIMITS NUMERICALLY AND GRAPHICALLY

In this section we use tables of values and graphs of functions to answer the question, What happens to the values $f(x)$ of a function f as the variable x approaches the number a?

Definition of Limit

We begin by investigating the behavior of the function f defined by

$$f(x) = x^2 - x + 2$$

for values of x near 2. The following table gives values of $f(x)$ for values of x close to 2 but not equal to 2.

x	$f(x)$	x	$f(x)$
1.0	2.000000	3.0	8.000000
1.5	2.750000	2.5	5.750000
1.8	3.440000	2.2	4.640000
1.9	3.710000	2.1	4.310000
1.95	3.852500	2.05	4.152500
1.99	3.970100	2.01	4.030100
1.995	3.985025	2.005	4.015025
1.999	3.997001	2.001	4.003001

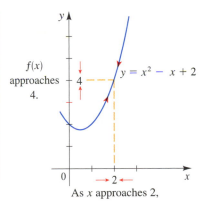

$f(x)$ approaches 4.

$y = x^2 - x + 2$

As x approaches 2,

FIGURE 1

From the table and the graph of f (a parabola) shown in Figure 1 we see that when x is close to 2 (on either side of 2), $f(x)$ is close to 4. In fact, it appears that we can make the values of $f(x)$ as close as we like to 4 by taking x sufficiently close to 2. We express this by saying "the limit of the function $f(x) = x^2 - x + 2$ as x approaches 2 is equal to 4." The notation for this is

$$\lim_{x \to 2} (x^2 - x + 2) = 4$$

In general, we use the following notation.

DEFINITION OF THE LIMIT OF A FUNCTION

We write

$$\lim_{x \to a} f(x) = L$$

and say

"the limit of $f(x)$, as x approaches a, equals L"

if we can make the values of $f(x)$ arbitrarily close to L (as close to L as we like) by taking x to be sufficiently close to a, but not equal to a.

Roughly speaking, this says that the values of $f(x)$ get closer and closer to the number L as x gets closer and closer to the number a (from either side of a) but $x \neq a$.

An alternative notation for $\lim_{x \to a} f(x) = L$ is

$$f(x) \to L \qquad \text{as} \qquad x \to a$$

which is usually read "$f(x)$ approaches L as x approaches a." This is the notation we used in Section 3.6 when discussing asymptotes of rational functions.

Notice the phrase "but $x \neq a$" in the definition of limit. This means that in finding the limit of $f(x)$ as x approaches a, we never consider $x = a$. In fact, $f(x)$ need not even be defined when $x = a$. The only thing that matters is how f is defined *near a*.

Figure 2 shows the graphs of three functions. Note that in part (c), $f(a)$ is not defined and in part (b), $f(a) \neq L$. But in each case, regardless of what happens at a, $\lim_{x \to a} f(x) = L$.

(a)

(b)

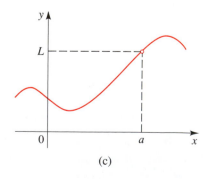
(c)

FIGURE 2

$\lim_{x \to a} f(x) = L$ in all three cases

Estimating Limits Numerically and Graphically

In Section 12.2 we will develop techniques for finding exact values of limits. For now, we use tables and graphs to estimate limits of functions.

EXAMPLE 1 ■ Estimating a Limit Numerically and Graphically

Guess the value of $\lim_{x \to 1} \dfrac{x-1}{x^2-1}$. Check your work with a graph.

SOLUTION

Notice that the function $f(x) = (x-1)/(x^2-1)$ is not defined when $x = 1$, but this doesn't matter because the definition of $\lim_{x \to a} f(x)$ says that we consider values of x that are close to a but not equal to a. The following tables give values of $f(x)$ (correct to six decimal places) for values of x that approach 1 (but are not equal to 1).

$x < 1$	$f(x)$		$x > 1$	$f(x)$
0.5	0.666667		1.5	0.400000
0.9	0.526316		1.1	0.476190
0.99	0.502513		1.01	0.497512
0.999	0.500250		1.001	0.499750
0.9999	0.500025		1.0001	0.499975

1

(1, 0.5)

0

FIGURE 3

2

On the basis of the values in the two tables, we make the guess that

$$\lim_{x \to 1} \frac{x-1}{x^2-1} = 0.5$$

As a graphical verification we use a graphing device to produce Figure 3. We see that when x is close to 1, y is close to 0.5. If we use the ZOOM and TRACE features to get a closer look, as in Figure 4, we notice that as x gets closer and closer to 1, y becomes closer and closer to 0.5. This reinforces our conclusion. ■

0.6

(1, 0.5)

0.9
0.4

1.1

FIGURE 4

EXAMPLE 2 ■ Finding a Limit from a Table

Find $\lim_{t \to 0} \dfrac{\sqrt{t^2+9}-3}{t^2}$.

SOLUTION

The table in the margin lists values of the function for several values of t near 0. As t approaches 0, the values of the function seem to approach $0.1666666\ldots$, and so we guess that

$$\lim_{t \to 0} \frac{\sqrt{t^2+9}-3}{t^2} = \frac{1}{6}$$

t	$\dfrac{\sqrt{t^2+9}-3}{t^2}$
± 1.0	0.16228
± 0.5	0.16553
± 0.1	0.16662
± 0.05	0.16666
± 0.01	0.16667

t	$\dfrac{\sqrt{t^2 + 9} - 3}{t^2}$
± 0.0005	0.16800
± 0.0001	0.20000
± 0.00005	0.00000
± 0.00001	0.00000

In Example 2 what would have happened if we had taken even smaller values of t? The table in the margin shows the results from one calculator; you can see that something strange seems to be happening.

If you try these calculations on your own calculator, you might get different values, but eventually you will get the value 0 if you make t sufficiently small. Does this mean that the answer is really 0 instead of $\frac{1}{6}$? No, the value of the limit is $\frac{1}{6}$, as we will show in the next section. The problem is that the calculator gave false values because $\sqrt{t^2 + 9}$ is very close to 3 when t is small. (In fact, when t is sufficiently small, a calculator's value for $\sqrt{t^2 + 9}$ is 3.000 . . . to as many digits as the calculator is capable of carrying.)

Something similar happens when we try to graph the function

$$f(t) = \frac{\sqrt{t^2 + 9} - 3}{t^2}$$

of Example 2 on a graphing device. Parts (a) and (b) of Figure 5 show quite accurate graphs of f and when we use the $\boxed{\text{TRACE}}$ feature, we can easily estimate that the limit is about $\frac{1}{6}$. But if we zoom in too far, as in parts (c) and (d), then we get inaccurate graphs, again because of problems with subtraction.

(a) $[-5, 5]$ by $[-0.1, 0.3]$ (b) $[-0.1, 0.1]$ by $[-0.1, 0.3]$ (c) $[-10^{-6}, 10^{-6}]$ by $[-0.1, 0.3]$ (d) $[-10^{-7}, 10^{-7}]$ by $[-0.1, 0.3]$

FIGURE 5

Limits That Fail to Exist

Functions do not necessarily approach a finite value at every point. In other words, it's possible for a limit not to exist. The next three examples illustrate ways in which this can happen.

EXAMPLE 3 ■ A Limit that Fails to Exist

The Heaviside function H is defined by

$$H(t) = \begin{cases} 0 & \text{if } t < 0 \\ 1 & \text{if } t \geqslant 0 \end{cases}$$

[This function is named after the electrical engineer Oliver Heaviside (1850–1925) and can be used to describe an electric current that is switched on at time $t = 0$.] Its graph is shown in Figure 6.

FIGURE 6

As t approaches 0 from the left, $H(t)$ approaches 0. As t approaches 0 from the right, $H(t)$ approaches 1. There is no single number that $H(t)$ approaches as t approaches 0. Therefore, $\lim_{t \to 0} H(t)$ does not exist. ∎

EXAMPLE 4 ■ A Function That Oscillates

Find $\lim\limits_{x \to 0} \sin \dfrac{\pi}{x}$.

SOLUTION

The function $f(x) = \sin(\pi/x)$ is undefined at 0. Evaluating the function for some small values of x, we get

$$f(1) = \sin \pi = 0 \qquad\qquad f\left(\tfrac{1}{2}\right) = \sin 2\pi = 0$$

$$f\left(\tfrac{1}{3}\right) = \sin 3\pi = 0 \qquad\qquad f\left(\tfrac{1}{4}\right) = \sin 4\pi = 0$$

$$f(0.1) = \sin 10\pi = 0 \qquad f(0.01) = \sin 100\pi = 0$$

Similarly, $f(0.001) = f(0.0001) = 0$. On the basis of this information we might be tempted to guess that

$$\lim_{x \to 0} \sin \frac{\pi}{x} = 0$$

 but this time our guess is wrong. Note that although $f(1/n) = \sin n\pi = 0$ for any integer n, it is also true that $f(x) = 1$ for infinitely many values of x that approach 0. (See the graph in Figure 7.)

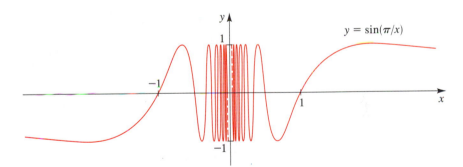

FIGURE 7

The broken lines indicate that the values of $\sin(\pi/x)$ oscillate between 1 and -1 infinitely often as x approaches 0. Since the values of $f(x)$ do not approach a fixed number as x approaches 0,

$$\lim_{x \to 0} \sin \frac{\pi}{x} \quad \text{does not exist} \qquad\qquad ■$$

Example 4 illustrates some of the pitfalls in guessing the value of a limit. It is easy to guess the wrong value if we use inappropriate values of x, but it is difficult to know when to stop calculating values. And, as the discussion after Example 2 shows, sometimes calculators and computers give the wrong values. In the next two sections, however, we will develop foolproof methods for calculating limits.

x	$\dfrac{1}{x^2}$
± 1	1
± 0.5	4
± 0.2	25
± 0.1	100
± 0.05	400
± 0.01	10,000
± 0.001	1,000,000

EXAMPLE 5 ■ A Function with a Vertical Asymptote

Find $\displaystyle\lim_{x \to 0} \frac{1}{x^2}$ if it exists.

SOLUTION

As x becomes close to 0, x^2 also becomes close to 0, and $1/x^2$ becomes very large. (See the table in the margin.) In fact, it appears from the graph of the function $f(x) = 1/x^2$ shown in Figure 8 that the values of $f(x)$ can be made arbitrarily large by taking x close enough to 0. Thus, the values of $f(x)$ do not approach a number, so $\lim_{x \to 0} (1/x^2)$ does not exist.

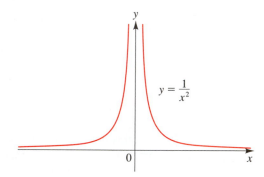

FIGURE 8

To indicate the kind of behavior exhibited in Example 5, we use the notation

$$\lim_{x \to 0} \frac{1}{x^2} = \infty$$

 This does not mean that we are regarding ∞ as a number. Nor does it mean that the limit exists. It simply expresses the particular way in which the limit does not exist: $1/x^2$ can be made as large as we like by taking x close enough to 0. Notice that the line $x = 0$ (the y-axis) is a vertical asymptote in the sense we described in Section 3.6.

One-Sided Limits

We noticed in Example 3 that $H(t)$ approaches 0 as t approaches 0 from the left and $H(t)$ approaches 1 as t approaches 0 from the right. We indicate this situation symbolically by writing

$$\lim_{t \to 0^-} H(t) = 0 \qquad \text{and} \qquad \lim_{t \to 0^+} H(t) = 1$$

The symbol "$t \to 0^-$" indicates that we consider only values of t that are less than 0. Likewise, "$t \to 0^+$" indicates that we consider only values of t that are greater than 0.

DEFINITION OF A ONE-SIDED LIMIT

We write

$$\lim_{x \to a^-} f(x) = L$$

and say the "left-hand limit of $f(x)$ as x approaches a" [or the "limit of $f(x)$ as x approaches a from the left"] is equal to L if we can make the values of $f(x)$ arbitrarily close to L by taking x to be sufficiently close to a and x less than a.

Notice that this definition differs from the definition of a two-sided limit only in that we require x to be less than a. Similarly, if we require that x be greater than a, we get "the **right-hand limit of $f(x)$ as x approaches a** is equal to L" and we write

$$\lim_{x \to a^+} f(x) = L$$

Thus, the symbol "$x \to a^+$" means that we consider only $x > a$. These definitions are illustrated in Figure 9.

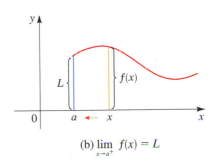

FIGURE 9

(a) $\lim_{x \to a^-} f(x) = L$

(b) $\lim_{x \to a^+} f(x) = L$

By comparing the definitions of two-sided and one-sided limits, we see that the following is true.

$$\lim_{x \to a} f(x) = L \quad \text{if and only if} \quad \lim_{x \to a^-} f(x) = L \quad \text{and} \quad \lim_{x \to a^+} f(x) = L$$

Thus, if the left-hand and right-hand limits are different, the (two-sided) limit does not exist. We use this fact in the next two examples.

EXAMPLE 6 ■ Limits from a Graph

The graph of a function g is shown in Figure 10. Use it to state the values (if they exist) of the following:

(a) $\lim_{x \to 2^-} g(x), \quad \lim_{x \to 2^+} g(x), \quad \lim_{x \to 2} g(x)$

(b) $\lim_{x \to 5^-} g(x), \quad \lim_{x \to 5^+} g(x), \quad \lim_{x \to 5} g(x)$

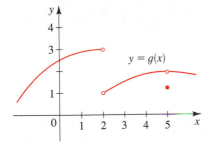

FIGURE 10

SOLUTION

(a) From the graph we see that the values of $g(x)$ approach 3 as x approaches 2 from the left, but they approach 1 as x approaches 2 from the right. Therefore

$$\lim_{x \to 2^-} g(x) = 3 \qquad \text{and} \qquad \lim_{x \to 2^+} g(x) = 1$$

Since the left- and right-hand limits are different, we conclude that $\lim_{x \to 2} g(x)$ does not exist.

(b) The graph also shows that

$$\lim_{x \to 5^-} g(x) = 2 \qquad \text{and} \qquad \lim_{x \to 5^+} g(x) = 2$$

This time the left- and right-hand limits are the same, and so we have

$$\lim_{x \to 5} g(x) = 2$$

Despite this fact, notice that $g(5) \neq 2$.   ■

EXAMPLE 7 ■ A Piecewise-Defined Function

Let f be the function defined by

$$f(x) = \begin{cases} x + 2 & \text{if } x < 1 \\ 2x^2 & \text{if } x \geq 1 \end{cases}$$

Graph f, and use the graph to find the following:

(a) $\lim_{x \to 1^-} f(x)$ (b) $\lim_{x \to 1^+} f(x)$ (c) $\lim_{x \to 1} f(x)$

SOLUTION

The graph of f is shown in Figure 11. From the graph we see that the values of $f(x)$ approach 3 as x approaches 1 from the left, but they approach 2 as x approaches 1 from the right. Thus, the left- and right-hand limits are not equal. So we have

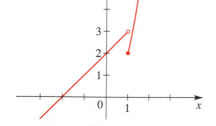

FIGURE 11

(a) $\lim_{x \to 1^-} f(x) = 3$ (b) $\lim_{x \to 1^+} f(x) = 2$ (c) $\lim_{x \to 1} f(x)$ does not exist.   ■

12.1 EXERCISES

1–6 ■ Complete the table of values (to five decimal places) and use the table to estimate the value of the limit.

1. $\lim_{x \to 4} \dfrac{\sqrt{x} - 2}{x - 4}$

x	3.9	3.99	3.999	4.001	4.01	4.1
$f(x)$						

2. $\lim_{x \to 2} \dfrac{x - 2}{x^2 + x - 6}$

x	1.9	1.99	1.999	2.001	2.01	2.1
$f(x)$						

3. $\lim\limits_{x \to 1} \dfrac{x - 1}{x^3 - 1}$

x	0.9	0.99	0.999	1.001	1.01	1.1
$f(x)$						

4. $\lim\limits_{x \to 0} \dfrac{e^x - 1}{x}$

x	-0.1	-0.01	-0.001	0.001	0.01	0.1
$f(x)$						

5. $\lim\limits_{x \to 0} \dfrac{\sin x}{x}$

x	± 1	± 0.5	± 0.1	± 0.05	± 0.01
$f(x)$					

6. $\lim\limits_{x \to 0^+} x \ln x$

x	0.1	0.01	0.001	0.0001	0.00001
$f(x)$					

7–12 ■ Use a table of values to estimate the value of the limit. Then use a graphing device to confirm your result graphically.

7. $\lim\limits_{x \to -4} \dfrac{x + 4}{x^2 + 7x + 12}$

8. $\lim\limits_{x \to 1} \dfrac{x^3 - 1}{x^2 - 1}$

9. $\lim\limits_{x \to 0} \dfrac{5^x - 3^x}{x}$

10. $\lim\limits_{x \to 0} \dfrac{\sqrt{x + 9} - 3}{x}$

11. $\lim\limits_{x \to 1} \left(\dfrac{1}{\ln x} - \dfrac{1}{x - 1} \right)$

12. $\lim\limits_{x \to 0} \dfrac{\tan 2x}{\tan 3x}$

13. For the function f whose graph is given, state the value of the given quantity, if it exists. If it does not exist, explain why.

(a) $\lim\limits_{x \to 1^-} f(x)$
(b) $\lim\limits_{x \to 1^+} f(x)$
(c) $\lim\limits_{x \to 1} f(x)$
(d) $\lim\limits_{x \to 5} f(x)$
(e) $f(5)$

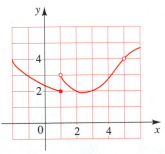

14. For the function f whose graph is given, state the value of the given quantity, if it exists. If it does not exist, explain why.

(a) $\lim\limits_{x \to 0} f(x)$
(b) $\lim\limits_{x \to 3^-} f(x)$
(c) $\lim\limits_{x \to 3^+} f(x)$
(d) $\lim\limits_{x \to 3} f(x)$
(e) $f(3)$

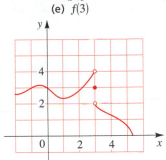

15. For the function g whose graph is given, state the value of the given quantity, if it exists. If it does not exist, explain why.

(a) $\lim\limits_{t \to 0^-} g(t)$
(b) $\lim\limits_{t \to 0^+} g(t)$
(c) $\lim\limits_{t \to 0} g(t)$
(d) $\lim\limits_{t \to 2^-} g(t)$
(e) $\lim\limits_{t \to 2^+} g(t)$
(f) $\lim\limits_{t \to 2} g(t)$
(g) $g(2)$
(h) $\lim\limits_{t \to 4} g(t)$

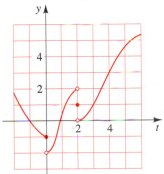

16. State the value of the limit, if it exists, from the given graph of f. If it does not exist, explain why.

(a) $\lim\limits_{x \to 3} f(x)$
(b) $\lim\limits_{x \to 1} f(x)$
(c) $\lim\limits_{x \to -3} f(x)$
(d) $\lim\limits_{x \to 2^-} f(x)$
(e) $\lim\limits_{x \to 2^+} f(x)$
(f) $\lim\limits_{x \to 2} f(x)$

17–22 ■ Use a graphing device to determine whether the limit exists. If the limit exists, estimate its value to two decimal places.

17. $\lim\limits_{x \to 1} \dfrac{x^3 + x^2 + 3x - 5}{2x^2 - 5x + 3}$

18. $\lim\limits_{x \to 2} \dfrac{x^3 + 6x^2 - 5x + 1}{x^3 - x^2 - 8x + 12}$

19. $\lim\limits_{x \to 0} \ln(\sin^2 x)$

20. $\lim\limits_{x \to 0} \dfrac{x^2}{\cos 5x - \cos 4x}$

21. $\lim\limits_{x \to 0} \cos \dfrac{1}{x}$

22. $\lim\limits_{x \to 0} \dfrac{1}{1 + e^{1/x}}$

23–26 ■ Graph the piecewise-defined function and use your graph to find the values of the limits, if they exist.

23. $f(x) = \begin{cases} x^2 & \text{if } x \leq 2 \\ 6 - x & \text{if } x > 2 \end{cases}$

(a) $\lim\limits_{x \to 2^-} f(x)$ (b) $\lim\limits_{x \to 2^+} f(x)$ (c) $\lim\limits_{x \to 2} f(x)$

24. $f(x) = \begin{cases} 2 & \text{if } x < 0 \\ x + 1 & \text{if } x \geq 0 \end{cases}$

(a) $\lim\limits_{x \to 0^-} f(x)$ (b) $\lim\limits_{x \to 0^+} f(x)$ (c) $\lim\limits_{x \to 0} f(x)$

25. $f(x) = \begin{cases} -x + 3 & \text{if } x < -1 \\ 3 & \text{if } x \geq -1 \end{cases}$

(a) $\lim\limits_{x \to -1^-} f(x)$ (b) $\lim\limits_{x \to -1^+} f(x)$ (c) $\lim\limits_{x \to -1} f(x)$

26. $f(x) = \begin{cases} 2x + 10 & \text{if } x \leq -2 \\ -x + 4 & \text{if } x > -2 \end{cases}$

(a) $\lim\limits_{x \to -2^-} f(x)$ (b) $\lim\limits_{x \to -2^+} f(x)$ (c) $\lim\limits_{x \to -2} f(x)$

DISCOVERY · DISCUSSION

27. A Function with Specified Limits   Sketch the graph of an example of a function f that satisfies all of the following conditions.

$$\lim\limits_{x \to 0^-} f(x) = 2 \qquad \lim\limits_{x \to 0^+} f(x) = 0$$

$$\lim\limits_{x \to 2} f(x) = 1 \qquad f(0) = 2 \qquad f(2) = 3$$

How many such functions are there?

28. Graphing Calculator Pitfalls
(a) Evaluate $h(x) = (\tan x - x)/x^3$ for $x = 1, 0.5, 0.1, 0.05, 0.01$, and 0.005.

(b) Guess the value of $\lim\limits_{x \to 0} \dfrac{\tan x - x}{x^3}$.

(c) Evaluate $h(x)$ for successively smaller values of x until you finally reach 0 values for $h(x)$. Are you still confident that your guess in part (b) is correct? Explain why you eventually obtained 0 values.

(d) Graph the function h in the viewing rectangle $[-1, 1]$ by $[0, 1]$. Then zoom in toward the point where the graph crosses the y-axis to estimate the limit of $h(x)$ as x approaches 0. Continue to zoom in until you observe distortions in the graph of h. Compare with the results of part (c).

12.2   FINDING LIMITS ALGEBRAICALLY

In Section 12.1 we used calculators and graphs to guess the values of limits, but we saw that such methods don't always lead to the correct answer. In this section we use the following properties of limits, called the *Limit Laws*, to calculate limits.

LIMIT LAWS

Suppose that c is a constant and that the following limits exist:

$$\lim\limits_{x \to a} f(x) \qquad \text{and} \qquad \lim\limits_{x \to a} g(x)$$

Then

1. $\lim\limits_{x \to a} [f(x) + g(x)] = \lim\limits_{x \to a} f(x) + \lim\limits_{x \to a} g(x)$ Sum Law

2. $\lim\limits_{x \to a} [f(x) - g(x)] = \lim\limits_{x \to a} f(x) - \lim\limits_{x \to a} g(x)$ Difference Law

3. $\lim\limits_{x \to a} [cf(x)] = c \lim\limits_{x \to a} f(x)$ Constant Multiple Law

4. $\lim\limits_{x \to a} [f(x)g(x)] = \lim\limits_{x \to a} f(x) \cdot \lim\limits_{x \to a} g(x)$ Product Law

5. $\lim\limits_{x \to a} \dfrac{f(x)}{g(x)} = \dfrac{\lim\limits_{x \to a} f(x)}{\lim\limits_{x \to a} g(x)}$ if $\lim\limits_{x \to a} g(x) \neq 0$ Quotient Law

These five laws can be stated verbally as follows:

Sum Law   **1.** The limit of a sum is the sum of the limits.

Difference Law   **2.** The limit of a difference is the difference of the limits.

Constant Multiple Law   **3.** The limit of a constant times a function is the constant times the limit of the function.

Product Law   **4.** The limit of a product is the product of the limits.

Quotient Law   **5.** The limit of a quotient is the quotient of the limits (provided that the limit of the denominator is not 0).

It's easy to believe that these properties are true. For instance, if $f(x)$ is close to L and $g(x)$ is close to M, it is reasonable to conclude that $f(x) + g(x)$ is close to $L + M$. This gives us an intuitive basis for believing that Law 1 is true.

If we use the Product Law repeatedly with $g(x) = f(x)$, we obtain the following Power Law. A similar law holds for roots.

LIMIT LAWS

6. $\displaystyle\lim_{x \to a} [f(x)]^n = \left[\lim_{x \to a} f(x)\right]^n$ where n is a positive integer   Power Law

7. $\displaystyle\lim_{x \to a} \sqrt[n]{f(x)} = \sqrt[n]{\lim_{x \to a} f(x)}$ where n is a positive integer   Root Law

$\left[\text{If } n \text{ is even, we assume that } \lim_{x \to a} f(x) > 0.\right]$

In words, these laws say:

Power Law   **6.** The limit of a power is the power of the limit.

Root Law   **7.** The limit of a root is the root of the limit.

EXAMPLE 1 ■ Using the Limit Laws

Use the Limit Laws and the graphs of f and g in Figure 1 to evaluate the following limits, if they exist.

(a) $\displaystyle\lim_{x \to -2} [f(x) + 5g(x)]$

(b) $\displaystyle\lim_{x \to 1} [f(x)g(x)]$

(c) $\displaystyle\lim_{x \to 2} \frac{f(x)}{g(x)}$

(d) $\displaystyle\lim_{x \to 1} [f(x)]^3$

FIGURE 1

SOLUTION

(a) From the graphs of f and g we see that

$$\lim_{x \to -2} f(x) = 1 \quad \text{and} \quad \lim_{x \to -2} g(x) = -1$$

Therefore, we have

$$\lim_{x \to -2}[f(x) + 5g(x)] = \lim_{x \to -2} f(x) + \lim_{x \to -2}[5g(x)] \qquad \text{Law 1}$$

$$= \lim_{x \to -2} f(x) + 5\lim_{x \to -2} g(x) \qquad \text{Law 3}$$

$$= 1 + 5(-1) = -4$$

(b) We see that $\lim_{x \to 1} f(x) = 2$. But $\lim_{x \to 1} g(x)$ does not exist because the left- and right-hand limits are different:

$$\lim_{x \to 1^-} g(x) = -2 \qquad \lim_{x \to 1^+} g(x) = -1$$

So we can't use Law 4. The given limit does not exist, since the left-hand limit is not equal to the right-hand limit.

(c) The graphs show that

$$\lim_{x \to 2} f(x) \approx 1.4 \qquad \text{and} \qquad \lim_{x \to 2} g(x) = 0$$

Because the limit of the denominator is 0, we can't use Law 5. The given limit does not exist because the denominator approaches 0 while the numerator approaches a nonzero number.

(d) Since $\lim_{x \to 1} f(x) = 2$, we use Law 6 to get

$$\lim_{x \to 1}[f(x)]^3 = [\lim_{x \to 1} f(x)]^3 \qquad \text{Law 6}$$

$$= 2^3 = 8$$

Applying the Limit Laws

In applying the Limit Laws, we need to use four special limits.

> **SOME SPECIAL LIMITS**
>
> 1. $\lim_{x \to a} c = c$
> 2. $\lim_{x \to a} x = a$
> 3. $\lim_{x \to a} x^n = a^n$ where n is a positive integer
> 4. $\lim_{x \to a} \sqrt[n]{x} = \sqrt[n]{a}$ where n is a positive integer and $a > 0$

Special Limits 1 and 2 are intuitively obvious—looking at the graphs of $y = c$ and $y = x$ will convince you of their validity. Limits 3 and 4 are special cases of Limit Laws 6 and 7.

EXAMPLE 2 ■ Using the Limit Laws

Evaluate the following limits and justify each step.

(a) $\lim_{x \to 5}(2x^2 - 3x + 4)$

(b) $\lim_{x \to -2} \dfrac{x^3 + 2x^2 - 1}{5 - 3x}$

SOLUTION

(a)
$$\lim_{x \to 5} (2x^2 - 3x + 4) = \lim_{x \to 5}(2x^2) - \lim_{x \to 5}(3x) + \lim_{x \to 5} 4 \qquad \text{Laws 2 and 1}$$

$$= 2 \lim_{x \to 5} x^2 - 3 \lim_{x \to 5} x + \lim_{x \to 5} 4 \qquad \text{Law 3}$$

$$= 2(5^2) - 3(5) + 4 \qquad \text{Special Limits 3, 2, and 1}$$

$$= 39$$

(b) We start by using Law 5, but its use is fully justified only at the final stage when we see that the limits of the numerator and denominator exist and the limit of the denominator is not 0.

$$\lim_{x \to -2} \frac{x^3 + 2x^2 - 1}{5 - 3x} = \frac{\displaystyle\lim_{x \to -2}(x^3 + 2x^2 - 1)}{\displaystyle\lim_{x \to -2}(5 - 3x)} \qquad \text{Law 5}$$

$$= \frac{\displaystyle\lim_{x \to -2} x^3 + 2 \lim_{x \to -2} x^2 - \lim_{x \to -2} 1}{\displaystyle\lim_{x \to -2} 5 - 3 \lim_{x \to -2} x} \qquad \text{Laws 1, 2, and 3}$$

$$= \frac{(-2)^3 + 2(-2)^2 - 1}{5 - 3(-2)} \qquad \text{Special Limits 3, 2, and 1}$$

$$= -\frac{1}{11} \qquad \blacksquare$$

If we let $f(x) = 2x^2 - 3x + 4$, then $f(5) = 39$. In Example 2(a), we found that $\lim_{x \to 5} f(x) = 39$. In other words, we would have gotten the correct answer by substituting 5 for x. Similarly, direct substitution provides the correct answer in part (b). The functions in Example 2 are a polynomial and a rational function, respectively, and similar use of the Limit Laws proves that direct substitution always works for such functions. We state this fact as follows.

LIMITS BY DIRECT SUBSTITUTION

If f is a polynomial or a rational function and a is in the domain of f, then

$$\lim_{x \to a} f(x) = f(a)$$

Functions with this direct substitution property are called **continuous at** a. You will learn more about continuous functions when you study calculus.

EXAMPLE 3 ■ **Finding Limits by Direct Substitution**

Evaluate the following limits.

(a) $\lim_{x \to 3} (2x^3 - 10x - 8)$

(b) $\lim_{x \to -1} \dfrac{x^2 + 5x}{x^4 + 2}$

SOLUTION

(a) The function $f(x) = 2x^3 - 10x - 12$ is a polynomial, so we can find the limit by direct substitution:

$$\lim_{x \to 3} (2x^3 - 10x - 12) = 2(3)^3 - 10(3) - 8 = 16$$

(b) The function $f(x) = (x^2 + 5x)/(x^4 + 2)$ is a rational function, and $x = -1$ is in its domain (because the denominator is not zero for $x = -1$). Thus, we can find the limit by direct substitution:

$$\lim_{x \to -1} \frac{x^2 + 5x}{x^4 + 2} = \frac{(-1)^2 + 5(-1)}{(-1)^4 + 2} = -\frac{4}{3} \qquad \blacksquare$$

■ Finding Limits Using Algebra and the Limit Laws

As we saw in Example 3, evaluating limits by direct substitution is easy. But not all limits can be evaluated this way. In fact, most of the situations in which limits are useful requires us to work harder to evaluate the limit. The next three examples illustrate how we can use algebra to find limits.

EXAMPLE 4 ■ Finding a Limit by Canceling a Common Factor

Find $\lim\limits_{x \to 1} \dfrac{x - 1}{x^2 - 1}$.

SOLUTION

Let $f(x) = (x - 1)/(x^2 - 1)$. We can't find the limit by substituting $x = 1$ because $f(1)$ isn't defined. Nor can we apply the Quotient Law because the limit of the denominator is 0. Instead, we need to do some preliminary algebra. We factor the denominator as a difference of squares:

$$\frac{x - 1}{x^2 - 1} = \frac{x - 1}{(x - 1)(x + 1)}$$

The numerator and denominator have a common factor of $x - 1$. When we take the limit as x approaches 1, we have $x \neq 1$ and so $x - 1 \neq 0$. Therefore, we can cancel the common factor and compute the limit as follows:

$$\lim_{x \to 1} \frac{x - 1}{x^2 - 1} = \lim_{x \to 1} \frac{x - 1}{(x - 1)(x + 1)} \qquad \text{Factor}$$

$$= \lim_{x \to 1} \frac{1}{x + 1} \qquad \text{Cancel}$$

$$= \frac{1}{1 + 1} = \frac{1}{2} \qquad \text{Let } x \to 1$$

This calculation confirms algebraically the answer we got numerically and graphically in Example 1 in Section 12.1. $\qquad \blacksquare$

Newton and Limits

In 1687 Isaac Newton (see page 854) published his masterpiece *Principia Mathematica*. In this work, the greatest scientific treatise ever written, Newton set forth his version of calculus and used it to investigate mechanics, fluid dynamics, and wave motion, and to explain the motion of planets and comets.

The beginnings of calculus are found in the calculations of areas and volumes by ancient Greek scholars such as Eudoxus and Archimedes. Although aspects of the idea of a limit are implicit in their "method of exhaustion," Eudoxus and Archimedes never explicitly formulated the concept of a limit. Likewise, mathematicians such as Cavalieri, Fermat, and Barrow, the immediate precursors of Newton in the development of calculus, did not actually use limits. It was Isaac Newton who first talked explicitly about limits. He explained that the main idea behind limits is that quantities "approach nearer than by any given difference." Newton stated that the limit was the basic concept in calculus, but it was left to later mathematicians like Cauchy to clarify these ideas.

EXAMPLE 5 ■ Finding a Limit by Simplifying

Evaluate $\displaystyle\lim_{h \to 0} \frac{(3+h)^2 - 9}{h}$.

SOLUTION

We can't use direct substitution to evaluate this limit, because the limit of the denominator is 0. So we first simplify the limit algebraically.

$$\lim_{h \to 0} \frac{(3+h)^2 - 9}{h} = \lim_{h \to 0} \frac{(9 + 6h + h^2) - 9}{h} \quad \text{Expand}$$

$$= \lim_{h \to 0} \frac{6h + h^2}{h} \quad \text{Simplify}$$

$$= \lim_{h \to 0} (6 + h) \quad \text{Cancel } h$$

$$= 6 \quad \text{Let } h \to 0 \qquad ■$$

EXAMPLE 6 ■ Finding a Limit by Rationalizing

Find $\displaystyle\lim_{t \to 0} \frac{\sqrt{t^2 + 9} - 3}{t^2}$.

SOLUTION

We can't apply the Quotient Law immediately, since the limit of the denominator is 0. Here the preliminary algebra consists of rationalizing the numerator:

$$\lim_{t \to 0} \frac{\sqrt{t^2 + 9} - 3}{t^2} = \lim_{t \to 0} \frac{\sqrt{t^2 + 9} - 3}{t^2} \cdot \frac{\sqrt{t^2 + 9} + 3}{\sqrt{t^2 + 9} + 3} \quad \text{Rationalize numerator}$$

$$= \lim_{t \to 0} \frac{(t^2 + 9) - 9}{t^2(\sqrt{t^2 + 9} + 3)} = \lim_{t \to 0} \frac{t^2}{t^2(\sqrt{t^2 + 9} + 3)}$$

$$= \lim_{t \to 0} \frac{1}{\sqrt{t^2 + 9} + 3} = \frac{1}{\sqrt{\lim_{t \to 0}(t^2 + 9)} + 3} = \frac{1}{3 + 3} = \frac{1}{6}$$

This calculation confirms the guess that we made in Example 2 in Section 12.1. ■

Using Left- and Right-Hand Limits

Some limits are best calculated by first finding the left- and right-hand limits. The following theorem is a reminder of what we discovered in Section 12.1. It says that *a two-sided limit exists if and only if both of the one-sided limits exist and are equal.*

$$\lim_{x \to a} f(x) = L \qquad \text{if and only if} \qquad \lim_{x \to a^-} f(x) = L = \lim_{x \to a^+} f(x)$$

When computing one-sided limits, we use the fact that the Limit Laws also hold for one-sided limits.

EXAMPLE 7 ■ Comparing Right and Left Limits

Show that $\lim\limits_{x \to 0} |x| = 0$.

SOLUTION

Recall that

$$|x| = \begin{cases} x & \text{if } x \geq 0 \\ -x & \text{if } x < 0 \end{cases}$$

The result of Example 7 looks plausible from Figure 2.

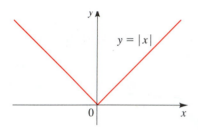

FIGURE 2

Since $|x| = x$ for $x > 0$, we have

$$\lim_{x \to 0^+} |x| = \lim_{x \to 0^+} x = 0$$

For $x < 0$, we have $|x| = -x$ and so

$$\lim_{x \to 0^-} |x| = \lim_{x \to 0^-} (-x) = 0$$

Therefore

$$\lim_{x \to 0} |x| = 0$$ ■

EXAMPLE 8 ■ Comparing Right and Left Limits

Prove that $\lim_{x \to 0} \dfrac{|x|}{x}$ does not exist.

SOLUTION

Since $|x| = x$ for $x > 0$ and $|x| = -x$ for $x < 0$, we have

$$\lim_{x \to 0^+} \frac{|x|}{x} = \lim_{x \to 0^+} \frac{x}{x} = \lim_{x \to 0^+} 1 = 1$$

$$\lim_{x \to 0^-} \frac{|x|}{x} = \lim_{x \to 0^-} \frac{-x}{x} = \lim_{x \to 0^-} (-1) = -1$$

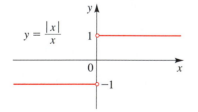

FIGURE 3

Since the right-hand and left-hand limits exist and are different, it follows that $\lim_{x \to 0} |x|/x$ does not exist. The graph of the function $f(x) = |x|/x$ is shown in Figure 3 and supports the limits that we found.   ■

EXAMPLE 9 ■ The Limit of a Piecewise-Defined Function

Let

$$f(x) = \begin{cases} \sqrt{x - 4} & \text{if } x > 4 \\ 8 - 2x & \text{if } x < 4 \end{cases}$$

Determine whether $\lim_{x \to 4} f(x)$ exists.

SOLUTION

Since $f(x) = \sqrt{x - 4}$ for $x > 4$, we have

$$\lim_{x \to 4^+} f(x) = \lim_{x \to 4^+} \sqrt{x - 4} = \sqrt{4 - 4} = 0$$

Since $f(x) = 8 - 2x$ for $x < 4$, we have

$$\lim_{x \to 4^-} f(x) = \lim_{x \to 4^-} (8 - 2x) = 8 - 2 \cdot 4 = 0$$

The right- and left-hand limits are equal. Thus, the limit exists and

$$\lim_{x \to 4} f(x) = 0$$

The graph of f is shown in Figure 4.   ∎

FIGURE 4

12.2 EXERCISES

1. Given that

$$\lim_{x \to a} f(x) = -3 \qquad \lim_{x \to a} g(x) = 0 \qquad \lim_{x \to a} h(x) = 8$$

find the limits that exist. If the limit does not exist, explain why.

(a) $\lim_{x \to a} [f(x) + h(x)]$

(b) $\lim_{x \to a} [f(x)]^2$

(c) $\lim_{x \to a} \sqrt[3]{h(x)}$

(d) $\lim_{x \to a} \dfrac{1}{f(x)}$

(e) $\lim_{x \to a} \dfrac{f(x)}{h(x)}$

(f) $\lim_{x \to a} \dfrac{g(x)}{f(x)}$

(g) $\lim_{x \to a} \dfrac{f(x)}{g(x)}$

(h) $\lim_{x \to a} \dfrac{2f(x)}{h(x) - f(x)}$

2. The graphs of f and g are given. Use them to evaluate each limit, if it exists. If the limit does not exist, explain why.

(a) $\lim_{x \to 2} [f(x) + g(x)]$

(b) $\lim_{x \to 1} [f(x) + g(x)]$

(c) $\lim_{x \to 0} [f(x)g(x)]$

(d) $\lim_{x \to -1} \dfrac{f(x)}{g(x)}$

(e) $\lim_{x \to 2} x^3 f(x)$

(f) $\lim_{x \to 1} \sqrt{3 + f(x)}$

3–8 ■ Evaluate the limit and justify each step by indicating the appropriate Limit Law(s).

3. $\lim_{x \to 4} (5x^2 - 2x + 3)$

4. $\lim_{x \to 3} (x^3 + 2)(x^2 - 5x)$

5. $\lim_{x \to -1} \dfrac{x - 2}{x^2 + 4x - 3}$

6. $\lim_{x \to 1} \left(\dfrac{x^4 + x^2 - 6}{x^4 + 2x + 3} \right)^2$

7. $\lim_{t \to -2} (t + 1)^9 (t^2 - 1)$

8. $\lim_{u \to -2} \sqrt{u^4 + 3u + 6}$

9–20 ■ Evaluate the limit, if it exists.

9. $\lim_{x \to 2} \dfrac{x^2 + x - 6}{x - 2}$

10. $\lim_{x \to -4} \dfrac{x^2 + 5x + 4}{x^2 + 3x - 4}$

11. $\lim_{x \to 2} \dfrac{x^2 - x + 6}{x + 2}$

12. $\lim_{x \to 1} \dfrac{x^3 - 1}{x^2 - 1}$

13. $\lim_{t \to -3} \dfrac{t^2 - 9}{2t^2 + 7t + 3}$

14. $\lim_{h \to 0} \dfrac{\sqrt{1 + h} - 1}{h}$

15. $\lim_{h \to 0} \dfrac{(2 + h)^3 - 8}{h}$

16. $\lim_{x \to 2} \dfrac{x^4 - 16}{x - 2}$

17. $\lim_{x \to 7} \dfrac{\sqrt{x + 2} - 3}{x - 7}$

18. $\lim_{h \to 0} \dfrac{(3 + h)^{-1} - 3^{-1}}{h}$

19. $\lim_{x \to -4} \dfrac{\dfrac{1}{4} + \dfrac{1}{x}}{4 + x}$

20. $\lim_{t \to 0} \left(\dfrac{1}{t} - \dfrac{1}{t^2 + t} \right)$

21–24 ■ Find the limit and use a graphing device to confirm your result graphically.

21. $\lim_{x \to 1} \dfrac{x^2 - 1}{\sqrt{x} - 1}$

22. $\lim_{x \to 0} \dfrac{(4 + x)^3 - 64}{x}$

23. $\lim_{x \to -1} \dfrac{x^2 - x - 2}{x^3 - x}$

24. $\lim_{x \to 1} \dfrac{x^8 - 1}{x^5 - x}$

25. (a) Estimate the value of

$$\lim_{x \to 0} \dfrac{x}{\sqrt{1 + 3x} - 1}$$

by graphing the function $f(x) = x/(\sqrt{1 + 3x} - 1)$.

(b) Make a table of values of $f(x)$ for x close to 0 and guess the value of the limit.

(c) Use the Limit Laws to prove that your guess is correct.

 26. (a) Use a graph of

$$f(x) = \frac{\sqrt{3 + x} - \sqrt{3}}{x}$$

to estimate the value of $\lim_{x \to 0} f(x)$ to two decimal places.

(b) Use a table of values of $f(x)$ to estimate the limit to four decimal places.

(c) Use the Limit Laws to find the exact value of the limit.

27–32 ■ Find the limit, if it exists. If the limit does not exist, explain why.

27. $\lim_{x \to -4} |x + 4|$

28. $\lim_{x \to -4^-} \frac{|x + 4|}{x + 4}$

29. $\lim_{x \to 2} \frac{|x - 2|}{x - 2}$

30. $\lim_{x \to 1.5} \frac{2x^2 - 3x}{|2x - 3|}$

31. $\lim_{x \to 0^-} \left(\frac{1}{x} - \frac{1}{|x|} \right)$

32. $\lim_{x \to 0^+} \left(\frac{1}{x} - \frac{1}{|x|} \right)$

33. Let

$$f(x) = \begin{cases} x - 1 & \text{if } x < 2 \\ x^2 - 4x + 6 & \text{if } x \geq 2 \end{cases}$$

(a) Find $\lim_{x \to 2^-} f(x)$ and $\lim_{x \to 2^+} f(x)$.

(b) Does $\lim_{x \to 2} f(x)$ exist?

(c) Sketch the graph of f.

34. Let

$$h(x) = \begin{cases} x & \text{if } x < 0 \\ x^2 & \text{if } 0 < x \leq 2 \\ 8 - x & \text{if } x > 2 \end{cases}$$

(a) Evaluate each of the following limits, if it exists.

(i) $\lim_{x \to 0^+} h(x)$ 　　(iv) $\lim_{x \to 2^-} h(x)$

(ii) $\lim_{x \to 0} h(x)$ 　　(v) $\lim_{x \to 2^+} h(x)$

(iii) $\lim_{x \to 1} h(x)$ 　　(vi) $\lim_{x \to 2} h(x)$

(b) Sketch the graph of h.

▣ DISCOVERY · DISCUSSION

35. Cancellation and Limits

(a) What is wrong with the following equation?

$$\frac{x^2 + x - 6}{x - 2} = x + 3$$

(b) In view of part (a), explain why the equation

$$\lim_{x \to 2} \frac{x^2 + x - 6}{x - 2} = \lim_{x \to 2} (x + 3)$$

is correct.

36. The Lorentz Contraction　In the theory of relativity, the Lorentz contraction formula

$$L = L_0 \sqrt{1 - v^2/c^2}$$

expresses the length L of an object as a function of its velocity v with respect to an observer, where L_0 is the length of the object at rest and c is the speed of light. Find $\lim_{v \to c^-} L$ and interpret the result. Why is a left-hand limit necessary?

37. Limits of Sums and Products

(a) Show by means of an example that $\lim_{x \to a} [f(x) + g(x)]$ may exist even though neither $\lim_{x \to a} f(x)$ nor $\lim_{x \to a} g(x)$ exists.

(b) Show by means of an example that $\lim_{x \to a} [f(x)g(x)]$ may exist even though neither $\lim_{x \to a} f(x)$ nor $\lim_{x \to a} g(x)$ exists.

12.3　TANGENT LINES AND DERIVATIVES

In this section we see how limits arise when we attempt to find the tangent line to a curve or the instantaneous rate of change of a function.

▮ The Tangent Problem

A *tangent line* is a line that *just* touches a curve. For instance, Figure 1 shows the parabola $y = x^2$ and the tangent line t that touches the parabola at the point $P(1, 1)$. We will be able to find an equation of the tangent line t as soon as we know its slope m. The difficulty is that we know only one point, P, on t, whereas we need two points to compute the slope. But observe that we can compute an approxima-

FIGURE 1

$Q(x, x^2)$ t

$y = x^2$ $P(1, 1)$

0   x

FIGURE 2

tion to m by choosing a nearby point $Q(x, x^2)$ on the parabola (as in Figure 2) and computing the slope m_{PQ} of the secant line PQ.

We choose $x \neq 1$ so that $Q \neq P$. Then

$$m_{PQ} = \frac{x^2 - 1}{x - 1}$$

Now we let x approach 1, so Q approaches P along the parabola. Figure 3 shows how the corresponding secant lines rotate about P and approach the tangent line t.

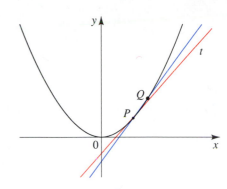

Q approaches P from the right

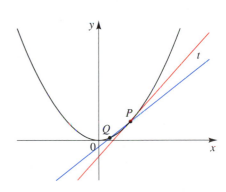

Q approaches P from the left

FIGURE 3

The slope of the tangent line is the limit of the slopes of the secant lines:

$$m = \lim_{Q \to P} m_{PQ}$$

So, using the method of Section 12.2, we have

$$m = \lim_{x \to 1} \frac{x^2 - 1}{x - 1} = \lim_{x \to 1} \frac{(x - 1)(x + 1)}{x - 1}$$

$$= \lim_{x \to 1} (x + 1) = 1 + 1 = 2$$

Now that we know the slope of the tangent line is $m = 2$, we can use the point-slope form of the equation of a line to find its equation:

$$y - 1 = 2(x - 1) \qquad \text{or} \qquad y = 2x - 1$$

The point-slope form for the equation of a line through the point (x_1, y_1) with slope m is

$$y - y_1 = m(x - x_1)$$

(See Section 1.10.)

We sometimes refer to the slope of the tangent line to a curve at a point as the **slope of the curve** at the point. The idea is that if we zoom in far enough toward the point, the curve looks almost like a straight line. Figure 4 illustrates this procedure for the curve $y = x^2$. The more we zoom in, the more the parabola looks like a line. In other words, the curve becomes almost indistinguishable from its tangent line.

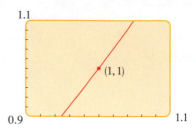

FIGURE 4

Zooming in toward the point $(1, 1)$ on the parabola $y = x^2$

If we have a general curve C with equation $y = f(x)$ and we want to find the tangent line to C at the point $P(a, f(a))$, then we consider a nearby point $Q(x, f(x))$, where $x \neq a$, and compute the slope of the secant line PQ:

$$m_{PQ} = \frac{f(x) - f(a)}{x - a}$$

Then we let Q approach P along the curve C by letting x approach a. If m_{PQ} approaches a number m, then we define the *tangent* t to be the line through P with slope m. (This amounts to saying that the tangent line is the limiting position of the secant line PQ as Q approaches P. See Figure 5.)

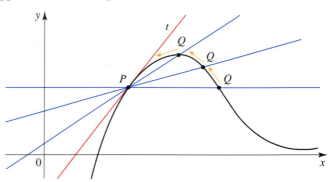

FIGURE 5

DEFINITION OF A TANGENT LINE

The **tangent line** to the curve $y = f(x)$ at the point $P(a, f(a))$ is the line through P with slope

$$m = \lim_{x \to a} \frac{f(x) - f(a)}{x - a}$$

provided that this limit exists.

EXAMPLE 1 ■ **Finding a Tangent Line to a Hyperbola**

Find an equation of the tangent line to the hyperbola $y = 3/x$ at the point $(3, 1)$.

SOLUTION

Let $f(x) = 3/x$. Then the slope of the tangent line at $(3, 1)$ is

$$m = \lim_{x \to 3} \frac{f(x) - f(3)}{x - 3} \qquad \text{Definition of } m$$

$$= \lim_{x \to 3} \frac{\dfrac{3}{x} - 1}{x - 3} \qquad f(x) = \frac{3}{x}$$

$$= \lim_{x \to 3} \frac{3 - x}{x(x - 3)} \qquad \begin{array}{l}\text{Multiply numerator and}\\ \text{denominator by } x\end{array}$$

$$= \lim_{x \to 3} -\frac{1}{x} \qquad \text{Cancel } x - 3$$

$$= -\frac{1}{3} \qquad \text{Let } x \to 3$$

Therefore, an equation of the tangent at the point $(3, 1)$ is

$$y - 1 = -\tfrac{1}{3}(x - 3)$$

which simplifies to

$$x + 3y - 6 = 0$$

The hyperbola and its tangent are shown in Figure 6.   ■

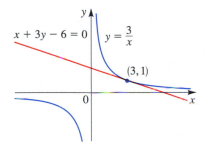

FIGURE 6

There is another expression for the slope of a tangent line that is sometimes easier to use. Let $h = x - a$. Then $x = a + h$, so the slope of the secant line PQ is

$$m_{PQ} = \frac{f(a + h) - f(a)}{h}$$

(See Figure 7 where the case $h > 0$ is illustrated and Q is to the right of P. If it happened that $h < 0$, however, Q would be to the left of P.)

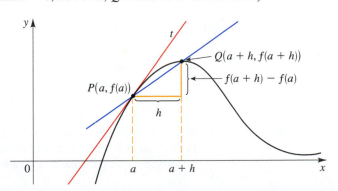

FIGURE 7

Notice that as x approaches a, h approaches 0 (because $h = x - a$), and so the expression for the slope of the tangent line becomes

$$m = \lim_{h \to 0} \frac{f(a + h) - f(a)}{h}$$

EXAMPLE 2 ■ Finding a Tangent Line

Find an equation of the tangent line to the curve $y = x^3 - 2x + 3$ at the point $(1, 2)$.

SOLUTION

If $f(x) = x^3 - 2x + 3$, then the slope of the tangent line where $a = 1$ is

$$m = \lim_{h \to 0} \frac{f(1 + h) - f(1)}{h} \qquad \text{Definition of } m$$

$$= \lim_{h \to 0} \frac{[(1 + h)^3 - 2(1 + h) + 3] - [1^3 - 2(1) + 3]}{h} \qquad f(x) = x^3 - 2x + 3$$

$$= \lim_{h \to 0} \frac{1 + 3h + 3h^2 + h^3 - 2 - 2h + 3 - 2}{h} \qquad \text{Expand numerator}$$

$$= \lim_{h \to 0} \frac{h + 3h^2 + h^3}{h} \qquad \text{Simplify}$$

$$= \lim_{h \to 0} (1 + 3h + h^2) \qquad \text{Cancel } h$$

$$= 1 \qquad \text{Let } h \to 0$$

So an equation of the tangent line at $(1, 2)$ is

$$y - 2 = 1(x - 1) \qquad \text{or} \qquad y = x + 1 \qquad ■$$

Derivatives

We have seen that the slope of the tangent line to the curve $y = f(x)$ at the point $(a, f(a))$ can be written as

$$\lim_{h \to 0} \frac{f(a + h) - f(a)}{h}$$

It turns out that this expression arises in many other contexts as well, such as finding velocities and other rates of change. Because this type of limit occurs so widely, it is given a special name and notation.

DEFINITION OF A DERIVATIVE

The **derivative of a function f at a number a**, denoted by $f'(a)$, is

$$f'(a) = \lim_{h \to 0} \frac{f(a + h) - f(a)}{h}$$

if this limit exists.

EXAMPLE 3 ■ Finding a Derivative at a Point

Find the derivative of the function $f(x) = 5x^2 + 3x - 1$ at the number 2.

SOLUTION

According to the definition of a derivative, with $a = 2$, we have

$$f'(2) = \lim_{h \to 0} \frac{f(2 + h) - f(2)}{h} \qquad \text{Definition of } f'(2)$$

$$= \lim_{h \to 0} \frac{[5(2 + h)^2 + 3(2 + h) - 1] - [5(2)^2 + 3(2) - 1]}{h} \qquad f(x) = 5x^2 + 3x - 1$$

$$= \lim_{h \to 0} \frac{20 + 20h + 5h^2 + 6 + 3h - 1 - 25}{h} \qquad \text{Expand}$$

$$= \lim_{h \to 0} \frac{23h + 5h^2}{h} \qquad \text{Simplify}$$

$$= \lim_{h \to 0} (23 + 5h) \qquad \text{Cancel } h$$

$$= 23 \qquad \text{Let } h \to 0 \qquad ■$$

We see from the definition of a derivative that the number $f'(a)$ is the same as the slope of the tangent line to the curve $y = f(x)$ at the point $(a, f(a))$. So the result of Example 2 shows that the slope of the tangent line to the parabola $y = 5x^2 + 3x - 1$ at the point $(2, 25)$ is $f'(2) = 23$.

EXAMPLE 4 ■ Finding a Derivative

Let $f(x) = \sqrt{x}$.
(a) Find $f'(a)$.
(b) Find $f'(1)$, $f'(4)$, and $f'(9)$.

SOLUTION

(a) We use the definition of the derivative at a:

$$f'(a) = \lim_{h \to 0} \frac{f(a+h) - f(a)}{h} \qquad \text{Definition of derivative}$$

$$= \lim_{h \to 0} \frac{\sqrt{a+h} - \sqrt{a}}{h} \qquad f(x) = \sqrt{x}$$

$$= \lim_{h \to 0} \frac{\sqrt{a+h} - \sqrt{a}}{h} \cdot \frac{\sqrt{a+h} + \sqrt{a}}{\sqrt{a+h} + \sqrt{a}} \qquad \text{Rationalize numerator}$$

$$= \lim_{h \to 0} \frac{(a+h) - a}{h(\sqrt{a+h} + \sqrt{a})} \qquad \text{Difference of squares}$$

$$= \lim_{h \to 0} \frac{h}{h(\sqrt{a+h} + \sqrt{a})} \qquad \text{Simplify numerator}$$

$$= \lim_{h \to 0} \frac{1}{\sqrt{a+h} + \sqrt{a}} \qquad \text{Cancel } h$$

$$= \frac{1}{\sqrt{a} + \sqrt{a}} = \frac{1}{2\sqrt{a}} \qquad \text{Let } h \to 0$$

(b) Substituting $a = 1$, $a = 4$, and $a = 9$ into the result of part (a), we get

$$f'(1) = \frac{1}{2\sqrt{1}} = \frac{1}{2} \qquad f'(4) = \frac{1}{2\sqrt{4}} = \frac{1}{4} \qquad f'(9) = \frac{1}{2\sqrt{9}} = \frac{1}{6}$$

These values of the derivative are the slopes of the tangent lines shown in Figure 8. ∎

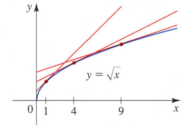

FIGURE 8

Instantaneous Rates of Change

In Section 2.4 we defined the average rate of change of a function f between the numbers a and x as

$$\text{average rate of change} = \frac{\text{change in } y}{\text{change in } x} = \frac{f(x) - f(a)}{x - a}$$

Suppose we consider the average rate of change over smaller and smaller intervals by letting x approach a. The limit of these average rates of change is called the instantaneous rate of change.

INSTANTANEOUS RATE OF CHANGE

If $y = f(x)$, the **instantaneous rate of change of y with respect to x** at $x = a$ is the limit of the average rates of change as x approaches a:

$$\text{instantaneous rate of change} = \lim_{x \to a} \frac{f(x) - f(a)}{x - a} = f'(a)$$

Notice that we now have two ways of interpreting the derivative:

- $f'(a)$ is the slope of the tangent line to $y = f(x)$ at $x = a$
- $f'(a)$ is the instantaneous rate of change of y with respect to x at $x = a$

In the special case where $x = t =$ time and $s = f(t) =$ displacement (directed distance) at time t of an object traveling in a straight line, the instantaneous rate of change is called the **instantaneous velocity**.

EXAMPLE 5 ■ Instantaneous Velocity of a Falling Object

If an object is dropped from a height of 3000 ft, its distance above the ground (in feet) after t seconds is given by $h(t) = 3000 - 16t^2$. Find the object's instantaneous velocity after 4 seconds.

SOLUTION

$h(t)$

After 4 s have elapsed, the height is $h(4) = 2744$ ft. The instantaneous velocity is

$$h'(4) = \lim_{t \to 4} \frac{h(t) - h(4)}{t - 4} \qquad \text{Definition of } h'(4)$$

$$= \lim_{t \to 4} \frac{3000 - 16t^2 - 2744}{t - 4} \qquad h(t) = 3000 - 16t^2$$

$$= \lim_{t \to 4} \frac{256 - 16t^2}{t - 4} \qquad \text{Simplify}$$

$$= \lim_{t \to 4} \frac{16(4 - t)(4 + t)}{t - 4} \qquad \text{Factor numerator}$$

$$= \lim_{t \to 4} -16(4 + t) \qquad \text{Cancel } t - 4$$

$$= -16(4 + 4) = -128 \text{ ft/s} \qquad \text{Let } t \to 4$$

The negative sign indicates that the height is *decreasing* at a rate of 128 ft/s. ■

EXAMPLE 6 ■ Estimating an Instantaneous Rate of Change

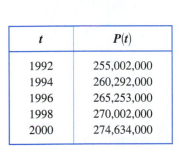

t	$P(t)$
1992	255,002,000
1994	260,292,000
1996	265,253,000
1998	270,002,000
2000	274,634,000

Let $P(t)$ be the population of the United States at time t. The table in the margin gives approximate values of this function by providing midyear population estimates from 1992 to 2000. Interpret and estimate the value of $P'(1996)$.

SOLUTION

The derivative $P'(1996)$ means the rate of change of P with respect to t when $t = 1996$, that is, the rate of increase of the population in 1996.

t	$\dfrac{P(t) - P(1996)}{t - 1996}$
1992	2,562,750
1994	2,480,500
1998	2,374,500
2000	2,345,250

Here we have estimated the derivative by averaging the slopes of two secant lines. Another method is to plot the population function and estimate the slope of the tangent line when $t = 1996$.

According to the definition of a derivative, we have

$$P'(1996) = \lim_{t \to 1996} \frac{P(t) - P(1996)}{t - 1996}$$

So we compute and tabulate values of the difference quotient (the average rates of change) as shown in the table in the margin. We see that $P'(1996)$ lies somewhere between 2,480,500 and 2,374,500. [Here we are making the reasonable assumption that the population didn't fluctuate wildly between 1992 and 2000.] We estimate that the rate of increase of the U.S. population in 1996 was the average of these two numbers, namely

$$P'(1996) \approx 2.4 \text{ million people/year} \qquad \blacksquare$$

12.3   EXERCISES

1–6 ■ Find the slope of the tangent line to the graph of f at the given point.

1. $f(x) = 3x + 4$ at $(1, 7)$

2. $f(x) = 5 - 2x$ at $(-3, 11)$

3. $f(x) = 4x^2 - 3x$ at $(-1, 7)$

4. $f(x) = 1 + 2x - 3x^2$ at $(1, 0)$

5. $f(x) = 2x^3$ at $(2, 16)$ **6.** $f(x) = \dfrac{6}{x + 1}$ at $(2, 2)$

7–12 ■ Find an equation of the tangent line to the curve at the given point. Graph the curve and the tangent line.

7. $y = x + x^2$ at $(-1, 0)$ **8.** $y = 2x - x^3$ at $(1, 1)$

9. $y = \dfrac{x}{x - 1}$ at $(2, 2)$ **10.** $y = \dfrac{1}{x^2}$ at $(-1, 1)$

11. $y = \sqrt{x + 3}$ at $(1, 2)$ **12.** $y = \sqrt{1 + 2x}$ at $(4, 3)$

13–18 ■ Find the derivative of the function at the given number.

13. $f(x) = 1 - 3x^2$ at 2

14. $f(x) = 2 - 3x + x^2$ at -1

15. $g(x) = x^4$ at 1 **16.** $g(x) = 2x^2 + x^3$ at 1

17. $F(x) = \dfrac{1}{\sqrt{x}}$ at 4 **18.** $G(x) = 1 + 2\sqrt{x}$ at 4

19–22 ■ Find $f'(a)$, where a is in the domain of f.

19. $f(x) = x^2 + 2x$ **20.** $f(x) = -\dfrac{1}{x^2}$

21. $f(x) = \dfrac{x}{x + 1}$ **22.** $f(x) = \sqrt{x - 2}$

23. (a) If $f(x) = x^3 - 2x + 4$, find $f'(a)$.
   (b) Find equations of the tangent lines to the graph of f at the points whose x-coordinates are 0, 1, and 2.
   (c) Graph f and the three tangent lines.

24. (a) If $g(x) = 1/(2x - 1)$, find $g'(a)$.
   (b) Find equations of the tangent lines to the graph of g at the points whose x-coordinates are -1, 0, and 1.
   (c) Graph g and the three tangent lines.

25. If a ball is thrown into the air with a velocity of 40 ft/s, its height (in feet) after t seconds is given by $y = 40t - 16t^2$. Find the velocity when $t = 2$.

26. If an arrow is shot upward on the moon with a velocity of 58 m/s, its height (in meters) after t seconds is given by $H = 58t - 0.83t^2$.
   (a) Find the velocity of the arrow after one second.
   (b) Find the velocity of the arrow when $t = a$.
   (c) When will the arrow hit the moon?
   (d) With what velocity will the arrow hit the moon?

27. The displacement (in meters) of a particle s moving in a straight line is given by the equation of motion $s = 4t^3 + 6t + 2$, where t is measured in seconds. Find the velocity of the particle s at times $t = a$, $t = 1$, $t = 2$, $t = 3$.

28. A spherical balloon is being inflated. Find the rate of change of the surface area $(S = 4\pi r^2)$ with respect to the radius r when $r = 2$ ft.

29. A roast turkey is taken from an oven when its temperature has reached 185°F and is placed on a table in a room where the temperature is 75°F. The graph shows how the temperature of the turkey decreases and eventually approaches room temperature. By measuring the slope of the tangent, estimate the rate of change of the temperature after an hour.

30. A cardiac monitor is used to measure the heart rate of a patient after surgery. It compiles the number of heartbeats after t minutes. When the data in the table are graphed, the slope of the tangent line represents the heart rate in beats per minute.

t (min)	36	38	40	42	44
Heartbeats	2530	2661	2806	2948	3080

(a) Find the average heart rates (slopes of the secant lines) over the time intervals $[40, 42]$ and $[42, 44]$
(b) Estimate the patient's heart rate after 42 minutes by averaging the slopes of these two secant lines.

31. A tank holds 1000 gallons of water, which drains from the bottom of the tank in half an hour. The values in the table show the volume V of water remaining in the tank (in gallons) after t minutes.

t (min)	5	10	15	20	25	30
V (gal)	694	444	250	111	28	0

(a) Find the average rates at which water flows from the tank (slopes of secant lines) for the time intervals $[10, 15]$ and $[15, 20]$.
(b) The slope of the tangent line at the point $(15, 250)$ represents the rate at which water is flowing from the tank after 15 minutes. Estimate this rate by averaging the slopes of the secant lines in part (a).

32. The table gives the world's population in the 20th century.

Year	Population (in millions)	Year	Population (in millions)
1900	1650	1960	3040
1910	1750	1970	3710
1920	1860	1980	4450
1930	2070	1990	5280
1940	2300	2000	6080
1950	2560		

Estimate the rate of population growth in 1920 and in 1980 by averaging the slopes of two secant lines.

DISCOVERY · DISCUSSION

33. Estimating Derivatives from a Graph   For the function g whose graph is given, arrange the following numbers in increasing order and explain your reasoning.

$$0 \qquad g'(-2) \qquad g'(0) \qquad g'(2) \qquad g'(4)$$

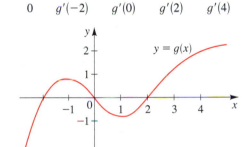

34. Estimating Velocities from a Graph   The graph shows the position function of a car. Use the shape of the graph to explain your answers to the following questions.
(a) What was the initial velocity of the car?
(b) Was the car going faster at B or at C?
(c) Was the car slowing down or speeding up at A, B, and C?
(d) What happened between D and E?

Applied Project

Designing a Roller Coaster

Suppose you are asked to design the first ascent and drop for a new roller coaster. By studying photographs of your favorite coasters, you decide to make the slope of the ascent 0.8 and the slope of the drop -1.6. You then connect these two straight stretches $y = L_1(x)$ and $y = L_2(x)$ with part of a parabola

$$y = f(x) = ax^2 + bx + c$$

where x and $f(x)$ are measured in feet. For the track to be smooth there can't be abrupt changes in direction, so you want the linear segments L_1 and L_2 to be tangent to the parabola at the transition points P and Q. (See the figure.)

1. To simplify the equations, you decide to place the origin at P. As a consequence, what is the value of c?

2. Suppose the horizontal distance between P and Q is 100 ft. To ensure that the track is smooth at the transition points, what should the values of $f'(0)$ and $f'(100)$ be?

3. If $f(x) = ax^2 + bx + c$, show that $f'(x) = 2ax + b$.

4. Use the results of Problems 2 and 3 to determine the values of a and b. That is, find a formula for $f(x)$.

5. Plot L_1, f, and L_2 to verify graphically that the transitions are smooth.

6. Find the difference in elevation between P and Q.

12.4   LIMITS AT INFINITY; LIMITS OF SEQUENCES

In this section we study a special kind of limit called a *limit at infinity*. We examine the limit of a function $f(x)$ as x becomes large. We also examine the limit of a sequence a_n as n becomes large. Limits of sequences will be used in Section 12.5 to help us find the area under the graph of a function.

Limits at Infinity

Let's investigate the behavior of the function f defined by

$$f(x) = \frac{x^2 - 1}{x^2 + 1}$$

as x becomes large. The table in the margin gives values of this function correct to six decimal places, and the graph of f has been drawn by a computer in Figure 1.

x	$f(x)$
0	-1.000000
± 1	0.000000
± 2	0.600000
± 3	0.800000
± 4	0.882353
± 5	0.923077
± 10	0.980198
± 50	0.999200
± 100	0.999800
± 1000	0.999998

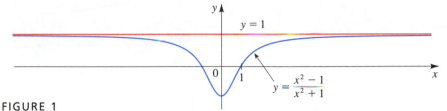

FIGURE 1

As x grows larger and larger, you can see that the values of $f(x)$ get closer and closer to 1. In fact, it seems that we can make the values of $f(x)$ as close as we like to 1 by taking x sufficiently large. This situation is expressed symbolically by writing

$$\lim_{x \to \infty} \frac{x^2 - 1}{x^2 + 1} = 1$$

In general, we use the notation

$$\lim_{x \to \infty} f(x) = L$$

to indicate that the values of $f(x)$ become closer and closer to L as x becomes larger and larger.

LIMIT AT INFINITY

Let f be a function defined on some interval (a, ∞). Then

$$\lim_{x \to \infty} f(x) = L$$

means that the values of $f(x)$ can be made arbitrarily close to L by taking x sufficiently large.

Another notation for $\lim_{x \to \infty} f(x) = L$ is

$$f(x) \to L \qquad \text{as} \qquad x \to \infty$$

The symbol ∞ does not represent a number. Nevertheless, we often read the expression $\lim_{x \to \infty} f(x) = L$ as

"the limit of $f(x)$, as x approaches infinity, is L"

or   "the limit of $f(x)$, as x becomes infinite, is L"

or   "the limit of $f(x)$, as x increases without bound, is L"

Limits at infinity are also discussed in Section 3.6.

Geometric illustrations are shown in Figure 2. Notice that there are many ways for the graph of f to approach the line $y = L$ (which is called a *horizontal asymptote*) as we look to the far right.

FIGURE 2

Examples illustrating $\lim\limits_{x \to \infty} f(x) = L$

Referring back to Figure 1, we see that for numerically large negative values of x, the values of $f(x)$ are close to 1. By letting x decrease through negative values without bound, we can make $f(x)$ as close as we like to 1. This is expressed by writing

$$\lim_{x \to -\infty} \frac{x^2 - 1}{x^2 + 1} = 1$$

The general definition is as follows.

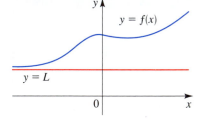

LIMIT AT NEGATIVE INFINITY

Let f be a function defined on some interval $(-\infty, a)$. Then

$$\lim_{x \to -\infty} f(x) = L$$

means that the values of $f(x)$ can be made arbitrarily close to L by taking x sufficiently large negative.

Again, the symbol $-\infty$ does not represent a number, but the expression $\lim\limits_{x \to -\infty} f(x) = L$ is often read as

"the limit of $f(x)$, as x approaches negative infinity, is L"

The definition is illustrated in Figure 3. Notice that the graph approaches the line $y = L$ as we look to the far left.

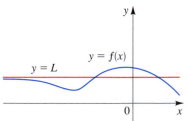

FIGURE 3

Examples illustrating $\lim\limits_{x \to -\infty} f(x) = L$

HORIZONTAL ASYMPTOTE

The line $y = L$ is called a **horizontal asymptote** of the curve $y = f(x)$ if either

$$\lim_{x \to \infty} f(x) = L \quad \text{or} \quad \lim_{x \to -\infty} f(x) = L$$

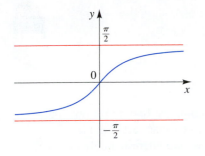

FIGURE 4

$y = \tan^{-1} x$

For instance, the curve illustrated in Figure 1 has the line $y = 1$ as a horizontal asymptote because

$$\lim_{x \to \infty} \frac{x^2 - 1}{x^2 + 1} = 1$$

As we discovered in Section 7.4, an example of a curve with two horizontal asymptotes is $y = \tan^{-1} x$ (see Figure 4). In fact,

$$\lim_{x \to -\infty} \tan^{-1} x = -\frac{\pi}{2} \quad \text{and} \quad \lim_{x \to \infty} \tan^{-1} x = \frac{\pi}{2}$$

so both of the lines $y = -\pi/2$ and $y = \pi/2$ are horizontal asymptotes. (This follows from the fact that the lines $x = \pm\pi/2$ are vertical asymptotes of the graph of tan.)

EXAMPLE 1 ■ Limits at Infinity

Find $\lim\limits_{x \to \infty} \dfrac{1}{x}$ and $\lim\limits_{x \to -\infty} \dfrac{1}{x}$.

SOLUTION

We first investigated horizontal asymptotes and limits at infinity for rational functions in Section 3.6.

Observe that when x is large, $1/x$ is small. For instance,

$$\frac{1}{100} = 0.01 \qquad \frac{1}{10,000} = 0.0001 \qquad \frac{1}{1,000,000} = 0.000001$$

In fact, by taking x large enough, we can make $1/x$ as close to 0 as we please. Therefore

$$\lim_{x \to \infty} \frac{1}{x} = 0$$

Similar reasoning shows that when x is large negative, $1/x$ is small negative, so we also have

$$\lim_{x \to -\infty} \frac{1}{x} = 0$$

It follows that the line $y = 0$ (the x-axis) is a horizontal asymptote of the curve $y = 1/x$. (This is a hyperbola; see Figure 5.)   ■

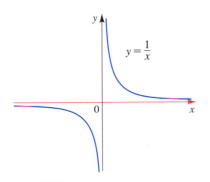

FIGURE 5

$\lim\limits_{x \to \infty} \dfrac{1}{x} = 0, \ \lim\limits_{x \to -\infty} \dfrac{1}{x} = 0$

The Limit Laws that were given in Section 12.2 also hold for limits at infinity. In particular, if we combine the Power Law with the results of Example 1, we obtain the following important rule for calculating limits.

> If k is any positive integer, then
>
> $$\lim_{x \to \infty} \frac{1}{x^k} = 0 \qquad \text{and} \qquad \lim_{x \to -\infty} \frac{1}{x^k} = 0$$

EXAMPLE 2 ■ Finding a Limit at Infinity

Evaluate $\displaystyle\lim_{x \to \infty} \frac{3x^2 - x - 2}{5x^2 + 4x + 1}$.

SOLUTION

To evaluate the limit at infinity of a rational function, we first divide both the numerator and denominator by the highest power of x that occurs in the denominator. (We may assume that $x \neq 0$ since we are interested only in large values of x.) In this case, the highest power of x in the denominator is x^2, so we have

$$\lim_{x \to \infty} \frac{3x^2 - x - 2}{5x^2 + 4x + 1} = \lim_{x \to \infty} \frac{3 - \dfrac{1}{x} - \dfrac{2}{x^2}}{5 + \dfrac{4}{x} + \dfrac{1}{x^2}} \qquad \text{\color{red}Divide numerator and denominator by } x^2$$

$$= \frac{\displaystyle\lim_{x \to \infty} \left(3 - \frac{1}{x} - \frac{2}{x^2}\right)}{\displaystyle\lim_{x \to \infty} \left(5 + \frac{4}{x} + \frac{1}{x^2}\right)} \qquad \text{\color{red}Quotient Law}$$

$$= \frac{\displaystyle\lim_{x \to \infty} 3 - \lim_{x \to \infty} \frac{1}{x} - 2\lim_{x \to \infty} \frac{1}{x^2}}{\displaystyle\lim_{x \to \infty} 5 + 4\lim_{x \to \infty} \frac{1}{x} + \lim_{x \to \infty} \frac{1}{x^2}} \qquad \text{\color{red}Sum and Difference Laws}$$

$$= \frac{3 - 0 - 0}{5 + 0 + 0} = \frac{3}{5} \qquad \text{\color{red}Let } x \to \infty$$

A similar calculation shows that the limit as $x \to -\infty$ is also $\frac{3}{5}$. Figure 6 illustrates the results of these calculations by showing how the graph of the given rational function approaches the horizontal asymptote $y = \frac{3}{5}$. ■

FIGURE 6

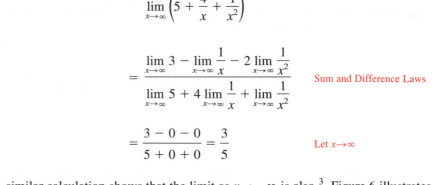

FIGURE 7

x	e^x
0	1.00000
−1	0.36788
−2	0.13534
−3	0.04979
−5	0.00674
−8	0.00034
−10	0.00005

EXAMPLE 3 ■ A Limit at Negative Infinity

Use numerical and graphical methods to find $\displaystyle\lim_{x \to -\infty} e^x$.

SOLUTION

From the graph of the natural exponential function $y = e^x$ in Figure 7 and the corresponding table of values, we see that

$$\lim_{x \to -\infty} e^x = 0$$

It follows that the line $y = 0$ (the x-axis) is a horizontal asymptote. ■

FIGURE 8

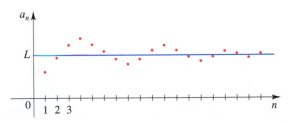

EXAMPLE 4 ■ A Function with No Limit at Infinity

Evaluate $\lim\limits_{x \to \infty} \sin x$.

SOLUTION

From the graph in Figure 8 and the periodic nature of the sine function, we see that, as x increases, the values of $\sin x$ oscillate between 1 and -1 infinitely often and so they don't approach any definite number. Therefore, $\lim_{x \to \infty} \sin x$ does not exist. ■

Limits of Sequences

In Section 10.1 we introduced the idea of a sequence of numbers $a_1, a_2, a_3, \ldots$. Here we are interested in their behavior as n becomes large. For instance, the sequence defined by

$$a_n = \frac{n}{n+1}$$

is pictured in Figure 9 by plotting its terms on a number line and in Figure 10 by plotting its graph. From Figure 9 or 10 it appears that the terms of the sequence $a_n = n/(n+1)$ are approaching 1 as n becomes large. We indicate this by writing

$$\lim_{n \to \infty} \frac{n}{n+1} = 1$$

FIGURE 9

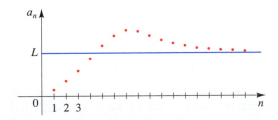

FIGURE 10

DEFINITION OF THE LIMIT OF A SEQUENCE

A sequence $a_1, a_2, a_3, \ldots$ has the **limit** L and we write

$$\lim_{n \to \infty} a_n = L \qquad \text{or} \qquad a_n \to L \text{ as } n \to \infty$$

if the nth term a_n of the sequence can be made arbitrarily close to L by taking n sufficiently large. If $\lim_{n \to \infty} a_n$ exists, we say the sequence **converges** (or is **convergent**). Otherwise, we say the sequence **diverges** (or is **divergent**).

This definition is illustrated by Figure 11.

FIGURE 11

Graphs of two
sequences with
$\lim\limits_{n \to \infty} a_n = L$

If we compare the definitions of $\lim_{n\to\infty} a_n = L$ and $\lim_{x\to\infty} f(x) = L$, we see that the only difference is that n is required to be an integer. Thus, the following is true.

> If $\lim_{x\to\infty} f(x) = L$ and $f(n) = a_n$ when n is an integer, then $\lim_{n\to\infty} a_n = L$.

In particular, since we know that $\lim_{x\to\infty} (1/x^k) = 0$ when k is a positive integer, we have

$$\lim_{n\to\infty} \frac{1}{n^k} = 0 \qquad \text{if } k \text{ is a positive integer}$$

We also note that the Limit Laws given in Section 12.2 also hold for limits of sequences.

EXAMPLE 5 ■ Finding the Limit of a Sequence

Find $\lim_{n\to\infty} \dfrac{n}{n+1}$.

SOLUTION

The method is similar to the one we used in Example 2: Divide the numerator and denominator by the highest power of n and then use the Limit Laws.

$$\lim_{n\to\infty} \frac{n}{n+1} = \lim_{n\to\infty} \frac{1}{1 + \dfrac{1}{n}} \qquad \text{Divide numerator and denominator by } n$$

$$= \frac{\displaystyle\lim_{n\to\infty} 1}{\displaystyle\lim_{n\to\infty} 1 + \lim_{n\to\infty} \frac{1}{n}} \qquad \text{Limit Laws 1 and 2}$$

$$= \frac{1}{1+0} = 1 \qquad \text{Let } n \to \infty$$

This shows that the guess we made earlier from Figures 9 and 10 was correct.

Therefore, the sequence $a_n = n/(n+1)$ is convergent.   ∎

EXAMPLE 6 ■ A Sequence That Diverges

Determine whether the sequence $a_n = (-1)^n$ is convergent or divergent.

SOLUTION

If we write out the terms of the sequence, we obtain

$$-1, 1, -1, 1, -1, 1, -1, \ldots$$

The graph of this sequence is shown in Figure 12. Since the terms oscillate between 1 and -1 infinitely often, a_n does not approach any number. Thus, $\lim_{n\to\infty} (-1)^n$ does not exist; that is, the sequence $a_n = (-1)^n$ is divergent.   ∎

FIGURE 12

EXAMPLE 7 ■ Finding the Limit of a Sequence

Find the limit of the sequence given by

$$a_n = \frac{15}{n^3}\left[\frac{n(n+1)(2n+1)}{6}\right]$$

SOLUTION

Before calculating the limit, let's first simplify the expression for a_n. Because $n^3 = n \cdot n \cdot n$, we place a factor of n beneath each factor in the numerator that contains an n:

$$a_n = \frac{15}{6} \cdot \frac{n}{n} \cdot \frac{n+1}{n} \cdot \frac{2n+1}{n} = \frac{5}{2} \cdot 1 \cdot \left(1 + \frac{1}{n}\right)\left(2 + \frac{1}{n}\right)$$

Now we can compute the limit:

$$\lim_{n\to\infty} a_n = \lim_{n\to\infty} \frac{5}{2}\left(1 + \frac{1}{n}\right)\left(2 + \frac{1}{n}\right) \qquad \text{Definition of } a_n$$

$$= \frac{5}{2} \lim_{n\to\infty}\left(1 + \frac{1}{n}\right) \lim_{n\to\infty}\left(2 + \frac{1}{n}\right) \qquad \text{Limit Laws 3 and 4}$$

$$= \frac{5}{2}(1)(2) = 5 \qquad \text{Let } n \to \infty \qquad ■$$

12.4 EXERCISES

1–2 ■

(a) Use the graph of f to find the following limits.
  (i) $\displaystyle\lim_{x\to\infty} f(x)$
  (ii) $\displaystyle\lim_{x\to-\infty} f(x)$

(b) State the equations of the horizontal asymptotes.

1.

2.

3–14 ■ Find the limit.

3. $\displaystyle\lim_{x\to\infty} \frac{6}{x}$

4. $\displaystyle\lim_{x\to\infty} \frac{3}{x^4}$

5. $\displaystyle\lim_{x\to\infty} \frac{2x+1}{5x-1}$

6. $\displaystyle\lim_{x\to\infty} \frac{2-3x}{4x+5}$

7. $\displaystyle\lim_{x\to-\infty} \frac{4x^2+1}{2+3x^2}$

8. $\displaystyle\lim_{x\to-\infty} \frac{x^2+2}{x^3+x+1}$

9. $\displaystyle\lim_{t\to\infty} \frac{8t^3+t}{(2t-1)(2t^2+1)}$

10. $\displaystyle\lim_{r\to\infty} \frac{4r^3-r^2}{(r+1)^3}$

11. $\displaystyle\lim_{x\to\infty} \frac{x^4}{1-x^2+x^3}$

12. $\displaystyle\lim_{t\to\infty} \left(\frac{1}{t} - \frac{2t}{t-1}\right)$

13. $\displaystyle\lim_{x\to-\infty} \left(\frac{x-1}{x+1} + 6\right)$

14. $\displaystyle\lim_{x\to\infty} \cos x$

15–18 ■ Use a table of values to estimate the limit. Then use a graphing device to confirm your result graphically.

15. $\displaystyle\lim_{x\to-\infty} \frac{\sqrt{x^2+4x}}{4x+1}$

16. $\displaystyle\lim_{x\to\infty} \left(\sqrt{9x^2+x} - 3x\right)$

17. $\displaystyle\lim_{x\to\infty} \frac{x^5}{e^x}$

18. $\displaystyle\lim_{x\to\infty} \left(1 + \frac{2}{x}\right)^{3x}$

19–30 ■ If the sequence is convergent, find its limit. If it is divergent, explain why.

19. $a_n = \dfrac{1 + n}{n + n^2}$

20. $a_n = \dfrac{5n}{n + 5}$

21. $a_n = \dfrac{n^2}{n + 1}$

22. $a_n = \dfrac{n - 1}{n^3 + 1}$

23. $a_n = \dfrac{1}{3^n}$

24. $a_n = \dfrac{(-1)^n}{n}$

25. $a_n = \sin(n\pi/2)$

26. $a_n = \cos n\pi$

27. $a_n = \dfrac{3}{n^2}\left[\dfrac{n(n + 1)}{2}\right]$

28. $a_n = \dfrac{5}{n}\left(n + \dfrac{4}{n}\left[\dfrac{n(n + 1)}{2}\right]\right)$

29. $a_n = \dfrac{24}{n^3}\left[\dfrac{n(n + 1)(2n + 1)}{6}\right]$

30. $a_n = \dfrac{12}{n^4}\left[\dfrac{n(n + 1)}{2}\right]^2$

31. (a) A tank contains 5000 L of pure water. Brine that contains 30 g of salt per liter of water is pumped into the tank at a rate of 25 L/min. Show that the concentration of salt after t minutes (in grams per liter) is

$$C(t) = \frac{30t}{200 + t}$$

(b) What happens to the concentration as $t \to \infty$?

32. The downward velocity of a falling raindrop at time t is modeled by the function

$$v(t) = 1.2(1 - e^{-8.2t})$$

(a) Find the terminal velocity of the raindrop by evaluating $\lim_{t\to\infty} v(t)$. (Use the result of Example 3.)

(b) Graph $v(t)$, and use the graph to estimate how long it takes for the velocity of the raindrop to reach 99% of its terminal velocity.

DISCOVERY · DISCUSSION

33. The Limit of a Recursive Sequence

(a) A sequence is defined recursively by $a_1 = 0$ and

$$a_{n+1} = \sqrt{2 + a_n}$$

Find the first ten terms of this sequence correct to eight decimal places. Does this sequence appear to be convergent? If so, guess the value of the limit.

(b) Assuming the sequence in part (a) is convergent, let $\lim_{n\to\infty} a_n = L$. Explain why $\lim_{n\to\infty} a_{n+1} = L$ also, and therefore

$$L = \sqrt{2 + L}$$

Solve this equation to find the exact value of L.

12.5   AREAS

We have seen that limits are needed to compute the slope of a tangent line or an instantaneous rate of change. Here we will see that they are also needed to find the area of a region with a curved boundary. The problem of finding such areas has consequences far beyond simply finding area. (See *Focus on Modeling*, page 955.)

The Area Problem

One of the central problems in calculus is the *area problem*: Find the area of the region S that lies under the curve $y = f(x)$ from a to b. This means that S, illustrated in Figure 1, is bounded by the graph of a function f (where $f(x) \geq 0$), the vertical lines $x = a$ and $x = b$, and the x-axis.

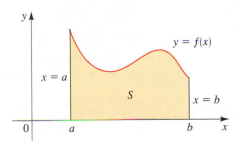

FIGURE 1

In trying to solve the area problem, we have to ask ourselves: What is the meaning of the word *area*? This question is easy to answer for regions with straight sides. For a rectangle, the area is defined as the product of the length and the width. The area of a triangle is half the base times the height. The area of a polygon is found by dividing it into triangles (as in Figure 2) and adding the areas of the triangles.

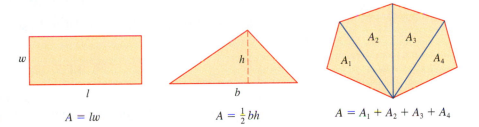

FIGURE 2 $A = lw$ $A = \frac{1}{2}bh$ $A = A_1 + A_2 + A_3 + A_4$

However, it is not so easy to find the area of a region with curved sides. We all have an intuitive idea of what the area of a region is. But part of the area problem is to make this intuitive idea precise by giving an exact definition of area.

Recall that in defining a tangent we first approximated the slope of the tangent line by slopes of secant lines and then we took the limit of these approximations. We pursue a similar idea for areas. We first approximate the region S by rectangles, and then we take the limit of the areas of these rectangles as we increase the number of rectangles. The following example illustrates the procedure.

EXAMPLE 1 ■ Estimating an Area Using Rectangles

Use rectangles to estimate the area under the parabola $y = x^2$ from 0 to 1 (the parabolic region S illustrated in Figure 3).

SOLUTION

We first notice that the area of S must be somewhere between 0 and 1 because S is contained in a square with side length 1, but we can certainly do better than that. Suppose we divide S into four strips S_1, S_2, S_3, and S_4 by drawing the vertical lines

FIGURE 3

(a)

(b)

FIGURE 4

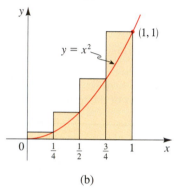

FIGURE 5

$x = \frac{1}{4}$, $x = \frac{1}{2}$, and $x = \frac{3}{4}$ as in Figure 4(a). We can approximate each strip by a rectangle whose base is the same as the strip and whose height is the same as the right edge of the strip [see Figure 4(b)]. In other words, the heights of these rectangles are the values of the function $f(x) = x^2$ at the right endpoints of the subintervals $[0, \frac{1}{4}]$, $[\frac{1}{4}, \frac{1}{2}]$, $[\frac{1}{2}, \frac{3}{4}]$, and $[\frac{3}{4}, 1]$.

Each rectangle has width $\frac{1}{4}$ and the heights are $(\frac{1}{4})^2$, $(\frac{1}{2})^2$, $(\frac{3}{4})^2$, and 1^2. If we let R_4 be the sum of the areas of these approximating rectangles, we get

$$R_4 = \tfrac{1}{4} \cdot (\tfrac{1}{4})^2 + \tfrac{1}{4} \cdot (\tfrac{1}{2})^2 + \tfrac{1}{4} \cdot (\tfrac{3}{4})^2 + \tfrac{1}{4} \cdot 1^2 = \tfrac{15}{32} = 0.46875$$

From Figure 4(b) we see that the area A of S is less than R_4, so

$$A < 0.46875$$

Instead of using the rectangles in Figure 4(b), we could use the smaller rectangles in Figure 5 whose heights are the values of f at the left endpoints of the subintervals. (The leftmost rectangle has collapsed because its height is 0.) The sum of the areas of these approximating rectangles is

$$L_4 = \tfrac{1}{4} \cdot 0^2 + \tfrac{1}{4} \cdot (\tfrac{1}{4})^2 + \tfrac{1}{4} \cdot (\tfrac{1}{2})^2 + \tfrac{1}{4} \cdot (\tfrac{3}{4})^2 = \tfrac{7}{32} = 0.21875$$

We see that the area of S is larger than L_4, so we have lower and upper estimates for A:

$$0.21875 < A < 0.46875$$

We can repeat this procedure with a larger number of strips. Figure 6 shows what happens when we divide the region S into eight strips of equal width. By computing the sum of the areas of the smaller rectangles (L_8) and the sum of the areas of the larger rectangles (R_8), we obtain better lower and upper estimates for A:

$$0.2734375 < A < 0.3984375$$

So one possible answer to the question is to say that the true area of S lies somewhere between 0.2734375 and 0.3984375.

FIGURE 6

Approximating S with eight rectangles

(a) Using left endpoints

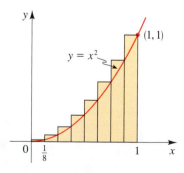

(b) Using right endpoints

n	L_n	R_n
10	0.2850000	0.3850000
20	0.3087500	0.3587500
30	0.3168519	0.3501852
50	0.3234000	0.3434000
100	0.3283500	0.3383500
1000	0.3328335	0.3338335

We could obtain better estimates by increasing the number of strips. The table in the margin shows the results of similar calculations (with a computer) using n rectangles whose heights are found with left endpoints (L_n) or right endpoints (R_n). In particular, we see by using 50 strips that the area lies between 0.3234 and 0.3434. With 1000 strips we narrow it down even more: A lies between 0.3328335 and 0.3338335. A good estimate is obtained by averaging these numbers: $A \approx 0.3333335$.   ■

From the values in the table it looks as if R_n is approaching $\frac{1}{3}$ as n increases. We confirm this in the next example.

EXAMPLE 2 ■ The Limit of Approximating Sums

For the region S in Example 1, show that the sum of the areas of the upper approximating rectangles approaches $\frac{1}{3}$, that is,

$$\lim_{n \to \infty} R_n = \tfrac{1}{3}$$

SOLUTION

R_n is the sum of the areas of the n rectangles shown in Figure 7. Each rectangle has width $1/n$, and the heights are the values of the function $f(x) = x^2$ at the points $1/n, 2/n, 3/n, \dots, n/n$. That is, the heights are $(1/n)^2, (2/n)^2, (3/n)^2, \dots, (n/n)^2$. Thus

$$R_n = \frac{1}{n}\left(\frac{1}{n}\right)^2 + \frac{1}{n}\left(\frac{2}{n}\right)^2 + \frac{1}{n}\left(\frac{3}{n}\right)^2 + \cdots + \frac{1}{n}\left(\frac{n}{n}\right)^2$$

$$= \frac{1}{n} \cdot \frac{1}{n^2}(1^2 + 2^2 + 3^2 + \cdots + n^2)$$

$$= \frac{1}{n^3}(1^2 + 2^2 + 3^2 + \cdots + n^2)$$

Here we need the formula for the sum of the squares of the first n positive integers:

$$1^2 + 2^2 + 3^2 + \cdots + n^2 = \frac{n(n+1)(2n+1)}{6}$$

Putting the preceding formula into our expression for R_n, we get

$$R_n = \frac{1}{n^3} \cdot \frac{n(n+1)(2n+1)}{6} = \frac{(n+1)(2n+1)}{6n^2}$$

Thus, we have

$$\lim_{n \to \infty} R_n = \lim_{n \to \infty} \frac{(n+1)(2n+1)}{6n^2}$$

$$= \lim_{n \to \infty} \frac{1}{6}\left(\frac{n+1}{n}\right)\left(\frac{2n+1}{n}\right)$$

$$= \lim_{n \to \infty} \frac{1}{6}\left(1 + \frac{1}{n}\right)\left(2 + \frac{1}{n}\right)$$

$$= \tfrac{1}{6} \cdot 1 \cdot 2 = \tfrac{1}{3}$$ ■

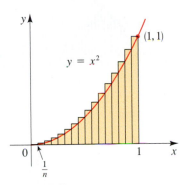

$y = x^2$

(1, 1)

$\frac{1}{n}$

FIGURE 7

This formula was discussed in Section 10.5.

It can be shown that the lower approximating sums also approach $\frac{1}{3}$, that is,

$$\lim_{n \to \infty} L_n = \tfrac{1}{3}$$

From Figures 8 and 9 it appears that, as n increases, both R_n and L_n become better and better approximations to the area of S. Therefore, we *define* the area A to be the limit of the sums of the areas of the approximating rectangles, that is,

$$A = \lim_{n \to \infty} R_n = \lim_{n \to \infty} L_n = \tfrac{1}{3}$$

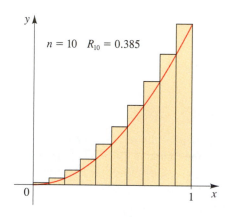

$n = 10 \quad R_{10} = 0.385$

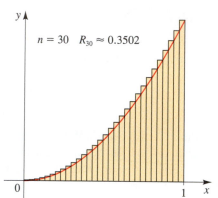

$n = 30 \quad R_{30} \approx 0.3502$

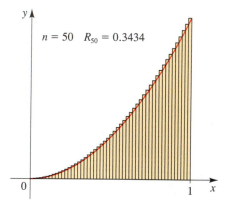

$n = 50 \quad R_{50} = 0.3434$

FIGURE 8

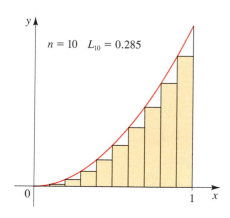

$n = 10 \quad L_{10} = 0.285$

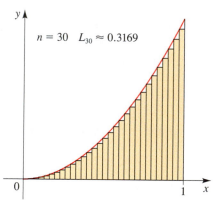

$n = 30 \quad L_{30} \approx 0.3169$

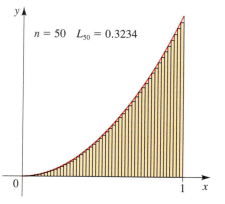

$n = 50 \quad L_{50} = 0.3234$

FIGURE 9

■ Definition of Area

Let's apply the idea of Examples 1 and 2 to the more general region S of Figure 1. We start by subdividing S into n strips $S_1, S_2, \ldots, S_n$ of equal width as in Figure 10.

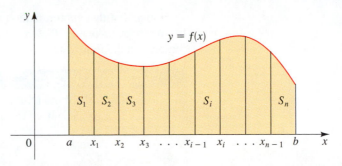

FIGURE 10

The width of the interval $[a, b]$ is $b - a$, so the width of each of the n strips is

$$\Delta x = \frac{b - a}{n}$$

These strips divide the interval $[a, b]$ into n subintervals

$$[x_0, x_1], \quad [x_1, x_2], \quad [x_2, x_3], \quad \ldots, \quad [x_{n-1}, x_n]$$

where $x_0 = a$ and $x_n = b$. The right endpoints of the subintervals are

$$x_1 = a + \Delta x, \quad x_2 = a + 2\,\Delta x, \quad x_3 = a + 3\,\Delta x, \quad \ldots, \quad x_k = a + k\,\Delta x, \quad \ldots$$

Let's approximate the kth strip S_k by a rectangle with width Δx and height $f(x_k)$, which is the value of f at the right endpoint (see Figure 11). Then the area of the kth rectangle is $f(x_k)\Delta x$. What we think of intuitively as the area of S is approximated by the sum of the areas of these rectangles, which is

$$R_n = f(x_1)\,\Delta x + f(x_2)\,\Delta x + \cdots + f(x_n)\,\Delta x$$

Figure 12 shows this approximation for $n = 2$, 4, 8, and 12.

FIGURE 11

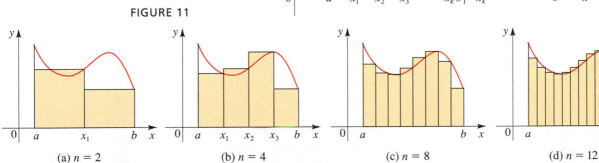

(a) $n = 2$ (b) $n = 4$ (c) $n = 8$ (d) $n = 12$

FIGURE 12

Notice that this approximation appears to become better and better as the number of strips increases, that is, as $n \to \infty$. Therefore, we define the area A of the region S in the following way.

> ## DEFINITION OF AREA
>
> The **area** A of the region S that lies under the graph of the continuous function f is the limit of the sum of the areas of approximating rectangles:
>
> $$A = \lim_{n\to\infty} R_n = \lim_{n\to\infty} \left[f(x_1)\,\Delta x + f(x_2)\,\Delta x + \cdots + f(x_n)\,\Delta x \right]$$
>
> Using sigma notation, we write this as follows:
>
> $$A = \lim_{n\to\infty} \sum_{k=1}^{n} f(x_k)\,\Delta x$$

In using this formula for area, remember that Δx is the width of an approximating rectangle, x_k is the right endpoint of the kth rectangle, and $f(x_k)$ is its height. So

Width: $\qquad \Delta x = \dfrac{b-a}{n}$

Right endpoint: $\qquad x_k = a + k\,\Delta x$

Height: $\qquad f(x_k) = f(a + k\,\Delta x)$

When working with sums, we will need the following properties from Section 10.1:

$$\sum_{k=1}^{n}(a_k \pm b_k) = \sum_{k=1}^{n}a_k \pm \sum_{k=1}^{n}b_k \qquad \sum_{k=1}^{n}ca_k = c\sum_{k=1}^{n}a_k$$

We will also need the following formulas for the sums of the powers of the first n natural numbers from Section 10.5.

$$\sum_{k=1}^{n}c = nc \qquad\qquad \sum_{k=1}^{n}k = \frac{n(n+1)}{2}$$

$$\sum_{k=1}^{n}k^2 = \frac{n(n+1)(2n+1)}{6} \qquad \sum_{k=1}^{n}k^3 = \frac{n^2(n+1)^2}{4}$$

EXAMPLE 3 ■ Finding the Area under a Curve

Find the area of the region that lies under the parabola $y = x^2$, $0 \le x \le 5$.

SOLUTION

The region is graphed in Figure 13. To find the area, we first find the dimensions of the approximating rectangles at the nth stage.

Width: $\qquad \Delta x = \dfrac{b-a}{n} = \dfrac{5-0}{n} = \dfrac{5}{n}$

Right endpoint: $\qquad x_k = a + k\,\Delta x = 0 + k\left(\dfrac{5}{n}\right) = \dfrac{5k}{n}$

Height: $\qquad f(x_k) = f\left(\dfrac{5k}{n}\right) = \left(\dfrac{5k}{n}\right)^2 = \dfrac{25k^2}{n^2}$

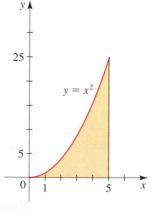

FIGURE 13

Now we substitute these values into the definition of area:

$$A = \lim_{n \to \infty} \sum_{k=1}^{n} f(x_k) \, \Delta x \qquad \text{Definition of area}$$

$$= \lim_{n \to \infty} \sum_{k=1}^{n} \frac{25k^2}{n^2} \cdot \frac{5}{n} \qquad f(x_k) = \frac{25k^2}{n^2}, \Delta x = \frac{5}{n}$$

$$= \lim_{n \to \infty} \sum_{k=1}^{n} \frac{125k^2}{n^3} \qquad \text{Simplify}$$

$$= \lim_{n \to \infty} \frac{125}{n^3} \sum_{k=1}^{n} k^2 \qquad \text{Factor } \frac{125}{n^3}$$

$$= \lim_{n \to \infty} \frac{125}{n^3} \cdot \frac{n(n+1)(2n+1)}{6} \qquad \text{Sum of squares formula}$$

$$= \lim_{n \to \infty} \frac{125(2n^2 + 3n + 1)}{6n^2} \qquad \text{Cancel } n \text{ and expand numerator}$$

$$= \lim_{n \to \infty} \frac{125}{6} \left(2 + \frac{3}{n} + \frac{1}{n^2} \right) \qquad \begin{array}{l}\text{Divide numerator and} \\ \text{denominator by } n^2\end{array}$$

$$= \frac{125}{6}(2 + 0 + 0) = \frac{125}{3} \qquad \text{Let } n \to \infty$$

We can also calculate the limit by writing

$$\frac{125}{n^3} \cdot \frac{n(n+1)(2n+1)}{6}$$

$$= \frac{125}{6} \left(\frac{n}{n} \right)\left(\frac{n+1}{n} \right)\left(\frac{2n+1}{n} \right)$$

as in Example 2.

Thus, the area of the region is $\frac{125}{3} \approx 41.7$.

EXAMPLE 4 ■ Finding the Area under a Curve

Find the area of the region that lies under the parabola $y = 4x - x^2$, $1 \le x \le 3$.

SOLUTION

We start by finding the dimensions of the approximating rectangles at the nth stage.

Width: $\qquad \Delta x = \dfrac{b - a}{n} = \dfrac{3 - 1}{n} = \dfrac{2}{n}$

Right endpoint: $\qquad x_k = a + k \, \Delta x = 1 + k\left(\dfrac{2}{n} \right) = 1 + \dfrac{2k}{n}$

Height: $\qquad f(x_k) = f\left(1 + \dfrac{2k}{n} \right) = 4\left(1 + \dfrac{2k}{n} \right) - \left(1 + \dfrac{2k}{n} \right)^2$

$$= 4 + \frac{8k}{n} - 1 - \frac{4k}{n} - \frac{4k^2}{n^2}$$

$$= 3 + \frac{4k}{n} - \frac{4k^2}{n^2}$$

Figure 14 shows the region whose area is computed in Example 3.

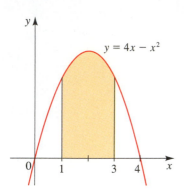

FIGURE 14

Thus, according to the definition of area, we get

$$A = \lim_{n \to \infty} \sum_{k=1}^{n} f(x_k) \, \Delta x = \lim_{n \to \infty} \sum_{k=1}^{n} \left(3 + \frac{4k}{n} - \frac{4k^2}{n^2} \right)\left(\frac{2}{n}\right)$$

$$= \lim_{n \to \infty} \left(\sum_{k=1}^{n} 3 + \frac{4}{n} \sum_{k=1}^{n} k - \frac{4}{n^2} \sum_{k=1}^{n} k^2 \right)\left(\frac{2}{n}\right)$$

$$= \lim_{n \to \infty} \left(\frac{2}{n} \sum_{k=1}^{n} 3 + \frac{8}{n^2} \sum_{k=1}^{n} k - \frac{8}{n^3} \sum_{k=1}^{n} k^2 \right)$$

$$= \lim_{n \to \infty} \left(\frac{2}{n}(3n) + \frac{8}{n^2}\left[\frac{n(n+1)}{2} \right] - \frac{8}{n^3}\left[\frac{n(n+1)(2n+1)}{6} \right] \right)$$

$$= \lim_{n \to \infty} \left(6 + 4 \cdot \frac{n}{n} \cdot \frac{n+1}{n} - \frac{4}{3} \cdot \frac{n}{n} \cdot \frac{n+1}{n} \cdot \frac{2n+1}{n} \right)$$

$$= \lim_{n \to \infty} \left[6 + 4\left(1 + \frac{1}{n}\right) - \frac{4}{3}\left(1 + \frac{1}{n}\right)\left(2 + \frac{1}{n}\right) \right]$$

$$= 6 + 4 \cdot 1 - \frac{4}{3} \cdot 1 \cdot 2 = \frac{22}{3}$$

12.5   EXERCISES

1. (a) By reading values from the given graph of f, use five rectangles to find a lower estimate and an upper estimate for the area under the given graph of f from $x = 0$ to $x = 10$. In each case, sketch the rectangles that you use.

  (b) Find new estimates using ten rectangles in each case.

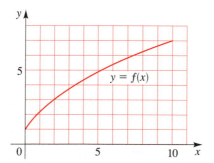

2. (a) Use six rectangles to find estimates of each type for the area under the given graph of f from $x = 0$ to $x = 12$.

   (i) L_6 (using left endpoints)
   (ii) R_6 (using right endpoints)

  (b) Is L_6 an underestimate or an overestimate of the true area?

  (c) Is R_6 an underestimate or an overestimate of the true area?

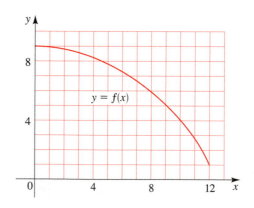

3–6 ■ Approximate the area of the shaded region under the graph of the given function by using the indicated rectangles. (The rectangles have equal width.)

3. $f(x) = \frac{1}{2}x + 2$

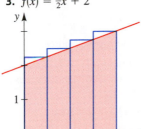

4. $f(x) = 4 - x^2$

5. $f(x) = \dfrac{4}{x}$

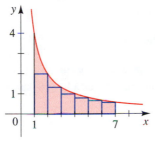

6. $f(x) = 9x - x^3$

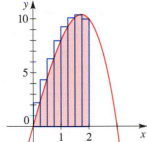

7. (a) Estimate the area under the graph of $f(x) = 1/x$ from $x = 1$ to $x = 5$ using four approximating rectangles and right endpoints. Sketch the graph and the rectangles. Is your estimate an underestimate or an overestimate?
  (b) Repeat part (a) using left endpoints.

8. (a) Estimate the area under the graph of $f(x) = 25 - x^2$ from $x = 0$ to $x = 5$ using five approximating rectangles and right endpoints. Sketch the graph and the rectangles. Is your estimate an underestimate or an overestimate?
  (b) Repeat part (a) using left endpoints.

9. (a) Estimate the area under the graph of $f(x) = 1 + x^2$ from $x = -1$ to $x = 2$ using three rectangles and right endpoints. Then improve your estimate by using six rectangles. Sketch the curve and the approximating rectangles.
  (b) Repeat part (a) using left endpoints.

10. (a) Estimate the area under the graph of $f(x) = e^{-x}$, $0 \le x \le 4$, using four approximating rectangles and taking the sample points to be
   (i) right endpoints
   (ii) left endpoints
   In each case, sketch the curve and the rectangles.
  (b) Improve your estimates in part (a) by using eight rectangles.

11–12 ■ Use the definition of area as a limit to find the area of the region that lies under the curve. Check your answer by sketching the region and using geometry.

11. $y = 3x$, $0 \le x \le 5$ **12.** $y = 2x + 1$, $1 \le x \le 3$

13–18 ■ Find the area of the region that lies under the graph of f over the given interval.

13. $f(x) = 3x^2$, $0 \le x \le 2$
14. $f(x) = x + x^2$, $0 \le x \le 1$
15. $f(x) = x^3 + 2$, $0 \le x \le 5$
16. $f(x) = 4x^3$, $2 \le x \le 5$
17. $f(x) = x + 6x^2$, $1 \le x \le 4$
18. $f(x) = 20 - 2x^2$, $2 \le x \le 3$

 DISCOVERY · DISCUSSION

19. Regions with Straight Versus Curved Boundaries
Write a short essay that explains how you would find the area of a polygon, that is, a region bounded by straight line segments. Then explain how you would find the area of a region whose boundary is curved, as we did in this section. What is the fundamental difference between these two processes?

12   REVIEW

CONCEPT CHECK

1. Explain in your own words what is meant by the equation

$$\lim_{x \to 2} f(x) = 5$$

Is it possible for this statement to be true and yet $f(2) = 3$? Explain.

2. Explain what it means to say that

$$\lim_{x \to 1^-} f(x) = 3 \quad \text{and} \quad \lim_{x \to 1^+} f(x) = 7$$

In this situation is it possible that $\lim_{x \to 1} f(x)$ exists? Explain.

3. Describe several ways in which a limit can fail to exist. Illustrate with sketches.

4. State the following Limit Laws.
(a) Sum Law
(b) Difference Law
(c) Constant Multiple Law
(d) Product Law
(e) Quotient Law
(f) Power Law
(g) Root Law

5. Write an expression for the slope of the tangent line to the curve $y = f(x)$ at the point $(a, f(a))$.

6. Define the derivative $f'(a)$. Discuss two ways of interpreting this number.

7. If $y = f(x)$, write expressions for the following.
(a) The average rate of change of y with respect to x between the numbers a and x.
(b) The instantaneous rate of change of y with respect to x at $x = a$.

8. Explain the meaning of the equation

$$\lim_{x \to \infty} f(x) = 2$$

Draw sketches to illustrate the various possibilities.

9. (a) What does it mean to say that the line $y = L$ is a horizontal asymptote of the curve $y = f(x)$? Draw curves to illustrate the various possibilities.
(b) Which of the following curves have horizontal asymptotes?
(i) $y = x^2$ (iv) $y = \tan^{-1} x$
(ii) $y = 1/x$ (v) $y = e^x$
(iii) $y = \sin x$ (vi) $y = \ln x$

10. (a) What is a convergent sequence?
(b) What does $\lim_{n \to \infty} a_n = 3$ mean?

11. Suppose S is the region that lies under the graph of $y = f(x)$, $a \le x \le b$.
(a) Explain how this area is approximated using rectangles.
(b) Write an expression for the area of S as a limit of sums.

EXERCISES

 1–6 ■ Use a table of values to estimate the value of the limit. Then use a graphing device to confirm your result graphically.

1. $\lim_{x \to 2} \dfrac{x - 2}{x^2 - 3x + 2}$

2. $\lim_{t \to -1} \dfrac{t + 1}{t^3 - t}$

3. $\lim_{x \to 0} \dfrac{2^x - 1}{x}$

4. $\lim_{x \to 0} \dfrac{\sin 2x}{x}$

5. $\lim_{x \to 1^+} \ln \sqrt{x - 1}$

6. $\lim_{x \to 0^-} \dfrac{\tan x}{|x|}$

7. The graph of f is shown in the figure. Find each limit or explain why it does not exist.
(a) $\lim_{x \to 2^+} f(x)$ (b) $\lim_{x \to -3^+} f(x)$
(c) $\lim_{x \to -3^-} f(x)$ (d) $\lim_{x \to -3} f(x)$
(e) $\lim_{x \to 4} f(x)$ (f) $\lim_{x \to \infty} f(x)$
(g) $\lim_{x \to -\infty} f(x)$ (h) $\lim_{x \to 0} f(x)$

8. Let

$$f(x) = \begin{cases} 2 & \text{if } x < -1 \\ x^2 & \text{if } -1 \le x \le 2 \\ x + 2 & \text{if } x > 2 \end{cases}$$

Find each limit or explain why it does not exist.
(a) $\lim_{x \to -1^-} f(x)$ (b) $\lim_{x \to -1^+} f(x)$
(c) $\lim_{x \to -1} f(x)$ (d) $\lim_{x \to 2^-} f(x)$
(e) $\lim_{x \to 2^+} f(x)$ (f) $\lim_{x \to 2} f(x)$
(g) $\lim_{x \to 0} f(x)$ (h) $\lim_{x \to 3} (f(x))^2$

9–20 ■ Use the limit laws to evaluate the limit, if it exists.

9. $\lim\limits_{x \to 2} \dfrac{x + 1}{x - 3}$

10. $\lim\limits_{t \to 1} (t^3 - 3t + 6)$

11. $\lim\limits_{x \to 3} \dfrac{x^2 + x - 12}{x - 3}$

12. $\lim\limits_{x \to -2} \dfrac{x^2 - 4}{x^2 + x - 2}$

13. $\lim\limits_{u \to 0} \dfrac{(u + 1)^2 - 1}{u}$

14. $\lim\limits_{z \to 9} \dfrac{\sqrt{z} - 3}{z - 9}$

15. $\lim\limits_{x \to 3^-} \dfrac{x - 3}{|x - 3|}$

16. $\lim\limits_{x \to 0} \left(\dfrac{1}{x} + \dfrac{2}{x^2 - 2x} \right)$

17. $\lim\limits_{x \to \infty} \dfrac{2x}{x - 4}$

18. $\lim\limits_{x \to \infty} \dfrac{x^2 + 1}{x^4 - 3x + 6}$

19. $\lim\limits_{x \to \infty} \cos^2 x$

20. $\lim\limits_{t \to -\infty} \dfrac{t^4}{t^3 - 1}$

21–24 ■ Find the derivative of the function at the given number.

21. $f(x) = 3x - 5$, at 4

22. $g(x) = 2x^2 - 1$, at -1

23. $f(x) = \sqrt{x}$, at 16

24. $f(x) = \dfrac{x}{x + 1}$, at 1

25–28 ■ (a) Find $f'(a)$. (b) Find $f'(2)$ and $f'(-2)$.

25. $f(x) = 6 - 2x$

26. $f(x) = x^2 - 3x$

27. $f(x) = \sqrt{x + 6}$

28. $f(x) = \dfrac{4}{x}$

29–30 ■ Find an equation of the tangent line shown in the figure.

29.

30.

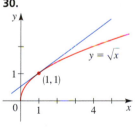

31–34 ■ Find an equation of the line tangent to the graph of f at the given point.

31. $f(x) = 2x$, at $(3, 6)$

32. $f(x) = x^2 - 3$, at $(2, 1)$

33. $f(x) = \dfrac{1}{x}$, at $\left(2, \dfrac{1}{2}\right)$

34. $f(x) = \sqrt{x + 1}$, at $(3, 2)$

35. A stone is dropped from the roof of a building 640 ft above the ground. Its height (in feet) after t seconds is given by $h(t) = 640 - 16t^2$.
(a) Find the velocity of the stone when $t = 2$.

(b) Find the velocity of the stone when $t = a$.
(c) At what time t will the stone hit the ground?
(d) With what velocity will the stone hit the ground?

36. If a gas is confined in a fixed volume, then according to Boyle's Law the product of the pressure P and the temperature T is a constant. For a certain gas, $PT = 100$, where P is measured in lb/in^2 and T is measured in kelvins (K).
(a) Express P as a function of T.
(b) Find the instantaneous rate of change of P with respect to T when $T = 300$ K.

37–42 ■ If the sequence is convergent, find its limit. If it is divergent, explain why.

37. $a_n = \dfrac{n}{5n + 1}$

38. $a_n = \dfrac{n^3}{n^3 + 1}$

39. $a_n = \dfrac{n(n + 1)}{2n^2}$

40. $a_n = \dfrac{n^3}{2n + 6}$

41. $a_n = \cos\left(\dfrac{n\pi}{2}\right)$

42. $a_n = \dfrac{10}{3^n}$

43–44 ■ Approximate the area of the shaded region under the graph of the given function by using the indicated rectangles. (The rectangles have equal width.)

43. $f(x) = \sqrt{x}$

44. $f(x) = 4x - x^2$

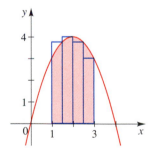

45–48 ■ Use the limit definition of area to find the area of the region that lies under the graph of f over the given interval.

45. $f(x) = 2x + 3$, $0 \le x \le 2$

46. $f(x) = x^2 + 1$, $0 \le x \le 3$

47. $f(x) = x^2 - x$, $1 \le x \le 2$

48. $f(x) = x^3$, $1 \le x \le 2$

1. (a) Use a table of values to estimate the limit.
   (b) Use a graphing calculator to confirm your answer graphically.

$$\lim_{x \to 0} \frac{x}{\sin 2x}$$

2. For the piecewise-defined function f whose graph is shown, find:

   (a) $\lim_{x \to -1^-} f(x)$ (b) $\lim_{x \to -1^+} f(x)$ (c) $\lim_{x \to -1} f(x)$

   (d) $\lim_{x \to 0^-} f(x)$ (e) $\lim_{x \to 0^+} f(x)$ (f) $\lim_{x \to 0} f(x)$

   (g) $\lim_{x \to 2^-} f(x)$ (h) $\lim_{x \to 2^+} f(x)$ (i) $\lim_{x \to 2} f(x)$

$$f(x) = \begin{cases} 1 & \text{if } x < -1 \\ 0 & \text{if } x = -1 \\ x^2 & \text{if } -1 < x \leq 2 \\ 4 - x & \text{if } 2 < x \end{cases}$$

3. Evaluate the limit, if it exists.

   (a) $\displaystyle\lim_{x \to 2} \frac{x^2 + 2x - 8}{x - 2}$ (b) $\displaystyle\lim_{x \to 2} \frac{x^2 - 2x - 8}{x + 2}$ (c) $\displaystyle\lim_{x \to 2} \frac{1}{x - 2}$

   (d) $\displaystyle\lim_{x \to 2} \frac{x - 2}{|x - 2|}$ (e) $\displaystyle\lim_{x \to 4} \frac{\sqrt{x} - 2}{x - 4}$ (f) $\displaystyle\lim_{x \to \infty} \frac{2x^2 - 4}{x^2 + x}$

4. Let $f(x) = x^2 - 2x$. Find:
   (a) $f'(x)$ (b) $f'(-1)$, $f'(1)$, $f'(2)$

5. Find the equation of the line tangent to the graph of $f(x) = \sqrt{x}$ at the point where $x = 9$.

6. Find the limit of the sequence.
   (a) $a_n = \dfrac{n}{n^2 + 4}$ (b) $a_n = \sec n\pi$

7. The region sketched in the figure in the margin lies under the graph of $f(x) = 4 - x^2$, above the interval $0 \leq x \leq 1$.
   (a) Approximate the area of the region with five rectangles, equally spaced along the x-axis, using right endpoints to determine the heights of the rectangles.
   (b) Use the limit definition of area to find the exact value of the area of the region.

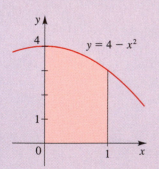

$y = 4 - x^2$

Focus on Modeling
Interpretations of Area

The area under the graph of a function is used to model many quantities in physics, economics, engineering, and other fields. That is why the area problem is so important. Here we will show how the concept of work (Section 7.8) is modeled by area. Several other applications are explored in the problems.

Recall that the work W done in moving an object is the product of the force F applied to the object and the distance d that the object moves:

$$W = Fd \qquad \text{work} = \text{force} \times \text{distance}$$

This formula is used if the force is *constant*. For example, suppose you are pushing a crate across a floor, moving along the positive x-axis from $x = a$ to $x = b$, and you apply a constant force $F = k$. The graph of F as a function of the distance x is shown in Figure 1(a). Notice that the work done is $W = Fd = k(b - a)$, which is the area under the graph of F [see Figure 1(b)].

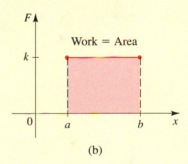

FIGURE 1
A constant force F

(a)

(b)

But what if the force is *not* constant? For example, suppose the force you apply to the crate varies with distance (you push harder at certain places than you do at others). More precisely, suppose that you push the crate along the x-axis in the positive direction, from $x = a$ to $x = b$, and at each point x between a and b you apply a force $f(x)$ to the crate. Figure 2 shows a graph of the force f as a function of the distance x.

FIGURE 2
A variable force

How much work was done? We can't apply the formula for work directly because the force is not constant. So let's divide the interval $[a, b]$ into n subintervals with endpoints $x_0, x_1, \ldots, x_n$ and equal width Δx as shown in Figure 3(a). The force at the right endpoint of the interval $[x_{k-1}, x_k]$ is $f(x_k)$. If n is large, then Δx is small, so the values of f don't change very much over the interval $[x_{k-1}, x_k]$. In other words f is almost constant on the interval, and so the work W_k that is done in moving the crate from x_{k-1} to x_k is approximately

$$W_k \approx f(x_k) \, \Delta x$$

Thus, we can approximate the work done in moving the crate from $x = a$ to $x = b$ by

$$W \approx \sum_{k=1}^{n} f(x_k) \, \Delta x$$

It seems that this approximation becomes better as we make n larger (and so make the interval $[x_{k-1}, x_k]$ smaller). Therefore, we define the work done in moving an object from a to b as the limit of this quantity as $n \to \infty$:

$$W = \lim_{n \to \infty} \sum_{k=1}^{n} f(x_k) \, \Delta x$$

Notice that this is precisely the area under the graph of f between $x = a$ and $x = b$ as defined in Section 12.5. See Figure 3(b).

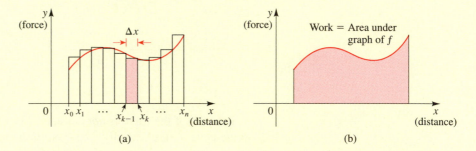

FIGURE 3
Approximating work

EXAMPLE ■ The Work Done by a Variable Force

A man pushes a crate along a straight path a distance of 18 ft. At a distance x from his starting point, he applies a force given by $f(x) = 340 - x^2$. Find the work done by the man.

SOLUTION

The graph of f between $x = 0$ and $x = 18$ is shown in Figure 4. Notice how the force the man applies varies—he starts by pushing with a force of 340 lb, but steadily applies less force. The work done is the area under the graph of f on the

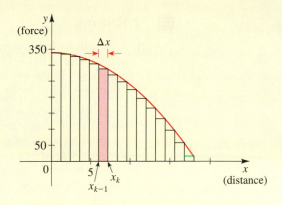

FIGURE 4

interval $[0, 18]$. To find this area, we start by finding the dimensions of the approximating rectangles at the nth stage.

Width:
$$\Delta x = \frac{b - a}{n} = \frac{18 - 0}{n} = \frac{18}{n}$$

Right endpoint:
$$x_k = a + k\,\Delta x = 0 + k\left(\frac{18}{n}\right) = \frac{18k}{n}$$

Height:
$$f(x_k) = f\left(\frac{18k}{n}\right) = 340 - \left(\frac{18k}{n}\right)^2$$

$$= 340 - \frac{324k^2}{n^2}$$

Thus, according to the definition of work we get

$$W = \lim_{n \to \infty} \sum_{k=1}^{n} f(x_k)\,\Delta x = \lim_{n \to \infty} \sum_{k=1}^{n} \left(340 - \frac{324k^2}{n^2}\right)\left(\frac{18}{n}\right)$$

$$= \lim_{n \to \infty} \left(\frac{18}{n} \sum_{k=1}^{n} 340 - \frac{(18)(324)}{n^3} \sum_{k=1}^{n} k^2\right)$$

$$= \lim_{n \to \infty} \left(\frac{18}{n}\, 340n - \frac{5832}{n^3}\left[\frac{n(n+1)(2n+1)}{6}\right]\right)$$

$$= \lim_{n \to \infty} \left(6120 - 972 \cdot \frac{n}{n} \cdot \frac{n+1}{n} \cdot \frac{2n+1}{n}\right)$$

$$= 6120 - 972 \cdot 1 \cdot 1 \cdot 2 = 4176$$

So the work done by the man in moving the crate is 4176 ft-lb.

■ Problems

1. A motorized winch is being used to pull a felled tree to a logging truck. The motor exerts a force of $f(x) = 1500 + 10x - \frac{1}{2}x^2$ lb on the tree at the instant when the tree has moved x ft. The tree must be moved a distance of 40 ft, from $x = 0$ to $x = 40$. How much work is done by the winch in moving the tree?

2. Hooke's law states that when a spring is stretched, it pulls back with a force proportional to the amount of the stretch. The constant of proportionality is a characteristic of the spring known as the **spring constant**. Thus, a spring with spring constant k exerts a force $f(x) = kx$ when it is stretched a distance x.

A certain spring has spring constant $k = 20$ lb/ft. Find the work done when the spring is pulled so that the amount by which it is stretched increases from $x = 0$ to $x = 2$ ft.

$f(x) = kx$

3. As any diver knows, an object submerged in water experiences pressure, and as depth increases, so does the water pressure. At a depth of x ft, the water pressure is $p(x) = 62.5x$ lb/ft². To find the force exerted by the water on a surface, we multiply the pressure by the area of the surface:

$$\text{force} = \text{pressure} \times \text{area}$$

Suppose an aquarium that is 3 ft wide, 6 ft long, and 4 ft high is full of water. The bottom of the aquarium has area $3 \times 6 = 18$ ft², and it experiences water pressure of $p(4) = 62.5 \times 4 = 250$ lb/ft². Thus, the total force exerted by the water on the bottom is $250 \times 18 = 4500$ lb.

The water also exerts a force on the sides of the aquarium, but this is not as easy to calculate because the pressure increases from top to bottom. To calculate the force on one of the 4 ft by 6 ft sides, we divide its area into n thin horizontal strips of width Δx, as shown in the figure. The area of each strip is

$$\text{length} \times \text{width} = 6\,\Delta x$$

If the bottom of the kth strip is at the depth x_k, then it experiences water pressure of approximately $p(x_k) = 62.5x_k$ lb/ft²—the thinner the strip, the more accurate the approximation. Thus, on each strip the water exerts a force of

$$\text{pressure} \times \text{area} = 62.5x_k \times 6\,\Delta x = 375x_k\,\Delta x \text{ lb}$$

(a) Explain why the total force exerted by the water on the 4 ft by 6 ft sides of the aquarium is

$$\lim_{n \to \infty} \sum_{k=1}^{n} 375x_k\,\Delta x$$

where $\Delta x = 4/n$ and $x_k = 4k/n$.

(b) What area does the limit in part (a) represent?

(c) Evaluate the limit in part (a) to find the force exerted by the water on one of the 4 ft by 6 ft sides of the aquarium.

(d) Use the same technique to find the force exerted by the water on one of the 4 ft by 3 ft sides of the aquarium.

NOTE Engineers use the technique outlined in this problem to find the total force exerted on a dam by the water in the reservoir behind the dam.

4. Since distance = speed × time, it is easy to see that a car moving, say, at 70 mi/h for 5 h will travel a distance of 350 mi. But what if the speed varies, as it usually does in practice?

   (a) Suppose the speed of a moving object at time t is $v(t)$. Explain why the distance traveled by the object between times $t = a$ and $t = b$ is the area under the graph of v between $t = a$ and $t = b$.

   (b) The speed of a car t seconds after it starts moving is given by the function $v(t) = 6t + 0.1t^3$ ft/s. Find the distance traveled by the car from $t = 0$ to $t = 5$ s.

5. If the outdoor temperature reaches a maximum of 90°F one day and only 80°F the next, then we would probably say that the first day was hotter than the second. Suppose, however, that on the first day the temperature was below 60°F for most of the day, reaching the high only briefly, whereas on the second day the temperature stayed above 75°F all the time. Now which day is the hotter one? To better measure how hot a particular day is, scientists use the concept of **heating degree-hour**. If the temperature is a constant D degrees for t hours, then the "heating capacity" generated over this period is Dt heating degree-hours:

$$\text{heating degree-hours} = \text{temperature} \times \text{time}$$

   If the temperature is not constant, then the number of heating degree-hours equals the area under the graph of the temperature function over the time period in question.

   (a) On a particular day, the temperature (in °F) was modeled by the function $D(t) = 61 + \frac{6}{5}t - \frac{1}{25}t^2$, where t was measured in hours since midnight. How many heating degree-hours were experienced on this day, from $t = 0$ to $t = 24$?

   (b) What was the maximum temperature on the day described in part (a)?

   (c) On another day, the temperature (in °F) was modeled by the function $E(t) = 50 + 5t - \frac{1}{4}t^2$. How many heating degree-hours were experienced on this day?

   (d) What was the maximum temperature on the day described in part (c)?

   (e) Which day was "hotter"?

Answers to Odd-Numbered Exercises and Chapter Tests

CHAPTER 1

Section 1.1 ■ page 12

1. Commutative Property for addition
3. Associative Property for addition   **5.** Distributive Property
7. $3x + 3y$ **9.** $8m$ **11.** $-5x + 10y$
13. (a) $\frac{7}{13}$ (b) $\frac{23}{30}$ **15.** (a) $\frac{4}{9}$ (b) $\frac{15}{2}$
17. (a) False   (b) True   **19.** (a) False   (b) True
21. (a) $x > 0$ (b) $t < 4$ (c) $a \geq \pi$ (d) $-5 < x < \frac{1}{3}$
(e) $|p - 3| \leq 5$
23. (a) $\{1, 2, 3, 4, 5, 6, 7, 8\}$ (b) $\{2, 4, 6\}$
25. (a) $\{1, 2, 3, 4, 5, 6, 7, 8, 9, 10\}$ (b) $\varnothing$
27. (a) $\{x \mid x \leq 5\}$ (b) $\{x \mid -1 < x < 4\}$
29. $-3 < x < 0$

31. $2 \leq x < 8$

33. $x \geq 2$

35. $(-\infty, 1]$

37. $(-2, 1]$

39. $(-1, \infty)$

41.

43.

45.

47. (a) 100 (b) 73 **49.** (a) 2 (b) -1
51. (a) 12 (b) 5 **53.** (a) 15 (b) 24 (c) $\frac{67}{40}$
55. (a) $\frac{7}{9}$ (b) $\frac{13}{45}$ (c) $\frac{19}{33}$ **57.** (a) Yes   (b) 170

Section 1.2 ■ page 23

1. $17^{-1/2}$ **3.** $\sqrt[3]{4^2}$ **5.** $\sqrt[5]{a^3}$ **7.** (a) 16 (b) -16 (c) 1
9. (a) $\frac{16}{25}$ (b) 1000 (c) 1024 **11.** (a) $\frac{2}{3}$ (b) 4 (c) $\frac{1}{2}$
13. (a) $\frac{3}{2}$ (b) 4 (c) $-\frac{1}{5}$ **15.** 5 **17.** 14 **19.** $\sqrt[3]{4}$
21. $2\sqrt{5}$ **23.** a^4 **25.** $6x^7y^5$ **27.** $16x^{10}$ **29.** $4/b^2$
31. $64r^7s$ **33.** $648y^7$ **35.** $\dfrac{x^3}{y}$ **37.** $\dfrac{y^2z^9}{x^5}$ **39.** $\dfrac{s^3}{q^7r^6}$
41. $x^{13/15}$ **43.** $16b^{9/10}$ **45.** $\dfrac{1}{c^{2/3}d}$ **47.** $y^{1/2}$ **49.** $\dfrac{32x^{12}}{y^{16/15}}$
51. $\dfrac{x^{15}}{y^{15/2}}$ **53.** $\dfrac{4a^2}{3b^{1/3}}$ **55.** $\dfrac{3t^{25/6}}{s^{1/2}}$ **57.** $|x|$ **59.** $x\sqrt[3]{y}$

61. $ab\sqrt[5]{ab^2}$ **63.** $2|x|$ **65.** (a) $\dfrac{\sqrt{6}}{6}$ (b) $\dfrac{\sqrt{3xy}}{3y}$ (c) $\dfrac{\sqrt{15}}{10}$
67. (a) $\dfrac{\sqrt[3]{x^2}}{x}$ (b) $\dfrac{\sqrt[5]{x^3}}{x}$ (c) $\dfrac{\sqrt[7]{x^4}}{x}$ **69.** (a) 6.93×10^7
(b) 2.8536×10^{-5} (c) 1.2954×10^8
71. (a) $319{,}000$ (b) 0.00000002670
(c) $710{,}000{,}000{,}000{,}000$ **73.** (a) 5.9×10^{12} mi
(b) 4×10^{-13} cm   (c) 3.3×10^{19} molecules
75. 1.3×10^{-20} **77.** 1.429×10^{19} **79.** 7.4×10^{-14}
81. 2.5×10^{13} mi   **83.** 1.3×10^{21} L
85. 4.03×10^{27} molecules   **87.** 41.3 mi

Section 1.3 ■ page 33

1. $5x^2 - 2x - 4$ **3.** $x^3 + 3x^2 - 6x + 11$
5. $9x + 103$ **7.** $-t^4 + t^3 - t^2 - 10t + 5$
9. $x^{3/2} - x$ **11.** $y^{7/3} - y^{1/3}$
13. $21t^2 - 29t + 10$ **15.** $3x^2 + 5xy - 2y^2$
17. $1 - 4y + 4y^2$ **19.** $4x^4 + 12x^2y^2 + 9y^4$
21. $2x^3 - 7x^2 + 7x - 5$ **23.** $x^4 - a^4$ **25.** $a - \dfrac{1}{b^2}$
27. $1 + 3a^3 + 3a^6 + a^9$ **29.** $x^5 + x^4 - 3x^3 + 3x - 2$
31. $1 - x^{2/3} + x^{4/3} - x^2$
33. $3x^4y^4 + 7x^3y^5 - 6x^2y^3 - 14xy^4$
35. $6x(2x^2 + 3)$ **37.** $3y^3(2y - 5)$ **39.** $(x - 4)(x + 2)$
41. $(y - 3)(y - 5)$ **43.** $(2x + 3)(x + 1)$ **45.** $9(x - 5)(x + 1)$
47. $(3x + 2)(2x - 3)$ **49.** $(2t - 3)^2$ **51.** $(r - 3s)^2$
53. $(x + 6)(x - 6)$ **55.** $(7 + 2y)(7 - 2y)$
57. $4ab$ **59.** $(x + 3)(x - 3)(x + 1)(x - 1)$
61. $(t + 1)(t^2 - t + 1)$ **63.** $(2x - 5)(4x^2 + 10x + 25)$
65. $(x^2 - 2y)(x^4 + 2x^2y + 4y^2)$ **67.** $x(x + 1)^2$
69. $x^2(x + 3)(x - 1)$ **71.** $(y + 2)(y - 2)(y - 3)$
73. $(2x^2 + 1)(x + 2)$ **75.** $3(x - 1)(x + 2)$
77. $y^4(y + 2)^3(y + 1)^2$ **79.** $x^{1/2}(x + 1)(x - 1)$
81. $x^{-3/2}(x + 1)^2$ **83.** $(x^2 + 3)(x^2 + 1)^{-1/2}$
85. $(a + 2)(a - 2)(a + 1)(a - 1)$ **87.** $16x^2(x - 3)(5x - 9)$
89. $(2x - 1)^2(x + 3)^{-1/2}\left(7x + \dfrac{35}{2}\right)$

Section 1.4 ■ page 42

1. $\dfrac{1}{x + 2}$ **3.** $\dfrac{x + 2}{x + 1}$ **5.** $\dfrac{y}{y - 1}$ **7.** $\dfrac{x(2x + 3)}{2x - 3}$

9. $\dfrac{1}{t^2 + 9}$ **11.** $\dfrac{x + 4}{x + 1}$ **13.** $\dfrac{(2x + 1)(2x - 1)}{(x + 5)^2}$

15. $x^2(x + 1)$ **17.** $\dfrac{x}{yz}$ **19.** $\dfrac{3x + 7}{(x - 3)(x + 5)}$

21. $\dfrac{1}{(x + 1)(x + 2)}$ **23.** $\dfrac{3x + 2}{(x + 1)^2}$ **25.** $\dfrac{u^2 + 3u + 1}{u + 1}$

27. $\dfrac{2x + 1}{x^2(x + 1)}$ **29.** $\dfrac{2x + 7}{(x + 3)(x + 4)}$ **31.** $\dfrac{x - 2}{(x + 3)(x - 3)}$

33. $\dfrac{5x - 6}{x(x - 1)}$ **35.** $\dfrac{-5}{(x + 1)(x + 2)(x - 3)}$ **37.** $-xy$

39. $\dfrac{c}{c - 2}$ **41.** $\dfrac{3x + 7}{x^2 + 2x - 1}$ **43.** $\dfrac{y - x}{xy}$ **45.** 1

47. $\dfrac{-1}{a(a + h)}$ **49.** $\dfrac{-3}{(2 + x)(2 + x + h)}$ **51.** $\dfrac{1}{\sqrt{1 - x^2}}$

53. $\dfrac{(x + 2)^2(x - 13)}{(x - 3)^3}$ **55.** $\dfrac{x + 2}{(x + 1)^{3/2}}$

57. $\dfrac{2x + 3}{(x + 1)^{4/3}}$ **59.** $\dfrac{3 - \sqrt{5}}{2}$ **61.** $\dfrac{2(\sqrt{7} - \sqrt{2})}{5}$

63. $\dfrac{-4}{3(1 + \sqrt{5})}$ **65.** $\dfrac{r - 2}{5(\sqrt{r} - \sqrt{2})}$ **67.** $\dfrac{1}{\sqrt{x^2 + 1} + x}$

69. True   **71.** False   **73.** False   **75.** True   **77.** False

79. (a) $\dfrac{R_1 R_2}{R_1 + R_2}$ (b) $\frac{20}{3} \approx 6.7$ ohms

Section 1.5 ■ page 57

1. (a) Yes (b) No   **3.** (a) Yes (b) No   **5.** 4   **7.** -9
9. -3 **11.** 12   **13.** $-\frac{3}{4}$ **15.** $\frac{32}{9}$ **17.** $-4, 2$
19. $-3, -\dfrac{1}{2}$ **21.** $-\dfrac{4}{3}, \dfrac{1}{2}$ **23.** $-1 \pm \sqrt{3}$ **25.** $-\dfrac{3}{2}, \dfrac{1}{2}$
27. $-2 \pm \dfrac{\sqrt{14}}{2}$ **29.** $-6 \pm 3\sqrt{7}$ **31.** $\dfrac{-3 \pm 2\sqrt{6}}{3}$
33. $\dfrac{1 \pm \sqrt{5}}{4}$ **35.** $\dfrac{8 \pm \sqrt{14}}{10}$ **37.** No real solution   **39.** $-\dfrac{7}{5}, 2$
41. $-50, 100$ **43.** -4 **45.** 4   **47.** 3   **49.** $\pm 2\sqrt{2}, \pm\sqrt{5}$
51. No real solution   **53.** $\pm 3\sqrt{3}, \pm 2\sqrt{2}$ **55.** $-1, 0, 3$
57. 27, 729   **59.** $-\dfrac{3}{2}, \dfrac{3}{2}$ **61.** 3.99, 4.01   **63.** $R = \dfrac{PV}{nT}$
65. $R_1 = \dfrac{RR_2}{R_2 - R}$ **67.** $x = \dfrac{2d - b}{a - 2c}$ **69.** $x = \dfrac{1 - a}{a^2 - a - 1}$
71. $r = \pm\sqrt{\dfrac{3V}{\pi h}}$ **73.** $b = \pm\sqrt{c^2 - a^2}$ **75.** 2   **77.** 1

79. 4.24 s   **81.** (a) After 1 s and $1\frac{1}{2}$ s   (b) Never
(c) 25 ft   (d) After $1\frac{1}{4}$ s   (e) After $2\frac{1}{2}$ s
83. (a) After 17 yr, on Jan. 1, 2009
(b) After 18.612 yr, on Aug. 12, 2010
85. 12 cm   **87.** 132.6 ft

Section 1.6 ■ page 70

1. $3n + 3$ **3.** $0.14x$ **5.** $4w + 10$ **7.** $\dfrac{25}{x + 3}$

9. 111, 112, 113   **11.** 19 and 36   **13.** $9000 at $4\frac{1}{2}$% and
$3000 at 4%   **15.** 7.5%   **17.** $7400   **19.** $45,000
21. Plumber, 70 h; assistant 35 h   **23.** 40 years old
25. 9 pennies, 9 nickels, 9 dimes   **27.** 45 ft
29. 120 ft by 120 ft   **31.** 9 cm   **33.** 48 cm
35. 60 ft by 40 ft   **37.** 2 ft by 6 ft by 15 ft   **39.** 4 in.
41. 18 g   **43.** 0.6 L   **45.** 35%   **47.** 10 m
49. 120 ft by 126 ft   **51.** 6.4 ft from the fulcrum
53. 37 min 20 s   **55.** 3 h   **57.** Henry, $4\frac{1}{2}$ h; Irene 3 h
59. $1\frac{1}{2}$ h   **61.** 4 h   **63.** 500 mi/h
65. 50 mi/h (or 240 mi/h)
67. 50   **69.** 16 mi; no   **71.** 7.52 ft   **73.** 120 ft
75. 35 yards   **77.** 169.1 in. (circle) and 190.9 in. (square)
79. 6 cubits

Section 1.7 ■ page 87

1. $\left\{0, \frac{1}{2}, \sqrt{2}, 2\right\}$ **3.** $\{-1, 2\}$ **5.** $\left(-\infty, \frac{7}{2}\right]$

7. $(4, \infty)$

9. $(-\infty, 2]$

11. $\left(-\infty, -\frac{1}{2}\right)$

13. $[1, \infty)$

15. $\left(\frac{16}{3}, \infty\right)$

17. $(-\infty, -1]$

19. $[-3, -1)$

21. $(2, 6)$

23. $\left[\frac{9}{2}, 5\right)$

25. $(-\infty, 2) \cup (5, \infty)$

27. $[-3, 6]$

29. $(-\infty, -1] \cup \left[\frac{1}{2}, \infty\right)$

31. $(-1, 4)$

33. $(-\infty, -3) \cup (6, \infty)$

35. $(-2, 2)$

37. $(-\infty, \infty)$

39. $(-\infty, -1) \cup [3, \infty)$

41. $\left(-\infty, -\frac{3}{2}\right)$

43. $(-\infty, 5) \cup [16, \infty)$

45. $(-2, 0) \cup (2, \infty)$

47. $[-2, -1) \cup (0, 1]$

49. $[-2, 0) \cup (1, 3]$

51. $\left(-3, -\frac{1}{2}\right) \cup (2, \infty)$

53. $(-\infty, -1) \cup (1, \infty)$

55. $(-7, 7)$

57. $[2, 8]$

59. $(-7, -3)$

61. $[1.3, 1.7]$

63. $(-4, 8)$

65. $(-6, 2)$

67. $68 \le F \le 86$ **69.** From 0 s to 3 s

71. Between 0 and 60 mi/h   **73.** More than 200 mi

75. (a) $-\frac{1}{3}P + \frac{560}{3}$ (b) From \$215 to \$290

77. ± 0.195 dollars   **79.** Between 20 and 40 ft

81. $-\frac{4}{3} \le x \le \frac{4}{3}$ **83.** $x < -2$ or $x > 7$

Section 1.8 ■ **page 100**

1.

3. (a)

(b) $\sqrt{10}$

(c) $\left(\frac{7}{2}, \frac{5}{2}\right)$

5. (a)

(b) $\sqrt{74}$

(c) $\left(\frac{5}{2}, \frac{1}{2}\right)$

7. (a)

(b) 10

(c) $(0, 0)$

9. 24

11. Trapezoid, area $= 9$

13.

15.

17.

19.

21.

23. $A(6, 7)$ **25.** $Q(-1, 3)$ **29.** (b) 10

33. 6.02, 5.22, 6.08 (correct to two decimals)

35. (a)

(b) $\left(\frac{5}{2}, 3\right), \left(\frac{5}{2}, 3\right)$

37. (a) $(8, 5)$ (b) $(a + 3, b + 2)$ (c) $(0, 2)$
(d) $A'(-2, 1), B'(0, 4), C'(5, 3)$ **39.** No, yes, no

41. No, yes, yes

43. x-intercept 0, y-intercept 0,
symmetry about origin

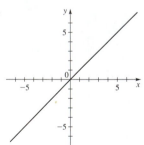

45. x-intercept 1,
y-intercept -1,
no symmetry

47. x-intercept $\frac{5}{3}$,
y-intercept -5,
no symmetry

49. x-intercepts ± 1,
y-intercept 1,
symmetry about y-axis

51. x-intercept 0,
y-intercept 0,
symmetry about y-axis

53. x-intercepts ± 3,
y-intercept -9,
symmetry about y-axis

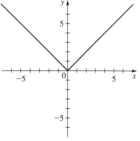

55. x-intercept 0,
y-intercept 0,
no symmetry

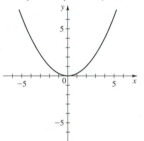

57. x-intercept 0,
y-intercept 0,
symmetry about y-axis

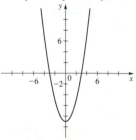

59. x-intercepts ± 4,
y-intercept 4,
symmetry about y-axis

61. x-intercept 0,
y-intercept 0,
symmetry about y-axis

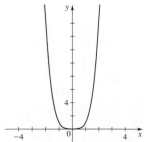

63. Symmetry about y-axis   **65.** Symmetry about origin
67. Symmetry about origin

69.   **71.**

73. $(x - 2)^2 + (y + 1)^2 = 9$ **75.** $(x + 1)^2 + (y - 5)^2 = 130$
77. $(x - 7)^2 + (y + 3)^2 = 9$ **79.** $(x + 2)^2 + (y - 2)^2 = 4$

81. $(1, -2), 2$ **83.** $(0, -3), \sqrt{7}$ **85.** $\left(-1, -\frac{1}{2}\right), \frac{1}{2}$

87.   **89.** 12π

19.   **21.**

23. No   **25.** Yes, 2   **27.** -4 **29.** $\dfrac{5}{14}$ **31.** $\pm 4\sqrt{2} \approx \pm 5.7$

33. $2.5, -2.5$ **35.** $5 + 2\sqrt[4]{5} \approx 7.99, 5 - 2\sqrt[4]{5} \approx 2.01$
37. $3.00, 4.00$ **39.** $1.00, 2.00, 3.00$ **41.** 1.62 **43.** $-1.00,$
$0.00, 1.00$ **45.** 2.55 **47.** $-2.05, 0, 1.05$ **49.** $[-2.00, 5.00]$
51. $(-\infty, 1.00] \cup [2.00, 3.00]$ **53.** $(-1.00, 0) \cup (1.00, \infty)$
55. $(-\infty, 0)$ **57.** $0, 0.01$

Section 1.10 ■ page 124

1. 2 **3.** $-\frac{1}{4}$ **5.** 4 **7.** $-\frac{9}{2}$

9. $-2, \frac{1}{2}, 3, -\frac{1}{4}$ **11.** $x + y - 4 = 0$ **13.** $3x - 2y - 6 = 0$

15. $x - y + 1 = 0$ **17.** $2x - 3y + 19 = 0$

19. $5x + y - 11 = 0$ **21.** $3x - y - 2 = 0$

23. $3x - y - 3 = 0$ **25.** $y = 5$ **27.** $x + 2y + 11 = 0$

29. $x = -1$ **31.** $5x - 2y + 1 = 0$ **33.** $x - y + 6 = 0$

35. (a)   (b) $3x - 2y + 8 = 0$

37. All lines pass through $(3, 2)$

39. $-1, 3$ **41.** $-\frac{1}{3}, 0$

Section 1.9 ■ page 112

1. (c)   **3.** (c)   **5.** (c)

7.   **9.**

11.   **13.**

15.   **17.**

43. $\frac{3}{2}, 3$

45. $0, 4$

71. (a)

(b) $(21.82, 13.82)$
(c) Price $21.82, amount 13.82

47. $\frac{3}{4}, -3$

49. $-\frac{3}{4}, \frac{1}{4}$

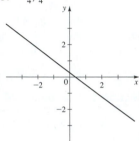

55. $x - y - 3 = 0$ **57. (b)** $4x - 3y - 24 = 0$
59. 16,667 ft
61. (a) 8.34; the slope represents the increase in dosage for a one-year increase in age.
(b) 8.34 mg
63. (a)

(b) The slope represents production cost per toaster; the y-intercept represents monthly fixed cost.

65. (a) $t = \frac{5}{24} n + 45$ **(b)** 76°F

67. (a) $P = 0.434d + 15$, where P is pressure in lb/in² and d is depth in feet   **(b)** 196 ft

69. (a) $C = \frac{1}{4} d + 260$
(b) $635
(c) The slope represents cost per mile.
(d) The y-intercept represents annual fixed cost.

Chapter 1 Review ■ page 129

1. Commutative Property for addition
3. Distributive Property
5. $-2 \leq x < 6$

7. $[5, \infty)$

9. 6   **11.** $\frac{1}{72}$ **13.** $\frac{1}{6}$ **15.** 11   **17.** 4   **19.** $16x^3$
21. $12xy^8$ **23.** x^2y^2 **25.** $\dfrac{x(2 - \sqrt{x})}{4 - x}$ **27.** $\dfrac{4r^{5/2}}{s^7}$
29. 7.825×10^{10} **31.** 1.65×10^{-32}
33. $3xy^2(4xy^2 - y^3 + 3x^2)$ **35.** $(x - 2)(x + 5)$
37. $(4t + 3)(t - 4)$ **39.** $(5 - 4t)(5 + 4t)$
41. $(x - 1)(x^2 + x + 1)(x + 1)(x^2 - x + 1)$
43. $x^{-1/2}(x - 1)^2$ **45.** $(x - 2)(4x^2 + 3)$
47. $\sqrt{x^2 + 2}\,(x^2 + x + 2)^2$ **49.** $6x^2 - 21x + 3$ **51.** $x - 7$
53. $2x^3 - 6x^2 + 4x$ **55.** $\dfrac{3(x + 3)}{x + 4}$ **57.** $\dfrac{x + 1}{x - 4}$
59. $\dfrac{1}{x + 1}$ **61.** $-\dfrac{1}{2x}$ **63.** $3\sqrt{2} - 2\sqrt{3}$ **65.** 5
67. No solution   **69.** $2, 7$ **71.** $-1, \frac{1}{2}$ **73.** $0, \pm \frac{5}{2}$
75. $\dfrac{-2 \pm \sqrt{7}}{3}$ **77.** -5 **79.** $3, 11$
81. 20 lb raisins, 30 lb nuts   **83.** $\frac{1}{4}(\sqrt{329} - 3) \approx 3.78$ mi/h
85. 110 min
87. $(-3, \infty)$ **89.** $(-\infty, -6) \cup (2, \infty)$

91. $(-\infty, -2) \cup (2, 4]$ **93.** $[2, 8]$

95. $-1, 7$ **97.** $[1, 3]$
99. (a)

(b) $\sqrt{193}$

(d) $y = -\frac{12}{7}x + \frac{24}{7}$

(c) $\left(-\frac{3}{2}, 6\right)$

(e) $(x-2)^2 + y^2 = 193$

101.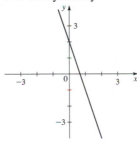

103. B

105. $(x+5)^2 + (y+1)^2 = 26$

107. Circle, center $(-1, 3)$, radius 1   **109.** No graph

111. No symmetry

113. No symmetry

115. Symmetry about y-axis

117. No symmetry

119.

121.

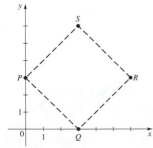

123. $2x - 3y - 16 = 0$ **125.** $3x + y - 12 = 0$

127. $x + 5y = 0$

129. $x^2 + y^2 = 169, 5x - 12y + 169 = 0$

Chapter 1 Test ■ page 132

1. (a)

(b) $(-\infty, 3], (0, 8]$ (c) 51

2. (a) 81   (b) -81 (c) $\frac{1}{81}$ (d) 25   (e) $\frac{9}{4}$ (f) $\frac{1}{8}$

3. (a) 1.86×10^{11} (b) 3.965×10^{-7}

4. (a) $6\sqrt{2}$ (b) $48a^5b^7$ (c) $\frac{x}{9y^7}$ (d) $\frac{x+2}{x-2}$ (e) $\frac{1}{x-2}$

(f) $-(x+y)$ **5.** $5\sqrt{2} + 2\sqrt{10}$ **6.** (a) $11x - 2$

(b) $4x^2 + 7x - 15$ (c) $a - b$ (d) $4x^2 + 12x + 9$

(e) $x^3 + 6x^2 + 12x + 8$

7. (a) $(2x-5)(2x+5)$ (b) $(2x-3)(x+4)$

(c) $(x-3)(x-2)(x+2)$ (d) $x(x+3)(x^2-3x+9)$

(e) $3x^{-1/2}(x-1)(x-2)$ (f) $xy(x-2)(x+2)$

8. (a) 6   (b) 1   (c) $-3, 4$ (d) $-1 \pm \frac{\sqrt{2}}{2}$

(e) No real solution   (f) $\pm 1, \pm\sqrt{2}$ (g) $\frac{2}{3}, \frac{22}{3}$ **9.** 120 mi

10. 50 ft by 120 ft

11. (a) $[-4, 3)$

(b) $(-2, 0) \cup (1, \infty)$

(c) $(1, 7)$

(d) $(-1, 4]$

12. Between $41°F$ and $50°F$ **13.** $0 \leqslant x \leqslant 6$

14. (a) $-2.94, -0.11, 3.05$ (b) $[-1, 3]$

15. (a) $S(3, 6)$ (b) 18

16. (a)

(b) x-intercepts $-2, 2$
y-intercept -4
(c) Symmetric about y-axis

17. (a) 20   **(b)** $(-1, -4)$ **(c)** $4x + 3y + 16 = 0$
(d) $3x - 4y - 13 = 0$ **(e)** $(x - 5)^2 + (y + 12)^2 = 169$
(f) $(x + 1)^2 + (y + 4)^2 = 100$

18. (a) Center $(0, 4)$; radius 3   **(b)** Center $(3, -5)$; radius 5

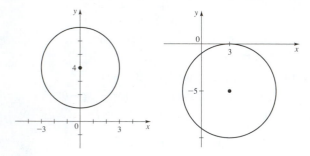

19. $y = \frac{3}{4}x + 6$, slope $\frac{3}{4}$, y-intercept 6

20. (a) $x + 3y - 7 = 0$ **(b)** $4x - y + 12 = 0$
21. (a) 35,000
(b) The slope represents the cost of producing one more
blender. The C-intercept represents the fixed costs.

Focus on Problem Solving ■ **page 138**

1. 37.5 mi/h   **3.** 150 mi   **5.** 2   **7.** 57 min
9. No, not necessarily   **11.** The same amount
13. (a) Uncle   **(b)** Father   **(d)** No   **15.** 8.49   **17.** 427
19. 40   **21.** $\frac{11}{2}$ **25.** Infinitely far

27.

CHAPTER 2

Section 2.1 ■ page 151

1. $f(x) = 7x + 2$ **3.** $f(x) = (x - 4)^2$ **5.** Divide by 2, then
add 7   **7.** Square, multiply by 3, then subtract 2
9.

input		output
0	square root	0
1	square root	1
4	square root	2

11.

x	$f(x)$
-1	3
0	1
1	3
2	9
3	19

13. $3, -3, 2, 2a + 1, -2a + 1, 2a + 2b + 1$
15. $-\dfrac{1}{3}, -3, \dfrac{1}{3}, \dfrac{1 - a}{1 + a}, \dfrac{2 - a}{a}$, undefined
17. $-4, 10, -2, 3\sqrt{2}, 2x^2 + 7x + 1, 2x^2 - 3x - 4$
19. $6, 2, 1, 2, 2|x|, 2(x^2 + 1)$ **21.** $4, 1, 1, 2, 3$
23. $8, -\dfrac{3}{4}, -1, 0, 1$ **25.** $x^2 + 4x + 5, x^2 + 6$
27. $x^2 + 4, x^2 + 8x + 16$ **29.** $3a + 2, 3(a + h) + 2, 3$
31. $5, 5, 0$
33. $3 - 5a + 4a^2, 3 - 5a - 5h + 4a^2 + 8ah + 4h^2,$
$-5 + 8a + 4h$
35. (a) $C(10) = 1532.1, C(100) = 2100$ **(b)** The cost of
producing 10 yd and 100 yd   **(c)** $C(0) = 1500$
37. (a) $D(0.1) = 28.1, D(0.2) = 39.8$ **(b)** 41.3 mi   **(c)** 235.6 mi
39. (a) $v(0.1) = 4440, v(0.4) = 1665$
(b) Flow is faster near central axis
(c)

r	$v(r)$
0	4625
0.1	4440
0.2	3885
0.3	2960
0.4	1665
0.5	0

41. (a) $T(5000) = 0, T(12,000) = 960, T(25,000) = 1975$
(b) The amount of tax paid on incomes of 5000, 12,000, and
25,000
43. $(-\infty, \infty)$ **45.** $[-1, 5]$ **47.** $\{x \mid x \neq 3\}$

49. $\{x \mid x \neq \pm 1\}$ **51.** $[5, \infty)$ **53.** $(-\infty, \infty)$ **55.** $\left[\frac{5}{2}, \infty\right)$
57. $[-2, 3) \cup (3, \infty)$ **59.** $(-\infty, 0] \cup [6, \infty)$
61. This person's weight increases as he grows, then continues to increase; the person then goes on a crash diet (possibly) at age 30, then gains weight again, the weight gain eventually leveling off.
63.

65. A won the race. All runners finished. Runner B fell, but got up again to finish second.
67.

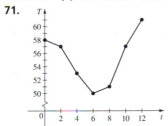

69. (a) 500 MW, 725 MW   (b) Between 3:00 A.M. and 4:00 A.M.   (c) Just before noon
71.

Section 2.2 ■ page 165

1. (a) $1, -1, 3, 4$ (b) Domain $[-3, 4]$, range $[-1, 4]$
3. (a) $f(0)$ (b) $g(-3)$ (c) $-2, 2$
5. (a) Yes   (b) No   (c) Yes   (d) No
7. Function, domain $[-3, 2]$, range $[-2, 2]$
9. Not a function
11. (a)

(b) Domain $(-\infty, \infty)$, range $(-\infty, \infty)$

13. (a)

(b) Domain $(-\infty, \infty)$, range $[-4, \infty)$

15. (a)

17. (a)

(b) Domain $[-4, 4]$, range $[0, 4]$

(b) Domain $[1, \infty)$, range $[0, \infty)$

19.

21.

23.

25.

27.

29.

31.

33.

35.

37.

39.

41. Yes
43. No
45. No
47. Yes
49. Yes
51. Yes

53. (a)

(b)

(c) If $c > 0$, then the graph of $f(x) = x^2 + c$ is the same as the graph of $y = x^2$ shifted upward c units. If $c < 0$, then the graph of $f(x) = x^2 + c$ is the same as the graph of $y = x^2$ shifted downward c units.

55. (a)

(b)

(c) If $c > 0$, then the graph of $f(x) = (x - c)^3$ is the same as the graph of $y = x^3$ shifted right c units. If $c < 0$, then the graph of $f(x) = (x - c)^3$ is the same as the graph of $y = x^3$ shifted left c units.

57. (a)   **(b)**

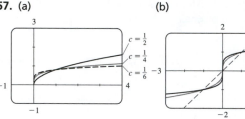

(c) Graphs of even roots are similar to $\sqrt{x}$; graphs of odd roots are similar to $\sqrt[3]{x}$. As c increases, the graph of $y = \sqrt[c]{x}$ becomes steeper near 0 and flatter when $x > 1$.

59.   **61.**

63.   **65.**

67.

69.

71.

73.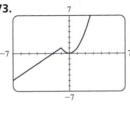

75.
$$C(x) = \begin{cases} 2 & 0 < x \leq 1 \\ 2.2 & 1 < x \leq 1.1 \\ 2.4 & 1.1 < x \leq 1.2 \\ \vdots \\ 4.0 & 1.9 < x < 2.0 \end{cases}$$

77. $f(x) = -\frac{7}{6}x - \frac{4}{3}, -2 \leq x \leq 4$ **79.** $f(x) = 1 - \sqrt{-x}$

Section 2.3 ■ page 172

1. $T = kx$ **3.** $v = k/z$ **5.** $y = ks/t$ **7.** $z = k\sqrt{y}$

9. $V = klwh$ **11.** $R = k\dfrac{i}{Pt}$ **13.** $y = 7x$

15. $M = 15x/y$ **17.** $W = 360/r^2$ **19.** $C = 16lwh$

21. $s = 500/\sqrt{t}$ **23.** (a) $F = kx$ (b) 8   (c) 32 N

25. (a) $C = kpm$ (b) 0.125   (c) \$57,500

27. (a) $R = kL/d^2$ (b) $0.002916\overline{6}$ (c) $R \approx 137 \, \Omega$

29. \$3.75   **31.** \$51,200   **33.** (a) $f = \dfrac{k}{L}$ (b) Halves it

Section 2.4 ■ page 182

1. 3   **3.** 5   **5.** 60   **7.** $12 + 3h$ **9.** $-1/a$ **11.** $\dfrac{-2}{a(a+h)}$

13. (a) $\frac{1}{2}$ **15.** $\frac{2}{3}$ **17.** $-\frac{4}{5}$

19. (a) 245 persons/yr   (b) -328.5 persons/yr
(c) 1990–1994   (d) 1994–1999

21. (a) 7.2 units/yr   (b) 8 units/yr   (c) -55 units/yr
(d) 1996–1997, 1997–1998

23. Increasing on $[-1, 1]$, $[2, 4]$; decreasing on $[1, 2]$

25. Increasing on $[-2, -1]$, $[1, 2]$; decreasing on $[-3, -2]$, $[-1, 1]$, $[2, 3]$

27. (a)

29. (a)
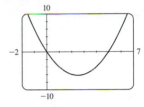

(b) Increasing on $[0, \infty)$; decreasing on $(-\infty, 0]$

(b) Increasing on $[2.5, \infty)$; decreasing on $(-\infty, 2.5]$

31. (a)

33. (a)

(b) Increasing on $(-\infty, -1]$, $[2, \infty)$; decreasing on $[-1, 2]$

(b) Increasing on $(-\infty, -1.55]$, $[0.22, \infty)$; decreasing on $[-1.55, 0.22]$

Section 2.5 ■ page 194

1. (a) Shift downward 3 units   (b) Shift right 3 units

3. (a) Stretch vertically by a factor of 3
(b) Shrink vertically by a factor of $\frac{1}{3}$

5. (a) Reflect in the x-axis and stretch vertically by a factor of 2
(b) Reflect in the x-axis and shrink vertically by a factor of $\frac{1}{2}$

7. (a) Shift right 4 units and upward $\frac{3}{4}$ unit   (b) Shift left 4 units and downward $\frac{3}{4}$ unit   **9.** (a) Shrink horizontally by a factor of $\frac{1}{4}$ (b) Stretch horizontally by a factor of 4

11. (a)   (b)

(c)

(d)

(e)

(f)

13. (a)

(b) (i)

(ii)

(iii)

(iv)

15. (a) Shift left 2 units   **(b)** Shift up 2 units

17. (a) Stretch vertically by a factor of 2

(b) Shift right 2 units, then shrink vertically by a factor of $\frac{1}{2}$

19. $g(x) = (x - 2)^2 + 3$ **21.** $g(x) = -5\sqrt{x + 3}$

23. $g(x) = 0.1 \left| x - \frac{1}{2} \right| - 2$

25.

27.

29.

31.

33.

35.

37.

39.

41.

For part (b) shift the graph in (a) left 5 units; for part (c) shift the graph in (a) left 5 units and stretch vertically by a factor of 2; for part (d) shift the graph in (a) left 5 units, stretch vertically by a factor of 2, and then shift upward 4 units.

43.

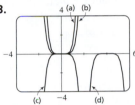

For part (b) shrink the graph in (a) vertically by a factor of $\frac{1}{3}$; for part (c) shrink the graph in (a) vertically by a factor of $\frac{1}{3}$ and reflect in the x-axis; for part (d) shift the graph in (a) right 4 units, shrink vertically by a factor of $\frac{1}{3}$, and then reflect in the x-axis.

45. (a)   (b)

47.   **49.**

51. Even   **53.** Neither

55. Odd   **57.** Neither

59. To obtain the graph of g, reflect in the x-axis the part of the graph of f that is below the x-axis.

61. (a)   (b)

   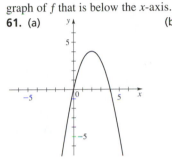

Section 2.6 ■ page 204

1. Vertex $(4, -16)$
x-intercepts 0, 8
y-intercept 0

3. Vertex $\left(\frac{3}{2}, -\frac{9}{2}\right)$
x-intercepts 0, 3
y-intercept 0

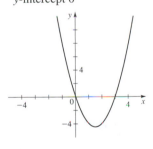

5. Vertex $(-2, -3)$
x-intercepts $-2 \pm \sqrt{3}$
y-intercept 1

7. Vertex $(-3, -1)$
x-intercepts $-4, -2$
y-intercept 8

9. Vertex $(-1, 1)$
no x-intercept
y-intercept 3

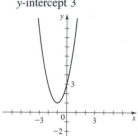

11. Vertex $(5, 7)$
no x-intercept
y-intercept 57

(b)

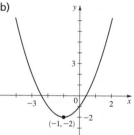

(c) Maximum $h\left(-\frac{1}{2}\right) = \frac{5}{4}$

25. Minimum $f\left(-\frac{1}{2}\right) = \frac{3}{4}$ **27.** Maximum $f(-3.5) = 185.75$
29. Minimum $f(0.6) = 15.64$ **31.** Minimum $h(-2) = -8$
33. Maximum $f(-1) = \frac{7}{2}$ **35.** $f(x) = 2x^2 - 4x$
37. $(-\infty, \infty), (-\infty, 1]$
39. 25 ft   **41.** \$4000, 100 units
43. 30 times   **45.** (a) -4.01 (b) -4.011025
47. Local maximum ≈ 0.38 when $x \approx -0.58$;
local minimum ≈ -0.38 when $x \approx 0.58$
49. Local maximum ≈ 0 when $x = 0$;
local minimum ≈ -13.61 when $x \approx -1.71$;
local minimum ≈ -73.32 when $x \approx 3.21$
51. Local maximum ≈ 5.66 when $x \approx 4.00$
53. Local maximum ≈ 0.38 when $x \approx -1.73$;
local minimum ≈ -0.38 when $x \approx 1.73$
55. 7.5 mi/h   **57.** 3.96°C   **59.** 7

13. Vertex $(-2, 19)$
x-intercepts $-2 \pm \frac{1}{2}\sqrt{19}$
y-intercept 3

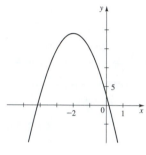

15. (a) $f(x) = -(x-1)^2 + 1$
(b)

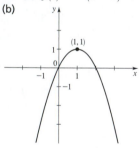

(c) Maximum $f(1) = 1$

19. (a) $f(x) = -\left(x + \frac{3}{2}\right)^2 + \frac{21}{4}$
(b)

(c) Maximum $f\left(-\frac{3}{2}\right) = \frac{21}{4}$

23. (a) $h(x) = -\left(x + \frac{1}{2}\right)^2 + \frac{5}{4}$

17. (a) $f(x) = (x+1)^2 - 2$
(b)

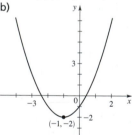

(c) Minimum $f(-1) = -2$

21. (a) $g(x) = 3(x-2)^2 + 1$
(b)

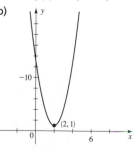

(c) Minimum $g(2) = 1$

Section 2.7 ■ page 214

1. $A(w) = 3w^2, w > 0$ **3.** $V(w) = \frac{1}{2}w^3, w > 0$
5. $A(x) = 10x - x^2, 0 < x < 10$ **7.** $A(x) = \left(\sqrt{3}/4\right)x^2, x > 0$
9. $r(A) = \sqrt{A/\pi}, A > 0$ **11.** $S(x) = 2x^2 + 240/x, x > 0$
13. $D(t) = 25t, t \geq 0$ **15.** $A(b) = b\sqrt{4 - b}, 0 < b < 4$
17. $A(h) = 2h\sqrt{100 - h^2}, 0 < h < 10$
19. (b) $p(x) = x(19 - x)$ (c) 9.5, 9.5
21. $-12, -12$
23. (b) $A(x) = x(2400 - 2x)$ (c) 600 ft by 1200 ft
25. (a) $f(w) = 8w + 7200/w$
(b) Width along road is 30 ft, length is 40 ft
(c) 15 ft to 60 ft
27. (a) $R(p) = -3000p^2 + 57{,}000p$ (b) \$19   (c) \$9.50
29. (a) $A(x) = 15x - \left(\dfrac{\pi + 4}{8}\right)x^2$
(b) Width ≈ 8.40 ft, height of rectangular part ≈ 4.20 ft
31. (a) $A(x) = x^2 + 48/x$
(b) Height ≈ 1.44 ft, width ≈ 2.88 ft
33. (a) $A(x) = 2x + 200/x$ (b) 10 m by 10 m
35. (a) $E(x) = 14\sqrt{25 + x^2} + 10(12 - x)$
(b) To point C, 5.1 mi from point B

Section 2.8 ■ page 225

1. $(f + g)(x) = x^2 + x + 2, (-\infty, \infty)$;
$(f - g)(x) = x^2 - x - 2, (-\infty, \infty)$;
$(fg)(x) = x^3 + 2x^2, (-\infty, \infty)$;
$(f/g)(x) = x^2/(x + 2), (-\infty, -2) \cup (-2, \infty)$
3. $(f + g)(x) = \sqrt{1 + x^2} + \sqrt{1 - x}, (-\infty, 1]$;
$(f - g)(x) = \sqrt{1 + x^2} - \sqrt{1 - x}, (-\infty, 1]$;
$(fg)(x) = \sqrt{1 - x + x^2 - x^3}, (-\infty, 1]$;
$(f/g)(x) = \sqrt{\dfrac{1 + x^2}{1 - x}}, (-\infty, 1)$
5. $(f + g)(x) = 8/(x(x + 4)), x \neq -4, x \neq 0$;
$(f - g)(x) = 4(x + 2)/(x(x + 4)), x \neq -4, x \neq 0$;
$(fg)(x) = -4/(x(x + 4)), x \neq -4, x \neq 0$;
$(f/g)(x) = -(x + 4)/x, x \neq -4, x \neq 0$
7. $[0, 1]$　**9.** $[4, \infty)$

11.

13. 　　**15.**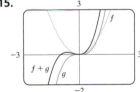

17. (a) 1　(b) -23　**19.** (a) -11　(b) -119
21. (a) $-3x^2 + 1$　(b) $-9x^2 + 30x - 23$
23. 4　**25.** 5　**27.** 4
29. $(f \circ g)(x) = 8x + 1, (-\infty, \infty)$;
$(g \circ f)(x) = 8x + 11, (-\infty, \infty)$;
$(f \circ f)(x) = 4x + 9, (-\infty, \infty)$;
$(g \circ g)(x) = 16x - 5, (-\infty, \infty)$
31. $(f \circ g)(x) = (x + 1)^2, (-\infty, \infty)$
$(g \circ f)(x) = x^2 + 1, (-\infty, \infty)$
$(f \circ f)(x) = x^4, (-\infty, \infty)$
$(g \circ g)(x) = x + 2, (-\infty, \infty)$
33. $(f \circ g)(x) = \dfrac{1}{2x + 4}, x \neq -2$
$(g \circ f)(x) = \dfrac{2}{x} + 4, x \neq 0$
$(f \circ f)(x) = x, x \neq 0$
$(g \circ g)(x) = 4x + 12, (-\infty, \infty)$

35. $(f \circ g)(x) = |2x + 3|, (-\infty, \infty)$
$(g \circ f)(x) = 2|x| + 3, (-\infty, \infty)$
$(f \circ f)(x) = |x|, (-\infty, \infty)$
$(g \circ g)(x) = 4x + 9, (-\infty, \infty)$
37. $(f \circ g)(x) = \dfrac{2x - 1}{2x}, x \neq 0$
$(g \circ f)(x) = \dfrac{2x}{x + 1} - 1, x \neq -1$
$(f \circ f)(x) = \dfrac{x}{2x + 1}, x \neq -1, x \neq -\tfrac{1}{2}$
$(g \circ g)(x) = 4x - 3, (-\infty, \infty)$
39. $(f \circ g)(x) = \sqrt[12]{x}, [0, \infty)$
$(g \circ f)(x) = \sqrt[12]{x}, [0, \infty)$
$(f \circ f)(x) = \sqrt[9]{x}, (-\infty, \infty)$
$(g \circ g)(x) = \sqrt[16]{x}, [0, \infty)$
41. $(f \circ g \circ h)(x) = \sqrt{x - 1} - 1$
43. $(f \circ g \circ h)(x) = (\sqrt{x} - 5)^4 + 1$
45. $g(x) = x - 9, f(x) = x^5$　**47.** $g(x) = x^2, f(x) = x/(x + 4)$
49. $g(x) = 1 - x^3, f(x) = |x|$
51. $h(x) = x^2, g(x) = x + 1, f(x) = 1/x$
53. $h(x) = \sqrt[3]{x}, g(x) = 4 + x, f(x) = x^9$
55. $A(t) = 3600\pi t^2$　**57.** $A(t) = 16\pi t^2$　**59.** (a) $f(x) = 0.9x$
(b) $g(x) = x - 100$　(c) $f \circ g(x) = 0.9x - 90$,
$g \circ f(x) = 0.9x - 100, f \circ g$: first rebate, then discount, $g \circ f$:
first discount, then rebate, $g \circ f$ is the better deal

Section 2.9 ■ page 236

1. No　**3.** Yes　**5.** No　**7.** Yes　**9.** Yes　**11.** Yes　**13.** No
15. No　**17.** (a) 2　(b) 3　**19.** 1
31. $f^{-1}(x) = \tfrac{1}{2}(x - 1)$　**33.** $f^{-1}(x) = \tfrac{1}{4}(x - 7)$
35. $f^{-1}(x) = 2x$　**37.** $f^{-1}(x) = (1/x) - 2$
39. $f^{-1}(x) = (5x - 1)/(2x + 3)$
41. $f^{-1}(x) = \tfrac{1}{5}(x^2 - 2), x \geq 0$
43. $f^{-1}(x) = \sqrt{4 - x}$　**45.** $f^{-1}(x) = (x - 4)^3$
47. $f^{-1}(x) = x^2 - 2x, x \geq 1$　**49.** $f^{-1}(x) = \sqrt[4]{x}$

51. (a) 　　(b)

(c) $f^{-1}(x) = \tfrac{1}{3}(x + 6)$

53. (a)

(b)

(c) $f^{-1}(x) = x^2 - 1, x \geq 0$
55. Not one-to-one

57. One-to-one

59. Not one-to-one

61. $x \geq 0, f^{-1}(x) = \sqrt{4 - x}$
63. $x \geq -2, h^{-1}(x) = \sqrt{x - 2}$

65.

67. (a) $f(x) = 500 + 80x$ (b) $f^{-1}(x) = \frac{1}{80}(x - 500)$, the number of hours worked as a function of the fee
(c) 9; if he charges $1220, he worked 9 h

69. (a) $v^{-1}(t) = \sqrt{0.25 - \dfrac{t}{18,\,500}}$

(b) 0.498; at a distance 0.498 from the central axis, the velocity is 30   **71.** (a) $F^{-1}(x) = \frac{5}{9}(x - 32)$; the Celsius temperature when the Fahrenheit temperature is x (b) $F^{-1}(86) = 30$; when the temperature is 86°F, it is 30°C

73. (a) $f(x) = \begin{cases} 0.1x & \text{if } x \leqslant 20{,}000 \\ 2000 + 0.2(x - 20{,}000) & \text{if } x > 20{,}000 \end{cases}$

(b) $f^{-1}(x) = \begin{cases} 10x & \text{if } x \leqslant 2000 \\ 10{,}000 + 5x & \text{if } x > 2000 \end{cases}$
If you pay x euros in taxes, your income is $f^{-1}(x)$
(c) $f^{-1}(10{,}000) = 60{,}000$

75. $f^{-1}(x) = \frac{1}{2}(x - 7)$ A pizza costing x dollars has $f^{-1}(x)$ toppings.

Chapter 2 Review ■ **page 240**
1. $-4, 8, -16, a^3 + 2a - 4, -a^3 - 2a - 4,$
$x^3 + 3x^2 + 5x - 1, 8x^3 + 4x - 4, 2x^3 + 4x - 10$
3. (a) $-1, 2$ (b) $[-4, 5]$ (c) $[-4, 4]$ (d) Increasing on $[-4, -2]$ and $[-1, 4]$; decreasing on $[-2, -1]$ and $[4, 5]$
(e) No   **5.** Domain $[-3, \infty)$, range $[0, \infty)$ **7.** $(-\infty, \infty)$
9. $[-4, \infty)$ **11.** $\{x \mid x \neq -2, -1, 0\}$
13. $(-\infty, -1] \cup [1, 4]$
15.

17.

19.

21.

23.

25.

27.

29.

31.

33. (iii)

35.

37.

39. $[-2.1, 0.2] \cup [1.9, \infty)$
41. $M = 8z$ **43.** (a) $I = k/d^2$ (b) 64,000   (c) 160 candles
45. 11.0 mi/h   **47.** 5   **49.** $\dfrac{-1}{3(3+h)}$

51.

Increasing on $(-\infty, 0], [2.67, \infty)$; decreasing on $[0, 2.67]$
53. (a) Shift upward 8 units   (b) Shift left 8 units
(c) Stretch vertically by a factor of 2, then shift upward 1 unit
(d) Shift right 2 units and downward 2 units
(e) Reflect in y-axis
(f) Reflect in y-axis, then in x-axis
(g) Reflect in x-axis   (h) Reflect in line $y = x$
55. (a) Neither   (b) Odd   (c) Even   (d) Neither
57. $f(x) = (x + 2)^2 - 3$ **59.** $g(-1) = -7$ **61.** 68 ft
63. Local maximum ≈ 3.79 when $x \approx 0.46$;
local minimum ≈ 2.81 when $x \approx -0.46$

65.

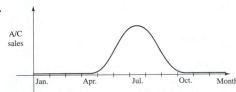

67. (a) $A(x) = 5\sqrt{3}x - \dfrac{\sqrt{3}}{2}x^2$ (b) 5 cm by $\dfrac{5\sqrt{3}}{2}$ cm

69. (a) $(f + g)(x) = x^2 - 6x + 6$
(b) $(f - g)(x) = x^2 - 2$
(c) $(fg)(x) = -3x^3 + 13x^2 - 18x + 8$
(d) $(f/g)(x) = (x^2 - 3x + 2)/(4 - 3x)$
(e) $(f \circ g)(x) = 9x^2 - 15x + 6$
(f) $(g \circ f)(x) = -3x^2 + 9x - 2$
71. $(f \circ g)(x) = -3x^2 + 6x - 1, (-\infty, \infty)$;
$(g \circ f)(x) = -9x^2 + 12x - 3, (-\infty, \infty)$;
$(f \circ f)(x) = 9x - 4, (-\infty, \infty)$;
$(g \circ g)(x) = -x^4 + 4x^3 - 6x^2 + 4x, (-\infty, \infty)$
73. $(f \circ g \circ h)(x) = 1 + \sqrt{x}$ **75.** Yes   **77.** No   **79.** No

81. $f^{-1}(x) = \dfrac{x + 2}{3}$ **83.** $f^{-1}(x) = \sqrt[3]{x} - 1$

85. (a), (b)

(c) $f^{-1}(x) = \sqrt{x + 4}$

Chapter 2 Test ■ page 244

1. (a) and (b) are graphs of functions, (a) is one-to-one
2. (a) $f(4) = \dfrac{2}{3}$, $f(6) = \dfrac{\sqrt{6}}{5}$, $f(a + 1) = \dfrac{\sqrt{a+1}}{a}$

(b) $[0, 1) \cup (1, \infty)$
3. (a) $M = kwh^2/L$ (b) 400   (c) 12,000 lb   **4.** $h + 7$
5. (a)   (b)

6. (a) Shift right 3 units, then shift upward 2 units
(b) Reflect in y-axis

7. (a) $f(x) = 2(x - 2)^2 + 5$

(b)

(c) $f(2) = 5$

8. (a) $-3, 3$ **(b)**

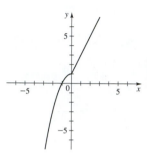

9. (a) $A(x) = 400x - 2x^2$ **(b)** 100 ft

10. (a) $(f \circ g)(x) = 4x^2 - 8x + 2$

(b) $(g \circ f)(x) = 2x^2 + 4x - 5$ **(c)** 2   **(d)** 11

(e) $(g \circ g \circ g)(x) = 8x - 21$

11. (a) $f^{-1}(x) = 3 - x^2, x \geqslant 0$ **(b)**

12. (a) Domain $[0, 6]$, range $[1, 7]$

(b)

(c) $\frac{5}{4}$

13. (a)

(b) No

(c) Local minimum ≈ -27.18 when $x \approx -1.61$; local maximum ≈ -2.55 when $x \approx 0.18$; local minimum ≈ -11.93 when $x \approx 1.43$

(d) $[-27.18, \infty)$ **(e)** Increasing on $[-1.61, 0.18] \cup [1.43, \infty)$; decreasing on $(-\infty, -1.61] \cup [0.18, 1.43]$

Focus on Modeling ■ page 250

1. (a)

(b) $y = 1.5125x - 2656.4$
(c) 365.6 ppm

3. (a)

(b) $y = -0.0995x + 13.9$, x in thousands of dollars
(c) 11.4 per 100 population   **(d)** 5.8 per 100 population

5. (a)

(b) $y = 2.6768x - 22.19$, $y = -3.5336x + 46.73$
(c) 0.77 is the correlation coefficient for the temperature-bites data, so a linear model is not appropriate; -0.89 is the correlation coefficient for the wind/bites data, so a linear model is more suitable in this case.

7. (a)

(b) $y = -3.9018x + 419.7$ (c) The correlation coefficient is -0.98, so a linear model is appropriate.   (d) 53%

9. (a)

(b) $y = 0.29083x - 501.8$ (c) 79.9 years

11. (a)

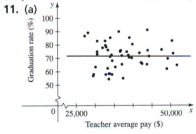

(b) $y = -0.000012x + 72.0$
(c) $r = -0.00812$; very little correlation; it appears that teachers' salaries do not affect graduation rates.

CHAPTER 3

Section 3.1 ■ page 269

1.

3.

5.

7.

9.

11.

13.

15.

17.

19.

21.

23.

25.

27.

29.

31.

33.

35.

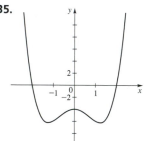

37. III

39. V

41. VI

43. $y \to \infty$ as $x \to \infty$, $y \to -\infty$ as $x \to -\infty$

45. $y \to \infty$ as $x \to \pm\infty$

47. $y \to \infty$ as $x \to \infty$, $y \to -\infty$ as $x \to -\infty$

49.

local maximum $(4, 16)$

51.

local maximum $(-2, 25)$
local minimum $(2, -7)$

53.

local minimum $(-3, -27)$

55.

local maximum $(-1, 5)$
local minimum $(1, 1)$

57. One local maximum, no local minimum

59. One local maximum, one local minimum

61. One local maximum, two local minima

63. No local extrema

65. One local maximum, two local minima

67.

Increasing the value of c stretches the graph vertically.

69.

Increasing the value of c moves the graph up.

71.

Increasing the value of c causes a deeper dip in the graph in the fourth quadrant and moves the positive x-intercept to the right.

73. (a)

(b) Three
(c) $(0, 2)$, $(3, 8)$, $(-2, -12)$
75. (g) $P(x) = P_O(x) + P_E(x)$, where $P_O(x) = x^5 + 6x^3 - 2x$ and $P_E(x) = -x^2 + 5$
77. (b) Domain $\{x \mid 0 < x < 18\}$
(c)

1728 in³

79. (a) 1380 rabbits, after 4.2 months
(b) After 8.4 months
81. (a) Three x-intercepts, two local extrema
(b) One x-intercept, no local extrema

Section 3.2 ■ page 278

In answers 1–23, the first polynomial given is the quotient and the second is the remainder.

1. $x + 1$, -11 **3.** $x^2 + 4x + 22$,   93
5. $x + 2$, $8x - 1$ **7.** $3x + 1$, $7x - 5$ **9.** $x^4 + 1$,   0
11. $x - 2$, -2 **13.** $3x + 23$,   138   **15.** $x^2 + 2$, -3
17. $x^2 - 3x + 1$, -1 **19.** $x^4 + x^3 + 4x^2 + 4x + 4$, -2
21. $2x^2 + 4x$,   1   **23.** $x^2 + 3x + 9$,   0
25. -3 **27.** 12   **29.** -7 **31.** -483 **33.** 2159   **35.** $\frac{7}{3}$
37. -8.279 **43.** $-1 \pm \sqrt{6}$ **45.** $x^3 - 3x^2 - x + 3$
47. $x^4 - 8x^3 + 14x^2 + 8x - 15$
49. $-\frac{3}{2}x^3 + 3x^2 + \frac{15}{2}x - 9$

Section 3.3 ■ page 287

1. $\pm1, \pm3$ **3.** $\pm1, \pm2, \pm4, \pm8, \pm\frac{1}{2}$ **5.** $-2, 1$ **7.** $-1, 2$
9. 2   **11.** $-1, 2, 3$ **13.** -1 **15.** $\pm1, \pm2$
17. $1, -1, -2, -4$ **19.** $\pm2, \pm\frac{3}{2}$ **21.** -2
23. $-1, -\frac{1}{2}, \frac{1}{2}$ **25.** $-\frac{3}{2}, \frac{1}{2}, 1$ **27.** $-1, \frac{1}{2}, 2$
29. $-3, -2, 1, 3$ **31.** $-1, -\frac{1}{3}, 2, 5$ **33.** $-2, -1 \pm \sqrt{2}$
35. $-1, 4, \dfrac{3 \pm \sqrt{13}}{2}$ **37.** $3, \dfrac{1 \pm \sqrt{5}}{2}$ **39.** $\dfrac{1}{2}, \dfrac{1 \pm \sqrt{3}}{2}$

41. $-1, -\frac{1}{2}, -3 \pm \sqrt{10}$
43. (a) $-2, 2, 3$

(b)

45. (a) $-4, 3$

(b)

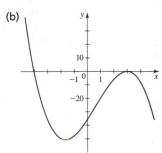

47. (a) $-1, 2$

(b)

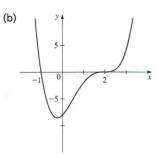

49. (a) $-1, 2$

(b)

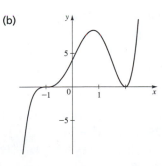

51. 1 positive, 2 or 0 negative; 3 or 1 real
53. 1 positive, 1 negative; 2 real
55. 2 or 0 positive, 0 negative; 3 or 1 real (since 0 is a zero but is neither positive nor negative)
61. $3, -2$ **63.** $3, -1$ **65.** $-2, \frac{1}{2}, \pm1$ **67.** $\pm\frac{1}{2}, \pm\sqrt{5}$

69. $-2, 1, 3, 4$ **75.** $-2, 2, 3$ **77.** $-\frac{3}{2}, -1, 1, 4$
79. $-1.28, 1.53$ **81.** -1.50 **83.** 11.3 ft
85. (a) It began to snow again.   **(b)** No
(c) Just before midnight on Saturday night
87. 2.76 m   **89.** 88 in. (or 3.21 in.)

Section 3.4 ■ page 298

1. Real part 5, imaginary part -7 **3.** Real part $-\frac{2}{3}$,
imaginary part $-\frac{5}{3}$ **5.** Real part 3, imaginary part 0
7. Real part 0, imaginary part $-\frac{2}{3}$
9. Real part $\sqrt{3}$, imaginary part 2
11. $5 - i$ **13.** $3 + 5i$ **15.** $12 + i$ **17.** $-19 + 4i$
19. $-4 + 8i$ **21.** $30 + 10i$ **23.** $-33 - 56i$ **25.** $27 - 8i$
27. $-i$ **29.** $\frac{8}{5} + \frac{1}{5}i$ **31.** $-5 + 12i$ **33.** $-4 + 2i$
35. $2 - \frac{4}{3}i$ **37.** $-i$ **39.** $-i$ **41.** 1   **43.** $5i$ **45.** -6
47. $(3 + \sqrt{5}) + (3 - \sqrt{5})i$ **49.** 2   **51.** $-i\sqrt{2}$ **53.** $\pm 3i$
55. $2 \pm i$ **57.** $-\frac{1}{2} \pm \frac{\sqrt{3}}{2}i$ **59.** $\frac{1}{2} \pm \frac{1}{2}i$ **61.** $-\frac{3}{2} \pm \frac{\sqrt{3}}{2}i$
63. $\dfrac{-6 \pm \sqrt{6}\,i}{6}$ **65.** $1 \pm 3i$

Section 3.5 ■ page 306

1. $-2, 1 \pm i\sqrt{3}$ **3.** $0, 1 \pm i$ **5.** $\pm i$ **7.** $\pm 2, \pm 2i$
9. $\pm 1, \frac{1}{2} \pm \frac{1}{2}i\sqrt{3}, \pm 1, -\frac{1}{2} \pm \frac{1}{2}i\sqrt{3}$

*In answers 11–27, the factored form is given first, then the
zeros are listed with the multiplicity of each in parentheses.*

11. $(x - 5i)(x + 5i)$; $\pm 5i$ (1)
13. $[x - (-1 + i)][x - (-1 - i)]$; $-1 + i$ (1), $-1 - i$ (1)
15. $x(x - 2i)(x + 2i)$; 0 (1), $2i$ (1), $-2i$ (1)
17. $(x - 1)(x + 1)(x - i)(x + i)$; 1 (1), -1 (1), i (1), $-i$ (1)
19. $16\left(x - \frac{3}{2}\right)\left(x + \frac{3}{2}\right)\left(x - \frac{3}{2}i\right)\left(x + \frac{3}{2}i\right)$; $\frac{3}{2}$ (1), $-\frac{3}{2}$ (1),
$\frac{3}{2}i$ (1), $-\frac{3}{2}i$ (1)
21. $(x + 1)(x - 3i)(x + 3i)$; -1 (1), $3i$ (1), $-3i$ (1)
23. $(x - i)^2(x + i)^2$; i (2), $-i$ (2)
25. $(x - 1)(x + 1)(x - 2i)(x + 2i)$; 1 (1), -1 (1),
$2i$ (1), $-2i$ (1)
27. $x(x - i\sqrt{3})^2(x + i\sqrt{3})^2$; 0 (1), $i\sqrt{3}$ (2), $-i\sqrt{3}$ (2)
29. $P(x) = x^2 - 2x + 2$
31. $Q(x) = x^3 - 3x^2 + 4x - 12$
33. $P(x) = x^3 - 2x^2 + x - 2$
35. $R(x) = x^4 - 4x^3 + 10x^2 - 12x + 5$
37. $T(x) = 6x^4 - 12x^3 + 18x^2 - 12x + 12$
39. $-2, \pm 2i$ **41.** $1, \dfrac{1 \pm i\sqrt{3}}{2}$ **43.** $2, \dfrac{1 \pm i\sqrt{3}}{2}$
45. $-\frac{3}{2}, -1 \pm i\sqrt{2}$ **47.** $-2, 1, \pm 3i$ **49.** $1, \pm 2i, \pm i\sqrt{3}$

51. 3 (multiplicity 2), $\pm 2i$ **53.** 1 (multiplicity 3), $\pm 3i$
55. (a) $(x - 5)(x^2 + 4)$ **(b)** $(x - 5)(x - 2i)(x + 2i)$
57. (a) $(x - 1)(x + 1)(x^2 + 9)$
(b) $(x - 1)(x + 1)(x - 3i)(x + 3i)$
59. (a) $(x - 2)(x + 2)(x^2 - 2x + 4)(x^2 + 2x + 4)$
(b) $(x - 2)(x + 2)[x - (1 + i\sqrt{3})][x - (1 - i\sqrt{3})]$
$[x + (1 + i\sqrt{3})][x + (1 - i\sqrt{3})]$
61. (a) 4 real   **(b)** 2 real, 2 imaginary   **(c)** 4 imaginary

Section 3.6 ■ page 321

1. x-intercept 2; y-intercept $-\frac{2}{3}$
3. x-intercepts $-1, 2$; y-intercept $\frac{1}{3}$
5. x-intercepts $-3, 3$; no y-intercept
7. x-intercept 3, y-intercept 3,   vertical $x = 2$; horizontal $y = 2$
9. Vertical $x = -2$; horizontal $y = 0$
11. Vertical $x = 3$, $x = -2$; horizontal $y = 1$
13. Horizontal $y = 0$
15. Vertical $x = -6$, $x = 1$; horizontal $y = 0$
17. Vertical $x = 1$

19.

21.

23.

25.

27.

29.

31.

33.

51.

53.

35.

37.

55.

57.

vertical $x = -3$

39.

41.

59.

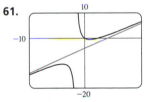

vertical $x = 2$

61.

vertical $x = -1.5$
x-intercepts 0, 2.5
y-intercept 0
local maximum $(-3.9, -10.4)$
local minimum $(0.9, -0.6)$
end behavior: $y = x - 4$

43.

45.

63.

vertical $x = 1$
x-intercept 0
y-intercept 0
local minimum $(1.4, 3.1)$
end behavior: $y = x^2$

47.

49.

65.

vertical $x = 3$
x-intercepts 1.6, 2.7
y-intercept -2
local maxima $(-0.4, -1.8)$,
$(2.4, 3.8)$
local minima $(0.6, -2.3)$
$(3.4, 54.3)$
end behavior $y = x^3$

67. (a)   (b) It levels off at 3000.

69. (a) 2.50 mg/L   (b) It decreases to 0.   (c) 16.61 h

71.

If the speed of the train approaches the speed of sound, then the pitch increases indefinitely (a sonic boom).

Chapter 3 Review ■ page 325

1.

3.

5.

7.

9.

x-intercepts -2.1, 0.3, 1.9
y-intercept 1
local maximum $(-1.2, 4.1)$
local minimum $(1.2, -2.1)$
$y \to \infty$ as $x \to \infty$
$y \to -\infty$ as $x \to -\infty$

11.

x-intercepts -0.1, 2.1
y-intercept -1
local minimum $(1.4, -14.5)$
$y \to \infty$ as $x \to \infty$
$y \to \infty$ as $x \to -\infty$

13. (a) $S = 13.8x(100 - x^2)$ (b) $0 \le x \le 10$
(c)   (d) 5.8 in.

In answers 15–21, the first polynomial given is the quotient and the second is the remainder.

15. $x - 1$, 3 **17.** $x^2 + 3x + 23$, 94
19. $x^3 - 5x^2 + 17x - 83$, 422 **21.** $2x - 3$, 12 **23.** 3
27. 8 **29.** (a) $\pm 1, \pm 2, \pm 3, \pm 6, \pm 9, \pm 18$
(b) 2 or 0 positive, 3 or 1 negative   **31.** $3 + i$ **33.** $8 - i$
35. $\frac{6}{5} + \frac{8}{5}i$ **37.** i **39.** 2 **41.** $4x^3 - 18x^2 + 14x + 12$
43. No; since the complex conjugates of imaginary zeros will also be zeros, the polynomial would have 8 zeros, contradicting the requirement that it have degree 4.
45. $-3, 1, 5$ **47.** $-1 \pm 2i$, -2 (multiplicity 2)
49. ± 2, 1 (multiplicity 3)   **51.** $\pm 2, \pm 1 \pm i\sqrt{3}$
53. $1, 3, \dfrac{-1 \pm i\sqrt{7}}{2}$
55. $x = -0.5, 3$ **57.** $x \approx -0.24, 4.24$
59.   **61.**

63.

65.

x-intercept 3
y-intercept -0.5
vertical $x = -3$
horizontal $y = 0.5$
no local extrema

67.

x-intercept -2
y-intercept -4
vertical $x = -1$, $x = 2$
slant $y = x + 1$
local maximum
$(0.425, -3.599)$
local minimum $(4.216, 7.175)$

69. $(-2, -28), (1, 26), (2, 68), (5, 770)$

Chapter 3 Test ■ page 328

1.

2. (a) $x^3 + 2x^2 + 2, 9$ (b) $x^3 + 2x^2 + \frac{1}{2}, \frac{15}{2}$
3. (a) $\pm 1, \pm 3, \pm\frac{1}{2}, \pm\frac{3}{2}$ (b) $2(x - 3)\left(x - \frac{1}{2}\right)(x + 1)$
(c) $-1, \frac{1}{2}, 3$
(d)

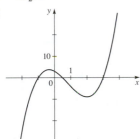

4. (a) $-1 - \frac{3}{2}i$ (b) $5 + i$ (c) $-1 + 2i$ (d) -1 (e) $6\sqrt{2}$
5. $3, -1 \pm i$ **6.** $(x - 1)^2(x - 2i)(x + 2i)$
7. $x^4 + 2x^3 + 10x^2 + 18x + 9$
8. (a) 4, 2, or 0 positive; 0 negative
(c) 0.17, 3.93

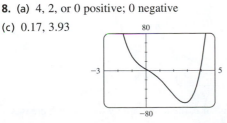

(d) Local minimum $(2.8, -70.3)$
9. (a) r, u (b) s (c) s

(d)

(e) $x^2 - 2x - 5$

Focus on Problem Solving ■ page 331

1. Not traversable   **3.** Traversable
5. Yes; tour is possible for odd n.   **7.** Inside
9. -36 **11.** No   **13.** $\frac{2}{3}$
15.

CHAPTER 4

Section 4.1 ■ page 343

1.

3.

5.

7.

31.

9. $f(x) = 3^x$ **11.** $f(x) = \left(\frac{1}{4}\right)^x$

13. III   **15.** I   **17.** II

19. $\mathbb{R}, (-\infty, 0), y = 0$ **21.** $\mathbb{R}, (-3, \infty), y = -3$

23. $\mathbb{R}, (4, \infty), y = 4$ **25.** $\mathbb{R}, (0, \infty), y = 0$

27.

29.

33. $y = 3(2^x)$

35. (a)

(b) The graph of g is steeper than that of f.

39. (a)

(i)

(ii)

(iii)

The graph of f ultimately increases much more quickly than g.

(b) 1.2, 22.4

41.

The larger the value of c, the more rapidly the graph increases.

43. (a) 13 kg   (b) 6.6 kg
45. (a) 0   (b) 50.6 ft/s, 69.2 ft/s
(c)   (d) 80 ft/s

47. (a) 5164   (b)
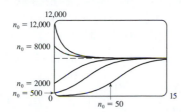
(c) 6000

49. (a) $16,288.95
(b) $26,532.98
(c) $43,219.42

51. (a) $4,615.87   (f) $4,704.93
(b) $4,658.91   (g) $4,704.94
(c) $4,697.04
(d) $4,703.11
(e) $4,704.68

53. (i)   **55.** (a) $7,678.96   (b) $67,121.04

57.   **63.**

65.   **67.**
Vertical asymptote $x = 0$,
horizontal asymptote $y = 1$

69. Local minimum $\approx (0.27, 1.75)$
71. (a) Increasing on $(-\infty, 1.00]$, decreasing on $[1.00, \infty)$
(b) $(-\infty, 0.37]$

Section 4.2 ■ **page 356**

1. (a) $5^2 = 25$ (b) $5^0 = 1$ **3.** (a) $8^{1/3} = 2$ (b) $2^{-3} = \dfrac{1}{8}$
5. (a) $e^x = 5$ (b) $e^5 = y$ **7.** (a) $\log_5 125 = 3$
(b) $\log_{10} 0.0001 = -4$ **9.** (a) $\log_8 \dfrac{1}{8} = -1$
(b) $\log_2 \dfrac{1}{8} = -3$ **11.** (a) $\ln 2 = x$ (b) $\ln y = 3$
13. (a) 1   (b) 0   (c) 2   **15.** (a) 2   (b) 2   (c) 10
17. (a) -3 (b) $\frac{1}{2}$ (c) -1 **19.** (a) 37   (b) 8   (c) $\sqrt{5}$
21. (a) $-\frac{2}{3}$ (b) 4   (c) -1 **23.** (a) 32   (b) 4
25. (a) 5   (b) 27   **27.** (a) 100   (b) 25   **29.** (a) 2   (b) 4
31. (a) 0.3010   (b) 1.5465   (c) -0.1761
33. (a) 1.6094   (b) 3.2308   (c) 1.0051
35. $y = \log_5 x$ **37.** $y = \log_9 x$ **39.** II   **41.** III   **43.** VI
45.

47. $(4, \infty)$, $\mathbb{R}$, $x = 4$ **49.** $(-\infty, 0)$, $\mathbb{R}$, $x = 0$

51. $(0, \infty)$, $\mathbb{R}$, $x = 0$ **53.** $(0, \infty)$, $\mathbb{R}$, $x = 0$

55. $(0, \infty)$, $[0, \infty)$, $x = 0$

57. $\left(-3, \infty\right)$ **59.** $(-\infty, -1) \cup (1, \infty)$ **61.** $(0, 2)$

63.

domain = $(-1, 1)$
vertical asymptotes
$x = 1$, $x = -1$
local maximum $(0, 0)$

65.

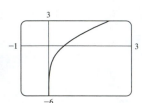

domain = $(0, \infty)$
vertical asymptote $x = 0$
no maximum or minimum

67.

domain = $(0, \infty)$
vertical asymptote $x = 0$
horizontal asymptote $y = 0$
local maximum
$\approx (2.72, 0.37)$

69. The graph of f grows more slowly than g.

71. (a)

(b) The graph of $f(x) = \log(cx)$ is the graph of $f(x) = \log x$ shifted upward $\log c$ units.

73. (a) $(1, \infty)$ **(b)** $f^{-1}(x) = 10^{2^x}$

75. (a) $f^{-1}(x) = \log_2\left(\dfrac{x}{1 - x}\right)$ **(b)** $(0, 1)$

Section 4.3 ■ page 363

1. $1 + \log_2 x$ **3.** $\log_2 x + \log_2 (x - 1)$ **5.** $10 \log 6$

7. $\log_2 A + 2 \log_2 B$ **9.** $\log_3 x + \frac{1}{2} \log_3 y$

11. $\frac{1}{3} \log_5(x^2 + 1)$ **13.** $\frac{1}{2}(\ln a + \ln b)$

15. $3 \log x + 4 \log y - 6 \log z$

17. $\log_2 x + \log_2(x^2 + 1) - \frac{1}{2} \log_2(x^2 - 1)$

19. $\ln x + \frac{1}{2}(\ln y - \ln z)$ **21.** $\frac{1}{4} \log(x^2 + y^2)$

23. $\frac{1}{2}\left[\log(x^2 + 4) - \log(x^2 + 1) - 2 \log(x^3 - 7)\right]$

25. $3 \ln x + \frac{1}{2} \ln (x - 1) - \ln (3x + 4)$

27. $\frac{3}{2}$ **29.** 1   **31.** 3   **33.** $\ln 8$ **35.** 16   **37.** $4 + \log 3$

39. $\log_3 160$ **41.** $\log_2(AB/C^2)$ **43.** $\log\left[\dfrac{x^4(x - 1)^2}{\sqrt[3]{x^2 + 1}}\right]$

45. $\ln[5x^2(x^2 + 5)^3]$

47. $\log\left[\sqrt[3]{2x + 1}\,\sqrt{(x - 4)/(x^4 - x^2 - 1)}\right]$

49. 2.321928   **51.** 2.523719   **53.** 0.493008

55. 3.482892

57.

Section 4.4 ■ page 372

1. 2.7726   **3.** 0.3495   **5.** -0.5850 **7.** 1.2040   **9.** 0.0767

11. 0.2524   **13.** 1.9349   **15.** -43.0677 **17.** 2.1492

19. 6.2126   **21.** -2.9469 **23.** -2.4423 **25.** 14.0055

27. ± 1 **29.** $0, \frac{4}{3}$ **31.** $\ln 2 \approx 0.6931, 0$

33. $\frac{1}{2} \ln 3 \approx 0.5493$ **35.** $e^{10} \approx 22026$ **37.** 0.01   **39.** $\frac{95}{3}$

41. $3 - e^2 \approx -4.3891$ **43.** 5   **45.** 5   **47.** $\frac{13}{12}$

49. 6   **51.** $\frac{3}{2}$ **53.** $1/\sqrt{5} \approx 0.4472$ **55. (a)** \$6435.09

(b) 8.24 yr   **57.** 6.33 yr   **59.** 8.15 yr   **61.** 8.30%

63. 13 days   **65. (a)** 7337   **(b)** 1.73 yr

67. (a) $t = -\frac{5}{13} \ln\left(1 - \frac{13}{60} I\right)$ **(b)** 0.218 s

69. 2.21   **71.** 0.00, 1.14   **73.** -0.57 **75.** 0.36

77. $2 < x < 4$ or $7 < x < 9$ **79.** $\log 2 < x < \log 5$

81. 101, 1.1   **83.** $\log_2 3 \approx 1.58$

Section 4.5 ■ page 386

1. (a) 500   **(b)** 45%   **(c)** 1929   **(d)** 6.66 h

3. (a) $n(t) = 18{,}000e^{0.08t}$ **(b)** 34,137

(c)

5. (a) $n(t) = 112{,}000e^{0.04t}$ (b) About 142,000   (c) 2008
7. (a) 20,000   (b) $n(t) = 20{,}000e^{0.1096t}$
(c) About 48,000   (d) 2010
9. (a) $n(t) = 8600e^{0.1508t}$ (b) About 11,600   (c) 4.6 h
11. (a) 2029   (b) 2049   **13.** 22.85 h
15. (a) $n(t) = 10e^{-0.0231t}$ (b) 1.6 g   (c) 70 yr
17. 18 yr   **19.** 149 h   **21.** 3560 yr
23. (a) 210°F   (b) 153°F   (c) 28 min
25. (a) 137°F   (b) 116 min
27. (a) 2.3   (b) 3.5   (c) 8.3
29. (a) 10^{-3} M   (b) 3.2×10^{-7} M
31. $4.8 \leqslant \text{pH} \leqslant 6.4$
33. $\log 20 \approx 1.3$ **35.** Twice as intense   **37.** 8.2
39. 6.3×10^{-3} W/m²   **41.** (b) 106 dB

Chapter 4 Review ■ page 390

1. $\mathbb{R}$, $(0, \infty)$, $y = 0$

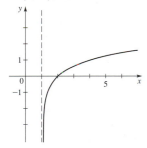

3. $\mathbb{R}$, $(-\infty, 5)$, $y = 5$

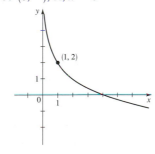

5. $(1, \infty)$, $\mathbb{R}$, $x = 1$

7. $(0, \infty)$, $\mathbb{R}$, $x = 0$

9. $\mathbb{R}$, $(-1, \infty)$, $y = -1$

11. $(0, \infty)$, $\mathbb{R}$, $x = 0$

13. $\left(-\infty, \frac{1}{2}\right)$ **15.** $(-\infty, -2) \cup (2, \infty)$ **17.** $2^{10} = 1024$
19. $10^y = x$ **21.** $\log_2 64 = 6$ **23.** $\log 74 = x$ **25.** 7
27. 45   **29.** 6   **31.** -3 **33.** $\frac{1}{2}$ **35.** 2   **37.** 92   **39.** $\frac{2}{3}$
41. $\log A + 2 \log B + 3 \log C$
43. $\frac{1}{2}\left[\ln(x^2 - 1) - \ln(x^2 + 1)\right]$
45. $2 \log_5 x + \frac{3}{2} \log_5(1 - 5x) - \frac{1}{2} \log_5(x^3 - x)$
47. $\log 96$ **49.** $\log_2\left[\dfrac{(x - y)^{3/2}}{(x^2 + y^2)^2}\right]$ **51.** $\log\left(\dfrac{x^2 - 4}{\sqrt{x^2 + 4}}\right)$
53. -15 **55.** $\frac{1}{3}(5 - \log_5 26) \approx 0.99$
57. $\frac{4}{3} \ln 10 \approx 3.07$ **59.** 3   **61.** $-4, 2$ **63.** 0.430618
65. 2.303600
67.

vertical asymptote
$x = -2$
horizontal asymptote
$y = 2.72$
no maximum or minimum

69.

vertical asymptotes
$x = -1$, $x = 0$, $x = 1$
local maximum
$\approx (-0.58, -0.41)$

71. 2.42   **73.** $0.16 < x < 3.15$
75. Increasing on $(-\infty, 0]$ and $[1.10, \infty)$,
decreasing on $[0, 1.10]$
77. 1.953445   **79.** $\log_4 258$
81. (a) $16,081.15   (b) $16,178.18   (c) $16,197.64
(d) $16,198.31   **83.** (a) $n(t) = 30e^{0.15t}$ (b) 55   (c) 19 yr
85. (a) 9.97 mg   (b) 1.39×10^5 yr
87. (a) $n(t) = 150e^{-0.0004359t}$ (b) 97.0 mg   (c) 2520 yr
89. (a) $n(t) = 1500e^{0.1515t}$ (b) 7940   **91.** 7.9, basic
93. 8.0

Chapter 4 Test ■ page 393

1.

$y = 2^x$
$y = \log_2 x$

2.

$(-2, \infty)$, $\mathbb{R}$, $x = -2$

3. (a) $\frac{3}{2}$ (b) 3 (c) $\frac{2}{3}$ (d) 2

4. $\frac{1}{2}\left[\log(x + 2) - 4 \log x - \log(x^2 + 4)\right]$

5. $\ln\left(\dfrac{x\sqrt{3 - x^4}}{(x^2 + 1)^2}\right)$ **6.** (a) 4.32 (b) 0.77 (c) 5.39 (d) 2

7. (a) $n(t) = 1000e^{2.07944t}$ (b) 22,627 (c) 1.3 h
(d)

8. (a) $A(t) = 12{,}000\left(1 + \dfrac{0.056}{12}\right)^{12t}$

(b) \$14,195.06 (c) 9.249 yr

9. (a)

(b) $x = 0, y = 0$
(c) Local minimum $\approx (3.00, 0.74)$
(d) $(-\infty, 0) \cup [0.74, \infty)$
(e) $-0.85, 0.96, 9.92$

Focus on Modeling ■ page 403

1. (a)

(b) $y = ab^t$, where $a = 4.041807 \times 10^{-16}$ and
$b = 1.021003194$, and y is the population in millions in
the year t.   (c) 457.9 million   (d) 221.2 million   (e) No
3. (a) Yes   (b) Yes, the scatter plot appears linear.

(c) $\ln E = 3.30161 \times 0.10769t$, where t is years since 1960 and
E is expenditure in billions of dollars.
(d) $E = 27.15633e^{0.10769t}$ (e) 1310.9 billion dollars
5. (a) $y = ab^t$, where $a = 301.813054$, $b = 0.819745$, and
t is the number of years since 1970
(b) $y = at^4 + bt^3 + ct^2 + dt + e$, where $a = -0.002430$,
$b = 0.135159$, $c = -2.014322$, $d = -4.055294$,
$e = 199.092227$, and t is the number of years since 1970
(c) From the graphs we see that the fourth-degree polynomial is
a better model.
(d) 202.8, 27.8; 184.0, 43.5

7. (a)

(b)

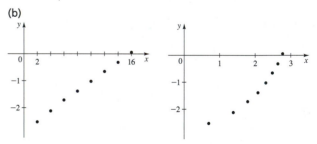

(c) Exponential Function
(d) $y = a \cdot b^x$ where $a = 0.057697$ and $b = 1.200236$
9. (a) $y = a + b \ln t$ where $a = -7154.888$, $b = 1016.007$,
and y is metric tons of coal produced in the year t
(b) 912 metric tons

CHAPTER 5

Section 5.1 ■ page 416

5. $P\left(\frac{4}{5}, \frac{3}{5}\right)$ **7.** $P\left(-\sqrt{5}/3, \frac{2}{3}\right)$ **9.** $P\left(-\sqrt{2}/3, \sqrt{7}/3\right)$
11. $t = \pi/4, (\sqrt{2}/2, \sqrt{2}/2)$; $t = \pi/2, (0, 1)$; $t = 3\pi/4$,
$(-\sqrt{2}/2, \sqrt{2}/2)$; $t = \pi, (-1, 0)$; $t = 5\pi/4, (-\sqrt{2}/2, -\sqrt{2}/2)$;
$t = 3\pi/2, (0, -1)$; $t = 7\pi/4, (\sqrt{2}/2, -\sqrt{2}/2)$; $t = 2\pi, (1, 0)$

13. $(0, 1)$ **15.** $\left(-\sqrt{3}/2, \frac{1}{2}\right)$ **17.** $\left(\frac{1}{2}, -\sqrt{3}/2\right)$
19. $\left(-\frac{1}{2}, \sqrt{3}/2\right)$ **21.** $(-\sqrt{2}/2, -\sqrt{2}/2)$
23. (a) $\left(-\frac{3}{5}, \frac{4}{5}\right)$ (b) $\left(\frac{3}{5}, -\frac{4}{5}\right)$ (c) $\left(-\frac{3}{5}, -\frac{4}{5}\right)$ (d) $\left(\frac{3}{5}, \frac{4}{5}\right)$
25. (a) $\pi/4$ (b) $\pi/3$ (c) $\pi/3$ (d) $\pi/6$
27. (a) $\pi/5$ (b) $\pi/6$ (c) $\pi/3$ (d) $\pi/6$
29. (a) $\pi/3$ (b) $\left(-\frac{1}{2}, \sqrt{3}/2\right)$ **31.** (a) $\pi/4$ (b) $(-\sqrt{2}/2, \sqrt{2}/2)$
33. (a) $\pi/3$ (b) $\left(-\frac{1}{2}, -\sqrt{3}/2\right)$
35. (a) $\pi/4$ (b) $(-\sqrt{2}/2, -\sqrt{2}/2)$
37. (a) $\pi/6$ (b) $\left(-\sqrt{3}/2, -\frac{1}{2}\right)$
39. (a) $\pi/3$ (b) $\left(\frac{1}{2}, \sqrt{3}/2\right)$
41. (a) $\pi/3$ (b) $\left(-\frac{1}{2}, -\sqrt{3}/2\right)$ **43.** $(0.5, 0.8)$
45. $(0.5, -0.9)$

Section 5.2 ■ page 426

1. $t = \pi/4$, $\sin t = \sqrt{2}/2$, $\cos t = \sqrt{2}/2$; $t = \pi/2$, $\sin t = 1$, $\cos t = 0$; $t = 3\pi/4$, $\sin t = \sqrt{2}/2$, $\cos t = -\sqrt{2}/2$; $t = \pi$, $\sin t = 0$, $\cos t = -1$; $t = 5\pi/4$, $\sin t = -\sqrt{2}/2$, $\cos t = -\sqrt{2}/2$; $t = 3\pi/2$, $\sin t = -1$, $\cos t = 0$; $t = 7\pi/4$, $\sin t = -\sqrt{2}/2$, $\cos t = \sqrt{2}/2$; $t = 2\pi$, $\sin t = 0$, $\cos t = 1$
3. (a) $-\sqrt{3}/2$ (b) $\frac{1}{2}$ **5.** (a) -1 (b) -1
7. (a) 1 (b) -1 **9.** (a) 0 (b) 0
11. (a) $\frac{1}{2}$ (b) 2 **13.** (a) $\frac{1}{2}$ (b) $\frac{1}{2}$
15. (a) $\sqrt{3}/3$ (b) $-\sqrt{3}/3$ **17.** (a) 2 (b) $-2\sqrt{3}/3$
19. (a) $\sqrt{2}/2$ (b) $\sqrt{2}$ **21.** (a) -1 (b) -1
23. $\sin 0 = 0$, $\cos 0 = 1$, $\tan 0 = 0$, $\sec 0 = 1$, others undefined
25. $\sin \pi = 0$, $\cos \pi = -1$, $\tan \pi = 0$, $\sec \pi = -1$, others undefined
27. $\frac{4}{5}, \frac{3}{5}, \frac{4}{3}$ **29.** $-\sqrt{11}/4, \sqrt{5}/4, -\sqrt{55}/5$ **31.** $\frac{9}{41}, \frac{40}{41}, \frac{9}{40}$
33. $-\frac{12}{13}, -\frac{5}{13}, \frac{12}{5}$ **35.** (a) 0.8 (b) 0.84147
37. (a) 0.9 (b) 0.93204 **39.** (a) 1 (b) 1.02964
41. (a) -0.6 (b) -0.57482 **43.** Negative
45. Negative   **47.** II   **49.** II
51. $\sin t = \sqrt{1 - \cos^2 t}$ **53.** $\tan t = (\sin t)/\sqrt{1 - \sin^2 t}$
55. $\sec t = -\sqrt{1 + \tan^2 t}$ **57.** $\tan t = \sqrt{\sec^2 t - 1}$
59. $\tan^2 t = (\sin^2 t)/(1 - \sin^2 t)$
61. $\cos t = -\frac{4}{5}$, $\tan t = -\frac{3}{4}$, $\csc t = \frac{5}{3}$, $\sec t = -\frac{5}{4}$, $\cot t = -\frac{4}{3}$
63. $\sin t = -2\sqrt{2}/3$, $\cos t = \frac{1}{3}$, $\tan t = -2\sqrt{2}$, $\csc t = -\frac{3}{4}\sqrt{2}$, $\cot t = -\sqrt{2}/4$
65. $\sin t = -\frac{3}{5}$, $\cos t = \frac{4}{5}$, $\csc t = -\frac{5}{3}$, $\sec t = \frac{5}{4}$, $\cot t = -\frac{4}{3}$
67. $\cos t = -\sqrt{15}/4$, $\tan t = \sqrt{15}/15$, $\csc t = -4$, $\sec t = -4\sqrt{15}/15$, $\cot t = \sqrt{15}$
69. Odd   **71.** Odd   **73.** Even   **75.** Neither

Section 5.3 ■ page 439

1.

3.

5.

7.

9.

11. $1, \frac{\pi}{2}$

13. $3, 2\pi/3$

15. $10, 4\pi$

17. $1, 6\pi$

19. $3, \frac{2}{3}$

21. $1, 2\pi, \pi/2$

23. $2, 2\pi, \pi/6$

25. $5, 2\pi/3, \pi/12$

27. $2, 3\pi, \pi/4$

29. $3, 2, -\frac{1}{2}$

31. $\frac{1}{2}, \pi, \pi/6$

33. $1, 2\pi/3, -\pi/3$

35. (a) $4, 2\pi, 0$ (b) $y = 4 \sin x$

37. (a) $3, 4\pi, 0$ (b) $y = 3 \sin \frac{1}{2} x$

39. (a) $\frac{1}{2}, \pi, -\pi/3$ (b) $y = -\frac{1}{2} \cos 2(x + \pi/3)$

41.

43.

45.

47.

49. (a) 20 s   (b) 6 ft   **51.** (a) $\frac{1}{80}$ min   (b) 80

(c)

(d) $140/90$; it is higher than normal

53.

55.

$y = x^2 \sin x$ is a sine curve that lies between the graphs of $y = x^2$ and $y = -x^2$

57.

$y = e^x \sin 5\pi x$ is a sine curve that lies between the graphs of $y = e^x$ and $y = -e^x$

59.

$y = \cos 3\pi x \cos 21\pi x$ is a cosine curve that lies between the graphs of $y = \cos 3\pi x$ and $y = -\cos 3\pi x$

61. Maximum value 1.76 when $x \approx 0.94$, minimum value -1.76 when $x \approx -0.94$ (The same maximum and minimum values occur at infinitely many other values of x.)
63. Maximum value 3.00 when $x \approx 1.57$, minimum value -1.00 when $x \approx -1.57$ (The same maximum and minimum values occur at infinitely many other values of x.)
65. 1.16   **67.** 0.34, 2.80
69. (a) Odd (b) $0, \pm2\pi, \pm4\pi, \pm6\pi, \ldots$
(c)

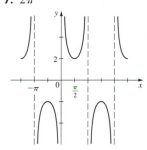

(d) $f(x)$ approaches 0
(e) $f(x)$ approaches 0

Section 5.4 ■ page 451

1. π

3. π

5. π

7. 2π

9. 2π

11. π

13. 2π

15. π

17. 2π

19. $\pi/2$

21. 1

23. π

25. π

27. $\pi/3$

29. $2\pi/3$

31. $\pi/2$

45. $\pi/2$

33. $\pi/2$

35. $\pi/2$

Chapter 5 Review ■ **page 452**

1. (b) $\frac{1}{2}, -\sqrt{3}/2, -\sqrt{3}/3$ **3.** (a) $\pi/3$ (b) $(-\frac{1}{2}, \sqrt{3}/2)$
(c) $\sin t = \sqrt{3}/2$, $\cos t = -\frac{1}{2}$, $\tan t = -\sqrt{3}$, $\csc t = 2\sqrt{3}/3$,
$\sec t = -2$, $\cot t = -\sqrt{3}/3$ **5.** (a) $\pi/4$ (b) $(-\sqrt{2}/2, -\sqrt{2}/2)$
(c) $\sin t = -\sqrt{2}/2$, $\cos t = -\sqrt{2}/2$, $\tan t = 1$, $\csc t = -\sqrt{2}$,
$\sec t = -\sqrt{2}$, $\cot t = 1$ **7.** (a) $\sqrt{2}/2$ (b) $-\sqrt{2}/2$
9. (a) 0.89121 (b) 0.45360 **11.** (a) 0 (b) Undefined
13. (a) Undefined   (b) 0 **15.** (a) $-\sqrt{3}/3$ (b) $-\sqrt{3}$
17. $(\sin t)/(1 - \sin^2 t)$ **19.** $(\sin t)/\sqrt{1 - \sin^2 t}$
21. $\tan t = -\frac{5}{12}$, $\csc t = \frac{13}{5}$, $\sec t = -\frac{13}{12}$, $\cot t = -\frac{12}{5}$
23. $\sin t = 2\sqrt{5}/5$, $\cos t = -\sqrt{5}/5$, $\tan t = -2$,
$\sec t = -\sqrt{5}$
25. $(16 - \sqrt{17})/4$ **27.** 3

29. (a) $10, 4\pi, 0$
(b)

31. (a) $1, 4\pi, 0$
(b)

37. 2

39. $2\pi/3$

33. (a) $3, \pi, 1$
(b)

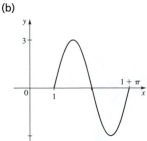

35. (a) $1, 4, -\frac{1}{3}$
(b)

41. $3\pi/2$

43. 2

37. $y = 5 \sin 4x$ **39.** $y = \frac{1}{2} \sin 2\pi\left(x + \frac{1}{3}\right)$

41. π

43. π

45. π

47. 2π

49. (a)

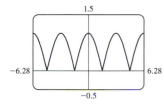

(b) Period π
(c) Even

51. (a)

(b) Not periodic
(c) Neither

53. (a)

(b) Not periodic
(c) Even

55.

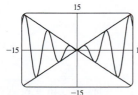

$y = x \sin x$ is a sine
function whose graph lies
between those of $y = x$
and $y = -x$

57.

The graphs are related
by graphical addition.

59. 1.76, -1.76 **61.** 0.30, 2.84

63. (a) Odd (b) 0, $\pm\pi$, $\pm 2\pi$, ...

(c)

(d) $f(x)$ approaches 0
(e) $f(x)$ approaches 0

Chapter 5 Test ■ page 454

1. $y = -\frac{5}{6}$ **2.** (a) $\frac{4}{5}$ (b) $-\frac{3}{5}$ (c) $-\frac{4}{3}$ (d) $-\frac{5}{3}$

3. (a) $-\frac{1}{2}$ (b) $-\sqrt{2}/2$ (c) $\sqrt{3}$ (d) -1

4. $\tan t = -(\sin t)/\sqrt{1 - \sin^2 t}$ **5.** $-\frac{2}{15}$

6. (a) 5, $\pi/2$, 0 (b)

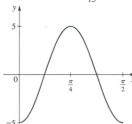

7. (a) 2, 4π, $\pi/3$ (b)

8. π

9. $\pi/2$

10. $y = 2 \sin 2(x + \pi/3)$

11. (a)

(b) Even
(c) Minimum value -0.11 when $x \approx \pm 2.54$, maximum value 1 when $x = 0$

Focus on Modeling ■ page 466

1. (a) $2, 2\pi/3, 3/(2\pi)$ (b)

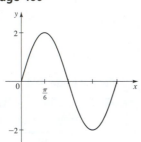

3. (a) $\frac{1}{4}, 4\pi/3, 3/(4\pi)$ (b)

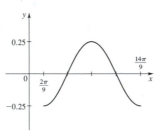

5. (a) 10 m/min
(b)

(c) 0.4 m

7. (a) 8900   (b) about 3.14 yr   **9.** $d(t) = 5 \sin 5\pi t$

11. $y = 21 \sin\left(\dfrac{\pi}{6} t\right)$

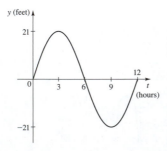

13. $y = 5 \cos 2\pi t$ **15.** $y = 11 + 10 \sin\left(\dfrac{\pi t}{10}\right)$

17. $y = 3.8 + 0.2 \sin\left(\dfrac{\pi}{5} t\right)$ **19.** $E(t) = 155 \cos 120\pi t$

21. (a) $45V$ (b) 40   (c) 40   (d) $V(t) = 45 \cos 80\pi t$

23. (a)

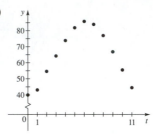

(b) $y = 22.9 \cos 0.52(t - 6) + 62.9$ where y is temperature (°F) and t is months (January $= 0$)

(c)

(d) $y = 23.4 \sin (0.48t - 1.36) + 62.2$

25. (a)

(b) $y = 30 \sin 0.52t + 50$ where y is the owl population in year t

(c)

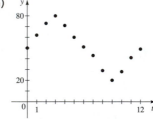

(d) $y = 25.8 \sin (0.52t - 0.02) + 50.6$

27. $f(t) = e^{-0.9t} \sin \pi t$ **29.** $c = \frac{1}{3} \ln 4 \approx 0.46$

CHAPTER 6

Section 6.1 ■ page 480

1. $\pi/5 \approx 0.628$ rad   **3.** $-8\pi/3 \approx -8.378$ rad   **5.** $\pi/3 \approx$ 1.047 rad   **7.** $-3\pi/4 \approx -2.356$ rad   **9.** $135°$ **11.** $150°$
13. $-270/\pi \approx 85.9°$ **15.** $-15°$ **17.** $410°, 770°, -310°,$ $-670°$ **19.** $11\pi/4, 19\pi/4, -5\pi/4, -13\pi/4$ **21.** $7\pi/4,$ $15\pi/4, -9\pi/4, -17\pi/4$ **23.** Yes   **25.** Yes   **27.** Yes
29. $13°$ **31.** $30°$ **33.** $280°$ **35.** $5\pi/6$ **37.** π **39.** $\pi/4$
41. $55\pi/9 \approx 19.2$ **43.** 4   **45.** 4 mi   **47.** 2 rad $\approx 114.6°$
49. $36/\pi \approx 11.459$ m   **51.** $330\pi \approx 1037$ mi
53. 1.6 million mi   **55.** 1.15 mi   **57.** 50 m^2 **59.** 4 m
61. 6 cm^2 **63.** $32\pi/15$ ft/s ≈ 6.7 ft/s
65. (a) 2000π rad/min   (b) $50\pi/3$ ft/s ≈ 52.4 ft/s
67. 39.3 mi/h   **69.** 2.1 m/s

Section 6.2 ■ page 489

1. $\sin \theta = \frac{4}{5}$, $\cos \theta = \frac{3}{5}$, $\tan \theta = \frac{4}{3}$, $\csc \theta = \frac{5}{4}$, $\sec \theta = \frac{5}{3}$, $\cot \theta = \frac{3}{4}$ **3.** $\sin \theta = \frac{40}{41}$, $\cos \theta = \frac{9}{41}$, $\tan \theta = \frac{40}{9}$ $\csc \theta = \frac{41}{40}$, $\sec \theta = \frac{41}{9}$, $\cot \theta = \frac{9}{40}$ **5.** $\sin \theta = 2\sqrt{13}/13$, $\cos \theta = 3\sqrt{13}/13$, $\tan \theta = \frac{2}{3}$, $\csc \theta = \sqrt{13}/2$, $\sec \theta = \sqrt{13}/3$, $\cot \theta = \frac{3}{2}$ **7.** (a) $3\sqrt{34}/34$, $3\sqrt{34}/34$ (b) $\frac{3}{5}, \frac{3}{5}$
(c) $\sqrt{34}/5$, $\sqrt{34}/5$ **9.** $\frac{25}{2}$ **11.** $13\sqrt{3}/2$ **13.** 16.51658
15. $x = 28 \cos \theta$, $y = 28 \sin \theta$
17. $\cos \theta = \frac{4}{5}$, $\tan \theta = \frac{3}{4}$, $\csc \theta = \frac{5}{3}$, $\sec \theta = \frac{5}{4}$, $\cot \theta = \frac{4}{3}$

19. $\sin \theta = \sqrt{2}/2$, $\cos \theta = \sqrt{2}/2$, $\tan \theta = 1$, $\csc \theta = \sqrt{2}$, $\sec \theta = \sqrt{2}$

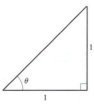

21. $\sin \theta = \sqrt{45}/7$, $\cos \theta = \frac{2}{7}$, $\tan \theta = \sqrt{45}/2$, $\csc \theta = 7\sqrt{45}/45$, $\cot \theta = 2\sqrt{45}/45$ **23.** $(1 + \sqrt{3})/2$ **25.** 1   **27.** $\frac{1}{2}$
29.   **31.**

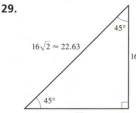

33. $\sin \theta \approx 0.45$, $\cos \theta \approx 0.89$, $\tan \theta \approx 0.50$, $\csc \theta \approx 2.24$, $\sec \theta \approx 1.12$, $\cot \theta \approx 2.00$
35. 1026 ft   **37.** (a) 2100 mi   (b) No   **39.** 19 ft
41. $38.7°$ **43.** 345 ft   **45.** 415 ft, 152 ft
47. 2570 ft   **49.** 5808 ft   **51.** 91.7 million mi
53. 3960 mi   **55.** 230.9   **57.** 63.7
59. $a = \sin \theta$, $b = \tan \theta$, $c = \sec \theta$, $d = \cos \theta$

Section 6.3 ■ page 501

1. (a) $30°$ (b) $60°$ (c) $60°$ **3.** (a) $\pi/4$ (b) $\pi/6$ (c) $\pi/3$
5. (a) $\pi/5$ (b) $\pi - \frac{11}{5} \approx 0.94$ (c) $3\pi/7$ **7.** $\frac{1}{2}$ **9.** $-\sqrt{2}/2$
11. $-\sqrt{3}$ **13.** 1   **15.** $-\sqrt{3}/2$ **17.** $\sqrt{3}/3$ **19.** $\sqrt{3}/2$
21. -1 **23.** $\frac{1}{2}$ **25.** 2   **27.** -1 **29.** undefined
31. III   **33.** IV   **35.** $\tan \theta = -\sqrt{1 - \cos^2\theta}/\cos \theta$
37. $\cos \theta = \sqrt{1 - \sin^2\theta}$ **39.** $\sec \theta = -\sqrt{1 + \tan^2\theta}$
41. $\cos \theta = -\frac{4}{5}$, $\tan \theta = -\frac{3}{4}$, $\csc \theta = \frac{5}{3}$, $\sec \theta = -\frac{5}{4}$, $\cot \theta = -\frac{4}{3}$
43. $\sin \theta = -\frac{3}{5}$, $\cos \theta = \frac{4}{5}$, $\csc \theta = -\frac{5}{3}$, $\sec \theta = \frac{5}{4}$, $\cot \theta = -\frac{4}{3}$
45. $\sin \theta = \frac{1}{2}$, $\cos \theta = \sqrt{3}/2$, $\tan \theta = \sqrt{3}/3$, $\sec \theta = 2\sqrt{3}/3$, $\cot \theta = \sqrt{3}$
47. $\sin \theta = 3\sqrt{5}/7$, $\tan \theta = -3\sqrt{5}/2$, $\csc \theta = 7\sqrt{5}/15$, $\sec \theta = -\frac{7}{2}$, $\cot \theta = -2\sqrt{5}/15$
49. (a) $\sqrt{3}/2$, $\sqrt{3}$ (b) $\frac{1}{2}$, $\sqrt{3}/4$ (c) $\frac{3}{4}$, 0.88967
51. 19.1   **53.** $66.1°$
55. (a)

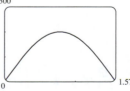

(b) 21.07
57. (a) $A(\theta) = 400 \sin \theta \cos \theta$
(b)

(c) width = depth ≈ 14.14 in.

Section 6.4 ■ page 510

1. 318.8   **3.** 24.8   **5.** $44°$ **7.** $\angle C = 114°$, $a \approx 51$, $b \approx 24$

9. $\angle C = 62°, a \approx 200, b \approx 242$

11. $\angle B = 85°, a \approx 5,$
$c \approx 9$

13. $\angle A = 100°, a \approx 89,$
$c \approx 71$

15. $\angle B \approx 30°, \angle C \approx 40°, c \approx 19$ **17.** No solution
19. $\angle A_1 \approx 125°, \angle C_1 \approx 30°, a_1 \approx 49;$
$\angle A_2 \approx 5°, \angle C_2 \approx 150°, a_2 \approx 5.6$
21. No solution   **23.** 219 ft
25. (a) 1018 mi   (b) 1017 mi   **27.** 155 m
29. (a) $d\dfrac{\sin \alpha}{\sin(\beta - \alpha)}$ (c) 2350 ft   **31.** 48.2°

Section 6.5 ■ page 518

1. 28.9   **3.** 47   **5.** 29.89°   **7.** 15
9. $\angle A \approx 39.4°, \angle B \approx 20.6°, c \approx 24.6$
11. $\angle A \approx 48°, \angle B \approx 79°, c \approx 3.2$
13. $\angle A \approx 50°, \angle B \approx 73°, \angle C \approx 57°$
15. $\angle A_1 \approx 83.6°, \angle C_1 \approx 56.4°, a_1 \approx 193;$
$\angle A_2 \approx 16.4°, \angle C_2 \approx 123.6, a_2 \approx 54.9$
17. No such triangle   **19.** 2   **21.** 25.4   **23.** 84.6°   **25.** 24.3
27. 2.30 mi   **29.** 23.1 mi   **31.** 2179 mi   **33.** (a) 62.6 mi
(b) S 18.2° E   **35.** 96°   **37.** 211 ft   **39.** 3835 ft
41. 3.85 cm² **43.** 14.3 m   **45.** $165,554

Chapter 6 Review ■ page 522

1. (a) $\pi/3$ (b) $11\pi/6$ (c) $-3\pi/4$ (d) $-\pi/2$
3. (a) 450°   (b) −30°   (c) 405°   (d) $(558/\pi)° \approx 177.6°$
5. 8 m   **7.** 82 ft   **9.** 0.619 rad $\approx 35.4°$
11. 18,151 ft²   **13.** (a) 300π rad/min ≈ 942.5 rad/min
(b) 7539.8 in./min $= 628.3$ ft/min   **15.** $\sin \theta = 5/\sqrt{74}$,
$\cos \theta = 7/\sqrt{74}, \tan \theta = \frac{5}{7}, \csc \theta = \sqrt{74}/5, \sec \theta = \sqrt{74}/7,$
$\cot \theta = \frac{7}{5}$ **17.** $x \approx 3.83, y \approx 3.21$
19. $x \approx 2.92, y \approx 3.11$
21.

23. $a = \cot \theta, b = \csc \theta$
25. 48 m   **27.** 1076 mi
29. $-\sqrt{2}/2$ **31.** 1

33. $-\sqrt{3}/3$ **35.** $-\sqrt{2}/2$ **37.** $2\sqrt{3}/3$ **39.** $-\sqrt{3}$
41. $\sin \theta = \frac{12}{13}, \cos \theta = -\frac{5}{13}, \tan \theta = -\frac{12}{5}, \csc \theta = \frac{13}{12},$
$\sec \theta = -\frac{13}{5}, \cot \theta = -\frac{5}{12}$
43. 60°   **45.** $\tan \theta = -\sqrt{1 - \cos^2\theta}/\cos \theta$
47. $\tan^2\theta = \sin^2\theta/(1 - \sin^2\theta)$
49. $\sin \theta = \sqrt{7}/4, \cos \theta = \frac{3}{4}, \csc \theta = 4\sqrt{7}/7,$
$\cot \theta = 3\sqrt{7}/7$
51. $\cos \theta = -\frac{4}{5}, \tan \theta = -\frac{3}{4}, \csc \theta = \frac{5}{3}, \sec \theta = -\frac{5}{4},$
$\cot \theta = -\frac{4}{3}$
53. $-\sqrt{5}/5$ **55.** 1   **57.** 5.32   **59.** 148.07   **61.** 77.82
63. 77.3 mi   **65.** 3.9 mi   **67.** 119.2 m   **69.** 14.98

Chapter 6 Test ■ page 526

1. $5\pi/3, -\pi/10$ **2.** 150°, 137.5°
3. (a) 400π rad/min ≈ 1256.6 rad/min   (b) 31,416 ft/min
4. (a) $\sqrt{2}/2$ (b) $\sqrt{3}/3$ (c) 2   (d) 1
5. $(26 + 6\sqrt{13})/39$ **6.** $a = 24 \sin \theta, b = 24 \cos \theta$
7. $(4 - 3\sqrt{2})/4$ **8.** $-\frac{13}{12}$ **9.** $\tan \theta = -\sqrt{\sec^2\theta - 1}$
10. 19.6 ft   **11.** 9.1   **12.** 250.5   **13.** 8.4   **14.** 19.5
15. (a) 15.3 m²   (b) 24.3 m   **16.** (a) 129.9°   (b) 44.9
17. 554 ft

Focus on Problem Solving ■ page 530

1. Tetrahedron 4, 6, 4; octahedron 8, 12, 6; cube 6, 12, 8;
dodecahedron 12, 30, 20; icosahedron 20, 30, 12
3. Hexagons   **5.** $P + 2\pi$ **7.** $2\left(\dfrac{\pi}{3} - \dfrac{\sqrt{3}}{4}\right) \approx 1.2$
9. 0   **11.** $\sqrt{a^2 + (b + c)^2}$
15. (a) $13 = 2^2 + 3^2, 41 = 4^2 + 5^2$
(c) $533 = 2^2 + 23^2 = 7^2 + 22^2$

CHAPTER 7

Section 7.1 ■ page 540

1. $\cos t$ **3.** $\sec x$ **5.** −1   **7.** $\csc u$ **9.** 1   **11.** $\cos y$
13. $\sin^2 x$ **15.** $\sec x$ **17.** $2 \sec u$ **19.** $\cos^2 x$
21. $\cos \theta$ **23.** LHS $= \sin \theta \dfrac{\cos \theta}{\sin \theta} =$ RHS
25. LHS $= \cos u \dfrac{1}{\cos u} \cot u =$ RHS
27. LHS $= \dfrac{\sin y}{\cos y} \sin y = \dfrac{1 - \cos^2 y}{\cos y} = \sec y - \cos y =$ RHS
29. LHS $= \sin B + \cos B \dfrac{\cos B}{\sin B}$
$= \dfrac{\sin^2 B + \cos^2 B}{\sin B} = \dfrac{1}{\sin B} =$ RHS

31. LHS $= -\dfrac{\cos\alpha}{\sin\alpha}\cos\alpha - \sin\alpha = \dfrac{-\cos^2\alpha - \sin^2\alpha}{\sin\alpha}$

$\quad = \dfrac{-1}{\sin\alpha} = $ RHS

33. LHS $= \dfrac{\sin\theta}{\cos\theta} + \dfrac{\cos\theta}{\sin\theta} = \dfrac{\sin^2\theta + \cos^2\theta}{\cos\theta\sin\theta}$

$\quad = \dfrac{1}{\cos\theta\sin\theta} = $ RHS

35. LHS $= 1 - \cos^2\beta = \sin^2\beta = $ RHS

37. LHS $= \dfrac{(\sin x + \cos x)^2}{(\sin x + \cos x)(\sin x - \cos x)} = \dfrac{\sin x + \cos x}{\sin x - \cos x}$

$\quad = \dfrac{(\sin x + \cos x)(\sin x - \cos x)}{(\sin x - \cos x)(\sin x - \cos x)} = $ RHS

39. LHS $= \dfrac{\dfrac{1}{\cos t} - \cos t}{\dfrac{1}{\cos t}} \cdot \dfrac{\cos t}{\cos t} = \dfrac{1 - \cos^2 t}{1} = $ RHS

41. LHS $= \dfrac{1}{\cos^2 y} = \sec^2 y = $ RHS

43. LHS $= \cot x \cos x + \cot x - \csc x \cos x - \csc x$

$\quad = \dfrac{\cos^2 x}{\sin x} + \dfrac{\cos x}{\sin x} - \dfrac{\cos x}{\sin x} - \dfrac{1}{\sin x} = \dfrac{\cos^2 x - 1}{\sin x}$

$\quad = \dfrac{-\sin^2 x}{\sin x} = $ RHS

45. LHS $= \sin^2 x\left(1 + \dfrac{\cos^2 x}{\sin^2 x}\right) = \sin^2 x + \cos^2 x = $ RHS

47. LHS $= 2(1 - \sin^2 x) - 1 = 2 - 2\sin^2 x - 1 = $ RHS

49. LHS $= \dfrac{1 - \cos\alpha}{\sin\alpha} \cdot \dfrac{1 + \cos\alpha}{1 + \cos\alpha} = \dfrac{1 - \cos^2\alpha}{\sin\alpha(1 + \cos\alpha)}$

$\quad = \dfrac{\sin^2\alpha}{\sin\alpha(1 + \cos\alpha)} = $ RHS

51. LHS $= \dfrac{\sin x - 1}{\sin x + 1} \cdot \dfrac{\sin x + 1}{\sin x + 1} = \dfrac{\sin^2 x - 1}{(\sin x + 1)^2} = $ RHS

53. LHS $= \dfrac{\sin^2 t + 2\sin t\cos t + \cos^2 t}{\sin t\cos t}$

$\quad = \dfrac{\sin^2 t + \cos^2 t}{\sin t\cos t} + \dfrac{2\sin t\cos t}{\sin t\cos t} = \dfrac{1}{\sin t\cos t} + 2$

$\quad = $ RHS

55. LHS $= \dfrac{1 + \frac{\sin^2 u}{\cos^2 u}}{1 - \frac{\sin^2 u}{\cos^2 u}} \cdot \dfrac{\cos^2 u}{\cos^2 u} = \dfrac{\cos^2 u + \sin^2 u}{\cos^2 u - \sin^2 u} = $ RHS

57. LHS $= \dfrac{\sec x}{\sec x - \tan x} \cdot \dfrac{\sec x + \tan x}{\sec x + \tan x}$

$\quad = \dfrac{\sec x(\sec x + \tan x)}{\sec^2 x - \tan^2 x} = $ RHS

59. LHS $= (\sec v - \tan v) \cdot \dfrac{\sec v + \tan v}{\sec v + \tan v}$

$\quad = \dfrac{\sec^2 v - \tan^2 v}{\sec v + \tan v} = $ RHS

61. LHS $= \dfrac{\sin x + \cos x}{\frac{1}{\cos x} + \frac{1}{\sin x}} = \dfrac{\sin x + \cos x}{\frac{\sin x + \cos x}{\cos x \sin x}}$

$\quad = (\sin x + \cos x)\dfrac{\cos x \sin x}{\sin x + \cos x} = $ RHS

63. LHS $= \dfrac{\frac{1}{\sin x} - \frac{\cos x}{\sin x}}{\frac{1}{\cos x} - 1} \cdot \dfrac{\sin x\cos x}{\sin x\cos x} = \dfrac{\cos x(1 - \cos x)}{\sin x(1 - \cos x)}$

$\quad = \dfrac{\cos x}{\sin x} = $ RHS

65. LHS $= \dfrac{\sin^2 u}{\cos^2 u} - \dfrac{\sin^2 u \cos^2 u}{\cos^2 u} = \dfrac{\sin^2 u}{\cos^2 u}(1 - \cos^2 u) = $ RHS

67. LHS $= (\sec^2 x - \tan^2 x)(\sec^2 x + \tan^2 x) = $ RHS

69. RHS $= \dfrac{\sin\theta - \frac{1}{\sin\theta}}{\cos\theta - \frac{\cos\theta}{\sin\theta}} = \dfrac{\frac{\sin^2\theta - 1}{\sin\theta}}{\frac{\cos\theta\sin\theta - \cos\theta}{\sin\theta}}$

$\quad = \dfrac{\cos^2\theta}{\cos\theta(\sin\theta - 1)} = $ LHS

71. LHS $= \dfrac{-\sin^2 t + \tan^2 t}{\sin^2 t} = -1 + \dfrac{\sin^2 t}{\cos^2 t} \cdot \dfrac{1}{\sin^2 t}$

$\quad = -1 + \sec^2 t = $ RHS

73. LHS $= \dfrac{\sec x - \tan x + \sec x + \tan x}{(\sec x + \tan x)(\sec x - \tan x)}$

$\quad = \dfrac{2\sec x}{\sec^2 x - \tan^2 x} = $ RHS

75. LHS $= \tan^2 x + 2\tan x\cot x + \cot^2 x = \tan^2 x + 2 + \cot^2 x$

$\quad = (\tan^2 x + 1) + (\cot^2 x + 1) = $ RHS

77. LHS $= \dfrac{\frac{1}{\cos u} - 1}{\frac{1}{\cos u} + 1} \cdot \dfrac{\cos u}{\cos u} = $ RHS

79. LHS $= \dfrac{(\sin x + \cos x)(\sin^2 x - \sin x\cos x + \cos^2 x)}{\sin x + \cos x}$

$\quad = \sin^2 x - \sin x\cos x + \cos^2 x = $ RHS

81. LHS $= \dfrac{1 + \sin x}{1 - \sin x} \cdot \dfrac{1 + \sin x}{1 + \sin x} = \dfrac{(1 + \sin x)^2}{1 - \sin^2 x}$

$\quad = \dfrac{(1 + \sin x)^2}{\cos^2 x} = \left(\dfrac{1 + \sin x}{\cos x}\right)^2 = $ RHS

83. LHS $= \left(\dfrac{\sin x}{\cos x} + \dfrac{\cos x}{\sin x}\right)^4 = \left(\dfrac{\sin^2 x + \cos^2 x}{\sin x\cos x}\right)^4$

$\quad = \left(\dfrac{1}{\sin x\cos x}\right)^4 = $ RHS

85. $\tan \theta$ **87.** $\tan \theta$ **89.** $3 \cos \theta$

91.

Yes

93.

No

Section 7.2 ■ page 548

1. $\dfrac{\sqrt{6} + \sqrt{2}}{4}$ **3.** $\dfrac{\sqrt{6} - \sqrt{2}}{4}$ **5.** $2 + \sqrt{3}$ **7.** $\dfrac{-\sqrt{6} - \sqrt{2}}{4}$

9. $\dfrac{\sqrt{6} + \sqrt{2}}{4}$ **11.** $\sqrt{2}/2$ **13.** $1/2$ **15.** $\sqrt{3}$

17. LHS $= \dfrac{\sin\left(\frac{\pi}{2} - u\right)}{\cos\left(\frac{\pi}{2} - u\right)} = \dfrac{\sin \frac{\pi}{2} \cos u - \cos \frac{\pi}{2} \sin u}{\cos \frac{\pi}{2} \cos u + \sin \frac{\pi}{2} \sin u}$

$= \dfrac{\cos u}{\sin u} =$ RHS

19. LHS $= \dfrac{1}{\cos\left(\frac{\pi}{2} - u\right)} = \dfrac{1}{\cos \frac{\pi}{2} \cos u + \sin \frac{\pi}{2} \sin u}$

$= \dfrac{1}{\sin u} =$ RHS

21. LHS $= \sin x \cos \frac{\pi}{2} - \cos x \sin \frac{\pi}{2} =$ RHS

23. LHS $= \sin x \cos \pi - \cos x \sin \pi =$ RHS

25. LHS $= \dfrac{\tan x - \tan \pi}{1 + \tan x \tan \pi} =$ RHS

27. LHS $=$

$\cos x \cos \frac{\pi}{6} - \sin x \sin \frac{\pi}{6} + \sin x \cos \frac{\pi}{3} - \cos x \sin \frac{\pi}{3}$

$= \frac{\sqrt{3}}{2} \cos x - \frac{1}{2} \sin x + \frac{1}{2} \sin x - \frac{\sqrt{3}}{2} \cos x =$ RHS

29. LHS $= \sin x \cos y + \cos x \sin y$
$- (\sin x \cos y - \cos x \sin y) =$ RHS

31. LHS $= \dfrac{1}{\tan(x - y)} = \dfrac{1 + \tan x \tan y}{\tan x - \tan y}$

$= \dfrac{1 + \frac{1}{\cot x} \frac{1}{\cot y}}{\frac{1}{\cot x} - \frac{1}{\cot y}} \cdot \dfrac{\cot x \cot y}{\cot x \cot y} =$ RHS

33. LHS $= \dfrac{\sin x}{\cos x} - \dfrac{\sin y}{\cos y} = \dfrac{\sin x \cos y - \cos x \sin y}{\cos x \cos y} =$ RHS

35. LHS $=$

$\dfrac{\sin x \cos y + \cos x \sin y - (\sin x \cos y - \cos x \sin y)}{\cos x \cos y - \sin x \sin y + \cos x \cos y + \sin x \sin y}$

$= \dfrac{2 \cos x \sin y}{2 \cos x \cos y} =$ RHS

37. LHS $= \sin((x + y) + z)$
$= \sin(x + y) \cos z + \cos(x + y) \sin z$
$= \cos z \left[\sin x \cos y + \cos x \sin y\right]$
$\quad + \sin z \left[\cos x \cos y - \sin x \sin y\right] =$ RHS

39. $2 \sin\left(x + \dfrac{5\pi}{6}\right)$ **41.** $5\sqrt{2} \sin\left(2x + \dfrac{7\pi}{4}\right)$

43. $f(x) = \sqrt{2} \sin\left(x + \dfrac{\pi}{4}\right)$

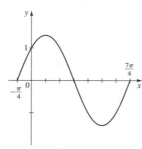

47. $\tan \gamma = \dfrac{17}{6}$

49. (a)

$\sin^2\left(x + \dfrac{\pi}{4}\right) + \sin^2\left(x - \dfrac{\pi}{4}\right) = 1$

51. (a)

(b) $k = 5\sqrt{2}$,
$\theta = \pi/4$

Section 7.3 ■ page 557

1. $\frac{120}{169}, \frac{119}{169}, \frac{120}{119}$ **3.** $-\frac{24}{25}, \frac{7}{25}, -\frac{24}{7}$ **5.** $\frac{24}{25}, \frac{7}{25}, \frac{24}{7}$

7. $-\frac{3}{5}, \frac{4}{5}, -\frac{3}{4}$ **9.** $\frac{1}{2}\left(\frac{3}{4} - \cos 2x + \frac{1}{4} \cos 4x\right)$

11. $\frac{1}{32}\left(\frac{3}{4} - \cos 4x + \frac{1}{4} \cos 8x\right)$

13. $\frac{1}{16}\left(1 - \cos 2x - \cos 4x + \cos 2x \cos 4x\right)$

15. $\frac{1}{2}\sqrt{2 - \sqrt{3}}$ **17.** $\frac{1}{2}\sqrt{2 + \sqrt{2}}$ **19.** $\sqrt{2} - 1$

21. $\frac{1}{2}\sqrt{2 - \sqrt{3}}$ **23.** (a) $\sin 36°$ (b) $\sin 6\theta$

25. (a) $\cos 68°$ (b) $\cos 10\theta$ **27.** (a) $\tan 4°$ (b) $\tan 2\theta$

29. $\sqrt{10}/10,\ 3\sqrt{10}/10,\ \frac{1}{3}$

31. $\sqrt{(3+2\sqrt{2})/6},\ \sqrt{(3-2\sqrt{2})/6},\ 3+2\sqrt{2}$

33. $\sqrt{6}/6,\ -\sqrt{30}/6,\ -\sqrt{5}/5$

35. $\frac{1}{2}(\sin 5x - \sin x)$ **37.** $\frac{1}{2}(\sin 5x + \sin 3x)$

39. $\frac{3}{2}(\cos 11x + \cos 3x)$ **41.** $2\sin 4x \cos x$

43. $2\sin 5x \sin x$ **45.** $-2\cos\frac{9}{2}x \sin\frac{5}{2}x$

47. $(\sqrt{2}+\sqrt{3})/2$ **49.** $\frac{1}{4}(\sqrt{2}-1)$ **51.** $\sqrt{2}/2$

53. LHS $= \cos(2\cdot 5x) =$ RHS

55. LHS $= \sin^2 x + 2\sin x \cos x + \cos^2 x = 1 + 2\sin x \cos x$
$=$ RHS

57. LHS $= \dfrac{(2\sin 2x \cos 2x)}{\sin x} = \dfrac{2(2\sin x \cos x)(\cos 2x)}{\sin x}$
$=$ RHS

59. LHS $= \dfrac{2(\tan x - \cot x)}{(\tan x + \cot x)(\tan x - \cot x)} = \dfrac{2}{\tan x + \cot x}$

$= \dfrac{2}{\frac{\sin x}{\cos x} + \frac{\cos x}{\sin x}} \cdot \dfrac{\sin x \cos x}{\sin x \cos x} = \dfrac{2\sin x \cos x}{\sin^2 x + \cos^2 x}$

$= 2\sin x \cos x =$ RHS

61. LHS $= \tan(2x + x) = \dfrac{\tan 2x + \tan x}{1 - \tan 2x \tan x}$

$= \dfrac{\frac{2\tan x}{1 - \tan^2 x} + \tan x}{1 - \frac{2\tan x}{1 - \tan^2 x}\tan x}$

$= \dfrac{2\tan x + \tan x(1 - \tan^2 x)}{1 - \tan^2 x - 2\tan x \tan x} =$ RHS

63. LHS $= (\cos^2 x + \sin^2 x)(\cos^2 x - \sin^2 x)$
$= \cos^2 x - \sin^2 x =$ RHS

65. LHS $= \dfrac{2\sin 3x \cos 2x}{2\cos 3x \cos 2x} = \dfrac{\sin 3x}{\cos 3x} =$ RHS

67. LHS $= \dfrac{2\sin 5x \cos 5x}{2\sin 5x \cos 4x} =$ RHS

69. LHS $= \dfrac{2\sin\left(\frac{x+y}{2}\right)\cos\left(\frac{x-y}{2}\right)}{2\cos\left(\frac{x+y}{2}\right)\cos\left(\frac{x-y}{2}\right)}$

$= \dfrac{\sin\left(\frac{x+y}{2}\right)}{\cos\left(\frac{x+y}{2}\right)} =$ RHS

73. LHS $=$

$\dfrac{(\sin x + \sin 5x) + (\sin 2x + \sin 4x) + \sin 3x}{(\cos x + \cos 5x) + (\cos 2x + \cos 4x) + \cos 3x}$

$= \dfrac{2\sin 3x \cos 2x + 2\sin 3x \cos x + \sin 3x}{2\cos 3x \cos 2x + 2\cos 3x \cos x + \cos 3x}$

$= \dfrac{\sin 3x(2\cos 2x + 2\cos x + 1)}{\cos 3x(2\cos 2x + 2\cos x + 1)} =$ RHS

75. (a)

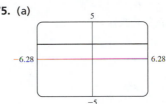

$\dfrac{\sin 3x}{\sin x} - \dfrac{\cos 3x}{\cos x} = 2$

77. (a)

(c)

The graph of $y = f(x)$ lies between the two other graphs.

83. (a) $P(t) = 8t^4 - 8t^2 + 1$
(b) $Q(t) = 16t^5 - 20t^3 + 5t$

Section 7.4 ■ page 567

1. (a) $\pi/6$ **(b)** $\pi/3$ **(c)** Not defined

3. (a) $\pi/4$ **(b)** $\pi/4$ **(c)** $-\pi/4$

5. (a) $\pi/2$ **(b)** 0   **(c)** π

7. (a) $\pi/6$ **(b)** $-\pi/6$ **(c)** Not defined

9. (a) 0.13889   **(b)** 2.75876   **11. (a)** 0.88998

(b) Not defined   **13.** $\frac{1}{4}$ **15.** 5   **17.** $\pi/3$ **19.** $-\pi/6$

21. $-\pi/3$ **23.** $\sqrt{3}/3$ **25.** $\frac{1}{2}$ **27.** $\pi/3$ **29.** $\frac{4}{5}$

31. $\frac{12}{13}$ **33.** $\frac{13}{5}$ **35.** $\sqrt{5}/5$ **37.** $\frac{24}{25}$ **39.** 1

41. $\sqrt{1 - x^2}$ **43.** $x/\sqrt{1 - x^2}$ **45.** $\dfrac{1 - x^2}{1 + x^2}$ **47.** 0

49. (a) $h = 2\tan\theta$ **(b)** $\theta = \tan^{-1}\left(\frac{h}{2}\right)$

51. (a) $\theta = \sin^{-1}\left(\frac{h}{680}\right)$ **(b)** $\theta = 0.826$ rad

53. (a)

Conjecture: $y = \pi/2$ for $-1 \le x \le 1$

55. (a) 0.28   **(b)** $(-3 + \sqrt{17})/4$

Section 7.5 ■ page 579

1. $\dfrac{\pi}{2} + 2k\pi$ **3.** $\dfrac{\pi}{3} + 2k\pi, \dfrac{5\pi}{3} + 2k\pi$

5. $\dfrac{4\pi}{3} + 2k\pi, \dfrac{5\pi}{3} + 2k\pi$

7. $\dfrac{\pi}{3} + 2k\pi, \dfrac{2\pi}{3} + 2k\pi, \dfrac{4\pi}{3} + 2k\pi, \dfrac{5\pi}{3} + 2k\pi$

9. $\dfrac{(2k+1)\pi}{4}$ **11.** $\dfrac{\pi}{2} + k\pi, \dfrac{7\pi}{6} + 2k\pi, \dfrac{11\pi}{6} + 2k\pi$

13. $-\dfrac{\pi}{3} + k\pi$ **15.** $\dfrac{\pi}{2} + k\pi$

17. $\dfrac{\pi}{3} + 2k\pi, \dfrac{5\pi}{3} + 2k\pi$ **19.** $\dfrac{3\pi}{2} + 2k\pi$

21. No solution   **23.** $\dfrac{7\pi}{18} + \dfrac{2k\pi}{3}, \dfrac{11\pi}{18} + \dfrac{2k\pi}{3}$

25. $\dfrac{1}{4}\left(\dfrac{\pi}{3} + 2k\pi\right), \dfrac{1}{4}\left(-\dfrac{\pi}{3} + 2k\pi\right)$ **27.** $\dfrac{1}{2}\left(\dfrac{\pi}{6} + k\pi\right)$

29. $4k\pi$ **31.** $4\left(\dfrac{2\pi}{3} + k\pi\right)$ **33.** $\dfrac{k\pi}{3}$

35. $\dfrac{\pi}{6} + 2k\pi, \dfrac{2\pi}{3} + 2k\pi, \dfrac{5\pi}{6} + 2k\pi, \dfrac{4\pi}{3} + 2k\pi$

37. $\dfrac{\pi}{8} + \dfrac{k\pi}{2}, \dfrac{3\pi}{8} + \dfrac{k\pi}{2}$ **39.** $\dfrac{\pi}{9}, \dfrac{5\pi}{9}, \dfrac{7\pi}{9}, \dfrac{11\pi}{9}, \dfrac{13\pi}{9}, \dfrac{17\pi}{9}$

41. $\dfrac{\pi}{6}, \dfrac{3\pi}{4}, \dfrac{5\pi}{6}, \dfrac{7\pi}{4}$ **43.** $\dfrac{\pi}{3}, \dfrac{2\pi}{3}, \dfrac{4\pi}{3}, \dfrac{5\pi}{3}$

45. $0, \dfrac{2\pi}{3}, \dfrac{4\pi}{3}$ **47.** (a) $1.15928 + 2k\pi, 5.12391 + 2k\pi$

(b) $1.15928, 5.12391$

49. (a) $1.36944 + 2k\pi, 4.91375 + 2k\pi$

(b) $1.36944, 4.91375$

51. (a) $0.46365 + k\pi, 2.67795 + k\pi$

(b) $0.46365, 2.67795, 3.60524, 5.81954$

53. (a) $0.33984 + 2k\pi, 2.80176 + 2k\pi$

(b) $0.33984, 2.80176$

55. $((2k+1)\pi, -2)$ **57.** $\left(\dfrac{\pi}{3} + k\pi, \sqrt{3}\right)$

59. $0.94721°$ or $89.05279°$ **61.** $44.95°$ or $135.05°$

63. $\dfrac{\pi}{8}, \dfrac{3\pi}{8}, \dfrac{5\pi}{8}, \dfrac{7\pi}{8}, \dfrac{9\pi}{8}, \dfrac{11\pi}{8}, \dfrac{13\pi}{8}, \dfrac{15\pi}{8}$

65. $\dfrac{\pi}{9}, \dfrac{2\pi}{9}, \dfrac{7\pi}{9}, \dfrac{8\pi}{9}, \dfrac{13\pi}{9}, \dfrac{14\pi}{9}$

67. $\dfrac{\pi}{2}, \dfrac{7\pi}{6}, \dfrac{3\pi}{2}, \dfrac{11\pi}{6}$ **69.** 0 **71.** $\dfrac{k\pi}{2}$

73. $\dfrac{\pi}{9} + \dfrac{2k\pi}{3}, \dfrac{\pi}{2} + k\pi, \dfrac{5\pi}{9} + \dfrac{2k\pi}{3}$

75. $0, \pm 0.95$ **77.** 1.92 **79.** ± 0.71

Section 7.6 ■ page 589

1. 4 **3.** $\sqrt{29}$

5. 2 **7.** 1

9.   **11.**

13.   **15.**

17.

19.

21.

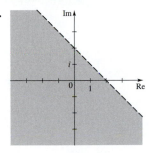

23. $\sqrt{2}\left(\cos\dfrac{\pi}{4}+i\sin\dfrac{\pi}{4}\right)$ **25.** $2\left(\cos\dfrac{7\pi}{4}+i\sin\dfrac{7\pi}{4}\right)$

27. $4\left(\cos\dfrac{11\pi}{6}+i\sin\dfrac{11\pi}{6}\right)$ **29.** $3\left(\cos\dfrac{3\pi}{2}+i\sin\dfrac{3\pi}{2}\right)$

31. $5\sqrt{2}\left(\cos\dfrac{\pi}{4}+i\sin\dfrac{\pi}{4}\right)$ **33.** $8\left(\cos\dfrac{11\pi}{6}+i\sin\dfrac{11\pi}{6}\right)$

35. $20(\cos\pi+i\sin\pi)$

37. $5\left[\cos\left(\tan^{-1}\tfrac{4}{3}\right)+i\sin\left(\tan^{-1}\tfrac{4}{3}\right)\right]$

39. $3\sqrt{2}\left(\cos\dfrac{3\pi}{4}+i\sin\dfrac{3\pi}{4}\right)$ **41.** $8\left(\cos\dfrac{\pi}{6}+i\sin\dfrac{\pi}{6}\right)$

43. $\sqrt{5}\left[\cos\left(\tan^{-1}\tfrac{1}{2}\right)+i\sin\left(\tan^{-1}\tfrac{1}{2}\right)\right]$

45. $2\left(\cos\dfrac{\pi}{4}+i\sin\dfrac{\pi}{4}\right)$

47. $z_1z_2=\cos\dfrac{4\pi}{3}+i\sin\dfrac{4\pi}{3}$

$\dfrac{z_1}{z_2}=\cos\dfrac{2\pi}{3}+i\sin\dfrac{2\pi}{3}$

49. $z_1z_2=15\left(\cos\dfrac{3\pi}{2}+i\sin\dfrac{3\pi}{2}\right)$

$\dfrac{z_1}{z_2}=\dfrac{3}{5}\left(\cos\dfrac{7\pi}{6}-i\sin\dfrac{7\pi}{6}\right)$

51. $z_1z_2=8(\cos150°+i\sin150°)$

$\dfrac{z_1}{z_2}=2(\cos90°+i\sin90°)$

53. $z_1z_2=100(\cos350°+i\sin350°)$

$\dfrac{z_1}{z_2}=\tfrac{4}{25}(\cos50°+i\sin50°)$

55. $z_1=2\left(\cos\dfrac{\pi}{6}+i\sin\dfrac{\pi}{6}\right)$

$z_2=2\left(\cos\dfrac{\pi}{3}+i\sin\dfrac{\pi}{3}\right)$

$z_1z_2=4\left(\cos\dfrac{\pi}{2}+i\sin\dfrac{\pi}{2}\right)$

$\dfrac{z_1}{z_2}=\cos\dfrac{\pi}{6}-i\sin\dfrac{\pi}{6}$

$\dfrac{1}{z_1}=\dfrac{1}{2}\left(\cos\dfrac{\pi}{6}-i\sin\dfrac{\pi}{6}\right)$

57. $z_1=4\left(\cos\dfrac{11\pi}{6}+i\sin\dfrac{11\pi}{6}\right)$

$z_2=\sqrt{2}\left(\cos\dfrac{3\pi}{4}+i\sin\dfrac{3\pi}{4}\right)$

$z_1z_2=4\sqrt{2}\left(\cos\dfrac{7\pi}{12}+i\sin\dfrac{7\pi}{12}\right)$

$\dfrac{z_1}{z_2}=2\sqrt{2}\left(\cos\dfrac{13\pi}{12}+i\sin\dfrac{13\pi}{12}\right)$

$\dfrac{1}{z_1}=\dfrac{1}{4}\left(\cos\dfrac{11\pi}{6}-i\sin\dfrac{11\pi}{6}\right)$

59. $z_1=5\sqrt{2}\left(\cos\dfrac{\pi}{4}+i\sin\dfrac{\pi}{4}\right)$

$z_2=4(\cos0+i\sin0)$

$z_1z_2=20\sqrt{2}\left(\cos\dfrac{\pi}{4}+i\sin\dfrac{\pi}{4}\right)$

$\dfrac{z_1}{z_2}=\dfrac{5\sqrt{2}}{4}\left(\cos\dfrac{\pi}{4}+i\sin\dfrac{\pi}{4}\right)$

$\dfrac{1}{z_1}=\dfrac{\sqrt{2}}{10}\left(\cos\dfrac{\pi}{4}-i\sin\dfrac{\pi}{4}\right)$

61. $z_1=20(\cos\pi+i\sin\pi)$

$z_2=2\left(\cos\dfrac{\pi}{6}+i\sin\dfrac{\pi}{6}\right)$

$z_2z_2=40\left(\cos\dfrac{7\pi}{6}+i\sin\dfrac{7\pi}{6}\right)$

$\dfrac{z_1}{z_2}=10\left(\cos\dfrac{5\pi}{6}+i\sin\dfrac{5\pi}{6}\right)$

$\dfrac{1}{z_1}=\tfrac{1}{20}(\cos\pi-i\sin\pi)$

63. -1024 **65.** $512(-\sqrt{3}+i)$ **67.** -1

69. 4096 **71.** $8(-1+i)$ **73.** $\dfrac{1}{2048}(-\sqrt{3}-i)$

75. $2\sqrt{2}\left(\cos\dfrac{\pi}{12}+i\sin\dfrac{\pi}{12}\right),$

$2\sqrt{2}\left(\cos\dfrac{13\pi}{12}+i\sin\dfrac{13\pi}{12}\right)$

77. $3\left(\cos\dfrac{3\pi}{8} + i\sin\dfrac{3\pi}{8}\right), 3\left(\cos\dfrac{7\pi}{8} + i\sin\dfrac{7\pi}{8}\right),$

$3\left(\cos\dfrac{11\pi}{8} + i\sin\dfrac{11\pi}{8}\right), 3\left(\cos\dfrac{15\pi}{8} + i\sin\dfrac{15\pi}{8}\right)$

79. $\pm 1, \pm i, \pm\dfrac{\sqrt{2}}{2} \pm \dfrac{\sqrt{2}}{2}i$

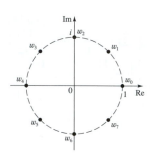

81. $\dfrac{\sqrt{3}}{2} + \dfrac{1}{2}i, -\dfrac{\sqrt{3}}{2} + \dfrac{1}{2}i, -i$

83. $\pm\dfrac{\sqrt{2}}{2} \pm \dfrac{\sqrt{2}}{2}i$

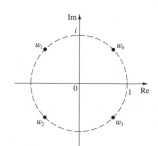

85. $\pm\dfrac{\sqrt{2}}{2} \pm \dfrac{\sqrt{2}}{2}i$

87. $2\left(\cos\dfrac{\pi}{18} + i\sin\dfrac{\pi}{18}\right), 2\left(\cos\dfrac{13\pi}{18} + i\sin\dfrac{13\pi}{18}\right),$

$2\left(\cos\dfrac{25\pi}{18} + i\sin\dfrac{25\pi}{18}\right)$

89. $2^{1/6}\left(\cos\dfrac{5\pi}{12} + i\sin\dfrac{5\pi}{12}\right), 2^{1/6}\left(\cos\dfrac{13\pi}{12} + i\sin\dfrac{13\pi}{12}\right),$

$2^{1/6}\left(\cos\dfrac{21\pi}{12} + i\sin\dfrac{21\pi}{12}\right)$

Section 7.7 ■ page 599

1.

3.

5.
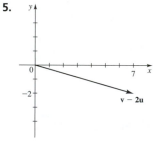

7. $\langle 3, 3 \rangle$ **9.** $\langle 3, -1 \rangle$ **11.** $\langle 5, 7 \rangle$ **13.** $\langle -4, -3 \rangle$
15. $\langle 0, 2 \rangle$ **17.** $\langle 4, 14 \rangle, \langle -9, -3 \rangle, \langle 5, 8 \rangle, \langle -6, 17 \rangle$
19. $\langle 0, -2 \rangle, \langle 6, 0 \rangle, \langle -2, -1 \rangle, \langle 8, -3 \rangle$
21. $4\mathbf{i}, -9\mathbf{i} + 6\mathbf{j}, 5\mathbf{i} - 2\mathbf{j}, -6\mathbf{i} + 8\mathbf{j}$
23. $\sqrt{5}, \sqrt{13}, 2\sqrt{5}, \frac{1}{2}\sqrt{13}, \sqrt{26}, \sqrt{10}, \sqrt{5} - \sqrt{13}$
25. $\sqrt{101}, 2\sqrt{2}, 2\sqrt{101}, \sqrt{2}, \sqrt{73}, \sqrt{145}, \sqrt{101} - 2\sqrt{2}$
27. $20\sqrt{3}\,\mathbf{i} + 20\mathbf{j}$ **29.** $-\dfrac{\sqrt{2}}{2}\mathbf{i} - \dfrac{\sqrt{2}}{2}\mathbf{j}$
31. $4\cos 10°\mathbf{i} + 4\sin 10°\mathbf{j} \approx 3.94\mathbf{i} + 0.69\mathbf{j}$
33. $15\sqrt{3}, -15$ **35.** $5, 53.13°$ **37.** $13, 157.38°$
39. $2, 60°$ **41.** $2\mathbf{i} - 3\mathbf{j}$ **43.** (a) $40\mathbf{j}$ (b) $425\mathbf{i}$
(c) $425\mathbf{i} + 40\mathbf{j}$ (d) 427 mi/h, N $84.6°$ E
45. 794 mi/h, N $26.6°$ W   **47.** (a) $10\mathbf{i}$ (b) $10\mathbf{i} + 17.32\mathbf{j}$
(c) $20\mathbf{i} + 17.32\mathbf{j}$ (d) 26.5 mi/h, N $49.1°$ E
49. (a) $22.8\mathbf{i} + 7.4\mathbf{j}$ (b) 7.4 mi/h, 22.8 mi/h
51. (a) $\langle 5, -3 \rangle$ (b) $\langle -5, 3 \rangle$ **53.** (a) $-4\mathbf{j}$ (b) $4\mathbf{j}$
55. (a) $\langle -7.57, 10.61 \rangle$ (b) $\langle 7.57, -10.61 \rangle$
57. $\mathbf{T}_1 \approx -56.5\mathbf{i} + 67.4\mathbf{j}, \mathbf{T}_2 \approx 56.5\mathbf{i} + 32.6\mathbf{j}$

Section 7.8 ■ page 609

1. (a) 2 (b) 45° **3.** (a) 13 (b) 56° **5.** (a) −1
(b) 97°   **7.** (a) $5\sqrt{3}$ (b) 30° **9.** Yes **11.** No
13. Yes **15.** 9 **17.** −5 **19.** $-\frac{12}{5}$ **21.** −24
23. (a) $\langle 1, 1 \rangle$ (b) $\mathbf{u}_1 = \langle 1, 1 \rangle$, $\mathbf{u}_2 = \langle -3, 3 \rangle$
25. (a) $\langle -\frac{1}{2}, \frac{3}{2} \rangle$ (b) $\mathbf{u}_1 = \langle -\frac{1}{2}, \frac{3}{2} \rangle$, $\mathbf{u}_2 = \langle \frac{3}{2}, \frac{1}{2} \rangle$
27. (a) $\langle -\frac{18}{5}, \frac{24}{5} \rangle$ (b) $\mathbf{u}_1 = \langle -\frac{18}{5}, \frac{24}{5} \rangle$, $\mathbf{u}_2 = \langle \frac{28}{5}, \frac{21}{5} \rangle$
29. −28 **31.** 25 **33.** 16 ft-lb **35.** 8660 ft-lb
37. 1164 lb **39.** 23.6°

Chapter 7 Review ■ page 612

1. LHS $= \sin\theta\left(\dfrac{\cos\theta}{\sin\theta} + \dfrac{\sin\theta}{\cos\theta}\right) = \cos\theta + \dfrac{\sin^2\theta}{\cos\theta}$
$\qquad = \dfrac{\cos^2\theta + \sin^2\theta}{\cos\theta} = \text{RHS}$

3. LHS $= (1 - \sin^2 x)\csc x - \csc x = \csc x - \sin^2 x\csc x - \csc x$
$\qquad = -\sin^2 x\,\dfrac{1}{\sin x} = \text{RHS}$

5. LHS $= \dfrac{\cos^2 x}{\sin^2 x} - \dfrac{\tan^2 x}{\sin^2 x} = \cot^2 x - \dfrac{1}{\cos^2 x} = \text{RHS}$

7. LHS $= \dfrac{\cos x}{\frac{1}{\cos x}(1 - \sin x)} = \dfrac{\cos x}{\frac{1}{\cos x} - \frac{\sin x}{\cos x}} = \text{RHS}$

9. LHS $= \sin^2 x\,\dfrac{\cos^2 x}{\sin^2 x} + \cos^2 x\,\dfrac{\sin^2 x}{\cos^2 x} = \cos^2 x + \sin^2 x = \text{RHS}$

11. LHS $= \dfrac{2\sin x\cos x}{1 + 2\cos^2 x - 1} = \dfrac{2\sin x\cos x}{2\cos^2 x} = \dfrac{2\sin x}{2\cos x} = \text{RHS}$

13. LHS $= \dfrac{1 - \cos x}{\sin x} = \dfrac{1}{\sin x} - \dfrac{\cos x}{\sin x} = \text{RHS}$

15. LHS $= \frac{1}{2}[\cos((x+y) - (x-y)) - \cos((x+y) + (x-y))]$
$\qquad = \frac{1}{2}(\cos 2y - \cos 2x) = \frac{1}{2}[1 - 2\sin^2 y - (1 - 2\sin^2 x)]$
$\qquad = \frac{1}{2}(2\sin^2 x - 2\sin^2 y) = \text{RHS}$

17. LHS $= 1 + \dfrac{\sin x}{\cos x} \cdot \dfrac{1 - \cos x}{\sin x} = 1 + \dfrac{1 - \cos x}{\cos x}$
$\qquad = 1 + \dfrac{1}{\cos x} - 1 = \text{RHS}$

19. LHS $= \cos^2\frac{x}{2} - 2\sin\frac{x}{2}\cos\frac{x}{2} + \sin^2\frac{x}{2} = 1 - \sin\!\left(2 \cdot \frac{x}{2}\right) = \text{RHS}$

21. LHS $= \dfrac{2\sin x\cos x}{\sin x} - \dfrac{2\cos^2 x - 1}{\cos x}$
$\qquad = 2\cos x - 2\cos x + \dfrac{1}{\cos x} = \text{RHS}$

23. LHS $= \dfrac{\tan x + \tan\frac{\pi}{4}}{1 - \tan x\tan\frac{\pi}{4}} = \text{RHS}$

25. (a)   (b) Yes

27. (a)   (b) No

29. (a) $2\sin^2 3x + \cos 6x = 1$

31. $0, \pi$ **33.** $\dfrac{\pi}{6}, \dfrac{5\pi}{6}$ **35.** $\dfrac{\pi}{3}, \dfrac{5\pi}{3}$ **37.** $\dfrac{2\pi}{3}, \dfrac{4\pi}{3}$
39. $\dfrac{\pi}{3}, \dfrac{2\pi}{3}, \dfrac{3\pi}{4}, \dfrac{4\pi}{3}, \dfrac{5\pi}{3}, \dfrac{7\pi}{4}$
41. $\dfrac{\pi}{6}, \dfrac{\pi}{2}, \dfrac{5\pi}{6}, \dfrac{7\pi}{6}, \dfrac{3\pi}{2}, \dfrac{11\pi}{6}$ **43.** $\dfrac{\pi}{6}$ **45.** 1.18
49. $\frac{1}{2}\sqrt{2 + \sqrt{3}}$ **51.** $\sqrt{2} - 1$ **53.** $\sqrt{2}/2$
55. $\sqrt{2}/2$ **57.** $\dfrac{\sqrt{2} + \sqrt{3}}{4}$ **59.** $2\dfrac{\sqrt{10} + 1}{9}$
61. $\frac{2}{3}\left(\sqrt{2} + \sqrt{5}\right)$ **63.** $\sqrt{(3 + 2\sqrt{2})/6}$ **65.** $\pi/3$
67. $\frac{1}{2}$ **69.** $2/\sqrt{21}$ **71.** $\frac{7}{9}$ **73.** $x/\sqrt{1 + x^2}$
75. $\theta = \cos^{-1}\dfrac{x}{3}$ **77.** (a) $\theta = \tan^{-1}\!\left(\dfrac{10}{x}\right)$ (b) none

79. (a)

(b) $4\sqrt{2}, \dfrac{\pi}{4}$ (c) $4\sqrt{2}\left(\cos\dfrac{\pi}{4} + i\sin\dfrac{\pi}{4}\right)$

81. (a)

(b) $\sqrt{34}$, $\tan^{-1}\left(\frac{3}{5}\right)$ **(c)** $\sqrt{34}\left[\cos\left(\tan^{-1}\frac{3}{5}\right) + i\sin\left(\tan^{-1}\frac{3}{5}\right)\right]$

83. (a)

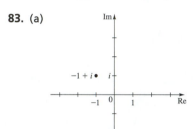

(b) $\sqrt{2}, \frac{3\pi}{4}$ **(c)** $\sqrt{2}\left(\cos\frac{3\pi}{4} + i\sin\frac{3\pi}{4}\right)$

85. $8(-1 + i\sqrt{3})$ **87.** $-\frac{1}{32}(1 + i\sqrt{3})$

89. $\pm 2\sqrt{2}(1 - i)$ **91.** $\pm 1, \pm\frac{1}{2} \pm \frac{\sqrt{3}}{2}i$

93. $\sqrt{13}, \langle 6, 4\rangle, \langle -10, 2\rangle, \langle -4, 6\rangle, \langle -22, 7\rangle$

95. $3\mathbf{i} - 4\mathbf{j}$ **97.** $(10, -2)$ **99. (a)** $(4.8\mathbf{i} + 0.4\mathbf{j}) \times 10^4$

(b) 4.8×10^4 lb, N 85.2° E   **101.** 5, 25, 60

103. $2\sqrt{2}, 8, 0$ **105.** Yes   **107.** No, 45°

109. (a) $17\sqrt{37}/37$ **(b)** $\langle \frac{102}{37}, -\frac{17}{37}\rangle$ **(c)** $\mathbf{u}_1 = \langle \frac{102}{37}, -\frac{17}{37}\rangle$,

$\mathbf{u}_2 = \langle \frac{9}{37}, \frac{54}{37}\rangle$ **111.** -6

Chapter 7 Test ■ page 615

1. (a) LHS $= \dfrac{\sin\theta}{\cos\theta}\sin\theta + \cos\theta = \dfrac{\sin^2\theta + \cos^2\theta}{\cos\theta} =$ RHS

(b) LHS $= \dfrac{\tan x}{1 - \cos x} \cdot \dfrac{1 + \cos x}{1 + \cos x} = \dfrac{\tan x(1 + \cos x)}{1 - \cos^2 x}$

$= \dfrac{\frac{\sin x}{\cos x}(1 + \cos x)}{\sin^2 x} = \dfrac{1}{\sin x} \cdot \dfrac{1 + \cos x}{\cos x} =$ RHS

(c) LHS $= \dfrac{2\tan x}{\sec^2 x} = \dfrac{2\sin x}{\cos x} \cdot \cos^2 x = 2\sin x\cos x =$ RHS

2. $\tan\theta$ **3. (a)** $\frac{1}{2}$ **(b)** $\dfrac{\sqrt{2} + \sqrt{6}}{4}$ **(c)** $\frac{1}{2}\sqrt{2 - \sqrt{3}}$

4. $(10 - 2\sqrt{5})/15$

5. (a) $\frac{1}{2}(\sin 8x - \sin 2x)$ **(b)** $-2\cos\frac{7}{2}x\sin\frac{3}{2}x$ **6.** -2

7.

Domain $\mathbb{R}$ Domain $[-1, 1]$

8. (a) $\theta = \tan^{-1}\dfrac{x}{4}$ **(b)** $\theta\cos^{-1}\dfrac{3}{x}$ **9. (a)** $\dfrac{2\pi}{3}, \dfrac{4\pi}{3}$

(b) $\dfrac{\pi}{6}, \dfrac{\pi}{2}, \dfrac{5\pi}{6}, \dfrac{3\pi}{2}$ **10.** 0.57964, 2.56195, 3.72123, 5.70355

11. $\frac{40}{41}$

12. (a)

(b) $2\left(\cos\dfrac{\pi}{3} + i\sin\dfrac{\pi}{3}\right)$ **(c)** -512

13. $-8, \sqrt{3} + i$

14. $-3i, 3\left(\pm\dfrac{\sqrt{3}}{2} + \dfrac{1}{2}i\right)$

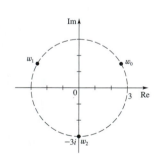

15. (a) $-6\mathbf{i} + 10\mathbf{j}$ **(b)** $2\sqrt{34}$
16. (a) $\langle 19, -3\rangle$ **(b)** $5\sqrt{2}$ **(c)** 0   **(d)** Yes
17. (a) $14\mathbf{i} + 6\sqrt{3}\mathbf{j}$ **(b)** 17.4 mi/h, N 53.4° E
18. (a) 45°   **(b)** $\sqrt{26}/2$ **(c)** $\dfrac{5}{2}\mathbf{i} - \dfrac{1}{2}\mathbf{j}$ **19.** 90

Focus on Problem Solving ■ page 618

1. 636,619   **3.** 16 cm   **7.** $\sqrt{1 + a}$ **9. (b)** 12 cm

(c) A flat disk   **11.** $\dfrac{\pi}{2}, \dfrac{5\pi}{4}, \dfrac{7\pi}{4}$ **13.** 5

CHAPTER 8

Section 8.1 ■ page 628

1. $(3, 1)$ **3.** $(4, 16), (-3, 9)$ **5.** $(2, -2), (-2, 2)$
7. $(-25, 5), (-25, -5)$ **9.** $(1, 2)$ **11.** $(-3, 4), (3, 4)$
13. $(-2, -1), (-2, 1), (2, -1), (2, 1)$
15. $(-1, \sqrt{2}), (-1, -\sqrt{2}), \left(\frac{1}{2}, \sqrt{\frac{7}{2}}\right), \left(\frac{1}{2}, -\sqrt{\frac{7}{2}}\right)$ **17.** $(4, 0)$
19. $(-2, -2)$ **21.** $(6, 2), (-2, -6)$ **23.** No solution
25. $(\sqrt{5}, 2), (\sqrt{5}, -2), (-\sqrt{5}, 2), (-\sqrt{5}, -2)$
27. $\left(3, -\frac{1}{2}\right), \left(-3, -\frac{1}{2}\right)$ **29.** $\left(\frac{1}{5}, \frac{1}{3}\right)$ **31.** $(-0.33, 5.33)$
33. $(2.00, 20.00), (-8.00, 0)$ **35.** $(-4.51, 2.17), (4.91, -0.97)$
37. $(1.23, 3.87), (-0.35, -4.21)$ **39.** $(-2.30, -0.70)$,
$(0.48, -1.19)$ **41.** 12 cm by 15 cm **43.** 15, 20
45. $(400.50, 200.25)$, 447.77 m **47.** $y = 3x + 5$
49. $\left(-\frac{1}{2}, -\frac{3}{2}\right), \left(\frac{1}{2}, \frac{3}{2}\right)$ **51.** $(\sqrt{10}, 10)$

Section 8.2 ■ page 635

1. $(2, 2)$

3. No solution

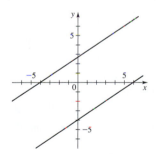

5. Infinitely many solutions

7. $(2, 2)$ **9.** $(3, -1)$ **11.** $(2, 1)$ **13.** $(3, 5)$ **15.** $(1, 3)$

17. $(10, -9)$ **19.** $(2, 1)$ **21.** No solution **23.** $\left(x, \frac{1}{3}x - \frac{5}{3}\right)$
25. $\left(x, 3 - \frac{3}{2}x\right)$ **27.** $(-3, -7)$ **29.** $\left(x, 5 - \frac{5}{6}x\right)$
31. $\left(-\dfrac{1}{a - 1}, \dfrac{1}{a - 1}\right)$ **33.** $\left(\dfrac{1}{a + b}, \dfrac{1}{a + b}\right)$ **35.** 22, 12
37. 5 dimes, 9 quarters
39. Plane's speed 120 mi/h, wind speed 30 mi/h
41. Run 5 mi/h, cycle 20 mi/h
43. 200 g of A, 40 g of B **45.** 25%, 10%
47. \$16,000 at 10%, \$32,000 at 6% **49.** 25
51. $(3.87, 2.74)$ **53.** $(61.00, 20.00)$

Section 8.3 ■ page 646

1. Linear **3.** Nonlinear **5.** $(1, 3, 2)$ **7.** $(4, 0, 3)$
9. $\left(5, 2, -\frac{1}{2}\right)$
11. $\begin{cases} x - 2y - z = 4 \\ -y - 4z = 4 \\ 2x + y + z = 0 \end{cases}$ **13.** $\begin{cases} 2x - y + 3z = 2 \\ x + 2y - z = 4 \\ 3y + 7z = 14 \end{cases}$
15. $(1, 2, 1)$ **17.** $(5, 0, 1)$ **19.** $(0, 1, 2)$
21. $(1 - 3t, 2t, t)$ **23.** No solution **25.** No solution
27. $(3 - t, -3 + 2t, t)$ **29.** $\left(2 - 2t, -\frac{2}{3} + \frac{4}{3}t, t\right)$
31. $(1, -1, 1, 2)$ **33.** $y = 2x^2 + 4x + 3$
35. $y = \frac{10}{3}x^2 + 5x - \frac{25}{3}$ **37.** \$30,000 in short-term bonds,
\$30,000 in intermediate-term bonds, \$40,000 in long-term
bonds **39.** Impossible

Section 8.4 ■ page 659

1. (a) Yes (b) Yes (c) $\begin{cases} x = -3 \\ y = 5 \end{cases}$
3. (a) Yes (b) No (c) $\begin{cases} x + 2y + 8z = 0 \\ y + 3z = 0 \end{cases}$
5. (a) No (b) No (c) $\begin{cases} x = 0 \\ 0 = 0 \\ y + 5z = 1 \end{cases}$
7. (a) Yes (b) Yes (c) $\begin{cases} x + 3y - w = 0 \\ z + 2w = 0 \\ 0 = 1 \\ 0 = 0 \end{cases}$
9. $(1, 1, 2)$ **11.** $(1, 0, 1)$ **13.** $(-1, 0, 1)$
15. $(-1, 5, 0)$ **17.** $(10, 3, -2)$ **19.** No solution
21. $(2 - 3t, 3 - 5t, t)$ **23.** No solution
25. $(-2t + 5, t - 2, t)$ **27.** $x = -\frac{1}{2}s + t + 6, y = s, z = t$
29. $(-2, 1, 3)$ **31.** $(-9, 2, 0)$ **33.** $(0, -3, 0, -3)$
35. $(-1, 0, 0, 1)$ **37.** $\left(\frac{7}{4} - \frac{7}{4}t, -\frac{7}{4} + \frac{3}{4}t, \frac{9}{4} + \frac{3}{4}t, t\right)$
39. $x = \frac{1}{3}s - \frac{2}{3}t, y = \frac{1}{3}s + \frac{1}{3}t, z = s, w = t$

41. 2 VitaMax, 1 Vitron, 2 VitaPlus

43. 5-mile run, 2-mile swim, 30-mile cycle

45. Impossible

Section 8.5 ■ page 669

1. No   **3.** $\begin{bmatrix} 1 & 3 \\ 1 & 5 \end{bmatrix}$ **5.** $\begin{bmatrix} 3 & 6 \\ 12 & -3 \\ 3 & 0 \end{bmatrix}$

7. Impossible   **9.** $\begin{bmatrix} 5 & 2 & 1 \\ 7 & 10 & -7 \end{bmatrix}$

11. $\begin{bmatrix} 5 & -2 & 5 \\ 1 & 1 & 0 \end{bmatrix}$ **13.** $\begin{bmatrix} -1 & -3 & -5 \\ -1 & 3 & -6 \end{bmatrix}$ **15.** $\begin{bmatrix} 13 & -\frac{7}{2} & 15 \\ 3 & 1 & 3 \end{bmatrix}$

17. $\begin{bmatrix} -14 & -8 & -30 \\ -6 & 10 & -24 \end{bmatrix}$ **19.** Impossible   **21.** $\begin{bmatrix} 3 & \frac{1}{2} & 5 \\ 1 & -1 & 3 \end{bmatrix}$

23. $\begin{bmatrix} 28 & 21 & 28 \end{bmatrix}$ **25.** $\begin{bmatrix} -1 \\ 8 \\ -1 \end{bmatrix}$ **27.** $\begin{bmatrix} 8 & -335 \\ 0 & 343 \end{bmatrix}$

29. Impossible   **31.** Impossible

33. $\begin{bmatrix} 2 & -5 \\ 3 & 2 \end{bmatrix}\begin{bmatrix} x \\ y \end{bmatrix} = \begin{bmatrix} 7 \\ 4 \end{bmatrix}$

35. $\begin{bmatrix} 3 & 2 & -1 & 1 \\ 1 & 0 & -1 & 0 \\ 0 & 3 & 1 & -1 \end{bmatrix}\begin{bmatrix} x_1 \\ x_2 \\ x_3 \\ x_4 \end{bmatrix} = \begin{bmatrix} 0 \\ 5 \\ 4 \end{bmatrix}$ **37.** $x = 2$, $y = -1$

39. $x = 2$, $y = 1$ **41.** $\begin{bmatrix} 3 & \frac{11}{2} \\ 2 & 5 \end{bmatrix}$ **43.** Impossible

45. (a) $\begin{bmatrix} 4,690 & 1,690 & 13,210 \end{bmatrix}$

(b) Total revenue in Santa Monica, Long Beach, and Anaheim, respectively

47. (a) $\begin{bmatrix} 105,000 & 58,000 \end{bmatrix}$ (b) The first entry is the total amount (in ounces) of tomato sauce produced, and the second entry is the total amount (in ounces) of tomato paste produced.

49. Only ACB is defined. $\begin{bmatrix} -3 & -21 & 27 & -6 \\ -2 & -14 & 18 & -4 \end{bmatrix}$ **51.** No

Section 8.6 ■ page 682

5. $\begin{bmatrix} 1 & -2 \\ -\frac{3}{2} & \frac{7}{2} \end{bmatrix}$ **7.** $\begin{bmatrix} 2 & -3 \\ -3 & 5 \end{bmatrix}$ **9.** $\begin{bmatrix} 13 & 5 \\ -5 & -2 \end{bmatrix}$

11. No inverse   **13.** $\begin{bmatrix} 1 & 2 \\ -\frac{1}{2} & \frac{2}{3} \end{bmatrix}$ **15.** $\begin{bmatrix} -4 & -4 & 5 \\ 1 & 1 & -1 \\ 5 & 4 & -6 \end{bmatrix}$

17. No inverse   **19.** $\begin{bmatrix} -\frac{9}{2} & -1 & 4 \\ 3 & 1 & -3 \\ \frac{7}{2} & 1 & -3 \end{bmatrix}$

21. $\begin{bmatrix} 0 & 0 & -2 & 1 \\ -1 & 0 & 1 & 1 \\ 0 & 1 & -1 & 0 \\ 1 & 0 & 0 & -1 \end{bmatrix}$ **23.** $x = 8$, $y = -12$

25. $x = 126$, $y = -50$ **27.** $x = -38$, $y = 9$, $z = 47$

29. $x = -20$, $y = 10$, $z = 16$ **31.** $x = 3$, $y = 2$, $z = 1$

33. $x = 3$, $y = -2$, $z = 2$ **35.** $x = 8$, $y = 1$, $z = 0$, $w = 3$

37. $\begin{bmatrix} 7 & 2 & 3 \\ 10 & 3 & 5 \end{bmatrix}$

39. (a) $\begin{bmatrix} 0 & 1 & -1 \\ -2 & \frac{3}{2} & 0 \\ 1 & -\frac{3}{2} & 1 \end{bmatrix}$ (b) 1 oz A, 1 oz B, 2 oz C

(c) 2 oz A, 0 oz B, 1 oz C   (d) No

41. (a) $\begin{cases} x + y + 2z = 675 \\ 2x + y + z = 600 \\ x + 2y + z = 625 \end{cases}$

(b) $\begin{bmatrix} 1 & 1 & 2 \\ 2 & 1 & 1 \\ 1 & 2 & 1 \end{bmatrix}\begin{bmatrix} x \\ y \\ z \end{bmatrix} = \begin{bmatrix} 675 \\ 600 \\ 625 \end{bmatrix}$

(c) $A^{-1} = \begin{bmatrix} -\frac{1}{4} & \frac{3}{4} & -\frac{1}{4} \\ -\frac{1}{4} & -\frac{1}{4} & \frac{3}{4} \\ \frac{3}{4} & -\frac{1}{4} & -\frac{1}{4} \end{bmatrix}$

He earns \$125 on a standard set, \$150 on a deluxe set, and \$200 on a leather-bound set.

43. $\frac{1}{2}\begin{bmatrix} e^{-x} & e^{-2x} \\ -e^{-2x} & e^{-3x} \end{bmatrix}$ **45.** $\begin{bmatrix} \cos\theta & -\sin\theta \\ \sin\theta & \cos\theta \end{bmatrix}$

Section 8.7 ■ page 697

1. 6   **3.** -4 **5.** Does not exist   **7.** $\frac{1}{8}$ **9.** 20, 20

11. $-12, 12$ **13.** 0, 0   **15.** 4, has an inverse   **17.** -6, has an inverse   **19.** 5000, has an inverse   **21.** -4, has an inverse

23. -18 **25.** 120   **27.** (a) -2 (b) -2 (c) Yes

29. $(-2, 5)$ **31.** $(0.6, -0.4)$ **33.** $(4, -1)$

35. $(4, 2, -1)$ **37.** $(1, 3, 2)$ **39.** $(0, -1, 1)$

41. $\left(\frac{189}{29}, -\frac{108}{29}, \frac{88}{29}\right)$ **43.** $\left(\frac{1}{2}, \frac{1}{4}, \frac{1}{4}, -1\right)$ **45.** $abcde$

47. 0, 1, 2   **49.** 1, -1 **51.** 21   **53.** $\frac{63}{2}$

Section 8.8 ■ page 703

1.

3.

17.

bounded

19.

bounded

5.

7.

21.

bounded

23.

bounded

9.

11.

25.

not bounded

27.

bounded

13.

not bounded

15.

not bounded

29.

bounded

31.

bounded

33.

bounded

35.

$(-1, 8)$ $(11, 8)$

37.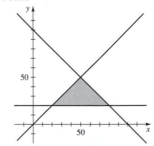

$(0.6, 3.4)$

$(6.4, -2.4)$

39. x = number of fiction books
y = number of nonfiction books

$$\begin{cases} x + y \le 100 \\ 20 \le y, x \ge y \\ x \ge 0, y \ge 0 \end{cases}$$

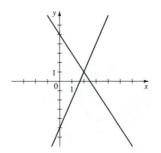

Section 8.9 ■ page 709

1. $\dfrac{A}{x-1} + \dfrac{B}{x+2}$ **3.** $\dfrac{A}{x-2} + \dfrac{B}{(x-2)^2} + \dfrac{C}{x+4}$

5. $\dfrac{A}{x-3} + \dfrac{Bx+C}{x^2+4}$ **7.** $\dfrac{Ax+B}{x^2+1} + \dfrac{Cx+D}{x^2+2}$

9. $\dfrac{A}{x} + \dfrac{B}{2x-5} + \dfrac{C}{(2x-5)^2} + \dfrac{D}{(2x-5)^3}$

$+ \dfrac{Ex+F}{x^2+2x+5} + \dfrac{Gx+H}{(x^2+2x+5)^2}$

11. $\dfrac{1}{x-1} - \dfrac{1}{x+1}$ **13.** $\dfrac{1}{x-1} - \dfrac{1}{x+4}$

15. $\dfrac{2}{x-3} - \dfrac{2}{x+3}$ **17.** $\dfrac{1}{x-2} - \dfrac{1}{x+2}$

19. $\dfrac{3}{x-4} - \dfrac{2}{x+2}$

21. $\dfrac{-\frac{1}{2}}{2x-1} + \dfrac{\frac{3}{2}}{4x-3}$ **23.** $\dfrac{2}{x-2} + \dfrac{3}{x+2} - \dfrac{1}{2x-1}$

25. $\dfrac{2}{x+1} - \dfrac{1}{x} + \dfrac{1}{x^2}$ **27.** $\dfrac{1}{2x+3} - \dfrac{3}{(2x+3)^2}$

29. $\dfrac{2}{x} - \dfrac{1}{x^3} - \dfrac{2}{x+2}$

31. $\dfrac{4}{x+2} - \dfrac{4}{x-1} + \dfrac{2}{(x-1)^2} + \dfrac{1}{(x-1)^3}$

33. $\dfrac{3}{x+2} - \dfrac{1}{(x+2)^2} - \dfrac{1}{(x+3)^2}$

35. $\dfrac{x+1}{x^2+3} - \dfrac{1}{x}$ **37.** $\dfrac{2x-5}{x^2+x+2} + \dfrac{5}{x^2+1}$

39. $\dfrac{1}{x^2+1} - \dfrac{x+2}{(x^2+1)^2} + \dfrac{1}{x}$

41. $x^2 + \dfrac{3}{x-2} - \dfrac{x+1}{x^2+1}$ **43.** $A = \dfrac{a+b}{2}, B = \dfrac{a-b}{2}$

Chapter 8 Review ■ page 711

1. $(2, 1)$ **3.** x = any number
$y = \frac{2}{7}x - 4$

5. No solution

7. $(-3, 3), (2, 8)$

9. $\left(\frac{16}{7}, -\frac{14}{3}\right)$

11. $(21.41, -15.93)$ **13.** $(11.94, -1.39), (12.07, 1.44)$

15. $(1, 1, 2)$ **17.** No solution

19. $x = -4t + 1,$
$y = -t - 1,$
$z = t$

21. $x = 6 - 5t,$
$y = \frac{1}{2}(7 - 3t),$
$z = t$

23. \$3000 at 6%, \$6000 at 7%   **25.** Impossible

27. $\begin{bmatrix} 4 & 18 \\ 4 & 0 \\ 2 & 2 \end{bmatrix}$ **29.** $\begin{bmatrix} 10 & 0 & -5 \end{bmatrix}$ **31.** $\begin{bmatrix} -\frac{7}{2} & 10 \\ 1 & -\frac{9}{2} \end{bmatrix}$

33. $\begin{bmatrix} 30 & 22 & 2 \\ -9 & 1 & -4 \end{bmatrix}$ **35.** $\begin{bmatrix} -\frac{1}{2} & \frac{11}{2} \\ \frac{15}{4} & -\frac{3}{2} \\ -\frac{1}{2} & 1 \end{bmatrix}$ **37.** $1, \begin{bmatrix} 9 & -4 \\ -2 & 1 \end{bmatrix}$

39. 0, no inverse   **41.** $-1, \begin{bmatrix} 3 & 2 & -3 \\ 2 & 1 & -2 \\ -8 & -6 & 9 \end{bmatrix}$ **43.** $(65, 154)$

45. $\left(\frac{1}{5}, \frac{9}{5}\right)$ **47.** $\left(-\frac{87}{26}, \frac{21}{26}, \frac{3}{2}\right)$

49.

bounded

51.

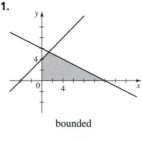

bounded

53. $x = \dfrac{b + c}{2}, y = \dfrac{a + c}{2}, z = \dfrac{a + b}{2}$ **55.** $2, 3$

57. $\dfrac{2}{x - 5} + \dfrac{1}{x + 3}$ **59.** $\dfrac{-4}{x} + \dfrac{4}{x - 1} + \dfrac{-2}{(x - 1)^2}$

Chapter 8 Test ■ page 713

1. Wind speed 60 km/h, airplane 300 km/h

2. $(3, -1)$; linear, neither inconsistent nor dependent

3. No solution; linear, inconsistent

4. $x = \frac{1}{7}(t + 1), y = \frac{1}{7}(9t + 2), z = t$;
linear, dependent

5. $(\pm 1, -2), \left(\pm\frac{5}{3}, -\frac{2}{3}\right)$; nonlinear

6. Peaches cost \$0.50, cherries cost \$2, pears cost \$1

7. Incompatible dimensions   **8.** Incompatible dimensions

9. $\begin{bmatrix} 6 & 10 \\ 3 & -2 \\ -3 & 9 \end{bmatrix}$ **10.** $\begin{bmatrix} 36 & 58 \\ 0 & -3 \\ 18 & 28 \end{bmatrix}$ **11.** $\begin{bmatrix} 2 & -\frac{3}{2} \\ -1 & 1 \end{bmatrix}$

12. B is not square   **13.** B is not square   **14.** -3

15. $(70, 90)$ **16.** $(5, -5, -4)$

17. $|A| = 0, |B| = 2, B^{-1} = \begin{bmatrix} 1 & -2 & 0 \\ 0 & \frac{1}{2} & 0 \\ 3 & -6 & 1 \end{bmatrix}$

18. (a)   (b)

19. $\dfrac{1}{x - 1} + \dfrac{1}{(x - 1)^2} - \dfrac{1}{x + 2}$ **20.** $(-0.49, 3.93), (2.34, 2.24)$

Focus on Modeling ■ page 719

1. 198, 195   **3.**

maximum 161
minimum 135

5. 3 tables, 34 chairs   **7.** 30 grapefruit crates, 30 orange crates

9. 15 Pasadena to Santa Monica, 3 Pasadena to El Toro,
0 Long Beach to Santa Monica, 16 Long Beach to El Toro

11. 90 standard, 40 deluxe   **13.** \$7500 in municipal bonds,
\$2500 in bank certificates, \$2000 in high-risk bonds

15. 4 games, 32 educational, 0 utility

CHAPTER 9

Section 9.1 ■ page 730

1. III **3.** II **5.** VI

Order of answers: focus; directrix; focal diameter

7. $F(1, 0)$; $x = -1$; 4

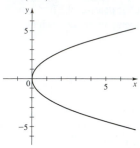

9. $F\left(0, \frac{9}{4}\right)$; $y = -\frac{9}{4}$; 9

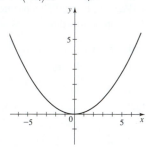

11. $F\left(0, \frac{1}{20}\right)$; $y = -\frac{1}{20}$; $\frac{1}{5}$

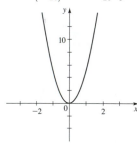

13. $F\left(-\frac{1}{32}, 0\right)$; $x = \frac{1}{32}$; $\frac{1}{8}$

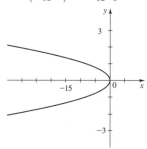

15. $F\left(0, -\frac{3}{2}\right)$; $y = \frac{3}{2}$; 6

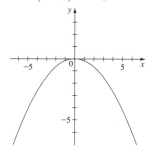

17. $F\left(-\frac{5}{12}, 0\right)$; $x = \frac{5}{12}$; $\frac{5}{3}$

19.

21.

23.

25. $x^2 = 8y$ **27.** $y^2 = -32x$ **29.** $y^2 = -8x$
31. $x^2 = 40y$ **33.** $y^2 = 4x$ **35.** $x^2 = 20y$
37. $x^2 = 8y$ **39.** $y^2 = -16x$ **41.** $y^2 = -3x$
43. $x = y^2$ **45.** $x^2 = -4\sqrt{2}\,y$
47. (a) $y^2 = 12x$ (b) $8\sqrt{15} \approx 31$ cm **49.** $x^2 = 600y$
51. (a) $x^2 = -4py$, $p = \frac{1}{2}$, 1, 4, and 8
(b) The closer the directrix to the vertex, the steeper the parabola.

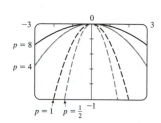

Section 9.2 ■ page 741

1. II **3.** I

Order of answers: vertices; foci; eccentricity; major axis and minor axis

5. $V(\pm 5, 0)$; $F(\pm 4, 0)$; $\frac{4}{5}$; 10, 6

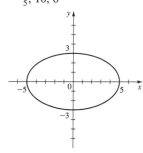

7. $V(0, \pm 3)$; $F(0, \pm\sqrt{5})$; $\sqrt{5}/3$; 6, 4

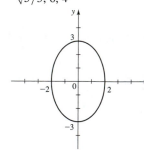

9. $V(\pm 4, 0)$; $F(\pm 2\sqrt{3}, 0)$; $\sqrt{3}/2$; 8, 4

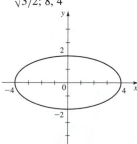

11. $V(0, \pm\sqrt{3})$; $F(0, \pm\sqrt{3/2})$; $1/\sqrt{2}$; $2\sqrt{3}$, $\sqrt{6}$

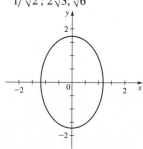

13. $V(\pm 1, 0)$; $F(\pm\sqrt{3}/2, 0)$; $\sqrt{3}/2$; 2, 1

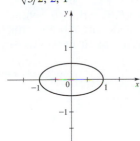

15. $V(0, \pm\sqrt{2})$; $F(0, \pm\sqrt{3}/2)$; $\sqrt{3}/2$; $2\sqrt{2}$, $\sqrt{2}$

43. $\dfrac{x^2}{2.2500 \times 10^{16}} + \dfrac{y^2}{2.2491 \times 10^{16}} = 1$

45. $\dfrac{x^2}{1{,}455{,}642} + \dfrac{y^2}{1{,}451{,}610} = 1$ **47.** $5\sqrt{39}/2 \approx 15.6$ in.

51. (a)

(b) Common major axes and vertices; eccentricity increases as k increases

17. $V(0, \pm 1)$; $F(0, \pm 1/\sqrt{2})$; $1/\sqrt{2}$; 2, $\sqrt{2}$

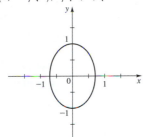

Section 9.3 ■ page 751

1. III   **3.** II

Order of answers: vertices; foci; asymptotes

5. $V(\pm 2, 0)$; $F(\pm 2\sqrt{5}, 0)$; $y = \pm 2x$

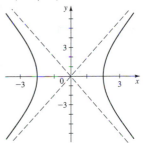

19. $\dfrac{x^2}{25} + \dfrac{y^2}{16} = 1$ **21.** $\dfrac{x^2}{4} + \dfrac{y^2}{8} = 1$ **23.** $\dfrac{x^2}{256} + \dfrac{y^2}{48} = 1$

25.

27.

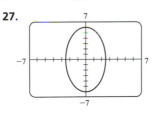

7. $V(0, \pm 1)$; $F(0, \pm\sqrt{26})$; $y = \pm\frac{1}{5}x$

29. $\dfrac{x^2}{25} + \dfrac{y^2}{9} = 1$ **31.** $x^2 + \dfrac{y^2}{4} = 1$ **33.** $\dfrac{x^2}{9} + \dfrac{y^2}{13} = 1$

35. $\dfrac{x^2}{100} + \dfrac{y^2}{91} = 1$ **37.** $\dfrac{x^2}{25} + \dfrac{y^2}{5} = 1$

39. $\dfrac{64x^2}{225} + \dfrac{64y^2}{81} = 1$

41. $(0, \pm 2)$

9. $V(\pm 1, 0)$; $F(\pm\sqrt{2}, 0)$; $y = \pm x$

11. $V(0, \pm 3)$; $F(0, \pm\sqrt{34})$; $y = \pm\frac{3}{5}x$

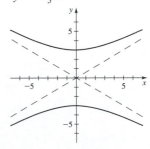

13. $V(\pm 2\sqrt{2}, 0)$;
$F(\pm\sqrt{10}, 0)$;
$y = \pm\frac{1}{2}x$

15. $V\left(0, \pm\frac{1}{2}\right)$;
$F\left(0, \pm\sqrt{5}/2\right)$;
$y = \pm\frac{1}{2}x$

17. $\dfrac{x^2}{4} - \dfrac{y^2}{12} = 1$ **19.** $\dfrac{y^2}{16} - \dfrac{x^2}{16} = 1$ **21.** $\dfrac{x^2}{9} - \dfrac{4y^2}{9} = 1$

23.

25.

27. $\dfrac{x^2}{9} - \dfrac{y^2}{16} = 1$ **29.** $y^2 - \dfrac{x^2}{3} = 1$

31. $x^2 - \dfrac{y^2}{25} = 1$ **33.** $\dfrac{5y^2}{64} - \dfrac{5x^2}{256} = 1$

35. $\dfrac{x^2}{16} - \dfrac{y^2}{16} = 1$ **37.** $\dfrac{x^2}{9} - \dfrac{y^2}{16} = 1$

39. (a) $x^2 - y^2 = c^2/2$

43. (a) 490 mi   (b) $\dfrac{y^2}{60{,}025} - \dfrac{x^2}{2475} = 1$ (c) 10.1 mi

45. (b)

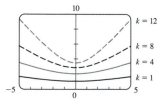

Section 9.4 ■ **page 760**

1. Center $C(2, 1)$;
foci $F(2 \pm \sqrt{5}, 1)$;
vertices $V_1(-1, 1)$,
$V_2(5, 1)$; major axis 6,
minor axis 4

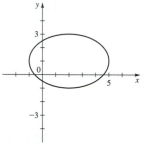

3. Center $C(0, -5)$;
foci $F_1(0, -1)$, $F_2(0, -9)$;
vertices $V_1(0, 0)$, $V_2(0, -10)$;
major axis 10, minor axis 6

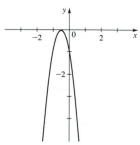

5. Vertex $V(3, -1)$;
focus $F(3, 1)$;
directrix $y = -3$

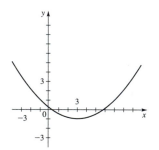

7. Vertex $V\left(-\frac{1}{2}, 0\right)$;
focus $F\left(-\frac{1}{2}, -\frac{1}{16}\right)$;
directrix $y = \frac{1}{16}$

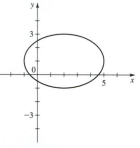

9. Center $C(-1, 3)$;
foci $F_1(-6, 3)$, $F_2(4, 3)$;
vertices $V_1(-4, 3)$, $V_2(2, 3)$;
asymptotes
$y = \pm\frac{4}{3}(x + 1) + 3$

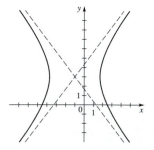

11. Center $C(-1, 0)$;
foci $F(-1, \pm\sqrt{5})$;
vertices $V(-1, \pm1)$;
asymptotes
$y = \pm\frac{1}{2}(x + 1)$

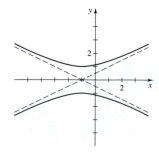

13. $x^2 = -\frac{1}{4}(y - 4)$

15. $\dfrac{(x - 5)^2}{25} + \dfrac{y^2}{16} = 1$

17. $(y - 1)^2 - x^2 = 1$

19. Ellipse; $C(2, 0)$;
$F(2, \pm\sqrt{5})$; $V(2, \pm 3)$;
major axis 6,
minor axis 4

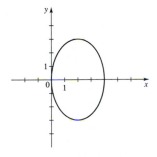

21. Hyperbola; $C(1, 2)$;
$F_1\left(-\frac{3}{2}, 2\right)$, $F_2\left(\frac{7}{2}, 2\right)$;
$V(1 \pm \sqrt{5}, 2)$;
asymptotes
$y = \pm\frac{1}{2}(x - 1) + 2$

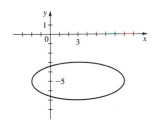

23. Ellipse; $C(3, -5)$;
$F(3 \pm \sqrt{21}, -5)$;
$V_1(-2, -5)$, $V_1(8, -5)$;
major axis 10,
minor axis 4

25. Hyperbola; $C(3, 0)$;
$F(3, \pm 5)$; $V(3, \pm 4)$;
asymptotes
$y = \pm\frac{4}{3}(x - 3)$

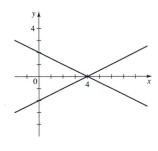

27. Degenerate conic
(pairs of lines),
$y = \pm\frac{1}{2}(x - 4)$

29. Point $(1, 3)$

31. **33.**

35. (a) $F < 17$ (b) $F = 17$ (c) $F > 17$

37. (a)

(c) The parabolas become narrower.

Section 9.5 ■ page 768

1. $(\sqrt{2}, 0)$ **3.** $(0, -2\sqrt{3})$ **5.** $(1.6383, 1.1472)$

7. $X^2 + \sqrt{3}\, XY + 2 = 0$

9. $7Y^2 - 48XY - 7X^2 - 40X - 30Y = 0$

11. $X^2 - Y^2 = 2$

13. (a) Hyberbola
(b) $X^2 - Y^2 = 16$
(c) $\phi = 45°$

15. (a) Parabola
(b) $Y = \sqrt{2}X^2$
(c) $\phi = 45°$

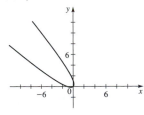

17. (a) Hyberbola

(b) $Y^2 - X^2 = 1$

(c) $\phi = 30°$

19. (a) Hyberbola

(b) $\dfrac{X^2}{4} - Y^2 = 1$

(c) $\phi \approx 53°$

21. (a) Hyberbola

(b) $3X^2 - Y^2 = 2\sqrt{3}$

(c) $\phi = 30°$

23. (a) Hyberbola

(b) $(X - 1)^2 - 3Y^2 = 1$

(c) $\phi = 60°$

25. (a) Ellipse

(b) $X^2 + \dfrac{(Y + 1)^2}{4} = 1$

(c) $\phi \approx 53°$

See graph at right.

27. (a) Parabola   (b)

29. (a) Hyperbola   (b)

31. (a) $(X - 5)^2 - Y^2 = 1$

(b) XY-coordinates: $C(5, 0)$; $V_1(6, 0), V_2(4, 0)$; $F(5 \pm \sqrt{2}, 0)$;
xy-coordinates: $C(4, 3)$; $V_1\left(\frac{24}{5}, \frac{18}{5}\right), V_2\left(\frac{16}{5}, \frac{12}{5}\right)$;
$F\left(4 \pm \frac{4}{5}\sqrt{2}, 3 \pm \frac{3}{5}\sqrt{2}\right)$

(c) $Y = \pm(X - 5)$; $7x - y - 25 = 0, x + 7y - 25 = 0$

33. $X = x \cos\phi + y \sin\phi$; $Y = -x \sin\phi + y \cos\phi$

Section 9.6 ■ page 782

1.

$\left(-3, \dfrac{3\pi}{2}\right), \left(3, \dfrac{5\pi}{2}\right)$

3.   **5.**

$\left(-1, -\dfrac{5\pi}{6}\right), \left(1, \dfrac{\pi}{6}\right)$ $(-5, 2\pi), (5, \pi)$

7. $(2\sqrt{3}, 2)$ **9.** $(1, -1)$ **11.** $(-5, 0)$ **13.** $\left(\sqrt{2}, \dfrac{3\pi}{4}\right)$

15. $\left(4, \dfrac{\pi}{4}\right)$ **17.** $\left(5, \tan^{-1}\frac{4}{3}\right)$ **19.** $\theta = \dfrac{\pi}{4}$

21. $r = \tan\theta \sec\theta$ **23.** $r = 4 \sec\theta$ **25.** $x^2 + y^2 = 49$

27. $x = 6$ **29.** $x^2 + y^2 = \dfrac{y}{x}$ **31.** $y - x = 1$

33. $x^2 + y^2 = (x^2 + y^2 - x)^2$ **35.** $x = 2$ **37.** $y = \pm\sqrt{3}\,x$

39. VI   **41.** II   **43.** I

45.   **47.**

49.

51.

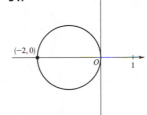

69. $0 \le \theta \le 4\pi$

53.

55.

71.

57.

59.

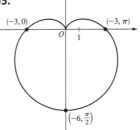

The graph of $r = 1 + \sin n\theta$ has n loops.

73. IV   **75.** III   **77.** (b) $\sqrt{10 + 6\cos(5\pi/12)} \approx 3.40$

61.

63.

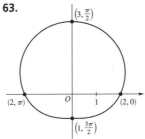

Section 9.7 ■ page 788

1. $r = 6/(3 + 2\cos\theta)$ **3.** $r = 2/(1 + \sin\theta)$

5. $r = 20/(1 + 4\cos\theta)$ **7.** $r = 10/(1 + \sin\theta)$

9. (a) 3, Hyperbola   **11.** (a) 1, Parabola

(b)

(b)

65.

67. $0 \le \theta \le 4\pi$

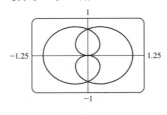

13. (a) $\frac{1}{2}$, Ellipse
(b)

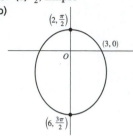

15. (a) $\frac{5}{2}$, Hyperbola
(b)

7. (a)

(b) $y = \dfrac{1}{x} + 1$

17. (a) $e = \frac{3}{4}$,   directrix $x = -\frac{1}{3}$

(b) $r = \dfrac{1}{4 - 3 \cos\left(\theta - \dfrac{\pi}{3}\right)}$

9. (a)

(b) $x^3 = y^2$

19. The ellipse is nearly circular
when e is close to 0 and becomes
more elongated as $e \to 1^-$. At $e = 1$,
the curve becomes a parabola.

11. (a)

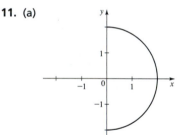

(b) $x^2 + y^2 = 4$

21. (b) $r = (1.49 \times 10^8)/(1 - 0.017 \cos\theta)$ **23.** 0.25

Section 9.8 ■ page 794

1. (a)

3. (a)

13. (a)

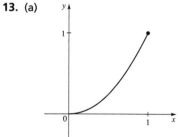

(b) $y = x^2$

(b) $x - 2y + 12 = 0$

(b) $x = (y + 2)^2$

5. (a)

(b) $x = \sqrt{1 - y}$

15. (a)

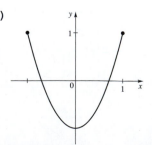

(b) $y = 2x^2 - 1$

17. (a) 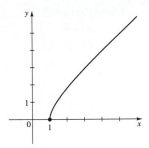   (b) $x^2 - y^2 = 1$

19. (a) 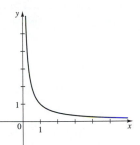   (b) $xy = 1$

21. (a) 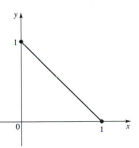   (b) $x + y = 1$

23. $x = 4 + t, y = -1 + \frac{1}{2}t$ **25.** $x = 6 + t, y = 7 + t$

27. $x = a \cos t, y = a \sin t$

31.   **33.**

37.   **39.**

41.

43. (a) $x = e^{t/12} \cos t, y = e^{t/12} \sin t$
(b)

45. (a) $x = \dfrac{4 \cos t}{2 - \cos t}, y = \dfrac{4 \sin t}{2 - \cos t}$
(b)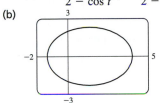

47. III   **49.** II

51. 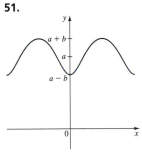   **53.** (b) $x^{2/3} + y^{2/3} = a^{2/3}$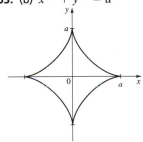

55. $x = a(\sin \theta \cos \theta + \cot \theta), y = a(1 + \sin^2 \theta)$

57. $y = a - a \cos\left(\dfrac{x + \sqrt{2ay - y^2}}{a}\right)$

Chapter 9 Review ■ page 797

1. $V(0, 0); F(0, -2);$
$y = 2$

3. $V(-2, 2); F\left(-\frac{7}{4}, 2\right);$
$x = -\frac{9}{4}$

5. $C(0, 0)$; $V(\pm 4, 0)$;
$F(\pm 2\sqrt{3}, 0)$; axes 8, 4

7. $C(0, 2)$; $V(\pm 3, 2)$;
$F(\pm\sqrt{5}, 2)$; axes 6, 4

23. Ellipse;
$F(1, 4 \pm \sqrt{15})$;
$V(1, 4 \pm 2\sqrt{5})$

25. Parabola;
$F\left(-\frac{255}{4}, 8\right)$;
$V(-64, 8)$

9. $C(0, 0)$; $V(\pm 4, 0)$;
$F(\pm 2\sqrt{6}, 0)$;

asymptotes $y = \pm\dfrac{1}{\sqrt{2}}x$

11. $C(-3, -1)$;
$V(-3, -1 \pm \sqrt{2})$;
$F(-3, -1 \pm 2\sqrt{5})$;
asymptotes $y = \frac{1}{3}x$,
$y = -\frac{1}{3}x - 2$

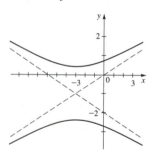

27. Ellipse;
$F(3, -3 \pm 1/\sqrt{2})$;
$V_1(3, -4)$, $V_2(3, -2)$

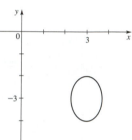

29. Has no graph   **31.** $x^2 = 4y$ **33.** $\dfrac{y^2}{4} - \dfrac{x^2}{16} = 1$

35. $\dfrac{(x-1)^2}{3} + \dfrac{(y-2)^2}{4} = 1$ **37.** $\dfrac{4(x-7)^2}{225} + \dfrac{(y-2)^2}{100} = 1$

39. $(x - 800)^2 = -200(y - 3200)$

41. (a) 91,419,000 mi   (b) 94,581,000 mi

43. (a)

13. $y^2 = 8x$ **15.** $\dfrac{y^2}{16} - \dfrac{x^2}{9} = 1$

17. $\dfrac{(x-4)^2}{16} + \dfrac{(y-2)^2}{4} = 1$

45. (a) Hyperbola
(b) $3X^2 - Y^2 = 1$
(c) $\phi = 45°$

47. (a) Ellipse
(b) $(X - 1)^2 + 4Y^2 = 1$
(c) $\phi = 30°$

19. Parabola;
$F(0, -2)$;
$V(0, 1)$

21. Hyperbola;
$F(0, \pm 12\sqrt{2})$;
$V(0, \pm 12)$

49. Ellipse

51. Parabola

53. (a)

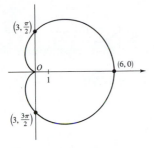

(b) $(x^2 + y^2 - 3x)^2 = 9(x^2 + y^2)$

55. (a)

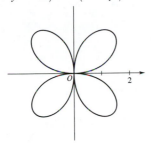

(b) $(x^2 + y^2)^3 = 16x^2y^2$

57. (a)

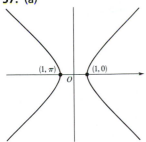

(b) $x^2 - y^2 = 1$

61. $0 \le \theta \le 6\pi$

59. (a)

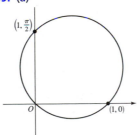

(b) $x^2 + y^2 = x + y$

63. $0 \le \theta \le 6\pi$

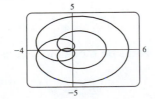

65. (a) $e = 1$, parabola

(b)

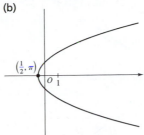

67. (a) $e = 2$, hyperbola

(b)

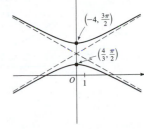

69. $x = 2y - y^2$

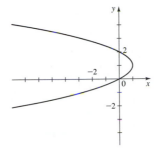

71. $(x - 1)^2 + (y - 1)^2 = 1$

73.

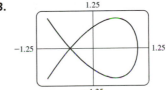

75. $x = \frac{1}{2}(1 + \cos \theta), \ y = \frac{1}{2}(\sin \theta + \tan \theta)$

Chapter 9 Test ■ page 800

1. $F(0, -3), \ y = 3$

2. $V(\pm 4, 0); \ F(\pm 2\sqrt{3}, 0); \ 8, 4$

3. $V(0, \pm 3)$; $F(0, \pm 5)$; $y = \pm\frac{3}{4}x$

4. $y^2 = -x$ **5.** $\dfrac{x^2}{16} + \dfrac{(y-3)^2}{9} = 1$

6. $(x-2)^2 - \dfrac{y^2}{3} = 1$

7. $\dfrac{(x-3)^2}{9} + \dfrac{(y+\frac{1}{2})^2}{4} = 1$

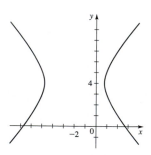

8. $9(x+2)^2 - 8(y-4)^2 = 72$

9. $(y+4)^2 = -2(x-4)$

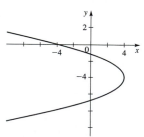

10. $\dfrac{y^2}{9} - \dfrac{x^2}{16} = 1$ **11.** $x^2 - 4x - 8y + 20 = 0$

12. $\frac{3}{4}$ in. **13.** (a) Ellipse (b) $\dfrac{X^2}{3} + \dfrac{Y^2}{18} = 1$

(c) $\phi \approx 27°$

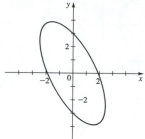

(d) $(-3\sqrt{2/5}, 6\sqrt{2/5})$, $(3\sqrt{2/5}, -6\sqrt{2/5})$

14.

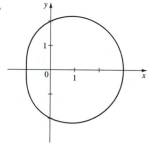

15. $(x-1)^2 + (y+2)^2 = 5$, circle

16. $r = \dfrac{1}{1 + 0.5\cos\theta}$

17. (a)

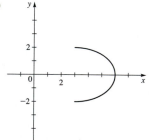

(b) $\dfrac{(x-3)^2}{9} + \dfrac{y^2}{4} = 1$

Focus on Modeling ■ **page 805**

1. $y = -\left(\dfrac{g}{2v_0^2\cos^2\theta}\right)x^2 + (\tan\theta)x$

3. (a) 5.45 s   **(b)** 118.7 ft   **(c)** 5426.5 ft

(d)

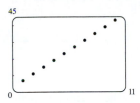

5. $\dfrac{v_0^2 \sin^2\theta}{2g}$ **7.** No, $\theta \approx 23°$

CHAPTER 10

Section 10.1 ■ page 816

1. 2, 3, 4, 5; 101   **3.** $\frac{1}{2}, \frac{1}{3}, \frac{1}{4}, \frac{1}{5}; \frac{1}{101}$ **5.** $-1, \frac{1}{4}, -\frac{1}{9}, \frac{1}{16}; \frac{1}{10,000}$

7. 0, 2, 0, 2; 2   **9.** 1, 4, 27, 256; 100^{100} **11.** 3, 2, 0, -4, -12

13. 1, 3, 7, 15, 31   **15.** 1, 2, 3, 5, 8

17. 7, 11, 15, 19, 23, 27, 31, 35, 39, 43

19. 12, 6, 4, 3, $\frac{12}{5}$, 2, $\frac{12}{7}$, $\frac{3}{2}$, $\frac{4}{3}$, $\frac{6}{5}$

21. $\frac{1}{2}, 2, \frac{1}{2}, 2, \frac{1}{2}, 2, \frac{1}{2}, 2, \frac{1}{2}, 2$

23. 2^n **25.** $3n - 2$ **27.** $(2n - 1)/n^2$ **29.** $1 + (-1)^n$

31. 1, 4, 9, 16, 25, 36   **33.** $\frac{1}{3}, \frac{4}{9}, \frac{13}{27}, \frac{40}{81}, \frac{121}{243}, \frac{364}{729}$

35. $\frac{2}{3}, \frac{8}{9}, \frac{26}{27}, \frac{80}{81}; S_n = 1 - \dfrac{1}{3^n}$

37. $1 - \sqrt{2}, 1 - \sqrt{3}, -1, 1 - \sqrt{5}; S_n = 1 - \sqrt{n + 1}$

39. 10   **41.** $\frac{11}{6}$ **43.** 8   **45.** 31   **47.** 385   **49.** 46,438

51. 22   **53.** $\sqrt{1} + \sqrt{2} + \sqrt{3} + \sqrt{4} + \sqrt{5}$

55. $\sqrt{4} + \sqrt{5} + \sqrt{6} + \sqrt{7} + \sqrt{8} + \sqrt{9} + \sqrt{10}$

57. $x^3 + x^4 + \cdots + x^{100}$ **59.** $\displaystyle\sum_{k=1}^{100} k$ **61.** $\displaystyle\sum_{k=1}^{10} k^2$

63. $\displaystyle\sum_{k=1}^{999} \dfrac{1}{k(k + 1)}$ **65.** $\displaystyle\sum_{k=0}^{100} x^k$ **67.** $2^{(2^n - 1)/2^n}$

Section 10.2 ■ page 822

1. (a) 5, 7, 9, 11, 13   **(b)** 2

(c)

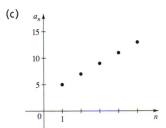

3. (a) $\frac{5}{2}, \frac{3}{2}, \frac{1}{2}, -\frac{1}{2}, -\frac{3}{2}$ **(b)** -1

(c)

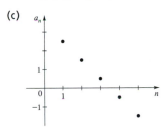

5. $a_n = 3 + 5(n - 1)$, $a_{10} = 48$

7. $a_n = \frac{5}{2} - \frac{1}{2}(n - 1)$, $a_{10} = -2$

9. Arithmetic, 3   **11.** Not arithmetic   **13.** Arithmetic, $-\frac{3}{2}$

15. Arithmetic, 1.7

17. 11, 18, 25, 32, 39; 7; $a_n = 11 + 7(n - 1)$

19. $\frac{1}{3}, \frac{1}{5}, \frac{1}{7}, \frac{1}{9}, \frac{1}{11}$; not arithmetic

21. $-4, 2, 8, 14, 21$; 6; $a_n = -4 + 6(n - 1)$

23. 3, 14, $2 + 3(n - 1)$, 299   **25.** 5, 24, $4 + 5(n - 1)$, 499

27. 4, 4, $-12 + 4(n - 1)$, 384

29. 1.5, 31, $25 + 1.5(n - 1)$, 173.5

31. $s, 2 + 4s, 2 + (n - 1)s, 2 + 99s$ **33.** $\frac{1}{2}$

35. $-100, -98, -96$ **37.** 30th   **39.** 100   **41.** 460

43. 1090   **45.** 20,301   **47.** 832.3   **49.** 46.75

51. $1250   **53.** $403,500   **55.** 20

57. (a) 576 ft   **(b)** $16n^2$ ft   **61.** Yes

Section 10.3 ■ page 830

1. (a) 5, 10, 20, 40, 80   (b) 2
(c)

3. (a) $\frac{5}{2}, -\frac{5}{4}, \frac{5}{8}, -\frac{5}{16}, \frac{5}{32}$ (b) $-\frac{1}{2}$
(c)

5. $a_n = 3 \cdot 5^{n-1}, a_4 = 375$ **7.** $a_n = \frac{5}{2}\left(-\frac{1}{2}\right)^{n-1}, a_4 = -\frac{5}{16}$
9. Geometric, 2   **11.** Geometric, $\frac{1}{2}$ **13.** Not geometric
15. Geometric, 1.1   **17.** 6, 18, 54, 162, 486; geometric,
common ratio 3; $a_n = 6 \cdot 3^{n-1}$
19. $\frac{1}{4}, \frac{1}{16}, \frac{1}{64}, \frac{1}{256}, \frac{1}{1024}$; geometric, common ratio $\frac{1}{4}$; $a_n = \frac{1}{4}\left(\frac{1}{4}\right)^{n-1}$
21. 0, ln 5, 2 ln 5, 3 ln 5, 4 ln 5; not geometric
23. 3, 162, $2 \cdot 3^{n-1}$ **25.** $-0.3, 0.00243, (0.3)(-0.3)^{n-1}$
27. $-\frac{1}{12}, \frac{1}{144}, 144\left(-\frac{1}{12}\right)^{n-1}$ **29.** $3^{2/3}, 3^{11/3}, 3^{(2n+1)/3}$
31. $s^{2/7}, s^{8/7}, s^{2(n-1)/7}$ **33.** $\frac{1}{2}$ **35.** $\frac{25}{4}$ **37.** 11th
39. 315   **41.** 441   **43.** 3280   **45.** $\frac{6141}{1024}$
47. 19 ft, $80\left(\frac{3}{4}\right)^n$ **49.** $\frac{64}{25}, \frac{1024}{625}, 5\left(\frac{4}{5}\right)^n$
51. (a) $17\frac{8}{9}$ ft   (b) $18 - \left(\frac{1}{3}\right)^{n-3}$ **53.** $\frac{3}{2}$ **55.** $\frac{3}{4}$ **57.** $\frac{1}{648}$
59. $-\frac{1000}{117}$ **61.** $\frac{7}{9}$ **63.** $\frac{1}{33}$ **65.** $\frac{112}{999}$ **67.** 3 m
69. (a) 2   (b) $8 + 4\sqrt{2}$ **71.** 1
73. $a \cdot \dfrac{1 - a^9}{1 - a} + 55b$

Section 10.4 ■ page 839

1. $13,180.79   **3.** $360,262.21   **5.** $5,591.79
7. $245.66   **9.** $2,601.59   **11.** $307.24
13. $733.76, $264,153.60
15. (a) $859.15   (b) $309,294.00   (c) $1,841,519.29
17. $341.24   **19.** 18.16%   **21.** 11.68%

Section 10.5 ■ page 846

1. Let $P(n)$ denote the statement $2 + 4 + \cdots + 2n = n(n + 1)$.
Step 1 $P(1)$ is true since $2 = 1(1 + 1)$.
Step 2 Suppose $P(k)$ is true. Then

$$2 + 4 + \cdots + 2k + 2(k + 1)$$
$$= k(k + 1) + 2(k + 1) \qquad \text{Induction hypothesis}$$
$$= (k + 1)(k + 2)$$

So $P(k + 1)$ follows from $P(k)$. Thus, by the Principle of Mathematical Induction, $P(n)$ holds for all n.

3. Let $P(n)$ denote the statement
$$5 + 8 + \cdots + (3n + 2) = \frac{n(3n + 7)}{2}.$$
Step 1 $P(1)$ is true since $5 = \dfrac{1(3 \cdot 1 + 7)}{2}$
Step 2 Suppose $P(k)$ is true. Then
$$5 + 8 + \cdots + (3k + 2) + [3(k + 1) + 2]$$
$$= \frac{k(3k + 7)}{2} + (3k + 5) \qquad \text{Induction hypothesis}$$
$$= \frac{3k^2 + 13k + 10}{2}$$
$$= \frac{(k + 1)[3(k + 1) + 7]}{2}$$

So $P(k + 1)$ follows from $P(k)$. Thus, by the Principle of Mathematical Induction, $P(n)$ holds for all n.

5. Let $P(n)$ denote the statement
$$1 \cdot 2 + 2 \cdot 3 + \cdots + n(n + 1) = \frac{n(n + 1)(n + 2)}{3}.$$
Step 1 $P(1)$ is true since $1 \cdot 2 = \dfrac{1 \cdot (1 + 1) \cdot (1 + 2)}{3}$.
Step 2 Suppose $P(k)$ is true. Then
$$1 \cdot 2 + 2 \cdot 3 + \cdots + k(k + 1) + (k + 1)(k + 2)$$
$$= \frac{k(k + 1)(k + 2)}{3} + (k + 1)(k + 2) \qquad \text{Induction hypothesis}$$
$$= \frac{(k + 1)(k + 2)(k + 3)}{3}$$

So $P(k + 1)$ follows from $P(k)$. Thus, by the Principle of Mathematical Induction, $P(n)$ holds for all n.

7. Let $P(n)$ denote the statement
$$1^3 + 2^3 + \cdots + n^3 = \frac{n^2(n + 1)^2}{4}.$$
Step 1 $P(1)$ is true since $1^3 = \dfrac{1^2 \cdot (1 + 1)^2}{4}$.

Step 2 Suppose $P(k)$ is true. Then

$$1^3 + 2^3 + \cdots + k^3 + (k + 1)^3$$

$$= \frac{k^2(k + 1)^2}{4} + (k + 1)^3 \qquad \text{Induction hypothesis}$$

$$= \frac{(k + 1)^2[k^2 + 4(k + 1)]}{4}$$

$$= \frac{(k + 1)^2(k + 2)^2}{4}$$

So $P(k + 1)$ follows from $P(k)$. Thus, by the Principle of Mathematical Induction, $P(n)$ holds for all n.

9. Let $P(n)$ denote the statement
$2^3 + 4^3 + \cdots + (2n)^3 = 2n^2(n + 1)^2$.
Step 1 $P(1)$ is true since $2^3 = 2 \cdot 1^2(1 + 1)^2$.
Step 2 Suppose $P(k)$ is true. Then

$$2^3 + 4^3 + \cdots + (2k)^3 + [2(k + 1)]^3$$

$$= 2k^2(k + 1)^2 + [2(k + 1)]^3 \qquad \text{Induction hypothesis}$$

$$= (k + 1)^2(2k^2 + 8k + 8)$$

$$= 2(k + 1)^2(k + 2)^2$$

So $P(k + 1)$ follows from $P(k)$. Thus, by the Principle of Mathematical Induction, $P(n)$ holds for all n.

11. Let $P(n)$ denote the statement
$1 \cdot 2 + 2 \cdot 2^2 + \cdots + n \cdot 2^n = 2[1 + (n - 1)2^n]$.
Step 1 $P(1)$ is true since $1 \cdot 2 = 2[1 + 0]$.
Step 2 Suppose $P(k)$ is true. Then

$$1 \cdot 2 + 2 \cdot 2^2 + \cdots + k \cdot 2^k + (k + 1) \cdot 2^{k+1}$$

$$= 2[1 + (k - 1)2^k] + (k + 1) \cdot 2^{k+1} \qquad \substack{\text{Induction}\\\text{hypothesis}}$$

$$= 2 + (k - 1)2^{k+1} + (k + 1) \cdot 2^{k+1}$$

$$= 2 + 2k2^{k+1} = 2[1 + k2^{k+1}]$$

So $P(k + 1)$ follows from $P(k)$. Thus, by the Principle of Mathematical Induction, $P(n)$ holds for all n.

13. Let $P(n)$ denote the statement $n^2 + n$ is divisible by 2.
Step 1 $P(1)$ is true since $1^2 + 1$ is divisible by 2.
Step 2 Suppose $P(k)$ is true. Now

$$(k + 1)^2 + (k + 1) = k^2 + 2k + 1 + k + 1$$

$$= (k^2 + k) + 2(k + 1)$$

But $k^2 + k$ is divisible by 2 (by the induction hypothesis) and $2(k + 1)$ is clearly divisible by 2, so $(k + 1)^2 + (k + 1)$ is divisible by 2. So $P(k + 1)$ follows from $P(k)$. Thus, by the Principle of Mathematical Induction, $P(n)$ holds for all n.

15. Let $P(n)$ denote the statement $n^2 - n + 41$ is odd.
Step 1 $P(1)$ is true since $1^2 - 1 + 41$ is odd.
Step 2 Suppose $P(k)$ is true. Now

$$(k + 1)^2 - (k + 1) + 41 = (k^2 - k + 41) + 2k$$

But $k^2 - k + 41$ is odd (by the induction hypothesis) and $2k$ is clearly even, so their sum is odd. So $P(k + 1)$ follows from $P(k)$. Thus, by the Principle of Mathematical Induction, $P(n)$ holds for all n.

17. Let $P(n)$ denote the statement $8^n - 3^n$ is divisible by 5.
Step 1 $P(1)$ is true since $8^1 - 3^1$ is divisible by 5.
Step 2 Suppose $P(k)$ is true. Now

$$8^{k+1} - 3^{k+1} = 8 \cdot 8^k - 3 \cdot 3^k$$

$$= 8 \cdot 8^k - (8 - 5) \cdot 3^k = 8 \cdot (8^k - 3^k) + 5 \cdot 3^k$$

which is divisible by 5 because $8^k - 3^k$ is divisible by 5 (by the induction hypothesis) and $5 \cdot 3^k$ is clearly divisible by 5. So $P(k + 1)$ follows from $P(k)$. Thus, by the Principle of Mathematical Induction, $P(n)$ holds for all n.

19. Let $P(n)$ denote the statement $n < 2^n$.
Step 1 $P(1)$ is true since $1 < 2^1$.
Step 2 Suppose $P(k)$ is true. Then

$$k + 1 < 2^k + 1 \qquad \text{Induction hypothesis}$$

$$< 2^k + 2^k \qquad \text{Because } 1 < 2^k$$

$$= 2 \cdot 2^k = 2^{k+1}$$

So $P(k + 1)$ follows from $P(k)$. Thus, by the Principle of Mathematical Induction, $P(n)$ holds for all n.

21. Let $P(n)$ denote the statement $(1 + x)^n \geqslant 1 + nx$
for $x > -1$.
Step 1 $P(1)$ is true since $(1 + x)^1 \geqslant 1 + 1 \cdot x$.
Step 2 Suppose $P(k)$ is true. Then

$$(1 + x)^{k+1} = (1 + x)(1 + x)^k$$

$$\geqslant (1 + x)(1 + kx) \qquad \text{Induction hypothesis}$$

$$= 1 + (k + 1)x + kx^2$$

$$\geqslant 1 + (k + 1)x$$

So $P(k + 1)$ follows from $P(k)$. Thus, by the Principle of Mathematical Induction, $P(n)$ holds for all n.

23. Let $P(n)$ denote the statement $a_n = 5 \cdot 3^{n-1}$.
Step 1 $P(1)$ is true since $a_1 = 5 \cdot 3^0 = 5$.
Step 2 Suppose $P(k)$ is true. Then

$$a_{k+1} = 3 \cdot a_k \qquad \text{Definition of } a_k + 1$$

$$= 3 \cdot 5 \cdot 3^{k-1} \qquad \text{Induction hypothesis}$$

$$= 5 \cdot 3^k$$

So $P(k + 1)$ follows from $P(k)$. Thus, by the Principle of Mathematical Induction, $P(n)$ holds for all n.

25. Let $P(n)$ denote the statement $x - y$ is a factor of $x^n - y^n$.
Step 1 $P(1)$ is true since $x - y$ is a factor of $x^1 - y^1$.
Step 2 Suppose $P(k)$ is true. Now

$$x^{k+1} - y^{k+1} = x^{k+1} - x^k y + x^k y - y^{k+1}$$

$$= x^k(x - y) + (x^k - y^k)y$$

But $x^k(x - y)$ is clearly divisible by $x - y$ and $x^k - y^k$ is divisible by $x - y$ (by the induction hypothesis), so their sum is divisible by $x - y$. So $P(k + 1)$ follows from $P(k)$. Thus, by the Principle of Mathematical Induction, $P(n)$ holds for all n.

27. Let $P(n)$ denote the statement F_{3n} is even.
Step 1 $P(1)$ is true since $F_{3 \cdot 1} = 2$, which is even.
Step 2 Suppose $P(k)$ is true. Now, by the definition of the Fibonnaci sequence

$$F_{3(k+1)} = F_{3k+3} = F_{3k+2} + F_{3k+1}$$
$$= F_{3k+1} + F_{3k} + F_{3k+1}$$
$$= F_{3k} + 2 \cdot F_{3k+1}$$

But F_{3k} is even (by the induction hypothesis) and $2 \cdot F_{3k+1}$ is clearly even, so $F_{3(k+1)}$ is even. So $P(k + 1)$ follows from $P(k)$. Thus, by the Principle of Mathematical Induction, $P(n)$ holds for all n.

29. Let $P(n)$ denote the statement
$F_1^2 + F_2^2 + \cdots + F_n^2 = F_n \cdot F_{n+1}$.
Step 1 $P(1)$ is true since $F_1^2 = F_1 \cdot F_2$ (because $F_1 = F_2 = 1$).
Step 2 Suppose $P(k)$ is true. Then

$$F_1^2 + F_2^2 + \cdots + F_k^2 + F_{k+1}^2$$
$$= F_k \cdot F_{k+1} + F_{k+1}^2 \qquad \text{\color{red}Induction hypothesis}$$
$$= F_{k+1}(F_k + F_{k+1})$$
$$= F_{k+1} \cdot F_{k+2} \qquad \text{\color{red}Definition of the Fibonnaci sequence}$$

So $P(k + 1)$ follows from $P(k)$. Thus, by the Principle of Mathematical Induction, $P(n)$ holds for all n.

31. Let $P(n)$ denote the statement $\begin{bmatrix} 1 & 1 \\ 1 & 0 \end{bmatrix}^n = \begin{bmatrix} F_{n+1} & F_n \\ F_n & F_{n-1} \end{bmatrix}$.

Step 1 $P(2)$ is true since $\begin{bmatrix} 1 & 1 \\ 1 & 0 \end{bmatrix}^2 = \begin{bmatrix} 2 & 1 \\ 1 & 1 \end{bmatrix} = \begin{bmatrix} F_3 & F_2 \\ F_2 & F_1 \end{bmatrix}$.

Step 2 Suppose $P(k)$ is true. Then

$$\begin{bmatrix} 1 & 1 \\ 1 & 0 \end{bmatrix}^{k+1} = \begin{bmatrix} 1 & 1 \\ 1 & 0 \end{bmatrix}^k \begin{bmatrix} 1 & 1 \\ 1 & 0 \end{bmatrix}$$
$$= \begin{bmatrix} F_{k+1} & F_k \\ F_k & F_{k-1} \end{bmatrix}\begin{bmatrix} 1 & 1 \\ 1 & 0 \end{bmatrix} \qquad \text{\color{red}Induction hypothesis}$$
$$= \begin{bmatrix} F_{k+1} + F_k & F_{k+1} \\ F_k + F_{k-1} & F_k \end{bmatrix}$$
$$= \begin{bmatrix} F_{k+2} & F_{k+1} \\ F_{k+1} & F_k \end{bmatrix} \qquad \text{\color{red}Definition of the Fibonnaci sequence}$$

So $P(k + 1)$ follows from $P(k)$. Thus, by the Principle of Mathematical Induction, $P(n)$ holds for all $n \geq 2$.

33. Let $P(n)$ denote the statement $F_n \geq n$.
Step 1 $P(5)$ is true since $F_5 \geq 5$ (because $F_5 = 5$).
Step 2 Suppose $P(k)$ is true. Now

$$F_{k+1} = F_k + F_{k-1} \qquad \text{\color{red}Definition of the Fibonnaci sequence}$$
$$\geq k + F_{k-1} \qquad \text{\color{red}Induction hypothesis}$$
$$\geq k + 1 \qquad \text{\color{red}Because } F_{k-1} \geq 1$$

So $P(k + 1)$ follows from $P(k)$. Thus, by the Principle of Mathematical Induction, $P(n)$ holds for all $n \geq 5$.

Section 10.6 ■ page 856

1. $x^6 + 6x^5y + 15x^4y^2 + 20x^3y^3 + 15x^2y^4 + 6xy^5 + y^6$

3. $x^4 + 4x^2 + 6 + \dfrac{4}{x^2} + \dfrac{1}{x^4}$

5. $x^5 - 5x^4 + 10x^3 - 10x^2 + 5x - 1$
7. $x^{10}y^5 - 5x^8y^4 + 10x^6y^3 - 10x^4y^2 + 5x^2y - 1$
9. $8x^3 - 36x^2y + 54xy^2 - 27y^3$

11. $\dfrac{1}{x^5} - \dfrac{5}{x^{7/2}} + \dfrac{10}{x^2} - \dfrac{10}{x^{1/2}} + 5x - x^{5/2}$

13. 15   **15.** 4950   **17.** 18   **19.** 32
21. $x^4 + 8x^3y + 24x^2y^2 + 32xy^3 + 16y^4$

23. $1 + \dfrac{6}{x} + \dfrac{15}{x^2} + \dfrac{20}{x^3} + \dfrac{15}{x^4} + \dfrac{6}{x^5} + \dfrac{1}{x^6}$

25. $x^{20} + 40x^{19}y + 760x^{18}y^2$
27. $25a^{26/3} + a^{25/3}$ **29.** $48{,}620x^{18}$
31. $300a^2b^{23}$ **33.** $100y^{99}$
35. $13{,}440x^4y^6$ **37.** $495a^8b^8$
39. $(x + y)^4$ **41.** $(2a + b)^3$
43. $3x^2 + 3xh + h^2$

Chapter 10 Review ■ page 858

1. $\frac{1}{2}, \frac{4}{3}, \frac{9}{4}, \frac{16}{5}, \frac{100}{11}$ **3.** $0, \frac{1}{4}, 0, \frac{1}{32}; \frac{1}{500}$
5. 1, 3, 15, 105; 654,729,075
7. 1, 4, 9, 16, 25, 36, 49
9. 1, 3, 5, 11, 21, 43, 85
11. (a) 7, 9, 11, 13, 15
(b)

(c) Arithmetic, common difference 2

13. (a) $\frac{3}{4}, \frac{9}{8}, \frac{27}{16}, \frac{81}{32}, \frac{243}{64}$

(b)

(c) Geometric, common ratio $\frac{3}{2}$

15. Arithmetic, 7   **17.** Arithmetic, $5\sqrt{2}$

19. Arithmetic, $t + 1$ **21.** Geometric, $\frac{4}{27}$ **23.** $2i$

25. 5   **27.** $\frac{81}{4}$ **29.** 64   **31.** 12,288   **35.** (a) 9   (b) $\pm6\sqrt{2}$

37. 126   **39.** 384   **41.** $0^2 + 1^2 + 2^2 + \cdots + 9^2$

43. $\dfrac{3}{2^2} + \dfrac{3^2}{2^3} + \dfrac{3^3}{2^4} + \cdots + \dfrac{3^{50}}{2^{51}}$ **45.** $\displaystyle\sum_{k=1}^{33} 3k$

47. $\displaystyle\sum_{k=1}^{100} k\, 2^{k+2}$ **49.** Geometric; 4.68559

51. Arithmetic, $5050\sqrt{5}$ **53.** Geometric, 9831

55. 13   **57.** 65,534   **59.** \$2390.27   **61.** $\frac{5}{7}$

63. $\frac{1}{2}\left(3 + \sqrt{3}\right)$

65. Let $P(n)$ denote the statement

$$1 + 4 + 7 + \cdots + (3n - 2) = \frac{n(3n - 1)}{2}.$$

Step 1 $P(1)$ is true since $1 = \dfrac{1(3 \cdot 1 - 1)}{2}$.

Step 2 Suppose $P(k)$ is true. Then
$$1 + 4 + 7 + \cdots + (3n - 2) + [3(k + 1) - 2]$$
$$= \frac{k(3k - 1)}{2} + [3k + 1] \quad \text{\color{red}Induction hypothesis}$$
$$= \frac{3k^2 - k + 6k + 2}{2}$$
$$= \frac{(k + 1)(3k + 2)}{2}$$
$$= \frac{(k + 1)[3(k + 1) - 1]}{2}$$

So $P(k + 1)$ follows from $P(k)$. Thus, by the Principle of Mathematical Induction, $P(n)$ holds for all n.

67. Let $P(n)$ denote the statement
$$\left(1 + \tfrac{1}{1}\right)\left(1 + \tfrac{1}{2}\right) \cdots \left(1 + \tfrac{1}{n}\right) = n + 1.$$

Step 1 $P(1)$ is true since $\left(1 + \tfrac{1}{1}\right) = 1 + 1$.

Step 2 Suppose $P(k)$ is true. Then

$$\left(1 + \frac{1}{1}\right)\left(1 + \frac{1}{2}\right) \cdots \left(1 + \frac{1}{k}\right)\left(1 + \frac{1}{k + 1}\right)$$
$$= (k + 1)\left(1 + \frac{1}{k + 1}\right) \quad \text{\color{red}Induction hypothesis}$$
$$= (k + 1) + 1$$

So $P(k + 1)$ follows from $P(k)$. Thus, by the Principle of Mathematical Induction, $P(n)$ holds for all n.

69. Let $P(n)$ denote the statement $a_n = 2 \cdot 3^n - 2$.

Step 1 $P(1)$ is true since $a_1 = 2 \cdot 3^1 - 2 = 4$.

Step 2 Suppose $P(k)$ is true. Then

$$a_{k+1} = 3a_k + 4$$
$$= 3(2 \cdot 3^k - 2) + 4 \quad \text{\color{red}Induction hypothesis}$$
$$= 2 \cdot 3^{k+1} - 2$$

So $P(k + 1)$ follows from $P(k)$. Thus, by the Principle of Mathematical Induction, $P(n)$ holds for all n.

71. Let $P(n)$ denote the statement $n! > 2^n$ for $n \geqslant 4$.

Step 1 $P(4)$ is true since $4! > 2^4$.

Step 2 Suppose $P(k)$ is true. Then

$$(k + 1)! = k!(k + 1)$$
$$> 2^k(k + 1) \quad \text{\color{red}Induction hypothesis}$$
$$> 2^{k+1} \quad \text{\color{red}Because } k + 1 > 2$$

So $P(k + 1)$ follows from $P(k)$. Thus, by the Principle of Mathematical Induction, $P(n)$ holds for all $n \geqslant 4$.

73. 255   **75.** 12,870

77. $16x^4 + 32x^3y + 24x^2y^2 + 8xy^3 + y^4$

79. $b^{-40/3} + 20b^{-37/3} + 190b^{-34/3}$

Chapter 10 Test ■ page 861

1. 0, 3, 8, 15; 99   **2.** -1 **3.** -4; 68, $a_n = 80 - 4(n - 1)$

4. $\frac{1}{5}, \frac{1}{25}$ **5.** (a) False   (b) True

6. (a) $S_n = \dfrac{n}{2}\left[2a + (n - 1)d\right]$ or $S_n = n\left(\dfrac{a + a_n}{2}\right)$

(b) 60   (c) $-\frac{8}{9}, -78$

7. (a) $S_n = \dfrac{a(1 - r^n)}{1 - r}$ (b) 58,025/59,049

8. $2 + \sqrt{2}$

9. Let $P(n)$ denote the statement
$$1^2 + 2^2 + \cdots + n^2 = \frac{n(n + 1)(2n + 1)}{6}.$$

Step 1 $P(1)$ is true since $1^2 = \dfrac{1(1 + 1)(2 \cdot 1 + 1)}{6}$.

Step 2 Suppose $P(k)$ is true. Then

$$1^2 + 2^2 + \cdots + k^2 + (k + 1)^2$$

$$= \frac{k(k + 1)(2k + 1)}{6} + (k + 1)^2 \qquad \text{Induction hypothesis}$$

$$= \frac{k(k + 1)(2k + 1) + 6\,(k + 1)^2}{6}$$

$$= \frac{(k + 1)[k(2k + 1) + 6(k + 1)]}{6}$$

$$= \frac{(k + 1)(2k^2 + 7k + 6)}{6}$$

$$= \frac{(k + 1)[(k + 1) + 1][2(k + 1) + 1]}{6}$$

So $P(k + 1)$ follows from $P(k)$. Thus, by the Principle of Mathematical Induction, $P(n)$ holds for all n.
10. (a) $(1 - 1^2) + (1 - 2^2) + (1 - 3^2) + (1 - 4^2) + (1 - 5^2)$
$= -50$
(b) 10
11. $32x^5 + 80x^4y^2 + 80x^3y^4 + 40x^2y^6 + 10xy^8 + y^{10}$
12. $\binom{10}{3}(3x)^3(-2)^7 = -414{,}720x^3$

Focus on Problem Solving ■ page 864

1. $1 + 3 + 5 + \cdots + (2n - 1) = n^2$ **5.** $\dfrac{n(n^2 + 1)}{2}$

7. (a) $\dfrac{2\sqrt{3}}{5}$

9. (a) 91; $\dfrac{n(n + 1)(2n + 1)}{6}$ (b) 441; $\left[\dfrac{n(n + 1)}{2}\right]^2$

CHAPTER 11

Section 11.1 ■ page 869

1. 12   **3.** (a) 64   (b) 24   **5.** 1024   **7.** 120
9. 144   **11.** 120   **13.** 32   **15.** 216   **17.** 480
19. No   **21.** 158,184,000   **23.** 1024   **25.** 8192
27. No   **29.** 24,360   **31.** 1050
33. (a) 16,807   (b) 2520   (c) 2401   (d) 2401   (e) 4802
35. 936   **37.** (a) 1152   (b) 1152   **39.** 483,840   **41.** 6

Section 11.2 ■ page 879

1. 336   **3.** 7920   **5.** 100   **7.** 2730   **9.** 151,200
11. 120   **13.** 24   **15.** 362,880   **17.** 997,002,000
19. 24   **21.** 60   **23.** 60   **25.** 15   **27.** 277,200

29. 2,522,520   **31.** 168   **33.** 56   **35.** 330
37. 100   **39.** 20   **41.** 210   **43.** 2,598,960
45. 120   **47.** 495   **49.** 2,035,800   **51.** 1,560,780
53. $22,957,480   **55.** (a) 56   (b) 256   **57.** 1024
59. 2025   **61.** 79,833,600   **63.** 131,072
65. 161,280   **67.** (a) 20,160   (b) 8640
69. 17,813,250   **71.** 182   **73.** 62,986

Section 11.3 ■ page 890

1. (a) $S = \{HH, HT, TH, TT\}$ (b) $\frac{1}{4}$ (c) $\frac{3}{4}$ (d) $\frac{1}{2}$
3. (a) $\frac{1}{6}$ (b) $\frac{1}{2}$ (c) $\frac{1}{6}$ **5.** (a) $\frac{1}{13}$ (b) $\frac{3}{13}$ (c) $\frac{10}{13}$
7. (a) $\frac{5}{8}$ (b) $\frac{7}{8}$ (c) 0   **9.** (a) $\frac{1}{3}$ (b) $\frac{5}{17}$
11. (a) $\frac{3}{16}$ (b) $\frac{3}{8}$ (c) $\frac{5}{8}$
13. $4C(13, 5)/C(52, 5) \approx .00198$
15. $4/C(52, 5) \approx 1.53908 \times 10^{-6}$
17. (b) $\frac{1}{16}$ (c) $\frac{3}{8}$ (d) $\frac{1}{8}$ (e) $\frac{11}{16}$ **19.** $\frac{9}{19}$
21. $1/C(49, 6) \approx 7.15 \times 10^{-8}$ **23.** (a) $\frac{1}{1024}$ (b) $\frac{15}{128}$
25. (a) $1/48^6 \approx 8.18 \times 10^{-11}$ (b) $1/48^{18} \approx 5.47 \times 10^{-31}$
27. $4/11! \approx 1.00 \times 10^{-7}$ **29.** (a) $\frac{3}{4}$ (b) $\frac{1}{4}$
31. (a) Yes   (b) No
33. (a) Mutually exclusive; 1
(b) Not mutually exclusive; $\frac{2}{3}$
35. (a) Not mutually exclusive; $\frac{11}{26}$
(b) Mutually exclusive; $\frac{1}{2}$
37. (a) $\frac{3}{4}$ (b) $\frac{1}{2}$ (c) 1   **39.** $\frac{21}{38}$ **41.** $\frac{31}{1001}$
43. (a) $\frac{3}{8}$ (b) $\frac{1}{2}$ (c) $\frac{11}{16}$ (d) $\frac{13}{16}$
45. (a) Yes   (b) $\frac{1}{36}$ **47.** (a) $\frac{1}{16}$ (b) $\frac{1}{32}$
49. $\frac{1}{12}$ **51.** $\frac{1}{1444}$ **53.** $\dfrac{1}{36^3} \approx 2.14 \times 10^{-5}$
55. (i)   **57.** $1 - \dfrac{P(365, 8)}{365^8} \approx 0.07434$
59. $600/P(40, 3) = 5/494$ **61.** 0.1

Section 11.4 ■ page 897

1. $1.50   **3.** $0.94   **5.** $0.92   **7.** 0   **9.** −$0.30
11. −$0.0526   **13.** −$0.50
15. No, she should expect to lose $2.10 per stock.
17. −$0.93   **19.** $1

Chapter 11 Review ■ page 898

1. 624   **3.** (a) 10   (b) 20   **5.** 120   **7.** 45
9. 17,576   **11.** 120   **13.** 5   **15.** 14
17. (a) 240   (b) 3360   (c) 1680
19. 40,320   **21.** (a) 31,824   (b) 11,760   (c) 19,448
(d) 2808   (e) 2808   (f) 6,683,040

23. (a) $\frac{3}{7}$ (b) $\frac{10}{21}$ (c) $\frac{19}{21}$ (d) $\frac{2}{21}$ (e) $\frac{1}{105}$ **25.** $\frac{2}{5}$
27. (a) $\frac{1}{13}$ (b) $\frac{2}{13}$ (c) $\frac{4}{13}$ (d) $\frac{1}{26}$ **29.** (a) $\frac{1}{6}$ (b) $\frac{5}{6}$
31. (a) $\frac{1}{1000}$ (b) $\frac{1}{200}$ **33.** 0
35. $0.00016 **37.** (a) 3 (b) .51
39. (a) 10^5 (b) 5^5 (c) $\frac{1}{32}$ (d) 75
41. (a) 144 (b) 126 (c) 84 (d) $\frac{7}{8}$

Chapter 11 Test ■ page 901

1. (a) 10^5 (b) 30,240 **2.** 60
3. (a) $P(30, 4) = 657,720$ (b) $C(30, 4) = 27,405$
4. 12 **5.** $4 \cdot 2^{14} = 65,536$ **6.** (a) $4! = 24$
(b) $6!/3! = 120$ **7.** $30 \cdot 29 \cdot 28 \cdot C(27, 5) = 1,966,582,800$
8. (a) $\frac{1}{2}$ (b) $\frac{1}{13}$ (c) $\frac{1}{26}$ **9.** (a) $\frac{5}{13}$ (b) $\frac{6}{13}$ (c) $\frac{9}{13}$
10. $C(5, 3)/C(15, 3) \approx 0.022$ **11.** $\frac{1}{6}$
12. $1 - 1 \cdot \frac{11}{12} \cdot \frac{10}{12} \cdot \frac{9}{12} \approx 0.427$ **13.** $0.65

Focus on Modeling ■ page 904

1. (b) $\frac{9}{10}$ **3.** (b) $\frac{7}{8}$ **7.** (b) $\frac{1}{2}$

CHAPTER 12

Section 12.1 ■ page 914

1. $\frac{1}{4}$ **3.** $\frac{1}{3}$ **5.** 1 **7.** -1 **9.** 0.51 **11.** $\frac{1}{2}$ **13.** (a) 2
(b) 3 (c) Does not exist (d) 4 (e) Not defined
15. (a) -1 (b) -2 (c) Does not exist (d) 2 (e) 0
(f) Does not exist (g) 1 (h) 3 **17.** -8 **19.** Does not
exist **21.** Does not exist
23. (a) 4 (b) 4 (c) 4

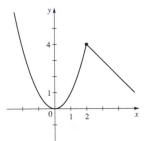

25. (a) 4 (b) 3 (c) Does not exist

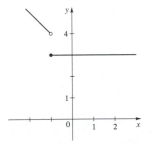

Section 12.2 ■ page 923

1. (a) 5 (b) 9 (c) 2 (d) $-\frac{1}{3}$ (e) $-\frac{3}{8}$ (f) 0
(g) Does not exist (h) $-\frac{6}{11}$ **3.** 75 **5.** $\frac{1}{2}$ **7.** -3 **9.** 5
11. 2 **13.** $\frac{6}{5}$ **15.** 12 **17.** $\frac{1}{6}$ **19.** $-\frac{1}{16}$
21. 4 **23.** $-\frac{3}{2}$

25. (a) 0.667

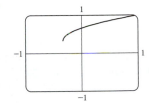

(b) 0.667 (c) $\frac{2}{3}$

x	$f(x)$
0.1	0.71339
0.01	0.67163
0.001	0.66717
0.0001	0.66672

x	$f(x)$
-0.1	0.61222
-0.01	0.66163
-0.001	0.66617
-0.0001	0.66662

27. 0 **29.** Does not exist **31.** Does not exist
33. (a) 1, 2 (b) Does not exist
(c)

Section 12.3 ■ page 932

1. 3 **3.** -11 **5.** 24 **7.** $y = -x - 1$ **9.** $y = -x + 4$
11. $y = \frac{1}{4}x + \frac{7}{4}$ **13.** $f'(2) = -12$ **15.** $g'(1) = 4$
17. $F'(4) = -\frac{1}{16}$ **19.** $f'(a) = 2a + 2$
21. $f'(a) = \dfrac{1}{(a + 1)^2}$

23. (a) $f'(a) = 3a^2 - 2$
(b) $y = -2x + 4$, $y = x + 2$, $y = 10x - 12$
(c)

25. -24 ft/s   **27.** $12a^2 + 6$ m/s, 18 m/s, 54 m/s, 114 m/s
29. $0.75°$/min   **31.** (a) -38.3 gal/min, -27.8 gal/min
(b) -33.3 gal/min

Section 12.4 ■ page 941

1. (a) $-1, 2$ (b) $y = -1$, $y = 2$ **3.** 0   **5.** $\frac{2}{5}$ **7.** $\frac{4}{3}$ **9.** 2
11. Does not exist   **13.** 7   **15.** $-\frac{1}{4}$ **17.** 0   **19.** 0
21. Divergent   **23.** 0   **25.** Divergent   **27.** $\frac{3}{2}$ **29.** 8
31. (b) 30 g/L

Section 12.5 ■ page 950

1. (a) 40, 52

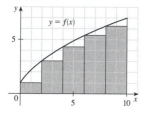

(b) 43.2, 49.2
3. 5.25   **5.** $\frac{223}{35}$

7. (a) $\frac{77}{60}$, underestimate

(b) $\frac{25}{12}$, overestimate

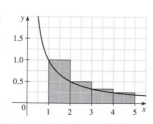

9. (a) 8, 6.875   (b) 5, 5.375

11. 37.5   **13.** 8   **15.** 166.25   **17.** 133.5

Chapter 12 Review ■ page 952

1. 1   **3.** 0.69   **5.** Does not exist   **7.** (a) Does not exist
(b) 2.4   (c) 2.4   (d) 2.4   (e) 0.5   (f) 1   (g) 2   (h) 0
9. -3 **11.** 7   **13.** 2   **15.** -1 **17.** 2   **19.** Does not exist
21. $f'(4) = 3$ **23.** $f'(16) = \frac{1}{8}$ **25.** (a) $f'(a) = -2$
(b) $-2, -2$ **27.** (a) $f'(a) = 1/(2\sqrt{a+6})$ (b) $1/(4\sqrt{2})$, 1/4
29. $y = 2x + 1$ **31.** $y = 2x$ **33.** $y = -\frac{1}{4}x + 1$
35. (a) -64 ft/s   (b) $-32a$ ft/s   (c) $\sqrt{40} \approx 6.32$ s
(d) -202.4 ft/s   **37.** $\frac{1}{5}$ **39.** $\frac{1}{2}$ **41.** Divergent   **43.** 3.83
45. 10   **47.** $\frac{5}{6}$

Chapter 12 Test ■ page 954

1. (a) $\frac{1}{2}$ (b)

2. (a) 1   (b) 1   (c) 1   (d) 0   (e) 0   (f) 0   (g) 4   (h) 2
(i) Does not exist   **3.** (a) 6   (b) -2 (c) Does not exist
(d) Does not exist   (e) $\frac{1}{4}$ (f) 2   **4.** (a) $f'(x) = 2x - 2$
(b) $-4, 0, 2$ **5.** $y = \frac{1}{6}x + \frac{3}{2}$ **6.** (a) 0   (b) Does not exist
7. (a) $\frac{89}{25}$ (b) $\frac{11}{3}$

Focus on Modeling ■ page 958

1. $57,333\frac{1}{3}$ ft-lb   **3.** (b) Area under the graph of
$p(x) = 375x$ between $x = 0$ and $x = 4$ (c) 3000 lb
(d) 1500 lb   **5.** (a) 1625.28 heating-degree hours   (b) $70°$F
(c) 1488 heating-degree hours   (d) $75°$F
(e) The day in part (a)

Index

Photo Credits

This page constitutes an extension of the copyright page. We have made every effort to trace the ownership of all copyrighted material and to secure permission from copyright holders. In the event of any question arising as to the use of any material, we will be pleased to make the necessary corrections in future printings. Thanks are due to the following authors, publishers, and agents for permission to use the material indicated.

CHAPTER 1

2: left, Jim Pickerell/ Stock Boston; right, © Corbis
77: American Heritage Center, University of Wyoming
106: National Portrait Gallery
135: Stanford University News Service
138: © Corbis

CHAPTER 2

142: left, Lori Adamski Peck/ Stone; right, Staffan Windstrand, Kevin Morris/ Corbis
163: Stanford University News Service
191: The Granger Collection
192: Courtesy of NASA
201: © USA Today, reprinted with permission
247: Kevin R. Morris/ Corbis
249: Eric & David Hosking/ Corbis
253: Phil Shemeister/ Corbis

CHAPTER 3

256: left, Brian Yarvin/ Photo Researchers Inc.; right, © Kerrick James/Stone
264: Courtesy of Ford Motor Co.
300: © Corbis
332: Michael Ng

CHAPTER 4

334: left, Joe McDonald/ Corbis; right, Owen Franken/ Corbis
343: Garry McMichael/ Photo Researchers Inc.

370: Joel W. Rogers/ Corbis
384: left, © Bettmann/ Corbis; right, © Hulton-Deutsch Collection/ Corbis
385: Roger Ressmeyer/ Corbis

CHAPTER 5

408: left, Michael Busselle/ Stone; right, Bob Krist/ Corbis
442: Jeff Lepore/ Photo Researchers Inc.

CHAPTER 6

472: left, Science Photo Library/ Photo Researchers Inc.; right, Alan Oddie/ PhotoEdit
509: Alan Oddie/ PhotoEdit

CHAPTER 7

534: left, © Corbis; right, J. L. Amos/ SuperStock

CHAPTER 8

620: left, Courtesy Ford Motor Co.; right, Tamara Munzer/ Stanford University, Eric Hoffman/ Ipsilon Networks, K. Claffy/ NLANR, and Bill Fenner/ Xerox PARC
626: Courtesy of NASA
663: The Archives of the National Academy of Sciences
667: Courtesy of CalTech
672: © Art Wolfe/ Stone
675: The Granger Collection
676: © Art Wolfe/ Stone

678: E.O. Hoppé/ Corbis
689: © Corbis
691: The Granger Collection

CHAPTER 9

722: left, © Corbis; right, Lew Long/ The Stock Market
754: Roger Ressmeyer/ Corbis
802: The Granger Collection

CHAPTER 10

806: left, Susan Van Etten/ PhotoEdit; right, © Cris Haigh/ Stone
814: The Granger Collection
815: © Corbis
818: Bill Ross/ Corbis
845: © Archivo Iconographico, S. A./ Corbis
854: Bill Sanderson/ Science Photo Library/ Photo Researchers Inc.

CHAPTER 11

866: left, Will and Denni McIntyre/ Photo Researchers Inc.; right, Kevin Morris/ Corbis
868: Courtesy of Math Dept., Stanford University
877: Ken Regan/ Camera 5

CHAPTER 12

906: left, Gar Ladd/ Photo Researchers Inc.; right, George Olson/ Photo 20-20

SEQUENCES AND SERIES

Arithmetic

$$a, a + d, a + 2d, a + 3d, a + 4d, \ldots$$

$$a_n = a + (n - 1)d$$

$$S_n = \sum_{k=1}^{n} a_k = \frac{n}{2}[2a + (n - 1)d] = n\left(\frac{a + a_n}{2}\right)$$

Geometric

$$a, ar, ar^2, ar^3, ar^4, \ldots \qquad a_n = ar^{n-1}$$

$$S_n = \sum_{k=1}^{n} a_k = a \frac{1 - r^n}{1 - r}$$

If $|r| < 1$, then the sum of an infinite geometric series is

$$S = \frac{a}{1 - r}$$

THE BINOMIAL THEOREM

$$(a + b)^n = \binom{n}{0}a^n + \binom{n}{1}a^{n-1}b + \cdots + \binom{n}{n-1}ab^{n-1} + \binom{n}{n}b^n$$

FINANCE

Compound interest

$$A = P\left(1 + \frac{r}{n}\right)^{nt}$$

where A is the amount after t years, P is the principal, r is the interest rate, and the interest is compounded n times per year.

Amount of an annuity

$$A_f = R \frac{(1 + i)^n - 1}{i}$$

where A_f is the final amount, i is the interest rate per time period, and there are n payments of size R.

Present value of an annuity

$$A_p = R \frac{1 - (1 + i)^{-n}}{i}$$

where A_p is the present value, i is the interest rate per time period, and there are n payments of size R.

Installment buying

$$R = \frac{iA_p}{1 - (1 + i)^{-n}}$$

where R is the size of each payment, i is the interest rate per time period, A_p is the amount of the loan, and n is the number of payments.

COUNTING

Fundamental Counting Principle

Suppose that two events occur in order. If the first can occur in m ways and the second can occur in n ways (after the first has occurred), then the two events can occur in order in $m \times n$ ways.

The number of **permutations** of n objects taken r at a time is

$$P(n, r) = \frac{n!}{(n - r)!}$$

The number of **combinations** of n objects taken r at a time is

$$C(n, r) = \frac{n!}{r!\,(n - r)!}$$

The number of **subsets** of a set with n elements is 2^n.

The number of **distinct permutations** of n elements, with n_i elements of the ith kind (where $n_1 + n_2 + \cdots + n_k = n$), is

$$\frac{n!}{n_1!\,n_2!\cdots n_k!}$$

PROBABILITY

If S is a sample space consisting of equally likely outcomes, and E is an event in S, then the probability of E is

$$P(E) = \frac{n(E)}{n(S)} = \frac{\text{number of elements in } E}{\text{number of elements in } S}$$

Complement of an event:

$$P(E') = 1 - P(E)$$

Union of two events:

$$P(E \cup F) = P(E) + P(F) - P(E \cap F)$$

Intersection of two independent events:

$$P(E \cap F) = P(E)P(F)$$

If a game gives payoffs of $a_1, a_2, \ldots, a_n$ with probabilities $p_1, p_2, \ldots, p_n$, respectively, then the **expected value** is

$$E = a_1 p_1 + a_2 p_2 + \cdots + a_n p_n$$

ANGLE MEASUREMENT

π radians $= 180°$

$1° = \dfrac{\pi}{180}$ rad $\qquad$ 1 rad $= \dfrac{180°}{\pi}$

$s = r\theta \qquad A = \frac{1}{2}r^2\theta \quad$ (θ in radians)

To convert from degrees to radians, multiply by $\dfrac{\pi}{180}$.

To convert from radians to degrees, multiply by $\dfrac{180}{\pi}$.

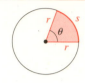

TRIGONOMETRIC FUNCTIONS OF REAL NUMBERS

$\sin t = y \qquad\qquad \csc t = \dfrac{1}{y}$

$\cos t = x \qquad\qquad \sec t = \dfrac{1}{x}$

$\tan t = \dfrac{y}{x} \qquad\qquad \cot t = \dfrac{x}{y}$

TRIGONOMETRIC FUNCTIONS OF ANGLES

$\sin \theta = \dfrac{y}{r} \qquad\qquad \csc \theta = \dfrac{r}{y}$

$\cos \theta = \dfrac{x}{r} \qquad\qquad \sec \theta = \dfrac{r}{x}$

$\tan \theta = \dfrac{y}{x} \qquad\qquad \cot \theta = \dfrac{x}{y}$

RIGHT ANGLE TRIGONOMETRY

$\sin \theta = \dfrac{\text{opp}}{\text{hyp}} \qquad \csc \theta = \dfrac{\text{hyp}}{\text{opp}}$

$\cos \theta = \dfrac{\text{adj}}{\text{hyp}} \qquad \sec \theta = \dfrac{\text{hyp}}{\text{adj}}$

$\tan \theta = \dfrac{\text{opp}}{\text{adj}} \qquad \cot \theta = \dfrac{\text{adj}}{\text{opp}}$

SPECIAL VALUES OF THE TRIGONOMETRIC FUNCTIONS

θ	radians	$\sin \theta$	$\cos \theta$	$\tan \theta$
0°	0	0	1	0
30°	$\pi/6$	$1/2$	$\sqrt{3}/2$	$\sqrt{3}/3$
45°	$\pi/4$	$\sqrt{2}/2$	$\sqrt{2}/2$	1
60°	$\pi/3$	$\sqrt{3}/2$	$1/2$	$\sqrt{3}$
90°	$\pi/2$	1	0	—
180°	π	0	-1	0
270°	$3\pi/2$	-1	0	—

SPECIAL TRIANGLES

GRAPHS OF THE TRIGONOMETRIC FUNCTIONS

SINE AND COSINE CURVES

$y = a \sin k(x - b) \quad (k > 0) \qquad y = a \cos k(x - b) \quad (k > 0)$

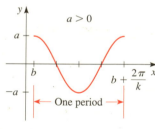

amplitude: $|a|$ $\qquad$ period: $2\pi/k$ $\qquad$ phase shift: b

GRAPHS OF THE INVERSE TRIGONOMETRIC FUNCTIONS

$y = \sin^{-1}x \qquad\qquad y = \cos^{-1}x \qquad\qquad y = \tan^{-1}x$